ADVANCED ACCELERATOR CONCEPTS

AIP
CONFERENCE
PROCEEDINGS 279

ADVANCED
ACCELERATOR
CONCEPTS

PORT JEFFERSON, NY 1992

EDITOR:

JONATHAN S. WURTELE

MIT

American Institute of Physics   New York

L.C. Catalog Card No. 93-71773
ISBN 1-56396-191-1
DOE CONF-9206193

Printed in the United States of America.

CONTENTS

PREFACE

The Third Advanced Accelerator Concepts Workshop was held in Port Jefferson, Long Island, from June 14–20, 1992. The format of the workshop closely followed that of the first two workshops held in August, 1986 in Madison, Wisconsin and in Lake Arrowhead in January, 1989. The workshop was sponsored by the American Physical Society, the U. S. Department of Energy, the Office of Naval Research, and Brookhaven National Laboratory.

The purpose of the workshop was to investigate new techniques for the production of ultra-high gradient acceleration and to address critical physics and engineering issues relating to all aspects of modern accelerators, including high brightness sources, microwave sources, accelerator configurations, and instrumentation requirements.

The meeting was divided between review presentations and subsequent working group sessions. Working groups were formed in physics and technology of high intensity electromagnetic waves, near field devices (e.g., wake field accelerators and semiconventional structures), plasma accelerators, physics, and technology of high brightness sources, far field devices (e.g., IFEL), instrumentation and diagnostics, and physics opportunities with advanced accelerators. The chairs of the various working groups reported on the substance of their deliberations at the plenary session. Summary talk papers for all but one of the groups are included in the proceedings.

The advanced accelerator research programs in both this country and abroad have matured considerably since the first workshop. Ideas are now replaced by experiments and new facilities will soon be operational. The interest of the participants was wide-ranging, from plasma acceleration to advances in slow-wave structures to muon colliders. We hope these proceedings give the reader a sense of the breadth and vitality of the field.

Special thanks go to the members of the organizing committee for their efforts. In addition, we wish to recognize particularly the considerable work carried out by Ilan Ben-Zvi, the local chair of the meeting, Kathleen Tuohy, who organized and ran the Scientific Secretariat, and Felicia Brady, who ably assisted in the preparation of these proceedings.

John A. Nation, Conference Chair
Jonathan S. Wurtele, Proceedings Editor

SCIENTIFIC ORGANIZING COMMITTEE

J. Nation (Chair, Cornell University)
A. Chao (SSC)
P. Chen (SLAC)
T. Katsouleas (USC)
C. Pellegrini (UCLA)
R. Siemann (SLAC)
P. Sprangle (NRL)
J. Wurtele (MIT)

I. Ben-Zvi (Local Chair, Brookhaven)
S. Chattopadhyah (LBL)
C. Joshi (UCLA)
F. Mills (Fermilab)
C. Roberson (ONR)
J. Simpson (Argonne)
D. Sutter (USDOE)

Dr. Stefano Alberti	MIT
Dr. Francois Amiranoff	Ecole Polytechnique
Dr. William Barletta	Lawrence Livermore National Laboratory
Mr. Nick Barov	UCLA
Dr. Stephen Benson	Duke University
Dr. Ilan Ben-Zvi	Brookhaven National Laboratory
Dr. D. Bernard	Ecole Polytechnique
Dr. Tarlochan Bhatia	U.S. Department of Energy
Professor Amitava Bhattacharjee	Columbia University
Dr. S. Alex Bogacz	Fermi National Accelerator Laboratory
Dr. Charles A. Braù	Vanderbilt University
Dr. Boris Briezman	University of Texas at Austin
Dr. K. C. Dominic Chan	Los Alamos National Laboratory
Dr. Paul J. Channel	Los Alamos National Laboratory
Dr. Chiping Chen	MIT
Dr. Pison Chen	Stanford Linear Accelerator Center
Dr. Shien-Chi Chen	MIT
Mr. Jack Sheng	University of Maryland
Dr. Eric Chojnacki	Argonne National Laboratory
Dr. Johannes Claus	Brookhaven National Laboratory
Dr. Christopher E. Clayton	UCLA
Dr. David B. Cline	UCLA
Dr. Manoel E. Conde	MIT
Dr. Richard Cooper	Los Alamos National Laboratory
Dr. Bruce Danly	MIT
Mr. Timothy Davis	Cornell University
Dr. Philip Debenham	U.S. Department of Energy
Mr. Chris Decker	UCLA
Dr. William Donaldson	University of Rochester
Dr. David H. Dowell	Boeing Physical Sciences Center
Dr. Eric Esarey	Naval Research Laboratory
Mr. Matthew Everett	UCLA
Dr. Richard C. Fernow	Brookhaven National Laboratory
Dr. Joachim Fischer	Brookhaven National Laboratory
Dr. Alan Fisher	Brookhaven National Laboratory
Dr. David Fisher	University of Texas at Austin
Professor Jorge R. Fontana	University of California at Santa Barbara
Dr. Henry Freund	SAIC
Dr. William E. Gabella	Fermi National Accelerator Laboratory
Dr. Wie Gai	Argonne National Laboratory
Dr. Juan Gallardo	Brookhaven National Laboratory
Dr. Luca Gianessi	ENEA-Frascati
Dr. Jermone Gonichon	MIT
Dr. Victor Granatstein	University of Maryland
Mr. William Graves	Fermi National Accelerator Laboratory
Dr. Gilbert F. Guinard	CERN

Dr. Bahman Hafizi	Naval Research Laboratory
Dr. Jacob Haimson	Haimson Research Corporation
Mr. Spencer C. Hartman	UCLA
Dr. H. Hirshfield	Omega P, Inc.
Dr. Ching-Hung Ho	Argonne National Laboratory
Dr. Mahir S. Hussein	Universidade de Sao Paulo
Dr. Gerhard Ingold	Brookhaven National Laboratory
Dr. Alan Jackson	Lawrence Berkeley Laboratory
Dr. Gerald P. Jackson	Fermi National Accelerator Laboratory
Dr. Robert A. Jameson	Los Alamos National Laboratory
Dr. Chan Joshi	UCLA
Dr. Yoon W. Kang	Argonne National Laboratory
Dr. Thomas C. Katsouleas	University of Southern California
Dr. Kwan-Je Kim	Lawrence Berkeley Laboratory
Dr. Wayne D. Kimura	STI Optronics
Dr. Harold G. Kirk	Brookhaven National Laboratory
Dr. George Kirkman	Integrated Applied Physics
Dr. Richard Konecny	Argonne National Laboratory
Dr. Jonathan Krall	Naval Research Laboratory
Dr. Samuel Krinsky	Brookhaven National Laboratory
Professor Norman M. Kroll	University of California at San Diego
Mr. Erjia Kuang	Cornell University
Dr. Andre Kudryavtsev	Institute of Nuclear Physics
Dr. Norman A. Kurnit	Los Alamos National Laboratory
Mr. Karl Kusche	Brookhaven National Laboratory
Mr. Chih-Hsiang Lai	University of Southern California
Dr. Peter Latham	University of Maryland
Dr. A. N. Lebedev	Phys. Inst. of Russian Academy of Science
Dr. Wim Leemans	Lawrence Berkeley Laboratory
Mr. Chia-Liang Lin	MIT
Mr. Ronglin Liou	University of Southern California
Mr. Robert Malone	Brookhaven National Laboratory
Dr. Ken Marsh	UCLA
Dr. Thomas C. Marshall	Columbia University
Mr. Henry W. Matthews	University of Maryland
Dr. Kirk T. McDonald	Princeton University
Mr. Achim Michalke	University of Wuppertal
Dr. Paolo Michelato	Univ. Degli Studi di Milano
Dr Frederick Mills	Argonne National Laboratory
Dr. Philippe Mine	Ecole Polytechnique
Dr. Alfred A. Mondelli	SAIC
Dr. Patrick Mora	Ecole Polytechnique
Dr. Warren B. Mori	UCLA
Dr. Gerard Mourou	University of Michigan
Professor John Nation	Cornell University
Dr. David Neuffer	CEBAF
Professor Yasushi Nishida	Utsunomiya University
Dr. Robert J. Noble	Fermi National Accelerator Laboratory
Dr. James H. Norem	Argonne National Laboratory
Dr. Patrick G. O'Shea	Los Alamos National Laboratory
Dr. Atsushi Ogata	KEK Natl. Lab. of High Energy Physics

Dr. Robert B. Palmer	Brookhaven National Laboratory
Dr. S. Park	UCLA
Dr. Zohreh Parsa	Brookhaven National Laboratory
Dr. Claudio Pellegrini	UCLA
Mr. Gerald J. Peters	U.S. Department of Energy
Dr. Igor Pogorelsky	Brookhaven National Laboratory
Mr. John Power	Argonne National Laboratory
Mr. Sankaranarayan Rajagopalan	Stanford Linear Accelerator Center
Dr. Govindan Rangarajan	Lawrence Berkeley Laboratory
Dr. Martin P. Reiser	University of Maryland
Dr. M. J. Rhee	University of Maryland
Dr. Sally Ride	University of California at San Diego
Dr. Charles W. Roberson	Office of Naval Research
Dr. James B. Rosenzweig	UCLA
Dr. Marc C. Ross	Stanford Linear Accelerator Center
Dr. Alessandro G. Ruggiero	Brookhaven National Laboratory
Dr. Robert Ryne	Brookhaven National Laboratory
Dr. Nicholas P. Samios	Brookhaven National Laboratory
Dr. Levi Schachter	Cornell University
Dr. Luca Serafini	INFN-Milan
Dr. Andrew M. Sessler	Lawrence Berkeley Laboratory
Dr. Robert Shafer	Los Alamos National Laboratory
Dr. Richard L. Sheffield	Los Alamos National Laboratory
Mr. Gennady Shvets	MIT
Dr. James Simpson	Argonne National Laboratory
Dr. Phillip Sprangle	Naval Research Laboratory
Dr. Triveni Srinivasan-Rao	Brookhaven National Laboratory
Dr. Loren Steinhauer	STI Optronics
Dr. Charles D. Striffler	University of Maryland
Dr. David Sutter	U.S. Department of Energy
Dr. Cha-Mei Tang	Naval Research Laboratory
Dr. Richard J. Temkin	MIT
Dr. Maury Tigner	Cornell University
Dr. Antonio C. Ting	Naval Research Laboratory
Mr. Gill Travish	UCLA
Dr. Donald Umstadter	University of Michigan
Dr. Arie van Steenbergen	Brookhaven National Laboratory
Dr. Alexander A. Verfolomeev	Kurchatov Institute
Dr. Changbaio Wang	Lawrence Berkeley Laboratory
Dr. Dunxiong Wang	University of Maryland
Mr. Xijie Wang	Brookhaven National Laboratory
Dr. Jie Wei	Brookhaven National Laboratory
Dr. Glen Westenskow	Lawrence Livermore Natl. Laboratory
Dr. Ronald Williams	UCLA
Dr. Jonathan Wurtele	MIT
Dr. David U. L. Yu	DULY Research, Inc.
Dr. L.-H.Yu	Brookhaven National Laboratory
Mr. Renshan Zhang	UCLA

IMPERATIVES FOR FUTURE HIGH ENERGY ACCELERATORS

M. Tigner
Cornell University, Ithaca, NY, 14853

ABSTRACT

The requirements for future electron positron linear colliders are examined. It is concluded that in the energy regime approaching 10 TeV the present single pass linear collider will probably be unsatisfactory due to high backgrounds, excessive power demand or both. Energy recovery and beam neutralization as ways to avoid this hiatus are discussed.

INTRODUCTION

More than half a century of experience with particle accelerators has amply shown that the science of the elementary material constituents and their interactions has been paced by progress in particle accelerator science and technology. The remarkable rate of this progress and its achievements are shown in the Livingston Chart, Fig. 1 showing the names, dates and center of mass (CM) energies reached. Most remarkable perhaps is the exponential increase in sophistication and power of all aspects of accelerator creation and operation that has been so demonstrated. These aspects include the ways in which we organize ourselves and our work as well as the means we develop for engineering with the electromagnetic interaction.

This success has been driven by the inherent excitement of winning new and deeper understanding of the universe around us. The results of this enterprise speak for themselves. More particularly, one may attribute the accomplishments of this science to the ability of its devotees to peer into the "seeds of time" and pave the way for the future to happen. For example, the outlines of the accelerators that will "happen" in the next two decades were already apparent in the proceedings of the ICFA workshops of 1978 and 1979.[1,2] The seeds of these ideas were already present and being discussed in the 60's and early 1970's.

So here we are today, again trying to see beyond the horizon. For the purposes of our discussion we shall assume the means for producing elementary interaction energies of a few TeV are in hand so that the TeV energy regime establishes our horizon. This assumption is not quite true of course but with the SSC under construction and LHC in the advanced planning stages it seems that the assumption is justifiable for hadronic colliders. On the lepton side too, there seems little doubt that the vehicle for TeV energy physics will be some type of linac using cavities and rf generation methods more or less similar to those now in use. The remaining challenge is to bring the cost within reason.

Our task this morning is to try and form some notion of what will be required of accelerators beyond this horizon.

GENERAL REQUIREMENTS

Energy   Schedule

As noted, getting beyond the horizon is herein defined as producing beams with elementary constituent energies well beyond a TeV. If we are to keep pace with the past, the Livingston chart tells us we need to take the next step by the year 2015 or so. Now you may well argue that this is not a hard and fast requirement and I would have to agree. We do, however, need to take into account the finiteness of human existence. Any curiosity driven enterprise needs to produce a significant enhancement in understanding on a time scale commensurate with a human generation or it will die. Thus while the pace set by our history chart is somewhat arbitrary, it does encompass a very basic fact which we cannot ignore.

Luminosity

To be of use as a physics vehicle our accelerator must produce a reaction rate at the energies of interest which is sufficient to reveal the underlying physics on a human time scale. While honest men may argue the details of the luminosity needed the theme is set by the very basic rule that the elementary constituent cross-sections will fall as s^{-1}. This is shown in Fig. 2 where we plot the $e^{+}e^{-}$ cross section versus collider center of mass (CM) energy. Thus we can say without much violence to the details that a successful collider of the future must be capable of a luminosity something like

$$L \geq 10^{33} E^2 (TeV) cm^{-2} s^{-1} \qquad (1)$$

$$L \equiv \text{Elementary Interaction Event Rate} / \sigma_{\text{elementary}}$$

Resources

While public support for this physics has been generous in the past and is clearly somewhat elastic, no one could argue that it is limitless. Thus a good part of the technical challenge will be in not only predicting the needed energy, but in reducing, by comparison with past efforts the unit energy and luminosity per dollar. Even with strong international collaborations, such as are likely to be the rule for the future, it is hard to imagine success at a cost exceeding 10^{10}\$. Likewise with energy usage in this increasingly energy conscious world. It seems highly unlikely that in the foreseeable future a facility consuming more than a few hundred megawatts in operation would be supportable. As we will see, this will be a major challenge.

Other   General   Requirements

The basic energy and luminosity requirements are relatively clear. There are others which are much less easy to define from our present viewpoint. The have to do with other aspects of using the accelerator for physics measurements and may be broadly classed as signal to noise ratio requirements. As such they involve not only the characteristics of the accelerator but also the robustness and selectivity of the detector(s). For example, while in hadronic colliders the elementary reactions of interest are characterized by s^{-1} dependent cross sections, the total cross-section continues to rise giving more and more unwanted reaction products to deal with as beam energies rise. Similarly in lepton machines the increasing radiation of electromagnetic and hadronic energy from the beam collision point as well as from upstream beam-matter collisions presents daunting problems. Before we can deal with these problems adequately in our designs, we will need the experience that only TeV colliders of the next generation can give. More on that presently.

Hadrons   and   Leptons

It would be possible to carry out a parallel analysis of hadron and lepton machines. The subject of proton accelerators beyond the SSC has been treated recently[3]. Since the focus of attention among those concerned with advanced accelerator concepts will likely, and rightly, be on linear lepton (photon) machines over the next few years, however, we shall concentrate on them here.

Particular Requirements for Lepton (Photon) Machines

For the sake of some definiteness, we will adopt here the single pass e^+e^- linear collider[4] as our paradigm. (It is, perhaps, possible that in the future conversion of the e beam energy to high energy photons could prove advantageous,[5,6] but such alternatives have their own severe complexity.) We begin by outlining the apparent constraints on single pass e^+e^- linear colliders in which the beams interact once and are spent. For simplicity, we will also assume, without much loss of generality, that the beams in the two arms have equal energies and equal currents with N particles per bunch with three dimensional Gaussian distributions and a macroscopic hitting frequency of f_c. The luminosity is then given by

$$L = \frac{N^2 R}{4\pi\sigma_x^2} f_c \cdot H_D(D_y) \qquad (2)$$

where σ_x is the *rms* horizontal beam size and R is $\sigma_x/\sigma_y \gg 1$. H_D is a luminosity enhancement factor representing the effect of the self pinching of the beams. For a number of reasons this enhancement will be rather small[7], less than a factor of two. Thus for clarity we will subsequently drop this factor.

From an engineering point of view this relation is highly constrained by our demand that L rise as γ^2 while at the same time maintaining desirable collision conditions. (γ is the beam lorentz factor in the lab.) These latter are affected by both single particle and coherent effects due to the large electromagnetic fields of the bunches at the interaction point(IP). The fields cause both self focusing of the beams and radiation called "beamstrahlung". Such effects in their totality are complex and their consequences are the subject of active study. Several excellent reviews exist[3,7,8]. Since the general features of these processes are relatively well known I will not repeat them here but merely employ the results derived in the references. These phenomena severely limit the useful luminosity that can be obtained from a single bunch and thus have important consequences for power consumption as the energy rises.

The parameters usually used to characterize the effects are D, Y, δ_E. D, the disruption parameter, measures the focusing of one beam by another. Formally it is the ratio of the rms beam length, σ_z, to the focal length of the beam-beam lens. In terms of beam parameters

$$D_y = \frac{2Nr_eR^2\sigma_z}{\gamma\sigma_x^2(1+R)} \tag{3}$$

D characterizes the stability of the beam bunches as they interpenetrate as well as that of streams of bunches as they pass near the IP. At the 1 TeV scale we will already be using the maximum allowed D it seems safe to say. Thus as we scale to higher energies we will be needing to prevent D from rising too.

Y is the primary factor governing the coherent radiation effects

$$Y \equiv \tfrac{2}{3}\frac{\hbar\omega_c}{E} \tag{4}$$

$\hbar\omega_c$ being the classically calculated critical energy of the synchrotron radiation emitted by a beam electron (positron) in the collective field of the oncoming beam. In terms of the commonly used beam parameters

$$Y_{av} \sim \frac{Nr_e\gamma R}{\alpha\sigma_z\sigma_x(1+R)} \quad \text{(Gaussian bunches)} \tag{5}$$

For small Y, radiation effects are relatively small, roughly speaking, and for Y large, they are large. For example, the number of $\gamma\gamma$ generated hadronic background events[9] per bunch crossing is proportional to n_γ^2 where

$$\eta_\gamma \cong \Gamma\frac{\sigma_z}{\gamma}Y_{av} \tag{6}$$

is the number of average energy photons emitted in a bunch-bunch collision. Γ is approximately constant. More dramatically, the number of coherently created pairs[7] is approximately

$$\eta_b \sim \tfrac{7}{128}\left[\frac{\alpha\sigma_z}{\gamma\lambda_e}Y\right]^2 e^{-16/3Y} \qquad Y<1 \tag{7}$$

The exponential factor changes by 9 orders of magnitude in the range $0.2 < Y < 1$, completely overwhelming other dependenceies.
Once a target energy and physics menu is selected, the final accelerator design must result from an optimization of the accelerator, final focus and detector considered together as a system. In these circumstances we may be permitted to vary Y and D somewhat to squeeze out the last bit of luminosity. For now, however, we are trying to look at the broad consequences of the push to energies well beyond our present horizons. For this purpose we must recognize the importance of background control

and we shall assume, accordingly, that we must hold Y approximately constant as E rises.

$\delta_E \equiv \langle \frac{\Delta E}{E} \rangle$ measures the average loss of energy by electrons (positrons) as they pass through opposing bunch at the IP. In terms of beam parameters

$$\delta_E \cong \Delta \frac{\sigma_z}{\gamma} Y_{av} \qquad (8)$$

where Δ changes little in $0.1 < Y < 1$. δ_E is only a rough average measure of the beamstrahlung effect. To get a better idea of what it means it is useful to look at the luminosity energy spectrum resulting from the action of beamstrahlung. Fig 3, adapted from ref.(8) shows this for a particular 1TeV+1TeV collider now being considered. Fig. 4, also adapted from ref.(8) gives a different view. It displays the relation between δ_E and the fraction of the interacting beam possessing energy fractions $x > x_{min}$.

There are three important consequences of this energy spreading. First, the machine is built to study interactions at its top energy. As shown in Figs. 3 and 4, the radiation can reduce significantly the effective luminosity at that top energy. Second, since the cross-section is rising rapidly as energy goes down, the low energy part of the luminosity will generate a disproportionate part of the events entering the detector. Being essentially uninteresting, they are unwanted background. Third, a great strength of e^+e^- colliders has been the so called beam constraint. We know that the energies of the particles in a particular event must add up to the beam energy since the primary particles are elementary. This is very useful in identifying wanted events, measuring energies of escaping neutrals and in suppressing some backgrounds. If the beam energy is spread significantly by beamstrahlung, this total energy constraint cannot be used. For some kinds of physics, of course, this may not be so important.

Consistent with our conservative approach, it would seem that limiting δ_E is also necessary as the energy rises.

These considerations indicate that, at least in the standard e^+e^-, single pass, linear collider adopted, controlling Y is imperative. We can express this constraint by combining the luminosity expression (2) with that for Y_{av} eliminating N. We obtain

$$L = KY_{av}^2 \frac{\sigma_z^2 R f_c}{\gamma^2} \text{ for } R \gg 1 \tag{9}$$

where K is a fundamental constant. As we must engineer the accelerator such that

$$L = L\gamma^2 \tag{10}$$

the net result is that the only "free" variables left, σ_z, R and f_c, must satisfy the relation

$$f_c R \sigma_z^2 = \frac{L}{KY_{av}^2}\gamma^4 \tag{11}$$

Replacing N in D_y and replacing σ_x by

$$\sqrt{\frac{\beta_x^* \varepsilon_N}{\gamma}} \tag{12}$$

where β_x^* is the horizontal focusing parameter at the IP and ε_N is the normalized emittance.

We find

$$D_y = \frac{2\alpha Y_{av}}{r_e} \frac{1}{\sqrt{\beta_x^* \varepsilon_N}} \cdot \frac{R\sigma_z^2}{\gamma^{3/2}} \tag{13}$$

Evidently, if we are able to keep Y constant, we should not have much difficulty in limiting D, δ_E and n_γ.

While the constraint of constant Y_{av}, equation (11), appears to be formidable, it may not be insurmountable. Although increasing σ_z is unlikely to be feasible[†] there may be some room for increasing R. It is not impossible to imagine, however, a very large increase in f_c. Even the highest f_c machines being discussed today for the 1 TeV regime have $f_c < 10$ kHz. In principle we could make $f_c \sim f_{rf}$ if we had an efficient enough accelerator. For an f_{rf} in the GHz range we could therefore get an enhancement of 10^5 more which could nicely support a factor of 10 more in energy.

Before concluding that all is well, however, we need to examine the implied beam power

---

[†] σ_z must be kept less than β_y^*, a parameter we shall be at some pains to decrease (or at least hold constant) as E rises. In Addition σ_z is constrained to be a small fractin of λ_{rf} to control energy spread out of the accelerator. Increasing λ_{rf} is unlikely to be practical.

$$P_b = 2Nef_cE \qquad (14)$$

Replacing N and f_c by the radiation constrained values above

$$P_b = A\frac{\sqrt{\beta_x^*\varepsilon_N}}{Y_{av}} \cdot \frac{\gamma^{1/2}}{R\sigma_z} \qquad (15)$$

A is a constant. Today, several concept designs for 0.5 to 2 TeV (CM) are under study[3,8]. Even those boasting the most economical use of beam power to produce luminosity have beam powers of more than 3 MW for 10^{33}cm^{-2}s^{-1} luminosity. If $R\sigma_z$ and ε_N are kept at values posited for 1 TeV regime linear colliders than were facing a beam power multiplication of 3000 for each tenfold energy increase according to this rule.

Even if we were clever enough to squeeze a factor of 10 out of the combination $\sqrt{\beta_x^*\varepsilon_N}/R$ the beam power for a tenfold energy increase would be unacceptable since by our paradigm the power is not recovered from the beam, but spent as unused heat.

Evidently then, the constraints derived from radiation considerations make is seem unlikely that the standard approach will suffice for energies much beyond 1 TeV.

POSSIBLE ALTERNATIVES

There appear to be three general paths which might enable us to avoid excessive input power. First, we might accept the rise in beam power by learning to recover it, keeping the total input power constant. Second, we might seek ways to mask out the effects of the coherent pairs while still controlling the energy spread degradation. Thirdly, we might seek to limit the radiation effects by neutralization. Of course we could also reduce radiation dramatically by accelerating muons instead of electrons but no practical means for producing muons with adequate intensity and brightness are known.

Energy Recovery

If the effects of disruption and energy spread degradation can be controlled then the beam will still be quite bright and stiff after interaction. Thus we might recover the beam power by reversing the phase of the wave particle interaction to extract monochromatic rf energy from the beam and feed back that energy for acceleration of new beam. One might even imagine slowing down the high energy beam in the same medium used to accelerate the oncoming beam[10]. This scheme presents major challenges. First, the acceleration-deceleration process must be very efficient. Second, the counter streaming beams in this simple version would disrupt themselves to destruction via the beam-beam interaction. Separate structures for acceleration and deceleration would thus be required, nearly doubling the cost of the facility. In addition, it must be noted that background control will still be a major problem due to the rising beam currents and the associated beam-gas and beam-wall interactions upstream. Nevertheless, it would seem that further study of this recovery option is merited.

Perhaps one should note parenthetically that even though the positron current would be rising with energy in this scheme, recent suggestions[11] for generating positrons from high energy electrons(positrons) combined perhaps with some particle recovery as well as energy recovery make the particle source problem soluble.

Coherent Pair Masking

Perhaps control of the unwanted electromagnetic background by masks can permit detection of wanted events in the face of large number of mostly forward going pairs. This technique is already under intense study for TeV regime colliders. Some real experience with this regime may show is how to accomplish the needed background reduction.

Neutralization

It has been pointed out many times that if the beam fields could be neutralized of the IP then the effects of beamstrahlung would be avoided. One possibility is to use a plasma[12] of just the right density at the IP. This scheme has at least two drawbacks. Since the density of plasma nuclei at the IP is at least of the order of the beam density the beam-nucleus interactions will produce intense backgrounds. In addition the wanted densities are likely to exceed practical limits for plasmas. Another possibility is the use of four beams in the collision, two counter streaming e^+ beams and two counter streaming e^- beams. While this idea was suggested many years ago[13], it has never been successfully been reduced to practice. Stability requirements[14] are severe in terms of the accuracy of the local charge neutralization and spatial overlap· In addition the extra beams mean doubling the accelerating medium and power supply.   Despite the technical difficulties involved the potential benefits of this scheme make its further study an important task.

Power   Constrained   Operation

If we were completely successful in the masking or neutralization approaches then we would be free from the radiation and pinching constraints.   There are then other constraints to face, particularly the constant power constraint.

Putting together the luminosity and power equations, (2) and (14) by eliminating N between them and by constraining luminosity to vary as γ^2 we find

$$P = B\sigma_x\sqrt{\frac{f_c}{R}}\,\gamma^2 \qquad (16)$$

where B is a constant.   If P is constrained to be a constant then we have the relation

$$\sigma_x^2\frac{f_c}{R} = \frac{const.}{\gamma^4} \qquad (17)$$

Before one can draw further conclusions one must have a model for the accelerating medium so that its constraints can be added. If we think of a conventional slow wave structure then our freedom to decrease f_c will be sharply limited by wakefield effects, leaving us with only σ_x and R to manipulate.   Supposing that we were clever enough to increase R another factor of ten beyond what will be achieved at 1 TeV.   We're then left with engineering a reduction in σ_x of a factor of 33 for each factor of ten gain in energy.   This ratio is not too much greater than the proposed horizontal beam size ratio of SLC and TeV range machines.   Thus by making further improvements in emittance and beam focusing technology, say channeling focusing with engineered crystals such improvements are not unthinkable.

CONCLUSION

We have seen that to be successful in our push to higher energies we are hemmed in on every side.   While increasing the energy we must raise the luminosity as the square, we have to overcome the radiation effects which tend to force us to higher beam power and, in addition, we have to hold the power constant. Evidently the standard model of the single pass, e^+e^-, linear collider is unable to meet these simultaneous requirements beyond the 1 TeV regime.   Some new paradigm which can meet the challenge awaits definition and development.

REFERENCES

1. Proceedings of the Workshop on Possibilities and Limitations of Accelerators and Detectors, FNAL, April 1979
2. Proceedings of the 2nd ICFA Workshop on Possibilities and Limitations of Accelerators and Detectors, CERN, June 1980
3. M. Tigner in Physics of the 100 GeV Mass Scale, SLAC 361, 1989
4. See for example R. Palmer in Annual Reviews of Nuclear and Particle Science 1990.40
5. V. Telnov, Nuclear Instruments and Methods, A294(1990)
6. P. Chen and V. Telnov, PRL 63(1989)
7. K. Yokoya and P. Chen, KEK Report 91-2, KEK Apr. 1991
8. K. Berkelman, CLIC Note 154, CERN, Feb. 1992
9. K. Berkelman, CLIC Note 164, CERN, May 1992
10. M. Tigner, Nuovo Cimento 37, 1965
11. See for example J. Rossbach, DESY 90-169, DESY, Dec. 1990
12. D. Whittum, A. Sessler and S. Yu, LBL preprint 25759, LBL July 1988
13. V. Balakin and A. Skrinsky, Op. Cit. 2

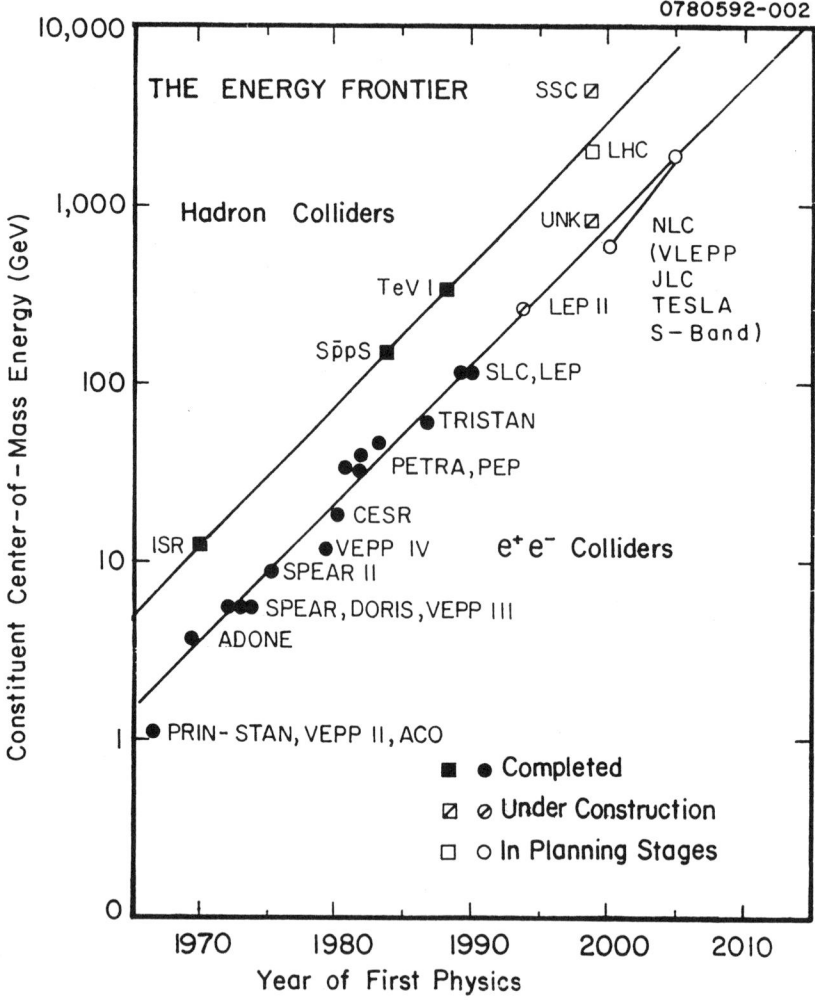

Fig. 1   The Livingston Chart for Colliders

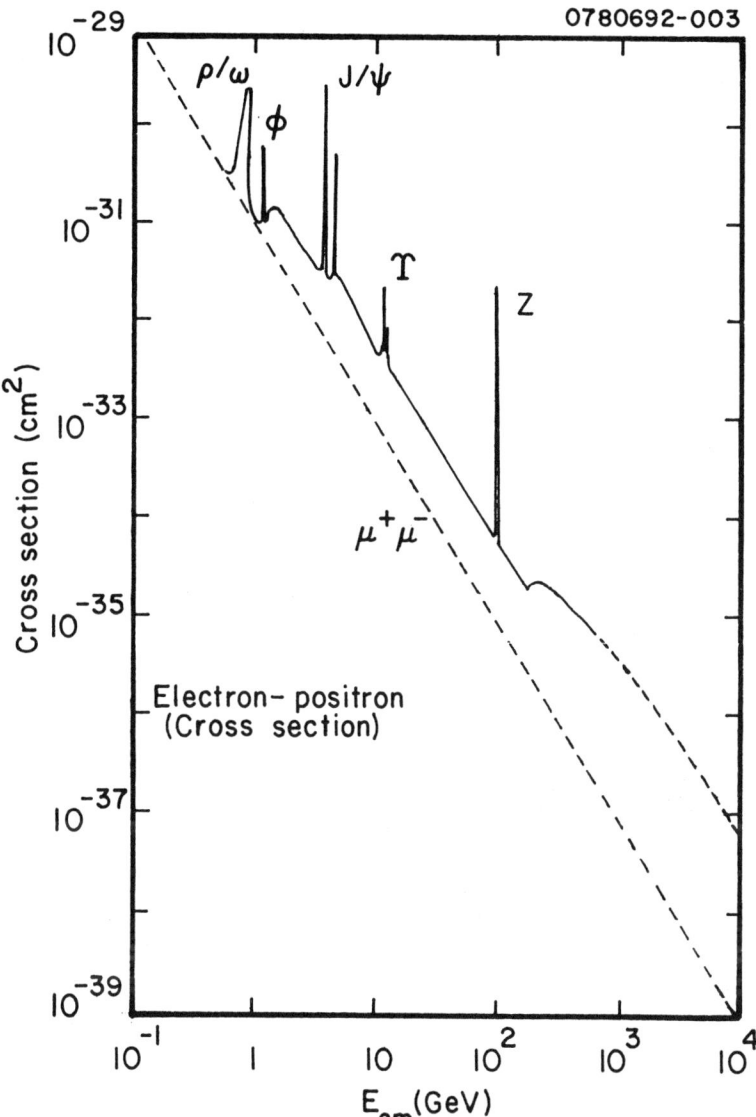

0780692-003

Fig. 2 The e⁺e⁻ annihilation cross section as a function of CM energy.

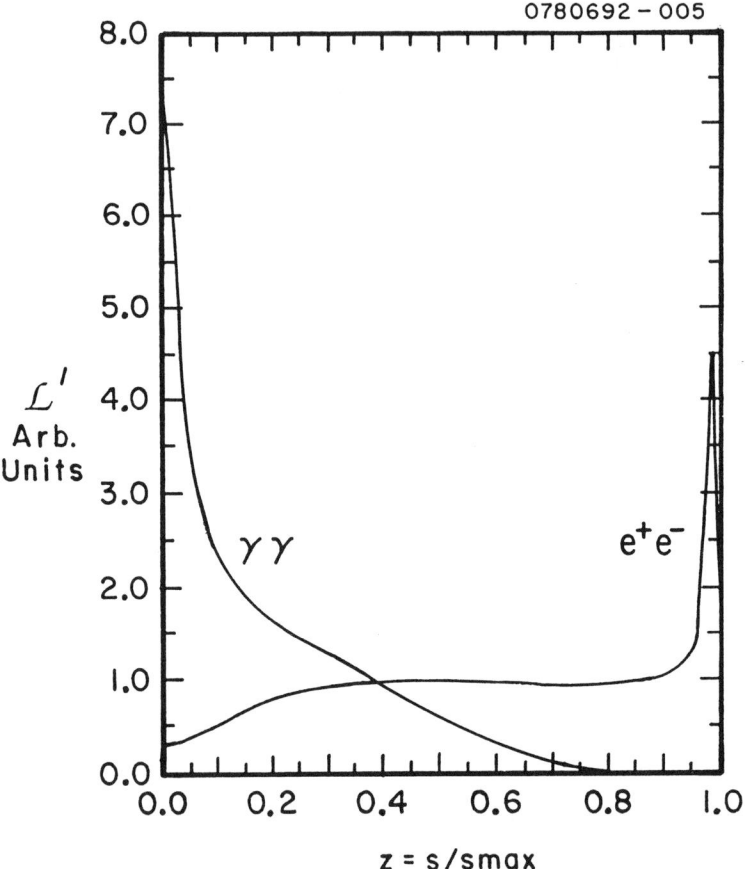

Fig. 3

Luminosity spectrum for a particular 2 TeV CM collider now under discussion.   The integral under the e⁺e⁻ curve is the luminosity calculated from the simple formula.   The curve marked γγ gives the two photon luminosity per energy   interval resulting   from   beamstrahlung.

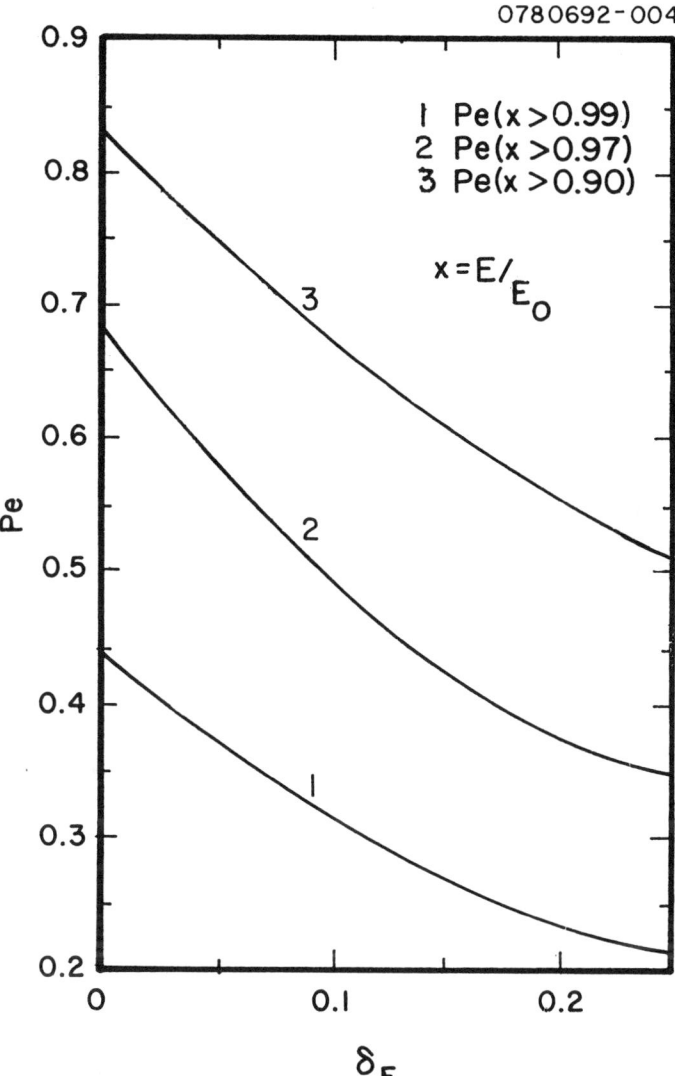

Fig. 4

The effect of δ_E on the width of the beam energy spectrum near the peak energy.   For example if we have a 10% beamstrahlung beamstrahlung parameter, δ_E, then the fraction of the beam particles having greater than 99% of the design energy is about 0.3.

SUMMARY REPORT: HIGH INTENSITY EM WAVES

V.L. Granatstein and C.D. Striffler
Laboratory for Plasma Research
University of Maryland, College Park, MD 20742

ABSTRACT

This is the Summary Report of the "High Intensity EM Waves" working group at the 3rd International Workshop on Advanced Accelerator Concepts held June 14-20, 1992 in Port Jefferson, Long Island, New York. The requirements for future linear colliders with regard to microwave amplifiers is delineated in Section I along with key parameters that allow for the determination of the present status of each research effort that was discussed at the conference. The present status of amplifiers with beam energies less than 800 keV are summarized in Section II. In Section III, we summarize the status of the more non-traditional schemes associated with the two-beam concept, in which the energy of the microwave-generating electron beam is much greater than 1 MeV.

I. INTRODUCTION

Future linear colliders will have a center-of-mass energy on the order of 1 TeV. To keep the accelerator length within acceptable bounds it has been proposed by researchers at the Stanford Linear Accelerator (SLAC)[1] that the accelerating gradient be on the order of 100 MV/m (compared with 17 MV/m in the presently operating Stanford Linear Collider).

To appreciate what demands will be made of the microwave amplifiers for these future colliders, it is useful to consider the relationship between peak microwave power required per unit accelerator length, p, the accelerator gradient, E_a, and the microwave wavelength, λ. Perry Wilson[2] has recently presented the result that for an accelerator structure consisting of either a chain of pillbox TM_{010} resonators or a disc-loaded, traveling-wave circuit with aperture radius $a = 0.175\ \lambda$, the microwave power per unit length is given by

$$p \approx 1.2 \times 10^{-7} E_a^2 \lambda^{1/2}. \tag{1}$$

(Throughout this paper, mks units are used unless otherwise noted so that p is in watts/m, E_a is in volts/m, and λ is in meters.)

The structure fill-time is given by[2]

$$t_f \approx 2.3 \times 10^{-5} \lambda^{3/2}. \tag{2}$$

Thus, the required microwave pulse energy per unit length is

$$u = p t_f \approx 2.8 \times 10^{-12} E_a^2 \lambda^2. \tag{3}$$

Then, a single microwave amplifier with peak output power P_p and pulse duration $\tau_p \geq t_f$ would be able to drive a length of accelerator structure

$$\ell_1 = \frac{P_p \tau_p \eta_c}{u} \approx 3.6 \times 10^{11} \frac{P_p \tau_p \eta_c}{E_a^2 \lambda^2}, \tag{4}$$

where we have used Eq. (3), and η_c is the efficiency of any pulse compression circuit that is used. If we estimate that each factor of 2 in pulse compression can be achieved with 90% efficiency,[3] then

$$\eta_c = 0.9^{\log_2(\tau_p/t_f)}. \tag{5}$$

As a practical limit, the useful pulse compression ratio $\tau_p/t_f \leq 16$.

It is interesting at this point to consider what value of E_a should be chosen to minimize the cost of an accelerator with a given final energy, U_f. The required overall accelerator length is given by

$$L = \frac{U_f}{e E_a} \tag{6}$$

while the total number of microwave tubes required, N_t, is obtained from Eqs. (4) and (6) as

$$N_t = \frac{L}{\ell_1} \approx 1.7 \times 10^7 \frac{U_f E_a \lambda^2}{P_p \tau_p \eta_c}. \tag{7}$$

Accelerator cost will increase both with the length of the required tunnel, L, and with the number of microwave tubes N_t. However, since $L \sim E_a^{-1}$ and $N_t \sim E_a$, the choice of E_a is not unambiguous, and involves a complicated analysis of such factors as tunnel costs vs. microwave tube manufacturing costs. For the purposes of this paper we will follow the lead of SLAC[1] and assume that a maximum acceptable value of L is about 10 km corresponding to $E_a = 100$ MV/m for a collider with $U_f = 1$ TeV.

Once E_a is chosen for a collider, Eq (4) together with Eqs. (2) and (5) may be used to evaluate ℓ_1, the length of accelerator that may be driven by a given microwave amplifier. The length ℓ_1 is a key parameter in evaluating the suitability of a specific microwave amplifier for driving the collider. As may be seen from Eq. (7), the number of microwave tubes required to achieve a given energy of the colliding particles will be inversely proportional to ℓ_1. For example, consider a 1 TeV collider with $E_a = 100$ MV/m; if $\ell_1 = 1$ m, then 10,000 microwave amplifiers would be required, whereas if $\ell_1 = 10$ m, only 1000 amplifiers would be necessary. The cost of the collider is strongly dependent on the number of microwave amplifiers. Each amplifier with its associated modulator, magnet and pulse compression circuit will cost hundreds of thousands of dollars. Thus, 10,000 amplifiers would be unacceptably expensive, and microwave amplifiers need to be developed with $\ell_1 \gg 1$m if TeV colliders with $E_a = 100$ MV/m are ever going to be built.

Note from Eq. (4) that ℓ_1 could be increased by choosing a higher microwave frequency if $P_p\tau_p$ decreased less rapidly than λ^2. In addition, high frequency has the advantages of increased limiting values of E_a as determined by rf breakdown,[4,5] smaller dark current at high values of E_a, and increased pulse repetition frequency[5] which diminishes problems caused by ground jitter. However, there is a practical upper limit on frequency that is currently estimated to be in the neighborhood of 35 GHz. At higher frequency, the difficulty of aligning the accelerator and/or deleterious wakefield effects might become unacceptable.

A number of participants in the 1992 Advanced Accelerator Workshop presented results and designs for microwave amplifiers intended for application to linear colliders; their contributions will be summarized below in part II. In addition to the ℓ_1 parameter, other microwave amplifier parameters are also of great importance in determining the cost of the rf driving systems for a collider. Salient among these parameters are amplifier efficiency and amplifier driving voltage. Generally, amplifier efficiency $\gtrsim 50\%$ is hoped for. It is the present 'consensus' that amplifier voltage should be less than 1 MV so that "standard" soft-tube-modulator/pulse transformer technology (with pulse length $\sim 1\mu\text{sec}$) could be used. At voltages above 1 MV, more exotic pulse power supplies such as induction linacs might be required and might be excessively costly, although this needs to be more carefully evaluated. For soft-tube modulators with pulse duration $\sim 1\mu\text{sec}$, the smaller the amplifier voltage, the less expensive will be both the modulators and the electron guns in the amplifier tube. However, at low voltages, pulse compression schemes are likely required and their cost and efficiency must be included in the overall system cost.

In addition to developing microwave amplifiers with $V \lesssim 1$ MV for driving colliders, the concept of two beam accelerators is also under study. In this concept, a relatively low energy, high current electron beam propagates alongside the main accelerator and continuously generates microwaves to be fed into the main accelerator structure. The beam generating the microwaves will likely be at an energy $\gg 1$ MeV; i.e., much larger than the beam energy in the individual microwave amplifiers discussed above. The status of two-beam accelerator research is described below in part III.

II. MICROWAVE AMPLIFIER RESULTS AND DESIGNS ($V \leq 800$ kV)

Status of currently operating microwave amplifiers

Experimental testing of only two new types of microwave amplifier with pulse duration equal to or longer than the structure fill-time (see Eq. (2)) were presented. These are the gyroklystron being studied at the University of Maryland[6] and the Free Electron Laser being studied at MIT.[7] Both types of amplifier are based on interaction of a spiralling or wiggling electron beam with fast electromagnetic waves ($v_{ph} > c$); and thus do not require slow wave periodic circuits or re-entrant cavities with small gaps. They therefore have the potential of utilizing circuits which have large transverse dimensions compared with the wavelength, and thus in

Table 1: Demonstrated amplifier performance ($\tau_p \geq t_f$); ℓ_1 is the length of accelerator which could be driven by a single amplifier at $E_a = 100$ MV/m.

Type of Amplifier/ (Research Institution)	f (GHz)	λ (cm)	P_p (MW)	τ_p (μs)	ℓ_1 (m)	V (kV)	η_A (%)	$\eta_A\eta_c$ (%)
SLC Klystron (SLAC)	2.856	10.5	65	3.5	0.59	350	45	36
X-Band Klystron (SLAC)	11.4	2.63	50	1.0	1.8	447	22	15
Gyroklystron (UMD)	9.85	3.05	27	1.4	1.0	425	32	22
FEL (MIT)	33	0.91	61	0.02	0.53	750	27	27
EIK (SRL/Haimson/MIT)	11.4	2.73	100	0.05	0.24	440	43	43
TWT (Cornell U.)	8.76	3.42	200	0.1	0.62	800	24	24

principle could be scaled to high power at high frequency. It should be noted that the FEL experiment used an emittance selector, and the amplifier efficiency figure, $\eta_A = 27\%$ is based on injected current only; based on total current, $\eta_A \sim 1\%$. The FEL experiment should therefore be regarded as preliminary with considerable research on beam formation required before an amplifier of 'engineering' interest could be demonstrated.

Experimental results on two advanced slow wave amplifiers were also presented; i.e. an Extended Interaction Klystron (EIK),[8] and a two section traveling wave tube (TWT).[9] The results were promising, although in both cases the pulse duration as limited by the available modulator fell somewhat short of the structure fill time given by Eq. (2).

In Table 1, the experimental results of the gyroklystron, FEL, EIK and TWT are displayed. For the purpose of comparison, the performance parameters of the S-band SLC klystron and the current results on an 11.4 GHz experimental SLAC klystron[10] are also listed in Table 1.

It should be noted that with both the X-band klystron and the gyroklystron progress has been made in extending the ℓ_1 parameter over what is possible with the SLC production klystrons. Presumably the FEL, EIK and TWT could also be made to operate with a longer pulse and an enhanced value of ℓ_1.

It should be appreciated that values of ℓ_1 in the range between 1 and 2 while an improvement in the state-of-the-art are not sufficient for driving a TeV collider, and we survey below the various amplifier types which have been proposed with much larger ℓ_1 values. We also note that none of the higher frequency amplifiers have yet equalled the SLC klystron in efficiency and improvement in this sense is generally called for.

phase bunching and power extraction as in the gyrotron; however, operation at a microwave frequency much larger than the electron cyclotron frequency is achieved via a Doppler shift. CARM amplifier experiments at 17.1 GHz are underway at MIT utilizing a 500 kV, 500 A, 50 ns electron beam. This project follows the study of a 35 GHz CARM amplifier in which $\sim$ 6% efficiency was realized.[15]

A preliminary design of a gyroresonance harmonic converter[16] was presented by Omega-P, Inc. It would use, for example, the output of a 65 MW, 2.856 GHz SLAC klystron to accelerate a 20-50 A electron beam from an initial energy of under 100 keV up to 1-3 MeV. This beam would then radiate coherently into a uniform cylindrical waveguide in a tapered magnetic field at a chosen harmonic of 2.856 GHz. Conversion efficiency for the fifth harmonic (14.3 GHz) has been calculated to be as high as 70%. The residual beam energy in this proposed device is expected to be efficiently recoverable using a single-stage depressed collector, since the spent beam is predicted to be nearly monoenergetic and phase coherent.

III. TWO-BEAM ACCELERATORS

Microwave generation with electron beams having energy $\gg 1$ MeV is being considered in two-beam accelerator schemes.[17] There are two major research efforts along these lines, one in California at Lawrence Livermore Laboratory in collaboration with the Lawrence Berkeley Laboratory and SLAC, and one at CERN. The group in California contemplates using induction linear accelerators for initial acceleration and re-acceleration of the microwave generating electron beam, while the CERN group will accelerate the electrons in rf superconducting cavities.

Two-beam accelerators based on induction linacs

Based in part on the successful demonstration of a 1 GW, 35 GHz, 40% efficient free electron laser[18] driven by the ETA induction linac, a two-beam accelerator scheme in which the microwaves would be generated by the FEL mechanism has been formulated.[19,20] The design presented at the 1992 Advanced Accelerator Workshop is based on free electron lasers operating at 17.1 GHz driven by a 12 MeV electron beam that will provide 200 MW/m to the collider accelerator.

The possibility of using the relativistic klystron[21] mechanism (probably in an extended interaction output circuit) to generate the microwaves in an induction linac based two-beam accelerator scheme is also under study. An output of 290 MW has been achieved using longitudinal bunching.[21] The design presented would use a 10 MeV electron beam to generate 500 MW/m in 11.4 GHz, 100 ns radiation pulses. A series of experiments employing transverse chopping has achieved an output power of 124 MW over a 30 nsec pulse. The stability of the power corresponded to an amplitude variation of less than 2% and a phase variation of less than 2°.[22,23] An ongoing experiment with a 5 MeV electron beam aims to generate 400 MW, 40 ns output power pulses at 11.4 GHz. Reacceleration experiments are also underway to demonstrate adding 250 keV of energy per induction cell. Beam instabilities and spurious mode control in an extended interaction output circuit are being studied using a prebunched (chopped) electron beam.

Two-beam accelerators based on superconducting rf cavities

A more radical design of a two-beam accelerator is embodied in the CERN linear collider (CLIC).[24,25] Prebunched electrons will be pre-accelerated to an energy of 3 to 6 GeV depending on the final main beam energy (250 GeV to 1 TeV) and re-accelerated as required in 350 MHz rf superconducting cavities. These electron bunches will then generate 30 GHz, 11 ns microwave pulses in a traveling wave circuit. The power transferred to the collider beam will be 140 MW/m to generate a gradient of E_a =80 MV/m.

Currently, experiments in a CLIC Test Facility are underway using a 60 MeV electron beam with the aim of generating a peak power of 60 MW at the exit of the transfer structure and 32 MW, 11 ns power pulses in a main accelerating section. This will be used to produce a gradient of 80 MV/m in a prototype of the collider structure which has been fabricated. Model work on the RF transfer structures and theoretical studies of beam dynamics in the drive linac are also underway.

AKNOWLEDGEMENTS

The authors are grateful to their colleagues who were kind enough to provide important comments on this manuscript in its draft form; they are as follows: B.G. Danly, G. Guignard, B. Hafizi, J. Hirshfield, P.E. Latham, J. Nation, A. Sessler, R. Temkin, G. Westenskow, and D. Yu.

REFERENCES

1. e.g., R.D. Ruth, "Progress on Next Generation Linear Colliders," AIP Conference Proceedings 184, ed. M. Month and M. Dienes, 1989, pp. 2209-2226.

2. P.B. Wilson, "Pulsed RF Technology for Future Linear Colliders, invited talk presented at the Washington, DC meeting of the American Physical Society, April 21, 1992.

3. Z.D. Farkas, G. Spalek, and P.B. Wilson, "RF Pulse Compression Experiment at SLAC," SLAC-PUB-4911, March 1989.

4. W.D. Kilpatrick, Rev. Sci. Instru. **28**, 824 (1957).

5. V.L. Granatstein and A. Mondelli, "Microwave Sources and Parameter Scaling for High Frequency Linacs," in Phys. Particle Accelerators, ed. M. Month and M. Dienes, AIP Conference Proceeding 153, New York, 1987, pp. 1506-1571.

6. W. Lawson, J.P. Calame, B. Hogan, P.E. Latham, M.E. Read, V.L. Granatstein, M. Reiser, and C.D. Striffler, Phys. Rev. Lett. **67**, 520 (1991); also, S.G. Tantawi, W.T. Main, P.E. Latham, G.S. Nusinovich, W. Lawson, C.D. Striffler, and V.L. Granatstein, IEEE Trans. Plasma Sci. **20**, 205 (1992).

7. M.E. Conde and G. Bekefi, Phys. Rev. Lett. **67**, 3082 (1991).

8. D.L. Goodman, D.L. Birx, and B.G. Danly, "Induction Linac Driven Relativistic Klystron and Cyclotron Autoresonance Maser Experiment," SPIE Vol. 1407, p. 217-225 (1991); J. Haimson and B. Mecklenburg, "Use of TW Output Structures for High Peak Power RF Generators," Proc. Linac Conf., Albuquerque, NM, 1990; J. Haimson and B. Mecklenburg, "Suppression of Beam Induced Pulse Shortening Modes in High Power RF Generator TW Output Structures," Proc. SPIE Symp. on Intense Microwave and Particle Beams II, January 1992.

9. D. Shiffler, J.A. Nation, L. Schacter, J.D. Ivers, and G.S. Kerslick, "A High Power, Two Stage, TWT AMplifier," J. Appl. Phys. **70**, 106 (1991), and L. Schacter, J.A. Nation, and D. Shiffler, "Theoretical Studies of High Power Cerenkov Amplifiers," J. Appl. Phys. **70**, 114 (1991).

10. G. Caryotakis, "Progress Report on SLAC NLC Klystrons," talk presented at the 1992 ECFA Workshop on e^+e^- Linear Colliders, Garmish-Partenkirchen, Germany, 25 July - 2 August 1992.

11. L. Schachter, T.J. Davis, and J.A. Nation, "A Proposed Extended Cavity for Co-axial Relativistic Klystrons," Proc. 9th Int. Conf. on High Power Particle Beams, Washington, DC, May 25-29, 1992 (to be published).

12. D.U.L. Yu, J.S. Kim, and P.B. Wilson, "Design of a High-Power Sheet Beam Klystron," Proc. of this Conf.

13. R.B. Palmer, W.B. Herrmansfeldt, and K.R. Eppley, "An Immersed Field Cluster Klystron," Proc. Int. Conf. on High Energy Acceleration, Tsukuba, Japan, August 22-26, 1989; SLAC-PUB-5026 (1989).

14. W.M. Manheimer, IEEE Trans. Plasma Sci. **18**, 632 (1990), and B. Hafizi, Y. Seo, S.H. Gold, W.M. Manheimer, and P. Sprangle, IEEE Trans. Plasma Sci. **20**, 232 (1992).

15. G. Bekefi, A.C. DiRienzo, C. Leibovitch, and B.G. Danly, Appl. Phys. Lett. **54**, 1302 (1990); B.G. Danly, F.V. Hartemann, T.S. Chu, D. Legorburu, W.L. Menninger, R.J. Temkin, G. Faillon, and G. Mourier, "Long-Pulse Millimeter-Wave FEL and Cyclotron Autoresonance Maser Experiments," Phys. Fluids **B4**, 2307 (1992).

16. J.L. Hirshfield, Phys. Rev. A **44**, 645 (1991).

17. A.M. Sessler in *Laser Acceleration of Particles*, ed. P.J. Channel, AIP Conf. Proc. **91**, 154 (1982).

18. T.J. Orzechowski, B.R. Anderson, J.C. Clark, W.M. Fawley, A.C. Paul, D. Prosnitz, E.T. Scharlemann, S.M. Yarema, D.B. Hopkins, A.M. Sessler, and J.S. Wurtele, Phys. Rev. Lett. **57**, 2172 (1986).

19. R.A. Jong, R.D. Ryne, G.A. Westenskow, and S.S. Yu, "A 17.1 GHz Free-Electron Laser as a Microwave Source for TeV Colliders," Proc. 11th Int. FEL Conf., Naples, FL, August 28-September 1, 1989, Nucl. Instrum. and Meth. in Phys. Res. A296, 776 (1990).

20. A.M. Sessler, D.H. Whittum, J.S. Wurtele, W.M. Sharp and M.A. Makowski, "Standing-Wave Free-Electron Laser Two-Beam Accelerator," Nucl. Instr. & Meth. in Phys. Res. **A306**, 592 (1991).

21. G.A. Westenskow, D.P. Aalberto, M.A. Allen, J.K. Boyd, R.S. Callin, G.A. Deis, H. Deruyter, K.R. Eppley, K.S. Fant, W.R. Fowkes, J. Haimson, H.A. Hoag, D.B. Hopkins, T.L. Houck, R.F. Koontz, T.L. Lavine, G.A. Loew, B. Mecklenburg, R.H. Miller, T.J. Orzechowski, R.D. Ruth, R.D. Ryne, A.M. Sessler, A.E. Vlieks, J.W. Wang, and S.S. Yu, "Relativistic Klystrons for High-Gradient Accelerators," Proc. 1990 Linear Accelerator Conference, Albuquerque, NM, September 10-14, 1990, pp. 192-194.

22. J. Haimson and B. Mecklenburg, "Design and Construction of a Chopper Driven 11.4 GHz Traveling Wave RF Generator," Proc. 1989 IEEE Particle Accelerator Conference, pp. 243-245.

23. T.L. Houck and G.A. Westenskow, "Status of the Choppertron Experiment," 16th Int. Linear Conf., Ottawa, Ontario, Canada, August 1992.

24. W. Schnell, "Status of CLIC," Proc. 1988 Linear Accelerator Conference, Williamsburg, VA, Oct. 3-7, 1988.

25. G. Guignard, "Status of CERN Linear Collider Studies," Proc. 1990, Linear Accelerator Conference, Albuquerque, NM, September 10-14, 1990, pp. 8-12.

EXPERIMENTAL GYROKLYSTRON RESEARCH AT THE UNIVERSITY OF MARYLAND FOR APPLICATION TO TEV LINEAR COLLIDERS

W. Lawson, V. L. Granatstein, B. Hogan, U.-V. Koc, P. E. Latham, W. Main,
H. W. Matthews, G. S. Nusinovich, M. Reiser, C. D. Striffler, and S. Tantawi
Laboratory for Plasma Research and Department of Electrical Engineering
University of Maryland, College Park, MD 20742 USA

ABSTRACT

X-Band and K-Band gyroklystrons are being evaluated for possible application to future linear colliders. So far we have examined ten different two- and three-cavity configurations. We have achieved a maximum peak power of 27 MW in ~1 μs pulses at a gain of 36 dB and an efficiency exceeding 32%. The nominal parameters include a 430 kV, 150-200 A beam with an average perpendicular to parallel velocity ratio near one. In this paper, we detail our progress to date and describe our plans for future experiments that should culminate in amplifier outputs in excess of 100 MW in 1 μs pulses.

INTRODUCTION

An international effort is underway to develop amplifiers in the 10 - 30 GHz range with peak powers in excess of 100 MW for driving future multi-TeV electron positron colliders [1]. At the University of Maryland, we are concentrating on the use of gyrotron amplifiers to achieve the required parameters. At the time our project first got underway, the state-of-the-art power for gyroklystrons hovered near 50 kW due in part to limitations imposed by spurious oscillations [2]. As an intermediate step toward the 100+ MW goal, we constructed and tested a system which was designed to surpass the 50 kW level by a factor of 500 [3]. We have achieved our goal [4], in part by increasing the power capability of the standard gyrotron electron gun [5] and by developing improved microwave absorbers to enhance stability [6]. This paper represents a summary of experimental work to date [4], [7]-[12].

A schematic of our experimental facility is shown in Fig. 1. A 1 - 2 μs, 500 kV, 400 A line-type modulator provides the required beam power with a maximum repetition rate of 4 Hz. A resistive divider shunts half the current and provides the intermediate voltage required for the double-anode magnetron injection gun (MIG). The MIG was designed to have a space-charge limiting current of 400 A and an axial velocity spread under 7% at the nominal current of 160 A with an average perpendicular to parallel velocity ratio of $\alpha = 1.5$. However, because of instabilities we have not achieved an α greater than about 1.2. Considerable flexibility in the magnetic field profile is achieved by using four independent supplies to power eight water-cooled pancake coils. The design magnetic field is 0.047 T at the cathode and 0.565 T $\pm$ 0.005 T for 25 cm in the circuit region, but the magnetic field can be varied along the axis to optimize performance.

An enlargement of the circuit region for the last two-cavity tube is shown in Fig. 2. The downtaper is lined with lossy dielectrics, which are indicated in black. The tapered ceramics are a non-porous mix of 20% SiC in 80% BeO. The other ceramics are made in-house from carbon-impregnated alumino-silicate. Most of the drift tube is also lined with lossy ceramics to suppress instabilities. A partially self-consistent code which includes realistic rf field profiles [13] and AC space-charge effects [14] is used to design the microwave circuit dimensions and predict amplifier performance. Power is injected from a 2 μs, 100 kW magnetron through a slit in the radial wall into the input cavity. Control over the quality factor (Q) in the input cavity is obtained from losses in a thin ceramic ring placed against the sidewall. The output cavity Q is predominantly due to diffractive losses from the cavity's output lip. The Q factors have spanned the range 125 - 500 in the various tubes, but the resonant frequencies have always been derived from an $m = 0$ TE mode at 9.85 GHz.

The output waveguide is shown in Fig. 1. A short 2° linear taper is followed by a non-linear taper which brings the waveguide radius to a value suitable for the copper beam dump. A cross-guide magnetic field at the end of the dump prevents high-energy electrons from traveling through the second non-linear taper and impinging on the five inch diameter half-wavelength BeO output window.

Two types of microwave diagnostics can be attached to the output waveguide. An anechoic chamber uses an open-ended piece of X-band waveguide as a receiving antenna to estimate output power and mode purity via far field measurements. The horn can be rotated by 90° and remotely swept transverse to the z-axis over one meter. The signal is split into several different size waveguides for instantaneous frequency estimates and is fed to an HP 8566B spectrum analyzer for time-averaged frequency resolution. Calibration measurements have been in good agreement with an uncertainty less than $\pm$ 1 dB. The second diagnostic involves an overmoded directional coupler and a liquid calorimeter. The coupler provides the microwave power envelope and gives an additional peak power estimate. The calorimeter consists of a methanol - water mixture flowing between two 17° conical pieces of polyethylene. Bench test measurements against a 20 W CW signal have given similar agreement for these diagnostics.

A total of six two-cavity and four three-cavity gyroklystron tubes have been tested. A summary of the main features of each tube is presented in Table 1. The table mainly denotes the features that were changed in going from one tube to the next. That is, the cavity Qs, the amount of loss in the various regions, and the output wall taper. In addition, the table displays the progression of the operating regime of the beam parameters. In Table 2, the main results from each tube are presented. These include the peak power, the peak efficiency, and the peak gain. The table also indicates the magnetic field profile employed and the factor that resulted in limiting the tube performance. In addition, the reference where more details can be obtained is given. A summary of the progress toward achieving high peak power is shown in Fig. 3. The saturated gain at each power level is also indicated. In the next two sections we will detail the results displayed in Table 1 and 2 and Fig. 3.

EXPERIMENTAL RESULTS–TWO CAVITY CIRCUITS

The first two tubes were plagued by a multitude of instabilities, produced power levels below the state of the art, and had signal gains less than 0 dB. These results occurred at a beam voltage of 183 kV, a current of 55 A, and a velocity ratio of $\alpha \approx 0.45$. The instabilities could be grouped into four classes. Modes in the first class existed mainly in the output waveguide in frequency ranges where the window was a good reflector. These modes required good post-output cavity beam quality and were suppressed by amplifier operation. The second class existed in the output waveguide adjacent to the output cavity and required significant reflections from the first non-linear taper. Whole tube modes comprised the third class and had their energy mainly in the drift tube with reflections provided by the cavities. The final mode class involves instabilities in the downtaper.

Instabilities in classes 3 and 4 were the most troublesome and usually involved modes with one azimuthal variation. These modes can be controlled by the introduction of loss into the downtaper and drift tube. Figure 4a shows the total single pass attenuation of the TE_{11} mode through these regions in Tube 1 in the most troublesome frequency range. The downtaper had only a few thin lossy rings and very little attenuation. The drift tube consisted of alternating rings of copper and ceramic (1% SiC in 99% MgO) in nearly equal amounts. Though there were some highly absorptive resonances, there was a large frequency range with negligible loss. In contrast, Fig. 4b shows the TE_{11} attenuation for Tube 6. This attenuation was sufficient to allow operation at the parameters listed in the abstract.

Tube 3 incorporated a downtaper and drift tube with attenuations near the final values of Fig. 4b. This allowed significant beam power to be used for the first time and resulted in a one and a half order of magnitude increase in output power. The input and output cavity Qs were 175 and 145, respectively. In a flat magnetic field of 0.452 T, a peak power above 1.5 MW was produced at 9.866 GHz when the beam voltage was 305 kV and the current was 108 A. Parameter space for amplification was limited by a class 2 instability near the amplifier frequency. We determined that the mode was the TE_{121} operating near cutoff in a constant radius waveguide section immediately following the output cavity. This mode could not be suppressed by amplifier operation but was eliminated in Tubes 4 and beyond by introducing a linear wall taper after the output cavity lip. With this modification, Tube 4 produced peak powers near 2.7 MW with a 427 kV, 130 A beam in a constant magnetic field of 0.537 T. Maximum efficiencies of both Tubes 3 and 4 were approximately 5%.

The final order of magnitude increase in power was achieved with Tube 5, which had additional loss in the downtaper and a higher Q in the output cavity (224). The primary performance increase came not from the circuit modifications

Table 1: Features of each Tube.

Tube	$V_{0,max}$ kV	I_0 A	Output cavity α	Input cavity Q	Buncher cavity Q	Output cavity Q	Output taper slant	Downtaper	Preliminary drift space	Drift tube
1	185	< 55	< 0.45	175	—	160	0°	Metal/lossy flat rings of C/AL-Sil.	Metal/lossy flat rings of C/AL-Sil.	Cu/lossy flat rings of 1% SiC, 99% MgO
2	325	< 150	< 0.49	175	—	160	"	Added loss	Added loss	"
3	375	40–140	< 1.0	175	—	145	"	Covered gate value. Added loss plus lossy flat rings of 20% SiC, 80% BeO	Metal/lossy flat rings of 20% SiC, 80% BeO.	Cu and nonperiodic configuration of tapered lossy rings of 20% SiC 80% BeO
4	425	<200	< 1.0	175	—	160	2°	"	"	"
5	440	<200	<1.0	214	—	224	"	Added loss and some tapered pieces	Added loss with all tapered pieces	"
6	440	<200	<1.0	500	—	224	"	Added loss	Added loss	Added loss
7	440	<225	.6–.7	250	270	200	2.2°	"	C/AL-Sil. everywhere flat	"
8	440	<225	.7–.8	"	"	350	"	C/AL-Sil.	"	"
9	440	<225	.7–.8	"	"	465 Long	> 2.2°	"	"	"
10	440	<225	.7–.8	"	"	700 TE_{012}	"	"	"	"

Table 2: Results of amplification studies for each Tube.

Tube (Ref)	Magnetic Field Configuration	Ouput Power – MW	Efficiency	Gain	Limiting Factor
1 (8)	Flat fields	$< P_{in}$	—	< 0 dB	Instabilities everywhere
2 (8)	//	—	—	—	//
3 (8)	// 0.452 T	1.5 MW [305 kV, 108 A]	5%	15-16 dB	Instability in region of output cavity/ waveguide
4 (8)	// 0.537 T	2.7 MW [427 kV, 130 A]	5%	19 dB	Instability in downtaper
5 (4)	Tapered fields; optimum 0.474 T in output cavity, 15% higher in input cavity	22 MW [425 kV, 150 A]	≤ 32%	≤ 26.5 dB	//
6 (9)	//	24 MW [425 kV, 190 A]	≤ 34%	≤ 34 dB	//
7 (10)	Tapered fields; strong & weak profiles, ∼ 20% and ∼ 30% tapers	23 MW [425 kV, 205 A]	≤ 36%	≤ 31 dB	//
8 (10)	//	27 MW 20 MW 16 MW	32% 28% 37%	36 dB 50 dB 33	//
9 (12)	//	22 MW	< 28%	< 44 dB	//
10 (12)	//	22 MW	< 27%	< 32 dB	//

but rather from the introduction of magnetic field tapering. A careful search of parameter space showed that the optimal input cavity field was 0.545 T while the best output cavity performance occurred at 0.474 T. Figure 5 reveals the dependence of peak power on input cavity magnetic field when the output cavity field is fixed at the optimal value. The beam voltage and current were 425 kV and 150 A, respectively. The simulated velocity ratio was $\alpha \approx 0.98$ in the output cavity and varied adiabatically with the magnetic field. At the optimum level, $\alpha \approx 1.15$ at the center of the input cavity. The efficiency at the maximum was 31% and the gain exceeded 26 dB.

The input power was limited to 45 kW due to breakdown at the input window and the drive curve indicated that Tube 5 was not saturated. An increase in the input cavity Q to ~500 in Tube 6 allowed saturation to be examined. Theory predicted this Q to be 80% of the required start oscillation value and oscillations were in fact observed in the input cavity at some points. The optimal amplified power occurred with the same 15% field profile. Two distinct points were examined. The first was the maximum efficiency point, where 22 MW was produced at 9.871 GHz with an efficiency of 34% and a gain of 34 dB. The beam voltage and current were 425 kV and 150 A, respectively. Figure 6 shows the time dependence of the cathode voltage and output power as measured by a calibrated crystal detector. The second point had a maximum power level of 24 MW at 9.875 GHz for the same field profile and beam voltage but with a current of 190 A.

EXPERIMENTAL RESULTS–THREE CAVITY CIRCUITS

The three-cavity tubes had three distinct departures from the two-cavity designs. A schematic of the three-cavity circuit is shown in Fig. 7. The first departure, of course, was the introduction of the buncher cavity whose Q (~270) was defined solely by an alumino-silicate ring on the outer wall. Two metal rods with rounded ends provided remote-controlled tunability of 120 MHz by adjustment of their insertion length. The second change was to use exclusively alumino-silicate in the drift tubes (Tubes 7 and beyond) and downtaper (Tubes 8 and beyond). The TE_{11} attenuation was always superior to their non-porous counterparts, the TE_{01} attenuation was adequate for cavity isolation at 9.85 GHz, and no outgassing problems were observed. The third difference was to use a lossy dielectric on the radial wall of the input cavity and modify the coupling slit. Both input cavities performed well; no noticeable advantages of either design were discerned during the amplifier studies.

All of the three cavity tubes had at least two distinct operating regimes controlled by the magnetic field taper. Figure 8 shows the axial variation of the magnetic guide field for these two regimes. The tapers which optimize each of the operating regimes vary slightly from tube to tube, so we give the range of tapers used. The steep taper has a 30-32% decrease and a field of 0.458 T in the output cavity, and the weak taper has a decrease of 17-22% and a field of 0.453-0.490 T in the output cavity. The two regimes were also affected differently by increased load reflections.

The primary difference in Tubes 7 to 10 is in the geometry of the output cavities. Figure 9 shows the output cavity cross-section of each of the tubes. Tube 7 operated well at high input power, producing 23 MW at 27% efficiency and 31 dB gain. At low input power, instabilities limited operation to lower α, or lower beam power. This prevented operation at higher gain and high power. We also discovered that the tube produced higher power when operated with the calorimeter (2% reflection) as opposed to the anechoic chamber ($< 0.1\%$ reflection).

We suspected that the downtaper instabilities observed in Tube 7 were suppressed with higher input power and thus we significantly increased the attenuation in the downtaper in Tube 8. We also suspected that the enhanced operation with a more reflective load indicated that a higher Q in the output cavity was necessary. At our optimum beam power (425 kV and 205 A) Tube 7 was operating at 5% of start oscillation current in the output cavity so we also increased the output cavity Q from 200 (Tube 7) to 350 (Tube 8). With these changes Tube 8 achieved 50 dB saturated gain at 20 MW, and 36 dB at the high power point of 27 MW where the efficiency was 32%. The tube was able to operate stably with input power near one hundred watts.

Ongoing numerical modeling of this device which considers only cavity modes has shown efficiencies from 21% to 38% depending on how the beam loading affects the cavity Qs and the value of the pitch angle. The theoretical results [15] are not consistent with the beam α predicted by the gun code [16]. Concerned that some of the interaction was occurring after the output cavity we decided to increase the length of the output cavity. The cavity could be lengthened while keeping the same operating frequency, either by reducing the radius, Tube 9, or by going to the second axial harmonic, TE_{012}, Tube 10. We also included a probe in the radial wall of the output cavity in Tubes 9 and 10 to directly monitor the cavity interaction.

For Tube 9, the best power and gain occurred at the same operating point and were 22 MW and 44 dB. In the weak taper regime the uncalibrated probe signal was directly proportional to the measurement made in the anechoic chamber. This suggests that the weak taper corresponds to an interaction in the output cavity. The probe indicated that for a strong taper the interaction occurs somewhere else.

The output cavity in Tube 10 had the same radius as both Tube 7 and 8 and an output cavity lip thickness the same as Tube 8. The cavity length, which was almost twice that of Tube 7 and 8 was intended to operate in the TE_{012} mode. The long length of the cavity caused the TE_{011} and similar lower order modes with one axial variation to have high Q, and hence low start oscillation currents (< 10 A). To suppress these modes, a thin ring of lossy dielectric was placed on the axial mid-plane of the cavity. Cold tests showed that this ring did not affect the Q of the TE_{012} mode while substantially reducing the Q of all single axial variation modes. As for Tube 9, initial observations indicate that Tube 10 has two regimes of operation. In one regime the output power is mostly due to interaction after the output cavity and in the other regime there is strong interaction in the output cavity. Tube 10 has a maximum output power of 22 MW.

To compare the operation of the four tubes, Fig. 10 shows plots of efficiency vs. beam current for each tube at the beam voltage of 425 kV. As expected from their similar designs, Tubes 7 and 8 gave similar operation. The increased Q in Tube 8 allows it to operate better in the current regime of 100-200 A. The efficiency of Tube 9 shows a strong dependence on beam current in the range of 150-200 A. Below 100 A, where large alphas could be achieved, the tube free-oscillated at the operating frequency. Tube 10, which had the longest output cavity, had significantly lower efficiency than Tubes 7 and 8.

FUTURE EXPERIMENTS

With the success of the intermediate experiments, we are now contemplating our approach to a 100+ MW device. The high power requirement places several constraints on the electron gun and the microwave circuit. To produce a higher power MIG, we must either increase the beam voltage, increase the peak electric field in the gun, enlarge the average beam radius, or decrease the applied magnetic field [17]. The new microwave circuit must continue to be stable to spurious modes, provide inter-cavity isolation, and dissipate a fair amount of average power. Our options will be explored by performing cold test studies, constructing and testing additional tubes with the existing beam parameters, and designing MIGs and microwave circuits for 300 - 400 MW beams.

Near term plans

While the low repetition rate of our facility doesn't allow an examination of the system's average power considerations, the only unique gyroklystron feature of potential concern is the power dissipated in the lossy ceramic of the penultimate cavity. We are initiating a study of the effects of temperature rise and means of cooling via a cold test mock-up of the buncher cavity. The available power in our 2.5 kW CW klystron is considerably higher than the current estimate of penultimate cavity energy deposition.

One way to lower magnetic field is to inject at the fundamental and extract at the second harmonic. We will test this concept by replacing the output cavity of Tube 6 with the 19.7 GHz TE_{02} cavity shown in Fig. 11. Several of the key dimensions are indicated in the figure. The small cavity radius insures that the TE_{01} cannot exist at 9.85 GHz. The adiabatic wall transitions are designed to minimize mode conversion to the TE_{01} at 19.7 GHz. A cold cavity design code indicates that the power flowing into the drift tube is 50 dB lower than the power extracted at the end of the system. Simulations of amplifier stability and operation have been performed assuming the nominal experimental beam parameters. The design quality factor of 780 exceeds the start oscillation requirement for magnetic fields above 0.545 T. Efficiencies of at least 25% have been predicted by the large signal code.

The output cavity has been constructed and cold tested. The measured parameters are f=19.64 GHz and Q=685. The existing non-linear tapers are inadequate in terms of TE_{01} – TE_{02} mode conversion, so new tapers with Dolphchebychev profiles and theoretical mode conversion under 45 dB have been designed and are being electroformed. Also under construction are a 60 dB overmoded TE_{02} directional coupler and mode converters from TE_{10} rectangular to TE_{01} circular and from TE_{01} to TE_{02} in circular waveguide.

Three additional experimental configurations are under consideration. The first is simply a shorter output cavity to be used if spurious oscillations in the output cavity become a problem. The only potential drawback is an increase in the 19.7 GHz TE_{01} mode content. A single short cavity filter before the output cavity can potentially increase isolation.

The other two approaches are aimed at improving efficiency. First, an inner conductor will be inserted into the second harmonic tube. The 2 mm radius will increase isolation between cavities. In the downtaper, the inner radius will be increased and lossy dielectrics will be added in an attempt to further suppress class 4 modes. Because of the electric field distribution of the TE_{11} mode, we anticipate a significant improvement in attenuation in the section closest to the gun. This should result in a higher attainable velocity ratio which the non-linear code predicts will translate into better efficiency.

If the velocity ratio cannot be improved, an alternative approach to higher efficiency and lower magnetic field is to operate at the fundamental with a large Doppler shift $k_z v_z$. We are currently designing a traveling wave output section (gyro-twistron) which would replace the output cavity of Tube 6. Though the design is not yet completed, initial indications are that efficiencies of 35% are possible with the existing experimental beam parameters.

Finally, two alternative concepts which could be tested on the existing facility are also being explored. A second harmonic output traveling wave section can be used to eliminate the backward TE_{01} component and insure isolation. The second approach is to use non-symmetric modes to enhance beam - microwave coupling at the second harmonic.

Long Term Plans

The approach to a 100+ MW system is dependent on the current and near term experiments and the selection of the optimal amplifier frequency. To conform to conventional modulator and emitter technology, we assume the new system's maximum beam voltage and cathode loading will be 500 kV and 8 A/cm², respectively. This implies that higher current, probably with a larger guiding center radius, gives the required approach to higher powers. As the guiding center is increased, one must either (1) use an extremely lossy lining (~15-20 dB per wavelength), (2) choose a mode with a higher cutoff, or (3) use a coaxial insert in order to maintain cavity isolation. Even if a material could be found that satisfies (1), the required heat load would probably be prohibitive. Approach (2) would require suppression of the lowest radial mode with the same azimuthal index (m) as the operating mode or a large increase in m. Either approach would probably entail a new input cavity scheme. The third approach is viable provided that TE_{0n} modes are used.

We have considered MIG designs at 11.424 GHz and 19.7 GHz for fundamental and second harmonic devices with 300 MW beams in coaxial geometry. All designs appear feasible with the fundamental 19.7 GHz design being the most difficult. In the following paragraphs, we detail only a scenario for a second harmonic experiment at 11.424 GHz. Other scenarios are receiving similar scrutiny.

Table 3. Nominal system parameters – 100+ MW System.

Beam Voltage	500 kV
Beam Current	600 A
Velocity Ratio	1.5
Magnetic Field	3.28 kG
Input Cavity	
Length	2.30 cm
Q	70
Output Cavity	
Length	4.20 cm
Q	368
Drift Tube	
Length	18 cm
Inner Radius	1.6 cm
Outer Radius	3.9 cm

A schematic for a system that utilizes a 5.712 TE_{01} input cavity and a second harmonic TE_{02} output cavity is shown in Fig. 12. Tentative parameters for such a system are given in Table 3. The current will be supplied by doubling the number of PFNs in the modulator and decreasing the repetition rate. The longer magnetic field region required for the interaction will be achieved by adding two field coils. The same supplies can be used as the magnet power requirement is considerably reduced at the lower field. Scaling the current beam from 200 A to 600 A requires a three-fold increase in guiding center radius. The inner conductor will be supported at the end of the beam dump and will include lossy ceramics to enhance stability. The output cavity radius is selected to preclude fundamental TE_{01} operation but may have to be increased if spurious modes become a problem. Initial large-signal modeling has predicted efficiencies above 27%; we expect this value to increase significantly when the design is optimized.

A MIG which provides the required beam parameters is shown in Fig. 13. Its single anode configuration reduces its overall size and complexity and eliminates compensation problems associated with the resistive divider. Simulations are performed with the EGUN [16] code. The average cathode radius is 7.13 cm and the magnetic field compression is 7.55 at $\alpha = 1.5$. The peak electric field is less than 90 kV/cm on the cathode and 30 kV/cm on the anode. The nominal current is only 26% of the space-charge limiting value. The simulated beam thickness allows only 1.1 mm clearance between the beam and the drift tube walls.

The emitter strip is slightly curved to reduce velocity spread with an average half-angle of 46°. The dependence of axial velocity spread on current is shown in Fig. 14 [18]. The magnetic field at the cathode is adjusted to maintain $\alpha = 1.5$. Near the design current, the beam is not fully mixed by the input cavity. This accounts for the difference in spread between the two cavities. The design can operate over an 800 A range with a velocity spread below 10% due to the laminar flow of electrons throughout much of the gun. This extended range will facilitate the study of high efficiency amplification in the 100+ MW regime.

ACKNOWLEDGMENT

This work was supported by the United States Department of Energy.

We wish to acknowledge the efforts of C. Bellamy, J. Cheng, O. Dajani, S. Demske, K. Lee, Q. Qian, M. Rimlinger and V. Specht.

REFERENCES

1. Proc. Int. Workshop on the Next Generation Collider, SLAC Report-335, Dec. 1988.

2. W. M. Bollen, et al., IEEE Trans. Plasma Sci. Vol. PS-13, p. 417 (1985).

3. K. R. Chu, et al., IEEE Trans. Plasma Sci. Vol. PS-13, p. 424 (1985).

4. W. Lawson, et al., Phys. Rev. Lett. Vol. 67, p. 520 (1991).

5. W. Lawson, et al., Int. J. Electron. Vol. 61, p. 969 (1986).

6. J. P. Calame and W. Lawson, IEEE Trans. Electron Dev. Vol. 38, p. 1538 (1991).

7. W. Lawson et al., AIP Conference Proceedings 193, Advanced Accelerator Concepts, Lake Arrowhead, CA, 1989, Editor, Chan Joshi, p. 274.

8. J. P. Calame, et al., J. Appl. Phys. Vol. 70, p. 2423 (1991).

9. W. Lawson, et al., IEEE Trans. Plasma Science Vol. 20, No. 3 (1992).

10. S. Tantawi, et al., IEEE Trans. Plasma Science, Vol. 20, No. 3 (1992).

11. W. Lawson, et al., Proceedings of the Beams '92 Conference, Washington, DC, May 1992.

12. W. Main, et al., Proceedings of the Beams '92 Conference, Washington, DC, May 1992.

13. J. Neilson, et al., IEEE Trans. Microwave Theory Tech. Vol. MTT-37, p. 1165 (1989).

14. P. E. Latham, IEEE Trans. Plasma Sci. Vol. 18, p.273 (1990).

15. P. E. Latham, et al., Proceedings of the Beams '92 Conference, Washington, DC, May 1992.

16. W. B. Herrmannsfeldt, SLAC Report-226, Nov. 1979.

17. W. Lawson, IEEE Trans. Plasma Sci. Vol. 16, p. 290 (1988).

18. W. Lawson and V. Specht, submitted to IEEE Trans. Electron Dev., May 1992.

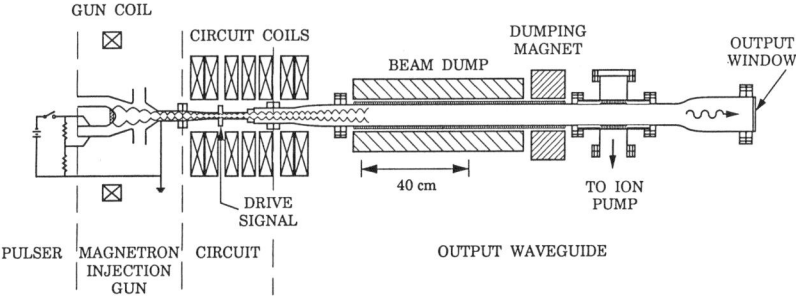

Figure 1. A schematic view of the gyroklystron experimental facility.

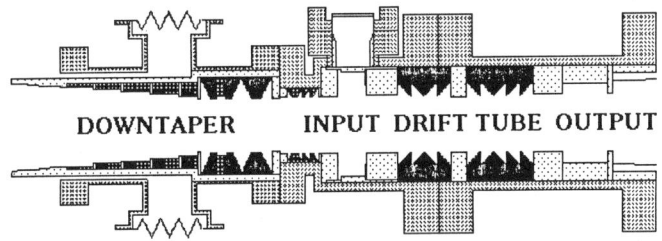

Figure 2. The two-cavity microwave circuit configuration.

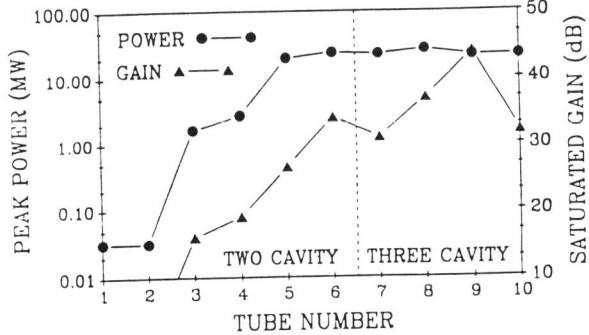

Figure 3. Summary of tube peak power performance.

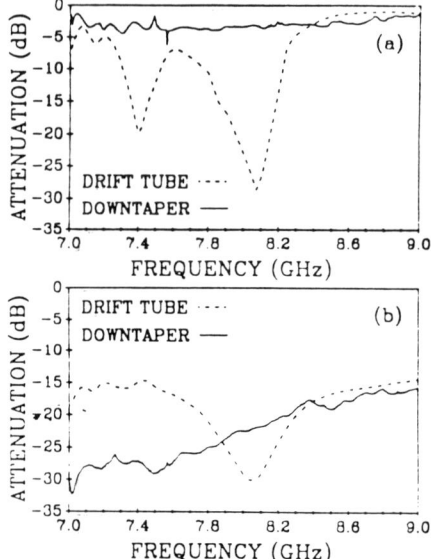

Figure 4. TE$_{11}$ attenuation vs. frequency in the downtaper and drift tube regions: (a) for Tube 1 and (b) for Tube 6.

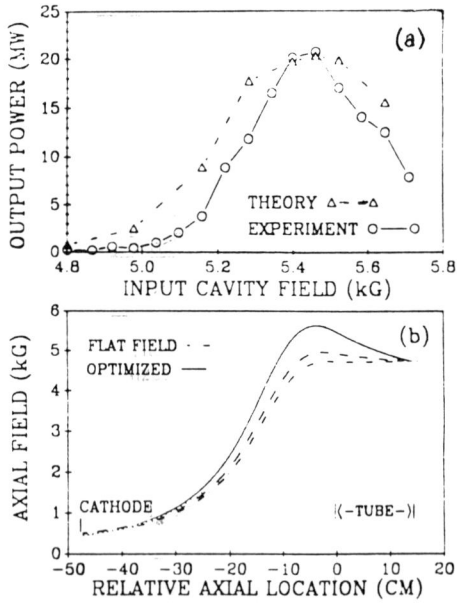

Figure 5. (a) Peak measured and theoretical output power vs. input cavity magnetic field, (b) axial field profile.

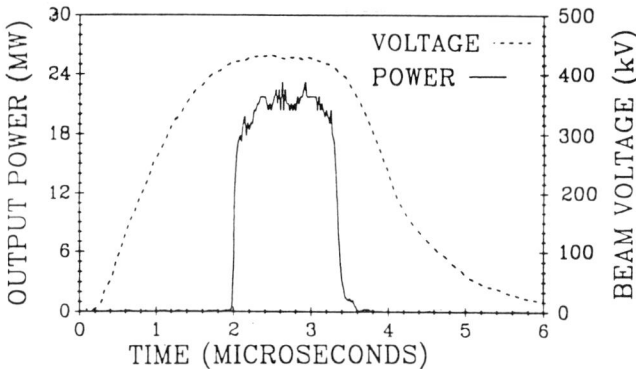

Figure 6. Time dependence of the output power and the beam voltage at the maximum efficiency point.

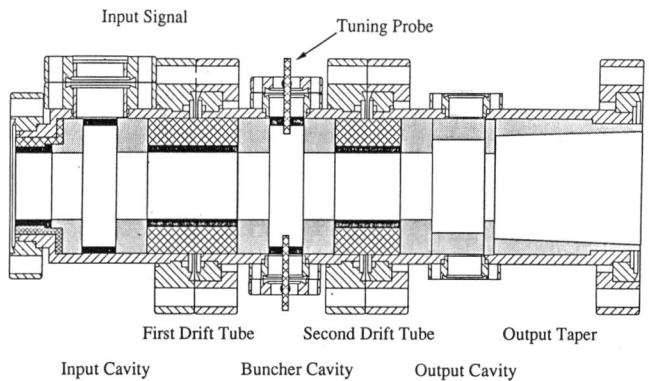

Figure 7. The three-cavity microwave circuit configuration.

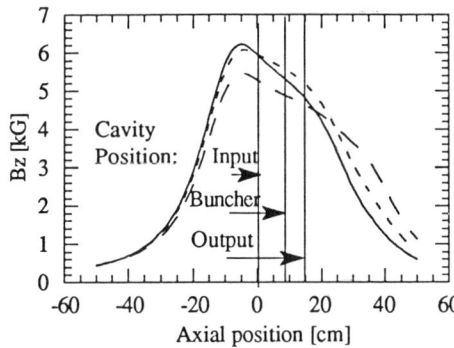

Figure 8. Axial variation of the guide field for the steep and weak tapers used in Tubes 8 and 9. Solid line ——— steep taper (30%), broken line – – – weak taper (17%) Tube 8, dotted line - - - - - weak taper (22%) Tube 9.

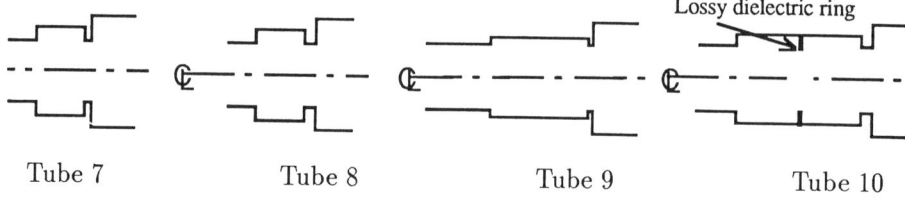

Figure 9. Cross-sectional diagrams of the output cavities of Tubes 7-10.

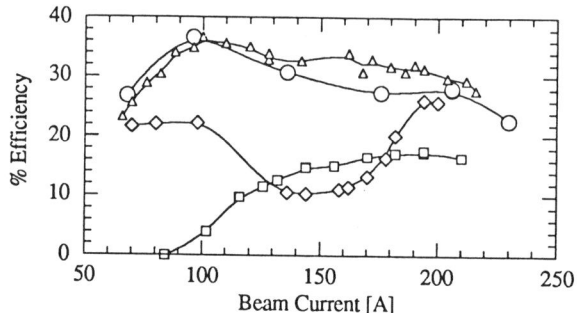

Figure 10. Dependence of efficiency on beam current for Tubes 7-10, at beam voltage 425 kV. The lines with circles (○) represent Tube 7, triangles (△) Tube 8, diamonds (◇) Tube 9, and squares (□) Tube 10.

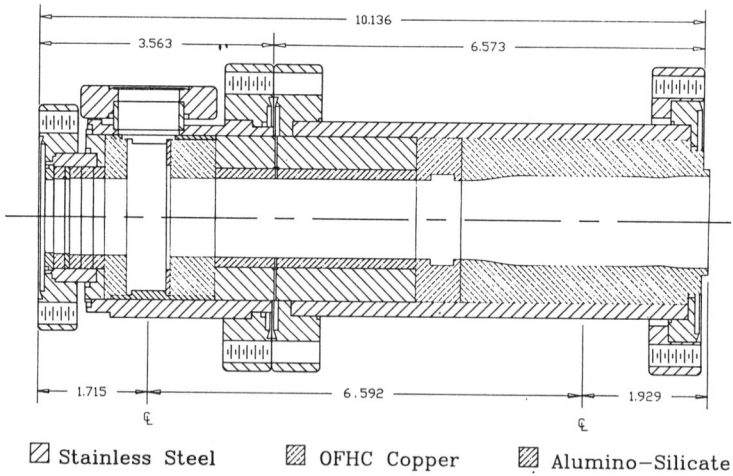

Figure 11. A second harmonic circuit.

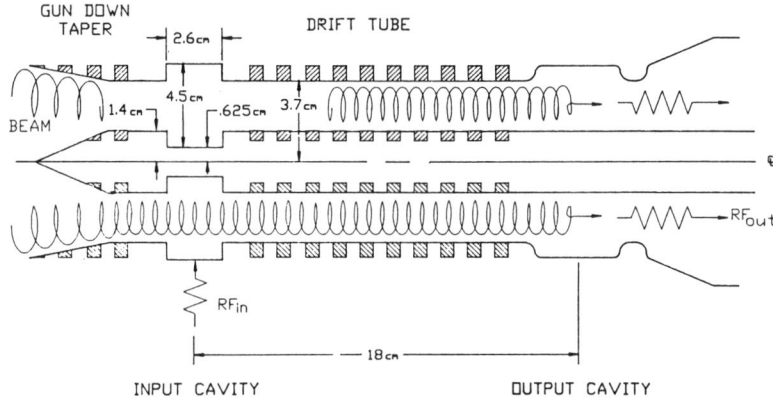

Figure 12. A coaxial second harmonic two-cavity circuit.

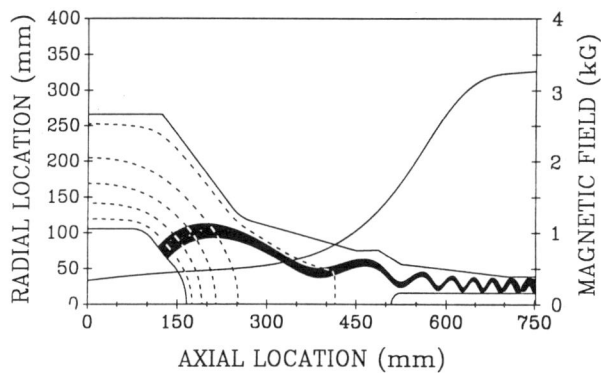

Figure 13. Electrode design and ray trajectories for the 300 MW MIG.

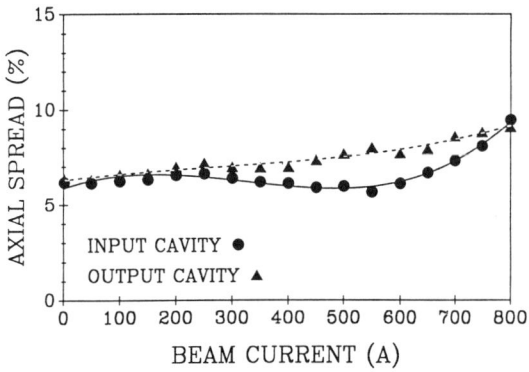

Figure 14. The dependence of axial velocity spread on current.

Modeling Relativistic Gyroklystron Amplifiers

P.E. Latham and G.S. Nusinovich

Laboratory for Plasma Research University of Maryland College Park, MD 20742

Abstract

Relativistic gyrotrons (electron kinetic energy on the order of the rest energy or greater) differ from their weakly relativistic counterparts in two important ways: the Larmor radius is large, which puts bounds on the allowable pitch angle for a given beam tunnel radius and magnetic field, and operation at Doppler upshifted frequencies, which allows for the extraction of energy from the axial motion, is feasible even in fast wave devices. Consequently, there are a number of issues that are not present at low voltage: computing the electromagnetic and AC space charge fields is not straightforward; achieving high efficiency while still maintaining the cutoff condition in the beam tunnels may be problematic, especially at cyclotron harmonics; and the use of slow wave structures to achieve very high electronic efficiency is a possibility. In this paper we consider how these issues affect modeling, and discuss the advantages and limitations of high voltage operation. We also present a specific example, the gyrotwistron, as a highly efficient device for operation in the high power regime.

I. Introduction.

There is currently a need for high power microwaves in a number of areas. Particle accelerators, plasma heating and diagnostics, space communication and material processing are some of the more conventional ones.[1] In addition, the removal of CFCs from the atmosphere and the powering of orbiting satellites from earth have been proposed.[2] For many of these applications, gyrotrons are a promising source. To accommodate the high power requirements, future gyrotrons would need to operate at relativistic voltages. At these voltages the Larmor radius is large, and the gyrotron cavities must be short compared to the cavities in weakly relativistic gyrotrons (for high efficiency operation, the number of cyclotron periods scales as[3] $\gamma_0/(\gamma_0 - 1)$ where γ_0 is the relativistic factor). These two trends raise a number of issues that are not present in weakly relativistic gyrotrons:

1. Because the Larmor radius is large, the beam tunnel must also be large; typically on the order of a free space wavelength. For harmonic and, to a lesser extent, Doppler upshifted operation, this may make it difficult to eliminate radiation in the beam tunnel. In addition, because of the high beam power, mode suppression in the beam tunnel is difficult.

2. With the shortening in cavity length, at a fixed operating frequency the cavity radius grows (see, for example, Fig. 1). This precludes operation in a pure TE_{mn} mode with a slowly varying axial profile, and makes modeling

the cold cavity fields non-trivial. On the plus side, however, because the cavities are short the axial wavenumber, k_z, can be large enough to extract significant energy from the axial motion.

3. Because of the large values of k_z that are typical of relativistic gyrotrons, modeling AC space charge effects, which can be important in gyroklystron amplifiers,[4] is difficult. As far as we know, there are no simple models for AC space charge when $k_z c/\omega$ is non-negligible (c is the speed of light and ω is the operating frequency).

In the remainder of this paper we discuss how these issues affect modeling. Our primary goal is to extract the essential physics of relativistic gyrotrons; this will allow us to quickly zero in on the parameters which produce the highest efficiency, and to construct simple models which can be implemented numerically. Both of these are important, as modeling a realistic device typically involves hundreds of simulations.

In Sec. II we examine in detail the consequences of the large Larmor radius, especially on the choice of mode and cyclotron harmonic. Section III covers the modeling of short cavities and Sec. IV discusses AC space charge. In Sec. V we consider operation with large Doppler upshift in both fast and slow wave structures and present an initial gyrotwistron design. Section VI contains our summary.

II. Large Larmor radius gyrotrons.

One of the fundamental requirements for gyrotrons is that no radiation propagate backwards from the output cavity into the beam tunnel. In conventional, weakly relativistic devices this is accomplished by operating the cavities near cutoff and decreasing the radius slightly in the beam tunnels. In relativistic gyrotrons, which generally employ short cavities far from cutoff, the tapers cannot be made smooth enough to eliminate conversion to lower order radial modes. Thus, a relativistic gyrotron operating in the TE_{mn} mode must have its beam tunnel radius small enough to cutoff the TE_{m1} mode, even when $n > 1$. If the drift tube radius is n_L Larmor radii thick but still small enough to cutoff the radiation at the operating frequency, ω, we arrive at the condition

$$\beta_{\perp 0} \frac{\omega}{\Omega_{c0}} < \frac{x_{m1}}{n_L[1 + \text{Im}\{k_z c/\omega\}^2]^{1/2}} \tag{1}$$

where k_z is the axial wavenumber of the cutoff TE_{m1} mode in the beam tunnel, $\beta_{\perp 0} = v_{\perp 0}/c$ is the perpendicular component of the beam velocity normalized to the speed of light, Ω_{c0} is the relativistic cyclotron frequency at the initial beam energy and x_{mn} follows from $J'_m(x_{mn}) = 0$ with J_m the mth order Bessel function, m the azimuthal mode number and n the radial mode number. In Eq. (1) $n = 1$. Note that harmonic operation ($s > 1$) makes this relation difficult to satisfy because the magnetic field, and thus Ω_{c0}, decreases with increasing harmonic number.

Equation (1) can, of course, always be satisfied by increasing the azimuthal mode number m until x_{m1} is large enough. However, at large m mode competition becomes more severe and the beam-wave coupling decreases. The decrease in coupling is a byproduct of the short cavities needed in relativistic gyrotrons. To see this, note that the beam-wave coupling is proportional to $J^2_{m-s}(k_\perp r_g)$ where $k_\perp$ is the perpendicular wavenumber, r_g is the guiding center radius and s is the harmonic number. This quantity peaks near $r_g \approx (m-s)/k_\perp$. Near cutoff, $k_\perp$ is close to ω/c and values of the guiding center for which $J^2_{m-s}(k_\perp r_g)$ is appreciable are small enough that the beam can fit inside a cutoff beam tunnel. For short cavities, say with a length on the order of a wavelength or less, the axial wavenumber is appreciable and $k_\perp$ is considerably smaller than ω/c. In that case, for guiding centers less than the radius of the cutoff beam tunnel the coupling $J^2_{m-s}(k_\perp r_g)$ is relatively small. Thus, except at high harmonics, high voltage gyroklystrons will probably be limited to reasonably small azimuthal mode numbers.

If we operate at low order azimuthal modes, then whether or not Eq. (1) can be satisfied at a reasonably large value of the pitch angle depends on the beam thickness; i.e. on n_L. The annular electron beam formed by a magnetron injection gun (MIG) usually has a thickness that is larger than twice the Larmor radius, so n_L is at least 2. In practice, it turns out that n_L can be made as small as 2.5 or 3. (Note that a MIG produces a hollow beam, so the cutoff frequency in the beam tunnel could be increased by operating in symmetric modes and using a coaxial insert. In that case the value of x_{m1} that appears in Eq. (1) would have to be modified appropriately.) If we take $n_L = 3$, a fairly reasonable estimate considering that clearance is needed between the beam and waveguide wall, for large enough magnetic field Eq. (1) is easily satisfied. At harmonic and Doppler upshifted operation, on the other hand, the magnetic field is reduced. In that case Eq. (1) may be difficult to satisfy without reducing $\beta_{\perp 0}$. However, the efficiency is highest when $\beta_{\perp 0}$ is relatively large and the magnetic field is low. To see this, note that for the averaged equations of motion (i.e. only a single harmonic present), there is a conserved quantity I given by

$$I = \frac{\gamma}{\gamma_0} - \frac{\beta^2_{\perp 0}}{2} \frac{\omega}{s\Omega_{c0}} \frac{P^2_\perp}{P^2_{\perp 0}} \qquad (2)$$

where $P_\perp$ is the perpendicular component of the canonical angular momentum. Equation (2) can be derived by noting that the averaged Hamiltonian depends only on the combination of coordinates $s\psi - \omega t$ where ψ is the gyrophase. Thus, the appropriate linear combination of conjugate momenta is conserved. That combination appears on the right hand side of Eq. (2). Using this conserved quantity, it is straightforward to show that the maximum single particle efficiency is given by the formula

$$\eta_{max} = \frac{\beta^2_{\perp 0}}{\beta^2_{\perp 0} + \beta^2_{z0}} \frac{\omega}{s\Omega_{c0}} \frac{\gamma_0 + 1}{2\gamma_0} \quad ; \qquad (3)$$

this is essentially the same as in Ref. 3. For conventional low-voltage gyrotrons, $\gamma_0 - 1 \ll 1$ and $\omega \simeq \Omega_{c0}$, Eq. (3) reduces to the usual expression $\eta_{\max} = \alpha_0^2/(1+\alpha_0^2)$ where $\alpha_0 = \beta_{\perp 0}/\beta_{z0}$ is the beam pitch angle. (Note that the averaged equations are not strictly valid when the cavities are short, but it turns out that for operation in the fundamental cyclotron harmonic, averaged equations represent a reasonably good approximation. Thus, Eq. (3), while not exact, does a fairly good job of predicting the maximum achievable efficiency.)

Equation (3) indicates that high efficiency occurs when both the Doppler upshift, $\omega/s\Omega_{c0}$, and the perpendicular velocity, $\beta_{\perp 0}$, are large. If we demand that the cyclotron resonance condition,

$$\omega - s\Omega_c - k_z v_z = 0 \ , \tag{4}$$

be approximately satisfied, then these two terms are related by $\omega/s\Omega_c = 1/(1 - k_z v_z/\omega)$. In that case, operation with fast waves ($k_z > \omega/c$) at a large Doppler upshift can be realized only at large axial, and thus small perpendicular, velocities. The optimum relation between pitch angle, α_0, and axial wavenumber, k_z, will be discussed in Sec. V. Note only that increasing $\beta_{\perp 0}$ and decreasing Ω_{c0} makes it difficult to satisfy Eq. (1), and this constraint becomes even more severe at harmonic operation. Thus, to access harmonic or Doppler upshifted regimes it may be necessary to replace the output cavity by a traveling wave section and operate in the gyrotwistron configuration;[5] in that configuration all the radiation is in the forward direction, there is no reason to have a cutoff beam tunnel, and we can drop the requirement given in Eq. (1). We discuss this option in Sec. V.

Another effect of the requirement that the beam tunnel be cutoff to the TE_{m1} mode is that we lose the option of placing the beam at the maximum for the mode of interest and near the minimum of competing modes. This is because the beam tunnel radius is on the order of a free space wavelength, so the beam-wave coupling terms, $J_{m-s}^2(k_\perp r_g)$, have broad maxima (relative to the size of the beam tunnel) as a function of guiding center radius r_g. Thus, there are always many modes that can be excited with more or less the same coupling strength. Nevertheless, eliminating mode competition in relativistic gyrotrons is, at least in principle, reasonably straightforward. Since the cavities are short, most of the mode competition occurs in the beam tunnel(s): in the beam tunnel before the first cavity and, in the case of gyroklystron amplifiers, in the drift sections between cavities. Because there is no radiation in the beam tunnels, they can be loaded with lossy dielectrics. We have had considerable success with this approach for the two[6] and three cavity[7] gyroklystron amplifiers built at the University of Maryland. By loading both the drift sections and gun downtaper regions with lossy dielectrics, we were able to achieve a pitch angle slightly greater than 1.0 at a voltage of 425 kV and currents up to 200 A (compared with $\alpha \approx 0.45$ at a voltage of 180 kV and current of 55 A with minimal loading[8]). Still, we did not reach our design value of $\alpha = 1.5$; to do that would probably require a redesign of the magnetron injection gun to include lossy dielectrics in the gun region.

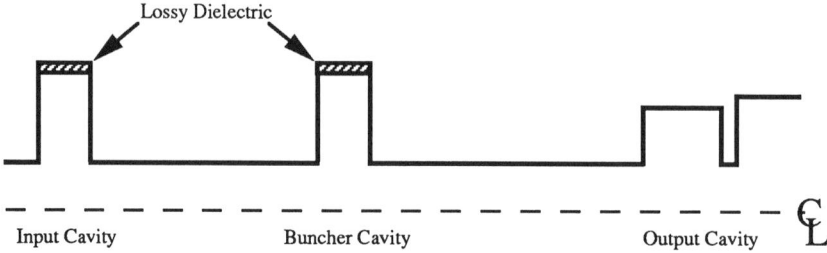

Figure 1: A typical three cavity gyroklystron drawn approximately to scale. The sharp boundaries on the cavities generate significant mode conversion, so the fields consist of a superposition of radial modes.

III. Numerical modeling of short cavity gyrotrons.

Conventional gyrotrons employ cavities that are many cyclotron periods long and operate in a pure TE_{mn} mode near cutoff. Consequently, a number of simplifying assumptions can be made when modeling these devices: the axial momentum, p_z, can be treated as constant, only a single harmonic needs to be included in the equations of motion, and a fixed Gaussian axial profile for the electromagnetic field can be used. In this regime, a simple set of averaged equations[9] provides a good description of the gyrotron.

Relativistic gyrotrons, on the other hand, employ short cavities (see Fig. 1) that do not operate in a single TE_{mn} mode, but instead consist of a sum of radial modes. For instance, in the cavity shown in Fig. 1, in each section (a section is defined as a region with constant wall radius) the z-component of the magnetic field may be written

$$B_z(r,\theta,z,t) = \text{Re}\left\{ \sum_n \left(f_n e^{ik_n z} + b_n e^{-ik_n z} \right) J_m(\gamma_{mn} r) e^{i(m\theta - \omega t)} \right\} \qquad (4)$$

where f_n and b_n are the forward and backward mode amplitudes, $\gamma_{mn} = x_{mn}/r_w$ is the perpendicular wavenumber, r_w is the wall radius, and k_n is related to γ_{mn} and ω via the usual dispersion relation

$$\frac{\omega^2}{c^2} = \gamma_{mn}^2 + k_n^2 \quad .$$

The mode amplitudes f_n and b_n can be found using the scattering matrix method.[10]

Because the fields are more complicated than for the weakly relativistic gyrotron, the equations are also more complicated. In addition, the cavities are too short to average over gyrophase, so the time derivatives of the particle coordinates consists of an infinite sum over harmonics. (In our numerical simulations,

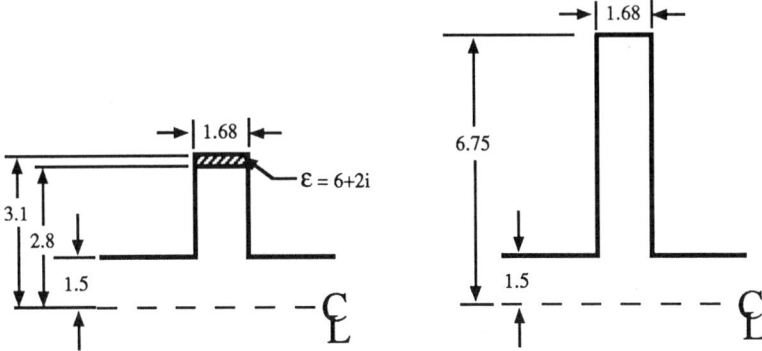

Figure 2: The input and buncher cavities used in the experiment (left) and the cavities used in our numerical modeling (right).

of course, we need only a few harmonics. In fact, it turns out that the dominant harmonic usually gives the correct answer to within 5% or less.)

These two additional complications – having many radial modes and more than one harmonic – are not too difficult to take into account, at least when we can assume that the cavity has perfectly conducting walls. What is a problem, however, is modeling the lossy dielectric that is needed in the input and intermediate cavities in gyroklystron amplifiers. This is primarily a technical issue: it is in principle possible to compute the fields in a cavity containing lossy dielectrics, but it is difficult. In addition, when the cavities are lossy the perpendicular wavenumber, γ_{mn}, is complex, and fast computation of Bessel functions with complex arguments is highly non-trivial. Speed is extremely important because many simulations are needed to come up with a realistic design.

For these reasons, when we model the intermediate cavities in our gyroklystron amplifiers we use perfectly conducting walls even when lossy dielectrics are present. For instance, Fig. 2 compares the input and buncher cavities used in the three cavity experiment and the ones used in our modeling. The cavity lengths are the same but the radius is increased to account for the lossy dielectric. This approach is far from ideal; because of the large disparity in radii, the amplitudes (f_n and b_n in Eq. 5) we used for the radial modes are not necessarily close to the actual experimental values. For instance, the start oscillation current in the cavity we use to model the three cavity experimental is about twice what it is in the experimental cavity. This indicates that we should, perhaps, model the cavities with a Q that is twice the value of the cold cavity Q. However, this may not take all effects into account; besides nonlinearities in the particle orbits, if the beam pulls any of the radiation away from the lossy dielectric and into the beam tunnel, the Q will increase even further.

In fact, there is some evidence for increased Qs; shown in Fig. 3 is a plot of

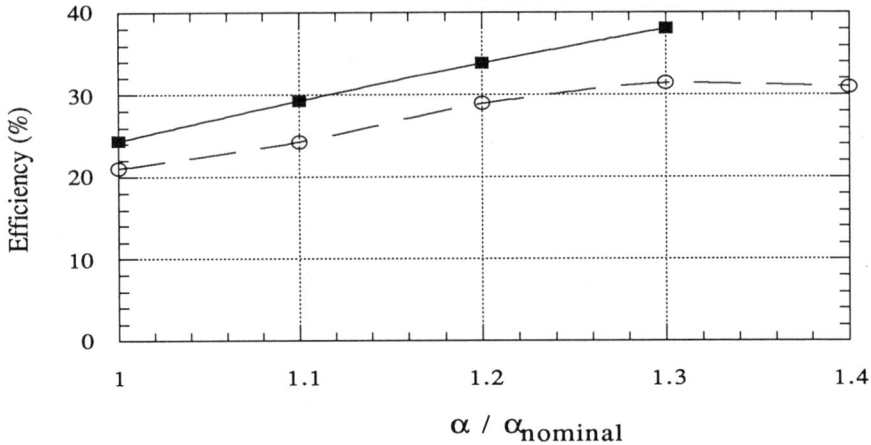

Figure 3: Efficiency versus $\alpha/\alpha_{\text{nominal}}$ where α_{nominal} corresponds to the predicted value of 0.82 in the output cavity.[11] For the lower curve (circles) the Qs were fixed at their cold cavity values. For the upper curve (squares) the Qs of the first and second cavity were chosen to maximize the efficiency.

efficiency versus $\alpha/\alpha_{\text{nominal}}$ (α_{nominal} is the value of the pitch angle predicted by the Herrmannsfeldt gun code[11]) for the three cavity experiment. Two cases are shown: the lower (dashed) curve corresponds to the experimental cold cavity Qs. In the upper (solid) curve, we set the Qs in the input and buncher cavities to the value which maximized the efficiency; typically around 4 times their cold cavity values. With fixed Qs we were unable to reach the experimental efficiency of 35%; only when we let the Qs increase were we able to achieve an efficiency consistent with the experiment, albeit at a larger pitch angle than was predicted by the Herrmannsfeldt code. We are currently investigating ways to model the effective Q analytically.

IV. AC space charge.

Numerical studies of weakly relativistic gyroklystron amplifiers[4] have shown that AC space charge has a significant effect on both gain and efficiency. The models used in these studies assume that the AC space charge forces act only in the direction transverse to the beam. This assumption is valid for weakly relativistic devices because the fastest growing space charge waves occur at an axial wavenumber that satisfies the resonance condition $k_z v_z = \omega - s\Omega_c$, and since ω is close to Ω_c, $k_z \approx 0$. However, as discussed above, relativistic gyrotrons operate with $k_z v_z$ a significant fraction of ω, so the assumption that the space charges forces act only in the transverse direction breaks down. As far as we know, there is no simple model for AC space charge valid at finite k_z.

Such a model is important for an accurate description of gyroklystron amplifiers. To get some idea of the effect of AC space charge on relativistic devices, we used the weakly relativistic model in numerical simulations of our 425 kV gyroklystron.[6] We found that the efficiency is slightly enhanced and the gain is greatly enhanced when AC space charge is included. Moreover, with AC space charge our simulations were in excellent agreement with the experiment, while without it we predicted a gain much lower than the observed experimental gain. Because the model we used was not strictly valid, it is hard to say how much of the result was fortuitous. One of our primary goals for the future is to construct a model of AC space charge that is valid for relativistic gyroklystron amplifiers with finite values of the axial wave number.

V. Large Doppler upshifted operation – gyrotwistrons.

To solve the problem of suppression of backward wave propagation in the beam tunnels, we can replace the output cavity by a traveling wave section. Then, since no radiation propagates backwards, there is no need for the radiation to be cutoff in the beam tunnel and the constraint implied by Eq. (1) no longer needs to be met. In this section we look at the design of a traveling wave device.

Equation (3) tells us that the highest efficiency occurs at the lowest magnetic field. Since amplifiers must operate close to the resonance condition given in Eq. (4), low magnetic field means large Doppler upshift. This is the cyclotron autoresonance maser (CARM) regime, and it has been known for some time that CARMs are capable, at least in principle, of high efficiency.

A natural operating point is the one that yields a maximum single particle efficiency, as given in Eq. (3), of 100%. An approximate value of the axial wavenumber, k_z, that produces this efficiency can be found by demanding that the amplifier operate exactly at resonance. Then, we may replace $s\Omega_{c0}/\omega$ in Eq. (3) by $1 - k_z v_{z0}/\omega$. Defining

$$h \equiv \frac{k_z c}{\omega} \quad ,$$

Eq. (3) becomes[12]

$$\eta_{max} = \frac{\alpha_0^2}{1 + \alpha_0^2} \frac{\gamma_0 + 1}{2\gamma_0} \frac{1}{1 - h\beta_{z0}} \tag{6}$$

where $\beta_{z0} = v_{z0}/c$ is the axial velocity normalized to the speed of light. The optimal value of h, h_{opt}, can now be found by solving Eq. (6) with $\eta_{max} = 1$. It is straightforward to show that

$$h_{opt} = 1 + \frac{(1 - \gamma_0^{-1} - \beta_{z0})^2}{2(1 - \gamma_0^{-1})\beta_{z0}} \quad . \tag{7}$$

This equation tells us that the optimum value of $k_z c/\omega$ is never less than 1. Thus, to achieve optimum efficiency, gyrotron amplifiers should use a slow wave structure. As was pointed out in Ref. 12, when $h = 1$ the corresponding value

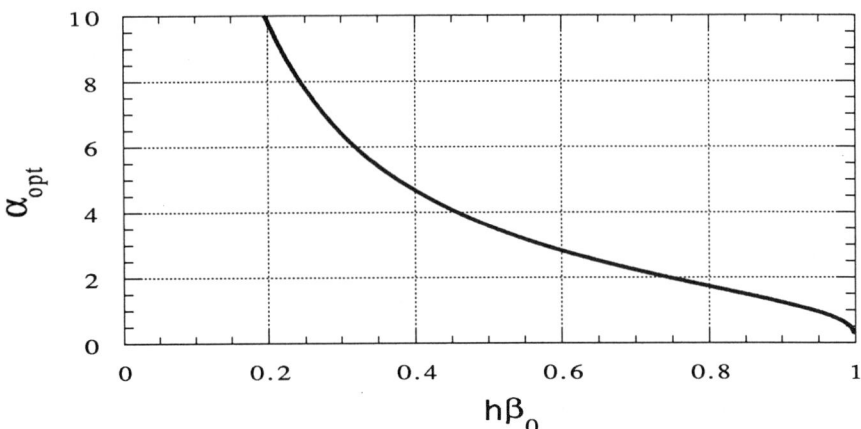

Figure 4: Optimum pitch angle (i.e. the pitch angle that maximizes the efficiency) versus $h\beta_0$ ($= k_z v_{z0}/\omega$).

of the pitch angle is $\alpha = \sqrt{2/(\gamma_0 - 1)}$. Except for highly relativistic velocities, a large pitch angle is required even when $k_z = \omega/c$. Note that this is contrary to the belief that CARMs work well at small pitch angles.

For fast wave devices k_z is less than ω/c, so h is less than 1 and it is not possible to achieve a single particle efficiency of 1. In that case, we can compute the optimum pitch angle for a given k_z. Maximizing η_{max} in Eq. (6) with respect to α_0 at a fixed beam energy, we find that the optimum pitch angle, α_{opt}, occurs at

$$\alpha_{opt} = \left[\frac{2\sqrt{1 - (h\beta_0)^2}}{1 - \sqrt{1 - (h\beta_0)^2}} \right]^{1/2} \tag{8}$$

where $\beta_0 = [1 - \gamma_0^{-2}]^{-1/2}$ is the total velocity normalized to the speed of light. The corresponding efficiency at the optimum value of α is

$$(\eta_{max})_{\alpha=\alpha_{opt}} = \frac{\gamma_0 + 1}{\gamma_0 + \sqrt{[1 - (h\beta_0)^2]/[1 - \beta_0^2]}} . \tag{9}$$

It is interesting to note that the optimal value of the pitch angle is relatively large, even when k_z is near ω/c. In Fig. 4 we plot α_{opt} versus $h\beta_0$, and we see that α_{opt} is above 1.5 until $h\beta_0$ is around 0.85. This is a relatively high value, especially considering that β_0 does not reach 0.85 until the beam voltage is 500 kV. Thus, the optimum pitch angle in fast wave traveling wave amplifiers will be large unless the beam is ultra-relativistic and $k_z c/\omega$ is near 1.

Although amplifiers can, in principle, achieve high efficiency, they are subject to severe mode competition, especially from backward waves. For this reason it

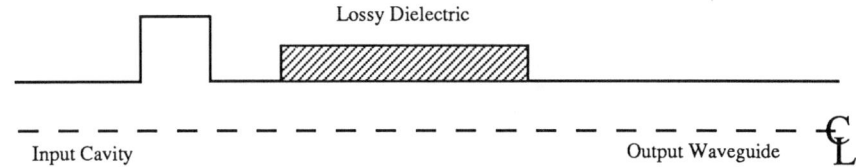

Figure 5: Gyrotwistron configuration.

is advantageous to operate in the gyrotwistron configuration,[5] in which the beam is prebunched in a cavity and then ballistically bunched in a cutoff or loaded section of waveguide. Tapering both the magnetic field and the radius of the output waveguide would also help decrease mode competition. We have done some preliminary studies of gyrotwistrons at 16 GHz. A schematic of the device we modeled is shown in Fig. 5. In this configuration, there is an input cavity, a drift section filled with lossy dielectric, and output waveguide. The purpose of the drift section is to allow the beam to bunch without significant interaction with the electromagnetic radiation. Because all the radiation is in the forward direction there is no need to have the drift tube cutoff.

In the drift section we impose a real and imaginary axial wavenumber k_z, and let the perpendicular structure be that of a perfectly conducting waveguide with perpendicular wavenumber given by $k_\perp = [\omega^2/c^2 - \text{Re}\{k_z\}^2]^{1/2}$. In the output section we assume perfectly conducting walls, so k_z is purely real and a function only of the radius. We fix the frequency, and in both the drift section and output waveguide we let the field amplitude evolve in z. Instead of simulating the input cavity numerically, we let the particles have an initial distribution (at the start of the drift tube) given by

$$\psi = \psi_0 \ \in [0, 2\pi] \qquad u_\perp = u_{\perp 0} + \frac{q}{L_d \psi_0'} \sin(\psi_0) \qquad u_z = u_{z0}$$

where ψ is the gyrophase, $u_\perp$ is the normalized perpendicular momentum, L_d is the drift tube length and ψ_0' is the value of $d\psi/dz$ evaluated at the beginning of the drift tube. Letting the initial value of u_z be independent of phase is consistent with an input cavity operating near cutoff; future modeling will use more realistic assumptions. With these initial values, at the end of the drift tube the particles will have a phase distribution corresponding to a bunching parameter equal to q.

The parameters for the device we simulated are listed in Table 1. Preliminary results are encouraging: as shown in Fig. 6, efficiency is on the order of 45-50%, more or less independent of the axial velocity spread out to a spread of 6% (the highest value we have simulated so far). However, the device is relatively long: the overall length, including the lossy drift section, varies between 100 and 200 cm. We suspect that the length can be decreased significantly by using a nonlinear rather than linear taper. In fact, a linear taper is probably a bad choice; it is

Table 1: Gyrotwistron parameters.

Parameter	Value
Beam voltage	425 kV
Beam current	160 A
Pitch angle ($\alpha_\perp = v_{\perp 0}/v_{z0}$)	1
Axial velocity spread ($\delta v_z/v_{z0}$)	6.5%
Frequency	16 GHz
Mode	TE_{01}
Lossy drift section: length	10 cm
$Re\{k_z c/\omega\}$	0.900
$Im\{k_z c/\omega\}$	0.025
Lossy drift section: radius	1.5 cm
$Re\{k_z c/\omega\}$	0.648
$Im\{k_z c/\omega\}$	0.000

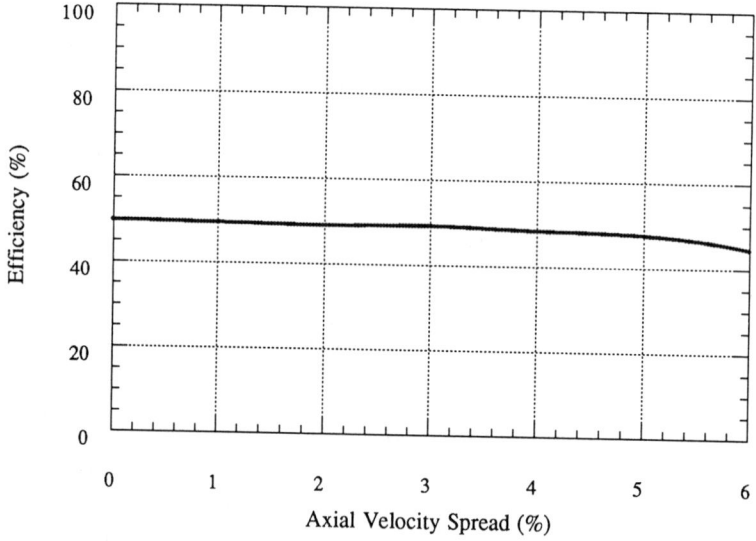

Figure 6: Efficiency versus axial velocity spread.

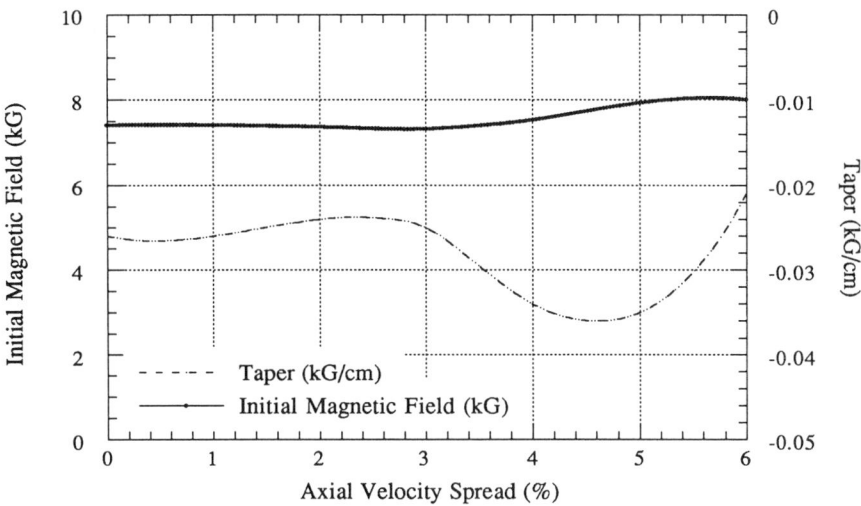

Figure 7: Initial magnetic field and magnetic field linear taper versus axial velocity spread.

likely that better results could be attained by adjusting the magnetic field for exact resonance in the beginning of the waveguide and then decreasing it when the field amplitude becomes large enough to trap a significant number of electrons. The peak magnetic field and taper are plotted in Fig. 7. The details of these plots should not be taken too seriously: they were created by optimizing efficiency with respect to magnetic field and magnetic field taper, but we did not perform a complete exploration of the parameter space. Thus, there is no guarantee that we found the global maximum; a more detailed analysis could yield slightly different results.

We also looked at the sensitivity of the phase of the output radiation to changes in the pitch angle and beam voltage. We found that the phase changed about 20 degree per percent change in pitch angle and about 40 degrees per percent change in beam voltage, with a weak dependence on velocity spread. This is comparable to conventional multi-cavity gyroklystron amplifiers operating at similar voltages. As in Ref. 13 we were able to almost eliminate the phase noise by adjusting the beam voltage and the pitch. For our parameters, the beam voltage was increased by 1/2% while the pitch angle was decreased by 1%. Such an adjustment could in principle be done with an appropriate active divider circuit on the magnetron injection gun.[14]

VI. Summary.

We have examined some of the issues relevant to relativistic gyrotrons. We found that high efficiency and the requirement that the radiation be cutoff in the beam tunnels at the operating frequency work in opposite directions, especially for harmonic devices. For gyro-TWT amplifiers, however, we can drop the cutoff condition; consequently, these devices have the potential to operate at high efficiency and low magnetic field. To decrease mode competition, it is likely that operation in the gyrotwistron configuration will be necessary. A preliminary gyrotwistron design achieved an efficiency of about 45%, more or less independent of axial velocity out to a spread of 6%. However, for a realistic device, the magnetic field and the overall length of the device need to be reduced. Both of these goals could probably be accomplished by using a nonlinear magnetic field taper rather than a linear one. In addition, we looked at the possibility of combining a gyrotwistron with a slow wave structure; this should increase the achievable efficiency.

Acknowledgements.

This work was supported by the United States Department of Energy.

References.

1. R.K. Parker and R.H. Abrams, Jr., "Radio-Frequency Vacuum Electronics: a Resurgent Technology for Tomorrow!," SPIE **791**, Millimeter Wave Technology IV and Radio Frequency Power Sources (1987).

2. J. Benford and J. Swegle, *Ninth International Conference on High-Power Particle Beams*, Washington, DC, May 1992; J. Benford and J. Swegle, *1992 IEEE International Conference on Plasma Science*, Tampa, FL, June 1992.

3. V.L. Bratman, N.S. Ginzburg, G.S. Nusinovich, M.I. Petelin and P.S. Strelkov, "Relativistic Gyrotrons and Cyclotron Autoresonance Masers," *Int. J. Electron.* **51**, 541 (1981).

4. P.E. Latham, "AC Space-Charge Effects in Gyroklystron Amplifiers," *IEEE Trans. Plasma Sci. Special Issue on High Power Microwave Generation* **18**, 273 (1991).

5. G.S. Nusinovich and H. Li, "Theory of the Relativistic Gyrotwistron," *Phys. Fluids B* **4**, 1058 (1992).

6. W. Lawson, J.P. Calame, B. Hogan, P.E. Latham, M.E. Read, V.L. Granatstein, M. Reiser, and C.D. Striffler, "Efficient Operation of a High-Power X-Band Gyroklystron," *Phys. Rev. Letters* **67**, 520 (1991).

7. S.G. Tantawi, W. Main, P.E. Latham, G.S. Nusinovich, W. Lawson, C.D. Striffler, and V.L. Granatstein, "High Power X-Band Amplification from an Overmoded Three-Cavity gyroklystron with a Tunable Penultimate Cavity," To be published in *IEEE Trans. Plasma Sci.* **20**, (1992).

8. J.P. Calame, W. Lawson, V.L. Granatstein, P.E. Latham, B. Hogan, C.D. Striffler, M. Reiser, and W. Main, "Experimental Studies of Stability and Amplification in Four Overmoded, Two-Cavity Gyroklystrons Operating at 9.87 GHz," *J. Appl. Phys.* **70**, 2423 (1991).

9. S.V. Gaponov, M.I. Petelin, and V.K. Yulpatov, "The Induced Radiation of Excited Classical Oscillators and Its Use in High-Frequency Electronics," *Radiophys. and Quantum Electron.* **10**, 794 (1967).

10. J. Neilson, P.E. Latham, M. Caplan, and W. Lawson, "Determination of the Resonant Frequencies in a Complex Cavity Using the Scattering Matrix Formulation," *IEEE Trans. Microwave Theory Tech.* **MTT-37**, 929 (1992).

11. W.B. Herrmannsfeldt, SLAC Report No. 226, 1979 (unpublished).

12. G.S. Nusinovich, "Cyclotron Resonance Masers with Inhomogeneous External Magnetic Fields," To be published in *Phys. Fluids B* **4**, July 1992.

13. W.L. Menninger, V.G. Danly, and R.J. Temkin, "Autophase cyclotron Autoresonance Maser Amplifiers," *Phys. Fluids B* **4**, 1077 (1992).

14. W. Lawson, private communication.

NARROW PASS-BAND HIGH POWER TWT AMPLIFIER

Levi Schächter and John A. Nation

Laboratory of Plasma Studies and School of Electrical Engineering
Cornell University, Ithaca NY

ABSTRACT

At power levels exceeding the $100MW$ level the spectrum of the output signal from a traveling wave tube amplifier contains a significant amount of power in frequencies other than the input frequency. The structure of the spectrum is determined by the noise in the system and the interference of the two waves bouncing due to finite reflections at the two ends of the waveguide. There are cases when the power in these selected frequencies is actually equal or even larger than in the initial frequency. Consequently, the device operates more like an oscillator than as an amplifier. In order to overcome this problem we have designed a narrow band periodic structure in which the electrons, whether they are bunched or not, can emit radiation only in a range of less than $50MHz$ - in comparison to $1.5GHz$ in the old structure. In this structure the group velocity is very low ($\beta_{gr} < 0.01$) and therefore the gain per unit length (for the same current) is about 5 times larger than in the previous structure. The low group velocity practically eliminates also the reflection problem since we can design the system such that by the time the first reflected wave reaches the input, the electron pulse has ended.

INTRODUCTION

Experiments indicate that power levels of more than $400MW$ in the X-band can be achieved with traveling wave tubes corresponding to more than 45% efficiency[1]. However the frequency content of these signals was broad ($> 300MHz$) and only about $200MW$ of this power spectrum actually overlaps, within $30MHz$, the input frequency. The remainder is amplified noise which occurs at selected frequencies. The interference of the two waves bouncing between the ends of the structure is responsible to this selection process[2]. Although the $400MW$ fits well the requirements of an acceleration system, the frequency content remains to be improved for a realistic application. In this paper we present suggested improvements in a TWT amplifier to ensure single frequency operation.

NARROW BAND STRUCTURE

The periodic structure which was used so far as the cold circuit of the amplifier consists of a corrugated waveguide with an average radius of $1.32cm$ and a corrugation amplitude of $0.4cm$. This system has a dispersion relation which can be approximated by

$$f(GHz) \approx 8.95 - 0.75cos(kL) ; \tag{1}$$

where $L = 0.7cm$ is the periodic length of the structure and k is the wave number in the first Brillouin zone. This corresponds to a pass band between 8.2 and 9.7 GHz. Typically the injected wave is at $8.76GHz$ and its bandwidth is about $1MHz$.

In the interaction process the electrons get bunched. The velocity modulation generates electrons in a relatively wide range of velocities. In the case of $100MW$ output we calculated a spread in the normalized electron velocity β of 0.8 to 0.97 (the average velocity at the input was 0.93 with less than 1% fluctuations). If we draw on the dispersion curve these two lines and check what frequency range they cover, we will find that this is more than $300MHz$ - Fig. 1. In other words we expect noise to occur in all this frequency range. As we mentioned above some of the frequencies will be selected in the reflection process and sidebands will ultimately occur[2].

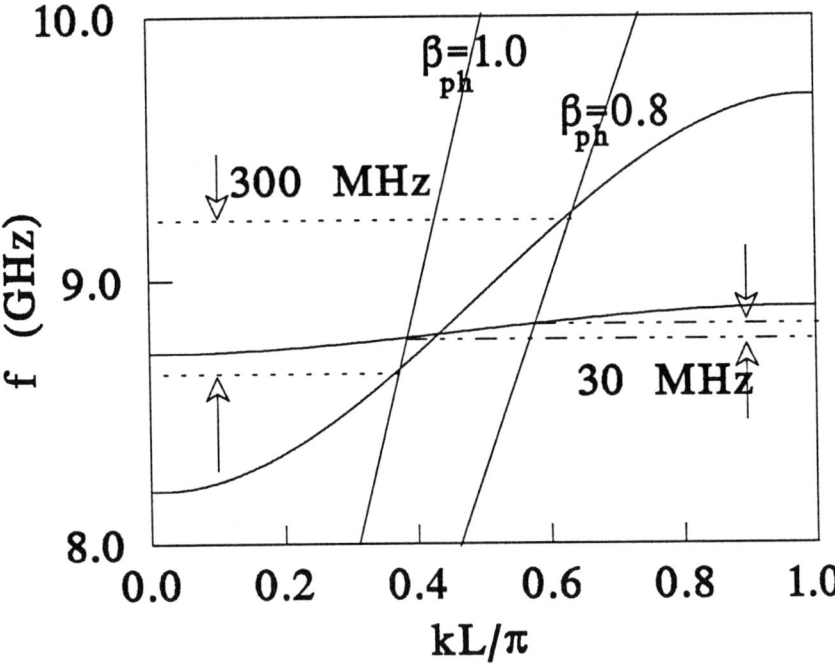

Fig. 1: The dispersion relation of the old and new structures. Note that in the old structure electrons with $0.8 < \beta < 1.0$ could radiate in a band of about $300MHz$ whereas in the new structure this band is only $30MHz$.

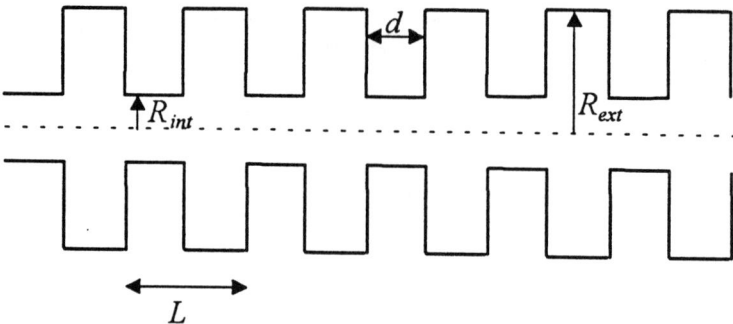

Fig. 2: The geometry of the new structure: $R_{ext} = 14.2mm$,
$R_{int} = 6.2mm$, $d=6mm$ and $L=12mm$.

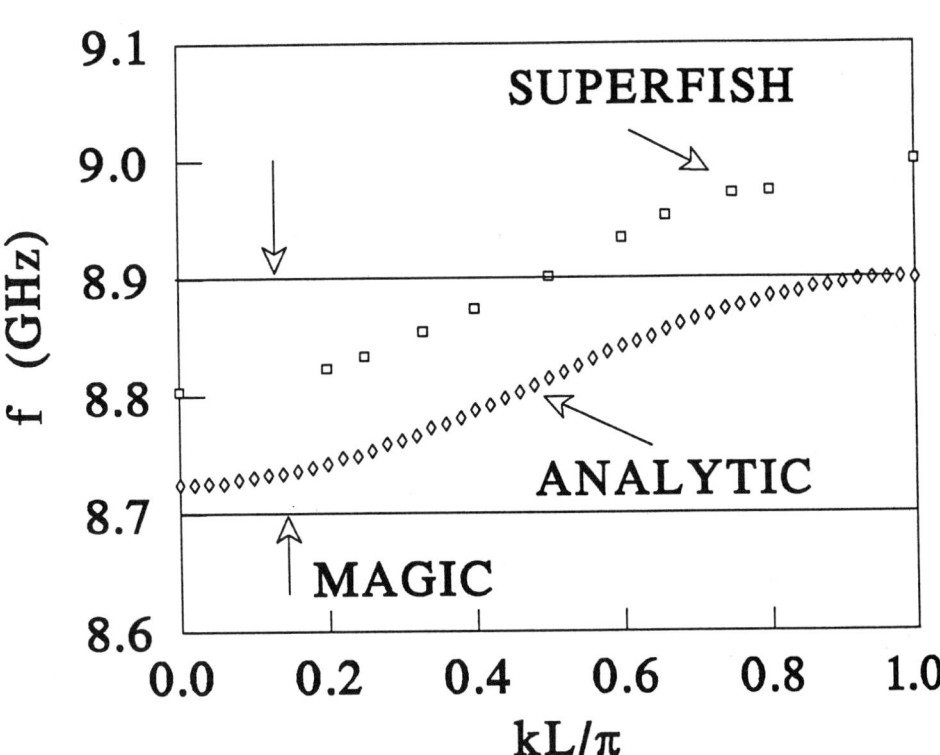

Fig. 3: Comparison of the results of three "cold" test on the narrow band structure.

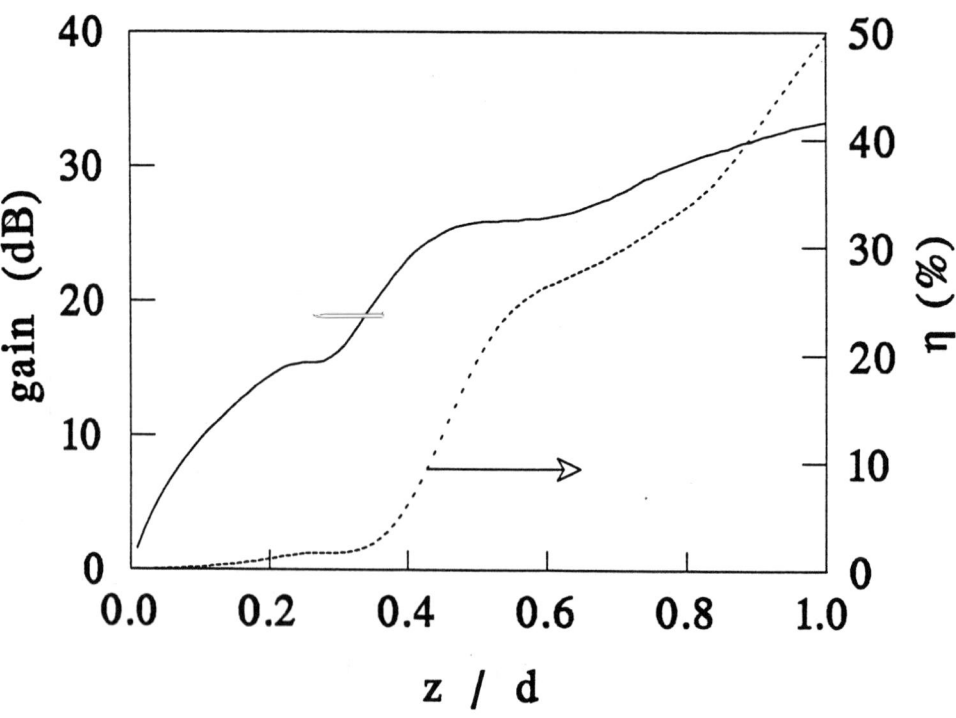

Fig. 4: The gain and the efficiency along the structure.

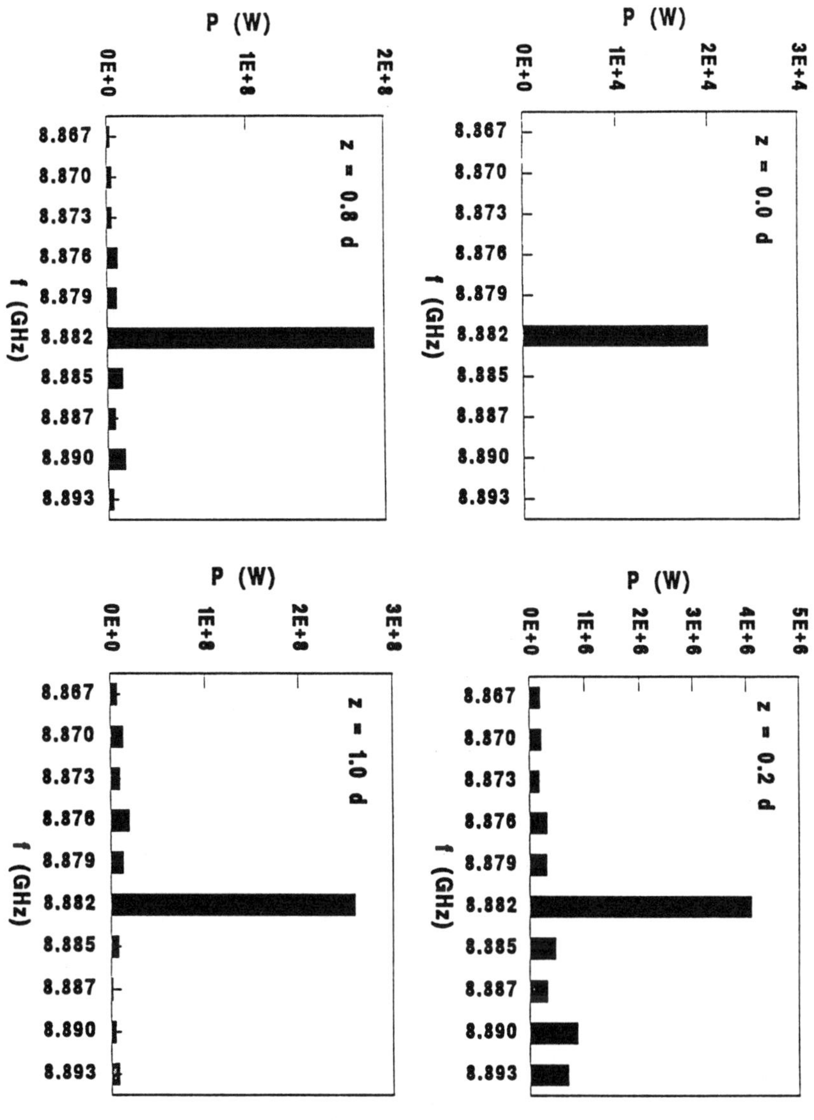

Fig. 5: The power in adjacent frequency at various points along the interaction region.

In order to avoid both the broad band noise problem and the effect of reflections we have designed a narrow band structure (NBS). It consists of a corrugated waveguide of $1.2cm$ periodicity, half of which is the groove; the internal radius is $0.62cm$ and the external is $1.42cm$ - see Fig. 2 . The dispersion relation was calculated analytically and the results were used for the design process. In Fig. 1 the dispersion curve of this structure is superimposed on the old one to illustrate the effect on the possible noise spectrum. For the same velocity spread as before we clearly observe that the noise spectrum is expected to be only $30MHz$. As in the case of the old structure (Eq.(1)) we can approximate the dispersion relation of this structure by

$$f(GHz) = 8.81 - 0.09cos(kL) . \tag{2}$$

This corresponds to a pass-band between $8.72GHz$ and $8.9GHz$ i.e. less than $200MHz$ compared to $1.5GHz$ in the previous case.

The dispersion relation of this structure has been calculated analytically, tested with MAGIC and SUPERFISH. The results are presented in Fig.3 . The results from SUPERFISH are systematically about $100MHz$ above the analytical results. With MAGIC we only determined the pass band of the structure and it fits well the analytical results. A 10 cells structure has been built and tested with a spectrum analyzer the cold test supports the SUPERFISH predictions. The reasons for discrepancy are yet to be investigated.

INTERACTION IN THE NARROW BAND STRUCTURE

In the frequency domain where the interaction occurs, we have lowered the group velocity of the structure by a factor of 20 namely, from 0.14 in the old structure to 0.01 in the NBS case. The coupling between the electrons and the wave can be shown to be inversely proportional to the energy velocity (which is the average power flow in one cell divided by the total energy stored in the same cell). The latter is closely related to the group velocity - in fact in non-highly dispersive electromagnetic structures the two are identical. In our case they seem to differ quite substantially $\beta_E \simeq 0.007$. The interaction has been analyzed using the following simplified set of equations:

$$\frac{d}{d\bar{z}}\gamma_i(\bar{z}) = -\frac{1}{2}\left[a_+(\bar{z})e^{j\chi(\bar{z})} + c.c.\right] , \tag{3}$$

$$\frac{d}{d\bar{z}}a_+(\bar{z}) = \alpha\langle e^{-j\chi_i(\bar{z})}\rangle , \tag{4}$$

$$\chi_i(\bar{z}) = \chi_i(0) + \int_0^{\bar{z}} d\xi[\frac{\omega}{c}d\frac{1}{\beta_i(\xi)} - kd] , \quad a \equiv eE_zd/mc^2 . \tag{5}$$

where α is the coupling coefficient and it is calculated analytically. This parameter in the NBS is about 4 times larger than in the old structure. Fig. 4 illustrates the gain and the efficiency in the structure for the following parameters: 10 periods, $P_{in} = 20kW$ at $8.882GHz$ and $800A$, $800kV$. The output gain is about $42dB$ corresponding to an efficiency of about 50% . The output power at the input frequency is almost $260MW$ from a total RF power of $320MW$ in the range between $8.867GHz$ and $8.893GHz$. Along the interaction region the frequency is stable as indicated in Fig.5 where we show the power at different adjacent frequencies.

In this high power, high efficiency device the electric field which develops on axis is very high - order of $170MV/m$ - and higher gradients are expected on the structure's surface. The proximity of the beam to the surface may be an advantage in this case since it may provide the "electrostatic insulation" to avoid breakdown.

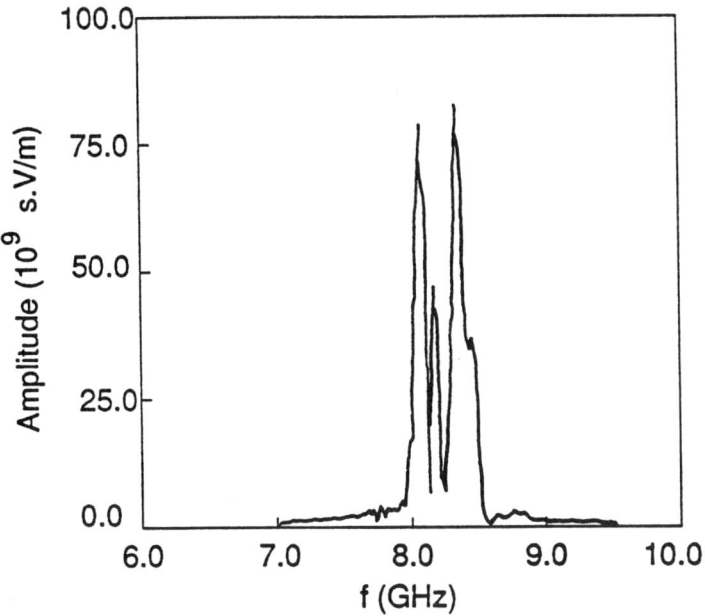

Fig. 6: The spectrum at the output of an old structure covers more than $500MHz$.

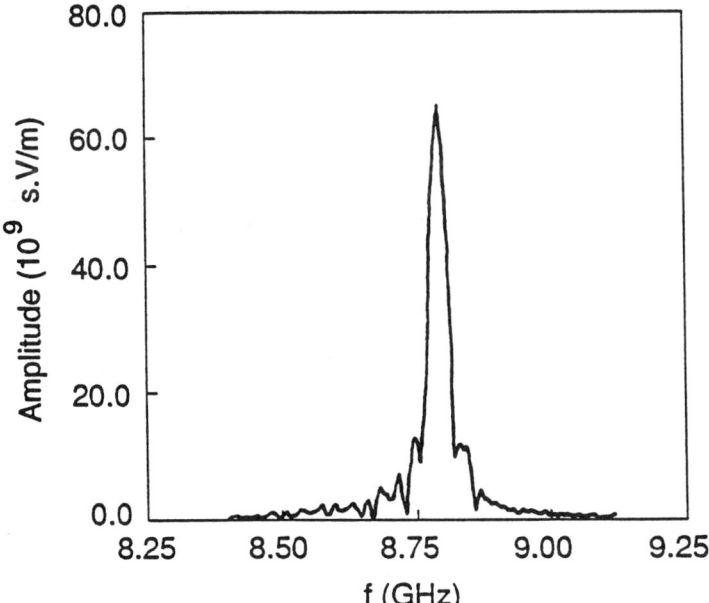

Fig. 7: The spectrum at the output of the new structure covers about $50MHz$.

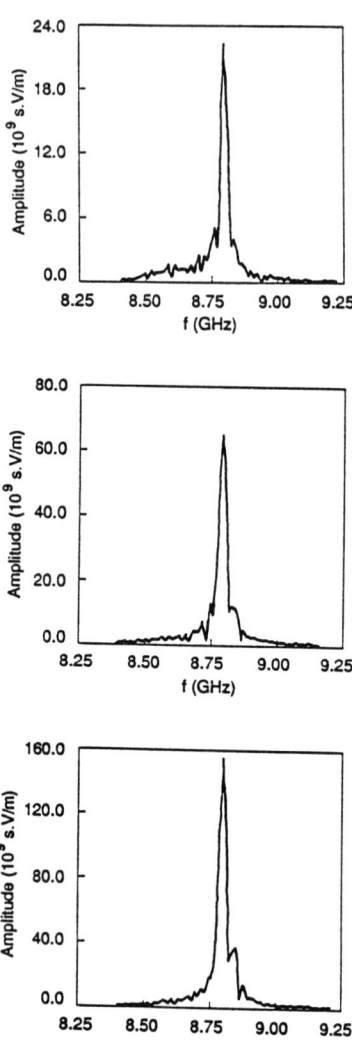

Fig. 8: The frequency is stable along the interaction region as indicated by these three frames each one representing the spectrum at
different points.

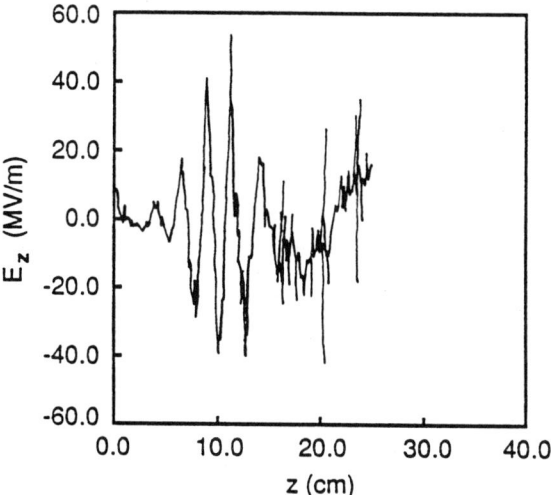

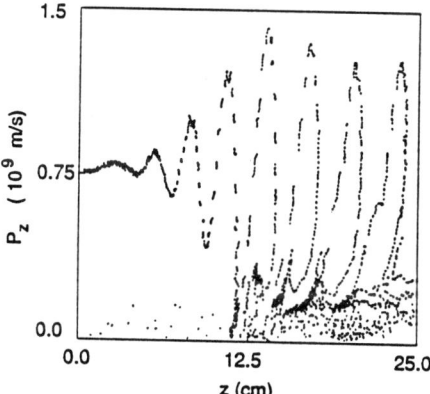

Fig. 9: Growth of the axial electric field in amplifier. The amplifier is 12cm long; the remainder is a uniform pipe. An electron phase-space distribution for the conditions shown above is illustrated in the lower frame.

MAGIC SIMULATIONS

The system has been simulated with the particle in cell code $MAGIC$. The three main features of the design namely, (i) narrow band, (ii) high gain and (iii) frequency stability have been confirmed by simulations. Figs. 6 and 7 illustrate a comparison of the bandwidth in the old structure compared to the new one; in the old case the signal spread was in a range of $500MHz$ whereas in the present case the indicated bandwidth is $50MHz$ which actually represents the temporal resolution of the simulation rather than the physical one which is expected to be even narrower. Fig. 8 shows the FFT of the signal at various points along the interaction region and clearly the frequency is the same in all three cases. Finally, the longitudinal electric field on axis is presented in order to show the spatial growth rate in the structure - Fig. 9. In the lower frame the corresponding phase-space distribution is illustrated. The amplifier ends at $z = 0.12m$ and is followed by a uniform waveguide. The loss of wave coherence and particle trapping is due to the dicontinuity at the transition between the two regions.

DISCUSSION AND CONCLUSIONS

In the design process of the NBS we have practically eliminated the possibility of sidebands. The way this was achieved is as follows: we have previously mentioned that the energy velocity is about $0.007c$. For a $12cm$ long structure the time the energy, which is reflected from the output, will take to reach the input, is of order of $60nsec$. This is of the same order of magnitude as the beam pulse which means that by the time the reflected wave reaches the input end there are no more electrons to interact with the wave.

The frequency content of the pulse is about $30MHz$ according to the analytical calculations and narrower than $50MHz$ as predicted by $MAGIC$. The gain per unit length in the NBS is at least 5 times larger than in the old structure. Based on the quasi-analytical calculations we should be able to achieve power levels of more than $250MW$ with an efficiency of about 45% and a frequency content of about $30MHz$.

This work was supported by the United States Department of Energy and by AFOSR.

REFERENCES

1.   D.A. Shiffler, J.A. Nation, L.Scahchter, J.D. Ivers and G.S. Kerslick; J. Appl. Phys. **70**(1), 106(1991).

2.   L. Schachter, J.A. Nation and D.A. Shiffler; J. Appl. Phys. **70**(1), 114(1991).

SECOND HARMONIC MAGNICON AMPLIFIER EXPERIMENT AT 11.4 GHz

S. H. Gold and C. A. Sullivan
Beam Physics Branch, Plasma Physics Division, Naval Research Laboratory
Washington, DC 20375–5000

B. Hafizi
Icarus Research, Bethesda, MD 20814

W. M. Manheimer
Senior Scientist for Fundamental Plasma Processes
Plasma Physics Division, Naval Research Laboratory
Washington, DC 20375–5000

ABSTRACT

We present a report on a program to develop a high–power X–band magnicon amplifier for linear accelerator applications. The goal of the program is to generate 50 MW at 11.4 GHz with 50% efficiency, using a 200 A, 500 keV electron beam produced by a cold–cathode diode on the NRL Long–Pulse Accelerator Facility. The initial experiment, designed to study the gain from the first (driven) deflection cavity to a second (passive) deflection cavity, is under way.

INTRODUCTION

The magnicon[1-3] is a "scanning beam" microwave amplifier tube, related to the gyrocon,[4] that is a potential successor to the klystron for powering future high–accelerating–gradient electron linear accelerators. Scanning beam devices modulate the insertion point of the electron beam into the output cavity in synchronism with the phase of a rotating rf wave. This synchronism creates the potential for an extremely efficient interaction in the output cavity, since every electron will in principle experience identical decelerating rf fields. In the magnicon, the output interaction is gyrotron–like, and requires a beam with substantial transverse momentum about the applied axial magnetic field. The transverse momentum is produced by spinning up the electron beam in a sequence of TM_{110} deflection cavities, the first of them driven by an external rf source. The output cavity employs an rf mode that rotates at the same frequency as the deflection cavity mode. As a result, the beam entering the output cavity is fully phase modulated with respect to the output cavity mode. The optimum magnetic field in the deflection cavities is twice the cyclotron resonant value at the drive frequency, while the output cavity operates as a first harmonic cyclotron device. These twin constraints lead naturally to the design of a second–harmonic amplifier, in which the output cavity operates at twice the frequency of the deflection cavities, and employs a TM_{210} mode. The overall design concept is shown in Fig. 1.

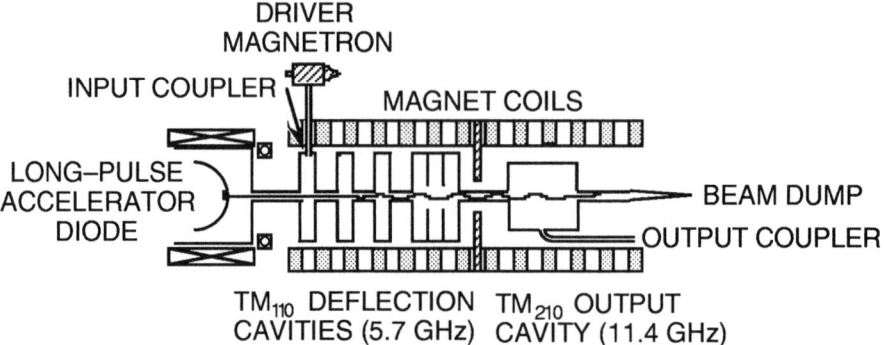

Fig. 1. NRL 11.4 GHz magnicon amplifier concept

In this paper, we discuss a preliminary experiment, employing only two 5.7 GHz deflection cavities, the first driven by an external source. We are performing parametric studies of the gain between these two cavities, preparatory to the design of a complete deflection system that will spin up an electron beam to high α for injection into an 11.4 GHz output cavity. Here, α is the ratio of perpendicular to parallel velocity.

APPARATUS

The first two–deflection–cavity magnicon gain experiment is under way on the NRL Long–Pulse Accelerator Facility.[5] It is shown schematically in Fig. 2. It employs a field–emission diode that was designed via the Herrmannsfeldt Code[6]

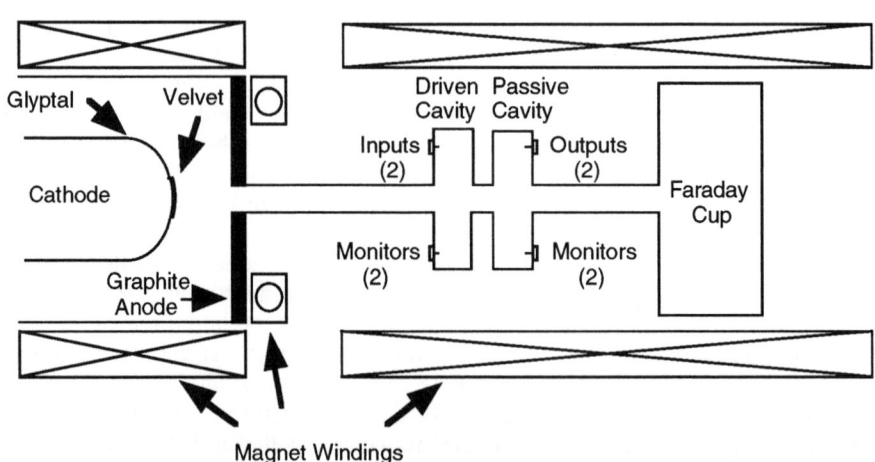

Fig. 2. Schematic of two–deflection–cavity gain experiment

to produce a high quality solid electron beam with minimum diameter and minimum velocity spread. The diode design is shown in Fig. 2. The measured beam parameters are ~200 A at 500 kV, with a beam diameter of ~5 mm at a final magnetic field of 8.1 kG. The Herrmannsfeldt Code predicts a mean α of ~.025 as the beam enters the first deflection cavity. The propagated current is measured by a Faraday cup, and beam cross sections are measured using witness plates.

A schematic of a single TM_{110}–mode deflection cavity is given in Fig. 3. There are four coupling ports spaced at 90° intervals in one end–wall of the cavity. Two adjacent pins are "long," for use in driving the cavity in each linear polarization, and the remaining pins are "short," in order to sample the cavity fields without significantly loading the cavity. The design value of the cavity quality factor, Q, is 1000, in each linear polarization, where the primary loading factor is the coupling pins. All the pins are coupled through type "N" connectors to external coaxial cables. The microwave cavities are fabricated from stainless steel, with a copper coating on interior surfaces to increase the ohmic Q. The cavities are of identical design. The first cavity is driven by a tunable C–band magnetron at approximately 5.7 GHz.

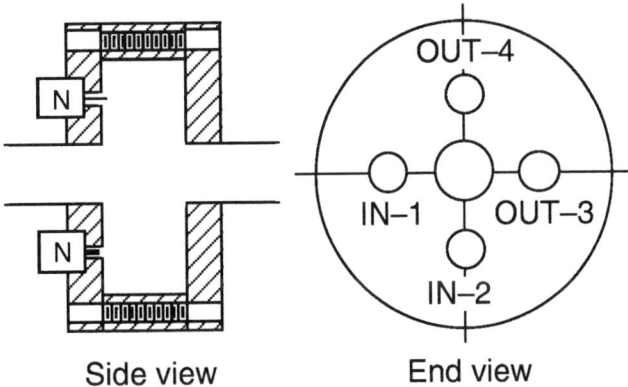

Fig. 3. Schematic of magnicon deflection cavity

The magnicon is designed to employ a circularly–polarized mode in the first cavity, and should generate a circularly–polarized mode in the second deflection cavity. The circularly–polarized signals are generated, and detected, using 3 dB hybrid couplers to drive (or sample) adjacent pins with a $\pi/2$ phase difference. However, the initial experiments are under way without 3 dB hybrids, and employ a linearly–polarized signal in the driven cavity. The coupling port in the orthogonal linear polarization is terminated in a matched load, in order to maintain the design value of Q. All of the coupling and sampling ports of each cavity are fully calibrated.

MAGNICON THEORY AND SIMULATION

The linear theory of the magnetized deflection cavities was first presented by Karliner, et al.,[1] and is developed in detail by Hafizi, et al.[7] The linear theory has been evaluated for a single on–axis electron, with no initial transverse momentum, and without finite beam radius and finite velocity spreads. Furthermore, it assumes that the electron energy is not changed by transit through the deflection cavities. In order to consider the use of more realistic beam parameters, such as are predicted by the diode simulation discussed above, and to model the variation of beam kinetic energy in the deflection cavities, a numerical simulation code for the deflection cavities was developed.[7] It is a self–consistent steady–state code that propagates particles through the TM_{110} fields of the first (driven) deflection cavity, through a drift space, and then through successive deflection cavities, followed by drift spaces. The rf field amplitudes are made (by iteration) self–consistent with the finite value of cavity Q and with the energy lost by the electron beam in transit through each cavity. The rf phase in each of the passive cavities is assumed to be the optimum phase to extract electron beam energy from an initially on–axis electron, since this should be a good approximation to the phase that is driven by a finite electron beam.

As regards the output cavity, we have performed both steady–state and time–dependent simulations. The time–independent simulations are an effective means of testing the sensitivity of the final cavity interaction to spreads in α or in guiding center radius. The time–dependent simulations are important in order to see whether the assumed final state is accessible and stable. This is especially important for a TM–mode interaction, where the oscillation in axial velocity has the effect of allowing more than one possible final state. Also, the time–dependent simulations are a simple means of determining the final phase relationship between the wave field and the particles. For a magnicon, the time–dependent simulations are particularly economical because, if spreads are neglected, all particles enter with the same phase between the gyro–orbit and the rotating rf fields in the cavity. Thus, a simulation can be performed using only one particle at each time step. Initial results of such a simulation are shown in Fig. 4 for an 11.4 GHz magnicon operating in a TM_{210} mode. These calculations employ a 6.2–cm–long, $Q=130$ output cavity driven by a 500 keV, 172 A, $\alpha=1.5$ electron beam with its guiding center radius equal to its Larmor radius of 0.38 cm in a magnetic field of 6.9 kG. Clearly, a high efficiency state ($\eta\sim60\%$) is accessible with fairly low rf electric field (~200 kV/cm).

EXPERIMENTAL RESULTS

In the first experimental test of microwave gain and beam frequency–pulling effects, the first deflection cavity was driven by a tunable C–band magnetron in a linearly–polarized TM_{110} mode through a single coupling pin. A sampling pin was used to diagnose the linearly–polarized signal in the driven cavity. The remaining pins of the first cavity led to matched loads. The signal in the second cavity was measured, in either linear polarization, by means of the sampling pins, with the larger coupling pins terminated to preserve the Q of 1000 seen in the cold tests.

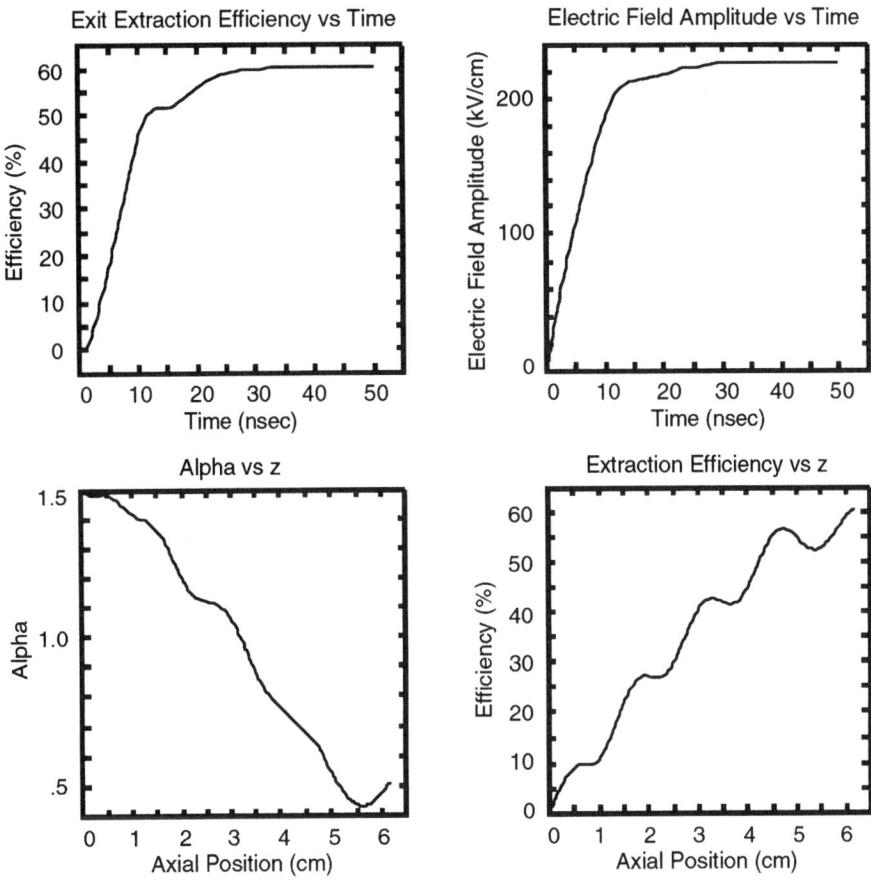

Fig. 4. Results of time–dependent simulation of output cavity operation

In the first set of measurements, the magnetron frequency was varied to measure the intercavity gain and bandwidth, and to observe the frequency pulling effect of the beam on the cold–cavity resonant frequency. A signal gain of ~6 dB was observed, into a single linear polarization of the output cavity, with a comparable signal in the orthogonal polarization, for a total gain of 9 dB. Theory suggests that only one circular polarization is effective in driving the interaction. Correcting for this as well leads to an apparent gain of ~12 dB, compared to ~13 dB predicted by theory and simulation. In addition, a frequency detuning of the first cavity was observed, due to the presence of the beam. This detuning was approximately −0.2%, in good agreement with the predictions of theory.

DISCUSSION

This first two–cavity gain experiment has demonstrated gain between a driven and a passive deflection cavity relevant to a future magnicon device. This gain is being studied as a function of macroscopic beam parameters (current, voltage) and magnetic field. The initial results compare well to predictions of theory. For the present experiment, a measurement of intercavity rf gain can be used to infer the amount of coherently phased transverse momentum (beam α) that has been produced. However, the driven α is still small, and cannot be measured directly without the addition of further gain cavities. In future experiments, additional deflection cavities will be employed, in order to study the coherent spin–up of the electron beam to high α for injection into a magnicon output cavity. The effect of velocity spread on the gain in the present experiment is predicted to be small. However, it will affect the quality of the final high α electron beam delivered to the output cavity. The real test of the final multicavity deflection system will be to produce a high α electron beam without excessive spreads in α or in gyrophase, and without employing excessively large rf electric fields in the final deflection cavity.

SUMMARY

The NRL X–band magnicon amplifier experiment has begun its initial experimental operation. The first experiments are examining only the deflection cavities, and are intended to measure the gain and beam loading in a pair of deflection cavities, the first driven and the second passive. At the same time, numerical simulation of both the deflection cavities and the output cavity is proceeding, in order to guide the future path of the experiment. The initial experimental results show an apparent gain of ~12 dB between a driven and a passive deflection cavity, accompanied by a frequency shift of approximately -0.2%, in good agreement with the predictions of theory and numerical simulation.

ACKNOWLEDGMENTS

This work was supported by the U.S. Department of Energy, under Interagency Agreement No. DE–AI05–91ER40638, and by the Office of Naval Research.

REFERENCES

1.   M. M. Karliner, E. V. Kozyrev, I. G. Makarov, O. A. Nezhevenko, G. N. Ostreiko, B. Z. Persov, and G. V. Serdobintsev, Nucl. Instrum. Methods Phys. Res. **A269**, 459 (1988).

2.   O. A. Nezhevenko, *Conference Record—1991 IEEE Particle Accelerator Conference*, edited by L. Lizama and J. Chew (IEEE, New York, 1991), pp. 2933–2942.

3.   W. M. Manheimer, IEEE Trans. Plasma Sci. **18**, 632 (1990).

4. G. I. Budker, M. M. Karliner, I. G. Makarov, S. N. Morosov, O. A. Nezhevenko, G. N. Ostreiko, and I. A. Shekhtman, Part. Accel. **10**, 41 (1979).

5. N. C. Jaitly, M. Coleman, S. Eckhouse, A. Ramrus, S. H. Gold, R. B. McCowan, and C. A. Sullivan, *Digest of Technical Papers—Eighth IEEE International Pulsed Power Conference*, edited by K. Prestwich and R. White (IEEE, New York, 1991), pp. 161–165.

6. W. B. Herrmannsfeldt, "Electron Trajectory Program," Stanford Linear Accelerator Report No. 226, 1979 (unpublished).

7. B. Hafizi, Y. Seo, S. H. Gold, W. M. Manheimer, and P. Sprangle, IEEE Trans. Plasma Sci. **20**, 232 (1992).

A GYRORESONANCE HARMONIC CONVERTER TO DRIVE NLC, THE NEXT LINEAR COLLIDER

J. L. Hirshfield
Omega-P, Inc., 2008 Yale Station, New Haven, CT 06520; and
Physics Department, Yale University, Box 6666, New Haven, CT 06511

A. K. Ganguly
Code 6840, U. S. Naval Research Laboratory, Washington, D. C. 20375

ABSTRACT

Current approaches are reviewed for developing an rf source for driving the Next Linear Collider (NLC) based on various gyroresonance interactions. An alternative mechanism is described, with the potential of converting 2.85 GHz power from a SLAC klystron to output at a harmonic ranging from 11.4 to 31.4 GHz. Preliminary calculations show that this new process can enjoy high conversion efficiency, strong resilience to axial momentum spread on the beam, and weak power flow into competing modes.

INTRODUCTION

Need is widely accepted for a novel rf driver for NLC (the Next Linear Collider) at a frequency above 10 GHz, to limit the machine length and its ac power consumption [1]. As a result, a considerable effort has been made recently to extend multi-cavity high-power klystron technology to 11.4 GHz [2], and to develop efficient gyroresonance sources for this application up to 35 GHz [3, 4, 5].

This paper will review the rationale for pursuing gyroresonance mechanisms as a solution to this problem, and will illustrate with four examples having decreasing levels of maturity. Reports have appeared recently on three of these promising mechanisms, so it is not necessary in this paper to present more than a brief overview of each. However, details are given of a novel idea [6, 7] that emerged in 1991 for exploiting the spatiotemporal nature of gyroresonance motion on the electron beam from a cyclotron autoresonance accelerator [8, 9]. Preliminary calculations show that fast wave radiation can be extracted from such a beam with high efficiency at a harmonic of the accelerator frequency [10]. It has furthermore been shown that conditions may be found under which the efficiency is not seriously diminished by moderate axial momentum spread on the beam, and that power flow from the beam into competing modes can be small. Thus it could be that an harmonic converter driven by an existing 60 MW, 2.85 GHz SLAC klystron will be technically and economically appealing as an alternative to the more conventional gyroresonance processes for generating the needed drive power for NLC.

THE RATIONALE FOR EMPLOYING GYRORESONANCE INTERACTIONS

Gyroresonance interactions have appeal for the development of high power microwave sources because of their intrinsic properties and because of difficulties encountered in designing conventional klystrons at short wavelengths.

Since gyroresonance interactions (e.g., gyrotrons, CARMS, gyroklystrons, gyro-TWTs, and gyroresonance harmonic converters) are by nature *fast wave* interactions, the microwave circuits employed are not constrained to dimensions much less than a free-space wavelength, as is customary for slow wave devices such as the klystron and TWT. In fact, whispering gallery gyrotron oscillators often have cavity diameters and lengths that extend to many free-space wavelengths [11]. Larger circuit dimensions will support greater stored rf energy without breakdown, and will usually distribute wall losses over a larger wall area; both of these factors increase the power-handling capacity of a fast wave circuit, in comparison with a conventional slow wave circuit operating at the same wavelength. Since a larger fast wave circuit can accommodate a larger electron beam (which is often annular in a gyroresonance device) demands are lightened on electron gun design, especially for cathode loading and accelerating electric field strength. For some gyroresonance interactions (such as gyro-TWT and CARM amplifiers, and gyroresonance harmonic converters) the microwave circuit is a smooth wall cylindrical waveguide; the absence of disk-loading or other discontinuities in the circuit avoids conditions that can lead to the beam breakup instability [12].

Despite clear problems with the multi-cavity klystron amplifier concept at frequencies above 10 GHz, impressive strides have recently been made. Reviews by Caryotakis [2], Lavine [13] and Wilson [14] detail advances in conventional klystron design and related advanced concepts to extend klystron operation to shorter wavelengths and higher average powers. Ingenious circuit designs for relativistic klystrons by Haimson and Mecklenburg [15] have recently been reported, in which 100 - 300 MW, 11.4 GHz power was generated in a 50 nsec pulse. But it remains to be proven if these strides can extend into the high average power regime required for NLC with efficiency values of 40% or more. Nevertheless, given the long and successful research and manufacturing history of high power klystron amplifiers, the available evidence suggests that conclusions as to their unsuitability for NLC may be premature.

A BRIEF REVIEW OF GYRORESONANCE DEVICE RESEARCH

Several gyroresonance interactions have received close attention over the past few years. Here a précis for three of these is given, in order of decreasing maturity. A proposal has also been made to develop a related gyroresonance interaction, the gyro-peniotron, as an accelerator driver at frequencies lower than those for NLC [16]. The mechanisms to be reviewed in this paper include the gyroklystron [3], the CARM amplifier [4], and the magnicon [5].

Beginning in about 1984, a group under the leadership of V. L. Granatstein at the University of Maryland undertook an intense program to develop the gyroklystron concept for driving NLC. Recent progress reports [17] show considerable theoretical and experimental achievement. Typical device operating parameters include a 1 - 2 microsec, 425 kV, 150-200 A beam from a space-charge limited magnetron injection gun with a 4 Hz repetition rate. Two- and three-cavity devices have been tested, with a variety of lossy dielectrics lining the drift tubes and cavities to suppress incipient beam instabilities. At the maximum instantaneous efficiency point 37% of the beam power was converted to 9.871 GHz radiation, while at the maximum output power point 27 MW of 9.875 GHz radiation was produced. The axial velocity spread on the beam of approximately 7%, and the need to operated the MIG gun to produce a beam with a velocity ratio of only about 1.1, can evidently explain why the efficiency is not higher. Near-term plans include second-harmonic operation to either lower the magnetic field at the gun to allow higher beam powers or to operate near 20 GHz, insertion of an inner conductor to increase isolation between cavities to suppress undesired oscillations, and use of a traveling-wave output section to enable further lowering of the gun magnetic field. The demonstrated energy per pulse of 39 J in these tests exceeds by over a factor-of-two what has been reported in advanced relativistic klystrons [15]. The challenge of increasing the device efficiency and average power is now being confronted aggressively.

Research on the cyclotron autoresonance maser (CARM) interaction intensified after early publications by Petelin [18], Hirshfield, Chu and Kainer [19], and Bratman, *et al* [20] pointed out that a considerable Doppler up-shift for radiation generated via the cyclotron maser instability could be realized, and that energy could be extracted from the axial as well as the transverse momentum components of the beam electrons. But early analyses also pointed out the severe sensitivity of the gain to axial momentum spread on the beam. A number of CARM oscillator experiments [21], but only one amplifier result [4] have been published. A group at MIT under the direction of B. G. Danly, J. Temkin, and J. Wurtele has undertaken to design an experiment to demonstrate CARM amplifier operation at 17.1 GHz for application as a driver for NLC. The beam for these experiments is produced with a 500 kV, 500 A, 50 nsec induction accelerator; transverse energy is imparted to the beam using a short wiggler. A beam with a velocity ratio of up to 3 has been produced in this manner, but axial momentum spread is believed to be too large to permit stable amplifier operation. Indeed high power oscillations are observed in several waveguide modes. The high axial momentum spread may be caused partially by scattering from a 45° mesh that is part of the rf input coupler; this design is to be modified to eliminated the mesh. Additional improvements planned in the near term include an improved output window to reduce internal circuit reflections and improved beam diagnostics to sharpen the beam spin-up by the wiggler. Theory predicts that a regime of moderate efficiency could be possible for a CARM amplifier, but values well in excess of the 6.3% reported at 35 GHz [4] will be needed before this interaction can be a practical candidate for driving NLC.

The magnicon, a modification described by Karliner, *et al* [22] of Budker's gyrocon [23], has been proven to operate efficiently at a frequency of 915 MHz, where 2.6 MW of power was generated with 85% electronic efficiency in a 30 microsec pulse. In 1991 [24] Nezhevenko described an experiment underway to demonstrate frequency doubling in a gyrocon, using a 420 kV, 240 A beam in a 2 microsec pulse that is expected to yield 60 - 70 MW of power at 7 GHz. Such an achievement would establish the magnicon to be a serious contender as a driver for NLC, assuming that similar results could be obtained at 10 GHz or higher. Such a goal has recently been set by a group at the U.S. Naval Research Laboratory under the leadership of S. H. Gold and W. M. Manheimer [5]. The NRL experiment is to be powered with a long-pulse electron beam accelerator using a cold field-emitter cathode. The beam parameters are intended to be 500 kV, 200 A, 1 microsec which, with a predicted device efficiency of 50%, is expected to generate 50 MW of power at 11.4 GHz. The experiment, like that described by Karliner, *et al* [22] is to be a doubler, with a multi-cavity deflection system driven at 5.7 GHz. A detailed analysis of this deflection system has recently been published [25]. Maintenance of good beam quality between the deflection system and the output cavity is recognized as a necessity that, if not observed, could result in a lower efficiency value. Success with the NRL experiment might lead to effort to build high-repetition rate version of the long-pulse accelerator, the efficiency of which will have to be taken into account in assessing the overall suitability of this approach to supply the driver for NLC.

THE GYRORESONANCE HARMONIC CONVERTER

The requirements for the rf driver for NLC include such basic issues as choice of operating frequency, overall device efficiency, and unit cost (for an accelerator installation requiring perhaps hundreds of devices). As the brief review given above indicates, it is not yet proven that any of the current approaches can satisfy all of these requirements. Based on preliminary study an alternative process, gyroresonance harmonic conversion, shows promise for satisfying these requirements. A conceptual sketch of a gyroresonance harmonic converter is shown in Fig. 1. In this device, the dominant power source is a 60 MW, 2.85 GHz SLAC klystron, a device that (together with its power supply and modulator) is already in production. Theoretical analysis shows that such a source can be used to generate a spatiotemporally modulated electron beam with high efficiency using cyclotron autoresonance acceleration [8, 9]. This beam is then decelerated in the course of generating coherent radiation at a harmonic of the 2.85 GHz drive frequency. The generation process can be highly efficient as well, since all the beam particles can remain in phase with the radiation during harmonic conversion. Calculations for fifth harmonic conversion described below show that conditions can be found where conversion into neighboring undesired harmonics can be suppressed, and where the efficiency of conversion is not dramatically diminished by a small axial velocity spread on the electron beam. It is not unreasonable to speculate that similar advantages would be enjoyed by an eighth or ninth harmonic converter, so that an efficient source in to 20 - 30 GHz frequency range might be produced in this manner as well. Needless to say, if this were a proven reality, it could have a dramatic effect on the design of NLC.

The work on this process published to date [6, 7] describes the linearized harmonic conversion from a spatiotemporally modulated beam of assumed properties into the fields of a rectangular waveguide. The objectives of this initial work were to find the parameter range where linearized power growth into a given harmonic could lead to a practical conceptual device design, where power growth into competing modes and competing harmonics could be minimized, and where the limits on beam quality needed to achieve these results could be ascertained. A major result of this initial work was the demonstration that mode competition at neighboring harmonics could be a serious issue in a rectangular waveguide, due to the dense mode spectrum and the relatively indiscriminate selection rules. However, it was also shown that use of a square waveguide supporting circularly polarized TE_{0m} modes could reduce this competition to tolerable levels. These results led to the realization that, not surprisingly, a circular waveguide would have harmonic power growth parameters similar to those of square waveguide, and that the governing selection rules are more favorable from the standpoint of mode competition than either rectangular or square waveguide. In what follows, a brief summary is given of the results of linear and nonlinear analysis for harmonic conversion in a cylindrical waveguide [26]. Examples from linearized theory are given for 50 MW fifth and eighth harmonic 14.25 and 22.80 GHz sources, and examples from a non-linear simulation codes using a 150 kV, 6.7 A beam for fifth harmonic conversion at 14.25 GHz are also given.

A class of electron beam equilibria is considered having a current density $J(r,t)$ characterized by the momentum-space integral

$$\mathbf{J}(r,\phi,z,\ t) = -e \int_{-\infty}^{\infty} du \int_{0}^{\infty} dw\, w \int_{0}^{2\pi} d\phi \left[\hat{\mathbf{e}}_\theta \frac{w}{\gamma} + \hat{\mathbf{e}}_z \frac{u}{\gamma} \right]$$

$$\times\ \delta\left(\theta - \theta_0 + \xi z - pt\right) f_0\left(u,w,\phi,\ r,\phi\right),$$

(1)

where r and ϕ are the radial and azimuthal coordinates, and u and w are components of a beam electron's momentum (divided by the electron rest mass m) along and across the z-axis. The relativistic Lorentz factor is related to the momenta by $\gamma^2 = (u^2 + w^2 + c^2)/c^2$. A uniform static magnetic field $\hat{\mathbf{e}}_z B_0$ is imposed. Unit vectors $\hat{\mathbf{e}}_\theta$ and $\hat{\mathbf{e}}_z$ are along the gyrating particles' angular and axial momenta. (The lowest-order particle orbits have no motion along their radii of gyration.) The distribution function for the beam electrons is given by $f_0(u,w,\phi,r,\phi)$, where ϕ is the azimuth angle in momentum space.

The delta-function in the integrand of Eq. 1 gives the equilibrium current density its spatiotemporal character: an individual particle moves on a helix of axial pitch number ξ, and the helix rotates with an angular frequency p. When the electrons are injected adiabatically along a slowly tapered magnetic field from an ideal cyclotron autoresonance accelerator driven at angular frequency p, it is shown in ref. 6 that $\xi = \gamma(p - \Omega)/u$, where $\Omega = eB/m\gamma$ is the gyrofrequency in the uniform region after the taper.

The rate of power transfer into TE_{mn} and TM_{mn} modes is calculated to first order. That is, any current density perturbations caused by the interaction with the vacuum fields of the waveguide are neglected. The fields are expanded in a guiding-center coordinate system to bring out the essential symmetries of the problem. From this, without loss of generality beyond the assumption of an axisymmetric beam distribution function where guiding center and momentum distributions are decoupled, the following significant results emerge.

(i)   Power growth at the m-th temporal harmonic is absent except for modes having an azimuthal mode index of m.

(ii) A beam with a uniform distribution of guiding centers will have a linearized m-th harmonic power growth rate no less than 90% that of a beam having common guiding centers if the ratio of the outer guiding center radius R_b to the waveguide radius R is less than $0.8944/j_m'$, where j_m' is the first zero of $J_m'(x)$. (For m = 5 this reduces to $R_b/R < 0.1394$.)

(iii) The diminution in linearized m-th harmonic power growth due to a uniform distribution of electron axial momentum will not be greater than 10% of the power growth rate for a cold beam if the normalized momentum spread is less than 48/N % where N is the number of interaction guide wavelengths in the harmonic converter.

(iv) Power transfer from a cold beam to TM modes is absent under conditions of phase matching, as a result of exact cancellation between contributions from the axial and transverse field components.

For fifth harmonic conversion, the lowest circular waveguide mode that can couple to the beam is TE$_{51}$. For this mode, linearized (i.e, first-order) theory gives for the spatial growth of power the result

$$P(L) = 1092.9 \left(\frac{\beta_\perp^2}{\beta_\parallel^3}\right) \left[J_5'(k_\perp\rho)\frac{L}{R}I_0\right]^2 \text{ watts,} \tag{2}$$

where $\beta_\perp$ and $\beta_\parallel$ are the transverse and axial electron velocity components (divided by c), $J_5'(k_\perp\rho)$ is the derivative of the fifth-order Bessel function with respect to its argument, $k_\perp$ is the cutoff wavenumber, ρ is the gyration radius, I_0 is the dc beam current in A, R is the waveguide radius, and L is the interaction length along the waveguide. To give an order-of-magnitude estimate of the parameters for a conceptual device, Eq. (2) has been evaluated for a 60 MW beam with current values in the range of 6 - 60 A. The relevant parameters for a fifth-harmonic converter at 14.25 GHz are shown in Table I.

beam voltage (MV)	beam current (A)	ρ (cm)	R (cm)	B (kG)	P(L)/L^2 (kW/cm^2)	L for P = 30 MW (cm)
1	60	1.714	2.370	2.477	118.4	15.9
3	20	1.843	2.397	5.624	11.85	50.3
5	12	1.861	2.401	8.801	4.18	84.7
10	6	1.870	2.403	16.759	1.03	170.4

Table I.   Parameters for conceptual fifth harmonic converters at 14.25 GHz for 60 MW beams with velocity ratio $\beta_\perp/\beta_\parallel = 2.0$. Waveguide mode is TE$_{51}$.

To illustrate how the parameters change as the choice of the harmonic number changes, the m = 8 counterpart of Eq. (2) has been evaluated for an eighth harmonic converter to produce power at 22.8 GHz. Table II gives the parameters for the same beams as in Table I.

beam voltage (MV)	beam current (A)	ρ (cm)	R (cm)	B (kG)	P(L)/L^2 (kW/cm^2)	L for P = 30 MW (cm)
1	60	1.714	2.227	2.477	142.77	14.5
3	20	1.843	2.253	5.624	12.49	49.0
5	12	1.861	2.256	8.801	4.41	82.5
10	6	1.870	2.258	16.759	1.09	165.7

Table II.   Parameters for conceptual eighth harmonic converters at 22.8 GHz for 60 MW beams with velocity ratio $\beta_\perp/\beta_\parallel = 2.0$. Waveguide mode is TE$_{81}$.

As is seen by comparing Tables I and II, linearized eighth harmonic power growth is slightly stronger than fifth harmonic, for the same beam parameters. This occurs since the harmonic coupling factor $J_m'(k_\perp \rho)$ falls only as $m^{-2/3}$ for relativistic beams, so long as $m \ll 3\gamma^3$; but the electric field amplitude is higher (by 29%) in the TE_{81} mode than in the TE_{51}, for the same power level and pipe radius; and the waveguide radius is smaller for the TE_{81} example than for the TE_{51}.

But beyond consideration of linearized harmonic power growth, issues such as saturated conversion efficiency, mode competition, and sensitivity to imperfect beam quality must be addressed before the mechanism can be established as a practical alternative to the other processes being studied for driving NLC. The study of these issues has recently begun, and preliminary results are cited here. These results are for fifth harmonic conversion, as considered for Table I above, but for a beam at 150 kV with only 1 MW of dc power. A down-tapered axial magnetic field has been shown to be essential for achieving significant harmonic conversion, in order to maintain phase synchrony during the beam deceleration process. Fig. 2 shows fifth harmonic conversion efficiency at 14.25 GHz as a function of axial distance for a linearly-tapered field; the efficiency is seen to grow to about 55% when the beam has no axial velocity spread. The form taken for the linearly tapered field is $B(z) = B_{gr} = 1.2102$ kG for $0 < z < z_0 = 0.5R$, and $B(z) = B_{gr} [1 - 0.003 (z - z_0)/R]$ for $z > z_0$. Also shown in Fig. 2 is the efficiency for a uniform magnetic field, which is seen to not exceed 10% due to dephasing. Fig 3 shows the effect of an addition of small axial velocity spread, for the same magnetic field profile as used for the cold beam example shown in Fig. 2. It is seen that the efficiency remains above 50% for spreads up to about 10%. The axial length for saturation L_s is also shown in Fig 3. One reason that the efficiency does not fall rapidly with increasing velocity spread is that the field taper chosen for the example in Figs. 2 and 3 is not the optimum for achieving maximum efficiency; thus a moderate loss of phase synchrony is already present for the cold beam, and is not worsened by a small velocity spread. To illustrate the effect of refining the magnetic field taper, results are shown in Fig 4 for a cold beam where $B(z) = B_{gr}$ for $0 < z < 2.2 R$ and thereafter the taper is adjusted to maintain exact phase matching along the interaction length. The fifth harmonic conversion efficiency is seen to increase monotonically to over 70%. Also shown in Fig. 4 are the growth patterns for the most serious competitors, the fourth harmonic TE_{41} mode, and the sixth harmonic TE_{61} mode. The simulations that led to this result were carried out on a multi-mode basis, so that the three modes were considered simultaneously [26]. As is seen, no serious mode competition develops in this case. At saturation, the individual particle motions remain in phase, but most of the motion is along the axial magnetic field and thus cannot drive the TE_{51} mode. However, the residual beam energy can be recovered with a single-stage depressed collector, since the particle energies are all identical. It should thus be possible, in principal, to recover nearly 100% of the initial beam power, either as fifth harmonic rf output or as dc power. When the magnetic field profile for highest fifth harmonic efficiency from a cold beam is used for a beam with finite axial velocity spread, the efficiency is found to diminish faster than for the case of the linear profile as shown in Fig. 2. This indicates that an optimum profile should be found for each given beam axial velocity distribution. Studies along such lines for beams capable of 50 MW level conversion at harmonics four through eleven are planned. These studies are part of an accelerated program to assess the potential of cyclotron autoresonance harmonic conversion as a mechanism for delivering rf power in the range 11.4 - 31.4 GHz to drive the next linear collider.

CONCLUSIONS

A brief review of progress on three established gyroresonance mechanisms for generating rf power to drive NLC has been presented. The University of Maryland gyroklystron has demonstrated 27 MW and 37% efficiency at 9.8 GHz, with an energy per pulse of 39 J. The MIT work on the CARM amplifier at 17.1 GHz has so far been plagued by oscillations, probably due to non-ideal beam quality. Work at NRL recently underway on a second harmonic magnicon at 11.4 GHz has progressed through the design of the critical beam deflection system, tests on which are expected before Spring 1993.

Preliminary theoretical work on cyclotron autoresonance harmonic conversion shows that device parameters are fairly modest for a 50 MW level source of 14.25 or 22.8 GHz radiation at either the fifth or eighth harmonic of a 2.85 SLAC klystron driver. Simulation studies of saturated efficiency, mode competition, and efficiency degradation due to imperfect beam quality for a fifth harmonic device at 14.25 GHz using a 1 MW, 150 kV beam have been presented. These show that efficiency values of 50% or more should be possible in the presence of up to 10% axial velocity spread, with a few percent or less of the rf power in competing modes. Extension of these studies to higher power beams and higher harmonics is planned during 1992.

Any gyroresonance device will be measured against the performance of advanced klystrons where decades of experience and continued development provide confidence within the accelerator community that an 11.4 GHz source can be produced. Issues such as overall system efficiency, high average power, rf phase stability, and operation up to 30 GHz have been discussed as favoring one or another gyroresonance device over a klystron. But until a laboratory demonstration is at hand, such claims will remain speculative.

ACKNOWLEDGEMENTS

Appreciation is extended to B. G. Danly, S. H. Gold, V. L. Granatstein, and P. Wilson for valuable discussions and for providing unpublished materials. This work was supported by the U.S. Office of Naval Technology and the Office of High Energy Research of the U.S. Department of Energy.

REFERENCES

1. A Mondelli, D. Chernin, V, Granatstein, P. Latham, W. Lawson, and M. Reiser, "RF Frequency Scaling and Gyroklystron Sources for Linear Supercolliders," in *Frontiers of Particle Beams*, Lecture Notes in Physics **296**, 533 (Springer-Verlag, Berlin, 1988).

2. G. Caryotakis, "Multimegawatt RF Power Sources for Linear Colliders," Proc. 1991 IEEE Particle Accelerator Conference, San Francisco, CA, p.2928.

3. V. L. Granatstein, W. Lawson, W. Main, and S. Tantawi, "Improved Prospects for Using Gyroklystrons to Drive Linear Colliders," in *The Physics of Particle Accelerators*, eds. M. Month and M. Dienes, AIP Conf. Proc. **249** (Amer. Inst. of Physics, New York, 1992) p. 1090; W. Lawson, J. P. Calame, B. Hogan, P.E. Latham, M.E. Read, V. L. Granatstein, M. Reiser, and C. C. Striffler, Phys. Rev. Lett. **67**, 520 (1991).

4. G. Bekefi, A. C. DiRienzo, C. Leibovitch, and B. G. Danly, Appl. Phys. Lett. **54**, 1302 (1990).

5. W. M. Manheimer, IEEE Trans. Plasma Science **18**, 632 (1990).

6. J. L. Hirshfield, Phys. Rev. A **44**, 645 (1991).

7. J. L. Hirshfield, Phys. Rev. A **46**, (Oct. 15, 1992).

8. R. Shpitalnik, C. Cohen, F. Dothan, and L. Friedland, J. Appl. Phys. **70**, 1101 (1991); R. Shpitalnik, J. Appl. Phys. **71**, 1583 (1992).

9. C. Chen, Phys. Fluids B **3**, 2933 (1991).

10. J. L. Hirshfield and A. K. Ganguly, "Coherent Fast Wave Radiation from Spatiotemporally Modulated Gyrating Relativistic Electron Beams, " Proc. 9th Intl. Conf. on High Power Particle Beams, 1992, Wash., D. C.

11. M. Blank, J. A. Casey, K. E. Kreischer, R. J. Temkin, and T. Price, Intl. J. Electronics **72**, 1093 (1992).

12. R. H. Helm and G. A. Loew, "Beam Breakup," in *Linear Accelerators*, eds. P. Lapostelle and A. Septier (North Holland Pub. Co., Amsterdam 1970), p. 173.

13. T. L. Levine, "Review of Pulsed RF Power Generation," in Proc. 3rd European Particle Accelerator Conf., Berlin, 1992.

14. P. Wilson, Bull. APS **37**, 941 (1992).

15. J. Haimson and B. Mecklenburg, "Suppression of Beam Induced Pulse Shortening Modes in High Power RF Generator TW Output Structures," in Proc. SPIE Symp. on Intense Microwave and Particle Beams II, Jan 1992.

16. A. K. Ganguly, S. Ahu, E. G. Zaidman, and A. S. Gilmore, Jr., IEEE Trans. Electron Devices **38**, 229 (1991).

17. S. G. Tantawi, W. T. Main, P. E. Latham, G. S. Nusinovich, W. G. Lawson, C. D. Striffler, and V. L. Granatstein, IEEE Trans. Plasma Science **20**, 205 (1992); W. Lawson, J. P. Calame, B. P. Hogan, M. Skopec, C. D. Striffler, and V. L. Granatstein, ibid, 216 (1992).

18. M. I. Petelin, Radiophysics and Quantum Elec. **17**, 686 (1974).

19. J. L. Hirshfield, K. R. Chu, and S. Kainer, App. Phys. Lett. **33**, 847 (1978).

20. V. L. Bratman and G. G. Denison, Intl. J. Electronics **72**, 983 (1992).

21. K. D. Pendergast, B. G. Danly, W. L. Menninger, and R. J. Temkin, Intl. J. Electronics **72**, 983 (1992).

22. M. M. Karliner, E. V. Kozyrev, I. G. Makarov, O. A. Nezhevenko, G. N. Ostreiko, B. Z. Persov, and G. V. Serdobintsev, Nucl. Instrum. Methods Phys. Res. A **269**, 459 (1988).

23. G. I. Budker, M. M. Karliner, I. G. Makarov, S. N. Morosov, O. A. Nezhevenko, G. N. Ostreiko, and I A. Shekhtman, Particle Accelerators **10**, 41 (1979).

24. O. A. Nezhevenko, "The Magnicon: A New RF Power Source for Accelerators," in IEEE Proc. Particle Accel. Conf. 1992, eds. L. Lizama and J. Chen (IEEE, New York, 1992), p. 2933.

25. B. Hafizi, Y. Seo, S. H. Gold, W. M. Manheimer, and P. Sprangle, IEEE Trans. Plasma Science **20**, 232 (1992).

26. J. L. Hirshfield and A. K. Ganguly (to be published).

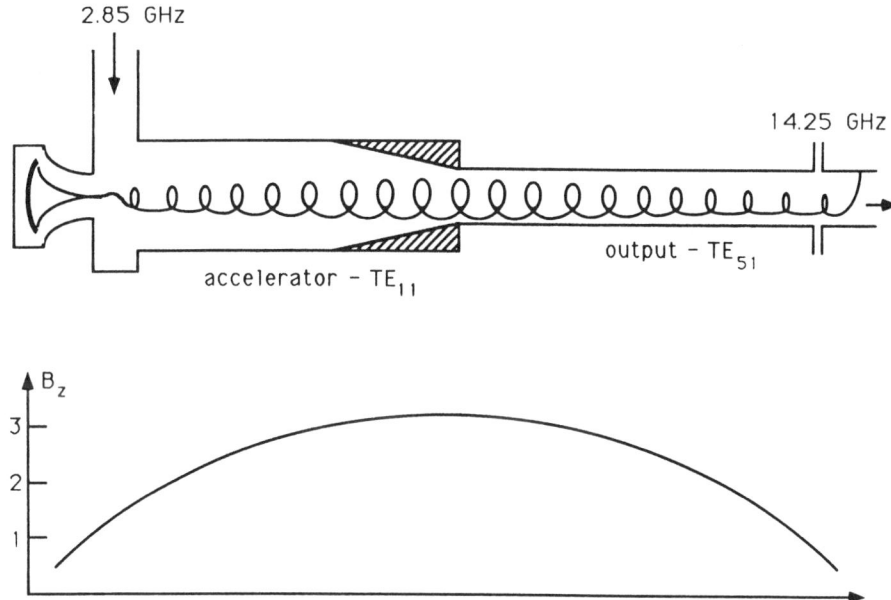

Fig. 1.   Schematic of a fifth harmonic converter (top), and of the static axial magnetic field profile required to maintain gyroresonance during beam acceleration and radiation (bottom).

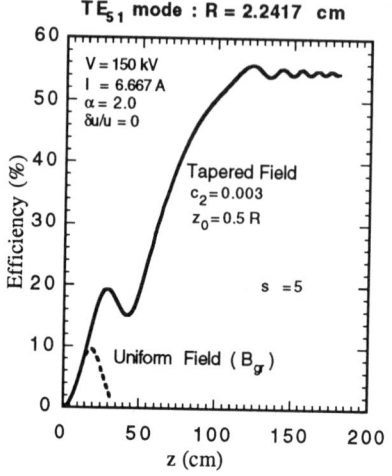

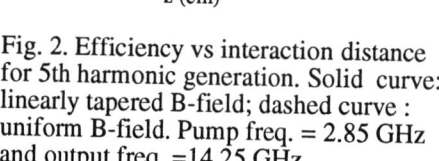

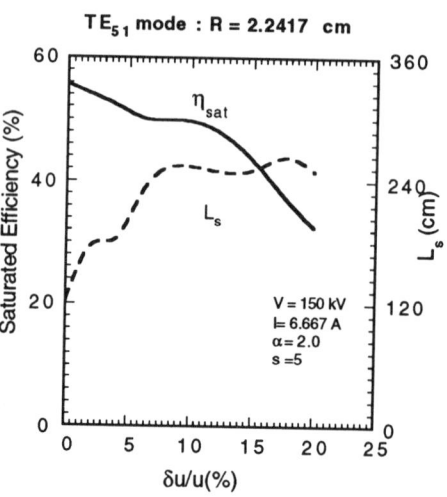

Fig. 2. Efficiency vs interaction distance for 5th harmonic generation. Solid curve: linearly tapered B-field; dashed curve : uniform B-field. Pump freq. = 2.85 GHz and output freq. =14.25 GHz

Fig. 3. Efficiency for the 5th harmonic interaction as a function of axial velocity spread in linearly tapered B-field.

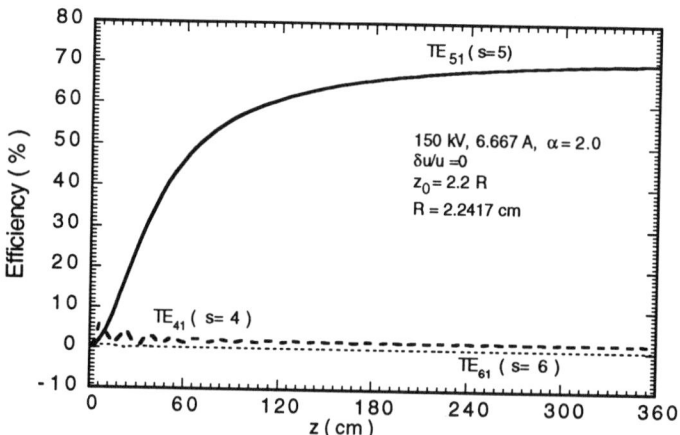

Fig. 4. Efficiency vs interaction distance for a 5th harmonic converter when three modes couple simultaneously to the beam. B-field taper is non-linear to preserve exact phase matching.

DESIGN OF A HIGH-POWER SHEET BEAM KLYSTRON

David U. L. Yu[*] and Jin Soo Kim[*]
DULY Research Incorporated, Rancho Palos Verdes, CA 90732
Perry B. Wilson[§]
Stanford Linear Accelerator Center, Stanford, CA 94305

INTRODUCTION

Future linear colliders require high-gradient (>100 MV/m) accelerators in order to maximize efficiencies and reduce accelerator lengths, thus minimizing capital and operating costs. To provide sufficient power to energize these accelerators, high-power microwave (HPM) sources operating at high frequencies with short pulse lengths ($\approx$100 ns) are needed. For next-generation linear colliders (up to 500 GeV per linac), gigawatt power sources at 11-14 GHz are being diligently sought worldwide. Such a source can be used to power a section of an accelerator about five meters long, producing accelerating gradients on the order of 100 MV/m. While lower frequencies are also being considered in other countries for less powerful colliders, still higher frequencies and gradients may be needed for second-generation colliders (2-5 TeV per linac). A single gigawatt HPM source is also sufficient to power an X-band, 500-600 MeV compact injector for commercial uses. For commercial applications as well as near-term linear colliders, the HPM source of choice is one which has the best performance per unit cost, and utilizes the most mature technology. In addition to reliability and cost, an important figure of merit is its efficiency.

There are two basic approaches to rf power production in meeting these requirements. In one approach, the pulse length of the power source matches directly the filling time of the accelerating structure. In recent years, high power sources such as the relativistic klystron and the free electron laser, both driven by the linear induction accelerator (LIA) which provides multi-gigawatt electron beams with short pulses, have been proposed[1] and actively pursued. Much progress has been made in the conceptual and technological development of these sources. The primary concerns for this approach are practical ones: The very high operating voltages (several megavolts) of the LIA's pose radiation shielding problems, and the system complexity and the large amount of ferrite used in the induction modules make the present-technology LIA inherently costly.

In another approach, rf power is first produced at a longer pulse length and lower peak power, and then, using modern pulse compression techniques,[2,3,4]

---
[*]Supported by U.S. DOE SBIR Grant No. DE-FG03-91ER81116.
[§]Supported by U.S. DOE Contract No. DE-AC03-76SF00515.

the pulse length is shortened while the power is increased. For example, a 1.2 microsecond-long, 100-MW pulse can be compressed in a twelve-to-one ratio to produce over 800 MW of power with a pulse length of 100 ns (a compression efficiency of about 70%). In the near future, a more modest compression factor of 5 or 6 will be used, with a power gain of about 4. The power requirements for a collider using such a pulse compression system are shown in Table I.

Table 1 RF Power Requirement for a Linear Collider at 11.4 GHz

Typical Filling Time	100 ns
Beam Pulse Length	
Short bunch train	15 ns
Long bunch train	100 - 150 ns
RF Pulse Length at Accelerator Input	
Short bunch train	125 ns
Long bunch train	200 - 250 ns
Klystron RF Pulse Length**	
Short bunch train	750 ns
Long bunch train	1.2 - 1.5 μs
Accelerating Gradient	
Stage I	40 - 50 MV/m
Stage II	80 - 120 MV/m
Peak Power per Meter at Accelerator Input	
Stage I	30 - 50 MW/m
Stage II	130 - 300 MW/m
Klystron Power for 5 m of Structure**	
Stage I	40 - 60 MW
Stage II	150 - 350 MW

**Assuming ×6 pulse compression at 67% efficiency, including transmission loss to the accelerator (×4 net power gain).

Conventional klystrons, which are the cornerstones of the rf power sources for low-energy linear accelerators, should not be overlooked for meeting the stage-I power requirements (40-60 MW). However, scaling conventional klystrons to higher stage-II power requirements (>150 MW), and possibly to higher frequencies for a multi-TeV collider, is severely limited by space charge effects due to high current densities, by electrical breakdown due to high voltages across smaller gaps, and by potential metal melting due to interception of the high-power-density beam. It would be of great benefit to overcome these difficulties by extending the concept of the conventional klystron, while still making good use of this economical, well-established technology.

The primary motivations for this second approach are higher efficiency, lower operating voltage and lower system costs. A low perveance klystron is preferred as it leads to higher efficiency. A low operating voltage (a few hundred kilovolts) is also highly desirable from an operational standpoint: it reduces radiation hazards, helps control the modulator cost, and increases moderator efficiency with a lower-turns-ratio pulse transformer. The sheet-beam klystron[5], which replaces a conventional round beam by a flat rectangular beam or by a hollow cylindrical beam, merits special attention as a HPM source in this regard.

Current round-beam klystron development programs face several limitations.[6] Space charge effects limit the output power to less than 100 MW and an efficiency less than 45% unless a super high beam voltage is used, e.g. 1 MV for klystron design at the Protvino, FSU, linear collider project. In this case, however, a conventional modulator is impractical and the klystron must be switched by a gridded gun, which imposes an additional burden on the complexity and reliability of the klystron design. Potential beam interception and metal melting pose a further danger for higher-power klystrons at X-band and higher frequencies.

The sheet-beam klystron (SBK) shown in Figure 1 significantly extenuates the space charge effects by flattening out the beam spatially. It can sustain a high total current, while keeping a low current density and a moderate voltage. Possibilities of breakdown are minimized by keeping a low voltage and by use of multiple output gaps. The serious potential problem of metal melting is avoided by keeping a low power density. As a function of frequency, the power output of a round-beam klystron falls off as the square of the wavelength. The power of a sheet-beam klystron, on the other hand, falls only linearly with wavelength. The SBK is thus more suitable for future linear colliders (or compact accelerators) operating at still higher frequencies and gradients. A sheet-beam klystron combined with pulse compression can produce the same peak power level as that produced by induction-linac driven power sources[1] such as relativistic klystrons and free electron lasers, but with higher efficiency and substantially lower cost.

In this paper we report on theoretical studies to demonstrate the conceptual feasibility of a high-power sheet-beam klystron operating at 11.4 GHz. Numerical simulations are used to investigate beam dynamics and rf issues. Specific problems with the gun design, magnetic focusing and rf coupling at the output gaps unique to the strip beam geometry of the SBK are considered. Very

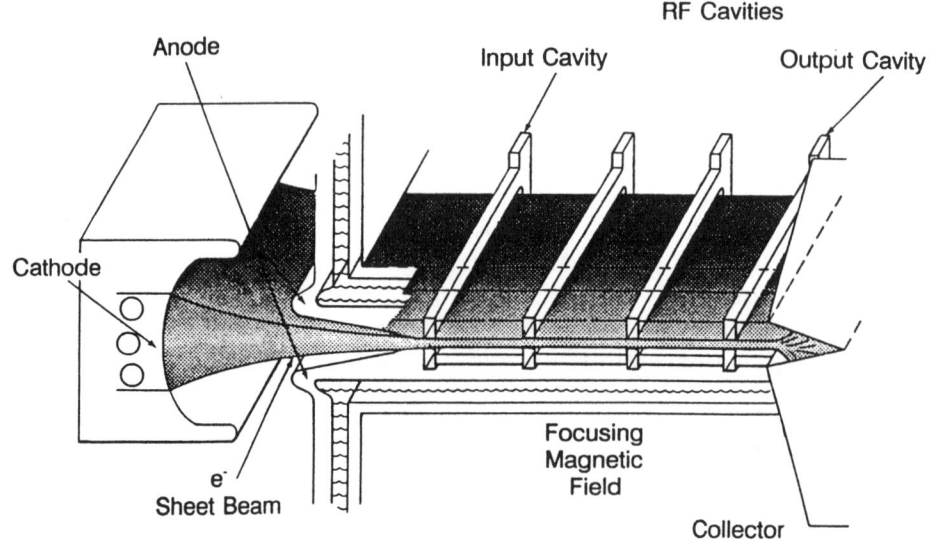

Figure 1   Schematic of Sheet Beam Klystron

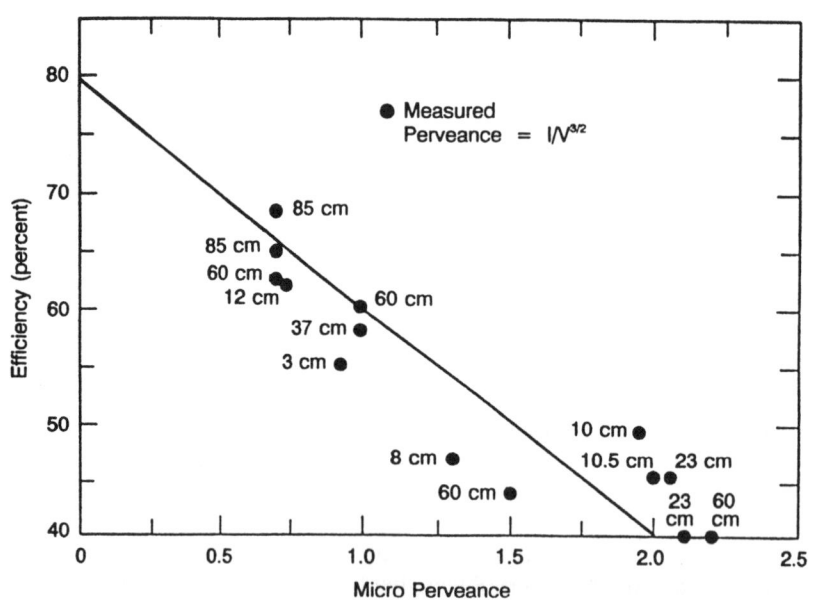

Figure 2   Klystron Efficiency vs Perveance

Table 2 Comparison of RF Power Sources Frequency Range 10-14 GHz

Device (Reference)	V_{beam} (kV)	Pulse Length (ns)	Micro-perv. (A/V$^{3/2}$ x10^{-6})	Peak Power (MW)	Eff. (%)	Freq. (GHz)	Note
Relativistic Klystron (LLNL Choppertron)	3000	10 30	0.2	450 150	25 15	11.4	(1)
Gyroklystron (U. Maryland)	500	1000	0.4	24	36	10	(2)
Conventional Klystron (Protvino, FSU) (SLAC)	1000 440	700 100 800	0.3 1.8	55 72 50	18 31 21	14 11.4	(3) (4)
Haimson Klystron (HRC)	500	50	1.4	100	40	11.4	(5)
Sheet-Beam Klystron (DULY/SLAC)	400	1000	0.15 per square	200	65	11.4	(6)
Cluster Klystron (BNL)	400		0.4	42x28	70	11.4	(6)

(1) Poor efficiency; longer pulse length is difficult
(2) Design limit; output pulse not flat
(3) Poor beam transmission; oscillation in output structure
(4) RF breakdown in output gap
(5) Behavior at longer pulse length is not known
(6) Paper design only

high efficiencies have been calculated for several values of perveance per square beam element of the SBK. We have found that a low-perveance, X-band, sheet-beam klystron is capable of producing over 200 MW of power at 65% efficiency, using a flat electron beam with a reasonable size of 0.4 cm x 16 cm, a moderate voltage of 400 keV, a strip gun compression ratio of 10:1, a cathode loading of 12 A/cm², a uniform focusing field of 3 kG, and a double-gap standing-wave output cavity. With only a constant longitudinal focusing field, the calculated efficiency of the SBK is among the highest of current or proposed HPM sources (see Table 2).

GENERAL DESIGN CONSIDERATIONS

Figure 1, adapted from ref. 5, shows a schematic of the sheet-beam klystron. Electrons thermionically emitted from a large hot cathode inside a vacuum tube are compressed by a specially designed strip gun into a flat sheet. The sheet beam is modulated by an rf signal at the input cavity, and subsequently bunched by idler and penultimate resonant cavities in the drift tube. The kinetic energy of the bunched electrons is partially extracted and converted to microwave power as the beam traverses the gap(s) in a traveling-wave or standing-wave output cavity. A longitudinal focusing field minimizes the transverse momentum spread of the beam in the drift tube by counteracting the space charge force with the Lorentz force. Finally, a collector following the output gap(s) collects the spent electrons.

A useful parameter which characterizes the klystron efficiency is the perveance. For a round beam, the perveance is defined as:

$$K = \frac{I}{V^{3/2}} \ ,$$

where I is the beam current and V is the beam voltage. With no confining field, a round beam spreads according to:

$$\frac{\Delta r}{2a} = 1.52 \times 10^{-4} K \left(\frac{2}{1+\gamma}\right)^{3/2} \left(\frac{z}{2a}\right)^2 \ ,$$

where Δr is the transverse spread at an axial distance z, a is the radius of the beam and γ is the Lorentz factor. The same expression applies to transverse beam spread for a sheet beam, with $2a$ replaced by t, the thickness of the sheet beam, and K replaced by $(\pi/2)K_\square$, where

$$K_\square = \frac{I_\square}{V^{3/2}}$$

and $I_\square$ is the current in a sheet-beam element with a square cross section of sides equal to the thickness of the beam[7]. The sheet beam can be thought of as a parallel array of beamlets, each having a current of $I_\square$. $K_\square$ for a sheet beam is typically many times smaller than K for a round beam with the same total current. Since the beam spread is proportional to the perveance, the transverse emittance is much better preserved for a sheet beam than a round beam with the same current. Conversely, for the same perveance, a sheet beam which is many "squares" in width can carry many times the current in a round beam.

Low perveance also improves longitudinal bunching of the beam. The ratio of the distance in the absence of space charge to a quarter of a reduced bunching plasma wavelength is on the order of the square root of the microperveance ($K/10^{-6}$):

$$\frac{z(ballistic)}{z(plasma)} = (factor) \times \sqrt{K/10^{-6}} \quad .$$

The factor in this expression ranges from about 2 for the input cavity of a typical klystron, to perhaps 1/2 for the penultimate cavity. Sharper longitudinal bunches with high rf current content can be obtained in the ballistic bunching regime. For the best efficiency, a microperveance less than one is desirable. The SBK designs described in detail later all have a perveance per square which is well below 10^{-6}, while round-beam klystrons typically have a perveance which is greater than this value. Figure 2 which plots experimental results of round-beam klystron efficiency vs perveance[8], shows this trend for better efficiency at lower perveance. Extrapolating the efficiency curve to low values of microperveance for a sheet beam (≤ 0.3), the tube efficiency can in principle be on the order of 70%.

There is still another reason why the sheet beam is preferred over the round beam. For the same total power, the power density (power per unit area) of the sheet beam is only a small fraction of that for the round beam, due to the lower perveance per square and lower operating voltage. Excessive power density of the e-beam could cause metal melting if there is beam interception in the drift tube. An electron beam impinging on copper will bring the metal within the electron penetration depth to the melting point at a current density of 50-70 A/cm^2 for a beam voltage of 300-600 kV. For a conventional round-beam klystron with a beam radius $a \approx \lambda/8$ (where λ is the operating wavelength), an efficiency of 50%, and a beam voltage of 500 kV, the output power at which the beam has the potential to melt the copper at the point of interception, assuming a hard-edge beam with a uniform current density, is $P(MW) \approx 0.85[\lambda(cm)]^2$. For a sheet beam, the corresponding expression for a beam voltage of 400 kV, an efficiency of 50%, and a beam thickness of $t \approx \lambda/6$, is $P(\text{MW/cm}) \approx 2\lambda(\text{cm})$. Thus, for an output power of 100 MW, the critical wavelength for potential beam melting is about 10 cm for a round-beam klystron, and 2.5 cm for a sheet-beam tube with a beam width of 20 cm.

   As mentioned earlier, the efficiency of the SBK can be characterized by the perveance per square, $K_\square$, defined as $I_\square/V^{3/2}$. The total beam current is equal to

$$I = I_\square \frac{w}{t} = K_\square V^{3/2} \frac{w}{t} \quad ,$$

where w and t are respectively the width and thickness of the beam. The current, and hence the output power, is limited by potential metal melting and by the cathode loading for a fixed beam area compression ratio in the gun design. Assuming the cathode is a strip of width w and height h, the transverse beam area compression ratio is:

$$R_c = \frac{h}{t} = \frac{I_\square}{I_A t^2} \quad ,$$

where I_A is the cathode loading current per unit area. The amount of current which can be carried by a sheet beam increases linearly with the allowable current loading at the cathode, other things being equal. Conversely, to reach a given current, the width of the sheet beam is inversely proportional to the allowable cathode loading. To assure long cathode life, I_A is limited to 10-15 A/cm^2 with current technology. The transverse beam area compression ratio, or the ratio of the cathode area to the beam cross section, is limited by a number of factors including alignment, thermal emittance, cathode roughness and edge emission, etc. For a round beam, an area compression ratio of 100 is consistent with good beam optics in the gun. The equivalent compression ratio for a sheet beam, where electrons from the cathode are compressed only in one dimension, is on the order of 10-12. In addition, the beam thickness must be less than about $t \approx \lambda/6$ to achieve good coupling at the rf gaps. Assuming $I_A = 12$ A/cm^2 and $R_c = 12$, the maximum current per square in the beam is

$$I_\square = R_c t^2 I_A \leq 4\lambda^2 \quad .$$

The limit on $I_\square$ due to potential copper melting is, from the previous paragraph,

$$I_o \leq 60t^2 \approx 1.7\lambda^2$$

In the above equations, the units of I_o and λ are in amp and cm, respectively. At 11.4 GHz ($\lambda=2.6$ cm), then $I_o \leq 28$ A for the beam compression/cathode loading limit, or $I_o \leq 12$ A for the melting limit. The melting limit is probably too conservative, since in practice beams are not hard-edged. Assuming that the melting limit can be exceeded by 60%, then a beam density of $R_c I_A = 100$ A/cm^2 is acceptable, corresponding to $R_c = 10$ and $I_A = 10$ A/cm^2. In this case, $I_o \leq 2.8 \lambda^2$, or $I_o \leq 20$ A at 11.4 GHz. The perveance per square will be less than 0.25×10^{-6} for any beam voltage greater than 180 kV. An SBK operating at 400 kV and an efficiency of 60% will then give an output power in MW per centimeter of beam width equal to $P/w = 4 \lambda$. At 11.4 GHz, the power is 10 MW/cm. Thus a beam width of 10/20 cm is needed for an output power of 100/200 MW.

SIMULATION OF SBK POWER AND EFFICIENCY

In an earlier design study of an X-band SBK, Eppley, Herrmannsfeldt and Miller (EHM)[5] simulated its properties using periodic permanent magnet focusing (ppm, or wiggler focusing). With a beam voltage of 200 keV, they found an efficiency of about 50%. However, the use of ppm focusing for a sheet beam klystron introduces an additional complexity in comparing the efficiency with that of a conventional klystron. We therefore felt it was of interest to calculate the properties of a sheet beam klystron using a conventional uniform, or possibly tapered, longitudinal magnetic focusing field. In particular, we will show that the efficiency is improved by this simpler focusing scheme.

Table 3 shows several sets of parameters for an SBK which can produce about 200 MW of power at 11.4 GHz. Columns 1-2 list two sets of parameters at an operating voltage of 400 keV. Columns 3-4 list two sets of parameters at an operating voltage of 300 keV. We have calculated the efficiencies with CONDOR, a 2-1/2 dimension particle-in-cell code, using constant longitudinal magnetic focusing. In comparison, the parameters used in a wiggler-focused SBK previously investigated by EHM are shown in the last column. For ease of comparison, we have chosen the same beam thickness for all cases. A cathode loading of 12 A/cm^2 and a gun compression ratio of 10:1, shown in columns 1 and 3, are consistent with the present manufacturing capabilities of round beam electron guns (area compression ratio about 100:1) and with long cathode life. A cathode loading of 20 A/cm^2 and a gun compression ratio of 12:1, shown in columns 2 and 4, on the other hand, should be within the reach of future manufacturing capabilities. However, the latter designs may possibly exceed the beam melting condition discussed in the previous section.

Table 3   Sheet-Beam Klystron Parameters

Voltage, keV	400	400	300	300	200
Cathode Loading, A/cm²	12	20	12	20	10
Compression Ratio	10:1	12:1	10:1	12:1	10:1
Beam Thickness, cm	0.4	0.4	0.4	0.4	0.4
Current per Square, A	19.2	38.4	19.2	38.4	16
Current per Unit Area, A/cm²	120	240	120	240	100
Microperveance/Sq, $\mu A/V^{3/2}$	.0759	.1518	.1168	.2337	.1790
Beam Width, cm	16	9	25	13	50
Beam Current, kA/m	4.8	9.6	4.8	9.6	4.0
Beam Current, kA	0.768	0.864	1.200	1.248	2.000
Focusing Field, kG	2.8	3.9	3.0	4.2	ppm
Efficiency, %	65	63	55	53	49
Power, MW	200	218	198	198	196

The higher beam voltages considered here result in lower perveances (at the same current per square) and improve efficiencies significantly. As a result of high efficiency and higher beam voltage, less total current is needed to achieve the same power. A lower required current results in a much narrower beam width than that considered by EHM.[5] A narrow beam considerably eases the tasks of designing rf output cavities and mechanical support of the drift tube. The beam width may be further reduced as the permitted cathode loading or the compression ratio increases.

In the CONDOR simulations, the sheet beam is assumed to be perfectly laminar as it enters a two-dimensional klystron, which consists of a rectangular drift channel, a drive cavity, two bunching cavities, and one or two output cavities. Each rf cavity is modelled as a port with a voltage and a phase, using a method developed by S. Yu[9]. For each value of beam voltage and current, we made a large number of computer runs, varying the cavity locations and voltages to obtain the parameters which would maximize the efficiency. Simulations were made for a beam voltage of 300-400 keV, and a beam current of 2-20 kA/m. A typical set of voltages, phases and locations of the cavities is shown in Table 4.

CONDOR calculates an electronic efficiency and a kinetic efficiency (see Figure 3). The electronic efficiency is considered to be more reliable and is shown in Table 3. The maximum efficiency for each case is determined using the following procedure. Initial runs include only the first two cavities. The voltage of the drive cavity is chosen so that the rf current peaks at a distance beyond the second cavity, about equal to one-quarter of the reduced plasma wavelength, the approximate location of the third cavity. The voltage of the second cavity is next varied until the rf current is about the same as the dc current. The location of the third cavity is chosen to be about 10% behind the peak position of the rf current. Its voltage is adjusted until the rf current reaches a maximum at the output port. The location of the output cavity coincides with the peak of the rf current. Its voltage is somewhat larger (by about 10%) than the input beam voltage. Using two output cavities in tandem lowers the voltage for each cavity, thus mitigating breakdown, and also increases the efficiency.

Table 4   Cavity Parameters in the Port Approximation (300 keV, 4.8 kA/m)

Cavity number	1	2	3	4	5
Location, mm	0	42	192	223	231
Gap width, mm	1.75	1.75	2.80	3.15	3.15
Voltage, V	6,165	54,000	200,000	250,000	250,000
Phase, rad	0	-1.57	-1.57	-0.70	+0.40

The longitudinal magnetic focusing field is modelled with an equivalent 2-D solenoid: two sheet currents flowing in opposite directions in planes above and below the beam, with the current flow perpendicular to the beam flow. The focusing field strength is varied in the range of 1-5 kG. We have found that a minimum focusing field strength of 3 kG (see Figure 4) is required to provide beam stability and good efficiency for a 300-keV, 1.2-A beam. The focusing strengths for other beam parameters are obtained by scaling the square of the focusing field strength with the product of the voltage, V, and the perveance per square, $I_\square$.

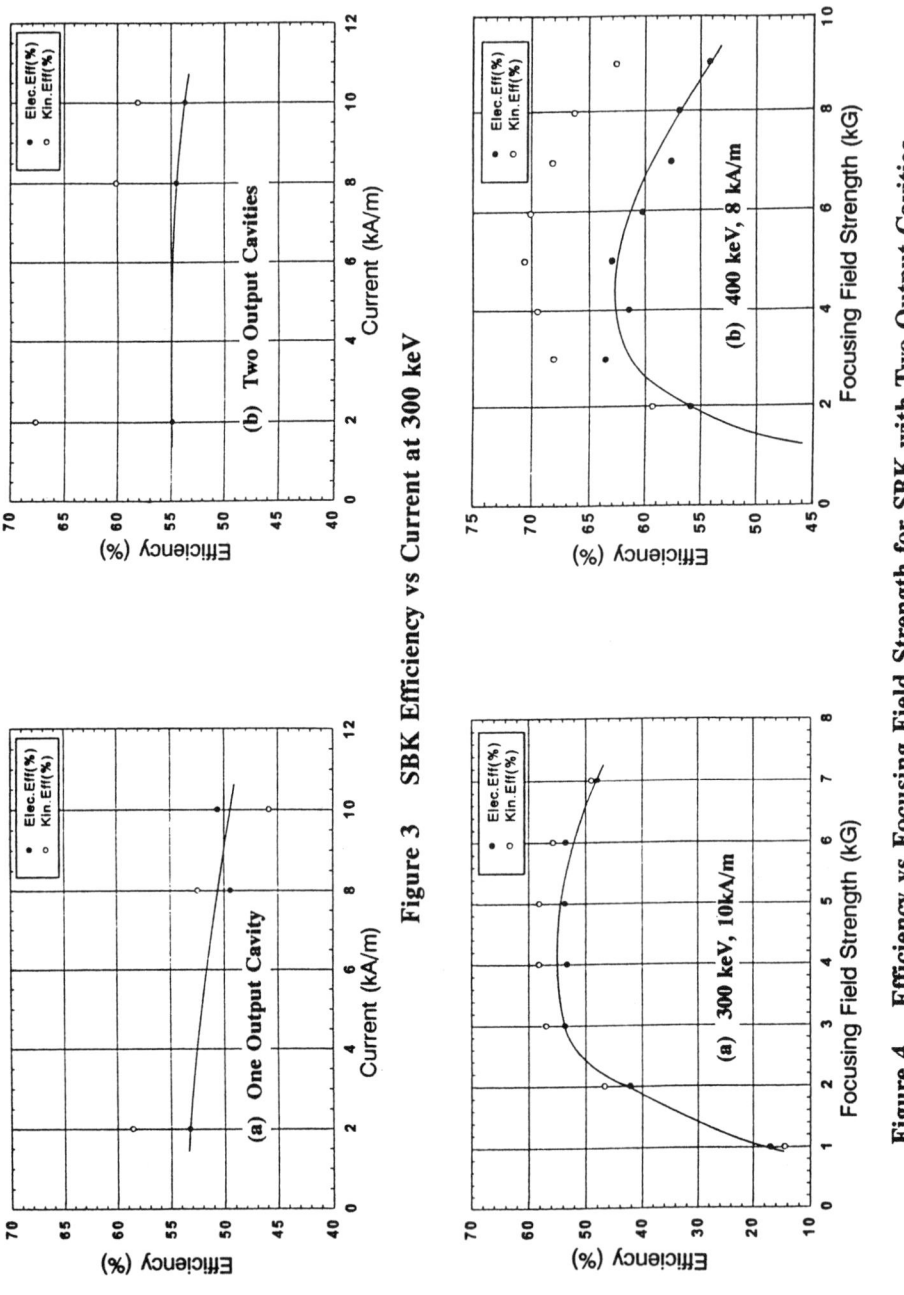

Figure 3   SBK Efficiency vs Current at 300 keV

Figure 4   Efficiency vs Focusing Field Strength for SBK with Two Output Cavities

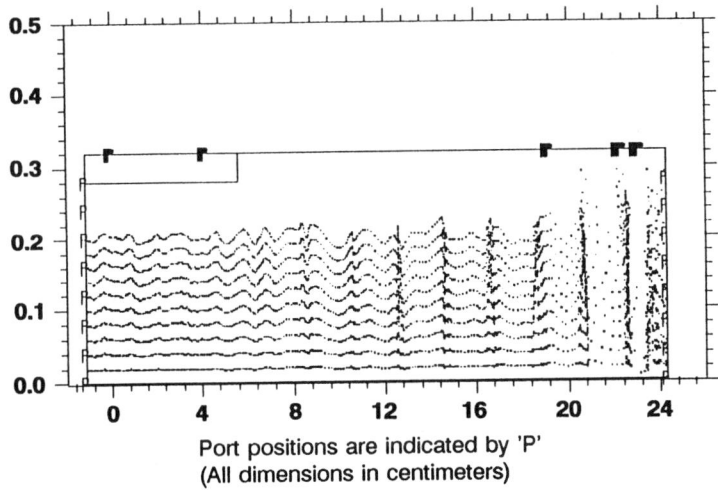

Port positions are indicated by 'P'
(All dimensions in centimeters)

Figure 5   **Electron Distribution for an SBK (300 keV, 8kA/m and 5 kG)**

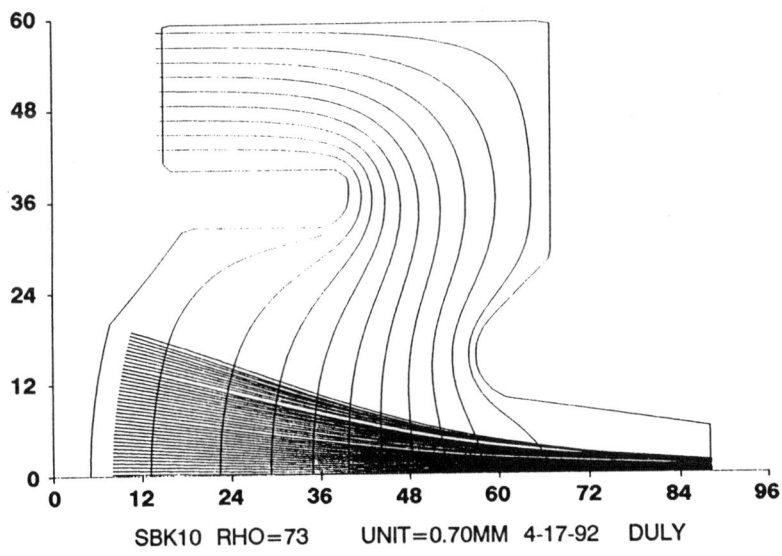

SBK10 RHO=73   UNIT=0.70MM 4-17-92   DULY

Figure 6   **EGUN Simulation of Strip-Beam Electron Trajectories at 400 keV**

Figures 3(a) and 3(b) plot efficiency vs current using one or two output cavities, respectively, to extract power from a 300-keV beam. For a given voltage, the efficiency is higher at a lower current. The efficiency vs current at 400 keV and 8 kA/m, using two output cavities, is calculated by CONDOR to be 63%. At a higher operating voltage, the SBK efficiency increases. By scaling the results at 300 keV, the efficiency at 400 keV and 4.8 kA/m is 65%. Figures 4(a) and 4(b) plot efficiency vs the focusing field strengths, respectively, for a 300-keV, 10-kA/m and a 400-keV, 8-kA/m SBK. As seen from these figures, there is a wide range of field strength over which the efficiency does not vary appreciably. It is desirable to use the lowest field strength to achieve beam stability and good efficiency, in order to minimize the cost of the magnets. Figure 5 shows a typical beam profile for a well-bunched, 300-keV, 4.8-kA/m sheet beam in the drift tube.

COMPONENT DESIGN CONSIDERATIONS

We have simulated the electron optics of a flat gun with the EGUN program in 2-D rectangular coordinates. Figure 6 shows trajectories of electrons emitted from a cylindrical cathode and focused by a 400-keV anode with a compression ratio of 12:1. In this simulation, the perveance of the gun is 10.2 microperveance/m, yielding a current of 2.58 kA/m. The parameters in this EGUN run are similar to those in Table 3. The Child-Langmuir law for thermionic emission in a plane diode should be valid for a cylindrical cathode-anode. The thermionic emission current density is proportional to the 3/2 power of the voltage, consistent with the concept of perveance. Optical aberrations due to the edge effects of a flat gun are in principle no different than those of a conventional round gun, which can be simulated with EGUN in 2-D. As an alternative to a flat sheet, a hollow cylindrical beam may also be considered.

A detailed sheet-beam gun design must also take into account the thermal and mechanical properties of the cathode and the heating elements behind it. The cathode should be heated with sufficient uniformity to avoid cool areas which would result in temperature limited emission and non-uniform current density in the beam, or hot spots which would result in shortened cathode life and non-uniform emission. As the cathode is raised to the operating point, differential thermal expansion can lead to distortion of the cathode surface with a consequent degradation in beam optics.

The dc beam can be focused with a solenoidal field of 600 G. The Brillouin field balances the space charge force and prevents the beam from spreading. After the rf is turned on, a 3-4 kG longitudinal magnetic focusing field is adequate to maintain beam stability in the drift tube as shown in the CONDOR simulations. A sheet-beam klystron with the same perveance per square as the round-beam perveance for a conventional klystron will require about the same focusing field to combat beam spreading due to space charge forces. However, we have shown earlier that the perveance per square for a typical SBK design will be a about a factor of ten lower than a round-beam klystron. As mentioned before, the lower perveance leads both to improved efficiency and to reduced danger from copper melting in the event of beam interception. An additional advantage of the lower perveance is that the required magnetic focusing field is lower--by a factor of about three (the square root of ten for the same beam voltage). This in turn leads to roughly a factor of five lower power consumption in the focusing coil.

We have found satisfactory results by imposing a constant longitudinal magnetic field in the CONDOR simulations. In addition to a constant magnetic field, the longitudinal profile of the focusing field can also be varied, for example, by separating the rectangular coil into a number of sections, each separately powered by different currents. A tapered field between the penultimate cavity and the output cavities should increase the efficiency of the SBK. The strength of the magnetic field needed to focus the sheet beam is proportional to the perveance. A low perveance beam is also less likely to have interception problems. The fringe field of the conventional focusing magnet tends to cancel transverse space charge force, although there remains a small residual shearing force which tends to distort the edges of a rectangular beam. The amount of beam shearing along the drift tube can be calculated from beam dynamics. The shearing is acceptable provided it is small compared with the beam width. If necessary, trim coils may be used to provide secondary edge focusing of the sheet beam. As a two-dimensional code, CONDOR cannot simulate the edge effects. In a finite-width sheet beam, as electrons are pushed sideways by the longitudinal magnetic field, they begin to execute betatron-like motion outside the main beam where the space charge force diminishes rapidly. After a maximum of a half betatron period, most fringe electrons rejoin the main beam and remain focused. A very small fraction of fringe electrons, however, may hit the walls of the drift tube. The detailed single-particle and plasma dynamics due to edge effects is complicated and is best studied with 3-D simulation programs. It is important to minimize the power consumption in the magnetic coils so that the overall efficiency of the SBK remains high. To focus a sheet beam with a pulse length of 1.5 μs, and with 200 MW peak and 120 kW average power, magnets should consume less than 5 kW of power.

Each rf cavity in the SBK has a different function and must be carefully designed in order to maximize the tube efficiency. The concerns of rf cavity design are mitigated by the very manageable beam width obtainable with our SBK parameters. It is much easier to achieve field uniformity across the rf cavities for a narrow beam. Unique to the sheet-beam klystron cavities is their relatively large transverse widths compared to the gap sizes. In order to provide good rf coupling to the sheet beam, it is important that the fields be approximately constant along the transverse width of each of these cavities, In general, it is not an easy task to design a cavity with flat field over a width of many wavelengths.

However, for the beam parameters which we have considered (see Table 3), the beam width can be as narrow as 9 cm at 400 keV, making this task more manageable. One design strategy is to operate the center portion near cut-off, with a widened portion about a quarter of a wavelength long at each end.

Another design issue is the potential rf back feedback problem if the non-cutoff TE mode is excited in the drift region, causing power to leak back upstream. This can occur, for instance, if there is asymmetry in the cavity gap as a result of manufacturing imperfections. Fortunately, this problem can be cured by the use of one or more quarter-wave chokes slots. For the output cavity, a number of design options are available for extraction of phase-stable, high-power microwaves. Like the other rf cavities, field uniformity across the transverse width is important. In addition, the gap voltage must not be too high to cause breakdown. Double-gap output cavity reduces the voltage and increases the efficiency. Other possible coupling methods include multi-gap standing wave or traveling wave output cavities.

Finally, the collector is one of the important design elements of a klystron. Since a klystron must sometimes operate with low or no drive power (and therefore no rf output power), the collector must be capable of absorbing the full dc beam power. In a correctly designed klystron, the beam begins to spread just following the output cavity as the focusing field begins to fringe. The sheet-beam klystron has a strong advantage over a conventional X-band klystron because the beam power density per unit area is below the critical value which would melt copper, even if the beam is not defocussed. In the conventional klystron, on the other hand, proper beam spreading before hitting the collector surface is of critical importance to avoid damage to the collector, and additional damage from pieces of molten copper falling on the cathode. Beam spreading is less critical in the collector region for a sheet-beam klystron.

CONCLUSIONS

Using the range of SBK parameters shown in columns 1-4 in Table 3, we have obtained the following Phase I simulation results:

- Good beam stability can be achieved by a low-perveance sheet beam with a reasonable transverse width, and with a current per square consistent with conventional gun cathode loading and cathode-to-anode convergence.

- Using a constant longitudinal magnetic field, a two-dimensional sheet beam can be focused in a drift tube with one drive cavity and two bunching cavities.

- A double-gap output cavity can extract high power with high efficiency from the sheet beam.

- The tube efficiency increases as the operating voltage increases. To avoid breakdown, however, probably the voltage should not be much higher than 400 keV.

- At a given operating voltage, there are considerable rf and mechanical design advantages in using a narrow beam, consistent with a high current per square. Beam interception must be avoided, however, to prevent potential copper melting.

The salient features of the SBK amply demonstrate its feasibility as a promising source of high power microwaves. While the calculated efficiency of the SBK has attained 65%, still higher efficiencies are potentially achievable by tapering the focusing field and by better output cavity designs.

ACKNOWLEDGMENT

We thank Dr. Ken Eppley and Dr. Bill Herrmannsfeldt, respectively, for their help with the CONDOR and EGUN programs for our application.

REFERENCES

1. A.M. Sessler, "The Free Electron Laser as a Power Source for a High-Gradient Accelerating Structure", in Laser Acceleration of Particles, P.J. Channel ed. (AIP Conf. Proc. No. 91, New York, 1982); S.S. Yu, "Physics of Relativistic Klystrons", Proceedings of Conf. on Advanced Accelerator Development, Orsay, France, July, 1987.

2. Z.D. Farkas, et al., "SLED: A Method of Doubling SLAC's Energy", Proceedings of the 9th International Conference on High Energy Accelerator", p.576, May, 1976.

3. T.L. Lavine, et al. "Binary Rf Pulse Compression Experiment at SLAC", SLAC-PUB-5277, June, 1990.

4. P.B. Wilson, Z.D. Farkas and R.D. Ruth, "SLED II: A New Method of Rf Pulse Compression", SLAC-PUB-5330, September, 1990.

5. K.R. Eppley, W.B. Herrmannsfeldt and R.H. Miller, "Design of a Wiggler-Focused, Sheet-Beam X-Band Klystron", SLAC-PUB-4221, February, 1987. Earlier references on the SBK can also be found herein.

6.   T. Lee and W. Herrmannsfeldt, private communications.

7.   P.B. Wilson, "Advances in Linear Collider Technologies", SLAC-AAS Note 48, September, 1989; and "Linear Collider Technologies: Accelerating Structures and Rf Sources", SLAC-AAS Note 57, November, 1990.

8.   R.B. Palmer and R.H. Miller, "A Cluster Klystron", SLAC-PUB-4706, September, 1988; also R.B. Palmer in this conference.

9.   S. Yu, "Particle-in-Cell Simulations of High Power Klystrons", SLAC-AP 34, September, 1984.

POWER EXCITATION BY THE
USE OF A RF WIGGLER*

A.G. Ruggiero

Brookhaven National Laboratory
Upton, NY 11973, USA

INTRODUCTION

It is well-known[1] that there are difficulties to obtain rf power sources of significant amount for frequencies larger than 3 GHz. Yet, rf sources in the centimeter/millimeter wavelength range would be very useful to drive, for example, high-gradient accelerating linacs for electron-positron linear colliders.[2] Ordinary conceived methods to produce radiation with short wavelengths is to let a bunch of electrons to travel on a circular orbit by the action of bending magnets[3] or along a magnetic structure with alternating field direction as in *wigglers* or in *undulators*[4]. Application of these devices to generate radiation in the centimeter-to-millimeter wavelength range, has often been proposed, and in a few cases also demonstrated[5-7].

We would like to propose an alternative method to produce such radiation. It makes use of a short electron bunch traveling along the axis of a waveguide which is at the same time excited by a TM propagating electromagnetic wave.[8] It is well known that radiation can be obtained by *wiggling* the motion of the electrons in a direction perpendicular to the main one.[4] The wiggling action can be induced by electromagnetic fields in a fashion similar to the one caused by wiggler magnets. We found that an interesting mode of operation is to drive the waveguide with an excitation frequency very close to the cut off. For such excitation, the corresponding e.m. wave travels with a very large phase velocity which in turn has the effect to increase the wiggling action on the electron bunch.

Our method, to be effective, relies also on the *coherence* of the radiation; that is the bunch length is taken to be considerably shorter than the radiated wavelength.[9,10] In this case, the total power radiated should be proportional to the square of the total number of electrons in the bunch.

The paper concludes with possible modes of operation, a list of performance parameters and a proposed experimental set-up.

---

* Work performed under the auspices of the U.S. Department of Energy.
Contribution to the Advanced Accelerator Concepts Workshop, Port Jefferson, NY, 14-20 June 1992.

FIELD DISTRIBUTION IN A WAVEGUIDE

We shall consider an infinitely long *waveguide*, straight, with rectangular cross-section of width w and height h. We shall introduce a rectangular coordinate system x, y and z; where x and y are the transverse distances from the upper left corner of the waveguide (see Fig. 1) and z is the longitudinal coordinate along the axis. In this section we describe the propagation of a TM traveling electromagnetic wave in the waveguide.[11] If we use the Lorentz representation, the fields can be derived from a scalar V and vector potential $\mathbf{A}$ satisfying the following equations

$$\nabla^2 V - \frac{1}{c^2}\frac{\partial^2 V}{\partial t^2} = 0 \tag{1}$$

$$\nabla^2 \mathbf{A} - \frac{1}{c^2}\frac{\partial^2 \mathbf{A}}{\partial t^2} = 0 \tag{2}$$

$$\operatorname{div}\mathbf{A} + \frac{1}{c}\frac{\partial V}{\partial t} = 0 \tag{3}$$

In cartesian coordinates the explicitly form of Eq. (1) is

$$\frac{\partial^2 V}{\partial x^2} + \frac{\partial^2 V}{\partial y^2} + \frac{\partial^2 V}{\partial z^2} - \frac{1}{c^2}\frac{\partial^2 V}{\partial t^2} = 0 \tag{4}$$

A solution of Eq. (4) is

$$V = V_0 \left(\sin \alpha_1 x + V_1 \cos \alpha_1 x\right)\left(\sin \alpha_2 y + V_2 \cos \alpha_2 y\right) e^{i(kz - \omega t)} \tag{5}$$

The nature of the traveling wave is described by the last factor where ω is the angular frequency and k the wave number which defines the longitudinal propagation mode. Insertion of Eq. (5) into Eq. (4) yields

$$k^2 = \frac{\omega^2}{c^2} - \alpha_1^2 - \alpha_2^2 \tag{6}$$

The horizontal and vertical propagation constants, respectively α_1 and α_2, are to be determined by specifying proper boundary conditions of the electric and magnetic fields at the walls of the waveguide. The same boundary conditions will be used to estimate V_1 and V_2 appearing at the right-hand side of Eq. (5).

According to the conventional waveguide terminology[11], a TM mode is defined as that traveling wave with vanishing magnetic field in the main direction of propagation, that is the z-axis of the waveguide. This mode is associated to solutions of the vector potential $\mathbf{A}$ which actually is completely directed along the z-axis

$$\mathbf{A} \equiv (0, 0, A) \tag{7}$$

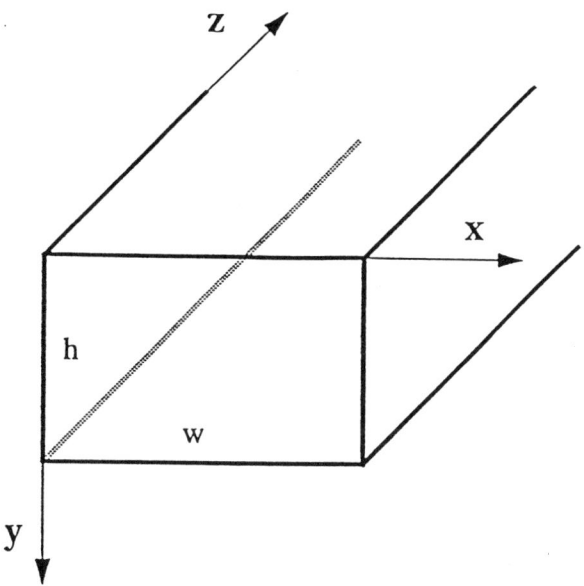

Fig. 1: Waveguide Geometry.

where A satisfies an equation similar to Eq. (4). Moreover, to satisfy the Lorentz condition represented by Eq. (3)

$$A = \beta_w V \tag{8}$$

where

$$\beta_w = \omega/kc \tag{9}$$

is the wave phase velocity, normalized to the speed of light. Thus the vector potential A is completely determined from the knowledge of the scalar potential V.

The electric E and magnetic B fields can be determined from the usual relations

$$E = -\mathrm{grad}\, V - \frac{1}{c}\frac{\partial A}{\partial t} \tag{10}$$

$$B = \mathrm{rot}\, A \tag{11}$$

We shall take the walls of the waveguide with the most general electromagnetic properties, described by the surface characteristic impedance ξ, a complex function of the angular frequency ω. As shown in ref. [8] we derive easily

$$V_j = -i\,\frac{\xi \beta_w \alpha_j}{k\left(\beta_w^2 - 1\right)} \tag{12}$$

where we let $j = 1$ or 2. Letting

$$tg\,\mu_j = \frac{2i\xi\beta_w\alpha_j k\left(\beta_w^2 - 1\right)}{k^2\left(\beta_w^2 - 1\right)^2 + \xi^2\beta_w^2\alpha_j^2} \tag{13}$$

it has been shown[8] that the eigenvalues of α_1 and α_2 are given by

$$\alpha_1 w - \mu_1 = \pi n \tag{14}$$

$$\alpha_2 h - \mu_2 = \pi m \tag{15}$$

with n, m integer real numbers. These equations can be used in conjunction to Eq. (6) to calculate the propagation constant k.

The constant V_0 in Eq. (5) determines the amplitude of the field potential and is related to the power flux in the waveguide. The following relation is obtained[8,11]

$$P = \frac{c\beta_w}{64}\pi V_0^2\left(\frac{h}{w}n^2 + \frac{w}{h}m^2\right) \tag{16}$$

where we have ignored the effect of the wall impedance.

PERFECTLY CONDUCTIVE WAVEGUIDE

A special case is a waveguide with perfectly conductive walls, that is $\xi = 0$. In this case it is easily seen that $V_1 = V_2 = 0$ and $\mu_1 = \mu_2 = 0$; moreover k^2 is real. Solving Eq. (6) gives the following dispersion relation

$$k = \sqrt{\frac{\omega^2}{c^2} - \frac{\omega_c^2}{c^2}} \tag{17}$$

where

$$\omega_c = \pi c\sqrt{\frac{n^2}{w^2} + \frac{m^2}{h^2}} \tag{18}$$

is the angular frequency at *cut-off*. It is convenient to introduce the form factor

$$q = \omega_c/\omega \tag{19}$$

It is seen from Eqs. (6 and 9) that

$$\beta_w = \frac{1}{\sqrt{1 - q^2}} \tag{20}$$

The range of values of the form factor fulfilling the *condition of propagation*, which corresponds to k positive, is

$$0 < q < 1 \tag{21}$$

that is $\omega > \omega_c$. It is then seen that β_w is always real and larger than 1; that is the wave phase velocity is always larger than the speed of light.

An inspection of the dispersion relation shows that below the cut-off, $\omega < \omega_c$, there is no propagation, and k assumes no real values. For large values of ω, k increases about linearly. An interesting plot, shown in Fig. 2, is the display of the wave phase velocity β_w versus the form factor q as given by Eq. (20). Observe that approaching the cut-off from below, $q \to 1$, the phase velocity β_w becomes infinitely large.

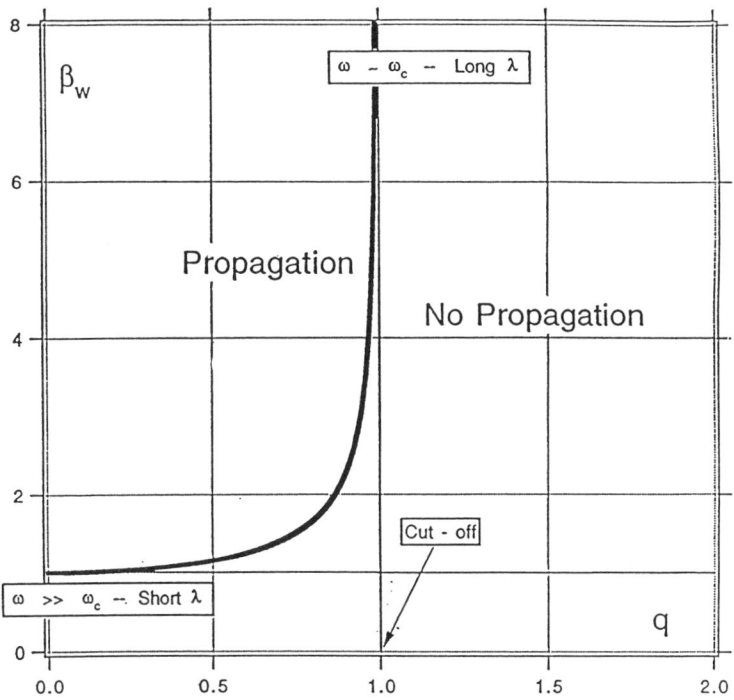

Fig. 2: Phase Velocity vs. Form Factor.

WAVEGUIDE WITH RESISTIVE WALLS

In the following we consider the case of *resistive* walls. For this case the surface characteristic impedance is

$$\xi = (1-i)\sqrt{\frac{\omega\mu}{8\pi\sigma}} \tag{22}$$

$$= (1-i)\mathcal{R}$$

where σ is the electric conductivity and μ the magnetic permeability of the wall material. The dispersion relation Eq. (6) can now be written

$$k^2 = \frac{\omega^2}{c^2} - \frac{\omega_c^2}{c^2} - \Delta^2 \tag{23}$$

where

$$\Delta^2 = \frac{\mu_1^2}{w^2} + \frac{\mu_2^2}{h^2} + 2\pi \left(\frac{n\mu_1}{w^2} + \frac{m\mu_2}{h^2} \right) \tag{24}$$

and k, Δ are complex quantities. Let us separate explicitly the real and imaginary parts; that is

$$k = k_r + i\, k_i \tag{25}$$

$$\Delta = \Delta_r + i\, \Delta_i \tag{26}$$

then

$$2k_r^2 = \frac{\omega^2 - \omega_c^2}{c^2} - \Delta_r^2 + \Delta_i^2 + \sqrt{\left(\frac{\omega^2 - \omega_c^2}{c^2} - \Delta_r^2 + \Delta_i^2 \right)^2 + 4\Delta_r^2 \Delta_i^2} \tag{27}$$

and

$$k_i = -\frac{\Delta_r \Delta_i}{k_r} \tag{28}$$

The last quantity k_i is the measure of the wave propagation attenuation per unit length, whereas k_r is the proper constant of propagation of the wave. The wave phase velocity is then given by

$$\beta_w = \frac{\omega}{ck_r} \tag{29}$$

from which we can derive the following relation to the form factor q

$$\beta_w^2 = \frac{2}{\left(1 - q_0^{-2} q^2 \right) + \sqrt{\left(1 - q_0^{-2} q^2 \right)^2 + 4c^4 \frac{\Delta_r^2 \Delta_i^2}{\omega_c^4} q^4}} \tag{30}$$

where

$$q_0^{-2} = 1 + \frac{c^2}{\omega_c^2} \left(\Delta_r^2 - \Delta_i^2 \right). \tag{31}$$

Both Δ_r and Δ_i depend on the angular frequency ω and the propagation constant k; nevertheless in the case of a good conductor, for instance copper with $\mu = 1$ and $\sigma = 5 \times 10^{17}$ s^{-1}, one can treat the contribution of the surface characteristic impedance $\mathcal{R}$, given by Eq. (22), as a perturbation to the field distribution. In proximity of the cut-off we can then let $\mathcal{R} = \mathcal{R}_c$ where $\mathcal{R}_c$ is $\mathcal{R}$ evaluated for $\omega = \omega_c$. Inspection of Eqs. (30 and 31) shows that β_w^2 is always a positive quantity for any value of q; the maximum occurs for $q = q_0$, which can be interpreted as a shift of the cut-off frequency. At cut-off the maximum is

$$\beta_{w\mathrm{max}}^2 \approx \frac{\omega_c^3}{2\pi^2 c^3 \mathcal{R}_c \left(\frac{n^2}{w^3} + \frac{m^2}{h^3} \right)} \tag{32}$$

COMPARISON OF DIFFERENT METHODS

For an electron moving along the longitudinal direction z, the field distribution in a waveguide is different from that the same electron would experience in the case of a wiggler magnet or a photon beam. To see this, we can operate a transformation of the field distribution in the frame where the electron is at rest. Let β and γ be the relativistic factors respectively for velocity and energy of the particle. Let us first consider the case of a planar electromagnetic wave moving in the z-direction. The electric and magnetic fields are perpendicular to each other and perpendicular to the direction of motion. The components of the fields of the plane wave in the laboratory frame are

$$E_y = E_z = B_x = B_z = 0 \tag{33}$$

$$E_x = B_y = E_0 \; e^{i(kz-\omega t)} \tag{34}$$

where E_0 is a constant amplitude, ω the angular frequency and k the propagation constant of the wave. After performing the relativistic transformation, the components of field distribution of the same wave, in the system where the electron is at rest, are

$$E'_y = E'_z = B'_x = B'_z = 0 \tag{35}$$

$$E'_x = B'_y = \gamma\,(1 + \beta)\,E_0 \; e^{i\left(k'z'-\omega't'\right)} \tag{36}$$

where we have denoted with prime the variables in the frame at rest. We have

$$k' = \gamma\left(k - \omega\frac{\beta}{c}\right) \tag{37}$$

$$\omega' = \gamma\,(\omega - \beta kc) \tag{38}$$

In the vacuum $k = \omega/c$ and the previous relations become

$$\omega' = \gamma\,(1 - \beta)\,\omega \tag{39}$$
$$k' = \omega'/c \tag{40}$$

Now let us consider the case of a wiggler magnet, that is a sequence of dipole magnets where the magnetic field is in the y-direction and changes periodically sign from one magnet to the next. Let L be the periodicity of the wiggler. In the laboratory frame the field components are

$$B_x = B_z = 0 \tag{41}$$
$$E_x = E_y = E_z = 0 \tag{42}$$
$$B_y = \pm B_w \tag{43}$$

where B_w is the wiggler field. There is no electric field in a wiggler. After performing the Lorentz transformation where the electron is at rest, one obtains a field distribution similar to that given by Eqs. (35,36) except that

$$E'_x = \beta\gamma B_w \tag{44}$$
$$B'_y = \gamma\,B_w \tag{45}$$

To a relativistic electron with $\beta \sim 1$, the wiggler is equivalent to a plane electromagnetic wave with amplitude $E_0 \simeq B_w/2$ and

$$\omega' = c\gamma k_w \tag{46}$$

where $k_w = 2\pi/L$. Thus the interaction of an electron with a planar electromagnetic wave or with a wiggler magnet is the same; if the electron moves against the wave, the consequence is a Compton scattering by which the electron loses energy and the photon field intensity increases.

Let us turn now our attention to the field distribution in the waveguide. After the Lorentz transformation to the frame where the electron is at rest is performed, the new field distribution is given by

$$E'_z = E_z \tag{47}$$
$$E'_x = \gamma\,(1 - \beta\beta_w)\,E_x \tag{48}$$
$$E'_y = \gamma\,(1 - \beta\beta_w)\,E_y \tag{49}$$

$$B'_z = 0 \tag{50}$$
$$B'_x = -\gamma\,(\beta_w - \beta)\,E_y \tag{51}$$
$$B'_y = \gamma\,(\beta_w - \beta)\,E_x \tag{52}$$

where k and ω are replaced by k' and ω' also given by Eqs. (37-38). Since ω and k are related to each other by the phase velocity given by Eq. (9), it is

$$k' = \gamma k\,(1 - \beta\beta_w) \tag{53}$$
$$\omega' = \gamma k c\,(\beta_w - \beta) \tag{54}$$

from which the phase velocity in the rest frame is

$$\beta'_w = \frac{\omega'}{k'c} = \frac{\beta_w - \beta}{1 - \beta\beta_w} \tag{55}$$

If we neglect the longitudinal component E'_z, it is seen that the two vectors $\mathbf{E}'$ and $\mathbf{B}'$ are perpendicular to the main direction of motion and to each other. Also, for a relativistic electron, $\beta \sim 1$ and the two vectors have about the same magnitude. This represents also an electromagnetic wave, but having nonplanar properties. Since $\beta_w > 1$, it is seen from Eq. (55) that in order for the wave to propagate in the positive direction of the z-axis also in the frame at rest, the electron should move in the opposite direction ($\beta < 0$) to start with. This will create an enhancement of the equivalent field as seen by the electron; otherwise, if the wave and the particle would move in the same direction there would be a cancellation, which is the most effective when $\beta_w \sim 1$ that is in the short wavelength regime, with frequencies ω well above the cut-off value ω_c. In the other regime with $\beta_w \gg 1$, in proximity of the cut-off, which corresponds to long wavelengths λ, the cancellation does not apply; a part from a sign, the amplitude of the wave is then proportional to

$\gamma\beta_w$ independently of the direction of motion of the electron with respect to the wave.

A case of interest is the following. The electron and the e.m. wave, as seen in the laboratory frame, are moving in the same direction, which is the positive direction of the z-axis of the waveguide, as shown in Fig. 3. Assume $\beta \sim 1$ and $\beta_w \gg 1$; then, from Eq. (55), $\beta'_w \sim -1$ and, in its own frame at rest, the electron sees a wave moving against its position; the consequence is again a forward Compton scattering. Let us also suppose that initially $x = y = 0$, then the only nonvanishing field components are, denoting $\phi = kz - \omega t$,

$$E'_x = 2\pi \frac{V_0}{w} \gamma \left(\beta\beta_w - 1 \right) \cos\phi \tag{56}$$

$$B'_y = -2\pi \frac{V_0}{w} \gamma \left(\beta_w - \beta \right) \cos\phi \tag{57}$$

which have an amplitude

$$\gamma E_w = 2\pi \frac{V_0}{w} \gamma\beta_w \tag{58}$$

that is to be compared to the value γB_w corresponding to a magnetic wiggler. Inspection of Eq. (58) shows that the field amplitude can reach very large values when driving the waveguide in proximity of the cut off.

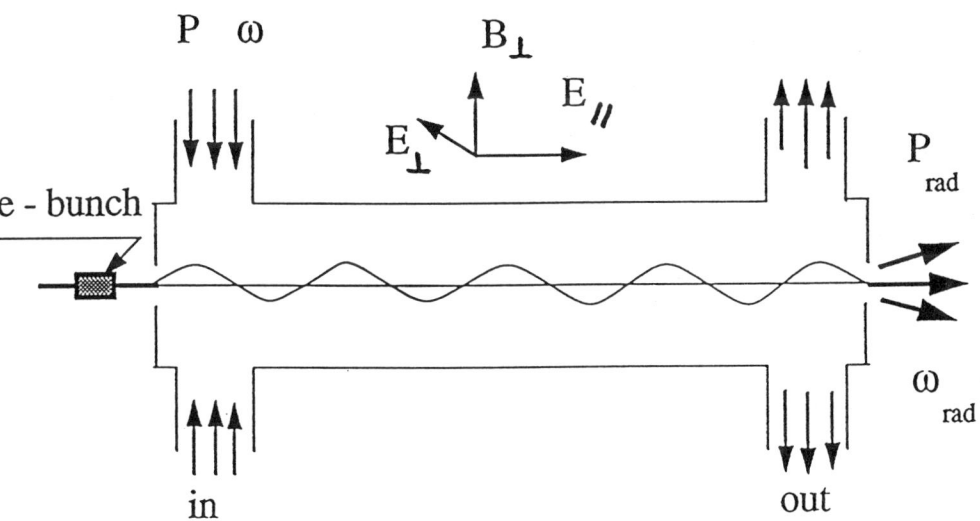

Fig. 3: The rf Wiggler Concept.

THE EQUATIONS OF MOTION

Consider now an electron with mass at rest m and electric charge e moving down the waveguide. The components of the equations of motion are

$$\frac{dp_x}{dt} = -e\left(\frac{\dot{z}}{c}\beta_w - 1\right) E_x \tag{59}$$

$$\frac{dp_y}{dt} = -e\left(\frac{\dot{z}}{c}\beta_w - 1\right) E_y \tag{60}$$

$$\frac{dp_z}{dt} = e\left(\frac{\dot{x}}{c}\beta_w E_x + \frac{\dot{y}}{c}\beta_w E_y + E_z\right) \tag{61}$$

where $\mathbf{p} \equiv (p_x, p_y, p_z)$ is the momentum and $\mathbf{v} \equiv (\dot{x}, \dot{y}, \dot{z})$ the velocity vector.

The equations of motion simplify considerably if we assume that the motion of the electron is confined in proximity of the $y = -h/2$ plane. In this case, if n is even and m is odd, $y = \dot{y} = 0$ is a solution of the equations of motion since $E_y = 0$. Moreover, if also x is very close to the $x = w/2$ axis, then in good approximation the equations of motion are

$$\frac{dp_x}{dt} \approx e\left(\frac{\dot{z}}{c}\beta_w - 1\right) V_0\alpha_1 \cos\phi \tag{62}$$

$$\frac{dp_y}{dt} \approx 0 \tag{63}$$

$$\frac{dp_z}{dt} \approx -e\frac{\dot{x}}{c}\beta_w V_0\alpha_1 \cos\phi \tag{64}$$

where we have taken the real part of the traveling wave exponential factor.

It is seen that the perturbation to the longitudinal motion is of first order in $\dot{x}$ and it can thus be neglected. At the same time

$$\frac{d\phi}{dt} = k\dot{z} - \omega$$
$$\approx kv - \omega \tag{65}$$
$$\approx ck\left(\beta - \beta_w\right) = -\Omega_0$$

Since $\beta_w > \beta$ this quantity is always negative. A phase slippage occurs when the particle and the electromagnetic wave are traveling in the same direction. In proximity of the cut-off, the phase slippage $\Omega_0 \approx \omega$. It is reasonable to assume that during the interaction with the electromagnetic wave, the velocity β of the electron does not change considerably and it remains close to unit.

In the approximation that the horizontal displacement remains small, that is $x \ll w/2$, Eq. (62) can be written as

$$\ddot{x} = \Omega^2 w \cos\phi \tag{66}$$

where

$$\Omega^2 = eV_0 \alpha_1 \frac{\beta\beta_w - 1}{m_0 \gamma w} \tag{67}$$

With a change of variables Eq. (66) becomes

$$\frac{d^2x}{d\phi^2} = \nu^2 w \cos\phi \tag{68}$$

with

$$\nu = \Omega/\Omega_0 \tag{69}$$

The solution of Eq. (68) can be easily derived to be

$$x = -a\cos\phi \tag{70}$$

that is an oscillation at the frequency equal to the phase slippage Ω_0 and amplitude

$$a = w\nu^2 \tag{71}$$

The solution given by Eq. (70) is correct only as long as the amplitude a of the oscillation is small compared to the width w of the waveguide, that is $\nu^2 \ll 1$, which sets a limit on the value of the voltage amplitude V_0.

ENERGY LOSS BY RADIATION

Consider an electron which is moving at relativistic velocity along the z-axis and at the same time is performing small amplitude oscillations at the angular frequency Ω_0. It is well known that the electron will lose energy by radiating electromagnetic waves moving forward in the same direction of the motion of the particle, within an angular aperture of about $1/\gamma$. In the approximation that the oscillatory motion has been occurring for an infinitely long period of time, the spectrum of the radiation is made of only one line at the angular frequency[4]

$$\omega_{\text{rad}} = 2\gamma^2\Omega_0 \tag{72}$$

The radiated power at that frequency by one electron can also be calculated

$$P_0 = \frac{1}{3}\frac{e^2}{c^3}\gamma^4 a^2 \Omega_0^4 \tag{73}$$

where a is the amplitude of the oscillation.

In the extreme case where the beam bunch is much longer than the wavelength of the radiation $2\pi c/\omega_{\text{rad}}$, each electron will radiate independently from the others and the total power radiated is $P_{\text{rad}} = NP_0$ where P_0 is the power from a single electron, given by Eq. (73), and N the total number of electrons in the bunch. On the other hand, when the bunch length is considerably smaller than the radiated wavelength it is conceivable that all the electrons are radiating *coherently* and in this case the total power is[9]

$$P_{\text{rad}} = N^2 P_0 \tag{74}$$

At the same time, though, in order to take advantage of this effect, it is also important that the transverse dimensions of the electron beam are made as small as possible. Indeed, they should not exceed the amplitude of the oscillations given by Eq. (71) and should be smaller than the bunch length itself.

APPLICATIONS

It is convenient to define two parameters that best summarize the interaction between the electron motion and the field in the waveguide. One is the *frequency transformer ratio* $r = \omega_{\text{rad}}/\omega$, that is the ratio of the *radiated* frequency to the *input* frequency to the waveguide. From Eqs. (65 and 72) we derive

$$r = 2\gamma^2 \frac{\beta_w - \beta}{\beta_w} \tag{75}$$

In proximity of the cut-off $\beta_w \gg 1$ and with good approximation $r \sim 2\gamma^2$.

The second parameter is the *power amplification factor* $\eta = P_{\text{rad}}/P$, that is the ratio of the power radiated by the beam bunch to the *input* power to the waveguide. Assuming that the condition of short bunches is satisfied, we derive

$$\eta = \frac{64 N^2 r_0^2 \gamma^2 \pi n^2 \left(\beta \beta_w - 1\right)^2}{3 h w^3 \left(\frac{n^2}{w^2} + \frac{m^2}{h^2}\right) \beta_w} \tag{76}$$

where $r_0 = 2.82 \times 10^{-15}$ m is the classical electron radius. An optimum case is given by a waveguide with a square cross-section, that is $h = w$, and by the lowest order of propagation, namely $m = 1$ and $n = 2$. In proximity of the cut-off then

$$\eta \simeq \frac{256}{15} \pi N^2 \gamma^2 \frac{r_0^2}{w^2} \beta_w \tag{77}$$

and from Eqs. (18 and 32)

$$\beta_w^2 \approx \frac{w \omega_c}{2c \, \mathcal{R}_c} \tag{78}$$

An application is a *frequency transformer*. In this mode of operation the power radiated by a short electron bunch is at a frequency larger than the one used in input to the waveguide. In this case it is sufficient that the power gain $\eta \sim 1$. An example of frequency transformer is shown in Table 1, where the waveguide material is taken to be warm temperature copper and the waveguide itself is driven in proximity of the cut-off where the phase velocity β_w is the largest.

An experimental demonstration of the method of generating power with the use of the rf wiggler is useful and feasible. It can be executed at the Accelerator Test Facility either at Brookhaven[12] or at Argonne National Laboratory[13]. A proposed experimental set up is shown in Fig. 4. Because of the relatively low energy of the electrons, the rf wiggler is immediately attached to the electron gun. It is followed by a dipole magnet to bend the electron beam out of the way and by a second waveguide properly terminated to trap and absorb the emitted radiation. The measurement and the experiment shall then be conducted as explained in ref. [10] which describes a similar experiment with a conventional wiggler (magnetic).

Table 1: An Example of Frequency Transformer

Kinetic Energy of Electrons	4 MeV
Number of Electrons	6×10^{10}
Bunch Length	1 mm
Input Frequency	1.3 GHz
Radiated Frequency	190 GHz
Frequency Transform Ratio	146
Power Amplification Factor	0.5
Phase Velocity, β_w	350
Cut-off Frequency	1.3 GHz
Waveguide Dimension, w	25.8 cm
Period of Oscillations	24.5 cm

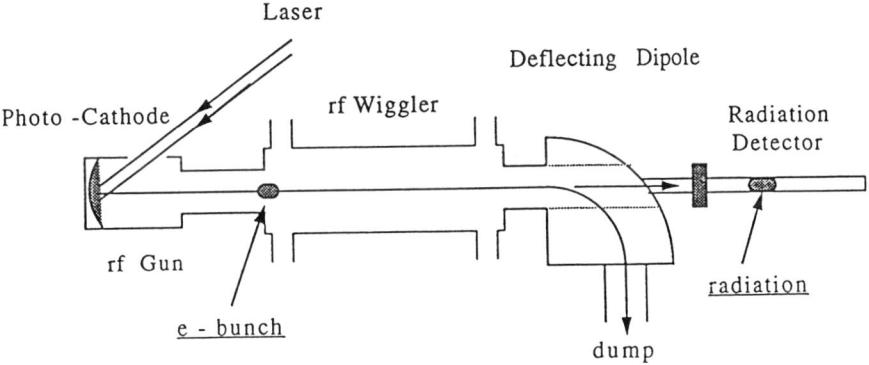

Fig. 4: Experimental Set-Up.

CONCLUSION

We have shown in this paper that it is possible to convert electromagnetic power from one frequency to another with reasonable efficiency by letting a short electron bunch interact with a waveguide driven by an electromagnetic wave in proximity of the cut-off. We have estimated the maximum power gain and the required electron bunch and dimensions; they are within reach of present state of the art of electron sources. Our method to be effective relies on the coherent radiation by which, if the wavelength radiated is larger than the bunch length, the power radiated is proportional to the square of the number of electrons in the bunch.

REFERENCES

1. M.A. Allen, et al., Proceedings of EPAC 90, Nice, France, June 1990, Vol. 1, p. 146.
2. H. Henke, Proceedings of EPAC 90, Nice, France, June 1990, Vol. 1, p. 174.
3. J. Schwinger, Phys. Rev. 75, No. 12 (1949), p. 1912.
4. J.M.J. Madey, J. Appl. Phys. 42, 1906 (1971).
5. R. Bonifacio, Proceedings of EPAC 90, Nice, France, June 1990, Vol. 1, p. 340.
6. A.M. Sessler, Proc. of the Workshop on Laser Acceleration of Particles, Los Alamos, New Mexico, 1982, AIP Conference Proceedings 91, 154 (1982).
7. T.J. Orzechowski, et al., Phys. Rev. Lett. 54, 889 (1985).
8. A.G. Ruggiero, BNL Internal Reports, AD/AP-35 and 41 (1992).
9. F.C. Michel, Phys. Rev. Lett. 48, No. 9, (1982), p. 580.
10. A.G. Ruggiero, contribution on this subject in these conference proceedings.
11. J.D. Jackson, Classical Electrodynamics, Chapter 8, John Wiley & Sons, 2nd Edition.
12. I. Ben-Zvi, 1991 Particle Accelerator Conference, IEEE 91CH3038-7, page 550.
13. P. Schoessow, et al., Proceedings of EPAC 90, Nice, France, June 1990, Vol. 1, p. 606.

RELATIVISTIC HELIX TRAVELING WAVE TUBE AMPLIFIERS

H.P. Freund,[†] N.R. Vanderplaats, and M.A. Kodis
Naval Research Laboratory, Washington, D.C. 20375

ABSTRACT

A relativistic field theory of a helix traveling wave tube (TWT) is described for the case in which a thin annular beam propagates through a sheath helix enclosed within a loss-free wall. The theory is applied to the study of a TWT with an intense relativistic electron beam. The analysis implicitly includes beam space-charge effects and is valid for arbitrary azimuthal mode number, and the coupled-wave Pierce theory is recovered in the *near-resonant* limit. The results indicate that impressive gains and efficiencies are possible in this regime. In addition, the interaction is relatively insensitive to the effects of a beam energy spread.

INTRODUCTION

The helix traveling wave tube (TWT) was first analyzed by Pierce and co-workers[1-3] based upon a coupled-wave analysis utilizing the vacuum modes of the helix and positive and negative energy space-charge waves. Theories based upon a more complete analysis of Maxwell's equations in the helix have also been developed[4-7]. Of these two approaches, the complete field theory is the more exact. Recently, a fully self-consistent and relativistic field theory of the helix TWT was developed by solution of the relativistic fluid equations and Maxwell's equations in a configuration which consists of the propagation of a thin annular electron beam down the axis of a sheath helix enclosed by a loss-free conducting wall[8]. This analysis is briefly described in the present paper, and used to analyze interaction for the case of an intense relativistic electron beam.

The general dispersion equation is compared with a Pierce analysis by the imposition of a *near-resonant* approximation. This shows how the self-consistent field theory reduces to the usual polynomial dispersion equation found in the Pierce analysis. The limiting cases of the coupled-wave theory which describe the ballistic and space-charge dominated regimes in which the gain scales as the cube and fourth root of the beam current respectively are explicitly addressed and compared with the results of the complete electromagnetic field theory. The field theory is then applied to the case of a relativistic helix TWT, and the results indicate that impressively high gains and efficiencies are possible, as is a remarkable insensitivity to beam thermal effects.

THE DISPERSION EQUATION

The equilibrium model we employ is of the propagation of a thin annular beam through a sheath helix-loaded cylindrical drift tube. Hence, the equilibrium is described via an azimuthally symmetric charge density $n_0(r) = n_b \Delta r_b \delta(r - r_b)$, where n_b denotes the ambient beam density, and $r_b \Delta r_b$ denote the radius and thickness of the annulus. The beam is assumed to propagate uniformly along the symmetry [i.e., z-] axis of the system at a velocity of $\mathbf{v}_0 = v_0 \, \hat{\mathbf{e}}_z$.

Permanent Address: Science Applications International Corp., McLean, VA 22102.

The circuit configuration describes a sheath helix located within a cylindrical waveguide of radius R_g. The sheath helix model assumes that the conducting wires of the helix are thin enough that the helix can be modeled by a thin conducting cylindrical sheet of radius R_h [< R_g], and with a pitch angle ϕ. Hence, the unit vector describing the pitch of the helix is $\hat{\mathbf{e}}_\phi = \hat{\mathbf{e}}_\theta \cos \phi + \hat{\mathbf{e}}_z \sin \phi$ in cylindrical coordinates.

The perturbed current density and beam velocity are obtained by a first order perturbation analysis in which we expand $n = n_0 + \delta n$ and $\mathbf{v} = \mathbf{v}_0 + \delta\mathbf{v}$. Hence

$$\left(\frac{\partial}{\partial t} + v_0 \frac{\partial}{\partial z}\right) \delta n + n_0 \nabla \cdot \delta\mathbf{v} = 0 \ ,$$

$$\left(\frac{\partial}{\partial t} + v_0 \frac{\partial}{\partial z}\right) \delta\mathbf{v} = -\frac{e}{\gamma_0 m_e}\left[\left(\mathbf{I} - \beta_0^2 \hat{\mathbf{e}}_z\hat{\mathbf{e}}_z\right) \cdot \delta\mathbf{E} + \beta_0 \,\hat{\mathbf{e}}_z \times \delta\mathbf{B}\right] \ ,$$

(1)

where e and m_e are the electron charge and rest mass, $\beta_0 \equiv v_0/c$, $\gamma_0 \equiv (1 - \beta_0^2)^{-1/2}$, $\mathbf{I}$ is the unit dyadic, and $\delta\mathbf{E}$ and $\delta\mathbf{B}$ denote the fluctuating electric and magnetic fields.

The source current and charge density are found after solution for the perturbed velocity and density subject to a Fourier analysis in which the perturbed quantities are assumed to vary as

$$\delta f(\mathbf{x},t) = \sum_{l=-\infty}^{\infty} \delta\hat{f}_l(r) \exp\left(ikz + il\theta - i\omega t\right) \ ,$$

(2)

where ω and k denote the angular frequency and wavenumber. Substitution of this form for the fluctuations into Eqs. (1) yields the solutions

$$\delta\hat{n}_l = \frac{n_0}{\Delta\omega}\left(k \, \delta\hat{v}_{z,l} - i \, \nabla_\perp \cdot \delta\hat{\mathbf{v}}_l\right) \ ,$$

$$\delta\hat{\mathbf{v}}_l = -\frac{ie}{\gamma_0 m_e \Delta\omega}\left[\left(\mathbf{I} - \beta_0^2 \hat{\mathbf{e}}_z\hat{\mathbf{e}}_z\right) \cdot \delta\hat{\mathbf{E}}_l + \beta_0\hat{\mathbf{e}}_z \times \delta\hat{\mathbf{B}}_l\right] \ ,$$

(3)

where $\Delta\omega \equiv \omega - kv_0$ describes the frequency tuning. As a result, the perturbed source current and charge density can be expressed as

$$\delta\hat{\mathbf{J}}_l \cong \frac{i}{4\pi} \frac{\omega_b^2}{\gamma_0 \Delta\omega} \Delta r_b \, \delta(r - r_b)\left[\frac{\omega}{\gamma_0^2 \Delta\omega} \hat{\mathbf{e}}_z\hat{\mathbf{e}}_z \cdot \delta\hat{\mathbf{E}}_l + \delta\hat{\mathbf{E}}_{\perp,l} + \beta_0\hat{\mathbf{e}}_z \times \delta\hat{\mathbf{B}}_l\right] \ ,$$

(4)

$$\delta\hat{\rho}_l \cong \frac{i}{4\pi} \frac{\omega_b^2}{\gamma_0^3 \Delta\omega^2} k \, \Delta r_b \, \delta(r - r_b)\hat{\mathbf{e}}_z \cdot \delta\hat{\mathbf{E}}_l \ ,$$

where $\omega_b^2 \equiv 4\pi e^2 n_b/m_e$ denotes the square of the plasma frequency.

Substitution of these sources into Maxwell's equations for the axial components of the fields yields

$$\left[\nabla_\perp^2 + \kappa^2\right]\delta\widehat{E}_{z,l} \cong \frac{\omega_b^2}{\gamma_0^3\Delta\omega^2}\kappa^2\Delta r_b\,\delta(r-r_b)\delta\widehat{E}_{z,l}\ ,$$

$$(5)$$

$$\left[\nabla_\perp^2 + \kappa^2\right]\delta\widehat{B}_{z,l} \cong 0\ .$$

for arbitrary charge and current densities, where $\kappa^2 \equiv \omega^2/c^2 - k^2$. The transverse components are

$$\delta\widehat{\mathbf{E}}_{\perp,l} = -\frac{i}{\kappa^2}\left[k\,\nabla_\perp\delta\widehat{E}_{z,l} - \frac{\omega}{c}\hat{\mathbf{e}}_z\times\nabla_\perp\delta\widehat{B}_{z,l}\right]\ ,$$

$$(6)$$

$$\delta\widehat{\mathbf{B}}_{\perp,l} = -\frac{i}{\kappa^2}\left[k\,\nabla_\perp\delta\widehat{B}_{z,l} + \frac{\omega}{c}\hat{\mathbf{e}}_z\times\nabla_\perp\delta\widehat{E}_{z,l}\right]\ .$$

Eqs. (5) are consistent with the physical mechanism in a TWT since the primary coupling is between the axial motion of the beam and the axial component of the electric field, and the axial component of the magnetic field is governed by the homogeneous wave equation. Note also that the source term for the axial electric field corresponds to a δ-function in radius; hence, there is a discontinuity in the axial electric field which can be expressed in terms of the following jump condition

$$\lim_{\varepsilon\to 0}\frac{d}{dr}\delta\widehat{E}_{z,l}\bigg|_{r=r_b-\varepsilon}^{r_b+\varepsilon} = \frac{\omega_b^2}{\gamma_0^3\Delta\omega^2}\kappa^2\Delta r_b\delta\widehat{E}_{z,l}(r_b)\ . \qquad (7)$$

The boundary conditions correspond to three regions:[8] (1) $0 \le r \le r_b$, (2) $r_b < r \le R_h$, and (3) $R_h < r \le R_g$. The boundary conditions at the wall require that the tangential component of the electric field and the normal component of the magnetic field must vanish. The boundary conditions at the beam require that the tangential components of the electric field and the axial and radial components of the magnetic field must be continuous. The tangential component of the electric field at the helix must be perpendicular to the direction of the helix. In addition, the tangential component of the electric field must be continuous. Finally, the tangential component of the magnetic field parallel to the helix must also be continuous.

The derivation of the dispersion equation involves finding the solution to Maxwell's equations subject to the jump and boundary conditions. To this end, we write the axial electric and magnetic fields in the three regions in the form

$$\delta\widehat{E}_{z,l} = \begin{cases} A_e J_l(\kappa r) & ;\,0\le r\le r_b \\ B_e J_l(\kappa r) + C_e Y_l(\kappa r) & ;\,r_b < r \le R_h \\ D_e J_l(\kappa r) + E_e Y_l(\kappa r) & ;\,R_h < r \le R_g \end{cases}\ ,$$

$$(8)$$

$$\delta\widehat{B}_{z,l} = \begin{cases} A_b J_l(\kappa r) & ;\,0\le r\le r_b \\ B_b J_l(\kappa r) + C_b Y_l(\kappa r) & ;\,r_b < r \le R_h \\ D_b J_l(\kappa r) + E_b Y_l(\kappa r) & ;\,R_h < r \le R_g \end{cases}\ ,$$

where J_l and Y_l denote the regular Bessel and Neumann functions respectively.

Application of the boundary and jump conditions results in a determinantal dispersion equation of the form[8]

$$\Lambda_l(\omega,k)\,\varepsilon_l(\omega,k) = -\frac{\omega_b^2}{4\gamma_0^3 c^2}\frac{\omega^2}{c^2}\kappa^2\sigma_b\frac{J_l^2(\kappa r_b)}{J_l(\kappa R_g)J_l(\kappa R_h)}\,W_l(R_g,R_h)\,,\tag{9}$$

where

$$\Lambda_l(\omega,k) \equiv \frac{\omega^2}{c^2}+\kappa^2\left(\frac{lk}{\kappa^2 R_h}-\tan\phi\right)^2\frac{J_l(\kappa R_h)J_l{}'(\kappa R_g)}{J_l(\kappa R_g)J_l{}'(\kappa R_h)}\frac{W_l(R_g,R_h)}{W_l{}''(R_g,R_h)}\,,\tag{10}$$

denotes the dispersion function for the cold-helix modes, [9] and

$$\varepsilon_l(\omega,k)\equiv\frac{\Delta\omega^2}{c^2}-\frac{\omega_b^2}{4\gamma_0^3 c^2}\kappa^2\sigma_b\frac{J_l(\kappa r_b)}{J_l(\kappa R_h)}\,W_l(r_b,R_h)\,,\tag{11}$$

denotes the dielectric function for the space-charge modes, $W_l(r_1,r_2)\equiv Y_l(\kappa r_1)J_l(\kappa r_2) - J_l(\kappa r_1)Y_l(\kappa r_2)$, and $W_l{}''(r_1,r_2)\equiv Y_l{}'(\kappa r_1)J_l{}'(\kappa r_2) - J_l{}'(\kappa r_1)Y_l{}'(\kappa r_2)$. As such, the space-charge mode dispersion equation, $\varepsilon_l(\omega,k)=0$, can be obtained from solution of Poisson's equation. This dispersion equation is valid for all azimuthal modes, for backward-propagating waves, and implicitly includes the RF space-charge effects. However, the dispersion equation has the same qualitative form as the Pierce TWT theory in that it is a product of the dispersion equations for the cold helix and the beam space-charge waves as well as a coupling term.

COMPARISON WITH PIERCE TWT THEORY

The determinantal dispersion equation (9) reduces to the Pierce TWT analysis in the *near-resonant* limit where it is assumed that the mode is azimuthally symmetric [i.e., $l = 0$], $|{\rm Im}\,k| \ll |{\rm Re}\,k|$, and $k \approx \omega/v_0$. In this limit, the dispersion equation can be expressed as a polynomial in k. In order to see how this is accomplished, we first note that the wavenumber in the cold helix can be approximated as

$$k_0^2(\omega)\equiv\frac{\omega^2}{c^2}\left[1+\cot^2\phi\,\frac{J_0(\kappa_e R_g)J_1(\kappa_e R_h)}{J_0(\kappa_e R_h)J_1(\kappa_e R_g)}\frac{W_0{}''(R_h,R_g)}{W_0(R_h,R_g)}\right]\,,\tag{12}$$

in the *near-resonant* regime, where κ has been replaced by $\kappa_e\equiv i\omega/\gamma_0 v_0$ [including in the W_0 and $W_0{}''$ functions]. Imposition of the near-resonant approximation on the space-charge contribution as well as on the coupling terms on the right-hand-side of (9) yields the familiar quartic polynomial dispersion from the Pierce Theory

$$\left(k^2-k_0^2\right)\left[\left(\omega-kv_0\right)^2-4QC^3k^2v_0^2\right]\cong 2C^3\frac{\omega}{v_0}k_0k^2v_0^2\,,\tag{13}$$

where

$$C^3=\frac{\omega_b^2\sigma_b}{8\gamma_0^5 c^2}\cot^2\phi\,\frac{J_0^2(\kappa_e r_b)J_1(\kappa_e R_h)}{J_0^2(\kappa_e R_h)J_1(\kappa_e R_g)}\,W_0{}''(R_h,R_g)\,,$$

$$\tag{14}$$

$$Q=\frac{\tan^2\phi}{2\beta_0^2}\frac{J_0(\kappa_e R_h)J_1(\kappa_e R_g)}{J_0(\kappa_e r_b)J_1(\kappa_e R_h)}\frac{W_0(R_h,r_b)}{W_0{}''(R_h,R_g)}\,.$$

There are two relatively simple limiting cases associated with this dispersion equation and may be termed the *ballistic* or *space-charge dominated* regimes.

Analogues to these regimes are found, for example, in free-electron lasers in which they are referred to as the Compton and Raman regimes.[10,11] The *ballistic* regime occurs when space-charge effects are negligible. The maximum growth rate occurs for $\Delta k = 0$. Noting that C^3 is typically negative, the solution is[8]

$$\text{Im } k \cong \pm \frac{\sqrt{3} k_0}{2} |C| . \tag{15}$$

Therefore, the growth rate scales as the cube root of the beam current. The *space-charge dominated* regime is found in the opposite limit. In this case, the dispersion equation is quadratic, and the maximum growth rate is, again, found when $\Delta k = 0$. This requires the choice of the negative-energy space-charge wave [i.e., $k = \kappa_- + \delta k$] since Q is typically negative as well. The solution for the maximum growth rate in this regime is

$$\text{Im } k \cong \pm \frac{k_0}{2} \sqrt[4]{\frac{C^3}{Q}} . \tag{16}$$

Hence, the growth rate scales as the fourth root of the current in this regime. There is an upper bound on the current in the space-charge dominated regime, however, since increasing beam currents modify the dispersion of the wave. For sufficiently high currents, the wave-particle resonance is altered and backward-propagating waves become important. It is important to recognize, however, that while the particular approximation which we have referred to as the space-charge dominated regime is invalid in this limit, the space-charge effects on the interaction are still important, but must be treated using the self-consistent field theory.

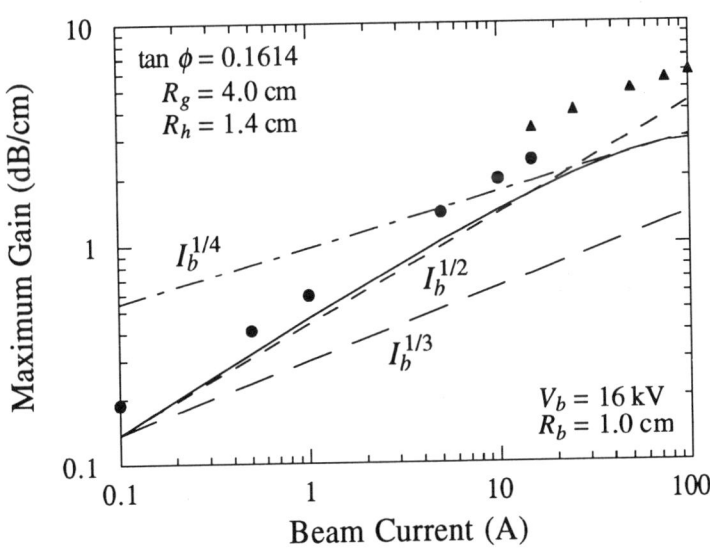

Fig. 1 Comparison of the maximum gain versus current with the ballistic (circles) and space-charge dominated (triangles) limits. Dashed lines show $I_b^{1/2}$, $I_b^{1/3}$, and $I_b^{1/4}$ scalings.

The specific behavior of the maximum gain from Eq. (9) with beam current is shown in Fig. 1 for a representative set of parameters. The ballistic (circles) and space-charge dominated (triangles) approximations are also shown, and the transition occurs for $I_b \approx 20$ A. There are several points of note. The first is that the scaling of the gain with $I_b^{1/3}$ in the ballistic regime is not found in the complete field theory. Rather, the gain is found to scale as $I_b^{1/2}$ to a good approximation up to the transition to the space-charge dominated regime. We take this to imply that space-charge forces are important even in the nominal ballistic regime *for this choice of parameters*. We do not maintain that this result will be found in general. The second point is that the gain from the complete field theory is found to scale as $I_b^{1/4}$ in the space-charge dominated regime, although the gain is lower than that predicted from (9).

THE RELATIVISTIC CASE

We now turn to the case of a helix TWT driven by an intense relativistic electron beam, and assume that the beam has a voltage of 500 kV and a current of 95 A. In order to obtain a resonant interaction at frequencies below approximately 20 GHz, we assume the helix has a radius of 0.8 cm, the wall has a radius of 1.2 cm, and the helix pitch is given by $\tan \phi = 1.4592$. In order to provide sufficient clearance to propagate the beam, we also assume that the beam radius is 0.35 cm. Note, however, that an axial guide magnetic field is still required in order to confine the beam.

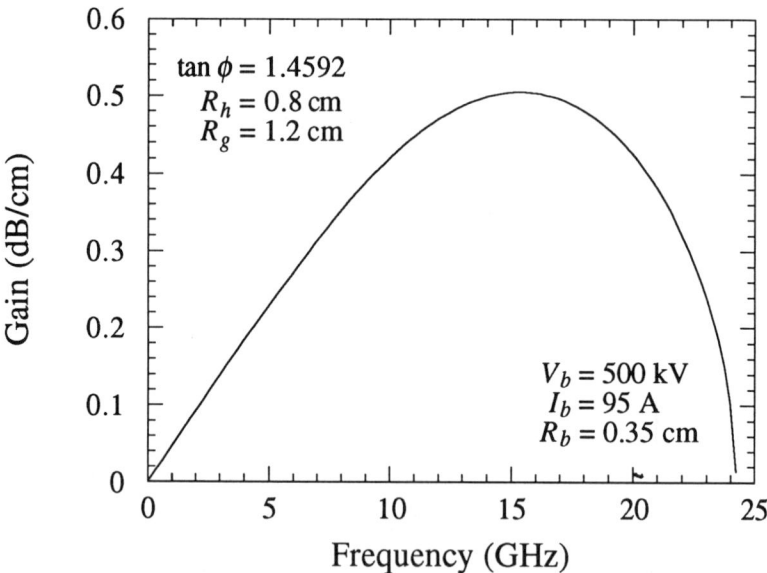

Fig. 2 Variation in the gain as a function of frequency for the relativistic TWT.

The variation in the gain per unit length as a function of frequency is determined by the imaginary part of the solution to Eq. (9). The results for the above-mentioned parameters are shown in Fig. 2. The maximum gain of approximately 0.5 dB/cm is found at a frequency in the neighborhood of 16 GHz. The bandwidth of the interaction is extremely broad, and the FWHM points extend over the range of approximately 5 - 23 GHz. It should be remarked in passing that the gain calculated from the Pierce theory (13) differs significantly from this result for frequencies greater than 8 - 9 GHz. The reason for this is that the Pierce theory is found under the assumption of a near-resonant approximation which breaks down in the relativistic regime due to the broad bandwidth of the interaction.

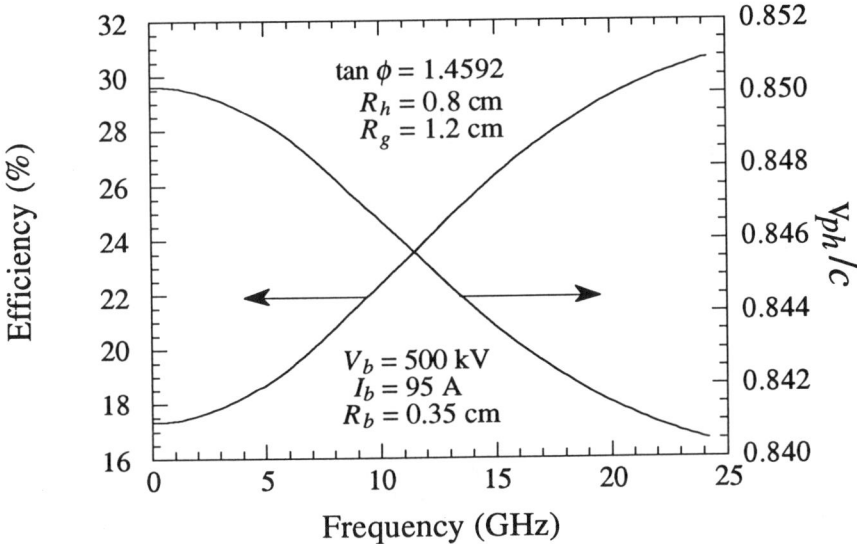

Fig. 3 Variation in the efficiency and phase velocity versus frequency for the relativistic TWT.

The interaction efficiency can be estimated by a phase trapping argument in which saturation occurs when the electron beam has decelerated by an amount $\Delta v = 2(v_0 - v_{ph})$, where v_{ph} denotes the phase velocity of the wave. The efficiency corresponding to this energy loss is

$$\eta \approx \frac{\gamma_0^3}{\gamma_0 - 1} \frac{v_0 \Delta v}{c^2} .$$ (19)

The phase velocity and, hence, the efficiency are determined by the real part of the solution for the wavenumber in the beam-loaded helix [note that the beam affects the dispersion both through the coupling to the space-charge wave and through the dielectric modification to the helix modes]. The results of this calculation are shown in Fig. 3 in which we plot both the efficiency and the phase velocity as functions of frequency. Observe that the phase velocity decreases and the efficiency increases with frequency over this band. It is clear that the resonant interaction occurs for phase velocities of $\approx 0.842 - 0.851$. The maximum efficiency is of the order of 30% at the high frequency end of this band, and is of the order of 28% at the maximum gain point.

Finally, we note that the sensitivity of the interaction to a beam energy spread can be estimated on the by noting that thermal effects begin to dominate the resonance condition when the thermal velocity Δv_z satisfies the condition $\Delta v_z/v_0 \approx \text{Im } k/\text{Re } k$. For this particular example, $\text{Im } kR_g \approx 0.059$ and $\text{Re } kR_g \approx 3.81$ at the frequency of maximum growth. This implies that $\Delta v_z/v_0 \approx 1.5\%$, which corresponds to an energy spread $\Delta\gamma_z/\gamma_0 \approx 4.2\%$.

SUMMARY AND DISCUSSION

A self-consistent relativistic field theory of a helix TWT has been presented which implicitly includes the effects of beam space-charge both in terms of the coupling to beam space-charge waves and the dielectric loading of the helix modes. The analysis is valid for arbitrary azimuthal mode numbers and for backward-propagating waves. The Pierce theory is recovered in the near-resonant approximation; however, that there is a discrepancy between the ballistic limit and the field theory even for low currents. The maximum gain from the self-consistent field theory is not found to scale as $I_b^{1/3}$ in this regime. Rather, it increases more rapidly than the cube root of the beam current, and we find that the maximum gain scales approximately as $I_b^{1/2}$ over a broad range of currents. This should not be interpreted, however, as a statement that the ballistic limit is never valid. Rather, the issue must be considered on an individual basis.

The results of the theory as applied to the case of a relativistic TWT driven by a high current electron beam yield impressive performance figures. For a 500 kV/95 A electron beam, peak gains in the K_u-band reach 0.5 dB/cm. In addition, projected efficiencies for this case reach up to 30%, and show a remarkable tolerance for beam energy spread. Since the wave-particle resonance is sensitively dependent upon the dimensions of the helix and the wall, it is expected that relativistic TWTs can be designed to operate over a wide frequency range ranging form the X-band through the K_a-band.

This level of performance must be considered in light of other recent work on relativistic intense beam slow-wave sources such as Cerenkov Masers and rippled-wall TWTs. In the case of Cerenkov Masers, a recent experiment conducted at General Dynamics demonstrated an efficiency of 11% at X-band frequencies using a 750 kV/3.1 kA electron beam.[13] However, two independent simulations of this experiment indicate that the device was too short to reach saturation, and that efficiencies as high as 30% would have been achievable with either a longer system or higher drive power.[14,15] Rippled-wall TWT experiments have also demonstrated efficiencies of theorder of 11% at X-Band frequencies using 850 kV/1 kA electron beams.[16] Finally, relativisitc intense beam BWO's using rippled-wall structures have demonstrated similar performance.[17] As a consequence, a variety of slow-wave devices using intense relativistic electron beams, included helix TWTs, have the potential for providing high power and high efficiency sources for a variety of applications including the next generation accelerator driver.

ACKNOWLEDGMENTS

This work was supported by the Office of Naval Research and the Office of Naval Technology.

REFERENCES

1. J.R. Pierce and L.M. Field, Proc. IRE **35**, 108 (1947).
2. J.R. Pierce, Proc. IRE **35**, 111 (1947).
3. J.R. Pierce, *Traveling-Wave Tubes* (Van Nostrand, New York, 1950).
4. O.E.H. Rydbeck, Ericsson Technics, No. 46 (1948).
5. L.J. Chu and J.D. Jackson, Proc. IRE **36**, 853 (1948).
6. A.H.W. Beck, *Space-Charge Waves* (Pergamon, New York, 1958).
7. R.G.E. Hutter, *Beam and Wave Electronics in Microwave Tubes* (Van Nostrand, New York, 1960).
8. H.P. Freund, M.A. Kodis, and N.R. Vanderplaats, IEEE Trans. Plasma Science (to appear in October 1992).
9. H.S. Uhm and J.Y. Choe, J. Appl. Phys. **53**, 8483 (1982).
10. H.P. Freund, P. Sprangle, D. Dillenburg, E.H. da Jornada, R.S. Schneider, and B. Liberman, Phys. Rev. A **26**, 2004 (1982).
11. H.P. Freund and A.K. Ganguly, Phys. Rev. A **28**, 3438 (1983).
12. H.P. Freund, Phys. Rev. A **37**, 3371 (1988).
13. E. Garate, H. Kosai, W. Peter, A. Fisher, W. Main, J. Weatherall, and R. Cherry, SPIE Proceedings on High Power Microwaves and Particle Beams, Los Angeles, CA January 1990.
14. W. Peter, E. Garate, W. Main, and A. Fisher, Phys. Rev. Lett. **65**, 2989 (1990).
15. H.P. Freund, Phys. Rev. Lett. **65**, 2993 (1990).
16. D. Shiffler, J.A. Nation, and G.S. Kerslick, IEEE Trans. Plasma Sci. **18**, 546 (1990).
17. Y. Carmel, K. Minami, R.A. Kehs, W.W. Destler, V.L. Granatstein, D. Abe, and W.L. Lou, Phys. Rev. Lett. **62**, 2389 (1989).

CONDITIONER FOR A HELICALLY TRANSPORTED ELECTRON BEAM*

Changbiao Wang**
Lawrence Berkeley Laboratory
University of California
Berkeley, CA 94720

ABSTRACT

The kinetic theory is developed to investigate a conditioner for a helically transported electron beam. Linear expressions for axial velocity spread are derived. Numerical simulation is used to check the theoretical results and examine nonlinear aspects of the conditioning process. The results show that in the linear regime the action of the beam conditioner on a pulsed beam mainly depends on the phase at which the beam enters the conditioner and depends only slightly on the operating wavelength. In the nonlinear regime, however, the action of the conditioner strongly depends on the operating wavelength and only slightly upon the entrance phase. For a properly chosen operating wavelength, a little less than the electron's relativistic cyclotron wavelength, the conditioner can decrease the axial velocity spread of a pulsed beam down to less than one-third of its initial value.

INTRODUCTION

In a beam conditioner, proposed by Sessler, Whittum and Yu,[1,2,3] a nearly monoenergetic beam has the shape of the beam's phase volume so changed that its axial velocity spread is improved. As a result, the beam conditioner can greatly reduce the spread in axial velocity of an electron beam, and hence it can be used, with advantage, on almost all fast wave devices. It is therefore natural that it has aroused considerable attention.[4,5]

There are different means for conditioning electron beams. The longitudinal electric field in a microwave cavity, as proposed by Sessler, et al, can be used to condition electron beams, and this is an efficient method. However, for low energy beams, transported by a solenoidal magnetic field, some other method of beam conditioning is required. The transverse electric field in an RF cavity can, conveniently, be used for this purpose.

In this paper, we present a kinetic formulation of a conditioner consisting of a microwave cavity operating in the TE_{011} mode while immersed in a uniform axial magnetic field. We treat analytically the linear problem of dependence of the axial velocity spread on the cavity length, and use simulation to examine non-linear aspects of the evolution of the spread with both the cavity length and the operating wavelength.

---

* Work supported by the Director, Office of Energy Research, Office of High Energy and Nuclear Physics, Division of High Energy Physics, of the U.S. Department of Energy under Contract Number DE-AC03-76SF00098
** On leave from University of Electronic Science and Technology of China, Chengdu, Sichuan, 610054, China

In a cavity operating in the TE_{011} mode, the electric field has only an azimuthal component with a radial distribution given by the first order Bessel function $J_1(k_c R)$, where k_c is the cutoff wave number, and R is the radial coordinate. For a single-energy electron beam with a sufficiently small beam radius and only one guiding center at the origin, the electrons with a larger gyration radius experience a stronger electric field decelerating force (for appropriate phase) than those with a smaller gyration radius. The larger the gyration radius is, the more energy the electron loses. By the coupling of energy with axial momentum (a relativistic effect), the axial velocity of the electron is increased if the effect of the time-dependent magnetic field is neglected. Therefore, as long as the beam pulse is sufficiently short the axial velocity spread will be improved.

For an actual electron beam, the guiding-center radius R_g ranges from zero to R_b, where R_b is the beam radius, and the gyration radius r_L ranges from zero to $(R_b - R_g)$. In such a situation, we can consider the azimuthal field as the sum of infinite cyclotron harmonics. Of all these harmonics only the zeroth one is important. So for those electrons with non-zero guiding center, the previous analysis holds. From this we can see that the axial velocity spread of an electron beam with multi-guiding centers also can be improved.

Generally speaking, increasing the cavity length increases the interaction time. In this case, however, non-linear effects become important. As it is well known, when the cavity operating frequency is slightly greater than the electron relativistic cyclotron frequency, the electron beam effectively interacts with cyclotron harmonics and, as a result, resonant emission appears, which is the basis of the maser instability.[6] At this time, most of the electrons lie in the decelerating electric field of the fundamental harmonic because of the negative mass effect[7,8,9] which results in particle bunching in the azimuthal direction as explained by Sprangle and Drobot.[10] So once the bunching appears, no matter whether the beam experiences a net transverse acceleration or deceleration at the beginning interaction (depending on the phase of the pulsed beam entrance of the cavity), the transverse velocities of the most electrons can be continuously reduced in the next interaction. Consequently, the axial velocity spread will be gradually improved until another nonlinear process (resonant absorption) arises, so that the transverse velocities of the electrons begin to increase, resulting in an increase in the axial velocity spread.

The beam conditioning presented here is different from the electron-beam cooling proposed by Hirshfield and Park.[11] In their proposal, the beam's distribution of energy is made narrower by use of both resonant emission and absorption. This process cannot be used to improve the axial velocity spread. For a single-energy electron beam with a spread in axial velocity, for example, it can do nothing because the width of the distribution in energy is null. The beam conditioner, in contrast, reduces the spread in axial velocity (instead of the energy spread) through the coupling of energy with axial momentum caused by resonant emission.

In Sec. II, a calculational model is set up to treat analytically a pulsed beam with the Vlasov theory. The perturbation distribution function of the pulsed beam conditioned by the RF cavity is derived and linear expressions for axial velocity spread are given. In Sec. III, numerical simulations are used to check the analytical results, and investigate the dependence of axial velocity spread on the cavity length and operating wavelength caused by the nonlinear interaction of the beam with the cavity field. Finally, in Sec. IV some conclusions are made.

LINEAR KINETIC THEORY OF THE BEAM CONDITIONER

In this section, based on linearized Vlasov equations, we will derive the perturbation distribution functions of a pulsed electron beam conditioned by the RF cavity and use them to obtain analytic expressions for the rms-normalized axial velocity spread.

In the model, we take the pulsed beam as a segment, which has a length L, of an infinitely long electron beam. We will first calculate the perturbation distribution function for the infinitely long beam, and then we use it to calculate the axial velocity spread of the considered segment. We assume that the electron beam is mono-energetic. The electron's transverse velocity is small compared with its axial velocity and variation in the pulse length is negligible when the pulsed beam goes from one end of the cavity to another. The beam pulse front is located at $z=0$ when $t=0$. At $t=d/v_0$ with d ($\geq L$) the cavity length and v_0 the total initial velocity, the beam pulse arrives at the front end of the cavity, and at $t=(d+L)/v_0$, the pulse beam has passed through the cavity, as shown in Fig. 1. For simplicity, the time-dependent magnetic field is neglected in the linear consideration.

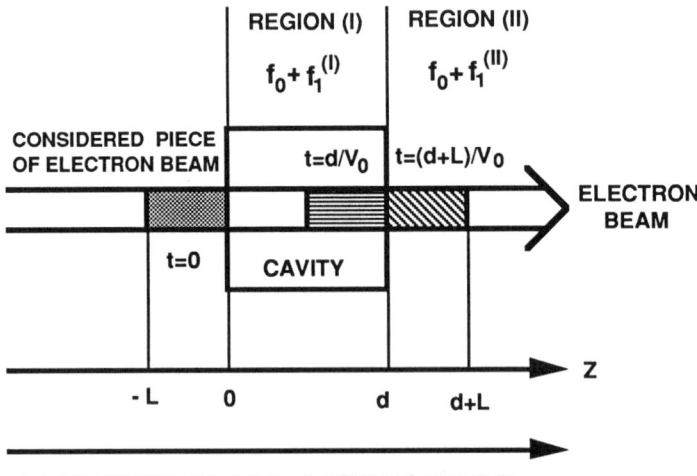

Fig. 1. Calculational model. The pulsed electron beam is taken as a segment of an infinitely long beam. When $t=0$, the front of the segment of beam is located at $z=0$. When $t=d/v_0$ and $t=(d+L)/v_0$, the front is at $z=d$ and $z=(d+L)$, respectively.

The Vlasov equation describing the beam conditioner is given by

$$\frac{\partial f}{\partial t} + \mathbf{v}.\frac{\partial f}{\partial \mathbf{x}} + e\,(\mathbf{E} + \mathbf{v} \times \mathbf{B}).\frac{\partial f}{\partial \mathbf{p}} = 0\,, \tag{1}$$

where

$$\mathbf{E} = \hat{\varphi}\,E_0 J_1(k_c R)\,\sin\frac{\pi}{d}z\,\sin \omega t\,, \tag{2}$$

$$\mathbf{B} = \hat{\mathbf{z}}\, B_0 \,. \tag{3}$$

In Eqs. (2) and (3), the cylindrical coordinates are used, and $\hat{\mathbf{R}}$, $\hat{\varphi}$ and $\hat{\mathbf{z}}$ are all unit vectors; E_0 is the TE_{011}-mode electric field amplitude, B_0 is the applied uniform axial magnetic field, and ω is the cavity operating frequency. According to the small signal assumption $|E_0/B_0 c| \ll 1$, where c is the light speed in free space, Eq. (1) can be linearized as

$$\left\{ \frac{\partial}{\partial t} + \mathbf{v} \cdot \frac{\partial}{\partial \mathbf{x}} + e\,(\mathbf{v} \times \mathbf{B}) \cdot \frac{\partial}{\partial \mathbf{p}} \right\} \left\{ \begin{array}{c} f_0 \\ f_1^{(II)} \end{array} \right\} = 0 \,, \tag{4}$$

and

$$\left\{ \frac{\partial}{\partial t} + \mathbf{v} \cdot \frac{\partial}{\partial \mathbf{x}} + e\,(\mathbf{v} \times \mathbf{B}) \cdot \frac{\partial}{\partial \mathbf{p}} \right\} f_1^{(I)} = -\,e\,\mathbf{E} \cdot \frac{\partial f_0}{\partial \mathbf{p}} \,, \tag{5}$$

where f_0 is the equilibrium distribution function for both regions (I) and (II), and $f_1^{(I)}$ and $f_1^{(II)}$ are, respectively, the perturbation distribution functions for the two regions. In region (I), there is a cavity field, whereas in region (II) there is no cavity field. So $f_1^{(I)}$ satisfies Eq. (5) and $f_1^{(II)}$ satisfies Eq. (4).

At $t = 0$ the electric field begins to condition the cavity-in part of the infinitely long electron beam so that it produces a perturbation of the distribution of the beam. Then the perturbation propagates with a velocity of v_z in the z-direction like a wave. So $f_1^{(I)}$ and $f_1^{(II)}$ are required to satisfy the following initial and boundary conditions

$$f_1^{(I)}(t \le 0) = f_1^{(II)}\!\left(t \le \frac{z-d}{v_z}\right) = 0 \,, \tag{6}$$

$$f_1^{(I)}(z = d) = f_1^{(II)}(z = d) \,. \tag{7}$$

Eqs. (4)-(7) are the basis of finding perturbation functions. Only after obtaining them can we calculate the axial velocity spread.

Equilibrium Distribution Function

To calculate perturbation distribution functions we first have to determine the equilibrium distribution function from Eq. (4). According to the first order partial differential equation theory, any combination of constants of motion from characteristic equations of Eq. (4) is a solution. So, if we find the constants of motion we can use them to construct equilibrium distribution functions in terms of a given electron beam. For convenience, we use cylindrical coordinates in the momentum space just as in the configuration space, that is, $p_x = p_\perp \cos\phi$, $p_y = p_\perp \sin\phi$, and $p_z = p_z$.

Calculations indicate that the characteristic equations of Eq. (4) have six independent constants of motion:

$$C_1 = p_\perp , \tag{8}$$
$$C_2 = p_z , \tag{9}$$

$$C_3 = \phi - \frac{|e|B_0}{p_z} z , \tag{10}$$

$$C_4 = R \cos\varphi - \frac{p_\perp}{|e|B_0} \sin\phi , \tag{11}$$

$$C_5 = R \sin\varphi + \frac{p_\perp}{|e|B_0} \cos\phi , \tag{12}$$

$$C_6 = \phi - \Omega t , \tag{13}$$

where e is the electron charge, and $\Omega = |e|B_0/(\gamma m)$ is the relativistic cyclotron angular frequency, with $\gamma = \left(p_\perp^2 + p_z^2 + m^2 c^2 \right)^{1/2}/(mc)$ the relativistic factor and m the electron rest mass.

Because the electrons gyrate in the axial magnetic field, it is more convenient to use those constants of motion characterizing guiding centers to construct equilibrium distribution functions, for this gives us a clear physical picture.

Setting $r_L = p_\perp/|eB_0|$ and $\phi = \theta + (B_0/|B_0|)\pi/2$, from Eqs.(11) and (12) we have

$$R_g \cos \varphi_g = R \cos\varphi - r_L \cos\theta , \tag{14}$$

$$R_g \sin \varphi_g = R \sin\varphi - r_L \sin\theta , \tag{15}$$

where R_g and φ_g are radial and azimuthal coordinates of the guiding center and they are all constants of motion.

When $B_0 > 0$, the electrons are right-rotated along the z-direction, and when $B_0 < 0$, the electrons are left-rotated. In the beam conditioner, unlike a gyrotron,[12,13] distinguishing the gyration direction is important because different gyration directions can result in different variations in velocity spread.

Suppose that the distribution of the guiding centers of the electron beam is uniform, so the equilibrium distribution function can be chosen as

$$f_0 = \delta(\gamma - \gamma_0) \, H(p_\perp) \, H[|eB_0|(R_b - R_g) - p_\perp] \, H(R_g) \, H(R_b - R_g) \, H(p_z) , \tag{16}$$

where γ_0 is the initial relativistic factor, and H(x) is a unit step function. Since γ, $p_\perp$, p_z, and R_g are all constants of motion, f_0 given by Eq. (16) is a solution of Eq. (4).

Perturbation Distribution Function

We will use the method of integration along characteristics to solve for $f_1^{(I)}$ and then directly determine $f_1^{(II)}$ by using $f_1^{(I)}$ and arguments involving constants of motion.

The perturbation distribution function $f_1^{(I)}$ can be expressed as

$$f_1^{(I)} = -e \int_0^t \mathbf{E}' \cdot \frac{\partial f_0}{\partial \mathbf{p}'} \, dt' . \tag{17}$$

To perform the above integration, we have to make local expansion of the electric field $\mathbf{E}$ in the guiding center (R_g, φ_g). Applying the Bessel function addition theorem

$$J_1(k_cR) \, e^{i(\varphi-\theta)} = \sum_{l=-\infty}^{+\infty} J_{1-l}(k_cR_g) \, e^{i(1-l)\varphi_g} J_l(k_cr_L) \, e^{i(l-1)\theta}, \tag{18}$$

we have

$$E_{r_L} = E_0 \sin \frac{\pi}{d} z \, \sin \omega t \sum_{l=-\infty}^{+\infty} (-1)^l J_l(k_cR_g) J_{l+1}(k_cr_L) \sin l(\varphi_g-\theta) , \tag{19}$$

$$E_\theta = E_0 \sin \frac{\pi}{d} z \, \sin \omega t \sum_{l=-\infty}^{+\infty} (-1)^l J_l(k_cR_g) J_{l+1}(k_cr_L) \cos l(\varphi_g-\theta) , \tag{20}$$

where E_{r_L} and E_θ are, respectively, the components of the r_L- and θ-directions in the guiding-center frame.

From Eqs. (19) and (20), we find that the electric field is expanded as a sum of infinite cyclotron harmonics. The amplitude of the l th harmonic is proportional to $J_l(k_cR_g)$. Because a small beam radius is used, k_cR_g is much less than unity. In addition, because the field of the first harmonic varies azimuthally, its effect on an electron tends to cancel when the electron makes a revolution in the linear limit. So, the effect of the zeroth one is dominant. It should be noted that the zeroth harmonic has only an azimuthal component of the electric field and it is axisymmetric in the guiding-center frame, just like the whole TE_{011}-mode electric field in the waveguide-axial frame. In fact, if we let R_g approach zero, Eqs. (19) and (20) go back to Eq. (2).

The equations describing the characteristics are given by

$$z' = z - v_z (t - t') , \tag{21}$$

$$\theta' = \theta - \Omega(t - t') \; , \tag{22}$$

where $v_z = p_z/(\gamma m)$ is the axial velocity and it is also a constant of motion.

Substituting Eqs. (19)-(22) into Eq. (17), after a tedious calculation we can obtain the first-region perturbation distribution function

$$f_1^{(I)} = \sum_{l=-\infty}^{+\infty} -\frac{B_0}{|B_0|}\frac{1}{4} e E_0 F_l G_l^{(I)} \; , \tag{23}$$

where

$$F_l = (-1)^l \left\{ J_l(k_c R_g) J_{l+1}(k_c r_L) \left[\frac{\partial f_0}{\partial p_\perp} + \frac{p_\perp}{\gamma(mc)^2} \frac{\partial f_0}{\partial \gamma} \right] - \frac{1}{|eB_0|} J_{l+1}(k_c R_g) J_l(k_c r_L) \frac{\partial f_0}{\partial R_g} \right\} \tag{24}$$

$$
\begin{aligned}
G_l^{(I)} = &-\frac{1}{\omega_{1l}} \left\{ \sin\left[l\varphi_g + \left(\frac{\pi}{d}z - l\theta\right) + \omega t\right] - \sin\left[l\varphi_g + \left(\frac{\pi}{d}z - l\theta\right) - \left(\frac{\pi}{d}v_z - l\Omega\right)t\right] \right\} \\
&+\frac{1}{\omega_{2l}} \left\{ \sin\left[l\varphi_g - \left(\frac{\pi}{d}z + l\theta\right) - \omega t\right] - \sin\left[l\varphi_g - \left(\frac{\pi}{d}z + l\theta\right) + \left(\frac{\pi}{d}v_z + l\Omega\right)t\right] \right\} \\
&-\frac{1}{\omega_{3l}} \left\{ \sin\left[l\varphi_g + \left(\frac{\pi}{d}z - l\theta\right) - \omega t\right] - \sin\left[l\varphi_g + \left(\frac{\pi}{d}z - l\theta\right) - \left(\frac{\pi}{d}v_z - l\Omega\right)t\right] \right\} \\
&+\frac{1}{\omega_{4l}} \left\{ \sin\left[l\varphi_g - \left(\frac{\pi}{d}z + l\theta\right) + \omega t\right] - \sin\left[l\varphi_g - \left(\frac{\pi}{d}z + l\theta\right) + \left(\frac{\pi}{d}v_z + l\Omega\right)t\right] \right\} \cdot
\end{aligned} \tag{25}
$$

In Eq. (25), ω_{1l}, ω_{2l}, ω_{3l}, and ω_{4l} are given by

$$\omega_{2l,1l} = \omega + \frac{\pi}{d}v_z \pm l\Omega \; , \tag{26}$$

$$\omega_{3l,4l} = \omega - \frac{\pi}{d}v_z \pm l\Omega \; . \tag{27}$$

On the basis of the perturbation distribution function in region (I), we can easily obtain the one in region (II). From Eq. (24) we can see that F_l is only a function of constants of motion and, of course, it is also a constant of motion. In Eq. (25), however, $G_l^{(I)}$ not only depends on the constants of motion φ_g, v_z, and Ω, but also depends on z, θ, and t, which are not constants of motion. So if we can use some constants of motion to take the place of them, then Eq. (4) is satisfied. To this end, setting $z = d$ in $G_l^{(I)}$ and then replacing t and θ by the following constants of motion:

$$t^* = t - \frac{1}{v_z}(z - d) \; , \tag{28}$$

$$\theta^* = \theta - \frac{\Omega}{v_z}(z - d) \; , \tag{29}$$

we obtain the second-region perturbation distribution function

$$f_1^{(II)} = \sum_{l=-\infty}^{+\infty} -\frac{B_0}{|B_0|}\frac{1}{4} e \, E_0 \, F_l \, G_l^{(II)} , \tag{30}$$

where

$$
\begin{aligned}
G_l^{(II)} &= \frac{H(t^*)}{\omega_{1l}} \left\{ \sin\left[l\left(\varphi_g - \theta^*\right) + \omega t^*\right] - \sin\left[l\left(\varphi_g - \theta^*\right) - \left(\frac{\pi}{d}v_z - l\Omega\right)t^*\right] \right\} \\
&\quad - \frac{H(t^*)}{\omega_{2l}} \left\{ \sin\left[l\left(\varphi_g - \theta^*\right) - \omega t^*\right] - \sin\left[l\left(\varphi_g - \theta^*\right) + \left(\frac{\pi}{d}v_z + l\Omega\right)t^*\right] \right\} \\
&\quad + \frac{H(t^*)}{\omega_{3l}} \left\{ \sin\left[l\left(\varphi_g - \theta^*\right) - \omega t^*\right] - \sin\left[l\left(\varphi_g - \theta^*\right) - \left(\frac{\pi}{d}v_z - l\Omega\right)t^*\right] \right\} \\
&\quad - \frac{H(t^*)}{\omega_{4l}} \left\{ \sin\left[l\left(\varphi_g - \theta^*\right) + \omega t^*\right] - \sin\left[l\left(\varphi_g - \theta^*\right) + \left(\frac{\pi}{d}v_z + l\Omega\right)t^*\right] \right\} . \tag{31}
\end{aligned}
$$

Since t^* and θ^* are all constants of motion, $f_1^{(II)}$ satisfies the equilibrium Vlasov equation. Indeed, it is easy to verify that $f_1^{(I)}$ and $f_1^{(II)}$ satisfy the initial and boundary conditions.

Axial Velocity Spread
We have obtained perturbation functions and now we can use them to calculate the axial velocity spread.

The rms-normalized axial velocity spread is defined by

$$\sigma_{\beta_z} = \sqrt{\langle \beta_z^2 \rangle - \langle \beta_z \rangle^2} , \tag{32}$$

where β_z is the axial velocity normalized to the light speed c, the averages $\langle \beta_z \rangle$ and $\langle \beta_z^2 \rangle$ are given by

$$\langle \beta_z, \beta_z^2 \rangle = A \int \left(\beta_z, \beta_z^2 \right) (f_0 + f_1) \, d^3p d^3x . \tag{33}$$

Here f_1 denotes $f_1^{(I)}$ or $f_1^{(II)}$, and

$$A = \frac{1}{\displaystyle\int (f_0 + f_1) \, d^3p d^3x} . \tag{34}$$

From Eqs. (32) and (33), we have

$$\langle \beta_z^2 \rangle - \langle \beta_z \rangle^2 = \left[\langle \beta_z^2 \rangle_0 - \langle \beta_z \rangle_0^2 \right] + \left[\langle \beta_z^2 \rangle_1 - \langle \beta_z \rangle_1^2 \right] - 2 \langle \beta_z \rangle_0 \langle \beta_z \rangle_1 , \tag{35}$$

where $\langle \ \rangle_0$ and $\langle \ \rangle_1$ denote taking an average with f_0 and f_1 respectively.
After a lot of calculations we can obtain[14]

$$\sigma_{\beta_z}^{(I)} = \sqrt{\frac{11}{1400}} \frac{1}{\beta_0} \left(\frac{R_b e B_0}{\gamma_0 mc}\right)^2 \left\{ 1 + \frac{4}{\pi} \frac{E_0}{B_0 c} \beta_0^2 \frac{d}{L} \frac{\sqrt{1-\left(\frac{\lambda}{2d}\right)^2}}{1-\left(\frac{\beta_0 \lambda}{2d}\right)^2} \right.$$

$$\left. \times \left[2\sin^2\left(\frac{\pi L}{2d}\right)\cos^2\left(\frac{\pi d}{\beta_0 \lambda}\right) + \frac{\beta_0 \lambda}{4d}\sin\frac{\pi L}{d}\sin\frac{2\pi d}{\beta_0 \lambda} \right] \right\}^{\frac{1}{2}}, \qquad (36)$$

$$\sigma_{\beta_z}^{(II)} = \sqrt{\frac{11}{1400}} \frac{1}{\beta_0} \left(\frac{R_b e B_0}{\gamma_0 mc}\right)^2 \left\{ 1 + \frac{4}{\pi} \frac{E_0}{B_0 c} \beta_0^2 \frac{d}{L} \frac{\sqrt{1-\left(\frac{\lambda}{2d}\right)^2}}{1-\left(\frac{\beta_0 \lambda}{2d}\right)^2} \right.$$

$$\left. \times \left[\sin^2\left(\frac{\pi L}{2d}\right) + \left(\frac{\beta_0 \lambda}{2d}\right)^2 \sin\frac{\pi L}{\beta_0 \lambda}\sin\frac{\pi(2d+L)}{\beta_0 \lambda} \right] \right\}^{\frac{1}{2}}, \qquad (37)$$

where λ is the cavity operating wavelength and $\beta_0 = v_0/c$.

Eq. (36) describes the axial velocity spread when the pulsed beam arrives at the front of the cavity and Eq. (37) describes the spread when it leaves the cavity. When the beam length approaches zero, the two formulas give the same result, as expected. Because the linear modification of the axial velocity spread is caused by the zeroth harmonic, it only depends on the cavity length normalized to an operating wavelength; that is, there is no dependence on what wavelength is used.

Taking $E_0 = 7.5 \times 10^4$ Volt/cm, $B_0 = 2500$ Gauss, ($E_0/B_0 c = 0.1$), $R_b = 1$ cm and $\gamma_0 = 2.47$,[15] from Eq. (37) we have drawn the dependence of the rms-normalized axial velocity spread on the normalized cavity length. As shown in Fig. 2, we can see that the maximum of the spread increases with the pulse length. For the pulsed beam with a length of 0.01 wavelength, the velocity spread is maximumly improved when the normalized cavity length is about 0.62. For the pulsed beam with a length of 0.5 wavelength, however, the spread is not improved and instead it is deteriorated. From this it can be inferred that the effect of the pulse length on velocity spread is important. From Fig. 2, we also can find that the spread varies quasi-periodically with the cavity length. The varying amplitude approaches zero as the cavity length increases infinitely. According to Eq. (37), the quasi-periodicity of the dependence of the spread on the cavity length is related to the electron's initial energy, the operating wavelength, and the pulse length.

It should be noted that the spread for 0.5 normalized cavity length, about 3.41%, is the same as that of the equilibrium beam. It seems that the beam is not affected at all when it passes through the cavity. This can be explained as follows: when the cavity length is equal to half an operating wavelength, the waveguide radius appproaches infinity and so the electric field within the electron beam vanishes. Accordingly, the beam cannot be conditioned.

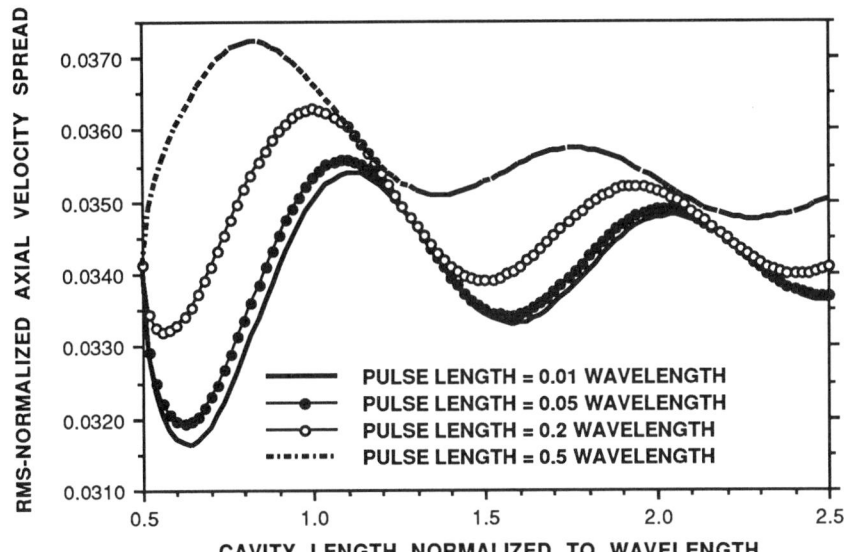

Fig. 2. Dependence of the rms-normalized axial velocity spread on the normalized cavity length. The entrance time of the pulsed beams is zero.

The linear theory indicates that the improvement on the spread is very small. Therefore, investigation of non-linear processes for the beam conditioner is necessary.

NONLINEAR EVALUATION OF THE BEAM CONDITIONER

In this section we will use the basic equations governing the nonlinear behavior of the beam conditioner to examine relations between the axial velocity spread and the cavity length.

In our procedure, the vacuum-cavity TE_{011}-mode fields are used and the contribution of the pulsed beam to the cavity fields is neglected. This is quite reasonable because the transverse velocities of the beam are rather small in the beam conditioner, unlike the cyclotron maser where an electromagnetic wave is efficiently amplified through the coupling between the wave and an electron beam with much larger averaged transverse velocity.[10] The electron orbits are related to the fields through the relativistic Lorentz force equations in the single-particle simulation. First, in order to check the previous linear kinetic theory we use only the TE_{011}-mode electric field and neglect its magnetic field to compute a single pulsed beam. Then we use both the electric and magnetic fields of the TE_{011} mode to compute the same pulsed beam and compare them with each other. This simulation reveals the nonlinear evolution of the rms-normalized axial velocity spread as a function of cavity length.

When only the electric field is used in the simulation, the axial momentum is a constant and it is examined to check the validity of the calculation. In the general case, all three checks have been passed by the code. When both the electric and magnetic fields are included, we use Liouville's theorem to check the code by computing the Jacobi determinant (time is taken as an independent variable) and, also, by reversing the computation and using the final values of a particle as initial conditions.

Lorentz Force Equations in the Guiding Center

In the previous linear theory, for the convenience of calculation, the TE_{011}-mode field is expanded as a sum of infinite harmonics. In the computation, however, it is more convenient to resolve directly the TE_{011}-mode field into components in the guiding-center frame without expansion into harmonics.

In the waveguide-axial frame, the TE_{011}-mode fields are given by

$$E_\varphi = E_0 J_1(k_c R) \sin \frac{\pi}{d} z \sin \omega t ,$$ (38)

$$B_R = - E_0 \frac{1}{\omega} \frac{\pi}{d} J_1(k_c R) \cos \frac{\pi}{d} z \cos \omega t ,$$ (39)

$$B_z = - E_0 \frac{k_c}{\omega} J_0(k_c R) \sin \frac{\pi}{d} z \cos \omega t .$$ (40)

In the guiding-center frame, E_φ and B_R are resolved into the following:

$$E_{r_L} = E_0 J_1(k_c R) \frac{R_g}{R} \sin \left(\theta - \varphi_g\right) \sin \frac{\pi}{d} z \sin \omega t ,$$ (41)

$$E_\theta = E_0 J_1(k_c R) \left[\frac{R_g}{R} \cos \left(\theta - \varphi_g\right) + \frac{r_L}{R} \right] \sin \frac{\pi}{d} z \sin \omega t ,$$ (42)

$$B_{r_L} = - E_0 \frac{1}{\omega} \frac{\pi}{d} J_1(k_c R) \left[\frac{R_g}{R} \cos \left(\theta - \varphi_g\right) + \frac{r_L}{R} \right] \cos \frac{\pi}{d} z \cos \omega t ,$$ (43)

$$B_\theta = E_0 \frac{1}{\omega} \frac{\pi}{d} J_1(k_c R) \frac{R_g}{R} \sin \left(\theta - \varphi_g\right) \cos \frac{\pi}{d} z \cos \omega t ,$$ (44)

where

$$R = \sqrt{R_g^2 + r_L^2 + 2 R_g r_L \cos \left(\theta - \varphi_g\right)} .$$ (45)

Here we use the same symbols as those in the linear theory. But it should be noted that some of them have different mathematical contents. For example, in the kinetic theory R_g and φ_g are functions of both the momentum variables and the configuration variables, whereas in this single-particle simulation they are fixed for a given guiding-center frame.

From Eqs. (41)-(44), the Lorentz force equations in the guiding-center frame can be written as

$$\frac{d\bar{r}}{dz} = \frac{\pi}{R_b} \sqrt{4\bar{d}^2 - 1} \frac{\beta_1}{\beta_3} ,$$ (46)

$$\frac{d\overline{\theta}}{d\overline{z}} = \frac{1}{2\overline{R}_b}\sqrt{4\overline{d}^2-1}\frac{1}{\overline{r}}\frac{\beta_2}{\beta_3}, \tag{47}$$

$$\frac{d\overline{t}}{d\overline{z}} = \frac{\overline{d}}{\beta_3}, \tag{48}$$

$$\frac{d\beta_1}{d\overline{z}} = \pi\sqrt{4\overline{d}^2-1}\frac{1}{\beta_3}\left\{\frac{\beta_2^2}{\overline{R}_b\overline{r}} - \frac{1}{\gamma}\left[(1-\beta_1^2)\overline{E}_r - \beta_1\beta_2\overline{E}_\theta - \beta_3\overline{B}_\theta + \beta_2\overline{B}_z\right]\right\}, \tag{49}$$

$$\frac{d\beta_2}{d\overline{z}} = \pi\sqrt{4\overline{d}^2-1}\frac{1}{\beta_3}\left\{-\frac{\beta_1\beta_2}{\overline{R}_b\overline{r}} + \frac{1}{\gamma}\left[\beta_1\beta_2\overline{E}_r - (1-\beta_2^2)\overline{E}_\theta - \beta_3\overline{B}_r + \beta_1\overline{B}_z\right]\right\}, \tag{50}$$

$$\frac{d\beta_3}{d\overline{z}} = \pi\sqrt{4\overline{d}^2-1}\frac{1}{\beta_3}\frac{1}{\gamma}\left[\beta_1\beta_3\overline{E}_r + \beta_2\beta_3\overline{E}_\theta + \beta_2\overline{B}_r - \beta_1\overline{B}_\theta\right], \tag{51}$$

where

$$\overline{E}_r = \alpha_1[J_0(\overline{R}_b\overline{R}) + J_2(\overline{R}_b\overline{R})]\overline{R}_g \sin 2\pi(\overline{\theta}-\overline{\varphi}_g)\sin\pi\overline{z}\sin 2\pi\overline{t} , \tag{52}$$

$$\overline{E}_\theta = \alpha_1[J_0(\overline{R}_b\overline{R}) + J_2(\overline{R}_b\overline{R})][\overline{R}_g\cos 2\pi(\overline{\theta}-\overline{\varphi}_g)+\overline{r}]\sin\pi\overline{z}\sin 2\pi\overline{t} , \tag{53}$$

$$\overline{B}_r = -\alpha_2[J_0(\overline{R}_b\overline{R}) + J_2(\overline{R}_b\overline{R})][\overline{R}_g\cos 2\pi(\overline{\theta}-\overline{\varphi}_g)+\overline{r}]\cos\pi\overline{z}\cos 2\pi\overline{t}, \tag{54}$$

$$\overline{B}_\theta = \alpha_2[J_0(\overline{R}_b\overline{R}) + J_2(\overline{R}_b\overline{R})]\overline{R}_g\sin 2\pi(\overline{\theta}-\overline{\varphi}_g)\cos\pi\overline{z}\cos 2\pi\overline{t}, \tag{55}$$

$$\overline{B}_z = \alpha_3 - \alpha_4 J_0(\overline{R}_b\overline{R})\sin\pi\overline{z}\cos 2\pi\overline{t} , \tag{56}$$

with $\alpha_1 = |e|R_bE_0/(2mc^2)$, $\alpha_2 = |e|R_bE_0/(4\overline{d}mc^2)$, $\alpha_3 = |e|B_0/(mk_cc)$,   and $\alpha_4 = |e|E_0/(m\omega c)$. The normalized quantities appearing in Eqs. (75)-(85) are defined by $\overline{z}=z/d$, $\overline{r}=r_L/R_b$, $\overline{\theta}=\theta/(2\pi)$, $\overline{t}=\omega t/(2\pi)$, $\beta_1=(dr_L/dt)/c$, $\beta_2=(r_Ld\theta/dt)/c$, $\beta_3=(dz/dt)/c$, $\overline{d}=d/\lambda$, $\overline{R}_b=k_cR_b$, $\overline{R}_g=R_g/R_b$, $\overline{\varphi}_g=\varphi_g/(2\pi)$, $\overline{R}=R/R_b$, and $\gamma=(1-\beta_1^2-\beta_2^2-\beta_3^2)^{-1/2}$.

Simulation Results

We used Eqs. (46)-(51) and made computations for a pulsed beam, immersed in a 2500 Gauss axial magnetic field, with a length of 0.5 cm, a radius of 1 cm, and an initial relativistic factor of 2.47.[15] The initial electron's relativistic cyclotron frequency is 2.83 GHz, corresponding to its relativistic cyclotron wavelength 10.6 cm in free space. Three layers of sample electrons are taken within the beam and each layer has six guiding centers with 209 electrons. Because the TE$_{011}$-mode fields are axisymmetrical, the six guiding centers are all placed at

$\varphi_g{=}0$. The guiding centers are distributed uniformly along the radial direction with the coordinates R_g/R_b = 0.0, 0.2, 0.4, 0.6, 0.8, and 1.0, and the distribution of the electrons on gyration orbits simulates the equilibrium distribution function, given by Eq. (16), of neglecting the gradient effect of the guiding center. The amplitude of the cavity electric field is taken as $7.5{\times}10^4$ Volt/cm.

First, let us examine the numerical simulation using only the electric field. Taking the operating wavelength as 10 cm, and the entrance time of the pulsed beam front as zero and 0.5T (T is the period of the cavity field), we find that the linear results agree qualitatively with the ones from the simulation, as shown in Fig. 3a and Fig. 3b, respectively. Both in the linear and simulation results, the axial velocity spread oscillates with the cavity length and the oscillation damps gradually. When the cavity length is larger than one wavelength, however, the nonlinear effect becomes very considerable. In the nonlinear interaction, the mean value of oscillation of the velocity spread evidently reduces with the cavity length, whereas in the linear result it keeps constant.

Then we made simulations for the same pulsed beam with the whole TE_{011}-mode field, including both electric and magnetic fields. Since the reduction in the mean value of oscillation is caused by resonant emission, it should not depend on the phase at which the pulsed beam enters the cavity. From Fig. 4 we can see, indeed, that these mean values are almost the same. The dependence of the rms-normalized energy spread on the normalized cavity length is shown in Fig. 5. From Fig. 4 and Fig. 5 we find that for short cavities no matter whether the axial velocity spread is increased or decreased, the energy spread is always increased.

To examine the dependence of the axial velocity spread on the operating wavelength and to find out at what wavelength the beam conditioner can best improve the beam's axial velocity spread, we made simulations for different wavelengths. The result indicates that the axial velocity spread strongly depends on the operating wavelength, as shown in Fig. 6. For a wavelength of 11 cm (2.73 GHz), the mean value of the axial velocity spread reduces most rapidly with the normalized cavity length. For too long, or short, a wavelength compared with 10.6 cm (corresponding to the initial electron's relativistic cyclotron frequency 2.83 GHz), the axial velocity spread cannot be improved. From Fig. 6, we also can find that in the linear regime the dependences of axial velocity spread on the cavity length normalized to different wavelengths are almost the same, which means that there is little dependence on what wavelength is used to normalize the cavity length. From this we can deduce that the effect of the zeroth harmonic is dominant and the effect of the first harmonic is negligible in the linear regime, which agrees with the previous linear theory.

Although the axial velocity spread rapidly reduces with the cavity length when the cavity operates at a wavelength of 11 cm, it very soon reaches its minimum value of 3.9%, only decreased by 2.1% compared with its initial value of 6%. If the cavity operates at 10 cm, the axial velocity spread decreases down to 1.8%, less than one third of its initial value. However, the cavity length is greater than that for the 11 cm case, as shown in Fig. 7.

From Fig. 7, we also can find that the time-dependent magnetic field plays such a role that the mean value of oscillation of the axial velocity spread is more rapidly decreased.

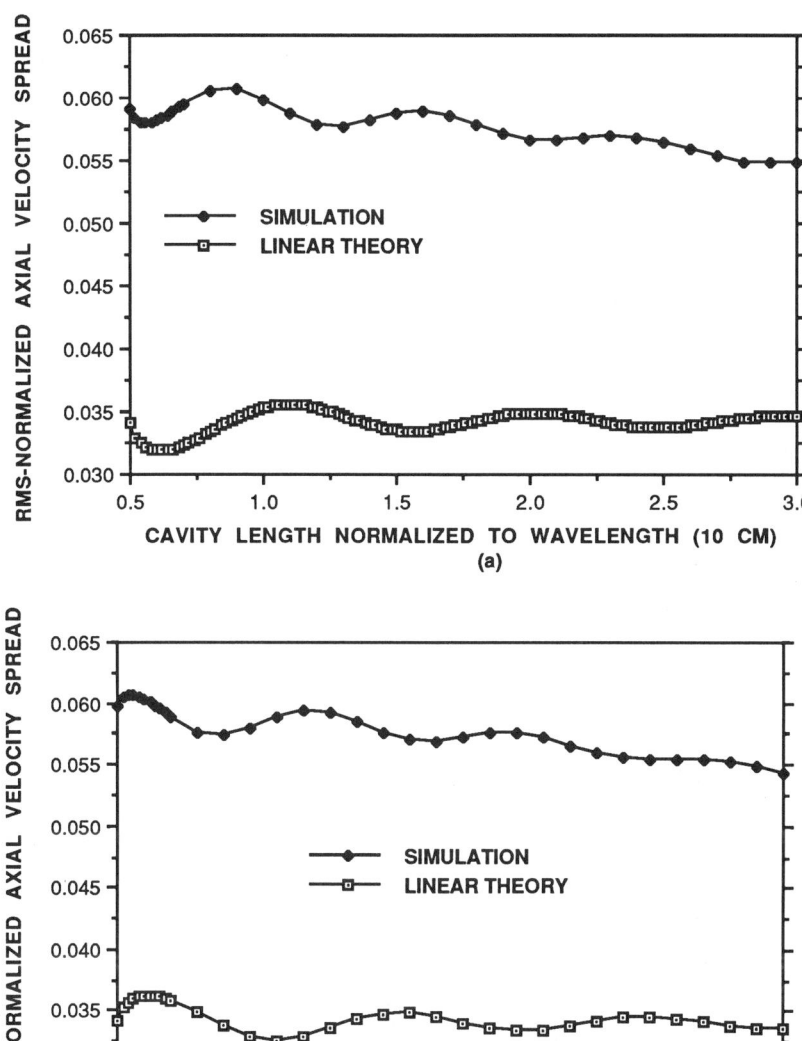

Fig. 3. Comparison of the linear result with that from the simulation using only the electric field. The operating wavelength is 10 cm. In the simulation, the gradient effect of the guiding center is neglected. (a) The entrance time is zero. When the normalized cavity length is equal to unity, the axial velocity spread is increased. (b) The entrance time is 0.5 T. When the normalized cavity length is equal to unity, the axial velocity spread is decreasd.

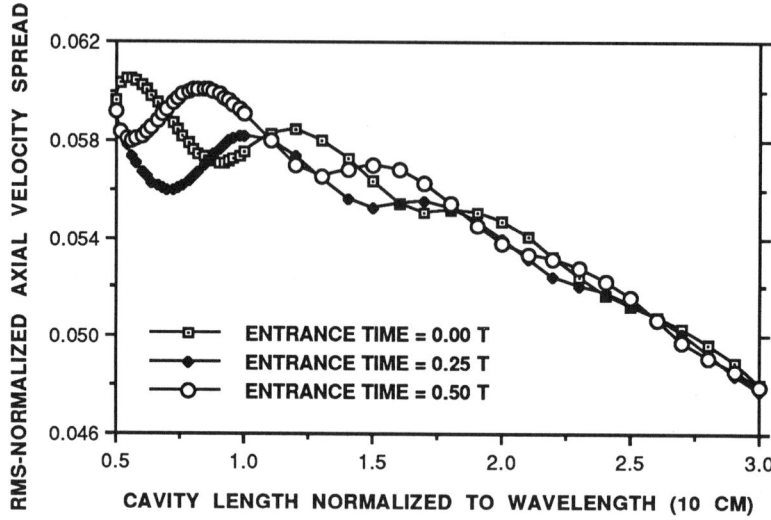

Fig. 4. Dependence of the rms-normalized axial velocity spread on the normalized cavity length when the pulsed beam enters the cavity at different times. Both the electric and the magnetic fields are included and the operating wavelength is 10 cm.

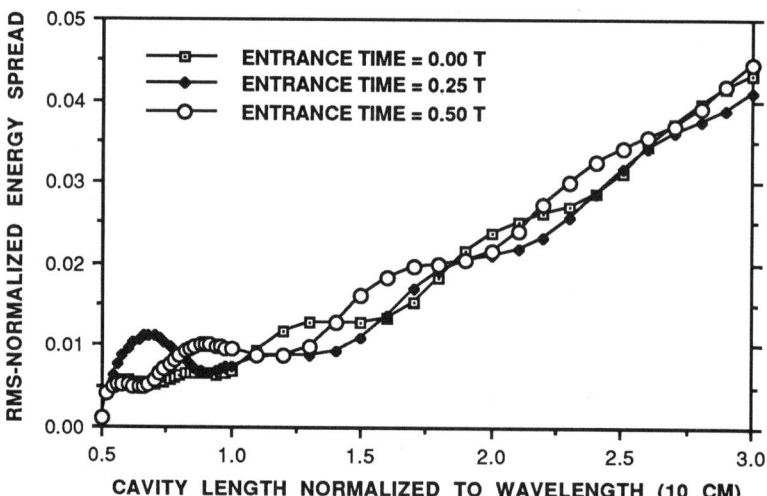

Fig. 5. Dependence of the rms-normalized energy spread on the normalized cavity length. The parameters are the same as those in Fig. 4.

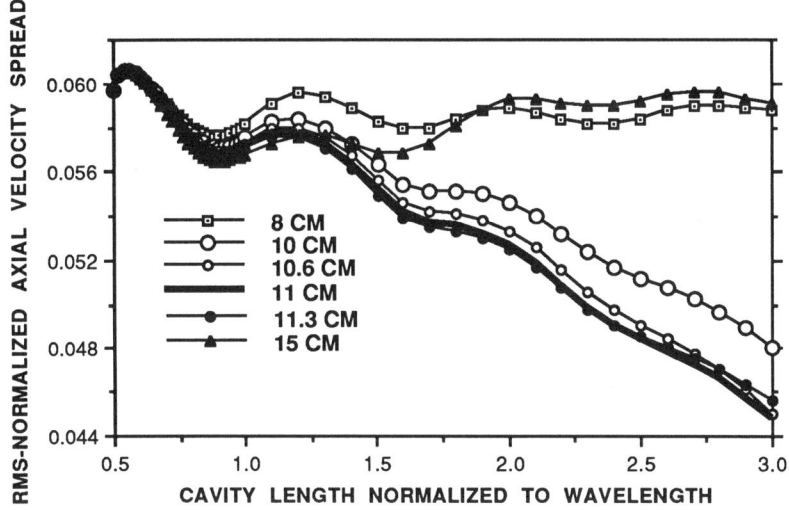

Fig. 6. Dependence of the rms-normalized axial velocity spread on the cavity length normalized to different operating wavelengths. Both the electric and magnetic fields are included and the entrance time is zero. In the linear regime, the dependences are almost the same.

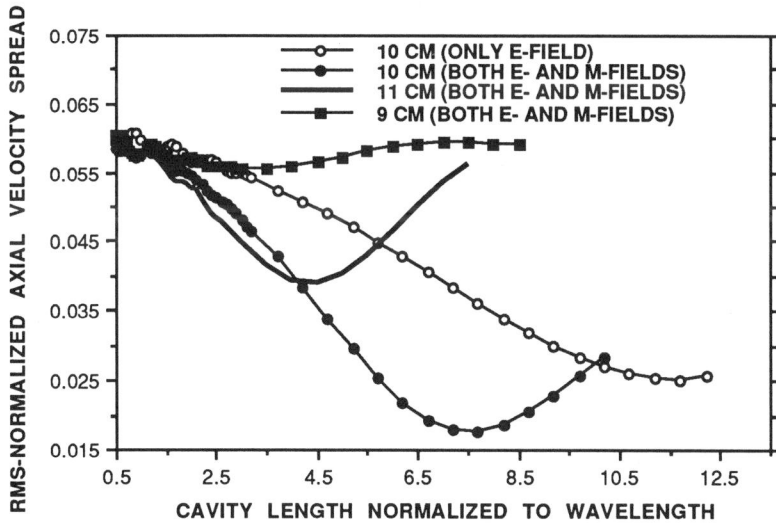

Fig. 7. Optimization of the operating wavelength (the entrance time is zero). When the operating wavelength is 10 cm, a little less than the initial electron's relativistic cyclotron wavelength, the beam conditioner best improves the axial velocity spread.

CONCLUSIONS

We have developed a linear kinetic theory to investigate a conditioner for a helically transported electron beam. The expressions for axial velocity spread of a pulsed beam conditioned by the RF cavity operating in the TE_{011} mode were derived. Numerical simulations were used to check the linear theory and it was found that the linear results are qualitatively in agreement with those from the simulations. We also have examined the nonlinear evolution of the axial velocity spread with the cavity length and the cavity operating wavelength due to the negative mass effect. In summary, we can make the following conclusions. In the linear regime, in which the cavity length is less than one operating wavelength, the modification of axial velocity spread is caused mainly by the interaction of the electrons with the zeroth harmonic, and hence whether the axial velocity spread is improved and this improvement mainly depends on the phase at which the pulsed beam enters the cavity and only slightly depends on the operating wavelength. In the nonlinear regime, the variation in axial velocity spread results from the interaction of the electrons with the fundamental harmonic based on the negative mass effect, and so it strongly depends on the operating wavelength and only slightly depends on the entrance phase of the pulsed beam. The simulation for a pulsed beam with a length of 0.5 cm, passing through a cavity operating at a wavelength of 10 cm, indicates that the rms-normalized axial velocity spread can be reduced down to 1.8%, less than one third of its initial value. From this we see that a beam conditioner can be used to decrease the spread in axial velocities for a low-energy electron beam.

ACKNOWLEDGMENTS

The idea was proposed by Andrew M. Sessler, to whom the author is indebted for his many good comments. The author wishes to thank Li-Hua Yu of BNL and Ming Xie of LBL for their helpful discussions. The author also wishes to thank D. Moretti, G. Rangarajan, and S. Krishnagopal for their assistance.

REFERENCES

1. A. M. Sessler, D. Whittum, and Li-Hua Yu, Phys. Rev. Lett. 68, 309 (1992).
2. A. M. Sessler, Lawrence Berkeley Laboratory ESG Tech Note 190, Mar. 10, 1992.
3. Li-Hua Yu, A. M. Sessler, and D. H. Whittum, Lawrence Berkeley Laboratory Report No. LBL-31198, 1991.
4. Y. H. Chin, K. J. Kim, and M. Xie, Lawrence Berkeley Laboratory Report No. LBL-30673, 1991.
5. K. -J. Kim, Lawrence Berkeley Laboratory Report No. LBL-31925, 1992.
6. J. L. Hirshfield and J. M. Wachtel, Phys. Rev. Lett. 12, 553 (1964).
7. C. E. Nielsen, A. M. Sessler, and K. R. Symon, in Proceedings of the International Conference on Accelerators (CERN, Geneva, 1959), p. 239.
8. V. K. Neil and W. Heckrotte, J. Appl. Phys. 36, 2761 (1965).
9. Y. Y. Lau and R. J. Briggs, Phys. Fluids 14, 967 (1971).
10. P. Sprangle and A. T. Drobot, IEEE MTT-25, 528 (1977).
11. J. L. Hirshfield and G. S. Park, Phys. Rev. Lett. 66, 2312 (1991); 68, 134 (1992).
12. K. R. Chu, Phys. Fluids 21, 2354 (1978).
13. Liu Shenggang, Scientia Sinica XXII, 901 (1979).
14. Changbiao Wang, Lawrence Berkeley Laboratory Report No. LBL-32222, 1992.
15. M. E. Conde and G. Bekefi, Phys. Rev. Lett. 67, 3082 (1991).

IMPEDANCE-BASED ANALYSIS
AND STUDY OF PHASE SENSITIVITY
IN SLOW-WAVE TWO-BEAM ACCELERATORS

Jonathan S. Wurtele[*]
Massachusetts Institute of Technology
Cambridge, MA 02139

David H. Whittum[**]
National Laboratory for High Energy Physics (KEK)
Tsukuba, Oho, Ibaraki, 305 Japan

Andrew M. Sessler[***]
Lawrence Berkeley Laboratory
University of California
Berkeley, CA 94720

ABSTRACT

This paper presents a new formalism which makes the analysis and understanding of both the relativistic klystron (RK) and the standing-wave free-electron laser (SWFEL) two-beam accelerator (TBA) available to a wide audience of accelerator physicists. A "coupling impedance" for both the RK and SWFEWL is introduced, which can include realistic cavity features, such as beam and vacuum ports, in a simple manner. The RK and SWFEL macroparticle equations, which govern the energy and phase evolution of successive bunches in the beam, are of identical form, differing only by multiplicative factors. Expressions are derived for the phase and amplitude sensitivities of the TBA schemes to errors (shot-to-shot jitter) in current and energy. The analysis allows, for the first time, relative comparisons of the RK and the SWFEL TBAs.

[*] Supported by the U.S. Department of Energy, Division of Nuclear and High Energy Physics, under contract No. DE-FG02-91ER40648.
[**] Supported by the Japan Society for the Promotion of Science, the U.S. National Science Foundation and the National Laboratory for High Energy Physics (KEK).
[***] Supported by the U.S. Department of Energy, Division of Nuclear and High Energy Physics, under contract No. DE-AC-03-SF-00098.

1. INTRODUCTION

The context and motivation for this work is the Two-Beam Accelerator (TBA) concept[1,2,3], which is, in essence, a high efficiency power converter, extracting energy from a low energy high-current electron "drive" beam and depositing it in a high energy electron or positron beam. In a TBA a drive beam of kiloampere current, bunched at centimeter wavelengths, passes through a periodic array of wiggler magnets, which extract the beam energy through a Relativistic Klystron (RK) or a Free-Electron Laser (FEL); at the same time, the beam passes through induction cells which replenish the beam energy, as seen in Figs. 1 and 2. The high power microwaves produced in the interaction region are periodically extracted and fed into a high gradient structure, where they accelerate the second beam.

The TBA configuration of present interest, the Standing-Wave Free-Electron Laser TBA (SWFEL/TBA), has grown out of a number of theoretical and conceptual refinements, including considerations of microwave extraction and phase and amplitude control.[4,5] In the original configuration, a large amplitude microwave signal propagates with the drive beam over the entire length of the accelerator. In each FEL section the microwave power was produced and extracted, by septa, in such a way that the total power remains roughly constant. This design allowed for the continuous longitudinal bunching of the electron beam through each FEL section. There were drawbacks to the scheme, however. The microwaves had to be transported across the induction unit gaps. Theoretical investigations discovered that while the longitudinal beam motion was stable, the rf phase shift produced by the FEL interaction had an undesirable sensitivity to shot-to-shot jitter in the induction units and in the beam current. It was proposed that the rf power be fully extracted at the end of each wiggler section, before re-acceleration, thereby limiting the total accumulated phase shift in the wave, and reducing the sensitivity to jitter in the system parameters.

Further considerations of the rf extraction mechanism led to the development of the SWFEL[6] in which the power is produced in a series of uncoupled cavities (the rf is cut off between the cavities), each of which is of order one wiggler oscillation in length. The FEL thus operates as a standing-wave device. The propagating beam provides the only coupling between the cavities. Numerical studies[7,8] have examined the phase sensitivity and longitudinal particle stability in the standing wave FEL in some detail.

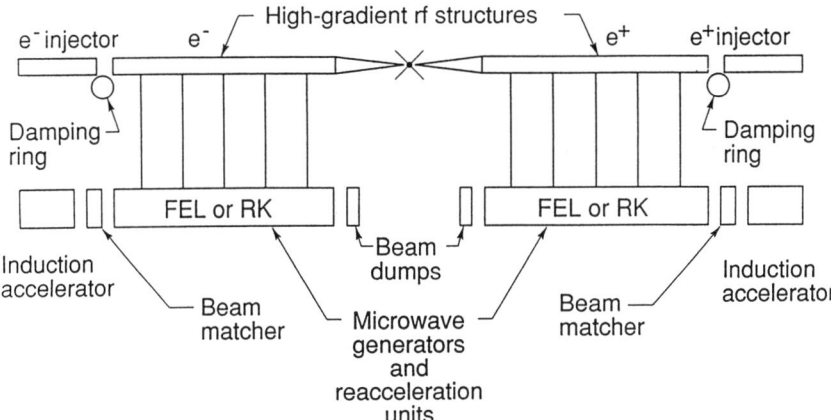

Fig. 1. A schematic of the structure of a TBA. The cavities can either be those of a relativistic klystron (RK), or those of a Standing-Wave FEL (SWFEL) in which case there is a wiggler magnetic field passing through the cavities.

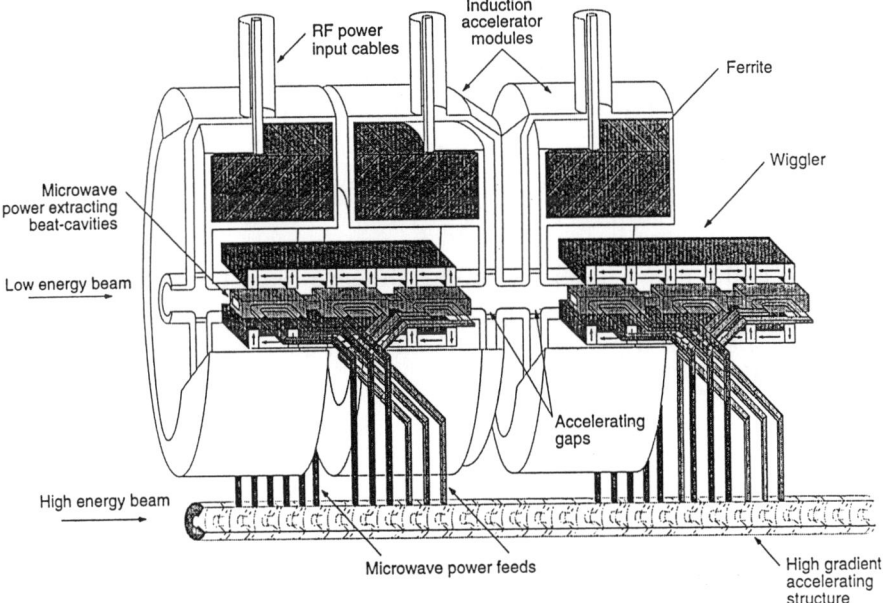

Fig. 2. A schematic of one superperiod of the Standing-Wave FEL TBA. Microwaves are produced by a low energy drive beam and fed into a high gradient structure to accelerate a high energy.

In all of the above work, the extraction units were taken as FELs. Alternatively, of course, it is possible to consider a Relativistic Klystron TBA (RK/TBA). This approach has been developed by the CERN Group.[9] Because it has been demonstrated experimentally that high power can be extracted from an RK[10], as well as from an FEL[11], both of these approaches are attractive. In fact, the standing wave FEL has many similarities to the relativistic klystron, the main difference between them being that the FEL produces power through the coupling of the transverse wiggle oscillation with the transverse electric field, while the klystron couples the longitudinal component of the electric field. Until now, no serious comparisons of these two approaches has been made. In fact, not even the formal framework in which such comparisons can be made has been developed. It is the purpose of this paper to set down such a framework.

In Section 2, we develop a formalism that allows comparisons of the RK and SWFEL TBAs. In Section 3 we evaluate phase sensitivities in the two approaches to a TBA. Section 4 contains discussion and conclusions.

2. FORMALISM

In this section we derive equations describing the coupling of beam electrons to cavity modes. We first decompose the vector potential in the Lorentz gauge,

$$\vec{A} = \frac{mc^2}{e} \sum_{\alpha} q_\alpha(t) \vec{a}_\alpha(\vec{r}),$$ (1)

where α is the mode index, q_α is the dimensionless mode amplitude and $\vec{a}_\alpha$ gives the spatial dependence of the mode. The electron mass is m, the speed of light is c and the electron charge is $-e$. The mode normalization is

$$\int_V d^3 r' \, \vec{a}_\alpha(\vec{r}\,) \bullet \vec{a}^*_\alpha(\vec{r}\,) = V$$ (2)

with V the cavity volume.

Maxwell's equations reduce to the well-known form

$$\left(\frac{\partial^2}{\partial t^2} + \frac{\omega_\alpha}{Q_\alpha} \frac{\partial}{\partial t} + \omega_\alpha^2 \right) q_\alpha(t) = \frac{4\pi e}{mc} \frac{1}{V} \int d^3 r' \, \vec{J}(\vec{r}\,',t) \bullet \vec{a}_\alpha(\vec{r}\,')$$ (3)

where the integral is over the cavity volume. We consider the interaction of the beam with a single cavity mode with a very high Q, and make an eikonal approximation, $q(t) = \Re\{b\, e^{i\varphi}\, e^{-i\omega t}\}$, where the phase φ and the amplitude b vary slowly on the time scale of the mode period. In terms of b, the energy stored per unit length is

$$U = \frac{1}{8\pi}\left[\frac{\omega^2 hw}{c^2}\right]\frac{m^2 c^4 b^2}{e^2}, \tag{4}$$

where h and w are the height and width of the cavity.

We will consider two cases: (1) coupling to a TE mode through the transverse current induced by a magnetic wiggler (FEL) and (2) coupling to a TM mode through the axial current (RK). In each case the coupling depends on the phase $\psi = \varphi + \theta$ of an electron's motion relative to the phase of the cavity fields. Here the phase θ is a particle variable. For an FEL this phase is given by

$$\theta = (k_w + k_z)z - \omega t, \tag{5}$$

where k_w is the wiggler wavenumber and k_z is the axial wavenumber for the forward-going component of the cavity mode. For a steady-state klystron this phase is

$$\theta = k_z z - \omega t - \theta_r, \tag{6}$$

where we have introduced the phase θ_r, that of a reference particle. Typically klystrons operate with $k_z = 0$, in a nearly single mode cavity. The SWFEL, on the other hand, operates in a highly overmoded cavity.

An important distinction between the SWFEL and the RK is that Eq. (5) defines a synchronous energy in terms of the system parameters, while Eq. (6), for the RK, only relates the phase of a particle to a reference phase and does not define a synchronous energy. The RK, therefore, can be operated at any energy (even GeV energies are possible), whereas the SWFEL requires a low energy (of order ten MeV) for resonance at microwave frequencies with reasonable wiggler parameters.

In terms of these variables the field equations in a given cavity may be written as

$$\frac{\partial}{\partial s} b e^{i\varphi} = i c \frac{1}{\eta} \frac{r}{Q} \frac{I}{I_A} \langle e^{-i\theta} \rangle, \tag{7}$$

where $s = v_z t - z$, with v_z the beam velocity. I is the average beam current, $I_A = mc^3/e \sim 17\text{kA}$, and the brackets indicate an average over a beam slice. The factor η depends on the kind of coupling. For an RK $\eta = 2$, while for an FEL, $\eta = a_w/2\gamma$, with γ the Lorentz factor, and a_w the wiggler parameter.

The shunt impedance per unit length r is given by [6]

$$\frac{r}{Q} = \frac{4\pi}{VL\omega} \left| \int_{-L/2}^{+L/2} dz \frac{\bar{v}(z)}{v_z} \cdot \bar{a}(z) \exp\left(-\frac{i\omega z}{v_z}\right) \right|^2, \tag{8}$$

where L is the cavity length. The SWFEL typically operates in the TE_{01p} mode of a rectangular cavity of width w and height h, so that

$$\frac{r}{Q} = \frac{Z_0}{8\pi} \frac{\lambda}{hw} \left(\frac{a_w}{\gamma}\right)^2 \left(\frac{\sin\chi}{\chi}\right)^2, \tag{9}$$

where $Z_0 = 4\pi/c$ (377Ω in MKS), λ is the free-space wavelength and $\chi = (\omega L/v_z - p\pi - k_w L)/2$ is the effective transit angle. For an RK operating in the TM_{m1p} mode,

$$\frac{r}{Q} = \frac{Z_0}{4\pi} \frac{\lambda}{hw} \left\{ \frac{k_x^2 + k_y^2}{k_x^2 + k_y^2 + k_z^2} \right\} \left(\frac{\sin\chi}{\chi}\right)^2, \tag{10}$$

where the transit angle is $\chi = (p\pi + \omega L/v_z)/2$. The coupling in the SWFEL is from the interaction of the wiggling velocity imparted to the beam by the wiggler and the transverse field of a TE mode, while the RK generates a shunt impedance from the axial coupling of the beam to the z-component of the electric field of a TM mode.

To complete the formulation, equations are required for the particle motion. It is convenient to linearize about the reference energy, so that the dynamical variables are θ and $\delta\gamma = \gamma - \gamma_r$, where γ_r is the resonant γ in the case of the FEL, or in the case of an RK, a reference γ. The phase evolution is found from Eqs. (5) and (6), so that[6]

$$\frac{d\theta}{dz} \approx 2\,\kappa\frac{\delta\gamma}{\gamma_r}\,,\tag{11}$$

$$\frac{d\delta\gamma}{dz} = -\eta\frac{\omega}{c}b\sin(\theta + \varphi) - \frac{eE_z}{mc^2}.\tag{12}$$

The constant $\kappa = \omega(1 + a_w^2/2)/2c\gamma^2$ for an FEL, while $\kappa = \omega/2c\gamma^2$ for the RK. Equations (7), (11), and (12) describe the self-consistent evolution of the beam and the cavity fields. *The SWFEL and RK are distinguished only through the values of η, κ and r/Q.* These equations can be obtained from a continuum limit of a discrete cavity analysis by assuming that the energy and phase change in a given cavity are small. Using Eq. (4) it is straightforward to check that these equations conserve energy.

We will consider an equilibrium (no z-dependence of the dependent variables) of a well-bunched constant-current beam, described by (fixed) parameters α, β (where the amount of detuning is characterized by β)

$$\theta_0(s) = \alpha + \beta s.\tag{13}$$

Substituting Eq. (13) in Eq. (7) one obtains the equilibrium field,

$$b_0(s)^2 - b_0(0)^2 =$$
$$2\frac{\varepsilon b_0(0)}{\beta}\left[\cos(\alpha + \varphi_0) - \cos(\alpha + \varphi_0 + \beta s)\right] + \left(\frac{2\varepsilon}{\beta}\right)^2\sin^2\left(\frac{\beta s}{2}\right).\tag{14}$$

Equation (12) and the equilibrium condition require a reaccelerating field given by

$$\frac{eE_z}{mc^2} = -\eta\frac{\omega}{c}\left[\tilde{b}_{0r}(0)\sin(\alpha + \beta s) + \tilde{b}_{0i}(0)\cos(\alpha + \beta s) + \frac{\varepsilon}{\beta}\sin(\beta s)\right].\tag{15}$$

Here we abbreviate $\tilde{b} = b_0 e^{i\varphi_0}$ and

$$\varepsilon = c\frac{1}{\eta}\frac{r}{Q}\frac{I}{I_0}.\tag{16}$$

The important results of this section are the equations describing an equilibrium (Eqs. 13 and 14), which allow us to study, as we shall in the next section, sensitivities to errors in beam current or energy, and the expression for the field (Eq. 7) in terms of a generalized coupling impedance defined in Eqs. (9) and (10).

3. RF AMPLITUDE AND PHASE SENSITIVITY TO JITTER

Next we consider the perturbation to the equilibrium resulting from an error in the initial beam energy $mc^2\Delta\gamma$ and an error ΔI in the beam current. The errors in amplitude and phase will be found by evaluating the perturbed field, $b_1(s,z) = b - b_0$.

The evolution of the perturbed phase $\theta(s,z) = \theta_0(s) + \theta_1(s,z)$ is governed by Eqs. (11) and (12), which combine to give

$$\frac{d^2\theta_1}{dz^2} + \Omega^2(s)\theta_1 = -\frac{2\eta\kappa\omega}{c\gamma}\{b_{i1}^\sim \cos\theta_0 + b_{r1}^\sim \sin\theta_0\}, \tag{17}$$

where the synchrotron period is

$$\Omega^2(s) = \frac{2\eta\kappa\omega}{\gamma c}b_0 \cos\psi_0 \approx \Omega^2(0) - \frac{4\varepsilon\kappa\eta\omega}{\gamma\beta c}\sin^2\frac{\beta s}{2}. \tag{18}$$

The perturbed eikonal, $\tilde{b}_1$, is determined from Eq. (6), to be:

$$\frac{\partial}{\partial s}\tilde{b}_1 = i\varepsilon_1 e^{-i\theta_0} + \varepsilon_0 e^{-i\theta_0}\theta_1. \tag{19}$$

Here $\varepsilon_1 = (cr\,\Delta I)/(\eta Q I_A)$, and $\varepsilon_0 = (crI_0)/(\eta Q I_A)$. Energy errors are included in the choice of initial condition for $d\theta_1/dz(z = 0)$. Errors in energy due to jitter in ε (due to loading of the induction cell circuit) are ignored.

Equation (19) may be solved up to quadrature and substituted in Eq. (17) to obtain

$$\frac{d^2\theta_1}{dz^2} + \Omega^2(s)\theta_1 =$$
$$-\int_0^s \mu_1 \cos\{\beta(s-s')\}ds' - \int_0^s \mu_0 \sin\{\beta(s-\check{s}')\}\theta_1(s',z)ds', \tag{20}$$

where we abbreviate $\mu_{\binom{1}{0}} = \dfrac{2\eta\kappa}{\gamma}\dfrac{\omega}{c}\varepsilon_{\binom{1}{0}}$. Note that the driving term

due to the current error (μ_1 term) is independent of z.

In order to solve Eq. (20), we set $\theta_1 = \delta + \tilde{\theta}_1$, where δ, the phase error due to current jitter, is independent of z, and evolves according to:

$$\Omega^2(s)\delta(s) =$$
$$-\int_0^s \mu_1 \cos\{\beta(s-s')\}ds' - \int_0^s \mu_0 \sin\{\beta(s-s')\}\delta(s')ds'. \tag{21}$$

As expected, $\delta(0) = 0$ and, when $\mu_1 = 0$, $\delta(s) = 0$ for all s. Equation (20) then reduces to

$$\frac{d^2\tilde{\theta}_1}{dz^2} + \Omega^2(s)\tilde{\theta}_1 = -\int_0^s \mu_0 \sin\{\beta(s-s')\}\tilde{\theta}_1(s',z)ds', \tag{22}$$

an equation of the "beam break-up" form. Note that the driving term on the right represents the feedback experienced by the perturbed beam.

To obtain an upper bound on the effect of this driving term, consider the problem with $\Omega(s) = \Omega(0)$. (This will lead to an upper bound, since $\Omega(s)$ increases with s, providing ever stronger focusing). The problem is then formally equivalent to the problem of "head-tail" beam break-up.[12] In the limit of large z, the asymptotic form is

$$\tilde{\theta}_1 \approx \exp\left\{i\Omega(0)z + \frac{3}{2}\left(\frac{\mu_0 \beta z s^2}{\Omega(0)}\right)^{1/3} e^{i\pi/6}\right\}. \tag{23}$$

Evidently, feedback is negligible when $\Omega(0) \gg \mu_0\beta z s^2$. For typical parameters, this driving term is indeed negligible, and the particle motion is well-described by a free synchrotron oscillation (i.e., the motion of a "test particle"). In this case, the solution to Eq. (23) is

$$\tilde{\theta}_1 = \frac{1}{\Omega(s)}\left(\frac{d\theta}{dz}\right)_0 \sin\Omega(s)z = 2\kappa\frac{\Delta\gamma}{\gamma_r\Omega(s)}\sin\Omega(s)z. \tag{24}$$

An analogous argument for $\delta(s)$ in Eq. (21) leads to:

$$\delta(s) = \frac{\mu_1 \sin \beta s}{\Omega^2(s)\beta}, \tag{25}$$

Employing Eqs (24) and (25), the error in the cavity field is obtained, from Eq. (19), as

$$\tilde{b}_1(s,z) = \varepsilon_0 \int_0^s ds' \theta_1(s',z)e^{-i\theta_0(s')} + i\varepsilon_1 \int_0^s ds' e^{-i\theta_0(s')}$$

$$= \varepsilon_0 \int_0^s ds' \left\{ \left[2\kappa \frac{\Delta\gamma}{\gamma} \frac{\sin(\Omega(s')z)}{\Omega(s')} \right] + \delta(s') \right\} e^{-i\theta_0(s')} + i\varepsilon_1 \int_0^s ds' e^{-i\theta_0(s')}. \tag{26}$$

Expressions for the amplitude (b_1) and phase (ϕ_1) errors caused by the energy error $mc^2\Delta\gamma$ and the current error ΔI may be obtained from Eq. (26). It suffices, in order to get a bound, to replace $\Omega(s)$ with $\Omega(0)$ in the κ-term. One finds after much algebra (and taking the equilibrium bunch length, s, such that $\beta s = -\pi$):

$$\frac{b_1}{b_0} = -\frac{\mu_1 g_r}{2\beta\Omega^2(0)} + \frac{\varepsilon_1}{\varepsilon_0}, \tag{27}$$

$$\phi_1 = -2\kappa \frac{\Delta\gamma}{\gamma} \frac{\sin \Omega_0(0)}{\Omega_0(0)} + \frac{\mu_1 g_i}{2\beta\Omega^2(0)}. \tag{28}$$

The factors g_r and g_i are

$$g_r = \int_0^\pi dy \frac{\sin y \cos y}{\left[1 + \Delta \sin^2 y / 2\right]},$$

$$g_i = \int_0^\pi dy \frac{\sin^2 y}{\left[1 + \Delta \sin^2 y / 2\right]}, \tag{29}$$

where

$$\Delta = \left[\frac{\Omega^2(s_F) - \Omega^2(0)}{\Omega^2(0)} \right]. \tag{30}$$

The factors g_r and g_i are of order unity; they may be easily evaluated numerically.

4. DISCUSSION AND CONCLUSIONS

The sensitivities of the RK/TBA and an SWFEL/TBA can now be compared using the results of the last section (Eqs. (27) through (30)). The first thing to observe is that the dependence upon current error, ΔI, is not excessive, nor is it very different for the RK and the SWFEL. This source of sensitivity must, and can, be controlled in either device.

The dependence upon energy errors, $mc^2\Delta\gamma$, is much more severe and it is different between the two devices. Note that it only affects φ_1 (and not the amplitude b_1). From Eq. (28) and the expressions for κ (just after Eq. (2)) we obtain, for the energy dependence,

$$\varphi_1 = -\left(\frac{\omega}{c\gamma^2}\right)\left(\frac{\Delta\gamma}{\gamma}\right)\frac{\sin\Omega(0)z}{\Omega(0)}; \qquad RK,$$

$$\varphi_1 = -\frac{\omega}{c\gamma^2}(1+a_w^2/2)\left(\frac{\Delta\gamma}{\gamma}\right)\frac{\sin\Omega(0)z}{\Omega(0)}; \qquad SWFEL. \tag{31}$$

The initial synchrotron period, given by Eq. (18), can be inserted to obtain the amplitude of phase error:

$$amplitude\ of\ \varphi_1 = -\left\{\frac{1}{(b_0(0)\cos\theta_0\varphi_0)}\right\}^{1/2}\left(\frac{\Delta\gamma}{\gamma}\right)\begin{cases}\left(\dfrac{1}{2\gamma}\right)^{1/2}; \ RK, \\[2ex] 2^{1/2}\left(\dfrac{1+a_w^2/2}{a_w}\right)^{1/2}; \ SWFEL.\end{cases} \tag{32}$$

Taking $b_0(0) \cos (\theta_0\varphi_0)$ the same for the two devices, we see that the factor $\left(\dfrac{1}{2\gamma}\right)^{1/2}$, for the RK, is replaced by $\left(2\sqrt{2}\right)^{1/2}$ for the SWFEL. (We have minimized the coefficient by taking $a_w = \sqrt{2}$.)

We see that, as a rule of thumb, the RK is roughly two times less sensitive to energy errors at a rather low energy than is the SWFEL. However, we must remember that the RK will have more severe wake-field effects than the SWFEL since it necessarily consists of smaller structures.

On the other hand, it is possible to operate the RK at a very high energy since no resonance condition must be satisfied (as in the

SWFEL). At large γ, the sensitivity to energy errors, $mc^2\Delta\gamma$, is very much less in the RK than in the SWFEL. Successful operation of an RK of high energy will, however, depend on acceleration of an intense bunched beam from a low energy, during which process phase errors may yet accumulate. Indeed, accelerating the drive beam of an RK to high energies, while maintaining its phase insensitivity is an important challenge for such a device, and remains to be analyzed.

In summary, we have developed formulas for the sensitivities of the SWFEL/TBA and the RK/TBA. The parameter which characterizes the sensitivity is displayed, and this has allowed us to compare the two schemes. It is shown that there is no vast difference between the two approaches, although the RK/TBA is less sensitive at high energies. The choice between them will probably be made on the basis of such issues as ease of construction, cost, and beam break-up limits (BBU).

In this last regard, note that the RK/FEL consists of single mode structures (as contrasted with the overmoded SWFEL/TBA), and, hence, will have a lower BBU limit or, equivalently, will be limited to operation at lower frequencies. Thus, high frequency operation (well above 10 GHz) will require the SWFEL/TBA; if one desires to operate at lower frequencies (well below 10 GHz) then the RK/TBA with its reduced sensitivities is the scheme of choice.

Finally, we note that the formalism which has been developed allows the input of a coupling impedance into the SWFEL and, therefore, the introduction of the features of a realistic cavity. It is interesting, but really obvious, that the properties of the cavity are included in just a single impedance function. We thus have the capability to analyze the performance of a SWFEL employing coupling impedances obtained by various electrodynamic codes such as SUPERFISH or MAFIA. In short, we have put the analysis of the standing-wave free-electron laser on the same footing as that of the relativistic klystron.

REFERENCES

1. A.M. Sessler, in *Laser Acceleration of Particles*, P.J. Channell ed., AIP Proceedings No. **91**, p. 163, New York, (1982).

2. D.B. Hopkins, A. M. Sessler, and J. S. Wurtele, Nucl. Instr. and Meth. in Phys. Res. **228**, 15 (1984).

3. E. Sternbach and A. M. Sessler, Nucl. Instr. and Methods in Phys. Res. **A250**, 464 (1986).

4. A.M. Sessler, E. Sternbach and J.S. Wurtele, Nucl. Instr. and Methods in Phys. Res. **B40/41**, 1064 (1989).

5. A.M. Sessler, D.H. Whittum and J.S. Wurtele, Proc. of the XIV Intern. Conf. on High Energy Accel., Tsukuba, Part. Accel. **31**, 69 (1990).

6. A.M. Sessler, et al., Nucl. Instr. and Methods in Phys. Res. **A306**, 592 (1991).

7. W.M. Sharp, et al., Linear Accelerator Conf, Albuquerque, Los Alamos National Laboratory Report LA-12004-C, 656 (1990).

8. W.M. Sharp, et al., Intense Microwave and Particle Beams II, O/E LASE' 91, Los Angeles, in Proc. Intern. Soc. Opt. Eng. (SPIE) (1991).

9. S. Van Der Meer, Particle Accelerators **30**, 127 (1990).

10. M.A. Allen et al., Phys. Rev. Lett. **63**, 2472 (1989).

11. T. Orzechowski et al., Phys. Rev. Lett. **57**, 2172 (1986).

12. A.W. Chao, B. Richter, and C.Y. Yao, XIth Int. Conf. on High Energy Accelerators, p. 597 (1980); Nucl. Instrum. Methods **178**, 1 (1980).

SENSITIVITY STUDIES OF A STANDING-WAVE FREE-ELECTRON LASER

Govindan Rangarajan and Andrew M. Sessler
Lawrence Berkeley Laboratory, University of California,
Berkeley, CA 94720

ABSTRACT

A standing-wave free-electron laser (SWFEL) has been proposed for use in a two-beam accelerator (TBA). We modify the previous one-dimensional discrete cavity model of the SWFEL by introducing drifts between cavities. We also vary the input beam energy as a function of the bunch number. In this new model, we obtain a stable equilibrium solution for a well-bunched beam (even after including all nonlinear terms). We obtain analytic expressions characterizing this equilibrium in the limit of small cavity lengths. We study the dependence of fluctuations in signal phase along the device as a function of detuning, input beam energy, beam length, current errors, and initial signal field amplitude. We are able to find an optimized set of parameters for which the output energy changes by less than 3% across the cavities for a 1% detuning. The maximum change in signal phase is less than 0.12 radians.

INTRODUCTION

A "two-beam accelerator" (TBA) has been proposed[1] as a device capable of achieving high accelerating gradients required for the next generation linear colliders. One possible configuration is to use a standing-wave free-electron laser (SWFEL)[2,3] as a power source for the high gradient structure in a TBA. In this device, irises are placed along the FEL wiggler to form a series of microwave cavities, and induction cells are placed between cavities to reaccelerate the beam (see Figure 1). The standing-wave signal that builds up in the cavities as the beam passes through is coupled to a parallel high-gradient radio-frequency accelerator.

The SWFEL has been studied in some detail in earlier papers[2-5]. In this paper, we study the discrete-cavity model introduced in Ref. [5] in greater detail.

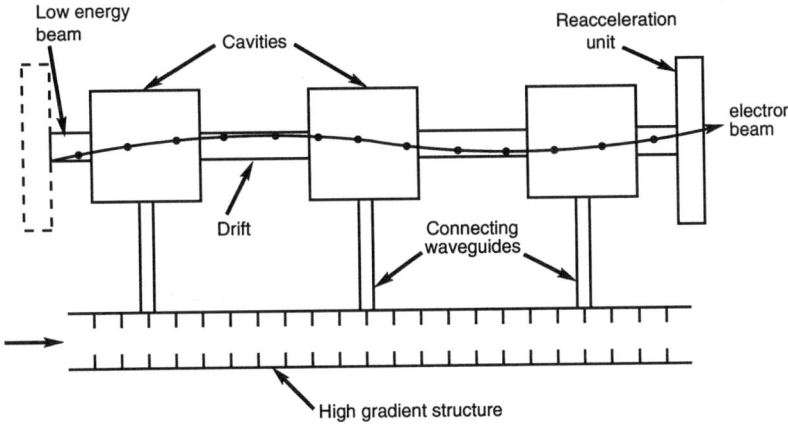

Figure 1: Conceptual layout of one section of a standing-wave TBA.

First, we introduce a new feature by putting drifts between cavities. Using this additional degree of freedom, we obtain an equilibrium solution for a well-bunched beam. Finally, we perform sensitivity studies around this equilibrium.

EQUILIBRIUM SOLUTION

Previously, an equilibrium solution was obtained[3] for a well-bunched beam in the continuous model of the SWFEL by linearizing the equations of motion. In this section, we obtain a general equilibrium solution (again for a well-bunched beam) using the full nonlinear set of equations. Moreover, this equilibrium solution does not assume small cavity lengths.

We start with the one-dimensional equations of motion in a discrete cavity model of the SWFEL. Since the model and the equations of motion have been discussed in detail elsewhere[5], we restrict ourselves to a brief description. The discrete-cavity model takes into account time-of-flight effects within the cavity and applies the reacceleration field only between cavities, where the ponderomotive force is absent. As in previous SWFEL models, only a single signal frequency is considered.

Within a cavity, we solve the conventional wiggle-averaged FEL equations. Denoting the particle phase $(k_s + k_w)z - \omega_s t$ by θ_j, the particle energy by γ_j, and taking $\tilde{z}$ to be the independent variable, the equations are given as follows

$$\frac{d\theta_j}{d\tilde{z}} = k_w + k_s - \frac{\omega_s}{c} - \frac{\omega_s}{2c\gamma_j^2}\left[1 + \frac{a_w^2}{2} - 2D_x a_w a_s \cos(\theta_j + \phi)\right], \qquad (1)$$

$$\frac{d\gamma_j}{d\tilde{z}} = -D_x \frac{\omega_s}{c} \frac{a_w}{\gamma_j} a_s \sin(\theta_j + \phi), \tag{2}$$

$$\frac{d\hat{a}}{d\tilde{z}} = i\eta \langle \frac{\exp(-i\theta_j)}{\gamma_j} \rangle. \tag{3}$$

Here a_s and ϕ denote the field amplitude and field phase respectively. The coupling coefficient D_x is given by

$$D_x = [J_o(\xi) - J_1(\xi)]/2, \tag{4}$$

where $\xi = \omega_s a_w^2 / (8ck_w \gamma_j^2)$. The coefficient η is given by

$$\eta = \frac{8\pi}{hw} \frac{eI_b}{m_e c^3} \frac{D_x a_w}{k_s}. \tag{5}$$

The full interaction of the electron beam with the SWFEL structure can be represented symbolically by the following recursion relations:

$$\theta_{k,l} = \theta_{k,l-1} + F_\theta(\theta_{k,l-1}, \gamma_{k,l-1}, (a_s)_{k,l-1}, \phi_{k,l-1}), \tag{6}$$

$$\gamma_{k,l} = \gamma_{k,l-1} + F_\gamma(\theta_{k,l-1}, \gamma_{k,l-1}, (a_s)_{k,l-1}, \phi_{k,l-1}) + \Delta\gamma_{k,l-1}, \tag{7}$$

$$(a_s)_{k,l} = (a_s)_{k-K',l} + F_a(\theta_{k-K',l}, \gamma_{k-K',l}, (a_s)_{k-K',l}, \phi_{k-K',l}), \tag{8}$$

$$\phi_{k,l} = \phi_{k-K',l} + F_\phi(\theta_{k-K',l}, \gamma_{k-K',l}, (a_s)_{k-K',l}, \phi_{k-K',l}). \tag{9}$$

Here, θ and γ are n-vectors where n is the number of particles in a bunch. The quantities $\theta_{k,l}$, $\gamma_{k,l}$, $(a_s)_{k,l}$, and $\phi_{k,l}$ represent values of θ, γ, a_s, and ϕ for the kth bunch at the beginning of the lth cavity. The functions F_θ, F_γ, F_a, and F_ϕ represent the FEL interaction within a cavity. The quantity $\Delta\gamma_{k,l}$ is the reacceleration field for kth bunch after the lth cavity. The bunch skip factor K' is equal to the number of bunches that pass through by the time the forward traveling radiation field makes a round trip within a cavity (see Ref. [5] for further details):

$$K' \approx 2L_c/\lambda_s. \tag{10}$$

Here, L_c is the length of the cavity and the signal wavelength λ_s is the average separation between electron bunches.

We are now in a position to derive an equilibrium solution. In a discrete-cavity model, an equilibrium solution is a solution where all the cavities behave in an identical fashion. The equilibrium is derived for a well-bunched beam i.e. θ and γ in Eqs. (6) and (7) are now scalars representing the average particle phase and average energy of an electron bunch. First, we modify the discrete-cavity model by introducing a drift space after each cavity. All drift spaces are taken to be identical to one another. The drift space is designed as follows. Consider the first bunch of electrons traversing the first cavity. By the time it exits the cavity, its θ would have changed by a finite amount, say $\Delta\theta$. The drift space is designed such that it exactly compensates for this change. That is, the particle phase changes by an amount $-\Delta\theta$ when passing through the drift. When the drift spaces are incorporated into our model, Eq. (6) is modified as follows

$$\theta_{k,l} = \theta_{k,l-1} + F_\theta(\theta_{k,l-1}, \gamma_{k,l-1}, (a_s)_{k,l-1}, \phi_{k,l-1}) - \Delta\theta. \tag{11}$$

Since we cannot change the drift length from bunch to bunch we have to ensure that all bunches undergo the same change $\Delta\theta$ in the first cavity. Only then will the θ correction scheme work for all bunches. This is achieved by choosing the input energy $\gamma_{k,1}$ such that the following condition is satisfied

$$F_\theta(\theta_{k,1}, \gamma_{k,1}, (a_s)_{k,1}, \phi_{k,1}) = \Delta\theta \quad \forall \; k. \tag{12}$$

Next, we choose the reaccleration field such that it restores the energy lost by a bunch in a cavity i.e.

$$\Delta\gamma_{k,l} = -F_\gamma(\theta_{k,l}, \gamma_{k,l}, (a_s)_{k,l}, \phi_{k,l}) \quad \forall \; k \text{ and } l. \tag{13}$$

Finally, the input radiation field amplitude and phase are taken to be the same for all cavities (i.e. $(a_s)_{1,l}$ and $\phi_{1,l}$ are independent of l). Once we satisfy all these conditions, we obtain a cavity-independent equilibrium. This can be seen as follows. From Eqs. (7) and (13), we find that $\gamma_{k,l} = \gamma_{k,1} \; \forall \; k$. From Eqs. (11) and (12), we find that $\theta_{k,2} = \theta_{k,1} \; \forall \; k$. Using these two equalities and the fact that $(a_s)_{1,2} = (a_s)_{1,1}$ and $\phi_{1,2} = \phi_{1,1}$, we find from Eqs. (8) and (9) that $(a_s)_{k,2} = (a_s)_{k,1}$ and $\phi_{k,2} = \phi_{k,1}$ for all k. Thus, the first two cavities behave in an identical fashion. Repeating this argument, one can easily show that the behaviour of any dynamical variable is independent of l. Since the drift spaces compensate for the change in θ and the reacceleration field compensates for the change in γ, the electron bunches see exactly the same initial conditions in all the cavities. Therefore, the resulting solution of the FEL equations is identical for all cavities.

The above equilibrium solution only fixes the input value of one variable – γ. The input values of the remaining variables are determined as follows. It is well known that multi-particle stability arguments favour a value of the equilibrium ponderomotive phase $\psi \; (= \theta + \phi)$ close to zero. However, other considerations like sensitivity to detuning (see the next section for further details) favour a ψ close to $\pi/2$. As a compromise, we typically take ψ to be a constant equal to $\pi/3$. This fixes $\theta_{k,1}$:

$$\theta_{k,1} = -\phi_{k,1} + \pi/3. \tag{14}$$

We still have to fix $\phi_{1,1}$ and $(a_s)_{1,1}$. Usually, $\phi_{1,1}$ is taken to be zero. To reduce sensitivity to detuning (see next section), $(a_s)_{1,1}$ is taken to be as high as is practically possible.

For long cavities, it is not possible to give an analytic expression for the equilibrium solution. However, this is possible for very short cavities where the continuous model is valid. This is the subject of our next subsection.

Equilibrium Solution in the Continuous Model

The continuous model is obtained by taking the cavity length to be so short that the Euler formula can be used to integrate the FEL equations within a cavity. From Eqs. (1)–(3) and (6)–(9), we obtain the following results (assuming a well-bunched beam)

$$F_\theta/L_c = k_w + k_s - \frac{\omega_s}{c} - \frac{\omega_s}{2c\gamma^2}\left[1 + \frac{a_w^2}{2} - 2D_x a_w a_s \cos(\theta + \phi)\right], \quad (15)$$

$$F_\gamma/L_c = -D_x \frac{\omega_s}{c}\frac{a_w}{\gamma}a_s \sin(\theta + \phi), \quad (16)$$

$$F_a/L_c = \eta \frac{\sin(\theta + \phi)}{\gamma}, \quad (17)$$

$$F_\phi/L_c = \eta \frac{\cos(\theta + \phi)}{a_s \gamma}. \quad (18)$$

Substituting the above expressions into Eqs. (7)–(9) and (11), we can convert them into the following differential equations:

$$\frac{d\theta}{dz} = k_w + k_s - \frac{\omega_s}{c} - \frac{\omega_s}{2c\gamma^2}\left[1 + \frac{a_w^2}{2} - 2D_x a_w a_s \cos(\theta + \phi)\right] - \overline{\Delta\theta}, \quad (19)$$

$$\frac{d\gamma}{dz} = -D_x \frac{\omega_s}{c}\frac{a_w}{\gamma}a_s \sin(\theta + \phi) - \frac{eE_z}{m_e c^2}, \quad (20)$$

$$\frac{\partial a_s}{\partial s} = \eta' \frac{\sin(\theta + \phi)}{\gamma}, \quad (21)$$

$$\frac{\partial \phi}{\partial s} = \eta' \frac{\cos(\theta + \phi)}{a_s \gamma}. \quad (22)$$

Here, z denotes the longitudinal distance along the device, E_z is the external reacceleration field, and s denotes the normalized distance from the beam head. The quantity $\overline{\Delta\theta}$ is the change in θ per unit length for the first electron bunch:

$$\overline{\Delta\theta} = \frac{D_x a_w \omega_s}{c\gamma_r^2}a_s(s=0)\cos\psi(s=0). \quad (23)$$

Here we have assumed that the first electron bunch comes in with the resonant energy γ_r and the ponderomotive phase $\psi(s=0)$. The field coupling coefficient η' is given as follows:

$$\eta' = \eta L_b/2, \quad (24)$$

where L_b is the length of the beam.

For an equilibrium solution, we require θ and γ to be z-independent. Denoting the equilibrium values by a subscript 0, we obtain the following result from Eq.

(19):

$$\gamma_0^2(s) = \frac{\left[\dfrac{\omega_s}{2c}\left(1 + \dfrac{a_w^2}{2}\right) - \dfrac{D_x a_w \omega_s}{c} a_{s0}(s)\cos\psi_0(s)\right]}{(k_w + k_s - \omega_s/c - \overline{\Delta\theta})}. \tag{25}$$

Typically, $\overline{\Delta\theta} \ll (k_w + k_s - \omega_s/c)$. Therefore, we get

$$\gamma_0^2(s) \approx \gamma_r^2 - \frac{D_x a_w \omega_s a_{s0}(s)\cos\psi_0(s)}{c(k_w + k_s - \omega_s/c)}. \tag{26}$$

From Eq. (20), we get

$$\frac{eE_z}{m_e c^2} = -D_x a_w \frac{\omega_s}{c} \frac{a_{s0}(s)\sin\psi_0(s)}{\gamma_0(s)}. \tag{27}$$

Before we can proceed further, we have to fix $\theta_0(s)$, which is a free parameter. To simplify calculations, we choose it as follows [cf. Eq. (14)]

$$\theta_0(s) = -\phi_0(s) + \pi/3. \tag{28}$$

That is, the equilibrium ponderomotive phase ψ_0 is a constant independent of s. Now, we can solve for $a_{s0}(s)$ and $\phi_0(s)$ from Eqs. (21) and (22). Assuming a constant current i.e.

$$\eta'(s) = \eta'_0, \tag{29}$$

we get the following expression for $a_{s0}(s)$ and $\phi_0(s)$:

$$a_{s0}(s) \approx a_{s0}(0) + \frac{\eta'_0 \sin\psi_0}{\gamma_r} s, \tag{30}$$

$$\phi_0(s) \approx \phi_0(0) + \cot\psi_0 \ln\left[1 + \frac{\eta'_0 \sin\psi_0}{\gamma_r a_{s0}(0)} s\right]. \tag{31}$$

By substituting these results into Eqs. (26)–(28), we can obtain analytic expressions for all quantities of interest. Similar expressions can be obtained for the linearly ramped current case.

NUMERICAL RESULTS

In this section, we numerically study the SWFEL model described in the previous section. We perform sensitivity studies around the new equilibrium described in Eqs. (12) — (14). First, we study a short cavity case (where $L_c = \lambda_s = 1.83$ cm) and then a long cavity case (where $L_c = 14.7$ cm). The simulation parameters are listed in Table 1. We set the initial signal level $|a_s(0)|$ by assuming some input power per unit length P_{in} and balancing this with cavity-wall losses, specified by an assumed cavity Q. The average beam current has been chosen to give an output energy per unit length of 10 J/m. For multiparticle simulations, we use a 1% spread in γ and a 10% spread in θ. Typically, we use 200 simulation particles.

Table 1: Simulation parameters for the standing-wave FEL

Parameter	Nominal value	Optimized value
average beam current (I_b)	1.8 kA	3.2 kA
beam length (L_b)	440.0 cm	440.0 cm
resonant energy (γ_r)	16.4	32.2
wiggler strength (a_w)	1.4	1.4
wiggler wavelength (λ_w)	37 cm	39.3 cm
wiggler length (L_w)	40 m	40 m
waveguide height (h)	3 cm	3 cm
waveguide width (w)	10 cm	10 cm
signal frequency $(\omega_s/2\pi)$	17.1 GHz	17.1 GHz
cavity quality (Q)	10^4	10^4
input power (P_{in})	8 kW/m	8 kW/m
output energy (W_{out})	10 J/m	10 J/m

Studies using short cavities

In this subsection, we numerically study a SWFEL using short FEL cavities. Length of each cavity is taken to be λ_s. In this case, our model of the SWFEL reduces to the continuous model introduced earlier[3] (other than for the drifts). All studies are performed around the equilibrium solution described in the previous section. This equilibrium is achieved in our numerical simulations by varying the input beam energy according to Eq. (25) and by fixing the drift lengths using Eq. (23). For the parameters given in Table 1, the variation in beam energy from the head to the tail of the beam is less than 4% (for a constant current, the input energy decreases linearly [cf. Eqs. (26) and (30)]). The reacceleration field compensates for the average energy lost in the previous cavity.

First, we study the effects of detuning on the output microwave energy per unit length W_{out} and signal phase ϕ. Detuning is achieved by offsetting the input energies of all particles by a fixed fraction $\Delta\gamma_0/\gamma_0$. Figure 2 shows the effects of 0, 0.5, and 1 percent detuning on W_{out} and ϕ for a well-bunched beam. As expected, for zero detuning we get a perfect equilibrium solution. In this case, we have verified that the analytic formulas given in Eqs. (30) and (31) agree quite well with the numerical simulations. However, for non-zero detuning, the amplitude of phase fluctuations exceeds the required tolerances (the maximum change in signal phase, $(\Delta\phi)_{max}$, needs to be less than 0.2 radians for the TBA to operate properly[3]).

We can get around this problem as follows. From earlier studies[4], we know that most of the phase ripple is introduced in the early part of the beam where $|a_0(s)|$ is small. This can be seen heuristically as follows. By linearizing around the equilibrium solution, we obtain the following dependence of the phase fluc-

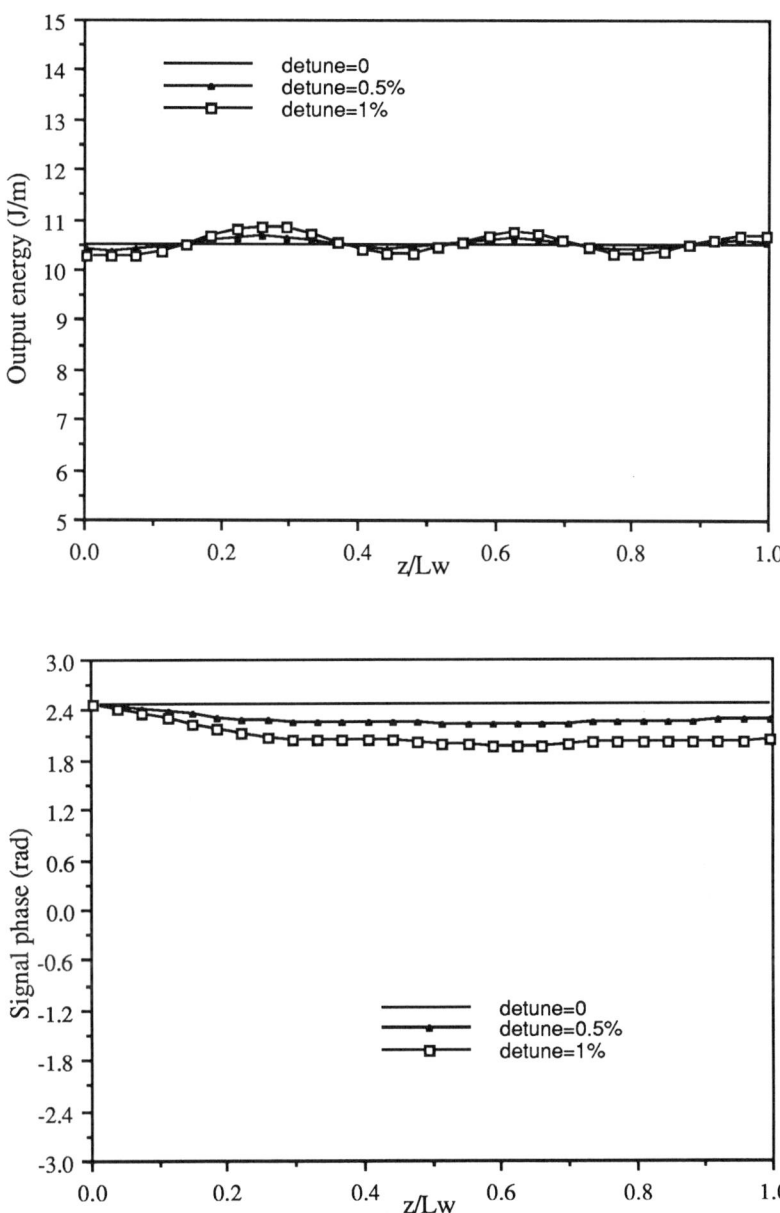

Figure 2: Single-particle simulations of the continuous model for the nominal parameters given in Table 1. Three values of $\Delta\gamma_0/\gamma_0$ are studied. Figure 2a shows the output energy per unit length W_{out} and Figure 2b shows the wave phase ϕ as functions of z.

tuation amplitude ϕ_1 on the detuning $\Delta\gamma_0/\gamma_0$:

$$\frac{\partial\phi_1}{\partial s} \approx -\eta' \, \frac{\cos\psi_0}{a_{s0}\gamma_0} \, \frac{\Delta\gamma_0}{\gamma_0}. \tag{32}$$

Thus, ϕ_1 builds up in amplitude when $a_{s0}(s)$ is small i.e. when s is small. Therefore, we can reduce the buildup by modifying the input beam energy as follows:

$$\gamma_{in}(s) = \gamma_0(s)\left[1 + \frac{\Delta\gamma_0}{\gamma_0}\frac{1}{1+\exp(\frac{s-s_0}{\delta s})}\right], \tag{33}$$

where we set $s_0/L_b = 0.2$ and $\delta s/L_b = 0.05$. With this modification, the magnitudes of phase fluctuations are within tolerance limits (cf. Figure 3). We repeat the calculation using many particles. In this case, the average input γ is given by Eq. (33). Again, the phase fluctuations are manageable (cf. Figure 4). These findings show that, in order to meet the tolerances on ϕ fluctuations, the beam energy needs to be near γ_0 for at most only the first 20% of the beam. This modulation of γ_0 should be achievable in practice beacuse of reduced beam loading near the head of the beam.

The above argument also suggests that we can reduce ϕ fluctuations by increasing the input power (thereby increasing $a_{s0}(0)$). This is found to be true. However, modifying the input beam energy as shown in Eq. (33) works better than increasing P_{in}. And once Eq. (33) is implemented, any increase in P_{in} has little effect. Therefore, in all simulations that follow, we will use γ_{in} as given by Eq. (33).

Using Eq. (32), we can justify the choice of $\pi/3$ for the equilibrium value of ψ_0. We see that the magnitude of ϕ_1 fluctuations is reduced as $\psi_0 \to \pi/2$. However, multiparticle stability requirements favour a ψ_0 value close to zero. Using numerical simulations, we find that $\psi_0 = \pi/3$ is a good compromise value.

Equation (32) also suggests that the magnitude of phase fluctuations can be reduced by going to a higher input γ. A higher input γ is achieved by keeping ω_s and a_w constant and increasing λ_w. To keep the output power level constant, we need to increase beam current as we increase γ. Figure 5 shows the variation of $(\Delta\phi)_{max}$ as a function of input γ for a 1% detuning. We see a significant reduction in $(\Delta\phi)_{max}$ as the input beam energy increases thus verifying the above hypothesis. However, we can not increase the input beam energy beyond a certain point. The initial cost of generating a very high energy beam would become too high (since the efficiency of a TBA decreases with a higher input beam energy) and the additional requirement of a higher beam current would lead to beam breakup problems. Keeping these considerations in mind, we have proposed an optimized set of parameters in Table 1. For these parameters, $(\Delta\phi)_{max}$ is only 0.12 radians for a 1% detuning.

If one increases the wiggler strength a_w while keeping the input γ fixed, $(\Delta\phi)_{max}$ increases. On the other hand, if λ_w is held fixed and a_w is increased (thus increasing input γ also), $(\Delta\phi)_{max}$ remains approximately constant. This is

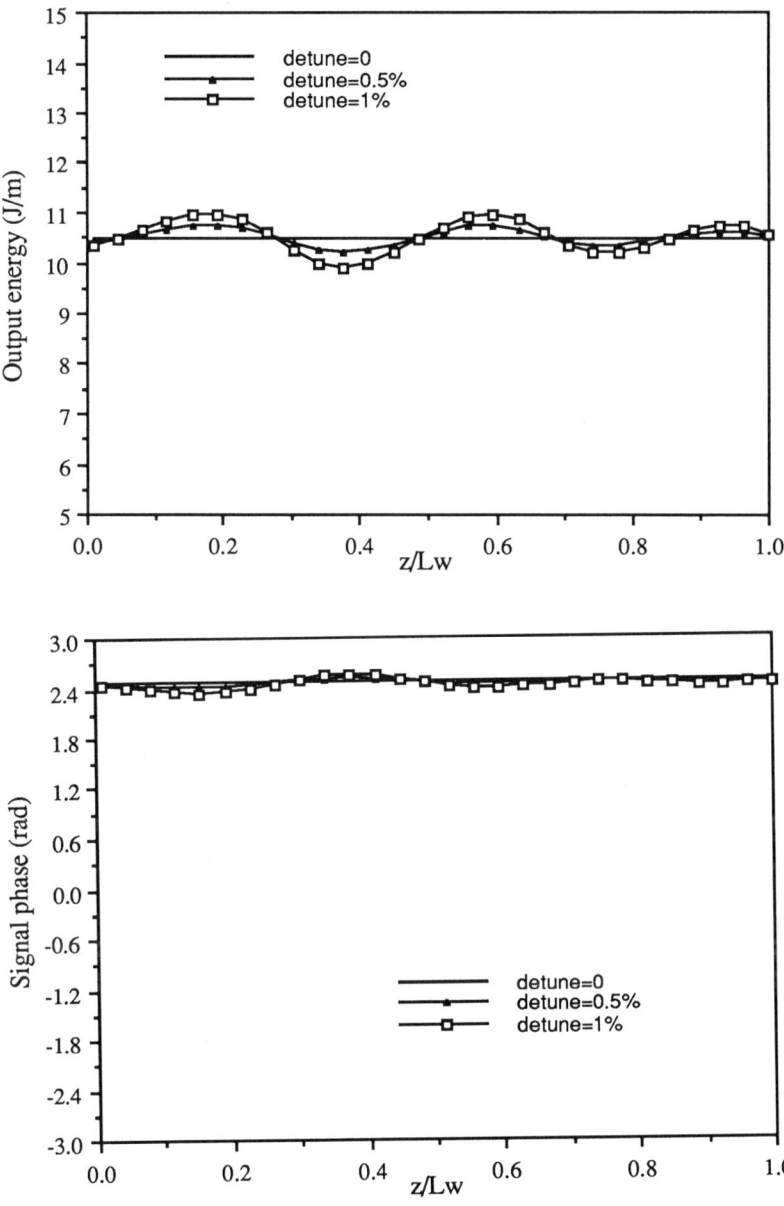

Figure 3: Single-particle simulations of the continuous model for the nominal parameters given in Table 1. Three values of $\Delta\gamma_0/\gamma_0$ are studied. In these simulations, $\Delta\gamma_0/\gamma_0$ is zero near the beam head and increases to the indicated value (0.5% or 1.0%) near $s/L_b = 0.2$. Figure 3a shows the output energy per unit length W_{out} and Figure 3b shows the wave phase ϕ as functions of z.

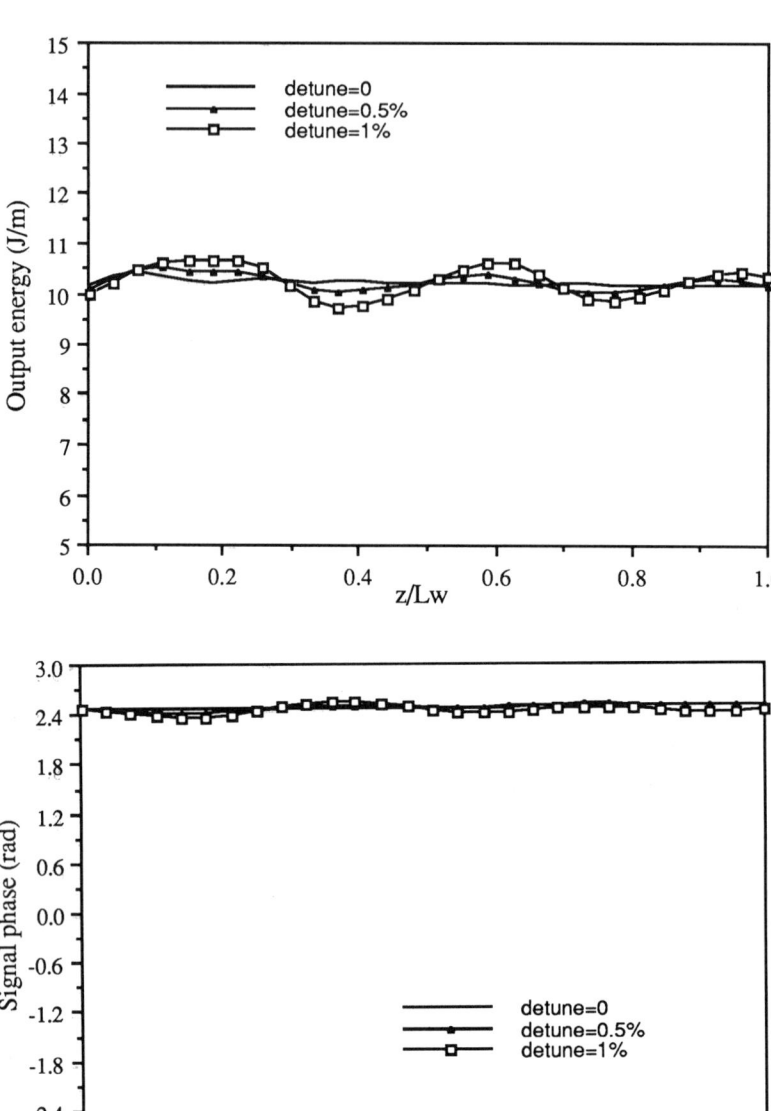

Figure 4: Multi-particle simulations of the continuous model for the nominal parameters given in Table 1. Three values of $\Delta\gamma_0/\gamma_0$ are studied. In these simulations, $\Delta\gamma_0/\gamma_0$ is zero near the beam head and increases to the indicated value (0.5% or 1.0%) near $s/L_b = 0.2$. Figure 4a shows the output energy per unit length W_{out} and Figure 4b shows the wave phase ϕ as functions of z.

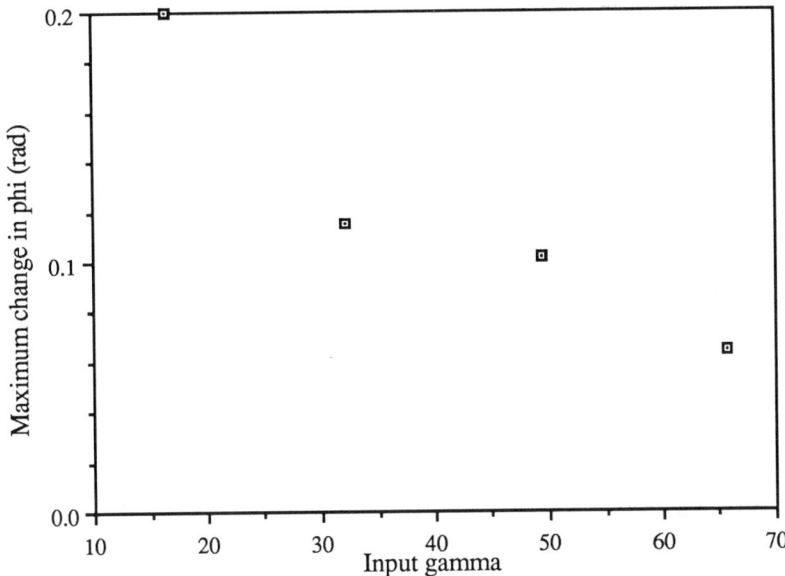

Figure 5: Single-particle simulations of the continuous model for the nominal parameters given in Table 1. This figure displays the variation of $(\Delta\phi)_{max}$ as a function of the input beam energy γ_0 for $\Delta\gamma_0/\gamma_0 = 1\%$. The input beam energy is varied by keeping ω_s and a_w fixed and varying only λ_w.

because increasing a_w increases $(\Delta\phi)_{max}$ whereas increasing γ decreases $(\Delta\phi)_{max}$. Therefore, if both are increased, these two effects cancel one another leading to a constant $(\Delta\phi)_{max}$.

Further, we find that a 2% error in beam current has no significant effect on $(\Delta\phi)_{max}$. Neither does a random 1% error in the magnitude of reacceleration field. For very short beams, $(\Delta\phi)_{max}$ tends to be higher. However, once the beam length exceed a critical value ($\sim$40 cms), $(\Delta\phi)_{max}$ settles down to a stable value independent of L_b.

Studies using long cavities

In this subsection, we perform sensitivity studies around the equilibrium solution using cavities of length 14.7 cm. This is done to check if long cavity lengths lead to new physical effects not present in the continuous model. We find that there are no significant deviations from the results obtained for the continuous model. As in the continuous model, specifying γ_{in} according to Eq. (33), increasing input power, and increasing input γ all help reduce ϕ fluctuations (see Figures 6 and 7). However, there is one case where the long cavity case appears (at first glance) to

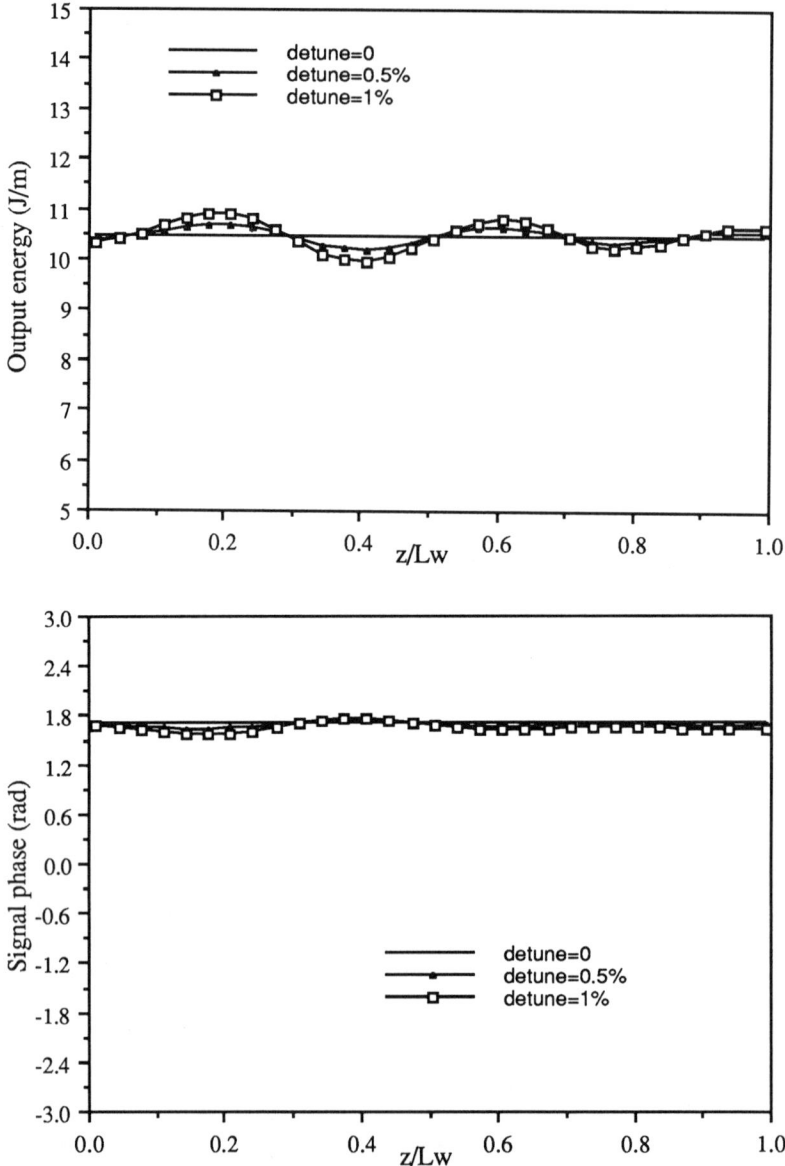

Figure 6: Single-particle simulations of the discrete cavity model ($L_c = 14.7$ cm) for the nominal parameters given in Table 1. Three values of $\Delta\gamma_0/\gamma_0$ are studied. In these simulations, $\Delta\gamma_0/\gamma_0$ is zero near the beam head and increases to the indicated value near $s/L_b = 0.2$. Figure 6a shows the output energy per unit length W_{out} and Figure 6b shows the wave phase ϕ as functions of z.

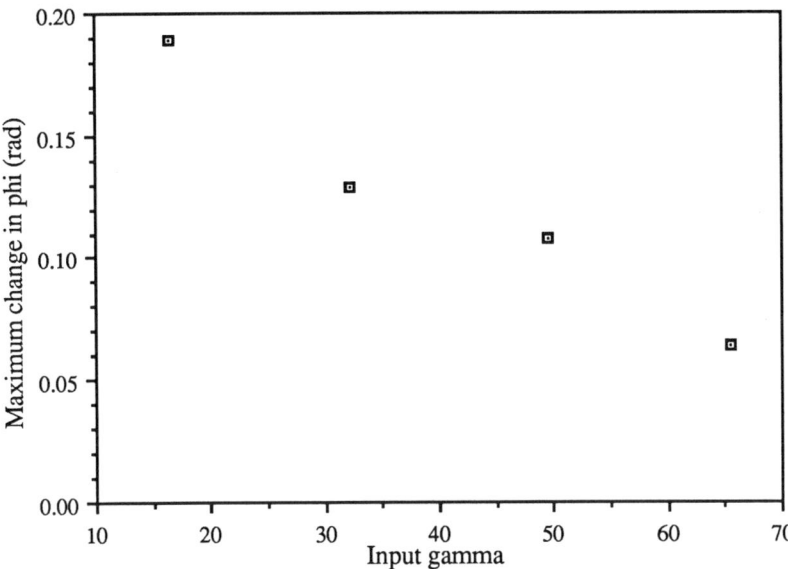

Figure 7: Single-particle simulations of the discrete cavity model ($L_c = 14.7$ cm) for the nominal parameters given in Table 1. This figure displays the variation of $(\Delta\phi)_{max}$ as a function of the input beam energy γ_0 for $\Delta\gamma_0/\gamma_0 = 1\%$. The input beam energy is varied by keeping ω_s and a_w fixed and varying only λ_w.

differ significantly from the continuous model. For a cavity of length 14.7 cm, we find that the beam length has to be greater than 300 cm before $(\Delta\phi)_{max}$ settles down to a stable value independent of L_b. This apparent discrepancy goes away when one analyses the situation more closely. If we look at the last bunch in a beam of length L_b, out of the L_b/λ_s bunches that have preceded it, it can interact only with $L_b/2L_c$ bunches [cf. Eq. (10)]. And this number is approximately equal for both the continuous model (where $L_c = 1.8$ cm and critical beam length is $\sim$40 cm) and the long cavity case (where $L_c = 14.7$ cm and critical beam length is $\sim$300 cm).

Finally, we find results similar to ones obtained above even if we increase the cavity length further — to 22 cm and further to 38.5 cm.

SUMMARY

We have developed a discrete cavity model of a SWFEL incorporating drifts in between cavities. A new equilibrium solution was found for this model (when the electrons are well bunched) by varying the input beam energy as a function of bunch number. We performed sensitivity studies around this new equilibrium. Remarkably similar sensitivities were observed irrespective of the cavity length. Keeping the beam energy close to the equilibrium value for the initial part of the beam was found to decrease sensitivity to detuning by a significant amount. A higher value of the input beam energy also led to decreased sensitivity. Errors in beam current and reacceleration field magnitudes did not lead to any significant increase in signal phase fluctuations. Using these results, we have been able to find a set of parameters for which the SWFEL has a tolerable sensitivity to detuning. One result that should help in future studies is the observation that the continuous model and the discrete cavity model behave in a similar fashion. Therefore, one needs to study only the more tractable continuous model in great detail.

ACKNOWLEDGEMENTS

This work was supported by the U.S. Department of Energy under contracts DE-AC03-76SF00098 (LBL) and W-7405-ENG-48 (LLNL). We would like to thank Dr. Bill Sharp and Dr. David Whittum for useful discussions.

REFERENCES

[1] A. M. Sessler, in *Laser Acceleration of Particles*, edited by P. J. Channell (AIP, New York, 1982), p.154.

[2] W. M. Sharp, A. M. Sessler, D. H. Whittum, and J. S. Wurtele, Proc. 1990 Linear Accelerator Conference, Albuquerque, LA-12004-C Conference, 656 (1991).

[3] A. M. Sessler, D. H. Whittum, W. M. Sharp, M. A. Makowski, and J. S. Wurtele, Nucl. Instr. and Meth. A306, 592 (1991).

[4] W. M. Sharp, G. Rangarajan, A. M. Sessler, and J. S. Wurtele, Proc. SPIE Conference 1407, 535 (1991).

[5] G. Rangarajan, A. Sessler, and W. M. Sharp, Proc. 12th Int. Free Electron Laser Conf., Nucl. Instr. and Meth. A (1992) (in press).

A   RF–linac, FEL Based Drive Beam Injector for CLIC

William A. Barletta
Lawrence Livermore National Laboratory , Livermore CA 94550 and
Dept. of Physics, UCLA, 405 Hilgard Avenue, Los Angeles, CA

and

Rodolfo Bonifacio
Universitá degli Studi Milano, Dipartimento di Fisica, Via Celoria 16, Milano, Italia

ABSTRACT

We describe a means of producing at train of 40 kA pulses of 3 ps duration as the drive beam for CLIC using an RF–linac driven free electron laser (FEL) buncher. Potential debunching effects are discussed. finally we describe a low energy test experiment.

INTRODUCTION

Among the many technological challenges of the CLIC two-beam accelerator scheme[1] to realize a TeV linear collider of electrons and positrons one of the most difficult is the generation of the drive beam. The CLIC drive beam is composed of 40 kA bunchlets of $\approx$ 3 ps duration ($2\sigma_0$) with a total charge of $\approx$160 nC. Eleven bunchlets are repeated at a frequency of 30 GHz to form a pulse train that is repeated 4 times at a frequency of 352 MHz to form a macropulse. The macropulses must be generated at a repetition rate of $\approx$ 2 kHz. These features are summarized in Table 1.

Table I. Features of the CLIC Drive Beam

Peak bunchlet current, I_{peak}	40 kA
Bunchlet rms duration	3 ps
Charge per bunchlet, Q	160 nC
Bunchlet spacing	1 cm
Bunchlet frequency	30 GHz
Bunchlets per pulse	11
Pulse frequency	352 MHz
Pulses per macropulse	4
Macropulse frequency	1500 Hz

Clearly the most difficult feature of producing the drive beam is the formation of such high charge, picosecond duration bunchlets at energies < 50 MeV. As it is unlikely that the bunchlets can be generated directly from a cathode with the desired waveform, one must consider other strategies, all of which involve one or more stages of bunch compression.

The consequences of the bunching properties of a high gain FEL have been investigated theoretically[2] and demonstrated experimentally[3], but the bunching of the beam current per se has never been measured. Recently Shay et al.[4] have proposed the use a high gain FEL driven by a 15 MeV, 2 – 3 kA linear induction accelerator to generate the CLIC drive beam. Here we extend to higher currents the ELFA high gain FEL[5] with wave guide slippage control driven by a 10 MeV superconducting RF linac operating at 352 MHz. In particular we propose to furnish the drive beam for CLIC drive beam with an amalgam of magnetic switching, rf-accelerator and FEL technologies. The use of a radio frequency linac facilitates the control of energy spread in the beam and makes it more practical to extend beam energy up 40 MeV if desirable. The use of a low frequency rf-accelerator avoids several technological difficulties related to voltage regulation in induction linacs[4,6]. Moreover, the 352 MHz linac is easily phased locked to the main drive beam linac. A schematic of our scheme using a FEL pre-buncher is shown in Fig. 1 and is described below.

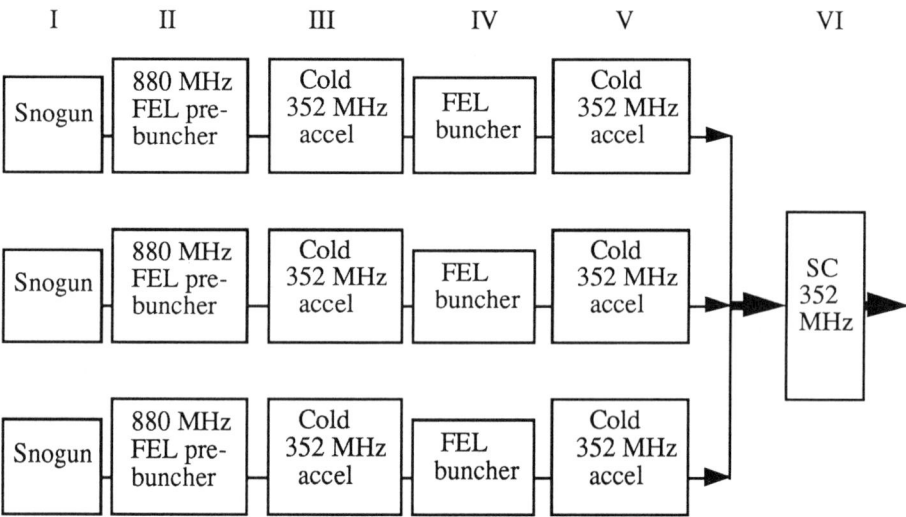

Figure 1. The proposed generator chain for the CLIC drive beam.. Section I - 1.25 MeV; 1.2 ns; 500 A; Section II - 1.25 MeV; 1 bunch @ 120 ps; 5 kA; Section III - 25 MeV; 1 bunch @ 120 ps; 5 kA; Section IV - 25 MeV; 4 bunches @ 3 ps; 50 kA; 30 ps spacing; Section V - 50 MeV; 4 bunches @ 3 ps; 50 kA; 30 ps spacing; Section VI - 3 GeV; 12 bunches @ 3 ps; 50 kA

At the entrance to the 30 GHz FEL buncher which compresses the beam current into 40 kA, 3 ps bunchlets, we require a beam of $\approx$ 4 kA at $\approx 20 - 30$ MeV. The beam train begins with a high voltage, high current electron gun. One such gun with suitable characteristics is the SNOGUN[6] presently under construction by Science Research Laboratory and UCLA. SNOGUN consists of a small thermionic cathode generating a 400 A, 1.5 ns beam in a 1.25 MeV induction stack driven with SNOMAD magnetic pulse compressors. Beam extraction is controlled with a grid driven by ferrimagnetic shock lines. At the exit of the electron gun, rf bunching cavities or an FEL pre-buncher can be used to compress the beam to a 4 kA, 120 ps pulse prior to its acceleration to 20 − 30 MeV in cryogenically cooled, high field 352 MHz cavities. The ≈20 MeV beam is injected with a 30 GHz signal into a waveguide inside a short, high field wiggler. With beam high currents the FEL process leads to high gain and rapid bunching. The termination of the FEL well before saturation maximizes compression of the beam current and minimizes energy spread to yield the desired peak currents and waveform. As the initial pulse can only yield 3 bunchlets the entire pulse train is produced by four lines running in parallel. The subsequent rapid acceleration prevents space charge debunching.

THE HIGH GAIN FEL AS A BUNCHER

When a beam traverses the FEL wiggler of period,λ_w, radiation is generated at the resonant wavelength λ_s. As the signal grows the pondermotive potential created by the wiggler field and the radiation, bunches the beam electrons periodically on the wavelength scale of the resonant (sometimes called optical) wavelength. The current compression due the FEL action can be computed from the bunching parameter, b, which is the ensemble average

$$b = \left| \left\langle e^{i\theta} \right\rangle \right| \tag{1}$$

where θ_i is the phase of the ith electron with respect to the radiation frequency. For an unbunched beam b = 0 while for a beam bunched to a delta function of current b = 1. The strong FEL bunching is evidenced that when the gain saturates the bunching factor reaches a maximum b_{max} = 0.8 regardless of the actual values of the initial current, λ_w, or λ_s. As an example we show in figure 2 a calculation of the bunching of a 10 MeV, 4 kA beam based on the 1–dimensional model[2] of the high gain FEL. One clearly sees the strong bunching action that derives from the dispersive path in the wiggler and the strong pondermotive potential.

The gain of the FEL and the speed of the bunching process is described by the BPN universal scaling parameter[2], ρ;

$$\rho = \frac{1}{\gamma} \left(\frac{a_w \omega_p}{4 c k_w} \right)^{2/3} \propto \frac{I^{1/3} B^{2/3} \lambda_w^{4/3}}{\gamma} \tag{2}$$

where γ is the usual relativistic factor, ω_p is the non-relativistic plasma frequency, a_w is the dimensionless vector potential of the wiggler and k_w is the wiggler spatial frequency.

$$\rho a_w \approx 0.66 \, B_w(\text{Tesla}) \, \lambda_w \, (\text{cm}) \tag{3a}$$

$$k_w \approx \frac{2 \pi}{\lambda_w} \tag{3b}$$

We note the experimentally verified feature of the high gain FEL, namely that the rate of the bunching action is proportional to $\rho\lambda_w$ and hence actually proceeds more rapidly as the input current increases. In fact, in the experiment of ref. 2 the bunched current inferred from the well–verified simulations of the experiment was ≈ 10 kA (though this feature was never measured). It is this characteristic that makes the FEL buncher so attractive in comparison with other bunching schemes when the final peak value of the bunched current must be extremely large as in the case of the CLIC drive bunches. If the initial current pulse is several optical wavelengths long, the output current from the FEL will be a train of high current bunchlets spaced at the optical wavelength. With respect to the requirements of the CLIC drive beam, achieving a precisely controlled bunchlet spacing at the desired interval corresponding to 30 GHz is a natural consequence of FEL action.

Another feature of the FEL that compares favorably with respect to other bunching schemes is that the length over which bunching occurs scales favorably (increases linearly with increased beam energy). The FEL action does, however, induce an energy spread in the beam that is proportional to the gain. By terminating wiggler before the FEL process saturates one can a) maximizes current multiplication, b) keep the induced energy variation small, and c)minimizes wiggler.

POTENTIAL LIMITING PHENOMENA

We now turn our attention to phenomena that can limit the performance of the FEL bunching process:

a) energy spread at the entrance to the wiggler,
b) loss of the radiation field via diffraction,
c) transverse and longitudinal space charge debunching.

Effects of energy spread

With respect to the intrinsic energy spread that is produced by the acceleration process upstream of the buncher the FEL compares well with other techniques. Quantitatively, we expect the FEL to be insensitive to energy spreads and variations as long as

$$\left(\frac{\Delta\gamma}{\gamma}\right)_{accelerator} < \rho. \tag{4}$$

For the simulation shown in Fig. 2, $\rho \approx 0.05$. As the FEL (or rf-cavity) pre-buncher in Fig. 1 is followed by an accelerator section that increases the mean beam energy by a factor of ≈ 5, this requirement is easily satisfied especially as the wiggler is actually truncated before saturation is reached. Moreover, in strong super-radiant regime to be used in our scheme, the sensitivity of the FEL is further reduced with respect to both the instantaneous energy spread or to the energy variation in beam .

Loss of the radiation field via diffraction,

In optical FELs using very low emittance beams the gain can be reduces and the bunching speed decreased if the radiation diffracts out of the electron beam too rapidly.

In our case this consideration does not apply as the radiation is confined to the region close to the beam by the waveguide that we use to provide slippage control. The effectiveness of the waveguide in eliminating the degradation of performance due to diffraction is displayed by the results of a full three dimensional simulation with the code, GINGER[8] that includes the effects of longitudinal space charge forces. Thus we are left with space charge as the only potentially deleterious effect.

In Fig. 2 and 3 we show GINGER simulations for parameters (Table II) with the output power and bunching as a function of the length of the wiggler (Figure 2) and also the histogram of the current and the electron longitudinal phase space in a wavelength (Fig. 3). Note in figure 2 the link between FEL gain and the bunching mechanism and the strong bunching action that derives from the combination of the dispersive path in the wiggler and the strong pondermotive potential.

Table II. Parameters for GINGER simulation

Peak bunchlet current	4 kA
Radiation wavelength	1 cm
Beam energy	9.5 MeV
Wiggler wavelength	24 cm
Wiggler field	2.5 kG
BPN parameter, ρ	0.13
Waveguide height	2.5 cm
Normalized emittance	1 π mm-mrad
Input power	1 KW
Output power	$\approx$ 5 GW
Saturation length	3.5 m

Transverse space charge effects

To begin our analysis we consider first the problem of the transverse space charge. Outside the bunched beam of radius R the radial electrostatic field, E_r, is

$$E_{r,\,peak}(r) = \frac{I_{peak}}{2\,\varepsilon_o\,\pi\,c\,\beta\,r} = \frac{60\,I_{peak}\,(A)}{r\,(m)}\,MV/m. \tag{5}$$

In eq. (5) c is the speed of light, ε_o is the permittivity of free space, and β is the velocity of the electrons, v, divided by c. For the 40 kA CLIC bunchlet at 20 MeV with an rms radius of 3 mm the radial electric field evaluated at the edge of the beam is $\approx$ 800 MeV/m. Thus transverse space charge would be disastrous were it not for the fact the a relativistic beam carries its own extremely strong, azimuthal focusing field, B_θ,

$$B_\theta(r) = \frac{\mu_o\,I_{peak}}{2\,\pi\,r} = \frac{I_{peak}\,(A)}{r\,(cm)}\,Gauss. \tag{6}$$

Once we include this pinch field in the Lorentz force equation we find that the relativistic electrons experience very little radial force.

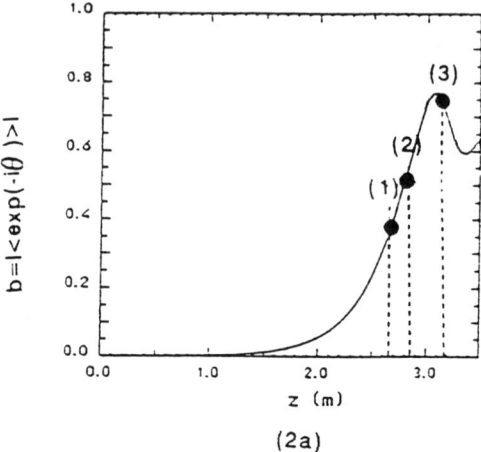

(2a)

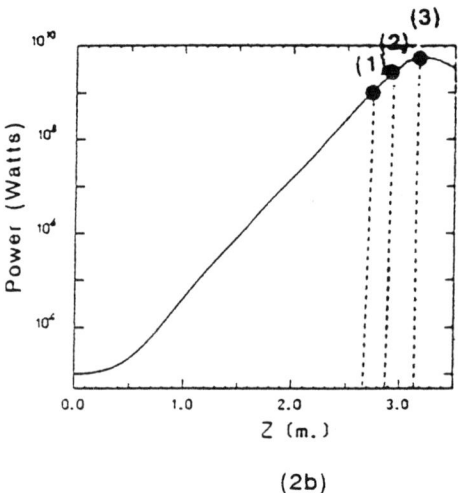

(2b)

Figure 2. GINGER simulation of the FEL for parameters of Table II. Panel 2a shows the bunching parameter as a function of the length of the wiggler. Panel 2b shows the output power in Watts as a function of the length of the wiggler.

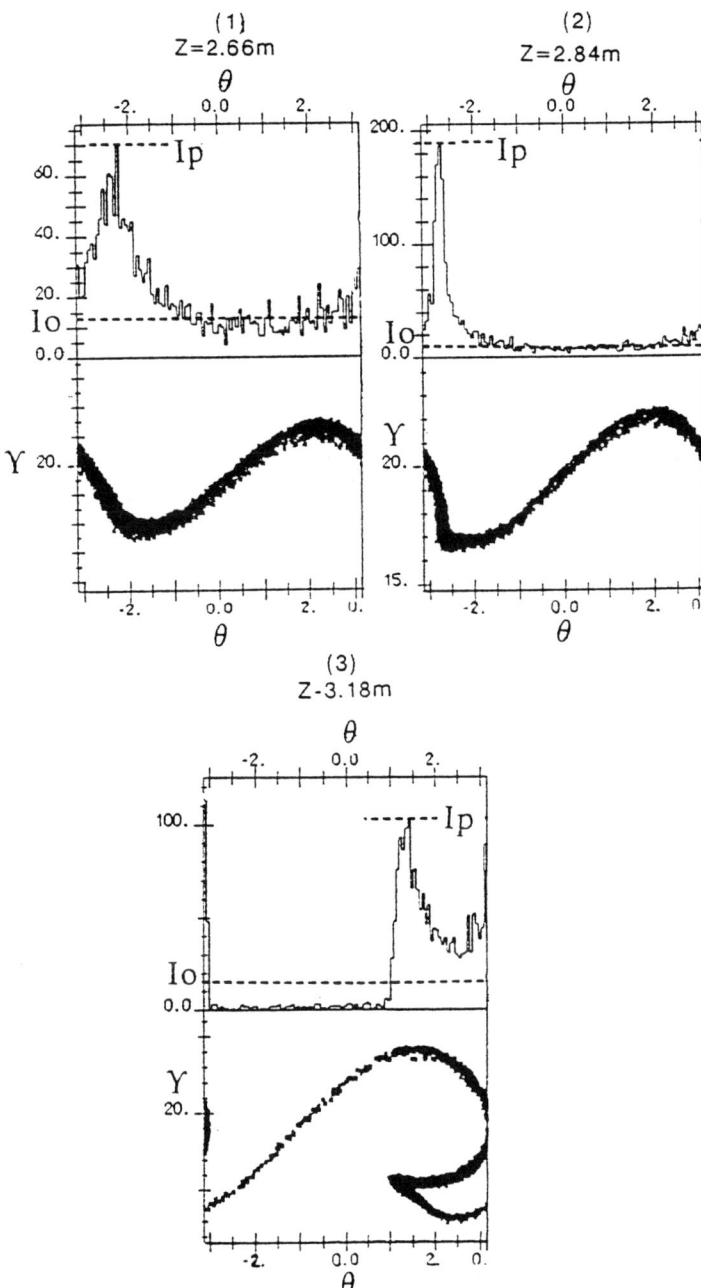

Figure 3. Phase histograms showing the electron current (top) and longitudinal phase space (bottom). I_0 is the average inpt current. I_p is the peak output current, The outputs are for the points marked in Fig. 2

$$F_r = e \left(E_r - v \, B_\theta \right).$$

(7)

Noting from eq. (5) and (6) that

$$B_\theta = \frac{v}{c} E_r \,,$$

(8)

we obtain the well known result

$$F_r = e \, E_r \left(1 - \frac{v^2}{c^2} \right) = \frac{1}{\gamma^2} e \, E_r.$$

(9)

Recall that 1 Gauss $\approx 3 \times 10^4$ V/m, we can see that outside of the wiggler at 20 MeV even a 40 kA bunchlet can be confined radially against space charge expansion by a magnetic field equivalent to a B_θ of only a few tens of Gauss. In fact, a larger field is actually needed to prevent the beam from expanding due to its finite emittance.

The relative effects of the emittance, E, and transverse space charge as the beam propagates in the z direction can be evaluated both outside of and within the wiggler through the use of the envelop equation for the rms radius of the beam,

$$\frac{d^2 R}{dz^2} - \frac{E^2}{R^3} + \frac{U}{R} + \frac{\left\langle K_\beta^2 R^2 \right\rangle}{R} = 0.$$

(10)

The transverse self-fields of the beam are described by potential U and the external focusing by the term K_β. If the external focusing can be described by an harmonic potential with betatron frequency k_β, the ensemble average in the last term is reduced to the simple product $k_\beta^2 R$. In the wiggler the betatron frequency is

$$k_\beta = \frac{2\pi}{\lambda_w} = \frac{2\pi a_w}{\sqrt{2}\, \lambda_w \gamma}\,.$$

(11)

In the absence of space charge neutralization from a background plasma the potential U may be expressed in terms of the Alfvén current as

$$U = -\frac{I}{\gamma^2 I_A} = -\frac{I}{17 \text{ kA } \gamma^3}\,.$$

(12)

The envelop equation implies a unique equilibrium radius for the beam, R_e,

$$R_e = R_0 \left[f + \sqrt{1 + f^2} \right]^{1/2}$$

(13)

with

$$R_0 = \sqrt{\frac{E}{K_\beta}} = 6 \times 10^{-2} \text{ (m)} \sqrt{\frac{E_n \text{ (m-rad)}}{B_w \text{(T)}}}$$

(14)

where R_0 is the equilibrium radius in the absence of space charge and

$$f = \frac{U}{2 E \, K_\beta} = \frac{U \gamma}{2 E_n \, K_\beta} = \frac{10^{-4} \, I \text{(kA)}}{\gamma \, B_w \text{(T) } E_n \text{(m-rad)}}.$$

(15)

In eq. (15) we have introduced the normalized emittance, E_n, which is equal to γE. From eq. (13) and (15) we can see the that transverse space charge effects can be ignored when f << 1, i.e., when

$$E_n > \frac{U\gamma}{2 k_\beta} = 10^{-6} \frac{I(kA)}{\gamma B_w(T)}. \tag{16}$$

To satisfy the high FEL conditions, the phase area of the electron beam should be smaller than that of the radiation; i.e.,

$$E_n < \frac{\lambda_s \gamma}{2\pi}. \tag{17}$$

For a 40 kA bunchlet at 20 MeV in a 0.66 T wiggler, both eq. (16) and (17) will be satisfied when

$$1.5 \times 10^{-4} \text{ m-rad} < E_n < 6 \times 10^{-2} \text{ m-rad} .$$

This condition is easily satisfied for our proposed injector for the CLIC drive beam; we choose a value, $E_n = 2 \times 10^{-3}$ m-rad, so that the beam has a convenient radius (≈ 3 mm) with respect to the height of the waveguide. From this discussion we conclude that the neglect of transverse space charge effects in the GINGER simulation upon which we would base a final design is justified.

Longitudinal space charge debunching

The potential difficulties arising from longitudinal space charge forces are more complicated to analyze as it depends on the geometry and charge distribution in the bunch and on the external boundary conditions. Fortunately with respect to the FEL action, in the wiggler all these dependences have been fully included in the GINGER simulation of a 10 MeV beam which was shown in Fig. 3. Therefore, we need on focus our attention on the debunching at the output of the wiggler. Following ref. 4 we quote the calculation[9] of the longitudinal electric field, E_z, of a relativistic beam with charge distribution, n_e, in a long conducting pipe of radius b. Up to terms of order ln (b/R)

$$E_z = -\alpha \frac{d n_e}{dz}. \tag{18}$$

where

$$\alpha = \frac{R^2}{4\pi\epsilon_o \gamma^2}. \tag{19}$$

Assuming the bunchlet to have a Gaussian charge distribution with width σ_0 and total charge Q,

$$n_e = \frac{Q}{\pi R^2} \frac{1}{\sqrt{2\pi} \sigma_o} \exp\left(-\frac{z^2}{2\sigma_o^2}\right). \tag{20}$$

One can easily evaluate the maximum value of E_z, which obtains when $z = \sigma_0$;

$$E_{z,\,max} = \frac{1}{4\,\pi\,\varepsilon_o}\,\frac{Q}{\sqrt{2\,\pi\,e}\,\,\gamma^2\,\sigma_o^2} \quad . \tag{21}$$

This formula can be readily understood as the field produced by a point charge of magnitude $Q/(2\pi e)^{1/2}$. The factor γ^2 comes from the Lorentz transformation from the beam frame into the laboratory frame; the factor $(2\pi e)^{1/2}$ derives from the Gaussian shape. In eq. (21) one can replace Q with $I\lambda_s$, where I is the current averaged over the interval λ_s at which the bunchlets are repeated, i.e., the optical wavelength of the FEL. Putting in physical quantities, we have

$$E_{z,\,max} = 8 \times 10^{-6}\frac{I(A)\,\,\lambda_s(m)}{\gamma^2\,\sigma_o^2\,(m)} \quad . \tag{22}$$

For our proposed scheme to produce a drive beam for CLIC, i.e., $I = 4$ kA, $\lambda_s = 10^{-2}$ m, $\sigma_0 = 5 \times 10^{-4}$ m, and $\gamma = 40$, $E_{z,\,max} \approx 0.4$ MV/m. Knowing $E_{z,\,max}$ we can estimate the energy spread induced by the space charge after the bunch travels a distance L out of the wiggler. From Newton's law

$$\left(\frac{\Delta\gamma}{\gamma}\right)_{sc} = \frac{e\,E_z\,L}{\gamma\,m\,c^2} \approx \frac{2\,E_z(MV/m)\,L\,(m)}{\gamma} \quad . \tag{23}$$

In our case,

$$\left(\frac{\Delta\gamma}{\gamma}\right)_{sc} \approx \frac{0.8}{40}\,L = 10^{-2}\,L\,(m) \quad . \tag{24}$$

The space charge induced energy spread must be added to that generated by the high gain FEL action, namely,

$$\left(\frac{\Delta\gamma}{\gamma}\right)_{FEL} \approx \rho, \tag{25}$$

where ρ is the FEL scaling parameter. For our parameters ρ is of order 10^{-2}. If the energy spread that can be tolerated in the input sections of the rf linac that raises the beam energy from 20 MeV to 3 GeV is in the range from 5 – 10 %, then eq. (24) implies that a drift of 5 to 10 m is tolerable. This distance is a reasonable length for the matching section from the wiggler to the front end of the drive linac. We must, of course, verify that the debunching produced in the length L is also acceptable.
    The change in bunch length is

$$\Delta\sigma = \Delta\beta\,L = \frac{\Delta\gamma}{\gamma^3}\,L = 2\,\frac{E_{z,\,max}}{\gamma^3}\,L^2 \quad . \tag{26}$$

The debunching length, L_D, is defined as the distance over the bunchlet doubles in length; i.e., $\Delta\sigma = \sigma_0$. Hence,

$$L_D = \sqrt{\frac{\sigma_o \, \gamma^3}{2 \, E_z}} \qquad (27)$$

For our proposed drive beam scenario for CLIC in which $E_z = 0.4$ MV/m, the debunching length is 8.8 m.

In an experiment to verify the longitudinal effects of space charge we should require that

$$\left(\frac{\Delta\gamma}{\gamma}\right)_{sc} \gg \left(\frac{\Delta\gamma}{\gamma}\right)_{FEL} \approx \rho \ . \qquad (28)$$

Inserting eq. (23) into eq. (26) we have

$$E_{z,\,max} \, L \gg \frac{\gamma\rho}{2} \ . \qquad (29)$$

For the CLIC buncher $L \gg 2$ m; as ρ scales as γ^1, $E_{z,max} L$ will be independent of γ. Hence, we can test the CLIC drive beam buncher and all the relevant physics in a low energy experiment as we discuss below.

First we apply the preceding analysis to the case of the 10 MeV GINGER simulation of Fig. 3. In that case

$$E_{z,\,max} \propto \frac{1}{\gamma^2} = 0.8 \text{ MV/m} \qquad (30)$$

and

$$\left(\frac{\Delta\gamma}{\gamma}\right)_{sc} \propto \frac{1}{\gamma^3} = 8 \times 10^{-2} \text{ L(m)} \ . \qquad (31)$$

Hence, $\Delta\gamma/\gamma \le 0.1$ if $L \approx 1.25$ m and $L_D = 2.3$ m.

If these values seem somewhat shorter than practical one can increase the energy from 10 to 20 MeV as in the previous example. A further increase of the beam energy to 40 MeV would eliminate space charge effects for all practical purposes. These features described above have already been confirmed in preliminary computations with the code, PARMELA, for a beam expanding against a solenoidal B field in otherwise free space. The PARMELA simulation assumes a hard edge distribution of charge inside the electron bunch. This assumption certainly increases the expected space charge effects at the beam boundary. Nonetheless at 20 MeV we find a debunching length of ≈ 3 m; at 40 MeV the debunching length exceeds 30 m. The results of a more detailed simulation including the effects of a Gaussian charge distribution and of conducting beam pipe boundaries will be presented elsewhere.

Finally we point out that extending the operating energy of the FEL from 20 to 40 MeV does not present any serious practical difficulty, because the resonance condition is such that an increase in γ is easily compensated for by increasing λ_w. Indeed increasing the wiggler period makes construction of the wiggler easier.

LOW ENERGY, LOW CURRENT TEST EXPERIMENT

All the bunching properties of the high gain RF-linac driven FEL with waveguide control of slippage can be tested in a low energy, low current experiment. Novel features of this experiment include the measurement of bunched current, slippage control in the FEL, and strong superadiant behavior of the FEL. With the same wiggler as we would use with the full 20 MeV, 4 kA beam, the laser gain will be nearly the same as the gain scales as $\rho \propto I^{1/3} \gamma^1$. If we decrease I by a factor of 400 from 4 kA to 10 A, we should simultaneously decrease the beam energy by a factor of 6.6, down to 2.5 MeV. Then the saturation length of the FEL amplifier will remain unchanged. The low energy experiment also allows us to test the relevant properties of space charge debunching.

Imposing the condition of eqn. (28) that the space charge induced energy spread exceeds the FEL induced energy spread, we conclude that we should make measurement of the energy spread and bunch length at a distance , L downstream of the wiggler, where

$$L(m) \gg \frac{\gamma^3 \sigma_o^2(m) \, \rho \, 10^5}{2 \, \lambda(m) \, I(A)} .$$

(32)

For the experimental parameters of $\gamma = 6$, $I = 10$ A, $\sigma_0 = 10^{-3}$ m, $\lambda = 10^{-2}$ m, the BPN parameter $\rho \approx 10^{-2}$; hence, for distances L exceeding 1 m space charge effects will dominate the debunching and will be readily distinguishable from debunching due to the FEL induced energy broadening. The parameters are summarized in Table III.

Table **III.** Parameters for of the low energy test experiment, ELFA-2

Peak bunchlet current	10 A
Radiation wavelength	1 cm
Beam energy	2.5 MeV
Wiggler wavelength	12 cm
Wiggler field	1.8 kG
BPN parameter, ρ	0.018
Waveguide height	1.7 cm
Normalized emittance	1 π mm-mrad
Input power	1 KW
Output power	≈ 0.5 MW
Saturation length	5 m

CONCLUSIONS

We have showed that an RF-linac injector with a FEL buncher appears to be a promising candidate to provide the CLIC drive beam. The required control of the beam energy (≈ 0.2 %) necessary to assure phase stability of the RF-current at 30 GHz imposed on the beam by the FEL is routinely achieved with care from RF-linacs unlike the comparable situation with induction linac produced beams. Issues related to space charge can be made negligible by an appropriate choice of beam energy at the entrance to the wiggler. Finally we have suggested a low energy, low current test experiment that would demonstrate all the relevant physics of our proposed scheme.

ACKNOWLEDGMENTS

We are indebted to Dr. Bellomo, Dr. Corsini, Dr. De Salvo, Dr. Pierini, Dr. Pullia and the rest of the ELFA theory group at Milano for the continuous discussion and for their analytical and computational support. We also acknowledge helpful discussions with Dr. Colin Johnson of CERN. One of us (WAB) wishes to thank the Milano section of INFN for its kind hospitality during the preparation of this report. His work has been partially supported by Lawrence Livermore National Laboratory under contract w-7405-eng 48 with the U. S. Dept. of Energy.

REFERENCES

1. W. Schnell, "The Drive Linac for a Two–stage RF Linear Collider", CERN–LEP-RF/88–59, June, 1988

2. R. Bonifacio, C. Pellegrini, and L. Narducci, Optical Communications, **50**, 373, 1984

3. T. J. Orzechowski et al. Phys. Rev. Lett., **54**, 889 (1985)

4. H. D. Shay, R. A. Jong, R. D. Ryne, E. T. Scharlemann, and S. S. Yu, "Use of a FEL as a Buncher for a TBA Scheme", Proc. of the Int. FEL Conference, (Paris, 1990)

5. R. Bonifacio et al., "The ELFA Project", Proc. of the European Particle Accelerator Conference, Nice, (June, 1990)

6. W. A. Barletta, "High Current Electron Accelerators using Pulsed Power Techniques", Proc. of the European Particle Accelerator Conference, Nice, (June, 1990).

7. W. A. Barletta and S. Hartmann , "SNOGUN, A Novel High-Repetition Rate, High Current Source of Electrons and Positrons", UCLA report, UCLA-CAA00070-1/91, January, 1991

8. W. Fawley and T. Scharlemann, Lawrence Livermore National Laboratory

9. R. J. Briggs and V. K. Neil, Plasma Physics, **8**, 255 (1966)

Report of the Working Group on Near-field Accelerators

J.B. Rosenzweig*

UCLA Department of Physics

405 Hilgard Ave., Los Angeles, CA 90024

February 12, 1993

Abstract

Advanced research on near-field acceleration, by the time of this meeting, has progressed into mature field of both experimental and theoretical investigation. This field has a number of subcomponents; a few of the more notable are wake-field acceleration, open structure laser acceleration, and novel rf structures. This paper will provide an overview of these subjects, and examine the most recent advances in near-field acceleration and the facilities developed for its study, as reported at the Port Jefferson workshop.

1 Wake-field Accelerators

Development of advanced high gradient radio-frequency (rf) linear accelerators requires that the problems of rf power generation and propagation in the linac structure be addressed anew. Historically, rf power for accelerators has been derived from electron tube devices such as klystrons. In these sources, high voltage electron beams interact with resonant rf cavities to produce the required electromagnetic power, which can then be coupled to an accelerating structure through a waveguide. It was recognized some years ago that this process could be simplified by allowing a high-current relativistic electron beam (the *driving beam*) to travel through the accelerating structure directly, radiating rf power into the accelerating mode which has a phase velocity $v_\phi = v_b \simeq c$. A trailing, lower current relativisitic beam can then be resonantly accelerated to high energy by this mode. This scheme is termed *wake-field acceleration*, as the coherent radiation of particles due to their passage through an accelerating structure is generally described as a wake-field.

*Work supported in part by U.S. Department of Energy Grant DE-FG-92ER40693.

A wake-field can be excited in structure by a beam of charged particles whenever the structure can support electromagnetic waves which have phase velocity less than c (slow waves). Seen in this light, one can in a generalized sense consider wake-fields to be coherent emission of Cerenkov radiation. Examples of wake-field structures can include linac-like periodic metallic waveguides, dielectric-lined waveguides, and (slightly expand the usage of the term structure) plasma. In these examples it is apparent that the radiation is generated by the polarization response to the beam's electromagnetic fields of resonator systems based microscopically on electronic – valence, atomic and free electrons, respectively – motion. Note that radiation generated by the interaction of of an electron beam with an externally applied magnetic field is excluded from the definition of wake-fields as we have given (*i.e.* a two-beam accelerator driven by an FEL would not be considered a wake-field accelerator).

The major advantage of wake-field acceleration over a more conventional approach is that the rf power is created inside of the accelerating structure itself, and thus considerations of how to efficiently excite the structure, which have played a dominant role in the design of high-gradient, high-frequency normal conducting linear colliders are mitigated. In a wake-field accelerator the accelerating rf is created, and then used for acceleration on time scale short compared to the power dissipation or transport in the structure. Thus the shunt impedance and group velocity of the structure have less importance in the design of a wake-field accelerator. In order for high gradients to be achieved in a wake-field accelerator, however, the longitudinal beam impedance $Z_\parallel$ should be maximized and the structure excited by a beam with very high peak current. Alternatively, one can relax the impedance requirements and drive the structure resonantly with a beam with current modulated at the rf frequency ω, as in a klystron.

1.1   Wake-fields: General Remarks

The wake-fields excited by a beam in a structure is generally defined to be only the field components with which a paraxial beam particle ($\mathbf{v_b} \simeq v_b \hat{z}$ interacts,[1] which are broken down into longitudinal and transverse parts,

$$W_z = E_z \tag{1}$$
$$\mathbf{W}_\perp = \mathbf{E}_\perp + v_b \hat{z} \times \mathbf{B}, \tag{2}$$

respectively. The longitudinal and transverse parts of the wake-field are not independent, but are related (in the approximation that the structure is translationally invariant in z, and thus the wake-field can be written as a function of $\zeta = z - v_b t$) by the Panofsky-Wenzel theorem[1][2]

$$\frac{\partial \mathbf{W}_\perp}{\partial \zeta} = \nabla_\perp W_z. \tag{3}$$

Since we are interested in large rates of energy transfer to the structure, it is useful to examine an upper limit, the energy loss per particle of a beam of N

electrons to Cerenkov radiation, in a polarizable medium with very large dielectric constant ϵ. This quantity is approximately[3]

$$eW_{z,-} \simeq \frac{1}{2}e^2 N k_0^2,$$ (4)

where now k_0 is the maximum wavenumber that can be excited by the beam, which is for practical purposes the inverse of the largest beam dimension, usually the rms beam length σ_z. The dependence of a single mode component of the wake-field (of wavenumber $k = \omega/v_b$) with linear response can be illustrated by writing

$$W_z(\zeta) = Z'(k)v_b \int_\zeta^\infty \lambda(\zeta') \cos\left(k(\zeta - zeta')\right) d(k\zeta'),$$ (5)

where $\lambda = I/v_b$ and $Z'(k)$ are the beam charge and the impedance per unit length of the structure at the wave-number k, respectively. This *convolution integral*, when evaluated for gaussian bunches, gives a wake-field amplitude proportional to $Z'(k) \exp\left(-k^2\sigma_z^2/2\right)$ (*i.e.* the response is proportional to the Fourier amplitude of the current at the wave-number k). These results illustrate the need for a short, high current beam, and a high frequency, high impedance medium or structure to generate large amplitude wake-fields from one bunch. In practice, the coupling of the beam to the structure does not approach this maximum for devices with vacuum holes for the beam, and the limiting coupling is approached only in plasma. Also, the creation and transport of the high current beams needed to obtain accelerating gradients over 100 MV/m is technically challenging, and thus use of a modulated current beam or pulse train to achieve the required current spectrum may be preferable in some schemes. In particular, since the Panofsky-Wenzel theorem in the states that the transverse and longitudinal wake-field amplitudes are proportional a lower impedance structure may be necessary in order to avoid beam-breakup instability (BBU).

In many wake-field acceleration schemes the accelerating beam and the driving beam are collinear, that is they travel nominally along the symmetry axis of the structure. In this case, if the electromagnetic response of the structure is linear, both beams experience the same impedance. this places constraints on both the efficiency of the scheme and the ratio of acceleration gradient in the driving beam's wake to the deceleration gradient inside of the driver. a common statement found in wake-field literature is termed the *fundamental theorem of beam loading*. This theorem states that, for point-like beams traversing a linear structure which supports only one resonant electromagnetic mode, that the ratio of the maximum accelerating wake-field felt by the trailing beam to the decelerating field felt driving beam (the *transformer ratio r*) is less than or equal to two[1]. This factor of two arises from the fact that the driving beam, because of the causal nature of the excitation, feels only half of its own the longitudinal field. In addition, it can be shown that the energy transfer efficiency is maximized when the accelerating beam has charge equal to the driving beam, but at the price of reducing R to unity.

This theorem has a very restrictive range of applicability, however, and there are a variety of ways of making $R > 2$ if its assumptions are violated. The first method is to shape the current profile[4], to give it a long rise ($\sigma_+ \gg k^{-1}$), and a short fall ($\sigma_- < k^{-1}$). In this way one can obtain $R \simeq 2k\sigma_+$. Physically, this method is similar to stretching a spring adiabatically, and then releasing it suddenly, allowing it to go in to large amplitude oscillation Maximizing R in this way is also equivalent to minimizing the energy spread in the driving beam. Another alternative is to use a medium (*e.g.* low density plasma[5]) with a nonlinear response. In this way the impedance seen by the beam is a falling function of field amplitude, allowing the possibility of enhanced R.

Still another method of transformer ratio enhancement is to allow the driving beam and accelerating beam to traverse different paths. An example of this method is found in the radial transformer[6] studied at DESY, in which a "smoke-ring" beam excites a disk loaded structure, and the wake-field amplitudes are enhanced by the geometric compression of the electromagentic pulse as it propagates inward towards the axis. The proposed CLIC linear collider scheme[7] also uses an off-axis beam, driven efficiently by a pulse train (modulated in time at the rf frequency of the structure) which is accelerated in a low frequency superconducting linac. The wake-producing "transfer structure" is relatively low impedance, since BBU instabiltiy must be suppressed over the entire length of the linear collider.

Other examples of the use of parallel driving and acclerating beam paths in wake-field accelerators are the relativistic klystron two-beam accelerator, and the dielectric wake-field two-beam accelerator[8]-[10]. Aspects of the latter device was presented in some detail at the workshop, and we will return to discuss it below.

1.2   Wake-fields in Disk-loaded Structures

The Advanced Accelerator Test Facility (AATF) at Argonne National Laboratory was constructed to test a variety of wake-field acceleration concepts, including acceleration in a standard disk-loaded structure[11]. The AATF consists of a 21 MeV, short pulse ($\sigma_z \simeq 2$ mm) driving beam which contains up to 4 nC of charge, and a weak (pC) "witness" beam at 15 MeV, which traverses an adjustable delay transport line. When the two beams are combined to pass through a wake-field device on parallel trajectories, the witness beam may be delayed up to a nanosecond and offset transversely by several millimeters. In this way, the wake-fields due to the driving beam can be mapped out by measuring the witness beam energy changes and transverse deflections as a function of delay and offset.

The initial experiments at the AATF were performed on a disk-loaded metallic waveguide structure, and provided a good test of the standard theoretical and computational models of wake-fields in these devices. Recently, further tests were carried out on scaled SLAC structures with transverse mode damping slots. These tests showed that the Q of the dipole modes in these structures can be drastically reduced. In addition to this work on transverse mode-damping, at the workshop E. Chojnacki presented promising results on new results concern-

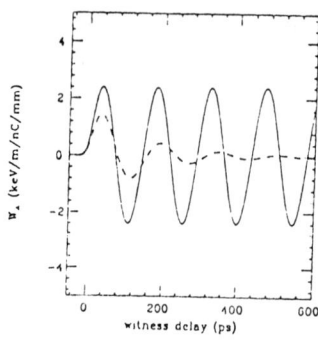

Calculated transverse wake fields. The broken
(solid) line refers to damped segmented (undamped uniform conductor) structure.

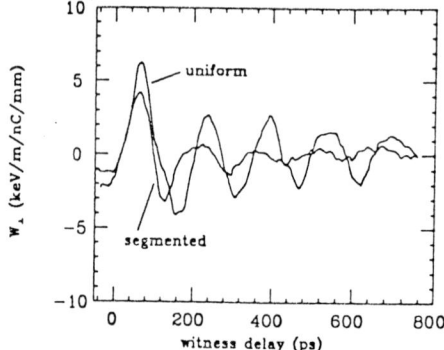

Measured transverse wake fields for both uniform outside conductor and segmented conductor dielectric
wake-field devices.

Figure 1: Longitudinal and transverse wake-fields in an undamped and damped
DWA tube.

ing mode-damping with absorbing material placed just behind narrow radial slots in nominally axisymmetric structures. Also, a new type of accelerating structure, the photonic band-gap structure (and its close cousin, the "onion-skin" structure), which should very effectively damp all higher order modes, was discused at the workshop. We will return to this subject in a later section.

1.3   Dielectric Wake-field Acceleration

In the Dielectric Wake-fied Accelerator (DWA) the electromagnetic wave is slowed down in a waveguide not by undulations in the wall, but by introduction of a dielectric liner. Initial tests of this scheme at the AATF verified the basic theory of this device, measuring a multi-mode wake-field maximum accelerating gradient of 0.5 MV/m with one picosecond resolutione[8].

The simplicity and flexibility of the DWA encouraged the Argonne accelerator research group to propose an upgraded facility, to make a 1 GeV staged (using up to six drive bunches derived from a single 150 MeV linac) demonstration device, the Argonne Wake-field Accelerator (AWA). For this device it is desired to achieve up to 100 MV/m acceleration gradient, and thus a short ($\sigma_z < 1.5$ mm), high current ($Q = 100$ nC) driving beam. The desired acceleration could be obtained in a collinear device, but the short range BBU would destroy the drive beam before its energy was expended, even assuming tight alignment tolerances.

The AWA design therefore calls for using a two-beam device, in which the drive beam traverses a parallel path in a dielectric lined pipe (wake tube) with relatively large inner diameter and group velocity of the desired mode. The rf power created in this wake tube is then transferred, through a quarter wavelength matching section, to an accelerating tube with smaller cross-sectional dimensions, and a higher dielectric constant. Compression of the transverese dimensions and lowering of the group velocity allows the electromagnetic energy density to be magnified in this section. Thus the accelerating gradient can be enhanced, and large transformer ratios achieved.

The dipole mode wake-fields generated in the wake tube section can be prevented from affecting the accelerating beam in this device in two ways. First, the matching section provides some rejection of modes other than the fundamental. Second, dipole modes can be damped, as suggested by Chojnacki[10], by making the outer boundary of the wake tube, instead of a solid metal wall, out of longitudinally oriented lengths of insulated wire, backed with microwave absorber. The wire allows the purely longitudinal currents associated with the monopole accelerating modes to travel unimpeded, while disrupting the azimuthal wall currents which support the dipole modes. The dipole modes radiate into the microwave absorber and are strongly damped. Some experimental results from the AATF are illustrated in Fig. 1, which shows damping of the dipole wake-field with almost no effect on the longitudinal wake-field.

The overall design and status of the AWA project was presented by J. Simpson in the plenary session at the workshop. The detailed design of the rf gun, preaccelerator, and the low energy beam dynamics were presented in the AWA was discussed by C. Ho in the working group. The results of this work give an appreciation of the intricacy of creating the source for an effective driving beam for a wake-field. The source for the CLIC two-beam wake-field accelerator, described by C. Johnson in the plenary session, also presents many interesting challenges in technological development. For more detail on these presentations, see the individual contributions to these proceedings. For further details on electron source development, also see the numerous contributions to these proceedings from the high-brightness source working group.

1.4   Plasma Wake-field Acceleration

The plasma wake-field accelerator (PWFA)[12] uses plasma as the wake-inducing medium. Since this provides a region of overlap with the plasma accelerator working group, a joint session was held with both the plasma and near field groups in attendance. In a PWFA[12], the beam excites the normal mode of the plasma, the electron plasma, or Langmuir, waves. These oscillations are *electrostatic* in the linear, small amplitude limit (they have no associated magnetic field), and have a frequency of $\omega_p = \sqrt{4\pi e^2 n_0/m_e}$, where n_0 is the unperturbed plasma density. Thus the plasma wake-fields are single-mode in this limit, where the perturbed electron density $n_1 \ll n_0$. Since the plasma electrons are in direct contact with the beam charge, the coupling of the beam to the plasma wave can approach the limit given by Eq. 4. The longitudinal wake-fields, unlike those in vacuum, have a strong variation in the transverse dimension. This nonlinear variation, which is very dependent on the exact configuration of the beam current, has a characteristic length scale of the beam size σ_r or the plasma skin depth c/ω_p, whichever is larger. From the Eq. 3, one can therefore deduce that the transverse wake-fields will be large and nonlinear. If the beam is long and narrow compared to c/ω_p, then the beam charge can be completely neutralized by the plasma electron response. Then the wake-fields are approximately equal to the magnetic self-fields of the beam, which is a very strong focusing force for a high current beam.

The initial tests of the PWFA were performed at the AATF, with a driving bunch of 2 nC, in a plasma of density $0.8 - 6 \times 10^{13}$ cm^{-3}. The wake-fields which were measured verified the predictions of linear theory, including the existence of strong transverse wake-fields within the driving beam[13]. Subsequent tests with twice the beam charge directly showed strong focusing of the driving beam within the plasma and concommitant enhancement of the plasma wave amplitude[14]. The amplitudes of the driven plasma waves in these experiments became nonlinear, and displayed characteristic steepened profiles. Experimentally measured nonlinear plasma wake-fields were found in these tests, and shown in Ref. [14]. This data provides a nice illustration of the Panofsky-Wenzel theorem, since the longitudinal derivative of the transverse wake-field was proportional to the longitudinal wake-field, even in the nonlinear regime. The maximum wake-field in

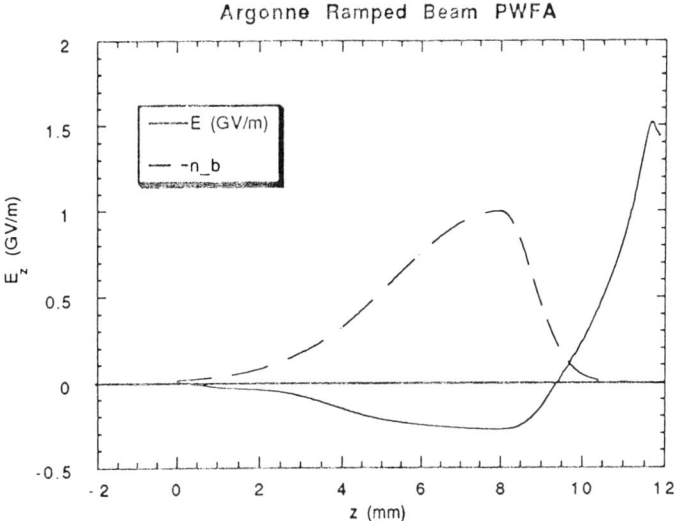

Figure 2: Plasma wake-fields the proposed Argonne nonlinear plasma wake-field acceleration experiment. Plasma density: $n_0 = 10^{14}$ cm^{-3}; beam parameters: $Q = 100$ nC, $\sigma_r = 130$ μm, rise, fall length $\sigma_{(+,-)} = 3$ mm.

these tests was approximately 6 MV/m.

Further tests of the PWFA were perfomred in Japan at KEK[15] using a pulse train of 6 bunches to resonantly excite wake-fields in a less than 10^{12} cm^{-3} plasma. No witness beam was used in these experiments – the wake-field response was scanned in frequency by varying the plasma density, and observing the energy change in each bunch. The maximum acceleration longitudinal field was about 20 MV/m in these experiments. These results were presented in the plenary session at the workshop, and more details can be found elsewhere in these proceedings.

In the future, the extremely high current (10 – 20 kA) pulse that the AWA is designed to provide is an ideal driver for a new regime of the plasma wake-field accelerator[16], in which the plasma electrons are completely ejected from the beam channel by the rising edge of the beam. Once the beam channel is rarefied of plasma electrons, the beam is contained in a cavity like nonlinear plasma wave. In this case the electromagnetic component of the wake-field gives a longitudinal field independent of radial position, and no net transverse wake-field, just as in a linac cavity. In addition, however, the electrostatic field of the ions provides a uniform, linear focusing force which is independent of longitudinal position. The acceleration and focusing characteristics of the PWFA in this limit are excellent, and not strongly dependent on the beam profile, in contrast to the linear regime. Also, this regime allows operation at lower plasma density, meaning a longer accelerating wavelength, and easing of source requirements and beam multiple scattering.

Entry into this regime requires that the focusing strength of the channel, the beam emittance and the current be such that the beam is self-consistently much more dense than the plasma over most of its length. This is the case for the AWA beam, which has a normalized emittance of 400 mm-mrad. In this extremely nonlinear regime of the PWFA, a ramped beam pulse can still be used to enhance the transformer ratio. An example of the expected performance of a proposed PWFA experiment which uses a ramped AWA pulse is shown in Fig. 2, for a plasma of density 10^{14} cm^{-3} and a beam rise/fall length of 3/1 mm. The transformer ratio in this case is $R > 5$, and the accelerating gradient is in excess of 1.5 GV/m, which is, remarkably, approximately the limit given by Eq. 4 assuming $k_0 = c/\omega_p$.

The general considerations concerning this nonlinear regime of the PWFA outlined above were presented by this author in the joint plasma/near-field session. A more detailed discussion of this subject can be found in Ref. [16]. Further work on the transverse dynamics of the driving beam in this regime of the PWFA was reported in the joint session by N.Barov. The next generation PWFA experiment at the AWA is expected to begin in earnest this year.

Additional experiments on the PWFA are planned at Novosibirsk, in which a modulated pulse train (similar in principle to the CLIC drive beam), derived from the BEPC storage ring, will be used to resonantly excite large amplitude plasma waves. This proposal was outlined at the workshop by B. Breizman, with details available eleswhere in these proceedings.

2   Novel Structures

Many new concepts in accelerating structures have been developed in recent years, and some of the most promising were discussed at the workshop. These concepts can be roughly divided into three categories: high frequency, high gradient structures, switched power devices, and structures with damped higher order modes. The high frequency structures have been investigated with an eye towards laser powering of the accelerator. The damped mode structures tend to be closer in frequency to standard linac structures, but may have unique geometrical attributes, and therefore electromagnetic mode characteristics, which set them apart.

2.1   High Frequency, High Gradient Structures

The only near-field laser-based acceleration concept to reported on at the workshop which has survived since the original workshop in this series is the "fox-hole" accelerating structure, an open structure which can be filled by a grazing incidence high power laser. Theoretical and computational results in support of the Brookhaven ATF experiment were reported by R. Fernow. This experiment will use a high power CO_2 ($\lambda = 10.6$ μm) laser to excite different types of fox-hole arrays, which act as low-Q (due to their open nature) high gradient accelerating structures. More details concerning these structures should be available elsewhere in these proceedings.

The use of such short wavelength structures raises, among other concerns, the issue of whether one can reliably fabricate these accelerators. F. Mills gave a presentation on application of the LIGA technique for microfabricating rf structures, showing that submicron tolerances may be met in this way.

Other high frequency structure experiments, performed at MIT, and reported by M. Conde, concerned filling a 33 GHz structure with a very short, high power pulse of rf derived from an FEL. Results were obtained on power matching, as well as dark current and x-ray production.

The use of very high gradients in accelerating structures also raises questions about the beam dynamics in such devices. For example, in open structures such as the fox-hole accelerator, there is often a field asymmetry which causes beam deflections. Even in axisymmetric devices, there can be strong effects induced by the transverse rf fields, which give very strong second order effects in the transverse motion. It is well known that for speed of light phase velocity structures that the accelerating field is independent of radius inside of the aperture, and thus (by the Panofsky-Wenzel theorem) the first order focusing vanishes. S. Hartman[17] showed in his presentation, however, that the second order focusing is very large in high gradient standing wave linacs, due to the alternating gradient force derived from the backward wave component of the rf. This effect can have major consequences in the rf photocathode sources, and in high gradient superconducting rf linear colliders. Along a similar line, A. Ruggiero gave a talk on use of backward travelling wave deflecting modes in cavities to create a strong electromagnetic undulator for generating radiation.

2.2   Damped Mode Structures

The subject of higher order mode control in linacs has attracted much attention, due to the constraints of multibunch BBU on accelerator performance. The solution to this problem has been two-fold: damping of higher order modes (HOMs), and frequency detuning of the HOMs along the structure. Only the direct damping of HOMs was discussed in this working group. As mentioned above, Chojnacki presented a fairly standard disk-loaded structure with HOM damping added as a perturbation. On the other hand some new structures were discussed which are much more exotic, and preclude high-Q HOMs by the very geometry.

Several new ideas of this sort utilize dielectrics to support a high-Q accelerating mode, while allowing the HOMs to travel away from the beam axis, hopefully minimizing the effects of BBU. N. Kroll spoke on the subject of a "photonic band-gap" accelerator structure. This device is shown in Fig. 3, and consists of an array of high ϵ dielectric cylinders which terminate on superconducting plates. In this structure, the electromagnetic wave spectrum has pass-bands in

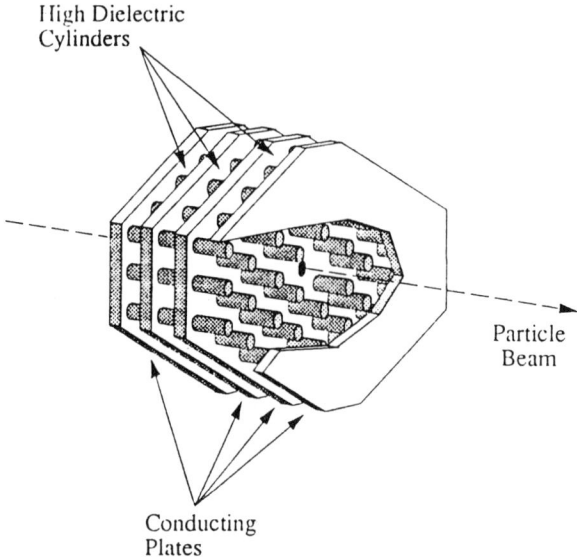

Figure 3: A photonic band-gap accelerating structure, with dielectric rods and superconducting plates.

transverse propagation which arise from constructive interference of the Bragg scattered waves from each dielectric cylinder. If one removes the on-axis cylinder, a "defect" accelerating mode can be trapped near the axis – the photonic band gap mode. This mode may, given sufficient ingenuity, be the only trapped mode, which means that effectively only the fundamental accelerating mode has a high Q. All other transverse and longitudinal modes will be damped, which has positive implications for multibunch operation.

Another accelerator structure based on interference effects from dielectric components is the "onion-skin" structure, which has a series of concentric dielectric shells of varying ϵ surrounding the beam hole. R. Cooper presented some work that had been done at Los Alamos on this structure, and pointed out the conceptual similarities to the photonic band gap structure.

Both of these structures are dependent on use of a small loss tangent dielectric such as sapphire to obtain high Q operation. Since the low frequency as well as the high frequency conductivity of these materials is tends to be low, charging from the beam spray could be a serious problem. This difficulty has been observed in wake-field experiments at Argonne. To alleviate this, a higher low frequency conductivity is introduced to the dielectric material. While the high frequency conductivity also suffers from this change, a wake-field structure can easily tolerate a lowered Q, since it is filled directly by the driving beam.

2.3   Switched Power Structures

The idea of using a switched DC voltage, which is transformed upwards by use of geometrical compression (*e.g.*a radial transmission line with a beam hole at the center) has been pursued for a number of years now. The proof-of-principle of such a scheme, using picosecond laser pulses, was performed at the University of Rochester. W. Donaldson reported on these technologically challenging experiments at the workshop, in which the azimuthally symmetric swithing was obtained to a good degree, and photoinjected electrons were accelerated.

While the results of the proof-of-principle experiment were impressive, it was also pointed out by Donaldson that the scheme is very complicated and hardware intensive. It is thus not a likely to be scaled up into a long, multi-cell linac structure without some major changes in the conceptual approach. More details on this work are available elsewhere in the proceedings.

References

[1] S.A. Heifets and S.A. Kheifets, *Rev. Mod. Phys.* **63**, 631 (1991).

[2] W.K.H. Panofsky and W. A. Wenzel, *Rev. Sci. Instr.* **27**, 967 (1956).

[3] Donald H. Perkins, *Introduction to High Energy Physics*, (Addison-Wesley, 1987).

[4] K.L.F. Bane, P. Chen, P.B. Wilson, *IEEE Trans. Nucl. Sci.* **32**, 3524 (1985)

[5] J. B. Rosenzweig, *Phys. Rev. Letters*, **58** (1987) 555.

[6] W. Bialowons, *et al.*, *Proc. Europ. Part. Accelerator Conf.* **1**, Ed. S. Tazzari, 490 (World Scientific, 1989).

[7] W. Schnell, CERN-LEP/88-59, CLIC Note 85 (Geneva, 1988).

[8] W. Gai, P. Schoessow, B. Cole, R. Konecny, J. Norem, J. Rosenzweig, and J. Simpson, *Phys. Rev. Lett.* **61**, 24 (1988).

[9] M. Rosing and W. Gai, *Phys. Rev. D* **42**,1829 (1990).

[10] E. Chojnacki, W. Gai, C. Ho, R. Konecny, S. Mtingwa, P. Schoessow, *J. Appl. Phys.* **69**, (1991).

[11] H. Figueroa, W. Gai, R. Konecny, J. Norem, P. Schoessow, and J. Simpson, Phys. Rev. Lett.,**60**, 2144 (1988).

[12] P. Chen, J. M. Dawson, R. W. Huff, and T. Katsouleas, *Phys. Rev. Lett.*, **54**, 693 (1985).

[13] J.B. Rosenzweig, D. Cline, B. Cole, H. Figueroa, W. Gai, R. Konecny, J. Norem, P. Schoessow, and J. Simpson, *Phys. Rev. Lett.*,**61,** 98 (1988).

[14] J. B. Rosenzweig, P. Schoessow, B. Cole, W. Gai, R. Konecny, J. Norem and J. Simpson, *Phys. Rev. A,* **39,** 1586 (1989).

[15] K. Nakanishi, *et al.*, *Nucl. Instr. Meth. A* **292,** 12 (1990).

[16] J.B. Rosenzweig, T. Katsouleas, and J.J. Su, *Phys. Rev. A* **44,** R6189 (1991).

[17] S.C. Hartman and J.B. Rosenzweig,to be published in *Phys. Rev. E,* March 1993.

PHOTONIC BAND GAP STRUCTURES:
A NEW APPROACH TO ACCELERATOR CAVITIES

N. Kroll,* D. R. Smith, and S. Schultz

Department of Physics, University of California, San Diego
9500 Gilman Drive, La Jolla, California 92093-0319

ABSTRACT

We introduce a new accelerator cavity design based on Photonic Band Gap (PBG) structures. The PBG cavity consists of a two-dimensional periodic array of high dielectric, low loss cylinders with a single removal defect, bounded on top and bottom by conducting sheets. We present the results of both numerical simulations and experimental measurements on the PBG cavity.

INTRODUCTION

We consider here a new class of resonant cavities which we are investigating in the context of what has come to be termed Photonic Band Gap (PBG) structures.[1,2] The concepts underlying the PBG condition, and the experimental and numerical confirmation of typical resonant modes, will be presented later in this paper. The configurations we have investigated are comprised of a periodic array of short dielectric cylinders, which are bounded on top and bottom by a metal sheet, and from which one cylinder has been removed at the central site. We find that the electromagnetic modes associated with these configurations are analogous to resonant cavity modes, with the E field polarized perpendicular to the metal sheets. Thus, if an aperture is placed in each metal sheet at the site of the missing cylinder, we can envision an electron beam entering through one plate, being accelerated by the parallel rf E field, and then emerging from the other plate. A suitable three-section coupled set of such structures could, for example, form the familiar 2π basic accelerator unit with $2\pi/3$ phase advance per section (Fig. 1).

The reasons why we believe this new class of structures may be particularly suited for accelerators are:

(i) They have a set of resonant frequencies which are entirely different from those associated with the usual cavity structures, and therefore may be much better suited to reducing the higher order mode problem. Indeed, it may be possible to arrange that there will be only one trapped mode.

(ii) The spatial distribution of the E field for the resonant mode falls off exponentially from the center of the active region, thus minimizing constraints on the boundary material.

(iii) The structures may be readily fabricated utilizing sapphire cylinders and superconducting niobium plates which will result in Q values $>10^6$. We note that the only superconducting material required is in the form of a flat sheet, with no bends, joints, or welds, a circumstance which may mitigate limitations of acceleration gradient associated with the superconducting surface.

*Also Stanford Linear Accelerator Center, Stanford University, Stanford CA 94309

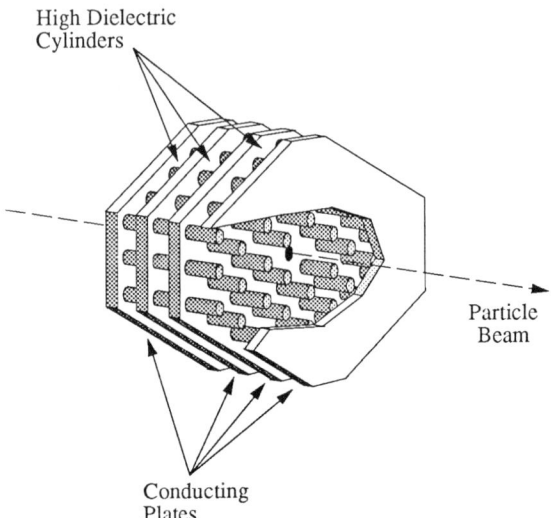

High Dielectric
Cylinders

Particle
Beam

Conducting
Plates

Fig. 1. A schematic view of the proposed 2π accelerator unit. In this example the unit consists of three triangular photonic lattices, separated by superconducting sheets. Each of the lattices has a cylinder removed to allow the formation of a defect mode with an electric field maximum in the center. Holes drilled through the conducting plates would allow a particle beam to be accelerated through the unit.

We envisage a number of potential problems with the proposed structures. In particular, there is the concern that dielectric material (such as pure sapphire) located inside the vacuum, and at a distance of ~1 cm from the electron beam will exhibit charging, multipaction, and/or dielectric breakdown. Furthermore, we have not studied problems which may be associated with junctions between the dielectric and the superconducting sheets. It will require further numerical simulation and the appropriate experimental testing to establish the optimum configurations and validity of this approach.

In Section I we present a physical explanation of the PBG concept. In Section II we discuss our experimental equipment and techniques. In Section III we discuss the methodology that has been formulated for the numerical determination of the electromagnetic "band structure", and also calculation of the defect states. In Section IV we present some of the experimental results which illustrate the key properties of selected PBG-defect structures. Some concluding discussion and remarks are given in Section V.

I. Introduction to the PBG Concept

It is simplest to introduce the PBG concept in the context of wave propagation in physical systems which have a periodic structure along some specified spatial direction. Well known examples are the wave function of a particle moving in a periodic potential, mode propagation in periodic linear accelerator structures, and a waveguide of uniform cross section filled with dielectric with permittivity varying periodically along the waveguide axis. For concreteness we fix our attention on the last of these and recall that such systems can be discussed in terms of modes which are fully characterized by the transverse variation of their electromagnetic fields along the waveguide cross section. Fixing our attention on a particular mode we note that such a structure will exhibit an alternating series of stop bands and pass bands as the frequency is varied.[3] In the language of the PBG community the stop bands are referred to as photonic band gaps; photonic because we are dealing with Maxwell's equations (not entirely logical because we are dealing with them classically) and bands gaps in analogy with the language used to describe electronic states in the periodic potentials of crystalline solids. A wave propagating in a dielectric free section of the waveguide which encounters a long section of the periodic structure will be totally reflected if its frequency lies within one of the band gaps. Thus, if one interrupts the periodicity by removing a section of the periodically varying dielectric, there is the possibility of a trapped mode in which a wave whose frequency lies in the band gap is reflected back and forth between the two dielectric filled sections; if the multiply reflected waves are in phase with one another then the possibility is realized. The resultant trapped mode is referred to as a defect state, associated with the fact that the gap in the periodicity is referred to as a defect, both in analogy with the terminology applied to electronic states in solids.

Before proceeding to a discussion of the case of periodicity in more than one dimension, we briefly mention an extension of the above discussion to the case of structures which are symmetric under rotation about an axis with dielectric constant which varies as a function of the radial distance r from the axis.[4] For example, one might consider a set of contiguous concentric tubes of alternating dielectric constant. It is easy to show that one can choose the radii of the cylinders so as to have bands of frequencies that are non-propagating for rotationally symmetric radial waves. By omitting an appropriate amount of the central section one can again have radial waves with frequency in the gap which experience multiple reflections from the surrounding structure. Again, if the dimensions are such that the multiple reflections are in phase, one has a trapped mode. One can readily envisage a mode of this sort which would be quite suitable for particle acceleration. We refer to such configurations as "radial band gap" structures with radial band gap modes.

We turn now to consideration of structures with two-dimensional periodicity of the sort discussed in the introduction. We first focus our attention on the case of an infinite periodic lattice. The solutions of Maxwell's equations can be chosen such that each component has the form $\exp(i\bar{k}\cdot\bar{R})F(\bar{R})G(z)$ where the functions F have the periodicity of the lattice. Here R is a two-dimensional position vector in the plane of periodicity and z the coordinate perpendicular to it. The two-dimensional vector k is referred to as the wave vector and may be thought of as residing in the reciprocal lattice space. For each value of k there are a discrete set of solutions with a discrete set of frequencies $\{\omega_n(k)\}$. The band index n refers to the particular discrete solution. The $\{\omega_n\}$, which may be thought of as two-dimensional surfaces in (k, ω) space, have the double periodicity of the reciprocal lattice. Each ω_n and the associated solutions represents an allowed frequency band of propagating waves. The frequency span associated with each band surface may or may not overlap that associated with other surfaces. Hence, in contrast with the one-dimensional case, there may or may not be "band gaps", that is, frequency ranges for which no propagating solutions exist. As the name suggests, a PBG system is an array which does contain band gaps.

A "defect", formed, say by removing one of the cylinders, presents the same possibility that we noted in the one-dimensional situation, namely a superposition of plane waves with frequency in the gap which are reflected into one another when they try to penetrate the surrounding lattice. For particular geometries and frequency these multiple reflections may be in-phase so as to allow them to combine to form a trapped mode, analogous to the localized electronic defect states familiar in semiconductor physics.

Experimental and theoretical studies of band structure and trapped localized defect modes in two-dimensional structures will be presented in the next sections.

II. EXPERIMENTAL APPARATUS

In order to experimentally study the band gap structure and associated defect modes, we utilize a microwave scattering chamber,[4] which enables us to make mappings of the standing wave defect modes, and also measure the microwave transmission through various lattices of dielectric cylinders. The interior of the scattering chamber is 1 cm high, 46 cm wide, and 51 cm long. The bottom and side walls of the chamber are machined out of a solid aluminum plate. On both ends of the chamber are standard 8-12 GHz waveguide fittings which can be used to detect or inject microwaves in the chamber via a tapered region integrally machined into the main plate. An aluminum cover plate, free to move laterally, completes the chamber. The scatterers inside the chamber are typically cylinders with height of 1 cm and a variety of radii and dielectric constants. Accurate lattices can be constructed by placing the cylinders into a precision drilled styrofoam (dielectric constant of 1.03) template. Finally, a thick layer of low density absorber is placed between the interior chamber walls and the styrofoam template, which serves to minimize reflection.

In conjunction with an HP network analyzer we are able to sweep the microwave frequency and make measurements of the power transmitted through the scattering region. We are also able to map the spatial structure of standing wave modes (e.g., defect modes) by weakly coupling to a tuned probe through any of a lattice of small holes in the cover plate. Using standard homodyne techniques, we can measure both the phase and amplitudes of the fields sampled by the probe. By mapping the Mie resonances associated with scattering from a single cylinder, we have found that the probe does not significantly perturb the system and that the chamber is adequately terminated.

III. THEORETICAL CALCULATIONS

A. Bandstructure

A great deal of work has recently been done to compute the band structure for a lattice of infinitely long dielectric cylinders.[5,6,7] These calculations have mainly been concerned with the specific case of waves propagating perpendicular to the cylinder axes; that is to say, modes for which the fields are independent of z, where z is the coordinate axis parallel to the cylinder axes. For this case modes with electric field polarization parallel and perpendicular to the cylindrical axis propagate independently. Equivalently, the modes can be characterized as transverse magnetic (TM) and transverse electric (TE) with respect to the z-axis. In order to discuss the full set of modes for the accelerator cavity which we have in mind, it is necessary to extend these calculations to include arbitrary propagation direction so that fields are assumed to have an exp(iqz) variation, q being the wavevector directed along the cylinder axes. This causes the TE and TM modes to hybridize and to lead to a coupled pair of second order 2-D partial differential equations instead of the uncoupled pair which occur for the q=0 case.

Maxwell's equations for the lattice of dielectric cylinders, taken to be infinite in the transverse directions, are

$$\vec{\nabla} \times \vec{E} = -i\frac{\omega}{c}\mu\vec{H} \tag{1}$$

$$\vec{\nabla} \times \vec{H} = i\frac{\omega}{c}\varepsilon\vec{E} \tag{2}$$

where $\varepsilon(x,y)$ is the periodic dielectric function representing the cylinders, and $\mu=1$ everywhere. We have assumed a time dependence of $e^{i\omega t}$. If we further assume an explicit dependence on the longitudinal (z) component such that

$$\vec{E} = \left(\vec{E}_t + \hat{z}E\right)e^{-iqz} \tag{3}$$

$$\vec{H} = \left(\vec{H}_t + \hat{z}H\right)e^{-iqz} \tag{4}$$

then Eqs. (1) and (2) yield

$$-iq\hat{z} \times \vec{E}_t - \hat{z} \times \vec{\nabla}E = -i\frac{\omega}{c}\mu\vec{H}_t \tag{5}$$

$$\vec{\nabla} \times \vec{E}_t = -i\frac{\omega}{c}\mu H\hat{z} \tag{6}$$

$$-iq\hat{z} \times \vec{H}_t - \hat{z} \times \vec{\nabla}H = i\frac{\omega}{c}\varepsilon\vec{E}_t \tag{7}$$

$$\vec{\nabla} \times \vec{H}_t = i\frac{\omega}{c}\varepsilon E\hat{z} \tag{8}$$

Eqs. (5) through (8) can be combined to yield

$$\vec{\nabla} \cdot \varepsilon\alpha\vec{\nabla}E + \varepsilon\kappa^2 = \beta\hat{z} \cdot \vec{\nabla} \times \alpha\vec{\nabla}H \tag{9}$$

$$\vec{\nabla} \cdot \alpha\vec{\nabla}H + \kappa^2 = -\beta\hat{z} \cdot \vec{\nabla} \times \alpha\vec{\nabla}E \tag{10}$$

where

$$\kappa^2 = \frac{\omega^2}{c^2} - q^2 \tag{11}$$

$$\beta = \frac{qc}{\omega} \tag{12}$$

$$\alpha = \frac{1 - \beta^2}{\varepsilon - \beta^2} \tag{13}$$

The transverse field components have been eliminated in Eqs. (9) and (10), and we are thus left with coupled linear equations for the z field components $E(x,y)$ and $H(x,y)$ with parameter β.

Since $\varepsilon(x,y)$ is invariant under translation by any distance composed of integral multiples of a lattice constant, we may expand it in an infinite sum over reciprocal lattice vectors as

$$\varepsilon(\vec{x}) = \sum_{\vec{G}} \varepsilon_{\vec{G}} e^{i\vec{G}\cdot\vec{x}} \tag{14}$$

where

$$\vec{G} \cdot \vec{R} = 2\pi n \qquad n = 0, \pm 1, \pm 2, \ldots \tag{15}$$

and

$$\tilde{\varepsilon}_{\vec{G}} = \delta_{\vec{G},0} + (\varepsilon_a - 1)f \frac{2J_1(|\vec{G}|a)}{|\vec{G}|a} \tag{16}$$

f is the filling factor, ε_a is the cylinder dielectric constant, a is the cylinder radius, and R is a translation vector. $\alpha(x,y)$ can be expanded similarly. (For future reference we note that $1/\varepsilon(x,y)$, which we will need later, may also be expanded in this way.) The translational invariance also restricts the field solutions of Eqs. (9) and (10) to satisfy

$$E(\vec{x} + \vec{R}) = e^{i\vec{k}\cdot\vec{R}} E(\vec{x}) \tag{17}$$

$$H(\vec{x} + \vec{R}) = e^{i\vec{k}\cdot\vec{R}} H(\vec{x}) \tag{18}$$

We may thus expand the field solutions in terms of reciprocal lattice vectors as

$$E(\bar{x}) = \sum_{\bar{G}} E_{\bar{G}} e^{i(\bar{k}+\bar{G})\cdot\bar{x}} \tag{19}$$

$$H(\bar{x}) = \sum_{\bar{G}} H_{\bar{G}} e^{i(\bar{k}+\bar{G})\cdot\bar{x}} \tag{20}$$

k is known as the wave vector, or crystal momentum in solid state physics; for an infinite system with translational symmetry, k indexes the allowed eigenmodes, which are known as Bloch states, or running wave states. (We are thinking of the states as Bloch states in 2-D modulated by exp(iqz).) Substituting all the above expansions into Eqs. (9) and (10), we arrive at the following coupled equations:

$$\sum_{\bar{G}'} \tilde{\alpha}_{\bar{G}-\bar{G}'} \left[\left(\bar{k}+\bar{G}\right)\cdot\left(\bar{k}+\bar{G}'\right)\tilde{H}_{\bar{G}'} + \beta\hat{z}\cdot\left(\bar{k}+\bar{G}\right)\times\left(\bar{k}+\bar{G}'\right)\tilde{E}_{\bar{G}'} \right] = \kappa^2 \tilde{H}_{\bar{G}} \tag{21}$$

$$\sum_{\bar{G}'} \varepsilon\alpha_{\bar{G}-\bar{G}'} \left(\bar{k}+\bar{G}\right)\cdot\left(\bar{k}+\bar{G}'\right)\tilde{E}_{\bar{G}'} - \beta\hat{z}\cdot\left(\bar{k}+\bar{G}\right)\times\left(\bar{k}+\bar{G}'\right)\tilde{\alpha}_{\bar{G}-\bar{G}'}\tilde{H}_{\bar{G}'} = \kappa^2 \sum_{\bar{G}'} \varepsilon_{\bar{G}-\bar{G}'}\tilde{E}_{\bar{G}'} \tag{22}$$

Eqs. (21) and (22) constitute a generalized eigenvalue equation of the form

$$\underline{M}\begin{pmatrix} \tilde{E} \\ \tilde{H} \end{pmatrix} = \kappa^2 \underline{N}\begin{pmatrix} \tilde{E} \\ \tilde{H} \end{pmatrix} \tag{23}$$

Equation (23) can be brought to a symmetric, standard eigenvalue equation by making the substitution

$$\tilde{E}_{\bar{G}} = \left[\frac{1}{\sqrt{\varepsilon}}\right]_{\bar{G}-\bar{G}'} \tilde{E}'_{\bar{G}'} \tag{24}$$

For the case when $\beta=0$, Eqs. (21) and (22) are uncoupled, and using Eq. (24) reduce to the matrix equations for the unform polarizations reported previously.

The results of calculations for a square lattice are shown in Fig. 2(a)-(d), where we include the bandstructures for a selection of different β values. For the $\beta=0$ case we plot the TM and TE modes separately. We note that for this particular example the $\beta=0$ TM mode band structure exhibits band gaps for all directions of k-vector, but that the $\beta=0$ TE mode band structure does not. Since the TE and TM modes hybridize for non-zero β, it is perhaps not surprising that none of the hybrid cases investigated show gaps. Even when both TE and TM bandstructures show gaps, the frequency range of the gaps may not overlap. Should this be the case it is likely that the gaps will be absent for non-zero β cases.

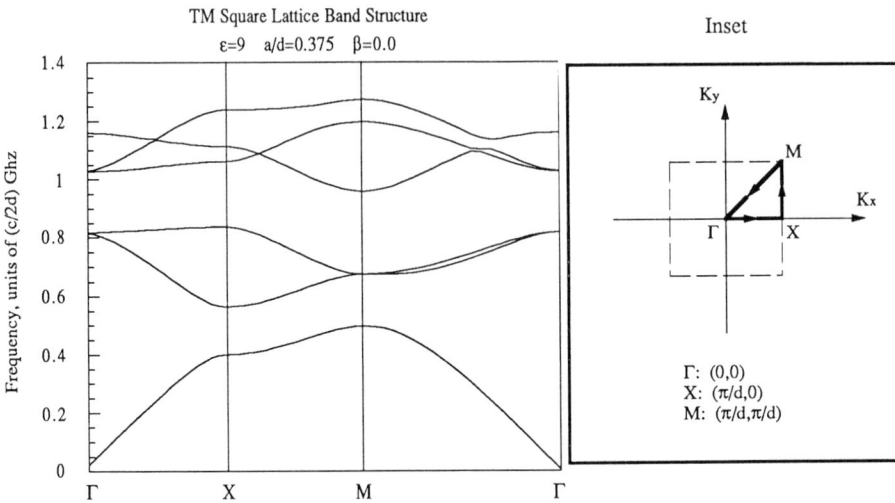

Fig. 2. (Inset) Because of the periodicity of the lattice, we may restrict our attention
to frequencies corresponding to a minimum set of k vectors. For a square lattice this
set, shown in the inset within the dashed lines, comprises also a square in reciprocal
lattice space. Furthermore, because the real lattice has fourfold rotational and
reflection symmetries, only the solutions for a single octant of this square are unique.
A plot of the frequencies corresponding to k vectors along the boundary of the octant
forms the "band structure". It is expected that each frequency span coincides with
that of the full $\omega_n(k)$ surface. (a) Band structure for the square photonic lattice. In
this case, the electric field is parallel to the cylinder axes (TM). There are two band
gaps for this polarization evident for the given frequency range. 109 reciprocal
lattice vectors were used to obtain the band structure plot.

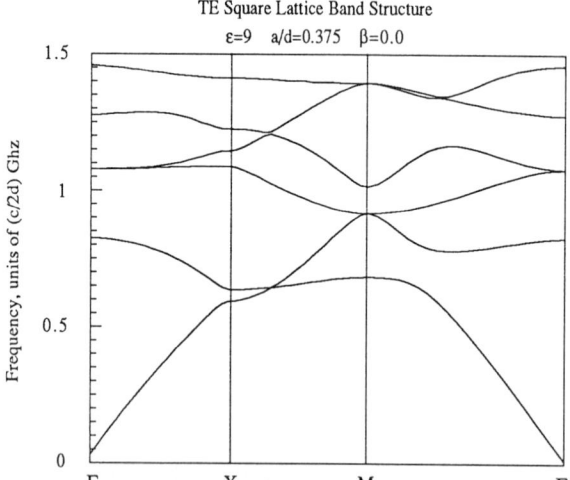

Fig. 2. (b) Here, the electric field is polarized perpendicular to the cylinder axes.
There are no gaps in this case.

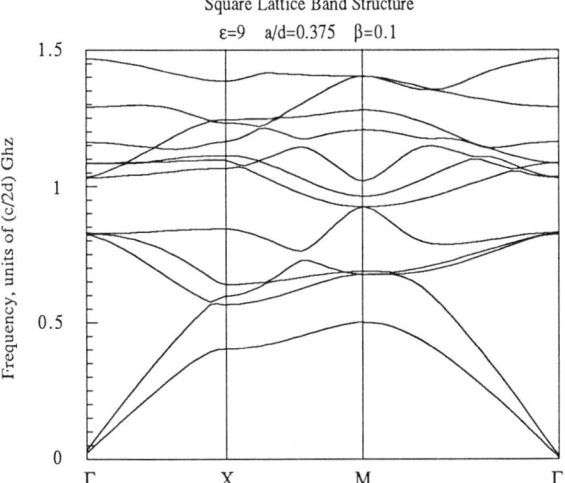

Fig.2 (c) The square lattice band structure computed with β = 0.1.

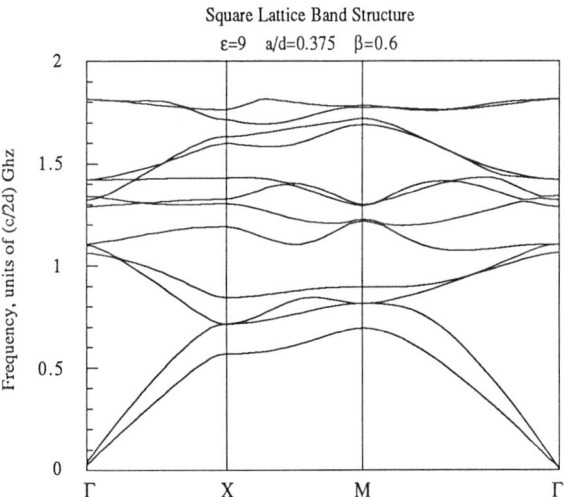

Fig. 2 (d) The square latticer band structure computed with β = 0.6.

B. Defect Studies

When the translational symmetry of the lattice is violated by, for example, the addition of some impurity, previously forbidden states with imaginary values of the Bloch vector k are now allowed. Thus, in addition to the extended periodic running wave states which comprise the band structures for infinite lattices, there may also be states which decay away exponentially from the perturbed site. If the perturbation consists of a single cylinder removal, the associated mode is a defect mode. In 1-D it is possible to analyze defect states by directly assuming that they are formed from Bloch waves with imaginary values of k.[3]

In higher dimensions the situation is more complicated, and we calculate the properties of impurity modes by expanding the new eigenfunctions in terms of the infinite lattice Bloch functions. Writing the dielectric function as

$$\varepsilon(\vec{x}) = \varepsilon_{per}(\vec{x}) + \varepsilon_{def}(\vec{x}) \tag{25}$$

where we have separated the periodic variation from the local defect function, we may write Eq.(9) as

$$\nabla^2 E(\vec{x}) + \frac{\omega^2}{c^2}\left(\varepsilon_{per}(\vec{x}) - 1\right)E(\vec{x}) + \frac{\omega^2}{c^2}E(\vec{x}) = -\frac{\omega^2}{c^2}\varepsilon_{def}(\vec{x})E(\vec{x}) \tag{26}$$

To arrive at Eq. (26) we set $\beta=0$, and concern ourselves only with the TM mode. We do this primarily because we do not have at hand a non-zero β case for which a gap is present. Since the TM mode has E parallel to the cylinder axis, this is the mode of primary interest for accelerator applications. Because of the finite spacing between plates for the accelerator cavity, only the TM modes exist for $\beta=0$ for relevent frequencies.

Expanding the solution in terms of the Bloch waves found previously,

$$\vec{E}(\vec{x}) = \sum_{\vec{k},n} f_{\vec{k},n} E_{\vec{k},n}(\vec{x}) \tag{27}$$

and using the orthogonality relation

$$\int E^*_{\vec{k},n}(\vec{x})\varepsilon_{per}(\vec{x})E_{\vec{k}',n'}dV = C_{\vec{k},n}\delta_{\vec{k}\vec{k}'}\delta_{nn'} \tag{28}$$

we find the eigenvalue equation

$$\sum_{\vec{k}',n'}\left[\int_{\substack{\text{defect}\\\text{volume}}}dV E^*_{\vec{k},n}(\vec{x})\varepsilon_{def}(\vec{x})E_{\vec{k}',n'}\right]f_{\vec{k}',n'} = \left[\frac{\omega^2_{\vec{k},n} - \omega^2}{\omega^2}\right]f_{\vec{k},n} \tag{29}$$

where n is the band index, and the normalization constant has been taken into the Bloch functions. Eq. (29) can be manipulated into a symmetric eigenvalue equation as

$$\sum_{\vec{k}'} \left[\frac{\delta_{\vec{k}\vec{k}'}}{\omega_{\vec{k}}^2} + \frac{M_{\vec{k}\vec{k}'}}{\omega_{\vec{k}}\omega_{\vec{k}'}} \right] \alpha_{\vec{k}'} = \frac{\alpha_{\vec{k}}}{\omega^2} \tag{30}$$

where

$$M_{\vec{k}\vec{k}'} = \sum_{\vec{G}} \sum_{\vec{G}'} \varepsilon_{\text{def}} f \left[\frac{J_1 \left(\left| \vec{G} - \vec{G}' + \vec{k} - \vec{k}' \right| a \right)}{\left| \vec{G} - \vec{G}' + \vec{k} - \vec{k}' \right| a / 2} \right] E_{\vec{G}}^{\vec{k}} E_{\vec{G}'}^{\vec{k}'} \tag{31}$$

and

$$\alpha_{\vec{k}} = \omega_{\vec{k}} f_{\vec{k}} \tag{32}$$

We can use Eq. (30) to numerically calculate the modified band structure due to the addition of a perturbation in dielectric constant to an otherwise perfect lattice. In order to follow this procedure, however, we must first choose a finite set of k values for the sum in Eq. (27). Choosing a set of N k values effectively constrains the size of the lattice to N sites; it is this sublattice, N sites plus the defect, that constitutes the new unit cell of an overall super-lattice. We expect that for bound modes in which the fields fall off exponentially from the center of a perturbation, reasonable numerical convergence may be achieved with a fairly small number N of sites.

To further reduce the computational task in the numerical calculation we can make use of any discrete rotational symmetry possessed by the impurity dielectric function. In the case of a defect in a square lattice, for example, there are three distinct mode types; monopole, dipole, and quadrupole. An accelerator mode must, of course, have monopole character.

We have completed calculations for defects in square and triangular lattices utilizing the above procedure. We find, however, that the calculated defect frequency is very sensitive to the number of reciprocal lattice vectors (RLV's) used in the expansions in Eqs. (14) and (19). While the band structure calculation exhibits reasonable convergence for the lower bands using only about one hundred RLV's, it appears that the equivalent convergence for the defect mode requires a much greater number of RLV's. Using an alternate technique, Meade et al.[8] have succeeded in calculating the frequency dependence of the defect modes in a square photonic lattice; their results agree very well with previously reported measurements.[5]

IV. EXPERIMENTAL RESULTS

The focus of our experimental investigations has been to verify the various theoretical predictions resulting from the numerical calculations. Once the calculations demonstrate the necessary accuracy, it will then be possible to search for the optimal parameters for an accelerator structure. In Fig. 3(a) we present transmission data for the triangular lattice, verifying the predicted second band gap. In Fig. 3(b) we present the transmission data for a triangular lattice with a defect (a single cylinder removed). Note the presence of the new peak in the forbidden region corresponding to the defect mode. The frequency width of the defect mode, zero in principle for an infinite lattice, in this case is set by the cylinder and boundary plate losses, in addition to the finite size of the lattice.

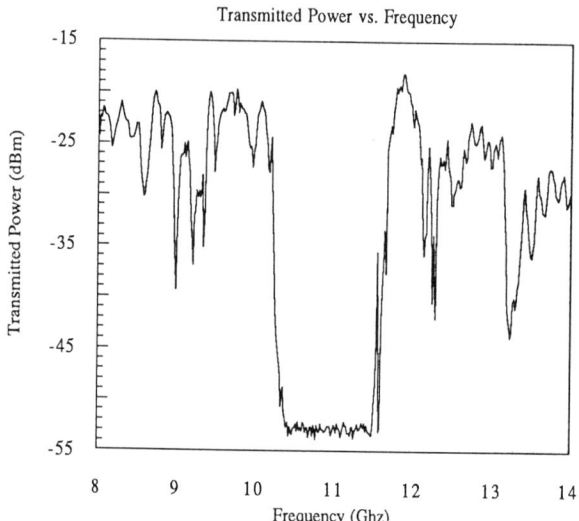

Fig. 3 (a): Transmitted power vs. frequency for a triangular photonic lattice. The lattice was composed of 200 dielectric cylinders with ε=9.0 in a styrofoam template with ε=1.03. The lattice spacing was 1.27 cm. In a transmission experiment, incident waves with frequencies corresponding to the forbidden band gap frequencies are exponentially attenuated across the lattice; thus, the sharp dip in transmitted power provides a measurement of a band gap from 10- 11.5 GHz.

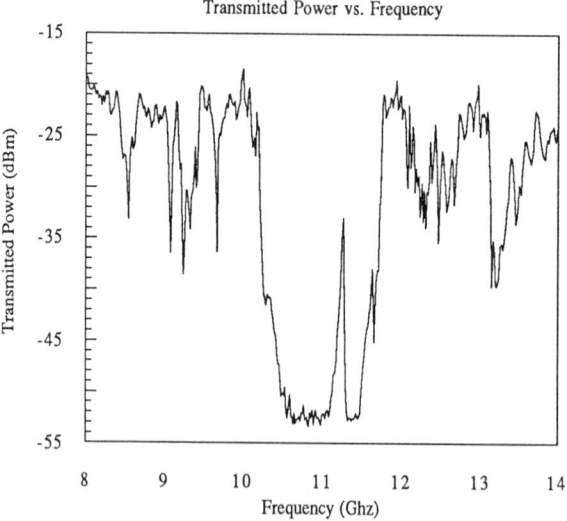

Fig. 3 (b) Transmitted power vs. frequency for a triangular lattice with a single cylinder removed. Note the new peak occurring in the band gap region corresponding to a defect mode.

In Fig. 4 we show spatial mappings of the defect modes for square and triangular lattices with the same dielectric and lattice constants. We see that the triangular mode falls off much more rapidly away from the defect center than does the square lattice mode, a feature suggested by our computer simulations. The defect modes in both cases are monopole in character.

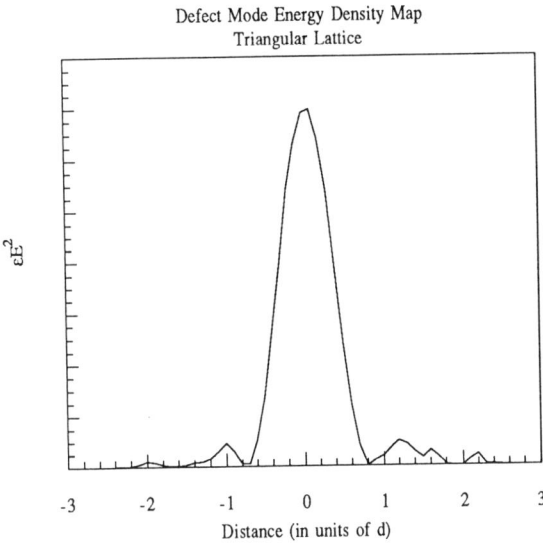

Defect Mode Energy Density Map
Triangular Lattice

Fig. 4 (a) A map of the electric energy density for a defect mode in the second gap of the triangular lattice (d = 1.27 cm).

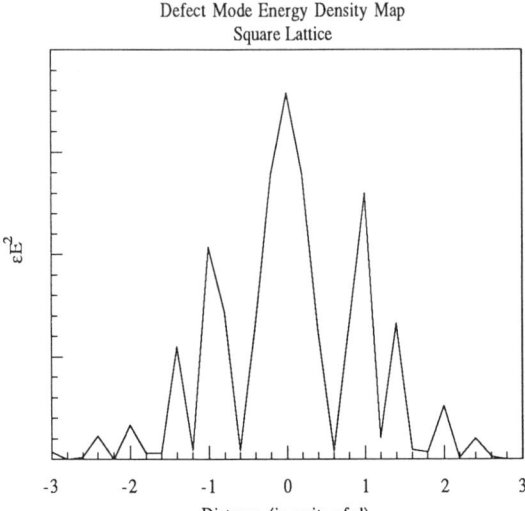

Defect Mode Energy Density Map
Square Lattice

Fig. 4 (b) A map of the electric energy density for a defect mode in the second gap of the square lattice (d = 1.56 cm).

V. CONCLUDING DISCUSSION AND REMARKS

In order to establish the utility of PBG structures as accelerator cavities, a great deal more work needs to be done. We discuss in this section a number of issues (in addition to those mentioned in the introduction) with which we are currently concerned.

While we have shown the existence of trapped modes whose configuration appears to be suitable for particle acceleration, we need to investigate the question of higher order modes. It appears from the work already done that the existence of band gaps is exceptional rather than typical. We are hopeful, therefore, that structures can be found which exhibit a single band gap. Because the frequency range of the gap is limited, it seems likely that the number of trapped modes present in any gap is small, and our results seem to imply that situations can be found in which the gap contains only a single such mode. Given that this is the case, there remains the possibility that there will also exist *quasi* localized modes in the allowed frequency ranges. Such modes would be the analogue of the damped resonances of waveguide loaded conventional cavities. To see how such modes might arise, recall that we found no band gaps for the finite q case for the square lattice. It is, however, very likely that a TM band gap mode is only slightly shifted and weakly coupled to propagating TE modes by a finite but small value of q, leading to a mode of finite but high Q_{ext}. Such modes may also arise under circumstances in which the overlap range of two gaps is small. In this case propagation may occur only over a very limited range of angles in wave vector space. Localized modes with finite Q_{ext} may well occur in such regions, the narrow angle directions of propagation acting like a set of waveguides loading the cavity

In the text above we have only considered the use of metallic end walls as terminations of the PBG cavity in the z direction. Mueller, et al.[9] have suggested using the Clogston[10] layered conductor scheme as a means of reducing end wall losses without resorting to superconductivity. We also plan to investigate the utility of the Zakowicz[11] dielectric coating scheme. We also need to investigate cavity-to-cavity coupling.

Apropos of the cavity coupling question, we are also considering the design of a band gap structure which avoids this issue. A smooth dielectric waveguide with $qc/\omega = 1$ would be a suitable structure for an inverse Cerencov accelerator. On the basis of the results reported above for non-zero q, it seems unlikely that a low loss structure could be formed from a lattice array of parallel cylindrical rods. On the other hand it seems clear that a low loss radial band gap structure, of the sort mentioned in the second section, can be designed.

ACKNOWLEDGEMENTS

We thank Dr. F. M. Mueller for informative conversations. This research has been supported by the National Science Foundation, grant DMR-89-15815, and the Department of Energy, contracts DE-AS0389ER40527 and DE-AC03-76SF00515.

REFERENCES

1. S. John, Phys. Rev. Lett. , 58, 2486 (1987).
2. E. Yablonovitch, Phys. Rev. Lett., 58, 2059 (1987).
3. D. Smith, R. Dalichaouch, N. Kroll, S. Schultz, S. L. McCall, P. M. Platzman, "*Photonic Band Structure and Defects in One- and Two-Dimensions*", submitted to J. Opt. Soc. Amer.B (1992).
4. See also W. C. Sailor, F. M. Mueller, and B. E. Carlsten, "Theory for a cylindrical pillbox accelerator cavity using layered structures for reducing skin-effect losses", preprint (submitted to IEEE TMTT); and the presentation of Dr. R. B. Palmer at the 3rd Workshop on Advanced Accelerator Concepts, Port Jefferson, NY June 14-20, 1992, entitled: "The Definition and Properties of Near Fields".
5. S. L. McCall, P. M. Platzman, R. Dalichaouch, D. Smith, S. Schultz, Phys. Rev. Lett., 67, 2017 (1991).
6. M. Plihal, A. Shambrook, A. A. Maradudin, Ping Sheng, Optics Communications, 80, 199 (1991).
7. M. Plihal, A. A. Maradudin, Phys. Rev. B, 44, 8565 (1991).
8. R. D. Meade, A. M. Rappe, K. D. Brommer, J. D. Joannopoulos, "*Accurate Theoretical Analysis of Photonic Band-Gap Materials*", preprint, submitted to Phys. Rev. Lett. (1992).
9. F. M. Mueller, private communication.
10. A. M. Clogton, Proc. of the I.R.E., pp. 767-782 (1951).
11. J. Zakowicz, J. Appl. Phys., 55, 3421 (1984).

PROPERTIES OF THE FOXHOLE ACCELERATING STRUCTURE*

R.C. Fernow & J. Claus
Brookhaven National Laboratory

Abstract

We examine some properties of a new type of open accelerating structure. It consists of a series of rectangular cavities, which we call foxholes, joined by a beam channel. The power for accelerating the particles comes from an external radiation source and enters the cavities through their open upper surfaces. Analytic and computer calculations are presented showing that the foxhole is a suitable structure for accelerating relativistic electrons.

Introduction

It is widely believed that the only practical means of pursuing accelerator-based particle physics after the SSC is through the construction of electron-positron linear colliders[1]. Cost optimization studies of these devices have indicated that the operating wavelength should be significantly shorter than the 10 cm used in the SLC, the first operating linear collider[2,3]. In addition, it has been proposed to investigate using the high peak power from lasers as the power source for the accelerator[4]. However, since the dimensions of the accelerating cavities scale with the wavelength, it becomes increasingly more difficult to construct iris-loaded, cylindrical cavities of the conventional type. This has lead to the proposal for an open, planar type of geometry for the accelerator cavities[5]. We are investigating the properties of one of these open structures, the foxhole structure[6,7], in order to judge its suitability for accelerators.

This investigation is somewhat prejudiced by the desire to design a structure suitable for $\lambda \approx 10~\mu$m that can be fabricated by methods that are conventional in the semiconductor industry. Foxholes deeper than $\lambda/2$ and "buried beam channels" may be difficult to make at $\lambda = 10~\mu$m. Such restrictions may be less severe at longer wavelengths, e.g. $\lambda \approx 1$ mm. There is evidence that deeper foxholes and buried channels would be advantageous.

Both the foxhole structure, shown in Fig. 1, and the conventional electron linac consist of chains of resonating cavities aligned along the axis of the beam path. The transverse walls between adjoining cavities have holes for beam passage. The excitation methods however are quite different. The cavities of the foxhole structure are individually and independently coupled to a common power source. The acceleration process is the same in the two structures. The cavities carry an axial electric field in the vicinity of the beam path. The phase difference $\Delta\phi$ between successive cavities must match the propagation time Δt of the particles. The tolerance on the average phase delay per cell, averaged over many cells, is very small.

---

* Work supported by the U.S. Department of Energy contract DE-AC02-76CH00016

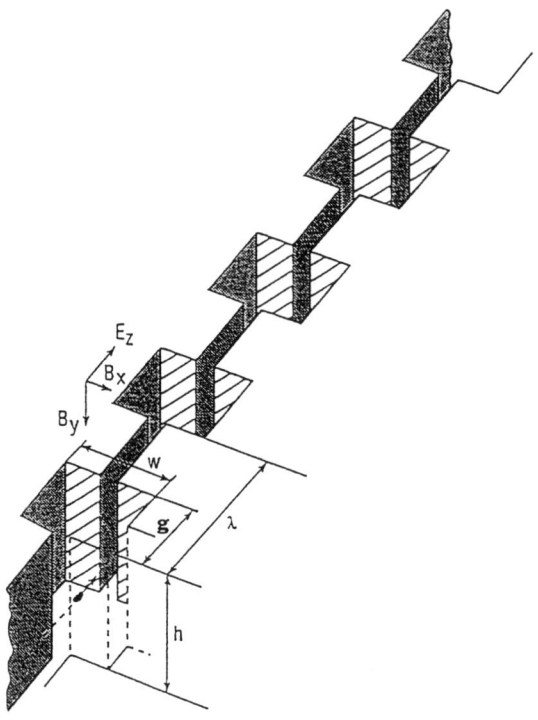

Figure 1: Schematic diagram of a foxhole accelerating structure.

The phase condition is relatively easy to satisfy in the foxhole structure. All the resonators are driven in parallel from a common power source that provides a common phase reference. The only critical parameter is the axial location of the centers of the resonators. The foxhole structure requires higher peak power than a conventional structure for the same frequency and accelerating gradient because all its resonators in a given section of accelerator are filled with energy simultaneously. The power must be provided during a time interval sufficiently long to guarantee that the first particle to enter and the last particle to leave the structure receive the same acceleration. However, if the whole accelerator is subdivided into n sections, the peak power per section becomes $1/n$ of what it would be with no subdivision.

The Q's of the resonators do not have to be particularly high (10- 20 would be good enough) and therefore the filling time in terms of RF cycles can be small.

However, the higher the operating frequency, the more the conventional structure resembles the foxhole in tolerances and power requirements. We conclude that for very high frequencies (e.g. $\lambda < 1$ cm) foxholes may be the way to go.

Foxhole structures can be designed at many levels of complexity. At this point in these studies, it seems prudent to forego sophistication in the interest of simplicity of fabrication and analysis. We consider the general design parameters for a simple foxhole structure suitable for accelerating 50 MeV electrons and for coupling to a 10 μm power source (CO_2 laser).

A section through the midplane of the proposed structure is shown in Fig.1. We choose to have one resonator per wavelength. The incident radiation sees the resonators as sections of rectangular waveguide with cross section w by g. We choose the accelerating gap $g \approx \lambda/2$. The width w and height h of the resonator are chosen to match the operating frequency. The beam moves from resonator to resonator via a slotted or buried channel with length $\beta\lambda - g$. These channels must have sufficient cross section to allow the beam to pass, but no more. The smaller this cross section the weaker the electromagnetic intercavity coupling. The cut off frequency of the channels, considered as waveguides, should be above the operating frequency.

The structure is excited by electromagnetic radiation that impacts perpendicularly onto the top surface and into the sections of waveguide. The electric component of the incident wave is directed along the electron beam. The structure resembles an Alvarez linac because it may be regarded as a collection of drift tubes (the channels or tunnels) that shield the electrons from the incident radiation field whenever that field is decelerating.

This shielding effect alone is not sufficient to make a useful accelerating structure because it leaves the travelling wave nature of the incoming radiation undisturbed. There is in the waveguide a magnetic field component of the travelling wave that is in phase with the accelerating electric field, perpendicular to it and to the direction of propagation. Electrons travelling along the beam axis are therefore not only accelerated longitudinally, but also transversely along the waveguide axis with nearly equal force. This transverse acceleration deflects the electrons off the axis and causes energy loss due to their emission of synchrotron radiation.

The problem with the deflecting magnetic field component can be solved by converting the travelling wave to a standing wave. The simplest way to accomplish this is to provide each resonator with a metallic floor at a point an odd number of quarter wavelengths below the particle beam axis. The electric and magnetic field components are shifted relative to each other by a quarter wavelength along the waveguide direction and by a quarter period in time. As a result particles passing through at places where and at times when the electric field is maximum will feel little magnetic field. Unless the characteristic impedance of the guide matches the radiation resistance of a row of equivalent waveguides radiating into free space the reflected radiation will be partially reflected again at the input end and the cavity will "fill" similar to a conventional one.

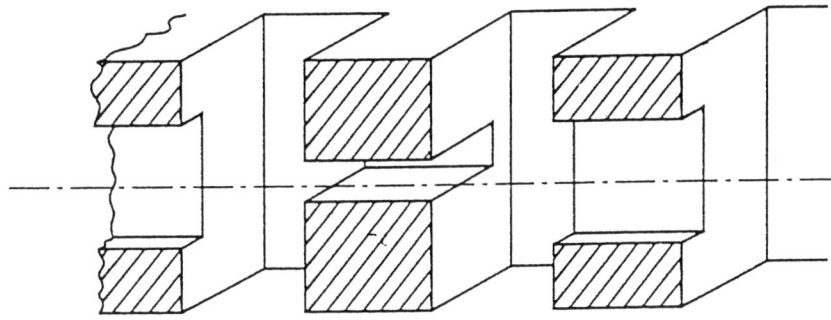

Figure 2: Vertical slice through an RF quadrupole foxhole structure.

At a later stage one might consider converting this structure into an RF quadrupole by modulating the channel cross section with axial position as shown in Fig. 2.

We should point out that the following analysis only considers the properties of isolated foxholes. Surface currents must exist in the metallic parts of the upper plate of the structure. If the currents originating in a given foxhole have not been significantly attenuated by the time they reach a neighboring cavity, the structure will behave like a system of coupled resonators. This coupling has been assumed to be small in this report.

Electromagnetic fields and forces

From conventional theory we find that a metallic waveguide of rectangular cross section $g \times w$ can support waves constant in the z direction described by

$$E_{Z\mp} = \tfrac{1}{2}E_A \cos\left[(2n-1)\frac{\pi x}{w}\right]e^{i(\omega t \mp ky)}$$

$$B_{X\mp} = \mp\frac{k}{\omega}\tfrac{1}{2}E_A \cos\left[(2n-1)\frac{\pi x}{w}\right]e^{i(\omega t \mp ky)} \tag{1}$$

$$B_{Y\mp} = \frac{i}{\omega}\frac{(2n-1)^2\pi^2}{w}\tfrac{1}{2}E_A \sin\left[(2n-1)\frac{\pi x}{w}\right]e^{i(\omega t \mp ky)}$$

These are identical waves travelling in opposite directions (upper and lower signs) along the y axis. The width w, radian frequency ω, and wavenumber k in the guide are coupled by the wave equation

$$\left[\frac{(2n-1)\pi}{w}\right]^2 + k^2 = \epsilon\mu\omega^2 \tag{2}$$

The two waves interfere to yield a standing wave

$$E_Z = E_A \cos\left[(2n-1)\frac{\pi x}{w}\right]\cos(ky)e^{i\omega t}$$

$$B_X = \frac{-k}{\omega}E_A \cos\left[(2n-1)\frac{\pi x}{w}\right]\sin(ky)e^{i(\omega t-\frac{\pi}{2})} \qquad (3)$$

$$B_Y = \frac{(2n-1)^2\pi}{\omega w}E_A \sin\left[(2n-1)\frac{\pi x}{w}\right]\cos(ky)e^{i(\omega t-\frac{\pi}{2})}$$

Inspection shows that E_Z is a maximum, while B_X and B_Y are zero at $x = 0, y = 0$. Note also that $E_Z=0$, $B_Y=0$ for all x at $ky = \pi/2$. This condition can be realized by placing a metallic floor in the guide at that point. We define $-h/2$ to be this particular value of y, so that the electron beam (system) axis is located $\Delta y = h/2$ above the floor. Substituting $k = (\pi/2)/(h/2)$ in the wave equation, one finds

$$\left(\frac{2n-1}{w}\right)^2 + \left(\frac{1}{h}\right)^2 = \left(\frac{2}{\lambda}\right)^2 \qquad (4)$$

where $\lambda = 2\pi c/\omega$ is the wavelength in free space. Since w is fixed by the construction of the guide, changes in λ require changes in h, i.e. the distance between the system axis and the foxhole floor depends on the frequency. The width w must be larger than $(2n-1)\lambda/2$ if h is to be real. Thus if $n = 1$, we must have $w > \lambda/2$.

The ratio between E_Z and H_X is independent of location:

$$\left|\frac{E_{Z\mp}}{H_{X\mp}}\right| = \frac{Z_o}{\sqrt{1 - [(2n-1)\frac{\lambda}{2w}]^2}} \qquad (5)$$

$$\equiv Z_C$$

where $Z_o = \sqrt{(\mu_o/\epsilon_o)}$ is the impedance of free space and Z_C is the familiar characteristic impedance. Z_C is also real for the case where $w > (2n-1)\lambda/2$, it is always larger than Z_o, increases with the mode number n, and decreases with increasing width w.

The conditions at the floor level ($E_Z = 0, B_Y = 0$) recur at levels

$$\Delta y = h, \ 2h, \ 3h, ...$$

above it. The field components B_X and E_Z in the incident field are independent of transverse location (x, z), while they vary like $cos[(2n-1)\pi x/w]$ at the top of the foxhole. The incident wave will therefore tend to excite a sum of modes with various mode numbers in such a way as to achieve independence of x. Choosing the width such that $3\lambda/2 > w > \lambda/2$ prevents the propagation of modes with $n > 1$. Although such modes do not propagate, their fields do reach inside the guide. Their magnitudes decrease as $exp[-k_n y]$ with distance y below the top the top surface. The B_X and B_Y components of these fields are in phase with E_Z. They are therefore non-zero on the system axis at the times that particles pass. Their effect can be minimized by making the k_n's of the troublesome modes large and the foxhole a multiple of the half guide wavelengths deep, rather than using the first $(n = 1)$ one.

If the characteristic impedance of the n=1 mode is not equal to the radiation resistance, $Z_r \neq Z_c$, the incident wave, returning to the top surface after reflection at the bottom, is partially reflected again at the interface of the foxhole with free space. The reflection coefficient is defined by

$$r \equiv \frac{Z_r - Z_C}{Z_r + Z_C}$$

$$= \frac{\frac{Z_r}{Z_o}\sqrt{1 - \left(\frac{\lambda}{2w}\right)^2} - 1}{\frac{Z_r}{Z_o}\sqrt{1 - \left(\frac{\lambda}{2w}\right)^2} + 1} \tag{6}$$

The choice of r controls the strength of the coupling between the internal and external fields, e.g. if $r \approx 0$, $Z_C \approx Z_r$ most of the energy incident on the interface is transmitted and little is reflected. We have as of now no analytic expression for Z_r/Z_o, but expect that $Z_o \leq Z_r < Z_C$. We use $Z_r = Z_o$ in the numerical calculations as a first approximation.

A surface current density flows along z on the horizontal sections and along y on the vertical sections in the figure. The magnitude and phase of these currents are determined by the local value of B_X. Thus they have a sinusoidal dependence on x across the foxhole. They are largest near the corners and zero at the midpoint of the vertical sections.

In terms of particle dynamics the foxhole structure behaves very much like a conventional linac. The net force on the particle can be found by integrating the time dependent force on a relativistic electron over one period of the structure, assuming that the fields are identically zero inside the beam channel. We find that

$$\langle F_X \rangle = \frac{q\lambda E_A}{2\pi w} \sin\frac{\pi x}{w} \cos\frac{\pi y}{h} \sin\Phi$$

$$\langle F_Y \rangle = \frac{q\lambda E_A}{2\pi h} \cos\frac{\pi x}{w} \sin\frac{\pi y}{h} \sin\Phi \tag{7}$$

$$\langle F_Z \rangle = \frac{q E_A}{\pi} \cos\frac{\pi x}{w} \cos\frac{\pi y}{h} \cos\Phi$$

where Φ is the relative phase. Note that the transverse forces cancel out exactly for particles moving along the axis $(x = 0, y = 0)$ and for phases corresponding to the zeros of the sine function. These equations are only valid in the open part of the structure and they do not include any contribution from the fringe field in the vicinity of the opening into the beam channel.

Wavelength response

Electromagnetic energy in the foxhole will be reflected back and forth between the upper and lower boundary surfaces. After a number of cycles the fields inside the foxhole cavity will approach an equilibrium value. The total field at any height y in the cavity can be found by summing the contributions of all the reflected waves. We find that the magnitude of E_Z relative to the field just below the open surface of the foxhole can be written as

$$\rho = \frac{2\cos ky}{\sqrt{1 - 2r\cos\alpha + r^2}} \tag{8}$$

where $\alpha = 2kh$. The phase relation can be written as

$$\tan\Phi = -\frac{(1+r)}{(1-r)}\cot\frac{\alpha}{2} \tag{9}$$

It is desirable to relate the field enhancement factor ρ to the strength of the incident wave E_o. In order to do this we assume that the dominant spatial Fourier component of the actual field that exists just above the upper surface of the foxhole has an amplitude comparable in magnitude to that of the incident field. Then we can write the field at the height y as

$$E_Z(y) \approx \rho(y)\tau E_o \tag{10}$$

This shows that the resultant field depends on the resonant properties of the cavity through the factor ρ, on the coupling efficiency through the transmission coefficient $\tau = 1 + r$, and on the strength of the incident field through the factor E_o.

The dependence of the field enhancement $\rho\tau$ and phase Φ on λ is shown in Fig. 3. We can use this curve to determine the dimensional tolerances required for building a foxhole accelerator.

RF cavity properties

Important radiofrequency (RF) properties of an accelerating cavity include the stored energy, power loss, quality (Q) factor, shunt impedance, and the fill time. The RF properties of the foxhole cavity are given in the table for a number of different values of w.

The total stored energy in the foxhole can be approximated by integrating the energy density over the geometrical dimensions of the foxhole. The total stored energy in the cavity is

$$U \approx \frac{1}{8}\epsilon_o E_A^2 gwh \tag{11}$$

There are two main sources of power loss in a foxhole. First there is the Joule heating loss due to the finite conductivity of the metallic walls. We have estimated this by the standard technique using the surface currents on the five

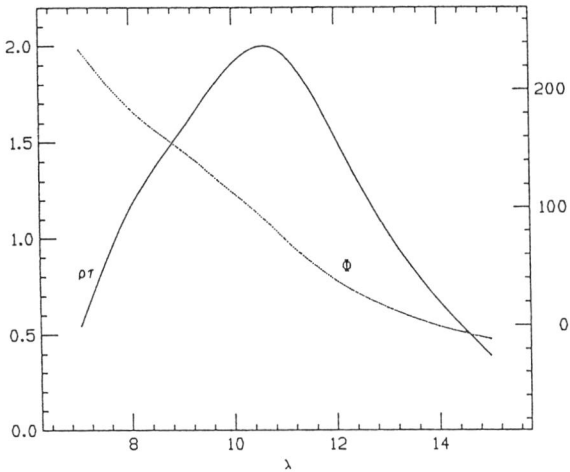

Figure 3: The magnitude (ρ,left scale) and phase (Φ,right scale) of the accelerating electric field in the foxhole as a function of the wavelength of the incident radiation. $w = 8$ and $g = 5.3$.

metallic walls. The second and dominant loss is the escape of radiation through the upper, open boundary. We found that the time averaged power flow into the cavity is

$$P = \frac{(1-r)^2 E_A^2 gw}{16 \ Z_C} \tag{12}$$

Some of the incoming power (P_d) is dissipated due to ohmic losses in the cavity walls and to interaction of the fields with the electron beam. If we assume the beam loading is negligible, the power that is not lost in the walls must be radiated back out of the cavity under steady state conditions. Thus we can write the radiated power (P_r) as $P_r = P - P_d$. The numerical examples given in the table show that the power loss will be dominated by the radiation losses through the open surface.

The Q of the foxhole cavity is

$$Q = \frac{2\pi n}{(1-r)^2} \left(\frac{\lambda_g}{\lambda}\right)^2 \tag{13}$$

where λ_g is the (guide) wavelength inside the foxhole. Note that since h is proportional to λ_g, the Q increases rapidly as the depth of the foxhole is increased.

We compute the fill time from the product of the value and the period (T) of the oscillation. The fill time is

$$t_F = \frac{-T}{\ln r} \tag{14}$$

The shunt impedance per unit length is

$$r_{sh} = \frac{4Z_o}{w} \left(\frac{1+r}{1-r}\right)^2 \tag{15}$$

RF cavity properties							
E_A=1 GV/m; $\lambda = 10.6\mu m$; $g = \lambda/2$							
w	h	U	P_{rad}	P_d	t_F/T	Q	r_{sh}
μm	μm	pJ	kW	W			MΩ/m
6	11.307	398	0.60	412	0.98	70.2	1144
8	7.076	332	3.59	277	0.51	15.3	336
10	6.250	366	5.99	292	0.40	10.4	210
12	5.907	416	8.13	328	0.34	8.7	156
16	5.617	527	12.1	420	0.28	7.5	106
20	5.497	645	15.8	520	0.25	7.0	81

Slotted channel structure

The analytic calculations given up to this point have disregarded the presence of channels between successive holes for the electron beam to pass through. We studied the effects caused by such channels numerically by means of the 3D EM field code MAFIA. This approach has its own limitations which must be taken into account in the interpretation of the results. They stem from the nature of the code and from the constraints on available memory and speed of the computer used.

Any actual accelerating structure must have a channel or tunnel for the beam to pass through. This channel must be big enough to encompass the tails of the particle transverse distribution, yet small enough so that it does not significantly disturb the field patterns in the cavity. At the 10 μm scale the slotted structure, shown in Fig. 1, is easier to build than a buried channel. It can be constructed by etching down along the y direction into a single piece of silicon. On the other hand, since the slot extends the full height of the cavity, it causes a greater disturbance in the cavity fields.

We next consider MAFIA calculations of the integrated forces on a particle travelling parallel to the z axis with velocity c and crossing a half cell of the structure (cavity plus channel). The calculated forces take into account the time variation of the fields. The F_Z force is shown in Fig. 4 as a function of x_p, the x position of a particle crossing the foxhole. The particles cross the cavity with y_p equal to half the cavity height. All particles travel with the phase that gives maximum acceleration. The slot widths are expressed in any system of length where $\lambda = 10$ units (in other words $s = 2$ means $s/\lambda = 0.2$). We see that F_Z drops rapidly as the slot width is increased, but it is relatively insensitive to the x_p position for non-zero slot dimensions.

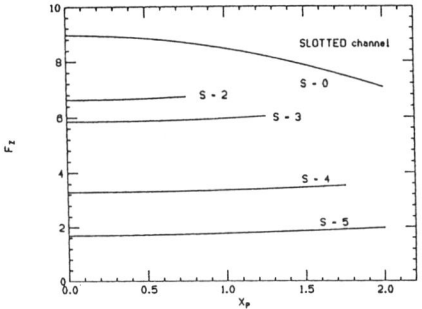

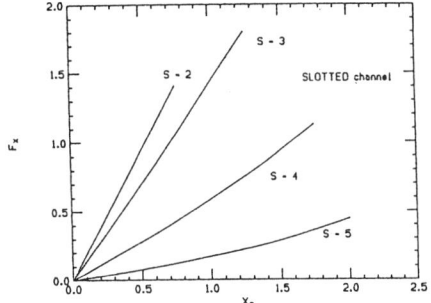

Figure 4: F_Z versus x_p.   **Figure 5:** F_X versus x_p.

Fig. 5 shows that the slot causes an x directed force that tends to become less significant for larger slot widths. Recall that the transverse forces are 90° out of phase with respect to F_Z. Thus there is no contribution to the transverse forces from the cavity fields in this case. This was verified in a MAFIA run with a slot width of zero. The transverse force contribution in Fig. 5 and in the following figures is due to the fringe field in the vicinity of the slot edges.

The dependence of the integrated force F_Z on the deviation of y_p from the half-height of the cavity is shown in Fig. 6. The particles cross the cavity with $x_p = 0$. F_Z has a significant dependence on y_p. Unless it is corrected, this effect would cause a correlation to develop between the momentum of particles in a beam bunch and their y position.

Fig. 7 shows that all particles in the slot see the same sign vertical force. Thus there is no axis with $F_Y = 0$, and F_Y is asymmetric for particles above and below the beam axis. Particles near the open surface of the foxhole experience the largest vertical forces, particularly for large slot widths. Because of symmetry there is no dependence of F_X on y_p for $x_p = 0$.

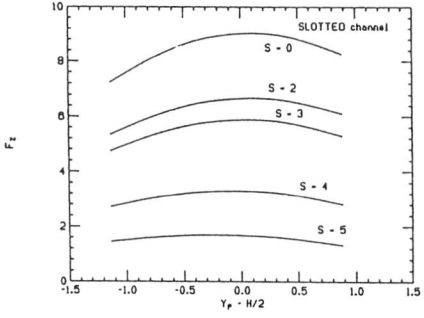

Figure 6: F_Z versus y_p.

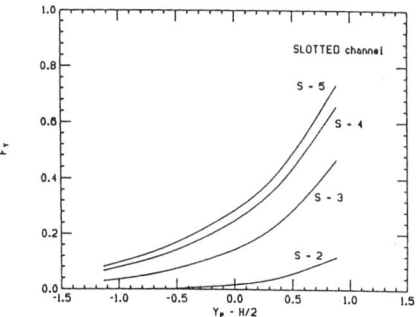

Figure 7: F_Y versus y_p.

Buried channel structure

It is more difficult to manufacture a buried channel structure than a slotted channel one at the 10 μm length scale. It requires etching along the x direction into two halves of the structure and then precisely aligning and joining the pieces together. Fig. 2 shows a MAFIA model of a buried channel structure. We use a square channel for the calculations.

MAFIA calculations of the integrated forces on a particle crossing a half cell of the structure (cavity plus channel) is shown in Fig. 8 as a function of x_p, the x position of a particle crossing the foxhole. The particles cross the cavity parallel to the z axis with velocity c and with y_p equal to half the cavity height. All particles travel with the phase that gives maximum acceleration. Fig. 8 shows that the integrated value of F_Z drops off rapidly with increasing channel width and is practically independent of x_p for channel widths less than 4. Comparing Fig. 8 with Fig. 4 shows that the net force in the z direction is significantly higher for the buried channel, despite the fact that the peak accelerating field in the open cavity is roughly the same for the two cases. This would seem to indicate that the field inside the channel contributes more effectively for the buried channel.

Fig. 9 shows that the channel causes an x directed force that grows with x_p. There is no tendency for F_X to decrease and flatten out for large s, as there is with the slotted channel. The dependence of F_Y on x_p is much weaker than for the slotted channel; the magnitude did not exceed 0.04 for the cases studied.

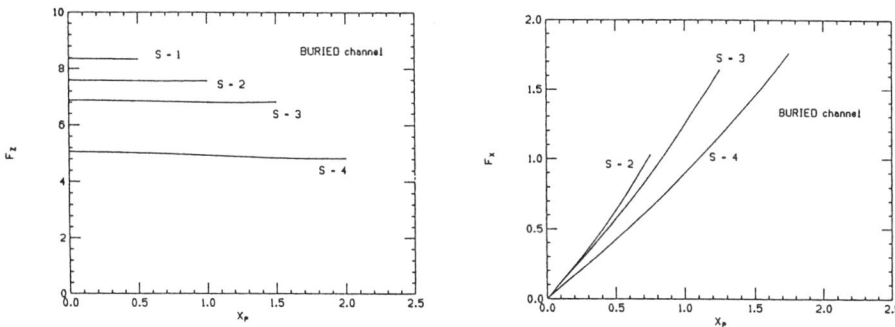

Figure 8: F_Z versus x_p.   **Figure 9**: F_X versus x_p.

The dependence of the integrated force on the deviation of y_p from the half-height of the cavity is shown in Fig. 10. The particles cross the cavity with $x_p = 0$. F_Z has a significant dependence on y_p. Unless it is corrected, this effect would cause a correlation to develop between the momentum of particles in a beam bunch and their y position.

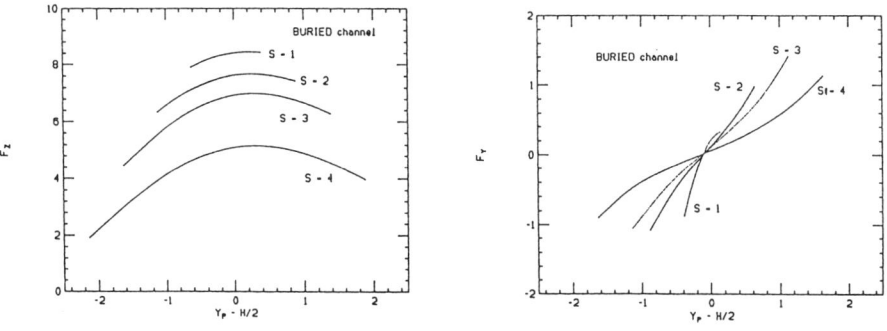

Figure 10: F_Z versus y_p.   **Figure 11**: F_Y versus y_p.

Fig. 11 shows that, unlike the slotted structure, the buried channel structure has no net F_Y on the beam axis. In this respect the forces more closely follow the ideal behavior described by Eqs. 7.

Conclusions

We have used simple analytic arguments and approximate computer modelling to show that the foxhole structure should be suitable for accelerating relativistic electrons. Reasonable geometries should give accelerating fields several times greater than the peak magnitude of the incident radiation. The technology exists now to make slotted channel structures that couple to optical radiation. Unfortunately the slotted channel foxhole has a relatively large, asymmetric vertical force that would have to be compensated. The buried channel foxhole on the other hand has a neutral axis where both transverse forces vanish. In this respect it acts similarly to the conventional disk-loaded structure.

Although we have discussed this structure in the context of laser driven acceleration, it is of course not restricted to use at optical wavelengths. It may be a reasonable choice for any wavelength between 10 μm and 1 cm, where building a conventional, closed cavity would be difficult or impossible.

References

1. See for example the review, R.H. Siemann, *Advanced accelerator concepts and electron-positron linear colliders*, Ann.Rev.Nucl.Part.Sci. 37:243-66, 1987, and the references mentioned there.

2. R.B. Palmer, *The interdependence of parameters for TeV linear colliders*, in S. Turner(ed), **New Developments in Particle Acceleration Techniques**, p. 80-121, CERN Report 87-11, 1987.

3. J. Claus, *Energy efficiency and choice of parameters for linear colliders*, in M. Puglisi et al(eds), **New Techniques for Future Accelerators**, p. 45-65, Ettore Majorana International Science Series 2g, Physical Sciences, Plenum Press, New York, 1987.

4. P.J. Channel(ed), **Laser Acceleration of Particles**, AIP Conf. Proc. No. 91, 1982; C. Joshi & T. Katsouleas(eds), **Laser Acceleration of Particles**, AIP Conf. Proc. No. 130, 1985.

5. R.B. Palmer, *Open accelerating structures*, in S. Turner(ed), op.cit., p. 633-41; N. Kroll, *General features of the accelerating modes of open structures*, in C. Joshi & T. Katsouleas(eds), op.cit., p. 253-70.

6. J. Claus, *Characteristics of foxhole resonator*, unpublished notes, 18 June 1986; J. Claus, *Laser linac: foxhole structure*, unpublished notes, Jan. 1990.

7. The properties of the foxhole structure are treated in greater detail in R. Fernow & J. Claus, *The foxhole accelerating structure*, Brookhaven National Laboratory report BNL-52336, 1992.

HIGH-GRADIENT LINAC STUDIES IN THE FREQUENCY RANGE OF CLIC

G. Guignard
Cern, Geneva, Switzerland

ABSTRACT

The present status of research and development for high-gradient linacs required for future linear colliders is described. The interval of RF frequency considered here ranges from about 10 GHz to 30 GHz and many studies made for NLC (Next Linear Collider) proposed at SLAC, KEK and IHEP as well as for CLIC (CERN Linear Collider) are covered. After a brief recall of the reason for operating at high frequency when high accelerating gradient is required, major issues are addressed. Effects of strong wake fields on the energy spread at the linac exit and on transverse beam stability, as well as possible compensations are summarized. Consequences for multibunch dynamics are also discussed, together with suggested remedies. The difficult question of the alignment tolerances is approached to underline the necessity of a very efficient scheme of trajectory correction, and the state of the art in building accelerating structure prototypes is given. Prospects for microwave klystrons as high-power RF sources for normal-conducting electron and positron linacs in NLC are briefly described. Finally, recent results concerning the CLIC scheme for generating the necessary power, based on a two-stage accelerator, are presented.

ACCELERATING GRADIENT AND RF FREQUENCY

The total average RF power that is required in one linac is given by

$$P_{RF} = \frac{P_b}{g^2 \eta} \tag{1}$$

where g^2 is the filling efficiency of the accelerating sections (i.e. the fraction of input energy left at the end of the fill time τ_0) and η the fractional energy extraction by the beam.[1] The beam power P_b is proportional to the final particle energy eU, the number N of particles in the beam and the repetition rate f_{rep},

$$P_b = eUNf_{rep} \tag{2}$$

while the fraction of stored energy extracted by a charge Ne is given by

$$\eta = \frac{NeZ_0\omega_0^2}{2\pi c\, E_0} \tag{3}$$

The quantity Z_0 is frequency-independent and only depends on the shape of the accelerating structure; it is the shunt impedance over Q factor per RF wavelength and is related to the shunt impedance over Q factor per unit length r_0' via

$$\omega_0 Z_0 = 2\pi c \, r_0'$$ (4)

The wave number ω_0 is equal to 2π times the RF frequency f_0, and E_0 stands for the accelerating gradient.

Since a high-energy extraction is desirable, the expression (3) advocates for high RF frequencies at a given beam charge. In particular, when aiming at very high gradients in order to limit the total length of the linac, very high frequencies f_0 are required. The highest possible value is eventually limited by the manufacturing and alignment tolerances as well as the wake fields, and this limit constrains the gradient E_0 actually attainable.

Other constraints on E_0 come from the difficulty in generating the required power. The peak power per unit length $\hat{P} / L_0$, in a structure of shunt impedance per unit length R_0' is indeed given by

$$\frac{\hat{P}}{L_0} = \frac{E_0^2}{g^2 \alpha R_0' \eta_2}$$ (5)

where α is the power flow attenuation constant and η_2 is the efficiency of the energy transfer from the source to the linac structure. The power requirements (5) related to the desired E_0 may be large and it is a challenge to develop power sources delivering the necessary energy and working at high frequency.

The scaling of E_0 with ω_0^2 suggested above keeps constant the fraction η and the average power P_{RF}, by virtue of Eqs. (1)–(3). In these conditions the peak power per unit length $\hat{P} / L_0$ is proportional to

$$\frac{\hat{P}}{L_0} \sim \omega_0^{-1/2} E_0^2 \sim \omega_0^{7/2}$$ (6)

which includes the fact that R_0' varies with $\sqrt{\omega_0}$.

Increasing E_0 in this way can be considered until a limit either on the manufacturing tolerances or on the development of power sources is reached, as mentioned already.

A third limitation may arise from the wake fields whose peak values depend on ω_0, the iris aperture a and the loss factor k_0 as follows,[1]

$$\hat{W}_T^\delta \sim \frac{k_0}{\omega_0 a^2} \sim \frac{R_0}{Q_0 a^2} \sim \frac{\omega_0}{a^2} \sim \omega_0^3$$

$$\hat{W}_L^\delta \sim k_0 \sim \frac{\omega_0 R_0}{Q_0} \sim \omega_0^2$$ (7)

The rapid increase of the point-charge wakes (mainly transverse) with ω_0, and the concomitant beam blow-up are the sources of this limitation.

Since both the achievable gradient and the detrimental effects increase with ω_0, the choice of the RF frequency results from a compromise. The frequency range considered in this study report goes from about 10 GHz to 30 GHz, while the accelerating gradient is supposed to be between 50 and 100 MV/m. This corresponds to linacs suitable for either CLIC (CERN Linear Collider using a Drive Linac) or NLC (Next Linear Collider using pulsed RF generators) as envisaged at SLAC, KEK or IHEP. Table I gives considered values of the parameters discussed above, for the different proposals.

Table I. Examples of parameters.

	E_0 (MV/m)	ω_0 (GHz)	$\hat{P} / L_0$ (MW/m)
CLIC	80–100	30.0	150
NLC: SLAC (NLC)	50	11.4	
KEK (JLC)	100	11.4	60–240
IHEP (VLEPP)	100	14.0	

WAKE FIELDS AND SINGLE-BUNCH DYNAMICS

As mentioned above, high-current beams induce in a high-gradient accelerator strong electromagnetic fields that increase rapidly with the RF frequency. These wake fields are responsible for energy loss, energy spread and transverse blow-up. Longitudinal wakes directly influence the distribution of energy inside the bunch, its contribution diminishing the accelerating field seen by the particles. Since these wakes are not uniform within a bunch, the energy loss is accompanied by an energy spread that induces variation of the focusing strength and dispersion of the trajectories inside the bunch in the presence of external magnetic quadrupoles. Transverse wakes deflect parts of the bunch and these deflections depend on the particle momentum and position with respect to the accelerator axis. Because of the energy spread and trajectory dispersion, dipole wakes produce kicks changing along the bunch and eventually producing an apparent beam blow-up (the "head" of the bunch deflecting the "tail"). Quadrupole wakes may also be present but they have been either considered as negligible or not yet studied in the different proposals.

Some compensation of the energy spread σ_E using wake potential versus RF sine wave is possible to high orders.[1,2] This is strongly desirable if the momentum acceptance at the exit of the linac is limited, as for instance in a final-focus system that typically accepts a $\Delta p/p$ of $\pm 2‰$ to $\pm 5‰$. The total accelerating gradient seen by a particle at position z in the bunch can be written:

$$G(z) = G_{RF} \cos(\omega_0 \frac{z}{c} - \phi_0) - W_L(z) \qquad (8)$$

and the balance between the RF wave and the longitudinal wake will obviously depend on the phase ϕ_0, but also on the bunch charge N and bunch length σ_z that enter in W_L. Knowing G(z) it is possible to find ϕ_0 and σ_z values which minimize the spread σ_E for a given N. Furthermore, the energy distribution $v(E)$ can be calculated[2] in order to study its properties and dependence on G_{RF}, ϕ_0, σ_z and N, by using

$$v(E) = \frac{1}{N}\frac{dN}{dE} = \frac{1}{N}\frac{dN}{dz}\frac{dz}{dE} = \frac{\rho(z)}{dE/dz} = \frac{\rho(z)}{e\,dG(z)/dz} \qquad (9)$$

where $\rho(z)$ is the charge distribution (Gaussian). Results concerning CLIC are given as examples.[2] Figure 1 shows G(z) for different charges N with $\phi_0 = 7$–$8.5°$ and $\sigma_z = 0.14$–0.17 mm. Figure 2 gives the corresponding energy distributions for $N = 5 \cdot 10^9$ and $6 \cdot 10^9$ per bunch. The flatter G(z), the smaller σ_E, and, in these two cases, the values obtained are about 4.6‰ and 7.7‰ respectively (in the absence of truncation of $v(E)$). Note also that this adjustment ends up with a minimal tail population and a somewhat narrow core size (Fig. 2).

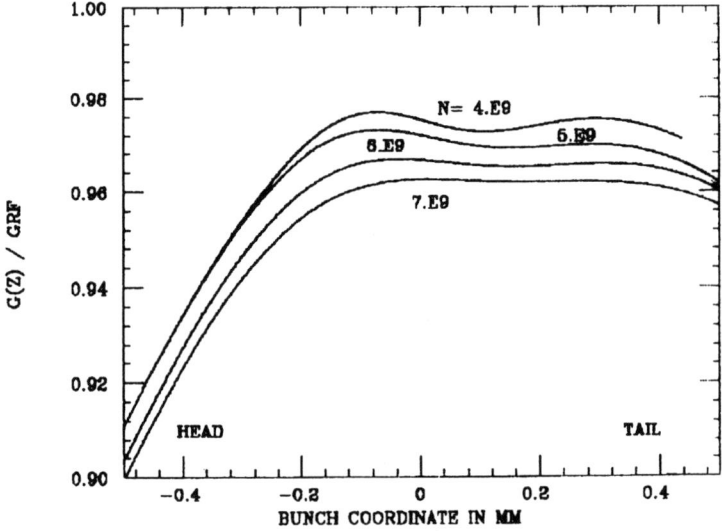

Fig. 1. Total gradient minimizing the energy spread σ_E.

Transverse instabilities due to dipole wakes are all the more critical the smaller the emittance. In the linac of colliders, the vertical emittance may be very small since the beam is usually flat. Having a flat beam or a large aspect ratio $R = \sigma_x/\sigma_y$ at the collision point reduces the average energy loss due to synchroton radiation. This energy degradation due to beamstrahlung induces a large energy spread, that should not, however, exceed the energy spreading related to background processes (imposing a limit of ~5% for tolerable σ_E from beamstrahlung). Furthermore, a large

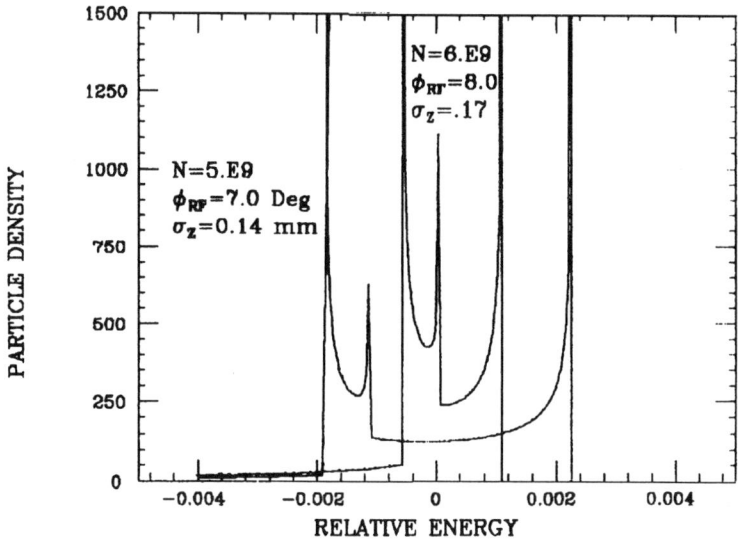

Fig. 2. Energy distribution minimizing σ_E for $N = 5 \cdot 10^9$ and $6 \cdot 10^9$.

ratio R allows the avoidance of an excessive repetition rate that varies for constant luminosity and beam–beam radiation σ_E as follows:

$$f_{rep}(R) = f_{rep}(R = 1)\frac{4}{RH_y} \tag{10}$$

where H_y is the vertical pinch enhancement factor (typically between 2 and 2.5). For these reasons, values of R and normalized vertical emittance $\gamma\varepsilon_y$ are respectively large and small in the proposals quoted (Table II).

Table II. Beam aspect ratio at final focus and V-emittance.

	R	$\gamma\varepsilon_y$ (rad m)
CLIC	10–20	$2 \cdot 10^{-7}$
NLC	100	$3 \cdot 10^{-8}$

The presence of strong dipole wakes (almost 20 times larger in CLIC than in NLC) implies large kicks originating from misalignments of the structure and off-centred trajectories. The corresponding equation of the transverse motion of single particles is given by

$$y'' + k_0^2 y = \left[k_0^2 - k^2(z,s)\right]y + \frac{r_0}{\gamma}\int_{-\infty}^{z}\rho W_T^{\delta}\left(z^* - z\right)y\left(z^*,s\right)dz^* \tag{11}$$

where k^2 characterizes the linac focusing (k_0^2 being its "average" value, independent of z) and the wake-field effect is integrated over the heading part of the bunch of charge distribution ρ. To counteract this effect, the idea consists of obtaining a coherent motion by imposing the same oscillation period to all particles. This condition, called autophasing by its author,[3] comes directly from inspecting Eq. (11):

$$k^2(z,s) = k_0^2 + \frac{r_0}{\gamma} \int_{-\infty}^{z} \rho W_T^\delta \left(z^* - z \right) dz^* \tag{12}$$

All the proposals for high-gradient, high-frequency linear accelerators strive to satisfy condition (12) or its linearized version[4]

$$\left. \frac{\partial k^2}{\partial z} \right)_{z=0} = \frac{r_0}{\gamma} N \left. \frac{\partial W_T}{\partial z} \right)_{z=0} \tag{13}$$

to limit beam blow-up. There are basically two ways for achieving this variation of the focusing strength with the position z inside the bunch:

1) Using the external focusing of a magnetic FODO lattice, the change of k^2 with z can be obtained via an imposed energy spread σ_{BNS}, since $k^2 = k_0^2 / p$ if p is the particle momentum. The required energy spread may come in turn from the dependence on z [Eq. (8)] of the accelerating gradient G combined with an adjustment of ϕ_0. A negative phase ϕ_0 is needed to ensure that the bunch tail is more focused than the head, according to Eq. (12). The subsequent σ_{BNS} ensuring stability is as large as 5% in CLIC, but only 0.2–0.6% in NLC proposals (in proportion to their wake fields). Note that this requirement conflicts with the minimization of σ_E discussed above and based on a positive phase ϕ_0, in particular for high-RF frequency linacs.

2) An elegant way to avoid the conflict with σ_E minimization and simultaneously create the spread in k^2 without the detour of a large energy spread does exist. It consists of generating part of the transverse focusing directly from RF fields oscillating at the frequency of the accelerating fields, in so-called microwave quadrupoles.[5] Since the radial electric field in a narrow slit vanishes in the mid-plane, the effective magnetic gradient due to the axial electric field and deduced from Maxwell's equations is given by[5]

$$G_m(T/m) = \frac{\pi}{c\lambda_{RF}} E_0(MV/m)\sin\phi_1 \tag{14}$$

where E_0 is the peak accelerating gradient, λ_{RF} the RF wavelength and ϕ_1 the RF phase angle measured from the top. In the direction perpendicular to the slit, the electric field is doubled (compared with circular aperture) and overcompensates the magnetic gradient by exactly a factor 2, thus forming a quadrupole. In practice an oval cavity with circular aperture (Fig. 3) is preferred to a circular cavity with slotted iris,[5] in order to have the required radius and surface finish at the aperture.

Fig. 3. Microwave quadrupole cell with flat cavity (CLIC).

In CLIC, where transverse wakes are large, it is proposed to generate the spread of k^2 with microwave quadrupoles using a phase ϕ_1 close to the phase ϕ_0 that minimizes σ_E, i.e. running near the maximum accelerating voltage. In this solution,[6] the main basic focusing k_0^2 is created by external magnetic quadrupoles, while the microwave quadrupoles are only responsible for the variation $k_z^2 - k_0^2$ (Fig. 4). Hence, transverse instabilities can be damped (Fig. 5) keeping the energy spread low.

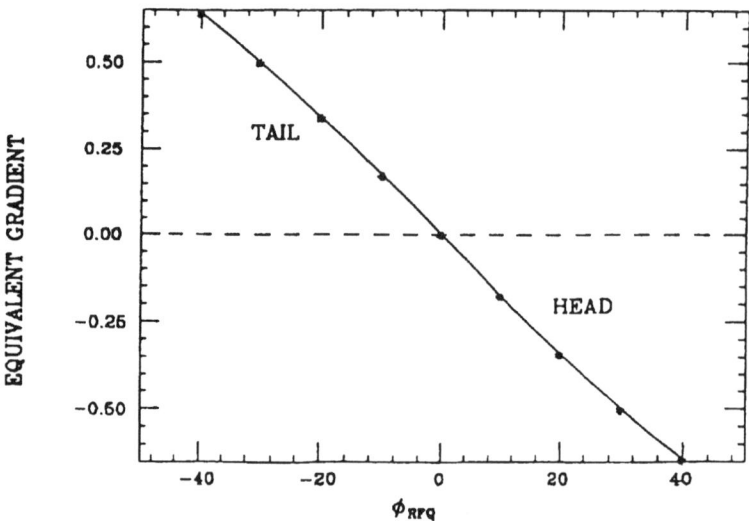

Fig. 4. Equivalent microwave gradient versus the phase.

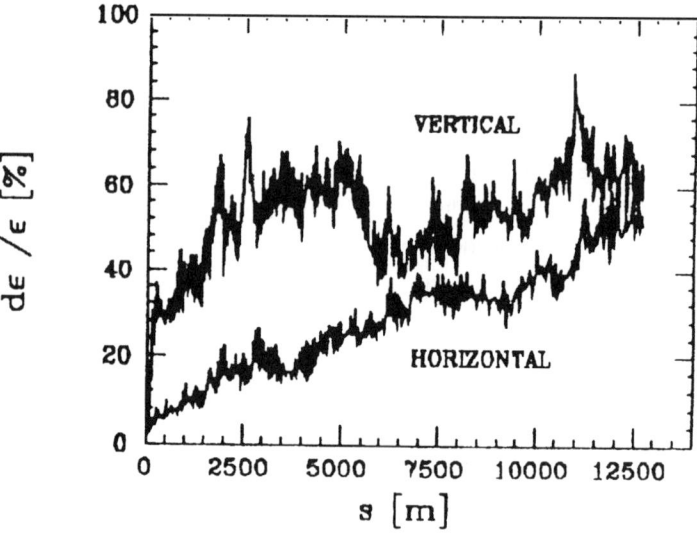

Fig. 5. Example of blow-up control with microwave quadrupoles and minimum σ_E.

WAKE FIELDS AND MULTIBUNCH DYNAMICS

Accelerating long bunch trains that extend over a period comparable to the filling time τ_0 of a cavity section may provoke instabilities of the whole train along the linac, due to interactions with RF fields. The shape, timing and modulation of the RF pulse, as well as long-range longitudinal wake field, are responsible for bunch-to-bunch energy variations. The subsequent energy spread can cause filamentation and emittance growth beyond the acceptance at the exit of the linac. Interactions with resonant transverse electromagnetic fields in disk-loaded waveguides, in particular with the so-called HEM_{11} dipole modes, produce increasing deflections along the bunch train that drive a transverse instability, called beam break-up.

The bunch-to-bunch energy variation produces effects that are more severe with long trains. The fundamental mode, as well as high-order modes of longitudinal wakes, is at the origin of inter-bunch beam loading and the actual RF pulse influences the energy spectrum. There is consequently a need to control bunch-to-bunch energy spreads and some compensation schemes have been studied:[7,8]

1) Matched filling, i.e. adjustment of the injection timing of the bunch train with respect to the RF pulse and appropriate choice of the bunch spacing. The idea is to have sufficient extra energy in the RF section fill between bunches to cope for the energy lost in accelerating the preceding bunches.

2) Staggered timing, i.e. delay of a subset of klystrons so that some accelerating sections are only partially filled during build-up of the beam-loading voltage to its steady-state value. The number of delayed klystrons is selected to produce a voltage equal to about twice the steady-state beam-loading voltage in the linac.

3) Modulation of RF input, i.e. phase adjustments or small klystron variations during the time when the bunch train is passing through a cavity section. This makes use of the propagation out of the section of the leading edge of the pulse while the train is passing over.

The results of such compensation schemes depend on the bunch length with respect to the section filling-time and on the bunch separation. In the NLC for instance,[7] the first method applies preferably to short trains (10 bunches of 10^{10} particles/bunch, lasting about 10% of τ_0). Figure 6 shows the energy deviation obtained with bunch separation of 16 RF wavelength, 5ns RF pulse rise-time and dispersion of RF frequency components. The fractional energy deviation remains below about 3‰ but ~75% of the maximum accelerating gradient is used. With long trains (70 bunches or more), the second method seems more appropriate and Fig. 7 shows results obtained in the same conditions of rise-time and dispersion. If the total energy deviation is about the same, only ~65% of the maximum gradient (assumed to be 50 MV/m) is available. Multibunch dynamics in CLIC has not yet been studied.

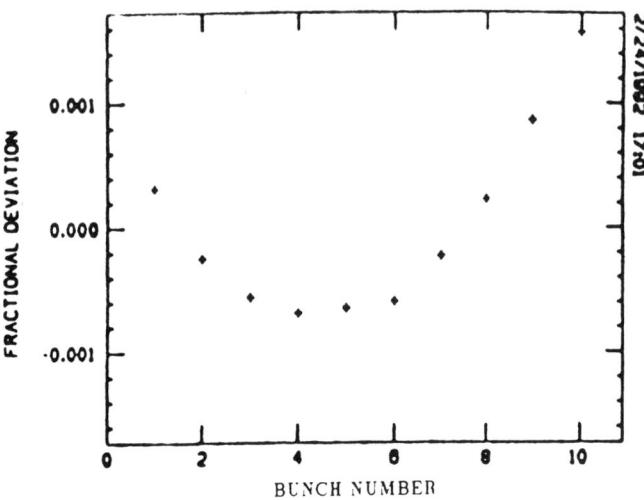

Fig. 6. Fractional energy deviation in a short train after matched filling.[7]

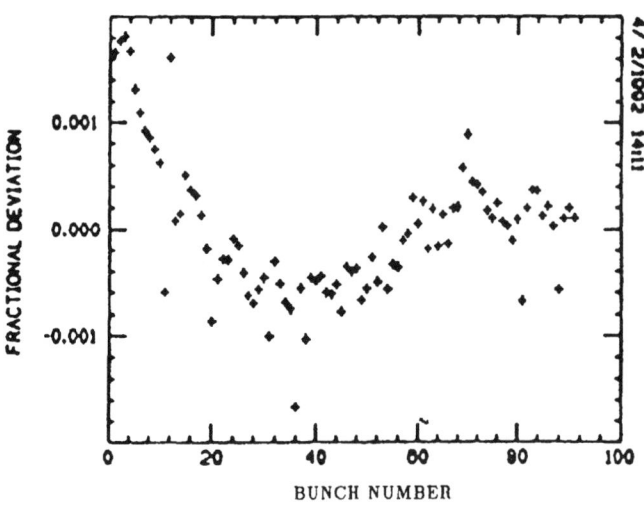

Fig. 7. Fractional energy deviation in a long train after staggered timing.[7]

Performance might probably require only 2–4 bunches per beam separated by 10–20 λ_{RF} ($\lambda_{RF}/c \cong 33$ ps) while the RF section filling-time is 11.1 ns. Hence, the first method seems applicable but this must still be checked.

Cumulative beam break-up due to long-range dipole modes of transverse wake fields can be severe. The transverse beam modulation is carried along the linac from accelerating section to accelerating section, through the beam. Blow-up then occurs and manifests itself along the linac as an amplitude growth from the head to the tail of the bunch train. Two possible cures have been investigated:[9,10,11]

1) damped structures; modified disk-loaded waveguides in which the power of the wake field modes is coupled out to lossy regions through radial slots in the disks and/or azimuthal rectangular waveguides, whereby the external Q-factor of the undesirable HEM mode is lowered to values typically below 20. This was studied in SLAC and KEK, the latter requiring Q-values between 15 and 70 for the predominant TM_{11} modes[12] (in order to limit the emittance growth within a factor $\sqrt{2}$ and alignment tolerances within 80 μm). The limitation of this method might come from the low Q-values required and the large number of cells involved, the practical difficulties increasing with the RF frequency.

2) staggered tuning;[9] variation in the cell dimensions in each accelerating section resulting in a cell-to-cell spread (by a few per cent) of the dipole mode frequencies. These modes are split into N_f frequency-components, whose distribution can be varied. The best frequency distribution seems to be a truncated Gaussian, since the initial roll-off of the wake is strong, with low partial recoherence within the length of a (short) bunch train.

Whilst the fabrication of damped structures has been tested,[11] the possibility of staggered tuning has been investigated by numerical simulations[13] and experimental measurements.[14] They concern a detuned 50-cavity disk-loaded structure with iris diameter ranging from 0.83 cm to 1.22 cm and Gaussian HEM frequency population centred at 14.45 GHz for a standard deviation of 1.07 GHz. The Advanced Accelerator Test Facility at Argonne makes it possible to measure the energy variation and the horizontal position of a witness bunch following the driving bunch in a time interval between 0 and 1 ns. The former yields the longitudinal wake potential and the latter gives the transverse potential as the structure is swept horizontally. Figure 8 compares calculation with experiment[14] and confirms roll-off expectation in this particular case, though it is not clear if recoherence takes place at larger distances z from the driving bunch (in NLC bunch separation is about 4 cm and a "short" train would cover ~ 3.5 m). No emittance blow-up simulations are known to the author for RF frequencies between 10 and 30 GHz, but the gain expected from staggered tuning can be illustrated by simulation results[15] obtained with 180 bunches, at lower frequency (3 GHz) and with $N_f = 25$ (Fig. 9). The effectiveness of staggered tuning is visible from the fact that without it about half of the bunches have too large amplitudes at the end of the linac, and from the difference in the scale used for the top picture (mm) and for the bottom one (μm).

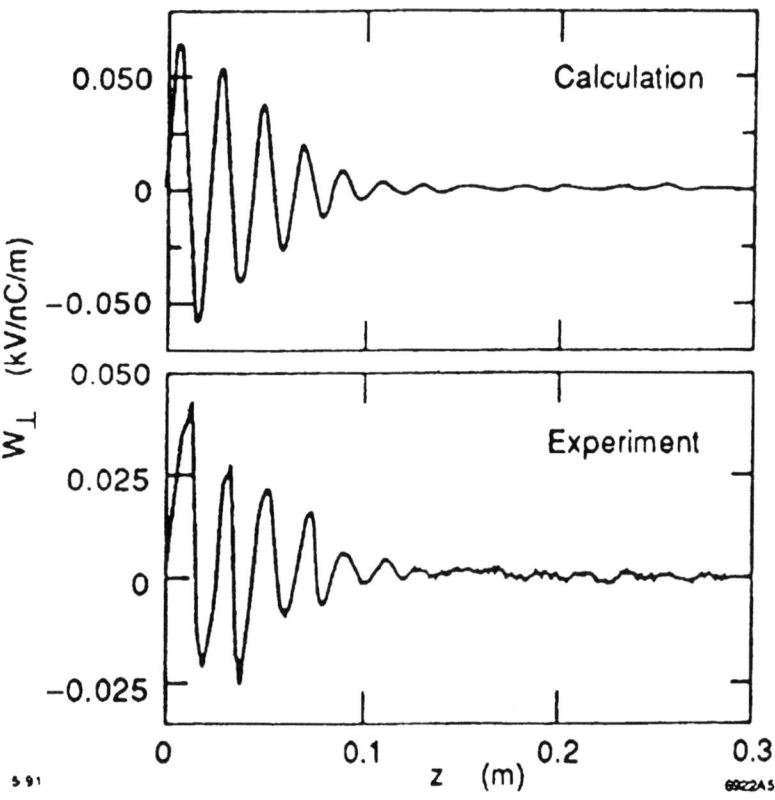

Fig. 8. Calculated and measured transverse wake potential
for a detuned 50-cavity structure.[14]

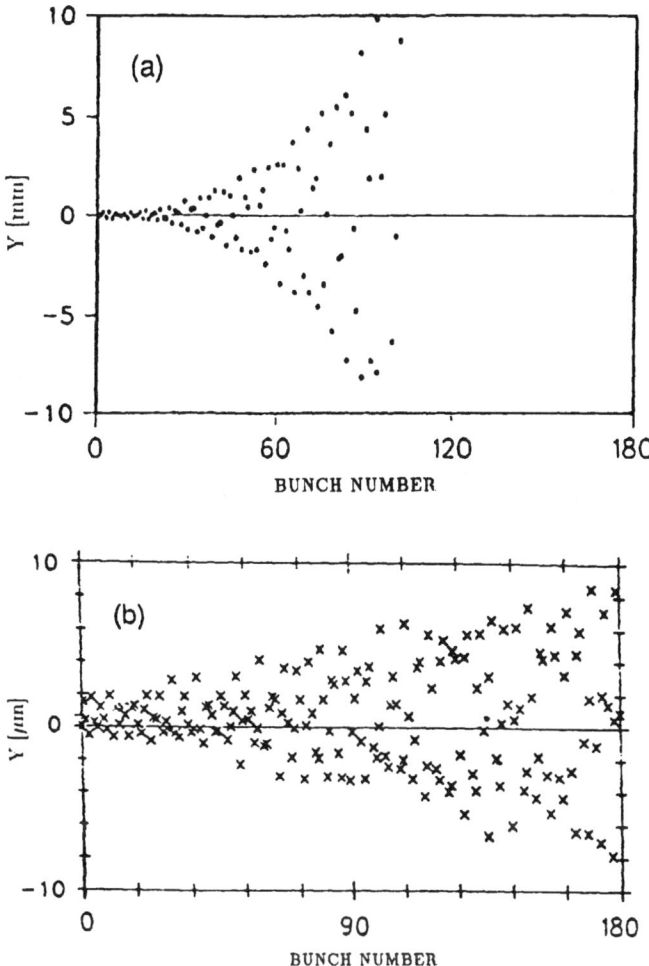

Fig. 9. Transverse bunch offsets at linac end, without (a) and with (b) staggered tuning.[15]

CONSTRAINTS ON THE FABRICATION OF ACCELERATING SECTIONS

The requirement for high gradient at high frequency calls for tight tolerances in the fabrication of the accelerating cavities. Approximately the same principles have been adopted in the design of the accelerator sections proposed at CLIC and at the JLC or VLEPP versions of NLC. The most promising manufacturing method, tested at CLIC, is the brazing of machined copper cups, and the construction of prototypes of CLIC structures[16] proved that this technique could be successfully acquired by industry.

For the CLIC 30-GHz structure, the tolerances on the cell dimensions (iris diameter of 4 mm and cell diameter of 8.7 mm) are of the order of 2–5 µm (in order to limit the total phase error to 5° over a section) and on the surface finish of about

0.05 μm (to obtain 95% of the nominal Q-value). The high-precision copper cups were made on Pneumo diamond-tool lathes, with laser interferometric feedback with 25 nm resolution. The machining accuracy consistently achieved by manufacturers would eliminate the need for dimple tuning, since the phase shift error was approximately 0.1°/cell. High-quality brazed joins were produced, with two diffusion-bonded annular surfaces at the inside and outer edges of the disks, to prevent braze leakage. Unacceptable frequency changes could be avoided during brazing operations. Complete structures (Fig. 10) with radial holes and vacuum manifolds for pumping and channels for cooling have been manufactured and measurements in the laboratory confirmed the expected parameter values.[16] In the JLC design, for example, damping slots could be incorporated in addition.

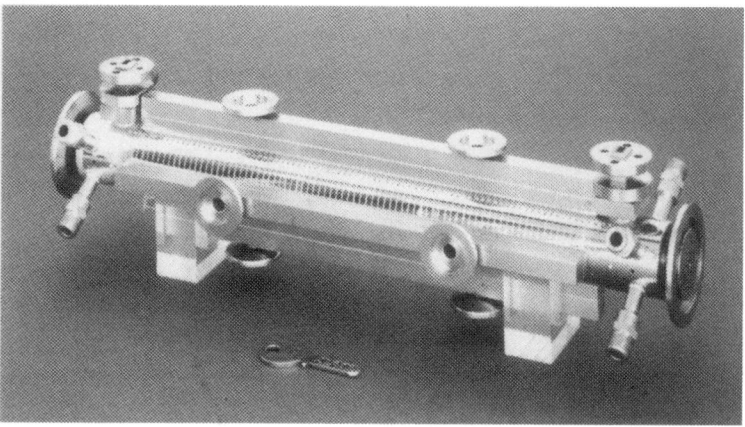

Fig. 10. Finished prototype of a 30 GHz accelerating section (CLIC).

Field-emitted electrons can create multipactoring resonance discharge at well defined field levels, and also dark currents due to the capture, bunching and acceleration of these electrons, eventually producing a parasitic beam. Recently, work was carried out on possible scaling laws[17] for these phenomena with the operating frequency. Basically, both phenomena scale with the frequency ω_0 for exact geometry scaling, but are influenced by the cavity shape near the axis for the same iris opening a. Working at high frequency with a large iris opening (or large a/λ_{RF} ratio) should be favourable in this respect. Numerical estimates[17] for CLIC cavities at 30 GHz indicate absence of electron capture and dark currents up to a gradient as high as 1400 MV/m. A possible limitation around 100 MV/m seems rather to come from second- or third-order multipactoring, in which emitted electrons lose energy by successive impacts, drift to the outer diameter, and eventually give rise to a breakdown of the structure. Further investigations are needed to better understand these phenomena that already look more critical at the lower end of the frequency interval considered.

TOLERANCE ON THE ALIGNMENT OF LINAC ELEMENTS

The dominant effect associated with misalignments is the blow-up of the projected emittance resulting from the incoherent motions due to transverse dipole wake fields (as already mentioned above). The acceptable growth of the normalized emittance $\gamma\varepsilon_y$ may vary from only 14% to a factor 3 or 4 depending on the proposal requirements. The subsequent alignment tolerances can be deduced from these numbers assuming first a simple one-to-one trajectory correction aiming at centring the beam in each position monitor (BPM) by moving the preceding lattice quadrupole. To give examples, the corresponding alignment tolerances for CLIC and NLC (SLAC version) are the following (r.m.s. values):

CLIC   3 µm on quadrupoles, 5 µm on structures,

NLC   7 µm on quadrupoles, 4 µm on structures.

In order to relax these tolerances while keeping the same acceptable emittance growth, a trajectory correction more efficient than the one-to-one method has to be applied. SLAC[18] proposed compensating for the dispersion while correcting the orbit and minimizing the wake-field dilutions caused by the corrected trajectory. The minimization procedure developed for achieving this is a weighted least-squares that minimizes the following sum:

$$\Phi = \sum_{j\in\{BPM\}} \frac{\left(\Delta y_{QF}\right)_j^2}{2\sigma_{prec}^2} + \frac{\left(\Delta y_{QD}\right)_j^2}{2\sigma_{prec}^2} + \frac{y_j^2}{\sigma_{al}^2 + \sigma_{prec}^2} = \Phi_{min} \qquad (15)$$

where y_j is the corrected trajectory amplitude, Δy_{QF} the difference trajectory resulting from both QF-variations and corrector adjustments (idem for Δy_{QD}, related to QD-variations however), σ_{prec} is the r.m.s. precision of the BPM readings and σ_{al} the r.m.s. BPM misalignments. Since the QF- and QD-fields are opposite and the QF- and QD-strengths are both supposed to be decreased when measuring Δy_j, the sum of these two variation terms mimics the effects of the dispersion (trajectory shift with momentum or quadrupole-strength deviation), while their difference mimics the effects of the wake field (sign depending only on the side where the trajectory is off-centred and not on the quadrupole field polarity). With this algorithm, the NLC (SLAC) alignment tolerances could be relaxed to 70 µm for both the quadrupoles and the accelerating structures. If all the quadrupoles are simultaneously detuned, only one difference Δy corresponding to the dispersion appears in the function Φ. Minimizing it was called dispersion-free correction while minimizing (15) was termed wake-free correction by its author.[18] Figure 11 shows beam distributions after one-to-one, dispersion-free and wake-free corrections in NLC. Plots on the left are projections onto the y-y´ phase plane, while right-hand plots are projections onto the y-z plane (z: longitudinal coordinate). One sees the strong dilution in (a), reduced to wake-field tail bend in (b) and minimized in (c).

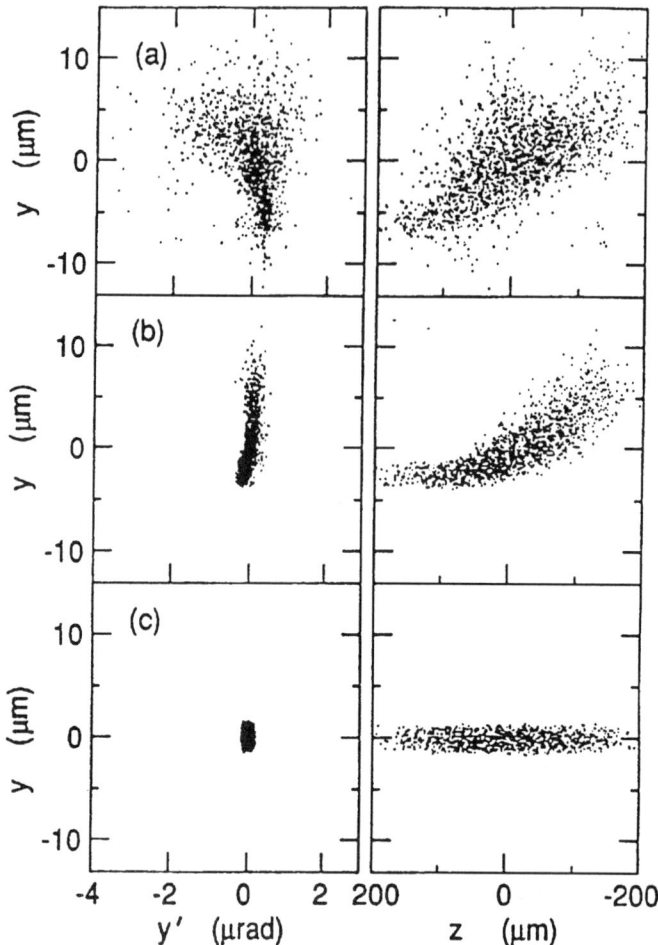

Fig. 11. Beam distributions after one-to-one (a), dispersion-free (b),
and wake-free (c) correction in NLC.[18]

CLIC started from this idea and focused on an achromatic trajectory correction[19] to higher order, developing the trajectory differences Δy_j (measured) and ΔY_j (due to corrections) in $\delta = \Delta p/p$,

$$\Delta y_j = \sum_n a_n^j \delta^n , \quad \Delta Y_j = \sum_n A_n^j \delta^n \qquad (16)$$

Calling y_j the measured trajectory and Y_j the one due to corrections, the function to be minimized becomes,

$$\Phi = \sum_{j\in\{BPM\}_y} \left\{ w_0 \frac{y_j + Y_j}{\sigma_{al}^2 + \sigma_{prec}^2} + \sum_{n\geq 1} w_n \delta^{2n} \frac{a_n^j + A_n^j}{2\sigma_{prec}^2} \right\} = \Phi_{min} \qquad (17)$$

where the sum applies to the quadrupoles focusing in the plane considered, in order to avoid too large a wake dilution. Each variation term of the second sum in Eq. (17) represents the dispersion order by order. In the applications, second-order corrections are actually implemented and the arbitrary weights w_n are used to optimize the results. With this algorithm, the CLIC alignment tolerances could probably be relaxed to about 5 μm or more for the quadrupoles and about 10 μm for the accelerating structures. The efficiency of this correction is illustrated in Fig. 12 showing a gain of about 3 in the emittance dilution in CLIC, after 12 km, for the same misalignments (the normalized initial emittance amounts to $0.5 \cdot 10^{-6}$ rad m).

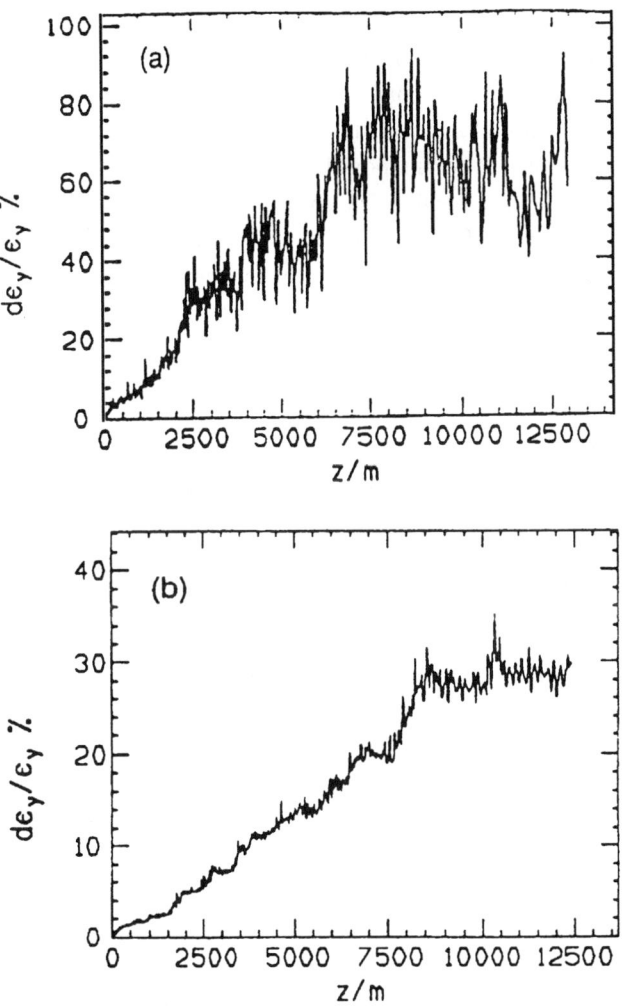

Fig. 12. Blow-up in CLIC main linac after one-to-one (a)
and achromatic (b) correction.

MICROWAVE KLYSTRONS AS RF POWER SOURCES

Conceptual designs of linacs for future colliders operating around 11 or 14 GHz call for microwave klystrons able to deliver as much as 100–200 MW in pulse lengths of the order of 1 μs. These requirements cannot be satisfied with existing microwave tubes and new klystron designs meet a certain number of challenges briefly recalled hereafter. The maximum power capability is limited by the area available to dissipate beam or RF losses and shrinking with the inverse of ω_0^2. Good power transfer efficiency from the beam to the output circuit and possible release of the intrabeam space charge forces favour high RF voltage V as well as low perveance defined by $I/V^{3/2}$; this implies a large RF gradient across the output gap. Permitting greater beam current I and power makes it necessary to achieve a higher beam convergence that involves better confinement and more precise beam optics. Finally, these high-current, high-voltage conditions increase the risk of failure mechanisms limiting the power; this concerns possible RF breakdowns mainly in the output circuit as well as intrapulse heating due to beam interception, the two mechanisms being interrelated. Typical figures considered are 40–50% for the power transfer efficiency, about 1 or 2 $\mu A/V^{3/2}$ for the perveance, beam convergence as high as 200, a gradient lower than 6 MV/cm in the output gap and a scraped beam fraction below 1%.

There are different means envisaged for trying to find solutions. Working at low perveance by increasing V is an important element for beam control, as empirically demonstrated for operational RF sources. Using excellent beam optics near the output circuit and reducing gap voltages by replacing the resonant cavity by multiple gaps or travelling-wave output are possible improvements. The klystron gun voltage can be divided by many intermediate anodes in order to provide the correct potential profile for beam formation and focusing, as in the VLEPP klystron design.[20] In this intricate design (Fig. 13), the beam is composed of many separate "beamlets" produced by different regions of the cathode and is switched by a non-intercepting control electrode (offering the possibility of pulsing the klystron using a quasi-d.c., high-voltage supply and a low-voltage modulator in series with the grid). In order to achieve very low perveance, a "cluster klystron" is proposed by a BNL–SLAC collaboration;[21] it is a collection of 42 separate beams, each comprising a 40-MW klystron and all sharing a common superconducting solenoid. Finally, the use of lumped shavers in order to control beam interception might be necessary.

Some characteristics of high-power klystron development projects, incorporating (separately) the features mentioned, are given in Table III. Achievements so far, as quoted in a recent review paper,[22] are briefly summarized below, together with the reasons for the limitation observed. The SLAC XC klystron achieved 40 MW, 0.8 μs or 72 MW, 0.1 μs pulses (Fig. 14), performance being

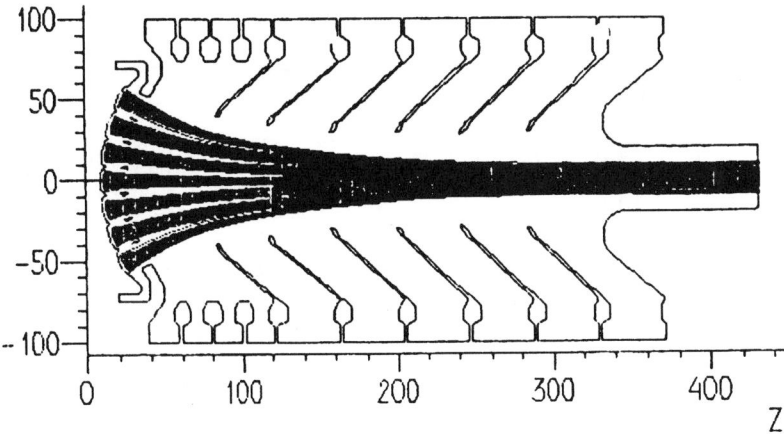

Fig. 13. The VLEPP X-band klystron's gun.

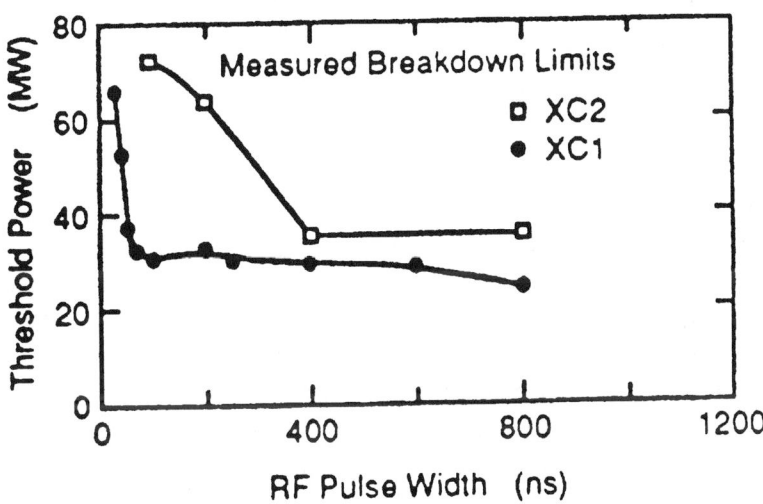

Fig. 14. Performance of the first X-band SLAC klystrons.[22]

limited by RF breakdown in the double-gap output circuit stimulated by beam
interception. In the KEK klystron, the output power was restricted to 22 MW by
failure of the gun ceramic insulator. In the VLEPP klystron, 50 MW has been
achieved, limited by a substantial beam loss that amounts to 70 out of 200 A. The
cluster klystron is still at the design stage.

Table III: High-power klystron projects.

Klystron	Wave-length (cm)	Power out (MW)	Pulse length (μs)	Voltage (kV)	Microper-veance (μA/V$^{3/2}$)	Frequency (GHz)
SLAC 5045	10.5	67	3.5	350	2.0	2.86
SLAC	10.5	150	1.0	450	2.0	2.86
SLAC XC	2.6	100	1.0	550	1.2	11.4
KEK	2.6	120	1.0	550	1.2	11.4
VLEPP	2.1	150	1.0	1000	0.3	14.2
Cluster	2.6	1680	1.0	400	0.4	11.4

DRIVE LINAC AS HIGH-FREQUENCY POWER SOURCE

Owing to difficulties met in the development of microwave klystrons operating in the frequency range of 11–14 GHz, it is unthinkable to use similar tubes to deliver the 30-GHz, ~150 MW/m peak power per unit length required in the CLIC linac structure. The CLIC scheme for generating the necessary power is based on a two-stage accelerator;[23] there is a drive beam that runs parallel to the main beam and ensures power flow to the main linac. The drive linac contains strings of travelling-wave transfer structures, in which short and intense bunches induce the required power that is then fed to the main linac, and sectors of superconducting (SC) cavities supplying energy to the beam when necessary. The use of SC cavities to reboost the drive beam as well as to accelerate it up to its initial energy is dictated by the concern of good extraction efficiency; LEP-type cavities operating at ~ 350 MHz with a gradient of ~ 6 MV/m are considered. The drive-beam energy should be of the order of a few GeV, which implies no longitudinal mixing inside the bunches and no phase slip with respect to the main beam. Owing to the unavoidable intermittent reacceleration, the drive beam has to be arranged in discrete trains of dense bunchlets that all contribute to the build-up of a decelerating field in the transfer structure and are separated by the 30-GHz wavelength. To generate a pulse of length equal to the main structure filling time τ_0 (11.1 ns), four such trains are needed, separated by the 350-MHz wavelength. The total charge needed per train is about 1.65 μC and the number of bunchlets per train depends on the matching of the decelerating field build-up to the SC cavity accelerating gradient. Simply minimizing the deviation of the linear build-up from the sinusoidal acceleration wave limits this number to 11, but the use of voltage harmonics schemes[24] (their sum giving almost a linear function as shown in Fig. 15) might allow for 43 bunchlets (with correspondingly lower charge per bunchlet).

This drive-linac conceptual design implies certain challenges which have been addressed, mainly the bunchlet generation, the transfer structure design and the beam

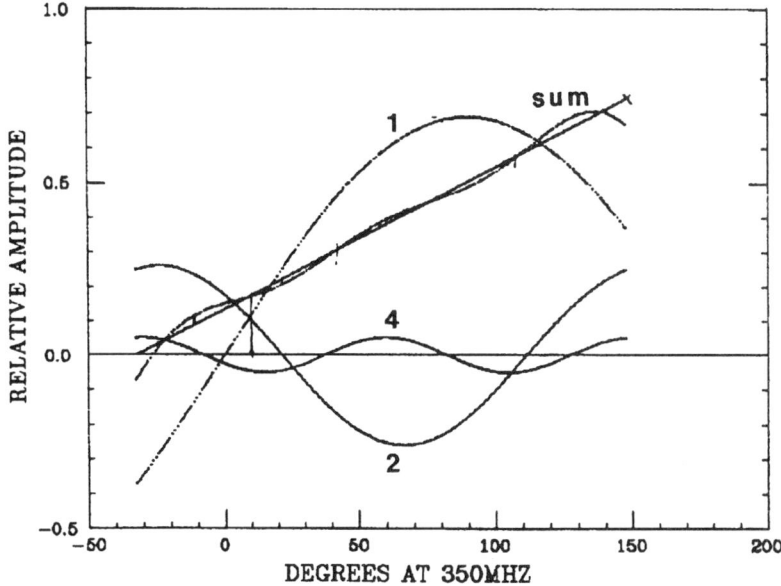

Fig. 15. Accelerating ramp obtained with 3 RF harmonics.

dynamics control. Owing to the difficulty of generating short (1 mm r.m.s.) and dense (up to 10^{12} particles) bunchlets, a test facility (CTF) has been built.[25] It includes an RF gun, a beam line acting as magnetic spectrometer, acceleration to 60 MeV at 3 GHz and RF power generation at 30 GHz (using prototype structures). A charge up to 30 nC per beam should be obtained using a laser-driven photocathode, synchronized with the RF. During first tests using a prototype of the main linac structure instead of an actual transfer structure, a 48-kW power pulse was extracted. The most recent transfer structure design proposed,[26] is based on power-collecting rectangular waveguides that run along the outside of the beam pipe (either on each side or above and below) and are coupled to the inside via slits about 0.5 m long (Fig. 16). The phase velocity in the waveguide is adjusted by periodic indentations. Numerical calculations and model work are being carried out in parallel to check the possibility of generating the required flat power pulse and the amplitude of the wake fields. This design must indeed provide the low impedance needed ($\sim$ 4.5 Ω/m for R´/Q) and the required decelerating field per bunchlet ($\sim$ 65 kV/m for a population of 10^{12}), while minimizing the undesirable wake-field modes that could compromise beam stability. The dynamics of such a beam includes special features: the energy differences between bunchlets are unusually large owing to the increasing decelerating field, the energy spread within each bunchlet is wide since the bunch length is not short with respect to λ_{RF}, and the amplitudes of the synchronous wake fields may be disturbing. The impact of these features on the beam transport and

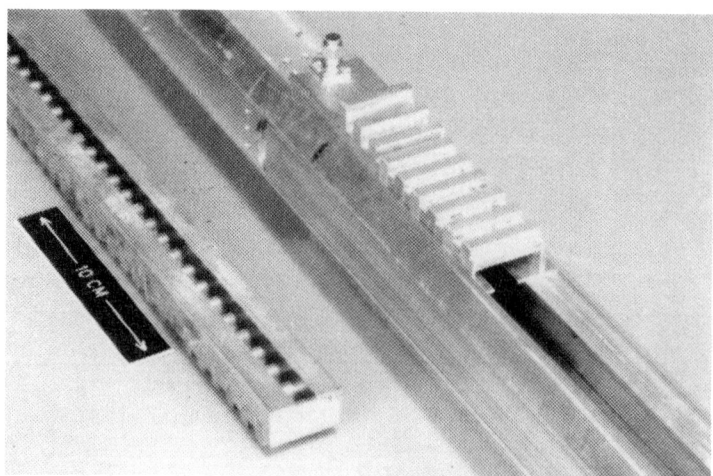

Fig. 16. Model of the most recent transfer structure design.

dynamics, in the presence of alignment imperfections, trajectory correction, variable wakes and magnetic focusing, is being investigated by numerical simulations.[27] Figure 17 gives an example of initial- and final-energy distributions in a train of 11 bunches travelling over ~ 3.5 km. Phase plots of the emittances (Fig. 18) show that all the bunchlets remain within the beam-pipe acceptance (circle tangent to the frame), with the assumptions retained for this calculation, in particular no synchronous, slowly damped transverse wakes. Further investigations are needed to check the feasibility of the scheme.

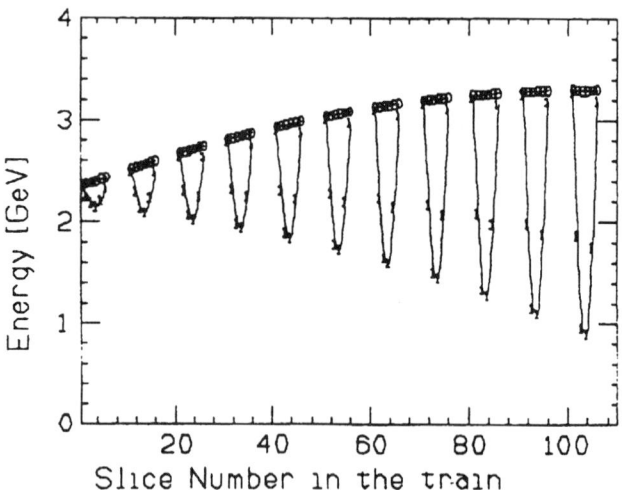

Fig. 17. Initial (0) and final (1) energy distributions in a drive-beam train.

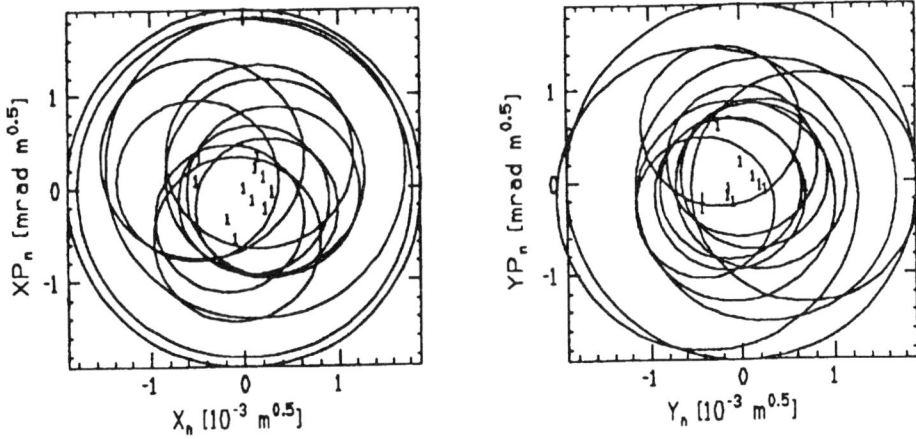

Fig. 18. Bunchlet emittances (H and V) at the drive-linac exit.

CONCLUSIONS

The choice of the RF-frequency in a high-gradient normal-conducting linac will probably fall in the 10–30 GHz interval and result from a compromise between the opposing requirements of saving power and minimizing harmful effects. A lot of studies and simulations improved the knowledge of the mechanisms involved in wake field effects and beam stability in the presence of one bunch or several bunches per beam. They revealed a number of promising correction possibilities aiming at low energy-spread and small transverse emittances at the exit of the linac. The question of the tight alignment tolerances required remains challenging, but looks solvable if good alignment strategies are defined. The idea of using microwave quadrupoles for stabilizing the beam in the presence of strong wave fields has been reinforced by numerical simulations. Model work on high-frequency accelerating structures proved that they could be manufactured by industry and recent studies indicate that the risk of dark currents decreases at higher frequency. Microwave klystrons for 250-GeV and 50-MV/m linacs seem feasible in the near future, as do the companion RF-pulse compression systems. Prospects for the peak power requirements at 500 GeV and 100 MV/m are however more distant, since technical limitations have to be overcome. For the CLIC scheme based on a two-stage accelerator, significant progress has been made on the study of the drive-beam dynamics and the design of the transfer structures. In this case, challenging issues are the generation of the required short and dense bunchlets, and the development of efficient transfer structures that are least harmful to the drive beam.

REFERENCES

1. R. B. Palmer, SLAC-PUB-4295, 1987;
   W. Schnell, SLAC/AP-61, 1987.

2. K. L. F. Bane, SLAC-AP-76, 1989;
   G. Guignard and C. Fischer, Proc. Part. Acc. Conf., San Francisco, 1991, Vol. 5, p. 3231.

3. V. E. Balakin, Inst. of Nucl. Phys., Novosibirsk, Preprint 88-100, 1988.

4. H. Henke and W. Schnell, CERN-LEP-RF/86-18, 1986.

5. W. Schnell and I. Wilson, Proc. Part. Acc. Conf., San Francisco, 1991, Vol. 5, p. 3237.

6. G. Guignard, CERN SL/91-19 (AP), 1991.

7. K. A. Thompson, SLAC AAS Note 71, 1992.

8. R. Ruth, SLAC-PUB-4541, 1988.

9. R. B. Neal (editor), The Stanford Two-Mile Accelerator
   (W. A.Benjamin Inc., New York, 1968).

10. R. B. Palmer, Proc. DPF Summer Study, Snowmass, SLAC-PUB-4542, 1988.

11. H. Deruyter et al., Proc. Linear Acc. Conf., Albuquerque, 1990, p. 132.

12. T. Higo et al., Proc. Linear Acc. Conf., Albuquerque, 1990, p. 147;
    T. Taniuchi et al., KEK Preprint 91–152, 1991.

13. K. A. Thompson and J. W. Wang, Proc. Part. Acc. Conf., San Francisco, 1991, Vol. 1, p. 431.

14. J. W. Wang et al., Proc. Part. Acc. Conf., San Francisco, 1991, Vol. 5, p. 3219.

15. N. Holtkamp, Private Communication;
    T. Weiland (spokesman) et al., DESY 91-153, 1991.

16. I. Wilson, W. Wuensh, C. Achard, Proc. EPAC 90, Nice, 1990, p. 943;
    I. Wilson and W. Wuensh, CERN-SL/90-103 (RFL), 1990.

17. R. Parody, Private communication, CERN, 1991.

18. T. Raubenheimer and R. Ruth, SLAC-PUB-5355, 1990;
    T. Raubenheimer, SLAC-387, UC-414, 1991.

19. C. Fischer and G. Guignard, Proc. EPAC 92, Berlin, 1992.

20. L. N. Arapov, V. E. Balakin, Y. Kazakov, Proc. EPAC 92, Berlin, 1992.

21. R. B. Palmer, W. B. Herrmannsfeldt, K. R. Eppley, Part. Acc. **30**, 197–209 (1990).

22. T. L. Lavine, Proc. EPAC 92, Berlin, 1992.

23. W. Schnell, CERN-LEP-RF/88-59, 1988.

24. L. Thorndahl, CLIC Note 152, CERN, 1991.

25. Y. Baconnier et al., Proc. Linear Acc. Conf., Albuquerque, 1990, p. 733.

26. G. Carron and L. Thorndahl, Proc. EPAC 92, Berlin, 1992.

27. G. Guignard, CERN SL/92-22 (AP), 1992.

ACKNOWLEDGEMENTS

The author is particularly grateful to N. Holtkamp, T. L. Lavine, K. A. Thompson and L. Thorndahl who have kindly made their most recent results available to him.

RECENT ACTIVITIES IN ACCELERATOR CODE DEVELOPMENT*

Richard K. Cooper and Robert D. Ryne
Accelerator Theory and FEL Technology Group, AT-7
Los Alamos National Laboratory, Los Alamos, NM 87545

ABSTRACT

In this paper we will review recent activities in the area of code development as it affects the accelerator community. We will first discuss the changing computing environment. We will review how the computing environment has changed in the last 10 years, with emphasis on computing power, operating systems, computer languages, graphics standards, and massively parallel processing. Then we will discuss recent code development activities in the areas of electromagnetics codes and beam dynamics codes.

INTRODUCTION

In this paper we will review recent activities in the area of code development as it affects the accelerator community. We cannot hope to be completely comprehensive in our coverage, but will try to give examples of development that reflect the trends in the field. In many cases these activities are driven by the changing computing environment. In other cases the activities are driven by the desire for more detailed knowledge of beam interactions in three dimensions, for greater spatial resolution, or for long time sequences. As computer speeds increase, and as computer memories get larger, we are able to perform calculations that we would not have thought of attempting even ten years ago.

The last ten years has been a most remarkable period of computer development. In August 1981 IBM introduced the PC. In 1982, IBM introduced the PC in Europe, Compaq introduced its IBM-compatible PC, and Sun Microsystems was founded and shipped their first workstation. The Macintosh[†] was introduced in early 1984.

Ten years ago the supercomputer market belonged to Cray and Control Data Corporation. The memories of these machines were typically one or two megawords, so that while the machines were fast, the size of problems that could be efficiently run was limited.

The past few years has seen a blurring of the boundaries between those applications requiring a supercomputer, those requiring a workstation and those requiring a PC. Today, if one is faced with running a one hour job on a workstation

---

*Work supported by U.S.D.O.E., Office of High Energy and Nuclear Physics, Office of Basic Energy Sciences, Office of Fusion Energy, Office of SSC, Office of Energy Research, and Office of Scientific Computing
†Macintosh is a registered trademark of Apple Computers, Inc.

or the equivalent two minute job on a Cray, one might just choose the workstation; since the Cray is shared among many users, one could possibly obtain his or her answers more quickly on the desk-top machine. Increasingly, workstation capabilities are being measured in significant fractions of a Cray-1. And workstation memories of 32 megabytes are increasingly common.

Along with hardware development we have seen the development of multitasking operating systems. Most notably, the UNIX[‡] operating system has become widespread among the scientific workstations and the supercomputer centers. The DOE-funded National Energy Research Supercomputer Center (NERSC) Crays are now running UNICOS, a UNIX-like operating system, as is also the DOE-funded Supercomputer Research Institute (SCRI) Cray Y-MP at Florida State University. The Crays at Los Alamos and at Argonne National Laboratory are also running UNICOS. The UNIX utilities such as **make** simplify the code maintenance effort a great deal, and add to the portability of many codes.

This commonality of operating systems across the scientific computing world has made it easier for the scientist to do his work on a variety of platforms and to begin to develop truly portable software. The adoption of the X-Window protocol[1] as an industry standard makes it possible to develop truly portable graphics applications as well. The widely adopted PostScript[§] page-description language has also made it possible to develop portable graphics applications.

The standardization of FORTRAN in FORTRAN 77[2], and more recently in FORTRAN 90[3], has been a real help to those who wish to produce portable software. The unfortunate omission of namelist and bit manipulation functions from FORTRAN 77 has limited its flexibility (and that of programmers) somewhat. The C programming language[4] has become a language with which to reckon. Like FORTRAN 77 and FORTRAN 90, C has an ANSI standard definition[5]. It may be that many feel that C is being forced down their throats by the computer scientists, but its popularity is due to the fact that it gives the programmer more flexibility and control than FORTRAN. To quote from the authors[6] of *Numerical Recipes in C*:

> In our view, C is now the only computer language that has a measurable chance of wresting the monopoly on scientific computing away from FORTRAN. C is superior to FORTRAN in a number of significant ways. Further, it has a "closeness to the machine" that is appealing to experienced FORTRAN programmers — and not shared by other high-level languages, notably Pascal. And it has a very high degree of portability from one machine to another.

UNIX operating systems are written in C, and C is often the only compiler that comes with a new workstation. The only current binding of the X-Window protocol is to the C language. In addition to traditional programming languages,

---

[‡]UNIX is a registered trademark of AT&T.
[§]PostScript is a trademark of Adobe Systems Incorporated.

there is now a great deal of interest in object-oriented techniques. This includes, for example, the C++ language ¶, which is an object-oriented extension[7] of C. Object-oriented techniques are already having a major impact in some areas, such as those involving graphical environments. However, the impact that these techniques will have on scientific computing is not clear.

The future (and it is now upon us) will include massively parallel computers, that is, computers carrying out the same calculation on many different data sets all at the same time. We have already seen[8] an implementation of the finite-difference time-domain algorithm for wakefield calculations programmed on the CM-2 computer, and it is to be anticipated that such applications will grow in number in the near term.

With the advent of massively parallel computers, such as the Thinking Machines Corporation Connection Machine (CM-2 or the newer, more powerful, CM-5), we have the opportunity to apply a new technology to the way we compute solutions to problems in the accelerator field. For some problems, such as the finite-difference time-domain calculation of charged-particle beam-generated fields in accelerating structures, the Connection Machine appears to be almost purpose-built. Other problems, such as the solution of Poisson's equation for space-charge fields, can benefit from parallel algorithms implemented on the CM-2 or the CM-5.

As an example of a massively parallel computer, consider the CM-2, which has 65,536 (single bit) processing elements. Each processor has its own memory for an aggregate of 8 gigabytes, with an additional disk storage of 20 gigabytes. Some of the features of the CM-2 which permit efficient parallel implementation of finite-difference codes include:

- **Virtual Machine Architecture.** One physical processor can simulate many virtual processors in a user-transparent manner. Each mesh cell can be assigned to a virtual processor, so the actual number of cells used may exceed the number of physical processors.

- **Fast Interprocessor Communication.** The CM-2 supports fast data transfer between nearest neighbors in the mesh geometry used for computing differences of fields on adjacent mesh cells. More general interprocessor data transfers are possible.

- **Flexible Problem Control.** Individual processors may be active or inactive depending on the state of the context flag, a 1 bit field on each processor. This feature can be used to mask off cells requiring special handling, as in the treatment of boundary conditions.

The slicewise model of computing presently employed on the CM-2 modifies the numbers above by clustering 32 or 64 single bit processors for efficient floating point operation. The Connection Machine remains a truly massively parallel

---

¶C++ is a trademark of AT&T.

machine with potential for significant speed increases over conventional supercomputers. The CM–5 will bring the latest technology developments to the field of massively parallel computing.

The CM–2 is a single-instruction, multiple-data machine, which implies that all processors run in lockstep, performing the same operation at the same time on different data streams.

The CM-5 purchased by Los Alamos is a massively parallel computer with 1024 processors. In its largest configuration (16 000) nodes, this computer can scale to a peak performance of a teraflop. Each node of the CM–5 is a 22-MIPS RISC Sun Sparc microprocessor containing four vector pipes.

The CM–5 will operate in a timesharing environment, and unlike the CM–2, can perform in multiple-instruction, multiple data (MIMD) mode.

CODE DEVELOPMENT

A. Electromagnetics Codes

A.1 POISSON/SUPERFISH

A suite of codes that has spanned the period of computer development from the middle ages ($\approx$ 1965) to the present is the POISSON group[9] of codes (SUPERFISH dates[10] from 1976). Release 3.0 of POISSON/SUPERFISH was officially distributed[11] in early 1992. In addition to major changes in the organization of the codes, Release 3 also contains a number of bug fixes and physics enhancements. These include corrected stored energy calculations and changes related to the built-in permeability tables. Release 3.0 is installed on the Crays at NERSC and at SCRI for the use of DOE researchers. This release comes in response to many requests for a "UNIX" version of the codes. We have developed a Makefile approach to compiling the codes in most UNIX systems, and have developed a script for changing the parameter MXDIM in all the codes with a single command.

In the old days, memory was scarce, and it was necessary to pack as much information into a computer word as possible in order to use all the available bits efficiently and permit the running of reasonable-size problems. This packing of information required unpacking, i.e. shifting and masking operations. Due to peculiarities of the coding, this packing also limited mesh sizes to 32767, namely $2^{15} - 1$. Over the years these bit operations have been a constant source of annoyance in moving the codes from one installation to another. In Release 4.0 of the POISSON/SUPERFISH codes (which is now undergoing pre-release testing), these shifting and packing operations have all been removed. Furthermore, the namelist input (for which there is no standard) has been replaced by FORTRAN 77 routines that encompass most implementations of namelist. The above changes make Release 4 more portable than previous releases, and able to handle problems with a very large number of mesh points. Also, Release 4 will contain several long overdue features. For example, we intend to provide arrow plots as well as the familiar (and confusing) plots of contours of constant values of rH_ϕ. Additionally, the popular FRONT program will be included in this release. We will also eliminate the confusing $x, y \rightarrow r, z$ in POISSON versus $x, y \rightarrow z, r$ in SUPERFISH inconsistency by allowing r, z input in addition to x, y input. It is worth pointing out that, in Release 4, the tape35 binary file used for communication between the various codes will not be compatible with older versions.

In response to the oft-expressed desire for portable graphics free from proprietary software, we have responded in two ways. First, starting with release 3.0, we have distributed a source file called pltgen.f which, when compiled, acts just like the old TEKPLOT, but which produces only a PostScript[12] file with the graphics plots. It does not produce screen graphics, and thus can be run from a dumb terminal and provides truly portable graphics. The pltgen.f file can also be compiled, after having been preprocessd by a program called FIXPLT, to produce graphics using a variety of graphics packages (GKS, DISSPLA, PLOT10). The second approach we have taken to provide portable no-cost graphics is to develop an X-Window version of the code earlier known as TEKPLOT (called PSFPLOT starting with Release 3). This version uses the XView∥ widget set[13], which is available via anonymous **ftp** from MIT. This plotting program (currently named XPP3 for Release 3 and XPP4 for Release 4) provides for mouse-selected menu-driven operation with no typing required. An optional PostScript plot file can be generated which incorporates only those views selected by the user by clicking on the Print button. This X-Window plotter is available to friendly users of Release 3 and will be released officially with Release 4. The program has been compiled and run on Sun-3s, Sun-4s, and UNICOS Crays at LANL, NERSC, and SCRI.

As alluded to in the introduction, many codes that formerly ran only on mainframes now run on scientific workstations and on PCs. The POISSON group is now available on all these platforms. The Los Alamos Accelerator Code Group maintains versions that run on mainframes and workstations. The PC field has been left to others. PC versions of POISSON and SUPERFISH have been available for several years from BNL (see the entry in the LAACG Compendium[14]). The POISSON code is commercially available from S. Humphries' Acceleration Consultants (again refer to Ref. 14) with a new name, EMP2, reflecting the extensive interfacing work that has been done to make the code both more user friendly and useful. Recently the Los Alamos AT–1 group has announced the availability of a PC version of SUPERFISH (for 386 and 486 computers only). The group will supply executable code and documentation to anyone on request**, but source code is not available.

A.2 MAFIA

∥XView is a trademark of Sun Microsystems, Inc.

**If interested in this version, contact Group AT–1, MS H817, Los Alamos, NM 87545, phone (505)667-6627.

The MAFIA family of codes[15] is one of the most widely used three-dimensional codes in the accelerator community. The original development of these codes was the result of a collaborative effort of groups at Los Alamos, DESY, and KFA-Jülich. Release 2.04 is available to users on the Crays at NERSC and is supported by the LAACG. A number of post-processors for release 2.04 have also been developed[16,17,18].

The latest release, MAFIA 3.1, runs on a number of platforms, including many workstations, and must be obtained in executable form from the principal author, T. Weiland. A new release, MAFIA 3.2, is in preparation. It is expected to include a number of enhancements, including: an X-Window interface, speed improvements, the ability to use $r - \phi - z$ coordinates, a shape editor and fully automatic optimization.

A.3 ARGUS

ARGUS, like MAFIA, is a family of three-dimensional simulation codes[19]. It was developed by Science Applications International Corporation (SAIC), and now, with support from the U.S. Department of Energy, it is being readied for widespread distribution. SAIC and LAACG will collaborate on release and user support of ARGUS. It will be free to DOE laboratories and contractors, and it will include source code and documentation. The official release is expected to occur early in 1993. Prior to that time, code enhancements are taking place, including a new (Motif-based) user interface, a complex eigenvalue solver, and rate equations for steady-state PIC calculations.

A.4 AMOS

AMOS is a $2\frac{1}{2}$-dimensional, finite-difference time-domain simulation code developed at LLNL[20]. It uses an interactive, window-based preprocessor (called DRAGON) for simple generation and editing of complex simulation models. It includes a postprocessor for the calculation of wake potentials and impedances, and a graphical postprocesser (AMOSGR) for $x-y$ plots and vector field plots. AMOS supports dielectric, permeable, and lossy materials. Furthermore, it can treat dispersive media, and it is often used to design accelerator components that include ferrite.

B. Beam Dynamics Coding

Recently there has been much activity in the development of tools for beam dynamics calculations, with an emphasis on particle tracking and analysis of maps. One of the earliest codes of this type was the program TRANSPORT[21]. Later, an alternate (though equivalent) approach based on Lie-algebraic tools was developed, and was the basis of the program MARYLIE[22]. The next major development was the introduction of tools by Berz that used Differential Algebra (DA)[23]. This work made it possible to perform calculations of essentially arbitrarily high order, as was implemented in the program COSY Infinity[24] . Today there are a number of codes available for performing a wide variety of beam optics calculations to very high order. Two of these codes are discussed below.

B.1 TRACY

In the area of beam optics calculations for circular machines, the work of Bengtsson, Forest and Nishimura (all at LBL) is notable. They have developed the program TRACY[25]. A major tenet of their approach is that the production of maps and the analysis of maps should be completely decoupled. (This view is especially valuable when dealing with nonideal beam transport systems, i.e. systems with errors). They rely heavily on Automatic Differentiation (of which DA is one implementation) and Symplectic Integrators. Their codes also treat radiative effects.

This group has also adopted a novel scheme to implement their programs. Previously, the major beam transport codes contained parsing subroutines (usually written by physicists) to interpret commands in the input file of these codes. In order to perform very complex tasks, these programs developed increasingly complex front ends. Instead, Bengtsson et al. have made use of PASCAL-S, a portable, self-contained subset of the PASCAL language that contains a compiler, interpreter, and most of the PASCAL language itself, all in about 5000 lines of code. Bengtsson extended PASCAL-S to contain not only most of the PASCAL language, but also a number of commands used in the analysis of beam transport systems. The input file to TRACY is in fact a subroutine written in this extended version of PASCAL-S. Clearly, one can write an input file to perform very complex tasks when one has a complete programming language at one's fingertips.

B.2 TLIE

In the area of very high order optics calculations, the work of van Zeijts (University of Maryland) and Neri (LANL) is notable. They have developed the program TLIE, which, like several other programs, can obtain maps to very high order. Additionally, they have incorporated powerful fitting and optimizing routines. For example, with TLIE one can choose to vary parameters to remove all third-order aberrations in a system while minimizing fifth order aberrations. They have also incorporated a flexible scheme for generating magnetic field profiles, and they have incorporated space charge to a limited extent. Like TRACY, the program TLIE is unorthodox in its implementation. The TLIE input language is based on a combination of the MAD[26] input language and the HOC6 YACC program[27]. The power and flexibility of the TLIE input scheme set a new standard for beam dynamics calculations.

REFERENCES

1. *X Protocol Reference Manual*, Volume Zero of The X Window System Series, O'Reilly & Associates, Inc., Sebastopol, CA 94572.

2. American National Standards Institute, Inc., "American National Standard Programming Language, FORTRAN," Standard ANSI X3.9-1978, (New York, 1978).

3. Report X3.198-1991 of X3J3, the Fortran Technical Committee of the American National Standards Institute; also report ISO/IEC 1539:1991.

4. Brian W. Kernighan and Dennis M. Ritchie, *The C Programming Language, Second Edition,* Prentice Hall Software Series, Prentice-Hall, Englewood Cliffs, New Jersey, 1988.

5. American National Standard for Information Systems – Programming Language – C, ANSI X3.159-1989, American National Standards Institute, 1430 Broadway, New York, NY 10018.

6. William H. Press, Brian P. Flannery, Saul A. Teukolsky, and William T. Vetterling, *Numerical Recipes in C, The Art of Scientific Computing*, Cambridge University Press, 1988, p. xi.

7. Bjarne Stroustrup, *The C++ Programming Language,* Addison-Wesley Publishing Company, 1987.

8. P. Schoessow, "Wakefield Calculations on Parallel Computers," *Proc. of the Conf. on Computer Codes and the Linear Accelerator Community,* LA-11857-C, Los Alamos, 1990, p. 377.

9. K. Halbach, "A Program for Inversion of System Analysis and Its Application to the Design of Magnets," Lawrence Berkeley Laboratory report UCRL-17436 (1967), CONF-670705-14.

10. See for example K. Halbach and R. F. Holsinger, "SUPERFISH" *Particle Accelerators* **7**, 213 (1976).

11. T. Barts and J. Merson, "Users' Notes for POISSON/SUPERFISH Release 3.0", LA-UR-91-4140, February 21, 1992.

12. Adobe Systems Incorporated, *PostScript Language Tutorial and Cookbook,* Addison-Wesley Publishing Company, Inc., 1985.

13. *XView Programming Manual,* Volume 7 of the X Window System Series, O'Reilly & Associates, Inc., Sebastopol, CA 94572.

14. Los Alamos Accelerator Code Group, "Computer Codes for Particle Accelerator Design and Analysis: A Compendium," , Los Alamos National Laboratory Unclassified Release LA-UR-90-1766, May 1990.

15. R. Klatt, F. Krawczyk, W.-R. Novender, C. Palm, T. Weiland, B. Steffen, T. Barts, M. J. Browman, R. Cooper, C. T. Mottershead, G. Rodenz,S. G. Wipf, "MAFIA- A Three-Dimensional Electromagnetic CAD System for Magnets, RF Structures, and Transient Wake-field Calculations," *1986 Linear Accelerator Conference Proc.*, Stanford Linear Accelerator Center Report SLAC-303, 276-278 (1986).

16. M. Jean Browman, "Special Mafia Postprocessors and Software Tools", Los Alamos National Laboratory Unclassified Release LA-UR-92-476.

17. M. Jean Browman, "Generating MAFIA Azimuthally Symmetric Cavities from a Two-Dimensional Cross Section," Los Alamos National Laboratory internal technical note AT-7:ATN-92-3.

18. M. Jean Browman, "Program POWER," Los Alamos National Laboratory Unclassified Release LA-UR-92-475.

19. A. Mondelli et al., "Application of the ARGUS Code to Accelerator Design Calculations," Proc. 1989 Particle Accelerator Conference, Chicago, IL.

20. J. DeFord et al., "The AMOS (Azimuthal MOde Simulator) Code," Proc. 1989 Particle Accelerator Conference, Chicago, IL.

21. K. L. Brown et al., "TRANSPORT–A Computer Program for Designing Charged Particle Beam Transport Systems," SLAC-91 (1977).

22. A. Dragt et al., "MARYLIE 3.0, A Program for Charged Particle Beam Transport Based on Lie Algebraic Methods," Univ. of Maryland (1987).

23. M. Berz, "Differential Algebraic Description of Beam Dynamics to Very High Orders," Particle Accelerators 24 (1989).

24. M. Berz, "COSY Infinity, a New Arbitrary Order Optics Code," Proceedings of the Conference on Computer Codes and the Linear Accelerator Community, LANL (1990).

25. J. Bengtsson et al., "TRACY User's Manual," Lawrence Berkeley Laboratory (in preparation).

26. F. C. Iselin, "The MAD Program Reference Manual," CERN-LEP-TH/85-15 (1985).

27. B. W. Kernighan and R. Pike, "The Unix Programming Environment," Prentice-Hall (1984).

REPORT OF THE WORKING GROUP ON

FAR FIELD ACCELERATORS

Cha-Mei Tang
Beam Physics Branch, Plasma Physics Division
Naval Research Laboratory, Washington, DC 20375-5320

ABSTRACT

This report describes the accomplishments of the Working Group on Far Field Accelerators. In addition to hearing presentations of current research, the group produced designs for "100 MeV" demonstration accelerators, "1 GeV" conceptual accelerators and a small electron beam source. Two of the "100 MeV" designs, an Inverse Free Electron Laser (IFEL) and an Inverse Čerenkov Accelerator (ICA), use the CO_2 laser and the 50 MeV linac at the Advanced Test Facility (ATF) at Brookhaven National Laboratory (BNL), requiring only modest changes in the current experimental setups. By upgrading the laser, an ICA design demonstrated 1 GeV acceleration in a gas cell about 50 cm in length. For high average power accelerators, examples based on the IFEL concept were also produced utilizing accelerators driven by high average power FELs. The Working Group also designed a small electron beam source based on the inverse electron cyclotron resonance concept. Accelerators based on the IFEL and ICA may be the first to achieve "100 MeV" and "1 GeV" energy gain demonstration with high accelerating gradients.

I. INTRODUCTION

Far field laser accelerating schemes utilize electromagnetic fields to accelerate electrons at a distance far from boundaries compared to the radiation wavelength. Two such schemes, which may be the first to achieve a 100 MeV or more energy gain with a high accelerating gradient, are the IFEL and ICA.

The Far Field Working Sessions heard reports on research by the participating members. Most papers described projects utilizing IFELs and ICAs. Some new concepts were presented, including vacuum acceleration without undulators, inverse-Bremsstrahlung electron acceleration and 1 MeV storage ring damped by lasers. Highlights from many of these papers will be outlined below in their appropriate sections.

Members of the Working Group were: W. Barletta, C. Brau, C. Chen, A. S. Fisher, J. R. Fontana, J. C. Gallardo, M. Hussein, W. D. Kimura, T. Marshall, C. Pellegrini, I. Pogorelsky, S. K. Ride, J. Sandweiss, G. Shvets, P. Sprangle, A. van Steenbergen, L. C. Steinhauer, D. Sutter, C. M. Tang, G. Travish and A. A. Varfolomeev

The working group also spent half of its time designing electron accelerators outlined by D. Sutter of DOE: i) "100 MeV" demonstration, ii) "1 GeV" baseline design and iii) small source for use as injector in existing laser accelerator experiments. The guidelines are: i) to provide a first pass parameter set, ii) to identify major research and development issues and iii) to obtain a rough cost estimate. The working group produced three "100 MeV" demonstration accelerator designs, two "1 GeV" baseline accelerator designs and a small electron beam source design. Since IFEL and ICA experiments on the ATF accelerator are close to demonstrating acceleration, modifications of those experiments to obtain 100 MeV energy gain can be accomplished with a modest budget in a short time.

II. INVERSE ČERENKOV ACCELERATOR

When laser light propagates at an angle with respect to an electron beam, as shown in Fig. 1, there is a component of the electric field in the axial direction. In order to utilize this component of the electric field to accelerate electrons in the axial direction to large energies with little phase slippage, it is necessary to slow the phase velocity of the laser beam to that of the electron velocity by changing the index of refraction, n. This can be accomplished by the introduction of a gas. The phase matching condition is

$$\theta_c = \cos^{-1}(1/n\beta),$$

where $\beta = v/c$, v is the velocity of the electron and θ_c is the Čerenkov angle.

The ICA concept has been demonstrated at Stanford.[1] STI Optronics reported on their ICA experiment[2] at the ATF with an improved laser configuration, which uses a radially polarized annular CO_2 laser beam formed through an axicon. For 100 GW peak laser power and a nominal H_2 gas pressure of 1.7 atm, the energy gain for electrons is predicted to be 38 MeV in 20 cm. STI also addressed the issues of multistaging and electron beam trapping,[3] which are important for large final acceleration energies. This analysis was helpful in assessing the multistaging issue in the working group designs.

A "100 MeV" demonstration experiment based on the ICA concept can be achieved by lengthening the gas cell of the STI experiment and using the existing CO_2 laser and accelerator at the ATF, shown in Fig. 2. Due to the change in β, the ICA requires two acceleration stages, each at the appropriate Čerenkov angle and each roughly 75 cm long. The accelerating gradient is approximately 67 MeV/m for a 50 GW CO_2 laser. The accelerated electrons will have a distribution of energies similar to the existing ICA experiment. More efficient acceleration of the electron bunch is possible using a prebuncher before the ICA cell. This "100 MeV" demonstration experiment can be achieved with low cost and within a short period of time.

A "1 GeV" baseline experiment based on the ICA concept, shown in Fig. 3, has an accelerating gradient of 2 GeV/m. This high gradient is possible with a 1 TW glass laser of wavelength 1 μm. The improvement in the acceleration gradient is due to the shorter wavelength and higher power. Gas breakdown is probably not a problem, even at this high power, because the peak laser power in the axicon configuration is distributed over a large area in a ring around the electron beam. Three acceleration stages are required, each at the appropriate Čerenkov angle, to adjust for the energy gain of the electron beam.

To achieve a few percent energy spread for the accelerated electron beam, a prebuncher has to be installed before the 3-stage ICA. The prebuncher can be a short ICA section (approximately 20 cm length) utilizing a 1 GW glass laser.

III. INVERSE FREE ELECTRON LASER

The field of a laser beam combined with a periodic magnetic field forms a ponderomotive traveling potential. When the phase velocity of the ponderomotive wave, v_{ph}, is matched to the axial electron beam velocity in the wiggler, v_z, electrons can be trapped and accelerated by varying the period and/or amplitude of the magnetic field, i.e.,

$$v_{ph} = \frac{\omega}{k + k_w} = v_z,$$

where ω is the frequency of the laser, $k = \omega/c$ is the wavenumber of the laser, $k_w = 2\pi/\ell_w$ is the wavenumber of the static magnetic field, $v_z/c = \sqrt{1 - 1/\gamma_z^2}$, $\gamma_z^2 = \gamma^2/(1 + a_w^2)$, γ is the relativistic mass factor before entering the magnetic field, $a_w = (eB_w/m_oc^2k_w)|_{rms}$ is the normalized vector potential of the static magnetic field, ℓ_w is the wavelength of the magnetic field and B_w is the magnitude of the magnetic field. The concept of the IFEL has been experimentally demonstrated[4] and experiments using improved configurations were presented in the working session.

Experimental results of the IFEL auto-accelerator at Columbia University[4] were reported by T. C. Marshall. This is a compact IFEL accelerator, where the radiation source is provided by an FEL. The same electron beam that generated the FEL enters a second wiggler that satisfies the IFEL acceleration condition. The results are in good agreement with simulation. In addition, Columbia proposed an IFEL Beat-Wave Accelerator,[5] in which the IFEL mechanism is used to bunch a dense but low-energy electron beam, and then uses the electric field generated between the bunches to accelerate another higher-energy electron beam.

Brookhaven National Laboratory gave a series of presentations on the development of IFEL experiments at the ATF.[6] The acceleration of electrons is accomplished by decreasing the wavenumber, k_w, of the wiggler. The period of the wiggler is 2.86 cm at the entrance and is increased to 4.32 cm at the exit.

The wiggler is 0.47 m long, with a peak magnetic field of 1.25 T. Computer calculations of a 200 GW CO_2 laser propagating in a low loss dielectric-coated waveguide yield acceleration of electrons to a final energy of 87.95 MeV for electrons injected into the wiggler at an energy of 48.9 MeV.

A B-factory design was presented by C. Pellegrini[7] based on the concept of an FEL driven IFEL. The generation of an FEL was necessary to provide a high average power laser source. Finally, A. A. Varfolomeev[8] presented analytical and numerical calculations on optical guiding effects in an IFEL.

The working group's designs of IFELs are extensions of the presentations. Two different approaches were used utilizing 1) the conventional laser and accelerator at ATF and 2) high average power IFEL accelerators driven by FELs.

A simple modification of BNL's IFEL experiment, lengthening the wiggler to 2 m, can lead to a "100 MeV" demonstration experiment, as shown in Fig. 4. The wiggler is divided into four parts for ease of construction. The parameters of the IFEL with four wiggler sections are listed in Table I. Similar to the "100 MeV" ICA demonstration, the "100 MeV" IFEL demonstration can be accomplished with a modest budget and in a short time.

An extension of the "100 MeV" design to "1 GeV" utilizing conventional lasers was not prepared in detail during the working session, due to time limitations. A guess at the configuration calls for 14 sections of wigglers with a total wiggler length of 8.4 m. The length of the accelerator would be about 10 m. The required laser power would vary with wavelength: i) $\lambda_L = 100$ μm, $P_L = 0.4$ TW, ii) $\lambda_L = 10.2$ μm, $P_L = 1.7$ TW and iii) $\lambda_L = 1$ μm, $P_L = 7$ TW, where λ_L is the laser wavelength and P_L is the laser power required to achieve the acceleration within the designed wiggler length. Due to the large laser power needed, a drive laser would have to be developed. This supports the concept of an FEL as a laser driver of an IFEL.

Working towards a high average power, high gradient accelerator, the UCLA group, headed by C. Pellegrini, proposed to develop IFELs driven by FELs rather than conventional lasers. These examples are similar to the B-factory design.

The parameters and schematic of the "100 MeV" IFEL are shown on the bottom half of Fig. 5. A 15 MeV gun with peak current of 100 A and 10 ps rms pulse length is injected into the undulator. The undulator length is 2 m with a uniform 16 cm period. The peak of the magnetic field increases from 337 G at the entrance to 7.6 kG at the exit. Since the wavelength of the injection laser is preferred to be long, the injected laser wavelength is picked to be 200 μm, and the required laser power is 40 GW. The propagation of the 200 μm laser beam in a 4 mm × 4 mm waveguide will reduce the slippage between the electron pulse and the laser pulse.

Currently, no conventional laser can produce the required power at 200 μm. An FEL is ideal for this purpose. Various conceptual FEL design configurations are possible to achieve the required power, provided the electron beam quality is good. The parameters for an FEL are shown in the top half of Fig. 5.

One example of a "1 GeV" setup, shown in Fig. 6, is achieved by connecting a series of modules similar to the one shown in Fig. 5. The period and amplitude of the magnetic field must be changed to match the accelerated beam energy. The input laser power is increased to 120 GW. The injected electron beam, 100 MeV at 300 A peak, is required to have a pulse length of 3.3 ps rms and an emittance of 3×10^{-6} m.

The FEL driven IFEL conceptual designs still have many issues to be addressed. The energy transfer between the laser beam and the electrons must be optimized. The beam quality requirement and the development of electron beams for the FEL and the injector are important. When $N\lambda \gtrsim \ell_b$, slippage control between the electron pulse and the laser pulse is important in the FEL as well as the IFEL. Here, N is the number of wiggler periods, λ is the wavelength of the laser and ℓ_b is the length of the laser pulse.

The electron beam and the high power FEL necessary to drive the IFEL are conceptually feasible, but require considerable development. The goal of high average accelerated beam power further requires the development of high repetition rate FELs and their drive accelerators.

IV. SMALL ACCELERATOR

An analysis of the scaling laws for the cyclotron resonance laser accelerator (CRLA)[9-12] was presented by C. Chen. The configuration is shown in Fig. 7. The resonance condition is

$$\gamma \left(1 - \frac{\beta_z}{\beta_{ph}}\right) \omega - \Omega_c = 0,$$

where $\Omega_c = eB_z/m_oc$, $\beta_{ph} = \omega/ck_z$. To overcome the dispersion of the laser field, the magnetic field has to be tapered. This mechanism is very simple and applicable to a small electron source.

Utilizing existing microwave and electron beam sources available in many major laboratories, this concept can produce a small electron beam, which may be used, for example, as an injector into other laser accelerators.

Two examples were constructed, based on a 33 GHz, 50 MW rf source in a 200 cm long 3 cm radius waveguide. The electron pulse duration is assumed to be 10 ps.

(i) Electrons can be accelerated from $\gamma = 1.5$ to $\gamma = 8.5$ by tapering the magnetic field from 7 kG to 14 kG. The peak current is 20-200 A with a total number of electrons of 1.2×10^9 -12×10^9.

(ii) Electrons can be accelerated from $\gamma = 3.0$ to $\gamma = 40$ by tapering the magnetic field from 2 kG to 4 kG. The number of electrons accelerated is 1×10^9-2×10^9 corresponding to a peak current of 16A-32 A. The laser energy is depleted by only 0.6%.

The beam would be extracted by magnetic cusp.

V. OTHER FAR FIELD ACCELERATION CONCEPTS

1. Vacuum Acceleration with Laser Field Only

Acceleration of electrons by a laser in free space without additional fields is possible with the appropriate configuration if the interaction length is finite. STI presented calculations based on the axicon configuration[13] originally suggested by D. Sutter. Electrons can be accelerated in a vacuum if the phase slippage between electrons and the light wave can be controlled. STI proposed minimizing the phase slippage by using the short interaction lengths provided by the axicon. High energies are attainable through multistaging.

Another configuration, presented by A. A. Varfolomeev,[14] consists of periodically spaced cavities. The optimal angle between the direction of electron propagation and the resonator axis must be small. Thus, resonator mirrors require holes for the electrons to enter and exit the resonator. Calculations indicated that a stable laser resonator mode is possible with holes, and far field laser driven accelerators without an undulator magnet are possible.

2. Nonlinear Amplification of Inverse-Bremsstrahlung Electron Acceleration (NAIBEA)

Acceleration of electrons with a laser and a small perpendicular static electric field[15] or magnetic field[16] was presented. Calculations of nonlinear equations show acceleration of the resonant electron ~ 150 MeV/m. However, many important issues have not yet been addressed, such as: the trapping efficiency, energy spread and emittance of the trapped beam.

3. Other Presentations

There were a few presentations that did not directly address high gradient acceleration with far fields. Nevertheless, the topics were somewhat related. They were: i) design of 1 MeV storage ring damped by lasers[17] ii) high brilliance, femtosecond x-ray source with FEL assist,[18] iii) SNōELF: a high current injector for CLIC[19] and iv) frequency upshifting in the FELs.[20]

VI. SUMMARY

The "100 MeV" demonstrations based on the ICA and IFELs utilizing the accelerator and the CO_2 laser at the ATF require only modest changes of current experimental setups. Thus, the time required for the demonstration would be short and costs would be low.

FEL driven IFELs may be necessary for "1 GeV" illustration experiment, both for single pulse or high repetition rate. There are many physics and design issues yet to be addressed.

The "1 GeV" baseline experiment based on the ICA concept and utilizing the ATF accelerator may be the top candidate for achieving high gradient acceleration (~ 2 GeV/m) using the laser acceleration concept. The proposed laser power has been reported.[21] Thus, the laser does not require long research development time and the cost would be moderate.

Acknowledgements

This work is supported by the U.S. Department of Energy and the Office of Naval Research.

Table I: Preliminary Parameters for A "100 MeV" IFEL Accelerator

Electron Beam		Modules		
	I	II	III	IV
Injection energy [MeV]	49	78	104	127
Exit energy [MeV]	78	104	127	149
Mean accel. gradient [MeV/m]	58	52	46	44
Number of electrons	10^9			
Energy Spread		$\lesssim$ few %		
Normalized emittance [m rad]			$\simeq 2 \times 10^{-5}$	

Wiggler

Module length [m]	0.5	0.5	0.5	0.5
Period λ_w [cm]	2.86-4.00	4.00-4.85	4.85-5.54	5.54-6.16
Field on axis (constant B_w) [T]	1.25			
Gap [mm]	4			

References

1. J. A. Edighoffer, W. D. Kimura, R. H. Pantell, M. A. Piestrup, and D. Y. Wang, Phys. Rev. A **23**, 1848 (1981).
2. W. D. Kimura, I. Pogorelsky, L. C. Steinhauer, S. C. Tidwell, G. H. Kim and K. P. Kusche, submitted to the AIP Proc. of the Advanced Accelerator Concepts Workshop, Port Jefferson, Long Island, NY, June 14-20, 1992.
3. L. C. Steinhauer and W. D. Kimura, submitted to the AIP Proc. of the Advanced Accelerator Concepts Workshop, Port Jefferson, Long Island, NY, June 14-20, 1992.

4. T. Marshall and I. Wernick, submitted to the AIP Proc. of the Advanced Accelerator Concepts Workshop, Port Jefferson, Long Island, NY, June 14-20, 1992.

5. F. Y. Cai and A. Bhattacharjee, Phys. Rev. A **42**, 4853 (1990).

6. A. Fisher, J. Gallardo, J. Sandweiss, A. van Steenbergen, submitted to the AIP Proc. of the Advanced Accelerator Concepts Workshop, Port Jefferson, Long Island, NY, June 14-20, 1992.

7. C. Pellegrini, submitted to the AIP Proc. of the Advanced Accelerator Concepts Workshop, Port Jefferson, Long Island, NY, June 14-20, 1992.

8. A. A. Varfolomeev and A. H. Hairetdinov, submitted to the AIP Proc. of the Advanced Accelerator Concepts Workshop, Port Jefferson, Long Island, NY, June 14-20, 1992.

9. W. B. Colson and S. K. Ride, Appl. Phys. **20**, 61 (1979).

10. P. Sprangle, L. Vlahos and C. M. Tang, IEEE Trans. on Nuclear Sci., **NS-30**, 3177 (1983).

11. C. Chen, Phys. Fluids B **3**, 2993 (1991).

12. C. Chen, Phys. Rev. A, in press, Nov. 1992.

13. L. C. Steinhauer and W. D. Kimura, "A New Approach for Laser Particle Acceleration in Vacuum", to be published in Journal of Applied Physics, Oct. 15, 1992.

14. A. A. Varfolomeev and A. H. Hairetdinov, submitted to the AIP Proc. of the Advanced Accelerator Concepts Workshop, Port Jefferson, Long Island, NY, June 14-20, 1992.

15. M. S. Hussein and M. P. Pato, Phys. Rev. Lett. **68**, 1136 (1992).

16. M. S. Hussein, submitted to the AIP Proc. of the Advanced Accelerator Concepts Workshop, Port Jefferson, Long Island, NY, June 14-20, 1992.

17. C. Brau, submitted to the AIP Proc. of the Advanced Accelerator Concepts Workshop, Port Jefferson, Long Island, NY, June 14-20, 1992.

18. W. A. Barletta and R. Bonifacio, submitted to the AIP Proc. of the Advanced Accelerator Concepts Workshop, Port Jefferson, Long Island, NY, June 14-20, 1992.

19. W. A. Barletta and R. Bonifacio, submitted to the AIP Proc. of the Advanced Accelerator Concepts Workshop, Port Jefferson, Long Island, NY, June 14-20, 1992.

20. J. Shvets and J. S. Wurtele, submitted to the AIP Proc. of the Advanced Accelerator Concepts Workshop, Port Jefferson, Long Island, NY, June 14-20, 1992.

21. C. Sauteret, D. Husson, G. Thiell, S. Seznec, S. Gary, A. Migus, and G. Mourou, Opt. Lett. **16**, 238 (1991).

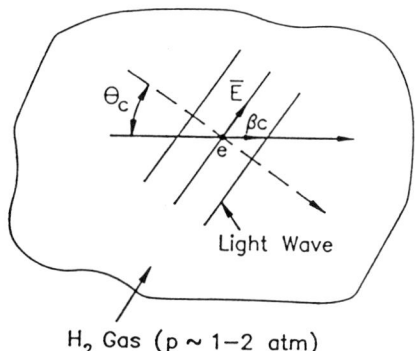

Fig. 1. Schematic illustrating the principle of the ICA concept.

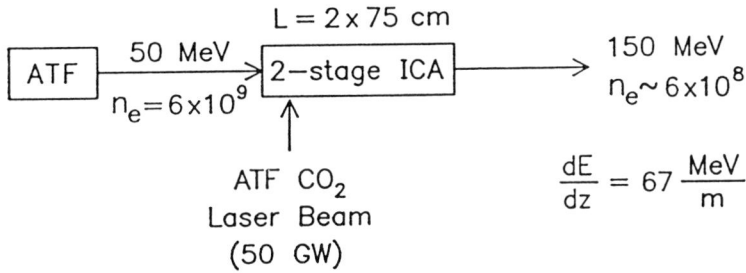

Fig. 2. Schematic of the "100 MeV" demonstration accelerator based on the ICA concept utilizing the CO_2 laser and accelerator at the ATF.

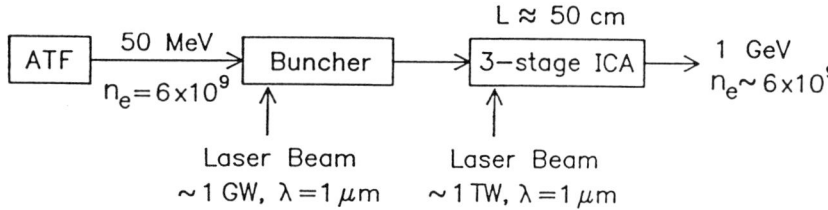

Fig. 3. Schematic of the "1 GeV" accelerator based on the ICA concept utilizing a solid state laser and the accelerator at the ATF.

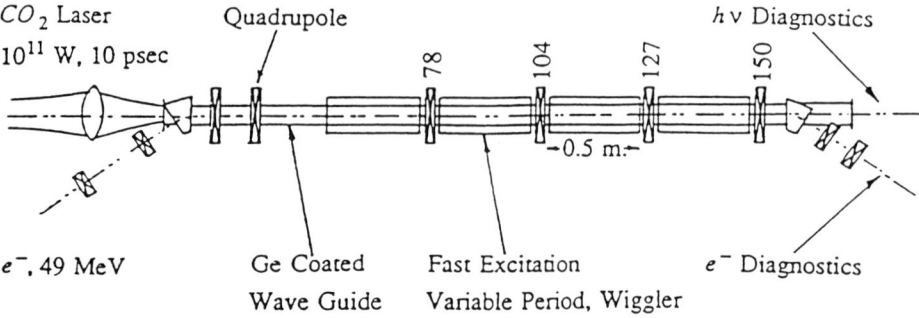

Fig. 4. Schematic of the "100 MeV" demonstration accelerator based on the IFEL concept utilizing the CO_2 laser and accelerator at ATF.

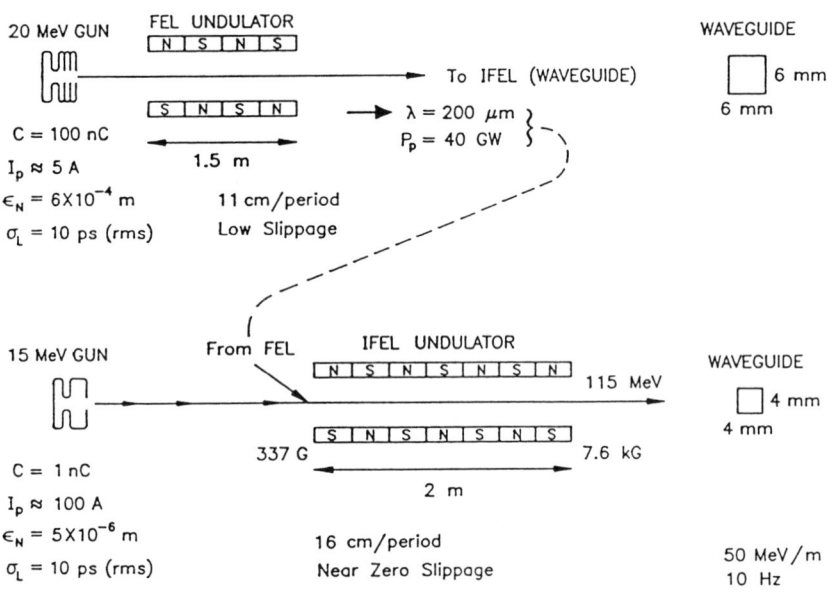

Fig. 5. Schematic of the "100 MeV" demonstration IFEL driven by FEL.

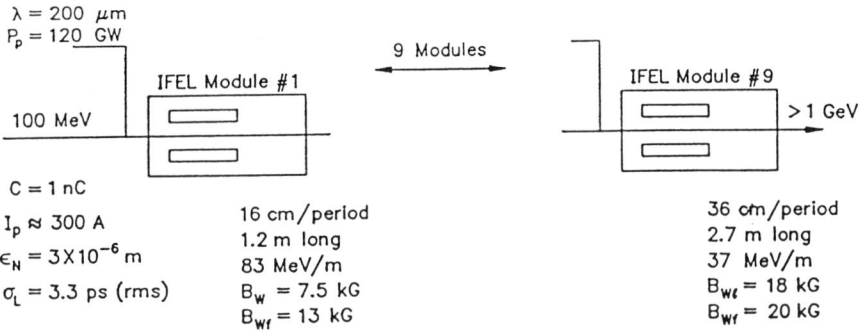

Fig. 6. Schematic of the "1 GeV" IFEL driven by FEL.

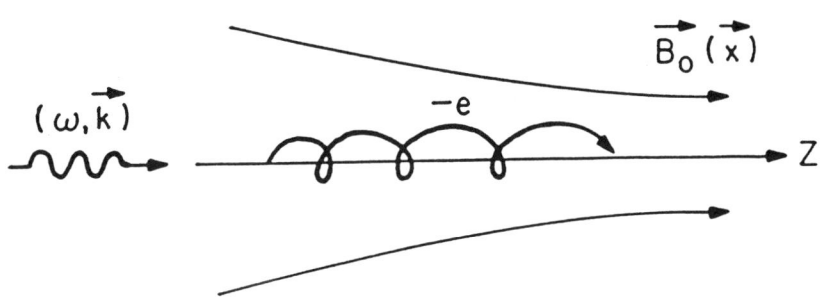

Fig. 7. Schematic of the CRLA concept.

Prospects and Limitations of
Cyclotron Resonance Laser Acceleration[†]

Chiping Chen
Plasma Fusion Center
Massachusetts Institute of Technology
Cambridge, Massachusetts 02139

ABSTRACT

The cyclotron resonance laser (CRL) accelerator is a novel concept of accelerating continuous charged-particle beams to moderately or highly relativistic energies. This paper discusses prospects and limitations of this concept. In particular, a three-dimensional, self-consistent theory is used to analyze the nonlinear interaction of an electron beam with an intense traveling electromagnetic wave in such an accelerator. The parameter regimes of experimental interest are identified on the basis of scaling calculations. The results of simulation modeling of a multimegavolt electron CRL accelerator are presented. The possibility of building continuous-wave (cw) CRL accelerators is discussed.

I. INTRODUCTION

There has been growing theoretical and experimental interest recently in the cyclotron resonance laser (CRL) accelerator [1]-[12], an advanced accelerator concept of producing charged-particle beams to moderately or highly relativistic energies using an intense coherent electromagnetic wave and a guide magnetic field configuration (Fig. 1). This novel accelerator has the following advantages over conventional radio-frequency (rf) accelerators: (i) acceleration of a continuous beam without microbunches; (ii) use of an oscillator, not necessarily an amplifier, as driver; (iii) use of a smooth-wall structure to avoid breakdown problems; and (iv) a high duty factor of acceleration. It is compact in comparison with electrostatic accelerators which are bulky but capable of accelerating continuous charged-particle beams. In addition to these intriguing features, there is a variety of potential applications of CRL accelerators. These include the following: (i) production of high-average-power charged-particle beams for material and chemical research; (ii) x-ray generation; and (iii) coherent millimeter wave generation, to mention a few examples. Proof-of-principle CRL accelerator experiments [3],[7],[11],[12] so far have had limited success, and have demonstrated the acceleration of electrons up to ∼ 0.5 MeV in the microwave regime. This paper discusses prospects and limitations of this concept.

---

†Research supported by the DOE under Grant No. DE-AC02-91ER40648.

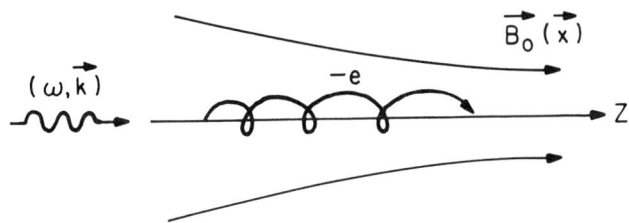

Figure 1: Schematic of a CRL accelerator.

II. SCALING PROPERTIES

Our earlier studies [9],[10] have shown that electron CRL accelerators with optimal magnetic taper possess the following scaling properties:

$$G \equiv \frac{\gamma_f - \gamma_i}{\gamma_i - 1} \propto \frac{1}{(\beta_{ph} - 1)^{0.5}} \quad \text{(independent of } a) , \tag{1}$$

$$k_{\parallel} z_m \propto \frac{1}{(\beta_{ph} - 1)^{0.6}} , \tag{2}$$

$$\frac{B_{0z}(z_m)}{B_{0z}(0)} = 2 - 3 , \tag{3}$$

accelerating gradient $\propto a$ (for a fixed β_{ph}) . $\qquad(4)$

In Eqs. (1)-(4), $\gamma_i mc^2$ and $\gamma_f mc^2$ are the average initial and final energies of the beam electrons, respectively, z_m is the maximum accelerating distance, $B_{0z}(z)$ is the magnetic field profile on the z axis, a is the normalized wave amplitude, and $\beta_{ph} - 1 \equiv \omega/ck_{\parallel} - 1$ is a measure of wave dispersion, where m is the electron rest mass, $\omega = 2\pi f$ and $k_{\parallel}$ are the frequency and axial wave number of the driving electromagnetic wave, respectively, and c is a speed of light in vacuum.

In the microwave regime, the normalized wave amplitude and the dispersion parameter are given by

$$a = 1.7 \times 10^{-4} \left(\frac{[cm]}{r_w} \right) \left(\frac{[GHz]}{f} \right) \left(\frac{P}{[W]} \right)^{1/2} , \tag{5}$$

$$\beta_{ph} - 1 = 40 \times \left(\frac{[cm^2]}{r_w^2} \right) \left(\frac{[GHz^2]}{f^2} \right) , \tag{6}$$

where $|\beta_{ph} - 1| \ll 1$ and the lowest-order transverse-electric (TE) cylindrical waveguide mode, TE_{11}, have been assumed, f and P are the frequency and power (or integrated axial Poynting flux) of the wave, respectively, and r_w is the waveguide radius. For microwave electron CRL accelerators, high energy gain, $G = 10 - 100$, and moderately high accelerating gradient, 1-10 MeV/m, can be achieved with presently available high-power microwave sources, [assuming the initial electron kinetic energy $(\gamma_i - 1)mc^2 \sim 100$ keV]. With a proper resonant cavity, a sizable wave amplitude $(a > 0.1)$ can be achieved at a moderate input rf power level $(P \sim 1$ MW). This shows prospects for the production of high-average-power $(> 100$ kW), relativistic $(> 1$ MeV) electron beams using continuous-wave CRL accelerators. Potential applications of high-average-power beams were mentioned in the Introduction.

In the optical regime, the normalized wave amplitude and the dispersion parameter are given by

$$a = 6 \times 10^{-6} \frac{\lambda}{[\text{cm}]} \left(\frac{\text{Laser Intensity}}{[\text{W/cm}^2]} \right)^{1/2}, \tag{7}$$

$$\beta_{ph} - 1 \approx \frac{\lambda^2}{2r_0^2}, \tag{8}$$

respectively. Here, $\lambda = 2\pi/k_{\parallel}$ is the laser wavelength and r_0 is the characteristic laser beam radius. For example, a 1 μm, 10^{16} W/cm^2 Nd:glass-based laser yields a respectable wave amplitude, $a = 0.06$, and a small dispersion parameter, $\beta_{ph} - 1 \sim 10^{-8}$ (assuming $r_0 = 0.5$ cm). With presently available high-intensity lasers, an accelerating gradient up to ~ 100 MeV/m may be achieved [4],[6]. However, the pulse length of such an ultra-intense laser is short (typically of order 1 ps), which imposes a limitation on the optical CRL accelerator concept in practice. There are other problems as well, such as wave diffraction, slippage between the wave and the beam, and the requirement of a strong magnetic field.

III. MODELING OF A MULTIMEGAVOLT CRL ACCELERATOR

The scaling calculations in Sec. II have revealed that at present, microwave CRL accelerators are most attractive from a practical point of view. The underlying self-consistent equations of motion describing the wave-beam interaction in the microwave CRL accelerator, which are identical to those describing the cyclotron autoresonance maser (CARM) [13]-[17], have been derived and presented elsewhere [10],[15],[16]. The scaling relations (1)-(4) and the self-consistent CRL accelerator equations in reference [10] are readily applicable for design modeling of CRL accelerators.

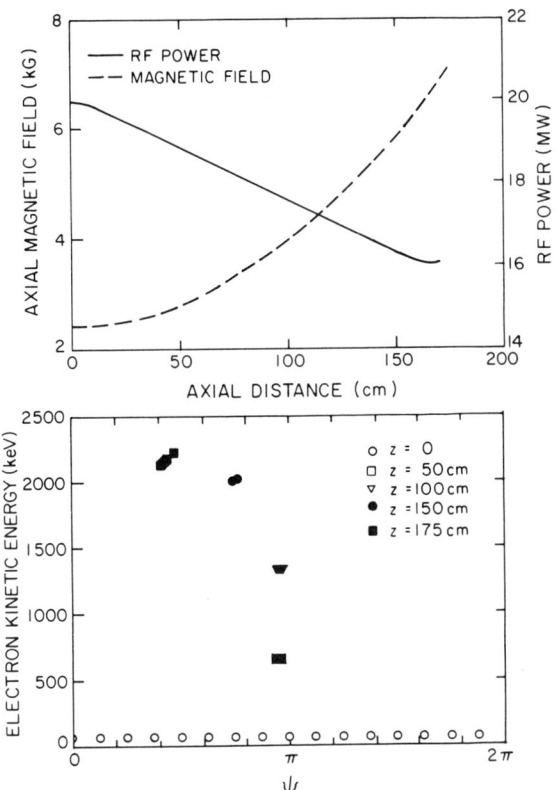

Figure 2: (a) Axial magnetic field and rf power as a function of the accelerating distance for an X-band CRL accelerator. (b) Phase space of beam electrons at several axial positions.

Table I. Parameters for a Proposed 2.3-MeV CRL Accelerator

Parameter	Symbol [Unit]	Value
Beam Current	I_b [A]	2
Initial Beam Voltage	V_i [kV]	75
Final Beam Voltage	V_f [kV]	2300
Acceleration Distance	z_m [cm]	170
Initial Axial Magnetic Field	$B_{0z}(0)$ [kG]	2.43
Final Axial Magnetic Field	$B_{0z}(z_m)$ [kG]	7.0
RF Frequency	f [GHz]	11.4
RF Power	P [MW]	20
Operating Mode		TE$_{11}$
Waveguide Radius	r_w [cm]	3.9

As an example, we discuss a moderately high-current, multimegavolt electron CRL accelerator powered by a pulsed 20 MW, 11.4 GHz rf source, (such as an X-band relativistic klystron amplifier). In this design study, the electron beam current is chosen to be sufficiently large so that effects of changes in both the wave amplitude and the axial wave number due to the *inverse* CARM interaction can be examined.

Figure 2 summarizes the design simulations which yield the electron energy gain $G = 30$, for the beam current $I_b = 2$ A and the initial beam voltage $V_i = 75$ kV. Effects of time-independent (DC) space-charge are not included in the simulation. The parameters of this proposed 2.3-MeV CRL accelerator are given in Table I. For a TE_{11} mode and the waveguide radius $r_w = 3.9$ cm, the normalized amplitude is $a = 0.017$ and the dispersion parameter is $\beta_{ph} - 1 = 0.02$, as evaluated from Eqs. (5) and (6).

The optimal axial magnetic field B_{0z} and rf power are plotted in Fig. 2(a) as a function of the accelerating distance z. The axial magnetic field exhibits approximately a quadratic dependence on the distance z. It increases no more than three-fold throughout the entire accelerating distance. Significant rf power is converted into electron beam power, corresponding to 20% efficiency.

Figure 2(b) shows the phase space of the electrons, where the vertical axis is the electron kinetic energy and the horizontal axis is the electron gyrophase with respect to the phase of the rf field. In Fig. 2(b), only sixteen macroparticles are plotted at a given axial distance z. The initial energy (or axial momentum) spread is assumed to be negligibly small, and the final energy spread from the simulation is $\sigma_\gamma / \langle \gamma \rangle = 0.9\%$, where $\sigma_\gamma = (\langle \gamma^2 \rangle - \langle \gamma \rangle^2)^{1/2}$.

IV. SUMMARY

The prospects and limitations of cyclotron resonance laser (CRL) accelerators have been discussed. Based on our scaling calculations and presently available intense lasers and microwave sources, microwave CRL accelerators have been identified as a practical concept in either pulse or continuous-wave (cw) operation, while optical CRL accelerators are so far of only theoretical interest. A multimegavolt microwave electron CRL accelerator experiment has been proposed.

Although our analyses have focused primarily on a traveling-wave configuration, the scaling laws are expected to be valid for a standing-wave configuration, a better representation of the driving wave in a resonant cavity. Finally, we believe that the scaling laws and theory presented in this paper will provide important guidelines for future experimental studies of CRL accelerators.

References

[1] A.A. Kolomenskii and A.N. Lebedev, Soviet Physics, Dokl. **7**, 745 (1963).

[2] C.S. Roberts and S.J. Buchsbaum, Phys. Rev. **135**, 381 (1964).

[3] H.R. Jory and A.W. Trivelpiece, J. Appl. Phys. **39**, 3053 (1967).

[4] W.B. Colson and S.K. Ride, Appl. Phys. **20**, 61 (1979).

[5] K.S. Golovanevsky, IEEE Trans. Plasma Sci. **PS-10**, 120 (1982).

[6] P. Sprangle, L. Vlahos, and C.M. Tang, IEEE Trans. Nuclear Sci. **NS-30**, 3177 (1983).

[7] D.B. McDermott, D.S. Furuno, and N.C. Luhmann, Jr., J. Appl. Phys. **58**, 4501 (1985).

[8] A. Loeb and L. Friedland, Phys. Rev. **A33**, 1828 (1986).

[9] C. Chen, Phys. Fluids **B3**, 2933 (1991).

[10] C. Chen, Phys. Rev. **A46**, 15 November (1992).

[11] R. Shpitalnik, C. Cohen, F. Dothan, and L. Friedland, J. Appl. Phys. **70**, 1101 (1991).

[12] R. Shpitalnik, J. Appl. Phys. **71**, 1583 (1992).

[13] V.L. Bratman, N.S. Ginsburg, G.S. Nusinovich, M.I. Petelin, and P.S. Strelkov, Int. J. Electron. **51**, 541 (1981).

[14] C. Chen and J.S. Wurtele, Phys. Rev. Lett. **65**, 3389 (1990).

[15] C. Chen and J.S. Wurtele, Phys. Fluids **B3**, 2133 (1991).

[16] H.P. Freund and C. Chen, Int. J. Electron. **72**, 1005 (1992).

[17] A.C. DiRienzo, G. Bekefi, C. Chen, and J.S. Wurtele, Phys. Fluids **B3**, 1755 (1991).

Strong Focusing for Planar Undulators*

G. Travish and J. Rosenzweig

Particle Beam Physics Laboratory
Department of Physics,
University of California at Los Angeles,
Los Angeles, CA 90024

Abstract

The range of parameters in which the free electron laser (FEL) is used has greatly expanded in recent years. The present trend towards short wavelength operation with long undulators places tight requirements on the electron beam quality. In particular, the need to maintain a well focused beam is critical to successful operation of such an FEL. This paper examines the use of alternating gradient (AG) sextupole focusing in planar undulators. After a brief review of various undulator focusing schemes, the equations of motion for an electron in an undulator field with a strong sextupole component are examined. It is shown that the mean electron longitudinal velocity can be kept constant through each focusing and defocusing section. Thus, the betatron velocity modulation associated with transverse focusing can be limited. Analytic as well as smooth approximation solutions are provided for AG sextupole focusing. Examples using the proposed SLAC 4nm FEL as well as the UCLA 10.6 µm FEL parameters are also discussed.

Introduction

Planar undulators are widely used in FEL applications. Permanent magnet planar undulators, especially those following the hybrid design,[1] offer a number of advantages over other schemes: high peak fields, simplicity, beamline accessibility and flexibility.[2] However, the lack of symmetric transverse focusing is a problem; planar, as opposed to helical, undulators only have "natural" focusing in one transverse dimension. Natural focusing preserves the phase relation between the optical field and the electron wiggle motion. This can be understood by noting that the average (over an undulator period) longitudinal velocity of the electron is constant over a betatron oscillation.

The need for electron beam focusing is a function of FEL parameters. Recent designs call for undulators tens of meters long.[3] Maintenance of small electron beam sizes through such distances is not feasible without a focusing mechanism. Other undulators operating at low beam energies (in the collective or Raman regime) require confinement of the beam even over short distances due to space charge induced divergence. Divergence of the electron beam in an FEL reduces efficiency through a lowering in beam density and reduction in the beam overlap with the

---

*Work performed under support from the U. S. Department of Energy, Grant DE-FG-92ER40693, Department of Education and the University of California.

radiation field. The need for focusing is also a function of the undulator period and the operating wavelength. Smaller scale lengths imply that the FEL is more sensitive to electron beam dimensions. These effects can be examined by using the dimensionless FEL parameter, ρ.[4] In SI units we may write[5]

$$\rho \cong 0.14 \frac{J^{1/3} B_u^{2/3} \lambda_u^{4/3}}{\gamma_r} \tag{1}$$

where J is the beam current density and the remainder of the notation is as given in Table 1.[6] In the high gain regime of an FEL, the power gain e-folding length is approximately

$$L_g = \frac{\lambda_u}{\sqrt{3}\,4\pi\rho}. \tag{2}$$

Thus a decrease in the beam density (J) will adversely affect gain. It is also clear that a high peak undulator field is desirable. Another requirement for efficient FEL operation is that the FEL parameter be greater than the energy spread (rms), σ_E:

$$\rho > \sigma_E. \tag{3}$$

A related requirement on the phase space of the electron beam is that it should overlap well with the output radiation.[7] For a beam with no external focusing the emittance must be smaller than the undulator period:

$$\varepsilon < \lambda_u / 4\pi. \tag{4}$$

This requirement can be mitigated somewhat through external focusing of the electron beam. A general statement can therefore be made, that many FELs can benefit from, and some cannot operate without, electron beam focusing.

Table 1: Notation used in this paper.

Undulator magnetic field	B_u
Normalized undulator field	$b_0 = (e/mc^2)B_u$
Electron charge	e
Electron mass	m
Speed of light	c
Undulator period	λ_u
Undulator wavenumber	$k_u = 2\pi/\lambda_u$
Electron Beam Energy (mc^2)	γ

FOCUSING SCHEMES

Various focusing schemes have been considered and presented in the literature. Quadrupole focusing has been used on a number of undulators.[8] Typically, external quadrupole magnets in a FODO lattice are superimposed over the undulator field. External magnets require that no permeable materials be used in the undulator; hybrid undulators are not compatible with external magnets. Alternatively, the quadrupoles can be interspersed with undulator sections. Regardless of the arrangement, quadrupole focusing differs from natural focusing in that it causes electron velocity modulation during betatron oscillations.[9] The effect is to modulate the phase of the electrons relative to the ponderomotive potential due to the undulator and optical beam. These oscillations degrade FEL performance by effectively debunching or detrapping the electrons.[10] The reduction in gain is especially undesirable in single pass amplifiers.

A method to achieve quadrupole focusing without external magnets is to *cant* the undulator poles (see Figure 1). By introducing a slight tilt to each undulator pole, alternating the tilt direction with each pole, it is possible to cause a quadrupole focusing field near the beam axis. However, this method suffers from the same problem as external quadrupole focusing. In addition, the canting can require more complicated undulator mechanics. It is of course possible to consider an alternating gradient scheme using canted poles.[11] While this would offer strong focusing without external magnets, it would still suffer from a phase modulation problem.

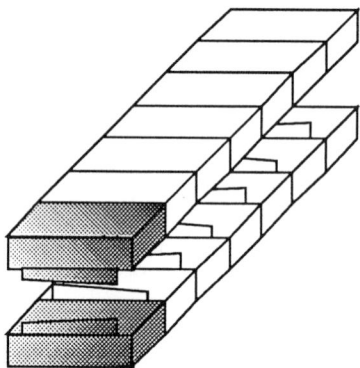

Figure 1: An undulator magnet with canted poles. The canting angle as well as the perspective are exaggerated for clarity.

Solenoidal confinement has been considered for high current (~kA) low energy (~MeV) beams.[12] Unfortunately, off axis electrons drift transversely. Rotational stabilization has been suggested as way of countering this problem.[13] Still, solenoidal confinement is not desirable for high field strength undulators or effective for high energy beams.

Ion focusing is a promising concept for undulator beam transport.[14] The idea of introducing a plasma into the beamline has its controversial points, but the potential benefits are impressive. An ion channel can offer extremely strong focusing; however, beam erosion, ion column collapse and instabilities (ion-hose, electron-hose, etc.) need to be avoided.[15] Ion focusing, like solenoidal focusing, provides both axisymmetric and continuous focusing. Both schemes, however, still cause longitudinal phase modulation of the electrons as they undergo betatron oscillations.

An elegant focusing scheme based on the converse to optical guiding has been theoretically derived.[16] Given a sufficiently strong optical field, the electron beam can be guided. The focusing is similar to natural focusing in its effects on the electron beam dynamics. The mutual focusing effect would be most pronounced in a high power system. Thus, an inverse FEL would be a candidate to observe this phenomenon.

The idea of shaping the poles of a planar undulator to provide focusing has been examined in detail by Scharlemann.[9] The scheme avoids the problems associated with pole canting by using sextupole focusing, which because the design orbit in the undulator is horizontally off-axis, gives rise to a quadrupolar field variation about the design orbit by *feed down* of multipole order. Proper shaping of the pole faces can produce a horizontal focusing up to the strength of the natural vertical focusing (see Figure 2). This concept has proven itself in application.[17] The major drawback of this scheme is that the strength of focusing is weak (as is discussed below); it is required to be less than the natural focusing of the undulator.

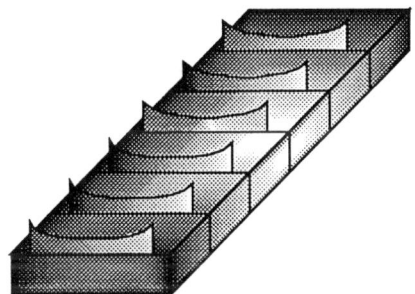

Figure 2: A sketch of the lower half of an undulator employing Sharlemann's shaped poles.

Following sections of this paper review Scharlemann's original scheme, and propose an extension to include alternating gradient or strong focusing. The theory which is developed is then applied to examples.

UNDULATOR POLE SHAPING

Scharlemann was first to derive the effect of parabolic pole face shaping on electron beam motion in a nominally planar undulator. A similar derivation is repeated here to remind the reader and elucidate the calculation in the next section. The geometry used in the calculations is given in Figure 3.

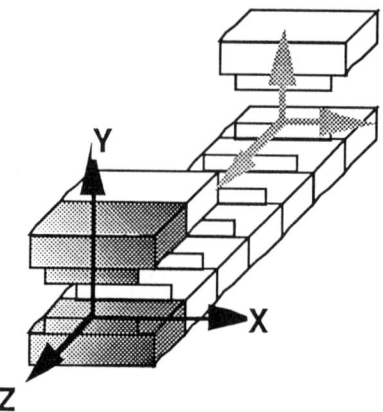

Figure 3: The geometry used in this paper. The beam would propagate along the positive z-axis (out of the page). Beam oscillations are assumed to be centered on the undulator axis.

Scharlemann begins with a normalized undulator magnetic field which satisfies the Maxwell equations and has a symmetric dependence on the transverse distance from the undulator axis:

$$\mathbf{b} = \hat{x}\frac{b_0 k_x}{k_y}\sinh(k_x x)\sinh(k_y y)\cos(k_u z)$$
$$+\hat{y}b_0 \cosh(k_x x)\cosh(k_y y)\cos(k_u z)$$
$$-\hat{z}\frac{b_0 k_u}{k_y}\cosh(k_x x)\sinh(k_y y)\sin(k_u z) \tag{5}$$

where the notation is as given in Table 1. The Maxwell equations additionally require that

$$k_u^2 = k_x^2 + k_y^2. \tag{6}$$

For k_x and k_y both real ($0 \le k_{x,y} \le k_u$) the above field is derivable from the scalar potential

$$\chi = -\frac{b_0}{k_y}\cosh(k_x x)\sinh(k_y y)\cos(k_u z). \tag{7}$$

A derivation of the field to arbitrary order is also possible, but analysis shows that the correction terms beyond second order are negligible (for known FEL parameters).[18] For small $k_x x$ and $k_y y$ the field may be therefore be approximated by

$$b_x = b_0 k_x^2 xy \cos(k_u z)$$

$$b_y = b_0 \left(1 + \frac{k_x^2 x^2}{2} + \frac{k_y^2 y^2}{2}\right) \cos(k_u z)$$
(8)

The electron equations of motion are straight forward to derive:

$$\ddot{x} = \frac{c}{\gamma}\left(\dot{z}b_y - \dot{y}b_z\right)$$

$$\ddot{y} = \frac{c}{\gamma}\left(\dot{x}b_z - \dot{z}b_x\right).$$
(9)

$$\ddot{z} = \frac{c}{\gamma}\left(\dot{y}b_x - \dot{x}b_y\right)$$

A dot is used to indicate a derivative with respect to time. A natural scale length of the problem is the undulator period. By separating the fast oscillations from the slow it is possible to simplify the equations. Following Scharlemann we define $r(x,y,z)=r_0+r_1$ where r_0 is constant over the undulator period (the slow betatron oscillation) and r_1 varies within a period (the fast undulator oscillation). Then, the equations of motion can be written, to leading order, as

$$\ddot{x}_0 = \frac{c\dot{z}_0}{\gamma}\langle b_y \rangle$$

$$\ddot{x}_1 = \frac{c\dot{z}_0}{\gamma} b_y$$
(10)

$$\ddot{y}_0 = \frac{c\dot{x}_1}{\gamma}\langle b_z \rangle - \frac{c\dot{z}_0}{\gamma}\langle b_x \rangle$$

where the brackets (<>) indicate averaging over an undulator period. The term $\ddot{y}_1$ can be neglected at this order in the analysis. It is easiest to integrate the expression for x_1 while inserting the expression for the magnetic field to yield

$$\dot{x}_1 = \frac{c}{\gamma k_u} b_0 \left(1 + \frac{k_x^2 x_0^2}{2} + \frac{k_y^2 y_0^2}{2}\right) \sin(k_u z).$$
(11)

A straight forward averaging and simplification gives the desired solution for the equations of motion:

$$\ddot{x}_0 + c^2 k_{\beta x}^2 x_0 \approx 0$$

$$\ddot{y}_0 + c^2 k_{\beta y}^2 y_0 \approx 0.$$
(12)

where the notation

$$k_{\beta(x,y)} = \frac{b_0}{\sqrt{2}\gamma k_u} k_{(x,y)} \tag{13}$$

has been introduced. Note that the focusing in the y plane has been reduced by a factor of k_y/k_u from the natural focusing strength. The above equations can be integrated (solved) by using the relation between the derivatives with respect to time and distance (z). Scharlemann has shown that the additional term coming from the longitudinal acceleration (velocity modulation) does not contribute to the average focusing. Thus, the relation $d/dt=v_z(d/dz)$ is a good approximation. Now it remains to evaluate the average transverse velocity,

$$\langle \beta_\perp^2 \rangle = \frac{1}{c^2}\left(\langle \dot{x}_1^2 \rangle + \dot{x}_0^2 + \dot{y}_0^2\right). \tag{14}$$

The x_1 term is averaged by noting that $<\sin^2(k_u z)>=1/2$. Then, eliminating higher order terms leaves

$$\langle \dot{x}_1^2 \rangle = \frac{c^2 b_0^2}{2\gamma^2 k_u^2}\left(1 + k_x^2 x_0^2 + k_y^2 y_0^2\right). \tag{15}$$

So, the average transverse velocity becomes

$$\langle \beta_\perp^2 \rangle = \frac{b_0^2}{2\gamma^2 k_u^2}\left(1 + k_x^2 x_\beta^2 + k_y^2 y_\beta^2\right), \tag{16}$$

where x_β and y_β are the amplitude of the transverse betatron oscillations (i.e. $x_0=x_\beta\sin[k_\beta x z+\phi_x]$). It is now clear that $<\beta_\perp^2>$ is a constant. That is, an electron's velocity averaged over an undulator period is constant through a betatron oscillation. This indicates that the (longitudinal) phase of the electrons (within a ponderomotive "bucket") is not modulated. Hence, sextupole focusing is not deleterious to electron bunching and FEL gain. In fact, as was discussed in the Introduction, the gain is expected to be higher since the beam density remains greater under focus.

While higher order corrections have been calculated in the literature,[18] Scharlemann's calculation gives the underlying physics and is accurate for FEL parameters. It is important to realize that Scharlemann's method provides weak, or constant gradient, focusing. A similar calculation, which points towards extension of Scharlemann's scheme to strong focusing, has been done to study the effects of the sextupole fields due to the finite width of undulator pole pieces. Dattoli and Renieri performed such a calculation using a smallness parameter to describe the effective horizontal defocusing sextupole strength.[19] They arrive at an expression similar to the one obtained by Scharlemann (Eq. 16, above) for $<\beta_\perp^2>$. Thus, the phenomenon of sextupole focusing and defocusing exists to some extent in all linear undulators.

Sextupole focusing can be implemented by machining a parabolic curve into the permeable metal pole pieces of a hybrid undulator (as was shown in Figure 2). It might also be possible to achieve an effective sextupole component by simpler methods. One scheme recently discussed is the use of side arrays of permanent magnets to shape the undulator field (see Figure 4a).[20] Another idea being considered is the use of moveable blocks within each pole piece (see Figure 4b).[21] This could allow for tunable field shaping, but may suffer from unacceptable large multipole moments.

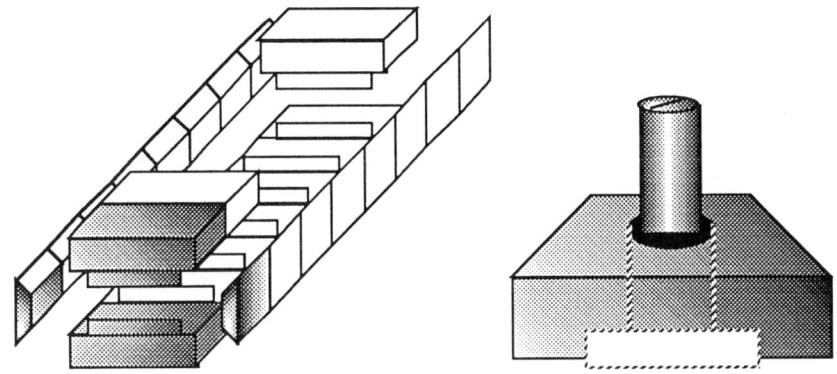

Figures 4a and 4b: Two alternatives to conventional pole shaping for producing field modifications. Figure 4a (left) shows an array of permanent magnets along the side of a planar undulator. Figure 4b (right) shows a single pole piece with an adjustable core.

Sextupole focusing satisfies the phase preservation requirement of an FEL; however, if it is constant gradient, it is weak focusing. Quadrupole focusing, on the other hand, can provide strong focusing (alternating gradient), but the phase modulation is detrimental to FEL action. The next section examines alternating gradient focusing using sextupole field components to provide strong focusing, while minimizing longitudinal phase modulation.

ALTERNATING GRADIENT SEXTUPOLE FOCUSING

Implementing strong focusing with sextupoles, which requires alternating from focusing to defocusing poles, can overcome the natural focusing strength limitation. A set of poles which focus in, say, x and defocus in y is followed by a set which defocuses in x and focuses in y. This is analogous to strong focusing with quadrupoles. Since sextupole fields are quadratic, the off-axis orbit taken by the design electron moves through a region with a linear gradient, giving rise to a quadrupole-like focusing. A schematic of strong sextupole focusing using pole shaping is given in Figure 5.

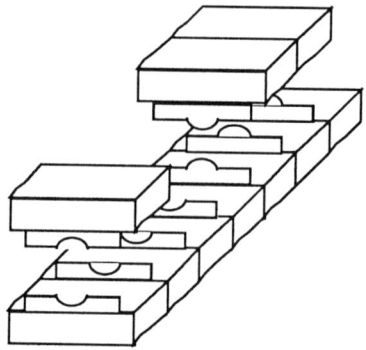

Figure 5: Sextupole AG focusing for planar undulators using pole shaping. A set of poles which focus (F) is followed by a set of poles which defocus (D) in order to form a FD lattice. This is repeated (FDFD...FD) the length of the undulator.

One constraint is that the design orbit of the electrons must match in the two types of undulator sections. It has already been verified that $<\beta_\perp^2>$ is constant in a weak focus (Scharlemann), and a weak defocus (Dattoli and Renieri). It is still necessary to show that the results are in fact valid for strong focusing. It is perhaps clearest to do this by extending the calculation in the previous section.[22]

We examine the case of strong defocusing in the undulation plane (x axis). By a simple extension, the analysis will hold for the focusing case. The constraint of Eq. 6 implies that when $k_y > k_u$ then k_x is imaginary, and likewise that when $k_x > k_u$ then k_y is imaginary. Recall that natural focusing in a planar undulator corresponds to the condition $k_x = 0$ and $k_y = k_u$. So, for defocusing in x (and hence strong focusing in y) $k_y > k_u$, and for strong focusing in x (and defocusing in y) $k_x > k_u$. For the strong *vertical* focusing case, we write $k_x = i\kappa_x$ where κ_x is real and non-negative. Then, the approximate normalized field in Eq. 8 can be rewritten as

$$b_x = -b_0\kappa_x^2 xy\cos(k_u z)$$

$$b_y = b_0\left(1 - \frac{\kappa_x^2 x^2}{2} + \frac{k_y^2 y^2}{2}\right)\cos(k_u z) \tag{17}$$

and the constraint $k_u^2 = k_x^2 + k_y^2 = k_y^2 - \kappa_x^2$ holds. For the strongly horizontally focusing case, the we can define $k_y = i\kappa_y$ where κ_y is real and non-negative. Then Eqs. 17 must be changed by substitution of the $-k_x^2$ for κ_x^2, and $-\kappa_y^2$ for k_y^2 to give the field components.

It is now straight forward to calculate the average transverse velocity as was done in the previous section. The equations of motion, Eq. 12, for the vertically focusing (horizontally defocusing) case become

$$\ddot{x}_0 - c^2 k_{\beta x}^2 x_0 \approx 0$$
$$\ddot{y}_0 + c^2 k_{\beta y}^2 y_0 \approx 0 \tag{18}$$

where, in analogy to the weak focusing case,

$$k_{\beta x} = \frac{b_0}{\sqrt{2}\gamma k_u} k_x . \tag{19}$$

Following the same argument as before it is possible to write the equations of motion with spatial (z axis) derivatives rather than time derivatives:

$$x_0'' - k_{\beta x}^2 x_0 = 0$$
$$y_0'' + k_{\beta y}^2 y_0 = 0 . \tag{20}$$

Then,

$$x_0 = x_\beta \cosh(k_{\beta x} z)$$
$$y_0 = y_\beta \cos(k_{\beta y} z) \tag{21}$$

where the phase terms have been suppressed. The above analysis holds in a similar fashion for the case when $k_x > k_y$ (strong focusing in x, defocusing in y). It is now straight forward to show that the average perpendicular velocity is given by

$$\left\langle \beta_\perp^2 \right\rangle^{\frac{F}{D}} = \frac{b_0^2}{2\gamma^2 k_u^2} \left(1 \, {}^+_- \, \kappa_x^2 x_\beta^2 \, {}^-_+ \, \kappa_y^2 y_\beta^2 \right) \tag{22}$$

where the top (bottom) signs hold for horizontal focusing (defocusing). Here the κ's are the real parts of the transverse undulator wavenumbers.

Equation 22 indicates two important points: the first is that $\langle \beta_\perp^2 \rangle$ is constant within individual focus and defocus sections. In addition, since one can show that the betatron amplitudes x_β and y_β are constant for each electron and remain the same across a lens boundary, one can see that $\langle \beta_\perp^2 \rangle$ is in general different in defocus and focusing sections. It is not feasible to make the velocities equal in the two types of sections for all electrons: any realistic beam will have a spread in betatron amplitude values. However, this shortcoming does not necessarily imply that strong sextupole focusing is problematic. Indeed, phase space mixing and possible detrapping is possible at the boundaries between a focus and defocus section. This situation inspires an analogy to tapered undulators. Theory indicates that the tapering should be performed gradually, but practical considerations can necessitate stepped tapering. Likewise, it is expected that if the focusing is not too strong the FEL gain will not be adversely affected. A calculation of the effective step difference in energy (due to a commensurate change in longitudinal velocity) across a focusing/defocusing section boundary, and a comparison of this quantity to the ponderomotive "bucket height" (the maximum energy deviation of the trapped electrons) should be performed for any case of interest, in order to determine whether or not detrapping may be a problem in this scheme.

If the average beta function is on the order of the gain length ($\beta \sim L_g$) then FEL operation (power output) is maximized for quadrupolar focusing.[23] Here the relevant length scale is not necessarily the beta function, but the length of each focusing section which need not be proportional to the average beta function. Perturbations caused by the focusing on a scale longer than the gain length should not be significant. The next section examines a solution to AG sextupole focusing and addresses the issue of focusing strength.

MATRIX DESCRIPTION OF AG FOCUSING

It is useful and straightforward to solve for the focusing effects of the AG sextupoles by using the transfer matrix description of the linear equations of motion. It is possible to use this method by substituting the equivalent quadrupole strength of the sextupole channel.

The transfer matrices for half of the focus (F) and defocus (D) section in a strong focusing lattice are defined by this prescription as follows

$$
F = \begin{bmatrix} \cos\dfrac{\theta}{2} & \dfrac{1}{k_\beta}\sin\dfrac{\theta}{2} \\ -k_\beta \sin\dfrac{\theta}{2} & \cos\dfrac{\theta}{2} \end{bmatrix}
$$

$$
D = \begin{bmatrix} \cosh\dfrac{\theta}{2} & \dfrac{1}{k_\beta}\sinh\dfrac{\theta}{2} \\ k_\beta \sinh\dfrac{\theta}{2} & \cosh\dfrac{\theta}{2} \end{bmatrix}
\tag{23}
$$

respectively, where $\theta = k_\beta L_q$ and L_q is the length of each section, and the focusing and defocusing strengths (k_β) are taken to be equal. Then, the total transfer matrix for one cell (one period of the focusing channel) is given by

$$
M_1 = \frac{F}{2}D\frac{F}{2} \quad \left(M_2 = \frac{D}{2}F\frac{D}{2} \right),
\tag{24}
$$

where a cell is started from the middle of a focus (defocus) section. Then,

$$
M_1 = \begin{bmatrix} \cos\theta\cosh\theta & \dfrac{1}{k_\beta}(\sin\theta\cosh\theta + \sinh\theta) \\ -k_\beta(\sin\theta\cosh\theta - \sinh\theta) & \cos\theta\cosh\theta \end{bmatrix}.
$$

$$
M_2 = \begin{bmatrix} \cos\theta\cosh\theta & \dfrac{1}{k_\beta}(\sin\theta + \sinh\theta\cos\theta) \\ -k_\beta(\sin\theta - \sinh\theta\cos\theta) & \cos\theta\cosh\theta \end{bmatrix}
$$

(25)

The parameter μ, the phase advance per cell, is then defined by $2\cos\mu = \mathrm{Tr}(M)$ = $2\cos\theta\cosh\theta$. For small angles ($\mu < \pi/4$) we may expand this transcendental expression to yield, in the smooth approximation,

$$
\mu \approx \frac{\theta^2}{\sqrt{3}}.
$$

(26)

The average beta function can be defined to be a geometric average of the minimum and maximum beta functions: $\beta_{max(min)}\sin\mu = [M_{1(2)}]_{1,2}$. This produces the relation

$$
\bar{\beta} = \frac{2\sqrt{3}}{k_\beta^2 L_q}.
$$

(27)

We note at this point the strong dependence on k_β, and that this beta function is $\sqrt{3}$ times larger than that for a thin lens FODO channel. Although this implies a larger beam (and so less dense), the variation of the beam size is smaller than in the thin lens case. This is advantageous in an FEL since large fluctuations in the beam size may be deleterious to gain and optical beam quality.

In order to make the strong focusing based on sextupole fields attractive, it must be clearly superior to weak focusing. This requires that

$$
\bar{\beta} < \beta_{\mathrm{weak}},
$$

(28)

where the quantity β_{weak} is defined as the beta function obtained for a round beam using Scharlemann's pole shaping scheme. Let the ratio between the strong and weak betatron wavenumber be $R = k_{\beta\mathrm{strong}}/k_{\beta\mathrm{weak}}$. Then $k_{\beta\mathrm{strong}} = R k_{\beta\mathrm{natural}}\sqrt{2}$ and Eq. 28 becomes[24]

$$
R > \left(\frac{2\sqrt{3}}{k_{\beta\mathrm{weak}} L_q}\right)^{1/2} = \left(\frac{4\sqrt{3}\gamma}{b_0 L_q}\right)^{1/2}.
$$

(29)

This requirement is an overestimate of the ratio R because of the previous requirement that the smooth approximation be valid. To show what happens when this requirement is lifted, consider the case of 90° phase advance per cell ($\mu=\pi/2$). Then,

$$\beta_{min} = \frac{1}{k_\beta} \quad \beta_{max} = \frac{1}{k_\beta} e^{\pi/2}$$

$$\bar{\beta} \approx \frac{1}{k_\beta} \sqrt{e^{\pi/2}} \approx \frac{2.2}{k_\beta} \tag{30}$$

Notice that the beta function is independent of the quadrupole length (assuming $L_q > \lambda_u$). In fact, the ratio R is independent of all parameters and it is only required that

$$R > 2.2. \tag{31}$$

The variation of the magnetic field with a strong sextupole component may be large enough to degrade the FEL synchronism condition, and this effect must be examined. The fractional variation of the magnetic field over the beam size can be expressed as

$$\left(\frac{\Delta B}{B}\right)_{beam} = \frac{(k_x \sigma_x)^2}{2} = \frac{(Rk_u \sigma_x)^2}{4} \tag{32}$$

where σ_x is the transverse beam size. Similarly, the variation of the magnetic field over the undulation orbit is

$$\left(\frac{\Delta B}{B}\right)_{undul.} = \frac{1}{2}\left(\frac{Rb_0}{\gamma k_u}\right)^2 \tag{33}$$

As the examples in the next section will show, requiring that these variations be small compared to the bandwidth of the FEL radiation ($\sim 1/N_{periods}$) is not unreasonable.

EXAMPLES

Two examples of high gain FEL amplifiers are now discussed, to show the potential usefulness of AG sextupole focusing in undulators. The UCLA Particle Beam Physics Laboratory is constructing a 10.6 μm FEL for the purpose of studying physics issues relevant to future short wavelength operation.[25] While the initial design calls for a single undulator section 60 cm long, future plans include adding a second section for a total length ~ 120-160 cm. The need for focusing might then become significant.

The short period modified hybrid undulator has flat poles. The natural vertical focusing has an average beta function of 10.5 cm, which if converted to the equivalent weak focusing round beam case would yield a beta function of 14.8 cm.

The strong focusing scheme proposed in this paper could be used to yield an average beta function as small as 3.8 cm with reasonable pole shapes (R=5.5). A plot of the pole shape is given in Figure 6. The reduced beta function would mean a beam density ~4 times greater than the round beam case, and over eight times greater than the actual situation of no horizontal focusing. The 1D linear FEL theory then gives at least a two-fold decrease in the power gain length.

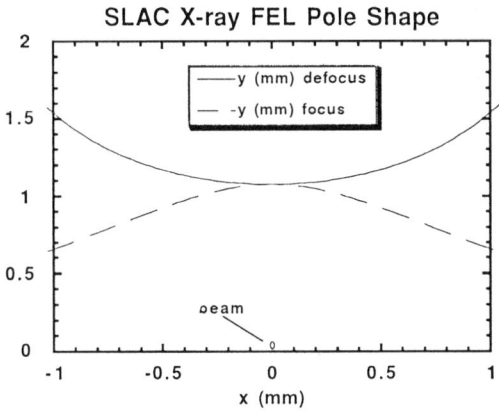

Figure 6: A plot of the pole face curvature required to implement a practical strong focus in the SLAC FEL. The solid line (top) is the convex extension to the pole face. The dashed line (bottom) is the concave indentation to the pole face. The beam (bottom, center) is barely visible on this scale.

The proposed SLAC based X-ray FEL poses a number of challenges including the need to propagate the electron beam along fifty meters of undulator.[3] Whereas the UCLA project can use focusing as an option to enhance performance, the SLAC FEL requires focusing. The natural round beam focusing beta function is 83 m whereas the design gain length requires a ~10 m beta function. So, weak focusing is insufficient to maintain the desired beam size, and the current density is such that the gain length would be approximately twice as large as the design value.

Strong focusing could be implemented in this undulator using curved pole pieces. An average beta function of 7.8 meters can be obtained with 90 degree phase advance per cell lattice, of half length 6 m, and R=21.4. Note that this section length is about equal to a gain length, and thus the deleterious effects of the sudden change in z-velocity of the electrons at the transitions should be mitigated. The radius of curvature of the poles would be small, and this might pose some technical problems. This issue will be investigated in future work. The alternative field shaping methods mentioned in the Introduction could also be used. It should be noted, however, that in this example, the fractional variation of the magnetic field over the beam cross section is small ($\sim 2 \times 10^{-4}$). The fractional variation of the undulator field over the undulating design orbit is even smaller ($\sim 4 \times 10^{-5}$).

CONCLUSIONS

Alternating gradient sextupole undulator focusing has been examined for use in free electron lasers. Sextupole fields appear to be an attractive option for strong focusing in that variations of the longitudinal velocity of the beam particles are minimized. This analysis indicates that small magnetic field variations over the beam cross section are sufficient to employ strong focusing. Field shaping alternatives might be superior to curved poles which have small radii of curvature, and thus tight machining and alignment tolerances. In addition, movable pole pieces could provide some tuning in this scheme, which might otherwise be viewed as unworkably rigid since the focusing in the device would be set at the time of the undulator construction and could not be subsequently changed. For a strong focusing channel, this may cause difficulty in tuning the beam optics over a range of energies.

Further work is needed to examine the effects of AG sextupole focusing on FEL gain. Ultimately, numerical simulations will be necessary because of a beam's inherent spread in the parameter space. Future work will also include simulations of field shaping schemes.

ACKNOWLEDGMENTS

The authors thank Claudio Pellegrini for many useful conversations.

REFERENCES

[1] K. Halbach, J. de Physique, Colloque C1 (1983).

[2] K. E. Robinson, et al.,*Hybrid Undulator Design Considerations*, Nucl. Inst. and Meth. A250, p. 100 (1986).

[3] C. Pellegrini, et al., A 2 to 4nm High Power FEL on the SLAC LINAC, Proc. Internatinal FEL Conf, Kobe, Japan, (To be published: 1991).

[4] R. Bonifacio, C. Pellegrini, andL. M. Narducci, Opt. Commun. 50, p373 (1984).

[5] R. Bonifacio, et al., *Physics of the High-Gain FEL and Superradiance*, La Rivista del Nuovo Cimento 13 9 (1990).

[6] For simplicity we ignore the Bessel function coefficient $JJ = J_0(x)-J_1(x)$ where $x = a_w^2/(1+2a_w^2)$.

[7] Here it is assumed that the electron and radiation beam sizes are equal. These relations hold for natural focusing and must, at least, be satisfied for other types of focusing. Another condition that must be met is that the Rayleigh range be greater than the gain length, but this is not directly relevant to focusing.

[8] T. J. Orzechowski, et al., *High Efficiency Extraction of Microwave Radiation from a Tapered Wiggler Free Electron Laser*, UCRL-94841 (1986).

[9] E. T. Scharlemann, *Wiggle plane focusing in linear wigglers*, J. Appl. Phys. 58 6, p. 2154 (1985).

[10] W. M. Fawley, D. Prosnitz, and E. T. Scharlemann, *Synchrotron-betatron resonances in free-electron lasers*, Phys. Rev. A 30 5, p2472 (1984).

[11] C. Pellegrini, J. Rosenzweig and G. Travish (unpublished).

[12] J. A. Pasour, F. Mako, and C. W. Roberson, *Electron drift in a linear magnetic wiggler with an axial guide field*, J. Appl. Phys. **53** 11, p. 7174 (1982).

[13] L. Friedland and R. E. Shefer, *Electron-beam confinment by rotational stabilization in a linear wiggler free electron laser*, J. Appl. Phys. **68** 10, p. 4958 (1990).

[14] L. H. Yu, A. Sessler, and D. H. Whittum, Free-Electron Laser Generation of VUV and X-Ray Radiation Using a Conditioned Beam and Ion-Channel Focusing, LBL-31198 and Proc. 1991 Free Electron Laser Conf. (Santa Fe, NM: 1991).

[15] H. Lee Buchanan, Electron beam propogation in the ion-focused regime, Phys. Fluids **30** 1, p221 (1987).

[16] V. G. Baryshevsky, I. Ya. Dubovskaya and O. N. Metelitsa, *Mutual focusing of an electron beam and an electromagnetic wave in a free electron laser*, Phys. Let. A **148** 5, p.272 (1990).

[17] For an early reference see W. M. Fawley, *Physics Design for the ATA Tapered Wiggler*, IEEE Trans. Nuc. Sci. **NS-32** 5, p.3424 (1985).

[18] For a complete derivation of the magnetic potential see John J. Barnard, *Anharmonic Betatron Motion in Free Electron Lasers*, Nuc. Inst. Meth. **A296**, p.508 (1990).

[19] G. Dattoli and A. Renieri, *The free electron laser*, <u>Laser Handbook 4</u>, eds. M. L. Stitch and M. S. Bass (Amsterdam: North-Holland).

[20] Private communication with A. A. Varfolomeev and A. S. Khlebnikov. See W. H. Urbanus, et al.,*A 1MW Free Electron Maser for Fusion Applications*, Proceedings 3rd Euro. Part. Accel. Conf., Berlin (1992), to be published.

[21] Private communication with A. Hairetdinov.

[22] It is interesting to note that the problem possesses four length scales: the undulator period, the AG period, the betatron orbit, and the synchrotron period. We eliminate the undulator period by averaging the electron motion on this length scale. And, we neglect the synchrotron period since it is typically very long.

[23] C. Pellegrini, Nucl. Instr. and Meth. **A272** p364 (1988).

[24] Here it is assumed that the weak transverse focusing is equal in the two planes. That is, $k_x^2 = k_y^2 = k_u^2/2$. The relation $\beta = 1/k_\beta$ holds throughout.

[25] A brief and up to date description of the UCLA project can be found in *The UCLA IR FEL Project*, Proc. International FEL Conf., Kobe, Japan, (To be published: 1992). A more detailed description is given in an article by J. W. Dodd, et al.. *The UCLA Infrared Free-Electron Laser*, Nucl. Instr. and Meth. **A304** p.155 (1991).

AN INVERSE FREE ELECTRON LASER ACCELERATOR EXPERIMENT

Iddo Wernick and T.C. Marshall
Department of Applied Physics, Columbia University, New York, NY 10027

ABSTRACT

A free electron laser was configured as an autoaccelerator to test the principle of accelerating electrons by stimulated absorption of radiation (λ=1.65mm) by an electron beam (750kV) traversing an undulator. Radiation is produced in the first section of a constant period undulator (l_{w1}=1.43cm) and then absorbed (~40%) in a second undulator, having a tapered period (l_{w2}=1.8-2.25cm),which results in the acceleration of a subgroup (~9%) of electrons to ~1MeV.

The principle of using free electron laser [FEL] physics to accelerate electrons was described by Palmer [1] in 1972, however despite the extensive development of the FEL there has been no demonstration of the stimulated absorption of a laser pulse accompanied by acceleration of a group of electrons while the electron beam is traversing an undulator [IFEL]. The idea has been re-examined in more detail [2-4] and appears to offer some promise to achieve an acceleration gradient ~ 1MV/cm in linear accelerators. In the electron rest frame, the magnetostatic field of the undulator is transformed into an electromagnetic wave which beats with the laser; acceleration occurs by keeping the phase of the electrons constant with respect to the beat wave, by varying the undulator period and/or magnetic field as the particle energy increases. We have devised a relatively simple experiment, done on the Columbia FEL facility [5], which demonstrates that acceleration does occur.

The Columbia FEL operates at a wavelength of ~1.6mm and produces about five MW from a 750kV electron beam. As there is no powerful laser source at this wavelength, we have configured the experiment as an "autoaccelerator"[IFELA] in which a subgroup of electrons is accelerated by the "inverse FEL" mechanism at the expense of the average energy of the entire beam. This is done by separating the undulator into two sections.

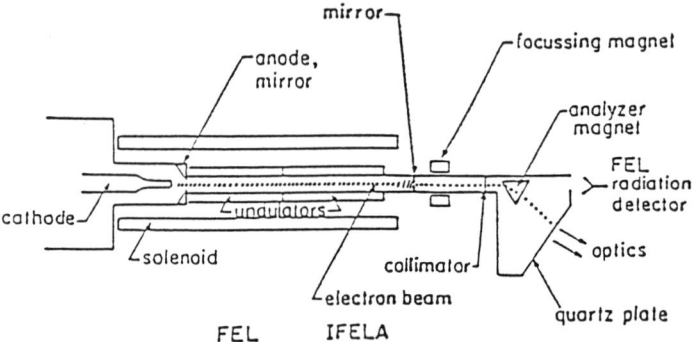

Figure 1: Schematic of the IFELA and the magnetic spectrometer. Electron emission occurs from a cold graphite cathode in a field-immersed diode, the beam is formed by a 4mm dia. aperture in a graphite anode, which also serves as the upstream mirror of the resonator.

The first section [Fig 1] together with a pair of 75% reflecting coaxial mirrors develops FEL radiation which grows from noise to saturated intensity and causes a bunching of the electrons. These particles then enter the second section of the undulator where the period is increased and then tapered along the axis so that a subgroup of electrons is accelerated as stimulated absorption of the wave occurs. We report measurements of this nonlinear absorption together with the electron energy spectrum. The accelerator section acts as a load for the oscillator, but its absorption is not high enough to prevent oscillation of the entire system.

Table I summarizes the parameters of the IFELA experiment. The undulator is a bifilar helical winding which provides a transverse field of order 600G following an adiabatic entry region.

Table I : Operating Conditions of the Columbia IFELA

Beam Energy	750-800kV
Beam Diameter	4mm
Beam current	150A
Pulse length	150nsec
Drift tube diameter	11mm
Undulator: first section	$l_{w1}= 1.4$cm
	$B\perp= 600$G
	length, 40cm
Undulator: second section	$l_{w2}= 1.8$-2.25cm
	taper$=\dfrac{1}{l_w}\dfrac{dl_w}{dz}=0.0067$cm^{-1}
	$B\perp=400$G
	length= 37.5cm

The beam is guided and focused along the drift tube by a uniform solenoidal field ~1T which causes "Group I" orbits [6]. The FEL power examined by a grating spectrometer shows a carrier wavelength of 1.6mm together with a pair of sidebands which carry about one-third of the total power; it was found that only the carrier was absorbed by the acceleration process [7]. The downstream mirror is polished graphite with a small hole on its axis followed by a collimator which forms the objective of the electron beam optics. A focusing solenoid guides the electrons beyond the fringe field of the solenoid. A dipole field using triangular polefaces deflects the beam and disperses the electrons onto a quartz viewing plate where the impact causes substantial Cerenkov light [8] (the quartz is painted with an opaque graphite film on the vacuum side). The light from the energetic electrons is directed to two lead-shielded photomultipliers located in a shielded room with the other electronics. The magnetic spectrometer is calibrated by using the electron beam with no undulator. The use of two photocells permits one photocell to monitor the principal group near the injection energy while the other scans the energy channels for the accelerated electrons.

The undulator in the IFELA section is designed using a numerical model to choose the appropriate field and taper. We use a set of equations [5] that models the electron motion in 1D along a single pass through the system and a self-consistent set of 2D field equations.

$$\frac{d\psi_i}{dz} = k_w + k_s - \frac{k_s}{\sqrt{1 - \dfrac{1 + a_w^2 + 2a_w\, a_s\, \cos\psi_i}{\gamma_i^2}}} + \frac{\partial\Phi}{\partial z} \tag{1}$$

$$\frac{d\gamma_i}{dz} = -\frac{a_w a_s k_s}{\gamma_i}\sin\psi_i + \frac{2\omega_p^2}{k_s c^2}[<\cos\psi>\sin\psi_i - <\sin\psi>\cos\psi_i] \tag{2}$$

$$(\nabla_\perp^2 + 2ik_s\frac{\partial}{\partial z} + 2i\frac{\omega}{c^2}\frac{\partial}{\partial t})\, u(r,t) = -F\frac{\omega_p^2 a_w}{c^2} <\frac{e^{-(\psi-\Phi)i}}{\gamma}> \tag{3}$$

Equation 1 is used to determine the phase Ψ of the electrons with respect to the signal wavefront as the electrons traverse the undulator. Equation 2 is the energy equation for the electrons and finally equation 3 is the 2D wave equation which describes the signal growth. The "Raman" term (i.e. terms $\propto \omega_p^2$) which accounts for the longitudinal space charge electric field from the electron bunches is included and is necessary to obtain the correct growth rate in the FEL section and the correct interaction in the IFELA section. The electron beam in the experiment has normalized parallel momentum spread $\sim 1\%$ [9], and is represented in the numerical study as a "cold" beam, since the trapping width, $\Delta\gamma/\gamma \propto \sqrt{a_s a_w}$ where a_s and a_w are the normalized vector potentials of the signal and undulator fields, is much larger than 1%.

The magnetic field of the FEL undulator section was chosen so that the power would grow roughly a factor of 25 in 40cm. This gain will sustain oscillation and results in a signal which reaches power saturation at the point where the electrons enter the IFELA section. The saturated power intensity on the beam axis, $\sim$ 10MW/cm^2, is consistent with the power output from an FEL oscillator device very similar to this one [10]. It is found numerically that the wave amplitude is reduced by one-half as it reaches the end of the IFELA undulator as a subgroup of the trapped electrons is accelerated to ~1MV. The code includes no slippage and therefore does not account for sideband radiation; hence the measured absorption of the FEL power by the accelerator module will be less than the model predicts. The taper of the undulator which optimizes the absorption and acceleration is found through trial and error of the numerical study and corresponds to an acceleration gradient of ~7kV/cm in this test experiment. The taper is used to generate a variable-period helix which is mapped onto a section of phenolic tube and then cut to specification "by hand". Measurements of the actual undulator period taper and magnetic field taper are then incorporated into the code to simulate the actual experimental situation. Once the taper is chosen, a series of simulations show that the acceleration is not very sensitive to variation in FEL power or undulator field, however, increasing the power input to the IFELA section will accelerate more electrons. The field in the IFELA undulator can be varied independently of the FEL undulator; both are powered by a capacitor bank discharge synchronized to the accelerator timing system.

A set of representative data is shown in Fig 2. Shots are selected for a relatively flat diode voltage history, with electron energy near resonance. A determining signature as the electron beam energy reaches the design value is a decrease of transmitted FEL power accompanied by an increase of light signal in the photomultipliers which respond to the accelerated electrons. The energy bins are separated by the resolution of the electron beam optics. Background light,obtained from operation of the apparatus with zero undulator field, is subtracted from the signal. The transmitted FEL power is monitored by a Schottky-barrier diode detector located

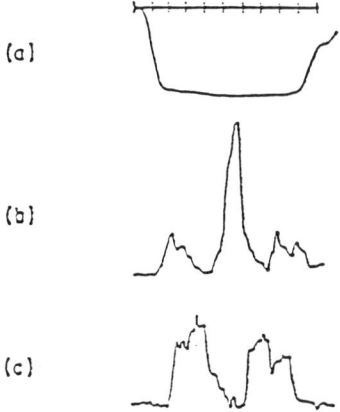

(a)

(b)

(c)

Figure 2. Representative signals obtained in the experiment. (a) Accelerator voltage, -800kV maximum; (b) signal from photocell monitoring 1MV electrons, showing a burst of electrons at middle of trace; (c) FEL power transmitted through the IFEL stage, showing absorption in middle of trace; 20nsec/div, horizontal scale.

along the beam axis. It is found that the fraction of power absorbed does not depend sensitively upon the power level itself. This behavior is to be expected from the model, which shows that increases in power absorbed result in more accelerated electrons but not necessarily higher electron energy.

Figure 3a displays results from a numerical simulation of the experiment showing the growth of the FEL signal, followed by the attenuation of the wave in the accelerator section. Fig 3b shows experimental data giving the FEL power transmitted through the accelerator section. This shows that there is a reduction of emitted power by ~40% in the vicinity of the resonant energy of the design. By reducing the undulator field in the accelerator section to 250G , the amount of observed absorption decreases to about 25% of the incident power. The study of the FEL spectrum [7] shows that the sidebands of the FEL power spectrum are not absorbed by the accelerator to a measurable amount, and therefore the reduction of the incident carrier intensity [~75%] indicated in the numerical simulation is larger than would be obtained if the sideband power is included in the measurement, as was the case.

Figure 4 shows the measured electron energy spectrum. The data is compared with the numerical simulation, run according to the experimental conditions. The computed spectrum is processed so that the ordinate corresponds to the number of simulation electrons contained in a bin having the same width as the experimental bin. The relative number of electrons accelerated and the acceleration energy are in good agreement with the numerical model. According to numerical simulation the smaller peak which occurs at $\gamma \approx 2.2$ does not result from the acceleration process.

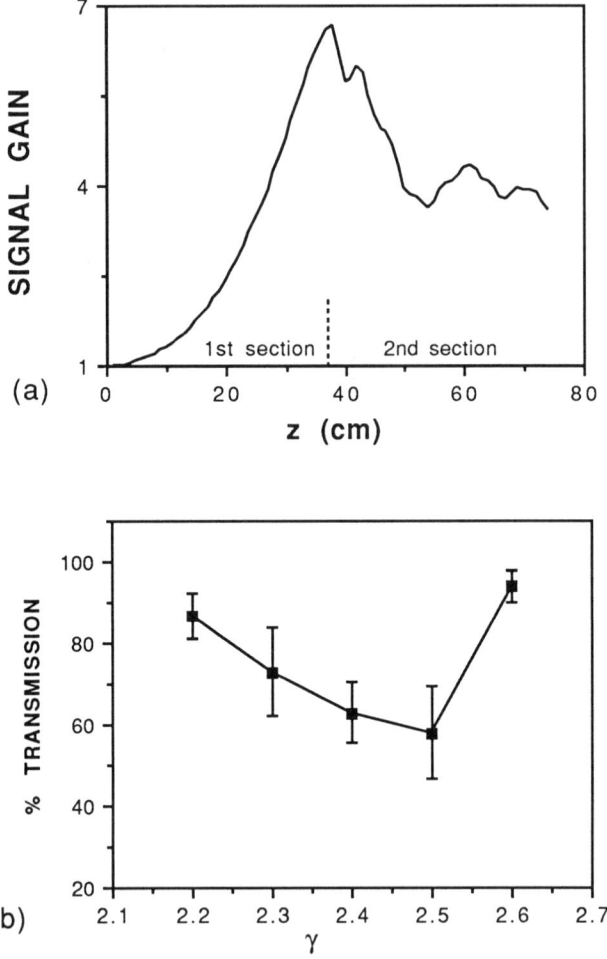

Figure 3. (a) Numerical result showing growth of FEL signal (initial intensity = 0.3MW/cm2) and attenuation in the IFELA section (begins at z = 37.5cm); (b) Radiation transmitted through the IFELA section as a function of initial electron beam energy. Parameters of simulation taken from Table I. Error bars indicate standard deviation of the data.

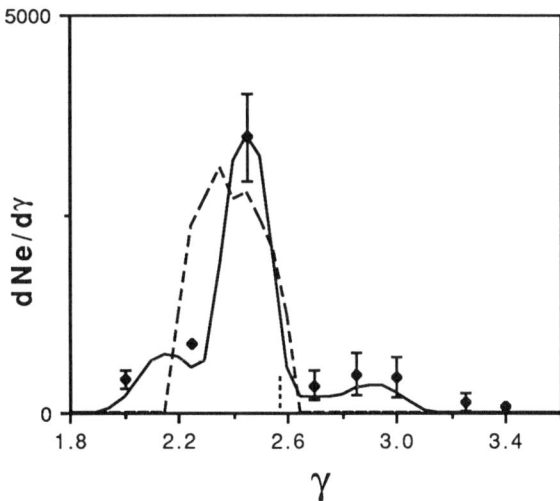

Figure 4. Measured electron energy spectrum(solid points);Electron energy spectrum obtained from simulation, at the exit from the IFELA(solid line); Electron energy spectrum obtained from simulation, at entry to the IFELA section(dashed line). The dotted vertical line denotes the injection energy.

For comparison, the dashed line on Fig 4 shows the electron spectrum computed (but not observed) at the end of the FEL section only; one notes there is a rapid cutoff of electrons having energy in excess of $\gamma = 2.7$, which is well below the energy of the accelerated group. Thus electrons at ~1MV have energy beyond the broadened distribution which results from the bunching and trapping of electrons from the FEL interaction.   Electrons above the injection energy in FELs have been observed[11, 12] however they are not "accelerated" by the IFEL mechanism. The number of accelerated electrons is about 9% of the total number, as can be estimated from Fig 4. The ratio of the power required to accelerate these electrons to the overall power of the electron beam is about 3%, which is less than the efficiency of the FEL (4-6%).

In an actual IFEL where the radiation is supplied by an external laser, the undulator field can be considerably larger (roughly a factor of ten greater than in our experiment) and the intensity of the laser wave can be higher by perhaps a factor of a thousand. Taking a 10.6μ laser wavelength as an example our observed acceleration gradient ~ 7kV/cm would scale up by a factor of ~100 for such a device. Techniques exist to control synchrotron radiation losses which are no longer negligible at high energy [3]. Our success with the experiment and its interpretation suggests that the IFEL is a promising technology for an "advanced accelerator" demonstration.

ACKNOWLEDGMENT

This research is supported by the NSF, grant ECS -89-13066, DOE grant 02-91ER40669, and ONR grant N00014-89-J-1652. The assistance of Y-P Chou in the early stages of this project is greatly appreciated.

REFERENCES

1. R. B. Palmer, J. Appl. Phys. $\underline{43}$, 3014 (1972).
2. P. Sprangle and C. M. Tang, IEEE Nucl. Sci. $\underline{NS-28}$, 3346 (1981).
3. E. D. Courant, C. Pellegrini, and W. Zakowicz, Phys. Rev. $\underline{A32}$, 2813 (1985).
4. A. van Steenbergen and C. Pellegrini, private commun.
5. A. Bhattacharjee, S. Y. Cai, S. P. Chang, J. W. Dodd, A. Fruchtman, and T. C. Marshall, Phys. Rev. A40, 5081 (1989).
6. L. Friedland, Phys. Fluids $\underline{23}$, p.2376 (1980).
7. T. C. Marshall, A. Bhattacharjee, S. Y. Cai, Y. P. Chou, and I. Wernick, Nuclear Instruments and Methods in Physics Research $\underline{A304}$, 683 (1991).
8. B. Kulke, M. J. Burns, and T. J. Orzechowski, LLNL Report UCRL-95598.
9. S. C. Chen and T. C. Marshall, IEEE J. Quantum Electronics $\underline{QE-21}$, 924 (1985).
10. F. G. Yee, T. C. Marshall, and S. P. Schlesinger, IEEE Trans. on Plasma Science $\underline{16}$, 162 (1988).
11. A. H. Lumpkin and R. B. Feldman, Nucl. Instruments and Methods in Physics Research, $\underline{A259}$, 19 (1987).
12. T. I. Smith, J. C. Frisch, R. Rohatgi, H. A. Schwettman, and R. L. Swent, Nucl. Instruments and Methods in Physics Research $\underline{A296}$, 33 (1990).

INVERSE FREE ELECTRON LASER ACCELERATOR[*]

A. Fisher, J. Gallardo, J. Sandweiss[**], A. van Steenbergen
National Synchrotron Light Source and Physics Department
Brookhaven National Laboratory, Upton, NY 11973

ABSTRACT

The study of the INVERSE FREE ELECTRON LASER, as a potential mode of electron acceleration, is being pursued at Brookhaven National Laboratory. Recent studies have focussed on the development of a low energy, high gradient, multi stage linear accelerator. The elementary ingredients for the IFEL interaction are the 50 MeV Linac e^- beam and the 10^{11} Watt CO_2 laser beam of BNL's Accelerator Test Facility (ATF), Center for Accelerator Physics (CAP) and a wiggler. The latter element is designed as a fast excitation unit making use of alternating stacks of Vanadium Permendur (VaP) ferromagnetic laminations, periodically interspersed with conductive, nonmagnetic laminations, which act as eddy current induced field reflectors. Wiggler parameters and field distribution data will be presented for a prototype wiggler in a constant period and in a $\approx 1.5\%$/cm tapered period configuration.

The CO_2 laser beam will be transported through the IFEL interaction region by means of a low loss, dielectric coated, rectangular waveguide. Short waveguide test sections have been constructed and have been tested using a low power cw CO_2 laser. Preliminary results of guide attenuation and mode selectivity will be given, together with a discussion of the optical issues for the IFEL accelerator.

The IFEL design is supported by the development and use of 1D and 3D simulation programs. The results of simulation computations, including also wiggler errors, for a single module accelerator and for a multi-module accelerator will be presented.

I. INTRODUCTION

The study of the Inverse Free Electron Laser (IFEL) as a potential mode for particle acceleration, has been pursued at Brookhaven National Laboratory for a number of years now.[1-4] More recent studies have focussed on the development of a low energy (few GeV) multistage linear accelerator.[5] Specifically also, the design of a short accelerator module has been pursued, which would make use of the ATF high brightness, 50 MeV, Linac beam and its high power, 10^{11} Watt, CO_2 laser beam. This would be used to carry out an accelerator demonstration stage experiment[6] with such parameters as to be of potential interest towards the further development of a "competitive" high gradient electron Linac with a wavelength

[*]Work performed under the auspices of the U.S. Department of Energy, under Contract No. DE-AC02-76CH00016.
[**]Yale University

parameter of 10 μm, based on the IFEL principle, for maximum operating energies of up to a few GeV.

In the IFEL accelerator, the electron beam is accelerated by the interaction with the laser radiation wave in the medium of a periodic wiggler field. The theoretical description of the interaction has been given by a number of authors. Here, in Section II, in order to simplify the parameterization for a single module IFEL accelerator stage, use is made of the basic formalism describing the IFEL accelerator as given by CPZ[3], with the further assumption that electron energy loss effects due to synchrotron radiation emission is taken to be zero and that the laser beam attenuation due to absorption by the accelerating electrons is negligible. The latter assumption is abandoned in further detailed treatment here and, in Section VI, following KMR[7], a self consistent system of Lorentz equations for the electrons and the wave equations for the input laser field is used, to form the basis of both a 1-D and 3-D IFEL simulation program.

As will be detailed in Section II, for the IFEL accelerator, maximum rate of acceleration is achieved with the use of the highest magnetic field, tapered period length wiggler, hence, substantial effort has been devoted to the design of a matched to the purpose, high field, variable period, fast excitation wiggler.[8-10] This is presented in Section III.

A limitation of the IFEL accelerator rate of acceleration is due to the varying E field magnitude with distance in a diffraction limited laser, free space, transport mode. Making use of a relatively short wiggler section (e.g. $\approx$ 0.6 m) it is possible to keep the Rayleigh length (defined as the distance over which the cross sectional area of a "Gaussian" laser beam doubles in value) to a small value (e.g. $\approx$ 0.3 m) thereby enhancing the effective use of the available laser power [$E_L \alpha \sqrt{(W/R)}$, where W = laser power, R = Rayleigh length; for a diffraction limited beam]. A disadvantage of this mode of laser transport is, however, that use must be made of periodically placed optical elements in a multistage accelerator system. Although this approach was studied in some detail and a possible conceptual approach was obtained[11] which preserved macroscopic synchronization of the electron bunches and the photon bunches in a multistage system, it was abandoned because of inherent complexity and difficulty in maintaining phase slip tolerances. Instead, the alternative of confinement of the laser radiation in a hollow optical waveguide has been adopted. As a matter of fact, the electron beam pipe passing through the IFEL wiggler acts as a (overmoded) wave guide for the CO_2 laser radiation, taking into account the relative magnitudes of the CO_2 laser wavelength and gap magnitude of the IFEL wiggler. In general, attenuation of CO_2 light in metallic guides can be significant for a long accelerator. The use of dielectric coated guides[12] can reduce the attenuation leading to calculated attenuation coefficients of 10^{-5} dB/m.

A description of square aperture, dielectric coated guides is given in Section III, together with early results of CO_2 laser attenuation measurements of both dielectric coated guide sections and uncoated comparable stainless steel guide sections.

The use of an overmoded metallic guide brings with it problems of mode coupling, matching of phase and group velocity and parasitic mode contamination. These optical issues are addressed in Section V.

As indicated above, for early parameterization of a single IFEL module, a 1-D simulation code was used to model the acceleration process. Subsequent to this, a 3-D code was further developed and adopted, specifically also to study transverse and longitudinal phase space distributions of the IFEL accelerator. Results from IFEL simulation computations are presented in Section VI. With the availability of these two programs, the parameterization of a multimodule IFEL accelerator was carried out together with the adoption of a sensible error spectrum, with the aim of not only gaining insight into the achievable parameters for a multimodule IFEL system, but also to optimize further the single first module for the accelerator demonstration stage. This is discussed in Section VII.

II. IFEL ACCELERATOR EXPERIMENT

From CPZ[3], for the case of zero synchrotron radiation emission and negligible attenuation of the laser beam due to energy transfer to the electron beam, the relevant IFEL accelerator equations are given by:

$$\left(\frac{d\gamma}{dz}\right) = \xi_n \, A \, \frac{K}{\gamma} \, \sin\psi$$

$$\left(\frac{d\psi}{dz}\right) = \kappa - [k/(2n\gamma^2)] \, (1 + \eta K^2 + K_L^2 + 2\eta^{1/2}\xi_n K \, K_L \cos\psi) \quad, \quad with$$

$$A = \left(\frac{e \, E_L}{m_o c^2}\right),$$

and E_L the electric field of the laser EM wave; $\xi_n = F_n$ for a planar wiggler. The F_n function (n = harmonic number) is given by (n = 1,3,5, ...):
$F_n = 1/2 \, \{J_{n-1}(x) - J_n(x)\}$ with $x = kK^2/8\kappa\gamma^2$ and J_i are Bessel functions.
Further: $K = eB\lambda_o/2\pi m_o c^2 = 93.4 \, B\lambda_o$ [T.m.] with λ_o the wiggler period length, and

$$K_L = \frac{eE_L\lambda}{2\pi m_o c^2}$$

$\eta = 1$, helical wiggler; $\eta = 1/2$ planar wiggler
Ψ = phase of the electron oscillation relative to the EM wave as seen by the electron.
$\Psi = (k + \kappa) \, z - \omega t$; $\omega = kc = (2\pi/\lambda)c$; $\kappa = 2\pi/\lambda_o$.

For electron acceleration it is necessary that $\Psi \approx$ constant, or $(d\Psi/dt) = 0$, from which the resonant condition follows as:

$$\kappa = \frac{k}{2n\gamma^2} (1 + \eta K^2) \quad for \ K_L < K$$

or, for $K >> 1$: $\lambda = (\eta \lambda_o K^2)/(2n\gamma^2)$.

With the synchronism condition, for a fixed value of λ, there remains one arbitrary function in the equations with which to optimize the accelerator, i.e. equivalent sets of accelerator parameter equations may be obtained for the case whereby either

$$\lambda_o(z) \quad or \quad B(z) \quad or \quad K(z) \propto B(z) \cdot \lambda_o(z)$$

is adopted as an arbitrary function in the equations. These equivalent sets of equations are given in Table I, where the rate of acceleration, $(d\gamma/dz)$, is given as a function of either $\lambda_o(z)$, $(\Omega/\omega) = C_1 \lambda B(z)$ or $K(z)$. In actual execution of designing an IFEL accelerator structure, limiting parameters are introduced by the maximum wiggler field value and minimum wiggler period length value. In effect, it is evident from the equations given in Table I that for a $\lambda_o(z) = $ const. accelerator, the maximum wiggler B value is encountered at the exit of the accelerator, but for a $K(z) = $ const. accelerator the maximum wiggler field magnitude is encountered at the injection end of the accelerator. In general, it can be shown[5] that for an IFEL accelerator, for which the synchrotron radiation loss term is negligible, and which has been optimized to the limit of the maximum wiggler field anywhere in the structure, the maximum rate of acceleration, averaged over the full accelerator structure length, is obtained for the case of a $B(z) = $ const. accelerator. This is further shown in Fig. 1 for the specific case of a 1 GeV IFEL accelerator for which a maximum (pulsed excitation) field value of 1.5 Tesla is adopted.

For the demonstration stage accelerator the parameters are developed based on the use of a constant maximum wiggler field magnitude of $B = 1.25$ T. The minimum value of the wiggler period length, in this case, with an injection energy of 50 MeV, is 2.87 cm ($n = 1$) which, taking a wiggler gap value of 0.4 cm into account, is technically acceptable. Recall also, however that in this case $\lambda_o \propto n^{1/3}$ $\gamma^{2/3}$, and that the harmonic mode of acceleration, $n = 3,5...$ could be employed in a short, low energy front end section, of the accelerator.

For the constant B accelerator structure, with an adopted value of $B_{max} = 1.25$ T and $n = 1$, to first order, the magnitudes of $\lambda_o(z)$, $K(z)$ and $(d\gamma/dz)$ are given by

$$\lambda_o = 0.0014 \ \gamma^{2/3} \ , \quad K = 0.167 \ \gamma^{2/3} \ , \quad \frac{1}{A}\left(\frac{d\gamma}{dz}\right) = 0.167 \ \gamma^{-1/3} \ F_1$$

with $A = A \sin\Psi$. The latter equation may be integrated to yield:

$$E(z) = \left[\frac{4}{3}(F_E \cdot 10^4)\left(\frac{2}{\eta}\right)^{1/3}(\xi \sin\psi)\left(\frac{\Omega}{\omega}\right)^{1/3}(m_o c^2)^{1/3}z + E^{4/3}(z=0)\right]^{3/4}$$

with $eE_O = F_E \cdot 10^4$ [MeV/m]; z[m]; $\xi = F_1$, and $F_E = (W_L/2 \ 10^{11})^{1/2}$.

For the case of the low loss waveguide CO_2 laser transport, the laser field magnitude follows from:

$$W = \left(\frac{1}{4Z_o}\right) \int\int E_i^2(0,0) \cos\left(\frac{\pi x}{2D}\right) \cos\left(\frac{\pi y}{2D}\right) dxdy = \left(\frac{D^2}{4Z_o}\right)\left(\frac{16}{\pi^2}\right) E_i^2(0,0) \text{ , } where$$

D is the half transverse aperture of the guide (1.4 mm). Hence, for a laser power of $W = 2.10^{11}$ Watts, a field magnitude of $E_\ell \approx 10^4$ MV/m is obtained.

The design parameters for an IFEL demonstration stage accelerator, with accelerator section length of 0.6 m can now be derived straightforwardly. An example of a self consistent set of parameters is tabulated in Table II. Energy and wiggler period length, as a function of distance are graphically given in Fig. 2. As is evident, with injection energy of $E_i = 50$ MeV, approximately a doubling of energy is achieved in a sector length of 0.6 m, i.e. with a given laser power of $W = 2 \times 10^{11}$ Watts and E field loss factor of $\alpha = 0.025$ ($E(z) = E_o$ exp $-\alpha z$) a mean accelerator field magnitude of ~ 90 MV/m may be achieved. The effectiveness of the IFEL accelerator, with the cited parameters, is further illustrated in Fig. 3 where the particle distribution in longitudinal phase space is shown at the exit of the IFEL module and where the population distribution is shown versus energy. Particle trapping in an accelerating "bucket" with explicit momentum separation of accelerated and non accelerated electrons is clearly in evidence.

Responding to the challenge posed during the Port Jefferson workshop (June 14-20, 1992), the parameters for a 100 MeV IFEL accelerator were derived based on projected near-term anticipated parameters for the electron beam and the CO_2 laser beam of the BNL ATF/CAP facility, i.e. laser parameters, $W = 10^{11}$ Watts, $\Delta t = 10$ psec, $J = 1$ Joule; and e⁻ beam parameters, $E = 50$ MeV, $I = 5$ mA, $(\Delta E/E) = \pm 3 \; 10^{-3}$, emittances 7 10^{-8} rad.m. With these parameters, in order to achieve an energy gain of 100 MeV, four accelerator modules similar to that enumerated in Table I are required. This is further shown in Fig. 4 where a four module test configuration is given. Evidently, with this relatively modest laser power a mean accelerating gradient of approximately 50 MeV/m is achievable.

III. FAST EXCITATION VARIABLE PERIOD WIGGLER

The IFEL electron accelerator as parameterized here makes use of a quasi-sinusoidal magnetic field, with constant maximum field amplitude, and varying wiggler period length. Related to the beam injection energy into the IFEL linear accelerator, this period length may vary from a few cm's in length to larger period length magnitudes. Such a structure could possibly be constructed using presently known techniques employing permanent magnet material. It would, however, be very high in cost because of the nonrepeat feature of the wiggler period length. Hence, for the present objective, a new design approach has been pursued, which makes use of easily stackable, geometrically alternating substacks of identical ferromagnetic material (Vanadium Permendur) laminations, which is driven in a fast excitation mode, and which makes use of interleaving of conductive, nonmagnetic, laminations, which act as eddy current induced "field reflectors".[8-10]

For the ferromagnetic laminations for this wiggler design a number of basic configurations have been studied by means of two dimensional mesh computations (POISSON) and by means of actual short wiggler model measurements. The adopted configuration is illustrated in Fig. 5. As shown, the magnetic material laminations are assembled in ($\lambda_o/4$) thickness substacks, and separated by nonmagnetic material laminations. Four straight current conductors, parallel to the axis of the composite assembly and interconnected only at the ends of the total assembly, constitute the current single excitation loop for the wiggler, permitting ease of stack assembly, compression of the stacks by simple tie rods, and ready adoption of either constant period length or sequentially varying period length.

Subsequent to early trials of a fast excitation driven wiggler, the use of eddy current induced "field reflectors" in the laminated wiggler core, was initiated, as also indicated in Fig. 5. This led to dramatic enhancement of maximum on axis field magnitude, for a specific wiggler period length and gap value, as shown by the experimental data given in Fig. 6. Field saturation is evident in these results for higher excitation current values. The onset of field saturation is clearly discernible with the onset of distortion of the magnetic measurement probe voltage versus time display. The field value corresponding to the onset of saturation, for a sequence of model measurements with different period length values, was obtained, both for the case of wiggler models without field reflection and with field reflection. This is summarized in Fig. 7. As is evident from these results the specification of 12.5 kG, for a 2.9 cm period length wiggler, with gap value of 4 mm, can readily be met for the fast excitation wiggler with field reflection.

The median plane field versus wiggler longitudinal coordinate was also measured for a number of wiggler models and full length prototypes,[13] including both constant period length wigglers and tapered period length wigglers. An example of this, with a 4% tapered period length is given in Fig. 8. Similarly, the results for a prototype wiggler, with constant period length of 3.76 cm is also shown in Fig. 8.

For selected cases, the harmonic content of the wigglers was obtained from measured field data and was found to be acceptably small, as shown, for example in Fig. 9 for the case of a constant period wiggler, $\lambda_o = 3.76$ cm. For this case also the first and second field integral values were obtained and are given in Fig. 10. No attempt was made as yet to optimize the latter results with modification of the wiggler end field taper or "external" field correction. Further studies, including also quantitative evaluation of the "phase integral", also for the tapered prototype wiggler case, are in progress.

IV. DIELECTRIC COATED WAVEGUIDE FOR CO_2 LASER TRANSPORT

The present design of the IFEL accelerator demonstration stage transports the CO_2 laser light through a low loss rectangular waveguide with selected wall thin film dielectric coating with a theoretical loss factor[12] of $\alpha = 10^{-4}$ m^{-1}, where α is defined by $P_\ell(z) = P_\ell(0) \exp(-2\alpha z)$. A development effort has been carried on aimed at the basic mechanical design of these guides, and to carry out loss measurements and mode spectrum measurement. The elementary Zakowicz configuration is shown in Fig. 11. Radiation wave attenuation occurs due to finite conductivity of the walls and the absorption in the dielectric coating. For $\lambda = 10$ μm, wall losses are much greater than absorption in the dielectric coating. The wall losses are proportional to $\oint_c |H_r|^2 d\ell$, , hence a reduction of losses is obtained with a reduction of $H_{\parallel}$ at the wall. This, for the E-M mode of relevance here, can be done by means of a dielectric layer of thickness given by $d = (\lambda/4) (\epsilon-1)^{-1/2}$ where $\epsilon = (an_o)^2$ and an_o is the refractive index of the coating. A number of waveguide sections were constructed with internal guide cross section of 2.8 x 2.8 mm^2 and external dimensions (matched to the wiggler aperture) of 3.8 x 19.0 mm^2. The configuration is shown in Fig. 12. For the basic body of the structure, stainless steel material was used with near optical finish on the "inside" surfaces. Either dielectric coated or uncoated guide sections were used in the guide measurement procedures. For the coated case, Ge coating to a thickness of 0.65 μm on a 0.1 μm Ag coating base was used on the "vertical" walls only. Calculations using the Zakowicz model (note Fig. 12), show that the Germanium coated guides should transmit, with small attenuation, CO_2 laser radiation over distances of > 1 km, where for the uncoated stainless steel guide case the power e-folding length should be of a magnitude of $\approx$ 10 cm only.

Waveguide test pieces so constructed were used to investigate coupling into the guides, transmission loss and end mode structure. Two lasers were used for these tests. The coupling measurement made use of a pulsed CO_2 oscillator of BNL's ATF/CAP facility, set for a wavelength of 10.2 μm. The pulse duration was 100 ns, and 1% of the 100 mJ beam was used. Subsequent measurements of transmission and mode used a cw CO_2 laser (courtesy of STI Optronics) set for 10.6 μm and various output powers from 100 to 500 mW. For both lasers, a HeNe alignment laser was directed along the path of the CO_2 beam, to provide a visible reference.

In both cases, the beam was focused into the waveguide using two ZnSe lenses with 25 cm focal lengths. The lenses were mounted on a 1.2 m optical rail, in order to allow rapid adjustment of their spacing. Positions were calculated to focus the TEM$_{oo}$ output of the lasers to a Gaussian waist with a radius ω_o between 0.8 and 1.2 mm at the entrance to the guide. These calculations were confirmed by measuring the beam profile at the location of the guide entry using a pyroelectric vidicon TV camera (Electrophysics 5400) and a digital frame grabber (Spiricon LBA-100 beam analyzer).

Measurements of mode coupling and guide attenuation are still in progress using progressively longer total guide sections. Preliminary measurement results may be summarized here, however, as follows:

* High laser power transmission, $\geq$ 80%, over lengths of guides of up to 0.8 m were measured. With individual guide sections, nominally 25 cm in length, in composite assemblies as indicated, the following power transmission values were obtained:

> SS (coupling), SS (sect.), Ge (Sect.); Transmission 90.3%
> SS (coupling), Ge (sect.), SS (Sect.); Transmission 86.1%
> SS (coupling), Ge (sect.), Ge (Sect.); Transmission 93.3%

*   Low loss power transmission of the coupling section is achieved with an entrance waist of the Gaussian beam at the guide entrance set at w/B = 0.71, or w = 1.0 mm.
*   Attenuation of the uncoated guide is far below the calculated e-folding length ($\approx$ 10 cm).
*   For the present experimental arrangements, in general, the Germanium coated guides (Zakowicz) enhanced the power transmission by about 7%, compared with the noncoated guides.

To measure the transverse mode in the waveguide, the pyroelectric vidicon was placed 7 cm downstream from three joined sections of waveguide totaling ~ 0.8 m. A Gaussian mode was obtained with careful alignment having a maximum transmission of 86% and exit beam cross section smaller than the input beam by about a factor of 2.

As indicated, further guide development is still actively being pursued. Based on the cited preliminary results it may be concluded now that the present guide configuration, even in the uncoated execution, may be suitable for the IFEL accelerator prototype experiment. Additional work is needed (and in progress) to study the benefits of dielectric coating in longer guide lengths, relevant to a multistage accelerator sequence.

V.   OPTICAL ISSUES IN THE IFEL ACCELERATOR

The use of a propagating E-M mode in a rectangular, partially dielectric coated, waveguide has been proposed by Zakowicz[12] for the IFEL accelerator, specifically for reasons of its theoretically predicted extremely low attenuation and its high efficiency in the use of laser power for the particle acceleration. As indicated above, this approach is pursued not only theoretically but also by means of extensive model development and testing.

The E-M mode adopted in the Zakowicz approach is the lowest possible mode of the low loss type. It has a sinusoidal variation of the amplitude of the component of the electric field in both transverse directions. The useful guide aperture is determined by the degree of field nonuniformity which is allowable in the accelerator operation. For example, if the amplitude seen by the beam is to be constant within 15%($\pm$7.5%), a diameter as low as 0.5 mm can be used.

Because the guide is highly "overmoded" i.e. it can support many higher order modes, both the phase and group velocity are close to the speed of light. Nevertheless, the (small) differences are important because of the tight phase tolerances with respect to the beam of electrons (phase velocity) and because of bunch length considerations (group velocity).

The relation between the propagation constant for the desired mode (k_{11}) and the free space propagation constant (k_o) is:

$$(k_{11} - k_o)/k_o = -(\pi/k_o B)^2$$

where B is the transverse dimension of the waveguide aperture. For the dimensions chosen, k_{11} differs from k_o by 2.00 x 10^{-2} radians per centimeter. To maintain a phase accuracy of 0.1 radian, the transverse dimension must hold to a tolerance of 70 μm.

The phase velocity does not depend on the dielectric thickness (within reasonable limits) but the propagation losses do so depend. As first pointed out by Zakowicz, the losses are surprisingly insensitive to the coating thickness. Typically, a 20% change in thickness leads to a 10% increase in the (very small) power loss. If 20% is taken as a reasonable target, the .645 μm thickness must be held to about 0.1 μm. This is a relatively easy tolerance for the film deposition technology used. The group velocity can be shown to be given by:

$$v_g = c \left(1 - 2[\pi/(k_o B)^2]\right)$$

where c is the free space velocity of light. For the chosen dimensions, the second term in the bracket is 6.38 x 10^{-6}. For comparison, an electron of 100 MeV has a fractional difference from light velocity of 1.25 x 10^{-5} so in 1.0 meter the laser pulse and such an electron would "slip" by 6.5 μm.

The laser power must be introduced into the guide in a manner that efficiently transforms the diffraction limited Gaussian output beam of the high power laser into the desired low loss accelerating mode. Zakowicz analyzed this "coupling" problem in a simple overlap integral fashion. The overlap between the input beam and the field pattern of the desired mode is taken to be proportional to the coupling efficiency into that mode. Zakowicz found that this method predicted that high efficiencies (over 95%) could be obtained if the waist in the Gaussian beam was located at the waveguide aperture and if the waist size was appropriately chosen (ω/B = 0.71).

This approach, while encouraging, is approximate and also does not give any information about the transition region over which the mode becomes established. Therefore, a series of calculations have been carried out in which the actual fields in the vicinity of the coupling aperture are computed in a diffraction type of approximation. The details of these calculations will be presented elsewhere and here only the results to date are summarized.

Although the formalism permits full three dimensional calculation, so far only two dimensional studies, neglecting polarization effects, have been carried out. Nevertheless these are informative and show features which are almost certainly general. It is found that the mode pattern transforms from the input Gaussian to a stable field distribution over a distance which is comparable to the Rayleigh length of the input beam. Thus, for a Rayleigh length of 30 cm, as contemplated here, the mode pattern requires about 50 cm to stabilize.

The calculation so far ignores all losses so that the energy which goes into the other modes manifests itself as amplitude and phase noise in the propagating light. For a particular choice of waist size, the best chosen so far, it is found that after the mode has stabilized ($\approx$ 50 cm as noted above) the amplitude typically fluctuates by $\pm$ 5% and the phase by $\pm$ .05 radian.

These values are consistent with the maximum amplitude due to the other modes of 0.05 of the amplitude of the desired mode. The phase fluctuation arises because the other modes have different phase velocities from that of the desired mode. These results indicate an efficiency in coupling the power into the desired mode of $\approx$ 90%, at least in the two dimensional case, neglecting polarization.

One important check on the calculation is given by the comparison of the phase velocity obtained from the diffraction propagation calculation and the analytic value obtained by Zakowicz. Fitting the calculated phase versus z for z = 50 cm to z = 200 cm a value of k - k_o = 1.07 x 10^{-2} is obtained. This is to be compared with 1.0 x 10^{-2} from the Zakowicz theory (adapted to a two dimensional case).

These results support the feasibility of efficiently coupling the laser beam into the desired mode. Further calculations are in progress to extend the studies to the full three dimensional case, to study the tolerances in waist size and alignment, and to include the effects of losses.

The above discussion is correct for the case where the metal walls can be treated as nearly (but not quite) perfect conductors and when the displacement current in the walls can be neglected. These assumptions are made at the outset in the work of Zakowicz. It was realized recently that at infra-red wavelengths these assumptions may not always be applicable and modifications to the detailed analyses may be needed. These corrections could account for the unexpectedly low attenuation of the uncoated waveguides. These modifications will be the subject of further studies.

VI. IFEL SIMULATION

A 1-D particle simulation code to model the acceleration process in a waveguide IFEL has been developed. This code incorporates in a self-consistent manner the longitudinal electron dynamics and the laser field; it also takes into account the properties of a realistic electron beam, i.e., finite radius, emittance and energy spread.

This 1-D numerical model has been used as a design tool for the demonstration stage experiment; in particular it provided the optimal tapered wiggler for the given input laser power, resonance angle and the peak wiggler field.

Subsequently, the multiparticle simulation Linac code PARMELA[14] (Phase And Radial Motion in Electron Linear Accelerators) was modified to simulate the full 3-D aspects of the IFEL interaction.

The electron orbits in a tapered linearly polarized wiggler are determined by numerically solving the relativistic Lorentz force equation. The following was included in the electron beam dynamics: a) An arbitrary initial electron distribution in "trace" space (x,x' = (dx/dz), y,y' = (dy/dz)) determined by the Twiss parameters α_x, β_x,α_y,β_y and the transverse emittances, ϵ_x and ϵ_y. b) An arbitrary longitudinal initial electron distribution in phase ϕ and energy W. c) A realistic piecewise constant tapered wiggler allowing for both horizontal and vertical focusing.

A small region of the center of the electron pulse of extent equal to the laser wavelength λ_s was modelled. Periodic boundary conditions are assumed and consequently, electrons slipping-out of the examined region are re-entered at the back.

The external laser is not a dynamical variable at this point; however, the transverse characteristics of the waveguide modes are incorporated in the simulations.

The physical model of an IFEL is described by the coupled system of equations for the i-th electron, as follows:

$$\frac{d\gamma^i}{d(ct)} = -\frac{e}{mc^2}\frac{K(x,y,z)}{2\gamma^i}[KK]\,E_0(x,y)\sin\left(\psi^i(z,t)+\phi_s\right)$$

$$\frac{d\psi^i(z,t)}{d(ct)} = \beta_z^i k_w(z) - \frac{k_s}{2\gamma^i}\left(1 + \left(\gamma^i\beta_x^i\right)^2 + \left(\gamma^i\beta_y^i\right)^2\right.$$

$$\left. + 2K(x,y,z)\,K_s\,[KK]\cos\left(\psi^i(z,t)+\phi_s\right)\right)$$

$$\frac{d\left(\gamma^i\beta_x^i\right)}{d(ct)} = \frac{e}{mc^2}\left(\beta_y^i B_z^w - \beta_z^i B_y^w\right)$$

$$\frac{d\left(\gamma^i\beta_y^i\right)}{d(ct)} = \frac{e}{mc^2}\left(\beta_z^i B_x^w - \beta_x^i B_z^w\right)$$

$$\frac{dx^i}{d(ct)} = \beta_x^i \qquad\qquad \frac{dy^i}{d(ct)} = \beta_y^i$$

where $E_0(x,y)$ and ϕ_s are the slowly varying laser field amplitude and phase; $K(x,y,z)$ is the wiggler constant, K_s is the corresponding laser constant and $[KK] = F_1$ as given in Section II. The relative phase of the electron wiggling motion to that of the laser oscillations is defined as:

$$\psi^i(z,t) = \int_0^z dz'\,k_w(z') + k_s z - k_s ct \ .$$

The laser field is assumed to be of the form,

$$E_s(x,y,z,t) = \overline{E}_o(x,y)\sin(k_s z - k_s ct + \phi_s) \qquad where$$

$\overline{E}_o(x,y)$ correspond to a TM_x waveguide mode as described in Ref. [12].

A few characteristic results of the application of the 3-D simulation code to the IFEL accelerator under study are given here.

The accelerated electrons oscillate about the axis with slowly increasing amplitude ($x \approx \sqrt{\lambda_w}(z)$; the nonaccelerated electrons on the other hand have an increasing amplitude ($x \approx \lambda_w^2(z)$) and some of them wander off axis. The velocity (x') component of typical electrons in the horizontal plane is shown in Fig. 13.

The transverse phase space at the end of the wiggler is given in Fig. 14, showing some emittance growth in the horizontal plane and two well defined group of electrons (accelerated and non-accelerated) in the vertical plane with different slopes; the emittances are comparable with the initial one. The initial phase space is included for comparison.

The energy spectrum at the end of the wiggler as shown in Fig. 15, clearly illustrates the fraction of accelerated electrons ($\approx$ % 50).

Further benchmarking of the code is in progress and an extension of the code to allow for dynamical variations of the laser amplitude and phase is also contemplated.

VII. IFEL MULTISTAGE ACCELERATOR, ERROR SIMULATION

Initially the 1-D simulation code has been used to define a sequence of optimally tapered wigglers for a given input laser power, resonance phase angle and peak wiggler field; and to calculate the particle capture (bucket acceptance) and losses (bucket leakage) of a single or multistage accelerator for variation of laser power, guide loss, energy spread and wiggler errors.   To this end, a basic multimodule structure 50-250 MeV IFEL accelerator was calculated, making use of a $W_L = 6.2 \ 10^{11}$ Watts laser power and a multiplicity of 8 elementary, $L_{sect} = 0.6$ m, units. The parameters of this structure are detailed in Table III.

A typical phase space (γ, Ψ) at the end of the 1th and the 8th module of this multistage accelerator is given in Fig. 16.   In addition to this, also the particle energy spectrum is shown, indicating the fraction of accelerated particles ($\sim 40\%$) and the fraction of unaccelerated electrons. The wiggler period $\lambda_w(z)$ as a function of distance is also given in this figure, both for the first stage and last stage of this accelerator.   This piecewise constant function is inherent in a periodic tapered wiggler and reflects a realizable wiggler that best fit the smooth theoretical curve obtained self-consistently to maintain the resonance condition.

The rate of acceleration $(d\gamma/dz)$ is proportional to $\gamma^{-1/3}$, this is further emphasized in Fig. 17 where (dE/dz) and $\gamma(N)$ are given as a function of accelerator module number.

The implicit vertical transverse focussing action of the planar wiggler is taken into account.   No effort has been made to shape the wiggler poles in order to achieve simultaneous horizontal transverse focussing. "External" focussing is added by means of an approximate FoDo channel, resulting in transverse optics for the 8 module accelerator as shown in Fig. 18.

A preliminary examination has been made of the sensitivity of the acceleration efficiency as a function of the laser power, loss coefficient and electron beam initial energy spread; for the case of a multistage ($N = 8$) accelerator. The results are shown in Fig. 19 where the 8 module accelerator net capture efficiency is shown versus resonant phase angle; in Fig. 20, where the beam capture efficiency is shown versus laser power and in Fig. 21, where the capture efficiency is shown as a function of guide attenuation coefficient α, for various laser power values. In addition, accelerator losses have been studied for the case of a wiggler with random pole-to-pole field errors. An example of this is given in Fig. 22 where the γ-Ψ plot is shown for the case of $\sigma(\Delta B_p/B_p) = 0.01$ and for the case of $\sigma(\Delta B_p/B_p) = 0.05$. Clearly, the latter case is unacceptable. With the present simulation program only phase errors are considered in this case. Aspects of electron beam "walk-off" have not been taken into account, as yet.   The predictions of the code have been compared with the linear theory as given in [3] and, in general, good agreement was obtained.   Note that the present code has been used as a design tool for the demonstration stage experiment. It is planned now to incorporate the transverse coordinates dependence of the laser and electron beam as well as the wiggler field with the use of the 3-D program cited in the foregoing.

CONCLUDING REMARKS

The study of the IFEL accelerator is continuing with near term emphasis on low loss guide development, IFEL accelerator transverse and longitudinal phase space transport and the further parameter optimization of a multimodule ≈ 1 GeV accelerator.

It is evident from the foregoing IFEL accelerator parameter study that for a single demonstration stage, aimed at approximately doubling the beam energy, a CO_2 laser power of $\approx 10^{11}$ Watts is satisfactory. For a cascaded IFEL accelerator system of final energy of ~ 1 GeV a laser power magnitude of $\geq 10^{12}$ Watts is required to make the overall device technically competitive.

Acknowledgement: The authors acknowledge the continuing support by the Advanced Technology R&D Branch, Division of High Energy Physics, U.S. Dept. of Energy. They also acknowledge engineering assistance from J. Sheehan, S. Ulc and M. Woodle. In addition, they wish to gratefully mention the participation in model/prototype measurements by T. Romano and J. Armendariz (wiggler) and by S. Coe and M. Weng (dielectric guides).

REFERENCES

1. R.B. Palmer, J. Appl. Phys. 43, 3014, 1972.
2. C. Pellegrini, P. Sprangle, W. Zakowicz, Intern. Conf. on High Energy Accelerators; 1983.
3. E. Courant, C. Pellegrini, W. Zakowicz, PR A, Vol. 32, No. 5, 1985.
4. E. Courant, J. Sandweiss, BNL 38915.
5. A. van Steenbergen, in Experimental Program, ATF, BNL 41664, 1988; and "IFEL Accelerator Demonstration Stage," in ATF, BNL 43702, 1989.
6. E. Courant, A. Fisher, J. Gallardo, C. Pellegrini (UCLA), J. Rogers, J. Sandweiss (Yale University), J. Sheehan, A. van Steenbergen, S. Ulc, M. Woodle, "Inverse Free Electron Laser (IFEL) Accelerator Development," ATF Users Meeting, BNL 47000, CAP 81-91P, p.235-278, October 1991; editor H.G. Kirk.
7. N. Kroll, P. Morton, M. Rosenbluth, IEEE of Quantum Electron. QE-17, 1436, 1981.
8. A. van Steenbergen, J. Gallardo, T. Romano, M. Woodle, "Fast Excitation Wiggler Development," Proc. Workshop 1 Angstrom FEL, Sag Harbor, NY, 1990, Editor J. Gallardo, p. 79, BNL 52273.
9. A. van Steenbergen, Patent Application 368618, June '89 (issued Aug. '90).
10. A. van Steenbergen, J. Gallardo, T. Romano, M. Woodle, "Fast Excitation Variable Period Wiggler," Proc. PAC San Francisco, IEEE Nucl. Sci., May 1991.
11. A. van Steenbergen, J. Gallardo, "IFEL Development, Parameterization 0.4 GeV Accelerator," Unpublished Notes, June 1990.
12. W. Zakowicz, J. Applied Physics 55, 9, 1984; also BNL 34347.
13. J. Armendariz, J. Gallardo, T. Romano, A. van Steenbergen, Fast Excitation Wiggler Field Measurement Results, BNL Report 47928 August 1992.
14. L. Young, "PARMELA", 1991 version, private communication.

Table II, IFEL Accelerator Experiment

Electron Beam			
	Injection energy	48.9	MeV
	Exit energy	87.95	MeV
	Mean accelerating field	89	MV/m
	Current, nominal	5	mA
	N(bunch)	$6\ 10^9$	e^-
	I(max)	100	A
	$\Delta E/E$ (one σ)	$\pm 3\ 10^{-3}$	
	Emittance (one σ)	$7\ 10^{-8}$	m.rad
	Beam radius	0.3	mm
Wiggler Parameters			
	Wiggler length	0.47	m
	Section length	0.6	m
	Period length	2.86-4.32	cm
	Gap	4	mm
	Field max.(Const.B mode)	1.25	T
	Beam oscillation, $a_{1/2}$	0.17-0.22	mm
Laser Parameters			
	Power, P(laser)	$2\ 10^{11}$	Watts
	Wavelength, λ	10.2	μm
	Max.field	$1.36\ 10^4$	MV/m
	Dielectric guide, loss par. α	0.025	m^{-1}
	Field attenuation/section	0.13	dB
	Pulse length (fwhm)	6	psec

Table I, Energy Dependence of Selected Parameters

$$\frac{d\gamma}{dz} = \left(\frac{2}{\eta}\right)^{\frac{1}{2}}.n.\xi\bar{A}\left(\frac{\lambda}{\lambda_o}\right)^{\frac{1}{2}}$$

$$\frac{d\gamma}{dz} = \left(\frac{2}{\eta}\right)^{\frac{1}{3}}\left(\frac{a}{3}\right)^{\frac{1}{3}}.n^{\frac{1}{3}}.\frac{1}{\gamma^{\frac{1}{3}}}.\xi\bar{A}$$

$$\frac{d\gamma}{dz} = \frac{K\,\xi\bar{A}}{\gamma}$$

Set: $\lambda_o = \lambda_o(z)$

$$\left(\frac{\lambda_o}{\lambda}\right) = \left(\frac{2}{\eta}\right)^{\frac{1}{3}}.\left(\frac{a}{3}\right)^{-\frac{2}{3}}.n^{\frac{1}{3}}.\gamma^{\frac{2}{3}}$$

$$\left(\frac{\lambda_o}{\lambda}\right) = \left(\frac{2}{\eta}\right).n.\frac{\gamma^2}{K^2}$$

$$\left(\frac{a}{3}\right) = \left(\frac{2}{\eta}\right)^{\frac{1}{2}}.\left(\frac{\lambda}{\lambda_o}\right)^{\frac{3}{2}}.n^{\frac{1}{2}}.\gamma$$

Set: $a = a(z)$

$$\left(\frac{a}{3}\right) = \left(\frac{2}{\eta}\right)\frac{K^3}{n\gamma^2}$$

$$K = \left(\frac{2}{\eta}\right)^{\frac{1}{2}}\left(\frac{\lambda}{2\pi}\right)\left(\frac{\lambda}{\lambda_o}\right)^{\frac{1}{2}}.n^{\frac{1}{2}}.\gamma$$

$$K = \left(\frac{2}{\eta}\right)^{\frac{1}{3}}\left(\frac{a}{3}\right)^{\frac{1}{3}}.n^{\frac{1}{3}}.\gamma^{\frac{2}{3}}$$

Set: $K = K(z)$

$$a_{\frac{1}{2}} = \left(\frac{\lambda}{2\pi}\right)\left(\frac{2}{\eta}\right)^{\frac{1}{2}}\left(\frac{\lambda}{\lambda_o}\right)^{\frac{1}{2}}.n^{\frac{1}{2}}$$

$$a_{\frac{1}{2}} = \left(\frac{2}{\eta}\right)^{\frac{2}{3}}\left(\frac{\lambda}{2\pi}\right)\left(\frac{a}{3}\right)^{-\frac{1}{3}}.n^{\frac{2}{3}}.\gamma^{\frac{1}{3}}$$

$$a_{\frac{1}{2}} = \left(\frac{\lambda}{2\pi}\right)\left(\frac{2}{\eta}\right)\frac{n\gamma}{K}$$

with: $\Omega = \frac{eB}{m_o c}$; $\omega = \frac{2\pi c}{\lambda}$; $\left(\frac{a}{3}\right) = 93.4\,B\lambda$ [T.m] ; $A\sin\psi = \bar{A}$, $A = \left(\frac{e\,E_\perp}{m_o c^2}\right)$

$K = \frac{eB\lambda_o}{2\pi m_o c^2} = \left(\frac{\lambda_o}{\lambda}\right)\left(\frac{\gamma\lambda_o}{2\pi\rho_o}\right)$; $\rho_o = \frac{\gamma m_o c^2}{eB}$; $\left(\frac{\lambda}{\lambda_o}\right) = \frac{nK^2}{2\pi\gamma^2}$; $\kappa c = (2\pi/\lambda)c$; $\kappa = 2\pi/\lambda_o$

$\xi_{3n}=1$, for a helical wiggler; $\xi_{3n}=F_n$ for a planar wiggler. n = harmonic number. n = 1,3,5...

$F_{1,3,5} = 1/2\ (J_{0,1,2}(x) - J_{1,2,3}(x))$; with $x = kK^2/8\kappa\gamma^2$ and J_i are Bessel functions

$\eta = 1$, helical wiggler; $\eta = 1/2$, planar wiggler

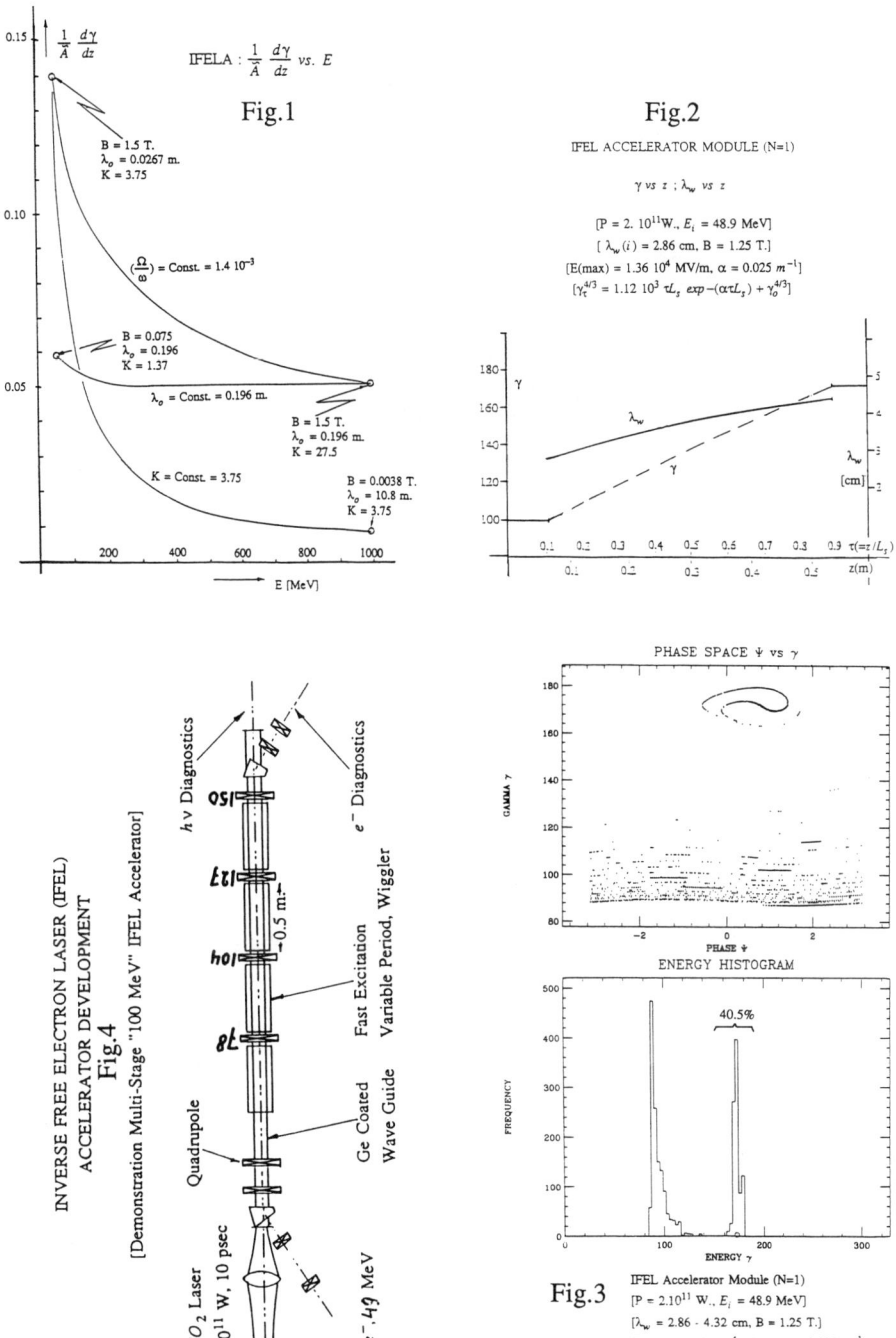

IFELA : $\frac{1}{A}\frac{d\gamma}{dz}$ vs. E

Fig.1

Fig.2

IFEL ACCELERATOR MODULE (N=1)

γ vs z ; λ_w vs z

[P = 2. 10^{11}W., E_i = 48.9 MeV]

[$\lambda_w(i)$ = 2.86 cm, B = 1.25 T.]

[E(max) = 1.36 10^4 MV/m, α = 0.025 m^{-1}]

[$\gamma_\tau^{4/3}$ = 1.12 10^3 τL_s $exp-(\alpha\tau L_s) + \gamma_0^{4/3}$]

PHASE SPACE Ψ vs γ

ENERGY HISTOGRAM

40.5%

Fig.3

IFEL Accelerator Module (N=1)

[P = 2.10^{11} W., E_i = 48.9 MeV]

[λ_w = 2.86 - 4.32 cm, B = 1.25 T.]

[E(max) = 1.36 10^4 MV/m, α = 0.025 m^{-1}]

INVERSE FREE ELECTRON LASER (IFEL)
ACCELERATOR DEVELOPMENT
Fig.4
[Demonstration Multi-Stage "100 MeV" IFEL Accelerator]

$h\nu$ Diagnostics

e^- Diagnostics

Fast Excitation Variable Period, Wiggler

Ge Coated Wave Guide

Quadrupole

CO_2 Laser 10^{11} W, 10 psec

e^- .49 MeV

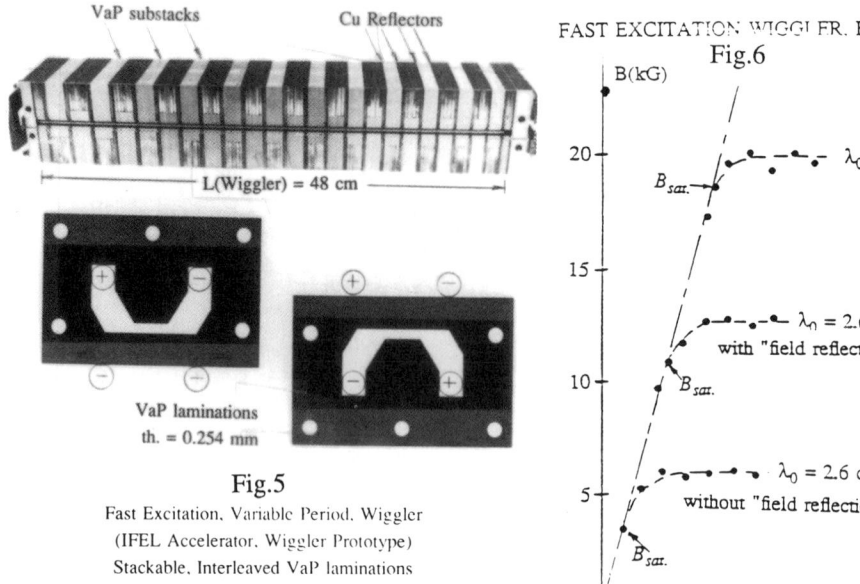

VaP substacks   Cu Reflectors

L(Wiggler) = 48 cm

VaP laminations
th. = 0.254 mm

Fig.5
Fast Excitation, Variable Period, Wiggler
(IFEL Accelerator, Wiggler Prototype)
Stackable, Interleaved VaP laminations
and Cu eddy current "field reflectors"

FAST EXCITATION WIGGLER, B vs I
Fig.6

$B(kG)$

20

$B_{sat.} \rightarrow$

$\lambda_0 = \infty$

15

$\lambda_0 = 2.6$ cm,
with "field reflection"

10

$B_{sat.}$

$\lambda_0 = 2.6$ cm.,
without "field reflection"

5

$B_{sat.}$

$I(kA)$

5   10

FAST EXCITATION WIGGLER, $B_{sat.}$ vs λ_0
(Gap = 4.0 mm)
Fig.7

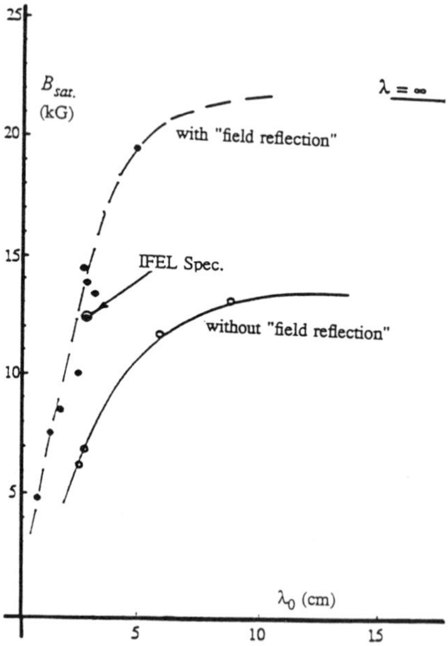

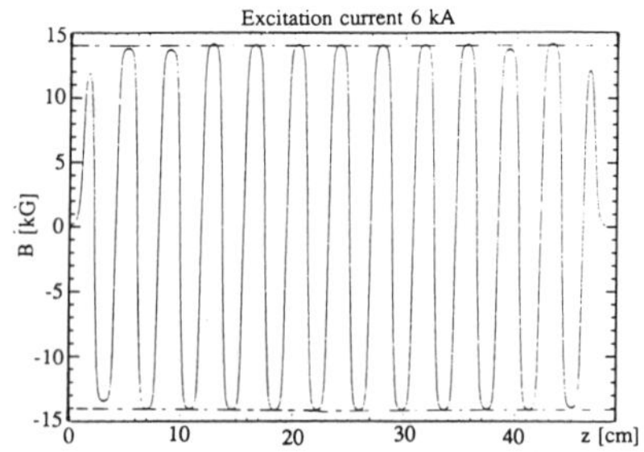

25

$B_{sat.}$
(kG)

$\lambda = \infty$

20

with "field reflection"

15

IFEL Spec.

without "field reflection"

10

5

λ_0 (cm)

5   10   15

Excitation current 6 kA

15
10
5
0
-5
-10
-15

B [kG]

0   10   20   30   40   z [cm]

Constant Period, Fast Excitation, Wiggler
$\lambda = 3.76$ cm, g = 0.40 cm

Fig.8

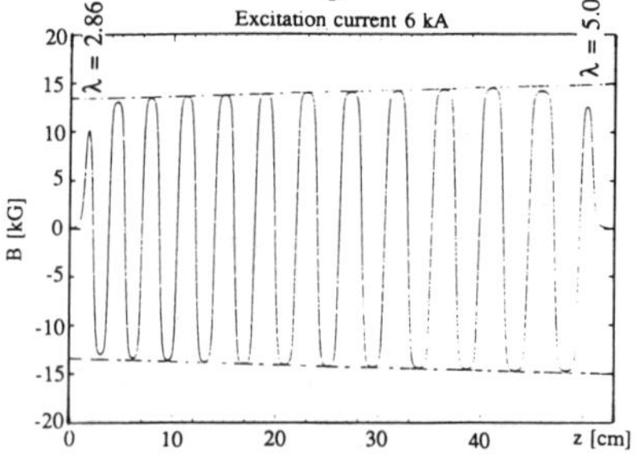

Excitation current 6 kA

$\lambda = 2.86$ cm $\lambda = 5.04$ cm

20
15
10
5
0
-5
-10
-15
-20

B [kG]

0   10   20   30   40   z [cm]

Variable Period, Fast Excitation, Wiggler
(IFEL Accelerator, Wiggler Prototype)
$\lambda = 2.98 - 5.04$ cm,   g = 0.40 cm

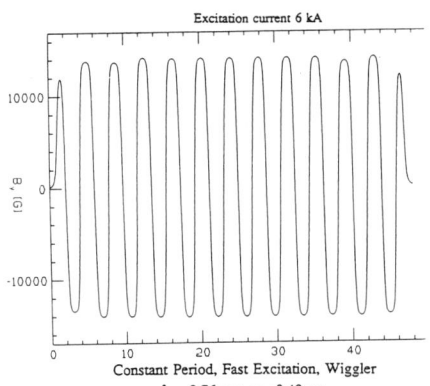

Excitation current 6 kA

Constant Period, Fast Excitation, Wiggler
λ = 3.76 cm, g = 0.40 cm

Fig.9

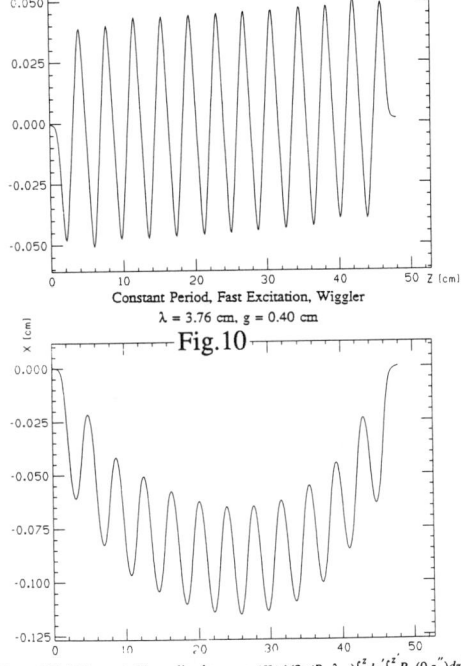

First Field Integral, Normalized : $x'(z) = (K/\gamma)(2\pi/B_w \lambda_w)\int_0^z B_y(0,z')dz'$

Constant Period, Fast Excitation, Wiggler
λ = 3.76 cm, g = 0.40 cm

Fig.10

Second Field Integral, Normalized : $x = (K/\gamma)(2\pi/B_w \lambda_w)\int_0^z dz'\int_0^z B_y(0,z'')dz''$

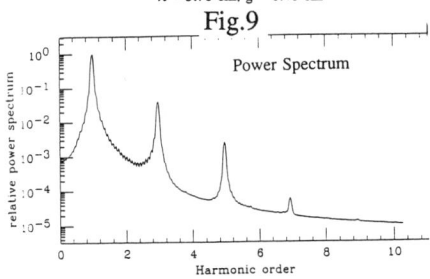

Power Spectrum

Power Spectrum, Constant Period Fast Excitation Wiggler
λ = 3.76 cm, g = 0.40 cm, I(exc) = 6 kA

LOW LOSS DIELECTRIC COATED WAVEGUIDE
[W.Zakowicz, J.Appl.Phys. 55 (9), 1984 and BNL 34347]

Fig.11

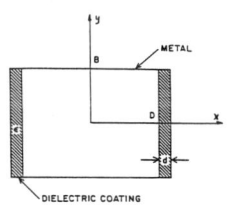

Minimum loss condition for TM_{11}^x hybrid mode:

$$d = (\lambda/4)(\varepsilon-1)^{-1/2}$$

where $\varepsilon = (an_0)^2$. an_0 is the refractive index of the dielectric coating.

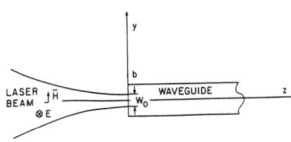

Transition Free Space - Guide Propagation

LOW LOSS DIELECTRIC COATED WAVEGUIDE
Fig.12

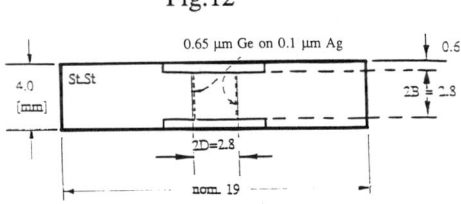

0.65 μm Ge on 0.1 μm Ag

Minimum loss condition for TM_{11}^x hybrid mode:

$$d = (\lambda/4)(\varepsilon-1)^{-1/2}$$

where $\varepsilon = (an_0)^2$. an_0 is the refractive index of the dielectric coating.

Attenuation constant α of the TM_{11}^x mode:

$$\alpha = \frac{c^2 \pi^{3/2}}{8\omega^{3/2}\sqrt{2\sigma}}\left(\frac{\varepsilon^2}{\varepsilon-1}\frac{1}{D^3}+\frac{1}{B^3}\right)+\frac{\pi^3 \varepsilon''\varepsilon}{16k_0^2 D^3 (\varepsilon-1)^{3/2}}$$

where D and B are guide dimensions, ω angular frequency, k_0 wave number in vacuum, σ wall conductivity, ε'' imag. part dielectric constant. Computed values, using $\sigma_{Cu} = 5. \cdot 10^{17}$ sec^{-1}, $\varepsilon(Ge) = 16$. $\varepsilon''(Ge) = 0.09$:

$$\alpha_w + \alpha_{diel} = 1.6 \cdot 10^{-4} + 1.4 \cdot 10^{-5} = 1.6 \cdot 10^{-4} \ [m^{-1}]$$

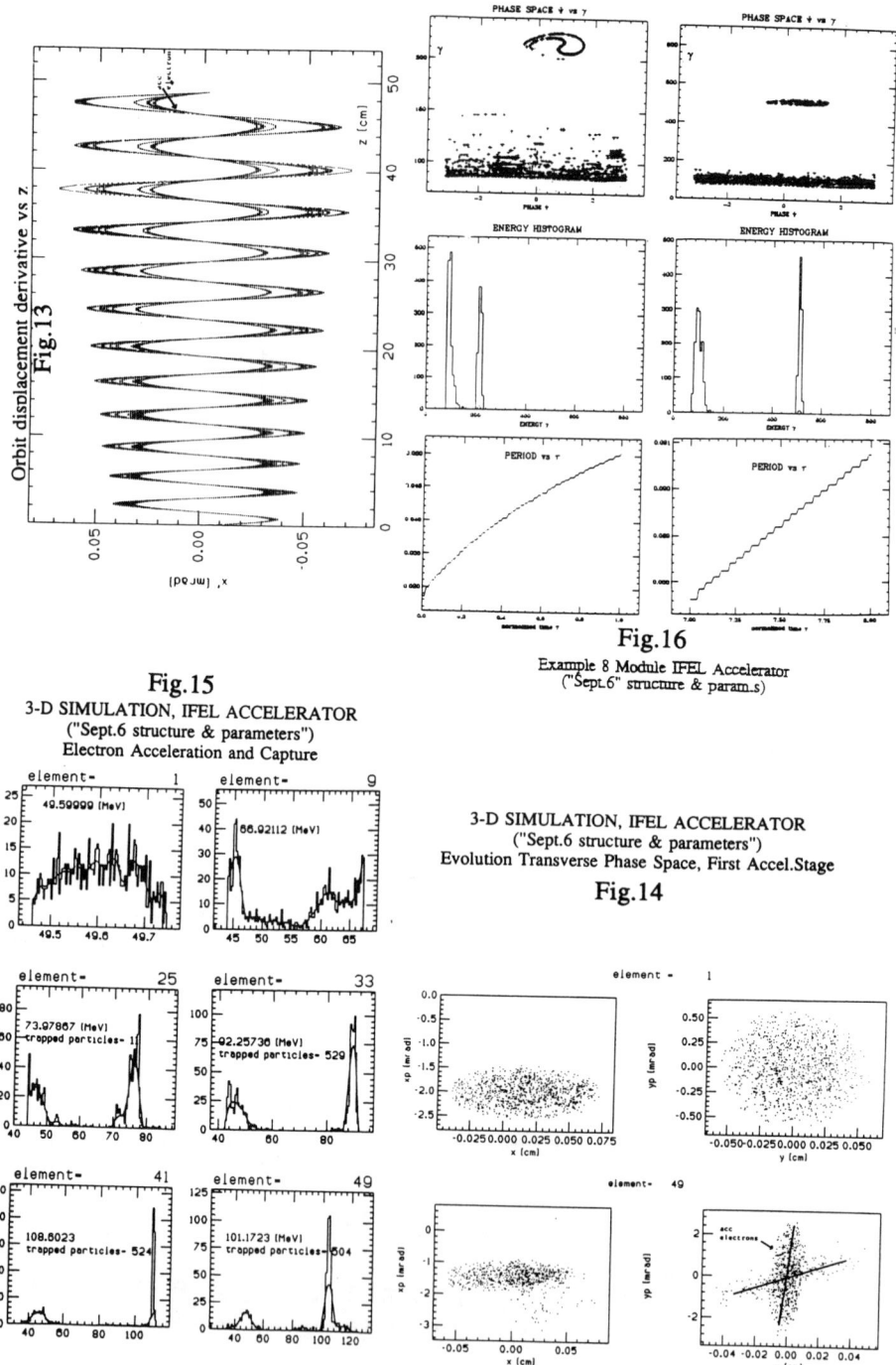

Fig.13

Fig.15

Fig.16

Fig.14

Example 8 Module IFEL Accelerator
("Sept.6" structure & param.s)

3-D SIMULATION, IFEL ACCELERATOR
("Sept.6 structure & parameters")
Electron Acceleration and Capture

3-D SIMULATION, IFEL ACCELERATOR
("Sept.6 structure & parameters")
Evolution Transverse Phase Space, First Accel.Stage

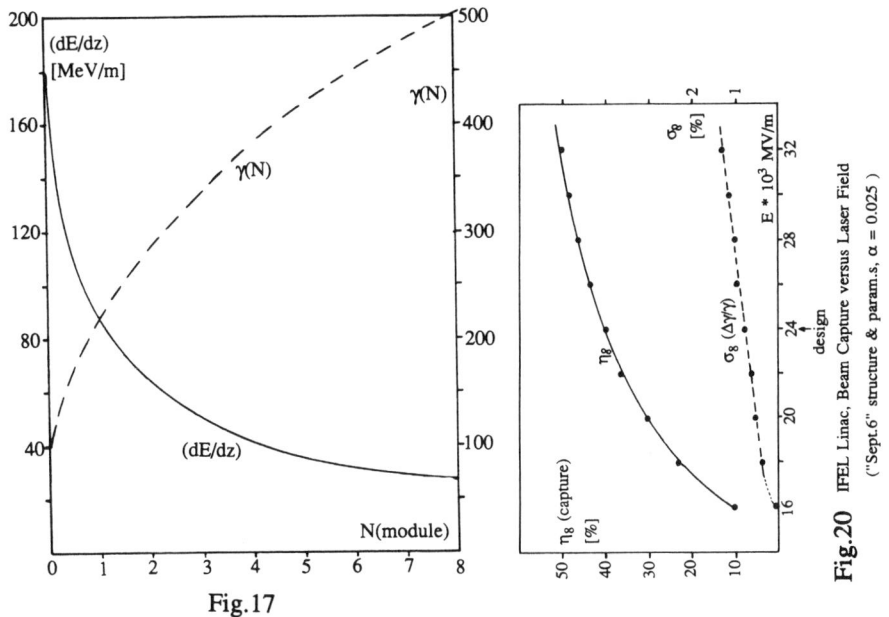

Fig.17

Example 8 Module IFEL Accelerator
("Sept.6" structure & param.s)

Fig.20 IFEL Linac, Beam Capture versus Laser Field
("Sept.6" structure & param.s, α = 0.025)

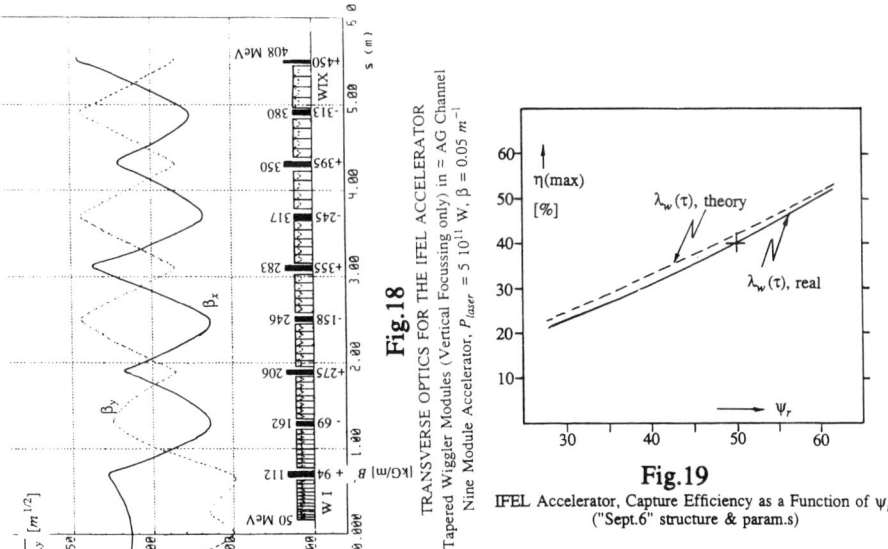

Fig.18

TRANSVERSE OPTICS FOR THE IFEL ACCELERATOR
Tapered Wiggler Modules (Vertical Focussing only) in = AG Channel
Nine Module Accelerator, P_{laser} = 5 10¹¹ W, β = 0.05 m^{-1}

Fig.19

IFEL Accelerator, Capture Efficiency as a Function of ψ_r
("Sept.6" structure & param.s)

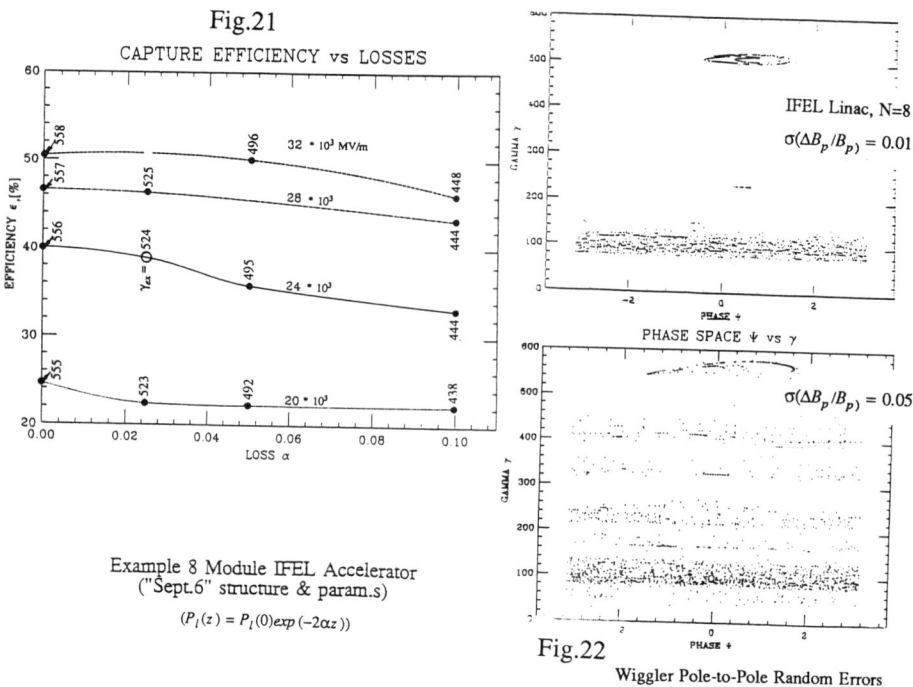

Fig.21

CAPTURE EFFICIENCY vs LOSSES

Example 8 Module IFEL Accelerator
("Sept.6" structure & param.s)

$(P_l(z) = P_l(0)exp\,(-2\alpha z))$

IFEL Linac, N=8

$\sigma(\Delta B_p/B_p) = 0.01$

PHASE SPACE ψ vs γ

$\sigma(\Delta B_p/B_p) = 0.05$

Fig.22

Wiggler Pole-to-Pole Random Errors

Table III, IFEL Multistage Accelerator

Electron Beam			Laser Parameters		
Injection energy	48.9	MeV	Power, P(laser)	$6.2\ 10^{11}$	Watts
Current, nominal	5	mA	Wavelength, λ	10.2	μm
N(bunch)	$6\ 10^9$	e^-	Max.field	$2.4\ 10^4$	MV/m
I(max)	100	A	Dielectric guide, loss α	0.025	m^{-1}
$\Delta E/E$ (one σ)	$\pm 3\ 10^{-3}$		Field attenuation/section	0.13	dB
Emittance (one σ)	$7\ 10^{-8}$	m.rad	Pulse length (fwhm)	6	psec
Beam radius	0.3	mm	Photon beam radius	0.8	mm

Accelerator: N = 8 sections; L(section) = 0.6 m.; Wiggler, constant B,

B(max) = 1.25 T.; Synchronous phase, ϕ_s = 50°, $\eta_{capt.}$ = 41%

Stage,N	L_W [m]	λ_W [cm]	Structure,m*	E(exit)[MeV]
1	0.48	2.36-5.05	23	108.56
2	0.50	5.05-6.12	17	144.28
3	0.45	6.12-6.89	13	171.50
4	0.49	6.89-7.50	13	193.83
5	0.53	7.50-7.99	13	212.82
6	0.48	7.99-8.40	11	229.35
7	0.50	8.40-8.76	11	243.94
8	0.52	8.76-9.08	11	256.94

*$L_W = [m\,(\lambda_W/2) + 2(\lambda_W/4)]$

LASER CAVITIES FOR FAR FIELD PARTICLE ACCELERATION WITHOUT RADIATION LIMITS TYPICAL FOR IFEL

A.A.Varfolomeev, A.H.Hairetdinov
Russian Science Center "Kurchatov Institute"
Moscow 123182, Russian Federation

ABSTRACT

Suggested earlier scheme of a periodical laser cavity for laser acceleration of high energy particles without using of any undulator is analyzed. It is shown by simulation that small nonaxial holes in cavity mirrors designed for beam transportation through the cavities don't change much the TEM_{00} mode configuration providing in appropriate way coupling of the accelerating particle with laser field what means that the scheme can be realized.

INTRODUCTION

Considering any new accelerator scheme one evaluate first its main potential factors as follows: the maximum acceleration rate, the maximum obtainable energy limit for accelerated particles, the types of particles which can be accelerated, the quality of the accelerated particle beams. The one of the most promising far field accelerator schemes is the inverse FEL scheme. In this scheme accelerating electrons and a laser beam propagate in average along the same axis. The coupling between the transversly polarized laser wave and the electrons appears due to small transverse deflections or oscillations of the electrons provided by undulator magnets. If optical cavity mirrors don't have holes for particle beam, then bending magnets also are to be used. Since that energy losses in magnetic fields are inevitable in that performances. The losses are increasing with energy of electrons at least as γ^2 along with decreasing acceleration rate and upper energy limit of particles being accelerated as well. In papers [1,2] it is shown that the acceleration rate is limited by some hundreds of MeV/m and upper energy limit by a hundred of GeV respectively.

So called noncollinear schemes [2-4] allow in principle to avoid the bending magnets and corresponding synchrotron radiation in them but the upper limits being caused by the undulator magnets are still valid. This very serious disadvantage of IFEL system can be avoided if the undulator magnets as well as bending magnets are excluded. Earlier we have suggested a laser driven far field acceleration scheme [5,6] which can work without using deflection magnets besides routine focusing ones. Now we would emphasize more on the possibility of construction of the appropriate periodical optical cavity providing the coupling of the driving laser beam with the accelerated particles.

LASER-DRIVEN FAR FIELD ACCELERATION IN VACUUM WITHOUT THE USE OF UNDULATORS

Conceptual design of the pure laser-driven accelerator without using undulator or other bending magnets is presented in Fig.1 (see [5,6]). A beam of accelerated high energy particles propagates in z direction without any induced transverse deviations. It crosses a row of separate optical cavities with their axises oriented at small angle to z axis. Each of these cavity sections is actually a confocal resonator where the TEM_{00} mode can be excited and maintained by an outer laser power source. For this mode the laser beam diameter in the focal area is defined as

$$d_0 = \sqrt{2 l \lambda_s / \pi} \tag{1}$$

where l is the waveguide section length, λ_s is the laser readiation wavelength. It can be seen from fig.1 that accelerating beam is crossing the laser beam. It goes through the small holes made in the mirrors on some distance from the cavity axis. We will define later how the hole parameters should be limited to provide Gaussian field distribution or mode shape.

The key point of the scheme is that it provides discrete time intervals for this beam interaction. It is switched on only for the moments when the interaction has the proper sign and switched off for other time intervals. We admit that the cavity section length is of the order of the radiation length i.e. $l \sim \lambda_s \gamma^2$ where γ is the particle energy in its rest mass units. The real size of interaction zone between accelerating particle and e.m. field of the laser mode is smaller (see the rectangulars in the fig.1) . If the e.m.wave is polarized in the drawing plane it has the electric field component $E_z = E_{s0} \sin\alpha$ accelerating (deaccelerating) particles moving along z axis.

The full strength beam interaction takes place in the focal regions near focuses z_{fi}. At larger distances from the central focal points z_{fi} the electric field component E_z is decreasing. It is caused by two reasons. Since the mode field amplitudes E_{s0} are decreasing along with increasing distances from the cavity axis and from the focus z_{fi} respectively and since at the longer distances from the focus the wave front becomes more spherical with diminished z-components of the electric field. As a result the particle travelling through the cavity along z axis will interact mainly in the central region with the wave field component of the type:

$$E_z = f(z - z_{fi}) E_{s0} \cos\left[\frac{2\pi}{\lambda}(z - ct) + \varphi_0\right] \tag{2}$$

where $f(z-z_{fi})$ is a function which describes field amplitude variation along z axis when particle is crossing Gaussian mode (Fig.2). More evident expression for the field profile along the particle axis within a cavity has the form:

$$f(s) = \cos\left(\frac{2\pi s}{1 + 1/s^2} - \text{arctg}(s)\right)(1 + s^2)^{3/2} \exp\left(-\frac{2\pi s^2}{1 + s^2}\right) \tag{3}$$

where $s = \dfrac{z - z_{fi}}{L_R}$ is the relative distance from the focus,

$L_R = \dfrac{\pi d_0^2}{4\lambda_s}$ is the Rayleigh length.

To maintain the resonant conditions of the interaction we should provide the function $f(z-z_{fi})$ to be periodical with period length

$$\lambda_0 = \frac{2\lambda_s \gamma_r^2}{1 + \alpha^2 \gamma^2} \tag{4}$$

where $\alpha \ll 1$, $\gamma \gg 1$. Apparently the definition is (4) exactly the same as one for the undulator period for noncollinear beam geometry with small noncollinearity angle and a vanishing undulator field.

Between neighbouring interaction zones the particles travel through small areas in the cavities with $E_z \sim 0$ and some drift spaces zone between the cavities. The relative phases of the laser intracavity beams should be fixed and should provide an integer number of 2π phase change for a particle traversing one space period of the design (the distance from one cavity cell entrance to the corresponding entrance point of the next cavity cell). It means that spatial synchronism will be conserved (for more details see [3,4]).

The formal similiarity with the noncollinear IFEL [2-4] is evident. The equations of motion can now be written in the form:

$$\frac{d(\gamma - \gamma_r)}{dz} = -\frac{\pi}{\lambda_0} K_s \aleph \sin(\Phi - \Phi_r)$$

$$\frac{d\Phi}{dt} = \frac{n\pi}{\lambda_0} \frac{\gamma - \gamma_r}{\gamma_r}$$

(5)

where the relative phase is given by

$$\Phi = \left(\frac{2\pi}{\lambda_0} + \frac{2\pi}{\lambda_s} \cos\alpha \right) z - \omega_s t + \Phi_0$$

Here Φ_r is the resonant phase, γ_r - is the resonant particle energy, $K_S = eE_{S0}\lambda_s/2\pi mc^2$ is the maximal relative wave amplitude, m is the particle mass, $\aleph$ is the average filling factor of the accelerator section. From comparison of the eq.(5) with that for the helical wiggler one can see that the considered laser-driven scheme is similiar to an IFEL with a helical wiggler. The formal difference is that the role of the wiggler constant $K_W = eB_{W0}\lambda_w/2\pi mc^2$ plays the expression

$$(K_W)_{eff} = \aleph \sin\alpha\gamma/2$$

(6)

On the basis of the above similarity and using the results of IFEL analysis [1-4] we can conclude that our accelerating scheme does work. The acceleration rate for the IFEL schemes for the case of the helical undulator and the circular polarized e.m.wave with the amplitude E_0 is given by:

$$mc^2 \frac{d\gamma}{dz} = eE_0 \sin\Phi_r \frac{K_w}{\gamma} - \frac{2}{3}\left(\frac{2\pi}{\lambda_w}\right)^2 e^2\gamma^2\left(K_w^2 + K_s^2 + 2K_w K_s\right) \qquad (7)$$

The first r.h. side term of the eq.(7) gives the energy pumping by the laser field. The second term on the right side relates to synchrotron radiation losses both in the undulator field with K_w amplitude and in the laser field with K_s amplitude respectively. These terms can be neglected for our scheme since we have no undulator field ($K_w=0$) and can compensate small deviations induced by laser field ($K_s \sim 0$). As a result we have for the acceleration rate provided by our scheme with using the linearly polarized laser beams the expression

$$mc^2 \frac{d\gamma}{dz} = eE_{s0} \sin\Phi_r \frac{\aleph}{2}\sin\alpha \qquad (8)$$

Unlike the IFEL scheme the rate (8) does not depend on enegy of accelerated particles. Formally we have no upper energy limit for particles being accelerated and no limitation on their mass. Heavy particles can be accelerated as well. One can easily satisfy the condition $\alpha\gamma \gg 1$. It will provide gain in the acceleration rate of the order $\aleph\alpha\gamma/2K_w$ in comparison with the IFEL scheme. The eq.(4) defines the length of the waveguide sections. Taking $\alpha\gamma \gg 1$ we recieve $\lambda_0 = 2\lambda_s/\alpha^2$ and $\lambda_0 \ll 2\lambda_s\gamma^2$. It means that the waveguide sections can be taken uniform with the equal length each. It will simplificate the cavity design.

LASER CAVITIES WITH NONAXIAL HOLES

It is supposed that with a laser oscillator being used one can excite similiar optical modes in all waveguide sections and control their phases to provide the relative coherence. For definition admit that TEM$_{00}$ modes are excited so that the cavity fields are described as the Gaussian beams with the beam diameters in the focal area equal to (1) . The r.m.s. radii of the Gaussian modes at the mirrors can be equall to $d_m = \sqrt{\frac{2}{\pi}\frac{\lambda_s}{\alpha}}$ (for the case $\alpha\gamma \gg 1$). This is nearly equal to the radial distanse of z-axis from the central mirror point $\propto \lambda_s/\alpha$. So we should test the stability of the Gaussian mode with the cavity mirror containing holes at the edge of the Gaussian mode i.e. at the edge of the first Fresnel zone.

To confirm the principal possibility of the realization of the above acceleration scheme we have made numerical calculation of the mode shapes for the optical resonator with holes in the spherical mirrors. As usually we have used the Fox-Li method [7]. The appropriate intergral equation was used for field distribution on the mirror surface

$$E = \frac{ik}{4\pi} \int E \frac{\exp(-ikR)}{R} (1 + \cos\theta) dS \qquad (9)$$

where k is the wave number, R is the distance between the position of a radiating element and a registration point of the radiation, θ is the angle between the vector $\vec{R}$ and the normal to the mirror surface (Fig.3).

A special 2D code was written. The field on the mirror was presented on a rectangular grid with the number of grid points $m > 16N^2$ where N is the Frenel number $N = a^2 / \lambda_s l$ and a is the mirror radius (see Fig.3). Parameters b and r define the position and the radius of the mirror holes respectively. Cavity modes with hole positions corresponding to N=1 as well as to N=2 were simulated.

Fig.4 presents the results showing the transition of the cavity field to a steady mode for N=1. Similiar results for N=2 are presented in Fig.5. For both cavities having small holes (but having radia$>>\lambda_s$) the e.m. waves do reach steady mode shapes after a few tens of reflections. These modes in effect are not distinguishable from the TEM_{00} mode defined for the conventional (with no holes) resonator. This conclusion is correct only for the parameter b comparable with the first Fresnel zone radius. It doesn't correct for smaller b values. For the smaller distances b increased diffraction losses prevent $TEM_{0\,0}$ construction. Results of this simulation can be considered as a principal approvement of the suggested optical cavities scheme for pure laser-driven far field acceleration.

REFERENCES

1. E.Courant, C.Pellegrini, W.Zakowics, Phys.Rev.32, 2813 (1985).
2. A.A.Varfolomeev,Yu.Yu.Lachin, Sov.Phys.Tech.Phys. 31(11),

1273   (1986).

3. A.A.Varfolomeev,Yu.Yu.Lachin, Noncollinear FEL and IFEL ,
   preprint IAE-4917114 (Atominform, Moscow, 1989).

4. A.A.Varfolomeev,Yu.Yu.Lachin, Nucl.Instr.Meth. A296,411
   (1990).

5. A.A.Varfolomeev, Sov.Phys.Tech.Phys.   35(6) 692 (1990).

6. A.A.Varfolomeev, A.H.Hairetdinov, Proc. of the workshop of
   new acceleration method, Erevan 1989. Voprosy atomnoi
   nauki i techniki, ser. jad.phys.issled, 6(14), 89   (1990).

7. A.G.Fox, T.Li, Bell Syst.Techn.J. 40, 453 (1960)

*) A full compensation of the transverce field components can be achieved
with the second   symmetrically spaced cavity row which can provide the
transverce E-field components of opposite sign to that of the first
cavity row.

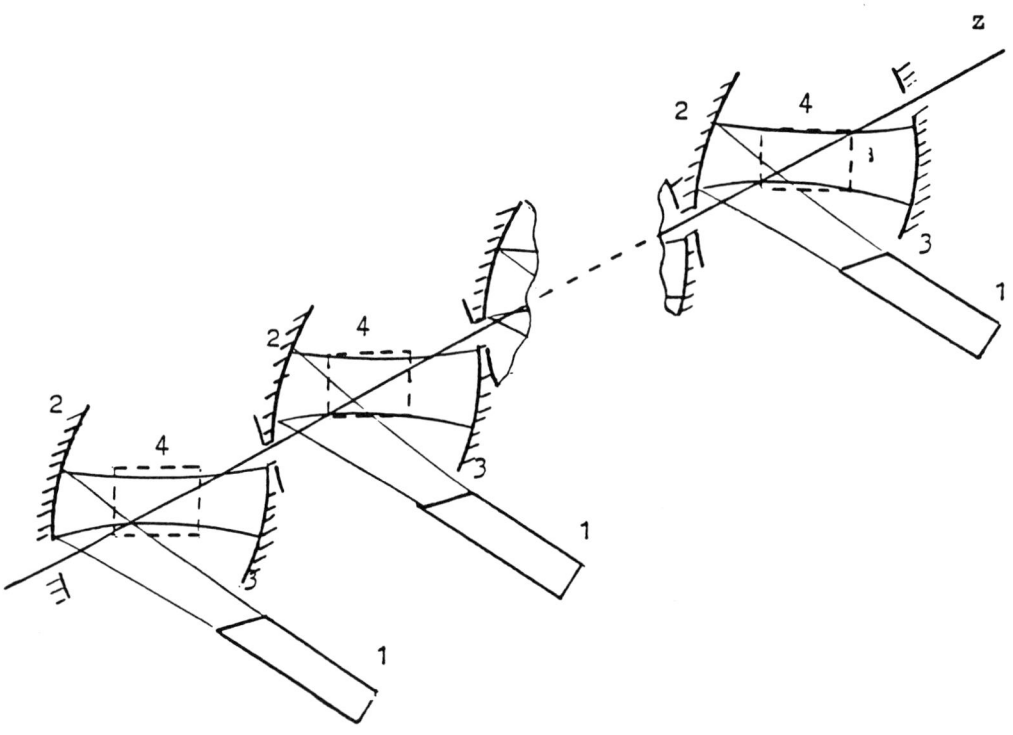

Fig.1. Scheme of laser-driven accelerator. 1 - laser power
sources; 2,3 - cavity mirrors ; 4 - interaction region

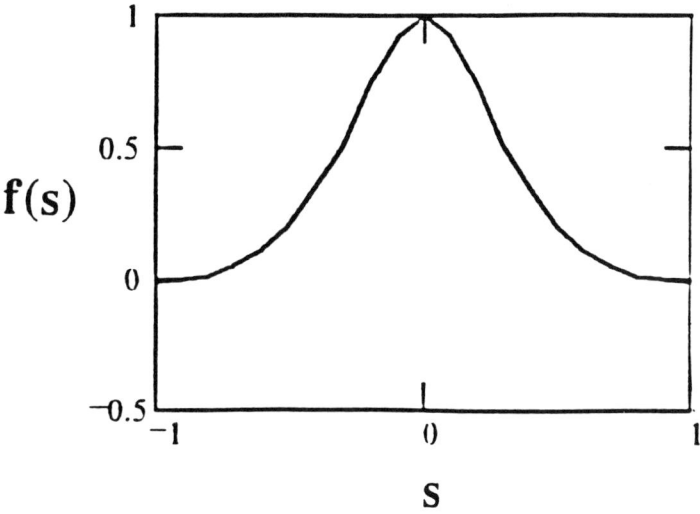

Fig.2.   The exact profile f(s) of E-field component parallel
to the particle beam axis within a cavity.

$$s = \frac{z - z_{fi}}{L_R}$$ -   relative   coordinate

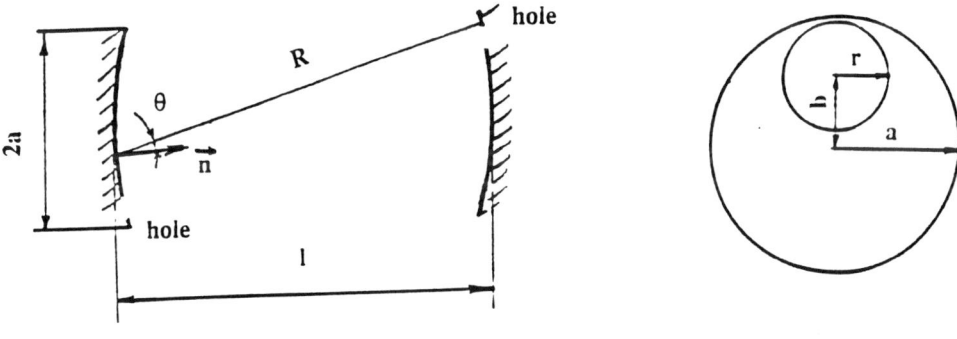

Fig.3.   The schematic picture of the cavity geometry used
in simulations.

Cavity without mirror holes

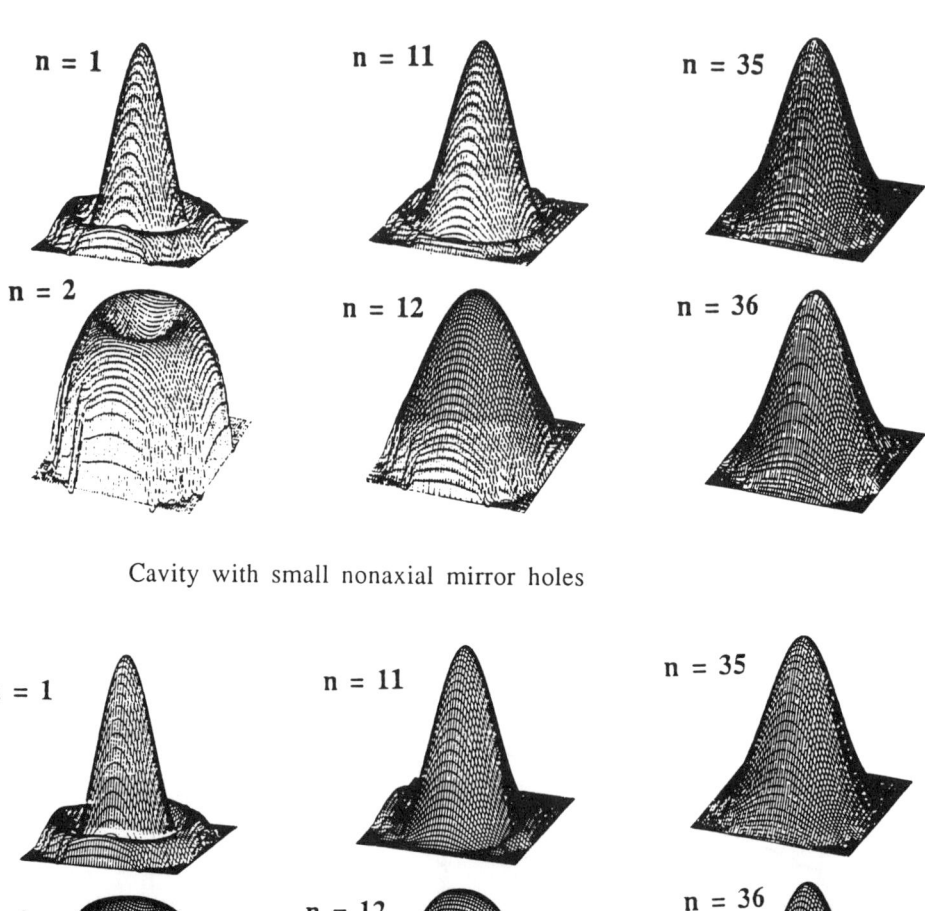

Cavity with small nonaxial mirror holes

Fig.4.   Field amplitude distribution as a function of
reflection number n , for N=1

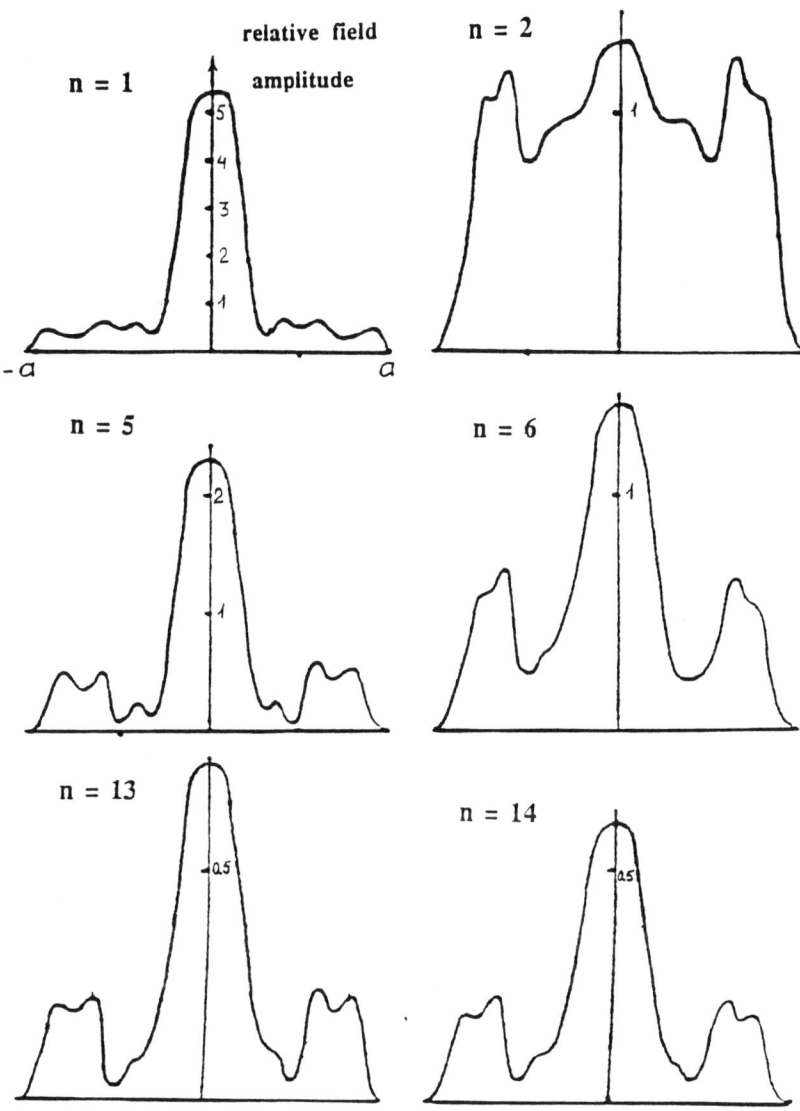

Fig.5.   Field amplitude distribution as a function of reflection number n , for N=2

OPTICAL GUIDING IN INVERSED FEL

A.A.Varfolomeev, A.H.Hairetdinov
Russian Science Center "Kurchatov Institute"
Moscow 123182, Russian Federation

ABSTRACT

The possibility of optical guiding phenomena in inversed FEL is investigated. An approximate analytical solution for the field transverse mode is obtained. Numerical calculations are presented.

INTRODUCTION

It is known that the 'optical guiding ' as phenomena does exist in the high gain FEL process. Physical reasons for the optical guiding in the FEL case appear to be rather evident. Diffractive spreading of the optical field is compensated by high gain of the fundamental mode of the field powered by a high quality electron beam and by refraction of the e.m. beam. The last one takes place due to dispersive properties of the electron beam refractive media. As a result the optical field is confined near the electron beam axis area.

'Optical guiding' means actually that a stable transverse profile f(x,y) of the field can be maintained along with its amplification over many Rayleigh ranges L_R along electron beam axis. The phenomena was investigated theoretically and in experiments (see, for example [1-3]).

The optical guiding for the inverse FEL process seems to be less evident. As we know there is only one publication relating to the problem [4]. From our point of view the optical guiding should exist as a phenomena in the IFEL process as well. Instead of field amplification we have now absorbtion of radiation by electron beam what can be formally considered as negative gain. The electron beam in the acceleration IFEL process also presents a refractive medium. So refractive processes should take place and act in the same way as in the FEL optical guiding process.

Another effects are connected with the absorbtion itself. The first is rather trivial one and relates to the case of small absorbtion length L_a . Diffraction losses can be neglected for short distances $L_a \ll L_R$.

An additional effect can exist also. Since the absorbtion takes place mainly in the central near z-axis area the diffraction from around area to the central area will be more significant for the IFEL than that for the FEL case. This mechanism can be easily understood with the extrapolation to the plane wave of high enough intensity. The field absorbtion in the central near axis area will be compensated by trivial diffraction from outside area if we consider long distances in comparison with the Rayleigh length L_R. This mechanism is actually similiar to that one which provides the so called 'diffraction-free' optical beams (see [4] and contained references). In what follows we will show that optical guiding of absorbing e.m. wave can exist in general case when the e.m. field absorbtion is caused by the inverse FEL acceleration process and the absorbtion length is not small. Analytical approach is used.

TRANSVERSE MODE EQUATION FOR IFEL

We will admit that paraxial approximation for e.m. field is valid and assume
1) e.m. wave can be considered as monochromatic with slowly varying amplitude and phase
2) transverse electron beam profile is independent on longitudinal distance
3) long e.b. bunch i.e. electron beam density doesn't depend on time.

This conditions are common to the FEL analysis. Since that we can use the following guided mode equation obtained and discussed by Moore [1] and Xie et al. [2] respectively. This equation can be written in the form

$$\nabla^2_\perp g + [\lambda + \nu - Pu(\vec{r}_\perp)/\lambda^2]g = 0 \tag{1}$$

Here following definitions are used. The function $g(\vec{r}_\perp)$ is the transverse profile function of the e.m. field.

$$a = g(\vec{r}_\perp)\exp(-i\lambda\tau) \tag{2}$$

λ is a propagation constant which we admit to be a parameter which can be found independently, $v = \dfrac{\partial \varsigma}{\partial \tau}$ is detuning parameter τ is the variable

$$\tau = \frac{z - ct(z)}{L} \qquad (3)$$

where L is the wiggler length. $P \equiv \dfrac{2\pi^2 e^2 K_w^2 [JJ]^2 L^2 N n_0}{mc^2 \gamma^3}$ is a numerical factor defined by K_w -wiggler strength, [JJ] - Bessel function difference for planar wiggler, n_0 - on-axis electron beam density, N- number of periods. For more details see [2].

Our purpose now is to show that the equation (1) can be solved in some analytical form with λ as a parameter. It can be found from IFEL equation which doesn.t take into account the optical guiding. Generally speaking it is some inconsistence to use λ parameters not from selfconsistent equations. But it can be approved as a first approximation since the optical guiding doesn't change much the λ itself. The electron beam profile will be taken in the Gaussian form.

$$u(r) = \frac{\exp(-r^2 / \sigma^2)}{2\pi\sigma} \qquad (4)$$

TRANSVERSE MODE FUNCTIONS

The eq.(1) was solved analytically by using the following approach. As a first approximation we neglected by $Pu(\bar{r}_\perp)/\lambda^2$ term and found g_0 function. As a next step we solved the equation

$$\nabla_\perp^2 g_1 + (\lambda + v)g_1 - (Pu(\bar{r}_\perp)/\lambda^2)g_0 = 0 \qquad (1')$$

The same procedure was used for the next step solution of g_2 term and so on. The final result is

$$g(r_\perp) = J_0(r_\perp \sqrt{\lambda + \nu}) + \frac{\pi P}{2\lambda^2} \left[Y_0(r_\perp \sqrt{\lambda + \nu}) \int_0^{r_\perp} x^3 u(x) J_0(r_\perp \sqrt{\lambda + \nu}) dx - \right.$$

$$\left. J_0(r_\perp \sqrt{\lambda + \nu}) \int_0^{r_\perp} x^3 u(x) J_0(r_\perp \sqrt{\lambda + \nu}) Y_0(r_\perp \sqrt{\lambda + \nu}) dx \right]$$

(5)

J_0 and Y_0 are the Bessel functions of the first and second kind respectively. We have calculated numerically some cases. The special program was written to calculate the value of Bessel functions of complex variable. The results presented in Fig.1-3.

RESULTS AND CONCLUSIONS

We have made one reasonable but very critical assumption that the transverse mode shape (2) can be real with λ parameter defined independently from our IFEL analysis without the use of this field modes and that can be expressed with only one λ parameter. Our results on the transverse mode function g(r) can be considered as possible if they exist as stable ones. For more realistic answer it is necessary to analize more carefully the influence of the real transverse size of the e.m. wave packet. From the presented results it can be concluded that the transverse size can be practically limited by some value which is not much larger than electron beam size. The most probable that we have here the effect similiar to the 'diffraction-free beam' effect [4]. In any case our result can be considered as a confirmation of a principal possibility of the optical guiding effect in the IFEL processes.

REFERENCES

1. G.T.Moore, Opt.Comm. 52, 46 (1984)
2. M.Xie, D.A.G.Deacon, J.M.J.Madey, Phys.Rev A (1989)
3. J.E.La Sala, D.A.G.Deacon, J.M.J.Madey et al. Nucl.Instr.Meth A272, 141 (1988)
4. S.Y.Cai, A.Bhattacharjee, T.C. Marshall, Nucl.Instr.Meth, A272, 481 (1988)
5. R.Bonifacio, F.Casagrande, G.Cerchioni et al. , Rivista del Nuovo Cimento, 13, n.9 (1990)

Table 1   Parameters used for Fig.1-3

wiggler period λ_w	6 cm
radiation wavelength λ_s	10 μm
wiggler strength K_w	1
total current I	100 A
number of periods N	100
electron energy (in rest mass units) γ_r	67.5
electron beam transverse distribution parameter	3 mm

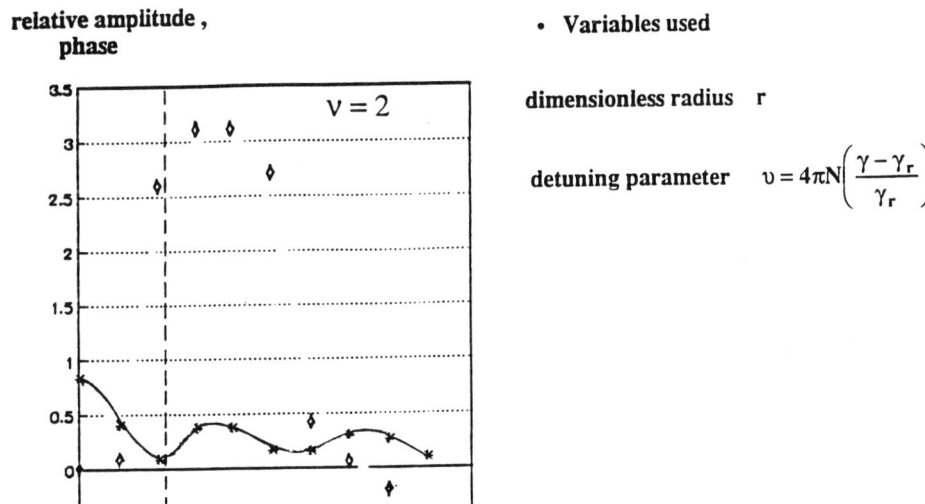

Fig.1   Transverse mode structure computated
with received analytical formulae for $\nu = 2$

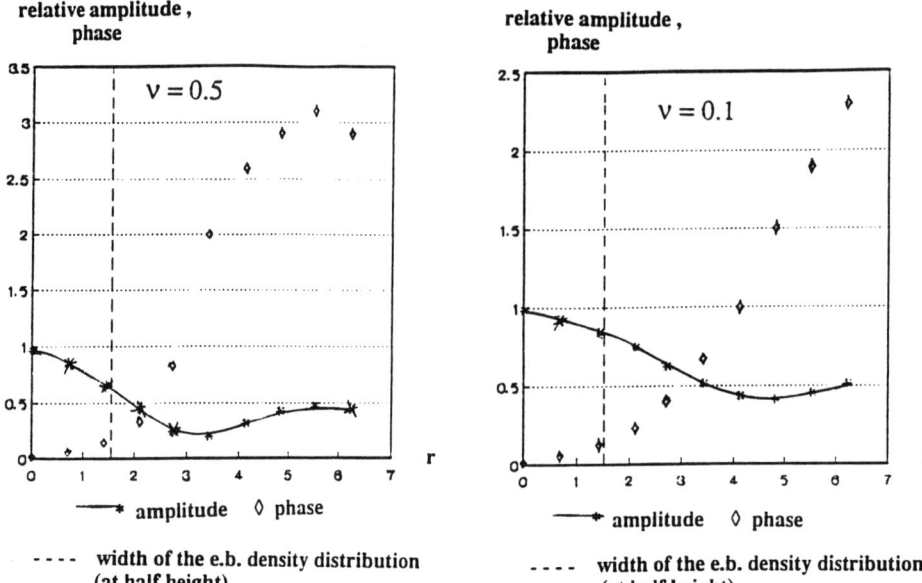

Fig.2   Transverse mode structure computated
with received analytical formulae for v=0.5 and v=0.1

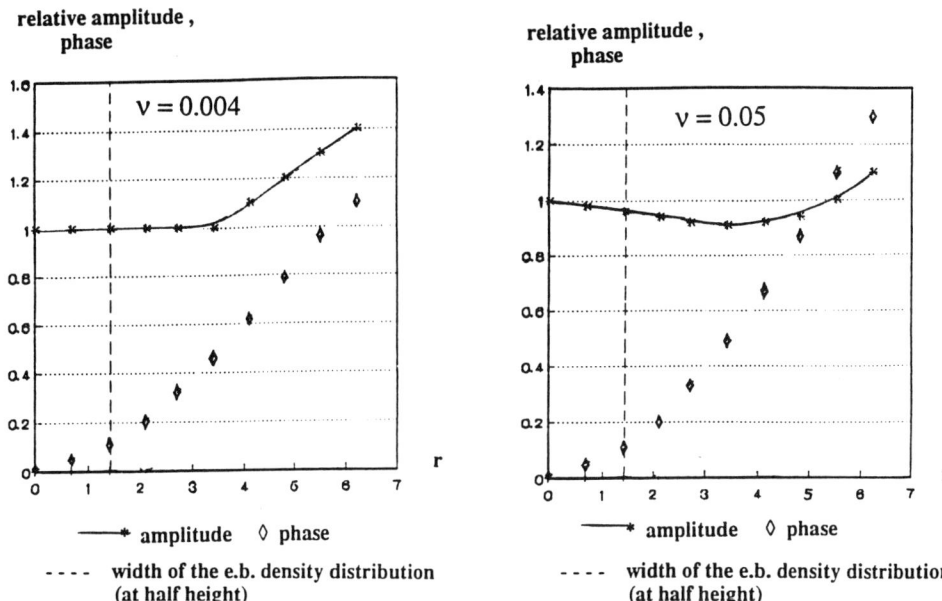

Fig.3   Transverse mode structure computated with
received analytical formulae for ν=0.05 and ν=0.004

USE ON AN INVERSE FREE ELECTRON LASER IN A LINEAR COLLIDER B FACTORY

C. Pellegrini[1], J. Sandweiss[2], and N. Barov[1]
1. Physics Dept., UCLA, Los Angeles, CA 90024
2. Physics Dept., Yale Univ., Hew Haven, CT 06511

ABSTRACT

We examine the possibility of using an IFEL as an accelerator in a linear collider B Factory. An IFEL is able to utilize a sizable fraction of the energy of the laser pulse used to accelerate the beams. It is also able to meet the stringent requirements imposed on the energy spread and luminosity at the interaction point. Two separate examples are considered, differing in the way the laser pulse energy is coupled to the electron beam. The first maximizes the slippage between the beam bunch and the radiation, in order to decrease the peak laser power. In the second example the slippage is minimized. This results in uniform beam loading and may in principle be run at higher efficiency and lower average power. We also address the laser required to drive this accelerator. The power and frequency requirements suggest the use of a FEL drive laser. Our design for this system includes the use of superconducting cavities to accelerate the drive beam, which is then propagated through an initially constant period undulator that is tapered after saturation.

1. INTRODUCTION

Although much work has been done recently to study new acceleration methods based on plasmas or lasers, not much work has been dedicated to study how these new accelerators could be used for a high energy physics program. Such work could illuminate what the strong or weak points of the new systems are when one asks not only for particles with a certain energy, but also for other beam characteristics like intensity, energy spread and emittance.

In this paper we discuss a linear collider B-factory based on an Inverse Free Electron Laser (IFEL). The main goal of the paper is to look at the feasibility of a useful high energy physics collider based on this novel acceleration technique. Hence we consider a system with a large luminosity, 10^{33} cm^{-2}s^{-1}, small beamstrahlung energy losses, and study the consequences of these initial choices on the IFEL accelerator. To explore the possibilities offered by the IFEL we consider several options, differing for the IFEL pulse propagation characteristics, and for the number of modules we use to reach the final energy. One example is based on a single acceleration module and a laser pulse group velocity greater than the beam velocity. Another example is based on the use of many accelerating modules and, in addition, a group velocity of the laser pulse which is modified to be equal to the beam velocity. As we will see, these different choices lead to very different requirements for the laser driving the IFEL, and for the design of the waveguide in which this laser pulse is propagating.

The laser driving the IFEL is assumed to be a Free Electron Laser (FEL). This seems to us a good choice because of the large laser power needed - in the MWatt range - and the large pulse repetition rate. In this power range, the FEL offers clear advantages. The FEL itself must be driven by an electron accelerator. Our choice of the accelerator type, a superconducting RF linac, has again been dictated by the high average power needed for the system, and by the high repetition rate, 10 kHz or larger, needed to obtain the large luminosity. Our model of the collider then consists of a superconducting RF linac, driving a high power FEL, followed by an IFEL, and the final focus system. In this paper we will discuss the main characteristics of all these components, with a goal of establishing the feasibility of such a system and identifying the major physical and technical problems.

The paper is divided into several parts. In section 2 we discuss the luminosity and other collider parameters, and find a set of parameters characterizing the electron and positron beams that must be produced by the IFEL. In section 3 we consider one of the two IFEL designs, based on a single accelerator module and a laser pulse with a group velocity larger than the beam velocity. In section 4 we discuss the second FEL example based on group velocity control and many modules. The difference between these two examples is twofold. The laser group velocity, when compared to the beam velocity, enters into the energy exchange between laser and accelerating beam. As we will see, this can lead to a difference in the energy required per laser pulse. The use of many accelerating modules is again done to reduce the requirement on laser energy per pulse and thus facilitate the FEL design.

The following two sections of the paper are dedicated to a discussion of the FEL, and of the superconducting driver linac.

2. DESIGN FORMULAE AND PARAMETER CHOICES

To facilitate the work and to make the essential features clear, we have made a number of simplifying assumptions, but we believe, not so many as to affect the comparison of this approach with other more conventional methods. Chief among these is the limitation to round beams ($\sigma_X = \sigma_Y$) and to an equal energy collider.

The basic IFEL equations are taken from reference[1] and the basic linear collider relations from reference[2].

The physics of a B factory determine some of the key parameters. However, even here there are a variety of approaches. For example, the B factory could operate on the $\Upsilon(4S)$ resonance with a corresponding gain in B production cross section, and with some, perhaps serious, constraints on studies of CP violation. Equal or unequal beam energies - at the same cm energy - could be used. The unequal beams offer the possibility of separating the decay vertices of the B pair, at the cost of some extra complexity.

Here we shall take the point of view that for an initial study we can consider the simpler case of a collider with equal beam energies, and operating at the $\Upsilon(4S)$ resonance. Operating at the $\Upsilon(4S)$ lessens (somewhat) the need for luminosity but places stringent limits on the beamstrahlung energy loss. All in all, we believe it is a useful example to illustrate the potential and problems of an IFEL as a B factory accelerator. We then have three basic parameters, vis.:

Luminosity - $\mathcal{L}$

Beam energy - E^*

Fractional Beamstrahlung Energy Loss - δ

We comment briefly on the choice of δ. The full width of the $\Upsilon(4S)$ is 24 MeV. Since δ is the average fractional energy loss, about $\delta/2$ is the fluctuation around the average which is relevant for determination of the collision energy. In addition, the energy loss of the two colliding particles is uncorrelated so the sum of their energies will fluctuate according to the quadrature of the individual fluctuations. Thus, the fluctuation in total collision energy will be $(\delta/2\sqrt{2})(2E^*)$ or ~ 7 MeV.

Another key parameter which will enter is the invariant emittance, ϵ_N, of the electron and positron beams. The Stanford SLC project has demonstrated a ϵ_N of 4×10^{-5} m-rad, and conceptual designs of damping systems aiming at $\epsilon_N = 10^{-6}$ m-rad have been presented[1,2]. We shall take $\epsilon_N = 2 \times 10^{-6}$ m-rad as a reasonable assumption for our IFEL designs. Following accepted practice, we define the additional quantities listed below:

f	the repetition rate of the linear collider system
N	the number of particles per bunch
D	the disruption parameter
H_D	the pinch luminosity enhancement factor
σ_t	the (rms) transverse beam dimension at the crossing
σ_z	the rms longitudinal beam dimension
β^*	the value of the β function at the crossing
γ	the relativistic factor (E^*/mc^2)

Reference[2] provides the following relations between these parameters.

$$\mathcal{L} = \frac{N^2 f H_D}{4\pi \sigma_t^2} \tag{1}$$

$$\sigma_t = \sqrt{\frac{\epsilon_N \beta^*}{\gamma}} \tag{2}$$

$$D = \frac{r_e N \sigma_z}{\gamma \sigma_t^2} \tag{3}$$

where r_e is the classical electron radius $= 2.82 \times 10^{-13}$ cm.

$$H_D = f(D) \tag{4}$$

where the function $f(D)$ is given in reference[2].

$$\delta = \frac{F r_e^3 N^2 \gamma H_D H_\gamma}{\sigma_t^2 \sigma_z} \tag{5}$$

where H_γ is a quantum correction term which can be taken equal to unity in all of this paper, and F is a constant approximately equal to 0.22.

The β^* at the I.P. and the bunch length, σ_z, must be comparable for maximum luminosity for a given value of N. We shall take:

$$\beta^* = \sigma_z \tag{6}$$

If, in addition to the "given" parameters, $\mathcal{L}$, E^*, and δ, we assume a repetition rate f, this system can be solved as follows. Eliminating N from equations (1) and (5), σ_z is given by:

$$\sigma_z = \frac{4\pi F r_e^3 \gamma \mathcal{L}}{f \delta} \tag{7}$$

One then examines the solutions obtained with various values of f to find the "most desirable". Of course, the power in the beam depends linearly on f via,

$$P_{Beam} = N E^* f \tag{8}$$

Given f, we can find σ_z, N, and all other parameters. The procedure described above produces a self-consistent set of parameters describing the colliding beams. This set of parameters is independent of the type of acceleration used to produce the electron and positron beams. To produce such beams with an IFEL, various design choices relative to the accelerator must be made.

According to the prescription presented in Section 3, we try various values of the repetition rate, f. We have found that f must be greater than about 10 KHz to obtain reasonable IFEL parameters. Mindful of the desirability of keeping f as low as possible we choose $f = 10^4$ Hz. The parameters of the final beam are summarized in Table 1.

3. A SINGLE MODULE IFEL WITH SLIPPAGE

In this section we discuss an IFEL based on a single accelerating module and no slippage control, which we will call case "A". Reference[1] describes the accelerator physics of an IFEL and we use the results in the following. As shown in[1] the wiggler tapering can be adjusted in many ways. In this section, we consider for simplicity only the case of a planar undulator with a constant period and a variable magnetic field. Use of a variable length wiggler system would permit a shorter accelerator but would not significantly change our conclusions.

We list the parameters which we shall use to describe the IFEL:

A	the effective cross sectional area of the light
λ	the laser wavelength
Λ	the wiggler period (constant)
E_0	the peak laser electric field strength
Ψ_r	the resonant phase of the accelerated particles. This is the phase of a particle which is accelerated without "Synchrotron" oscillation
B	the maximum wiggler magnetic field strength
P	the peak laser power density
l_L	the laser pulse length
W	the total laser pulse energy
η	the fraction of light energy extracted by the beam
L_A	acceleration length

The theory of the IFEL provides certain key relations between some of these parameters. In our application to B factory energies several simplifications apply. First, in all cases of interest, the wiggler parameter a_w, $\left(a_w = \frac{eB\Lambda}{2\pi mc^2}\right)$, is much larger than the "wiggler parameter" for the laser field a_L, $\left(a_L = \frac{eE_0\lambda}{2\pi mc^2}\right)$. And, secondly, the asymptotic energy - at which synchrotron radiation losses exactly balance acceleration - is much larger than the final energy of the accelerator.

We now consider the interaction of the light wave pulse with the particle bunch. We remind the reader that, if it's group velocity is c, the light wave "outruns" the particle motion by one light wavelength λ per wiggler period Λ, as required by the synchronism condition. The light pulse must have a length, l_L, large enough so that at the end of the accelerator there is still some light overlapping the particle bunch. This leads to the condition,

$$l_L \gg \text{particle bunch length} \tag{9}$$

We shall take

$$l_L = \frac{\lambda L_A}{\Lambda} + 4\sigma_Z \tag{10}$$

where, in practice, the light pulse would be made somewhat longer than the minimum required so as to make timing between the light pulse and the particle bunch less critical. We also need to confine the laser pulse with a waveguide. Unlike the following section, we assume that the modification to the group velocity due to the presence of the waveguide can be neglected.

With these approximations, and assuming $a_w \ll 1$, the synchronism condition of the IFEL becomes,

$$a_w = 2\gamma\sqrt{\frac{\lambda}{\Lambda}} \tag{11}$$

and for a fixed period, the magnetic field then changes as:

$$B = 2\gamma\sqrt{\frac{\lambda}{\Lambda}}\left(\frac{2\pi mc^2}{\Lambda e}\right) \tag{12}$$

The linear rate of increase of the relativistic factor γ with distance Z along the beam direction is,

$$\frac{d\gamma}{dZ} = JJ\frac{eE_o\sin\Psi_r}{mc^2}\sqrt{\frac{\lambda}{\Lambda}} \tag{13}$$

where the Bessel function factor approaches $JJ \simeq .69$ for large a_w. From equation (13) we see that for constant E_o and Λ, the rate of energy gain is constant along the accelerator. The accelerator length L_A is given by

$$L_A = \frac{E^* - E_i}{mc^2 d\gamma/dZ} \tag{14}$$

where the beam energy at injection has been assumed to be, $E_i = 500$ MeV.

The particles remove a fraction, η, of the energy of the part of the light pulse which overlaps them during the acceleration process. The beam loading relation for the IFEL - the number of particles, N, which can be accelerated having fixed η - is derived in Appendix 1,

$$N = \frac{1}{JJ}\frac{E_o A\eta}{8\pi e\sin\Psi_r}\sqrt{\frac{\lambda}{\Lambda}} \tag{15}$$

From the point of view of power economy, we would like a large value for the efficiency η. Too high an efficiency leads to particle detrapping due to the variation in the electric field over the beam. The problem of maximizing the efficiency has not been analyzed in detail for the IFEL. A computer simulation would be useful for this purpose. We simply choose a value of η ($\eta = 0.1$) which seems reasonable.

We next consider the choice of laser parameters. We have chosen a laser wavelength $\lambda = 10\,\mu$m as a reasonable choice between the advantages and disadvantages of 'small' wavelengths. In addition, because of CO_2 lasers a great deal of technology exists for handling 10 μm light.

For the wiggler a key parameter is the maximum magnetic field. We choose a B_{max} of 50 KG - a value which should be well within the range of superconducting designs. Having λ and B_{max}, we may use equation (12) to find the (constant) wiggler period, $\Lambda = 26.5$ cm. This value should not be a problem for the wiggler design.

The laser electric field required to keep the beam loading parameter $\eta = 0.1$ can be obtained from equation (15). To use this equation we must choose Ψ_r and an effective cross sectional area for the laser beam. We choose, $\sin(\Psi_r) = 1/\sqrt{2}$, which gives a good compromise between a large accelerating field and a large accelerating bucket area.

A number of studies have been done on possible guide structures for the 10 μm laser beam[3,4,5]. These indicate that a guide cross sectional area of $\simeq 1$ cm^2 is feasible. Choosing the dimensions of the wave guide to be .8 × .8 cm, an effective area of $A = 0.32$ cm^2 is applicable because the field distribution is approximately sinusoidal in the transverse cross section. With these choices, equation (15) gives $E_o = 1.04 \times 10^8$ V/cm, and a laser power density of, $P = 1.45 \times 10^{13}$ W/cm^2

The length of the wiggler is determined from equations (13) and (14):

$$L_A = 146 \, \text{m} \tag{16}$$

thus, the effective accelerating gradient is 31 MeV/m.

From (10), (14), and (16) we calculate the length of the laser pulse:

$$l_L = 6.62 \, \text{cm} \tag{17}$$

The total energy in the laser pulse, W. is:

$$W = \frac{PAl_L}{c} = 104 \, \text{Joules} \tag{18}$$

The properties of the IFEL linear collider at the $\Upsilon(4S)$ are summarized in Table 2, column 1.

4. AN IFEL WITH REDUCED SLIPPAGE AND MANY MODULES

In this section we present a different approach for the IFEL, which we call case "B". The operation of a FEL near zero slippage is examined in reference[8]. The advantage for this IFEL design is a reduction of the laser energy per pulse for any given peak electric field. Conceptually, the reduction in slippage is brought about by a combination of decreasing the waveguide transverse dimension and increasing the undulator period. We consider a planar undulator and only the case of zero slippage, which will demonstrate the general features of operating with reduced slippage. In the interest of power economy, we want the light pulse length to be close to the bunch length. We take

$$\sigma_L = \sigma_Z \tag{19}$$

where σ_Z is the bunch length and σ_L characterizes the light pulse intensity profile. At this σ_L, laser pulse lengthening due to dispersion limits the use of wavelengths much longer than tens of microns. We choose 10 microns for this example, noting that it is harder to modify the group velocity of shorter wavelengths without restricting the vertical beam aperture. We choose a laser of 10 microns, which satisfies both of the above conditions.

Given the transverse waveguide dimension, b, and the wavelength of the light, λ, we state the result from reference[8] for the undulator period, Λ at zero slippage:

$$\Lambda = \frac{(2b)^2}{\lambda} \tag{20}$$

The need for keeping the undulator period as short as possible is discussed in Appendix 2. We choose $b = 2$ mm which gives a favorable Λ ($\Lambda = 1.6$ m), but requires external focusing of the beam greater than the natural focusing. One possibility for this is the focusing scheme suggested in reference[7] for the proposed SLAC FEL.

The synchronism condition for zero slip from reference[8] differs from the ordinary synchronism condition by a factor of $\sqrt{2}$

$$a_w = \sqrt{2}\gamma\sqrt{\frac{\lambda}{\Lambda}} \tag{21}$$

which sets the magnetic field to $B = .22$ kG at the beginning of the acceleration and $B = 2.3$ kG at the end.

We want to relax the requirements on the drive laser by having multiple accelerator sections. This imposes the difficulty of having to match the phase of the laser to the bunching in the electron beam. Using 18 accelerating sections and a power efficiency of $\eta = .1$, the peak electric field in the waveguide is computed from the final energy of the beam.

$$E_0^2 = 8\pi N m c^2 \frac{\gamma^* - \gamma_i}{\sqrt{2\pi}\sigma_L A \eta} \frac{1}{R} \tag{22}$$

where R is the number of sections, and A is the effective waveguide cross sectional area, $A = ab/2$. For a waveguide with $a = 2$ mm, and $b = 4$ mm (the shorter dimension being transverse to the wiggle plane), we arrive at $E_0 = 1.84 \times 10^{10}$ V/m. The resulting acceleration is,

$$\frac{d\gamma}{dZ} = \frac{1}{2}\frac{E_0 a_w}{\gamma_r m c^2}\sin\left(\Psi_r\right) \tag{23}$$

where γ_r is the relativistic factor of the resonant particle, and $\Psi_r = \frac{\pi}{4}$ is the resonant phase. This equation can be integrated to obtain the length of the accelerator,

$$L_{acc} = 2mc^2 \frac{\gamma_f - \gamma_i}{E_0 a_w \sin(\Psi_r)} \qquad (24)$$

which gives an accelerator length of 282 meters and an acceleration gradient of 16 MeV/m.

The drive laser must produce 4.9 Joule pulses with a 4.5×10^{13} Watt/cm^2 peak intensity and must produce a train of 18 bunches 10^4 times per second.

One advantage of using multiple sections in the IFEL is the possibility of using a single drive laser. A switching technology such as germanium windows illuminated by UV light can direct the laser pulse to the appropriate section.

5. FEL DRIVER

The drive laser is characterized by its high peak power and large repetition frequency, as shown in Table 2. This characteristic makes it difficult to use a conventional atomic or molecular laser, like a CO_2 system. We have therefore studied a FEL driver, using a superconducting linac to produce the electron beam.

The FEL operates starting with a low power signal from a CO_2 laser; this signal is amplified to saturation in a constant period wiggler, and further amplified in a tapered wiggler, to reach the required power.

In this section we will indicate with $E_D = mc^2\gamma$ and N_D the energy and number of electrons per bunch of the superconducting linac. If η_D is the efficiency of energy transfer from the electron beam to the laser radiation, the laser pulse energy is

$$W_L = \eta_D N_D E_D \qquad (25)$$

The laser wavelength λ is related to the FEL wiggler parameter K_D and period λ_D, by

$$\lambda = \frac{\lambda_D}{2\gamma_D^2}(1 + \frac{K_D^2}{2}) \qquad (26)$$

To describe the FEL we use a simple one dimensional model. In the constant period part of the wiggler the gain length is given by

$$L_G = \frac{\lambda_D}{2\sqrt{3}\pi\rho} \qquad (27)$$

with the FEL parameter ρ given by

$$\rho = \left(\frac{K_D \lambda_D}{4\gamma} \frac{\Omega_p}{2\pi c} \right)^{2/3} \qquad (28)$$

where $\Omega_P = (4\pi r_e c^2 \eta_D / \gamma_D)^{1/2}$ is the electron beam plasma frequency and η_D its density. The saturation power level is

$$P_{\text{sat.}} = \rho I_D E_D \tag{29}$$

I_D being the driver beam current, and the saturation length, for a small input signal,

$$L_{\text{sat.}} \simeq \frac{\lambda_D}{\rho} \tag{30}$$

After reaching saturation we taper the FEL wiggler to reach an efficiency of about 40%. We assume the tapering to be done in a constant period, variable magnetic field undulator. In this case the beam energy decrease is still described by (13), but changing Ψ_r to $-\Psi_r$. Some of the main FEL and driver beam parameters are given in Tables 3 and 4.

6. A SUPERCONDUCTING LINAC DRIVER FOR THE FEL

The electron beam needed to drive the FEL has a very large peak and average current, and it's characteristics are beyond the present state of the art. However, beams similar to this are being considered for other applications in the area of linear collider, and might become a reality in the near future. One way to produce such a beam is by using a superconducting linac, as we will discuss now. Given the long bunch length and the large number of particles per bunch, it is convenient to use a low frequency linac. We consider here a 350 MHz system based on the superconducting cavities developed at CERN. If the accelerating voltage gradient is E_{sc}, the energy per unit length in the linac is

$$W_{sc} = \frac{E_{sc}^2}{\frac{R_{sc}}{Q_{sc}} \omega_{sc}} \tag{31}$$

Where R_{sc} and Q_{sc} are the shunt impedance and the quality factor. The beam loading determines an energy spread, given approximately by

$$\left(\frac{\Delta E}{E}\right)_{BL} = \frac{1}{2} \frac{e N_D E_{sc}}{W_{sc}} \tag{32}$$

The bunch length, σ_D, also produces an energy spread

$$\left(\frac{\Delta E}{E}\right)_\sigma = \frac{1}{2} \left(\frac{\sigma_D \omega_{sc}}{c}\right)^2 \tag{33}$$

The two effects of beam loading and RF voltage curvature can be partly compensated against each other by running the linac off crest, resulting in slower acceleration and a longer linac.

The power needed to operate the linac is partly needed for cooling, P_{cooling}, and partly to accelerate the beam, P_{RF}; the cooling power is

$$P_{\text{cooling}} = \eta_{\text{cooling}}^{-1} W_{sc} L_{sc} \frac{\omega_{sc}}{2\pi Q_{sc}} \tag{34}$$

where $\eta_{\text{cooling}} = 1/1500$ and the beam power is

$$P_{\text{RF}} = f N_D E_D \qquad (35)$$

with f the collision frequency.

A possible list of parameters is given in Table 5.

7. CONCLUSIONS

An examination of the results obtained shows that while the proposed system is feasible, it puts stringent requirements on the laser driving the IFEL, and on the electron superconducting linear accelerator powering the FEL. Some of these requirements, like the number of particles per bunch in the superconducting linac become easier in case B, than in A.

The work presented in this paper is only a preliminary investigation, and the two examples that we have discussed are not optimized systems. Extending this investigation to more cases might produce other system configurations with lower peak and average power requirements. In addition we need to study in more details the beam dynamics in the IFEL, to investigate the final beam emittance and energy spread and to optimize the wiggler design, including the initial and final undulator sections, which can be used to match the beam to the interaction region.

TABLE 1. BEAM PARAMETERS

Beam Energy (GEV)	5
Luminosity $(\text{cm}^{-2}\text{s}^{-1})$	10^{33}
Average Bmstlg. Fract. ΔE	2×10^{-3}
No. of Particles/Bunch	1.14×10^{10}
Pulse repetition rate (sec^{-1})	10^4
σ_t (μm)	0.249
β^* at I.P. (mm)	0.31
σ_z (mm)	0.31
Disruption Parameter D	16
Pinch Enhancement	6
Invariant Emittance (m-rad)	2×10^{-6}

TABLE 2. IFEL CHARACTERISTICS

Parameter:	Case A	Case B
Laser Wavelength (μm)	10	10
Injection Energy (MeV)	500	500
Dimensions of Laser Waveguide (cm)	.8 × .8	.2 × .4
Peak Laser Electric Field (V/m)	1.04×10^{10}	1.84×10^{10}
Laser Power Density (Watts/cm^2)	1.45×10^{13}	4.49×10^{13}
Laser Pulse Energy (Joules)	104	4.9
Bunch Train Repetition Rate (Hz)	10^4	10^4
Number of Bunches per Train	1	18
Average Laser Power (Watts)	1.04×10^6	8.8×10^5
Max. Wiggler Magnetic Field(Gauss)	5×10^4	2.3×10^3
Wiggler Period (cm)	26.5	160
Total Length of Wiggler (m)	146	282
Accelerating Gradient(MV/m)	31	16

TABLE 3. FEL PARAMETERS

Parameter:	Case A	Case B
Wiggler Period, cm	20	20
Wiggler Parameter	13.74	13.74
FEL Parameter	2.6×10^{-2}	1.9×10^{-2}
Saturation Length, m	7.63	10.4

TABLE 4. ELECTRON BEAM PARAMETERS

Parameter:	Case A	Case B
Electron Energy, MeV	500	500
Pulse Length, cm	0.66	.031
Peak Current, KA	23.5	9.1
Normalized Beam Emittance, m-rad	400	400
Average Beam Size in Wiggler, cm	0.14	0.14

TABLE 5. SUPERCONDUCTING LINAC

Parameter:	Case A	Case B
Beam Energy, MeV	500	500
Frequency, MHz	350	350
Shunt Impedance, GΩ/m	1.3×10^3	1.3×10^3
Quality Factor	5×10^9	5×10^9
Electric Fields, MV/m	15	15
Stored Energy, J/m	394	394
Linac Length, m	33	33
Total Stored Energy, KJ	13	13
Electrons/Bunch	3.2×10^{12}	1.53×10^{11}
Bunches/Bunch Train	1	18
Bunch Repetition Rate, KHz	10	10×18
RF Power, MW	2.6	2.2
Cooling Power, MW	1.3	1.3
Energy/Bunch, J	260	12
Beam Loading	1%	.8%
Bunch Length, cm	0.66	.031
Energy Spread Due to Bunch Length	0.1 %	-

APPENDIX A. DERIVATION OF THE BEAM LOADING EQUATION

In this appendix, we derive a beam loading equation for the case of Section 3. In this case, the bunch slips relative to the laser pulse an amount λ every undulator period Λ. During a distance dz, the bunch is exposed to fresh electromagnetic energy of amount,

$$dW_i = \frac{E_0^2 A}{8\pi} \frac{\lambda}{\Lambda} dz \qquad (A.1)$$

The number of particles present will determine how much of this radiation is used up. The energy used up by the beam is governed by the IFEL relation from equation (13) of Section 3.

$$\frac{d\gamma}{dz} = JJ \frac{eE_0 \sin \Psi_r}{mc^2} \sqrt{\frac{\lambda}{\Lambda}} \qquad (A.2)$$

Having traveled the same distance, dz, the beam will gain an energy,

$$dW_b = JJNeE_0 \sin \Psi_r \sqrt{\frac{\lambda}{\Lambda}} dz \qquad (A.3)$$

Dividing this by the incoming energy from equation (1) will give the efficiency, η.

$$\eta = \frac{dW_b}{dW_i} \qquad (A.4)$$

The number of particles which can be accelerated at a given amount of beam loading is then,

$$N = \frac{E_0 A \eta}{JJ 8\pi e \sin\Psi_r} \sqrt{\frac{\lambda}{\Lambda}} \qquad (A.5)$$

APPENDIX B. THE FINAL BEAM ENERGY SPREAD

This appendix will address the contribution to the energy spread at the interaction point brought about by the IFEL. Reference[1] explains in detail the longitudinal phase space dynamics of an IFEL. The motion relative to the resonant particle can be described by an effective Hamiltonian,

$$H = \frac{k_w}{\gamma_r}\gamma_1^2 - JJ\frac{eE_0}{mc^2}\sqrt{\frac{\lambda}{\Lambda}}\left(\cos\Psi + \Psi\sin\Psi_r\right) \qquad (B.1)$$

where $\gamma_1 = \gamma - \gamma_r$.

This has the feature that the buckets become very small as Ψ_r approaches $\pi/2$.

The full height of the phase space bucket described by (B.1) is,

$$\delta\gamma = 2\sqrt{\frac{2\gamma eE_0\Lambda JJ}{2\pi mc^2}\left(\frac{\lambda}{\Lambda}\right)^{1/2}\left(\cos\Psi_r - (\Psi_r - \pi/2)\sin\Psi_r\right)} \qquad (B.2)$$

Having fixed Ψ_r, the quantity inside the square root is proportional to $E_{acc}\gamma_r\Lambda$. In the interest of increasing the acceleration gradient E_{acc}, we must minimize the undulator period Λ.

If the energy spread of the injector is assumed to be much less than this value, a filamentation of the longitudinal phase space will result. Let us assume that we can achieve a state where the buckets are only half filled, so that the most energetic particle has only half of the energy needed to escape from the bucket. In this case, γ changes slowly enough in equation (B.1) to be considered adiabatic. The resultant energy increase is only proportional to $\gamma^{1/4}$.

In order to take advantage of this, the beam can be prebunched before entering the main accelerator. This step will not only minimize the final energy spread but also improve the initial acceptance of the accelerator. For the parameters of case "A", and assuming half filled buckets, this model predicts that all of the particles will lie in a 15 MeV band at the beginning of acceleration, (ignoring phase,) and a 28 MeV band at the end. The width of this distribution will be less than that of the resonance.

REFERENCES

1. E.D. Courant, C. Pellegrini, and W. Zakowicz, Phys. Rev A, 32, 2813 (1989).
2. U. Amaldi, Proc. US,CERN School on Part. Acc. at So. Padre Island, Texas, October 1986 ed. M. Month (Springer Verlag, 1988).
3. R. Palmer, SLAC-PUB-44295, (1987).
4. W. Zakowicz, J.Appl. Phys. 55, 9, (1984).
5. J. Sandweiss, BNL 35444, August 1984.
6. J. Sandweiss, BNL 40310, August 1987.
7. G. Travish, Submitted to Proc. Adv. Acc. Concepts at Port Jefferson NY,1992.
8. S.K. Ride, Appl. Phys. Lett. 57 1283 (1990)

ROLE OF BEAM QUALITY IN A FREE-ELECTRON LASER IN THE GAIN-FOCUSING REGIME

B. Hafizi[1,a] and C. W. Roberson[2]

[1] *Icarus Research*
7113 Exfair Road, Bethesda, MD 20814
[2] *Physics Division*
Office of Naval Research, Arlington, VA 22217

Abstract

We discuss the effect of beam emittance and energy spread in a free-electron laser operating in the gain-focusing regime. The variation of growth rate, radius of curvature of wavefronts, filling factor and efficiency with emittance and energy spread is derived. The analysis is based on the Vlasov-Maxwell system of equations and results are obtained by minimizing a variational functional. When plotted as a function of emittance, the efficiency at maximum growth rate peaks at a nonzero value of emittance. For small values of energy spread, the efficiency at maximum growth rate increases with energy spread, in contrast to intuitive expectations.

The results of an analytical study of the effect of emittance and energy spread[1-16] on a free-electron laser (FEL) in the gain-focusing regime[17-20] of operation are presented. Based on the Vlasov-Maxwell equations a differential eigenvalue equation for the wavenumber k of the radiation is derived and solved by a variational technique. The variational parameter furnishes the spot size and the radius of curvature of radiation wavefronts.[19,20] The stationarity condition imposed on the variational functional implies that our results are insensitive to the choice of the trial function.[21] The variation of growth rate, radius of curvature of wavefronts, filling factor and efficiency with emittance and energy spread is displayed graphically.

The model consists of a matched electron beam and a matched radiation beam propagating along the z axis through a planar wiggler. The wiggler vector potential is given by $\mathbf{A}_w = A_w \cosh(k_w y) \sin(k_w z)\mathbf{e}_x$, where A_w is the amplitude, $2\pi/k_w$ is the period and $\mathbf{e}_x$ is the unit vector along the x axis. The vector potential of the optical beam is given by $\mathbf{A}_s = \frac{1}{2}A_s(y)\exp[i(kz - \omega t)]\mathbf{e}_x + c.c.$, where ω is the angular frequency and A_s is the amplitude.

[a] Work performed at Plasma Physics Division, Naval Research Laboratory, Washington, DC 20375-5000

The equations of motion of an electron are derived from the Hamiltonian function $-p_z(y, p_y; t, -E; z)$:[22]

$$p_z = E/c - (m^2c^3/2E)$$
$$\times \left\{ 1 + \frac{p_y^2 + P_x^2}{m^2c^2} + \frac{a_w^2}{2}[1 + (k_w y)^2] + \frac{a_w a_s}{2i} f_B \exp[i(k + k_w)z - i\omega t] + c.c. \right\},$$

where $E \equiv \gamma mc^2$ is the energy of an electron of rest mass m and charge $-|e|$, t is the time, (P_x, p_y) are the momenta conjugate to the coordinates (x, y), $a_{w,s} = |e|A_{w,s}/mc^2$, $f_B = J_0(\xi) - J_1(\xi)$ is the usual difference of Bessel functions and $\xi = (a_w/2)^2/(1 + a_w^2/2)$.

In the absence of the optical field electrons perform betatron oscillations in the $y - p_y$ plane. The area in this plane is the action $I \equiv \int\int dy\, dp_y/2\pi = H/k_\beta$, where $H = cp_y^2/2E + Ek_\beta^2 y^2/2c$ is the Hamiltonian for the transverse motion and $k_\beta = a_w k_w/\sqrt{2}\gamma$ is the betatron wavenumber.

The electron distribution function evolves according to the Vlasov equation. For the equilibrium distribution we choose

$$F(E, P_x, I) = n_{b0} \frac{\exp[-(\gamma - \gamma_0)^2/\sigma_\gamma^2]}{\sqrt{\pi}\,\sigma_\gamma\, mc^2} \delta(P_x) \frac{\exp(-\sqrt{2}\,I/a_w k_w mc\sigma_b^2)}{\sqrt{\pi}\,a_w k_w mc\sigma_b}, \quad (1)$$

where $E_0 = \gamma_0 mc^2$ is the mean energy, $\sigma_\gamma mc^2$ is energy spread and $n_b(y) = n_{b0}\exp(-y^2/2\sigma_b^2)$ is the spatial density, with peak value n_{b0} and width σ_b.

In the Coulomb gauge the wave equation for the optical field takes the form

$$\frac{d^2 a_s}{dy^2} + [(\omega/c)^2 - k^2]a_s - \frac{if_B^2}{n_b}\left(\frac{\omega_b a_w}{2c}\right)^2 \int \frac{d\Gamma}{\gamma^2} \frac{\partial F}{\partial\gamma} \int_{-\infty}^0 d(\omega z'/c)\, a_s[y(z')] \exp[i\phi(z')] = 0,$$
$$(2)$$

where $\omega_b = (4\pi n_b|e|^2/m)^{1/2}$ and $d\Gamma = dP_x dp_y dE$. From the Hamiltonian functions, $-p_z$ and H, it follows that $\phi(z') = (k + k_w - \omega/c\beta_z)z'$ and $y(z') = y\cos(k_\beta z') + (cp_y/k_\beta E)\sin(k_\beta z')$, where

$$\beta_z^{-1} = 1 + [1 + a_w^2/2 + (P_x/mc)^2 + \sqrt{2}a_w k_w I/mc]/2\gamma^2, \quad (3)$$

and y and p_y denote the location of an electron in the transverse phase space at $z' = 0$.

Equation (2) may be written as $\mathcal{L}a_s(y) = 0$, where $\mathcal{L}$ is a linear operator. The eigenvalue k may be obtained by using the variational principle $\delta J = 0$, where $J \equiv \int_{-\infty}^\infty dy\, a_s(y)\mathcal{L}\tilde{a}_s(y)$, $\tilde{a}_s(y) = \exp(-k_s y^2/2\zeta_R)$ is the trial function, $k_s = 2\gamma_0^2 k_w/(1 + a_w^2/2)$, $\zeta_R \equiv k_s(2/\sigma_s^2 + ik_s/R)^{-1}$ is the variational parameter, σ_s is the spot size of the optical beam and R is the radius of curvature of the wavefronts. This is evidently a reasonable form for the trial function in view of the transverse profile of the electron beam density indicated following Eq. (1).

Evaluating the functional J and equating the result to zero leads to

$$\mu - \frac{k_{\beta 0}/k_w}{8k_{\beta 0}\zeta_R} + \frac{k_{\beta 0}/k_w}{\sqrt{2k_{\beta 0}\zeta_R}}\left(\frac{\gamma_0 D}{\sigma_\gamma}\right)^2 \sum_{r=-\infty}^{\infty}\int_0^\infty dx\left(1+\frac{2\sigma_\gamma}{\gamma_0}\xi\right)[1+\xi Z^*(\xi^*)]$$

$$\times I_r^2\left(\frac{k_s\epsilon}{2k_{\beta 0}\zeta_R}x\right)\exp\left[-\left(1+\frac{k_s\epsilon}{k_{\beta 0}\zeta_R}\right)x\right] = 0, \qquad (4)$$

where $x = cI/k_{\beta 0}E_0\sigma_b^2$, $\mu k_w = \omega/c - k$, $\epsilon = k_{\beta 0}\sigma_b^2$ is the unnormalized emittance,

$$D = \left[\sqrt{\frac{2\pi}{k_{\beta 0}k_s}}\frac{\tilde{I}_b}{I_A}\frac{a_w^2}{(1+a_w^2/2)}\right]^{1/2} f_B,$$

$\xi = \gamma_0[-\mu + (1 - \omega/\omega_s) - 2rk_{\beta 0}/k_w - (k_{\beta 0}/k_w)k_s\epsilon x]/2\sigma_\gamma$, $\omega_s = ck_s$, $I_A = 1.7 \times 10^4\gamma_0\beta_{z0}$ A is the Alfvén current, $\tilde{I}_b$ is the current per unit length, I_r is the modified Bessel function of order r, $Z(\xi)$ is the plasma dispersion function[14] and $k_{\beta 0}$ is the betatron wavenumber evaluated at γ_0.

Equation (4) and an equation obtained by equating to zero the derivative of Eq. (4) with respect to the variational parameter, determine μ and ζ_R. Taking $1 + (2\sigma_\gamma/\gamma_0)\xi \approx 1$ in Eq. (4), the scaled parameters are μ/D, $(1 - \omega/\omega_s)/D$, $k_{\beta 0}\zeta_R$, $k_s\epsilon$, $k_{\beta 0}/k_w D$ and $\sigma_\gamma/\gamma_0 D$.[11,23]

Figure 1 shows the scaled spatial growth rate, Im μ/D, the scaled radius of curvature, $k_{\beta 0}R$, the filling factor, σ_b/σ_s and the scaled efficiency, η/D, as functions of $k_s\epsilon$. In this example the electron beam is initially monoenergetic, i.e., $\sigma_\gamma = 0$, the emittance being due to pitch-angle scattering.

All numerical results presented here correspond to the particular value of 'detuning', $1 - \omega/\omega_s$, that gives the maximum growth rate. In Fig. 1, for example, the detuning varies from point to point as the emittance changes.

The nonlinear efficiency, η, may be determined from the energy lost by electrons when they are trapped in the ponderomotive buckets. It may be shown that $\eta = 2\gamma_0^2 < \beta_z - \beta_{ph} > /(1 + a_w^2/2)$, where $\beta_{ph} = \omega/(k_w + \text{Re } k)$ is the phase velocity of the buckets and $<>$ indicates an average over the distribution in Eq. (1).[1] Making use of Eq. (3), we find $\eta = -\text{Re}\,\mu + (1 - \omega/\omega_s) - (k_{\beta 0}/k_w)k_s\epsilon$. It is shown elsewhere that this expression is in good agreement with numerical simulations as the detuning is varied.[24] Figure 1 (d) shows that as $k_s\epsilon$ increases the efficiency at first increases, reaches a maximum (for both $k_{\beta 0}/k_w D = 1$ and 0.1) and eventually tends towards zero. We find that as $k_s\epsilon$ increases from small values the difference between the velocities of the *fastest growing* ponderomotive wave and the beam increases and this accounts for the initial increase in efficiency in Fig. 1 (d).[8] Eventually, however, the slowing down of the beam with increasing emittance predominates and the curve for the efficiency turns over and tends towards zero.

Our expression for efficiency can be used if the ponderomotive wave 'sees' the electrons as a cold beam. An estimate of when this is valid is provided by the relative magnitude of the axial thermal velocity, $\beta_{z,th} \equiv < (\beta_z - <\beta_z>)^2 >^{1/2}$, and $<\beta_z - \beta_{ph}>$. Making use of Eq. (3) one finds

$$\frac{\beta_{z,th}}{<\beta_z - \beta_{ph}>} = \left[2\left(\frac{\sigma_\gamma}{\gamma_0 D}\right)^2 + \left(\frac{k_{\beta 0}}{k_w D}k_s\epsilon\right)^2\right]^{1/2}\left(\frac{\eta}{D}\right)^{-1}. \qquad (5)$$

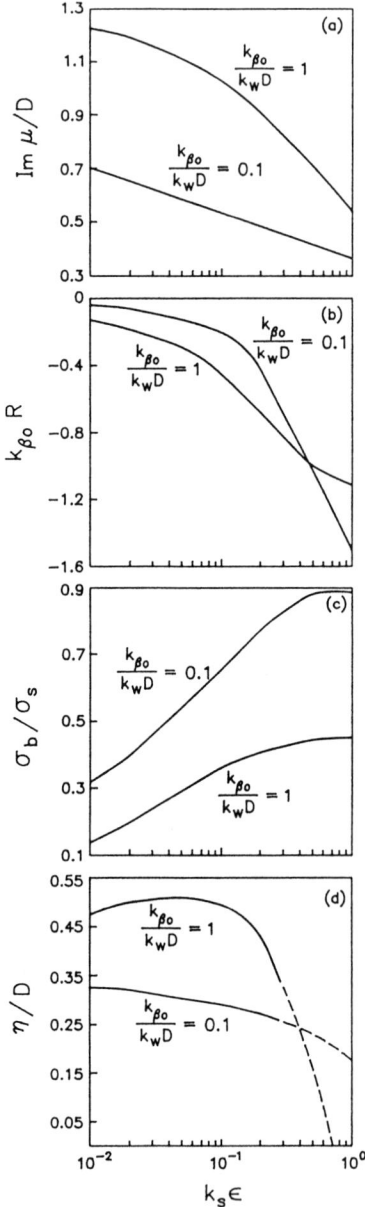

Figure 1: Plot of $\mathrm{Im}\,\mu/D$, $k_{\beta0}R$, σ_b/σ_s and η/D versus $k_s\epsilon$ for a monoenergetic beam. The dashed portion of the curves in (d) lie in the regime where kinetic effects are expected to modify the efficiency.

Equation (5) indicates that for the dashed portion of the curves in Fig. 1 (d) the ponderomotive wave is resonant with thermal electrons and the efficiency is expected to be significantly modified by *kinetic effects in the nonlinear stage of the interaction.*

To study the effect of energy spread, Fig. 2 shows the results, for $k_s\epsilon = 0.1$, as a function of the energy spread $\sigma_\gamma/\gamma_0 D$. Surprisingly, Fig. 2 (d) shows that the efficiency is a monotonically increasing function of $\sigma_\gamma/\gamma_0 D$. Equation (5) indicates that the dashed portion of the curves in Fig. 2 (d) is expected to be significantly modified by *kinetic effects in the nonlinear stage of the interaction.* In their region of validity, both curves in Fig. 2 (d) indicate a significant increase in the maximum efficiency with increasing energy spread, in contrast to intuitive expectations.

We have examined the effect of emittance and energy spread on the spatial growth rate, radius of curvature of radiation wavefronts and filling factor. The analysis is based on the Vlasov-Maxwell system of equations and a variational solution of the eigenvalue equation, with a trial function that depends on a variational parameter. We have found that i) As a function of beam emittance, the efficiency peaks at a nonzero value of emittance and ii) For small values of energy spread, the efficiency increases with energy spread on the beam.

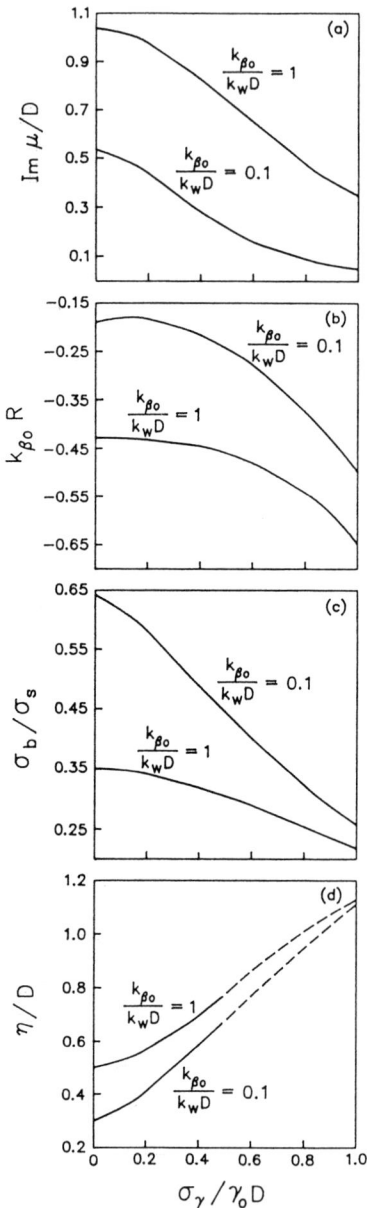

Figure 2: Plot of $\operatorname{Im}\mu/D$, $k_{\beta 0}R$, σ_b/σ_s and η/D versus $\sigma_\gamma/\gamma_0 D$ for $k_s\epsilon = 0.1$. The dashed portion of the curves in (d) lie in the regime where kinetic effects are expected to modify the efficiency.

Acknowledgment
The authors are grateful to Dr. D. G. Colombant for valuable help in the numerical solution of the dispersion and the variational relations. This work was supported by the Office of Naval Research and SDIO-IST.

References

[1] C. W. Roberson and P. Sprangle, Phys. Fluids B1, 3 (1989).

[2] D. C. Quimby and J. C. Slater, IEEE J. Quantum Electron., QE-21, 988 (1985).

[3] W. B. Colson, J. C. Gallardo and P. M. Bosco, Phys. Rev. A 34, 4875 (1986).

[4] J. C. Goldstein, T. F. Wang, B. E. Newnam and B. D. McVey, in *IEEE Proc., Particle Accel. Conf.*, 1987, E. R. Lindstrom and L. S. Taylor, eds., (IEEE, Washington, DC, 1987), p. 202.

[5] J. C. Goldstein and B. D. McVey, Nucl. Instrum. Methods Phys. Res. A 259, 203 (1987).

[6] J. K. Boyd, W. B. Colson and E. T. Scharlemann, Nucl. Instrum. Methods Phys. Res. A 272, 590 (1988).

[7] E. Jerby, Nucl. Instrum. Methods Phys. Res. A 272, 457 (1988).

[8] C. W. Roberson, Y. Y. Lau and H. P. Freund, in *High Brightness Accelerators*, A. K. Hyder, M. F. Rose and A. H. Guenther, eds., (Plenum, New York, 1988), p. 627.

[9] Y. Seo, V. K. Tripathi and C. S. Liu, Phys. Fluids B1, 221 (1989).

[10] C. W. Roberson and B. Hafizi, Nucl. Instrum. Methods Phys. Res. A 296, 477 (1990).

[11] L. H. Yu, S. Krinsky and R. L. Gluckstern, Phys. Rev. Lett. 64, 3011 (1990).

[12] C. W. Roberson and B. Hafizi, IEEE J. Quantum Electron. QE-27, 2508 (1991).

[13] H. P. Freund, R. C. Davidson and D. A. Kirkpatrick, IEEE J. Quantum Electron. QE-27, 2550 (1991).

[14] R. C. Davidson, *Physics of Nonneutral Plasmas*, (Addison-Wesley, CA, 1990), chap. 7.

[15] A. M. Sessler, D. H. Whittum and L. H. Yu, Phys. Rev. Lett. 68, 309 (1992).

[16] B. Hafizi and C. W. Roberson, Phys. Rev. Lett. 68, 3539 (1992).

[17] G. T. Moore, Opt. Commun. 52, 46 (1984).

[18] E. T. Scharlemann, A. M. Sessler and J. S. Wurtele, Phys. Rev. Lett. 54, 1925 (1985).

[19] M. Xie and D. A. G. Deacon, Nucl. Instrum. Methods Phys. Res. A 250, 426 (1986).

[20] P. Sprangle, A. Ting and C. M. Tang, Phys. Rev. Lett. 59, 202 (1987).

[21] P. M. Morse and H. Feshbach, *Methods of Theoretical Physics, part II,* (McGraw-Hill, New York, 1953), sec. 9.4.

[22] N. M. Kroll, P. L. Morton and M. N. Rosenbluth, IEEE J. Quantum Electron. QE-17, 1436 (1981).

[23] The role of the parameter $(k_w D/k_\beta)$ in an FEL in the gain-focusing regime was first pointed out in Ref. 10. It is shown in Refs. 10 and 12 that the wavelength λ and the emittance ϵ in an FEL in the gain-focusing regime are related by $\lambda = (k_w D/k_\beta) G(\sigma_b/\sigma_s) \pi\epsilon$, where G is a function of the filling factor, σ_b/σ_s. This expression is a generalization of the relation $\lambda = \pi\epsilon$, which is applicable to the case of a freely-expanding electron beam and a freely-diffracting radiation beam.

[24] B. Hafizi, A. Ting, P. Sprangle and C. M. Tang, Phys. Rev. A38, 197 (1988).

THE "NAIBEA" ACCELERATOR

M.S. Hussein
Nuclear Theory and Elementary Particle
Phenomenology Group, Instituto de Física
Universidade de São Paulo, C.P. 20516
01498, São Paulo, São Paulo, Brazil

ABSTRACT

The recently developed Nonlinear Amplification   of
Inverse Bremsstrahlung Electron Acceleration Concept   is
given adetailed qualitative description here.

INTRODUCTION

In a recent paper[1], M.P. Pato and I have developed
the NAIBEA concept of accelerating relativistic charged
particle to Gev or Tev[2] energies using lasers as the prin
cipal supplier of energy, and an alternating static electric field as a modulator of the transverse motion.   The
resulting machines were found more than a factor 20 smaller
than conventional ones. Generalization of the ideas   of
Ref.1) to the use of alternating static magnetic   field
and arbitrary state of polarization of the laser was made
by Hussein, Pato and Kerman[3]. In this communication   we
supply the qualitative considerations of NAIBEA[1].

GENERAL CONSIDERATION

The motivation for seeking alternatives to   conventional acceleration concepts is clear. The available elec
tric field usually employed to accelerate particles   is
about 10 MeV/m. Thus to reach e.g. the CEBAF energy (4
Gev) one needs a tube of about at 400 meters in length
(reduced when using superconductivity).

It is clear that to reach higher energies bigger and
bigger machines are required and one naturally   starts
seeking alternatives to the conventional concept.

A laser with a power of say P (W/cm²) supplies   an
electric field intensity of

$$eE_{laser} = \left[\frac{4\pi e^2}{c} P\right]^{1/2} \tag{1}$$

If we take for P, say, 10^{16} W/cm² , we obtain

$$eE_{laser} \cong 2.0 \times 10^5 \text{ Mev/m} = 0.2 \text{ Tev/m} \tag{2}$$

If one were to utilize even as little as 1% of the elec-

tric field intensity of the laser, one would end up with
an accelerator which is 100 times smaller than the conven
tional ones. The problem one faces here in that direction
of the electric field, to which the particle velocity must
couple, is perpendicular to the direction of propagation
of both the laser (the Poynting vector) and eventually
the particle.

To be able to accelerate the particle along the laser
Poynting vector, one must have a very small component of
the particle velocity along the laser electric field and
be sure to have the particle motion transverse to the
laser Poynting vector well contained: oscillatory. The
applied alternating static electric or magnetic field is
the needed degree of freedom to guarantee a transverse
oscillatory motion of the particle which must be within
the transverse extention of the laser pulse.

The above considerations allow us to make some useful
estimates.

Let me call the wavelength of the laser λ_0. The wave
length seen in the particle rest frame is obtained from
Doppler shift argument to be

$$\lambda = \lambda_0 \left(\frac{1 + \beta_0}{1 - \beta_0}\right)^{1/2} \quad , \quad \beta_0 = \frac{v_0}{c} \tag{3}$$

where v_0 is the initial velocity of the particle. The ve-
locity of the particle in the laboratory is γv_0, $\gamma = \sqrt{1 - \beta_0^2}$.
Thus the distance in the Lab. traveled by the particle
during a time lapse of $\frac{\lambda}{c}$ is

$$\Delta Z = \frac{\lambda}{c} \gamma v_0 = \lambda_0 \frac{\beta_0}{1 - \beta_0} \tag{4}$$

Therefore if λ_0 is several microns, ΔZ could be macroscopic
(several tens of cm's) if β_0 is close enough to unity
(relativistic injected particles). The above observations
allow the laser accelerator to be macroscopic.

If, for simplicity, we take ΔZ to be $n \lambda$, where n is
the number of cycles within the laser pulse (in fact, because of the acceleration, $\Delta Z > n \lambda$), and call the diameter
of the laser d_0 (which is not Doppler shifted), then after
$n \lambda$ encounters with the laser, the particle will have traversed (through the action of the applied alternating
static field) the laser n times. The gain in energy, $\Delta\epsilon$,
of the particle after traveling a distance of $n \lambda$ is then
given by

$$\Delta\epsilon \cong e \, E_{laser} \, (n \, d_0)$$

To be able to make sensible comparison with conventional accelerators, it is useful to introduce[4] an effective laser electric field intensity $E_{eff.}$, such that

$$\Delta \epsilon \equiv e\ E_{eff}(n\lambda) \tag{6}$$

We therefore obtain

$$eE_{eff} = e\ E_{laser}\ \frac{d_0}{\lambda}$$

or

$$e\ E_{eff} = e\ E_{laser}\ (\frac{1 - \beta_0}{\beta_0})\ (\frac{d_0}{\lambda_0}) \tag{7}$$

If we take e.g. $d_0 = 1$ mm, $\lambda_0 = 10^{-2}$ mm, $\beta_0 = 0.9999$, then, for the laser of Eq.1), we find

$$e\ E_{eff} = 2\ Gev/m$$

More realistic calculation, to be described bel ow, gives a smaller value for $e\ E_{eff}$.

Thus, we can say that Eq.(7) supplies an upper limit to the effective laser electric field intensity. Multiplying this intensity by the tube length supplies us with an upper limit to the gain in energy. It is an upper limit since: a) the distance $n\lambda$ is much smaller than the actual distance traveled by the particle during n encounters with the laser, and b) the laser field intensity E_{laser} has its maximum value in the center of the laser beam. It decays in the transversal plane, usually, as $\exp[-|x^2 + y^2|/w_0^2]$, where w_0 is the spot size of the presumed Gaussian beam[5]. We now turn to a general formula tion of NAIBEA[3].

FORMAL DEVELOPMENT

To simplify the presentation, we consider the units of mass in mc^2, the vector potential A in mc/e, the distance x in $1/k$, where k is the wave number and time, t in $1/\omega$; ω being the frequency. We take the direction of wave propagation and particle, acceleration to be along z. The Hamiltonian of the system, particle + laser + applied field is (note that we are working in the temporal gauge)

$$H = [1 + P_z^2 + (P_x - A_x)^2 + (P_y - A_y)^2]^{1/2} \equiv \gamma \tag{8}$$

in our units.

In Eq.(8) $\vec{A} = \vec{A}^{(0)}(t-z) + \vec{A}_{app}(t,z)$. Then $\vec{E}_{app} = -\dot{\vec{A}}_{app}$ and $\vec{B}_{app} = \vec{\nabla} \times \vec{A}_{app}$ are the external applied fields which add to those of the traveling laser pulse $\vec{A}^{(0)}(t-z)$.

Hamiltons equations follow from Eq.(8), i.e.

$$\dot{P}_z = -\frac{\partial \gamma}{\partial z} \ , \ \dot{P}_x = 0 \ , \ \dot{P}_y = 0 \tag{9}$$

$$\dot{\gamma} = \frac{\partial \gamma}{\partial t} = -\frac{\vec{P} - \vec{A}}{\gamma} \cdot [\frac{\partial \vec{A}^{(0)}}{\partial t} + \frac{\partial \vec{A}_{app}}{\partial t}]$$

Further reduction of $\dot{P}_z$ gives

$$\dot{P}_z = \frac{\vec{P} - \vec{A}}{\gamma} \cdot \left[\frac{\partial \vec{A}^{(o)}}{\partial \not{z}} - \frac{\partial \vec{A}_{app}}{\partial z} \right] \tag{10}$$

Combining $\dot{\gamma}$ and $\dot{P}_z$ we obtain

$$\dot{\gamma} - \dot{P}_z = \frac{1}{\gamma} \left[(\vec{A} \times \vec{B}_{app})_z - \vec{A} \cdot \vec{E}_{app} \right] \tag{11}$$

which, when integrated yields

$$\gamma = P_z + u \; ; \; u = u_o + \int_{-\infty}^{t} \frac{[(\vec{A} \times \vec{B}_{app})_z - \vec{A} \cdot \vec{E}_{app}]}{\gamma} dt' \tag{12}$$

$$u_o = \frac{1}{\gamma_o} (1 + \beta_o)^{-1} = \left[\frac{1 - \beta_o}{1 + \beta_o} \right]^{1/2}$$

At this point, we remark, as done in Ref. 1), that since $\vec{A}^{(0)}$ will be the dominant field, it is more convenient to use the phase $\varphi = t - z$ as an integration variable. Then, sice $\dot{\varphi} = 1 - \dot{z}$ and $\dot{z} = \partial \gamma / \partial P_z = P_z / \gamma$ and from (6), $1 - \dot{z} = u / \gamma$, we have $d\varphi / u = dt / \gamma$. Further, since $\dot{u} = (1 - \dot{z}) du / d\varphi = (u / \gamma) du / d\varphi$, we obtain from Eq. (12) the following

$$u^2(\varphi) = u_o^2 + 2 \int_{-\infty}^{\varphi} \left[(\vec{A} \times \vec{B}_{app})_z - \vec{A} \cdot \vec{E}_{app} \right] d\varphi' \tag{13}$$

and from $\gamma^2 = (P_z + u)^2 = 1 + P_z^2 + \vec{A}^2$

$$\gamma(\varphi) = \frac{1 + \vec{A}^2 + u^2}{2u} \; ; \; P_z = u \frac{dz}{d\varphi} = \frac{1 + \vec{A}^2 - u^2}{2u} \tag{14}$$

$$z(\varphi) = \int_{-\infty}^{\varphi} \left[\frac{1 + \vec{A}^2 - u^2}{2u^2} \right] d\varphi' = \int_{-\infty}^{\varphi} \left[\frac{\gamma(\varphi')}{u(\varphi')} - 1 \right] d\varphi' \tag{15}$$

The x and y coordinates of the particle are determined from the equations $\dot{P}_x = 0$ and $\dot{P}_y = 0$, which yield for the canonical momenta, $P_x = 0$ and $P_y = 0$, and thus the physical momenta are given by

$$P_x = \gamma \dot{x} = -A_x \quad ; \quad P_y = \gamma \dot{y} = -A_y \tag{16}$$

or

$$X(\varphi) = -\int_{-\infty}^{\varphi} \frac{A_x(\varphi')}{u(\varphi')} \, d\varphi' \tag{17}$$

$$y(\varphi) = -\int_{-\infty}^{\varphi} \frac{A_y(\varphi')}{u(\varphi')} \, d\varphi' \tag{18}$$

The set of equations (13-15) constitutes the Generalized NAIBEA equations of Ref.3. These equations are important generalizations of the NAIBEA equations of Ref. 1 in that: 1) the laser could be in any state of polariza tion, and have any pulse shape, and 2) the applied EM field E_{app} and B_{app} is quite general. Note that since $\vec{A}$ is defined to within an arbitrary constant, we have here full freedom in giving the electron non zero initial value of P_x and/or P_y (Eq. 16). The trajectory parameter, Q, which was introduced in Ref. 1 is here generalized to be a vector in the x-y plane and is defined by the equation

$$\frac{d\vec{Q}}{d\varphi} = -\vec{A} \tag{19}$$

It is a simple matter to show that the second derivative of $\vec{Q}$ can be written as

$$\frac{d^2\vec{Q}}{d\varphi^2} = -\frac{d\vec{A}^{(0)}}{d\varphi} + \frac{\gamma}{u}\vec{E}_{app} + \left[\frac{\gamma}{u}-1\right]\vec{\tilde{B}}_{app} \tag{20}$$

$$\vec{\tilde{B}} = \hat{3} \times \vec{B}$$

and the rate of change of γ with respect to φ

$$\frac{d\gamma}{d\varphi} = \frac{\vec{P} \cdot \vec{E}}{u} = -\frac{1}{u}\vec{A} \cdot \vec{E} = \frac{1}{u}\frac{d\vec{Q}}{d\varphi} \cdot \vec{E} \tag{21}$$

Note that Eq.(20) is a nonlinear second-order differential equation for $\vec{Q}$, since u and γ depend on $\frac{d\vec{Q}}{d\varphi}$ and $\frac{d\vec{Q}^2}{d\varphi^2}$. The solution of this equation completely determines the trajectory of the particle.

For γ to increase with φ, $\frac{d\vec{Q}}{d\varphi} \cdot \vec{E}(\varphi)$ must always   be positive (notice that $u(\varphi) > 0$)

$$\frac{d\vec{Q}}{d\varphi} \cdot \vec{E} > 0 \tag{22}$$

The above is the condition that guarantees that when $E_i(\varphi_j) = 0$, $P_i(\varphi_j) = \frac{dQ_i}{d\varphi_j}$ is also zero, then $\frac{d\gamma(\varphi_j)}{d\varphi} = 0$ and $\frac{d^2\gamma(\varphi_j)}{d\varphi^2} = 0$. Having such an inflection point in γ at φ_j (instead of a maximum) guarantees that γ keeps increasing for $\varphi > \varphi_j$. If B_{app} is taken to be zero as Ref. 1,   then $\vec{p} \cdot \vec{E} = \vec{p} \cdot \vec{E}^{(0)} + \vec{p} \cdot \vec{E}_{app}$. If $\vec{E}$ is taken in the y-direction, then we have $p_y E_y^{(0)} + p_y E_{app} > 0$. since $E_y^{(0)}$ is the dominant field except when passing through zero, the above   condition says that we use E_{app} to "fine tune" the sign of P_y so that it is always the same as that of $E_y^{(0)}$. The fundamental role of the applied field is to guarantee the validity of Eq.(22). This can happen even if $E_{app}/E^{(0)}$ or $B_{app}/B^{(0)}$ or both are much smaller than unity. The injection of electrons with a non-zero $P_x(0)$ or $P_y(0)$, albeit very small, is very important to set the machine to work. This is so since in the x-y plane the motion of the electron must be oscillatory in accordance with Eq.(12). The idea behind the NAIBEA acceleration is to optimally determine the applied field so that: a) the transversal motion of the particle is well confined to be within the transversal dimension of the laser beam, and b) the acceleration coefficient u (Eq.(13)) is made as close to zero as possible.

SPECIFIC CHOICES

We turn now to specific choices of the accelerator. We consider a linearly polarized pulse with $\vec{A}$ taken to be along the y direction. We first consider a   constant applied electric field, E_{app}, then

$$\vec{A} = \left[A_y^{(0)}(\varphi) - E_{app}t - P_y(0) \right] \hat{y} \tag{23}$$

With the $\vec{A}$ above used in Eqs. (20) and (21) we recover the NAIBEA equation of Ref. 1. Inversions of E_{app} are made   at appropriate values of z to assure the validity of Eq.(22), namely $p_y(n\pi) = 0$. This means that the applied field is inverted at φ_j's such that $\left.\dfrac{dp_y(\varphi)}{d\varphi}\right|_{\varphi_j = \frac{n+1}{2}\pi} = 0.$

We now replace E_{app} by a constant magnetic field along x. Then

$$\vec{A} = \left[A_y^{(0)}(\varphi) - B_{app}z - P_y(0)\right]\hat{y} \qquad (24)$$

The resulting NAIBEA equations are almost identical to those of Ref. 1 except for a change in sign of the second term in Eq.(8) of that reference (with E_{app} replaced by B_{app}). Further, Eq.(13) reduces to

$$u^2 = u_o^2 - 2\,B_{app}\,Q \qquad (25)$$

The y-component of the momentum is given by (Eq. 18).

$$P_y(\varphi) = P_y(0) - A_y^{(0)}(\varphi) + B_{app}z \qquad (26)$$

The continuous acceleration of the particle results if B_{app} is chosen so that $P_y(n\pi) = 0$. This requires changing the sign of B_{app} at appropriate places $\left(\varphi \simeq \frac{n+1}{2}\pi\right)$.

We consider the following numerical example. The initial value of γ, $\gamma_0 = 70$, $P = \frac{3}{2}\nu_o^2 10^{15}$ W/cm^2 for $\lambda_o = 10^{-3}$ cm. The parameter $\nu_o = 1$ refers to the maximum value of electric field of the pulse in our units. We also take nine cycles within the pulse. The shape of the pulse is taken to be a Gaussian, $A^2 = \exp(-\varphi^2/\Delta^2)$, with $\Delta = 3\pi$. The applied field intensity is taken to be 2.34 teslas which corresponds to $\sim 5 \times 10^{-5}$ that of the laser. The electrons are injected at an angle of $0.6°$ with respect to the z-axis $\left(P_y(0) \approx \frac{0.6}{180}\pi\gamma_o\right)$. We consider, as an example nine changes in the sign of the modulated applied magnetic field. Since we have not self consistently chosen the position of the field reversal we have not actually optimized the decrease of u in Eq.(13). In figure 1 we show the change of γ vs. z obtained by solving Eqs.21,20 and 15. The gain in energy is a factor of 35 over a distance (accelerator length) of seven meters! The accelerator length could be

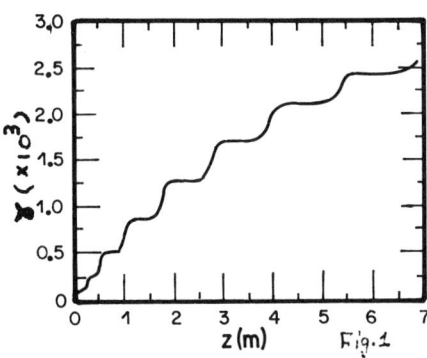

Fig.1

made smaller if full optimization is accomplished. It is interesting to calculate the effective field for this case. From Eqs. 1,5 and 6 we obtain the value e $E_{eff} \cong 150$ Mev/m.

The correspomding trajectory of the particle, confined in the (z,y)-plane is shown in figure 2.

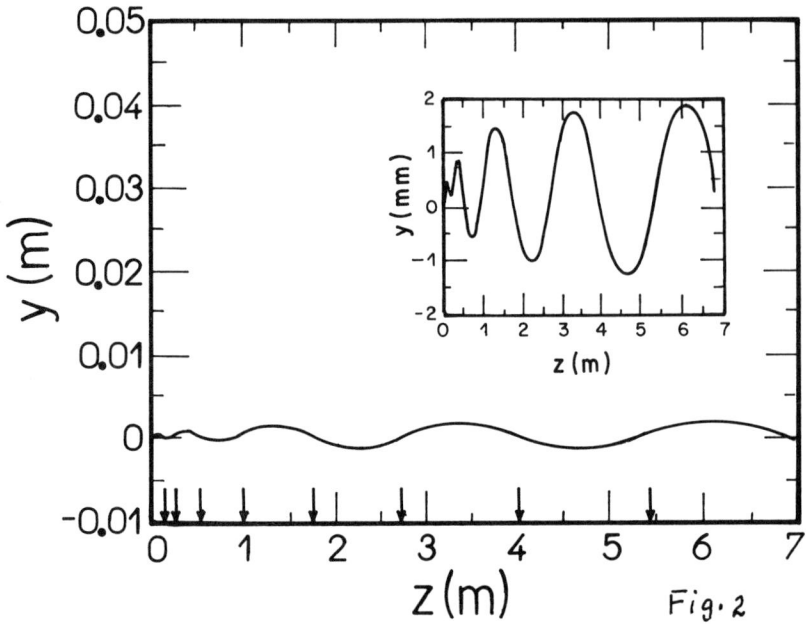

Fig. 2

The arrows indicate the positions where the applied magnetic field is inverted. These positions (in meters) are given in table 1.

1	2	3	4	5	6	7	8	9
0.15	0.24	0.51	1.0	1.75	2.75	4.0	5.4	6.9

Table 1

As a second example, we consider proton acceleration using our concept. We thus take for the initial energy of the protons, the value 0.5 Tev. This corresponds to an initial velocity of $V_O = 0.999998$ c. The laser power is taken here to be $P = 6.6 \times 10^{20}$ W/cm² with $\lambda_o = 5 \times 10^{-3}$ cm. The laser electric field intensity is e $E_{laser} = 49$ Tev/m. T initial injection angle is $\theta_o = 0.02°$. A Gaussian laser pulse with the above maximum intensity is considered with a width of 3π. The applied magnetic field intensity was taken to be 10 teslas, which is 5.14×10^{-7} E_{laser}. During the acceleration the sign of the applied field was changed

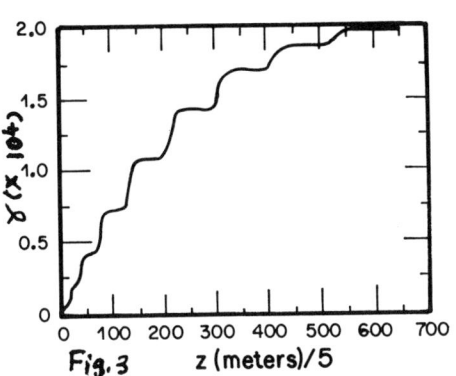

Fig. 3   z (meters)/5

11 times. The result are shown in figure 3.   The proton reaches an energy of 20 Tev, within a length of 3 km.   The effective laser electric field intensity is found here to be $e\,E_{eff} = 6.6$ Gev/m.   The trajectory of the proton in the y-z plane is shown in fig.4. The arrows indicate the positions where the last seven inversions of the field are made

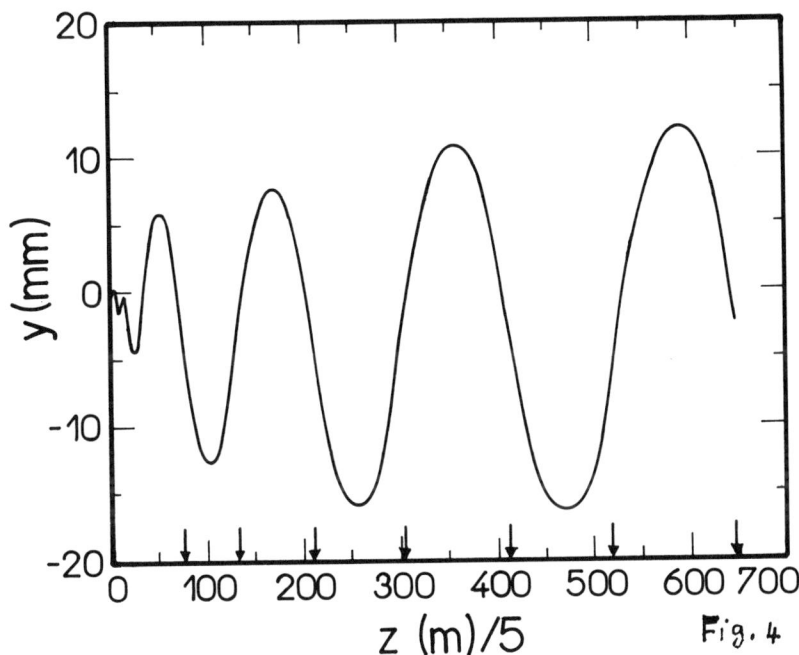

z (m)/5   Fig. 4

These positions as well as the first six of them are given in the table 2 below (in km).

1	2	3	4	5	6	7	8	9	10	11
0.031	0.05	0.085	0.18	0.375	0.665	1.045	1.515	2.052	2.63	3.23

Table 2

It is clear from Table 2 that there is ample space along the tube to further fine tune the particle trajectory.

As a final remark we mention that our NAIBEA accelerator can be further optimized by better determining the values of the available fine tune variables: θ_0 , B_{app} , and the inversion positions. The final aim is to have a rather "small", very high energy accelerator with the lowest possible value of B_{app} for a given laser power.

This work was done in collaboration with M.P. Pato and A.K. Kerman. The CNPq-Brasil and IBM do Brasil supplied a partial support.

REFERENCES

1. M.S. Hussein and M.P. Pato, Phys. Rev. Lett. <u>68</u>, 1136 (1992) and references there in.

2. M.S. Hussein and M.P. Pato, Modern Physics Letters B, in press (1992).

3. M.S. Hussein, M.P. Pato and A.K. Kerman, submitted to Phys. Rev. Lett. (1992).

4. M.S. Hussein and M.P. Pato, to appear in Modern Physics B (1992).

5. P.W. Milonni and J.H. Eberly, "Lasers" (John Wiley & Sons, New York, 1988) Chapter 14.

SUMMARY OF THE WORKING GROUP ON PLASMA ACCELERATORS

T. Katsouleas
University of Southern California
Los Angeles, CA 90089-0484

It has been nearly a decade since the workshops in this series began in Los Alamos. This year's workshop was the first in which experimental results of ultra high-gradient acceleration of injected electrons were presented. The new results set a tone that was maintained by the high-level of presentations made throughout the working group.

The plasma working group activities and this summary divide into two parts. In the first part, the group heard experimental and theoretical presentations of progress since the last Advanced Accelerator Concepts Workshop in Lake Arrowhead. In the second part, the working group addressed the charge: to identify an optimized design for a 100MeV plasma accelerator demonstration and a 1GeV demonstration.

I. PROGRESS REPORTS SINCE LAKE ARROWHEAD

A. Experimental Presentations

The working group first heard more detailed presentations of the experimental results that were given in the invited talks to the general workshop. Highlights are briefly summarized below.

1. UCLA Beatwave Acceleration Results (C. Clayton, K. Marsh, M. Everett)

Electrons at 2 MeV were injected into a plasma driven by a CO_2 laser beating on two lines (λ_0, $\lambda_1 = 10.6 \mu$, 10.3μ) that were resonant with a plasma at a density of 8.6×10^{15} cm^{-3}. Accelerated electrons have now been detected to energies above 9 MeV by a double apertured cloud chamber as well as surface barrier detectors. Up to 10^5 electrons were accelerated (approximately 1-10% of the injected number). The acceleration length is not measured directly, but Rayleigh length estimates and images of the laser-ionized plasma indicate it to be of the order 1-2 cm suggesting an accelerating gradient of over 500 MeV/m. The number of accelerated electrons and the amplitude of the plasma wave indicate that two key factors to the success of the experiment were (1) the uniformity of the plasma produced by laser ionization and (2) the shortening of the effective rise time of the lasers (in plasma periods) by lowering the plasma density. The shorter effective rise time increases the saturation amplitude of the plasma wave, reduces the density homogeneity requirement and delays the onset of ion motion and competing instabilities.

M. Everett also presented results obtained in a gas jet rather than a static gas-fill plasma source. Diagnostics of the plasma density showed that the uniformity was less in the gas jet ($\pm 15\%$) than in the static fill.

2. The Ecole Polytechnique Beat Wave Experiment (F. Amiranoff, P. Miné)

In the French experiment, a high phase-velocity plasma wave is driven by a YAG laser beating at $\lambda = 1.06\mu$ and 1.05μ in a 10^{17} cm^{-3} density plasma. Time-resolved measurements of the plasma wave amplitude are performed by laser-scattering. These demonstrated a gradient of $.3 - 1$GeV/m and are consistent with predictions of their model that saturation is due to ion instabilities. These results confirm that short laser pulses less than $1 - 2$ ion periods duration are needed. Their results also show the effective use of multi-photon ionization for controlling plasma density to within 1%. F. Amiranoff also applied his calculations of hydrodynamic ion motion to the case of the UCLA parameters. These showed that the laser pondermotive pressure pushes out on the order of 10% of the plasma density which was consistent with the observation that the density of neutral gas used in the UCLA experiment was about 10% higher than the resonant plasma density.

3. Japanese Experiments (A. Ogata, Y. Nishida)

A number of experimental results are described by A. Ogata in these proceedings. Particularly noteworthy are the highest-gradient particle-driven plasma wakefield acceleration experiments to date. Using a train of pulses from the 500 MeV linac at KEK, electrons in a later bunch of the train were accelerated by 10MeV over .5m by the plasma wakes produced by the earlier bunches. The experiment was prevented from achieving higher energy by the inability to control the phasing of the bunches and by the relatively low density plasma source ($\sim 10^{11}$ cm^{-3}).

Also reported was the first proof-of-principle thin plasma lens experiment which was performed at U. of Tokyo. A 20MeV beam was observed to focus by about 20% by passing it through a pre-formed plasma of density $\sim 10^{11}$ cm^{-3} and length ~ 30 cm. This reduction, though small due to their beam parameters, was in agreement with plasma lens theory.

At U. of Utsunomiya, Y. Nishida's group has performed demonstration experiments of surfatron-type acceleration ($E \times B$) in two devices. One is a plasma driven by microwave radiation and the other a slow-wave structure. The acceleration of 50 KeV electrons was mapped as a function of applied B-field in agreement with theory.

4. Tunnel-Ionization Experiments (W. Leemans)

Detailed scaling laws for the plasma densities and temperatures produced by a laser via tunneling ionization were presented. These show that short laser wavelengths are needed at high density in order to avoid a density "clamping" due to plasma refraction/diffraction of the laser.

5. Lawrence Livermore Laboratory Experiments with "High-Brightness Laser" (W. B. Mori)

W. B. Mori presented first results of laser-plasma interaction experiments with the Table-top Ten Terawatt (T^3) class laser at Livermore. Production of fully-ionized plasmas of densities up to 5×10^{18} cm^{-3} and 10 Rayleigh lengths ($\sim$ 3cm) were demonstrated by a 1μ laser.

6. U. of Michigan Experiments (D. Umstadter)

D. Umstadter presented results showing periodic self-focusing of a laser in a gas/plasma. The physics is rich and not completely identified, but it appears that the focusing is due to a Kerr effect in the gas and not due to a plasma focusing mechanism.

B. New Theoretical Results Presented

The Working Group heard progress on a number of theoretical fronts including the development of new simulation codes, new parameter regimes for old schemes and some entirely new schemes. Mechanisms of optically guiding laser pulses stimulated the greatest discussion. Some highlights are mentioned below.

1. New Simulation Codes (J. Krall, B. Breizman, C. Decker)

The development of 2-D non-linear fluid models is enabling the rapid exploration of the laser, beam and plasma parameter space of interest for plasma accelerators. J. Krall presented two new codes, one a fluid model which models self-consistently laser evolution and wake excitation and the other a Vlasov code for modeling beam-driven wakefields. B. Breizman presented results of a 2-D fluid code for modeling nonlinear wakes. It is nicknamed Novocode and is available to the community. C. Decker presented results of a PIC code with a "cyclic mesh" that follows a laser pulse enabling the study of laser propagation over many Rayleigh lengths with PIC simulations.

2. Optical Guiding (E. Esarey, P. Sprangle, J. Krall, P. Mora, C. Decker, N. Barov)

Simulations by the NRL group and P. Mora confirmed that relativistic self-focusing is not sufficient to guide lasers for the laser wakefield accelerator. For short pulses it was shown shortly after the Arrowhead workshop by Sprangle's group and verified in simulations at UCLA that the head of a laser pulse is not self-guided. At this meeting new results by NRL and Mora showed that longer pulses tend to focus into beamlets (separated by a wakefield wavelength) rather than to focus as a whole beam. These beamlets resemble the beats of the beatwave accelerator and resonantly excite a large wake. Alternative means of guiding lasers were presented including specially tailoring the laser pulse to be wide at the front and narrow at the back while maintaining a constant power throughout (Sprangle, Krall), and the use of performed channels of various types to confine the laser (NRL, UCLA groups).

Similar problems of guiding the head of short particle beams are encountered in the particle-driven plasma wakefield scheme. N. Barov presented a DC magnetic field as a means to overcome this.

3. Very Non-Linear Wakes [the "blowout regime"] (J. Rosenzweig, W. B. Mori)

It has recently been shown that wakefields driven by very intense particle beams ($n_b > n_0$) or lasers ($V_{osc}/c > 1$) exhibit unique accelerating properties. In such cases the driver is so intense it blows nearly all the electrons out leaving a nearly uniform ion column. After the pulse passes, the electrons rush back in to produce a large accelerating field E_z. As was demonstrated by Rosenzweig for particle drivers and Mori for laser drivers, the accelerating field E_z is independent of r and the focusing force W_r is linear in r. Both of these have positive implications for the quality of the beam accelerated by such a scheme.

Mori also presented results on the non-linear group velocity of laser pulses in 1-D. Rather paradoxically they find that the non-linear group velocity exceeds the linear group velocity, but due to distortion of the pulse the wake velocity is less than the linear group velocity. These results show that dephasing of accelerated particles can be the limiting process in the LWFA once optical guiding is achieved.

4. Plasma Lenses (S. Rajagopalan, P. Lai, R. Williams)

Investigations of plasma lenses included comparisons of uniform (Rajagopalan) adiabatic (Williams) and optimally tapered lenses (Lai). The simulations by Rajagopalan included cases with pre-ionized and beam ionized plasmas for SLC parameters. These are promising for a possible lens demonstration at the Stanford Final Focus Test Beam in 1994.

5. Ionization Wakes (P. Mori, D. Fisher)

A novel method of producing wakes from **long** pulses was proposed by the UCLA and the U. of Texas groups. The sudden ionization of a gas when a laser reaches an ionization threshold intensity gives an "effective" laser rise time equal to the ionization time (≤ 100 f sec). This affords a means of circumventing the need for very short lasers in order to study wakefield accelerators.

II. EXPERIMENTAL PLANS

With the growing body of experimental and theoretical background on plasma accelerators and the emergence of new laser and beam technologies, the possibility for exciting experimental progress has never been greater. Not surprisingly a number of proposals for experiments were presented. Ideas from many of these were incorporated into the Designs of the Working Group for 100MeV and 1GeV demonstration accelerators.

Among the planned experiments already in progress or under construction are those at Ecole Polytechnique, Osaka and UCLA. At Ecole Polytechnique a cw 10 MeV accelerator is being added to the beatwave experiment as an injector for particle acceleration experiments in late 1993. The Institute for Laser Engineering in Osaka will commit their main laser for one year to a laser wakefield experiment that is predicted to reach approximately 40 MeV. The initial experiments will be somewhat hampered by the lack of an injector, and $1 - 2$ MeV electrons must be generated and injected by hitting a solid target with a laser. At UCLA a collaborative effort between C. Joshi and C. Pellegrini has built a laser-driven 4.5

MeV injector that will be used for plasma lens and wakefield accelerator experiments.

New experiments were also described by A. Ting as part of an NRL/U. Michigan collaboration on laser wakefield acceleration. These would employ a T^3 class laser recently purchased at NRL and a .5 - 1 MeV injector. A. Kudryatsev described an intriguing plasma wakefield experiment that could be done in Novosibirsk with a modest level of outside support. This would employ the existing BEP storage ring beam as a driver, and could in principle accelerate electrons over a meter length by 100 MeV with the present beam if it were properly chopped and by 1GeV if it were also longitudinally compressed. J. Rosenzweig described the Argonne National Laboratory plans for wakefield acceleration to 100 MeV and 1 GeV, and their design is included in the next section.

III. THE WORKING GROUP CHARGE: 100 MEV AND 1 GEV DEMONSTRATION DESIGNS

The following designs are a product of the working group. The criteria for the designs were (1) they should be based on the availability of technology that is already existing or reasonably expected to exist within three years, and (2) they should involve a minimum of undemonstrated concepts. For example, optical guiding is **not** assumed in the laser-driven designs, although if it can be successfully accomplished it will greatly reduce the laser power requirements and increase the final energy possible. The designs considered are for the laser wakefield, beat wave and plasma wakefield schemes.

A. *100 MeV Laser Wakefield Accelerator (LWFA)*

The design equation summarizing the energy gain per Rayleigh length (Z) in the LWFA is

$$\frac{\Delta W}{Z_R} = \frac{.95 \text{ MeV}}{\pi} \frac{\lambda_\mu}{\tau_{FWHM}} P_{TW}$$

where λ_μ is the laser wavelength in microns, τ_{FWHM} is the full-width at half maximum of the laser pulse in psecs and P_{Tw} is the laser power in terawatts. To reach 100 MeV we choose a T^3 class laser based on the following parameters given by G. Morou and expected to exist within two years:

laser energy	1.5 Joules
pulse length	.1 psec
wavelength	.8 μ (Ti-Saphire)

From these we calculate the following:

plasma density	$n_0 = 3 \times 10^{17}$ cm^{-3}
plasma wave amplitude ($\delta n/n$)	.11
Rayleigh length $\times$ π	1.7 cm
Max. energy gain	110 MeV

Critical Issues: At 100 fsec, pulse compression of the laser pulse may need to be done in vacuum to avoid dispersion in the window of the plasma vacuum chamber. The plasma source should not be an issue as plasmas of this size and density have already been produced using laser ionization of a static gas fill. Finally, we note that a 1GeV LWFA with present/near term laser technology is possible **if** some form of optical guiding (channel or laser shaping) is successful (see J. Krall in these Proceedings for a sample design).

B. Beat Wave Designs

Guided by the experience gained in recent experiments, the working group opted for short pulse beat wave designs based on T^3 class short-pulse lasers operating on two lines. These take advantage of many of the best properties of previous beat wave and wakefield designs: Short pulses driving plasma waves over a few (e.g., N = 5 – 10) plasma cycles relax the power requirement (by approximately N) compared to LWFA while remaining short enough to avoid ion motion and relax plasma homogeneity requirements.

	100 MeV	1GeV
laser wavelengths	1.05μ, 1.06μ	
plasma density	10^{17} cm^{-3}	
laser pulselength	1.4 psec	4 psec
laser power	7 TW	14 TW
laser spot size (2σ)	100 μ	200 μ
Rayleigh length (Z_R)	.8 cm	3.1 cm
plasma homogeneity	± 20%	±7%
nomalized plasma wave amplitude ($\delta n/n$)	.35	.5
Gradient	100 MeV/cm	160 MeV/cm
Length (2Z_R)	1 cm	6.3 cm
Final Energy	100 MeV	1 GeV

Critical Issues: The technologies are in hand for these experiments. The 100 MeV laser parameters already exist, and the GeV example is an extension to two frequencies of the parameters of a laser currently operating at U. of Lemail in France (50 TW at 1 psec). The conversion of some of the laser power to a second frequency should be straight forward using Raman cells but needs to be demonstrated (see C. Joshi, et al., in these Proceedings). Self-focusing of the beating lasers (see P. Gibbon in the Proc. Lake Arrowhead Workshop, AIP Conf. Proc. No. 193, p. 126, 1989) may enable both cases to reach much higher energy.

C. Plasma Wakefield Accelerator Designs

The working group adopted the ANL designs as representative examples. The numbers in paranthesis are for the GeV design:

Drive Beam

Energy	50 MeV (200 MeV)
particle number (N)	5×10^{11}
n. emittance (ε_N)	400 mm - mrad
bunch length (σ_z)	1.5 mm
equilibrium spot size (σ)	230 μ

Plasma

density (n_0)	10^{14} cm^{-3}
n_b/n_0 (beam to plasma density)	6
gradient	10 MeV/cm
length	10 cm (1 m)
max. energy gain: Gaussian bunches (Shaped bunches @200 MeV)	100 MeV (1 GeV)

Critical Issues: Both of these designs drive the plasma toward the blow out regime ($n_b > n_0$) which has been simulated but not yet studied experimentally. Evolution of the beam is a key question: the head of the beam may need to be guided with a magnetic field to prevent erosion over meter lengths. The elctron hose instability has been predicted (D. Whittum, et al., Phys. Rev. Lett. **67**, 991, 1991) in the blowout regime and remains an open question.

D. Injector Requirements

Beam-Driven Designs:
For the electron-driven wakefield accelerators, the injected beam is assumed to be of the same type as the driver but with fewer particles. A witness beam with 1% of the drive beam number would not heavily beam load and would exceed the design goal of 5×10^8 electrons.

Laser Designs:
Approximate design criteria for the injector beam are given by the following considerations. The fraction of transmitted electrons is approximately $T \approx T_{11} \cdot T_\perp$, where $T_\perp \approx (a/\varepsilon)^2$, T_{11} is a longitudinal transmission factor ($\leq 1/2$ for an unphased

beam), ε is the beam emittance and a the plasma wave aceptance $a \le .2 \ [(\delta n/n)/\gamma]^{1/2}$ w_o, where w_o is the wave width, $\delta n/n$ the amplitude and γ the beam Lorentz factor. The maximum number of accelerated electrons is given by beam loading considerations: $N_{MAX} \approx 10^5 \ (\delta n/n) \ \sqrt{n_o} \ \pi w_o^2$ (for a 20% beam load; T. Katsouleas, et al., Part. Accel. **22**, 81 1987).   Based on the above and a design goal of 5×10^8 transmitted electrons (.5 $\times 10^8$ electrons/bucket $\times$ 10 buckets), we calculate that $\varepsilon_N \le$ a $\gamma \sim 6$ mm-mrad $\sqrt{\gamma}$ for the laser designs in Sec. B. With some safety factors ($\varepsilon \approx$ 1/2 a, T $\approx$ .1) we obtain the following injector requirements.

$$\varepsilon_N = 10\pi \text{ mm - mrad}, \qquad\qquad Q \approx 1nC$$
$$\text{pulse length} < 10 \ \lambda_p \approx 4 \text{ psec}, \qquad \gamma \ge 12$$

We note that the beam must be focused down to 100μ to match into the plasma wave. To safely trap and accelerate electrons to the design energy (i.e., to prevent T_{11} from becoming small), P. Mora has estimated that $\gamma \ge 4 - 8$ is required.

RESULTS OF THE EXERCISE

As a result of the exercise to develop 100 MeV/1GeV demonstration designs the following observations become apparent:

- 100 MeV laser and beam-driven plasma accelerators are in good position to be demonstrated within a 3 year time frame.

- Plasma sources are no longer a critical issue and appear to be solved by laser ionization of gasses.

- The design goal of 5×10^8 electrons is possible with a reasonable injector.

Finally, although costs have not been considered in detail, comparison of the 100 MeV and 1 GeV designs suggests that their relative cost is incremental (perhaps a factor of two). The current cost of T^3-class lasers is on the order of one million dollars as is the cost of a high-current electron injector. This suggests that the cost of 100 MeV/GeV plasma accelerators would eventually compare favorably to conventional approaches.

Acknowledgement

The assistance of several members of the working group in refining the design calculations are gratefully acknowledged as is B. Breizman for serving as co-chair.

Acceleration of Injected Electrons by the Plasma Beat Wave Accelerator

C. Joshi, C.E. Clayton, K. A. Marsh, A. Dyson,

M. Everett, A. Lal, W.P. Leemans[a], R. Williams[b],

T. Katsouleas[c], and W. B. Mori

Electrical Engineering Department, UCLA, Los Angeles, CA 90024
(a) Lawrence Berkeley Laboratory, Berkeley, CA 94720
(b) Florida A & M University, Tallahassee, FL 32307
(c) University of Southern California, Los Angeles, CA 90089

Abstract

In this paper we describe the recent work at UCLA on the acceleration of externally injected electrons by a relativistic plasma wave. A two frequency laser was used to excite a plasma wave over a narrow range of static gas pressures close to resonance. Electrons with energies up to our detection limit of 9.1 MeV were observed when 2.1 MeV electrons were injected in the plasma wave. No accelerated electrons above the detection threshold were observed when the laser was operated on a single frequency or when no electrons were injected. Experimental results are compared with theoretical predictions, and future prospects for the plasma beat wave accelerator are discussed.

I. Introduction

In the Plasma Beat Wave Accelerator[1] (PBWA), two co-propagating laser beams with frequencies and wavenumbers (ω_1, k_1) and (ω_2, k_2) resonantly drive a plasma wave with frequency $\omega_p = \omega_2 - \omega_1$ and wavenumber $k_p = k_2 - k_1$. The amplitude modulated electromagnetic wave envelope of the laser pulse exerts a periodic nonlinear force, the ponderomotive force, on the plasma electrons, causing them to bunch. The resulting space-charge wave has a phase velocity $v_\phi = \omega_p/k_p$ that is nearly equal to the speed of light c if $\omega_1 \approx \omega_2 \gg \omega_p$. If an electron bunch is now injected with a velocity close to this, it can be trapped and accelerated much in the same way as a surfer riding an ocean wave.

The major attraction of the PBWA is that extremely large acceleration gradients can be produced in the plasma wave. The longitudinal electric field associated with such a wave is given by $E = 0.96\varepsilon\sqrt{n_0}$ V/cm, where ε is the density modulation (n_1/n_0) and n_0 is the plasma electron density in cm^{-3}. Thus, for $\varepsilon = 0.1$ and $10^{15} < n_0(cm^{-3}) < 10^{17}$, accelerating fields of $\simeq 0.3 < E(GeV/m) < 3$ are possible. It is the purpose of our current experimental program to demonstrate such ultra-high gradient acceleration over a reasonable distance. This paper is organized as follows:

I.   Introduction
II.   Experimental Parameters
III.   Numerical Modeling of the Experiments
IV.   3-D Particle Trajectory Simulations
V.   Experimental Set-Up
VI.   Results
VII.   Plasma Beat Wave Accelerator in the 100 MeV - 500 MeV Range
VIII. Conclusions

II. Experimental Parameters

Our experiment consists of four major components: the CO_2 laser driver to act as the electromagnetic energy source; the plasma to convert the transverse oscillating field of the laser into the longitudinal oscillating field of a plasma wave to accelerate particles; a pre-accelerated bunch of electrons for injection into the plasma wave; and diagnostics to detect the accelerated electrons, plasma wave and plasma characteristics. In Table 1 we give experimental parameters for the laser, the plasma, and the electron beam (injector). These parameters will be used throughout the rest of the paper unless otherwise noted.

1. Laser

Wavelengths	10.59 μm, 10.29 μm
Energy per Line	60 J, 15 J
Electron Quiver Velocities/c	0.17, 0.07
Spot radius	150 μm
Rayleigh range $2z_o$	1.3 cm
Risetime	150 ps
FWHM	300 ps

2. Plasma

Density	8.6×10^{15} cm^{-3}, Hydrogen
Plasma Frequency ω_p	5.2×10^{12} s^{-1}
Plasma wave wavelength	360 μm
Lorentz Factor γ_{ph}	34
Gradient for a 10% Wave	1 GeV/m
Source	Tunnel ionization

3. Injector

Energy	2.1 MeV ($\gamma = 5.2$)
Emittance	6π mm-mrad
Micropulse Frequency	9.3 GHz
Electrons per micropulse	1.7×10^7
Spot Radius	125 μm

Table 1

III. Numerical Modeling of the Experiment

Before we describe the experimental set-up and the results, we present the results from a simple one-dimensional model that includes plasma formation and the beat wave growth and saturation via relativistic detuning[2]. The plasma is assumed to be formed via tunnel ionization[3] of the neutral gas which in our model is atomic hydrogen. In this model the plasma density increases as

$$\frac{dn(t)}{dt} = \omega(E,t)[n_0 - n(t)], \qquad (1)$$

where n(t) is the time dependent electron plasma density, n_0 is the initial equivalent monoatomic neutral gas density and $\omega(E,t)$ is the laser dependent Keldysh tunneling ionization rate given by[4]

$$\omega(E,t) = \frac{4me^4}{\hbar^3} \left[\frac{E_i}{E_H}\right]^{5/2} \xi \, \exp\left[-\frac{2}{3}\left[\frac{E_i}{E_H}\right]^{3/2}\xi\right]. \qquad (2)$$

Here E_H and E_i are the ionization energies (potential) of hydrogen and the atom in question, $\xi = E_a/E(t)$ where $E_a = m^2 e^5/\hbar^4 \simeq 5.21 \times 10^{11}$ V/m is the atomic unit of electric field, and E(t) is the amplitude of the applied electric field. We also assume that only a small fraction of the total number of electrons leave the ionized volume and that the resultant space charge is sufficient to confine the remaining electrons.

Figure 1 shows the calculated evolution of the plasma density (normalized to neutral gas density) as a function of time assuming atomic hydrogen gas. The laser pulse had a 150 ps (linear) risetime, a full width of 300 ps, and a peak intensity of 1.2×10^{14} W/cm^2. Figure 1(a) shows the growth of the laser intensity, in terms of its normalized electric field, $v_{osc}/c = eE/m\omega c$. As shown in Fig. 1(b), ionization begins at $\simeq 13$ ps and a fully ionized plasma is formed within 10 ps after this. The peak density of the fully ionized gas in this example is 2% less than the exact resonant density. Here resonant density is that density at which the plasma frequency $\omega_p = (4\pi n_0 e^2/m)^{1/2}$ is equal to the difference frequency $\Delta\omega \equiv \omega_2 - \omega_1$. Once the density is such that the beat frequency is within the bandwidth of this harmonically driven oscillator the plasma wave grows rapidly. The amplitude of the plasma wave is given by solving the equation[5]

$$\frac{d}{dt}\left(\gamma\frac{d\xi}{dt}\right) + \omega_p^2\xi = \alpha_1\alpha_2\frac{c^2\Delta k}{2}\sin\left(\Delta\omega t - \Delta k\xi\right). \tag{3}$$

Here $\Delta k = k_2 - k_1$, ξ is the Lagrangian displacement variable describing the wave oscillating quantity, $\gamma = (1 - (\dot{\xi}/c)^2)^{-1/2}$ is the relativistic Lorentz factor, and $\alpha_{1,2} = eE_{1,2}/m\omega_{1,2}c$ are the normalized quiver velocities in the laser fields of amplitudes $E_{1,2}$. As can be seen from Eq. (3) the plasma responds to the driving force as an anharmonic oscillator, the anharmonicity coming from the mass dependence of ω_p. As the amplitude of the plasma oscillation grows the natural frequency (ω_p) decreases due to the relativistic increase in the electron mass. Consequently, the driver and the plasma wave gradually shift in phase and saturation occurs when the two are $\pi/2$ out of phase with one another. In Fig. 1(c) we see that for the conditions of Fig. 1 (a,b), the plasma wave begins to grow once the ionization is complete at 23 ps and saturates at 75 ps, well before the peak of the laser pulse. The saturated amplitude is $\varepsilon = n_1/n_o = 0.13$. The accelerating electric field is related to the amplitude by $eE_{ac} = \varepsilon mc\omega_p$ which in this case (see Table 1 for experimental parameters) is 1.2 GeV/m even though the plasma density was 2% less than the resonant density.

Figure 2 shows the saturated value of the plasma wave amplitude and the time to saturation versus density, normalized to resonant density. The time to saturation includes the time it takes to ionize the gas. It can be seen that the optimum plasma density is $\approx 3\%$ above the resonance value. In this case the plasma wave builds up to a peak amplitude of 37% in about 120 ps. In terms of ion plasma periods saturation occurs in $< 2\nu_{pi}^{-1}$. Thus any instabilities involving ions do not have a chance to grow significantly. Therefore, damping of the beat-driven plasma wave by mode coupling[6] to the ion-acoustic wave

(SBS) is probably not too severe, and neither self-focusing[7] nor the modulational instability[8] are very important. Thus, this simple 1-D model is very useful for narrowing the parameter space for the experiment. From Fig. 2 it can be seen that for our conditions a density mismatch of $\pm 4\%$ is still allowable for ε to be greater than 5% for our laser intensities. It should be noted that while exact density resonance is not critical, plasma homogeneity is. Spatially inhomogeneous plasmas will produce phase mixing and the acceleration process will be degraded.

IV. 3-D Particle Trajectory Simulations

We have carried out computer simulations of trajectories of injected particles in a finite diameter electron plasma wave to see the expected number and energy spectra of the accelerated electrons in our experiment. The model has been extensively tested and documented in a previous publication[9]. For our experimental conditions, it is reasonable to assume from Equation (2) that several Rayleigh ranges will be fully ionized by the laser. The laser intensity relative to the peak is assumed to vary as $1/(1 + z^2/z_o^2)$, assuming gaussian optics. In this case, the axial envelope of the plasma wave amplitude just prior to saturation is given by

$$E_{accl}(z) = E_{max} / \left(1 + \frac{z^2}{z_o^2}\right), \tag{4}$$

where $E_{max} = \varepsilon m c \omega_p / e$ as before. The radial profile of the accelerating field is assumed to be gaussian and the radial field component $E_r(r, z, t)$ has a form given in Ref (9), equation 19. The resonant density plasma is assumed to exist over a length $\pm 3z_o$ where $z_o = 8$ mm. We inject 2.5×10^5 electrons with energy $\gamma = 5$ and an emittance of 6π mm-mrad and collect electrons exiting from the plasma wave in a f/10 cone. In Fig. 3 we plot a histogram of the number of electrons accelerated versus energy (γ) for a 15% wave. The spectrum peaks at the injected energy of $\gamma = 5$ and exponentially decreases to an energy of approximately $\gamma = 16$. The distribution then shows a long tail, with energies up to $\gamma = 25$ possible for this condition. In addition to the accelerated electrons, electrons have also been decelerated by the plasma wave to energies of $\gamma = 3$ or 4. Also note that in the simulations 2.5×10^5 electrons were injected as opposed to in the experiment where a microbunch contains up to 1.7×10^7 electrons. Thus, each electron in Fig. 3 represents 70 accelerated electrons in that energy bin in the experiment.

V. Experimental Set-Up

i) Laser System
The experimental apparatus is shown schematically in Fig. 4. The CO_2 laser system used in this experiment was specifically designed for the beat-wave

iii) Electron Beam Transport System and Acceptance Considerations

Figure 6 shows the various components of the beam transport line and the calculated trajectory of the electron beam spatial envelope. The emittance of the beam at the final focus was estimated by measuring the beam spot size (x_o) and divergence (x_o'). The emittance is given by $\epsilon = \pi x_o x_o'$, and was measured to be 6π mm-mrad. In our experiment the laser beam diameter (and therefore the plasma wave diameter) and the plasma wave wavelength have comparable dimensions. This means that the radial field $E_r(r,z)$ of the plasma wave is on the same order of magnitude as the longitudinal field. Our 3-D particle trajectory simulations (Section IV) discussed earlier have shown that these radial fields can cause a large net loss of injected electrons for finite emittance beams. To quantify the coupling efficiency, the well known concepts of accelerator acceptance and beam matching can be applied to our experiment[12]. To minimize the electron loss, the beam emittance should be smaller than the PBWA acceptance. It can be shown that the electron transmission by the plasma wave can be expressed as

$$T = \left(\frac{I_{trans}}{I_{in}}\right) = \left(\frac{a}{\epsilon}\right)^2 \tag{5}$$

for $T < 1$ where a is the acceptance given by

$$a = \sqrt{\frac{n_1}{n_0}\frac{1}{\gamma}\frac{2R}{9}}. \tag{6}$$

Here n_1/n_o is the plasma wave amplitude, γ is the injector energy and R is the spot size (radius) of the laser. Thus, for n_1/n_o of 10% for example, $a \simeq 4.6\pi$ mm mrad and $T \simeq 0.60$.

Even if the beam emittance is small, the electron beam must be properly focused into the accelerator or some electrons will be lost. Therefore, in the experiment the electron beam is focused to a spot size (x_o) somewhat smaller than the radius of the plasma wave. The optimum beam divergence (x_o') is given by $x_o' = a/x_o$. For our conditions the optimum value of $x_o' = 40$ mrad. The final focusing of the electron beam is f/10 which gives $x_o' = 50$ mrad. This condition is close to the optimum but can vary with n_1/n_o and errors in the calculated acceptance value.

iv) Electron Detection

As shown in Fig. 4, both the accelerated and the non-accelerated electrons enter a variable-field imaging electron spectrometer. The non-accelerated electrons are dumped onto low density plastic. Lead shielding reduces the flux of background x-rays reaching the electron detectors thereby reducing the background or noise levels to a value negligibly small compared to the signal levels ultimately obtained. The accelerated electrons exit the vacuum through a 25 μm thick Mylar window and are detected either electronically by one or more silicon surface barrier detectors (SBD) or photographically by the tracks they leave in a cloud chamber. The SBD has a 600 μm copper window which is "light tight" to soft x-rays but still "transparent" to energetic electrons with greater than 3 MeV energy. Along with a charge-sensitive preamplifier, the SBD produces about 20 mV per electron in the range 1-10 MeV. The preamplifier saturates at around 2.5 V thus limiting the number of detectable electrons to about 125 before saturation.

The electron track detector is a diffusion cloud chamber[13] which uses su-persaturated methanol vapor in 1 atm of air to form visible tracks as electrons ionize the molecules along their path. The lead-shielded chamber has a 6 μm thick Mylar window over a 3 mm entrance hole located about 5 cm from the vacuum window. Electron scattering in the two windows and the intervening air is calculated to be less than 1° for 5 MeV electrons. The tracks are recorded with a CCD camera. A uniform, 260 G magnetic field can be applied to the active region of the cloud chamber from coils located outside the lead shielding. The cloud chamber has a much wider dynamic range than the SBD. At low electron fluxes, one can count individual electron tracks while at high fluxes, the brightness of the cloud chamber image was calibrated (with the 2.1 MeV linac beam) to be roughly proportional to the flux of electrons, at least up to 400 electrons/mm^2.

v) Optical Diagnostics

The plasma wave was probed with three optical diagnostics as well as with an electron beam. The density fluctuations associated with the beatwave were probed directly with near-forward scattering of the CO_2 laser itself into the Stokes sideband[14]. This light is collected in a forward annulus with a half-angle range of f/10 and sent to a spectrograph/pyroelectric array combination for time-integrated spectral measurements. The other two optical diagnostics, Thomson scattering and backscatter, do not probe the beatwave directly, but rather its mode-coupled daughter wave[6] at around $k_s \simeq 2k_1$ due to the presence of an ion acoustic wave at $2k_1$ from stimulated Brillouin scattering (k_1 is the wavevector of the stronger 10.591 μm pump). In Thomson scattering, a frequency-doubled YAG probe beam of 5 nsec duration is focused to a 50 μm diameter spot within the 10 mm core of the plasma mentioned earlier. The geometry has been chosen to k-match to waves with $k = (2.0 \pm 0.5)\, k_1$. Thus, in addition to mode coupled waves, the diagnostic will pick up scattered light from stimulated Compton[15] and Brillouin[16] driven density fluctuations. The scattered light is sent to a spectrograph/streak camera combination with 0.5 Å and 25 psec resolution. The third optical diagnostic is the time-integrated spectrum of the backscattered CO_2 light captured on a frame-grabbed pyroelectric camera.

VI. Results

i) Plasma Formation

As discussed previously, one of the most crucial requirements for the beat wave acceleration experiment is the formation of a homogeneous, magnetic field free plasma of the required density (8.6×10^{15} cm^{-3}) that is at least 1 cm long. Such plasmas were produced by tunnel ionization of a static gas in the interaction chamber. That the plasmas were indeed produced by tunnel ionization and the physical properties of such plasmas are documented in various publications[3,17]. Here we show evidence of plasma size, density and temperature. The visible emission from the plasma was imaged onto a CCD camera on every data shot. One such image is shown in Fig. 7(a). This shows the overall plasma size to be more than 20 mm long in the axial direction with a core region of the plasma emitting uniform intensity visible radiation that is about 10 mm long. We point out that while this visible image is not a conclusive measure of either plasma density or homogeneity (being dependent on both density and temperature), it is reasonable to assume that in this core region (that is less than $2z_o$ long) both are fairly constant. Thus we will take the interaction length to be 10 mm.

The density and temperature of the plasma were estimated from the Compton scattered spectrum part of the Thomson scattering diagnostic[17]. One example is shown in Fig. 7(b). The fitting of the observed scattered light spectrum to the theoretical spectrum is only possible at early times and gives in this case a density of $n_e = 6 \pm 1 \times 10^{15}$ cm^{-3} and a parallel (or longitudinal) temperature of $T_\parallel = 75 \pm 10$ eV. Therefore, $k\lambda_d \simeq 1.0$ in this case. The measured density is in reasonable agreement with the fill pressure of the hydrogen gas (100 mT in this case). Clearly in this strongly damped regime[17] the scattering diagnostic cannot be expected to give the plasma density with any accuracy greater than this. We also found that strong refraction of the laser beam due to plasma lensing effects takes place for fill pressures corresponding to fully ionized densities of $\simeq 2 \times 10^{16}$ cm^{-3}. Thus we assume that fully ionized plasmas with densities less than 10^{16} cm^{-3} can be produced simply by varying the static pressure of the fill gas in the chamber, with plasma densities between 6.6×10^{15} cm^{-3} and 1.3×10^{16} cm^{-3} corresponding to pressures in the range 100-200 mT.

ii) Electron Acceleration

Since the object of this experiment was the demonstration of acceleration of externally injected electrons by the beat-excited relativistic plasma wave, we carried out several important control experiments to ensure that what we were expecting to observe was a real PBWA effect. We carefully calibrated the SBDs for distribution of noise signals from background x-rays as shown in Fig. 8(a). Here the SBD was set to observe 3.5 MeV electrons. At higher energy settings of the electron spectrometer magnet the noise level spectrum shifts to lower mean values of the signals than shown in Fig. 8(a). Figure 8(b) shows the cloud chamber image when the injected electrons are dumped into the beam dump plastic but the cloud chamber is set to look at 5.2 MeV electrons. In this picture only low energy electrons produced by photoionization by x-ray noise photons can be seen. No energetic electrons were observed on either detector when none were injected into plasmas produced by either single or dual frequency beams. This implies that in our experiments there is no contribution to the observed signal due to self-trapping of the background plasma electrons[18]. This is in contrast to recent Osaka experiments[19] where high energy electrons were reportedly observed with single and dual frequency laser-plasma interactions. In addition, for single frequency illumination no acceleration of electrons was observed even with injection of 2.1 MeV electrons. This implies that with dual frequency illumination and injection, accelerating fields associated with waves generated by Raman forward scattering have a negligible contribution to the observed signal[20]. Finally, no signal was detected when the laser and the electron beam were simultaneously fired into an evacuated chamber (no plasma) as expected.

In our experiments, injected electrons were seen to gain energy only when a two frequency laser beam was fired in a static gas over a narrow range of pressures. Fig. 9(a) and (b) show tracks from two laser shots for which the plasma density was nearly resonant (143 mT fill pressure), with the laser operating on both wavelengths, and with external electrons injected at 2.1 MeV. For these shots the electron spectrometer was set to direct 5.2 MeV ± 0.5 MeV electrons into the cloud chamber and 5.9 MeV electrons into the SBD. In Fig. 9(a), the magnetic field of the cloud chamber is turned off and the image shows hundreds of accelerated electrons entering the cloud chamber. On this shot, the SBD signal was saturated. In order to estimate the number of electrons in this particular image, we calibrated the brightness of the image against other images produced by using the 2.1 MeV injector as the source. This was done in the following way: First the electron linac current was measured vs. the filament current (cathode temperature). Then the electron spectrometer magnet current was adjusted to send the linac electrons directly into the cloud chamber (Fig. 9(c,d)), and the cathode temperature of the linac's gun was varied to obtain cloud chamber images in the brightness range of those measured (as in Fig. 9(a)). The image in Fig. 9(c) shows an electron flux of 60 electrons/mm^2, while Fig. 9(d) corresponds to a flux of 480 electrons/mm^2. Using this method we estimate that the image shown in Fig. 9(a) is produced by 150 electrons/mm^2 flux at 5.2 MeV with an error bar of ±30%. From knowledge of the dispersion and vertical imaging of the electron spectrometer, we estimate that this flux corresponds to 2×10^4 electrons in the 5.2 ± 0.25 MeV energy bin.

In Fig. 9(b), we show at this same energy setting of the electron spectrometer the effect of applying a 260 G external magnetic field to the cloud chamber. The magnetic field of the cloud chamber is used to bend these electrons inside the cloud chamber. The solid line superimposed on this figure is the calculated trajectory of a 5.2 MeV electron using the relativistic gyroradius formula

$$r_e = \left(\gamma^2 - 1\right)^{1/2} mc^2/eB. \tag{7}$$

It can be seen that the trajectories of the central particles match closely with the calculated trajectory and therefore confirms that these electrons have indeed gained energy from the beat wave - plasma interaction. The other trajectories have the same radius of curvature but are due to electrons entering the cloud chamber at various angles due to the imaging property of the dipole magnet of the electron spectrometer and the finite emittance of the beam. The SBD signal on this shot was 360 mV or about 18 electrons (assuming a SBD sensitivity of $\simeq 20$ mV/electron). This agrees closely with the number of tracks seen in approximately the same solid angle in the cloud chamber image, confirming that the SBDs are indeed "seeing" electrons. This diagnostic, we believe, gives irrefutable proof that the externally injected electrons have gained substantial energy in our experiment.

The experiment was repeated over a range of fill pressures with the electron spectrometer set to observe various energies. Fig. 10(a) shows a summary of the signals obtained on the SBDs at various acceleration energies. After each shot with a high electron signal, a noise spectrum such as the one shown in this figure was taken by firing up to 80 linac shots. The noise spectrum in Fig. 10(a) was taken at a spectrometer setting of 3.5 MeV. As mentioned earlier, at higher energy settings the noise level shifts to even lower values. The measured electron signals are clearly many standard deviations larger than the noise, and seven shots saturated the detector. Fig. 10(b) shows the signal levels on the SBD's at various energies versus the fill pressure. As stated previously, at these laser intensities, fully ionized plasmas are formed up to a plasma density of $n_e \leq 2 \times 10^{16}$ cm^{-3}. Therefore we assume that the desired density up to this limit can be obtained by changing the measured gas pressure. From Fig. 2(b) the calculated pressure range over which a 5% amplitude or greater plasma wave is expected is 128 - 137 mT, with the exact resonance being at 131.5 mT and the optimum pressure (for compensating the relativistic detuning) at 136 mT. Figure 10(b) shows that the signals are essentially in the noise below 131 mT, rising rapidly around 140 mT. However, there are not enough shots above 148 mT to confirm the exact location of the peak of the electron signal. The data points bracketed with an arrow indicated that on these shots the SBD signals were saturated. Thus these data points are indicative of the lower bound of the electron signal. However, the experiment appears to work best with about 5-10 mT higher pressure than the expected optimum. This may be a result of hydrodynamic expansion of the hydrogen plasma which could lower the density by $5 - 10\%$ during the 70-100 ps growth time of the plasma wave. The maximum magnetic field of the electron spectrometer limited the highest observable electron energy to 9.1 MeV. Even at this energy, we were still able to saturate the SBD. Thus many electrons gained at least 7 MeV in traversing long plasma wave, implying an accelerating gradient of more than 0.7 GeV/m which corresponds to a wave amplitude of at least 8%.

iii) Estimate of Number of Accelerated Electrons

In our experiment the injected electron microbunch is not deterministically synchronized to any particular point in the plasma wave with an accuracy (jitter) better than ± 100 ps. As a result, the numbers of accelerated electrons can vary greatly from one laser shot to the next. However, by taking a large number of shots one can estimate the maximum number of electrons observed in a particular energy bin and thereby crudely calculate the number of accelerated electrons in a particular energy interval. From the cloud chamber data, we found 2×10^4 electrons at 5.2 MeV within a 0.5 MeV energy bin. The saturated signal at 9.1 MeV indicates that we had at least 125 electrons (2.5 electrons/mm^2 on the SBD) at that energy. Assuming that the accelerated electron spectrum between 5.2 MeV and 9.1 MeV is exponentially decreasing from 2×10^4 to 125, and using the known dispersion of the spectrometer, we estimate that the total number of electrons in this part of the accelerated spectrum is $4 \pm 2 \times 10^4$. Our micropulse contained about 4.25×10^6 electrons in both accelerating and focusing phases of the buckets of the plasma wave. Thus approximately 1% of the available electrons were observed to be accelerated. A better estimate based on the single shot measurement of the complete spectrum of the accelerated electrons is currently under way.

iv) Correlation with Optical Diagnostics: Thomson and Raman Scatter

The measured electron signals were correlated with optical diagnostics to further confirm that the acceleration observed was associated with the relativistic plasma waves. The time resolved Thomson scattered spectrum is shown in Fig. 11(a). It shows a broad band of scattered frequencies between $0 < (\omega_o - \omega_s)/\Delta\omega < 1.5$ which are due to SCS[21] or SRS. Here, ω_o and ω_s are the scattered and incident frequencies of the Thomson probe respectively, and $\Delta\omega \equiv \omega_2 - \omega_1$ is the beat frequency. There is in addition a narrower but much more intense feature at a frequency shift corresponding to $\Delta\omega$. This feature generally shows two temporal bursts. The first has a typical growth time of 50-70 ps and is thought to be due to the mode coupling of the relativistic plasma wave[6] from the still growing ion acoustic wave. As expected, the strongest electron signals were observed on shots when the first burst at $\omega_o - \omega_s \equiv \Delta\omega$ was intense while SCS was still occurring. This is shown in Fig. 11(c). Not all laser shots with a strong burst at $\Delta\omega$ produced electron signals because of the 60% probability of the synchronization of the electrons and the plasma wave mentioned earlier. The second peak, which persists after SCS is over, is thought to arise from counter-propagating optical mixing which excites a slow phase velocity plasma wave. Such a wave cannot accelerate relativistic electrons significantly.

The backscatter spectrum, Fig. 11(b), typically shows two distinct peaks on two frequency laser shots. The first peak is shifted from the laser frequency by $\approx 0.8\Delta\omega$ (10.80-10.86 μm), and is present on all two frequency shots. Its location varies 5-10% from shot to shot, and its origin is still being studied. The second peak is shifted from the laser frequency by $\approx \Delta\omega$ (10.9 μm). It appears only on shots in which a plasma beat wave is excited. As with the Thomson scattering diagnostic, the strongest electron signals were observed when the feature at ≈ 10.9 μm was relatively intense (see Fig. 11(c)). The correlation of the electron signal with strong scattering signals in the Thomson and the backscatter spectra further supports the notion that the electrons are accelerated by the relativistic plasma wave excited by collinear optical mixing.

v) Correlation with Forward Scatter

Beat wave excitation of a relativistic plasma wave is a four wave process. In addition to the pump waves ω_1 and ω_2 coupling to Stokes sideband at $\omega_1 - \omega_p$ and anti-Stokes sideband at $\omega_2 + \omega_p$ must be considered[14]. For small plasma wave amplitudes the generation of Stokes and anti-Stokes sidebands can be considered as Bragg scattering of the incident pump waves from the plasma wave, which acts as a moving phase grating. Thus, the observation of the Stokes and anti-Stokes sidebands in the same direction as the pump waves is diagnostically important because not only does it confirm that the relativistic plasmon has been excited but it also can give an independent estimate of the amplitude length product, $(n_1/n_o \cdot L)$, of this wave. This same quantity is also responsible for the energy gain of electrons since $n_1/n_o \propto E_{acc}$. In our experiments we monitored the Stokes sideband at $\omega_1 - \omega_p$. However, because of the choice of the particular wavelengths and the high laser intensities, we found that rotational Raman scattering in the 100 meters of air (between the laser output window and the vacuum chamber) produced Stokes signals that were comparable to the Stokes radiation generated by the plasma.

To try to separate the rotational Raman light from the Stokes light generated from the plasma wave, we blocked a central f/14 cone in the forward scattered light and collected radiation in an annulus out to f/10.5. Since the Stokes radiation should originate in a smaller spot size than the incident radiation, it should therefore be diverging at a larger angle. The resulting Stokes shifted signal had a similar correlation with electrons to that observed with the backscatter and Thomson scatter (Fig. 11(c)), suggesting that the Stokes radiation did originate from the plasma, but more work is needed to conclusively resolve this issue. At any rate, large Stokes signals always accompanied any electron signals indicating at least that we have a two frequency laser pulse where both frequencies are temporally overlapped with one another.

vi) Estimate of Plasma Wave Amplitude From Thomson Scattering Measurements

As seen in a previous section, the electron energy gain measurements imply that plasma waves of amplitude n_1/n_o greater than 8% were excited over a 10 mm length in this experiment. Since the maximum energies observed were detection system limited, this is likely to be an underestimate. Estimates of the relativistic plasma wave amplitude, however, can also be made from three features of the Thomson scattered spectra: harmonic components, the mode coupling feature at $\Delta\omega$; and sub harmonic components. We shall briefly discuss each of these.

When the relativistic plasma wave becomes large it can steepen. Its steepened waveform can be decomposed into Fourier harmonics of the plasma frequency as[22]

$$\frac{n_e - n_o}{n_o} = \sum_{m=1}^{\infty} \left(\frac{n_1}{n_o}\right)^m \frac{m^m}{2^{m-1}m!} \tag{8}$$

where n_1/n_o is the amplitude of the fundamental component. Thus

$$n_2/n_1 \simeq n_1/n_o \simeq \frac{n_e - n_o}{n_o}, \tag{9}$$

$$n_2/n_o \simeq (n_1/n_o)^2. \tag{10}$$

In other words, the ratio of the second harmonic to the fundamental gives a measure of the absolute density pertubation of the fundamental.

In the experiment, the Thomson scatter does not measure the Fourier components of the relativistic plasma wave directly but its coupling to the ion acoustic wave due to stimulated Brillouin scattering. However, if the relativistic wave contains harmonics ($n\omega_p, nk_p = n\Delta k$), then since $n\Delta k \ll 2k_{1,2} \simeq k_{ion}$ the coupled mode or the quasi-mode will also show the harmonic components and the amplitude ratio of these harmonic components of the mode-couple waves will reflect the amplitude ratio of the harmonic components of the relativistic wave. Figure 12(a) shows a scan through the frequency axis of a Thomson scattered spectrum. This shows a clear large peak at $\Delta\omega$ and a much smaller peak at $2\Delta\omega$. The square-root of the ratio of the scattered power in the second harmonic to the fundamental is proportional to n_1/n_o. Using this technique we estimate n_1/n_o between $15 - 30\%$ at the point probed by the Thomson scatter diagnostic.

A second method to determine the plasma wave amplitude is the absolute amplitude of the coupled mode itself[6]. To obtain the saturated amplitude of the coupled mode we solve the modified version of Eq. (3) as follows:

$$\ddot{E} + \omega_p^2 \left(1 + \varepsilon \sin k_i x\right) E + c_e^2 E'' = \frac{1}{2}\omega_p^2 \alpha_1 \alpha_2 \sin\left(k_p x - \omega_p t\right) \tag{11}$$

where E is the wave electric field normalized to the cold plasma wave-breaking amplitude $E_{cold} = mcw_p/e$; εn_o is the depth of the periodic ion pertubation, k_i is the ion ripple wavenumber, and $c_e^2 = 3kT_e/m_e$. If we assume that the maximum amplitude of the plasma wave is much less than that due to relativistic detuning we can neglect relativistic effects and obtain the saturated amplitude of the relativistic plasma wave $E(\Psi_o)$ and its coupled mode $E(\Psi_1)$ as

$$E_{sat}\left(\Psi_o\right) = \alpha_1\alpha_2 f\left(p\right)/\varepsilon \tag{12}$$

and

$$E_{sat}\left(\Psi_1\right) = \alpha_1\alpha_2/\varepsilon, \tag{13}$$

where

$$f(p) = P + \left(1 + p^2\right)\left(2 + p^2\right)^{-1/2} \simeq 2p \tag{14}$$

for $p^2 \gg 1$ and $p = (3/\varepsilon)\left(k_i\lambda_D\right)^2$. Thus

$$E_{sat}\left(\Psi_0\right) = 2pE_{sat}\left(\Psi_1\right). \tag{15}$$

Taking $\varepsilon = 0.2\% - 0.4\%$, $T = 180$ eV, $n_e = 8.5 \times 10^{15}$ cm^{-3} and $E_{sat}\left(\Psi_1\right) = 3 \times 10^{-4}$ we obtain $E_{sat}\left(\Psi_1\right)$ or $(n_1/n_o)_{sat} = 22 - 44\%$. Clearly, this model is only valid for n_1/n_o smaller than 37% which is the relativistically limited maximum amplitude. However, the point is that the amplitude one obtains using this method is comparable to the harmonic ratio method and is greater than 10%.

Finally, we have observed sub-harmonic components $\left(\Delta\omega/2, \frac{3}{2}\Delta\omega\right)$ in the Thomson scattered spectrum (Fig. 12(b)). We have previously shown in a theoretical paper that a large amplitude relativistic plasma wave can develop such subharmonic components in the presence of an ion ripple[23]. Although we do not infer the plasma wave amplitude from these components we can assume that their observation implies a large amplitude plasma wave.

VII. Plasma Beat Wave Accelerator in the 100 MeV - 500 MeV Range

In this section we would like to propose the next phase of experimental program on the Plasma Beat Wave Accelerator. The main goals of such a program are:

1) The demonstration of substantial energy gain for the injected electrons. The design goal is a 100 MeV to 500 MeV accelerator.

2) Acceleration of a substantial number of electrons, perhaps greater than 10^8.

3) Demonstrate that the injected electrons are deeply trapped; consequently the microbunches that are formed have a small energy spread.

Our proposal uses a ≈ 1 μm laser as a driver. In order to keep the plasma length small, we employ a higher plasma density than has been hitherto employed in 1 μm beat wave experiments at the Rutherford Laboratory (U.K.)[24] and at Ecole Polytechnique (France)[25]. This avoids the need for employing any laser beam guiding techniques to keep the laser beam focused over a length much greater than that in vacuum. The parameters of the laser beam and the plasma as well as those of the plasma wave are summarized in Table 2.

1. Laser

Wavelengths	1.053 μm, 1.086 μm
Wavelength difference	285 cm^{-1}: Vibrational Raman
Energy per line	10 J
Pulse duration (FWHM)	4 ps
Peak intensity per line	10^{16} W/cm^2
Electron Quiver Velocities/c	0.1
Spot radius	90 μm
Rayleigh range z_o	2.4 cm

2. Plasma

Plasma wave wavelength	35 μm
Density	9×10^{17} cm^{-3}, Hydrogen
Source	multi-photon ionization

3. Plasma Wave

Final Energy	> 500 MeV
γ_{ph}	33
n_1/n_o	0.25
Accelerating gradient	22.8 GeV/m
Acceleration length	$\approx$ 2.4 cm

Table 2

The two laser lines needed to resonantly drive the plasma wave are generated by Raman shifting the driver (1.053 μm) wavelength on a vibrational transition of 285 cm^{-1}. The oscillator for this laser will probably be a Ti-sapphire oscillator working on 1.053 μm wavelength. Such an oscillator has been shown to generate sub 100 fs pulses[26]. The oscillator pulses are stretched on a grating to a few hundred picosecond long injection pulse. Such a pulse can then be amplified in a series of Nd:phosphate amplifiers to give an output energy of up to 30 J. This amplified pulse can then be Raman shifted using an oscillator-amplifier Raman downshifting technique[27]. Conversion efficiency of up to 30% into the first Stokes sideband seems possible. If the laser output at 1.053 μm is not of the desired spatial quality, it may be necessary to use the first and the second Stokes to do the beat wave excitation[27]. These frequency shifted lines should be focusable to a nearly diffraction limited spot size.

Fig.13(a) shows the amplitude (n_1/n_0) of the excited beat wave versus the laser pulse width for a constant laser energy of 10 J per line and a plasma density of 9×10^{17} cm^{-3}. The laser pulse is assumed to be triangular with a risetime equal to the full-width at half-maximum. It can be seen that for a 4 ps pulse the saturated plasma wave amplitude is in excess of 0.25, limited by relativistic detuning at the exact resonance. If one precompensates for the relativistic detuning by introducing an initial density mismatch (as shown in Fig.13(b)) one can obtain plasma wave amplitudes in excess of 30%. Assuming a plasma wave of 25%, the accelerating field is 22.8 GeV/m, which implies that an energy gain of greater than 500 MeV can be obtained over an acceleration length in a uniform field of only ≈ 2.2 cm. As can be seen from Table 2, for a laser spot diameter of 180 μm, the Rayleigh range is 2.4 cm. Thus for a peak intensity of 10^{16} W/cm^2 at the beam focus we expect a fully ionized plasma over $\approx$ two Rayleigh lengths due to multi-photon ionization.

A deterministically synchronized short pulse, high current electron injector is necessary for this experiment. The ideal injector is a photoinjector driven RF gun giving an output bunch containing 1 nc or greater in a < 4 ps long pulse. Recently such an injector has been developed at UCLA[28] and is available for the experiment. Preliminary measurements show that up to 1.5 nC of charge with energy of $\approx$ 4 MeV can be extracted in an approximately 10 ps long bunch from a gun that is only about 10 cm long.

We are currently studying all the relevant design issues associated with carrying out such an experiment.

VIII. Conclusions

In summary, high-gradient acceleration of externally injected electrons by a relativistic plasma wave excited by a two frequency laser beam has been demonstrated. No accelerated electrons were observed when none were injected or when the laser was operated on a single frequency. However, electrons up to our detection limit of 9.1 MeV were observed when 2.1 MeV electrons were injected in a plasma wave excited (over a narrow range of static gas pressures close to the resonance) by a dual frequency laser beam. The accelerated electron signal was found to be correlated with indirect measurements of the amplitude of the plasma wave using Thomson scattering, Raman backscatter, and Raman forward scatter. The energy gain of the electrons suggests plasma wave amplitudes of at least 8% over a 10 mm interaction length. Thomson scattering measurements indicate plasma wave amplitudes up to 15-30%, offering the possibility of measuring even greater energy gains in future experiments.

These experiments have shed light on what is important in future experiments. First, that tunnel or multi-photon ionized plasmas are homogeneous enough for coherent, macroscopic acceleration. Second, the laser pulse should be short to reduce the ion effects (typically $\tau < 3\nu_{pi}$) and the modulational instability. Third, the peak laser intensity should be such that $I\lambda^2 \sim 2 \times 10^{16}$

W/cm^2 · μm^2 in order to get substantial beat wave amplitudes. These considerations will play an important role in the designs of a 100 MeV and a 1 GeV plasma beat wave accelerator experiment. The present experiment has the potential to show energy gains of 15-20 MeV. The maximum energy gain in this experiment will be limited by the length of the plasma available for acceleration due to the laser intensity profile. By going to a 1 μm laser and a two orders of magnitude higher plasma density, it may be possible to reach hundreds of MeV energy gain.

Acknowledgments

The authors would like to acknowledge useful discussions with Drs. W. B. Mori and P. Mora, and Professors J. M. Dawson and T. Katsouleas and thank M. T. Shu and D. Gordon for their technical assistance. This work is supported by the U. S. Department of Energy under grant no. De-FG03-92ER40727.

References

1. T. Tajima and J.M. Dawson, Phys.Rev.Lett. 43, 267 (1979).

2. The 1-D simulation is a Fortran code that integrates the differential equation (3) by the Runge-Kutta method. It includes tunnel ionization and relativistic saturation effects.

3. W.P. Leemans et al., Phys.Rev.Lett. 68, 321 (1992).

4. L.D. Landau and E.M. Lifshitz, Quantum Mechanics (Pergamon Press), 1978.

5. M.N. Rosenbluth and C.S. Liu, Phys.Rev.Lett. 29, 701 (1972).

6. C. Darrow et al., Phys.Rev.Lett. 56, 2629 (1985).

7. C.E. Max et al., Phys.Rev.Lett 33, 209 (1974).

8. D. Pesme et al., Laser and Particle Beams 6, 199 (1988).

9. R.L. Williams et al., Lasers and Particle Beams 8, 427 (1990).

10. E. Yablonovitch and J. Goldhar, App.Phys.Lett 25, 580 (1974).

11. C. Clayton and K. Marsh, Rev.Sci.Inst., to be published, March 1993.

12. K. Marsh, UCLA PPG-1469, 1992.

13. A. Langsdorf, Jr., Rev.Sci.Inst. 10, 91 (1939).

14. B. Cohen et al., Phys.Rev.Lett 29, 581 (1972).

15. W.P. Leemans et al., Phys.Rev.Lett 67, 1434 (1991).

16. C.E. Clayton et al., Phys.Rev.Lett. 51, 1656 (1981).

17. W. P. Leemans, Ph.D Thesis, UCLA, 1992.

18. D. W. Forslund et al., Phys.Rev.Lett 54, 558 (1985).

19. Y. Kitagawa et al, Phys.Rev.Lett 68, 48 (1992).

20. C. Joshi et al., Phys.Rev.Lett 47, 1285 (1981).

21. W.P. Leemans et al.,Phys.Rev.A (1992).

22. D. Umstadter et al., Phys.Rev.Lett 59, 292 (1987).

23. H. Figueroa and C. Joshi, Phys.Fluids 30, 2294 (1987).

24. A.E. Dangor et al., Phys.Scrip. T30, 107 (1990).

25. F. Amironov et al., Phys.Rev.Lett 68, 3710 (1992).

26. J. Kafka et al., IEEE J.Quant.Elec 28, 2151 (1992).

27. Y.R. Shen, The Principles of Nonlinear Optics (John Wiley and Sons), 1984.

28. S. Park et al, same proceedings.

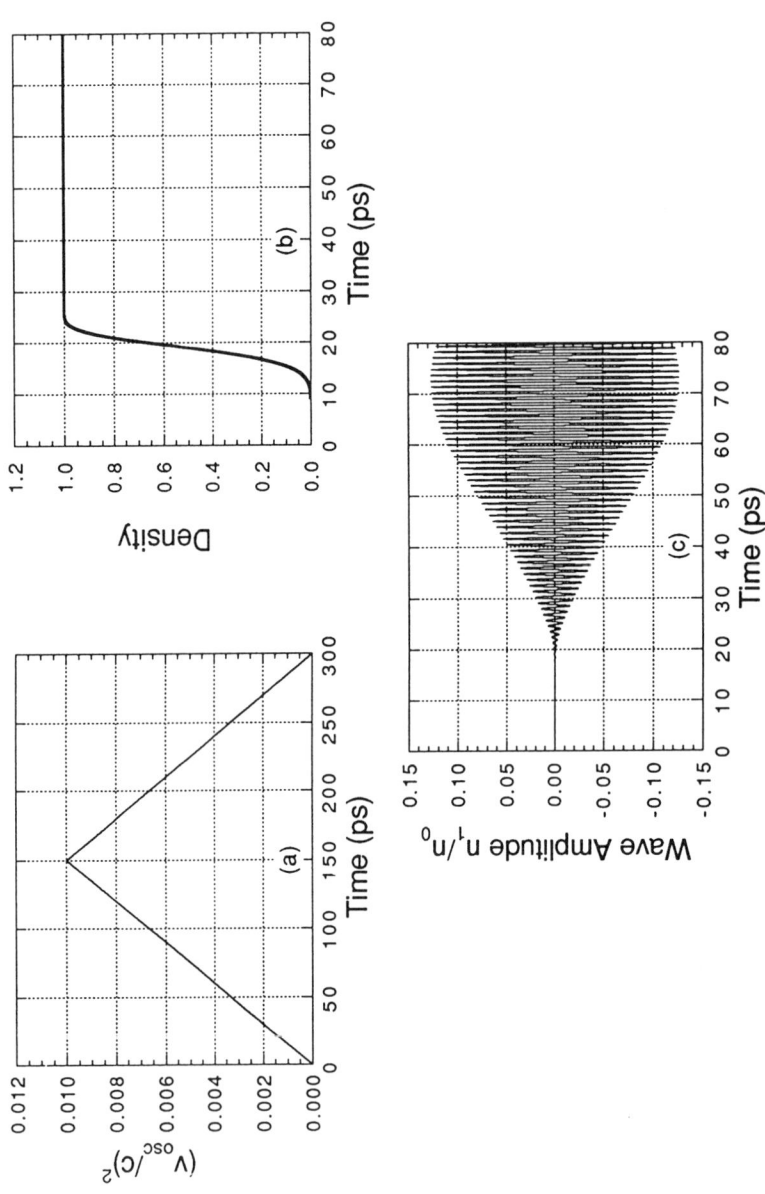

Fig. 1. Simulation of ionization and plasma wave growth. a) The intensity of one laser ($v_o/c = 0.1$) versus time. b) Plasma density due to tunnel ionization by the electric field of the laser. The density saturates at 23 ps due to 100% ionization of the neutrals. c) Growth and saturation of the relativistic plasma wave. In this example the fully ionized plasma has a density 2% below the resonant value where $\Delta\omega \equiv \omega_p$.

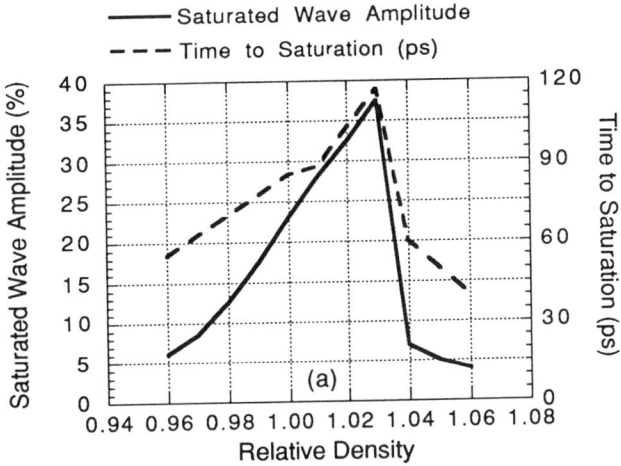

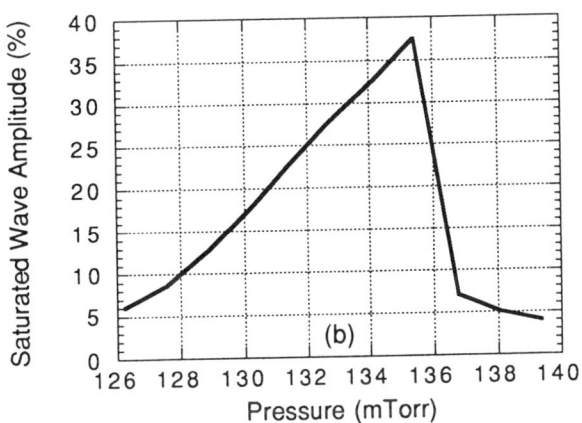

Fig. 2. Saturated value of the plasma wave amplitude $\varepsilon_{sat} = (n_1/n_o)$ and the time in psec to reach saturation versus plasma density relative to the exact resonant density. The conditions are the same as in Figure 1.

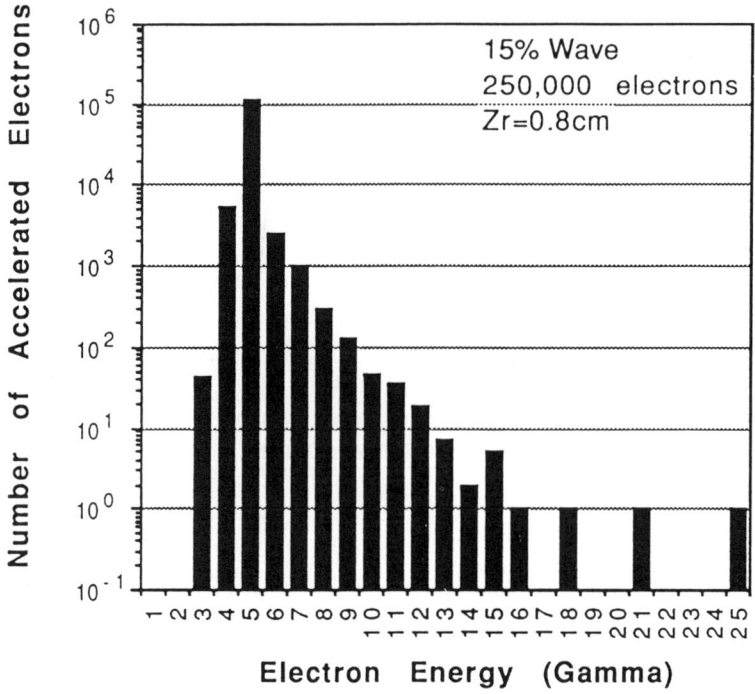

Fig. 3. Results of the 3D particle trajectory simulations which include longitudinal and radial fields of the plasma wave and finite emittance of the injected electron beam. Histogram showing number of particles in an energy bin with $\Delta\gamma = 1$ versus γ for a $n_1/n_o = 0.15$.

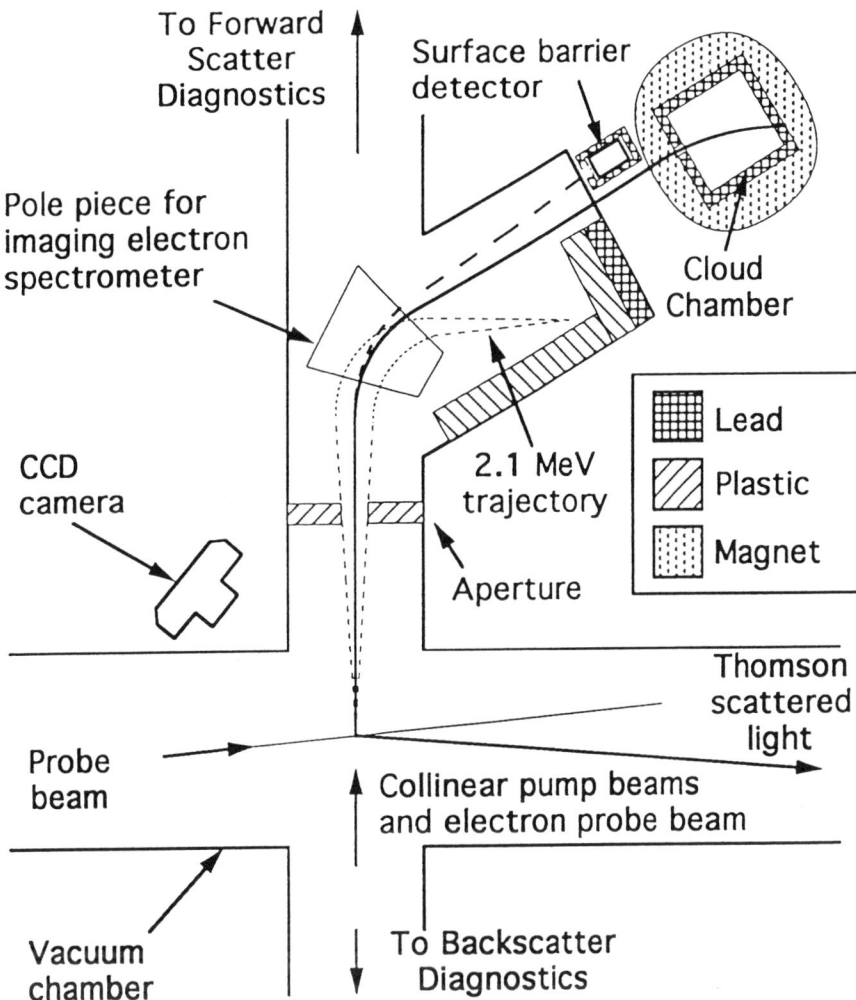

To Forward
Scatter
Diagnostics

Surface barrier
detector

Pole piece for
imaging electron
spectrometer

Cloud
Chamber

CCD
camera

2.1 MeV
trajectory

Lead

Plastic

Magnet

Aperture

Thomson
scattered
light

Probe
beam

Collinear pump beams
and electron probe beam

Vacuum
chamber

To Backscatter
Diagnostics

Fig. 4. Schematic arrangement of the electron acceleration experiment.

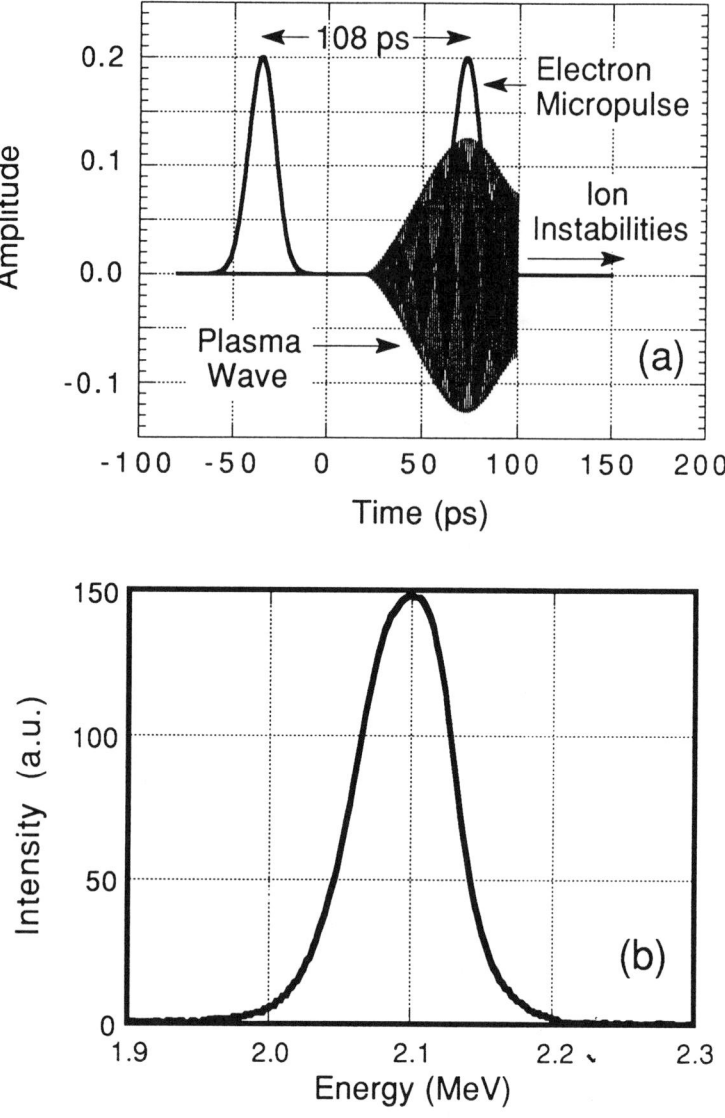

Fig. 5. a) Schematic of the micropulses (20ps wide), separated by 108 ps, and the plasma beat wave; b) the energy spectrum of the injected electrons.

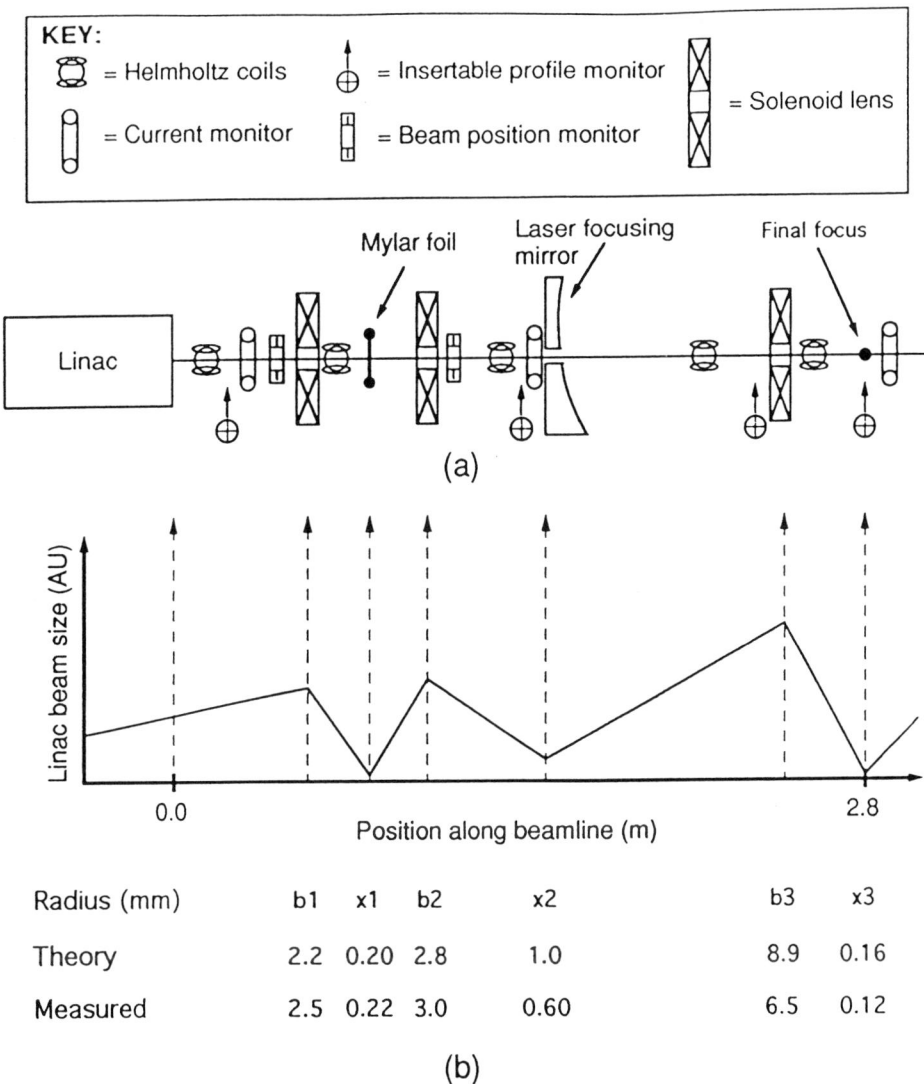

Fig. 6. a) The beam transport line and diagnostics. b) The measured and theoretical beam envelope size.

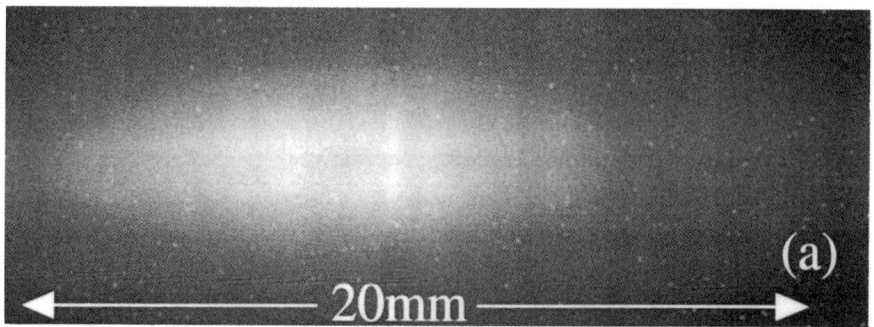

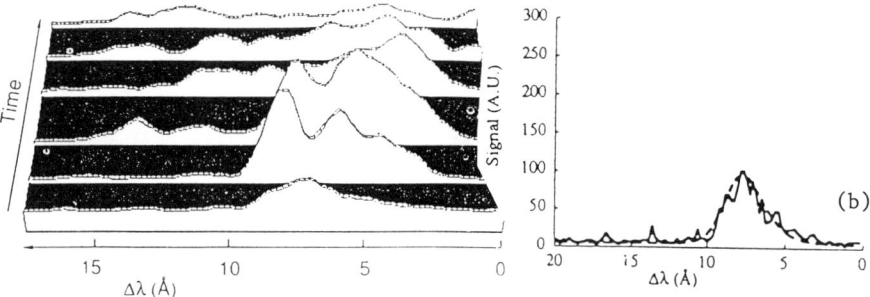

Fig. 7. a) The image of the visible radiation emitted by the plasma b) Line outs from Thomson scattering (single frequency) at $2k_1$ from electron fluctuations due to stimulated Compton scattering and a theoretical fit to this measured spectrum. The theoretical fit gives the plasma density $n_o = 6 \times 10^{15}$ and plasma longitudinal temperature $T_\parallel = 75$ eV.

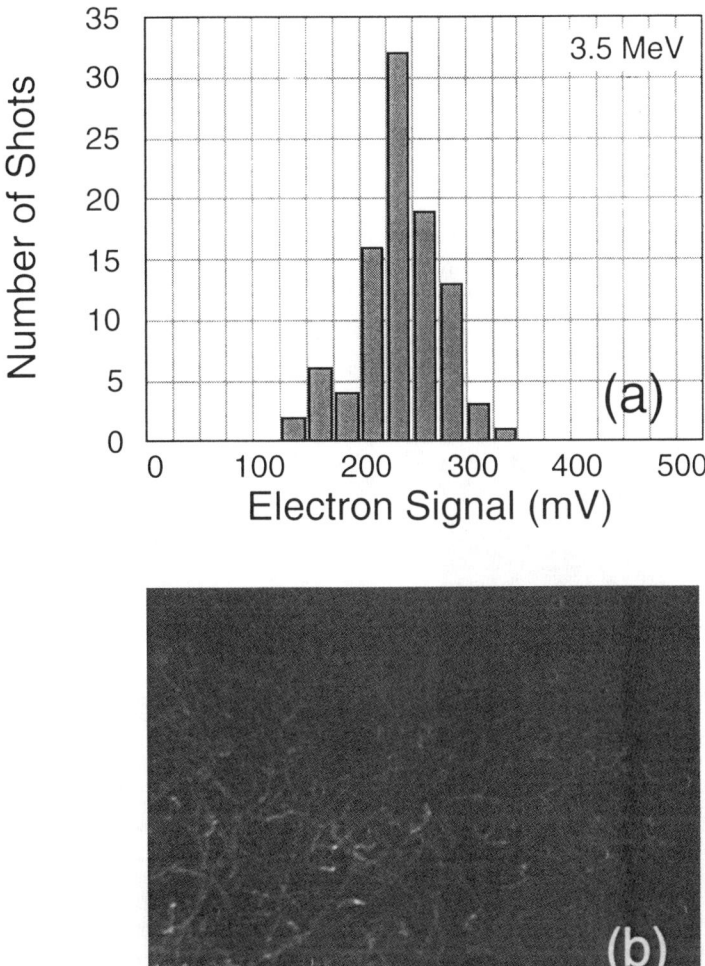

Fig. 8. a) The surface barrier detector signal levels due to statistical noise from x-rays. b) Cloud chamber image showing only signal produced by low energy electrons produced in the cloud chamber by photoionization by x-ray photons.

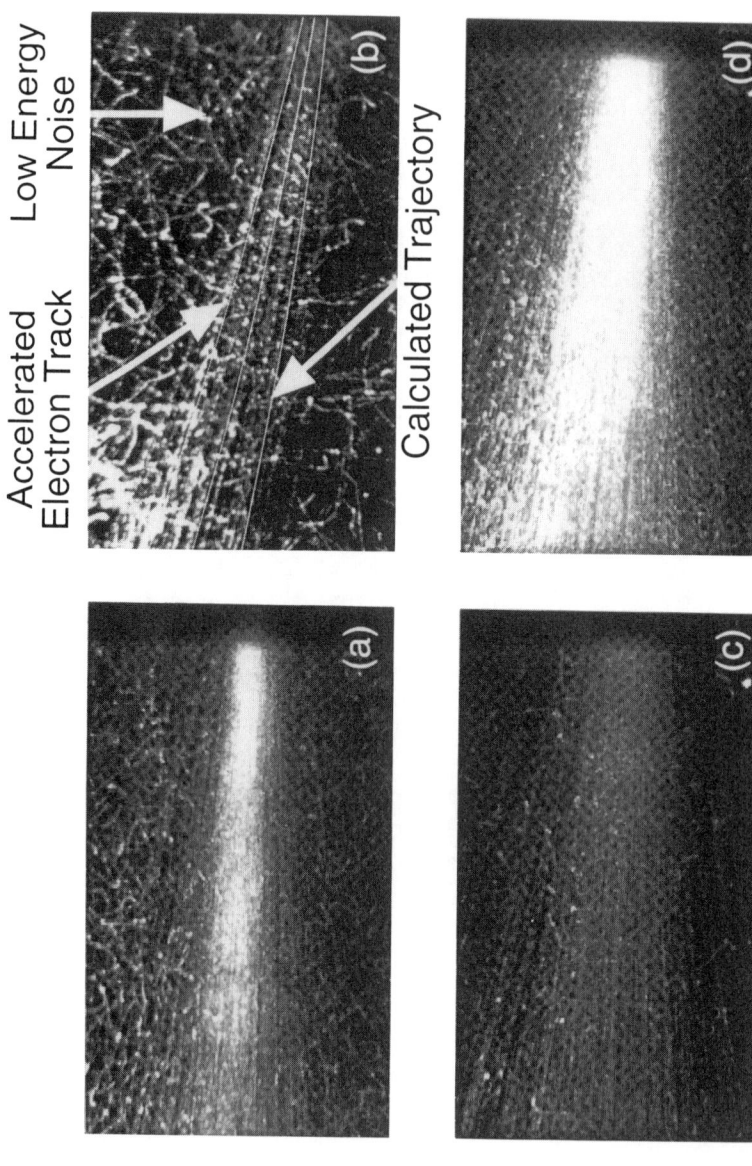

Fig. 9. a) Accelerated electrons trajectories at 5.2 MeV in the cloud chamber with no applied cloud chamber magnetic field ; b) Curved accelerated electrons trajectories at 5.2 MeV when a 260 G field is applied to the cloud chamber. In this image, the length is 7.25cm and the height is 3.4cm. c) The trajectories of the injected 2.1 MeV electrons in the cloud chamber (no B field), with a flux of 60 electrons/mm², and d) 480 electrons/mm².

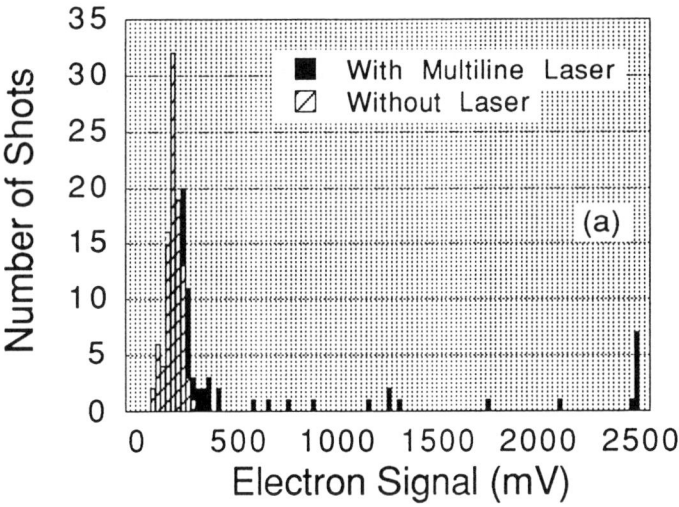

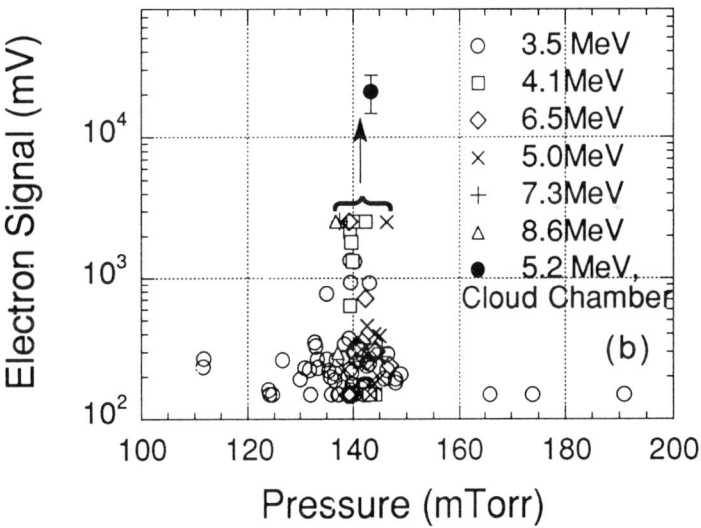

Fig. 10. a) Surface barrier detector signal levels (raw data) and noise histogram (shown in hatched bar). b) Surface barrier detector signal levels versus fill pressure of neutral gas.

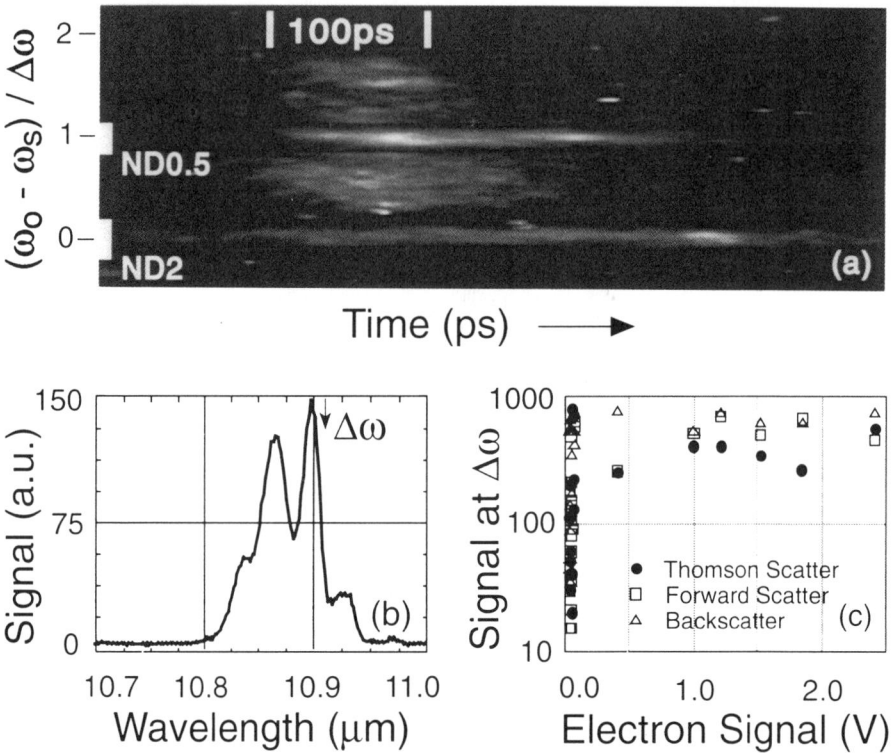

Fig. 11. a) Time resolved Thomson scattered spectrum (dual frequency). b) Time integrated Raman backscattered spectrum; c) correlation of the electron signal with the peak of the first burst in the Thomson scattered spectrum at $(\omega_o - \omega_s) = \Delta\omega$, the backscatter spectrum near $\Delta\omega$, and the Stokes peak at $\Delta\omega$ in the near forward direction.

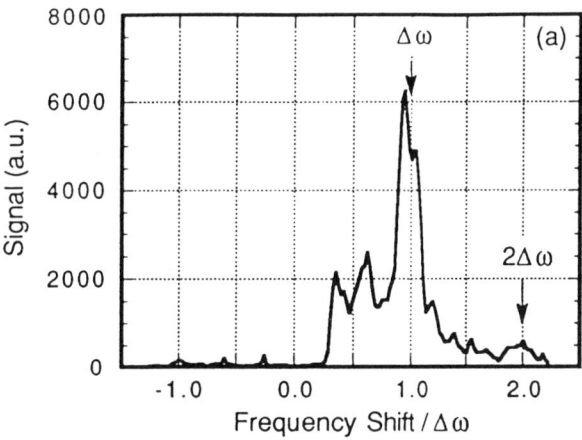

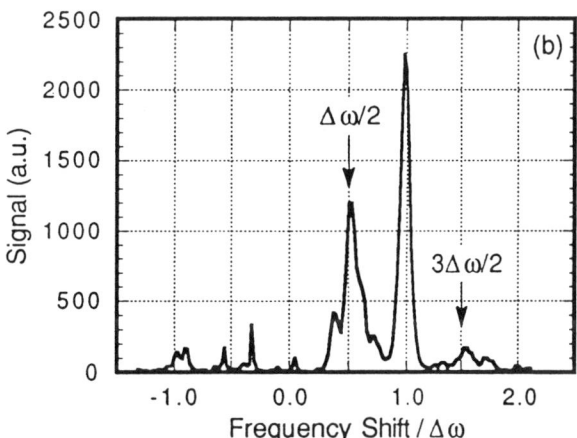

Fig. 12. a) A scan through a Thomson scattered spectrum at $k_s \simeq 2k_1$ showing a large scattered peak at $\Delta\omega$ and a smaller but distinct peak at $2\Delta\omega$. b) Subharmonic components at $\Delta\omega/2$ and $3/2\,\Delta\omega$ in a Thomson scattered spectrum at $k_s \simeq 2k_1$.

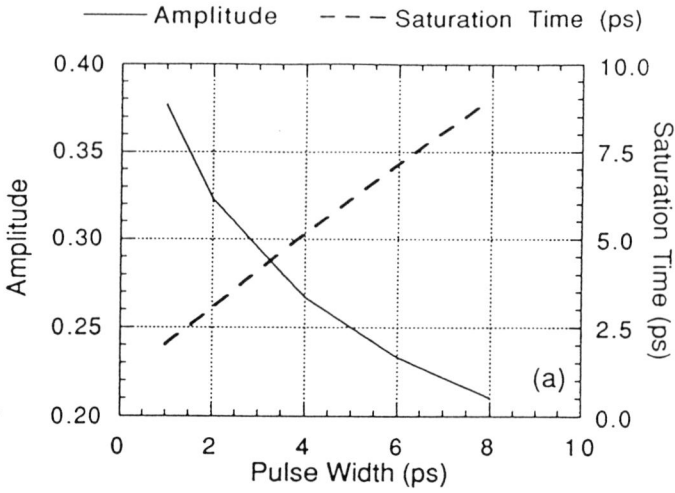

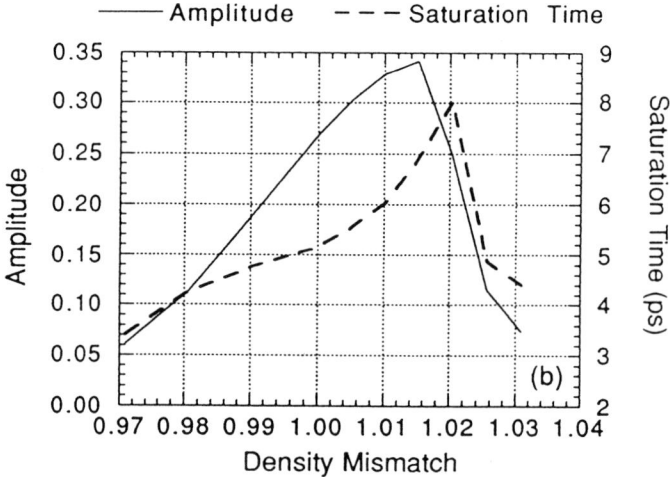

Fig. 13. a) The amplitude n_1/n_0 of the plasma beat wave versus laser pulse width for a laser energy of 10J per line, a density $n_0 = 9 \times 10^{17}$ cm^{-3}, and wavelengths $\lambda_1 = 1.053\mu$m and $\lambda_2 = 1.086\mu$m. b) The plasma wave amplitude versus density mismatch, for a laser energy of 10J, pulse width of 4 ps, and a density $n_0 = 9 \times 10^{17}$ cm^{-3}.

BEAT-WAVE EXPERIMENTS AT ECOLE POLYTECHNIQUE

F. Amiranoff, A. Dyson, M. Laberge,
J.R. Marquès, F. Moulin E. Fabre
Laboratoire d'Utilisation des Lasers Intenses*
Ecole Polytechnique 91128 Palaiseau France

P. Benkheiri, F. Jacquet, J. Meyer and Ph. Miné
Laboratoire de Physique Nucléaire des Hautes Energies*
Ecole Polytechnique 91128 Palaiseau France

B. Cros and G. Matthieussent
Laboratoire de Physique des Gaz et des Plasmas*, Université Paris-Sud, 91405 Orsay France

P. Mora
Centre de Physique Théorique*, Ecole Polytechnique 91128 Palaiseau France

C. Stenz
Groupe de Recherche sur l'Energétique des Milieux Ionisés*
Université d'Orléans 45000 Orléans France

* Laboratoires associés au Centre National
de la Recherche Scientifique

ABSTRACT

We summarize the results obtained in the context of the beat-wave experiments performed at Ecole Polytechnique in France. We first study the generation of long and homogeneous plasmas by multiphotoionization of low density gases and observe the density variation with time due to the ponderomotive force of the ionizing laser beam. In the Nd-laser beat-wave experiments we observe in the resonant conditions the generation of intense electron plasma waves. We also measure some daughter ion and electron waves generated by the decay of these primary electron waves by modulational instability (coupling with ion motions). These results show the importance of the ion hydrodynamic for the generation and the evolution of the accelerating electric fields in beat-wave experiments.

INTRODUCTION

The use of high intensity laser beams and plasmas has been proposed to generate high accelerating gradients [1,2]. A plasma is capable of transforming a part of an intense electromagnetic wave (laser beam) into an electrostatic wave with a relativistic phase velocity. The most studied mechanism at present is the beat-wave : the charge separation between ions and electrons in the plasma is induced by the beating of two laser beams with sligthly different frequencies [3-10]. If this frequency difference $\delta\omega$ is close to the natural oscillation frequency of the electrons in the plasma ω_p, the amplitude of the electron oscillations, and thus the amplitude of the accelerating electric field, will grow secularly in time

411

until it saturates because of other mechanisms [11-15]. The goal of the beat-wave program developed at Ecole Polytechnique is to demonstrate (if it is true) the possibility of accelerating electrons in a relativistic plasma wave generated by beat-wave. The program also includes the study of all physical mechanisms relevant to the growth and saturation of these accelerating plasma waves. In this paper we will describe the first two steps of the experiments.

The resonant condition between the plasma frequency and the difference frequency $\delta\omega$ makes it necessary to generate a long and homogeneous plasma with a very precise electron density. The required precision depends on the laser parameters and typical values range from a few percent with CO_2 lasers at $\lambda \approx 10$ μm and I $\approx 10^{14}$ W/cm^2 to less than 1% for Nd-lasers at $\lambda \approx 1$ μm and I $\approx 10^{14}$ W/cm^2. Thus we studied in a first step the generation of homogeneous plasmas by multiphotoionization of low density gases. In particular we looked very carefully at the density variations with time after the initial full ionization and demonstrated the role of the ponderomotive force of the laser itself in this evolution.

In a second step we performed beat-wave experiments with Nd-lasers in these plasmas. We measured the electron waves directly generated by the beat-wave and some ion and electron waves resulting from the decay of these primary high-amplitude waves by the modulational instability.

MULTIPHOTOIONIZATION   EXPERIMENTS

The principle of these experiments is to focus a high power laser beam into a gas containment vessel. Near the focus of the lens the laser intensity is high enough to exceed the ionization threshold : the gas is thus very rapidly fully ionized [16-18] and, in the case of hydrogen or deuterium, the electron density is equal to the initial atomic density. The electron density is measured by small angle Thomson scattering of the ionizing beam itself.

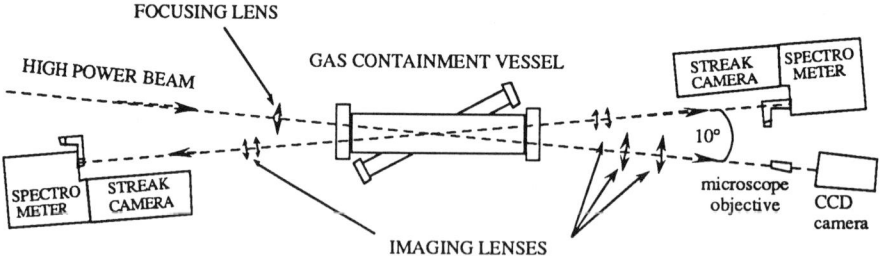

Fig 1. Experimental set-up for the multiphotoionization experiments

The experiments were performed with one beam of the neodymium-glass laser facility at L.U.L.I. After frequency doubling, it delivers up to 20 J in 600 ps (FWHM) pulses or 10 J in 200 ps pulses at $\lambda = 0.53$ μm. The 90 mm diameter beam is focused by a 1-m focal length doublet into the vessel filled with either H_2 or D_2 gas at pressures of a few Torr leading to electron densities of the order of $10^{17} e^-/cm^3$ (cf fig 1). The laser light transmitted through the plasma is imaged with a high magnification onto a CCD camera thus giving an image of the focal spot with a resolution of 10 μm. The intensity profile consists of an enveloppe of 50±20 μm FWHM with local hot spots of 2 to 3 times higher intensity and FWHM close to the diffraction limit. The maximum intensity is then close to 10^{16} W/cm² while the intensity averaged on the focal spot at half maximum is of the order of 10^{15} W/cm². To measure the density, the scattered light is collected by f/10 optics at 10°. The image of the plasma is formed with a magnification unity into a spectrograph streak-camera combination with an overall spectral resolution of 1Å. All the details on the gas vessel and the alignment sytem using a new technique are given in ref [17].

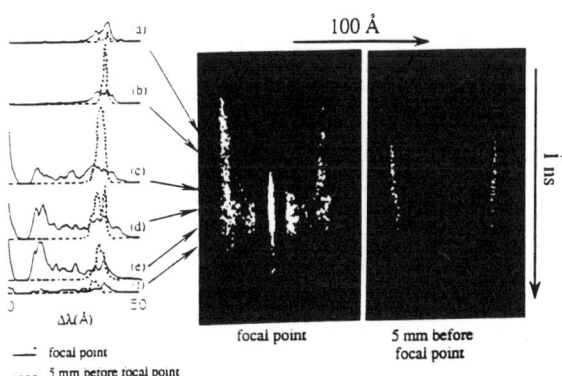

Fig 2. Deuterium low angle spectra at focus and 5 mm before the focal point. P = 3 Torr, λ = 0.53 μm, I = 1.4 10^{15}W/cm². The lineouts correspond to a) -320 ps, b) -240 ps, c) -80ps, d) 0 ps, e) 80 ps, f) 160 ps from the maximum of the scattered electron signal. Plain lines correspond to the focal point, dashed lines to 5 mm before the focal point. The central line is attenued by a neutral density filter.

Typical spectra measured in 3 Torr of deuterium are shown in fig 2. At such low scattering angles [19] the spectrum consists of a central feature - blocked in our case - and two narrow satellites at ± ω_p from the laser frequency. The distance between the two satellites is thus a direct measurement of the electron density. At the beginning of the pulse, the satellites are very narrow and indicate a very homogeneous plasma with an electron density corresponding to a fully ionized gas. As time increases the satellites broaden indicating the presence of different densities. The detailed evolution of these spectra mainly depends on the intensity profile of the laser beam. We can see in fig 2 that the plasma evolves more slowly far from the focal spot where the laser intensity is lower. Typical results show a broadening of the density profile of a few percent per 100 ps. The variation of the "mean density", defined as the density where the emission is maximum, is also typically 3% per 100ps. The last observation is that, despite the difference in the ion mass, there is no clear difference between spectra measured in hydrogen or deuterium.

We explain these density evolutions with a simple model. The plasma is assumed to be initially homogeneous and fully ionized with a radius r_p. It then evolves because of two effects : the thermal pressure generates a radial expansion in vacuum and the ponderomotive pressure associated with the laser intensity gradients tends to bore holes in the high intensity regions (fig 3).

Fig 3. Density evolution with time : numerical results for a gaussian laser intensity profiles in a hydrogen plasma. P = 3.75 Torr, r_p = 125 μm, T_e = 12 eV, T_e/T_i = 6 , λ =0.53μm, a gaussian pulse of 600 ps (FWHM), E = 22 J, ρ = 55 μm. The lineouts correspond to t = − 600 ps to +400 ps from the maximum of the laser intensity with 200 ps steps.

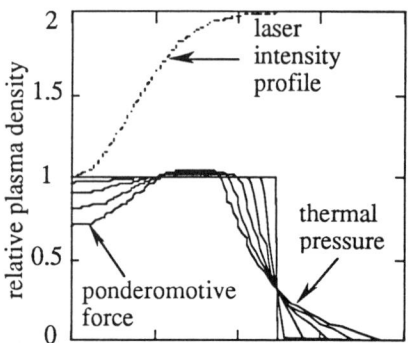

From the usual fluid equations we obtain the evolution of the plasma density :

$$M n \frac{dv}{dt} = - \left(Z T_e + T_i \right) \frac{dn}{dr} - \frac{Z n}{n_c} \frac{\nabla I}{2c}$$ [1]

$$\frac{dn}{dt} + n \nabla . v = 0$$ [2]

A characteristic time τ_{pond} for the density evolution due to the ponderomotive force is obtained by neglecting the thermal pressure. Assuming a constant pulse in time with a gaussian laser intensity profile: $I(r) = I_M \exp(-r^2/\rho^2)$, an analytic solution of the problem is easily obtained on the axis of the cylinder and for small density variations − i.e near the beginning of the ionization −. The on axis density is given by

$$n = n_0 \left[1 - (t/\tau_{pond})^2 \right]$$ [3]

where

$$\tau_{pond} = \left[\frac{M n_c c}{Z} \frac{\rho^2}{I_M} \right]^{1/2}$$ [4]

This shows that the density evolution is a very rapid function of the focal spot radius ρ ; at fixed laser energy I_M scales as $1/\rho^2$

so that the density evolution scales as $1/\rho^4$. For a constant pulse, the evolution on the axis of the plasma is parabolic in time, i.e slow at the beginning. Typical parameters for hydrogen : $Z=1$, $M=m_p$, $\lambda=0.53\mu m$, $I_M =10^{15}$ W/cm^2, $\rho=30\mu m$, give $\tau_{pond} = 420$ ps and a relative density variation $\delta n/n = 1$ % (3%) after 40 ps (70 ps). We can see from [4] that the variation is slower for deuterium. After 100 ps the density on the axis changes respectively by 5.7% and 2.8% for H_2 and D_2.

On the other hand, neglecting the ponderomotive term, we obtain the usual rarefaction characteristic velocity c_s given by : $c_s = [(ZT_e + T_i)/M]^{1/2}$. Thus, for a given plasma radius r_p, the rarefaction wave starting from the edge of the plasma will reach the axis after a time delay $\tau_{dif} = r_p/c_s$. With a plasma radius $r_p = 50$ μm, equal electron and ion temperatures $T_e = T_i = 10$ eV, we obtain $\tau_{dif} = 1.1$ ns so that near the beginning of the pulse the thermal diffusion only modifies the edge of the plasma while the ponderomotive force is active in the center as observed in fig 3.

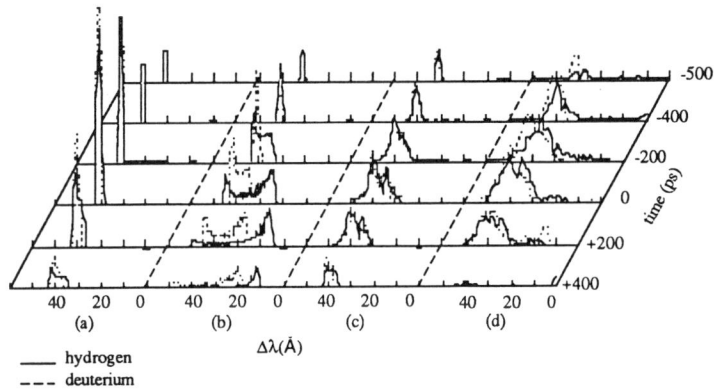

hydrogen
deuterium

Fig 4. Comparison between experimental (4.d) and calculated spectra. P=3.75 Torr, I=1.3 10^{15}W/cm^2. Calculated spectra with gaussian beam and radius 55μm (4a) and 25μm (4b). Calculated spectra with a modulated beam $I(r)=(I_M/(1+\beta))(1+\beta\cos(m\pi r/\rho))$ $\exp(-(r/\rho)^2)$ .with $\rho=55\mu m$, $\beta=0.8$ and $m=8$ (fluctuation wavelength $=10\mu m$).

We numerically solved equations [1] and [2] for our laser conditions in the case of purely gaussian (cf fig 3) and modulated beams. The comparison between experiments and calculations is shown in fig 4. Plain lines correspond to hydrogen and dotted lines to deuterium. The spectra in fig 4.a and 4.b are calculated with gaussian profiles and respectively $\rho = 55$ μm and 25 μm, whereas the case of the modulated intensity profile is shown in fig 4.c. One satellite of the experimental spectra is shown in fig 4.d. It is clear in this figure that the agreement between calculated and experimental spectra is very bad for gaussian intensity profiles ; with the measured radius of the laser envelope (55 μm) the spectra are much too narrow and with a smaller radius the shape of the spectra is quite different from the measured one. Moreover, in both cases, the calculations predict a large difference between hydrogen and deuterium. On the opposite, a fairly good agreement is obtained when we introduce intensity modulations in the beam : the shape of the spectra reproduce the experimental one and the predicted difference between hydrogen and deuterium is quite small as observed experimentally.

BEAT - WAVE   EXPERIMENTS

The experimental set-up for beat-wave experiments is shown in fig 5. Two synchronized Nd-YAG (λ=1.064 μm, τ=160ps (FWHM)) and Nd-YLF (λ=1.053 μm, τ=90ps) pulses are amplified in the laser chain of the LULI up to 7J with an output diameter of 90 mm. These two pulses are colinearly focussed by a 1m focal-length bichromatic doublet into the gas vessel filled with deuterium. They first ionize the gaz and then generate electron plasma waves by the beat-wave mechanism. These waves' can couple with the ions and decay into daughter ion and electron waves in almost every direction [14,15]. A 300 mJ, τ=160 ps, 30 mm diameter green probe beam is focussed in the plasma by the same lens as the two infrared beams. The relativistic on-axis plasma waves diffract a part of this beam at nearly 0°, whereas ion and electron waves almost perpendicular to the laser axis and with $k_p \lambda_D \approx 0.18$ (the Debye length λ_D is 0.1 μm in the resonant conditions) diffract the probe beam at 10°.

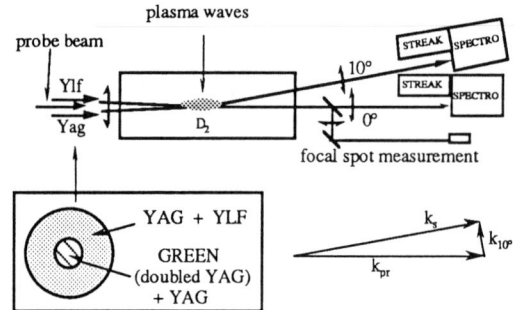

Fig 5 Experimental set-up. The three wavelengths (YAG, YLF and doubled YAG) are spatially distributed as shown and focussed in the middle of the gas chamber. k_{pr} is the probe beam wave-number and $k_{10°}$ the wave-number of the wave measured at 10°. $k_{0°}$ generated by beat-wave is on the axis and $\approx k_{pr}/191$

A typical signal observed at 0° is shown in fig 6. Two electron satellites appear at $\pm\delta\omega$ on each side of the "masked" probe beam wavelength. The beat-wave being a resonant process, these satellites should only appear near the resonant density ($n_e \approx 1.07$ $10^{17} e^-/cm^3$) that is a D_2 pressure of 1.65 Torr at 23°C. The "resonance" curve in fig 6 shows a large signal at any gas pressure. This is in fact due to a signal generated in the gas surrounding the plasma with an amplitude comparable to the expected scattered signal. At the resonant pressure, this signal would correspond to a product of the plasma wave density perturbation by the length of the wave of $(\delta n/n).L \approx (0.4-4)$%.mm, that is, for instance, a field of the order of 1GV/m over 1mm.

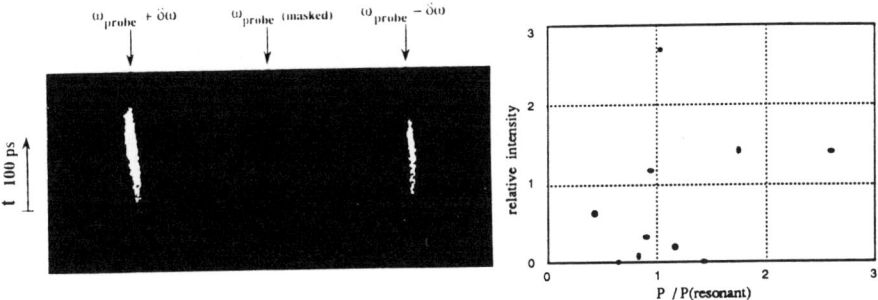

Fig 6. Experimental spectrally resolved signal measured at 0° and scattered signal at 0° and + δω vs normalized pressure

Taking into account the aperture of the focussing lens, the direct waves generated by beat-wave are enclosed in a cone centered on the laser axis with a half-angle of 83°. When reaching a high amplitude, these waves are unstable with respect to the coupling with ions. A whole spectrum of waves is generated and we observe some of them on the 10° diagnostic.

Fig 7.A scattered spectrum measured at 10° in D_2, 1.95 Torr, I_{max}(YAG) = I_{max}(YLF) $=10^{14}$W/cm². The central line (ω_{probe}) is attenuated by 100 and the first satellites ($\omega_{pr} \pm \delta\omega$) by 10. Lineouts at the maximum of the emission are shown on the right

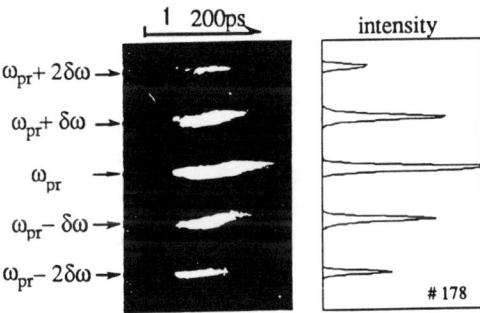

Fig 7 shows an example of scattered signal close to the resonant pressure. The central line whose frequency is equal to the probe beam frequency is an evidence of low frequency ion waves. On each side of this line appear satellites at ±δω and ±2δω. The ±δω lines are due to electron waves at frequency δω, whereas the non-resonant satellite is an evidence that the δω waves have a large amplitude and can generate harmonics by nonlinear coupling. The resonance curves are shown in fig 8. All these lines only appear near the resonant pressure and disappear far from resonance or when we use only one laser frequency.

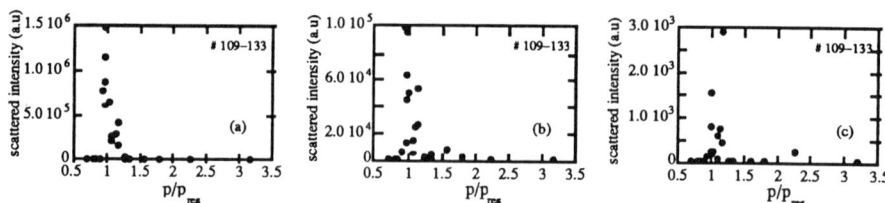

Fig 8. Intensity at 10° as function of normalized fill pressure.
(a): central line (ω_{pr}) ; (b): first satellite $(\omega_{pr} + \delta\omega)$; (c):
second satellite $(\omega_{pr} + 2\delta\omega)$. The three vertical scales use the
same unit. The points on the x axis are respectively 5, 4 and 2
orders of magnitude lower than the maximum signal.

From the intensity of the scattered signals, it is possible to
estimate the amplitude of the waves ; we obtain density perturbations related to the waves detected at 10° of $(\delta n/n)_{10°} \geq (0.4-1.2)\%$
for the electrons and $(\delta n/n)_{10°} \geq (1.2-3.7)\%$ for the ions. From the
ratio of the amplitudes of the two satellites at $\pm\delta\omega$ and $\pm2\delta\omega$ we
can get an estimate of the amplitude [10] of the direct waves of
$(\delta n/n)_{\approx 0°} = (1-5)\%$.

We made a theoretical estimate of the saturation level of the
relativistic plasma wave due to modulational instability (M.I).
Assuming a homogeneous initial density we calculate with the use of
usual fluid equations the growth of the relativistic plasma wave.
At each time step we compute the growth rate of the M.I and assume
that the wave saturates when the time integral of the growth rate
reaches a value of 5 as suggested by particle simulations [14]. At
this value for our experimental conditions the saturation value of
$\delta n/n = 1.5\%$ is reached 50 ps before the maximum of the laser pulses.
For comparison ,we note that in the absence of M.I, the saturation
due to relativistic detuning would lead to a maximum value of 14%
near the maximum of the pulses.

CONCLUSIONS

In conclusion we observed two effects due to the ion hydrodynamic which affect the growth and saturation of electron plasma
waves generated by beat-wave. Multiphotoionization generates the
long and homogeneous plasmas necessary for beat-wave experiments,
but the ponderomotive force due the the laser intensity gradients
very rapidly modifies the density profiles. We observed typical
density variations of (1-3)% per 100ps. In beat-wave experiments,
the modulational instability destroys the accelerating plasma waves
and generates a whole spectrum of low phase-velocity daughter
waves.

These results show the importance of using short laser pulses
of a few ps and "beautiful" focal spots in order to avoid any
hydrodynamic problems in beat-wave experiments.

ACKNOWLEDGMENTS

We greatly acknowledge the help of B. Montes, P. Poilleux, and the staff of the LULI during these experiments. We also thank A.E. Dangor, A. Dymoke-Bradshaw and C. Gouédard for their collaboration. This work was partially supported by the DRET under contract n° 86-196 and the EEC.

REFERENCES

1. John M. Dawson ; Scientific American. **260**, 54 (March 1989)
2. *Advanced Accelerator Concepts*, edited by C. Joshi,
   AIP Conf. Proc. n° 193 (AIP, New York, 1989)
3. M. Rosenbluth and C.S. Liu , Phys. Rev. Lett. **29**, 701 (1972);
   T. Tajima and J.M. Dawson , Phys. Rev. Lett. **43**, 267 (1979);
4. C.E. Clayton, C. Joshi, C. Darrow and D. Umstadter, Phys. Rev.
   Lett. **54**, 2343 (1985) and F. Martin and T.W. Johnston, Phys.
   Rev. Lett. **55**, 1651 (1985).
5. C. Darrow *et al.*, Phys. Rev. Lett. **56**, 2629 (1986).
6. N.A. Ebrahim *et al.*, SLAC Report 303 (1986) Available from the
   National Technical Information Service, U.S Department of
   Commerce, 5285 Port Royal Road, Springfield, VA 22161.
7. F. Martin, J.P. Matte, H. Pepin and N.A Ebrahim, New
   Developments in Particle Acceleration Techniques (Orsay 1987)
   CERN 87-11, available at CERN Service d'Information
   Scientifique.
8. Y. Kitagawa *et al.*, Phys. Rev. Lett. **68**, 48 (1992)
9. A.E Dangor, A.K.L. Dymoke-Bradshaw and A.E Dyson, Physica
   Scripta. **T30**, 107 (1990).
10. F. Amiranoff *et al.*, Phys. Rev. Lett. **68**, 3710 (1992)
11. T. Katsouleas and J.M Dawson; Phys. Rev. Lett. **51**, 392 (1983)
12. P. Sprangle, E. Esarey, A. Ting and G. Joyce ; Appl. Phys.
    Lett. **53**, 2146 (1989)
13. P.Mora, Revue. Phys. Appl. **23**, 1489 (1988)
14. P. Mora, D. Pesme, A. Héron, G. Laval and N. Silvestre,
    Phys. Rev. Lett. **61**, 1611 (1988) ;
15. D. Pesme, S.J. Kartuneen, R.R.E Salomaa , G. Laval and N.
    Silvestre , Lasers and Particle Beams. **6**, 199 (1988).
16. A.E Dangor, A.K.L. Dymoke-Bradshaw and A.E Dyson, J. Appl.
    Phys. **64**, 6182 (1988).
17. F.Amiranoff *et al.*, Rev. Sci. Inst. **61**, 2133 (1990).
18. J.R.Marquès *et al.* , submitted to Physics of Fluids B.
19. J. Sheffield, *Plasma Scattering of Electromagnetic Radiation*.
    (Academic Press New York 1975).

Plasma Lens and Wake Experiments in Japan

Atsushi Ogata

KEK, National Laboratory for High Energy Physics
Oho,Tsukuba 305 Japan

Abstract

Plasma lens and wake experiments performed and planned in Japan
are reviewed. Overdense plasma lens experiments were conducted at the
University of Tokyo on a 18MeV linac. The change in energy distribution
of linac beams caused by the plasma wakefield was measured at KEK on
a 500MeV linac. Laser wakefield acceleration experiments are planned at
Osaka University using a 30TW Nd:glass laser.

1. Introduction

This paper reviews plasma lens and wake experiments performed and planned
in Japan. These include plasma lens experiments, plasma wakefield acceleration
(PWA) experiments and laser wakefield acceleration (LWA) experiments.

The plasma lens experiments were conducted at the University of Tokyo on
a 18MeV linac during 1990-1991. Single-bunched beams were introduced into an
overdense plasma and the transverse beam sizes were measured at three different
points downstream of the plasma as a function of the plasma density. The profiles
in the horizontal-longitudinal plane were also measured using a streak camera.
The experimental results in the vertical direction agree well to a prediction based
on linear theory.

The plasma wakefield acceleration experiments which started in 1989 at
KEK on a 500MeV linac have recently been terminated. The linac provided a sequence of several bunches which generated a wakefield in a plasma. The wakefield,
in turn, caused an energy change in the following bunches. By analysing the energy
of each bunch, we could observe the energy tansfer between the bunches through
plasma waves without using a test beam. Data processing taking into account
the finite bunch length which was comparable, or even longer than, the plasma
wavelength provided information that the observed energy change, if converted
into the values at the bunch center, amounted to more than 20MeVm^{-1}.

Laser wakefield acceleration experiments are planned at Osaka University
using a Nd:glass laser with a 1.05μm wavelength, 1psec pulse duration and 30TW
peak power. The same laser will also be used to ionize a gas into a plasma and
to produce a test beam. Particle acceleration will be demonstrated by injecting a
test beam comprising a few MeV electrons emitted from a solid target irradiated
by a branch beam of the laser. The acceleration gradient of the wakefield can

exceed 2GeVm^{-1}. Since the acceleration length will be limited by the diffraction, the total energy gain will be $\sim 45\text{MeV}$.

In addition to the three results reported here, the results of two experiments have already been published. One involves the observation of a plasma wakefield buildup caused by a train of bunches; this experiment was conducted at the University of Tokyo on the 18MeV linac.[1] The idea of obtaining a high acceleration gradient in a plasma wakefield accelerator using a train of bunches is therefore examined. The amplitude of the plasma wave, which was detected using a coaxial diode detector, could not grow infinitely, but reached saturation. However, the wakefield at saturation could be 400-times greater than that caused by a single bunch. The other involves a study of ion-wave wakefields conducted at Utsunomiya University.[2] Plasma wakefields in the ion-wave regime were excited by injecting ion bunches with a variety of shapes, including the door-step shape. An excited-wave amplitude $(\delta n/n_0)$ of up to 17% was observed.

The present paper is divided into four sections. Sections 2, 3 and 4 describe the plasma lens experiments on a 18MeV linac, plasma wakefield acceleration experiments on a 500MeV linac, and plan of laser wakefield experimets, respectively. PWA and LWA are compared at the end of the last section.

2.   Observation of the Plasma Lens Effect

A plasma lens based on self-focusing due to the shielding of space charge of a particle beam by a quiescent plasma has been proposed as a final focus device for the next generation of linear colliders.[3] This concept was experimentally examined. Although the results agree well with the theoretical predictions, certain aspects of the results remain unexplained.

The plasma lens was first demonstrated by the ANL group.[5] Because their plasma was dense and long, the focal point fell inside the plasma. Following them, we have reported the effect of a thin overdense plasma lens.[6] The present experiments were conducted in order to verify the previous results, in which the ratio of the plasma density to the beam density ranged up to $n_e/n_b \sim 10$. A higher plasma density region, up to $n_e/n_b \sim 60$, was also explored in the present experiments. The results have shown that (1) the observed plasma lens effect in the vertical direction agrees well with a calculation based on linear theory;[3] (2) indication of an instability was found in the region $n_e/n_b > 50$; and (3) the present results seems to reproduce the emittance reduction discovered in the previous experiments.

2.1.   Experimental apparatus

The experiments were conducted at the University of Tokyo on a 18MeV linac.[4] Single-bunch beams were introduced into a plasma chamber, which was separated from the linac duct by the differential pumping technique. Transverse beam profiles were obtained on three phosphor screens (Desmarquest AF995R) which were located 880, 1380 and 1880mm from the center of the plasma chamber.

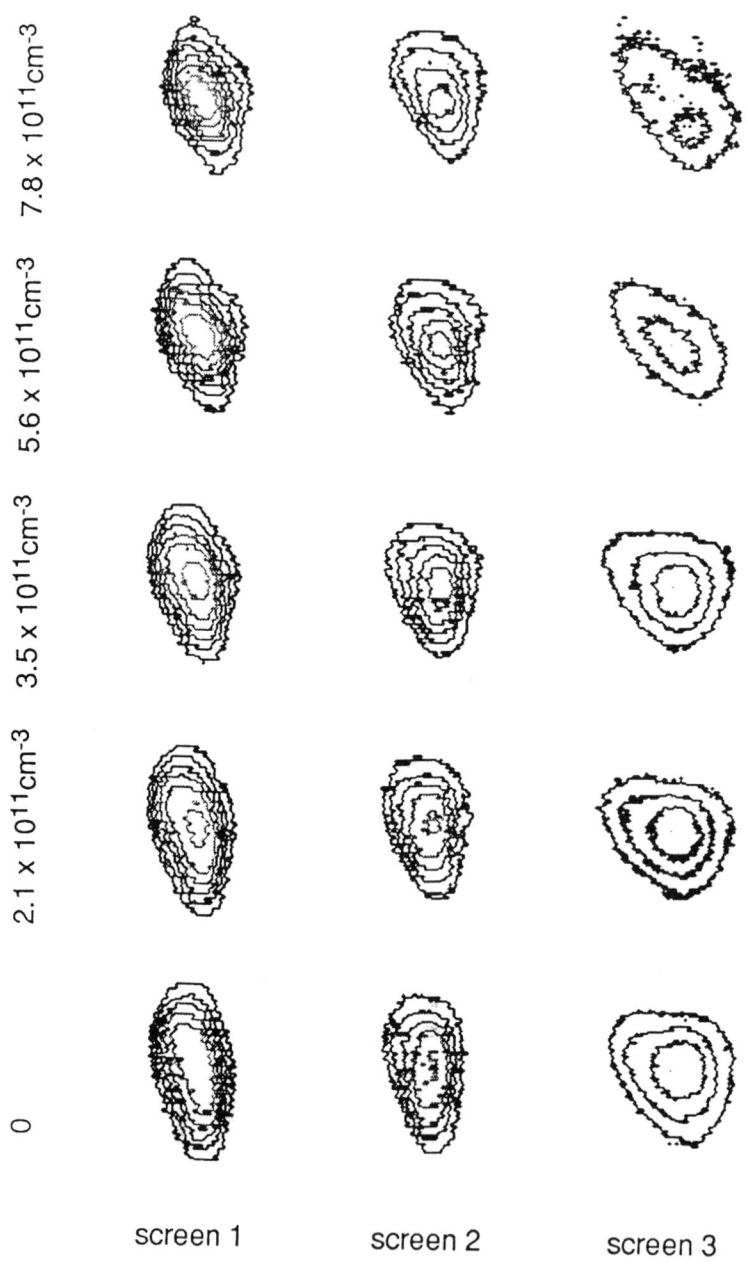

Figure 1: Typical transverse profiles obtained at three screens. The vertical mgnification is about 1.4-times higher than the horizontal magnification.

We call them the first, second and third screens in this paper, counting from the one nearest to the plasma. CCD TV cameras were used to observe images on the three screens, which were triggered in synchronism with the linac beams. A carbon block was set at the end of this beamline in order to measure the beam current. The fluctuation of the current was within $\pm 5\%$ throughout the data taking. We found the bunch charge to be 512pC with a repetition rate of 6.25Hz.

The plasma was produced in a chamber (147mm in inner diameter and 360mm in length) by a discharge between the LaB_6 cathodes and the chamber in synchronism with the linac bunch. The plasma pulse width was about 2msec. It was confined by the multidipole field of permanent magnets placed around the chamber periphery. The magnetic field had a maximum value (700G) at the chamber wall. One of the features of this confinement is that there was no magnetic field along the beam transport. The argon plasma density ranged from $.5 - 10 \times 10^{11} cm^{-3}$ and the temperature ranged from $2.5 - 4eV$, as measured by a Langmuir probe. The plasma length along the beam transport was about 150mm.

A streak camera was placed so as to observe the same first Desmarquest screen that was used for the transverse profile measurement. The horizontal slit in front of the lens introduced light only from the vertical beam center into the camera. The phosphorescent mechanism is described elsewhere.[7] In order to obtain good statistics, in spite of the poor light intensity, we added 512 streak pictures before computer-image processing.

The beam parameters at the plasma center in the absence of a plasma were as follows: $\sigma_{y0} = 2.26mm, \sigma'_{y0} = 0.837mrad, \rho_{y0} = -0.713, \sigma_{x0} = 2.77mm, \sigma'_{x0} = 2.25mrad$, and $\rho_{x0} = -0.956$. Their derivations are described in the following section. The streak-camera measurement gave the rms bunch length as being $\sigma_{l0} = 4.18mm$; this, however, is the value calculated using a superposition of 512 streak pictures. We thus use the value obtained from a single-bunch measurement several years ago, $\sigma_{l0} = 3mm$,[4] giving an average electron density inside the bunch of $n_b = 1.23 \times 10^{10} cm^{-3}$.

These two measurements using a streak camera give the resolution of the camera in the longitudinal measurements as $((4.18mm)^2 - (3mm)^2)^{1/2} = 2.91mm$. This must be mainly due to the time jitter of the camera triggerging.

2.2. Experimental results

Figure 1 shows typical transverse profiles on the three screens, recorded by the CCD cameras and contoured by a computer. As the plasma density increases, the image changes mainly in the vertical direction. Above $7 \times 10^{11} cm^{-3}$ or $n_e/n_b > 50$, two peaks are distinguishable on the third screen.

The intensity distribution on each screen was integrated both vertically and horizontally. The horizontal and vertical beam sizes were calculated from the resultant one-dimensional distributions. The so-called rms beam sizes were derived from the width of the distribution, to give $\exp(-1/2)$ of the peak. Figure 2 shows

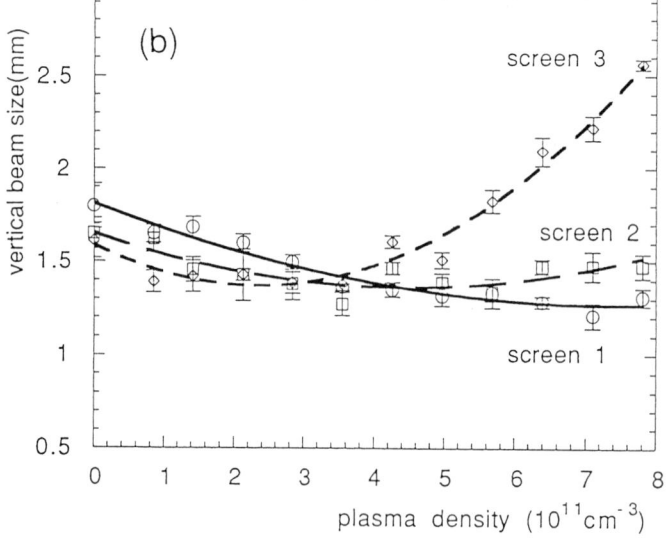

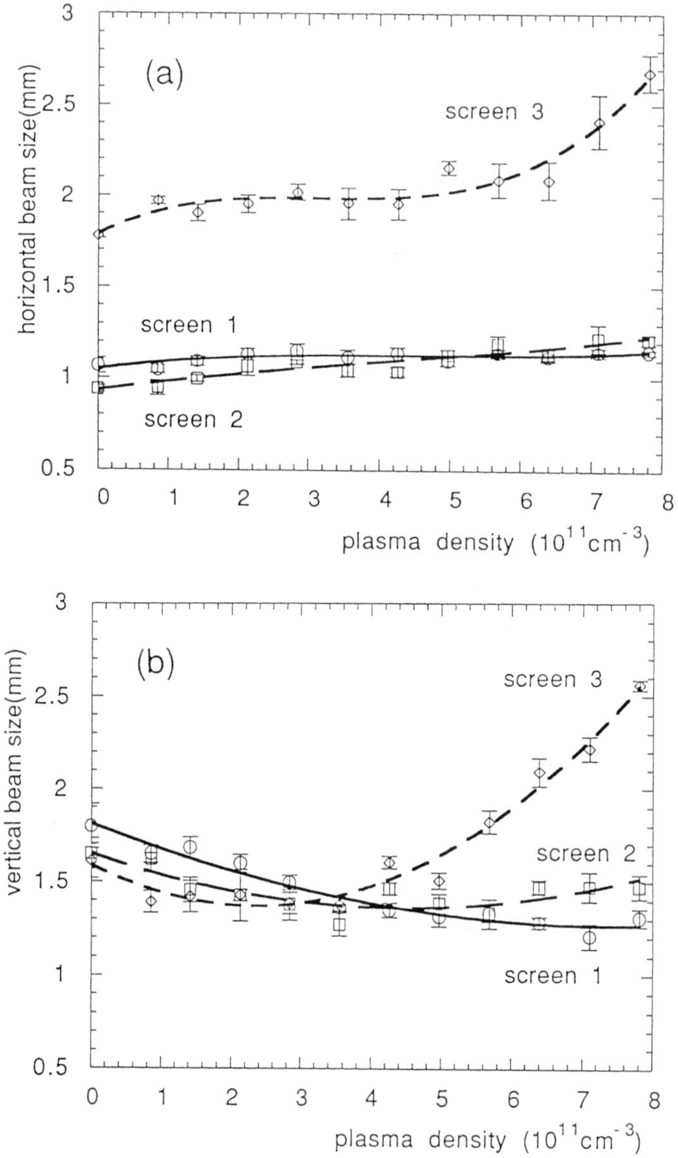

Figure 2: Horizontal(a) and vertical(b) beam sizes observed on three screens. The error bars were obtained from 10 measurements. The lines indicate approximations using third-order(a) and second order(b) polynominals.

the horizontal and vertical sizes as a function of the plasma density. The three bold lines indicate approximations using third-order (Fig.2(a)) and the second order(b) polynominals. Since no remarkable change was found in the horizontal size, except in the high-density region, the following analysis was made only concerning the vertical size.

Because free space existed between the plasma and the phosphor screens, we could derive three parameters (two Twiss parameters and the emittance) at the plasma as a function of the plasma density from a set of three data at the three screens. We can approximate the transverse beam profile at the plasma position by a Gaussian characterized by three parameters: σ, σ' and ρ. Specifically, the beam in the transverse phase space at the plasma is expressed by the matrix

$$\Sigma = \begin{pmatrix} \sigma^2 & \rho\sigma\sigma' \\ \rho\sigma\sigma' & \sigma'^2 \end{pmatrix}, \tag{1}$$

or the contour equation of the beam is written as

$$\frac{1}{1-\rho^2}\left(\frac{y^2}{\sigma^2} - \frac{2\rho yy'}{\sigma\sigma'} + \frac{y'^2}{\sigma'^2}\right) = 1. \tag{2}$$

The beam at the screen i (where $i = 1, 2, 3$) is then written by the matrix

$$\Sigma_i = F_i \Sigma F_i', \tag{3}$$

where F_i is the transfer matrix of free space with a distance s_i between the plasma and the i-th screen. F_i' denotes the transpose of F_i. Beam sizes $\sigma_1, \sigma_2, \sigma_3$ at the three screens then become

$$\sigma_i^2 = \sigma^2 + 2s_i\sigma\sigma' + s_i^2\sigma'^2. \tag{4}$$

Upon solving the three simultaneous equations with $i = 1, 2, 3$ we obtain σ, σ', and ρ. The twiss parameters (β and γ) and the emittance (ϵ) were derived using the relations

$$\beta = \sigma^2/\epsilon, \gamma = \sigma'^2/\epsilon, \epsilon = \sigma\sigma'(1 - \rho^2)^{1/2}. \tag{5}$$

The vertical Twiss parameters and the emittance were calculated in two ways using the experimental data. We first calculated them directly from a trio of data: σ_1, σ_2 and σ_3. The results are plotted in Fig.3; the dependence on the plasma density was approximated using third-order polynominals (indicated by thin solid lines). The second calculation used a method described in the previous report,[6] in which the density dependence of the beam size was first approximated by quadratic curves, as shown in Fig.2. The Twiss parameters and the emittance were then calculated as continuous functions of the plasma density from the approximated, but continuous, beam sizes. The thick solid lines in Fig.3 indicate the results.

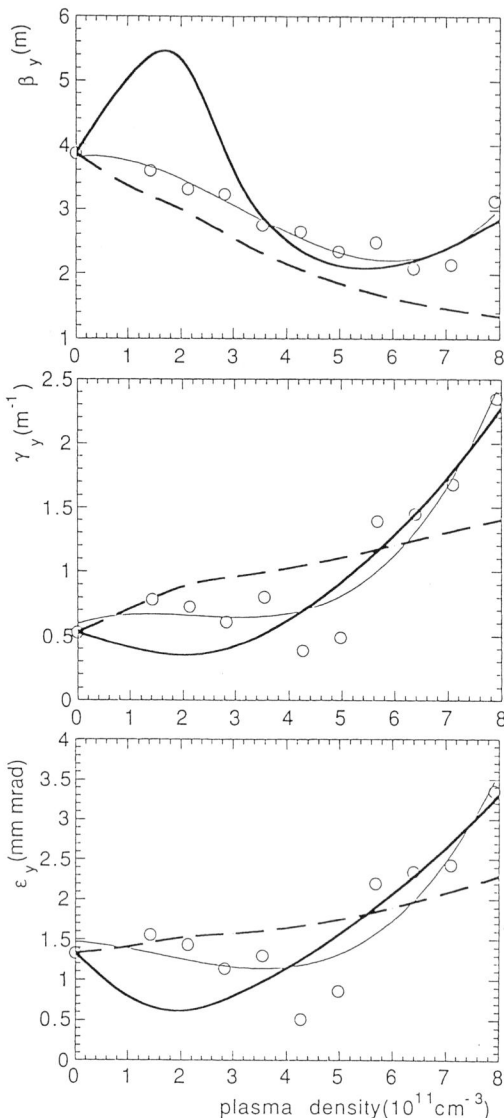

Figure 3: Dependence of vertical Twiss parameters and emittnce on the plasma density. The data points and the thin solid lines were calculated directly from the measured beam sizes, while the thick solid lines were calculated from a quadratic approximation. The dashed lines were calculated based on a round-beam model.

We now try to explain the density dependence theoretically. In the range $n_e > 0.5 \times 10^{11} \text{cm}^{-3}$, Chen's conditions for the round-beam limit, $1/(4\pi r_e a^2) \gg n_e \gg 4Nk_p^2 b/(\pi a^2)$, are satisfied,[3] where the parabolic profiles in both the transverse and longitudinal directions are assumed to be approximations to a Gaussian, with a and b denoting the bunch radius and half of the bunch length, respectively; i.e., the transverse distribution is approximated by $f(r) = 1 - r^2/a^2$, and the longitudinal distribution is approximated by $g(\zeta) = 1 - (\zeta + b)^2/b^2$, where ζ $(-2b < \zeta < 0)$ denotes the longitudinal position inside a bunch. What we observed was the rms size. We consider that the parameters characterizing the parabola distribution (a and b) should be determined so as to give the same area as does a Gaussian distribution with the same peak intensity. We then have $a = (9\pi/8)^{1/2}\sigma$, and $b = (9\pi/8)^{1/2}\sigma_l$. Under these conditions, the focusing strength is proportional to ζ^3:

$$K(\zeta) = \frac{e^2 k_p^2 N \zeta^3}{\gamma m c^2 a^2 b^2}. \tag{6}$$

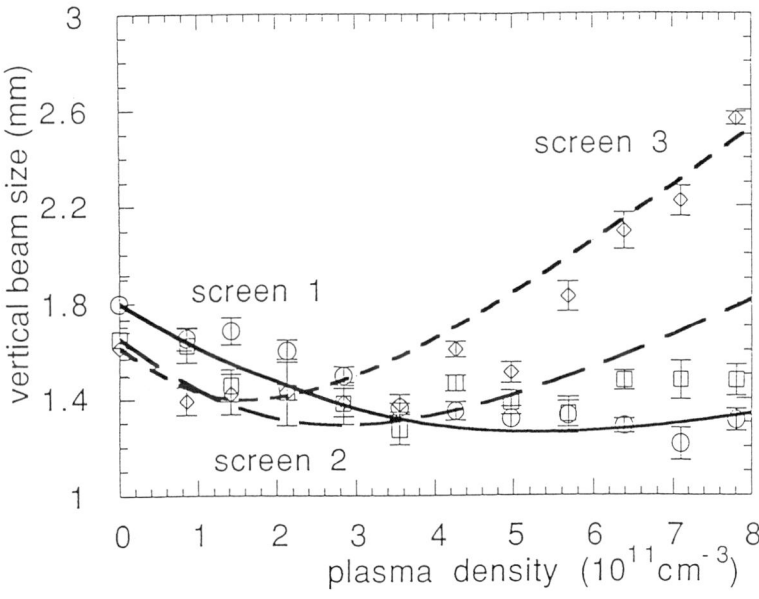

Figure 4: Theoretical dependences of the vertical beam sizes at three screens on the plasma density. Experimental data are also given.

Let us now express the beam at the plasma position in the absence of a plasma by a matrix Σ_0. Using a familiar transfer matrix of a thick lens with length l,

$$P(\zeta) = \begin{pmatrix} \cos(K(\zeta)l)^{1/2} & \sin(K(\zeta)l)^{1/2}/K(\zeta)^{1/2} \\ -K(\zeta)^{1/2}\sin(K(\zeta)l)^{1/2} & \cos(K(\zeta)l)^{1/2} \end{pmatrix}, \tag{7}$$

we can express the beam in the presence of a plasma by

$$\Sigma(\zeta) = P(\zeta)\Sigma_0 P'(\zeta). \tag{8}$$

By inserting $\Sigma(\zeta)$ into eq.(3) we obtain the beam parameters at a screen. Here, we write the result as $\Sigma_i(\zeta)$, because it is a function of ζ. The beam size at a screen, $\sigma_i(\zeta)$, is the square root of the $(1,1)$ element of the $\Sigma_i(\zeta)$. The beam size that we have observed was the average of $\sigma_i(\zeta)$ weighted by the longitudinal distribution: $i.e.$,

$$(\sigma_{obs})_i = \int_{-2b}^{0} g(\zeta)\sigma_i(\zeta)d\zeta / \int_{-2b}^{0} g(\zeta)d\zeta. \tag{9}$$

The thick lines in Fig.4 give the theoretical density dependencies of the vertical sizes at the three screens, where we set the lens length at 15cm. As it shows, the experimental and theoretical sizes agree fairly well.

Because of the nonlinear operation of eq.(9), the resultant longitudinal distribution is no longer Gaussian. This operation does not conserve the emittance, but increases it, if it is perfunctionaly calculated from the lefthand side of eq.(9) using eq.(5). The dashed lines in Fig.3 show the apparent dependence of the emittance on the plasma density, together with those of the Twiss parameters, β and γ. The figure shows that the experimental emittance reduction overcomes the apparent emittance increase in the low-density region, though the density dependences of the experimentally-obtained emittance are somewhat different due to two ways of data-processing.

From the streak pictures, longitudinal profiles along the beam center were derived. The longitudinal bunch sizes, the barycenter shifts and the peak intensities were then calculated. Here, the bunch size was again defined by using the full-width-exp(-1/2)-maximum. Figure 5 shows them together with thin lines approximating their density dependence by third-order polynominals. The longitudinal line distribution should be proportional to $g(\zeta)/[\sigma_{x1}(\zeta)\sigma_{y1}(\zeta)] = [1 - (\zeta + b)^2/b^2]/[\sigma_{x1}(\zeta)\sigma_{y1}(\zeta)]$.

The thick lines in Fig.5 give the width, peak position, and peak internsity calculated from this equation. The peak intensities were normalized by the value in the absence of a plasma. Only the plasma density dependence of the vertical size $(\sigma_{y1}(\zeta))$ was taken into account in the calculatiòn, while the horizontal size $(\sigma_{x1}(\zeta))$ was regarded as being constant. The finite resolution of the streak camera was taken into account in deriving both the width and the peak intensity. Though the tendency agrees between the experiments and theory, the experiments show

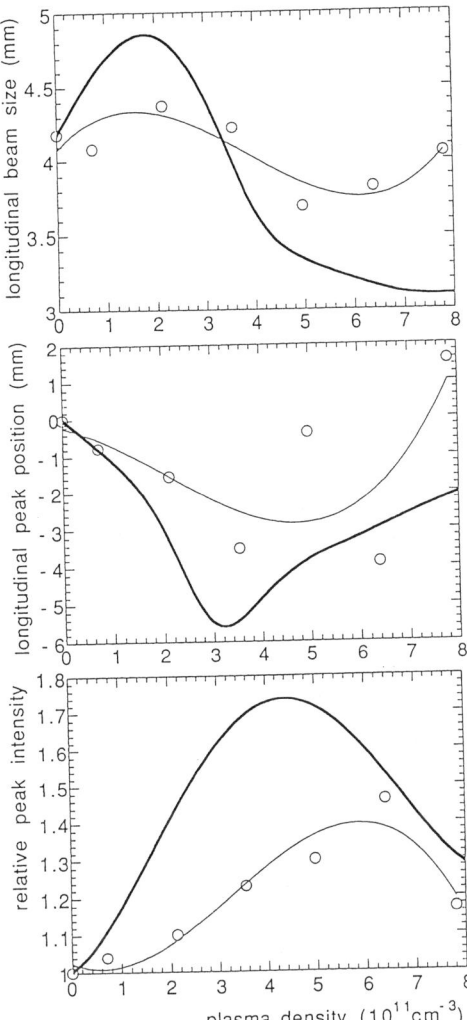

Figure 5: Dependence of the longitudinal bunch size, longitudinal peak position and peak intensity on the plasma density neasured by a streak camera. The peak intensity was normalized by the value in the absence of a plasma. The thick solid lines give the theoretical values based on a round beam model, while the thin lines show approximations of experimental data using third-order polynominals.

weaker density dependences for all three curves. It should be noted that the theoretical calculation is not strict, since it adopted a parabolic distribution as an approximation of a Gaussian distribution.

2.3.   Discussion and conclusions

One comment should be made concerning the peak split appearing on the third screen at the high plasma density region of Fig.1. We found that this structure was quite robust and reproducible. It must contribute to the emittance increase at the high-density region shown in Fig.3. One of the possible mechanisms causing this peak split would be the Weibel instability.[8] However, this instability is possible only for wide beams in which $k_p a > \pi$. Our plasma was not sufficiently dense to fulfil this condition.

We cannot explain the phenomena given in Fig. 2(b), that the plasma lens has little effect in the horizontal direction. Figure 1 shows that the vertical distribution was always broder than the horizontal distribution. This is most remarkable at the first screen. One possibility is the existence of an obstacle upstream of the plasma. If it scraped both the right and the left sides of the beam, the beam could not have a Gaussian distribution in the (x, x') phase space when entering the plasma. In this case, the lens could not be effective horizontally.

Figure 3 suggests the possibility that the emittance calculation adopted previously[6] might exaggerate its dependence on the plasma density. However, the same figure still shows the existence of a modest emittance decrease in the density region $n_e/n_b < 40$. A previous report[6] suggested the possibility that the transverse-emittance decrease is compensated for by a longitudinal-emittance increase. The hypothesis is as follows. The beam particles experience not only a transverse wakefield which causes the lens effect, but also a longitudinal wakefield which causes deceleration. It is null at the head of the beam and maximum at the tail. A substantial energy spread is thus introduced which increases the longitudinal-emittance. This increase in turn decreases the transverse-emittance.

Though this hypothesis is plausible, no specific mechanism has yet been designated which mixes the particles in six-dimensional phase space to enable transverse-longitudinal coupling. In addition, a fact counter to this hypothesis is suggested in the present experiments: the bunch length shortening shown in Fig. 5. The energy spread could be compensated for by bunch shortening, so that the longitudinal emittance would be conserved independently of the transverse emittance.

We also tried to study the underdense region, finding only a large statistical variance in the experimental data due to a weak lens effect. A longer plasma is necessary to study this region using this linac.

In conclusion, these experiments have verified the previous results in one aspect. The two-dimensional linear theory of a plasma wakefield still explains the results, in spite of the fact that the measurement was also carried out in the high-density plasma region $n_e/n_b \gg 10$. However, a new phenomenon was found which caused two peaks on the transverse profile and an emittance increase in the density region above $n_e/n_b > 50$. An emittance reduction was observed below $n_e/n_b < 40$. Its mechanism still remains to be analysed.

3.   Observation of a Beam Energy Shift Caused by a Plasma Wakefield

A plasma wakefield accelerator (PWA)[9] is the first and the only type that has demonstrated the acceleration of bunched electrons.[10,11] Our experiments concerning PWA were conceived by the use of a high-intensity electron beam for positron production in the KEK linac.[11] The linac provides a sequence of multiple bunches which generate wakefields in a plasma to either accelerate or decelerate trailing bunches. By analysing the energy of each bunch, we can observe the energy transfer between the bunches through plasma waves without the necessity of a test charge beam. Theory tells us that plasma wakefields are enhanced at certain plasma frequencies which are resonant with the linac bunch frequency. Because the plasma frequency is determined by the plasma density, we can probe the resonances by controlling the plasma density. Preliminary experiments have produced an energy shift of 12MeV at the maximum in a low-density plasma of the order of 10^{11}cm^{-3}.[11,12] The obtained energy shifts are consistent with, or even larger than, the values predicted by linear theory.

This section describes experiments in a plasma with a density of the order of 10^{12}cm^{-3}, which is optimum for the given linac beam. The next subsection describes the experimental apparatus, i.e., the linac and plasmas. The results of experimens are then given. The two major findings are: 1) energy shifts larger than the prediction were observed in the first three bunches, and 2) energy shifts in the latter bunches decreased, but were still comparable to the prediction.

3.1.   Experimental apparatus

In the KEK PF electron linac for positron generation, the beam emitted by a gun is compressed to less than 2nsec in a sub-harmonic buncher. A 2856MHz rf buncher then separates the 2nsec pulse into a train of 6 bunches with a 350psec (0.104m) spacing. They are then accelerated up to $\sim$ 500MeV. Bunches with a total charge of $5 - 10$nC are focused on a plasma by a quadrupole triplet at the end of the linac. The rms radius and length of the bunches are around 1 and 3mm, respectively.

The plasma chamber, made of Pyrex glass, has a diameter of 50mm and a length of 1m with a $0.5 - 1$kG solenoid magnetic field. The plasma of unit length gives the plasma wakefield directly in MeV/m units. Ionization was realized by a helicon wave,[13] which was excited by a $5 - 10$MHz, 1kW rf wave fed through a helical antenna. The discharge pulse had a duration of 10msec with a rate of 0.5Hz, equal to the linac beam rate. Argon gas was fed through a gas-flow controller in order to maintain a neutral gas pressure of $4 - 8 \times 10^{-4}$torr for a plasma density of $2 - 8 \times 10^{12}$cm^{-3}.

A Langmuir probe was used to measure the plasma temperature and density at the longitudinal and radial center of the plasma. In addition to the standard Langmuir probe, the current to a titanium end plate of the plasma chamber was also used to diagnose the plasma temperature and density. A PCD array combined with a 488nm filter measured the transverse plasma profile. An antenna combined to a waveguide and a diode detector was sometimes used to detect plasma waves. The plasma temperature ranged from 2 to 5eV, and the rms plasma radius was around 5mm. The plasma density was controlled by changing the gas flow rate, the solenoidal magnetic field, the rf power, and/or the timing: the bunches could interact with a plasma of lower density if we controlled the timing so that they could pass the after-glow plasma region.

The combination of a bending magnet and a streak camera made it possible to measure the energy spectrum of each bunch. The bunches which were analyzed in the bending magnet traveled in air over a length of 0.5m to a mirror while radiating Cherenkov radiation. The mirror reflected only the radiation, while transmitting the electron beam. The reflected radiation was finally focused onto the streak camera. It was found that the energy aperture of the magnet (15MeV) could not cover the energies of all the particles in a bunch. We therefore inserted a slit in front of the streak camera and swept the analyzing field. The two-dimensional particle distribution was also measured in the phase-space of the longitudinal co-ordinates and particle energy, i.e., (z, E) phase-space, by this method.

Integration of each energy spectrum gives the bunch intensity of each bunch. We found that the total charge was distributed among the six bunches by ratios of approximately 0.07 : 0.29 : 0.26 : 0.23 : 0.10 : 0.01 in the present experiment. The standard deviation of each bunch intensity was about 25%, calculated from 40 energy spectra. The reproducibility was not assured if an intermission, usually longer than 30min, was taken during linac operation.

3.2. Predictions of linear theory

Let us first examine what linear theory predicts.[11] If all of the bunches and the plasma are on the axis, the resultant wakefield is a linear summation of the individual wakefields, which is a function of both the plasma density and the position. The position dependence is given in Fig. 6, and the density dependence is given by lines in Figs. 7 and 8. In the calculations of these figures, it was assumed that a total charge of 10nC is distributed by the ratio already described, and that each bunch has a longitudinal Gaussian distribution with a standard deviation of 3mm, and a transverse parabolic distribution with a radius of 1.4mm.

Figure 6 shows the predicted amplitude of the longitudinal wakefield on the axis at $n_e = 5.05 \times 10^{12} \text{cm}^{-3}$ as a function of the longitudinal position. One can consider this to be the time evolution of the wakefield. The bunch positions and their intensities are also given here. The lines in Fig.7 indicate the prediction of the amplitude of the longitudinal wakefield at each barycenter of a bunch as being a function of the plasma density.[11] At resonant densities, where $\omega_p = n\omega_{rf}$ (n is an integer), all of the bunches are decelerated so as to produce the maximum amplitude of the wakefield behind the bunch train. Figure 6 gives a case in which the resonant condition $n = 7$ is satisfied. If we could inject a test bunch when the wakefield is maximum-positive, we could demonstrate a most effective acceleration, though it is impossible in the present setup.

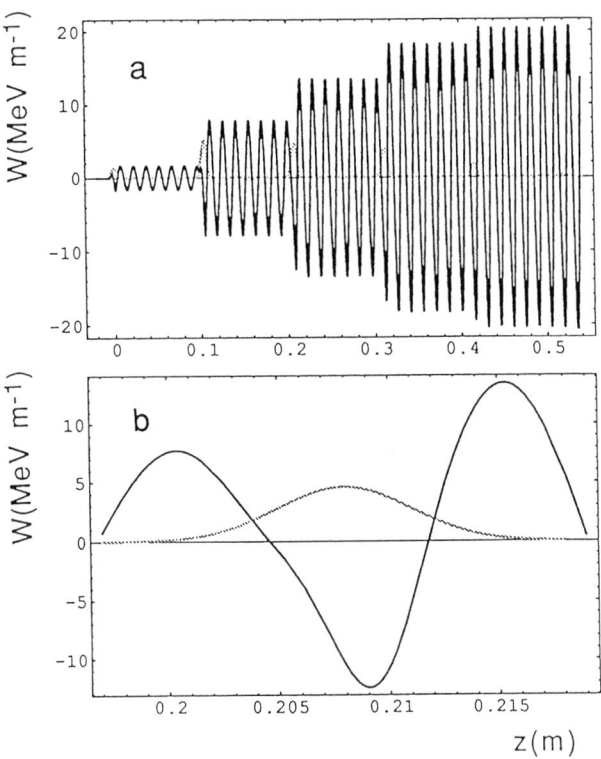

Figure 6: (a)Amplitude of the longitudinal wakefield at $n_e = 5.05 \times 10^{12}\text{cm}^{-3}$ on the axis as a function of the longitudinal position. (b)Enlargement of (a) around the third bunch.

In a plasma with a density of $\sim 10^{11}\text{cm}^{-3}$ we have directly observed barycenter energy differences.[11] However, in a denser plasma, as in the present experiments, we must take account of the fact that the bunch length is comparable to, or even longer than, the plasma wavelength. Figure 6(b) shows the plasma wakefield around the third bunch, together with the longitudinal bunch shape, when

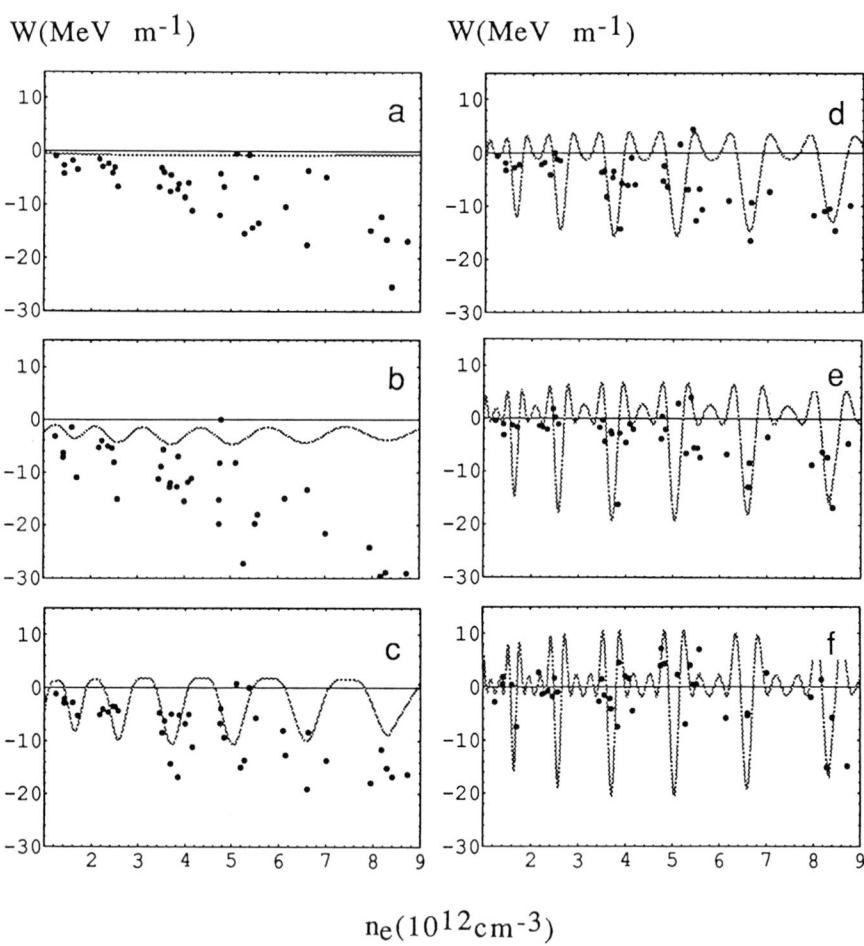

Figure 7: Amplitude of the longitudinal plasma wakefield at the bunch centers. The points indicate the observed values: (a) first bunch, (b) second bunch, (c) third bunch, (d) fourth bunch, (e) fifth bunch and (f) sixth bunch.

W(MeV m^{-1})

W(MeV m^{-1})

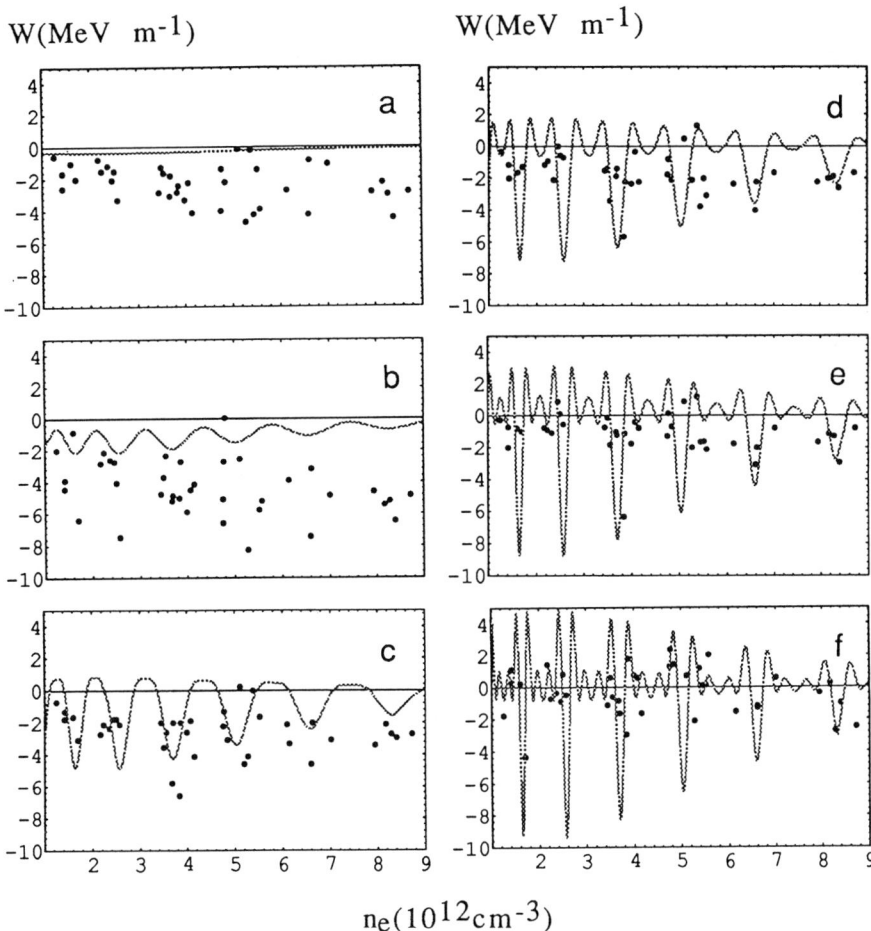

$n_e(10^{12}cm^{-3})$

Figure 8: Amplitude of the longitudinal plasma wakefield averaged over the bunch volume. The points indicate the observed values: (a) first bunch, (b) second bunch, (c)third bunch, (d) fourth bunch, (e) fifth bunch and (f) sixth bunch.

the density is $5.05 \times 10^{12} \mathrm{cm}^{-3}$. Though the main body experiences a decelerating field, the head and tail experience an accelerating field. In addition, bunches also have finite size in the radial direction. The observed barycenter energy shifts should fit the averaged wakefield over the bunch volume.

This averaged wakefield was calculated under the following assumptions and are shown by the lines in Fig. 8. It is assumed that: 1) the energy distribution in the absense of a plasma is Gaussian and is independent of the longitudinal distribution, and that 2) a plasma modifies the energy distribution without changing the longitudinal distribution. Comparing Figs. 7 and 8, we find that the averaged wakefield is a decreasing function of the plasma density in this region, in spite of the fact that the barycenter wakefield has an obtuse peak at around $n_e = 5 \times 10^{12} \mathrm{cm}^{-3}$.

3.3.   Experimental results

3.3.1.   energy shift

The energy spectra were measured alternately with and without a plasma. Examples are given in Fig. 9. The barycenter differences between the spectra with and without a plasma were calculated for all bunches as a function of the plasma density; they are also shown in Fig. 8. Because a Langmuir probe cannot indicate the absolute plasma density correctly, a conversion coefficient from the probe voltages to the plasma densities is adjusted so that the resultant plasma frequencies of the third bunch show resonances where theory predicts. The scattering of the data points is mainly due to the timewise non-uniformity of the plasma density. Though the variation during the time for the bunches to pass the plasma is negligible, the variation during the scanning of the magnetic field of the energy analyser, which requires about 50 linac beams, often amounts to $10^{11} \mathrm{cm}^{-3}$.

The agreement between the calculation and experiments given in Fig. 8 is poor. The observed energy shifts are larger than the calculated values (gray lines) in the first three bunches, comparable to the calculation in the latter bunches. Let us multiply the experimental data by the ratio of the wakefield at the bunch center to that averaged over the bunch volume in order to obtain the experimental wakefield at the bunch center. The results are given in Fig. 7. Some data suggest that the wakefield at the bunch center exceeds $30 \mathrm{MeVm}^{-1}$ in the second bunch.

The buildup of a wakefield in a train of bunches is considered to be useful in order to increase the transformer ratio.[15] If the wakefield were built up at the pace of first three bunches, the wakefield would amount to nearly $100 \mathrm{MeVm}^{-1}$ at the last bunch. However, Fig. 7 shows that the observed energy shifts do not exceed the prediction regarding the latter bunches. This must be partly because the buildup was not very suuccessful, at least in the third through the sixth bunches. A similar phenomenon has already been found,[12] in which a bunch with the maximum intensity showed an appreciable energy shift, while the bunch following just after this maximum made no energy change. Possible technical problems include the

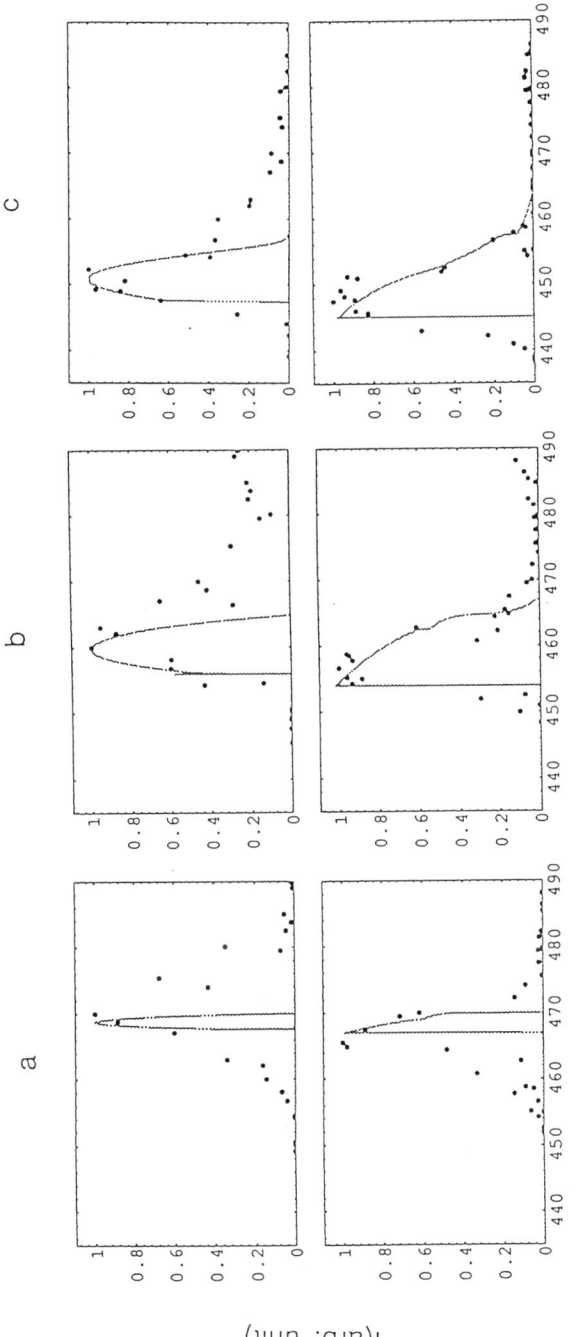

Figure 9: Energy spectra of the first three bunches in the absence of a plasma (upper) and in the presence of the plasma (lower) with a density of $n_e = 5.2 \times 10^{12} \mathrm{cm}^{-3}$. The data points show the experimentally-obtained values, while the gray lines show the theoretical values: (a) first bunch, (b) second bunch and (c) third bunch.

fact that the bunches were not on the same axis, due to the transversal wakefield.[14] The small plasma radius ($\sim$ 5mm) made this problem more serious.

3.3.2.   energy spectra

Let us examine the experimentally-obtained energy spectra in detail. We first consider the barycenter energies in the absence of a plasma. It was found that the latter bunches have smaller barycenter energies. This is because of the wakefield caused by the interaction between bunches and the linac structure; the latter bunches are more decelerated by the wakefield built up by the preceding bunches. This wakefield in the absence of a plasma was calculated and is shown in Fig. 10 as a function of the longitudinal position.[16] The experimantally-obtained barycenter energy of each bunch is also given. In the calculation, only the fundamental mode was taken into account, and the loss factor was assumed to be $3.75 \times 10^{13} \text{VC}^{-1} \text{m}^{-1}$.[14] The linac length is 80m and the initial beam energy is assumed to be 470MeV.

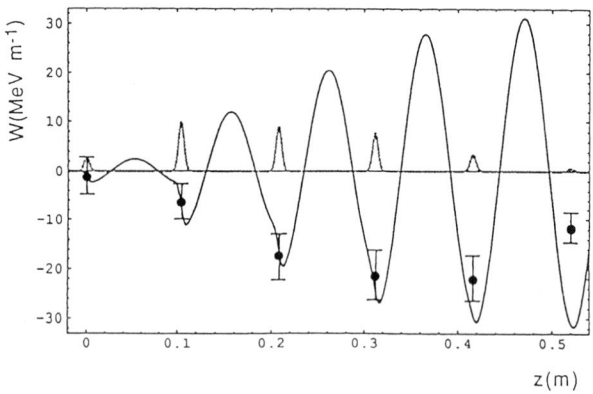

Figure 10: Longitudinal wakefield caused by the linac structure and the observed barycenter energies of the bunches.

The agreement between the experiments and the theory is excellent for the first three bunches, but gradually deteriorates in the latter bunches. This is probably because the latter bunches are off-centered due to the transverse wakefield, a characteristic of the linac structure.[14] This must contribute to the fact shown in Figs. 7 and 8 that a wakefield does not build up remarkably in the latter bunches.

Figure 9 gives three pairs of energy spectra both with and without a plasma for the first three bunches at $n_e = 5.2 \times 10^{12} \text{cm}^{-3}$. The data points show the observed values, while the gray lines are the results of a trial to explain these distributions. Let us begin with the case of no plasma. We first assumed that

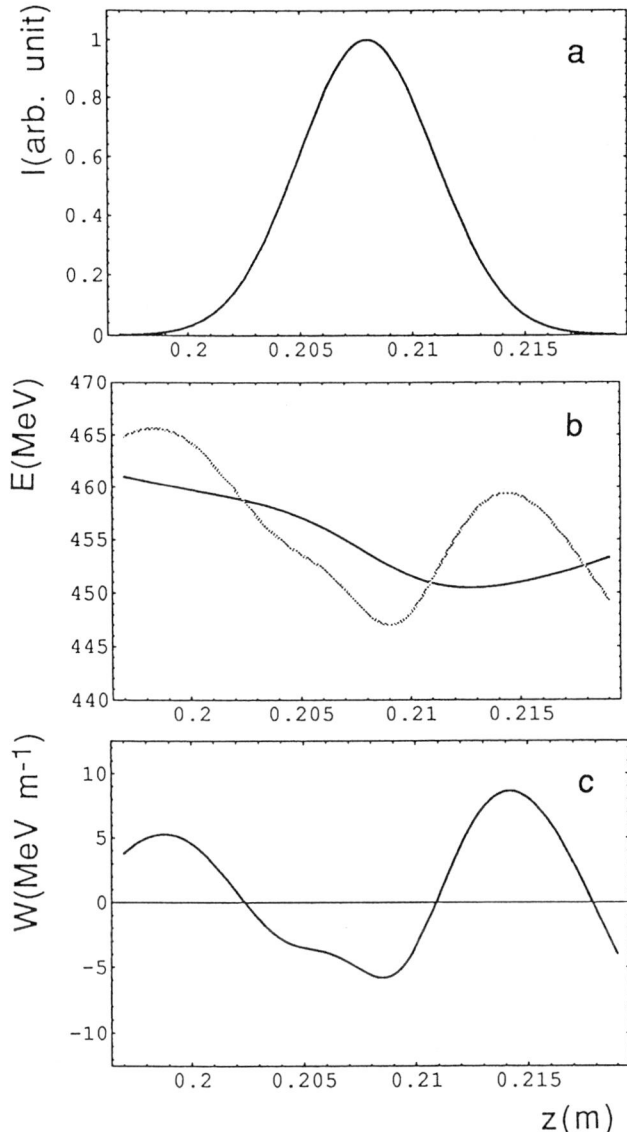

Figure 11: (a) Longitudinal particle distribution of the third bunch. (b) Beam energy dependence on the distance along the beam-axis. The black line indicates the case without a plasma, while the gray line indicates the case with the plasma of $n_e = 5.2 \times 10^{12} \text{cm}^{-3}$. (c)Wakefield change caused by a plasma of $n_e = 5.2 \times 10^{12} \text{cm}^{-3}$.

the longitudinal distribution of a bunch was Gaussian, as shown in Fig. 11(a). Measurements using a streak camera support this assumption. It was then assumed that the dependence of the beam energy upon the longitudinal position was governed solely by the wakefield caused by the linac structure (already shown in Fig. 10, but enlarged and shown again here for the third bunch by the black line in Fig. 11(b), as the beam energy dependence on the distance along the beam axis). We can imagine that Fig. 11(b) shows a three-dimensional diagram in (z, E) phase-space, the third axis of which shows the population; the mapping of this diagram onto the z-axis is given in Fig. 11(a), the Gaussian distribution. If we map it onto another axis, we can generate the energy spectrum of the third bunch, which is given by the gray line in the upper box of Fig. 9(c).

The plasma excites its own wakefield in the third bunch, which was calculated and is shown in Fig. 11(c). The bunch feels both the wakefield of the structure (the black line of Fig. 11(b)) and this plasma wakefield. The resultant trace on the (z, E) space is shown by the gray line in Fig. 11(b). From this gray line and the assumption of a Gaussian distribution, we can reconstruct the energy spectrum, which is shown by the gray line in the lower box of Fig. 9(c). Figures 9 (a) and (b) were also derived by the same procedure.

The first bunch has a broader spectra than that predicted. The second bunch has a high-energy tail in the absence of a plasma, which is wiped out by a plasma. This is a phenomenological reason why the barycenter difference with and without a plasma is larger than predicted. Aside from these observations, the calculation well predicts both the absolute positions and widths of the spectra macroscopically in the second and third bunches. Microscopically, we find that the calculated energy spectra have steep cut-off edges at the low-energy side. This is based on the assumption that the distributions in (z, E) phase space are given by lines, which were not unique-valued, but often folded in the low-energy side. The actual distribution should have a width around the lines to broaden the spectrum.

3.3.3.   distribution in the phase space

Measurement of two-dimensional particle distribution in (z, E) phase space was attempted in order to verify the analysis described above; the longitudinal profiles of the third bunch were measured as a function of the energy selected by the energy analyser. The results are given in Fig. 12. The jitter of the triggering and the poor reproducibility of beam intensities make any quantitative discussion difficult concerning the dependence of the energy spectra on the longitudinal coordinates. It is, however, evident that the plasma widens the energy spectrum in both high- and low-energy directions.

3.4.   Discussions and conclusion

The previous section reports three findings. The first one is that the observed wakefields were always comparable to or larger than the calculated values. They

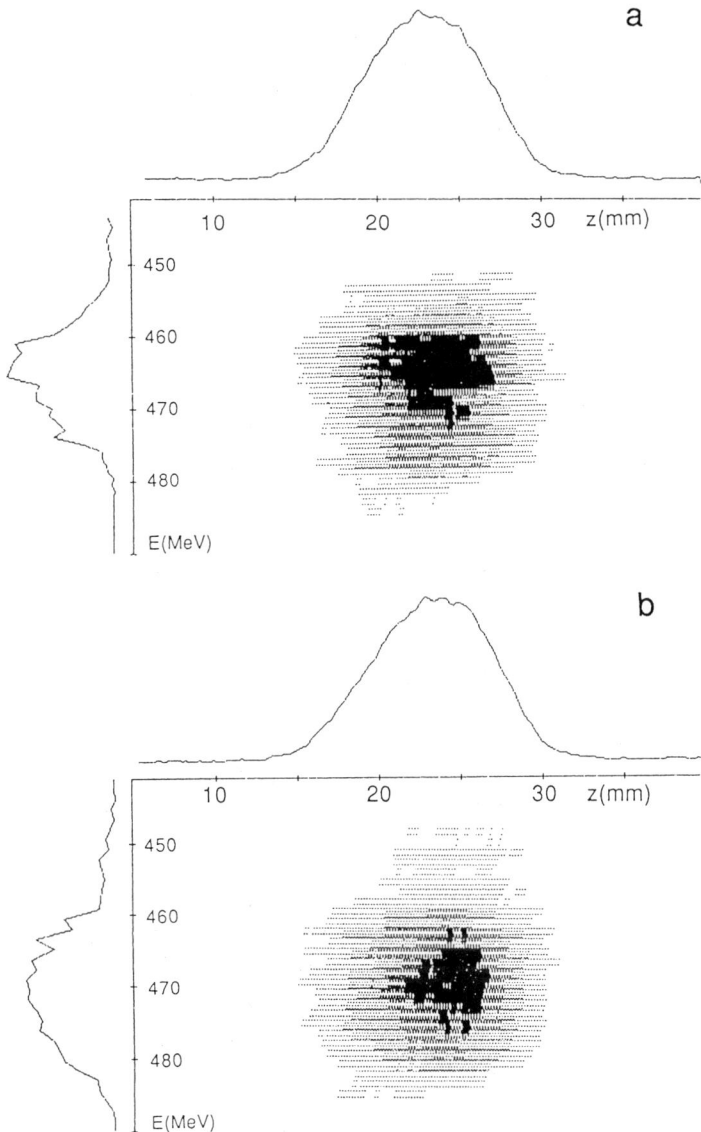

Figure 12: Two-dimensional distribution of the third bunch in (z, E) phase space in the absence of a plasma(a) and in the presence of the plasma(b) with a density of $n_e = 8 \times 10^{12} \mathrm{cm}^{-3}$.

certainly exceed 20MeVm^{-1}, and sometimes exceed 30MeVm^{-1}.

The second is that the buildup of a wakefield was not remarkable at the third through sixth bunches. This is probably because all the bunches were not on the same axis. It has been observed in this linac that the initial uniform transverse shift randomizes the transverse positions of the bunches at the end of the accelerator structure, and that the first two or three bunches still remain on the same axis even in such cases.[14]

The larst finding of large energy shifts in the preceding two bunches is puzzling. The following speculation excludes the possiblity of a nonlinear mechanism: The perturbed electron density must be approximately equal to the average electron density in a bunch, which is $\sim 5 \times 10^{11}\text{cm}^{-3}$ if the charge of the bunch is 1nC. This value is smaller than the background density by an order of magnitude in these experiments. Although self-focusing of the driving bunch can lead to an increase in the longitudinal wakefield, this self-focusing is weak in our high-γ bunches. One may suspect the existence of bunches preceding to the first one, whose energy shifts are extraordinary large. Certainly, we have not scanned the streak camera once after we found the bunches. The existence of large preceding bunches is, however, not probable, because of the following facts: 1) the linac is equipped with a subharmonic buncher which compresses the width of the bunch train to less than 2nsec; and 2) although the observed plasma-density dependence (Fig. 8 (a)) is noisy, no resonant structure was found, as in other bunches.

In conclusion, we have measured the plasma wakefiled caused by a sequence of bunches in plasmas with a density on the order of 10^{12}cm^{-3}, where the plasma wavelength is comparable, or even shorter than, the bunch length (3mm). Data processing which took into account the bunch length has shown that the experimentally-obtained wakefield was often larger than that predicted by linear theory; it certainly exceeded 20MeVm^{-1} at the bunch center. The observed wakefields were much larger than those calculated in the first three bunches, and comparable to those calculated in the last three bunches. The relatively-small energy shifts in the latter bunches were probably due to the fact that the trailing bunches were displaced when they entered the plasma.

4.   Plan of Laser Wakefield Acceleration

It is known that an ultrashort laser pulse is capable of exciting a plasma wave with a large amplitude and a relativistic phase velocity.[18,19] Rrecent progress regarding ultrashort super-intense lasers allows us to test the principle of this acceleration scheme, called laser wakefield acceleration (LWA). The intense short laser pulse with a peak power of 30TW and the pulse width of 1psec is provided by a Nd:glass laser system at Institute of Laser Engineering, Osaka University.[23] By using this laser we should be able to achieve an intensity of $10^{17} - 10^{18}\text{Wcm}^{-2}$. The intensity is sufficiently strong to create a highly-ionized plasma within an ultrashort time scale due to the multiphoton ionization and/or tunneling ionization processes. At an appropriate gas pressure, a large-amplitude wakefield is generated behind a laser pulse propagating through the plasma due to the ponderamotive force. According to a fluid model of plasma dynamics, since the phase velocity of the plasma wave is highly relativistic, the wakefield can accelerate charged particles trapped by plasma oscillation. The experiment will be able to diagnose the acceleration of relativistic electrons produced by intense laser irradiation.

4.1.   Experimental setup

4.1.1.   laser

Figure 13 shows the experimental setup. The laser will be used in three ways: to ionize gas; to produce a wakefield; and to make probing beams.

A 1.052μm Nd:glass laser pulse of 130psec duration is coupled to a single mode fiber of 1.85km length. A chirped pulse of 150psec duration and 1.8nm bandwidth at the exit of the fiber is amplified to an energy of 41J with a beam diameter of 14cm. The beam is then split in two ways. A 10J branch is used to produce test electron beams, while the main 30J pulse is compressed to a width of 1psec by a pair of gratings. The output from the compression stage is focused into a vacuum chamber containing helium or hydrogen gas with a focal spot of 100μm for gas ionization and wake excitation.

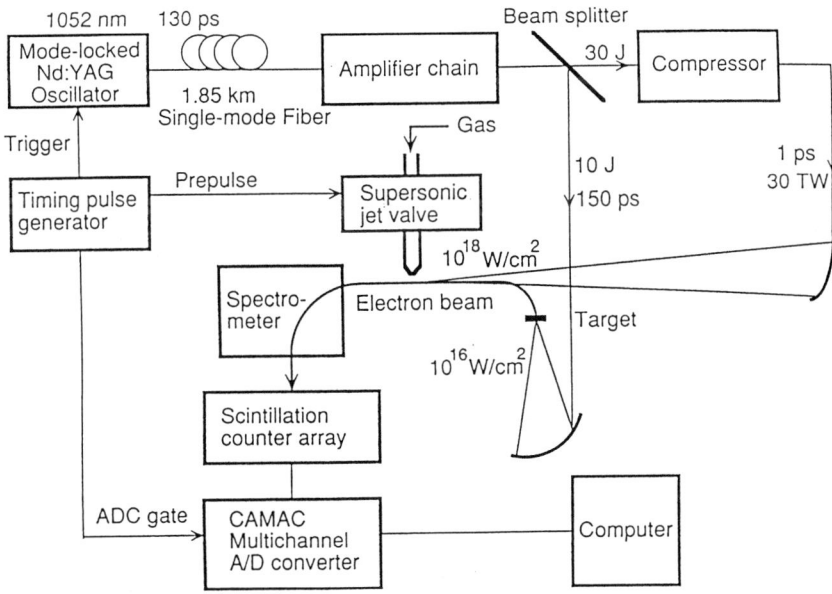

Figure 13: Diagram of the setup of the LWA experiments.

4.1.2.   gas ionization

A high-power laser can cause tunneling ionization of atoms on an ultrafast time scale.[20] The onset of tunneling ionization is predicted by a simple Coulomb-barrier model. The threshold intensity for the production of charge state Z is given by[21]

$$I_{th}(\text{Wcm}^{-2}) = 2.2 \times 10^{15} Z^{-2} (U_i(\text{eV})/27.21)^4, \tag{10}$$

where U_i is the ionization potential. It gives $1.37 \times 10^{14} \text{Wcm}^{-2}$ for the hydrogen atom. The ionization rate for hydrogen atom is given by[22]

$$W_H = 1.61\omega_{au}(10.87 E_{au}/E_0)^{1/2} \exp(-2E_{au}/(3E_0)), \tag{11}$$

where ω_{au} is the atomic unit of frequency $(4.1 \times 10^{16} \text{ sec}^{-1})$ and E_{au} is the atomic field strength $(5.1 \times 10^9 \text{ Vcm}^{-1})$. A laser with an intensity of $4 \times 10^{14} \text{Wcm}^{-1}$ can produce a fully ionized hydrogen plasma in $\sim$ 1fsec.

The appropriate plasma density required for excitation of the plasma wave will be produced by adjusting the filling pressure of gases without any electric discharge device or pre-ionizing laser pulses in these experiments. A supersonic jet valve will control the gas pressure, as shown in Fig.13.

4.1.3.   test beam production and beam diagnostics

For acceleration experiments it is necessary to use electrons with an energy that satisfies the trapping condition corresponding to the amplitude of plasma waves. An electron characterized by γ' in the wave frame is trapped in the wave of field E_z' if $eE_z'/(m_e c\omega_p) \geq \gamma' - 1$. The trapping condition in the laboratory frame thus becomes[25]

$$eE_z/(m_e c\omega_p) \geq \gamma(1 - \beta_\phi\beta) - 1/\gamma_\phi, \tag{12}$$

where β_ϕ is the phase velocity of the plasma wave and γ_ϕ is its relativistic factor, defined as

$$\beta_\phi = \frac{v_p}{c} = (1 - \frac{\omega_p^2}{\omega_0^2})^{1/2}, \gamma_\phi = \frac{1}{(1 - \beta_\phi^2)^{1/2}} = \frac{\omega_0}{\omega_p}. \tag{13}$$

The minimum threshold kinetic energy trapped by the plasma potential is about 40keV for an excitation of 10^{18} Wcm^{-2} intensity.

It is known that a large amount of electrons with energies in the MeV range are created in plasmas produced by the irradiation of an intense laser pulse on solid targets. The electron generation is explained by the Raman instability or the resonance absorption of the laser radiation. The energies of such electrons can satisfy the above-mentioned trapping condition. According to experiments involving CO_2 laser-produced plasmas, electrons with energies of up to 1.4MeV were observed in a 5°-wide cone about the target normal during the 300psec risetime of the laser pulse with an intensity of 10^{14}Wcm^{-2}.[24] The simulation of Raman forward scattering for a Nd:glass laser gives an instability threshold intensity of

$\sim 10^{16} \mathrm{Wcm}^{-2}$. The typical fluence was $\sim 10^7 \mathrm{electrons}$ $\mathrm{keV}^{-1}\mathrm{str}^{-1}$ at an energy of 1MeV.

In order to produce an electron probing beam part of the laser pulse before compression is split and 10 J out of 41J is focused onto an aluminum target with a focal spot of $30\mu\mathrm{m}$ in diameter. In order to inject electrons emitted from the target into the laser wakefield in the waist of the laser beam, a dipole magnet is used to select the electron energy in the range $0.2 - 3$ MeV. This spectrograph is placed between the solid target and the image point of the electrons so that horizontal and vertical focusing is achieved by an appropriate choice of the magnetic edge angle. Adjustment of the collimator provides an electron beam with an image diameter of $40\mu\mathrm{m}$ and an energy spread of 10% at 1MeV. The typical intensity of pulsed probing leads to $\sim 10^6$ electrons at 1MeV.

The electrons must be injected along the axis of the main laser beam at the time delay within 50psec behind the laser pulse. The optimum delay is achieved by adjusting the optical path length of two laser pulses. Electron acceleration occurrs at the waist of the laser beam characterized by the Rayleigh length in the plasma chamber. The accelerated electrons are bent by angle of 90° in the dipole field of the spectrometer placed at the exit of the plasma chamber, as shown in Fig.13. The spectrometer covers the energy range $10 - 45$MeV at a dipole field of 4.3kG. The electron detector comprises an the array of 32 scintillation counters, each of which is assembled with 1cm wide scintillators and a 1/2-in photomultiplier. The pulse heights of the detector array outputs are measured by fast multichannel CAMAC ADCs gated in coincidence with the laser pulse. The energy resolution of the spectrometer is 1.3MeV/channel.

4.2.   Estimation of acceleration

Let us assume a bi-Gaussian profile of the laser pulse with an rms pulse length σ_z and an rms radius σ_r. The laser intensity is expressed by

$$I(r,z) = \frac{2P}{\pi w^2(z)} \exp(-\frac{2r^2}{w^2(z)}), \qquad (14)$$

where P is the peak power of the laser pulse. The spot size $w(z)$ of the laser beam is[26]

$$w(z) = w_0(1 + (z/z_R)^2)^{1/2}, \qquad (15)$$

where w_0 is the radius of the beam waist, and

$$z_R = \pi w_0^2/\lambda_0, \qquad (16)$$

is the Rayleigh length in the vacuum; λ_0 is the wavelength of the laser. The longitudinal wakefield excited by a Gausssian laser pulse is written as

$$eE_z = \frac{m_e c^2 \epsilon_0}{z_R(1 + (z/z_R)^2)} \exp(-\frac{r^2}{w_0^2(1 + (z/z_R)^2)}) \cos\psi, \qquad (17)$$

where $\psi = k_p z - \omega_p t$ and

$$\epsilon_0 = \frac{\Omega_0 P \lambda_0 k_p \sigma_z}{\pi^{1/2}(m_e c^2)^2 \lambda_p} \exp(-\frac{k_p^2 \sigma_z^2}{4}). \tag{18}$$

Here, $\Omega_0 = (\mu_0/\epsilon_0)^{1/2} = 377\Omega$ is the vacuum resistivity, λ_0 the laser wavelength and λ_p the plasma wavelength.

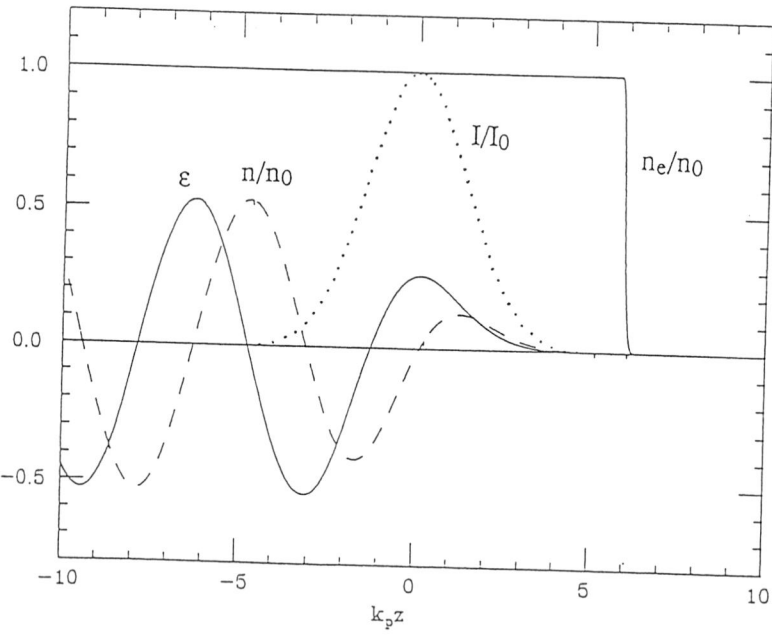

Figure 14: Evolution of the electron density of a hydrogen plasma, the density perturbation, and electric field excited by a 1 psec, 10^{18} Wcm^{-2} Nd:glass laser. The initial atom density is 2.415×10^{15} cm^{-3}.

Figure 14 shows what linear theory predicts; i.e., evolution of the electron density of a hydrogen plasma, its density perturbation and the electric field caused by a 1psec, 10^{18}Wcm^{-2} Nd:glass laser. The initial atom density is assumed to be 2.415×10^{15}cm^{-3}.

The maximum accelerating gradient is achieved at a plasma wavelength of $\lambda_p = \pi\sigma_z$. As an example, the maximum of the plasma wave excited by a 1.052μm laser with an intensity of 10^{18}Wcm^{-2} and 1 psec pulse duration leads to 2.5GeVm^{-1} at a plasma density of 2.415×10^{15}cm^{-3}. Note that the gas is ionized long before the laser peak arrives.

The maximum energy gained by a synchronized electron with a velocity equal to the phase velocity of the plasma wave is obtained by integrating the longitudinal wakefield along the laser beam axis:

$$(\Delta E)_{max} = \int_{-\infty}^{+\infty} E_z(z)dz = \pi m_e c^2 \epsilon_0 \cos \psi_s. \tag{19}$$

Here, ψ_s is the synchronous phase of the electron captured by the wave potential. A laser pulse of wavelength $\lambda_0 = 1.052 \mu m$ should be able to produce the maximum energy gain,

$$(\Delta E(\text{MeV}))_{max} = 1.49 P(\text{TW})/\tau_\text{L}(\text{psec}), \tag{20}$$

where τ_L is the pulse width in fwhm, $c\tau_L = 2\sigma_z \ln 2$. A 30TW, 1psec laser will provide an energy gain of $\sim$ 45MeV. It should be noted that the energy gain is proportional to the laser power (P).

4.3.   Discussion and conclusions

There are some phenomena which can be verified by these experiments. First is a nonlinear wave. Figure 14 gives the prediction of linear theory. However, the high density-perturbation must invoke a nonlinear steeping and periodic lengthening of the plasma wave.[27] Second is the effect of the plasma temperature. It has been reported that tunnel-ionized plasmas have high temperature.[28] It is doubtful if the prediction based on the assumption of cold plasmas is applicable. Two calculations of the influence of the temperature on wake generation are contradictory to each other.[29]

In conclusion, experiments of the LWA experiments are planned using a 1psec, 30TW Nd:glass laser. The laser is useful both in order to ionize gasses by the tunneling effect and to excite a wakefield in the resultant plasma. Particle acceleration can be demonstrated by injecting a few MeV electrons emitted from a solid target irradiated by a laser beam branched from the main laser. The expected accelerating gradient of the wakefield exceeds 2GeVm^{-1}, and the expected energy is 45MeV since the acceleration length is limited by the diffraction.

We now compare the two acceleration methods, PWA and LWA, though their acceleration mechanism is the same once a plasma is generated. The acceleration gradient of PWA is proportional to the density of the driving bunch. A longer plasma gives a greater energy gain. To the contrary, the laser power determines the energy gain of LWA, which is the product of the acceleration gradient and the accleration length. In PWA, the plasma density should be proportional to the beam density, so that the gradient is consequently proportional to n_e. On the other hand, the plasma density of LWA should agree with the laser-pulse length, so that the energy gain is proportional to $n_e^{1/2}$. One advantage of LWA over PWA is that gas ionization is providced by the tunneling process of the laser. One problem which LWA must solve is the limit of the acceleration length. In our plan using a 30TW laser, the expected energy gain is only $\sim$ 45MeV, in spite of the high acceleration gradient, 2.5GeVm^{-1}, since the acceleration length is limited by diffraction.

If we bear in mind the realization of a wakefield accelerator, the main issue concerning a comparison is the high-power short-pulse laser $v.s.$ the high-density driving beams. The production of driving beams requires a dedicated accelerator. Contrarily, the recent development of the laser has made a high-power ($\sim 10\text{TW}$), high-repetition ($\sim 10\text{Hz}$), table-top laser commertially available. This fact makes studies of LWA more attractive than that of PWA, though the hitherto achievement of PWA will make the realization of $\sim 100\text{MeVm}^{-1}$ probable within a couple of years.

References

1. A. Ogata, Y. Yoshida, N. Yugami, Y. Nishida, H. Nakanishi, K. Nakajima, H. Shibata, T. Kozawa, T. Kobayashi and T. Ueda, *Conf. Record IEEE Part. Accel. Conf.,* San Francisco, 1991, p622; to be published in **Part. Accel.**.
2. Y. Nishida, T. Ozaki, N. Yugami and T. Nagasawa, **Phys. Rev. Lett.** 66(1991) 2328.
3. P. Chen, **Part. Accel. 20**(1987) 171.
4. H. Kobayashi, T. Ueda, T. Kobayashi, S. Tagawa and Y. Tabata, **Nucl. Instr. and Meth. B10/11**(1985) 1004.
5. J. B. Rosenzweig, R. Schessow, B. Cole, C. Ho, W. Gai, R. Konecny, S. Mtingwa, J. Norem, M. Rosing and J. Simpson, **Phys. Fluids B2**(1990) 1376.
6. H. Nakanishi, Y. Yoshida, T. Ueda, T. Kozawa, H. Shibata, K. Nakajima, T. Kurihara, N. Yugami, Y. Nishida, T. Kobayashi, A. Enomoto, T. Oogoe, H. Kobayashi, B. S. Newberger, S. Tagawa, K. Miya and A. Ogata, **Phys. Rev. Lett. 66**(1991) 1870.
7. S. Tagawa, *Proc. 7th Symp. Accel. Sci. Tech.,* Wako, Japan, 1991, p293.
8. J. J. Su, T. Katsouleas, J. M. Dawson, P. Chen, M. Jones and R. Keinigs, **IEEE Trans. Plasma Sci. PS-15**(1987) 192.
9. P. Chen, J. M. Dawson, R. W. Huff and T. Katsouleas, **Phys. Rev. Lett. 54** (1985) 693.
10. J. B. Rosenzweig, D. B. Cline, B. Cole, H. Figueroa, W. Gai, R. Konecny, J. Norem, P. Schoessow and J. Simpson, **Phys. Rev. Lett. 61**(1988) 98.
11. K. Nakajima, A. Enomoto, H. Kobayashi, H. Nakanishi, Y. Nishida, A. Ogata, S. Ohsawa, T. Oogoe, T. Shoji and T. Urano, **Nucl. Instr. and Meth. A292**(1990) 12.
12. A. Enomoto, H. Kobayashi, K. Nakajima, H. Nakanishi, Y. Nishida, A. Ogata, S. Ohsawa, T. Oogoe, Y. Suetsugu and T. Urano, *Proc. 2nd. Eur. Part. Accel. Conf.,* Nice, 1990, p634.
13. R. W. Boswell, **Plasma Phys. Controlled Fusion 26**(1984) 1147.
14. Y. Ogawa, T. Shidara and A. Asami, **Phys. Rev. D43**(1991) 258.
15. K. Nakajima, **Part. Accel. 32**(1990) 209.

16. K. L. F. Bane, P. B. Wilson and T. Weiland, in *Phys. High Energy Particle Accelerators,* AIP Conf. Proc. No.127, (Am. Inst. Phys., New York, 1983) p875.

17. M. Takao, Y. Ogawa, T. Shidara and A. Asami, *Conf. Record IEEE Part. Accel. Conf.,* San Francisco, 1991, p497.

18. T. Tajima and J. M. Dawson, **Phys. Rev. Lett. 43**(1979) 267.

19. P. Sprangle, E. Esarey, A. Ting and G. Joyce, **Appl. Phys. Lett. 53**(1988)2146.

20. L. V. Keldysh, **Zh. Eksp. Teor. Fiz. 47**(1964) 1945.

21. N. M. Penetrante and J. N. Bardsley, **Phys. Rev. A43**(1991) 3100.

22. M. V. Ammosov, N. B. Delone and V. P. Krainov, **Zh. Eksp. Teor. Fiz. 91**(1986) 2008.

23. K. Yamakawa, H. Shiraga, Y. Kato and C. P. J. Barty, **Opt. Lett. 16**(1991) 1593.

24. C. Joshi, T. Tajima and J. M. Dawson, **Phys. Rev. Lett. 47**(1981) 1285; S. Aithal, P. Lavigne, H. Pepin, T. W. Johnston and K. Estabrook, **Phys. Fluids 30**(1987) 3825.

25. N. A. Ebrahim, P. Lavigne and S. Aithal, **IEEE Trans. Nucl. Sci. NS-32** (1985) 3539.

26. K. Nakajima, A. Enomoto, H. Nakanishi, A. Ogata, Y. Suetsugu, Y. Kato, Y. Kitagawa, K. Mima, H. Shiraga, K. Yamakawa, T. Shoji, Y. Nishida, N. Yugami, M. Downer, W. Horton, B. Newberger and T. Tajima, to be published in *Proc. 3rd Eur. Part. Accel. Conf.,* Berlin, 1992.

27. R. Bingham, U. De Angelis, M. R. Amin, R. A. Cairns and B. McNamara, **Plasma Phys. Controlled Fusion 34**(1992) 557.

28. W. P. Leemans, C. E. Clayton, W. B. Mori, K. A. Marsh, A. Dyson and C. Joshi, **Phys. Rev. Lett. 68**(1992) 321.

29. T. Katsouleas and W. B. Mori, **Phys. Rev. Lett. 61**(1988) 90; A. Ts. Amatuni, S. S. Elbakian and E. V. Sekhpossian, **Part. Accel. 36**(1992) 241.

SELF-GUIDING OF HIGH-INTENSITY LASER PULSES FOR LASER WAKE FIELD ACCELERATION

D. Umstader and X. Liu

Center for Ultrafast Optical Science
University of Michigan, Ann Arbor, MI 48109-2099

ABSTRACT

A means of self-guiding an ultrashort and high-intensity laser pulse is demonstrated both experimentally and numerically. Its relevance to the laser wake field accelerator concept is discussed. Self-focusing and multiple foci formation are observed when a high peak power ($P > 100$ GW), 1 μm, subpicosecond laser is focused onto various gases (air or hydrogen). It appears to result from the combined effects of self-focusing by the gas, and de-focusing both by diffraction and the plasma formed in the central high-intensity region. Quasi-stationary computer simulations show the same multiple foci behavior as the experiments. The results suggest much larger nonlinear electronic susceptibilities of a gas near or undergoing ionization in the high field of the laser pulse. Although self-guiding of a laser beam by this mechanism appears to significantly extend its high-intensity focal region, small-scale self-focusing due to beam non-uniformity is currently a limitation.

INTRODUCTION

With the recent development of compact ultra-short pulse and high intensity lasers[1], proof-of-principle tests of proposed novel laser-plasma electron accelerators[2][4] are now feasible. A practical laser accelerator, however, will require a long high-intensity laser-plasma interaction region, which has yet to be produced. Thus a means of self-guiding intense lasers must be found. Towards this end, self-guiding mechanisms involving laser-plasma effects[6], such as relativistic self-focusing and cavitation, have been investigated[3]. Recently, however, it has been shown theoretically[4] that the conditions required for self-focusing a pulse with a Gaussian temporal profile by these methods are incompatible with those required for laser wake field acceleration.

Laser wake field acceleration requires high-intensity ultra-short laser pulses with laser pulse widths (τ_l) satisfying the condition $\tau_l \sim \tau_p$, where $\tau_p = 2\pi/\omega_p$ is the plasma period, the inverse of the plasma frequency ($\omega_p = 4\pi n_f^2/m$). This places a restriction on the density of free electrons (n_f) for a given laser pulse width. For instance, for a picosecond laser, the suitable plasma density is $n_f = 10^{16}$ cm^{-3}. Because of collective plasma effects, however, in order for the above methods of self-guiding to work, several plasma periods are required in a pulse width. In other words relativistic self-focusing will only work at densities that are higher than are suitable for laser wake field acceleration.

We report in this paper the observation, both experimentally and numerically, of self-focusing and multiple-foci formation of subpicosecond laser pulses, by a method that does not require several plasma periods in a pulse width. It involves both plasma and, in addition, neutral gas effects on the beam propagation. Although multiple sparks produced by longer laser pulses at atmospheric pressures were observed in gas breakdown experiments[7], and self-focusing was identified as the mechanism, here we concentrate on self-guiding under conditions relevant to laser wake field acceleration, i.e., with ultra-short laser pulses and lower gas densities.

Self-focusing or self-defocusing, in either atomic or ionized media, originates from a radial intensity-dependent refractive index gradient. Self-focusing occurs when a negative gradient acts as a positive lens, and vice versa for self-defocusing. The mechanism responsible for the index change depends on the laser pulse duration and intensity. For cw and long pulses, thermal[8], ponderomotive[5], and molecular orientational nonlinear effects[9] are dominant. For pulses of picosecond and shorter duration, only the electronic nonlinear susceptibility and ionization are important[9]. Finally, at the highest laser intensities, relativistic effects must be considered[6].

Self-guiding by the mechanism that we will be discussing arises naturally when an intense laser is focused into a gas. A plasma will be formed wherever the intensity exceeds the ionization threshold. In the gas region, a negative radial index gradient will be created by the intensity-dependence of the nonlinear susceptiblity, focusing the beam. In the plasma region, a positive radial index gradient will be created by the intensity-dependence of the ionization, defocusing the beam. A focused beam will also defocus due to diffraction even without an index gradient. A combination of these three effects may result in multiple self-focusing.

EXPERIMENTAL ARRANGEMENT

A Nd:glass laser system[10], based on the principle of chirped-pulse amplification[11], was used in the experiment. Laser pulses of 1.05-μm wavelength, 400-fs duration, and 10^5-intensity contrast ratio were focused into a chamber that was backfilled with either air or hydrogen. Lenses of either of two focal lengths were used, 20 cm (f/9), or 40 cm (f/18). Pressures of 1 - 500 Torr in the chamber were measured to within an accuracy of 0.5% with a capacitance manometer. When focused into an evacuated chamber using the f/9 lens, the laser beam was measured to have a two-and-a-half times diffraction-limited spot size of about 40 μm, a peak intensity up to 10^{16} W/cm^2, and an energy per pulse in the range of 20 - 140 mJ. The laser power used in the experiment was kept below the threshold for relativistic self-focusing[6].

Without considering the plasma defocusing effect, the power level of the laser pulses in these experiments exceeds the critical power, $P_{cr} = \lambda^2 c/16\pi^2 n_2$[12], for self-focusing to occur in the medium with nonlinear refractive index n_2. This self-focusing would lead to catastrophic intensities in the focused region. However, plasma defocusing prevents such a process from occuring, and actually leads to multiple foci formation, as shown in both the experiments and numerical simulations.

In order to image the location of the foci, light emitted from the laser-produced plasma was recorded with a digital CCD camera looking perpendicular to the laser axis. An infrared filter was placed in front of the CCD camera to block any stray laser light.

EXPERIMENTAL RESULTS

At low gas densities ($p \leq 25$ Torr) and low laser energies, no evidence for self-focusing was observed. Images recorded under these conditions (using the f/9 lens) indicate the formation of a plasma column of length roughly twice the Rayleigh range, as shown in Fig. 1(a). The midpoint of this plasma column serves as reference point indicating the location of the best focus in the case when no significant self-focusing or self-defocusing occurs. Beam propagation was investigated under various conditions, by varying either the laser intensity, polarization state, focusing geometry, or the gas density.

As the gas density was increased to about $p = 50$ Torr, the plasma column moved backward from the reference point toward the incident laser beam, and a second focus developed (Fig. 1(b)). As the gas density increased even further yet, three or more foci appeared (Fig. 1(c)–(d)). The energy threshold for $p = 50$ Torr appears to be about 40 mJ, which corresponds to a peak power of about 100 GW. The number of self-focused foci and their positions varied from shot to shot as the energy in the laser pulse fluctuated. The $f = 40$ cm lens produced similar effects, with correspondingly longer focal region (Fig. 2(a) – (d)). The general behavior of the focus moving toward the incident laser beam with increasing density is expected because as the gas density increases, so does the electronic nonlinear refractive index, and the resulting positive lens effect causes the beam to focus earlier.

The propagation was sensitive to the laser intensity and beam quality, but not polarization. Linear polarization, both perpendicular and parallel to the CCD camera axis, as well as circular polarization, produced similar results. When the pressure was held constant at some value ($p > 200$ Torr), self-focusing was observed for laser energies varying between 30 – 120 mJ, generally with higher energies producing more foci. Some nonuniformity of the laser beam profile was present, as evidenced by the on-axis image of the laser focal spot made when the laser was focused into vacuum, shown in Fig. 3(a). Because the incident laser beam has spatial nonuniformity, and as such is unstable against small scale self-focusing when propagated in a nonlinear medium, the beam will eventually break-up. An on-axis image made with the chamber filled with 50 Torr of hydrogen (Fig. 3(b)) confirmed this. The trajectory of some of the beam foci appear to be curved in Figs. 1 and 2. We believe this is caused by nonlinear bending also due to the nonuniformity of the incident laser beam[13].

THEORETICAL MODEL

We modeled the laser beam propagation taking into consideration the above-mentioned effects. A focused high-intensity laser pulse may produce a plasma near the center of the beam by a number of ionization mechanisms. For our laser parameters, tunneling ionization dominates. We use only hydrogen with a single electron in the model for simplicity. Because of the short laser pulse duration, we also neglect any thermodynamic motion of the plasma during the interaction. The medium is assumed to be a mixture of free and bound electrons. If we follow a particular point in the pulse's temporal envelope (or assume the quasi-stationary condition), the free electron density as a function of the laser field can be approximated by $n_f = n_t \exp[-E_s/E(\mathbf{r})]$ and the bound electron density by

$$n_b = n_t - n_f = n_t(1 - \exp[-E_s/E(\mathbf{r})]), \tag{1}$$

where $E = E(\mathbf{r})$ is the laser electric field, E_s the saturation field for ionization, and n_t the total electron density, which is proportional to the original gas density (pressure).

The bound electrons, in the high laser field, contribute to the nonlinear refractive index of the medium. Usually the nonlinear index can be written as $\eta_{NL}(\mathbf{r}) = n_2|E(\mathbf{r})|^2$, where n_2, the nonlinear refractive index constant, is positive and proportional to the bound electron density. The nonlinear index, n_2, is related to the third order nonlinear susceptibility $(\chi^{(3)})$ by,

$$n_2 = \frac{12\pi n_b}{\eta}\chi^{(3)}_{1111}(-\omega;\omega,\omega,-\omega) = n_2' \times n_b. \tag{2}$$

The free electrons, on the other hand, contribute to the plasma index of refraction. This is given by $\eta_{plasma} = \sqrt{1 - n_f/n_c} \simeq 1 - (1/2)(n_f(\mathbf{r})/n_c)$ when the density of free electrons is much less than the critical density, or, $n_f(\mathbf{r}) \ll n_c$. The combined index of refraction is then given by $\eta = 1 + n_2|E(\mathbf{r})|^2 - (1/2)(n_f(\mathbf{r})/n_c)$. In the paraxial approximation, we can thus write the nonlinear Schroedinger equation for the complex amplitude of the laser field $E = A\exp[i(k_o z - \omega t)]$ as

$$i\frac{\partial A}{\partial z} = -\frac{1}{2k_o}\nabla_\perp^2 A - k_o\left(n_2|A|^2 - \frac{1}{2}\frac{n_f}{n_c}\right)A. \tag{3}$$

Substituting n_2 and n_f as functions of the laser field amplitude, we get:

$$i\frac{\partial A}{\partial z} = -\frac{1}{2k_o}\nabla_\perp^2 A - k_o\left\{n_2'n_t\left(1 - \exp\left[-\frac{E_s}{|A(\mathbf{r})|}\right]\right)|A|^2 - \frac{n_t}{2n_c}\exp\left[-\frac{E_s}{|A(\mathbf{r})|}\right]\right\}A. \tag{4}$$

The first term on the right hand side represents beam diffraction, the second, self-focusing, and the third, plasma defocusing.

SIMULATION RESULTS

Rather than attempting an analytical solution of an equation as nonlinear as Eq. (4), we instead solved it numerically. A cylindrically symmetric radial beam profile was assumed. Some computer simulation results are presented in Fig. 4(a) – (d), in which it was further assumed that the initial beam profile was Gaussian. The ionization saturation intensity, $I_s = (c/8\pi)E_s^2$, is taken to be $I_s = 10^{15}$ W/cm^2[14][15]. I_s can also be estimated from the tunneling ionization rate[16]

$$\nu = 3.8 \times 10^{11}\epsilon_I^{1/2}\left(\frac{I}{\epsilon_I^3}\right)^{1/4}\exp\left[-2.4 \times 10^6\left(\frac{\epsilon_I^3}{I}\right)^{1/2}\right],$$

where ϵ_I is the ionization energy, expressed in eV, of the gas molecule (~ 14 eV for hydrogen). For subpicosecond pulses ν must be on the order of 10^{13} sec^{-1} for a significant portion of the gas to be ionized, which also gives I_s to be on the order of 10^{15} W/cm^2. It can be seen from these figures that for certain range of values of $n_2 = n_2'n_t$ in Eq. (4), the simulation results show the same behavior as the experiments, i.e., self-focusing, backward movement of the focus, and multiple foci formation.

It was found from these simulations that, by using values of n_2 that were measured at intensities much lower than the ionization threshold[17], the laser never reaches its vacuum spotsize or peak intensity (Fig. 4(b)). Clamping of the peak intensity, and hence maximum plasma density, has been observed both experimentally and numerically in 10.6-μm laser interactions[18]. However, the self-guiding effects observed in our experiments are reproduced in the simulations only when much larger values of n_2 are used, about an order of magnitude larger. This may be explained by the fact that when the gas is close to ionization, the bound electrons become highly anharmonic, and higher order nonlinear effects may become important. In this case, higher order terms in the intensity should be included in the nonlinear index, η_{NL}, as

$$\eta_{NL} = (n_2 + n_4|E|^2 + \cdots)|E|^2.$$

On the other hand, the perturbative approach to the nonlinear susceptibility in the high laser field can totally break down, as evidenced in above threshold ionization studies[19]. Therefore the above power expansion in terms of $|E|^2$ may not be applicable in the high field limit. The nonlinear properties of the gas may be very different compared to the lower field situations, and warrant further investigations. Also, the large nonlinear current predicted to result when a gas is undergoing ionization—specifically tunneling ionization[20]—may also be important for the laser beam propagation. Finally, to better simulate the experiment, attenuation, and both the spatial and temporal shape of the laser pulse must be considered, all of which limit the distance the beam can be guided.

RELEVANCE TO LASER WAKE FIELD ACCELERATION

In this section we discuss the relevance of the above results to the laser wake field accelerator concept. The question that must be answered is whether or not self-guiding by this mechanism will occur at densities suitable for wakefield generation.

As mentioned in the introduction, the laser pulse width used in this concept determines the required density of free electrons. The converse is also true; higher density requires a shorter pulse width. This may be made more quantitative as follows. From the relationship between the laser pulsewidth and the plasma period, $\tau_l = \tau_p$, required for wake field acceleration, it can be seen that the resonant density is inversely proportional to the square of the pulse width, $\tau_l = 2\pi/\omega_p \propto n_e^{-1/2}$. It should also be noted that the amplitude of the wake field increases linearly with the density, as given by Poison's equation: $\nabla \cdot E = 4\pi\rho = 4\pi(n_i - n_e)e$. Thus for the highest resonant density, and also wake field amplitude, the shortest laser pulse width should be used, assuming a fixed intensity. For the shortest pulsewidths currently attainable, 400 fs, still maintaining the intensity required for wakefield generation (see article on laser technology by Mourou and Umstadter in these Proceedings), the resonant density is $n_e = 5 \times 10^{16}$ cm^{-3}. We must now address issue of whether or not self-guiding at this density is possible.

The results shown above indicate that self-guiding is possible at densities of ... for a laser intensity of .... At higher intensities ....

SUMMARY

In summary, the observations reported here, of self-focusing with multiple foci by a high-intensity subpicosecond laser, may provide a solution to the important problem of extending the focusing region in high-intensity laser wake field interactions. This multi-focusing mechanism may only be understood by considering both gas and plasma contributions to the index of refraction. However, a better understanding of the nonlinear properties of a gas in a high laser field is required. Finally, small-scale self-focusing due to beam non-uniformity remains the outstanding issue limiting self-guiding by this mechanism.

ACKNOWLEDGEMENTS

This work was partially funded by the National Science Foundation Center for Ultrafast Optical Science, contract #PHY8920108. The authors thank C. Y. Chien for helpful assistance with the laser operation, and G. Mourou for his encouragement during the experiment.

REFERENCES

[1] G. Mourou and D. Umstadter, *Development and Applications of Compact High-Intensity Lasers*, Phys. Fluids B **4**, 2315.

[2] T. Tajima and J. M. Dawson, Phys. Rev. Lett **43**, 267 (1979); L. M. Gorbunov and V.I. Kirsanov, Zh. Eksp. Teor. Fiz. **93**, 509 (1987) [Sov. Phys. JETP **66**, 290 (1987)].

[3] A. B. Borisov, A. B. Borovskiy, B. B. Korobkin, A. M. Prokhorov, O. B. Shiryaev, X. M. Shi, T. S. Luk, A. McPherson, J. C. Solem, K. Boyer, and C. K. Rhodes, Phys. Rev. Lett. **68**, 2309 (1992); A. B. Borisov, A. B. Borovskiy, O. B. Shiryaev, B. B. Korobkin, A. M. Prokhorov, J. C. Solem, T. S. Luk, K. Boyer, and C. K. Rhodes, Phys. Rev. A **45**, 5830 (1992).

[4] P. Sprangle, E. Esarey and A. Ting, Phys. Rev. Lett., 23 Apr (1990); P. Sprangle, E. Esarey, A. Ting and G. Joyce, Appl. Phys. Lett. **53**, 2146 (1988); A. Ting, E. Esarey, and P. Srangle, Phys. Fluids B **2**, 1390 (1990); P. Sprangle, E. Esarey, and A. Ting, Phys. Rev. A **41**, 4463 (1990); P. Sprangle, C. M. Tang and E. Esarey, IEEE Trans. Plasma Sci. PS-15, 145 (1987); E. Esarey, A. Ting, P. Sprangle, and G. Joyce, Comments on Plasma Phys. Controlled Fusion **12**, 191 (1989).

[5] C. Max, "Physics of the Coronal Plasma in Laser Fusion Targets," in *Laser-Plasma Interaction,* R. Balian and J. C. Adam, eds. (North Holland, Amsterdam, 1982).

[6] C. Max, J. Arons and A. B. Langdon, Phys. Rev. Lett. **33**, 209 (1974); G. Z. Sun, E. Ott, Y. C. Lee, and P. Guzdar, Phys. Fluids **30**, 526 (1987); P.

Sprangle, C. M. Tang, and E. Esarey, IEEE Trans. Plasma Sci. **PS-15**, 145 (1987). J. Solem, T. S. Luk, and K. Boyer, IEEE Jour. of Quant. Elect. **25**, 2423 (1989).

[7] V. V. Korobkin and A. J. Alcock, Phys. Rev. Lett. **21**, 1433 (1968). A. J. Alcock, C. DeMichelis, V. V. Korobkin, and M. C. Richardson, Appl. Phys. Lett. **14**, 145 (1969). F. N. Bunkin, I. K. Krasyuk, V. M. Marchenko, P. P. Pashinin, and A. M. Prokhorov, Soviet Phys. JETP **33**, 717 (1971).

[8] T. Afshar-rad, L. A. Gizzi, M. Desselberger, F. Khattak, O. Willi, and A. Giulietti, Phys. Rev. Lett. **68**, 942 (1992); P. E. Young, H. A Baldis, T. W. Johnston, W. L. Kruer and K. G. Estabrook, Phys. Rev. Lett. **63**, 2812 (1988); P. E. Young, R. P. Drake, E. M. Campbell and K. G. Estabrook, Phys Rev. Lett. **61**, 2336 (1988). M. J. Herbst, J. A. Stamper, R. R. Whitlock, R. H. Lehmberg, and B. H. Ripin, Phys. Rev. Lett. **46**, 328 (1981); R. Craxton and R. L. McCrory, J. Appl. Phys. **56**, 108 (1984).

[9] Y. R. Shen, *Principles of Nonlinear Optics* (Wiley, New York, 1984).

[10] Y. Beaudoin, C. Y. Chien, S. Coe, J. Tapie, and G. Mourou, submitted to Optics Letters (1992).

[11] P. Maine, D. Strickland, P. Bado, M. Pessot and G. Mourou, IEEE J. Quantum Electron. **QE-24**, 398 (1988).

[12] O. Svelto, in *Progress in Optics*, E. Wolf, editor, North-Holland, Amsterdam, 1974, Vol. XII, pp. 1–51.

[13] I. Golub, Y. Beaudoin, and S. L. Chin, J. Opt. Soc. Am. **B5**, 2490 (1988)

[14] X. Liu, D. Umstadter, S. Coe, C. Chien, E. Esaray, and P. Sprangle, in *OSA Proceedings on Short Wavelength Coherent Radiation: Generation and Applications, 1991*, P. Bucksbaum and N. Ceglio, eds. (Optical Society of America, Washington, D.C., 1991), pp. 7–11.

[15] R. Dewhurst, J. Phys. D. Appl. Phys., **11**, 191 (1978)

[16] G. M. Weyl, in *Laser-Induced Plasmas and Applications*, L. J. Radziemski and D. A. Cremers, editors, M. Dekker, New York, 1989, pp. 1–67.

[17] Y. Shimoji, A. Fay, R. Chang, and N. Djeu, J. Opt. Soc. Am. **B6**, 1994 (1989)

[18] W. P. Leemans, *Topics in High Intensity Laser Plasma Interaction*, PhD. Thesis, (UCLA, Los Angeles, 1991) p. 59.

[19] J. Eberly, J. Javanainen, K. Rzazewski, Phys. Reports, **204** 331-383 (1991)

[20] F. Brunel, J. Opt. Soc. Amer. B **7**, 521 (1990).

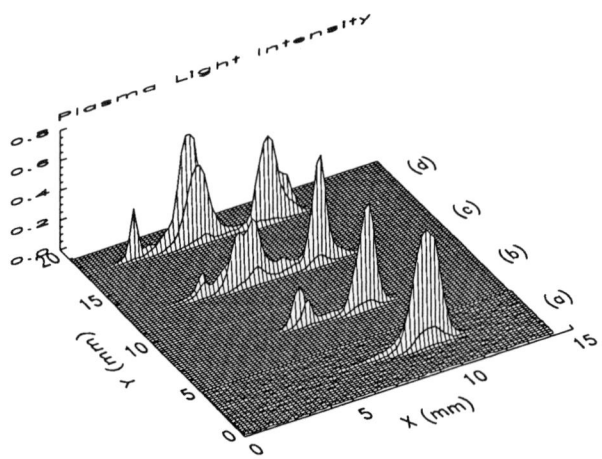

Figure 1: Normalized plasma column images with $f = 20$ cm lens in hydrogen showing multiple foci formation as pressure increases. The laser peak power for (b)–(d) was about 200 GW (80 mJ per pulse). (a) $p = 25$ Torr, Incident laser pulse energy ~ 20 mJ, focus position reference shot; (b) $p = 50$ Torr; (c) $p = 200$ Torr; (d) $p = 350$ Torr. The beam propagates from left to right.

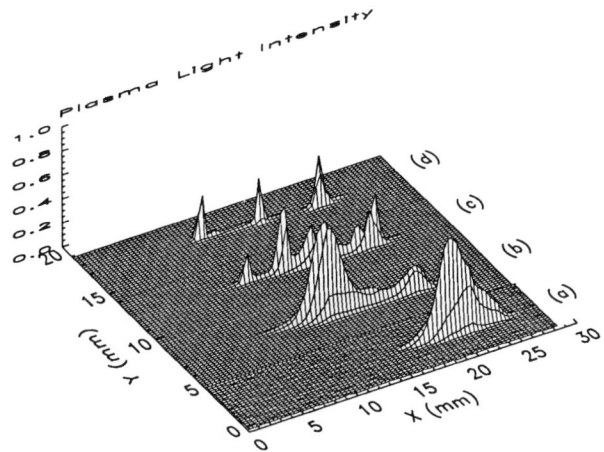

Figure 2: Normalized plasma column images with $f = 40$ cm lens in hydrogen. The laser peak power for (b)–(d) was again about 200 GW. (a) $p = 30$ Torr, 40 mJ; (b) $p = 100$ Torr; (c) $p = 250$ Torr; (d) $p = 500$ Torr. The beam propagates from left to right.

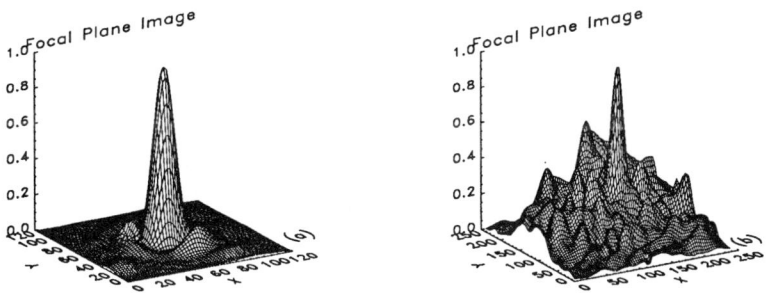

Figure 3: On-axis image of the laser focal plane showing beam break-up: (a) in vacuum, laser energy 10 mJ; (b) $p = 50$ Torr, 40 mJ, the intensity in vacuum would be 7×10^{15} W/cm^2.

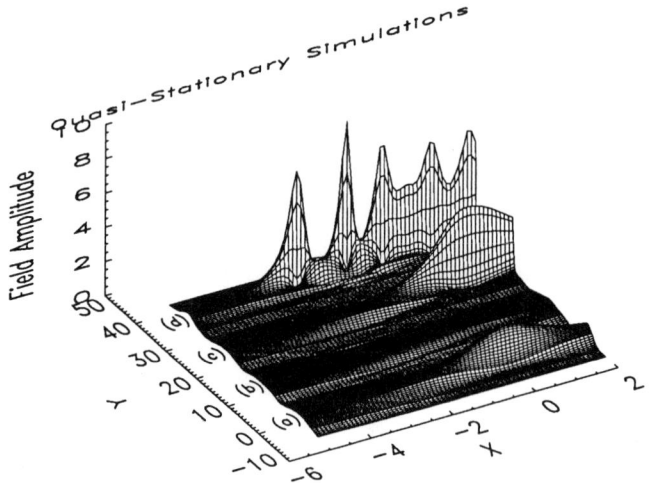

Figure 4: Numerical solution of Eq. (4). An $f/10$ lens is used for 1 μm laser wavelength. The x-axis is in units of vacuum Rayleigh range, the y-axis is in units of focused vacuum spot size. The laser field is in units of the ionization saturation field E_s, which is taken to be 2.9×10^6 esu (corresponding to saturation intensity $I_s = 10^{15}$ W/cm^2). For all cases, the focused vacuum field strength is $E = 6.5 \times 10^6$ esu ($I = 5 \times 10^{15}$ W/cm^2), and values of $n_2 = n_2' n_t$ are given for atmospheric pressure. For (b)–(d) the density is 100 Torr of hydrogen. (a) in vacuum, the beam comes to a focus at $x = 0$; (b) $n_2 = 1.5 \times 10^{-16}$ esu (ref. [17]), the peak field strength is 3.5 times lower than in vacuum; (c) $n_2 = 7.1 \times 10^{-16}$ esu; (d) $n_2 = 1.0 \times 10^{-15}$ esu.

Development and Study of an Excitation System for a Superatmospheric CO2 Amplifier with a Large Discharge Gap

G.A. Baranov, A.A. Kuchinsky, V.P. Tomashevich,
A.G. Maslennikov

D.V. Efremov Scientific Research Institute
of Electrophysical Apparatus

Metallostroy, 189631 St. Petersburg, Russia

Abstract

The results of the development of an excitation system for an ultra-short optic pulse CO amplifier at 10.6µm wavelength are given. Volumetric self-sustained discharge with a preliminary gas ionization by soft X-ray radiation was used for an active medium excitation. The discharge was formed in a system of ring-shaped electrodes with a 4 cm discharge gap in a volume of 300 cm^3. An electron gun with a bremsstrahlung anode operating at 60 kV and discharge current of 1500 ns was used. A six stage pulse generator supplied up to 400 kV at the electrodes and specific energy deposition in the discharge was approximately 150 J/1*atm.

The conditions necessary for formation of a volume discharge at active medium pressures of 1-10 atm were studied. Stable discharge in the excitation system with an interelectrode distance of 4 cm at the active medium (CO2/N2/He=1/1/8) pressure of 8 atm and energy deposition of 150 J/1* atm was realized.

Finally, the possibilities and ways for further increase of the discharge stability at high gas pressures, and the calculated characteristics of a laser amplifier with a discharge volume of 5*5*60 cm^3 and a working pressure up 10 atm are discussed.

1.   Introduction

It is well known that for the solution of a series of fundamental physical problems and, in particular, for practical implementation of different laser accelerator projects, we need laser systems with a high peak power and a beam pulse duration of < 10 ps[1-3]. Obtaining ultrashort laser pulses is a rather complicated problem and has been focused upon durgin the recent years[4].

The most encouraging successes achieved in the IR-range of spectrum 10.6µm may be attributed to the application of the double-stage semiconductor Nd/YAG laser beam-controlled shutters[4-6]. However, specific features, inherent for pulse compression systems, set an upper bound on a value of an output generator energy, approximately 0.01 mJ. Therefore, an increase of a laser pulse energy (its duration being unchanged) can be obtained only as a result of its further amplification. For this purpose, a high-pressure (7-10 atm) amplifier can be used.

Development and construction of the above said C02-amplifiers with an aperture of >10 cm% is a high-lighted task, depending strongly on the solution of both fundamental and engineering problems. Here the main obstacle is the realization of a volume discharge in a high-pressure gas mixture in systems with large interelectrode gaps.

2.   Experimental Set Up

This paper presents the results of development and studies of an excitation system applied in high-pressure CO2-amplifiers with the discharge gap (DG) of 4 cm. In these researches an experience gained in development of large discharge gap atmospheric pressure lasers pumped by a volumetric self-sustained discharge (VSD)[7-9] was extensively used.

Soft X-ray radiation (SXR) was used for preliminary ionization of an active medium. The efficiency and expediency of the SXR may be attributed to its higher penetration in comparison with UV-radiation and, in addition, to higher structural reliability and the simplicity of the X-ray ionizer, compared to an e-beam device. An increasing dependence of initial electron density (n) upon molecular gas concentration in the active mixture and its total pressure is also of significant value.

VSD was excited in the discharge gap of 4 cm height and 300 cm in volume. A high-voltage specially-shaped disk electrode made of aluminum with a plane part 10 cm in diameter was used.   A ground electrode through which SXR penetrated the active volume was manufactured of brass grid. A plate or a film separating the high pressure active medium from an X-ray source was positioned under this grid.

An explosive-emission diode with a thick bremsstrahlung anode (A1, 0.2 mm)[10] served as an SXR source. Maximum energy of bremsstrahlung radiation was E = 6 keV, and the SXR pulse duration was 1500 ns.

3.   Experimental Results and Discussion

Two methods were used for VSD formation: the first consisted of a fast switching on of a discharge supply source (high rate of the DG voltage rise) - called the "short front" regime.   In this case, a low-inductive six-stage pulse voltage generator (GIN) with the following parameters was used: L = 150 nH, the capacity - C = 16.7nF, the voltage pulse front duration -t(f) = 30 ns, the RC decay constant = 20 ms, the time delay between excitation and SXR pulses -t(d) = 500 ns. The second method, called "slow front" regime, involved the Hovel-Fitch generator instead of the GIN with t(f) = 1000 ns. An SXR pulse was applied to the active volume 500 ns after voltage was supplied. The active gas mixture was CO2:N2:He = 1:1:8.

The "short front" regime allowed us to excite VSD at a gas pressure in the DG up to Pmax = 1.5 atm. Minimum electric field strength in the DG was (E/P) min = 14 KV/cm*atm, where P is gas pressure. Under these conditions, SXR penetrated the active volume through 1 mm thick aluminum plate; changing the polarity of the voltage applied to the DG (we mean here the use of photoemissive properties of a cathode) made it possible to increase Pmax up to 3 atm and to reduce (E/P min to 123 Kv/cm*atm) without any special efforts. Our next step was a 500 ns pulse delay relative to the moment of the Fitch generator[9] which gave us a chance for VSD realization at Pmax and (E/P) min of 7 atm and 10.5 kV/cm*atm, respectively. And, at last, the replacement of aluminum separating window for a 0.01 mm maylor film allowed us to increase an ultimate pressure up to 8 atm and produce a (E/P) min value to 9 kV/cm*atm (at energy deposition v150 J/1*atm). Obtained results are likely to be commented as density of electrons (produced under SXR) simultaneously reducing their lifetime because of recombination and attachment processes. Therefore, VSD formation in the "slow front" regime appeared to be preferable. In this case, an SXR pulse was applied to the DG at the moment when voltage at the electrodes reached the values at which electrons were effectively multiplied by means of collisional ionization; as to recombination and attachment, they are no more so highly important as in the "short front" regime. A density increase in the near to the cathode[11] due to photoemission, also contributes to higher VSD stability. Shift of the SXR spectrum in the DG to the long wavelength range (which is achieved by replacing of the aluminum plate with maylor film) will also make a pronounced effect[12].

4. Conclusion

In conclusion, let us consider some ways and opportunities to attain higher VSD stability, which can be also used to increase the maximum pressures of the active mixture, higher content of molecular gases and specific energy deposition into the discharge.

1. One of these ways is an application of electrodes (namely, cathode) exhibiting in anisotropic-resistive properties[13]. Experiments on the atmospheric pressure CO2-lasers have shown that the above cathodes provide higher VSD stability, not less than twice (in terms of specific energy deposition and duration of stable burning)[14].

2. Introduction of easily ionizing impurities into the active mixture in the "slow front" regime also results in higher VSD stability marging [15-16]. However, there are definite obstacles for the use of similar impurities in the devices with the SXR-preionization because of the possible plasma-chemical reactions, causing degradation of the active mixture.

3. Highly interesting are applications of SXR sources based on the runaway electron effect. These sources are characterized by lower long-wavelength boundary of X-ray quanta, approximately 10 keV[8].

4.   For high pressure VSD realization, a combined excitation scheme involving a high-voltage nanosecond transformer for fast multiplication of original electrons may appear to be promising[11]. It is evident that for the development of the super-atmospheric C02-amplifier with a large discharge gap, both all known methods for VSD stabilization and new advanced technical solutions are to be used in combination.

At present, research on the above mentioned trends is being done at Efremov Institute (St. Petersburg). At the end of 1992 the start up of an experimental prototype of similar laser with the discharge volume of 5*5*60 cm 3 is planned. At the active mixture pressure up to 10 atm, the VSD excitation system should provide formation of stable discharge with an energy deposition not less than 150 J/1* atm at the molecular gas content of approximately 20%.

References

1. "Laser-driven electron positron colliders", *Phys. Today*, 1983, **36, N 2**, 19-21.
2. Batchelor et al., *Proceedings of the 1989 Particle Acceleration Conference*, March 20-23, 1989, 273.
3. W.D. Kimira, D.Y. Wang, M.A. Piestrup, A. Fauchet, J.A. Eddinghoefer and R.H. Pantell, *IEEE J. Quant. Elec.*, 1982, **QE-19**, 239.
4. Pogorelsky, A.S. Fisher and J. Veligdan, P. Russell, "Spatial Dynamics of Picosecond CO2 Laser Pulses Produced by Optical Switching in GE." *SPIE QE/Laser '91*, 1991, Los Angeles, CA, January 20-25, 240.
5. P.B. Corkum, *IEEE J. Quant. Elec.*, 1985, **QE-21**, 216.
6. S.A. Jamison and A.V. Nurmico, *Appl. Phys. Letts.*, 1978, **33**, 598.
7. V.A. Burtsev, A.G. Gordeichik, A.A. Kuchinsky, V.A. Rodichkin, V.A. Smirnov, V.P. Tomashevich, "Kvantovaia elektronika," 1988, **15** 1376.
8. S.L. Kulakov, A.A. Kuchinsky, A.G. Maslennikov, U.V. Rybin, V.A. Smirnov, V.P. Tomashevich, I.V. Shestakov, "ZHTF", 1990, 60, N12, 43.
9. A.G. Gordeichik, A.G. Maselnnikov, A.A. Kuchinsky, V.A. Rodichkin, V.A. Smirnov, V.P. Tomashevich, I.V. Shestakov, E.G. Yankin, "Kvantovaia elektronika', 1991, 18, 1173.
10. M.A. Vasilevsky, V.A. Rodichkin, I.M. Roiofe, E.G. Yankin, "ZHTF", 1985, 55, 1118.
11. R.S. Taylor, *Appl. Phys. B.*, 1986, 41, N 1, 1.
12. A.V. Kozyrev, T.D. Korolev, G.Z. Mesyats, Y.N. Novoselov, A.M. Prokhorov, V.S. Skakun, B.F. Tarasenko, S.A. Genkin, "Kvantovaia elektronica," 1984, 11, 524.
13. A.A. Velikin, M.A. Kanatenko, V.A. Pivovar, I.V. Podmoshensky, T.D. Sidorova, "TVT", 1988, 26, 37.
14. A.D. Berezin, M.A. Kanatenko, A.A. Kuchinsky, V.P. Tomashevich. Inter. Conf., "Physics of glow temperature of plasma", Minsk, 1991, part 2, 127.
15. V.V. Apollonov, G.G. Baitsur, A.M. Prokhorov, K.N. Firsov, "Pisma ZHTF", 21978, 12, 135.
16. V.V. Apollonov, G.G. Baitsur, V.P. Minenkov, A.M. Prokhorov, B.V. Semkin, K.N. Firsov, B.G. Shubin, A.V. Yushin, "Kvantovaia elektronika," 1987, 14, 220.

ON THE POSSIBILITY FOR EXPERIMENTS ON PLASMA WAKE-FIELD ACCELERATION IN NOVOSIBIRSK

A.A. Bechtenev, B.N. Breizman, P.Z. Chebotaev, I.A. Koop, A.M. Kudryavtsev, V.M. Panasyuk, Yu.M. Shatunov, A.N. Skrinsky, I.B. Vasserman
Budker Institute of Nuclear Physics, Novosibirsk, Russia 630090

ABSTRACT

We propose to build a new research facility to study a Plasma Wake-Field Acceleration (PWFA) where an accelerating gradient of several hundred MeV/m exists over a macroscopic distance. The Institute has an electron beam source with parameters that are uniquely suitable for this experiment. An essential element of the facility will be a novel technique for beam modulation at submillimeter wavelength. Our numerical simulations show that the modulated high energy beam can drive a quasistationary nonlinear plasma wave with a controllable phase of the accelerating field.

INTRODUCTION

The development of particle acceleration technology to higher accelerating gradients is a key issue for the next generation of linear colliders. In a conventional accelerating structure, it is difficult to obtain a gradient much higher than 100 MeV/m. Therefore, it is highly desirable to pay special attention to the development of the alternative schemes that have yet to be extensively explored and may even appear somewhat exotic at the moment. An option of particular interest is the plasma wake-field accelerator (PWFA). The basic physical principle of this concept[1,2] has been tested experimentally in 1988 in the USA at the Argonne National Laboratory.[3,4] More recently, a similar experiment has been performed in Japan[5] with somewhat improved parameters. Though the experiments[3-5] present a highly convincing proof of the feasibility of the PWFA, there are still a number of further problems to solve for practical application of this concept. At the moment, an important step would be the demonstration of the PWFA with an accelerating gradient of several hundred MeV/m over a macroscopic distance. This would actually open an area in parameter space that is inaccessible to conventional accelerators. To pursue this goal we propose to build a new research facility at the Budker Institute of Nuclear Physics in Novosibirsk. Such an experiment takes advantage of the Institute's extensive expertise in accelerator and in plasma physics and can use existing hardware.[6]

Instead of using a driving electron beam from a linac we propose to have a driver from the storage ring, BEP, which is capable of producing a very low emittance beam of high energy and current. An important advantage of such a high energy driving beam is that when the beam is properly shaped it is essentially the equivalent of a high precision accelerating structure. It can keep the quasistationary traveling wave in the plasma with a controllable phase of the accelerating field. We also propose a new technique for submillimeter modulation of the driving beam. With such modulation, the driver matches to a high density plasma that is required for the generation of a large amplitude plasma wave. It follows from our numerical simulations[6,7] that the nonlinear wake-field excited by the driver is suitable for particle acceleration for a reasonably long time.

Along with the advances in developing the PWFA, the proposed facility has some other attractive options, such as the experimental study of collective beam-plasma interaction in a new range of parameters, beam focusing by a plasma lens, the testing of various vacuum and dielectric cavities as accelerating structures, etc. By being the facility at a "desktop" scale compared to that needed for typical accelerator experiments, and by invoking ideas from different fields of physics, the proposed facility can be very convenient for an active and productive collaboration of different research groups. It will also provide valuable educational opportunities.

The feasibility of a PWFA-based collider is strongly dependent not only on the results of the proposed experiment, but also on whether other technological problems are solved such as the problem of staging. However, it is obviously necessary for the design of a large-scale device to first have a single accelerating section working efficiently. This particular problem can be fully and appropriately addressed with the proposed experimental facility. It will provide a relevant physical and technological background for further steps at a very moderate cost.

BASIC PHYSICAL ISSUES

The concept of the PWFA stems from the idea of using a plasma as a resonator in the accelerating structure to avoid using a vacuum resonator with walls exposed to strong electric field. The required accelerating field can be produced by charge separation in a plasma. An upper limit of the field is determined by plasma nonlinearity being of the order of $(4\pi nmc^2)^{1/2}$, where n is the plasma density. It should be emphasized that high plasma density prevents the wake-field from breaking nonlinearly. The characteristic wavelength of the field is $2\pi c/\omega_p = (\pi mc^2/ne^2)^{1/2}$. To provide a high efficiency of wave excitation, the current of the driving beam must change considerably over the scale $2\pi c/\omega_p$ (which is a rather small scale in a dense plasma). The technical problem of producing either a sharply cut or modulated beam is one of the major difficulties for the PWFA. A reasonable compromise can be achieved when the plasma density is of the order of 10^{15} cm^{-3}. Then the limiting gradient is 3 GeV/m for a modulation wavelength of 1mm.

It is easier to excite the wake-field with a train of electron bunches (all spaced one plasma wavelength apart) than with a single bunch of higher density.[6-7] To reach the wave breaking limit with a single bunch, the density of the bunch must be comparable to the plasma density, whereas with N bunches in the train, the density can be reduced to n/N. However, it might be difficult to use efficiently more than 10 to 20 bunches in the driving beam. This is because to gain from a larger number of bunches would require a very uniform plasma density, and also because the motion of plasma ions can break the synchronism between the driver and the excited wave.

To simulate the excitation of the wave numerically we have developed a code which solves the problem for the "rigid" (high energy) driving beam with a fully nonlinear and three-dimensional fluid description of plasma electrons. This code provides a convenient frame for optimizing beam parameters in the experiment and interpreting experimental data.

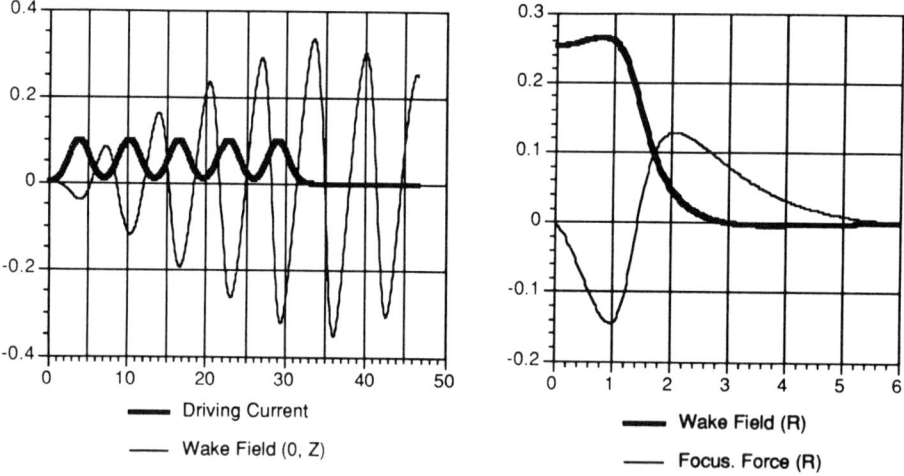

Fig. 1   Excitation of the wake-field by five electron bunches spaced one plasma wavelength apart, each with peak density of $0.1n$. 1a - normalized driving beam current and wake-field. 1b - normalized radial profiles of the wake-field and focusing force at the right end point on the wake-field longitudinal profile. The wake-field and the focusing force are normalized to $(4\pi nmc^2)^{1/2}$.

The results of a sample computational run for $n = 10^{15}\,\text{cm}^{-3}$, and a modulated driving beam with a Gaussian radial profile and with a peak density $n_b = 10^{14}\,\text{cm}^{-3}$, are shown in Fig. 1. Figure 1a shows the longitudinal profiles of the beam density and the axial electric field E_z on the beam axis. The beam moves to the left. The radial profiles of E_z and the focusing force, F, at the right endpoint of the wake-field's longitudinal profile ($z\omega_p/c = 46$) are shown in

Fig. 1b. Both radius and axial coordinate are normalized to $c/\omega_p = 0.168\,\text{mm}$. The field is normalized to $(4\pi nmc^2)^{1/2} = 3.08\,\text{GeV/m}$ and the beam density is normalized to the plasma density. The calculations show that, in spite of non-linear effects, the wake-field has a regular structure over several periods behind the driving beam. Also, there is a focusing force acting on the trailing particles near the maximum of the accelerating field (for the electrons, a negative sign of F corresponds to focusing). Similar focusing and defocusing forces act on the driving beam. They impart transverse momenta to beam electrons, whereupon the longitudinal velocity of the beam shifts downward from the velocity of light. As a result, the phase of the accelerating field changes gradually, which limits the total energy gain. To study this dephasing, one needs a more sophisticated kinetic code (now under development).

SCENARIO OF THE EXPERIMENT

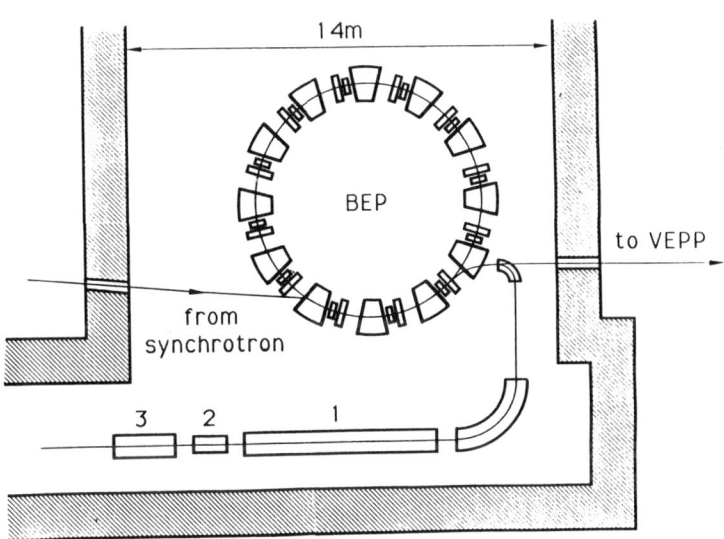

Fig. 2 Layout of the experiment. 1 - beam modulator; 2 - plasma; 3 - energy analyzer

The overview of the experiment is shown in Fig. 2. The source of the driving beam will be the electron-positron booster, BEP,[8] of the VEPP-2M collider, which is now in operation in the Budker Institute of Nuclear Physics. The booster operates in a single bunch regime. The electron beam extracted from BEP will be transferred to the PWFA channel with two 90° bending magnets. There is an option of having a bending bunch compressor in the transfer line. The BEP and PWFA facilities are located in the same shielded experimental room. After having passed the deflecting magnets and quadruple lenses the beam will enter the modulating section. The section consists of two sets of resonant cavities with a chopper in between. The RF magnetic field in the first set of cavities imparts transverse momenta to the electrons. The cavity dimensions and the excited mode are chosen to meet the condition that the "head" and the "tail" of the beam achieve opposite momenta. Then the beam will be rotated inertially in a free space over an angle of about 45° and cut into several equally spaced bunches with a one-dimensional cutting grate (chopper). The chopper is combined with the lenses which reverse the direction of beam rotation. Then the head-tail difference in the transverse momenta of beam electrons will be eliminated by a set of compensating cavities to have the modulated beam aligned as it goes through the plasma section. By choosing a proper width and position of the last chopper slit the witness particles will be cut from the driving beam with the same chopper. An energy and angular distribution of the electrons that pass through the plasma will be measured to reveal the wake-field effects.

TECHNICAL COMPONENTS

BEP

The BEP storage ring design and construction were specialized for two goals:

1. To have a large acceptance for high efficiency of positron production for VEPP-2M experiments.

2. To produce and study a high intensity beam ($N \sim 10^{12}$) with low emittance as a source for the future linear colliders.

The design parameters of the BEP electron beam are:

Energy $E = 120 - 850\,\mathrm{MeV}$
Maximum current $I = 2A(N = 10^{12})$
RMS beam length $\sigma_y = 10\,\mathrm{cm}$
Vertical emittance $\varepsilon_z = 10^{-8}\,\mathrm{cm\ rad}$
Horizontal emittance $\varepsilon_x = 6 \times 10^{-6}\,\mathrm{cm\ rad}$

The measured beam parameters are plotted in Fig. 3 versus beam current. The energy spread $\Delta E/E$, the total horizontal beam size x_t and the betatron beam size x_β are in good agreement with the values calculated from the quantum fluctuations of synchrotron radiation and intra-beam multiscattering. The related horizontal betatron emittance, ε_x, is $5 \times 10^{-6}\,\mathrm{cm\ rad}$ at a beam current of 0.5 A.

The vertical emittance is at least one order of magnitude less than the horizontal one, but it still requires precise measurement.

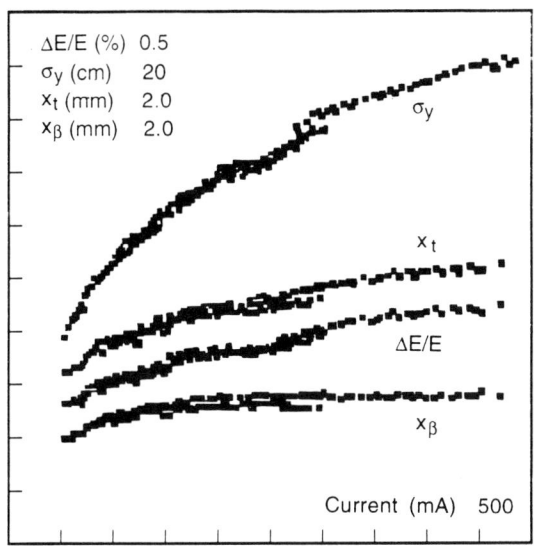

Fig. 3 Measured BEP-beam parameters at 360 MeV beam energy. The values in the upper left corner correspond to the upper endpoint on the vertical axis; $\Delta E/E$ — energy spread, σ_y — RMS beam length x_t — total horizontal size, x_β — betatron size.

The actual RMS beam length inferred from Fig. 3 is longer than the designed one. This results from the distortion of the longitudinal potential well with the voltage induced by the beam at the inductive impedance of the vacuum chamber. To reduce the length of the beam to the design value or even smaller value, we plan to install an additional superconducting resonator with a 30 times increase of qU, where q is the harmonic number and U is the RF-amplitude. An appropriate resonator is being constructed.

Beam transfer line

As shown in Fig. 2, this line starts from the channel that goes from the BEP to the VEPP-2M storage ring. The line has two 90° bending magnets with a straight section in between. The first magnet has a strong field (up to 5T) and relatively small radius (43 cm). The second, conventional, magnet, has a 150 cm radius. This scheme of the line allows the construction of the PWFA facility within the biologically shielded room of the BEP.

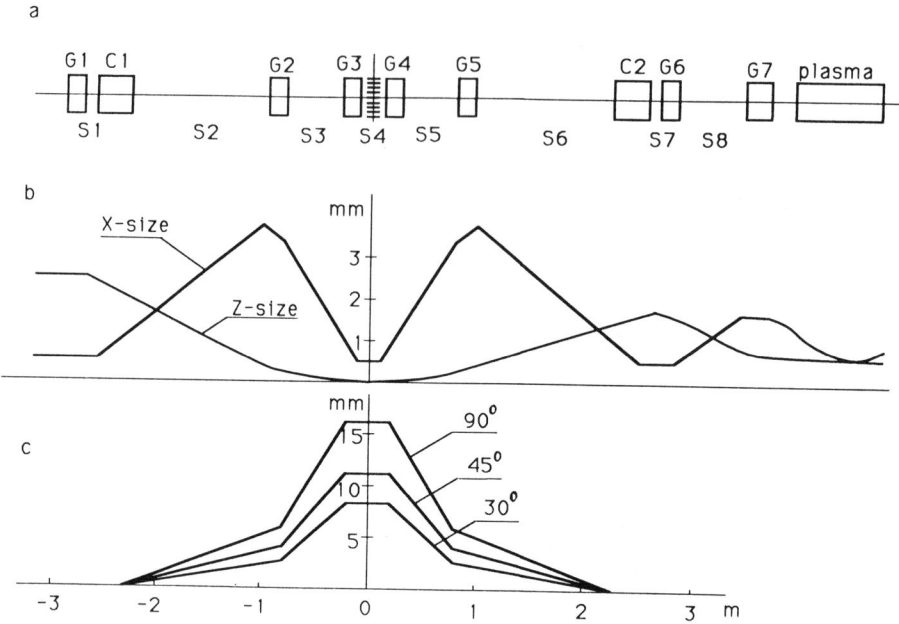

Fig. 4. Beam modulator and plasma section. 4a - elements of the optical system. 4b - evolution of beam transverse dimensions. 4c - trajectories of beam electrons for different phases $\phi = 360°l/\lambda, l$ - coordinate of the electron in the beam frame. The calculations performed for the beam energy 700 MeV, RF power per cavity 0.15 MWt with 8 cavities in a set, RF frequency about 3 GHz.

The electron optical system of the beam modulator shown in Fig. 4a is designed to meet the following requirements:

1. To have the cavities deflect the beam in z-direction.
2. To provide deep modulation, which requires the thickness of the beam (in z-direction) be less than the spacing of the chopping grate.
3. To keep the transverse beam dimensions small all along its path through the plasma.
4. To have a pair of defocusing lenses in order to decrease the distance from the deflecting cavities to the chopping grate.

For the components of the electron optical system we use the following notations: G_i for the quadruple lenses, S_i for the straight sections, and C_i for the resonant cavities.

Element	Parameters	
	Gradient (kG/cm)	Length (m)
G1	-0.89	0.2
S1	—	0.1
C1	—	0.4
S2	—	1.3
G2	1.43	0.2
S3	—	0.5
G3	-0.74	0.2
S4	—	0.2
G4	-0.74	0.2
S5	—	0.5
G5	1.43	0.2
S6	—	1.3
C2	—	0.4
S7	—	0.1
G6	-1.15	0.2
S8	—	0.5
G7	1.28	0.3

Table 1: The components of the electron optical system

The evolution of beam transverse dimensions as the beam passes the optical system is shown in Fig. 4b (for the BEP beam currently available). The beam thickness at the chopper, σ_z, is 5×10^{-3} cm. The cross section of the beam in the plasma does not exceed 1×1 mm^2.

The sets of cavities which are designed to make the beam rotate in the z-y-plane are shown in Fig. 5. Each cavity is a simple cylindrical resonator of length $\lambda/2$.

The cavities are excited at the E_{010} mode. The beam passes through the cavity near the wall where the magnetic field has its maximum value. The transverse momentum is given to the electrons by the Lorentz force. The RF oscillations in neighboring cavities of the set have opposite phases. The deflection angle, α, of the electron at the output is given by

$$\alpha = \Delta P_z / P = 1.2 \times 10^{-3} (W/N)^{1/2} N (\lambda/10)^{1/4} (800/E) \sin \phi ,$$

where W is the total feeding power in megawatts, N is the number of cavities in the set, λ is the vacuum wavelength of the RF field in centimeters, E is the beam energy in megaelectronvolts, and ϕ is the phase related to the coordinate, l, of the electron in the beam frame ($\phi = 2\pi l/\lambda$). The Q factor of the cavity is taken to be 17000. The trajectories of beam electrons with different phases are shown in Fig. 4c.

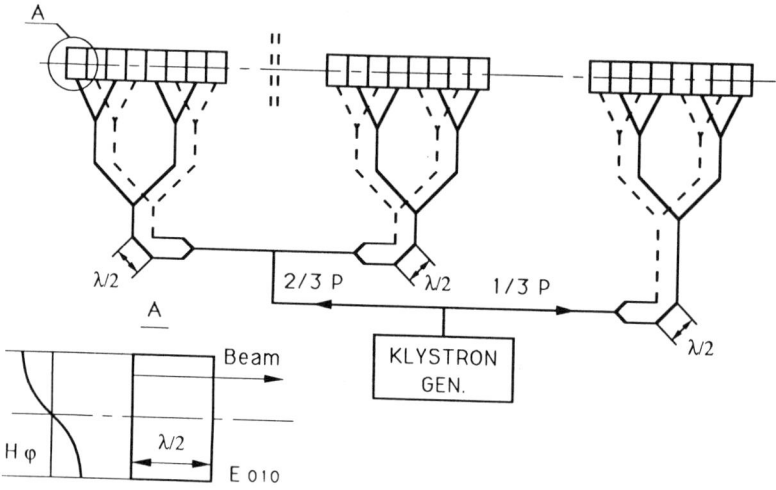

Fig. 5.   Feeding of modulator and analyzer cavities

Chopping grate

The width of the molybdenum grate plate in the direction of beam propagation is chosen to be of the order of the radiative length, i.e. 3-5 mm. For the beam rotated over 45°, the thickness of the plates, as well as plate spacing, must be equal to half a plasma wavelength. An electron that is slowed down or scattered by the grate will be absorbed by the collimating diaphragms. All this will make the beam deeply modulated at the output. The last slit of the grate will be movable in order to have a trailing witness electron bunch at a controllable distance from the driving beam.

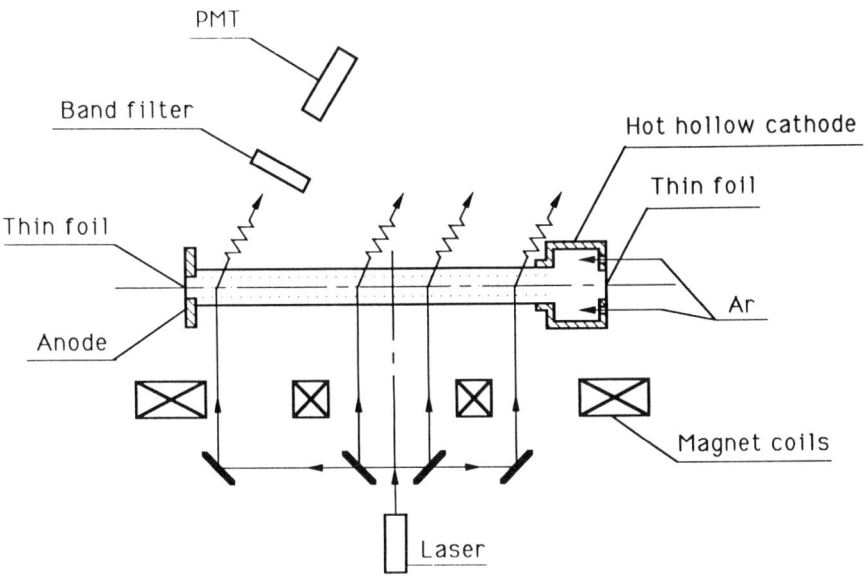

Fig. 6. Principal scheme of the plasma section with Thomson scattering diagnostic.

The proposed experiment requires a plasma with an electron density n in a range of $10^{15}\,\mathrm{cm}^{-3}$. An appropriate length of the plasma column is about 1m, with a diameter of 2 cm. To provide the conditions for the resonant excitation of the wave by a modulated beam with 10 to 20 driving electron bunches, the longitudinal profile of the plasma density must be sufficiently uniform and reproducible from shot to shot; an accuracy of 5% is sufficient. A relevant technique for creating such a plasma is a discharge with a hot hollow cathode in argon in a strong longitudinal magnetic field (about 1 T). This type of discharge has been studied in Ref. 9. It can provide the required plasma parameters at a rather moderate power (about 30 kWt) and gas flux (about $10l$ tor/s). The proposed scheme of the plasma section is shown in Fig. 6. The profile of the plasma density will be measured with a Thomson scattering technique and Langmuir probes.

Energy analyzer

The beam passed through the plasma will be collimated to an angular spread of less than 5×10^{-4} and a radius of 0.2 mm. The electrons will then be swept horizontally with the set of cavities similar to that used in the beam modulator. The electrons will also be deflected vertically by a stationary magnetic field. A high resolution two-coordinate secondary emission detector will be used to measure the beam charge distribution in the x-z-plane. From the results of these measurements, an electron energy can be found as a function of the electron position, l, in the beam frame. This will give a distribution of the wake-field along the beam axis.

REFERENCES

1. P. Chen, J.M. Dawson, R.W. Huff, and T. Katsouleas, Phys. Rev. Lett. **54**, 693 (1985).

2. R.D. Ruth, A. Chao, P.L. Morton, and P.B. Wilson, Particle Accelerators **20**, 171 (1985).

3. J. Rosenzweig, D. Cline, B. Cole, *et al.*, Phys. Rev. Lett. **61**, 98 (1988).

4. J. Rosenzweig, P. Schoessow, B. Cole, *et al.*, Phys. Rev. A **39**, 1586 (1989).

5. A. Ogata, Proceedings of International Topical Conference on Research Trends in Coherent Radiation Generation and Particle Accelerators, February 11-13, 1991, La Jolla, California.

6. B.N. Breizman, P.Z. Chebotaev, A.M. Kudryavtsev, *et al.*, Proceedings of the VIII International Conference on High-Power Particle Beams (BEAMS-90), July 2-5, 1990, Novosibirsk.

7. B.N. Breizman, D.L. Fisher, P.Z. Chebotaev, and T. Tajima, Proceedings of International Topical Conference on Research Trends in Coherent Radiation Generation and Particle Accelerators, February 11-13, 1991, La Jolla, California.

8. V.V. Anashin, I.B. Vasserman, V.G. Vestcherevich, *et al.*, Proceedings of the XIII International Conference on High Energy Particle Accelerators, vol .1, p. 155, 1987, Novosibirsk.

9. B.F.M. Pots, J.J.H. Coumans, D.C. Schram, Phys. Fluids, **24**, 517 (1981).

Wakefield Excitation by a Short Laser Pulse

D.L. Fisher and T. Tajima
Institute for Fusion Studies
University Of Texas at Austin
Austin,Texas 78712

M.C. Downer
Physics Department
University of Texas at Austin
Austin, Texas 78712

The generation of a wakefield in a dense plasma allows us to obtain accelerating electric fields well in excess of the typical breakdown field limit. Since the original suggestion[1] using a short laser pulse to excite a wakefield, a number of investigations have been carried out including the laser beatwave excitation, charged particle excitation[2], short laser pulse[1,3] and others. The short laser excitation has become experimentally plausible only recently because of rapid development of short pulse lasers[4]. In the present paper we summerise various new effects associated with the production of a wakefield. An intense enough laser pulse (or charged particle beams) quickly ionize the gas at or near the front of the pulse, and thereby set up an ionization front. This we find carries a substantial chemical potential. The ionization of a gas (or metal) will produce a density gradient in the ionization front and this ionization front may be seen as a pressure force in the moving frame of the front. This force will induce the plasma oscillation in its wake with phase velocity that of the ionization front which is approximently equal to c for a laser or relativistic particle beam.

For a short ionization front, $l_i << \frac{2\pi c}{\omega_p}$, analysis and simulation show that the peak amplitude of the electric wakefield due to the ionizations (chemical) potential for the one dimensional case gives $E_{peak} = \frac{\omega_p}{ec} T_e$ where T_e is the temperature of the electrons (ions are fixed). For a low intensity laser pulse, the ionization effect can under certain circumnstances induce a greater wakefield than that due to the ponderomotive force of the laser pusle. For a density of $10^{16} cm^{-3}$ and a longitudinal electron temperature of 5eV, the peak wakefield is $E_{peak} = 0.1 MV/m$. The intensity at which the ponderomotive potential and the ionization potential are equal is $I = 9.4 \omega_{laser}^2 T_e$ ergs/s-cm^2 for the one dimensional case where ω_{laser} is the frequency of the laser.

Below this intensity the ionization potential dominates. Since the wakefield is proportional to the temperature when the plasma is heated quickly after the ionization process, the amplitude of the wakefield increases. However, heating after the ionization process takes place mainly in the radial direction and a two dimensional effect is important to consider.

We have found that the wakefield in the longitudinal direction can be enhanced by the presence of the radial charge separation which may therefore produce substantial enhancement of the longitudinal field due to only the radial heating. The charge seperation comes from the radial plasma oscillations of the electron charge density. This radial effect can overcome the reduction of the longitudinal field due to the normal longitudinal charge seperation when the pulse radius is decreased. Three dimensional (axially symmetric) simulations show the radial effect alone in both the ionization and ponderomotive effects can induce large longitudinal wakefields. For a gaussian radial ionization profile with $\sigma_r = 1.4 \frac{c}{\omega_p}$, radial temperature of 56eV and a longitudinal temperature of 0eV the peak wakefield induced is $\frac{eE}{m\omega_p c} = 0.00053$ due only to the radial effect.

M.C. Downer's lab will be used to investigate the ionization and ponderomotive potenitals induced by a short laser pulse. In the low intensity regime ($I = 10^{14}$ W/cm^2) differentation between between the ponderomotive and the stronger ionization potential will be studied. With a density of 10^{16}cm^{-3} we expect to see an ionization induced wakefield of $E_{peak} \sim 0.1$ MV/m without heating. In the high intensity regime ($I \sim 10^{18}$ W/cm^2 and $\lambda_{laser} = 800$nm) simulations show a peak longitudinal field for the one dimensional case of $E_{peak} \sim 1$GV/m due to the ponderomotive potential.

The experimental diagnoses of the plasma oscillations will be in the reference frame of the primary laser pulse. This will be accomplished by injecting a probe pulse, which copropagates behind the primary pulse, then measuring frequency and phase shifts induced on the probe by the plasma density oscillations. The frequency shifts occure due to the rapid modulation of the index of refraction by the plasma oscillations. This technique (called "blue-shifting"[5-7] or "photon acceleration"[8]) has been used to diagnose ionization dynamics in intense femtosecond laser pulse fields. The probe accumulates a blue shift as long as it remains in the increasing density half-cycle and is red shifted while in the decreasing density half-cycle. By varying the pump-probe time delay, the spectral shifts of the probe alternates from blue to red to blue, etc. allowing individual cycles to be observed.

This work was supported by USDoE and TNRLC.

REFERENCES

1. T. Tajima and J.M. Dawson, Phys. Rev. Lett. 43, 267 (1979)

2. P. Chen et al., Phys. Rev. Lett. 54, 2343 (1985)

3. P. Sprangle, E. Esarey, A. Ting, and G. Joyce, Appl. Phys. Lett. 53, 1266,2146 (1988)

4. P. Maine, D. Strickland, P. Bado, M. Pesot, and G. Mourou, IEEE J. Wuant. Electron. QE-24, 398 (1988); J. Squier, F. Salin, E. Mourou, and D. Harter, Opt. Lett. 16, 324 (1991)

5. W.M. Wood, G. Focht, and M.C. Downer, Opt. Lett. 13, 984 (1988)

6. M.C. Downer, W.M. Wood, and J.I. Trisnadi, Phys. Rev. Lett 65, 2832 (1990)

7. M.C. Downer, J.I. Trisnadi, W.M. Wood, and W.C. Banyai, in Spectral Line Shapes, Vol.6,L., eds., Frommhold and J.W. Keto, American Institute of Physics Conference Proceddings 216, pp. 107-126 (1990).

8. S.C. Wilks, J.M. Dawson, W.B. Mori, T. Katsouleas, and M.E. Jones, Phys. Tev. Lett. 62, 2600 (1989); E. Esarey, A. Ting, and P. Sprangle, Phys. Rev. A, 42,3526 (1990).

Laser Wakefield Acceleration & Optical Guiding in a Hollow Plasma Channel

T. Katsouleas, T. C. Chiou

Dept. of Electrical Engineering-Electrophysics
University of Southern California
Los Angeles, CA 90089-0484

C. Decker , W. B. Mori

University of California, Los Angeles, CA 90024

J. S. Wurtele, G. Shvets

Massachusetts Institute of Technology, Cambridge, MA 02139

J. J. Su , National Central Univ. , Taiwan

Abstract

The accelerating and focusing wake fields that can be excited by a short laser pulse in a hollow underdense plasma are examined. The evacuated channel in the plasma serves as an optical fiber to guide the laser pulse over many Rayleigh lengths. Wake fields excited by plasma motion at the edge of the channel extend to the center where they may be used for ultra-high gradient acceleration of particles over long distances. The wake field and equilibrium laser profiles are found analytically and compared to 2-D PIC simulations. Laser propagation is simulated over more than ten Rayleigh lengths. The accelerating gradients on the axis of a channel of radius c/ω_p are of order of one fourth of the gradients in a uniform plasma. For present high-power lasers, multi-GeV/m gradients are predicted.

I. Introduction

Recent advances in short pulse laser technology[1,2] offer the possibility of ultra-high gradient acceleration of particles in laser-driven plasma wakes [3,4]. Sub-picosecond ten-terawatt pulses can generate acceleration gradients of order 10

GeV/m in a plasma of density $\sim 10^{17} cm^{-3}$; however Rayleigh lengths of such pulses are typically only a few millimeters. Thus it has been widely recognized that some form of optical guiding is needed. Several approaches to this problem have been previously discussed. In this article we propose a new scheme based on propagating a short laser pulse in a hollow channel embedded in an otherwise homogeneous underdense plasma. This scheme provides optical guiding of the laser pulse and nice accelerating properties for particles.

Several approaches to optical guiding have been previously discussed. First is self-focusing due to relativistic quiver motion of plasma electrons in the laser field[5]. Above a threshold power this mechanism provides optical guiding of the main body of the pulse. However, 1-D theory[6] and 2-D simulations[7,8] have shown that the leading edge of the pulse is not well guided. As a result, roughly the front c/ω_p of the pulse etches or diffracts away each Rayleigh length. To overcome this problem one can use a laser pulse that is very wide in the front and narrow in the back. This makes the Rayleigh length at the front sufficiently long.

A second approach to optical guiding is to confine the laser in a pre-formed channel in the plasma. A version of this was proposed several years ago by T. Tajima[9] known as the plasma fiber accelerator. In this scheme a long laser pulse propagates down a rippled vacuum channel in an overdense plasma. Since the laser is evanescent in overdense plasma, this was supposed to act as a slow wave conducting structure. Subsequent simulations showed that resonance absorption at each reflection of the laser would quickly degrade the laser[10]. More recently, optical guiding in a tailored density channel has been investigated by P.Sprangle and coworkers[8]. The density on axis is taken to be the usual density for wake excitation (i.e., n_0 is chosen such that $\pi c/\omega_p$ matches the laser pulse length), and the density rises parabolically with radius to confine the laser.

Here we propose a scheme for wakefield acceleration in a hollow underdense plasma. Since the index of refraction $\sqrt{\varepsilon}$ in underdense plasma is less than the index in the evacuated central region ($\sqrt{\varepsilon} = 1$), the laser can be confined as if it were in an optical fiber. The short laser pulse drives a plasma wake in the plasma at the edge of the hollow channel. The electrostatic fringe fields of the wake extend to the center of the channel, enabling focusing and acceleration of particles down the axis. This scheme differs from the plasma fiber accelerator in two important aspects. First, since the plasma is underdense resonance absorption at the channel walls is avoided. Second, the accelerating fields arise from a plasma wake rather than from a TM component of the laser mode.

In Sec. II we analyze the plasma mode excited in a hollow plasma. In Sec. III we analyze the laser mode. The wake amplitudes are estimated in Sec. IV and compared to 2-D PIC simulations in Sec. V.

II. Plasma Surface Mode - Fringe Field Accelerator

We now consider the normal wake field modes that can be excited in the hollow channel. Assume that the laser pulse propagates down the z-direction and the plasma slab extends infinitely in the x-direction. The channel supports a surface plasma wave. The mode structure follows from Maxwell's equations

$$\nabla \times E = -\frac{1}{c}\frac{\partial B}{\partial t} \tag{1a}$$

$$\nabla \times B = \frac{4\pi}{c}j + \frac{1}{c}\frac{\partial E}{\partial t} \tag{1b}$$

$$j = -en_0 v_1 \tag{2}$$

where n_0 is the unperturbed plasma density, v_1 is the perturbed electron velocity. Combining Eqs. (1a) and (1b) and using Eqs. (2) we can get

$$\nabla (\nabla \bullet \vec{E}) - \nabla^2 \vec{E} = -\frac{\omega_p^2}{c^2}\vec{E} - \frac{1}{c^2}\frac{\partial^2 \vec{E}}{\partial t^2} \tag{3}$$

where $\omega_p^2 = 4\pi n_0 e^2/m$.

Inside the plasma, we look for solutions of the form $\vec{E} = \vec{E}e^{i(k_z z - \omega t)}e^{-k_\perp(y-a)}$, with $\vec{E} = E_y\hat{y} + E_z\hat{z}$ and Eqs. (3) we can obtain

$$-k_\perp^2 E_z + ik_z(ik_z E_z - k_\perp E_y) = \left(-\frac{\omega_p^2}{c^2} + \left(\frac{\omega^2}{c^2} - k_z^2\right)\right)E_z \tag{4a}$$

$$-k_\perp^2 E_y - k_\perp(ik_z E_z - k_\perp E_y) = \left(-\frac{\omega_p^2}{c^2} + \left(\frac{\omega^2}{c^2} - k_z^2\right)\right)E_y \tag{4b}$$

Assuming $\frac{\omega}{c} \approx k_z$ and equating the determinant of the coefficients in Eqs. (4) to 0 yields

$$(k_\perp^2 - k_p^2)(k_z^2 - k_p^2) = 0 \tag{5}$$

where $k_p \equiv \frac{\omega_p}{c}$. It will turn out that $k_z = k_p$ will never satisfy the boundary conditions. Thus we have $k_\perp = k_p$ and

$$E_y = i \cdot \frac{k_z}{k_p}E_z \tag{6}$$

We note that these solutions give $\nabla \bullet \vec{E} = 0$, so that the plasma density is perturbed only at the surface of the channel.

Inside the channel, assuming $\vec{E} = \vec{E}e^{i(k_z z - \omega t)}$ and using $\omega_p = 0$ and $\nabla \cdot \vec{E} = 0$, Eqs. (3) become

$$\frac{\partial^2 \vec{E}}{\partial y^2} - \Delta^2 \vec{E} = 0 \tag{7}$$

where $\Delta^2 = k_z^2 - \dfrac{\omega^2}{c^2} \approx k_z^2 \left(1 - \left(\dfrac{v_g^{light}}{c} \right)^2 \right)$, the solutions are

$$E_z(y, z, t) = A\cosh(\Delta y)\, e^{i(k_z z - \omega t)} \tag{8a}$$

$$E_y(y, z, t) = C\sinh(\Delta y)\, e^{i(k_z z - \omega t)} \tag{8b}$$

with $C = -i \cdot \dfrac{k_z}{\Delta} A$ \hfill (9)

The dispersion relation follows from Eqs. (6), (8), (9) and the continuity of E_t and $D_n = \varepsilon E_n$ across the boundary, where t and n are the surface tangential and normal, respectively and $\varepsilon = 1 - \dfrac{\omega_p^2}{\omega^2}$:

$$-\frac{\tanh(\Delta a)}{\Delta} k_p = 1 - \frac{\omega_p^2}{\omega^2} \tag{10}$$

For high frequency lasers ($\dfrac{\omega_0}{\omega_p} \gg 1$), we expect $\dfrac{\omega}{k_z} \approx c$ and $\Delta \approx 0$, so that the dispersion relation simplifies to

$$\omega \approx \frac{\omega_p}{\sqrt{1 + k_p a}} \tag{11}$$

and $E_z(y, z, t) = Ae^{-i\omega(t - \frac{z}{c})}$, $E_y(y, z, t) = -i \cdot \dfrac{\omega}{c} \cdot Aye^{-i\omega(t - \frac{z}{c})} \approx B_x$. Thus the accelerating field is nearly uniform across the channel.

III. Laser Mode

The mode structure of the laser in the hollow plasma is well approximated by the mode structure in an optical fiber with refractive index $n_1 = 1$ inside and

$n_2 = \sqrt{1 - \dfrac{\omega_p^2}{\omega_0^2}}$ outside. Because the laser field has maximum at the channel center, we look for the fundamental TE mode solution of Maxwell's equations. In slab geometry, this is[11]

$$E_x = \begin{bmatrix} E_0 \cos(k_y y) e^{-jkz} & |y| \le a \\ E_0 \cos(k_y a) e^{-p(|y|-a)} e^{-jkz} & |y| \ge a \end{bmatrix} \tag{12}$$

where $k_y^2 = k_0^2 - k^2$, $p^2 = k^2 - k_0^2 n_2^2$ and $k_0 = \omega_0/c$, ω_0 is the frequency of the laser pulse. Applying the continuity of H_y gives the characteristic equations

$$p = k_y \tan(k_y a) \tag{13a}$$

$$p^2 + k_y^2 = k_p^2 \tag{13b}$$

In cylindrical geometry, the appropriate solutions of Maxwell's equations are the so-called HE_{11} mode[12] in an optical fiber which in this case, because $n_1 - n_2 \ll 1$, can be approximated by linearly polarized mode LP_{01}. Choosing the laser pulse to be polarized in the x-direction, the field is

$$E_x = \begin{bmatrix} E_0 J_0(hr) e^{-jkz} & r < a \\ E_0 \cdot \dfrac{J_0(ha)}{K_0(qa)} \cdot K_0(qr) e^{-jkz} & r > a \end{bmatrix} \tag{14}$$

where $h^2 = k_0^2 - k^2$, $q^2 = k^2 - k_0^2 n_2^2$. and J_0 is the Bessel function of first kind. The characteristic equation becomes

$$h \cdot \frac{J_1(ha)}{J_0(ha)} = q \cdot \frac{K_1(qa)}{K_0(qa)} \tag{15a}$$

$$(ha)^2 + (qa)^2 = (k_p a)^2 \tag{15b}$$

Eqs. (13) and Eqs. (15) can be solved numerically for k_y and h as functions of the channel radius. The results are plotted in Fig. 1. These results allow us to determine the equilibrium laser profile (Eqs.(12) or (14)) for diffraction-free propagation of the laser down the channel (Fig. 2).

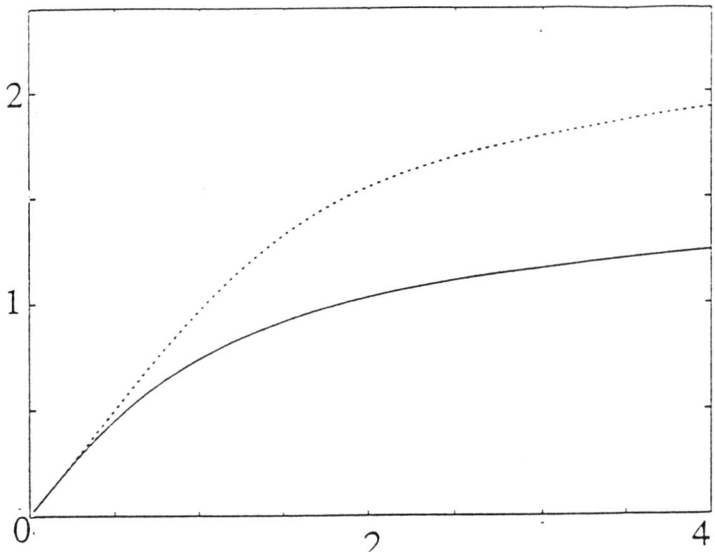

Figure 1. Dispersion relation of the electromagnetic wave (i.e. the laser) in the hollow plasma. Dashed curve : ha vs. $k_p a$ for cylindrical geometry; solid curve : $k_y a$ vs. $k_p a$ for slab geometry.

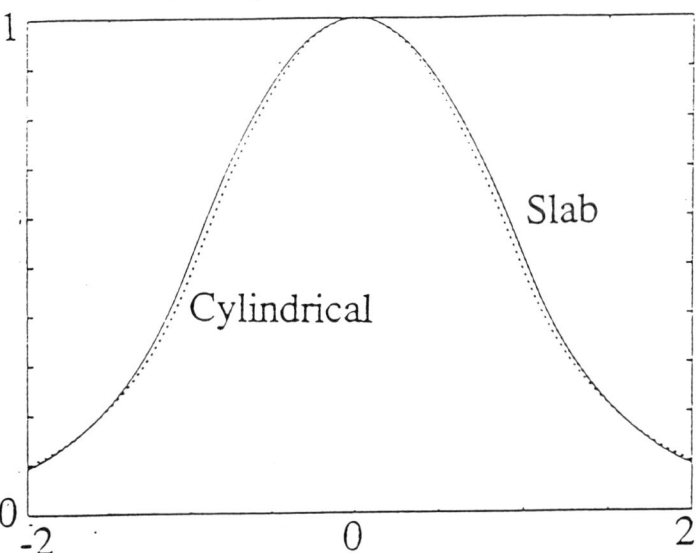

Figure 2. Equilibrium laser profile in the hollow plasma channel. Cylindrical - E_x vs. r/a; slab - E_x vs. y/a for $k_p a = 2$

IV. Amplitude of the Accelerating Gradient

We now estimate the amplitude of the accelerating gradient by the following simple model :

From Eqs. (3) and the assumption that $\dfrac{\omega}{k_z} \approx c$ we obtain

$$(\nabla_\perp^2 - k_p^2)\vec{E} = \nabla(\nabla \bullet \vec{E}) \tag{16}$$

As noted after Eqs. (6), $\nabla \bullet \vec{E}$ is zero except at the surface of the channel. The density perturbation at the surface is estimated by considering the Lagrangian displacement ξ of an electron fluid element in the y-direction : $n_1 = n_0 \xi \delta(y - a)$. So,

$$\nabla \bullet \vec{E} = -4\pi e n_0 \xi \delta(y - a) \tag{17}$$

The displacement ξ is caused by the pondermotive force of the laser pulse. Thus,

$$\ddot{\xi} + \omega^2 \xi = \frac{F_p}{m} = -\frac{1}{4}\nabla_\perp \left(\frac{v_{osc}}{c}\right)^2 \Bigg|_{y=a} \tag{18}$$

where $v_{osc} = \dfrac{eE_0}{m\omega_0}$ and v_{osc}/c is the normalized laser amplitude on axis.

Evaluating Eqs. (18) for a square laser pulse of duration $\pi c/\omega$ and substituting the result into Eqs. (16) and integrating across the delta - function gives

$$\frac{eE_z}{m\omega_p c} = -\frac{1}{4}\left(\frac{v_{osc}}{c}\right)^2 \sqrt{1 + k_p a} \cdot \frac{k_y}{k_p}\sin(2k_y a)\sin\omega(t - z/c) \tag{19}$$

For example, for $k_p a = 1$ and $v_{osc}/c = 0.3$, the normalized accelerating gradient is 0.023. This is about one half of the wakefield amplitude that would be excited in a homogeneous plasma by the same laser.

V. Simulations

To test the simple model just presented, we perform 2-D fully electromagnetic PIC simulations of the Hollow Channel Laser Wakefield Accelerator. The parameters used in the simulations are given in the caption to Fig. 3. Sample results are shown in Figs. 3-6. Fig. 3 shows the real space of the electrons in the hollow plasma. A slight rippling of the channel due to radial plasma motion is visible. Fig. 4 shows the laser profiles at t=0 (left) and after propagating about 13 Rayleigh lengths. The axial plasma wake is shown in Fig. 5. The peak accelerating field is $eE/m\omega_p c \approx 0.011$ compared to 0.015 predicted by Eq. (19). [Eq. (19) needs to be multiplied by 0.69 to compare to the polynomial pulse shape used in the simulations rather than the square pulse assumed for obtaining Eq. (19)]. The mode structure of the accelerating field is apparent in the radial slice of 3-D (Fig. 5b). The wavelength in Fig. 5a is clearly not $2\pi c/\omega_p$ as in a homogeneous plasma mode. It is $2\pi c/0.7\omega_p$ compared to $2\pi c/0.707\omega_p$ predicted from Eq.(11). The focusing field of the plasma wake shown in Fig.6 is in qualitative agreement with Eqs.(8b)

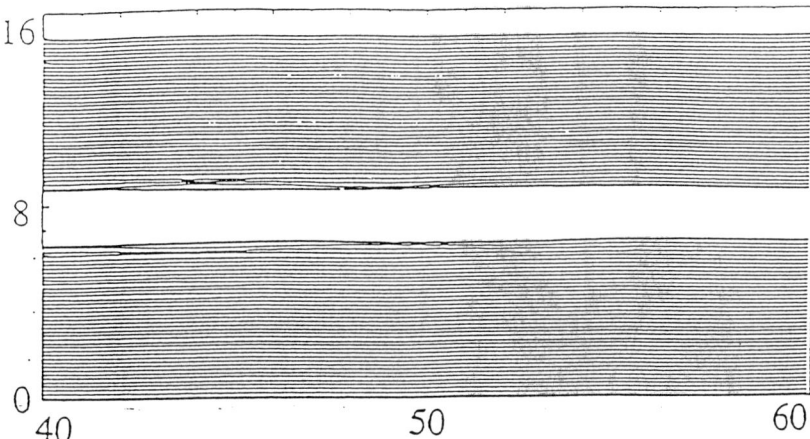

Figure 3 Real space of hollow plasma electrons ($\omega_p y/c$ vs. $\omega_p z/c$). The channel extends from $\omega_p y/c = 6.5$ to 8.5 ($k_p a = 1$); the cell size is $0.1 c/\omega_p$ in z and $0.15 c/\omega_p$ in y, 8 particles per cell, time step $= 0.25\omega_0^{-1}, \omega_0/\omega_p = 5$.

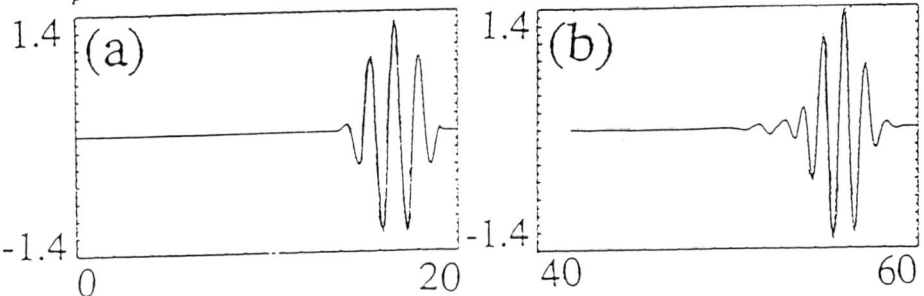

Figure 4 Laser electric field $eE_x/m\omega_p c$ vs. $\omega_p z/c$ at (a) t = 0 and, (b) after 13 Rayleigh lengths.

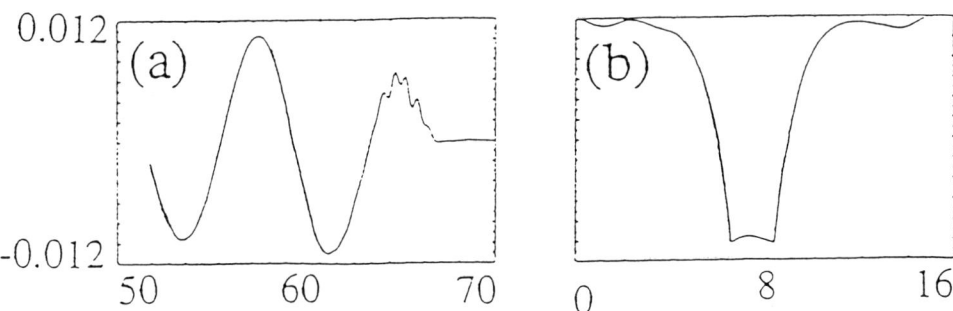

Figure 5 Longitudinal plasma wakefield $eE_z/m\omega_p c$ (a) vs. $\omega_p z/c$ on axis and (b) vs. $\omega_p y/c$ at $z = 60.8 c/\omega_p$.

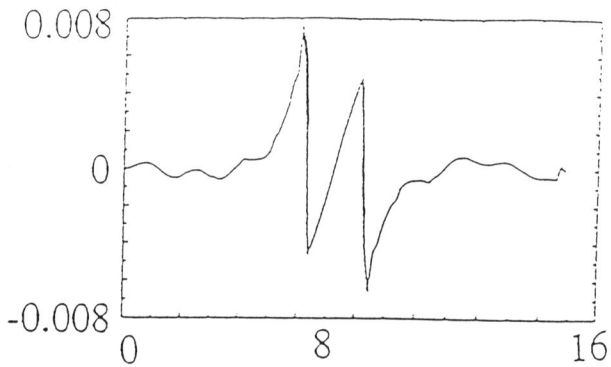

Figure 6   Radial plasma wake field $eE_y/m\omega_p c$ vs. $\omega_p y/c$ at $z = 59.8c/\omega_p$.

VI. Discussion

Naturally, the promise of the scheme analyzed here depends on the ability to form a hollow channel in a plasma. Several suggestions for doing this are currently being explored. These include ionization of a gas jet with a fine wire to block the central flow of gas[13] or channel formation by a pre-cursor pulse of laser or beam energy[14]

Acknowledge

We gratefully acknowledge useful input from David Whittum and from Bob Bingham. Work supported by USDOE - A/C # DE - FG03 - 92ER40745

References

1. G. Morou, These proceedings.
2. C. B. Darrow, et al., Advanced Acceleration Concepts., C. Joshi, ed., AIP Conf. Proc. 193, 50 (AIP, NY. 1989).
3. T. Tajima, J. M. Dawson, Phys. Rev. Lett. 43. 267 (1979).
4. P. Sprangle, et al. in Ref. [2], 376; T. Katsouleas, W. B. Mori and C. Darrow in Ref. [2], 165.
5. C. Max., J. Arons and A. B. Langdon, Phys. Rev. Lett. 33, 209 (1974). G. Schmidt and W. Hortou, Comments Plasma Phys. 9, 85 (1985). P. Sprangle, et al. IEEE Trans. on Plasma Science PS-15 , 145 (1987).
6. P. Sprangle, E. Esarey and A. Ting, Phys. Rev. Lett. 64, 2011 (1990).
7. T. Katsouleas, et al., in Nonlinear Dynamics and Particle Acceleration, AIP. Conf. Proc. 134 (AIP, NY. 1990). W. B. Mori, T. Katsouleas, Proc. of 2nd

EPAC, Nice, France; (June 12 -16, 1990)

8. P. Sprangle, et al. at in these Proc.

9. D. C. Barnes, T. Kurki-Suonio, T. Tajima, IEEE Trans. on Plasma Science, <u>PS-15</u> , 154, (1987).

10. W. B. Mori, et al. IEEE. Trans. on Nuclear Science, <u>NS -32</u>, 3555 (1985).

11. "<u>Electromagnetic Principles of Integrated Optics</u>", Donald L. Lee, (Wiely, New York, 1986).

12. "<u>Optical Electronics</u>", 3rd ed. Amnon Yariv, (Holt, Rinehart and Winston, New York, 1985).

13. M. Everett, private communication.

14. W. B. Mori, et al. Phys. Rev. Lett. <u>60</u>, 1299 (1988).

ELECTRON ACCELERATION AND OPTICAL GUIDING
IN THE LASER WAKEFIELD ACCELERATOR

P. Sprangle, E. Esarey, J. Krall, G. Joyce and A. Ting
Beam Physics Branch, Plasma Physics Division
Naval Research Laboratory, Washington, DC 20375-5000

ABSTRACT

The laser wakefield acceleration concept is studied using a general 2D formulation based on relativistic fluid equations. Simulations of an intense laser pulse propagating in a plasma address both optical guiding and wakefield generation issues. The ponderomotive forces associated with the laser pulse envelope expels virtually all of the plasma electrons, resulting in the generation of large amplitude accelerating and focusing fields. It is shown that relativistic guiding is ineffective in preventing the diffraction of short pulses and long pulses are broken up into beamlet segments due to wakefield effects. The use of preformed plasma density channels or tailored pulse profiles allows intense laser propagation over many Rayleigh lengths. Large amplitude wakefields can be generated over long distances in these cases and are suitable for electron trapping, acceleration and focusing.

INTRODUCTION

The possibility of utilizing the fields of intense laser beams to accelerate particles to high energies have attracted a great deal of interest.[1-4] The fundamental motivation for studying laser driven accelerators is the ultra-high fields associated with high intensity pulsed lasers. The laser transverse electric field in units of TV/m is given by

$$E_L(\text{TV/m}) = 2.6 \times 10^{-9} I_L^{1/2} [\text{W/cm}^2] = 3a_L/\lambda[\mu m],$$

where I_L is the laser intensity in W/cm^2, λ is the laser wavelength in μm and a_L is the unitless laser strength parameter. Recent developments in compact laser systems[5-8] have resulted in laser intensities of $I_L \sim 10^{18}$ $\text{W/cm}^2 (a_L \gtrsim 1)$ for $\lambda = 1$ μm. The transverse electric field associated with these lasers is typically, $E_L \gtrsim 3$ TV/m. The various laser acceleration concepts involve transforming a small fraction of the transverse laser field into an effective longitudinal accelerating field.

A laser driven plasma wave accelerator that has a number of attractive features is the laser wakefield accelerator[9-17] (LWFA).

In the LWFA, a short (~ 1 psec) intense (a_L ~ 1) laser pulse propagates through an underdense plasma, $\lambda^2/\lambda_p^2 \ll 1$, where $\lambda_p = 2\pi c/\omega_p$ is the plasma wavelength and $\omega_p = (4\pi e^2 n_0/m_0)^{1/2}$ is the plasma frequency. The ponderomotive force associated with the laser pulse envelope, $F_p \sim \nabla a^2$, generates a trailing plasma wave (wakefield) by expelling electrons from the region of the laser pulse, i.e., the laser pulse acts like a negative charge. If the laser pulse is sufficiently intense virtually all of the plasma electrons will be expelled. The accelerating gradient and transverse focusing field associated with the wakefield can accelerate and focus a trailing electron beam. If the laser pulse length, $c\tau_L$, is approximately equal to the plasma wavelength, $c\tau_L \simeq \lambda_p$, the ponderomotive force excites large amplitude plasma waves (wakefields) with phase velocities equal to the laser pulse group velocity. Figure 1 shows the wakefields generated behind the laser pulse.

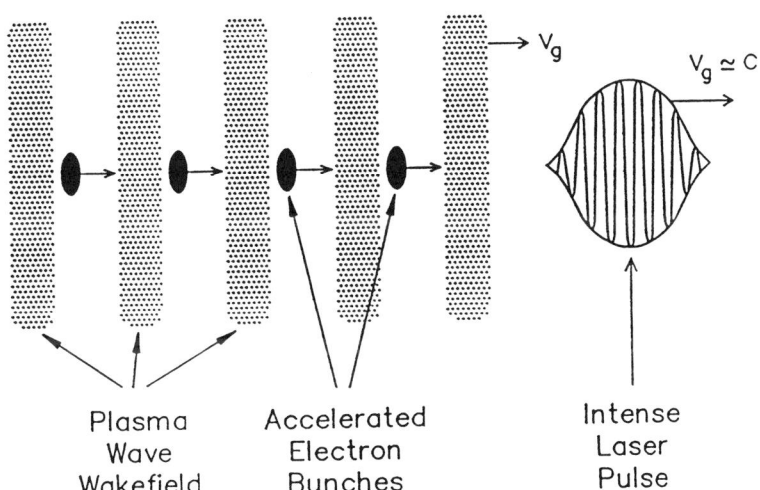

Plasma	Accelerated	Intense
Wave	Electron	Laser
Wakefield	Bunches	Pulse

Fig. 1.   Schematic showing wakefields generated by an intense laser pulse. The ponderomotive forces expel the plasma electrons, leaving behind a positive charge which in effect sweeps through the plasma at the speed of light.

The ratio of the accelerating field to the laser field in the LWFA is given by[13-16]

$$E_{acc}/E_L = 1.6 \times 10^{-11}(n_o[cm^{-3}])^{1/2}\lambda[\mu m] \frac{a_L}{(1+a_L^2/2)^{1/2}},$$

where n_o is the ambient plasma density in cm^{-3}. For $n_o = 10^{17}$ cm^{-3}, $a_L = 1$ and $\lambda = 1$ μm, the accelerating field is $E_{acc} \simeq 12$ GeV/m, which is 0.4% of the laser field ($E_L = 3$ TV/m). In the LWFA, the plasma serves a dual purpose, it acts as a medium which i) transforms a fraction ($\sim$ 1%) of the laser field into an accelerating field and ii) modifies the refractive index to optically guide the laser pulse.

In the absence of optical guiding, the interaction distance, L_{int}, will be limited by diffraction. Diffraction limits the interaction distance to $L_{int} \simeq \pi Z_R$, where $Z_R = \pi r_{Lo}^2/\lambda$ is the vacuum Rayleigh length and r_{Lo} is the minimum waist. The maximum energy gain of the electron beam in a single stage is $\Delta W \simeq E_{max}L_{int}$ which, in the limit $a_L^2 \ll 1$, may be written as

$$\Delta W[MeV] \simeq 580(\lambda/\lambda_p)P_L[TW].$$

As an example, consider a $\tau_L = 1$ psec linearly polarized laser pulse with $P_L = 10$ TW, $\lambda = 1$ μm and $r_L = 30$ μm ($a_L = 0.72$). The requirement that $c\tau_L \simeq \lambda_p$ implies a plasma density of $n_o = 1.2 \times 10^{16}$ cm^{-3}. The wakefield amplitude is $E_{acc} = 2.0$ GeV/m and the interaction length is $L_{int} = 0.9$ cm. Hence, a properly phased tailing electron bunch would gain an energy of $\Delta W = 18$ MeV in a single stage without optical guiding.

The major limitation of laser acceleration schemes is the limited acceleration distance, due to diffraction.[11,12] The interaction length and consequently the electron energy gain, may be greatly increased by optically guiding the laser pulse in the plasma. In plasmas, nonlinear and relativistic effects associated with intense laser fields can significantly modify the propagation characteristics of the laser.[11-28] Optical guiding in plasmas can be achieved by relativistic effects,[11-26] preformed density channels[15-17] or tailored laser pulse profiles.[15,17]

NONLINEAR FORMULATION

The large differences between the laser wavelength and other characteristic longitudinal lengths in the system, i.e., laser propagation distance, laser pulse length and plasma wavelength, make the direct numerical integration of the dynamical equations over extended distances impractical.

In the following, a fully nonlinear, relativistic, two-dimensional axisymmetric laser-plasma propagation model is formulated and numerically evaluated for laser pulses of ultra-high intensities and arbitrary polarizations.[17] The formulation has a number of unique features which allow numerical simulations to be carried out over extended laser propagation distances. The appropriate Maxwell-Fluid equations are recast into a convenient form by i) performing a change of variables to the "speed of light" coordinates, ii) applying the quasi-static approximation[13,14] (QSA), iii) expanding in two small parameters (which are independent of the laser intensity) and iv) averaging over the fast spatial scale length, i.e., the laser wavelength. The resulting equations are used to study the i) failure of relativistic focusing for short laser pulses,[13-17] ii) modulation of long laser pulses by wakefield effects,[15-17,27] iii) optical guiding of tailored laser pulses,[15,17] iv) use of plasma density channels to guide intense laser pulses,[15-17] and v) the generation of large amplitude wakefields.[13-17]

Field and Fluid Equations

The plasma is modelled using relativistic cold fluid equations. The fields associated with the intense laser-plasma interaction may be described by the normalized scalar ϕ and vector $\underset{\sim}{a}$ potentials, $\phi = |e|\Phi/m_o c^2$ and $\underset{\sim}{a} = |e|\underset{\sim}{A}/m_o c^2$. In the following, the Coulomb gauge is used, $\nabla \cdot \underset{\sim}{a} = 0$, the ions are assumed stationary and thermal effects are neglected. It is convenient to introduce the normalized fluid momentum $\underset{\sim}{u} = \gamma \underset{\sim}{v}/c$ and the fluid quantity $\rho = n/(n_o \gamma)$, where v is the electron fluid velocity, $\gamma = (1+u^2)^{1/2}$ is the relativistic factor of the electron fluid, n is the electron fluid density and n_o is the ambient density. In the fluid limit, the electron current is given by $\underset{\sim}{J} = -|e|n\underset{\sim}{v}$ and the normalized wave equation may be written as,

$$\left(\nabla^2 - \frac{1}{c^2}\frac{\partial^2}{\partial t^2}\right)\underset{\sim}{a} = k_p^2\rho\underset{\sim}{u} + \frac{1}{c}\frac{\partial}{\partial t}\nabla\phi, \tag{1}$$

where $k_p^2 = \omega_p^2/c^2 = 4\pi|e|^2n_o/m_oc^2$. The electrostatic potential is determined from Poisson's equation,

$$\nabla^2\phi = k_p^2\left(\gamma\rho - \rho^{(o)}\right), \tag{2}$$

where $\rho^{(o)}$ is the initial ion density profile, which may have the form of a pre-existing plasma density channel, i.e., $\rho^{(o)} = n_i(r)/n_o$. In the cold fluid limit, the electron response is determined by the momentum equation,

$$\frac{1}{c}\frac{d}{dt}\underset{\sim}{u} = \nabla\phi + \frac{1}{c}\frac{\partial}{\partial t}\underset{\sim}{a} - \frac{1}{\gamma}\underset{\sim}{u} \times (\nabla \times \underset{\sim}{a}), \tag{3}$$

and the continuity equation,

$$\frac{1}{c}\frac{\partial}{\partial t}(\rho\gamma) + \nabla\cdot(\rho\underset{\sim}{u}) = 0. \tag{4}$$

The momentum equation may be written in the form,

$$\frac{1}{c}\frac{d}{dt}(\underset{\sim}{u} - \underset{\sim}{a}) = \nabla\phi - \frac{1}{\gamma}(\nabla\underset{\sim}{a})\cdot\underset{\sim}{u}, \tag{5a}$$

which, after some manipulations, can be written as

$$\frac{1}{c}\frac{\partial}{\partial t}(\underset{\sim}{u} - \underset{\sim}{a}) = \nabla(\phi - \gamma) + \frac{1}{\gamma}\underset{\sim}{u} \times [\nabla \times (\underset{\sim}{u} - \underset{\sim}{a})]. \tag{5b}$$

Using the relation $\gamma^2 = 1 + \underset{\sim}{u}^2$, the momentum equation may be used to derive the relation,

$$\frac{d}{dt}(\gamma - \phi) = -\frac{\partial}{\partial t}\phi + \frac{1}{\gamma}\underset{\sim}{u}\cdot\frac{\partial}{\partial t}\underset{\sim}{a}. \tag{6}$$

Combining Eq. (5a) and Eq. (6) gives

$$\frac{1}{c}\frac{d}{dt}(\gamma - u_z - \phi + a_z) = -\left(\frac{\partial}{\partial z} + \frac{1}{c}\frac{\partial}{\partial t}\right)\phi + \frac{1}{\gamma}\underset{\sim}{u}\cdot\left(\frac{\partial}{\partial z} + \frac{1}{c}\frac{\partial}{\partial t}\right)\underset{\sim}{a}. \tag{7}$$

Change of Independent Variables

The full set of equations are recast into speed of light coordinates by changing variables from z,t to $\zeta = z-ct$ and $\tau = t$, where z and t are laboratory frame variables denoting the distance along the laser propagation axis and time, respectively. In these variables, the field equations (the wave equation and Poisson's equation) become,

$$\left(\nabla_\perp^2 + \frac{2}{c}\frac{\partial^2}{\partial\tau\partial\xi} - \frac{1}{c^2}\frac{\partial^2}{\partial\tau^2}\right)\underset{\sim}{a} = k_p^2\rho\underset{\sim}{u} + \left(\frac{1}{c}\frac{\partial}{\partial t} - \frac{\partial}{\partial\xi}\right)\nabla\phi, \tag{8a}$$

$$\left(\nabla_\perp^2 + \frac{\partial^2}{\partial\xi^2}\right)\phi = k_p^2(\gamma\rho - 1). \tag{8b}$$

The fluid response is determined by the momentum equations,

$$\left(\gamma\frac{\partial}{\partial\xi} - \underset{\sim}{u}\cdot\nabla\right)(\underset{\sim}{u} - \underset{\sim}{a}) = -\gamma\nabla\phi + (\nabla\underset{\sim}{a})\cdot\underset{\sim}{u} + \frac{\gamma}{c}\frac{\partial}{\partial\tau}(\underset{\sim}{u} - \underset{\sim}{a}), \tag{9a}$$

which can also be written as

$$\frac{\partial}{\partial\xi}(\underset{\sim}{u} - \underset{\sim}{a}) = \nabla(\gamma - \phi) - \frac{1}{\gamma}u \times [\nabla \times (\underset{\sim}{u} - \underset{\sim}{a})] + \frac{1}{c}\frac{\partial}{\partial\tau}(\underset{\sim}{u} - \underset{\sim}{a}), \tag{9b}$$

the continuity equation,

$$\frac{\partial}{\partial\xi}[\rho(\gamma - u_z)] = \nabla_\perp\cdot(\rho\underset{\sim}{u}_\perp) + \frac{1}{c}\frac{\partial}{\partial\tau}(\rho\gamma), \tag{10}$$

along with the equation,

$$\left(\gamma\frac{\partial}{\partial\xi} - \underset{\sim}{u}\cdot\nabla\right)(\gamma - u_z - \phi + a_z) = \frac{\gamma}{c}\frac{\partial}{\partial\tau}(\gamma - u_z + a_z) - \underset{\sim}{u}\cdot\frac{1}{c}\frac{\partial}{\partial\tau}\underset{\sim}{a}. \tag{11}$$

Quasi-Static Approximation (QSA)

The fluid response may be greatly simplified by using the quasi-static approximation.[13,14] The quasi-static approximation assumes that the laser pulse evolution time is long compared to the transit time of the electrons through the pulse. Here, the electron transit time through the laser pulse, which is equal to the laser pulse duration, τ_L, is assumed to be short compared to the laser pulse evolution time, τ_e, which is determined by the pulse diffraction time, $\sim Z_R/c$, or by the pulse dispersion time, $\sim \omega/\omega_p^2$, where ω is the laser frequency. In the QSA, the electrons experience essentially static fields, allowing the $\partial/\partial\tau$ derivatives to be neglected in the fluid equations, but not in the wave equation. It can be shown that in the quasi-static approximation, the quantity $\gamma - u_z - \phi + a_z$ is an invariant which is set equal to unity, i.e., its value prior to the arrival of the laser pulse, and hence,

$$\gamma - u_z = 1 + \phi - a_z. \tag{12}$$

Consequently, the normalized longitudinal momentum u_z and the relativistic factor γ are given by

$$u_z = \frac{(1 + u_\perp^2) - (1 + \phi - a_z)^2}{2(1 + \phi - a_z)}, \tag{13a}$$

$$\gamma = \frac{(1 + u_\perp^2) + (1 + \phi - a_z)^2}{2(1 + \phi - a_z)}. \tag{13b}$$

In the quasi-static approximation, the electron fluid momentum is given by

$$\frac{\partial}{\partial \xi} (\underset{\sim}{u} - \underset{\sim}{a}) = \nabla(\gamma - \phi) - \frac{1}{\gamma} \underset{\sim}{u} \times [\nabla \times (\underset{\sim}{u} - \underset{\sim}{a})] = \nabla(\gamma - \phi), \tag{14}$$

where we note that $\nabla \times (\underset{\sim}{u} - \underset{\sim}{a}) = 0$ is a consequence of axial symmetry, i.e., conservation of canonical angular momentum and shows the irrotational nature of the ponderomotive flow.

Slow Time and Space Scale Equations

The resulting equations are expanded to first order in the parameters $\varepsilon_1 = 1/kr_L \ll 1$ and $\varepsilon_2 = k_p/k \ll 1$, where $k = 2\pi/\lambda$, $k_p = 2\pi/\lambda_p = \omega_p/c$, and r_L is the laser spot size. All the fluid and field quantities are expanded in slow and fast terms, i.e., $\underset{\sim}{Q} = \underset{\sim}{Q}_s + \underset{\sim}{Q}_f$. The fast quantities are of the general form $\underset{\sim}{Q}_f = \hat{Q}_f(r, \zeta, \tau)\exp(imk\zeta)/2 + c.c.$, where $m = 1,2,3,\ldots$ and $\hat{Q}_f$ is complex and slowly varying in ζ. Within this representation, the nonlinear fluid equations are averaged over the laser wavelength in the (ζ, τ) frame. The ζ-averaging allows for all the laser-plasma response quantities to be evaluated on the slow spatial scale, i.e., λ_p or $c\tau_L$, permitting solutions over extended propagation distances.

The resulting equations describe the slowly varying components of the fluid and field quantities:

$$\nabla_{\perp}^2 \underset{\sim}{a} = k_p^2 \rho \underset{\sim}{u} - \partial(\nabla\phi)/\partial\zeta, \tag{15a}$$

$$\nabla_{\perp}^2 \phi + \partial^2\phi/\partial\zeta^2 = k_p^2(\gamma\rho - \rho^{(o)}), \tag{15b}$$

$$\partial(\underset{\sim}{u} - \underset{\sim}{a})/\partial\zeta = \nabla(\gamma - \phi), \tag{15c}$$

$$\partial(\rho(1 + \psi))/\partial\zeta = \nabla_{\perp}\cdot(\rho\underset{\sim}{u}_{\perp}), \tag{15d}$$

where the subscript s, denoting the slow component of the quantity, has been dropped, $\rho^{(o)} = n^{(o)}/n_o$, $n^{(o)}(r)$ is the initial plasma density profile prior to the laser interaction which may be a function of radial position and $\psi = \phi - a_z$ is the wake potential. Equations (15a-d) represent, respectively, the slow components of the wave, Poisson's, momentum and continuity equations. The slowly varying component of the relativistic factor is

$$\gamma = (1 + \psi)^{-1}\left(1 + \underset{\sim}{u}_{\perp}^2 + |\underset{\sim}{\hat{a}}_f|^2/2 + (1 + \psi)^2\right)/2, \tag{16}$$

where a linearly polarized laser pulse with amplitude $|\underset{\sim}{\hat{a}}_f|$ is assumed throughout this paper. The transverse component of the laser radiation field is $\underset{\sim}{a}_f = \underset{\sim}{\hat{a}}_f(r,\zeta,\tau)\exp(ik\zeta)/2 + \text{c.c.}$, where $\underset{\sim}{\hat{a}}_f$ is the complex, slowly varying amplitude which satisfies the parabolic (reduced) wave equation,

$$\left(\nabla_{\perp}^2 + 2c^{-1}k\partial/\partial\tau\right)\underset{\sim}{\hat{a}}_f = k_p^2\rho\underset{\sim}{\hat{a}}_f. \tag{17}$$

Within the QSA, the self-consistent, slowly varying equations in the (ζ,τ) variables, describing the laser-plasma interaction, to first order in ε_1 and ε_2, are given by Eqs. (15)-(17).

Equations (15a-d) can be combined to yield a single equation for ψ in terms of $|\underset{\sim}{\hat{a}}_f|^2$ of the form $\partial^2\psi/\partial\zeta^2 = G(\psi, |\underset{\sim}{\hat{a}}_f|^2)$, where G is an involved function. The equation for ψ is obtained by noting that $\rho = (\rho^{(o)} + k_p^{-2}\nabla_{\perp}^2\psi)/(1 + \psi)$ and $\underset{\sim}{u}_{\perp} = \rho^{-1}k_p^{-2}\nabla_{\perp}\partial\psi/\partial\zeta$. Note also that the refractive index is solely a function of ψ through ρ, i.e., $n_R = 1 - k_p^2\rho/2k^2$. Equation (17) together with $\partial^2\psi/\partial\zeta^2 = G$ completely describe the 2D-axisymmetric laser-plasma interaction. The wake potential, ψ, is related to the axial electric field, E_z, of plasma response (wakefield) by $\hat{E}_z = -\partial\psi/\partial\zeta$, where $\hat{E}_z = |e|E_z/m_oc^2$.

Limiting Cases

In the quasi-static approximation, the self-consistent laser plasma interaction is described by Eqs. (15)-(17). In general, these equations must be solved numerically. However, it is possible to examine the laser-plasma interaction analytically in two limits: (a) the broad pulse limit and (b) the narrow pulse limit.

a.   Broad pulse limit

Consider the broad pulse limit[13,14] in which $|\nabla_\perp| \ll |\partial/\partial\xi|$. In this limit, the $\nabla_\perp$ derivatives may be neglected in the quasi-static model and one finds $u_\perp \simeq a_z \simeq a_\perp \simeq 0$ and $\rho \simeq 1/(1 + \phi)$. In this limit, the laser-plasma interaction is completely described by

$$\left(\nabla_\perp^2 + \frac{2ik}{c}\frac{\partial}{\partial\tau}\right) \tilde{a}_\perp = k_p^2 \frac{\tilde{a}_\perp}{(1 + \phi)}, \tag{18a}$$

$$\frac{\partial^2}{\partial\xi^2}\phi = \frac{1}{2} k_p^2\left[\frac{(1 + a^2)}{(1 + \phi)^2} - 1\right], \tag{18b}$$

where $(1 + a^2)^{1/2} \simeq (1 + |\hat{a}_f|^2)^{1/2}$, for a linearly polarized laser. References 13 and 14 showed theoretically that relativistic guiding, which requires laser powers $P_L \geq P_{crit} = 17(\lambda_p/\lambda)^2[GW]$, does not occur for short pulses, $c\tau_L < \lambda_p/(1 + |\hat{a}_f|^2/2)^{1/2}$. For short pulses the plasma can not collectively respond to modify the refractive index.

b.   Narrow pulse limit

Consider the narrow pulse limit[23-26] in which $|\nabla_\perp| \gg |\partial/\partial\xi|$. In this limit, the $\partial/\partial\xi$ derivatives may be neglected and one finds $u_z \simeq u_\perp \simeq a_z \simeq a_\perp \simeq 0$ and $\gamma \simeq (1 + \phi) \simeq (1 + a^2)^{1/2}$. In this limit the plasma density response is given by

$$\rho \simeq (1 + a^2)^{-1/2}\left[k_p^{-2}\nabla_\perp^2(1 + a^2)^{1/2} + 1\right]. \tag{19}$$

where $\rho > 0$ has been assumed. The laser pulse evolution is determined by

$$\left(\nabla_\perp^2 + \frac{2ik}{c}\frac{\partial}{\partial\tau}\right)a_\perp = k_p^2 a_\perp(1 + a^2)^{-1/2}\left[k_p^{-2}\nabla_\perp^2(1 + a^2)^{1/2} + 1\right]. \tag{20}$$

As an example, it is possible to determine the conditions necessary to optically guide a narrow pulse. Consider a linearly polarized pulse of the form $a_{\perp} = |a| \cos k\xi$, such that the slowly varying part of $(1 + a^2)^{1/2}$ is $(1 + |a|^2/2)^{1/2}$. An envelope equation governing the laser spot size r_L may be obtained by applying the source dependent expansion method[29] to the wave equation. Assuming that the radiation field is adequately described by the lowest order Gaussian mode, $|a| = (a_0 r_{Lo}/r_L) \exp(-r^2/r_L^2)$, where a_0 is the peak amplitude and r_{Lo} is the minimum spot size, the laser spot size can be shown to evolve according to

$$\frac{d^2}{d\tau^2} r_L = \frac{4c^2}{k^2 r_s^3} \left[1 - \alpha \left(1 + \frac{8}{k_p^2 r_L^2} \right) \right], \tag{21}$$

where the limit $|a|^2/2 \ll 1$ has been assumed. Here $\alpha = k_p^2 a_0^2 r_{Lo}^2/32 = P/P_c$, where $P = 21.5(a_0^2 r_{Lo}^2/\lambda^2)$ GW is the laser power (in units of GWs) and

$$P_c = 17.4(k^2/k_p^2) \text{ GW}, \tag{22}$$

is the critical power (in units of GWs) necessary for the onset of relativistic focusing as obtained from conventional theory. Also, $\rho > 0$ implies $4a_0^2/(k_p^2 r_L^2) < 1$, which is easily satisfied for cases of interest. The first term on the right of Eq. (21) represents vacuum diffraction. The terms proportional to α are the focusing terms due to the plasma, the second of which is due to the density perturbation and is typically small. Hence, optical guiding of a narrow pulse requires

$$P/P_c \geq 1/(1 + 8/k_p^2 r_L^2) \simeq 1, \tag{23}$$

where it has been assumed $8/k_p^2 r_L^2 \ll 1$, as is typically the case. Thus, when $P \geq P_c$, the relativistic focusing forces exceed the effects of vacuum diffraction and the radiation beam may propagate at a constant spot size.

LWFA SIMULATIONS RESULTS

a)   Short Pulsed Propagation

Simulations[15,17] of short pulse propagation confirm the predictions of Refs. 13 and 14.   The results are shown in Fig. 2

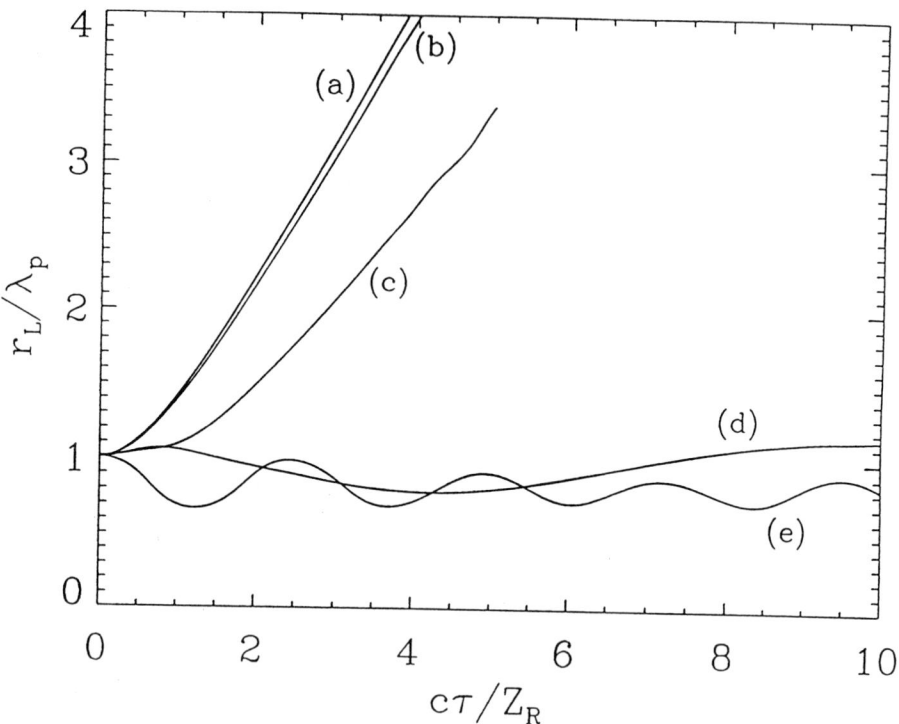

Fig. 2.   Laser spot size r_L (at $\zeta = -L/2$) versus propagation distance normalized to the Rayleigh length, $c\tau/Z_R$, for (a) vacuum diffraction, (b) an ultra-short pulse with L = $\lambda_p/4$, (c) a short pulse with L = λ_p, (d) a shaped pulse and (e) a channel-guided pulse. The spot size is normalized to the plasma wavelength λ_p = 0.03 cm.

for the parameters $\lambda_p = 0.03$ cm ($n_o = 1.2 \times 10^{16}$ cm^{-3}), $r_L = \lambda_p$ (Gaussian radial profile), $\lambda = 1$ μm ($Z_R = 28$ cm) and $P_L = P_{crit}$. The initial axial laser profile is given by $|\hat{\underset{\sim}{a}}_f(\zeta)| = a_L \sin(-\pi\zeta/L)$ for $0 < -\zeta < L = c\tau_L$, where $a_L = 0.9$ for the above parameters. Simulations are performed for two laser pulse lengths, $L = \lambda_p$ ($\tau_L = 1$ ps) and $L = \lambda_p/4$ ($\tau_L = 0.25$ ps). The initial normalized laser intensity, $|\hat{\underset{\sim}{a}}_f|^2$, is shown in Fig. 3 for $L = \lambda_p$. The spot size at the pulse center versus propagation distance $c\tau$ is shown in Fig. 2 for (a) the vacuum diffraction case, (b) the $L = \lambda_p/4$ pulse and (c) the $L = \lambda_p$ pulse. The $L = \lambda_p/4$ pulse diffracts almost as if in vacuum. The $L = \lambda_p$ pulse experiences a small amount of initial guiding before diffracting.

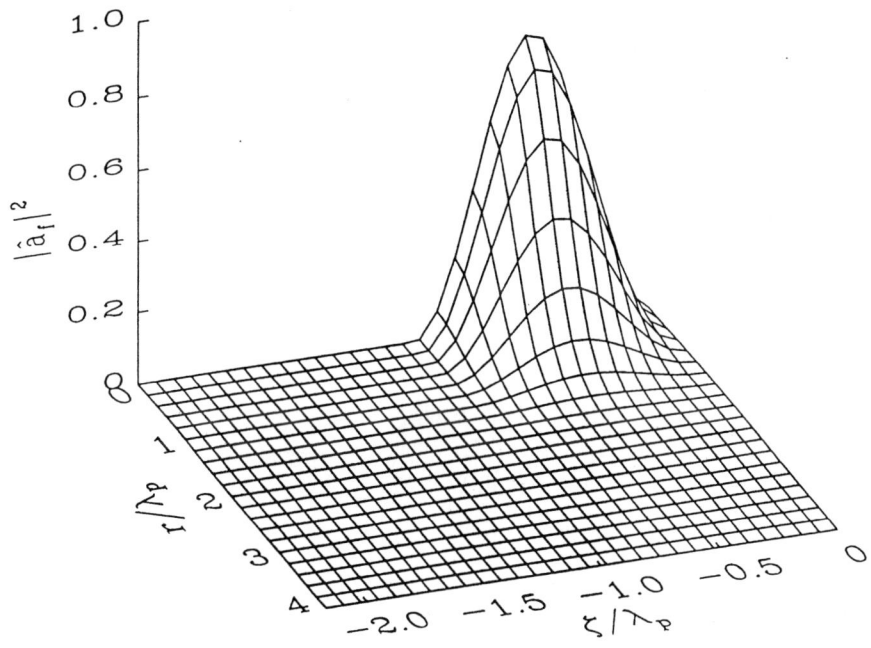

Fig. 3.   Normalized laser intensity $|\hat{\underset{\sim}{a}}_f|^2$ in the speed of light frame (ζ,τ) at $\tau = 0$ for the parameters $a_o = 0.9$, $L = r_L = \lambda_p = 0.03$ cm and $\lambda = 1$ μm. In the (ζ,τ) frame, the plasma flows from right to left.

b)   Laser Pulse Modulation

The wakefield generated by the finite rise time of a long pulse can modulate the pulse structure.[15,17] Consider a long laser pulse in which the body of the pulse has a constant amplitude with $P_L = P_{crit}$ and, therefore, should be relativistically guided. The amplitude of the wakefield generated by the front of the pulse is determined by the rise time. The wakefield, which consists of a plasma density modulation of the form $\delta n = \delta n_0(r)\cos(k_p\zeta)$, modifies the plasma's refractive index.[27] In regions of a local density channel, i.e., where $\partial\delta n/\partial r > 0$, the radiation focuses. In regions where $\partial\delta n/\partial r < 0$, diffraction is enhanced.

Pulse modulation is illustrated by a simulation of a long pulse (a long rise, $L_{rise} = 5\ \lambda_p$, followed by a long flattop region, $L_{flat} = 5\ \lambda_p$) with $P_L = P_{crit}$ ($a_L = 0.09$, $r_L = 10\ \lambda_p$, $\lambda_p = 0.03$ cm and $\lambda = 1\ \mu m$). Simulations indicate that for $P_L \geq P_{crit}$, an unstable wakefield is excited at the front of the pulse which rapidly modulates the pulse profile. Figure 4 shows the pulse

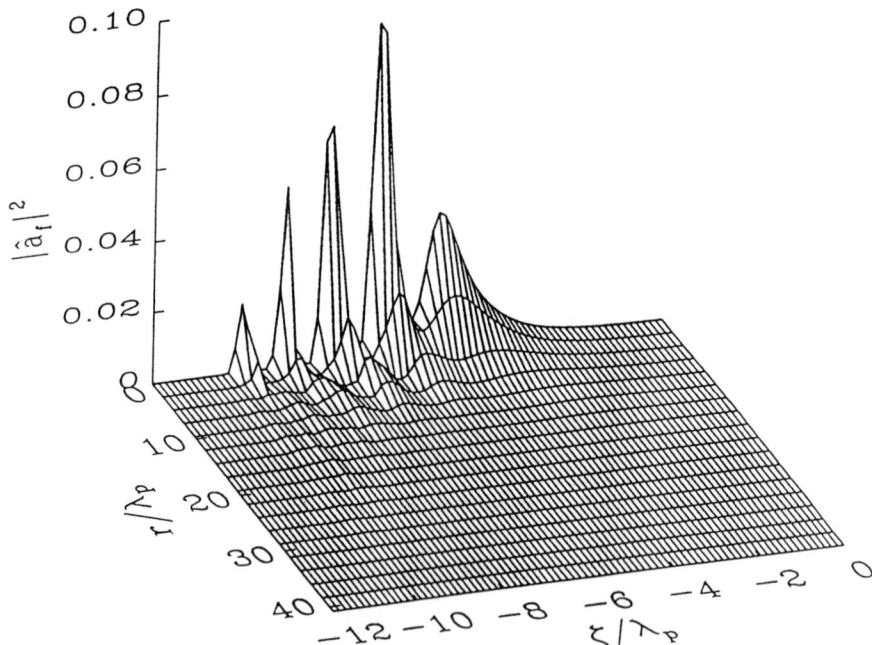

Fig. 4.   Normalized laser intensity $|\hat{a}_f|^2$ at $c\tau = 2\ Z_R$ for a long pulse showing modulation.

modulation, where $|\hat{a}_f|^2$ is plotted at $c\tau = 2Z_R$ for the above initial parameters. At high intensities, i.e., $a_L^2 > 1$, the modulation is reduced.

c)   Tailored Laser Pulse

A tailored laser pulse can propagate many Rayleigh lengths without significantly altering its original profile.[15,17] Consider a long laser pulse, $c\tau_L > \lambda_p$, in which the spot size is tapered from a large value at the front to a small value at the back, so that the laser power, $P_L \sim r_L^2 |\hat{a}_f|^2$, is constant throughout the pulse and equal to P_{crit}. The leading portion ($\ll \lambda_p$) of the pulse will diffract as if in vacuum.[13,14] However, since r_L is large at the front of the pulse, the Rayleigh length is also large. Hence, the locally large spot size allows the pulse front to propagate a long distance, whereas the body of the pulse will be relativistically guided. Also, since $|\hat{a}_f|^2$ increases slowly throughout the pulse, detrimental wakefield effects are reduced.

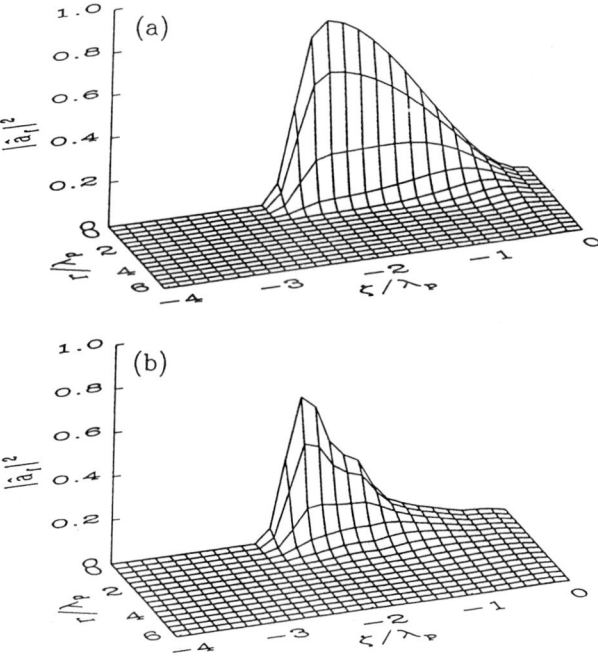

Fig. 5.   Normalized laser intensity $|\hat{a}_f|^2$ at (a) $\tau = 0$ and at (b) $c\tau = 24\ Z_R$ for a tailored pulse.

Figure 5(a) shows the initial profile of a tailored pulse in which $|\hat{a}_f|$ increases from 0.09 to 0.9 over a length $L_{rise} = 2\ \lambda_p$. Here, $P_L = P_{crit}$ throughout the pulse, $|\hat{a}_f| r_L = 0.9\lambda_p$ ($\lambda_p = 0.03$ cm, $\lambda = 1$ µm), which implies a decrease in r_L from $10\lambda_p$ to λ_p. At peak intensity the vacuum diffraction length is $Z_R = 28$ cm. The effectiveness of pulse tailoring can be seen by the $r_L(c\tau)$ plot in Fig. 2(d) and in Fig. 5(b), where a plot of $|\hat{a}_f|^2$ at $c\tau = 24\ Z_R$ demonstrates that the pulse is distorted but largely intact. At $c\tau = 24\ Z_R = 6.8$ m, the peak accelerating gradient of the wakefield behind the pulse is $E_{acc} = 1.3$ GeV/m.

d)   Pulse Propagation in a Plasma Density Channel

A preformed plasma density channel can guide short, intense laser pulses.[15-17] In the weak laser pulse limit, $|\hat{a}_f|^2 \ll 1$, the index of refraction is given by $\eta_R = 1 - k_p^2 \rho^{(o)}/2k^2$. Optical guiding requires $\partial \eta_R/\partial r < 0$, hence, a preformed density channel, $n^{(o)}(r) = n_o \rho^{(o)}(r)$, may prevent pulse diffraction. Analysis of the wave equation in the weak pulse limit indicates that a parabolic density channel will guide a Gaussian laser beam provided that the depth of the density channel[15-17] is $\Delta n = 1/\pi r_e r_L^2$, where $\Delta n = n^{(o)}(r_L) - n^{(o)}(0)$ and r_e is the classical electron radius, see Fig. 6.

Plasma Density/Laser, Profile

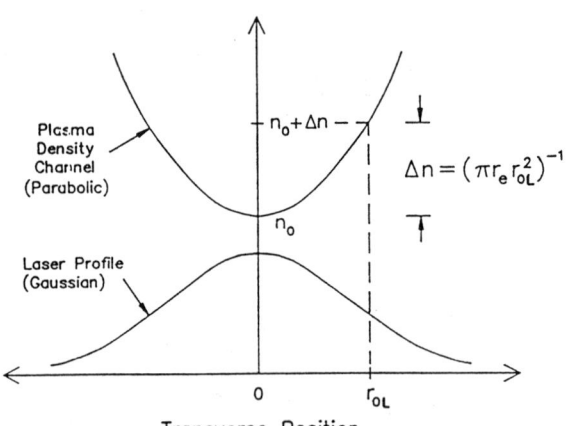

Plasma Density Channel (Parabolic)

Laser Profile (Gaussian)

$n_o + \Delta n$

n_o

$\Delta n = (\pi r_e r_{oL}^2)^{-1}$

0

r_{oL}

Transverse Position

Fig. 6.   Profile of the preformed plasma density channel necessary to optically guide a laser pulse. The required plasma density profile is quadratic in the radial coordinate and the density channel depth is Δn.

A simulation of channel guiding is shown in Figs. 7-9 for a laser pulse with the initial conditions of Fig. 3 propagating in a parabolic density channel with $\Delta n = 1.3 \times 10^{15}$ cm^{-3} and $n^{(o)}(0) = 1.2 \times 10^{16}$ cm^{-3}. Figure 6 shows the laser intensity at $c\tau = 10$ Z_R. The laser pulse shows some distortions but remains essentially guided. Guiding is confirmed by the $r_L(c\tau)$ plot in Fig. 2(e). The $r_L(c\tau)$ oscillations indicate a slight mismatch between the laser and channel parameters. This is caused by the laser pulse further reducing the density in the region of peak intensity as can be seen in Fig. 8. The axial wakefield E_z, plotted in Fig. 9, shows a peak accelerating gradient of 4.6 GeV/m. Simulations have shown that the large wakefields generated by a guided pulse can longitudinally and transversely trap and accelerate a trailing electron bunch to high energies.

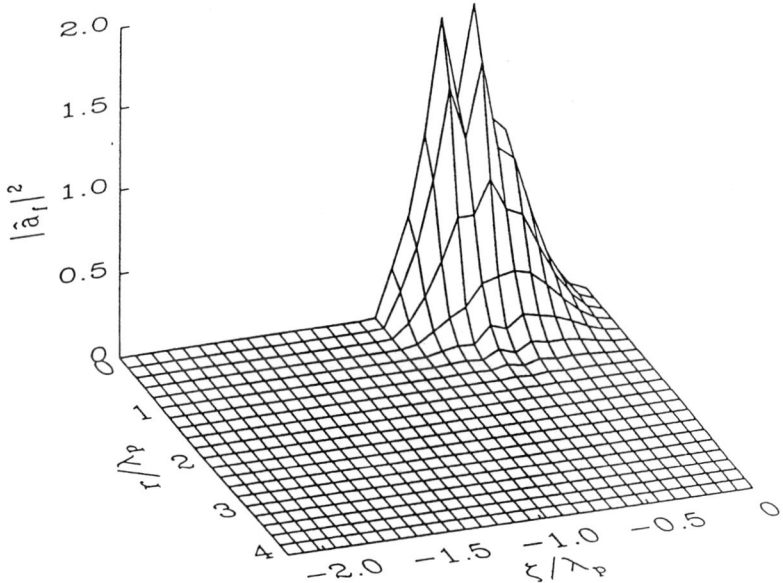

Fig. 7.   Normalized laser intensity $|\hat{a}_f|^2$ at $c\tau = 10$ Z_R for the laser pulse of Fig. 2 propagating in a plasma density channel.

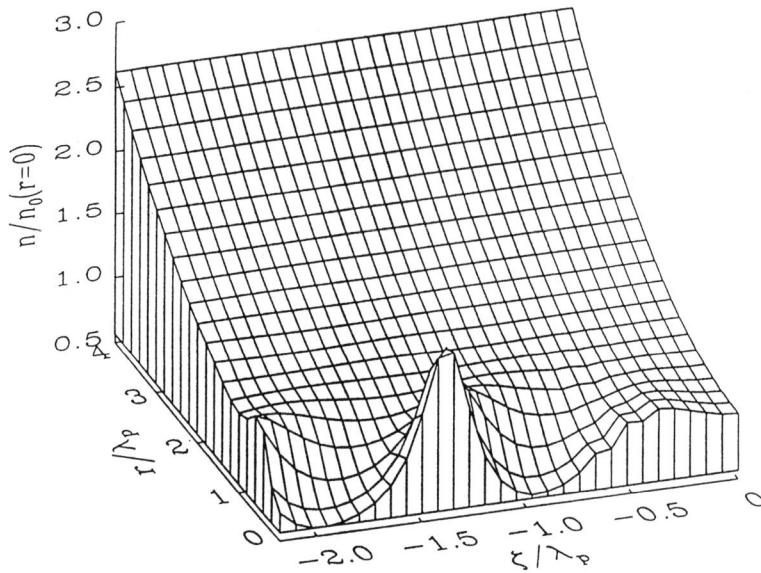

Fig. 8.   Plasma electron density n/n_0 at $c\tau = 10\ Z_R$ for the channel-guided case.

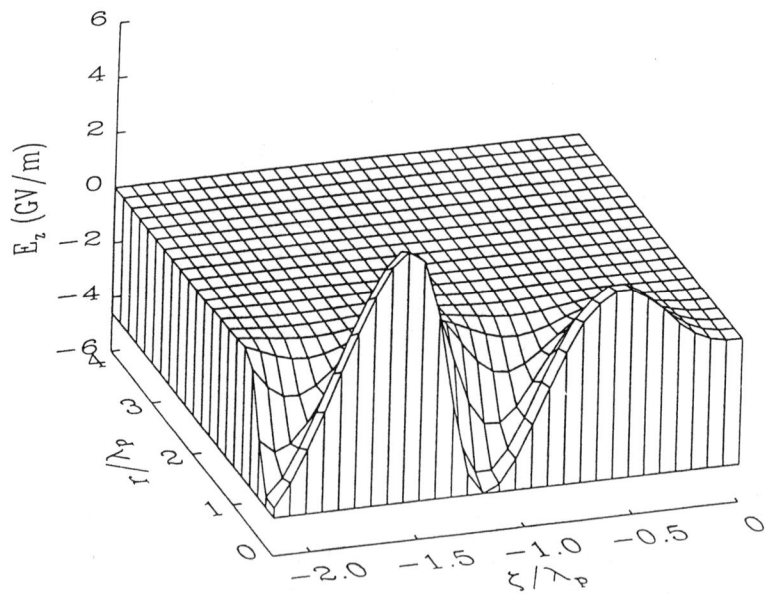

Fig. 9.   Axial electric field E_z at $c\tau = 10\ Z_R$ for the channel-guided case.

e)   Electron Trapping, Acceleration and Focusing

The large amplitude wakefield which can be generated over extended distances in a preformed plasma density channel can trap, focus and accelerate electrons to high energy. In the following, a laser pulse is propagated in a preformed plasma density channel with $\Delta n = 3.16 \times 10^{16}$ cm^{-3} and $n^{(o)}(0) = 7.8 \times 10^{16}$ cm^{-3}. The laser pulse parameters are $P_L = 40$ TW, $a_L = 0.72$, $r_L = 0.5 \lambda_p$ and $\lambda_p = 0.012$ cm. The laser pulse is propagated in the density channel over a distance equal to $c\tau = 20 \, Z_R = 23$ cm. In this example, the electrons are nearly totally expelled from the vicinity of the laser pulse. Figures 10 and 11 show the accelerating gradient and transverse focusing field associated with the wakefield at $c\tau = 20 \, Z_R$. The maximum accelerating gradient is $E_{acc} = 5.5$ GV/m and the maximum transverse field at $r = r_L/2$ is $E_r = 3.0$ GV/m. Figure 12 shows the configuration space for the cylindrically symmetric injected electron beam. The initial energy and radius of the continuous electron beam is $E_{inj} = 2.0$ MeV and $r_b = 10 \, \mu m$, respectively. Figure 13 shows the beam after being accelerated to 1.0 GeV in a distance of $c\tau = 23$ cm. The initially continuous beam becomes bunched and focused during the acceleration process. Approximately 70% of the initial beam electrons are trapped and accelerated. Figure 14 shows the energy distribution of the injected electron beam and Fig. 15 shows the final beam energy distribution of the accelerated beam. In this example, only three periods of the accelerating wakefield are shown. Figure 16 shows the peak energy (solid curve) and average energy (dashed curve) of the trapped electrons as a function of acceleration distance.

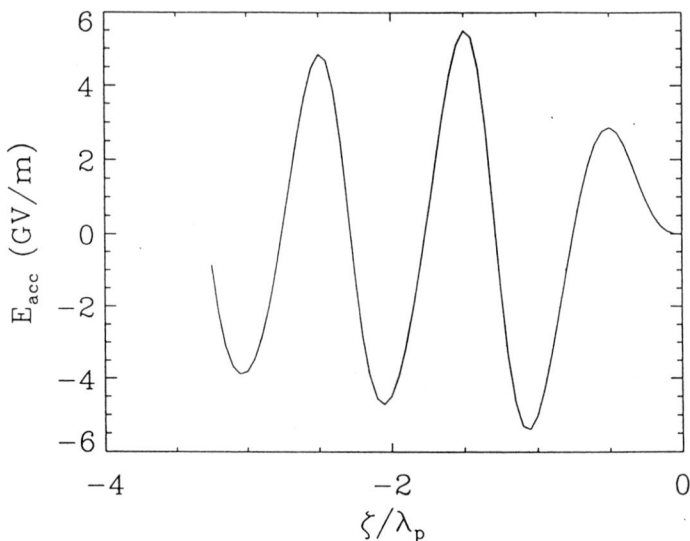

Fig. 10.   Accelerating gradient in the speed of light frame at $c\tau = 20$ $Z_R = 23$ cm.

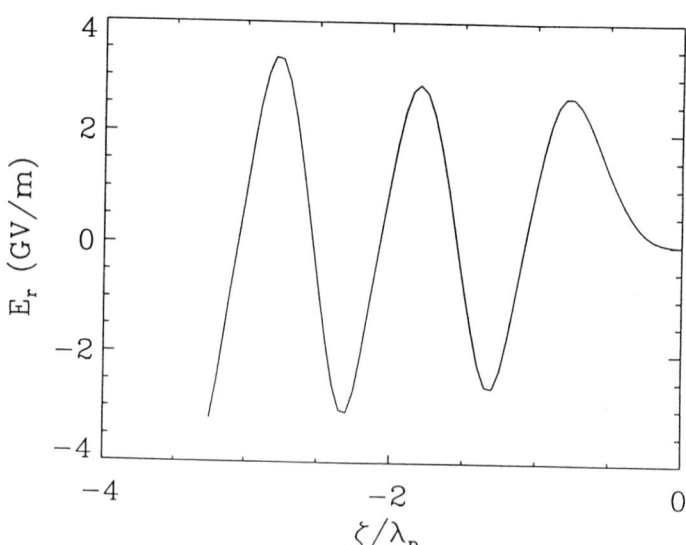

Fig. 11.   Transverse field in the speed of light frame at $c\tau = 20$ $Z_R = 23$ cm.

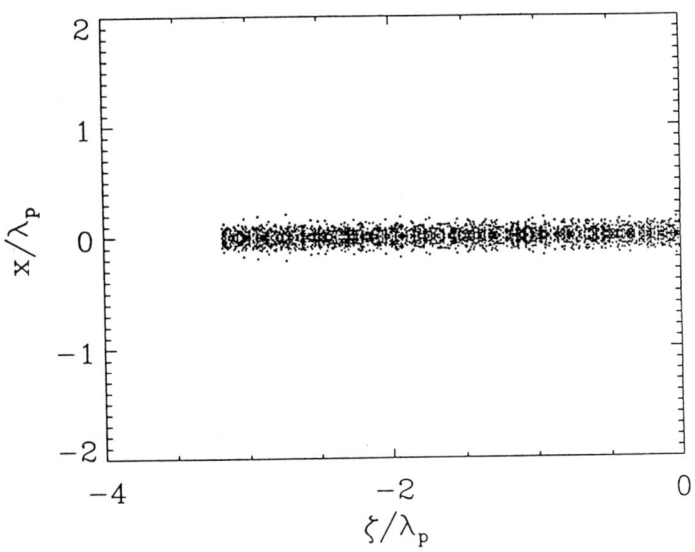

Fig. 12. Transverse and longitudinal spatial distribution of injected electron beam.

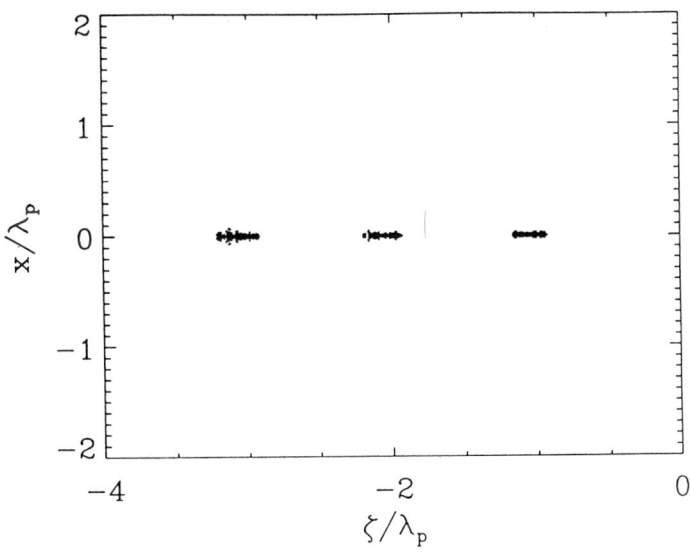

Fig. 13. Transverse and longitudinal spatial distribution of electron beam accelerated to 1 GeV in a distance of 20 cm.

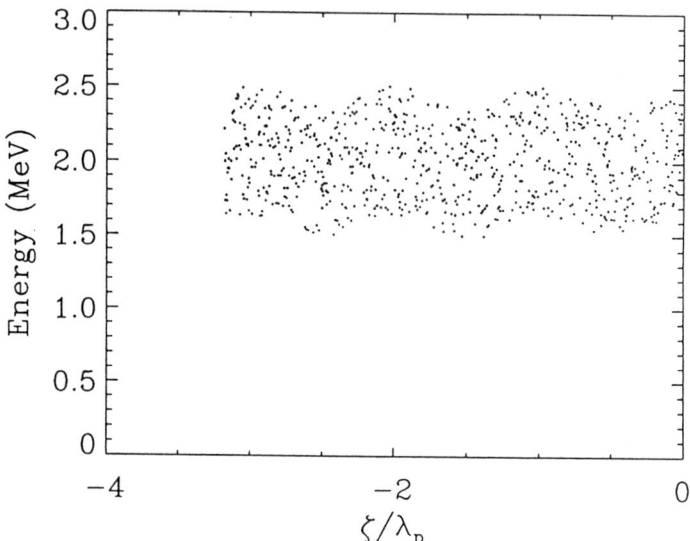

Fig. 14.   Energy distribution of injected electron beam.

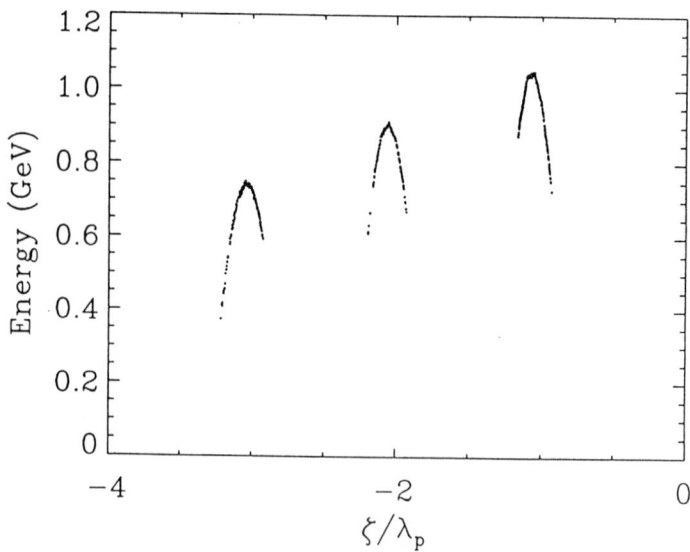

Fig. 15.   Energy distribution of accelerated beam.

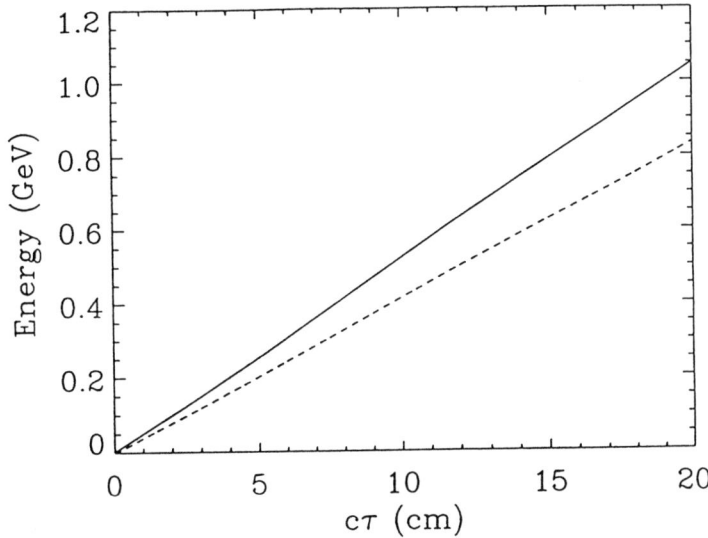

Fig. 16. Peak energy of trapped electrons (solid curve) and average energy of trapped electrons (dashed curve) as a function of acceleration distance.

CONCLUSION

The laser wakefield accelerator is capable of generating ultra-high accelerating and focusing fields over extended distances (many Rayleigh lengths). The ultra-high accelerating gradients and focusing fields are generated by the ponderomotive forces associated with the envelope of the laser pulse. These forces can totally expel electrons from the vicinity of the laser pulse, leaving behind an effective positive charge which sweeps through the plasma at the speed of light. It has been demonstrated that relativistic effects cannot prevent short laser pulses ($c\tau_L \lesssim \lambda_p$) from diffracting. Preformed plasma density channels or tailored laser pulse profiles can provide a means of propagating laser pulses many Rayleigh lengths. We have shown, using present-day laser technology, that the LWFA can produce accelerating gradients in excess of of 5.5 GV/m over distances greater than 20 cm. In this example a continuous injected 2 MeV electron beam was accelerated to 1 GeV in a distance of 20 cm and the accelerated beam becomes modulated at the plasma wavelength.

ACKNOWLEDGMENT

This work was supported by the Department of Energy and the Office of Naval Research.

REFERENCES

1.   Laser Acceleration of Particles, Edited by P. Channell, AIP Conf. Proc. No 91. (AIP, NY 1982).
2.   Laser Acceleration of Particles, Edited by C. Joshi and T. Katsouleas, AIP Conf. Proc. No. 130 (AIP, NY, 1985).
3.   Advanced Accelerator Concepts, Edited by F. Mills, AIP Conf. Proc. No. 156 (AIP, NY, 1986).
4.   Advanced Accelerator Concepts, Edited by C. Joshi, AIP Conf. Proc. No. 193 (AIP, NY, 1986).
5.   D. Strickland and G. Mourou, Opt. Commun. 56, 216 (1985).
6.   P. Maine, D. Strickland, P. Bado, M. Pessot, and G. Mourou, IEEE J. Quantum Electron. QE-24, 398 (1988).
7.   M. D. Perry. F. G. Patterson, and J. Weston, Opt. Lett. 15, 1400 (1990).
8.   G. Mourou and D. Umstadter, Phys. Fluids B, July (1992).
9.   T. Tajima and J. M. Dawson, Phys. Rev. Lett. 43, 267 (1979).
10. L. M. Gorbunov and V. I. Kirsanov, Sov. Phys. JETP 66, 290 (1987).
11. P. Sprangle, E. Esarey, A. Ting, and G. Joyce, Appl. Phys. Lett. 53, 2146 (1988).
12. E. Esarey, A. Ting, P. Sprangle, and G. Joyce, Comments Plasma Phys. Controlled Fusion 12, 191 (1989).
13. P. Sprangle, E. Esarey, and A. Ting, Phys. Rev. Lett. 64, 2011 (1990); Phys. Rev. A 41, 4463 (1990)
14. A. Ting, E. Esarey, and P. Sprangle, Phys. Fluids B 2, 1390 (1990).
15. P. Sprangle, E. Esarey, J. Krall and G. Joyce, presented at the Ultra High Field Symposium, APS-DPP Meeting, Tampa, FL, Nov, 1991.
16. P. Sprangle and E. Esarey, Phys. Fluids B, July (1992).
17. P. Sprangle, E. Esarey, J. Krall and G. Joyce, submitted for publication in Phys. Rev. Lett. (1992).
18. A. G. Litvak, Sov. Phys. JETP 30, 344 (1970).
19. C. Max, J. Arons, and A. B. Langdon, Phys. Rev. Lett. 33, 209 (1974).
20. P. Sprangle, C. M. Tang, and E. Esarey, IEEE Trans. Plasma Sci. PS-15, 145 (1987).
21. E. Esarey, A. Ting, and P. Sprangle, Appl. Phys. Lett. 53, 1266 (1988).
22. W. B. Mori, C. Joshi, J. M. Dawson, D. W. Forslund, and I. M. Kindel, Phys. Rev. Lett. 60, 1298 (1988).
23. G. Z. Sun, E. Ott, Y. C. Lee, and P. Guzdar, Phys. Fluids 30, 526 (1987).
24. T. Kurki-Suonio, P. J. Morrison, and T. Tajima, Phys. Rev. A 40, 3230 (1989).
25. P. Sprangle, A. Zigler, and E. Esarey, Appl. Phys. Lett. 58, 346 (1991).

26. A. B. Borisov, A. V. Borovskiy, O. B. Shiryaev, V. V. Korobkin, A. M. Prokhorov, J. C. Solem, T. S. Luk, K. Boyer, and C. K. Rhodes, Phys. Rev. A 45, 5830 (1992).
27. E. Esarey and A. Ting, Phys. Rev. Lett. 65, 1961 (1990).
28. X. Liu and D. Umstadter, submitted to Phys. Rev. Lett.
29. P. Sprangle, A. Ting, and C. M. Tang, Phys. Rev. Lett. 59, 202 (1987) and Phys. Rev. A 36, 2773 (1987).

NUMERICAL SIMULATION OF A 450 MeV SINGLE-STAGE
LASER WAKEFIELD ACCELERATOR

J. Krall, A. Ting, P. Sprangle, E. Esarey and G. Joyce

Beam Physics Branch, Plasma Physics Division
Naval Research Laboratory, 4555 Overlook Avenue, SW
Washington, DC 20375-5000

ABSTRACT

In the laser wakefield accelerator, a short ($\tau_L < 1$ ps), high power ($P > 10^{12}$ W) laser pulse propagates in plasma to generate a large amplitude ($E > 1$ GV/m) wakefield. We present two laser wakefield accelerator simulations, each driven by a 13 TW, 1.3 J laser pulse, demonstrating acceleration of particles to 90 MeV and 450 MeV respectively. In the latter case, the acceleration is enhanced by relativistic effects, but limited by phase detuning due to the reduced group velocity of the laser pulse.

INTRODUCTION

In the laser wakefield accelerator (LWFA) concept[1,2], an intense laser pulse drives a plasma wave which, in turn, accelerates a trailing electron bunch. A limitation on the acceleration distance is imposed by the diffraction length, or Rayleigh length, of the laser pulse:

$$z_R = \frac{k}{2} r_{L,0}^2 \qquad (1)$$

where k is the laser wavenumber and $r_{L,0}$ is the laser spotsize at the point of minimum focus. We have shown[3,4] that it is possible to overcome this limitation through the use of a plasma density channel or by "tailoring" a laser pulse with power equal to the critical power for relativistic optical guiding:[5]

$$P_c[GW] \simeq 17.4 \left(\frac{\lambda_p}{\lambda} \right)^2 \qquad (2)$$

where $\lambda = 2\pi/k$, $\lambda_p = 2\pi c/\omega_p$ is the plasma wavelength, $\omega_p = (4\pi n_0 e^2/m)^{1/2}$, n_0 is the plasma density, m is the electron mass, c is the speed of light in vacuum and e is the elementary charge. This paper was motivated by our interest in designing a simple proof-of-principle experiment and thus we will work within the limitations imposed by laser diffraction.

Even within the context of Eq. (1), current laser technology[6] is advanced enough that energies in the 100 MeV range are obtainable in a single stage, without relying on nonlinear or relativistic effects. We find, however, that relativistic effects can enhance considerably the peak accelerating gradient.

In some cases, an additional limitation on the peak energy that may be obtained in a single-stage LWFA is imposed by the group velocity of the laser pulse. This is $v_g = c(1 - 1/\gamma_g^2)^{1/2}$, where γ_g is given approximately by[7]

$$\gamma_g \simeq \frac{\omega}{\omega_p} \left(1 + \frac{a_0^2}{2} \right)^{1/2} . \tag{3}$$

With $v_g < c$, phase detuning limits the maximum acceleration to

$$\Delta\gamma_{max} \simeq \frac{\pi}{2} \left(\frac{\omega}{\omega_p} \right)^2 a_0^2 \left(1 + \frac{a_0^2}{2} \right)^{1/2} . \tag{4}$$

In Eqs. (3) and (4), $\omega = c/k$ and $a_0 = eA_0/mc^2$ is the normalized amplitude of the laser vector potential field which is assumed to be linearly polarized throughout this paper. Note that this limitation may be overcome by tapering the plasma density versus distance. This, however, is a greater complication than we wish to consider here.

In the following sections we present the results of two LWFA simulations. The first is a simple case which is optimized according to linear theory ($L_{laser} = \lambda_p/2$, where L_{laser} is the full-width-at-half-maximum length of the laser intensity profile on axis). In the second case, which is nearly identical to the first (only the plasma density is changed), relativistic and phase detuning effects become important.

These simulations are performed by numerically integrating a modified version of the laser-plasma fluid model described in Refs. 3 and 4. That model takes advantage of the fact that the laser envelope evolves on a slow time scale relative to a plasma period, $Z_R/c \gg 1/\omega_p$. In deriving the model, a coordinate transformation is made from (r,z,t) to (r, ζ=z-ct,τ=t). The laser pulse moves in the positive z direction such that the front of the laser pulse remains at or near $\zeta = 0$ in these coordinates. The physical region of interest extends from $\zeta = 0$, where the plasma is unperturbed, to $\zeta < 0$. The quasi-static approximation is then used, in which it is assumed that the length scale over which the laser field evolves is long compared to the laser pulse length ($Z_R \gg L_{laser}$). In this limit derivatives with respect to τ are dropped relative to derivatives with respect to ζ in the plasma fluid equations, but not in the wave equation. The model equations are then averaged over the laser frequency such that only the leading order terms in the small parameters $1/kr_L$ and ω_p/ω are retained, where $\omega \simeq kc$. Thus the simulation proceeds on a time scale dictated by the laser envelope, rather than its frequency, and on a space scale dictated by λ_p and/or L_{laser}.

The effects of phase detuning due to $v_g < c$ are included by using the exact wave equation to describe the laser pulse, rather than the reduced wave equation of Refs. 3 and 4. This can be written as:

$$\left(\nabla_\perp^2 + \frac{1}{c}\frac{\partial^2}{\partial\tau^2} + \frac{2}{c}\frac{\partial}{\partial\tau}\frac{\partial}{\partial\zeta} + \frac{2ik}{c}\frac{\partial}{\partial\tau}\right)\hat{a}_f = k_p^2\rho_s\hat{a}_f \,. \tag{5}$$

where $\hat{a}_f$ is the slowly varying laser amplitude of the normalized vector potential of the laser pulse ($a_f = \hat{a}_f e^{ik\zeta}$), $k_p = \omega_p/c$, $\rho_s = n/\gamma_s n_0$, n is the plasma electron number density, n_0 is the unperturbed density and γ_s is the relativistic factor for the plasma electrons.

EXAMPLE 1: A 90 MeV ACCELERATOR

In these runs, we will consider a Gaussian laser pulse with $\lambda = 1$ μm, $a_0 = 0.79$, $r_{L,0} = 31$ μm and $L_{laser} = 30$ μm (100 fs). In this case, the laser power is P = 21.5 $(a_0 r_{L,0}/\lambda)^2$ GW = 13 TW, the energy per pulse is 1.3 J and the Rayleigh length is $Z_R = 0.3$ cm. These laser parameters are within the bounds of present technology. The simulation begins at $\tau = 0$ with the converging laser pulse entering the plasma. The plasma reaches full density at $c\tau = 2 Z_R$. The laser pulse reaches its focus at approximately $c\tau = 4 Z_R$ and the simulation continues until $c\tau = 10 Z_R = 3.0$ cm. The plasma density profile is shown in Fig. 1. The point of minimum focus of the laser pulse is indicated by a dashed line in the figure.

According to linear theory[2] the optimum wakefield will be obtained at a plasma density for which $\lambda_p = 2 L_{laser} = 60$ μm, or $n_0 = 3.0 \times 10^{17}$ cm^{-3}. At this density, $P < P_c \simeq 63$ TW such that relativistic guiding effects are unimportant. Figure 2(a) shows the laser intensity $|\hat{a}_f|^2$ of the converging laser pulse as it enters the plasma at $c\tau = 2 Z_R$ The electric field on axis is shown in Fig. 2(b). As the laser reaches the point of minimum focus the wakefield is maximum. This is shown in Fig. 3. It is clear in this case that the presence of the plasma does not strongly effect the evolution of the laser pulse. This is verified by Fig. 4, where the peak accelerating field, plotted versus time, is symmetric about $c\tau = 4 Z_R$.

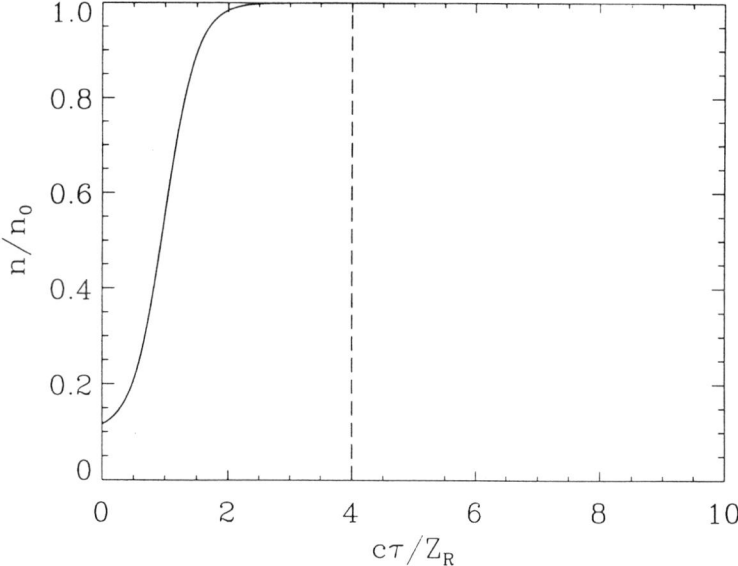

Fig. 1 Plasma density profile. The dashed line indicates the point at which the laser would reach its minimum spotsize in vacuum.

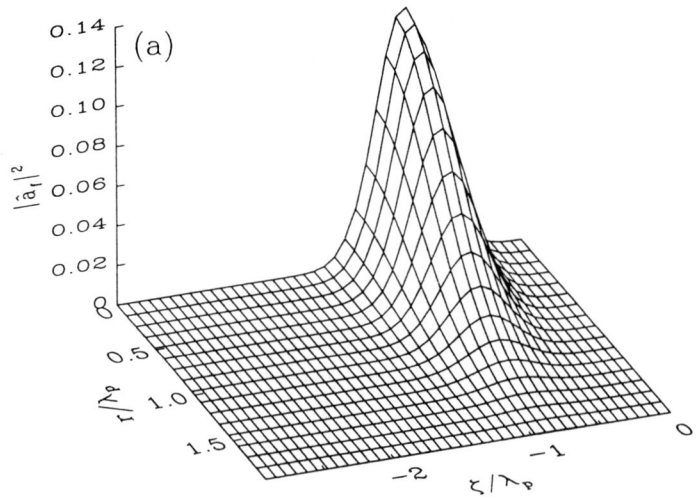

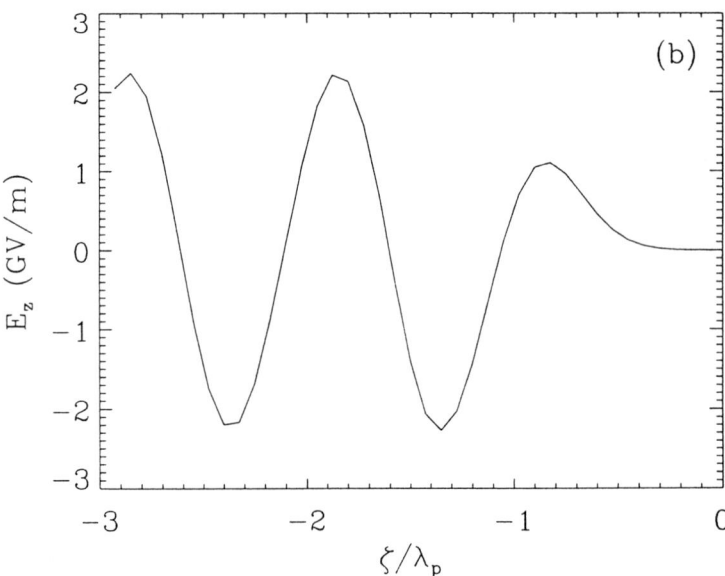

Fig. 2   Plots at $c\tau = 2\,Z_R$ showing (a) a surface plot of the laser intensity $|\hat{a}_f|^2$, sampled over a coarse grid (the numerical grid is much finer), and (b) axial electric field E_z versus ζ. The laser pulse is moving towards the right.

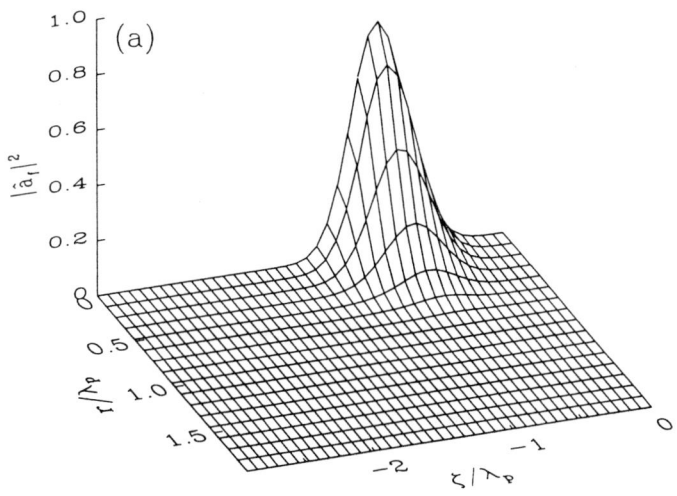

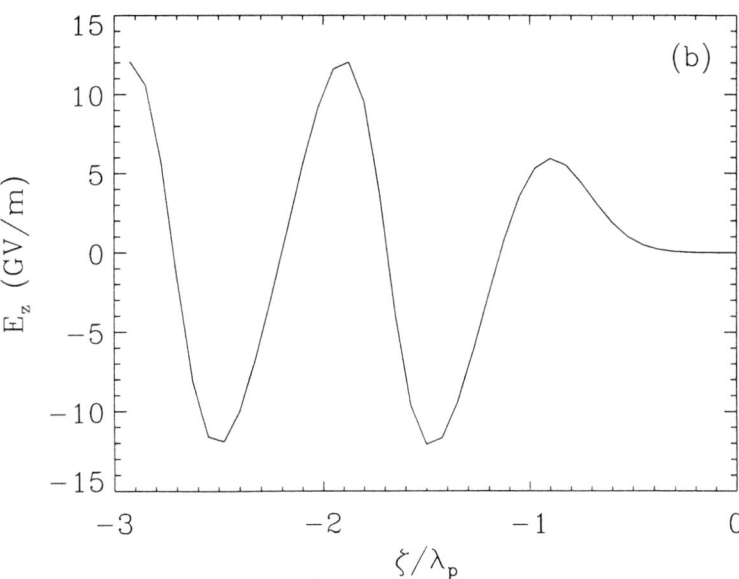

Fig. 3   Plots at $c\tau = 4\,Z_R$ showing (a) laser intensity $|\hat{a}_f|^2$ and (b) axial electric field E_z.

To study the acceleration and trapping of electrons by the high-gradient wakefield, we have modified an existing particle code to accelerate a distribution of non-interacting test particles in the time-resolved electric and magnetic fields. Here, we consider a continuous beam of 3.0 MeV electrons with emittance ϵ_n = 130 mm-mrad. The initially converging beam is focused to a minimum RMS radius r_b = 200 μm at $c\tau = 4\,Z_R$. After $c\tau = 10\,Z_R$ = 3.0 cm a small fraction ($\approx$ 0.5 %) of the original particle distribution has been trapped and accelerated (this fraction can be improved by using a lower emittance beam). Figure 5 shows a single bunch of particles which has been trapped at a position corresponding to the leading accelerating peak in E_z. The peak energy plotted versus time, Fig. 6, indicates a peak energy of 90 MeV over 3 cm or 3 GeV/m.

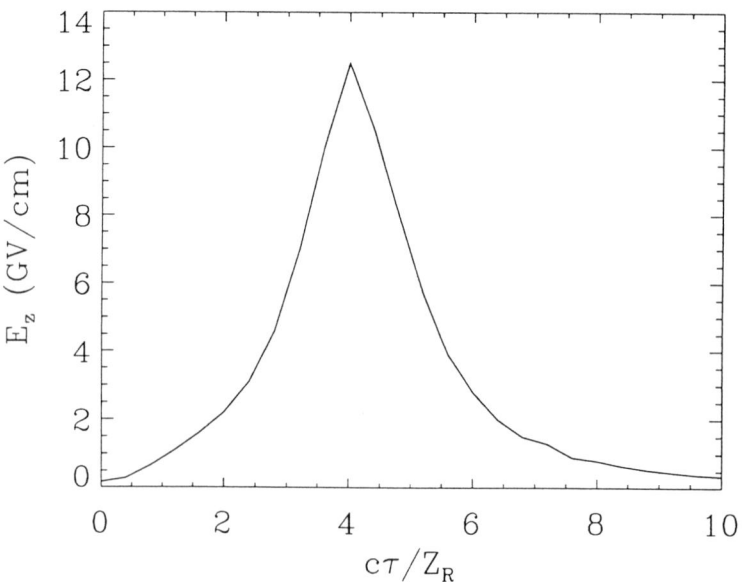

Fig. 4   Peak accelerating field versus $c\tau$.

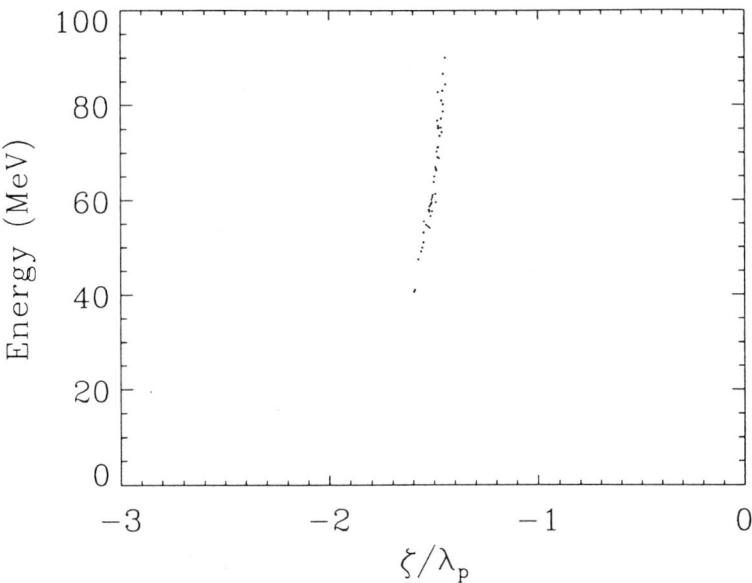

Fig. 5   Test-particle energy distribution at $c\tau = 3.0$ cm $= 10\,Z_R$.

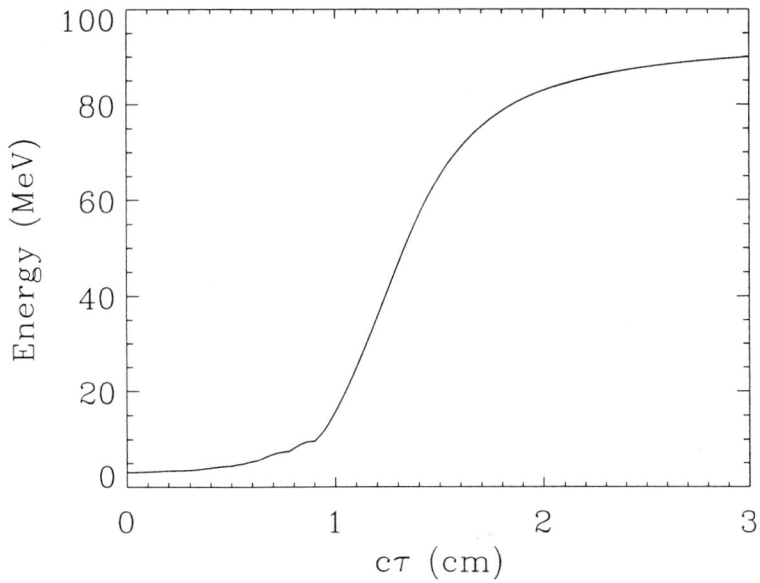

Fig. 6   Peak particle energy versus time.

EXAMPLE 2: A 450 MeV ACCELERATOR

We now consider a simulation nearly identical to that presented above. In this case, the plasma density is increased to $n_0 = 3.0 \times 10^{18}$ cm^{-3} such that the plasma wavelength is reduced to $\lambda_p = 20$ μm. More significantly, the critical power for relativistic guiding has been reduced to $P_c = 7$ TW, such that $P \approx 2P_c$. Since the laser parameters have not been changed, the laser pulse now extends over several plasma wavelengths and, according to linear theory, is far from optimum. This can be seen in Fig. 7, which shows the laser intensity of the converging laser pulse as it enters the plasma at $c\tau = 2$ Z_R and the corresponding accelerating electric field on axis.

Two important physical effects come into play in this case. Firstly, because $P > P_c$ and $L_{laser} > \lambda_p$, relativistic guiding effects, combined with the density modulation of the plasma wake, will create alternating regions of focusing and defocusing in the laser pulse, axially modulating the laser pulse at the plasma wavelength.[3,8] The modulated pulse resonantly excites the plasma wave which, in turn, further increases the modulation of the laser pulse. Secondly, Eq. (3) gives $\gamma_g \approx 23$ for this case, limiting the theoretical maximum energy to approximately 230 MeV, assuming no modification of the laser intensity due to self-focusing.

Figure 8 shows the laser intensity and axial electric field at $c\tau = 4$ Z_R. The modulation of the laser pulse and the nonlinear excitation of the plasma wave are clearly observable. Figure 8(a) also shows that relativistic optical guiding effects have focused the laser to obtain a much higher intensity than was observed in the 100 MeV case (see Fig. 3). The peak accelerating field as a function of time, Fig. 9, is in striking contrast to the simpler 100 MeV case (see Fig. 4).

As before, a simulated beam of noninteracting test particles is allowed to evolve in the time-resolved wakefield. In this case, approximately 2% of the particles are trapped and accelerated. The peak particle energy of 450 MeV is observed at $c\tau = 1.6$ cm $= 5.33$ Z_R, as shown in Fig. 10. At $c\tau = 3.0$ cm $= 10$ Z_R, however, the peak particle energy has dropped to 360 MeV due to the reduced group velocity which causes the electrons to become dephased (see Fig. 11). A plot of peak particle energy versus time, Fig. 12, shows acceleration to 450 MeV over 1.6 cm or 29 GeV/m.

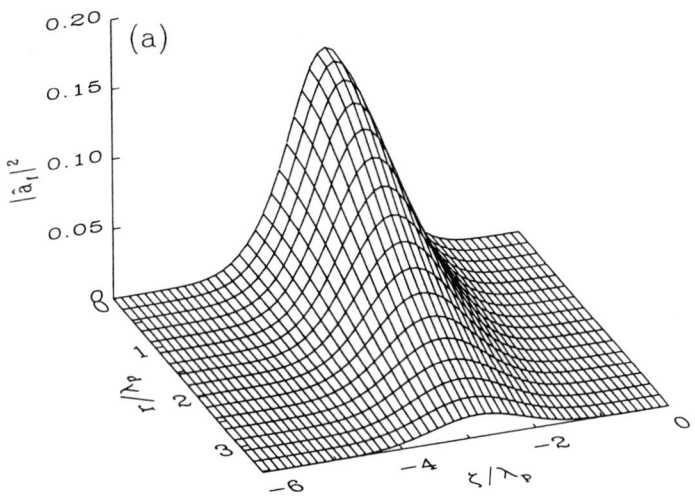

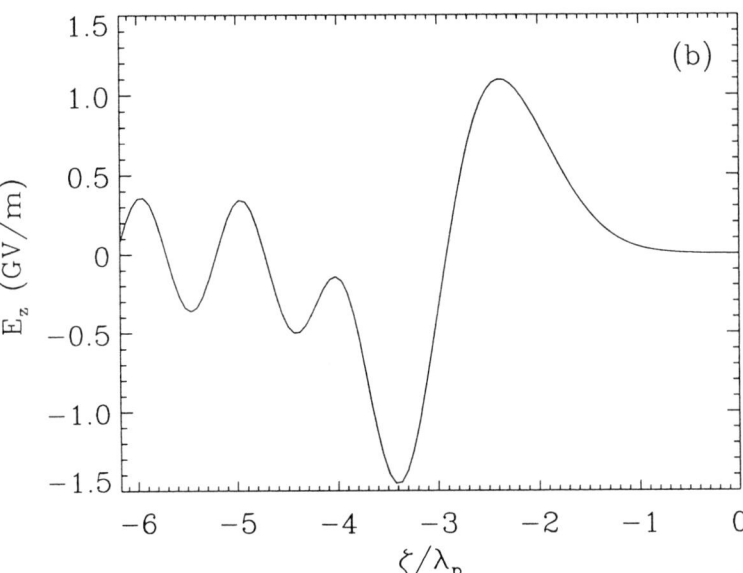

Fig. 7 Plots at $c\tau = 2\,Z_R$ showing (a) laser intensity $|\hat{a}_f|^2$ and (b) axial electric field E_z.

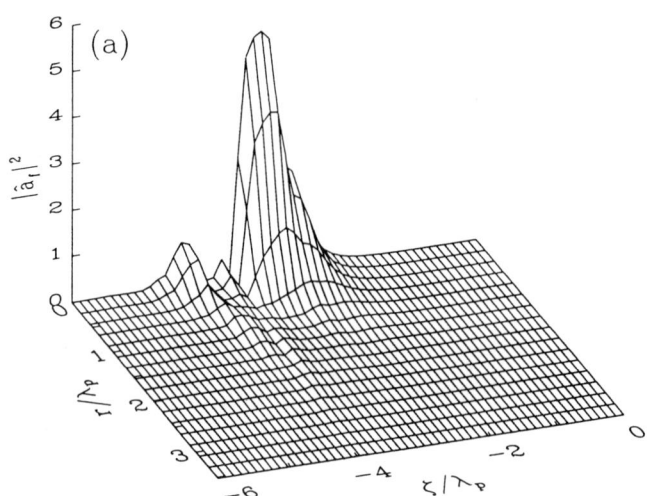

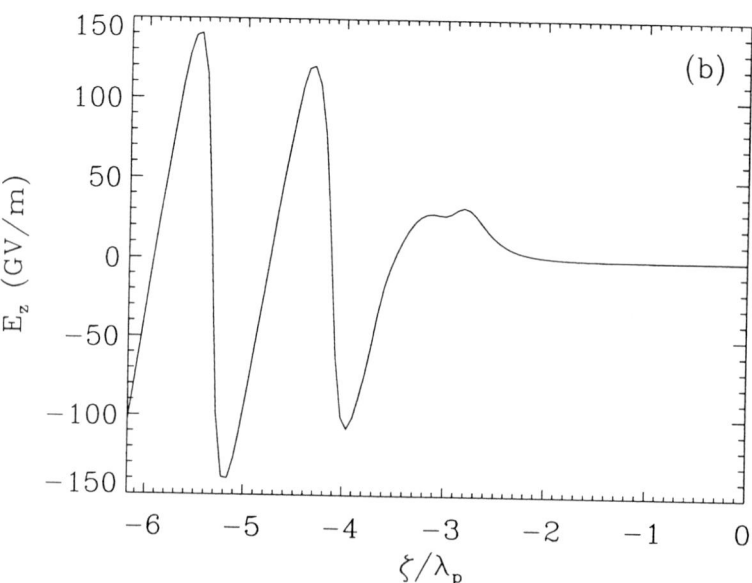

Fig. 8   Plots at $c\tau = 4\,Z_R$ showing (a) laser intensity $|\hat{a}_f|^2$ and (b) axial electric field E_z.

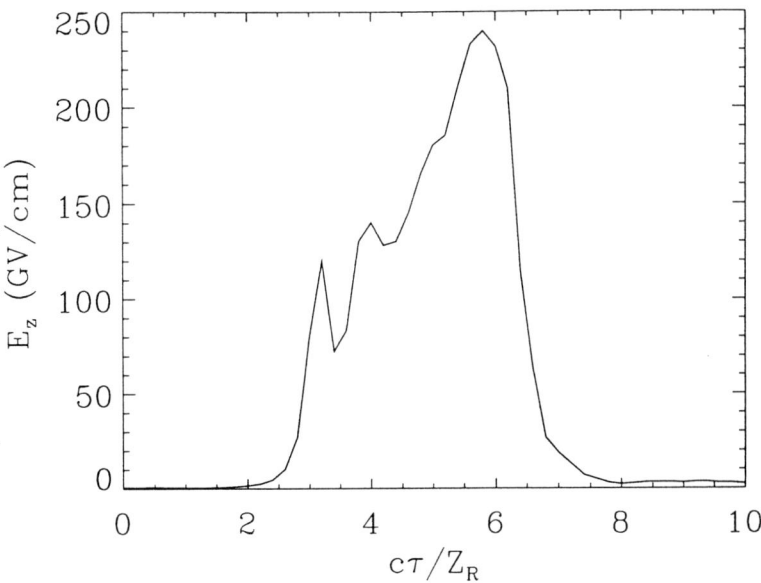

Fig. 9   Peak accelerating field versus cτ.

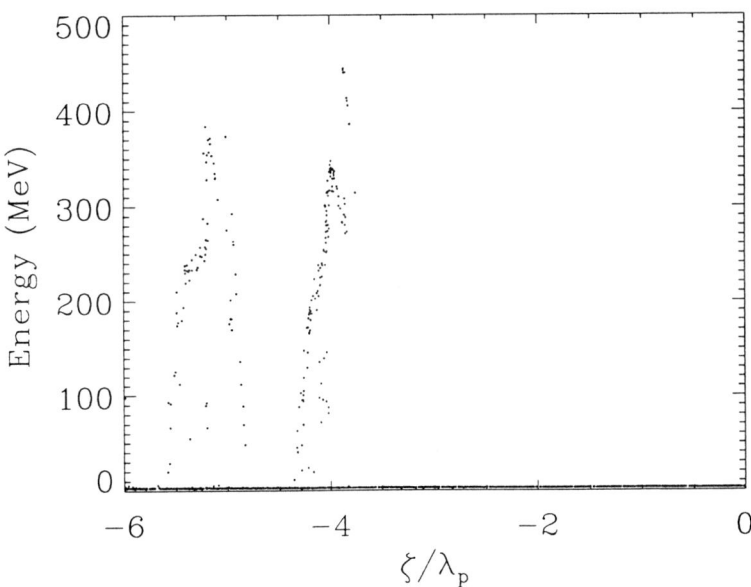

Fig. 10   Test-particle energy distribution at cτ = 1.6 cm = 5.33 Z_R.

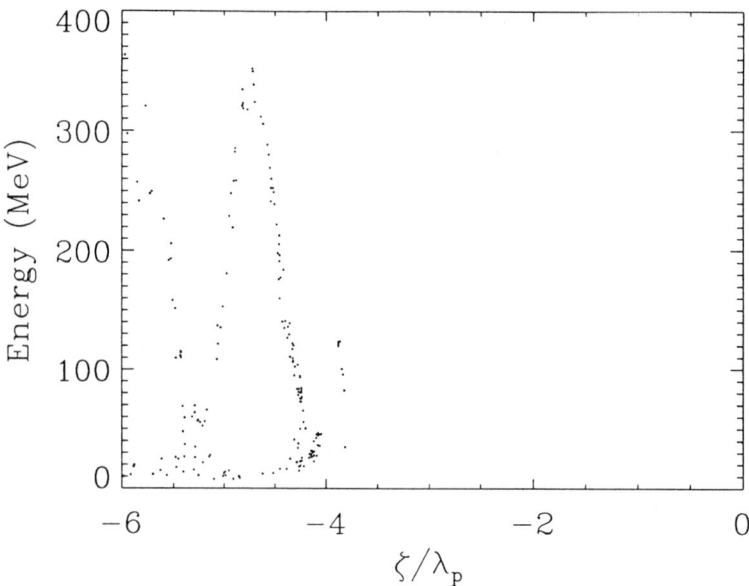

Fig. 11   Test-particle energy distribution at $c\tau = 3.0\,\text{cm} = 10\,Z_R$.

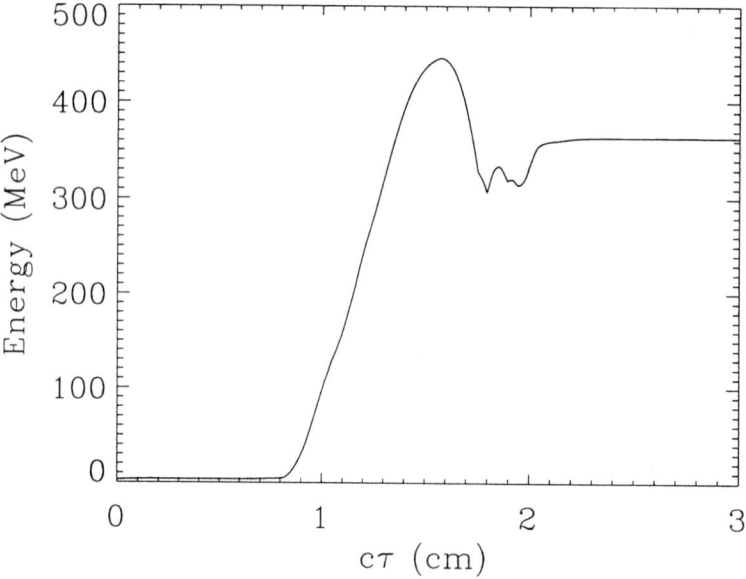

Fig. 12   Peak particle energy versus time.

CONCLUSIONS

We have presented simulations of two sets of parameters, demonstrating acceleration of particles to 90 MeV and 450 MeV respectively. In the latter case, the acceleration was enhanced significantly by relativistic effects, but limited by the reduced group velocity of the laser pulse.

The most notable aspect of these simulations is the fact that, by simply increasing the plasma density, which is easily adjusted experimentally, one can test both the simple linear theory and the highly nonlinear, highly relativistic regime.

ACKNOWLEDGMENTS

This work was supported by the U. S. Department of Energy and the Office of Naval Research.

REFERENCES

1.   T. Tajima and J. M. Dawson, Phys. Rev. Lett. 43, 267 (1979); L. M. Gorbunov and V. I. Kirsanov, Sov. Phys. JETP 66, 290 (1987).

2.   P. Sprangle, E. Esarey, A. Ting and G. Joyce, Appl. Phys. Lett. 53, 2146 (1988); E. Esarey, A. Ting, P. Sprangle and G. Joyce, Comments Plasma Phys. Controlled Fusion 12, 191 (1989).

3.   P. Sprangle, E. Esarey, J. Krall and G. Joyce, submitted to Phys. Rev. Lett.; P. Sprangle, E. Esarey, J. Krall, G. Joyce and A. Ting, these proceedings.

4.   J. Krall, G. Joyce, P. Sprangle and E. Esarey, these proceedings.

5.   G. Z. Sun, E. Ott, Y. C. Lee, and P. Guzdar, Phys. Fluids 30, 526 (1987); P. Sprangle, C. M. Tang, and E. Esarey, IEEE Trans. Plasma Sci. PS-15, 145 (1987); E. Esarey, A. Ting, and P. Sprangle, Appl. Phys. Lett. 53, 1266 (1988); W. B. Mori, C. Joshi, J. M. Dawson, D. W. Forslund and R. M. Kindel, Phys. Rev. Lett. 60, 1298 (1988).

6.   G. Mourou and D. Umstadter, Phys. Fluids B 4, 2315 (1992).

7.   P. Sprangle and E. Esarey, Phys. Fluids B 4, 2241 (1992).

8.   E. Esarey and A. Ting, Phys. Rev. Lett. 65, 1961 (1990); A. Ting, E. Esarey and P. Sprangle, Phys. Fluids. B 2, 1390 (1990).

NUMERICAL SIMULATIONS OF OPTICAL GUIDING OF LASER PULSES IN A PLASMA

J. Krall, G. Joyce, P. Sprangle and E. Esarey
Beam Physics Branch, Plasma Physics Division
Naval Research Laboratory, 4555 Overlook Avenue, SW
Washington, DC 20375-5000

ABSTRACT

In the laser wakefield accelerator, a short ($\tau_L < 1$ ps), high power ($P > 10^{12}$ W) laser pulse propagates in plasma to generate a large amplitude ($E > 1$ GV/m) wakefield. We present an axisymmetric nonlinear fluid model that allows simulation of laser pulse propagation through a plasma on the plasma time scale. We find that a laser pulse will propagate through a plasma for many vacuum diffraction lengths if either of two conditions are met: 1) an appropriately shaped plasma density channel can be obtained or 2) an ultra-high power tailored laser pulse can be created.

INTRODUCTION

In the laser wakefield accelerator (LWFA) concept[1], an intense laser pulse drives a plasma wave which, in turn, accelerates a trailing electron bunch. The problem of laser pulse propagation in a plasma over long distances must be solved in order to demonstrate the viability of this concept. By long distances, we mean $z \gg Z_R$, where

$$Z_R = \frac{k}{2} r_{L,0}^2 \qquad (1)$$

is the vacuum diffraction length, or Rayleigh length, of the laser pulse, k is the laser wavenumber and $r_{L,0}$ is the initial spotsize of the laser pulse.

For high intensity laser pulses, which are characterized by $a_0 \gtrsim 1$, where $a_0 = eA_0/mc^2$ is the normalized amplitude of the laser vector potential field, m is the electron mass, c is the speed of light in vacuum and e, assumed positive, is the elementary charge, plasma wakefields can be obtained with amplitudes at or near the nonrelativistic wavebreaking field:

$$E_{WB} \equiv mc\omega_p/e , \qquad (2)$$

where $\omega_p = (4\pi n_0 e^2/m)^{1/2}$ and n_0 is the plasma density. For example, with $n_0 = 10^{18}$ cm^{-3}, $E_{WB} \simeq 100$ GeV/m.

We have shown, via numerical simulation, that laser diffraction in a plasma can be overcome by either of two methods: plasma channel guiding or relativistic optical guiding of a tailored pulse.[2] In this paper, we describe the numerical model by which these results were obtained and give a further example. Because we make use of the fact that the plasma evolves on a slow time scale relative to that corresponding to the laser frequency, $\omega_p << \omega \simeq kc$, the resulting code has tremendous advantages over conventional particle simulations, which must resolve the fast laser frequency.

MODEL EQUATIONS FOR NUMERICAL SOLUTION

Model equations giving the axisymmetric plasma response to a given laser field in the quasi-static approximation have been recently derived.[2] In deriving these equations, a coordinate transformation was made from (r,z,t) to $(r,\zeta=z-ct,\tau=t)$. The laser pulse moves in the positive z direction such that the front of the laser pulse remains at $\zeta = 0$ in these coordinates. The physical region of interest extends from $\zeta = 0$, where the plasma is unperturbed, to $\zeta < 0$. In the quasi-static approximation, it is assumed that the scale length over which the laser field evolves is long compared to the laser pulse length ($Z_R >> L_{laser}$). In this limit derivatives with respect to τ are dropped relative to derivatives with respect to ζ in the plasma fluid equations, but not in the wave equation. The model equations, after averaging over the laser frequency, to leading order in the small parameters $1/kr_L$ and ω_p/ω, are given by

$$(\nabla_\perp^2 + \frac{\partial^2}{\partial\zeta^2})\phi_s = k_p^2[\gamma_s\rho_s - \rho^{(0)}] , \tag{3}$$

$$\nabla_\perp^2\psi_s - k_p^2\rho_s\psi_s = k_p^2[\gamma_s - \rho^{(0)}] , \tag{4}$$

$$\nabla_{\perp}^2 a_{\perp,s} = k_p^2\rho_s u_{\perp,s} - \nabla_\perp\frac{\partial\phi_s}{\partial\zeta} , \tag{5}$$

and

$$\frac{\partial u_{\perp, s}}{\partial \zeta} = \frac{\partial a_{\perp, s}}{\partial \zeta} + \nabla_{\perp}(\gamma_s - \phi_s) , \tag{6}$$

where the Coulomb gauge has been used and the subscript s indicates quantities associated with the slow time-scale plasma response: $\phi_s = e\Phi_s/mc^2$ is the normalized electrostatic potential, $\psi_s = \phi_s - a_{z,s}$, a_s is the normalized vector potential, $u_s = v_s/c$ is the normalized plasma electron fluid velocity, $\rho_s = n/\gamma_s n_0$, n is the plasma electron number density and n_0 is the unperturbed density at r = 0. The definitions $\rho^{(0)} = \rho_s(\zeta=0)$ and $k_p = \omega_p/c$ are also used. In this model, the plasma ions are assumed to be immobile. The laser field enters through the relativistic factor for the plasma electrons:

$$\gamma_s = 1 + \frac{u_{\perp, s}^2 + \hat{a}_f \cdot \hat{a}_f^* /2 + \psi_s^2}{2(1 + \psi_s)} , \tag{7}$$

where a_f, which varies on a fast time-scale, is the normalized vector potential of the laser pulse and $\hat{a}_f$ is the slowly varying laser amplitude. This form of the γ_s equation assumes a linearly polarized laser field. For circular polarization, the factor of 1/2 in the second term of the numerator above is dropped. The laser field, which is assumed to have a phase velocity, $v_{ph} = c$, is updated via a reduced wave equation:

$$(\nabla_{\perp}^2 + \frac{2ik}{c}\frac{\partial}{\partial \tau})\hat{a}_f = k_p^2 \rho_s \hat{a}_f . \tag{8}$$

Equations (3-7) are numerically intractable in the sense that it is not clear which equation should be solved first, second, etc. or if an iterative method of solving the equations can be expected to converge on a meaningful solution. We avoid this numerical quagmire by recasting the problem into a single equation for the quantity ψ_s. From Eqs. (3) and (4), we have

$$\frac{\partial^2 \psi_s}{\partial \zeta^2} = (\frac{\gamma_s - \psi_s - 1}{\psi_s + 1})(k_p^2 \rho^{(0)} + \nabla_{\perp}^2 \psi_s) - \nabla^2 a_{z,s} . \tag{9}$$

From Eq. (6) and the gauge condition we can write $\nabla^2(a_{z,s})$ in terms of ψ_s and $u_{\perp,s}$:

$$\nabla^2 a_{z,s} = \nabla_\perp^2 (\gamma_s - \psi_s) - \frac{\partial}{\partial \zeta}(\nabla_\perp \cdot \underline{u}_{\perp,s}) \ . \tag{10}$$

From Eq. (5) and the gauge condition, we write $\underline{u}_{\perp,s}$ in terms of ψ_s:

$$\underline{u}_{\perp,s} = \frac{1}{k_p^2 \rho_s} \frac{\partial}{\partial \zeta}(\nabla_\perp \psi_s) \ . \tag{11}$$

For completeness, we point out that Eq. (4) can be recast as an equation for ρ_s in terms of ψ_s:

$$\rho_s = \frac{1}{1 + \psi_s} (\rho^{(0)} + \frac{1}{k_p^2} \nabla_\perp^2 \psi_s) \ . \tag{12}$$

The ψ_s equation can be written in the following form:

$$\left[1 - \frac{\nabla_\perp^2}{k_p^2 \rho_s} + \left(\frac{\nabla_\perp \rho_s}{k_p^2 \rho_s^2}\right) \cdot \nabla_\perp\right] \frac{\partial^2 \psi_s}{\partial \zeta^2}$$

$$= \left[1 - \frac{\nabla_\perp^2}{k_p^2 \rho_s} + \left(\frac{\nabla_\perp \rho_s}{k_p^2 \rho_s^2}\right) \cdot \nabla_\perp\right] [(\frac{\gamma_s - \psi_s - 1}{1 + \psi_s}) (k_p^2 \rho^{(0)} + \nabla_\perp^2 \psi_s)]$$

$$+ \nabla_\perp \cdot \{[\frac{1}{k_p^2} \frac{\partial}{\partial \zeta}(\frac{1}{\rho_s})] [\frac{\partial}{\partial \zeta} \nabla_\perp \psi_s] + \frac{1}{\rho_s} (\gamma_s - \psi_s - 1) \nabla_\perp \rho_s\} \ . \tag{13}$$

Equation (13) can be integrated numerically in ζ in a straightforward manner by starting at the $\zeta = 0$ boundary, where the plasma is unperturbed and where we assume $\psi_s = \nabla_\perp \psi_s = \nabla_\perp^2 \psi_s = \partial \psi_s / \partial \zeta = 0$. Radial boundary conditions are dictated by axisymmetry at $r = 0$ and by the imposition of a metallic wall at $r = r_w \gg r_{L,0}$. Equations (13) and (8), along with the auxiliary Eqs. (7), (11) and (12), provide a complete description of the laser-plasma interaction on the plasma time scale.

SIMULATION RESULTS

As an example, we consider a tailored laser pulse with modest energy and power (i.e., within the bounds of present technology). A tailored pulse is one in which the laser spotsize is tapered from a large value to a small value over the length of the laser pulse while keeping the laser power constant throughout the pulse and equal to the critical power[3] for relativistic optical guiding. This condition is met when

$$a_0(\zeta)\, r_{L,0}(\zeta) = 0.9\ \lambda_p \ . \tag{14}$$

Figure 1 shows the laser intensity $|\hat{a}_f|^2$ and laser spotsize r_L for a tailored pulse with $a_{0,peak} = 0.9$ and $r_{L,0,min} = \lambda_p$ tapered over a length of $2\lambda_p$. With $n_0 = 1.24 \times 10^{18}$ cm^{-3} ($\lambda_p = 30$ μm) and laser wavelength $\lambda = 1$ μm, the laser power is ≈ 16 TW and the pulse energy is ≈ 3 J. Figure 2 shows the laser intensity and spotsize at $c\tau = 36\ Z_R = 10.2$ cm. The pulse is distorted but largely intact. Figure 3, a plot of the axial electric field at $c\tau = 36$ Z_R, shows a peak accelerating gradient of 15 GV/m. Because the pulse has expanded somewhat, this is smaller than the average value of 18 GV/m over the length of the simulation. Figure 4, laser spotsize at the point of peak laser intensity plotted versus τ, confirms that the laser pulse remains focused throughout the simulation. For comparison, the curve corresponding to vacuum expansion of the laser pulse is also shown.

To study the acceleration and trapping of electrons by the high-gradient wakefield, we have modified an existing particle code to accelerate a distribution of non-interacting test particles in the time-resolved electric and magnetic fields. For our present example, we consider a continuous beam with average energy 2.0 MeV, rms radius $r_b = 5$ μm, and emittance $\epsilon_n = 1.0$ mm-mrad, shown in Fig. 5 (a). After 10 cm, Fig. 5 (b), the particles have been trapped into two bunches corresponding to the two accelerating phases of $E_z(\zeta)$ as shown in Fig. 3. In Fig. 5 (b), the average energy is 1.6 GeV, $r_b = 0.75$ μm, and $\epsilon_n = 1.4$ mm-mrad. Approximately 56% of the particles shown in Fig. 5 (a) remain in Fig. 5 (b). A phase-space plot at $c\tau = 10$ cm, Fig. 6, shows that the majority of the particles are close to the peak energy of 1.8 GeV. Plots of peak and average particle energy versus time, Fig. 7, reflect the varying amplitude of the plasma wake, which varies with laser intensity. The highest acceleration occurs at $c\tau = 1.4$ cm $\approx 5\ Z_R$ which corresponds to the minimum spotsize shown in Fig. 4.

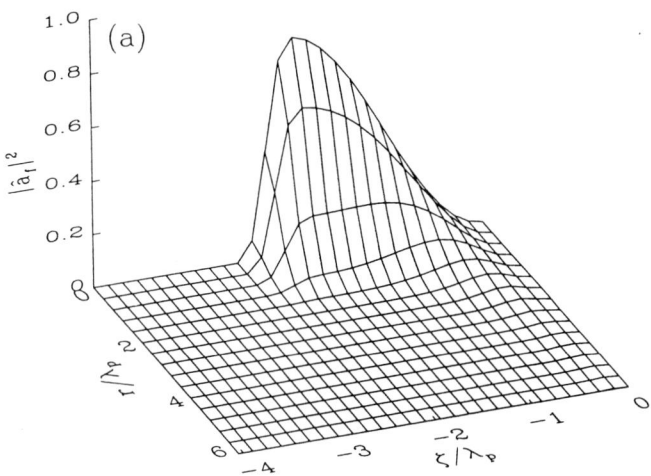

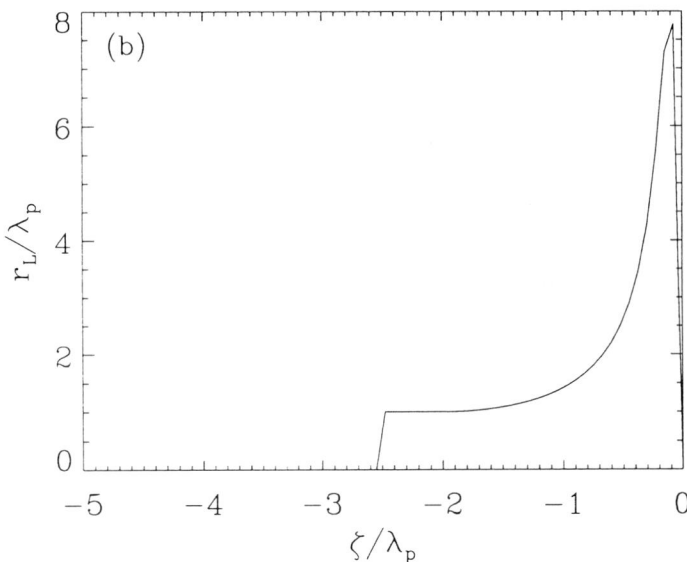

Fig. 1 Plots showing (a) laser intensity $|\hat{a}_f|^2$ and (b) laser spotsize r_L at $\tau = 0$.

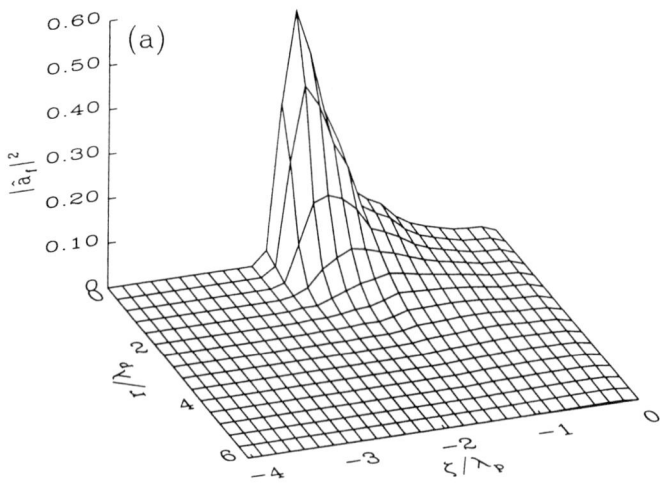

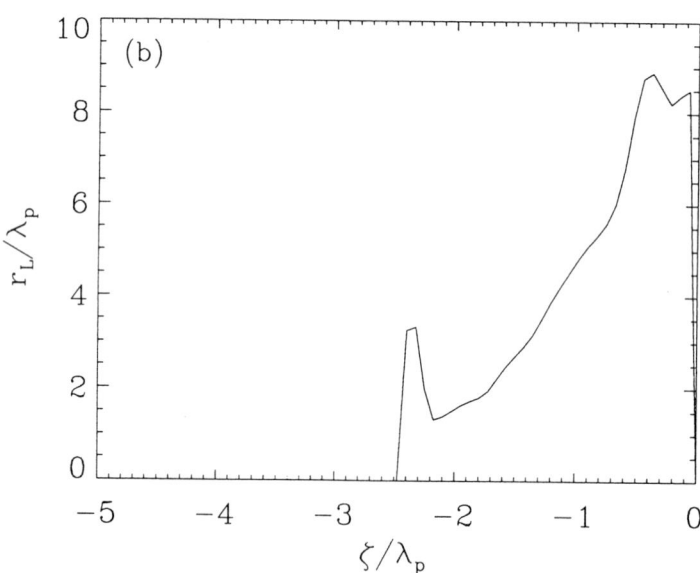

Fig. 2   Plots showing (a) laser intensity $|\hat{a}_f|^2$ and (b) laser spotsize r_L at $c\tau = 36\,Z_R$.

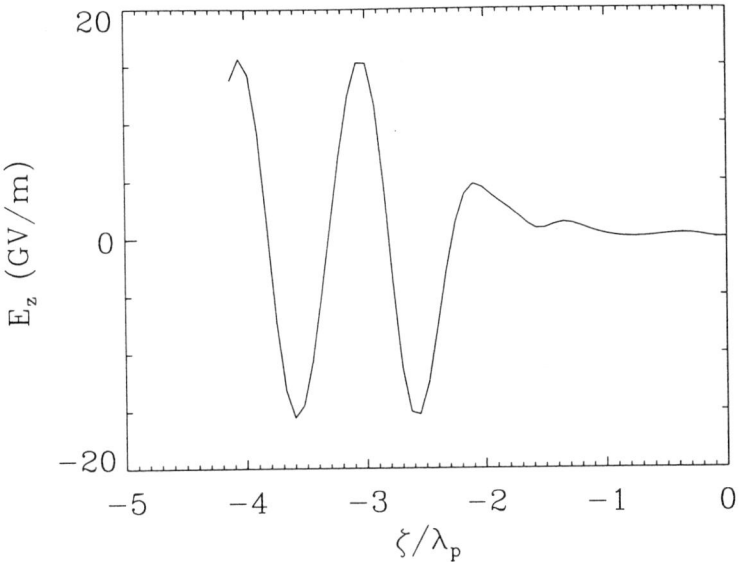

Fig. 3   Axial electric field E_z versus ζ at $c\tau = 36\,Z_R$.

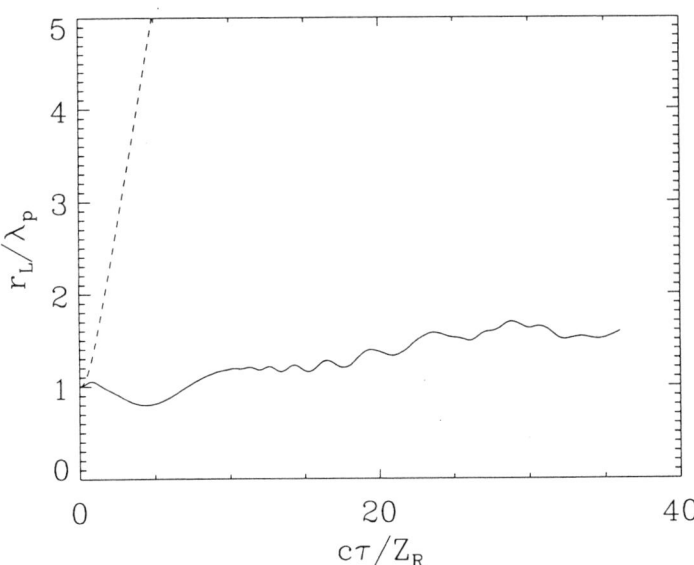

Fig. 4   Laser spotsize r_L, at point of peak intensity in the pulse, versus time (solid line) and vacuum diffraction curve (dashed line).

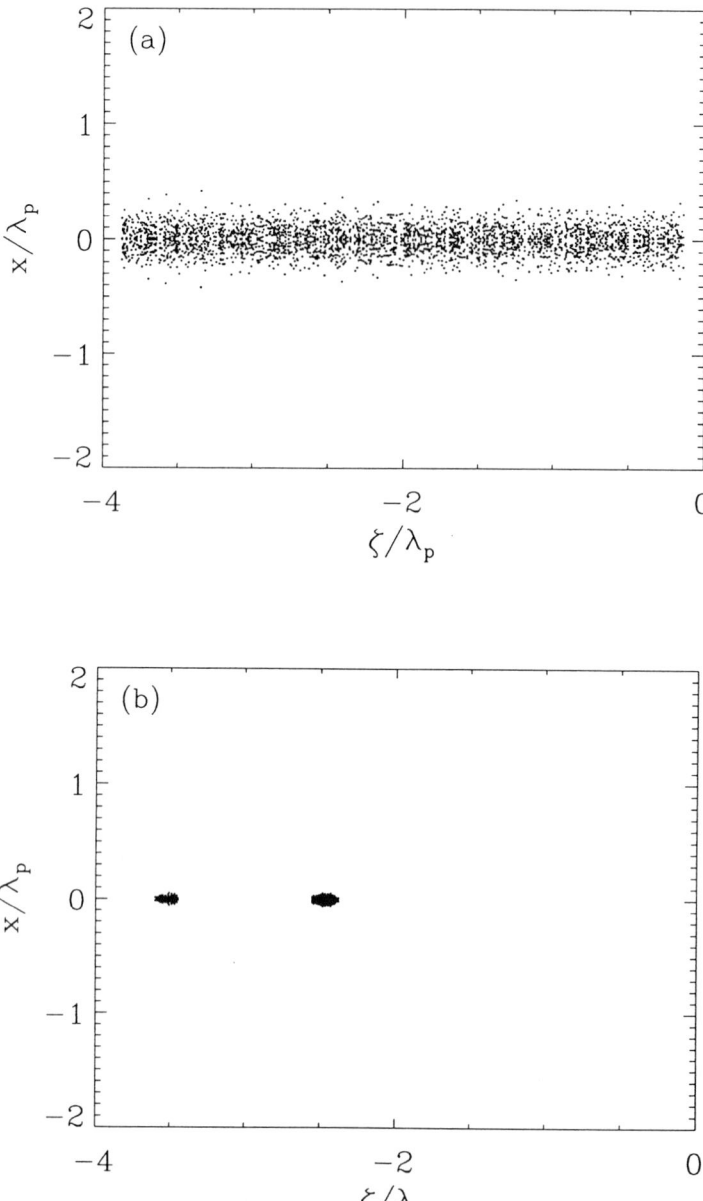

Fig. 5   Plots of (x, ζ) phase-space showing the test particles at (a) $\tau = 0$ and (b) $c\tau = 10$ cm.

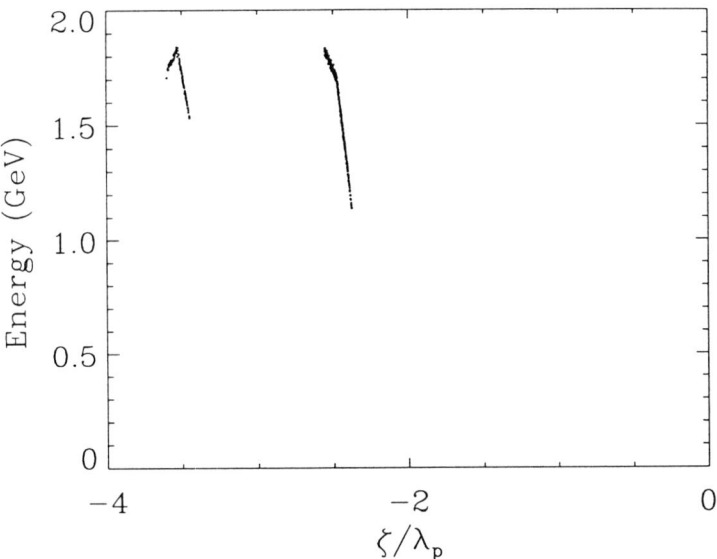

Fig. 6 Phase-space plot showing the test particle energy distribution at $c\tau = 10$ cm.

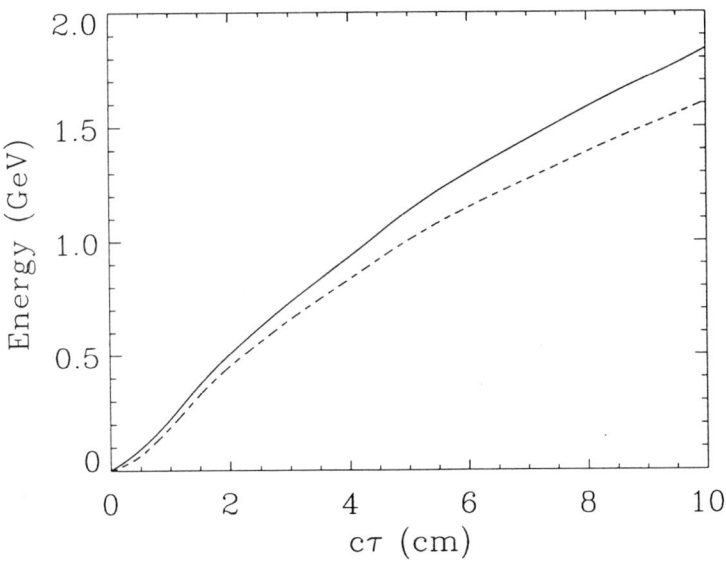

Fig. 7 Peak particle energy (solid line) and average energy (dashed line) versus time.

CONCLUSIONS

We have presented an axisymmetric nonlinear fluid model that allows us to simulate laser pulse propagation through a plasma on the plasma time scale, $\omega_p \ll \omega$. Using this code, we have demonstrated that a "tailored" pulse can propagate for distances in excess of 36 vacuum diffraction lengths with only minimal distortion. We have also simulated particle acceleration, demonstrating that a 16 terrawatt, 3 Joule laser pulse can drive a wake such that 1.8 GeV electrons are produced in a single 10 cm accelerating stage. These laser parameters (16 TW, 3 J) are consistent with present technology.

We have also performed simulations of plasma density channel guiding over similarly large distances with even less distortion of the laser pulse than is observed for the tailored pulse shown here. We conclude that a laser pulse will propagate through a plasma for many vacuum diffraction lengths if either of two conditions are met: 1) an appropriately shaped plasma density channel can be obtained or 2) an ultra-high power tailored laser pulse can be created.

ACKNOWLEDGMENTS

We wish to acknowledge many helpful discussions with A. Ting of N.R.L. This work was supported by the U. S. Department of Energy and the Office of Naval Research.

REFERENCES

1.   T. Tajima and J. M. Dawson, Phys. Rev. Lett. <u>43</u>, 267 (1979).; L. M. Gorbunov and V.I. Kirsanov, Sov. Phys. JETP <u>66</u>, 290 (1987); P. Sprangle, E. Esarey, A. Ting and G. Joyce, Appl. Phys. Lett. <u>53</u>, 2146 (1988); E. Esarey, A. Ting, P. Sprangle and G. Joyce, Comments Plasma Phys. Controlled Fusion <u>12</u>, 191 (1989).
2.   P. Sprangle, E. Esarey, J. Krall and G. Joyce, submitted to Phys. Rev. Lett.
3.   G. Schmidt and W. Horton, Comments Plasma Phys. Controlled Fusion <u>9</u>, 85 (1985); G. Z. Sun, E. Ott, Y. C. Lee, and P. Guzdar, Phys. Fluids <u>30</u>, 526 (1987); P. Sprangle, C. M. Tang, and E. Esarey, IEEE Trans. Plasma Sci. <u>PS-15</u>, 145 (1987); E. Esarey, A. Ting, and P. Sprangle, Appl. Phys. Lett. <u>53</u>, 1266 (1988).

MULTISTAGING AND *e*-BEAM TRAPPING IN LASER PARTICLE ACCELERATORS

L. C. Steinhauer and W. D. Kimura
STI Optronics, Inc.
2755 Northup Way, Bellevue, WA 98004-1495

ABSTRACT

Any practical laser particle acceleration (LPA) scheme will necessarily require a series of acceleration sections arranged in tandem. The phase of the electrons with respect to the laser field can be readjusted at the beginning of each stage, thereby determining that stage's contribution to the acceleration and/or focusing of the *e*-beam. Multistaging, in fact, offers a natural way to control the *e*-beam/laser interaction by tuning each successive stage to achieve a high acceleration gradient while preserving or enhancing the quality of the *e*-beam (e.g., emittance and energy spread). The stage-to-stage evolution of the beam can be treated somewhat independently of specific details in each accelerator section. Therefore, the laser/*e*-beam interaction within the sections is treated using a generic approach. The basic conditions for beam trapping are established for a multistaged system. Calculations of the longitudinal and transverse beam dynamics are performed to determine the evolution of a finite-emittance *e*-beam through successive stages. It is found that maintaining the *e*-beam quality requires only that the trapping condition be satisfied.

INTRODUCTION

Research in laser particle accelerators (LPA) has concentrated primarily on the issue of achieving high acceleration gradients. Some work has also looked at focusing of *e*-beams by the laser beam. A practical area which has received little attention is that an actual LPA would probably consist of a number of laser-driven acceleration sections in a manner analogous to microwave-driven accelerators. These sections can affect the quality of the *e*-beam, in particular the emittance. This issue is further complicated when laser-induced focusing or defocusing also occurs within these sections. Multistaging is important because it introduces a new degree of freedom in the design of an LPA system, namely the phase of the electron bunches relative to the laser wave as they enter each acceleration section. Since this "entrance phase" can be controlled, it becomes the chief tuning parameter for controlling the beam trapping and evolution. We present in this paper the preliminary results of our analysis of multistaging, and show that it is possible to select the entrance phase point of each acceleration section to assure beam trapping <u>and</u> preserve the transverse (emittance) and longitudinal (bunching) quality of the *e*-beam.

BASIC FIELD EQUATIONS

The stage-to-stage evolution can be treated somewhat independently of the details within each section. However, a reasonable laser focal structure with self-consistent longitudinal and transverse electromagnetic fields should be used. For convenience, we adopt the laser beam focusing geometry that is associated with inverse Čerenkov acceleration,[1-2] and has been proposed as a new vacuum laser accelerator geometry.[3] This geometry, shown in Fig. 1, features a radially polarized laser beam and an axicon mirror producing a linear focus that is aligned with the *e*-beam axis. Although we have selected a fairly specific geometry, its primary function is to assure self-consistency between the transverse and longitudinal fields acting upon the particles. Though details of the accelerator section differ from one LPA concept to another, they share the common feature that the relative laser/*e*-beam phase at the entrance of each section determines the degree of acceleration and focusing within that section. In that sense then the approach taken here is generic to all multistaged LPAs.

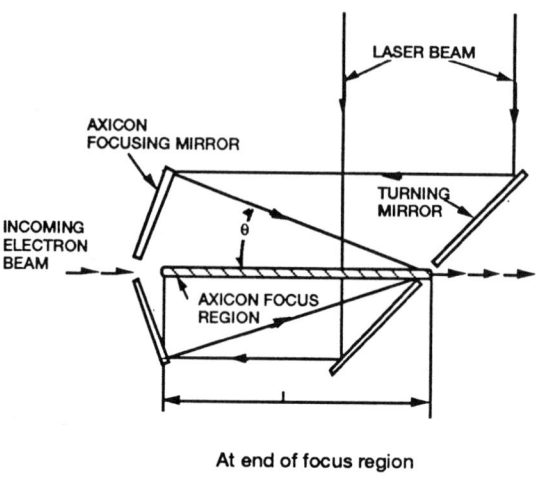

At end of focus region

91 19542

Fig. 1. Model for laser/*e*-beam interaction geometry. The laser beam consists of a radially polarized field that passes through an axicon and converges at an angle θ onto the *e*-beam. L is the interaction-region length.

Any electromagnetic field structure can be represented as the superposition of basis functions that employ Bessel functions.[4] In general, the superposition takes the form of an integral over the basis functions. For axisymmetric radially-polarized electromagnetic fields, the basis functions have the common form,

$$E_r = E_0 J_1(\alpha r) \exp[i(\omega t - \kappa z)], \tag{1a}$$

$$E_z = E_0 \frac{\alpha}{\kappa} J_0(\alpha r) \exp[i(\omega t - \kappa z - \frac{\pi}{2})], \tag{1b}$$

$$B_\theta = E_0 \frac{\omega}{c^2 \kappa} J_1(\alpha r) \exp[i(\omega t - \kappa z)], \tag{1c}$$

where E_0 = constant is the reference electric field. The parameters α and κ are related to the crossing angle θ (the angle at which the rays cross the axis) by,

$$\alpha = \frac{\omega}{c} \sin\theta, \tag{2a}$$

$$\kappa = \frac{\omega}{c} \cos\theta. \tag{2b}$$

The fields in Eqs. 1 and 2 resemble those for radially-polarized modes in a waveguide (in which case there would be discrete values of θ, one for each mode).

The longitudinal extent of the laser/particle interaction zone within a given accelerator section is L. It is determined by the spacing between the axicon mirror at the entrance and the turning mirror at the exit of the section (see Fig. 1). The natural reference scale for L is the *slip distance*, defined as the longitudinal distance over which the particle phase, $\psi \equiv \omega t - \kappa z(t)$, [$z(t)$ is the particle position as a function of time], slips by π. Assuming that the wave propagates in a vacuum:

$$L_\pi = \frac{\lambda}{2} (1 - \beta\cos\theta)^{-1} \approx \frac{\lambda}{\theta^2} , \tag{3}$$

where $\lambda = 2\pi c/\omega$ is the wavelength. The approximation assumes small crossing angle and high energy particles (specifically $\gamma \gg 1/\theta$). In analyzing the laser/e-beam interaction within each section we will assume that the phase slips by π over the interaction length, i.e., $L = L_\pi$. In an actual vacuum accelerator[3] L may be several times L_π which merely reduces the net acceleration by the factor $(L/L_\pi)^{1/2}$. The case L $= L_\pi$ corresponds roughly to an inverse Čerenkov accelerator where phase slip is eliminated by introducing a gas into the interaction region. Note that L_π is twice the Rayleigh range, the minimum natural length of the focal region.

It is useful to determine the ideal electron energy gain in one section of the accelerator. This allows E_0 (the reference electric field in Eq. 1) to be related to the energy gain $\Delta\gamma$. As derived in the Appendix,

$$E_0 = \frac{\pi}{2} \frac{m_e c^2}{-e} \frac{<\gamma'>_I}{\theta} , \tag{4}$$

where $-e$ and m_e are the electron charge and rest mass, respectively, and $<\gamma'>_I \equiv \Delta\gamma/L_\pi$, i.e., the average acceleration gradient over a single-section interaction length of L_π. The derivation assumed an electron traveling along the axis and the optimal entrance phase.

TRANSVERSE e-BEAM DYNAMICS

Consider now the transverse forces exerted on the electrons by the laser. The radial component of the Lorentz force for the fields in Eq. 1 is

$$F_r = -eE_0 \left[1 - \frac{\beta_z}{\cos\theta} \right] J_1(\alpha r) \cos\psi , \tag{5}$$

where $\beta_z \equiv (1/c)dz/dt$, and $\psi \equiv \omega t - \kappa z$ is the phase.

The phase ranges for acceleration and focusing are illustrated in Fig. 2. Also shown is an example electron with an entrance phase slightly less than π. The entrance phase ψ_0 is the value of ψ at the entrance to a given section. In slipping by $\Delta\psi = \pi$, this particular electron at first, briefly, experiences a decelerating force (negative F_z), but is accelerated over most of its path. At the same time it experiences a focusing force (negative F_r) over the early part of its path, and a defocusing force toward the end. Fig. 3 shows the effect on electron trajectory for two specific cases. In the first, ψ_0 is chosen for maximum acceleration. In this case the electron path is first bent inward (focusing) and then bent outward (defocusing) with little net focusing effect. In the second case, ψ_0 is chosen for maximum focusing. This case produces no acceleration.

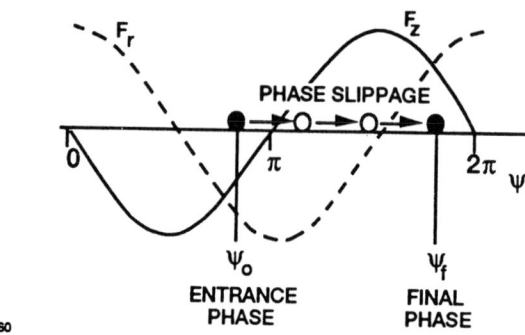

Fig. 2.   Phase ranges for acceleration and focusing.

(a) Initial Phase Tuned For Maximum Acceleration: ψ_0 =

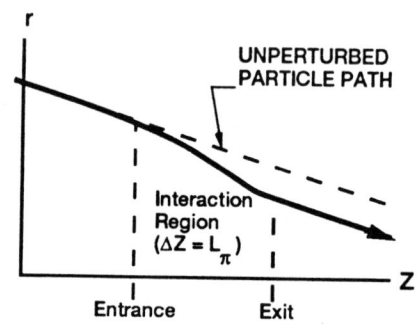

(b) Initial Phase Tuned For Maximum Focusing: ψ_0 =

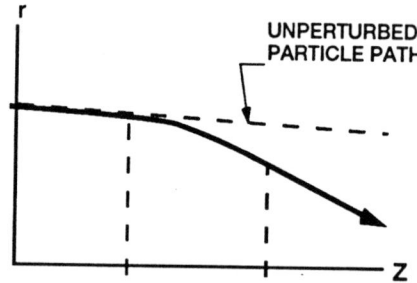

Fig. 3.   Limiting cases of transverse *e*-beam dynamics when governed by the phase relationships given in Fig. 2.

These examples illustrate the role of the entrance phase ψ_0 in *tuning* the interaction. Depending on the choice of ψ_0 the interaction can lead to net acceleration or deceleration, net focusing or defocusing, or some combination of the two. That is, both the longitudinal and transverse dynamics are controlled by the choice of ψ_0. As will be shown later, in a multistaged system, the choice of ψ_0 will also affect the e-beam trapping and quality.

LASER BEAM EFFECT ON e-BEAM TRAPPING

The quality of the beam (e.g., the emittance) depends on the ability to trap individual particles. Trapped electron motion is taken here to mean that the electron remains near the axis ($x = y = 0$) and its phase ψ (a measure of longitudinal position) remains close to that of the other electrons. Consider the case $L = L_\pi$; then from Eqs. A4 and A5 (using $\psi_f = \psi_0 + \pi$ since $L = L_\pi$), the energy gain for an electron close to the axis is

$$\Delta\gamma \approx -L_\pi <\gamma'>_I \cos\psi_0 . \tag{6}$$

Recall that $<\gamma'>_I$ is the acceleration gradient for the optimum ψ_0 electron. Hence, net acceleration occurs within the entrance phase range $\pi/2 < \psi_0 < 3\pi/2$, with maximum acceleration at $\psi_0 = \pi$. Similarly, it can be shown that the change in transverse momentum goes as

$$\Delta\frac{dx}{dz} \approx -\frac{\pi}{2} x_0 \frac{<\gamma'>_I}{\gamma_0} \sin\psi_0 , \tag{7}$$

where x_0 is the entrance x position. Here, for simplicity, the electron trajectory is assumed to lie in the $y = 0$ plane. Focusing (negative $\Delta dx/dz$) occurs within the entrance phase range, $0 < \psi_0 < \pi$. Thus, the twin requirements of acceleration *and* focusing restrict the entrance phase to the range $\pi/2 < \psi_0 < \pi$.

Consider now a treatment of electron trapping that accounts for both transverse and longitudinal motion. In order to place these calculations in context, the individual electrons considered are referenced to an electron beam with a finite spread of particle positions and velocities, i.e., an e-beam with finite emittance. For this purpose it is assumed that the particles initially belong to a Kaphinskij-Vladimirskij (K-V) distribution,[5] for which all electrons lie on an ellipse in x-p_x space, where $p_x \equiv (\gamma/c)dx/dt$ is the dimensionless momentum in the x direction (the dimensional factor is $m_e c$). At the entrance to the first section all electrons are assumed to have the same initial energy, $\gamma = \gamma_0$, and same phase, $\psi = \psi_0$. The K-V distribution can be represented parametrically as follows:

$$x(u) = \Delta x \cos(u); \quad p_x = \Delta p \sin(u); \quad 0 < u < 2\pi , \tag{8}$$

where u is the parameter that specifies a particular electron in the distribution. Employing the root-mean-square definition of the emittance ε_{Nx},

$$\varepsilon_{Nx} \equiv 4\left[<x^2><p_x^2> - (<x><p_x>)^2\right]^{1/2}, \tag{9}$$

the normalized emittance of the K-V distribution (Eq. 8) is

$$\varepsilon_{Nx} = 2\Delta x \Delta p_x . \tag{10}$$

Fig. 4 illustrates the K-V distribution for $\varepsilon_N = 7\pi$ mm-mrad. The arrows at $\alpha x = \pm 2.405$ identify the location (if $y = 0$) where the J_0 Bessel function is zero, i.e., the accelerating force falls to zero. The points marked A, B, C, and D represent particular electrons in this distribution. For these the parameter u in Eq. 8 is 0, $\pi/4$, $\pi/2$, and $3\pi/4$, respectively. These are the <u>initial</u> positions of the electrons (i.e., before the first stage of the accelerator); the evolution of their position in phase space through successive sections is examined next.

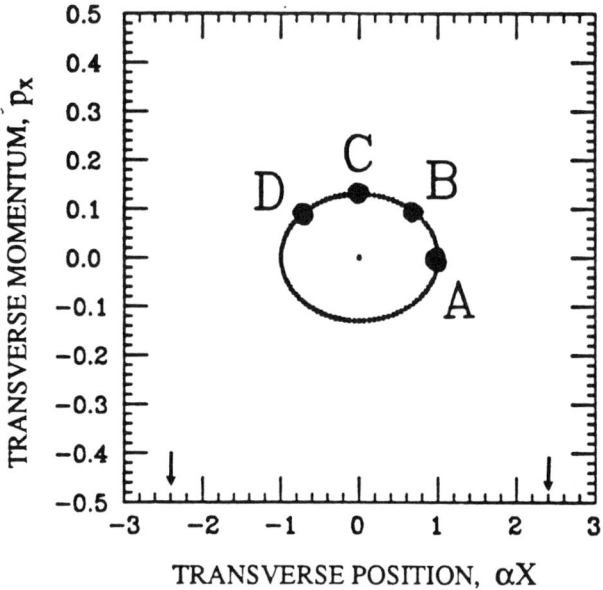

Fig. 4.   Example of a Kaphinskij-Vladimirskij (K-V) distribution with $\varepsilon_N = 7\pi$ mm-mrad, $\theta = 20$ mrad, and $\lambda = 10.6$ μm.

The trapping of the electron labeled A is illustrated in Fig. 5 for a 100-stage LPA where each acceleration stage is separated by a drift section (with no forces on the electrons) also of length L_π. Four pairs of plots are shown, each pair for a different entrance phase ψ_0. The transverse phase space behavior is displayed in the left plots, in which the αx-p_x space evolution is traced from the initial position (marked by a "+" symbol). The dots indicate the position at the entrance to each successive stage, leading to the final position after 100 stages (marked by an "x" symbol). The longitudinal phase space behavior is displayed in the right plots where the ψ - p_z space evolution is traced. Here $p_z = (\gamma/c)dz/dt$ is the normalized longitudinal momentum ($p_z \approx \gamma$ for a paraxial beam). The actual quantities displayed, $\Delta p_z/p_z$ and $\Delta\psi$, reference the momentum and phase to the ideal electron that runs exactly along the z axis. Clearly, the cases $\psi_0 = 0.7\pi$ and 0.9π are trapped in both the transverse and longitudinal phase planes. The cases $\psi_0 = 1.1\pi$ and 1.3π are untrapped in both planes. Transverse trapping is taken to mean that the electron remains near the axis (specifically, inside the zero point of J_0). Longitudinal trapping is taken to mean that the electron remains near the reference particle in either ψ or p_z coordinates.

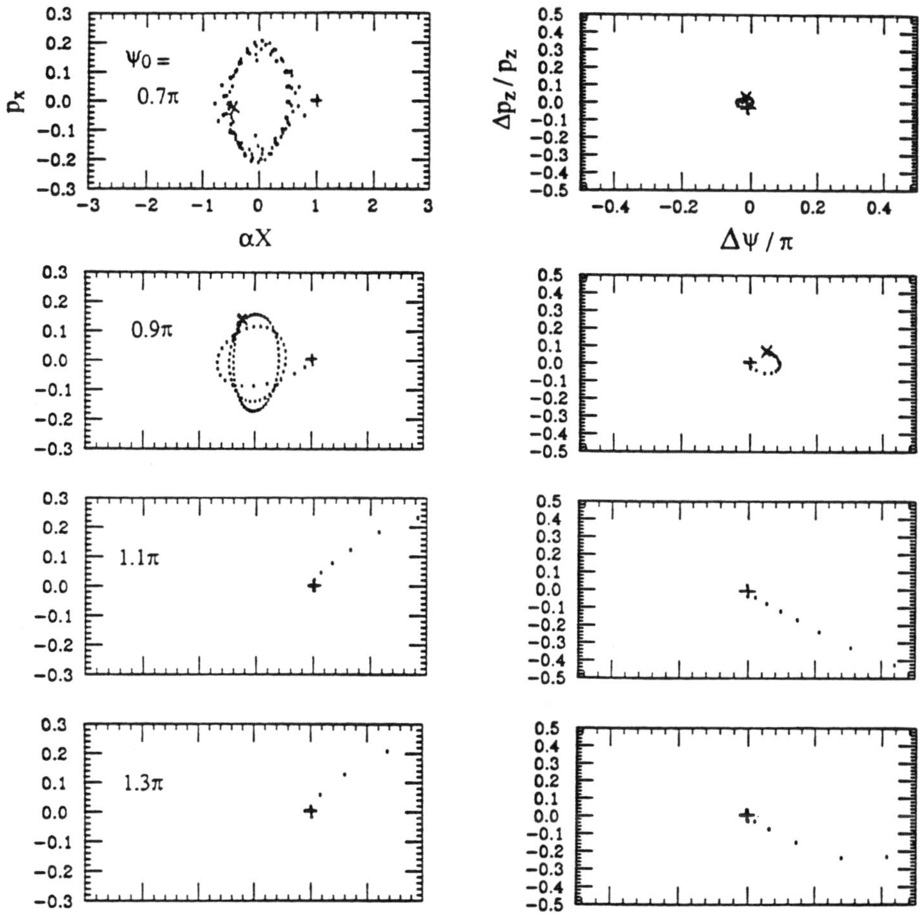

Fig. 5.   Particle trapping as a function of entrance phase for the electron labeled A in Fig. 4. The initial position is indicated by a "+" symbol and the final position after 100 acceleration stages is indicated by a "x" symbol.

Shown in Fig. 6, for an initial emittance of 7π mm-mrad, is the stability of the four electrons with initial positions labeled A, B, C, and D in Fig. 4. Electrons with larger initial outward momentum tend to be untrapped. Here electron C, which has the largest $|p_x|$, is lost the fastest. Electron B is also lost, while electrons A and D are trapped. Note that transverse trapping and detrapping match up with longitudinal trapping and detrapping, respectively.

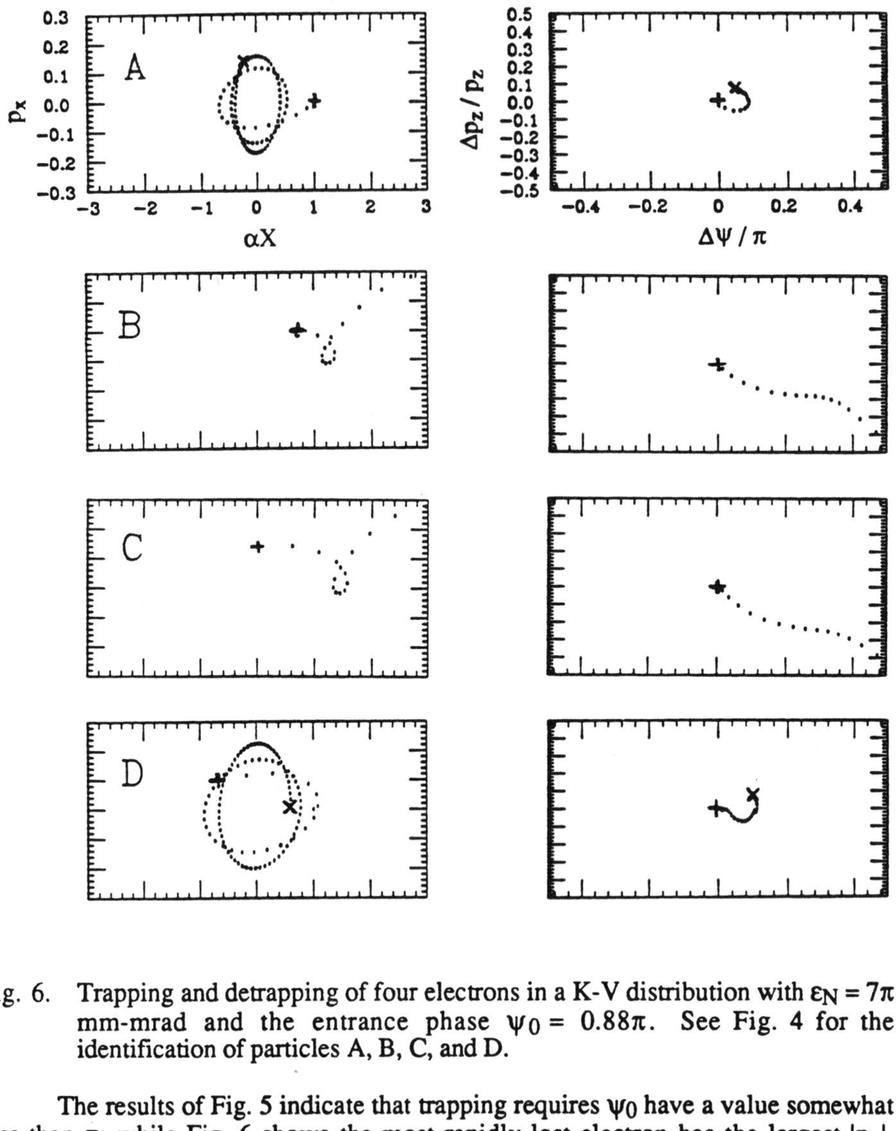

Fig. 6.   Trapping and detrapping of four electrons in a K-V distribution with $\varepsilon_N = 7\pi$ mm-mrad and the entrance phase $\psi_0 = 0.88\pi$. See Fig. 4 for the identification of particles A, B, C, and D.

The results of Fig. 5 indicate that trapping requires ψ_0 have a value somewhat less than π; while Fig. 6 shows the most rapidly lost electron has the largest $|p_x|$. Therefore, there must be a marginal entrance phase $(\psi_0)_{st}$ that marks the trapping boundary, such that if $\psi_0 \leq (\psi_0)_{st}$ then all particles in the K-V distribution will be trapped. This marginal entrance phase for trapping is shown in Fig. 7 as a function of the emittance of the K-V distribution. Also shown is a curve for $\gamma_0 = 200$ indicating that trapping becomes easier at higher energies.

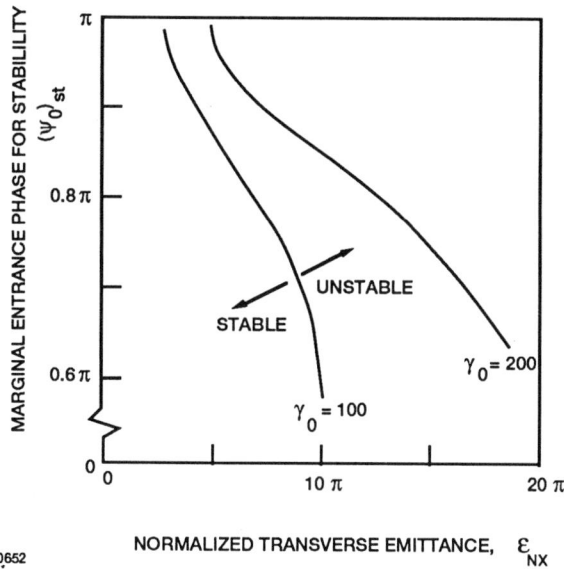

92 20652

NORMALIZED TRANSVERSE EMITTANCE, ε_{NX}

Fig. 7.   Marginal conditions for trapping of electrons in a K-V distribution.

EFFECT ON e-BEAM QUALITY

In general, a given emittance describes only the beam quality with respect to two of the six dimensions in phase space (x, p_x, y, p_y, z, p_z), and until now we have been dealing primarily with the transverse components. The longitudinal emittance (z-p_z plane) is also an important beam quality index since the phase (a relative measure of longitudinal position) clearly has a profound influence on acceleration and focusing. The longitudinal emittance is defined here as follows,

$$\varepsilon_{Nz} \equiv \left[<(\psi - \psi_R)^2><(p_z - p_{zR})^2> - (<\psi - \psi_R><p_z - p_{zR}>)^2 \right]^{1/2}. \quad (11)$$

The longitudinal emittance is defined relative to the reference electron designated by the subscript R, i.e., the electron traveling exactly along the axis. This choice is important since in absolute terms ψ may drift considerably within a single accelerator section; however, since the electrons tend to slip together, the relevant question regarding longitudinal beam quality is whether the electrons remain bunched in ψ relative to each other. Equation 11, which uses the differences from the reference electron, reflects this important distinction.

The most basic principle governing beam quality is Liouville's theorem, which effectively is a continuity equation for the particle distribution function (density of particles in phase space). In brief, the theorem states that the density of particles in phase space remains constant. A consequence is that an interaction that reduces ε_{Nx} and ε_{Ny} (i.e., improves the transverse beam quality), will necessarily increase ε_{Nz} (i.e., degrades the longitudinal quality).

The complementary linkage of transverse and longitudinal beam quality is evident from an example calculation for a particular distribution. Fig. 8 shows the evolution of ε_{Nx} and ε_{Nz} for a K-V distribution in a 100-stage accelerator with $\gamma = 100$. For this example the initial transverse emittance (5π mm-mrad) and entrance phase (0.8π) were chosen low enough that all electrons are trapped. Since the K-V distribution used (Eq. 8) has no initial spread in ψ and γ, the initial longitudinal emittance is zero. The oscillations of the transverse and longitudinal emittances display complementary behavior — a downswing in ε_{Nx} (improved transverse beam quality) is simultaneous with an upswing in ε_{Nz} (degraded longitudinal beam quality). In the later stages corresponding to higher γ, the beam quality factors change little. On the whole, the relative changes in either emittance are relatively modest.

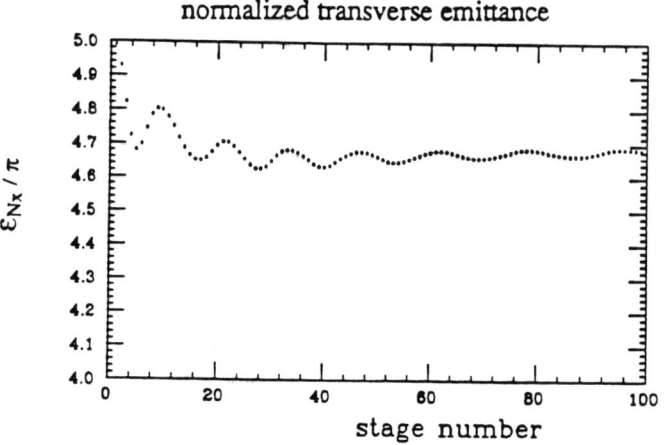

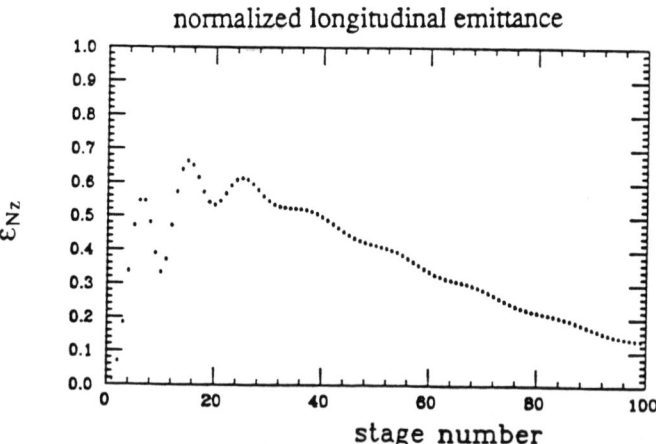

Fig. 8. Evolution of transverse and longitudinal emittance in a 100-stage LPA based upon a K-V distribution with 8 particles, $\varepsilon_{Nx} = 5\pi$ mm-mrad (initial), $\Delta\gamma_I = 12.4$, $\gamma_0 = 100$, and $\psi_0 = 0.8\pi$.

CONCLUSION

Multistaging, besides being necessary in any practical laser accelerator, also introduces an important flexibility that can help in the overall laser acceleration process. By proper selection of the phase at the entrance of each section, it is possible to accelerate and trap the electrons over many sections. Importantly, it is shown that achieving trapping is sufficient to preserve the beam quality. Beam quality includes both the transverse emittance and the bunching in phase and energy (longitudinal emittance).

Another important conclusion concerns laser acceleration schemes in which significant phase slippage ($\geq\pi$) occurs, such in the proposed vacuum accelerator mentioned earlier.[3] Our analysis shows that the total phase slippage need not be limited to a small fraction of π in order to preserve the beam quality.

ACKNOWLEDGMENTS

The authors wish to thank J.R. Fontana for his helpful comments. This work was supported by the U.S. Department of Energy, Contract No. DE-AC06-83ER40128.

APPENDIX

It is important to calculate the ideal electron acceleration in order to relate the acceleration to the reference electric field E_0, which appears in the definition of the generic electromagnetic fields in Eq. 1. The longitudinal Lorentz force for these fields is

$$F_z = eE_0\,\theta\left[-J_0(\alpha r)\sin\psi - \frac{1}{\theta c}\frac{dr}{dt}J_1(\alpha r)\cos\psi\right]. \tag{A1}$$

The second (magnetic) term can be neglected as relatively small for the conditions considered in this paper. Then, with the approximations $\beta_z \approx 1$ and $J_0(\alpha r) \approx 1$, and considering an electron running exactly along the z-axis, the longitudinal equation reduces to

$$\frac{d\gamma}{dz} = \frac{eE_0\theta}{m_ec^2}(-\sin\psi)\,, \tag{A2}$$

where z has replaced t as the independent variable. Considering that $dz = -(L_\pi/\pi)d\psi$, Eq. A2 can be integrated from initial ψ_0 to final ψ_f values,

$$\Delta\gamma = \frac{eE_0\theta}{m_ec^2}L_\pi\frac{1}{\pi}(\cos\psi_f - \cos\psi_0)\,. \tag{A3}$$

The optimum acceleration occurs for an entrance phase $\psi_0 = \pi$, and a final phase $\psi_f = 2\pi$, i.e., an interaction over one slip distance $\Delta z = L_\pi$ (Eq. 3). The ideal acceleration gradient is $\langle\gamma'\rangle_I \equiv \Delta\gamma/L_\pi$. Then,

$$\langle\gamma'\rangle_I = \frac{2}{\pi}\left[\frac{eE_0\theta}{m_ec^2}\right]\,, \tag{A4}$$

which relates the electric field E_0 to the ideal acceleration gradient $\langle\gamma'\rangle_I$. This leads to Eq. 4 in the main text.

REFERENCES

1. J. R. Fontana and R. H. Pantell, J. Appl. Phys. **54**, 4285 (1983).
2. W. D. Kimura, I. Pogorelsky, L. C. Steinhauer, S. C. Tidwell, G. H. Kim, and K. P. Kusche, "Inverse Čerenkov Acceleration Experiment," in AIP Proceedings of Advanced Accelerator Concepts Workshop, Port Jefferson, Long Island, NY, Jun. 14-20, 1992.
3. L. C. Steinhauer and W. D. Kimura, "A New Approach for Laser Particle Acceleration in Vacuum," accepted for publication in Journal of Applied Physics.
4. R. D. Romea and W. D. Kimura, Phys. Rev. D **42**, 1807 (1990).
5. I. M. Kapchinskij and V. V. Vladimirskij, in Proceedings of the International Conference on High Energy Accelerators, (CERN, Geneva, 1959), p. 274.

Optimal Density Taper for Plasma Lenses

T. Katsouleas and C. H. Lai

Dept. of Electrical Engineering-Electrophysics
University of Southern California
Los Angeles, CA 90089-0484

PACS Nos.: 52.40Mj, 41.85-p; 52.75Di; 29.27-a

Abstract

The optimal plasma density profile for a plasma lens is investigated. It is suggested that the optimal plasma lens for focusing electron beams is one in which the plasma density increases in proportion to the density of the beam as the beam pinches in the lens. In this way the focusing strength is the maximum possible while remaining in the underdense plasma regime in which the spherical aberrations are small. The minimum spot size possible with such a lens is calculated analytically for both round and flat beams and compared to the corresponding limits for uniform and adiabatic plasma lenses. The luminosity enhancement from such a tapered plasma lens can be several times that of a uniform lens. The form of the density taper is found numerically for sample sets of beam parameters.

I. Introduction

Plasma lenses are under active consideration for providing the final spot sizes required in high-energy linear colliders,[1-4] including the Stanford Linear Collider and associated Final Focus Test Beam[5]. Proof-of-principle demonstrations of the mechanism of a plasma lens have recently been performed in overdense plasmas,[6,7] and the mechanism of the underdense plasma lens[4,3] is well known as the IFR (ion focused regime) beam transport scheme[8]. Most recent plasma lens proposals operate in the underdense plasma lens regime in which the beam density is larger than the plasma density. In this case focusing of an electron beam occurs when the space charge of the beam electrons blows out the plasma electrons, leaving the positive ions behind. The ion space charge focuses the beam. This underdense regime is attractive because the radial dependence of the net focusing force ($F = eE_r$) is very linear and is independent of the radial beam profile[4]. On short time scales, the ion density is just the background plasma density (n_0), so that the lens strength ($K \equiv \dfrac{F/r}{\gamma mc^2}$) is given simply by Gauss' law:

$$K = 2\pi \, n_0 \, e^2/\gamma \, mc^2 \qquad (1)$$

Previously, the spot size compression possible with a passive plasma slab placed in the path of a relativistic charged particle beam had been found. In previous treatments, the density of the slab was considered to be either uniform or adiabatically rising. Here we consider a lens profile that is tapered such that the plasma density experienced by the beam increases in proportion to the beam density as the beam pinches. In this way the beam experiences the maximum possible pinching force, thereby obtaining the shortest possible focal length and smallest spot size (since the spot size σ is crudely the focal length f times the angular spread of the beam $\theta \approx \varepsilon/\sigma$ where ε is the beam emittance). At the same time, tapering the plasma density in proportion to the beam density keeps the lens in the underdense plasma regime where aberrations are small[4]. In this paper we find (a) the density profile corresponding to the above assumption and (b) the improvement possible with this density profile.

II.   Spot size limits of uniform and adiabatic lenses

Though the focusing strength of plasma lenses can be incredibly high (e.g. 3 gigaGauss/cm for beam and plasma densities of order 10^{18} cm^{-3}), the final spot size from plasma lenses is limited by several effects. For uniform underdense lenses the competition between the need for small spot size at the lens entrance to obtain maximum focusing strength and the tendency of such a narrow beam to diverge quickly limits the final spot size (σ^*). In terms of the beam beta function ($\beta^* = \sigma^{*2}/\varepsilon$, where ε is the beam emittance[10] $= \frac{1}{\pi}$ times the area of the beam in transverse phase space) this limit is approximately[3,4]

$$\beta^* \sim \frac{1}{K\beta_o} \sim \frac{\sqrt{2\pi}\,\sigma_z\,\varepsilon_n}{Nr_e} \tag{2}$$

where σ_z is the length of the beam, N the number of beam particles, $\varepsilon_n = \gamma\varepsilon$ and $r_e = e^2/mc^2$. Moreover, synchrotron radiation losses from the focusing force of a lens cause an uncorrectable chromatic aberration known as the Oide limit[13]. The Oide limit becomes apparent only if β^* approaches a value as small as

$$\beta^* \geq \left(\frac{275}{3\sqrt{6\pi}}\, r_e\, \lambda_c\, F \right)^{2/7} \gamma(\varepsilon_n)^{3/7} \tag{3}$$

for flat beams where F is a function of lens length and strength and λ_c is the Compton wavelength ($\hbar/mc$). For parameters typical of present beams and plasmas and F of order unity, inequality (2) is more restrictive than (3) for beam energies less than about 1TeV.

The spot size from adiabatic plasma lenses is limited for other reasons. First, the spot size compression is proportional to the ratio of final to initial plasma density to the one-fourth power. (For example, a four order of magnitude increase in plasma density is needed to reduce the spot size by a factor of ten). Since the plasma's density increases much faster than the beam's, the adiabatic lens will be limited at the point when the plasma density reaches the beam density (since beyond this point no further increase in focusing strength is obtained by increasing the plasma density). Coincidentally, this consideration gives a spot size limitation that is essentially identical to Eq. (2). However, the adiabatic lens does have the advantages that it is not subject to the Oide limit,[13] nor is it sensitive to plasma inhomogeneities or beam energy spread.

The scheme we propose here overcomes the limitation given by Eq. (2). Thus, it is appropriate when Eq. (2) rather than (3) is the limit to spot size as is typically the case for present and next generation collider parameters.

III.   Optimally tapered lens analysis

To begin to analyze the optimally tapered plasma lens profile, we consider the well known equation for the beam beta function (β) from beam optics[10]:

$$\beta''' + 4K(s)\, \beta' + 2\,K'\,(s)\,\beta = 0 \tag{4}$$

where primes (′) denote derivatives with respect to s, the coordinate along the direction of beam propagation. The beam spot size is related to β by

$$\sigma^2 = \varepsilon\beta \tag{5}$$

The idea of the optimally tapered plasma lens is to increase the plasma density (n_0) in proportion to the peak beam density (n_{bo}):

$$\frac{n_{bo}}{n_o} \equiv \alpha = \text{a constant} > 1 \tag{6}$$

Choosing α near to 1 maximizes the focusing strength while keeping the lens in the underdense regime. For a bi-Gaussian beam of the form $n_b = n_{bo} \exp\,[- (r^2/2\sigma^2 + z^2/2\sigma_z^2)]$, n_{bo} is given by

$$n_{bo} = \frac{N}{(2\pi)^{3/2}\, \sigma^2\, \sigma_z} \tag{7}$$

where N is the number of beam electrons. Substituting Eqs. (6), (7) and (5) into (1) gives

$$K(s) = \frac{C}{\beta(s)} \tag{8}$$

where $C = Nr_e/\,\alpha\sqrt{2\pi}\,\gamma\,\sigma_z\,\varepsilon = \text{constant}$.

Substituting Eq. (8), Eq. (4) simplifies to

$$\beta''' + \frac{2C}{\beta}\,\beta' = 0. \tag{9}$$

This equation was studied in Ref. 2 in a different context; namely, to find the thick lens correction to the focal length of an overdense plasma lens. The problems associated with an overdense lens will be discussed further in the final section of this paper.

Integrating twice yields

$$1/2\,\beta'^2 - 1/2\,\beta_0'^2 = \beta_0''\,(\beta - \beta_0) - 2C\,(\beta\ln\beta/\beta_0 - \beta + \beta_0) \tag{10}$$

Here β_0, β_0', β_0'' refer to the beta function and its derivatives just inside the lens entrance $(s = 0+)$. At the lens entrance β_0 and β_0' are continuous, but β_0'' satisfies a jump condition given by integrating Eq. (4) across the lens boundary:

$$\beta_0'' = \beta''\,(s = 0+) = \beta''(0-) - 2\,K\,(0)\,\beta(0). \tag{11}$$

Prior to the lens, the beam satisfies Eq. (4) in free space $(K = 0)$, namely

$$\beta\,(s) = \beta_0^*\,(1 + \frac{(s - s_0)^2}{\beta_0^{*2}}) \tag{12}$$

where β_0^* is the value of β at the beam's natural waist $(s = s_0)$. From (11) and (12) then $\beta_0' = -2s_0/\beta_0^* = -2\,\sqrt{\beta_0/\beta_0^* - 1}$ and $\beta_0'' = 2/\beta_0^* - 2C$.

To solve for the motion (i.e., $\beta(s)$) we must integrate Eq. (10) once more:

$$s(\beta) = -\int_{\beta_0}^{\beta} \frac{d\beta}{\sqrt{\beta_0'^2 + 2\beta_0''\,(\beta - \beta_0) - 4C\,(\beta\ln\beta/\beta_0 - \beta + \beta_0)}}$$

$$= \beta_0 \int_{1}^{\beta/\beta_0} \frac{dx}{2\sqrt{Rx - K_0\beta_0^2\,x\,\ln x - 1}} \tag{13}$$

where $R \equiv \beta_0/\beta_0^*$ and $K_0 = C\beta_0^{-1}$ is K at the lens entrance.

Eq. (13) may be solved numerically for $\beta(s)$, from which we can determine the plasma density taper profile from Eqs. (8) and (1):

$$\frac{n_0(s)}{n_0} = \frac{\beta_0}{\beta(s)} \tag{14}$$

To find the minimum spot size (i.e. minimum β) we return to Eq. (10) and solve for β at the turning point $(\beta' = 0)$:

$$\frac{\beta_{min}}{\beta_0^*} = \frac{1}{1 + K_0\beta_0^*\beta_0 \ln \beta_0/\beta_{min}} \tag{15}$$

For comparison in a uniform lens the corresponding minimum beta is

$$\beta_{min}/\beta_0^* = \frac{1}{1 + K\beta_0^* (\beta_0 - \beta_{min})} \tag{16}$$

Although Eqs. (15) and (16) are transcendental we can see that ramping the plasma density improves the plasma lens by a factor of approximately $\ln \beta_0/\beta_{min} > \ln \beta_0/\beta_0^*$.

For example, for $R = 3$ and $K_0 \beta_0^{*2} = 2$, the β reduction possible with a uniform lens is $\beta^*/\beta_0^* \approx 1/7$; while that with an optimally tapered lens is $\beta^*/\beta_0^* \approx 1/28$. The β function and plasma density profile from Eqs. (13) and (14) are shown for this example in Fig. 1. An example for parameters similar to the Stanford Linear Accelerator Center (SLAC) beam is given in Table 1. For the beam parameters given in Table 1, uniform and adiabatic lenses reduce the final spot size from 1.5μ to approximately 0.6μ (See Eq. 2). Using a tapered plasma lens with $R = 3$ and plasma density ramped from 10^{17} to 10^{19} cm^{-3} over 5 mm gives a final spot size from Eq. (15) of 0.3μ. Adiabatic, Uniform, and Tapered Lenses are compared in Table 1.

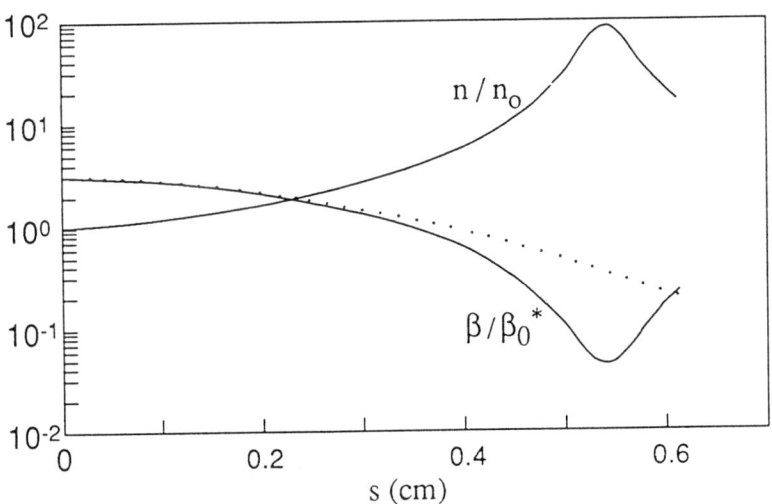

Fig. 1. Beam β reduction and plasma density vs. distance from Eq,(13) and (14) for the example of Table 1; $\beta_0/\beta_0^* = 3$, $K\beta_0^{*2} = 2$. (Solid line:optimally tapered lens, dotted line: uniform lens.)

IV.   Effect of density taper errors

In order to test the sensitivity of the present scheme to errors in the density profile, we solve Eq. (4) numerically for several density profiles that approximate the optimal profile. The approximate profiles considered were (a) an exponential power law best fit, (b) an optimal profile truncated after one order of magnitude increase in density, (c) a simple exponential, and (d) a linear profile. For all cases the initial beam and plasma parameters correspond to the example in Fig. 1 and Table 1. The following exponential power law (case a) gives a good approximation of the optimal profile as seen in Fig. 2: $n = n_0 \exp(ax^2 + bx^3)$, $a = 4.64/\beta_0^2$, $b = 299.4/\beta_0^3$, and $x = 0$ to $.24 \beta_0$. The beta reduction for this approximate profile is quite close to the optimal case. $[\beta^*/\beta_0^* = 1/32$, slightly better than the exact case. However, the plasma becomes slightly overdense, so Eq. (1) and this result are only approximately valid.] The spot size reduction for the other cases is summarized in Table 2. For case (b), the plasma lens was truncated at the point where the density n reached ten times its initial value n_0. From there, the beam propagates in vacuum to its waist. Cases (c) and (d) are shown in Figs. 3 and 4. For the exponential profile the slope of the density rise at the lens entrance was taken to match the optimal case: $n = n_0 e^{ax}$ and $a = -\beta_0'/\beta_0$. From Table 2 we see that all of the approximate profiles give smaller spot sizes than a uniform lens. Moreover, **any** underdense density profile that rises will give spot sizes somewhere between the uniform and optimal lens cases.

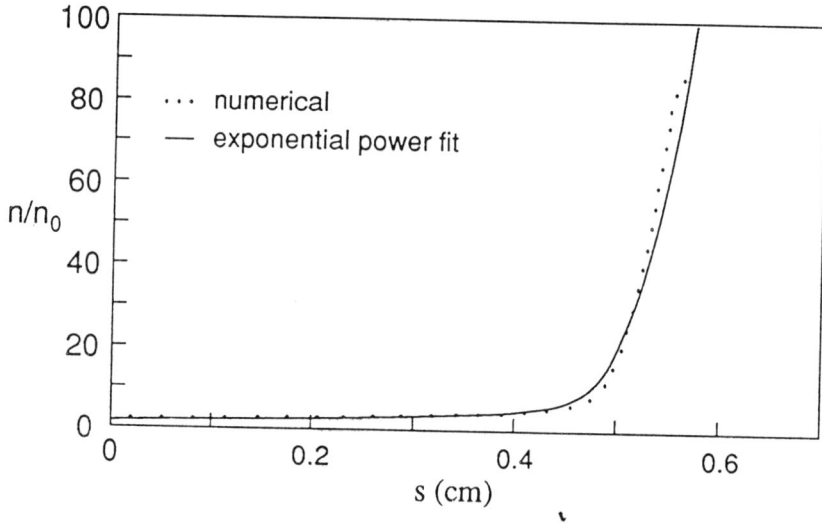

Fig. 2. Optimal plasma density profile from Eqs. (13) and (14) for
the example of Table 1 and approximate fit given by
$n/n_0 = \exp(4.64\, x^2/\beta_0^2 + 288.4\, x^3/\beta_0^3)$; $\beta_0/\beta_0^* = 3$, $K\beta_0^{*2} = 2$.

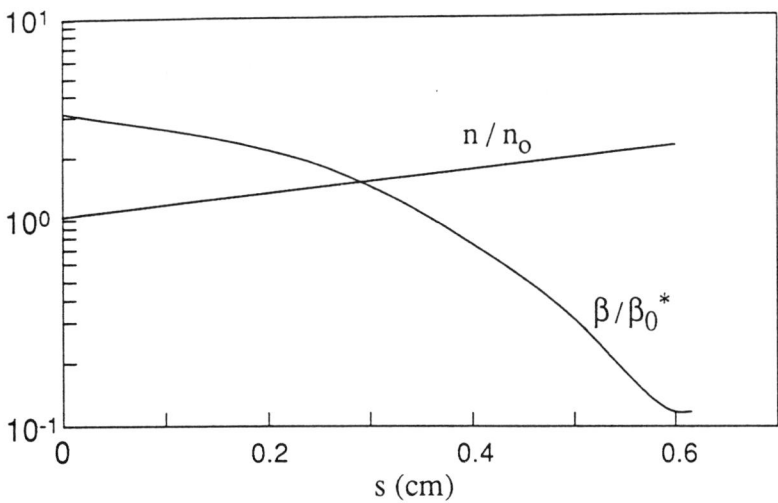

Fig. 3. Beam β reduction for an exponential profile for the
example of Table 2.

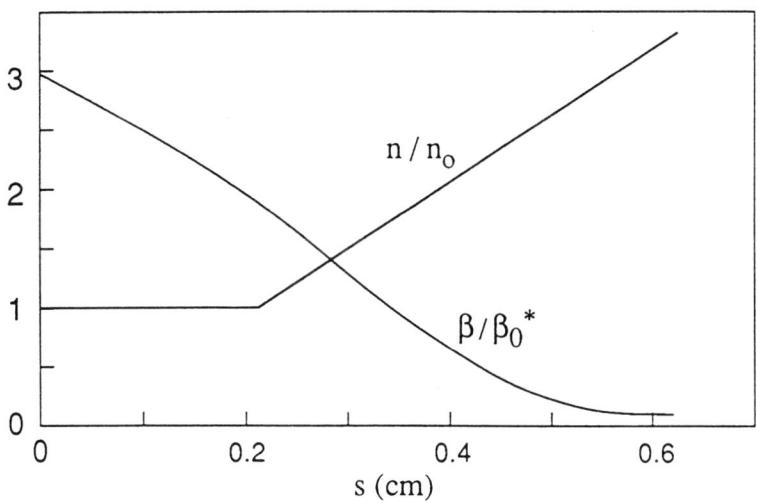

Fig. 4. Beam β reduction for a piecewise linear density profile for
the example of Table 2.

V.   Flat Beams

Many future collider designs call for flat beams (i.e. beams that are very much wider in one direction (x) than the other (y)). For flat beams, the counterparts to Eqs. (1) and (8) for the narrow dimension are

$$K_y = 4\pi \, n_0 \, e^2/\gamma \, mc^2 \tag{1'}$$

$$K_y = \frac{C_y}{\sqrt{\beta_y}} \tag{8'}$$

where $C_y = 2N \, r_e/\sqrt{2\pi\varepsilon_y\varepsilon_x\beta_x} \cdot \alpha \cdot \sigma_z$. Substituting Eq. (8'), Eq. (4) can be simplified to:

$$\beta_y''' + 3\frac{C_y}{\sqrt{\beta_y}} \beta_y' = 0 \tag{17}$$

Integrating twice and putting $\beta' = 0$, we obtain the minimum spot size β^*:

$$\frac{\beta^*}{\beta_0^*} = \frac{1}{1 + 2 \, K_0\beta_0^* \, (\beta_0 - \sqrt{\beta_0\beta^*})} \tag{18}$$

Neglecting β^* on the right side yields

$$\frac{\beta^*}{\beta_0^*} \approx \frac{1}{1+2 \, K_0 \, \beta_0^*\beta_0} \tag{19}$$

or

$$\frac{\beta^*}{\beta_0^*} \approx \frac{1}{1 + 2 \, C_y \, \sqrt{R} \, \beta_0^{*3/2}} \tag{20}$$

where $R = \dfrac{\beta_0}{\beta_0^*}$

Comparing (19) with (16), which also applies to flat beams, we note that the improvement is approximately a factor of 2.

Also note from (20), $\dfrac{\beta^*}{\beta_0^*}$ is inversely proportional to $\sqrt{R}$, that is, we get better spot size reduction by matching larger R into the lens. However, the neglect of β_x variations in the analysis limits R. The neglect of β_x variations is justified for $s_0 \ll \beta_{xo}$, or from Eq. (12)

$$R = \frac{\beta_0}{\beta_0^*} = 1 + \left(\frac{s_0^2}{\beta_0^{*2}}\right) << 1 + \left(\frac{\beta_{xo}^*}{\beta_0^{*2}}\right)^2 \tag{21}$$

When (21) does not hold, we have to consider variations in the x-direction.

Table 3 gives a comparison between the uniform lens and tapered lens for different R (R = 100 & 50). In these examples based on expected parameters of the SLAC Final Focus Test Beam, the final spot sizes reach values comparable to the Oide limit[13] due to quantum synchrotron radiation. The beam beta functions computed numerically from Eq. (17) for the examples in Table 3 are shown in Fig. 5 and the required density profile is shown in Fig. 6.

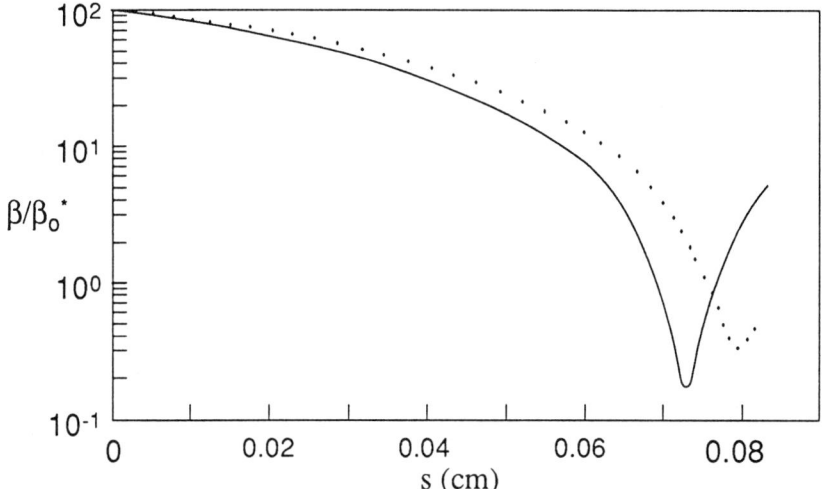

Fig. 5. β functions ($\beta(s)/\beta_0^*$) vs. distance in uniform (dotted line) and optimally tapered (solid line) plasma lenses for the flat beam example of Table 3 (R=100).

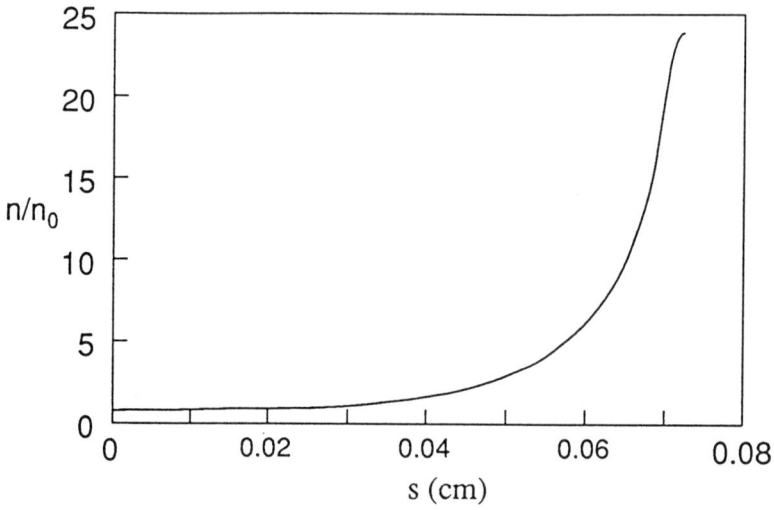

Fig. 6. Optimal plasma density profile corresponding to Table 3 example for R=100.

VI.   Discussion

Thus far we have neglected chromatic and spherical aberrations which we would expect to limit the spot size to $\sigma^* \geq \sigma_0 \, \Delta\gamma/\gamma$ or $l \, \Delta K/K$, respectively, where l is the length of the lens determined from Eq. (13). Ultimately, the final spot size is limited by chromatic aberrations arising from quantum synchrotron radiation - the Oide limit.[13] Other lens considerations include the contribution of the plasma to background noise events in the detectors due to beam-plasma scattering. These will be an order of magnitude more in the tapered lens due to its higher density.

Finally, we mention the overdense plasma lens regime. In this regime, Eq. (8) is automatically satisfied (taking $\alpha = 1$) even without tailoring the plasma density. Thus one could apply the present scheme without the need to carefully control the plasma density. However, the overdense regime is subject to large spherical aberrations[3,4] not considered in this analysis. Similarly, the analysis here is approximately valid for positron beams; however the spherical aberrations for these are large in both under- and overdense plasmas[4].

In summary we have found the minimum β of an electron beam passing through a plasma lens that increases in density as the beam density increases. The minimum β is a few times smaller than it would be in a uniform density plasma lens for round beams [Eq. (15)] and approximately two times smaller for flat beams [Eq. (18)].

Work supported by U.S.D.O.E., ONR, and LLNL-PPRI. Discussions with C. Clayton, R. Liou, M. Gundersen, J. M. Dawson,W. B. Mori, R. Williams, J. Rosenzweig, and W. Leemans are appreciated. Thanks to John Dawson's Physics 222 class for asking the question that led to this calculation.

Table 1. Comparison of Uniform, Adiabatic, and Optically Tapered Plasma Lenses

	Uniform Lens	Adiabatic Lens	Tapered
Initial Beam Parameters[†]			
σ_o^*	1.5μ	1.5μ	1.5μ
β_o^*	.75 cm	.75 cm	.75 cm
β_o	2.25 cm	1.5 cm	2.25 cm
Lens Parameters			
Position (s_o)	1.06 cm	.75cm	1.06 cm
Thickness (l)	.64 cm	.68 cm ($\alpha = 1$)[††]	.545 cm
Initial Density (cm^{-3})	1.9×10^{17}	4.5×10^{16}	1.9×10^{17}
Final Density (cm^{-3})	1.9×10^{17}	5.8×10^{18}	1.5×10^{19}
β^*	.11cm	.14 cm	.027 cm
σ^*	570 nm	630 nm	280 nm

[†] Note: All cases assume N = 2 x 10^{10}, ε= 3 x 10^{-8} cm - rad, γ = 10^5, σ_z = 1mm.

[††] Assumes a profile of the form in Ref. 9: $n(s) = n_o (1 - 2 \alpha s/\beta_o)^{-2}$.

Table 2. Comparison of Optimal and Approximate Lens Profiles

		Uniform	Optimal	Cut Off $n/n_0 = 10$	Exponential Density	Piecewise linear
Initial Beam Parameters	β_0^* [cm]	.75	.75	.75	.75	.75
	β_0 [cm]	2.25	2.25	2.25	2.25	2.25
Lens parameter	Thickness [cm]	0.64	0.545	0.45	0.60	0.62
	Initial Density [cm^{-3}]	1.9×10^{17}	"	"	"	"
	Final Density	1.9×10^{17}	1.5×10^{19}	1.9×10^{18}	4×10^{17}	6.27×10^{17}
Final Beam Parameters	β^* [cm]	0.11	0.027	0.05	0.082	0.074
	β^*/β_0^*	0.15	0.036	0.067	0.11	0.1

Table 3. Parameters of Plasma Lenses for Flat Beam Cases

	Uniform Lens		Tapered Lens	
Initial Beam Parameters				
β_o^* (cm)	$\beta_{ox}^* = .33$ $\beta_{oy}^* = .012$		$\beta_{ox}^* = .33$ $\beta_{oy}^* = .012$	
N	2×10^{10}		2×10^{10}	
σ_z	.5 mm		.5 mm	
γ	10^5		10^5	
ε(cm-rad)	$\varepsilon_x = 3 \times 10^{-8}$ $\varepsilon_y = 3 \times 10^{-9}$		$\varepsilon_x = 3 \times 10^{-8}$ $\varepsilon_y = 3 \times 10^{-9}$	
β_o (cm) [†]	$\beta_{ox} = 4.7$ $\beta_{oy} = 1.2$ (R = 100)	$\beta_{ox} = 4.7$ $\beta_{oy} = 0.6$ (R = 50)	$\beta_{ox} = 4.7$ $\beta_{oy} = 1.2$ (R = 100)	$\beta_{ox} = 4.7$ $\beta_{oy} = 0.6$ (R = 50)
Lens Parameters				
Density (cm^{-3})	4.2×10^{18}	6.0×10^{18}	4.2×10^{18} $\rightarrow$ 9.7×10^{19}	6.0×10^{18} $\rightarrow$ 8.4×10^{19}
Final Beam Parameters				
Focal distance (cm) = plasma thickness	0.078	0.061	0.072	0.057
β_y^* (cm)	0.00376	0.00468	0.0023	0.0031
β_y^*/β_o^*	0.313	0.39	0.19	0.26
σ_y^* [††]	34 nm	37 nm	26 nm	31 nm

---

† Note: Inequality (20) for this case is $R \ll 1 + \left(\dfrac{\beta_{xo}^*}{\beta_{yo}^*}\right)^2 \approx 750$

†† Note: Synchrotron radiation has been neglected (see Ref. 13)

References

1. P. Chen, Part. Accel. 20, 171 (1987).

2. J. B. Rosenzweig and P. Chen, Phys. Rev. D. 39, 2039 (1989).

3. P. Chen, S. Rajagopalan, J. Rosenzweig, Phys. Rev. D 40, 923   (1989).

4. T. Katsouleas, J.J. Su, W. B. Mori, J. M. Dawson, Phys. Fluids B 2 (6), 1384 (1990); J. J. Su, T. Katsouleas, J. M. Dawson, R. Fedele, Phys. Rev. A 41, 3321 (1990).

5. D. Betz, et al., Proc. IEEE PAC Conf., San Francisco, May 6-9, 1991.

6. H. Nakanishi, et al. Phys. Rev. Lett. 66, 1870 (1990).

7. J. B. Rosenzweig, et al., Phys. Fluids B 2 (6), 1990.

8. R. J. Briggs, Phys. Rev. Lett. 54, 2588 (1985).

9. P. Chen, K. Oide, A. M. Sessler, S. S. Yu, Phys. Rev. Lett. 64, 1231 (1990); R. Williams, U.C. Los Angeles Ph.D. Thesis (1992).

10. D. Cary, The Optics of Charged Particle Beams (Harwood   Academic, New York, 1987).

11. D. Whittum, U.C. Berkeley Ph.D. Thesis (1990).

12. R. Erickson, Stanford Linear Accelerator Center Report No. SLAC-Pub-4479, 1987 (unpublished).

13. K. Oide, Phys. Rev. Lett. 61, 1713 (1988); K. Hirata, B. Zotter and K. Oide, Phys. Lett. B 224, 437 (1989). For the example in Table 2, Column 1, the Oide limit is below 22nm.

Numerical Studies on the
Ramped Density Plasma Lens

R. L. Williams
Department of Physics, and CeNNAs
Florida A. & M. University, Tallahassee, Florida 32307

T. Katsouleas
Department of Electrical Engineering-Electrophysics
University of Southern California, Los Angeles, CA 90089-0484

ABSTRACT

We consider the so-called adiabatic plasma lens when the plasma density is ramped too quickly to be considered adiabatic. The lens length can be much shorter in such a case, but the final spot size is shown to be larger by a factor of $\sqrt{1 + \alpha^2}$ than for a slowly ramped plasma lens with the same initial and final density (where $\alpha = -\beta'/2$ is proportional to the plasma density gradient). We find that the final spot size is the same whether or not the Courant-Snyder parameters of the beam (α and β) are matched to the lens. However, matched beams allow the plasma density to be lower while unmatched beams allow the lens to be shorter (for the same α and for the same final to initial plasma density ratio). Finally, we find that a smaller spot size can be obtained for a given lens length and density ratio by starting at smaller α and increasing α along the lens.

INTRODUCTION

Plasma lenses have been proposed [1] and studied recently as a way to reduce e^-e^+ beam spot sizes in future TeV linear colliders. Small spot sizes increase the luminosity, $\mathcal{L}$, at the collision point, and spot dimensions on the order of 1 nm × 190 nm and smaller will be desired in the future [2] [3] [4]. The luminosity is given by [4]

$$\mathcal{L} = \frac{N^2 b f H_D}{4\pi\sigma_x\sigma_y} \approx \mathcal{O}\left(10^{33} \text{ cm}^{-2} \text{ sec}^{-1}\right) \tag{1}$$

where N is the number of particles per bunch, b is the number of bunches per linac pulse, f is the pulse repetition rate, H_D is the luminosity enhancement factor due to disruption, σ_x is the bunch width in the x direction, and σ_y is the bunch width in the y direction. Disruption is the additional pinching of one beam by the space charge of the other beam during the collision. In this study we examine in detail one method that has been proposed for decreasing σ_x and σ_y, which is called the adiabatic, or ramped density, or continuous focus plasma lens.

Strong focusing fields will be required to obtain small spot sizes. Using thin lenses of any kind, there is a fundamental limit (the Oide limit [5]) on the final spot size due to the energy spread resulting from the synchrotron radiation emitted during strong focusing. The "adiabatic focuser" has been proposed [6] as an alternative to the thin lens since it would not be as severely restricted by the Oide limit. In an adiabatic plasma focuser beam electrons propagate into an increasing density plasma and eject the background electrons which creates an ion channel [7][8]. The resulting ion space-charge field provides an increasing focusing force that continuously and, as proposed, adiabatically squeezes the beam to smaller spot sizes. Because there is a continuous focus rather than a single focal point, chromatic aberrations are reduced. In an adiabatic plasma lens the spot size compression (ratio of minimum spot size to the spot size at the lens entrance) is equal to $(n_o/n_f)^{1/4}$ where $_{o,f}$ refer to the initial and final plasma density or focusing strength of the lens.

In studying the adiabatic focuser we have found that, as expected [9], the same result is obtained even if the ramping of the plasma is not adiabatic as long as the beam and plasma are matched. However, the matching requires that the beam be converging as it enters the plasma. Since the convergence would produce a smaller spot size even without the plasma, the ratio of final beam size to minimum beam size without the plasma is larger, $(n_o/n_f)^{1/4}\sqrt{1+\alpha^2}$, and is in fact the same for both matched and unmatched beams. Thus the choice of how to match the beam to the plasma changes the lens length and density (for a given beam and ratio of n_o/n_f but not the final spot size. We also find that smaller spot sizes can be obtained by using ramped profiles that become less adiabatic (i.e. letting $\partial\beta/\partial s$ increase rather than be constant as assumed in previous work).

Our tool for examining the adiabatic focuser is a 2-D numerical simulation code that calculates the trajectories of electrons injected into a region of space that has an increasing focusing force [10]. The focusing force models the ion space-charge force, however, the results of this study would pertain to any device or medium that gives an increasing focusing force, not necessarily a plasma.

BACKGROUND

In the adiabatic plasma lens concept an electron bunch passes through an underdense plasma such that $n_o < n_b$, where n_o is the plasma electron density and n_b is the electron beam density. (Particle-in-Cell simulations have shown that a more accurate inequality would be $n_o/2 \leq n_b$ [11].) The leading electrons in the bunch repel all of the background plasma electrons, leaving a region of immobile and unshielded positively charged ions with density n_{io}. This is sometimes called the ion focusing regime [7]. The repulsive electrostatic forces of the electrons in the beam are canceled by the attractive $\bar{v} \times \bar{B}$ forces, where $\bar{v}$ is the electron velocity and $\bar{B}$ is the magnetic field due to the beam current. The unshielded ions of the background plasma produce a space-charge force, F_y, which from Gauss' law is

$$\frac{F_y}{y} = -2\pi n_{io}e^2 = 300\,\frac{\text{MegaGauss}}{\text{cm}} \times \frac{n_{io}(\text{cm}^{-3})}{10^{17}} \tag{2}$$

where y is the transverse displacement of the electron and e is the electron charge. This is the force that focuses the electrons. We note that in magnetic lenses, the focusing force is on the order of 10 kiloGauss/cm.

The equation of motion of a particle in the lens is

$$\frac{d^2y(s)}{ds^2} + K(s)y(s) = 0 \tag{3}$$

where s is the coordinate along the axis of the lens and the lens focusing strength is

$$K(s) = \frac{F_y(s)}{y(s)}\frac{1}{\gamma m_o c^2} = \frac{2\pi n_i(s)e^2}{\gamma m_o c^2} = \frac{\omega_p^2(s)}{2\gamma c}, \tag{4}$$

$\omega_p(s)$ is the plasma frequency given by $\omega_p(s)^2 = 4\pi n_i(s)e^2/m_o$, the electron mass is m_o, γ is the relativistic factor, and c is the speed of light. This has solution $y(s) = A\beta_l(s)^{1/2}\cos(\Psi(s) + \phi)$ where A and ϕ are integration constants, $\Psi(s)$ is the phase advance, and $\beta_l(s)$ is the amplitude function, or beta-function.

In the originally proposed adiabatic lens scheme [6] $\beta_l(s)$ varies linearly with s as $\beta_l(s) = \beta_{lo} - 2\alpha_l(s)s$ with $\beta_l(0) = \beta_{lo}$ and $\alpha_l(s) = -\frac{1}{2}d\beta_l(s)/ds = -\frac{1}{2}\beta_l'(s)$, where we define $\alpha_l(s)$ as the amount of lens "adiabaticity". In the uniformly ramped lens $\alpha_l(s) = \alpha_l(0) = \alpha_{lo} = $ constant, which was the case considered in reference [6]. The adiabatic condition is that $\alpha_{lo} < 1$. For the uniformly ramped adiabatic lens equation 4 becomes

$$K(s) = \frac{1 + \alpha_{lo}^2}{\beta_l(s)^2} = \frac{1 + \alpha_{lo}^2}{(\beta_{lo} - 2\alpha_{lo}s)^2}. \tag{5}$$

The value of $\beta_l(s)$ at the entrance of the lens, ($s = 0$), is determined by the beam. The plasma frequency and thus plasma density at the entrance to the lens is obtained using $\beta_l(0) = \beta_{lo} = \sqrt{2\gamma(1 + \alpha_{lo}^2)}(c/\omega_{po})$ and inverting. For the uniformly ramped plasma lens we have the important relations

$$K(s) \propto \frac{1}{\beta_l^2(s)} \propto n(s) \propto \frac{1}{y^4(s)}. \tag{6}$$

The coordinates of the beam as it propagates through the lens are related by the beam ellipse equation (an exact constant of the motion)

$$\frac{\left[y(s)^2 + (\alpha_b(s)y(s) + \beta_b(s)y'(s))^2\right]}{\beta_b(s)\epsilon} = 1 \tag{7}$$

where $y'(s) = dy(s)/ds$, the ellipse skewness factor is $\alpha_b(s) = -\frac{1}{2}d\beta_b(s)/ds = -\frac{1}{2}\beta_b'(s)$, $\beta_b(s)$ is the beam betatron wavelength, and ϵ is the beam emittance.

Given ϵ, the waist betatron wavelength, β_b^*, and the waist width, σ^*, we can determine $\alpha_b(s)$, $\beta_b(s)$, and $\sigma(s)$ versus s for the beam envelope in free space using $\beta_b(s) = \beta_b^*(1 + s^2/\beta_b^{*2})$, $\sigma^2(s) = \beta_b(s)\epsilon$, and $\beta_b'(s) = -2\alpha_b(s) = 2s/\beta_b^*$, where $\sigma(s)$ is the beam envelope width at s.

In the unmatched beam and lens case the entrance to the lens is placed at the waist of the beam, as illustrated in figure 1(a). For the unmatched beam and lens case we have

$$\alpha_b(0) = 0, \quad \alpha_l(0) > 0, \quad \text{and} \quad \beta_l(0) = \beta_b(0) \neq 0. \tag{8}$$

To match the lens to the beam we have

$$\alpha_b(0) = \alpha_l(0) \neq 0 \quad \text{and} \quad \beta_l(0) = \beta_b(0) \neq 0. \tag{9}$$

Matching is accomplished by first fixing a value for the lens α_{lo} at its entrance, then moving the lens entrance upstream from the beam waist to the position s where the beam's $\alpha_b(s)$ equals α_{lo}. The value of the beam's $\beta_b(s)$ is calculated and then the lens β_{lo} is set equal to it. Figure 1(b) shows that for the matched beam and lens case there is a continuous transition of the slope of the beam envelope as it enters the lens from free space.

The electron transverse displacement and beam envelope width are related to the lens parameters by

$$\frac{y(s)}{y(0)} = \frac{\sigma(s)}{\sigma(0)} = \left[\frac{\beta_l(s)}{\beta_{lo}}\right]^{1/2} \frac{\cos \Psi(s)}{\cos \Psi(0)} \leq \left[\frac{n_{io}}{n_i(s)}\right]^{1/4} \frac{1}{\cos \Psi(0)} \tag{10}$$

where we have used $\cos \Psi(s) \leq 1$. The ratio of the maximum to minimum beam envelope widths gives the spot size compression factor, $\sigma(0)/\sigma(s)$, where $\sigma(0)$ is the maximum width measured at the entrance to the lens and $\sigma(s)$ is the minimum width measured at the lens exit. For the unmatched beam-lens case equation 10 becomes

$$\frac{\sigma(s)}{\sigma(0)} \leq \left[\frac{n_{io}}{n_i(s)}\right]^{1/4} \sqrt{1 + \alpha_{lo}^2} = \left[\frac{\beta_l(s)}{\beta_{lo}}\right]^{1/2} \sqrt{1 + \alpha_{lo}^2}. \tag{11}$$

(a) Beam and Lens are Unmatched:

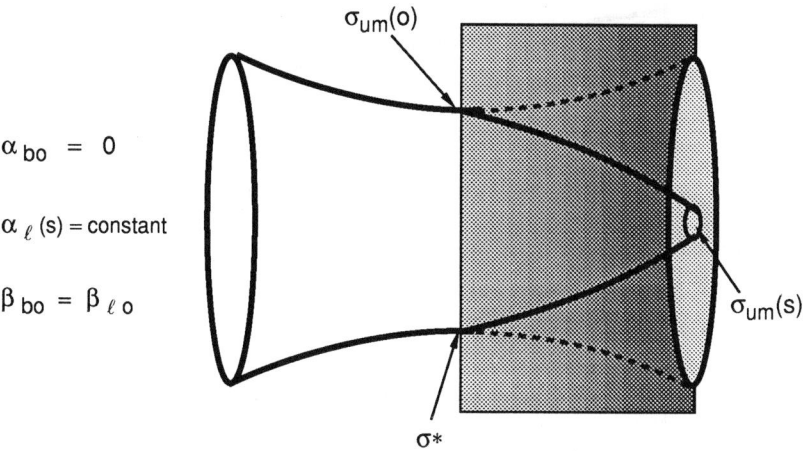

$\alpha_{bo} = 0$

$\alpha_{\ell}(s) = $ constant

$\beta_{bo} = \beta_{\ell o}$

(b) Beam and Lens are Matched:

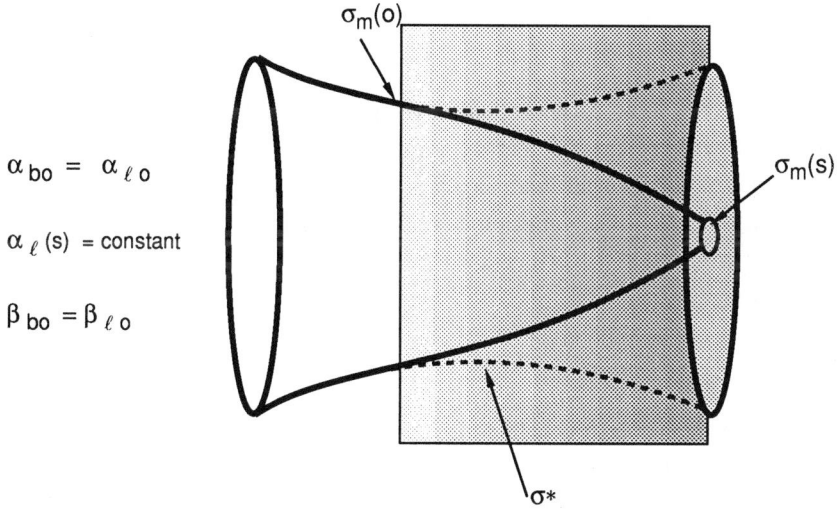

$\alpha_{bo} = \alpha_{\ell o}$

$\alpha_{\ell}(s) = $ constant

$\beta_{bo} = \beta_{\ell o}$

Figure 1: Geometry of beam and continuous focus plasma lens showing the (a) unmatched and (b) matched schemes.

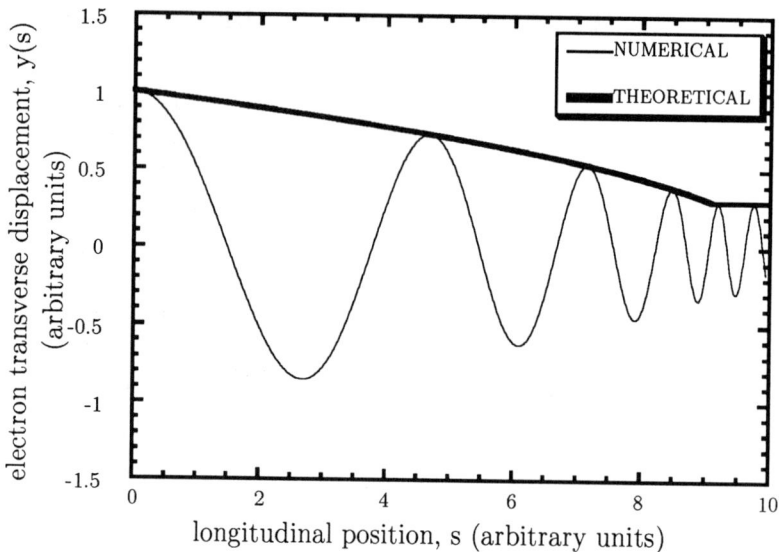

Figure 2: Electron trajectory and beam envelope in ramped density plasma lens.

For the matched beam-lens case, equation 10 becomes

$$\frac{\sigma(s)}{\sigma(0)} \leq \left[\frac{n_{io}}{n_i(s)}\right]^{1/4}. \tag{12}$$

In figure 2 we show a theoretical envelope of the electron beam along with one representative oscillatory electron trajectory in an adiabatic lens. The theory curve is a plot of equation 11, and the trajactory is a plot of the numerical solution to the equation of motion, equation 3, for $K(s)$ given by equation 5. In figure 2 the focusing strength $K(s)$ is increased by a factor of 10^4, i.e. until the following condition is met

$$\frac{1}{(\beta_{lo} - 2\alpha_{lo}s)^2} \geq \frac{10000}{\beta_{lo}^2}. \tag{13}$$

Then the focusing strength is held constant (starting at $s \approx 9$) so that we can display and measure the trajectories which then have constant betatron wavelengths and the beam has a constant envelope width.

NUMERICAL RESULTS

The variation of beam spot size compression as a function of α_{lo} and of matching is shown in figure 3. For small values of α_{lo}, less than approximately 0.3, the reduction due to unmatching is negligible. But for larger values of α_{lo} up to at

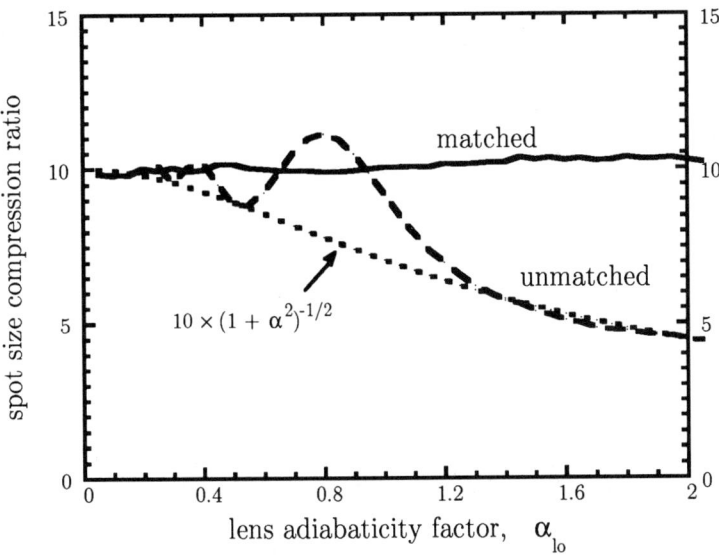

Figure 3: Spot size compression vs adiabaticity factor for matched and unmatched beam and lens.

least 2.0, the matched beam compression remains constant and the unmatched beam compression decreases. Note that for $\alpha_{lo} > 1$ the adiabatic condition does not hold. Therefore we conclude that for matched beam and lens, it is not necessary to limit the plasma lens by the adiabatic condition. We have also plotted the theoretical reduction factor which follows the numerical data for the unmatched case. The oscillation in the unmatched compression curve in figure 3 can be explained as follows. The unmatched compression is proportional to $\cos \Psi(s)$ and oscillates as the length of the lens varies.

In figure 4 we have plotted numerical simulation results for the beam envelope widths at the entrance and exit of the lens for the matched and unmatched cases, versus α_{lo}. This figure shows an increasing width at the entrance to the matched lens, a constant width at the entrance to the unmatched lens located at the beam waist, and equal widths at the exit of the lenses which increases as $\sqrt{1 + \alpha_{lo}^2}$. Note that the compression ratio of the matched lens is greater than that for the unmatched lens, but the final spot sizes for both cases are nearly equal.

An advantage of the matched lens is that it requires a smaller focusing force to obtain the same final envelope width. Figure 5 shows the variation in plasma density at the entrance versus α_{lo} for the matched and unmatched cases for the beam at the SLAC End Station. This clearly shows that the matched lens requires

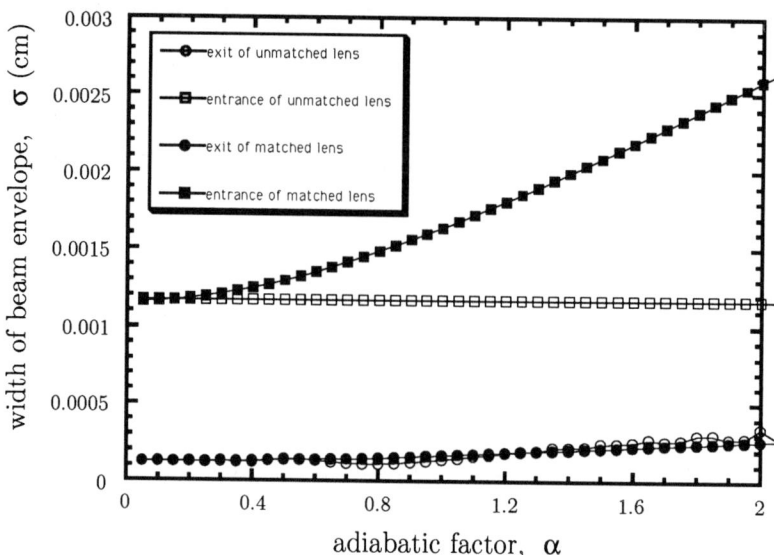

Figure 4: Variation of beam envelope width at the lens entrance and exit with adiabaticity factor for matched and unmatched beam and lens.

a lower plasma density. The length, L, of the matched and unmatched lens were calculated in the numerical simulations and are shown in figure 6.

For the unmatched lens we find that $L_{um} = \beta_b^*/2\alpha_{lo}$ and for the matched lens we find $L_m = \beta^*(1 + \alpha_{lo}^2)/2\alpha_{lo}$. We find that there is a minimum at $\alpha_{lo} = 1/\sqrt{2}$ for the matched lens.

Possible disadvantages of the matched lens are that it will be wider since it intercepts the beam at a larger envelope width, and its entrance must be moved upstream of the focus as α_{lo} increases. As α_{lo} increases the length shortens and thus the exit of the lens can be upstream of the original free space waist. The lens shifts the interaction point upstream as illustrated in figure 7. There the longitudinal positions of the matched lens entrance and exit are plotted versus α_{lo}. The free space beam waist is located at $s = 0$ and the beam propagates from $-s$ to $+s$. The location of the unmatched lens entrance and exit can be inferred from figure 6 if its vertical axis is relabeled to "longitudinal position, s" and it is recalled that the unmatched lens entrance is always at $s = 0$.

Thus far we have assumed that the plasma density ramps uniformly with the same degree of adiabaticity (i.e, $\alpha_l(s) =$ constant). Next we consider deviations from this assumption. We include the effect of nonuniform ramping in $\alpha_l(s)$ by setting

$$\alpha_l(s) = \alpha_{lo} \left(1 - 2\delta_l s \right) \tag{14}$$

so that

$$\beta_{\delta l}(s) = \beta_{\delta lo} - 2\alpha_l(s)s = \beta_{\delta lo} - 2\alpha_{lo} \left(s - 2\delta_l s^2 \right) \tag{15}$$

where δ_l is the nonuniformity factor for the lens. Thus $\delta_l < 0$ ($\delta_l > 0$) corresponds to a lens that is becoming less (more) adiabatic with distance into the lens. We are concerned with $\delta_l < 0$ which increases the focusing strength. For this case we find that

$$K_\delta(s) = \frac{1 + \alpha_{lo}^2 - 2\alpha_{lo}\beta_{\delta lo}\delta_l}{(\beta_{\delta lo} - 2\alpha_{lo}s + 2\alpha_{lo}\delta_l s^2)^2} = \frac{\text{constant}}{\beta_{\delta l}(s)} , \tag{16}$$

and that

$$\beta_{\delta lo} = \left[-\frac{\alpha_{lo}\gamma c}{\omega_{po}}\delta_l + \sqrt{\left(\frac{\alpha_{lo}\gamma c}{\omega_{po}}\right)^2 \delta_l^2 + 2\gamma \left(1 + \alpha_{lo}^2\right)} \right] \frac{c}{\omega_{po}} . \tag{17}$$

Spot size compression obtained by numerical simulations versus the nonuniformity factor is plotted in figure 8 for $\alpha_{lo} = 1/\sqrt{2}$ and for the matched and unmatched cases.

The upper curve in figure 8 shows that spot size compression is constant for the matched case, where we are using the same definition of matching and unmatching as before. The next lower curve in the figure shows that the compression eventually becomes reduced by approximately the factor $1/\sqrt{1 + \alpha_{\delta lo}^2}$ as δ_l gets more negative. Also shown is the theoretical value of the nonuniform unmatched compression.

The length of the lens can be reduced even more than as shown in figure 6 by varying δ_l as illustrated by the numerical simulation results of figure 9. In figure 9 we take the starting point as the minimum of the matched curve shown in figure 6. Therefore $\delta_l = 0$ in figure 9 corresponds to the point $\alpha_{lo} = 1/\sqrt{2}$ in figure 6 where the matched curve is minimum. This shows that the length of the lens is greatly reduced as δ_l goes negative.

The effect of beam energy spread is shown in figure 10, in which spot size compression is plotted versus the fractional reduction in beam energy and α_{lo}. In this simulation, all of the electrons in the beam were reduced in energy by the same amount before injection into the plasma lens. The beam energy was varied from 100% to 70%. The figure shows that the compression remains constant for small α_{lo} but increases slightly for large α_{lo} as the energy is reduced. These curves imply that electrons that lose energy will be focused to a spot size at least as small as that of the highest energy particles. The aberrations due to energy spread (due to synchrotron radiation, for example) should be very small in this scheme. Therefore the ramped density plasma lens would not be severely restricted by the Oide limit.

SUMMARY

We have examined in detail the beam spot size compression for matched and unmatched beam-lens cases, and for uniformly and nonuniformly varying focusing forces in a plasma lens. We have found that the focusing need not be adiabatic to obtain small spot sizes, a finding which reduces the length requirements of the lens. The results are applicable to any strong continuously focusing scheme, not just a plasma lens.

Spot size compression, as a function of adiabaticity factor, α_{lo}, was constant in the matched case and was reduced in the unmatched case. The ratio of the final envelope width to the width of the free space beam waist was equal in the matched and unmatched cases. Thus we conclude that the two cases are equally effective. The strength of the focusing force required to obtain the same final envelope width was smaller in the matched case than in the unmatched case. This is the primary advantage of the matched case. On the other hand, the unmatched lens has the advantage that the lens is shorter. The matched lens has a minimum width at $\alpha_{lo} = 1/\sqrt{2}$ but is always longer than the unmatched lens. The entrance to the matched lens moves upstream of the beam's free space waist and the lens width increases as α_{lo} increases. The unmatched lens is always located at the waist and has a constant width. The exit of the matched lens can move upstream of the original free space waist as α_{lo} gets large.

One of many possible nonuniform ramping profiles was studied. As the nonuniformity factor was increased negatively, the matched lens compression remained constant and the unmatched lens compression was reduced. Thus a matched density profile that starts at modest α and increases α along the lens offers a smaller spot size for the same lens length and initial to final plasma density ratio. The beam energy spread, which could result from synchrotron radiation, was found not to be a cause of chromatic aberrations.

Future investigations could address nonlinearities in the focusing force. The radiation emitted by the strongly focusing particles, in both the classical regime and the quantum regime, could be calculated and the Oide limit could be tested directly. Possible improvements to the code could be to include a third dimension, to add space charge, and to include arbitrary E and B fields. The code could eventually model colliding beams of various shapes and various focusing

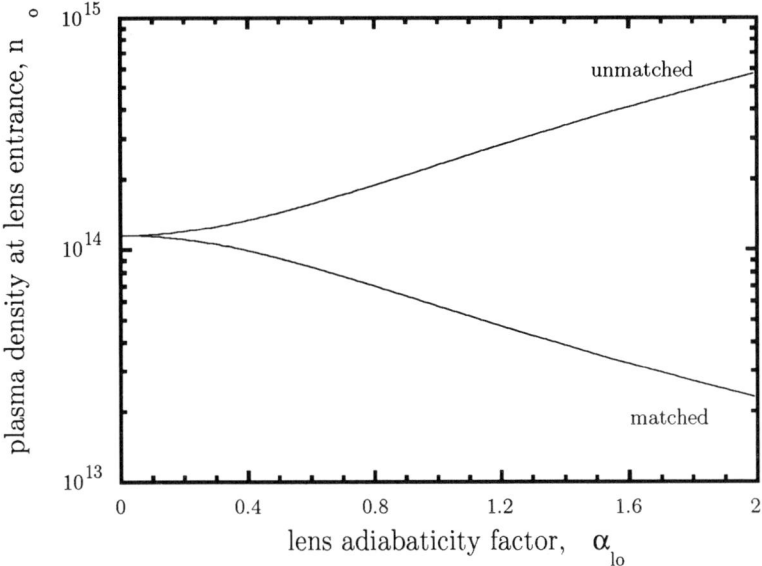

Figure 5: Variation of plasma density at lens entrance vs adiabaticity factor for matched and unmatched beam and lens. For this example $\mathcal{E} = 15$ GeV and $\beta^* = 12$ cm.

configurations.

This work is supported by U.S.D.O.E.. Partial support for R.L.W. is provided by NASA Grant No. NAGW-2930.

References

[1] P. Chen. A Possible Final Focusing Mechanism for Linear Colliders. *Part. Accel.* **20**, 171 (1987).

[2] B. Richter. Very High Energy Colliders. *IEEE Tran. Nucl. Sci.* **NS-32**, 3828 (1985).

[3] R. B. Palmer. The Interdependence of Parameters for TeV Linear Colliders. SLAC **Pub. 4295**, 1987.

[4] P. B. Wilson. Linear Accelerators for TeV Colliders. SLAC **Pub. 3674 (Rev.)**, 1985.

[5] K. Oide. Synchrotron-Radiation Limit on the Focusing of Electron Beams. *Phys. Rev. Lett.* **61**, 1713 (1988).

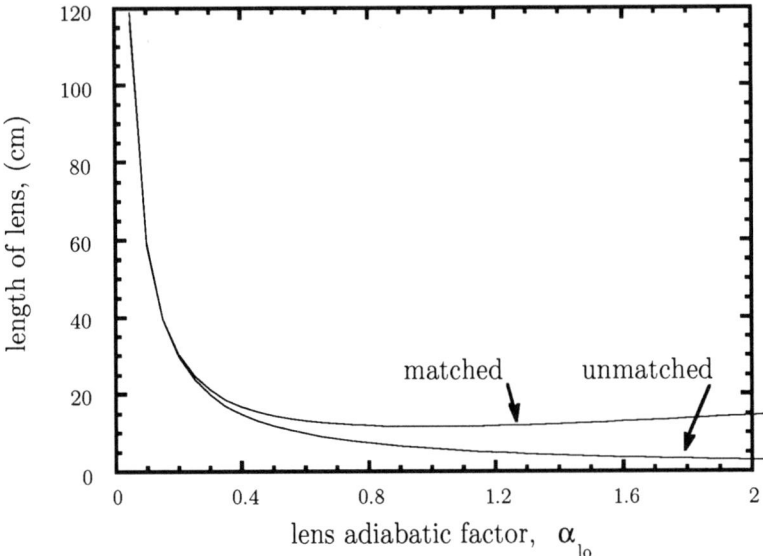

Figure 6: Variation of beam length with adiabaticity factor for matched and unmatched beam and lens.

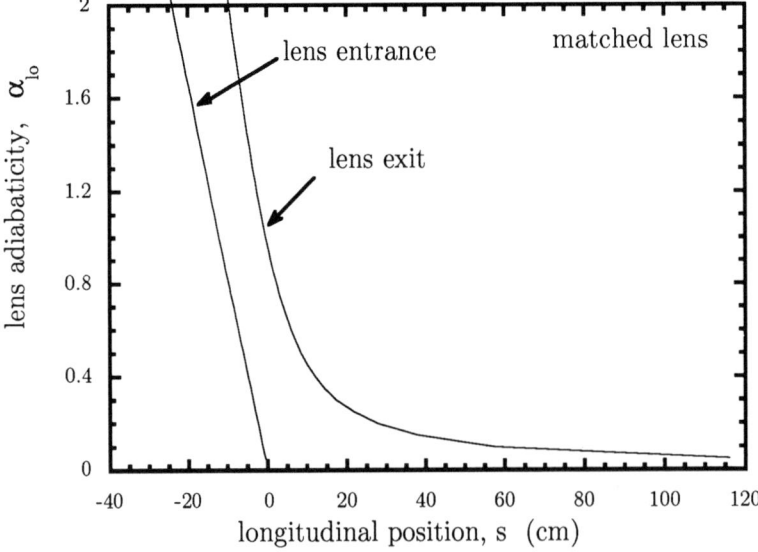

Figure 7: Variation of position, s, of matched lens entrance and exit versus adiabatic factor.

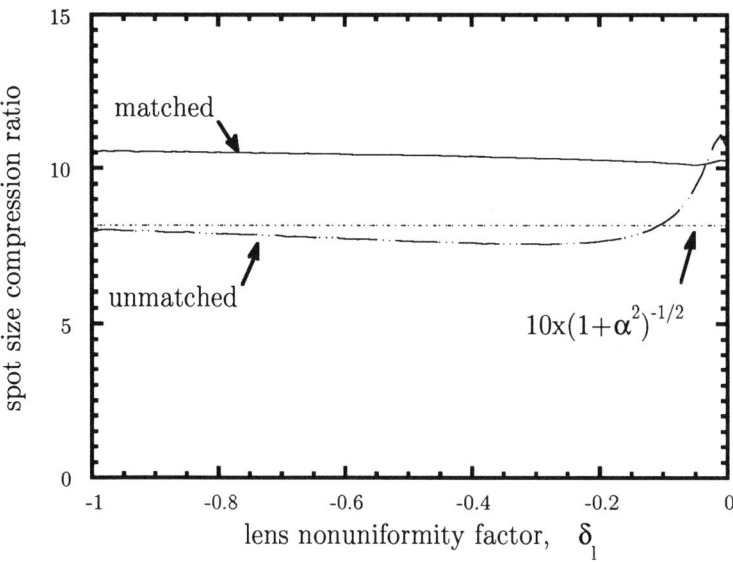

Figure 8: Spot size compression vs nonuniformity for matched and unmatched beam-lens.

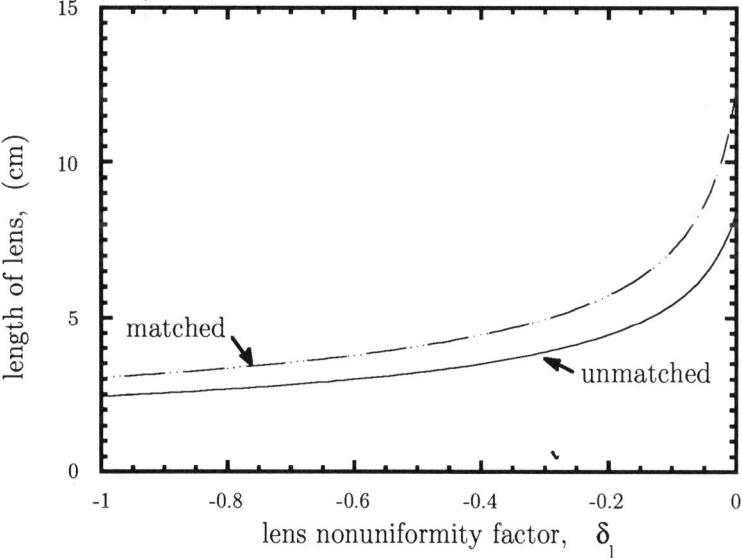

Figure 9: Length of lens versus nonuniformity factor. The adiabaticity factor equals $1/\sqrt{2}$.

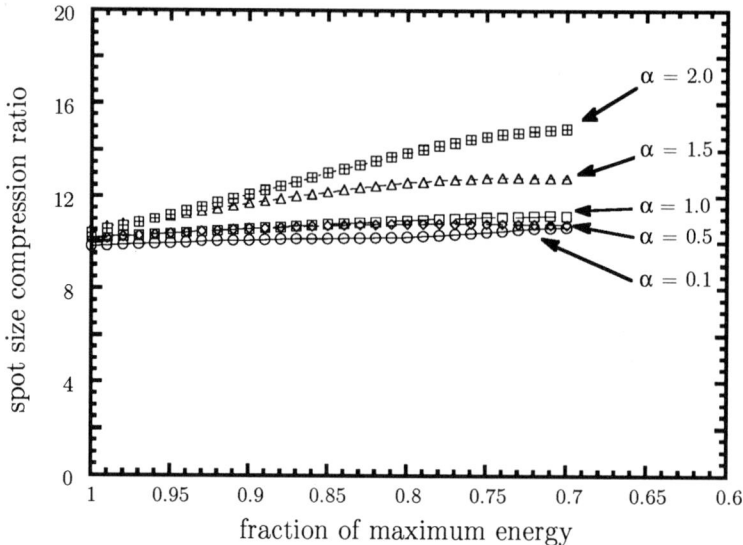

Figure 10: Spot size compression vs electron energy and adiabaticity factor.

[6] P. Chen, K. Oide, A. M. Sessler, and S. S. Yu. An Adiabatic Focuser. *Part. Accel.* **31**, 7 (1990).

[7] D. H. Whittum. Theory of the Ion-Channel Laser. Ph.D. Thesis. UC Berkeley, 1990 and LBL Report **LBL-29720**.

[8] D. H. Whittum. A Continuous Plasma Final Focus. Proceedings of the Topical Conference on *Research Trends in Nonlinear and Relativistic Effect in Plasmas*, edited by V. Stefan, (LaJolla, 1990) and LBL report **LBL-28644 (LBL-27965 Rev.)**.

[9] A. Sessler, private communication.

[10] R. L. Williams. Interactions of relativistic electrons with relativistic plasma waves. Ph.D. Thesis, UCLA, 1992, unpublished.

[11] J. J. Su, T. Katsouleas, J. M. Dawson, and R. Fedele. Plasma Lens for Focusing Particle Beams. *Phys. Rev.* **A 41**, 3321 (1990).

REPORT ON THE HIGH-BRIGHTNESS SOURCE WORKING GROUP*

Patrick O'Shea
MS J579
Accelerator Technology Division
Los Alamos National laboratory
Los Alamos NM 87545

Martin Reiser
Laboratory for Plasma Studies
University of Maryland
College Park MD 20742

Introduction

The charges of the high-brightness source working group were two-fold:

- Report the state of the art of electron sources and promising new concepts
- Develop an outline design for an inexpensive electron injector suitable for use in advanced acceleration concept experiments.

There were twenty-four formal presentations in the working sessions. The topic can de divided into three categories as indicated in the table below.

Category	Number of papers
Photocathode sources	12
Other sources	6
Beam transport	6

A complete list of all the papers presented is includes at the end of this report.

In the first session of the working group there was some debate with regard to the appropriate definitions for emittance and brightness. We agreed that both emittance (ε) and brightness (B) would be would be presented as normalized root-mean-square (rms) quantities according to the following definitions:

$$\tilde{\varepsilon}_n = \beta\gamma(<x^2><x'^2> - <x.x'>^2)^{\frac{1}{2}} \quad \text{units: } \pi \text{ mm-mrad}$$

*Work performed under the auspices of the US Department of Energy

$$B_n = \frac{2I}{[\pi \tilde{\varepsilon}_n]^2} \qquad \text{units: A/(m-rad)}^2$$

Note that the factor of π is explicitly used in the calculation of B_n. A detailed discussion of this topic can be found in the paper by Lejeune and Aubert[1]. Our use of these definitions and units should not be considered to be an attempt to legislate definitions. Our aim is to ensure some degree of consistency and to minimize confusion in the comparison of the various electron sources discussed in the working session and in this report.

Photocathodes
It is evident that a strong trend exists toward the use of photocathode sources in applications that require both high-brightness and high current. Photocathodes are now a reliable source in regular use at a number of accelerator facilities[2]. The demonstrated brightness of electron beams from photocathode sources has reached unprecedented levels. Beams with $\tilde{B}_n \approx 3 \times 10^{12}$ A/(m-rad)^2 at currents in excess of 100 A have been accelerated to high energy in rf linacs[3,4].
Two photocathode approaches have been adopted:
- High quantum efficiency (> 1%), usually multi-alkali semiconductors using a drive laser operating in the visible (527 nm). The advantages of this approach are that the drive laser construction is straight forward and suitable for high duty operation. The disadvantage is that multi-alkali cathodes are easily contaminated by poor vacuum conditions, with typical operating life-times being on the order of 10-20 hours before refurbishment is required

- Low quantum efficiency (<0.1%), usually metal or LaB_6 using a drive laser operating in the ultra-violet (260 nm). The advantage of this approach is that the cathode is very robust and has a very long life-time. The disadvantage is that the drive laser is more difficult because of the high power required in the UV and the difficulty in scaling to high-duty operation.

It was agreed that the ideal photocathode would combine the advantages of the above two approaches, having a quantum efficiency in the range 0.1 - 0.5 % at 527 nm and be insensetive to vacuum conditions. While no such cathode was reported on in the working session, there is work under way at BNL to develop a field-enhanced metal photocathode that may approach the ideal.

Extreemly high current densities have been extracted from photocathodes. Peak values of close to 7 kA/cm^2 from a multi-alkali cathode at LANL[3] and 21 kA/cm^2 from a metal cathode at BNL[5] have been reported. Operation at high duty (>10%) using a multi-alkali photocathode in a room-temperatureh rf linac has been reported from Boeing[6] at peak currents in excess of 100 A and true average currents approaching 0.1 A and $\tilde{\varepsilon}_n < 10\ \pi$ mm-mrad.

There ara at present two methodes of minimizing the emittance growth resulting from space-charge forces near the photocathode. One method is to have a very high accelerating gradient e.g. 100 MV/m at Brookhaven. The other is to use a linear focusing lens to compensate for the linear space-charge forces. Because of the fact that photocathode sources in rf cavities produce bunched beams in which the transverse displacement of particles resulting from linear space-cahrge forces is correlated with axial position in the bunch, emittance correction schemes using various focusing techniques are possible. The results from the LANL group show that the use of an appropriately placed solenoidal lens can be used to unwind the transverse phase-space along the bunch and to correct for the linear space-charge forces. This scheme works well for accelerating gradients of a few tens of MV/m. Such a scheme will not work for high gradient photoinjectors such as at Brookhaven[7]. In the case of low duty operation where high gradients do not cause thermal management problems the Brookhaven scheme is applicable. For high-duty operation where a low accelerating gradient is required the LANL scheme is more applicable.

The use of rf focusing has been proposed for the correction of rf induced emittance growth. One variant uses a decoupled unsymmetrical rf cavity that can be phased independently of the photocathode cell to recover the minimum emittance condition even for electron injection at high rf phases[8]. The other scheme[9] corrects for rf induced emittance blowup by running the photocathode cavity with both the accelerating and a higher harmonic mode. The presence of the higher order mode has been shown in simulations to cancel the emittance growth resulting from time-dependent accelerating rf fields. No report was received of either of these techniques being used in practice.

Work is under way at the University of Wuppertal[10] and the University of Rochester[11] on superconducting photoinjectors. Superconducting photoinjectors should produce high-quality beams similiar to room-temperature devices. Their usefullness will be for appplications that require a high duty factor with modest average current. We estimate that the maximum average current from a superconducting photoinjector will be on the order of 20 mA because with a multi-alkali photocathode(5%

quantum efficiency). This estimate is bsaed on the fact that the maximum average drive-laser power that can be dissipated before the photocathode substrate goes normal is on the order of 1 W/cm^2.

Other Sources

Ferro-electric cathodes

Work is under way at Cornell[12] on the use of electrically stressed ferro-electric materials to produce bright electron beams. These dwevices show promise of current densities in excess of 1 kA/cm^2 with pulse lengths on the order of 100 ns.

Field-Emission Arrays

The Naval Research Laboratory[13] is studying field-emission arrays to produce gated pulses as short as 100 ps with current densities of 1 kA/cm^2 and a projected $\acute{B}_n \approx 10^{12}$ A/(m-rad)^2.

Pseudospark

The pseudospark hollow cathode discharge is underinvestigation at Integrated Applied Physics Inc., MIT, the University of Southern California[14] and the University of Maryland[15]. The source is very simple to build and operate. Currents in excess of 3 kA have been generated at 35 kV with $\tilde{B}_n \approx 3 \times 10^{12}$ A/(m-rad)^2. These measurements were made while the electron beam was still in the gaseous medium near the source. Acceleration to higher energies before extraction of the beam into vacuum will be necessary in order to avoid severe space-charge induced emittance growth.

Inexpensive Electron Injectors

Traditionally high-brightness electron injectors have not been inexpensive. The working group was asked to develop an outline design for a simple inexpensive injector suitable for use as a source of electrons for advances acceleration schemes.

The design guidelines are given in the table below

Energy	10 MeV
Current (peak)	16 A
Charge per bunch	10^9

Pulse length	10 ps
Repetition rate	10 Hz
Emittance	< 10 π mm-mrad
Construction time	1 year
length	< 1m

The candidate designs considered reflect the opinions and biases of those present at the working session. It is beyond the scope of this report to provide significant design or costing details. However we have provided cost estimates for the major components that would have to be procured from outside vendors. We have not estimated the labor costs involved in assembly, however the designs outlined below could readily be assembled by student labor. Because of the short pulse length and low emittance rf photocathode sources were deemed to be superior to rf thermionic guns. Rf photocathode sources can produce beams that exceed the quality required directly from the photoinjector, whereas thermionic rf injectors require bunching and momentum filtering. In addition the rf photocathode offers a wide range of options for the spatial and temporal tailoring of the pulse shape. Such flexibility may be of importance to the users of the injectors. In two of the options described below, the Grumman-BNL gun and the microtron, the photocathode may be replaced by a thermionic cathode if so desired.

Grumman-Brookhaven Photoinjector

The G-BNL photoinjector[16] is a 3.5 cell photoinjector with a copper UV-photocathode. Its performance in terms of the criteria is given below:

Energy	10 MeV
Current (peak)	>100 A
Charge per bunch	>1 nC
Pulse length	<10 ps
Repetition rate	>10 Hz
Emittance	< 2 π mm-mrad
Construction time	1 year
Length	< 1 m

The cost estimate for the G-BNL gun is given below.

Accelerator cavities + cathode   200 k$

Rf driver (S-band klystron and modulator)	330
Drive-laser	100
Magnets	10
Vacuum equipment	20
Diagnostics, controls	40
Total	700

Microtron

Microtron accelerators have been used for many years with thermionic cathodes in the rf cavity. The typical output current is a few amps. This current could the significantly increases by the use of a photocathode to produce electron beams of up to 20 MeV in a simple system.

Energy	20 MeV
Current (peak)	>20 A
Charge per bunch	>0.2 nC
Pulse length	<10 ps
Repetition rate	>10 Hz
Emittance	< 10 π mm-mrad
Construction time	1 year
Diameter	1 m

The sample cost estimate below is based on the actual cost plus drive-laser cost for a microtron installed at Frascati[17] in the late 1980s.

Accelerator cavities +cathode	40 k$
Rf driver (magnetron) + associated hardware	500
Drive-laser	100
Magnet, vacuum chamber	83
Vacuum equipment	50
Diagnostics, controls, cooling	40
Total	813

Superconducting photoinjector

At first glance it might appear that the use of superconducting technology would be inappropriate for an inexpensive injector. Examination of the cost estimates for the systems described above shows that the major expense is associated with the rf driver. Because of the low average beam power required for the injector the superconducting option becomes attractive because of the low cost of the rf system. As can be seen below the performance of this system would be similar to the G-BNL injector.

Energy	10 MeV
Current (peak)	>100 A
Charge per bunch	>1 nC
Pulse length	<10 ps
Repetition rate	>10 Hz
Emittance	< 2 π mm-mrad
Construction time	1 year
length	2 m

This cost estimate below is based on work at the University of Wuppertal[18]

Accelerator cavities +cathode	100 k$
Rf driver (solid state) + associated hardware	20
Drive-laser	100
Cryostat	150
Vacuum equipment	10
Magnets	20
Diagnostics, controls	40
Liquid helium per year	25
Total	465

Pseudo spark with induction linac

One unifying characteristic of the three options mentioned above is the requirement for the purchase of a large quantity of sophisticated customized equipment from outside vendors. We felt it appropriate to consider an option that could in principle be assembled from simple readily available components. The option of a pseudospark source with post acceleration with an induction linac is ideal for the adventurous researcher who likes to build all components in-house with student labor. As indicated previously the pseudospark source has the demonstrated brightness to meet our requirements, however it is not yet clear that the beam can be accelerated to high energy and extracted form the gaseous medium of the source while preserving the beam brightness. The inherent pulse width of such a device is about 1 ns at a current in excess of 1 kA. The cost of building a 35 kV pseudospark source is less than 10 k$[19]. Induction modules can be easily fabricated at low cost, and be installed on

Option	Advantages	Disadvantages	Cost (k$)
G-BNL Injector	Proven, compact, upgradable, very high quality electron beam	rf cost	700
Microtron	proven, compact	Emittance and current not as good as G-BNL	800
Superconducting Injector	Simple rf system, compact, upgradable, very high quality electron beam	Handling of LiHe, delicate hardware, complete system unproven	470
Pseudospark with induction linac	Build in stages, no rf, low-tech, parts easy to obtain, high rep. rate (1 kHz)	Unproven, questions concerning emittance	?

Comparison of the options for a low cost injector

a piece-wise basis for concept demonstration. An accurate estimate for an injector with an output energy in excess of 10 MeV in not possible at this time.

The advantages and disadvantages of each option are summarized in the attached table.

Summary

As indicated in the introduction photocathodes sources are well on their way to becoming the source of choice for high brightness electron beam generation. their performance and shortcomings are well understood theoretically. Advanced system designs are now being built that offer a significant increase in brightness over the present state of the art in the next year[20,21]. The fact that photocathodes have been shown to emit currents >> 1 kA/cm^2 indicates that a brightness in excess of 10^{14} A/(m-rad)^2 is possible for source temperatures less than 1 eV. The difficulty is not so much in the generation of a bright high-current electron beam but in its transport through the accelerator system while preserving the brightness. Drive lasers for photocathodes continue to be expensive and relatively complex devices. The advent of diode-pumped technology will enhance the laser stability and efficiency and reliability.

The alternative source concepts presented offer the advantage of being simpler to operate than photocathode systems, however they remain unproven in their ability to generate beams that can be accelerated without significant emittance growth.

Whether the inexpensive injectors described in the report can be constructed awaits more detailed design studies.

K.J. Kim	LBL	Beam dynamics of rf photocathode gun
P.G. O'Shea	LANL	APEX 40MeV photoinjector Llnac
S. Park	UCLA	UCLA compact linac
S.C. Hartman	UCLA	Results form the UCLA photoinjector
R.L. Sheffield	LANL	Physics of AFEL linac
K.C.D. Chan	LANL	The perfect joint: wakefields
I.S. Lehrman	Grumman	Grumman-Brookhaven photoinjector
S.C. Chen	MIT	17 GHz photocathode gun
A. Michalke	Wuppertal	Superconducting photoinjector
W. Donaldson	Rochester	Superconducting rf gun
N.A. Kurnit	LANL	Electromagnetic pulses for high-brightness sources
T.S-Rao	BNL	Phoroemission from Au and Cu films
L. Giannessi	Frascati	Synchrotron radiation triggered rf gun
P. Michelato	Milano	High-efficiency photocathodes at Milano
L. Schachter	Cornell	Ferro-electric cathodes
C.M. Tang	NRL	Field-emission arrays
S. Benson	Duke	Thermionic microwave guns
M.J. Rhee	Maryland	Pseodospark electron source
G. Kirkman	IAP	Hollow-cathode electron source
C. Wang	LBL	RF conditioning of helically transported beam
L. Schachter	Cornell	Injector based on an x-band TWT
P. Sprangle	NRL	Beam conditioner for an FEL
B. Hafizi	NRL	Emittance and energy spread in an FEL
C.A. Brau	Vanderbilt	Compton-damped compact storage ring

Papers presented at the working session

References

1. C. Lejeune and J. Aubert, in Advances in Electronics and Electron Optics, Supplement 13A, edited by A. Septier, published by Academic Press, page 159, (1980)
2. C. Travier, Particle Accelerators, 36, 33, (1991)
3. P.G. O'Shea et al. these preceedings
4. K. Batchelor et al., Nucl. Instr. and Meth ., A318, 372, (1992)
5. T. Srinvasan-Rao et al. , J. Appl. Phys. 69,3219, (1991)
6. D. Dowell, Boeing, personal communication
7. K.J. Kim, these proceedings
8. L. Serafini et al. Nucl. Instr. and Meth ., A318, 275, (1992)
9. L. Serafini et al. Nucl. Instr. and Meth ., A318, 301, (1992)
10. A. Michalke these proceedings
11. W. Donaldson these proceedings
12. L. Schachter these proceedings
13. C. M. Tang these proceedings
14. G. Kirkman these proceedings
15 M.J. Rhee these procedings
16 I.S Lehrman these proceedings
17 L. Giannessi personal communication
18 A. Michalke personal communication

19 M.J. Rhee personal communication
20. R.L. Sheffield personal communication
21. I.S. Lehrman personal communication

THE BNL ACCELERATOR TEST FACILITY AND EXPERIMENTAL PROGRAM[1]

Ilan Ben-Zvi

NSLS, Brookhaven National Laboratory, Upton NY 11973

and

Physics Department, SUNY Stony Brook, NY 11794

ABSTRACT

The Accelerator Test Facility (ATF) at BNL is a users' facility for experiments in Accelerator and Beam Physics. The ATF provides high - brightness electron beams and high-power laser pulses synchronized to the electron beam, suitable for studies of new methods of high-gradient acceleration and state-of-the-art Free-Electron Lasers.

The electrons are produced by a laser photocathode rf gun and accelerated to 50 MeV by two traveling wave accelerator sections. The lasers include a 10 mJ, 10 ps Nd:YAG laser and a 500 mJ, 10 to 100 ps CO_2 laser. A number of users from National Laboratories, universities and industry take part in experiments at the ATF. The experimental program includes various laser acceleration schemes, Free-Electron Laser experiments and a program on the development of high-brightness electron beams.

The ATF's experimental program commenced in early 1991 at an energy of about 4 MeV. The full program, with 50 MeV and the high-power laser will begin operation this year.

THE ACCELERATOR TEST FACILITY

Introduction. The Accelerator Test Facility (ATF) is operated as a user facility for accelerator and beam physicists by the Brookhaven National Laboratory (BNL) for the US Department of Energy. The National Synchrotron Light Source department at BNL has responsibility for the

---

[1] Work performed under the auspices of the U.S. Department of Energy.

construction, operation, safety and administration of the ATF. The experimental program is coordinated by the BNL Center for Accelerator Physics. Support for the ATF is provided by the Advanced Technology R&D Branch, Division of High-Energy Physics of the DOE.

The central theme of the ATF research is the interaction of high-brightness electron beams with electromagnetic energy. For this purpose the ATF is equipped with a laser-photocathode rf-gun, an electron linac, high-power short pulse lasers synchronized with the electron beam and a variety of diagnostic equipment. To provide a wide range of beam parameters for the users, the electron beam charge may be varied from 0.1 pC at a normalized rms emittance of 10^{-8} πm-rad to 1 nC at a normalized rms emittance of 7×10^{-6} πm-rad.

The macropulse repetition rate is variable from 1.5 to 6 pulses per second. The length of a macropulse is 3.5 μS. In each macropulse one may produce from 1 to 100 micropulses with a FWHM pulse length of 6 ps, at a 24.5 ns period. The electron beam energy from the gun is variable up to 4.6 MeV and the linac energy, initially at 50 MeV, will be upgraded to 100 MeV by providing additional rf power.

The Nd:YAG laser radiation at 1.06 μm with a pulse energy of about 10 mJ in 10 ps and the CO_2 radiation at 10.6 μm with a pulse energy of about 0.5 J, in 100 ps (10 ps in the future), both synchronized to the electron beam, will be available at the experimental area.

The rf gun. The BNL laser-photocathode rf-gun is a resonant π-mode 1½ cell cavity operating at 2856 MHz [1]. The cavity, shown in Fig. 1, is 83.08 mm inner diameter and 78.75 mm long, with a beam aperture of 20 mm. It has a Q of 11900 and a calculated shunt impedance of 57 MΩ/m, which corresponds to a beam energy of 4.65 MeV at a structure peak power of 6.1 MW. At this power, the peak surface electric field is 119 MV/m and the cathode field is 100 MV/m.

The gun power is derived from the linac klystron using a high - power hybrid junction, and fed through a power attenuator and phase shifter so that the power level and phase can be varied from a few KW up to a maximum of 12.5 MW. For a high micropulse charge the emittance growth due to space charge effects is minimized by operating the gun in the highest possible electric accelerating field.

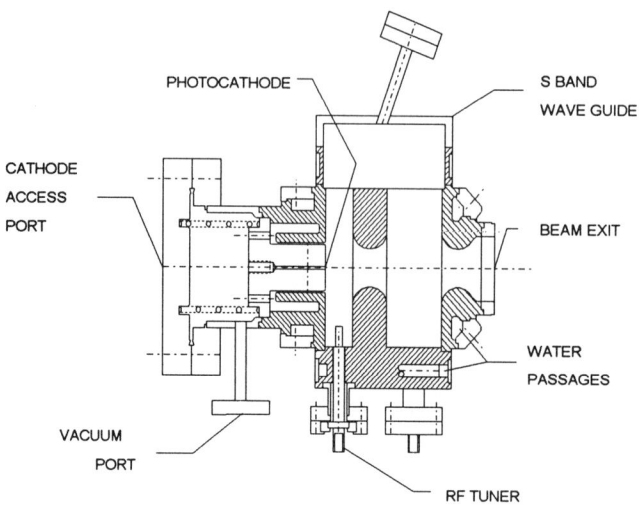

FIGURE 1. RF GUN.

The rf field contribution to the emittance is minimized in two ways. The first of these involves providing a nearly linear dependence of the transverse fields with beam radius by a suitable cell design. The other uses cancellation of the strong radial fields near the aperture through operation in π-mode and crossing of the boundary between the two cells when the electric fields vanish. This last condition, in conjunction with the requirement for a high cathode electric field due to the space charge forces, determines the optimal (charge dependent) phase of the laser pulses relative to the rf wave for minimum emittance at the gun output. Since there is no compensation for the transverse deflection at the exit of the gun, the greater part of the rf induced emittance growth takes place there.

The photocathode. A study of various materials [2] for the photocathode has shown that certain metals have a good combination of quantum efficiency, high damage thresholds, a work function close to the photon energy of the laser and good mechanical and chemical stability. Copper and yttrium metal cathodes proved particularly robust.

We are using a frequency quadrupled Nd:YAG laser, with photon energy of 4.65 eV. Yttrium has a work function of about 3.1 eV and a quantum efficiency of up to 10^{-3}. Thus the photoelectrons can emerge with up to 1.5 eV energy. We may assume that the electrons are emitted with an isotropic angular distribution at the cathode. For a Gaussian laser power distribution with $\sigma_r=3$ mm this would lead to an initial invariant emittance of 3.5π mm-mrad.

A copper cathode with a work function of 4.3 eV contributes about half the initial emittance but only offers a quantum efficiency of only 10^{-4}. We have operated both copper and yttrium cathode for prolonged periods in the gun, with vacuum cycling and through power conditioning of the gun, with a negligible loss in the cathode quantum efficiency.

RF gun research. An important aspect of the research at the ATF is the continued development of high-brightness electron guns. In the design of the current version of the ATF gun, the beam dynamics in the gun has been calculated with a version of PARMELA, modified to include ejection of low-energy electrons from a photocathode by a laser pulse as well as the image charge effects at the cathode [1]. The rf fields were calculated by SUPERFISH and transferred to PARMELA as Fourier coefficients.

The rms invariant emittance ε_n has been optimized for a charge of 1 nC at a cathode field of 100 MV/m. In this simulation the laser pulse length is $\sigma_z=0.6$ mm and its spot radius is $\sigma_r=3$ mm. The optimal ε_n is 7.3 mm-mrad, which is mostly due to space charge effects (6.2 π mm-mrad), while the cathode and the rf field contribute the remaining emittance. The energy spread is $\sigma_E=17$ keV. This small value is the result of the short laser pulse and is not due to an energy selection slit.

In recent work [3] time dependent correlations were found between the transverse displacement and the divergence of the electrons exiting the gun. In calculating the emittance for slices that are one tenth of the bunch length, a value that is five times smaller has been found. This suggests that an emittance correction method may be found to result in much improved emittance.

A collaboration between BNL and Grumman Aerospace Corporation has been set up to develop the next generation photocathode gun for the ATF. The MAGIC particle-in-cell (PIC) code is being used for this research. This code is also being applied to the present gun for confirmation of the modeling. This code calculates the rf fields of the cavity internally and provides for a complete and accurate description of wake fields and space charge forces. The MAGIC code simulation [4] includes the fields in the exit region of the gun.

The resolution of the MAGIC simulations was increased beyond that previously available to properly resolve the bunch and suppress grid heating [5]. The increased resolution, coupled with a field algorithm that damps artificial particle noise, results in emittance values significantly

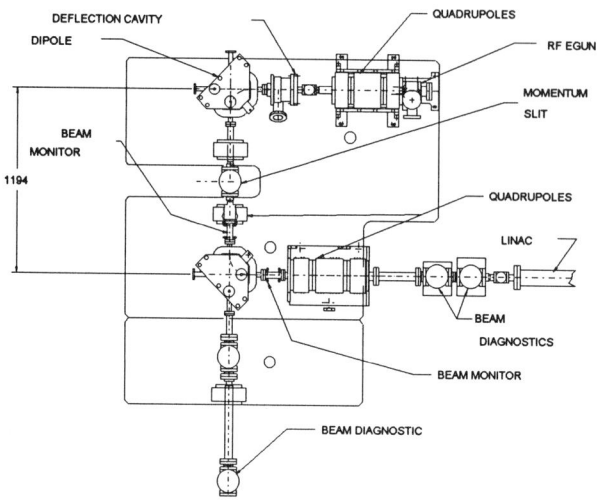

FIGURE 2. THE ATF FRONT-END.

lower (up to a factor of 2 lower) than those predicted by our PARMELA simulations. The results that are forthcoming from the ATF experimental program should resolve the discrepancy.

Low-energy beam transport. The ATF low-energy beam transport, shown in Fig. 2, is designed for easy illumination of the cathode by the laser, longitudinal and transverse phase space diagnostics, bunch compression and availability of the beam for experiments at low-energy.

The LEBT consists of two quadrupole triplets and a 180° achromatic double bend. It is equipped with a momentum selection slit at the maximum horizontal dispersion point (4 mm per %), a vertical rf deflection cavity for streak measurements, phosphorous screen beam profile monitors with CCD cameras, fast strip line wall current monitors and Faraday cups.

The momentum selection slit has a remote control, beam charge measurement capability and a phosphorous layer monitored by a CCD camera. The rf deflection cavity makes it possible to measure the longitudinal phase space of the 6 ps electron bunches.

The laser optics of the front end provides a remote control of the UV light spot on the cathode: intensity, position and size. The laser power is monitored by an energy meter and a fast photodiode and the spot shape and size by a CCD camera.

The LEBT has been modeled with a version of TRANSPORT with linear space charge added to study the effect of this on the beam profiles [6]. The first triplet is operated in an asymmetric mode to compensate for certain high order aberrations. For the 1 nC beam described above, the beam pipe scrapes about 25% of the beam at the centers of the first quadrupole triplet. The geometric horizontal emittance ε growths from 0.8 π mm-mrad to about 2.2 π mm-mrad and the vertical emittance is preserved in passing through the transport line. Further modelling, done with PARMELA, indicates an emittance growth due to non-linear space charge effects. Consequently a direct injection path is being constructed in which the beam is not bent by dipoles before acceleration by the linac. With this layout and with a solenoid focussing lens following the gun the normalized rms emittance at the linac entrance is about 5 π mm-mrad.

The linac. The ATF linac consists of two SLAC linac sections, ($2\pi/3$ mode, disc loaded, travelling wave, 3.05 meters long) produced at IHEP Academia Sinica. The linac and gun are driven by a single XK5 klystron delivering about 25 MW of peak power at a frequency of 2856 MHz and with a pulse duration of 3.5 µs. The rf system is designed to run at repetition rates up to 6 Hz so average power and heating are minimal.

A phase locked oscillator operating at the 35th harmonic of 81.6 MHz is used to drive a series of solid state and triode amplifier stages to obtain the up to 1 KW of drive power for the klystron. The fundamental also drives and synchronizes the Nd:YAG laser. The linac and waveguide system are temperature stabilized to within ±0.1°C. The gun has an independent regulation system.

Calculations of the linac beam dynamics have been done numerically, including the effects of space charge and wake fields. We find that for a properly matched transverse beam no increase in transverse emittance occurs in the acceleration process through the linac.

An increase of the longitudinal emittance takes place for a 1 nC, 6 ps beam microbunch due to the combined effect of the short range longitudinal wake field and the curvature of the accelerating waveform. The effect is minimized in the usual way by a proper choice of the bunch phase.

A second modulator and klystron are proposed for future operation of the ATF, so that the gun and the linac will be driven by two independent rf amplifiers. This will improve the amplitude and phase control and increase the ATF energy to about 70 MeV. Upgrade to 100 MeV can be achieved by providing a klystron for each accelerator section.

Adaptive control. While the macropulse-to-macropulse energy of the linac is very stable, various effects contribute to the energy spread of the beam within a single macropulse, which will contain up to 100 microbunches. Beam loading effects would produce an energy variation of about 8% during the macropulse. The modulator voltage ripple of ±0.3% contributes ±0.6% to the rf power and an estimated ±2° to the phase error with the klystron saturated.

Our approach to this problem is the use of an adaptive feed-forward amplitude and phase control system [7]. An attenuator and phase shifter in the low-level drive system vary the amplitude and phase of the klystron unsaturated output. These control elements are driven by arbitrary function generators (AFGs). A PC computer reads a digitizer that samples the cavity field and phase values and sends data to the AFGs.

In a test of the system we have simultaneously stabilized the RF amplitude and phase of the gun cavity. A stability of better than ± 0.5% of the amplitude and ±1° of the phase of the cavity were obtained over a 3 µs pulse. Regulation of just the cavity's amplitude resulted in an amplitude fluctuation of less than ± 0.2%.

Beam line diagnostics. Part of the ATF high-energy beam transport line and its diagnostic system are shown in Fig. 3.

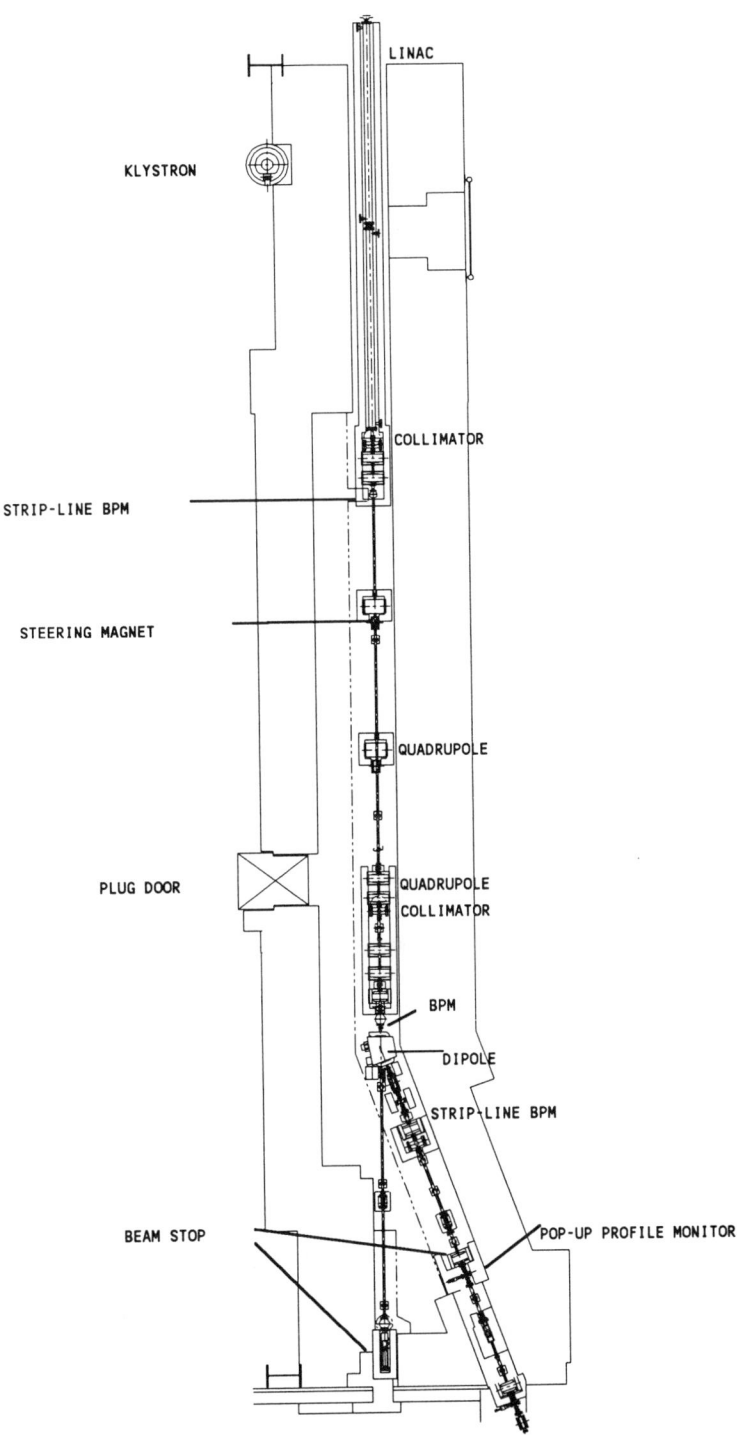

FIGURE 3. THE ATF LINAC & HIGH ENERGY BEAM TUNNEL

The ATF electron beam diagnostics incorporate non destructive, strip-line beam position monitors with a wide band response, to allow for micropulse-by-micropulse measurements of the beam position (or energy) and current. Faraday cups are provided at a few locations for exact beam current measurements.

Remotely controlled destructive beam profile monitors based on phosphor screens and CCD cameras are used for emittance measurements [8]. A large number of pneumatically operated monitors will give beam position and size information. Momentum analyses at both 4.5 MeV and 50 MeV are made by measuring the beam profile (using a phosphor screen) and position (using a strip line detector) at the maximum dispersion points.

Beam profile measurement of the 50 MeV beam in the time domain will be done by means of a fast magnetic septum kicker system and the phosphor screen - CCD camera monitor.

An emittance selection system is incorporated in the post-linac beam transport line, comprising two collimators with a variable aperture. The collimators are phosphor coated and monitored by a CCD camera. The emittance selection system is designed to define the extremely small emittance beams of the laser acceleration experiments [9].

A fast beam profile detector is being developed [10], based on induced conduction in a sapphire substrate on which a meander strip line is deposited. This detector is designed to convert the beam intensity distribution in a given dimension to a pulse height distribution in time for each microbunch (24.5 ns period). A silicon semiconductor detector comprising microstrips 1 μm wide will provide an ultra-high-resolution beam position monitor for positioning the extremely small beams used by the laser acceleration experiments.

Laser systems. A Nd:YAG laser system is used for exciting the gun photocathode and controlling a semiconductor switch in the picosecond CO_2 laser. This system includes a Spectra-Physics CW oscillator (wavelength 1064 nm), mode locked to the 40.8 MHz, a sub-harmonic of the 81.6 MHz rf reference source.

A Lightwave Electronics series 1000 timing stabilizer [11] is used to phase-lock the oscillator to the reference, reducing pulse-to-pulse timing jitter, or phase noise, of the laser to better than 1 ps. The oscillator pulse (about 80 ps long) is then chirped in a 200 m optical fiber and amplified in a pulsed regenerative amplifier. [12] The amplifier bandwidth chops the chirped pulse to about 10 ps.

The output is frequency doubled and then transported about 30 m to the gun hutch where a second doubling takes place. At this point there is an energy of 150 μJ in a 6 ps pulse at the operating wavelength of 266 nm. This is sufficient to produce 1 nC of electron charge from a copper photocathode.

A modification of the Nd:YAG system will provide a pulse train mode, in which up to 100 microbunches, separated by 24.5 ns may be switched by a Pockels cell.

Part of the output from the Nd:YAG at 1064 nm is used to slice a short, synchronized CO_2 laser pulse of 10 to 300 psec duration out of a 100 ns pulse from a CO_2 oscillator by using germanium plates that change from transmitters to reflectors when hit by 1064 nm light [13].

A broadband, 4 Atmosphere isotopic mix CO_2 amplifier, is being used to amplify the sliced pulse. The amplifier, which has a 2 m long active area and a UV pre-ionized TE discharge, has been developed in collaboration with Los Alamos National Laboratory. The amplifier has demonstrated a pulse energy of several hundred mJ in 100 to 300 ps pulse. For pulses shorter than 30 ps, an isotope mix will be used, $^{12}C^{16}O_2/^{12}C^{18}O_2/^{12}C^{16}O^{18}O$ at ratios of 1/1/2. The gas will be used in a closed loop longitudinal gas circulation with a warm (100°C) catalytic converter. Since the timing of the CO_2 pulse is determined by the Nd:YAG pulse, it is synchronized to the electron beam as well.

Control system. The control system is based on a VAX 4000 series system. A Microvax-II/GPX is used off-line for software development and testing.

Control and monitoring of the facility's devices (magnet power supplies, beam position monitors, timing system, etc.) is through a Kinetic Systems Corporation CAMAC byte-serial highway driver connected to 4 CAMAC crates. From the CAMAC, communication to local devices is via industry-standard hardware interfaces and protocols such as EIA-RS-232 and IEEE-488.

Both machines are equipped with Ethernet interfaces and are connected by DECnet to HEPNET and Internet. The computers operate under version 5.4 of DEC's VMS operating system. Support for software development in both C and FORTRAN programming languages is provided.

In addition, a commercial control system software package, marketed by Vista Control Systems, Inc., is used to build window-based operator interfaces. All windowing operations are done using DECwindows, Digital's implementation of the X-windows standard. Operators interact with the control system through "point and click" pull-down menus that graphically display controls, overviews of the facility's status and alarm conditions.

By employing X-windows technology, these detailed graphic presentations will be available throughout the ATF. The Vista package also includes a database generator, various report writers, a line sequencer and a library of program development routines.

THE EXPERIMENTAL PROGRAM

General. One experimental beam line at the gun energy is available as a branch of the LEBT and three high-energy beam lines are available at the experimental hall, shown in Fig. 4.

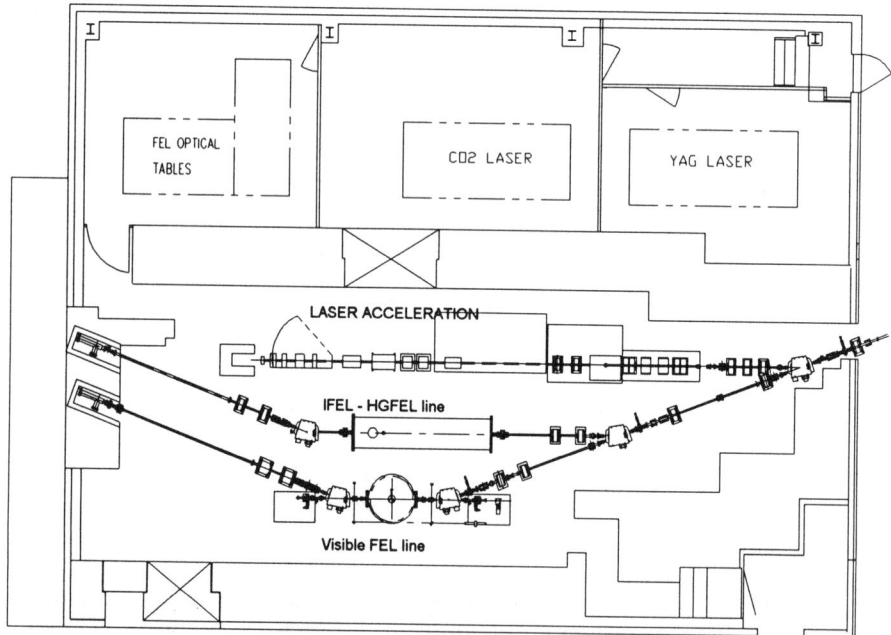

Figure 4. The ATF experiment hall and laser rooms.

The high-energy beam lines are equipped with emittance selection as well as diagnostics for the longitudinal and transverse emittance, beam profile,

energy, beam position and current. The CO_2 laser radiation may intercept two of the lines to perform experiments on laser acceleration or photon-electron scattering. The third line is dedicated to FEL experiments.

Proposals for experiments or letters of intent are considered for approval by the Steering Committee of the BNL Center for Accelerator Physics. The currently approved experiments at the ATF are:

1. Visible FEL (BNL, Stony Brook, Rocketdyne)
2. Grating Acceleration (BNL, LANL, Princeton, UCLA).
3. Nonlinear-Compton Scattering (Princeton, BNL, LANL).
4. Inverse Cerenkov Acceleration (STI, BNL, LANL).
5. Study of Spiking Phenomena in FELs (Columbia).
6. Room temperature, pulsed Microwiggler (MIT).
7. Inverse FEL (BNL, LANL, Yale, UCLA).
8. Cerenkov and Metal Grating FELs (Dartmouth).
9. High Gain Harmonic Generation FEL (BNL, Grumman).

A user's meeting and simultaneous CAP committee meeting are scheduled normally about October every year or two at BNL, depending on the status of the program. In the following we shall describe briefly the experiments that are currently approved for the ATF.

Visible FEL. (Spokesperson I. Ben-Zvi, BNL and SUNY Stony Brook). This FEL oscillator experiment [14] is designed to explore the short wavelength limit imposed by emittance in compact accelerators.

A novel superferric microundulator [15] with a period of 8.8 mm, gap of 4.4 mm and axial magnetic field of 4.7 kG will produce radiation at a wavelength of 500 nm with the 50 MeV ATF beam. This FEL will be used as a test-bed for the Columbia and MIT experiments.

The FEL interaction and resonator design were studied in detail [16]. An output power of 10 MW peak and a gain of about 25% are expected at the ATF beam parameters but with only 50 A peak current.

Grating Accelerator. (Spokesperson R.C. Fernow, BNL). This experiment is designed to investigate new methods of particle acceleration using a short-pulse CO_2 laser as the power source and grating-like structures as accelerator 'cavities' [17].

The CO_2 laser will be focussed to a 3 mm long line on the surface of the microstructure. With 10 mJ of laser energy in a 10 ps pulse, the electric field in the spot will be around 1 GV/m.

Various accelerating microstructures are being developed [18] as well the techniques to guide and diagnose micron sized beams in structures of the same scale. The structures and diagnostics rely on advanced silicon microfabrication techniques that are being explored at BNL.

Nonlinear Compton Scattering. (Spokesperson K.T. McDonald, Princeton U.). This investigation of the nonlinear Compton scattering is the first experiment in a program to study the nonlinear quantum electrodynamics of electrons and photons in an intense electromagnetic wave [19].

The 10 μm CO_2 laser beam with a peak intensity of $>10^{15}$ Watts/cm^2 will be brought into a head-on collision with a single bunch of 50 MeV electrons at the ATF. At such intensities it is probable that an electron absorbs several laser photons before emitting a single photon of higher energy.

In addition, the transverse oscillations of an electron inside the laser beam are relativistic. This leads to a shift in the effective mass of the electron that is discernible in the spectrum of radiated photons.

Inverse Cerenkov Acceleration. (Spokesperson W.D. Kimura, STI Optronics). The goal of this program is to build upon earlier Stanford experiments [20] and demonstrate Inverse Cerenkov Acceleration (ICA), using radially polarized laser beam and axicon focussing [21].

Preliminary scaling and modeling have shown the viability of this method for electron acceleration to high energies [22]. A computer modeling for the ATF design parameters (50 MeV beam, 10 GW peak CO_2 laser power etc.) was performed. A radially polarized laser beam in a medium of low-pressure hydrogen gas was assumed. The result was an acceleration of over 25 MeV in a 20 cm interaction length [23].

Study of Spiking Phenomena in FELs. (Spokesperson T.C. Marshall,

Columbia Univ.) In this research project, spiking phenomena (very short pulses of high-power radiation) will be investigated.

Spiking has been observed at Columbia [24] in mm wavelength and at Stanford [25] in the infra-red. At the ATF spiking will be explored in the visible range. The spike is an energy-compression feature that extracts energy from a group of electrons in the slippage distance and concentrates it into a shorter interval [26].

The short spikes of high-power radiation may have useful applications. Alternatively the spiking can be eliminated by careful adjustment of the resonator detuning.

Pulsed Electromagnet Microwiggler- Driven FEL. (Spokesperson G. Bekefi, MIT). In this research program an FEL experiment will be carried out on the ATF using a 70 period, 8.8 mm/period pulsed electromagnet wiggler with 4.4 mm gap [27].

This wiggler has a planar geometry and separately tunable half periods. A 20 period sub-section of this wiggler has been tested recently [28]. The measured peak axial magnetic field was 4.2 kG at a current (per coil) of 48 amperes. The field uniformity without tuning is 1% rms.

The MIT study of this class of microwiggler has demonstrated 0.4% rms random field error and repetition rates of 1 to 2 Hz with 1 ms half sine waves. This performance is obtained using a very simple technology (water cooling and Microsil ferromagnetic cores).

Cerenkov and Smith-Purcell FEL. (Spokesperson J. Walsh, Darthmouth). This experimental program will be carried out at the low energy beam transport port at energies of less than 4.5 MeV. Various gratings and dielectrics will be used to generate slow waves in the structure that will be synchronous with the electron beam, leading to gain and lasing action at FIR wavelengths.

High Gain Harmonic Generation FEL. (Spokesperson L.H. Yu, BNL). The motivation for this experiment is to present a proof-of-principle for the proposed UV-FEL User's Facility at BNL. In the experiment the frequency of a CO_2 seed laser will be tripled by utilizing two superconducting wigglers and a dispersive
section.

The first wiggler will be used in conjunction with the CO_2 seed laser to generate a ponderomotive force that will bunch the electron beam. The bunching will then be enhanced by the dispersion section. The second wiggler, tuned to the third harmonic of the seed laser will follow. In the beginning of the second wiggler the bunched beam will produce super-radiant emission (characterized by a quadratic growth of the radiated power), then the radiation will be amplified exponentially. The last part of the wiggler will be tapered.

The plan is to study the evolution of the various radiation growth mechanisms as well as the coherence of the tripled and exponentially amplified radiation.

Operational Status. All the electron gun and linac systems have been installed and operated at or close to design levels. The gun has been operated at incident power levels over 6 MW and has reached a cathode surface electric field of 98 MV/m. The measured shunt impedance is 50 M Ω/m.

The Nd:YAG laser has been operated for extended periods. The UV beam (265 nm) on the gun cathode has reached pulse energies of 180 µJ, well above the design value. Photo-ejected electrons were accelerated to 4.2 MeV and the charge and emittance were measured at various energies. The laser pulse width was derived by measuring the energy spread at the low-energy beam transport center. This procedure results in a FWHM laser pulse width of 5 ps and a centroid jitter of 0.8 ps rms over a period of 10 minutes. With a measured charge is about 1.0 nC, this results in a peak photoemitted current of 150 Amps. There is yet an uncertainty concerning the value of the electron pulse length. A streak camera measurement of the electron pulse length will be done shortly.

The measured normalized rms emittance is 4.0 π mm-mrad at a charge of 1.0 nC. A quantum efficiency for the copper cathode was measured to be about 7×10^{-5} [30]

All the ATF transport elements and diagnostics are in place and are operational from the central control computer. The magnetic field quality of all the transport quadrupoles has been measured using a rotating coil and spectrum analyzer to establish the multipole content, magnetic center and excitation curve. The beam optical elements have been surveyed with a tolerance of ±0.1 mm. The linac is ready to accelerate electron beams of up to 50 MeV down to beam stops in the beam tunnel.

The CO_2 laser oscillator and amplifier systems are operational and up to 0.5 Joule pulses have been delivered to experiments at the Experimental Hall.

The user program has started with a measurement of Smith-Purcell radiation at the LEBT beam line and Phase I of the Inverse Cerenkov Accelerator program, which uses the high power CO_2 laser. Currently work is in progress on the beam dump shielding and the experimental hall beam transport elements.

REFERENCES

1. K. Batchelor, H. Kirk, J. Sheehan, M. Woodle and K. McDonald, Proceedings of the 1988 European Particle Accelerator Conference, Rome, Italy, June 2-12, 1988.

2. J. Fischer and T. Srinivasan-Rao, Proceedings of the 4th Workshop on Pulse Power Techniques fo Future Accelerators, Erice, Italy March 4-9, 1988.

3. J.C. Gallardo and R.B. Palmer, IEEE J. of Quantum Electronics **26 no.8** 1328 (1990).

4. I.S. Lehrman, I.A. Birnbaum, S.Z. Fixler, R.L. Heuer, S. Siddiqi, I. Ben-Zvi, K. Batchelor, J.C. Gallardo, H.G. Kirk, T. Srinivasan-Rao, Proceedings of the 1991 International Free-Electron Laser Conference, Santa Fe NM.

5. C.K. Birdsall and A.B. Langdon, Plasma Physics via Computer Simulation, McGrow-Hill, New York (1985) p. 176.

6. X.J. Wang, H.G. Kirk, C. Pellegrini, K.T. McDonald and D.P. Russell, Proceedings of the 1989 IEEE Particle Accelerator Conf. March 20-23, 1989, p.307.

7. R. Zhang, I. Ben-Zvi and J. Xie , to be submitted to Nucl. Instr. and Meth. A.

8. D.P. Russell, K.T. McDonald and B. Miller, Proceedings of the 1989 Particle Accelerator Conference, March 20-23, 1989, p.1510.

9. X.J. Wang and H.K. Kirk, Proceedings of the 1991 Particle Accelerator Conference, San Francisco, May 6-9 1991.

10. J. Rogers and A. Grey, BNL, private communication.

11. M.J.W. Rodwell, D.M. Bloom and K.J. Weingarten, IEEE   J. of Quantum Electronics, **25 No. 4**, April 1989.

12. T. Shimada, I.J. Bigio, R.F. Harrison and N.A. Kurnit, CLEO'89, Baltimore, MD, April 24-28, 1989.

13. I. Pogorelsky, A.S. Fisher, J. Veligdan and P. Russell, Proceedings of the SPIE OE/Lasers 91, Los Angeles CA, January 20-25 1991.

14. K. Batchelor, I. Ben-Zvi, R. Fernow, J. Gallardo, H. Kirk, C. Pellegrini, A. van Steenbergen and A. Bhowmik, Nucl. Instr. and Meth. **A296**, 239 (1990);

K. Batchelor, I. Ben-Zvi, R.C. Fernow, A.S. Fisher, A. Friedman, J. Gallardo, G. Ingold, H. Kirk, S. Kramer, L. Lin, J.T. Rogers, J.F. Sheehan, A. van Steenbergen, M. Woodle, J. Xie, L.H. Yu and R. Zhang, Proceedings of the 1991 International Free-Electron Laser Conference, Santa Fe NM.

15. I. Ben-Zvi, Z.Y. Jiang, G. Ingold, L.H. Yu and W.B. Sampson, Nucl. Instr. and Meth. **A297**, 301 (1990);

I. Ben-Zvi, R. Fernow, J. Gallardo, G. Ingold, W. Sampson, M. Woodle, Proceedings of the 1991 International Free-Electron Laser Conference, Santa Fe NM.

16. A. Bhowmik, N. Lordi, I. Ben-Zvi and J. Gallardo, Proceedings of the International Conference Lasers'90, Dec. 10-14 1990 San Diego Ca.

17. R.B. Palmer, Particle Accelerators **11**, 81 (1980).

18. J. Warren, Proceedings of Material Research Society Sym. **76**, 129 (1987).

19. K.T. McDonald, DOE/ER/3072-38, Princeton University (Sept. 1986).

20. W.D. Kimura, D.Y. Wang, M.A. Piestrup, A. Fauchet, J.A. Eddinghofer and R.H. Pantell, IEEE Journal of Qauntum Electronics **QE-18**, 239

(1982).

21. J.R. Fontana and R.H. Pantell, J. Appl. Phys. **54**, 4285 (1983).

22. R.D. Romea and W.D. Kimura, Phys. Rev. D. **42**, 1807 (1990).

23. W.D. Kimura, STI Optronics, private communication.

24. J.M. Dodd and T.M. Marshall, IEEE Trans. Plasma Sci. **18 No. 3**, 447 (1990)

25. B.A. Richman, J.M.J. Madey and E. Szarmes, Phys. Rev. Lett. **63**, 1682 (1989).

26. S.Y. Cai and A. Bhattacharjee, to be published in Phys. Rev. A.

27. R. Stoner, S.C. Chen and G. Bekefi, IEEE Trans. Plasma Sci. **18 No. 3**, 387 (1990).

28. R. Stoner, MIT, private comunication.

29. I. Ben-Zvi, A. Friedman, C.M. Hung, G. Ingold, S. Krinsky, L.H. Yu, I. Lehrman and D. Weissenburger, Proceedings of the 1991 International Free-Electron Laser Conference, Santa Fe NM.

30. K. Batchelor, I. Ben-Zvi, R.C. Fernow, J. Fischer, A.S. Fisher, J. Gallardo, G. Ingold, H. Kirk, L. Lin, R. Malone, K. McDonald, I. Pogorelsky, D. Russel, T. Srinivasan-Rao, J.T. Rogers, J.F. Sheehan, T. Tsang, J. Sheehan, S. Ulc, X.J. Wang, M. Woodle, J. Xie, and R. Zhang, Proceedings of the 1991 International Free-Electron Laser Conference, Santa Fe NM.

HIGH POWER PICOSECOND CO_2 LASER SYSTEM
for ATF ELECTRON ACCELERATOR PROJECT

I. Pogorelsky

STI Optronics, Brookhaven National Laboratory

Upton, NY 11973

ABSTRACT

A picosecond high power CO_2 laser system is under the operation at the BNL Accelerator Test Facility. The present paper describes its components and the research efforts to meet the requirements for novel laser acceleration techniques to be tested at the ATF.

1. INTRODUCTION

The Accelerator Test Facility[1] at Brookhaven National Laboratory is preparing to test several laser-acceleration schemes. In all cases, the high-power CO_2 laser will interact with a 50 MeV high-brightness e-beam produced by a conventional RF linear accelerator. One program investigates electron acceleration by the inverse Čerenkov effect, first demonstrated at Stanford University.[2] Computer modeling of the ATF experiment predicts that a radially polarized CO_2 laser beam with 10-GW peak power, focused in an axicon-like geometry on the electron beam may result in an acceleration of up to 10 MeV in a 20-cm interaction length.[3] Other methods to be tested using the same CO_2 laser are acceleration on micrograting structures[4] and an inverse free-electron laser.[5]

The high power picosecond CO_2 laser system, being assembled at the ATF (Fig.1) includes: TEA CO_2 laser oscillator delivering 100-ns, 100-mJ linearly polarized gaussian beam of the 10 μm wavelength-tunable radiation; a semiconductor switching system for cutting out a 10...300-ps, $\approx$0.5 MW pulse from the oscillator's output, driven by the picosecond Nd:YAG laser; and a high-pressure, multipass CO_2 amplifier to raise peak power up to 10 GW range.

Oscillator and amplifier were delivered to BNL by Los Alamos National Laboratory. The main activity inside the ATF towards the completion of the picosecond CO_2 laser system includes research on the semiconductor picosecond switching system and multipass amplification through the gain cell. The operation, design and test results for the main components of the ATF CO_2 laser system are described in the following chapters.

2. CO_2 OSCILLATOR AND Nd:YAG LASER

The hybrid single-longitudinal zero-transverse mode TEA CO_2 laser oscillator is the source of the 10 μm beam. Fig.2 presents its principal diagram. The UV-preionized main discharge with a $60\times2\times2$ cm^3 volume is sustained by application of a 40-kV pulsed voltage between two optimally profiled electrodes. A diffraction grating, used as an end reflector in the laser cavity, selects predetermined line in the CO_2 rotational spectrum. A low-pressure auxiliary discharge tube serves to specify a narrow high-gain spectral zone in the center of the selected rotational line and create favorable conditions for single longitudinal mode selection. It is called the "smoothing" tube, because it helps to exclude high-frequency modulation over the output pulse profile due to the mode beating. Final fine cavity length adjustment is performed by the piezo-electric drive of the output mirror. Two intracavity on-axis iris diaphragms are used to select a single low order transverse mode. The output laser pulse energy is 100 mJ, about 50% of which is contained in the leading main pulse with a FWHM duration of 50-75 ns. The maximum pulse repetition rate is 0.3 Hz due to the limitations of the present HV power supply. The beam diameter at FWHM of energy profile is about 3 mm.

ATF CO_2 LASER SYSTEM

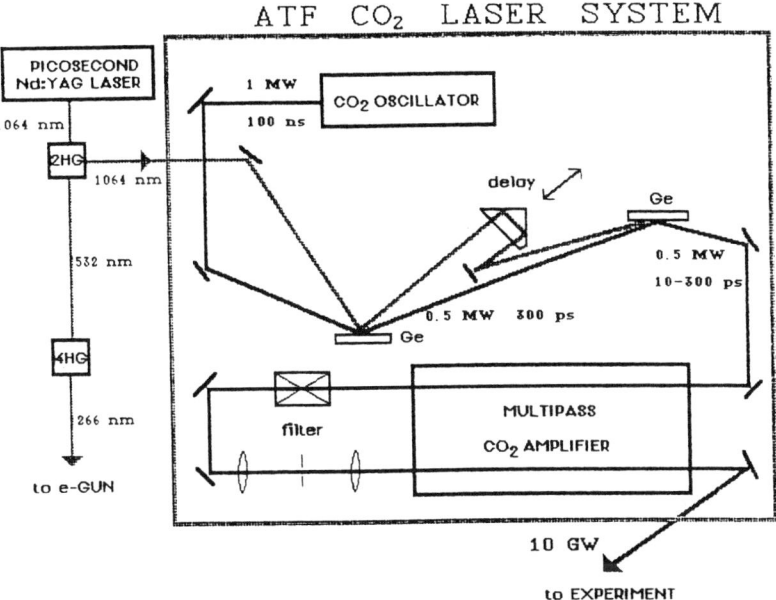

Figure 1: ATF Picosecond CO_2 laser system.

The control beam for ultra-fast amplitude modulation of the output CO_2 pulse on semiconductor switches is provided by actively-mode-locked CW Nd:YAG oscillator, delivering a train of 200-ps 1.06-μ pulses, followed by a fiber for pulse chirping and a regenerative amplifier to select, compress and amplify a single pulse, resulting in 6-ps 10-mJ pulses of linearly polarized light.[10] To drive the photocathode of the linac's electron gun, this radiation is frequency quadrupled to 266 nm. The remaining infrared, about 60% of the initial laser pulse energy, is deflected to the switching system.

Controlling the electron beam and the picosecond CO_2 beam with the same source allows the synchronization essential to laser accelerator experiments. The relative delay times to trigger CO_2 oscillator and amplifier were adjusted so that the few-picosecond 1.06-μm pulse struck the semiconductor switch at the moment of maximum 10-μm radiation intensity at the same point, and the sliced 10-μm pulse passes through the amplifier within the maximum gain period of the discharge.

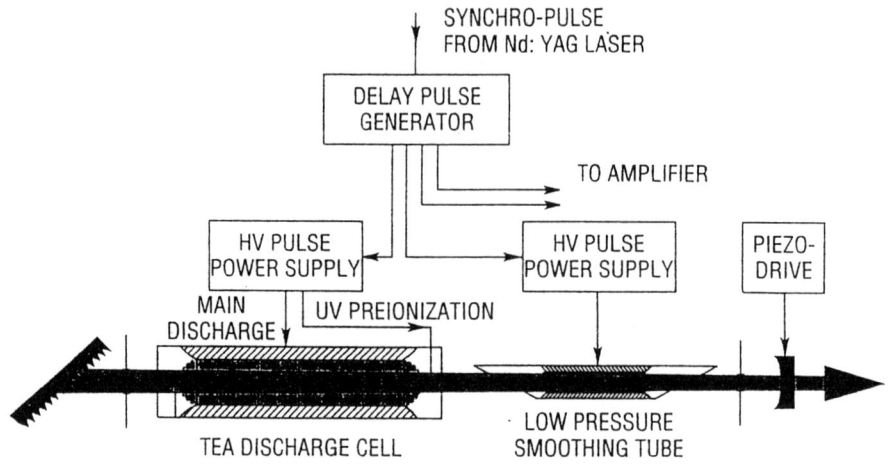

Figure 2: Single mode grating tuned CO_2 oscillator.

3. PICOSECOND PULSE SLICING SYSTEM

3.1. Principles of the switching

The picosecond switching method is based on modulating the reflective and transmissive properties of a semiconductor by optically controlling the free charge-carrier density. A short-wavelength picosecond laser pulse with a photon energy above the band gap of the semiconductor rapidly creates a highly reflective subsurface electron-hole plasma in a semiconductor, such as germanium, which is normally transparent to 10-μm radiation. N is linearly proportional to the absorbed control-pulse energy E by the relation

$$N = \alpha E / (h\nu), \tag{1}$$

where α is the absorption coefficient and $h\nu$ is the photon energy. For an Nd:YAG control laser, $\alpha^{-1} = 1$ μm, and, for a characteristic pulse fluence of ≈ 1 mJ/cm^2, the density of excess free carriers created in Ge is more than 2×10^{19}/cm^3. This is sufficient to "metallize" the Ge so that it switches from a window at 10 μm to a highly reflective mirror. Fig.3a illustrates a dinamics of the semiconductor switching. A Ge plate is illuminated by p-polarized CO_2 laser light at the Brewster angle (76°), to insure a low background reflection.

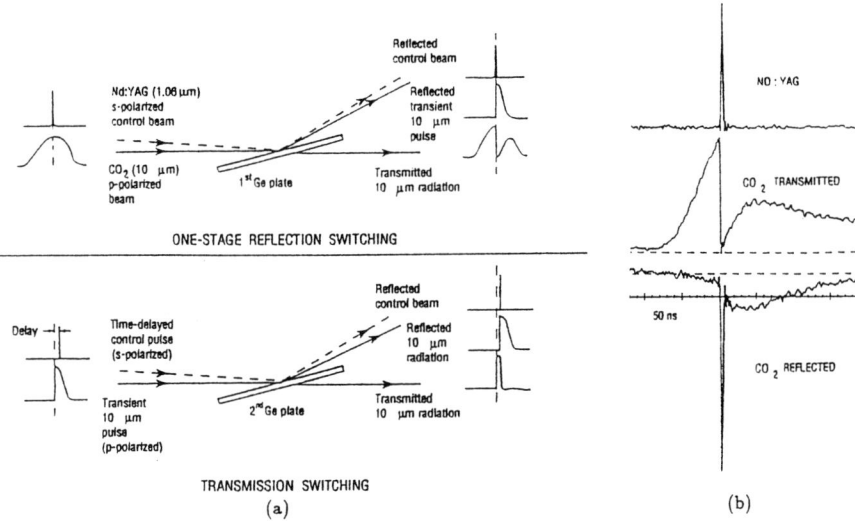

Figure 3: Principle and observations of the CO_2 pulse semiconductor slicing with the Nd:YAG picosecond laser pulse.

(a) - Principal scheme for semiconductor optical switching of a picosecond 10-μm pulse with the 1.06-μm control pulse.

(b) - Oscilloscope traces of the Nd:YAG (upper trace), transmitted CO_2 (middle trace), and reflected CO_2 (lower trace) radiation observed after a reflection switch. Sharp Nd:YAG ($\approx$ 6 ps) and transient reflected CO_2 ($\approx$ 300 ps) pulses are not properly resolved in time.

After the termination of the control pulse, the main process governing the time evolution of the excess free carriers, and hence the reflectivity, will be ambipolar diffusion:[8]

$$N\left(r,t\right) = N\left(r,0\right)\exp\left(\alpha^2 Dt\right)\mathrm{erfc}\left[\alpha\left(Dt\right)^{1/2}\right], \tag{2}$$

where the ambipolar diffusion constant $D = 65$ cm^{2}sec^{-1}, and $N(r,0)$ is the free-carrier density at the end of the control pulse according to Eq.(1). The characteristic time, $\alpha^{-2}D^{-1}$, is then 150 ps. Electron-hole recombination is much slower, with a typical time of $\approx$ 50 ns.[9]

To define the trailing edge of the pulse, shortening it to a few picoseconds, the complement to reflection switching, transmission switching, is used for a second stage. An optically delayed control pulse cuts off the trailing edge of the transient pulse by initiating reflection and absorption with a cross-section[11] $\sigma = 6 \times 10^{-16}$ cm^2. The resulting "sliced" transmitted pulse has the desired few-picosecond length.

3.2. Experimental set up and performance of the picosecond slicer

A detailed analysis and experimental test of the different slicing schemes[12] resulted in a development of the optimized optical set up illustrated by Fig.4. One of the distinctive features of this set up is in a polarization rotation of the control YAG beam. This provides an effective way of utilizing the control pulse with maximum efficiency by optimally redistributing its energy between the reflection and transmission switches. Since the reflection switch is at Brewster angle, by properly polarizing the Nd:YAG beam before each switch, we could arrange to absorb a portion of the energy at the reflection switch and, after an additional rotation, absorb the remainder at the transmission switch.

The semiconductor switches are composed of polished intrinsic Ge slabs placed on high resolution rotary stages. The reflection switch has been placed on two perpendicularly oriented rotary stages giving the alignment in vertical and horizontal planes with the accuracy 0.3 arc-sec. Mirror M_1, placed on the same stage as Ge plate, automatically conserved overall alignment of the slicing system during angular adjustment of the reflection switch. A half-wave plate in the path of the incident control beam regulates Nd:YAG energy consumption at the first Ge plate between 20% and 100%. Reflected by the Ge plate s-polarized Nd:YAG beam is transformed into a p-polarized beam after two passes through the quarter-wave plate. After a variable time-delay introduced by the movable mirror M_4, the Nd:YAG pulse is directed to the second Ge switch, onto the spot irradiated by the transient CO_2 pulse.

The control system makes it possible a monitoring of the CO_2 and Nd:YAG radiation at the various stages of the slicing process within a time resolution of 300 ps (limited by the oscilloscope), 50 nJ in energy, and 0.1 mm over a laser beam's cross-section. Spatial energy distributions in the laser beams are monitored in a single shot by an Electrophysics 5400-00Z pyroelectric camera which is read by Spiricon LBA-100 Laser Beam Analyzer.

The onset of reflection was monitored on an oscilloscope, as shown in Fig. 3b, which shows the Nd:YAG sampled by a photodiode (upper trace), the transmitted CO_2 recorded by a photon-drag detector (middle trace), and the reflected CO_2 pulse recorded by a photoelectromagnetic HgCdTe detector (lower trace).

Experiments were performed with control 1.06-μm energy consumption on the reflection switch varying between 2.5 and 3.5 mJ. After a presize angular alignment of the reflection switch at the Brewster angle, a background reflection of a 100-ns CO_2 insident pulse has not exceeded 1 μJ.

Alignment and performance of the transmission switch is not so much critical for the slicing process. Oscillograms of the Ge transmission at different control energies are shown in Fig. 5. This shows the same dip as that shown in Fig. 3b (middle trace), but with better time-resolution. Fig. 6 displays the experimental dependence of the transmitted CO_2 energy upon the delay time of the control pulse coming to the transmission switch. This curve was obtained with an energy absorption on the reflection and transmission switches of 3 mJ each. Under these conditions, the total energy contained in the transient pulse (infinite delay) was $\approx$100 μJ. Differentiation of this curve gives the transient reflected-pulse shape. Its 200 ps plateau is due to a saturation effect, when the initial free-carriers density immediately after termination of the control pulse is about twice that required for 100% reflection at Ge.

By proper positioning mirror M_4, the 10 μm pulse can be sliced to various picosecond pulse widths with a peak power 0.3-0.5 MW. The shortest length for efficient slicing depends on the transient pulse's rise-time, which is, from these observations, shorter than 20 ps.

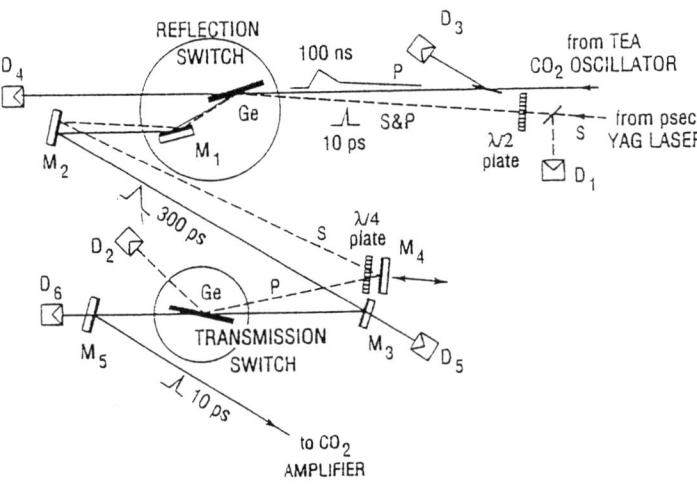

Figure 4: Picosecond pulse slicing with a polarization rotation of the control beam.

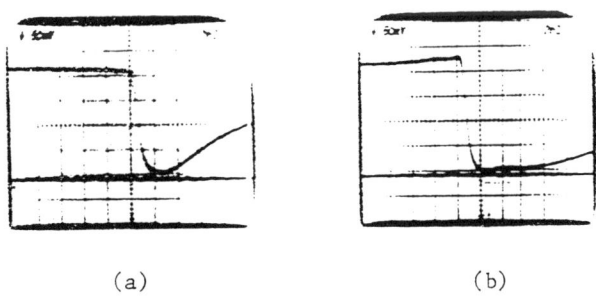

(a)   (b)

Figure 5: Oscillograms of the transmission of the Ge switch obtained with a) 2 mJ and b) 4 mJ absorbed energy of the Nd:YAG laser pulse, observed after the first reflection switch with full-aperture collimated laser beams (2 ns/div).

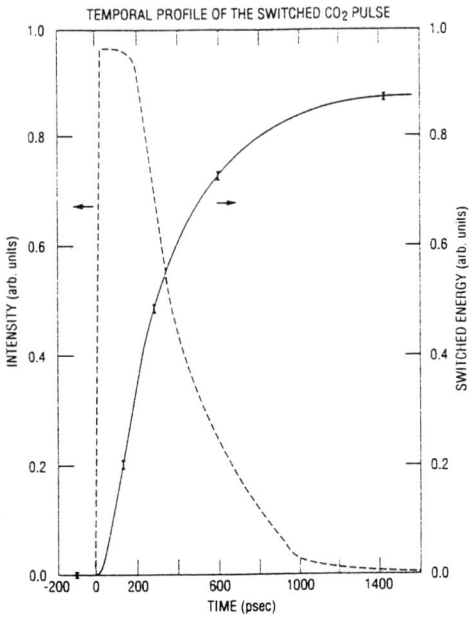

Figure 6: Sliced CO_2 laser energy vs.the arrival-time delay of the Nd:YAG pulse at the transmission switch, relative to the transient CO_2 pulse (right), and the shape of the CO_2 pulse (left) calculated by differentiating the experimental curve.

4. MULTIPASS CO_2 AMPLIFIER

4.1. Specifics of the picosecond pulse amplification in a multiatmosphere CO_2 laser

The small-signal gain spectrum of CO_2 laser is strongly modulated due to its vibrational-rotational (V-R) nature. Near the most intense 10.6 μm P(20) transition, lines are separated by 55 GHz. As a result, the spectrum width of a short laser pulse (less than $\approx$20 ps) may cover several descrete transition lines. The electric field of such an input pulse excites a polarization in the CO_2 molecules, which are in various V-R states. Since molecules in different states are characterized by different frequencies, these polarization components become eventually dephased. As a result, the spectral and time structure of the induced radiation will not be equal any more to those of the initial pulse. It will be revealed in certain distortions of the amplified pulse.

Fig.7, 8 illustrate some results of the picosecond CO_2 pulse amplification computer modeling[13]. Let us give qualitative explanations to these results.

When a short laser pulse propagates in a resonant-amplifying molecular medium, two characteristic time constants and corresponding them spectral constants should be taken into consideration. The first is a time, δt, corresponding to a frequency interval between the centers of V-R lines, which is 18 ps for the P-branch of the 10-μm band. Other is a collisionally induced dipole dephasing time, T_2, related to a Lorenzian line shape of a V-R transition by the ratio

$$\delta\nu^{-1} = \pi T_2,$$

where T_2 is inversely proportional to the gas pressure and, for 1 atm CO_2 gas mixture, is approximately 200 ps.

The spectral width of the individual V-R transition at atmospheric pressure is much less than the spectral separation between lines, and gain spectrum may be considered as discrete. If the input laser pulse width, τ_0, is longer than δt, then just one rotational line is involved in the amplification. During the amplification, the spectrum of the input pulse is filtered by the gain spectrum and eventually pulse duration increases. In the case that the input pulse width is shorter than δt, the laser pulse interacts with several rotational lines. The discrete gain spectrum transforms the spectrum of the input pulse from continuous to discrete, and its Fourier transform corresponds to a periodic with δt short pulse train (Fig.7(b)).

At 10 atm, the pressure broadening effect smooths the discrete gain spectrum fAs a result, or the same pulse width τ_0 =3 ps, a 10-atm case shows smaller pulse splitting as compared with the 1-atm case (see Fig.7(c)). Because T_2 is inversely proportional to the gas pressure, it becomes $\approx$20 ps at 10 atm. Therefore, a 30-ps pulse may propagate without appreciable distortions (Fig.8(b)). In general, if τ_0 is much longer than T_2 (in other words, the spectrum of the input pulse is narrower than the individual V-R line), the pulse can propagate without distortion up to a certain energy level at which a saturation effect shortens the pulse width. The saturation energy level depends upon pulse length and molecular relaxation time constants. In 10-atm, it is close to 0.5 J/cm^2 for both 3-ps and 30-ps pulses.

4.2. Apparatus

A high-pressure multipass CO_2 amplifier pumped by UV preionized high-voltage transverse electric discharge serves to raise the peak power of the sliced picosecond 10 μm pulse from $\approx$0.5 MW up to a required level of $\approx$10 GW. The laser head consists of a cylindrical plexiglas vessel designed to house the main discharge electrodes and the preionizing spark-gap arrays. Serially placed, two sets of main copper electrodes define a discharge volume 2x5x60 cm^3 (x2) = 1,200 cm^3. Preionization spark-gap arrays are arranged along the side walls of the discharge volume. AR coated ZnSe wadge windows have apertures of 65x25 mm^2. Plexiglas cell is nested on an oil tank containing the high-voltage pulse forming circuit. Two sets of 2-stage Marx generators supply up to 140 kV pulses to the main discharge with a specific energy loading up to 650 J/l. UV preionization discharge is loaded with 35 kV pulse voltage. The triggering of the preionization and two main discharge sections is accomplished through five air spark-gaps. By the proper time adjustment of the air spark-gap triggering, we are able to create conditions for the maximum gain in the amplifier just when the picosecond laser pulse propagates through it. Closed loop gas supply manifold includes a 2-stage diaphragm-type compressor with a productivity of 0.8 c.f./min, and a catalytic regenerator. Low rate gas exchange inside the amplifier volume is arranged through two symmetrical axial gas flows directed from the end flanges towards the 10-mm exhaust port in the central portion of the amplifier cell.

Small signal gain up to 2.4 %/cm at 10.6 μm have been reported for this device when operated with gas mixture $CO_2/N_2/He$: 1/1/18 at a total pressure of 4 atm (LANL result).

To avoid the unwanted pulse splitting and simplify the experimental procedure, all the amplification tests were performed with a 200-300 ps pulse delivered to the amplifier just after the reflection switching.

From the modeling results we can see Fig.7(d) that a transfer to a multyizotopic mixture of $^{12}C^{16}O_2$ $:^{12} C^{16}O^{18}O$ $:^{12} C^{18}O_2$ helps to suppress a short picosecond pulse splitting. We plan to implement this approach after the optimization of the multipass amplifier, improvement of the gas purification system, and construction of a single-shot 10-μm autocorrelator. This work is presently in progress.

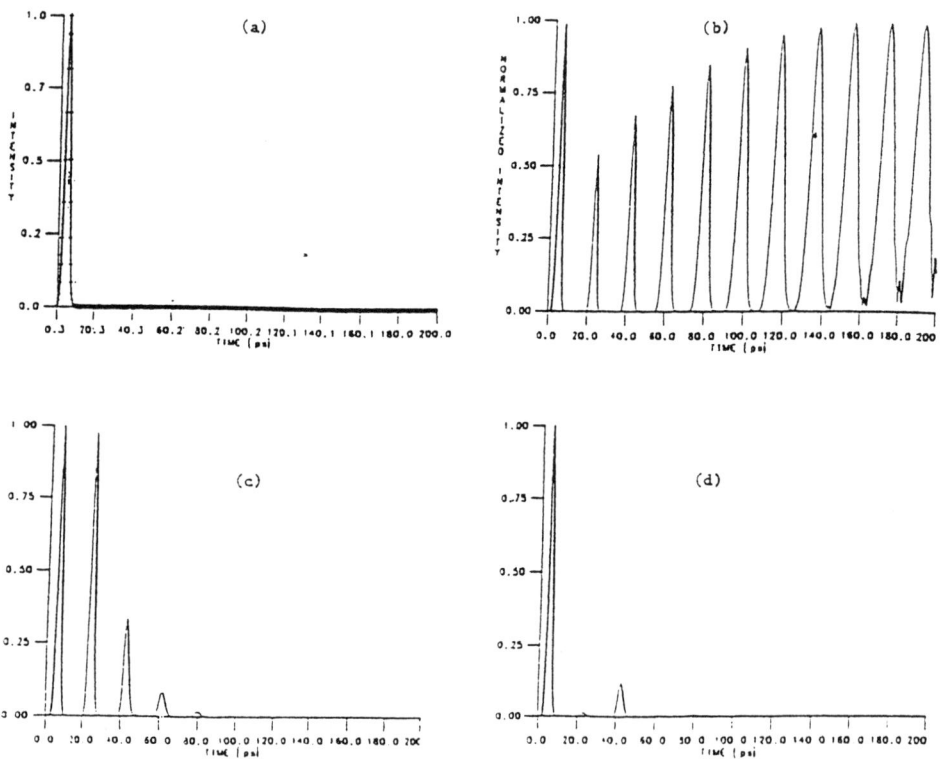

Figure 7: 3-ps CO_2 pulse propagation in the amplifier. Initial intensity 1 MW/cm²; a) initial 3-ps pulse shape; b) after amplification with a total gain $g_0 l = 10$ in a regular 1-atm mixture; c) same in a 10-atm regular mixture; d) same in a 4-atm isotopic mixture.

4.3. Optical set up for the multipass amplifier

The important problems related to the multipass amplifier optical design are: how to satisfy conditions of minimal distortions and losses during the laser beam propagation through the amplifier, and how to prevent parasitic self-lasing under a high energy loading into discharge.

The experimental tests of different optical arrangements confirm the choice of its ultimate version presented by Fig.9.

The specifics of this arrangement are as follows:

a) Laser beam is directed close to the plane of symmetry between the electrodes in order to avoid its distortion and diffraction.

b) A concave mirror with a 20 m radius of curvature is used after 2 passes to compensate beam divergence.

c) The roof mirror positioning at the opposite side helps to prevent the creation of stable cavity conditions. 4 passes may be arranged in this way without self-lasing.

d) Due to the limited aperture of the amplifier windows, no additional mirrors may be placed without creation of a stable cavity and self-lasing. Consequently, the same set of mirrors is used to arrange up to 8 passes.

e) To suppress parasitic feedback through the optics directing the beam for the next 4 passes, an optical polarization isolator is placed on that way.

The isolator consists of two crossed polarizers (Ge Brewster plates) with a Pockels cell between them. By applying the 10-ns $\lambda/2$-voltage pulse synchronously with the picosecond pulse arrival, we open the isolator to transmit the pulse to the next course of amplification. Without the HV activation, the isolator blocks any CO_2 radiation with the extinction ratio 1000:1, which is enough to prevent self-lasing build-up. A spatial filter, placed after the isolator, helps to telescope the laser beam before the final passes and to purify the output spatial and angular intensity distributions.

The aforementioned multipass amplifier configuration is presently under optimization. We have demonstrated so far up to 600 mJ output after 6 passes through the amplifier. Output beam has close to the Gaussian spatial intensity distribution with FWHM ≈ 1 cm. Meeting the requirements for a laser excelerator, we are aiming to the 10 GW output level. The potentials for that are: in reducing the sliced pulse width by the transmission switch, an increase of pass number up to 8, and an optimization of the spatial filter.

5. CONCLUSIONS

1. A high-power CO_2 laser system is assembled and tested at the ATF.

2. CO_2 picosecond pulse slicing system with polarization rotation of the control pulse is designed. The slicing system is capable of delivering ≈ 0.5 MW peak power in pulses of variable duration between 10...300 ps.

3. 8-pass amplification optical set up is assembled.

4. A 10-ns polarization isolator is installed into the multipass amplifier, in order to suppress self-lasing without attenuation of the amplified picosecond pulse.

5. ≈ 3 GW of a peak power in ≈ 200 ps 10.6 μm laser pulse was obtained in 6 passes through the amplifier.

6. Upon the completion of proposed modifications, the CO_2 laser system will be available for the application experiments with output peak power ≈ 10 GW in ≈ 100-ps pulse.

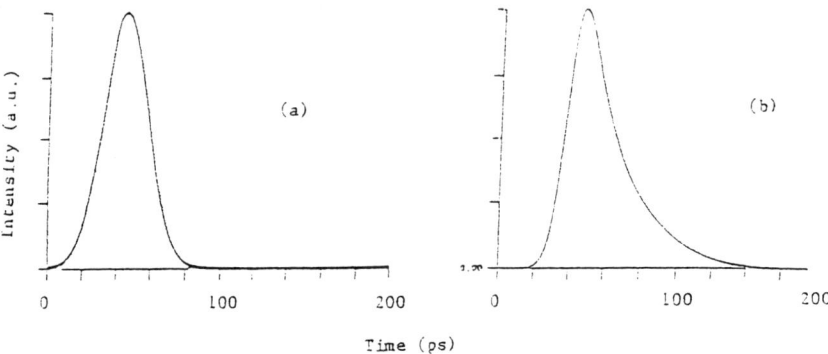

Figure 8: 30-ps CO_2 pulse propagation in the amplifier. Initial intensity 1 MW/cm²; a) initial pulse shape; b) the same after amplification in a 10-atm regular mixture, $g_0 l = 10$.

6. REFERENCES

1. K. Batchelor et al., *Proceedings of the 1989 Particle Accelerator Conference*, March 20-23, 1989, p.273

2. W.D. Kimura, D.Y. Wang, M.A. Piestrup, A. Fauchet, J.A. Eddinghofer, and R.H. Pantell, *IEEE J. Quant. Electron.*, vol.QE-18, 239 (1982).

3. W.D. Kimura, R.D. Romea, S.C. Tidwell, and D.H. Ford, *Advanced Accelerator Concepts, AIP Conf. Proc.*, vol.193, 203 (1989).

4. W. Chen et al., *BNL Report* No. 43465, 1989

5. A. van Steenbergen, J. Gallardo, T. Romano, and M. Woodle, *"1 Angstrom FEL" Workshop*, Sag Harbor, NY, April 22-27, 1990

6. P.B. Corkum, *IEEE J. Quant. Electron.*, vol.QE-21, 216 (1985).

7. S.A. Jamison and A.V. Nurmikko, *Appl. Phys. Letts.*, vol.33, 598 (1978).

8. S.A. Jamison, A.V. Nurmikko, and H.J. Gerritsen, *Appl. Phys. Letts.*, vol.29, 640 (1976).

9. A.J. Alcock and P.B. Corkum, *Can. J. Phys.*, vol.57, 1280 (1979).

10. T. Shimada, I.J. Bigio, R.F Harrison, and N.A. Kurnit, *CLEO '89*, Baltimore, MD, April 24-28, 1989

11. A.F. Gibson, C.A. Rosito, C.A. Raffo, and M.J. Kimmit, *Appl. Phys. Letts.*, vol.21, 356 (1972).

12. I. Pogorelsky et al, *OE/LASE'91*, Los Angeles, CA, January 20-25, 1991; I. Pogorelsky et al, *BNL Tecnical Paper*, BNL-45802 (1991).

13. F. Kanary, *STI Optronics Technical Note*, Draft (1988).

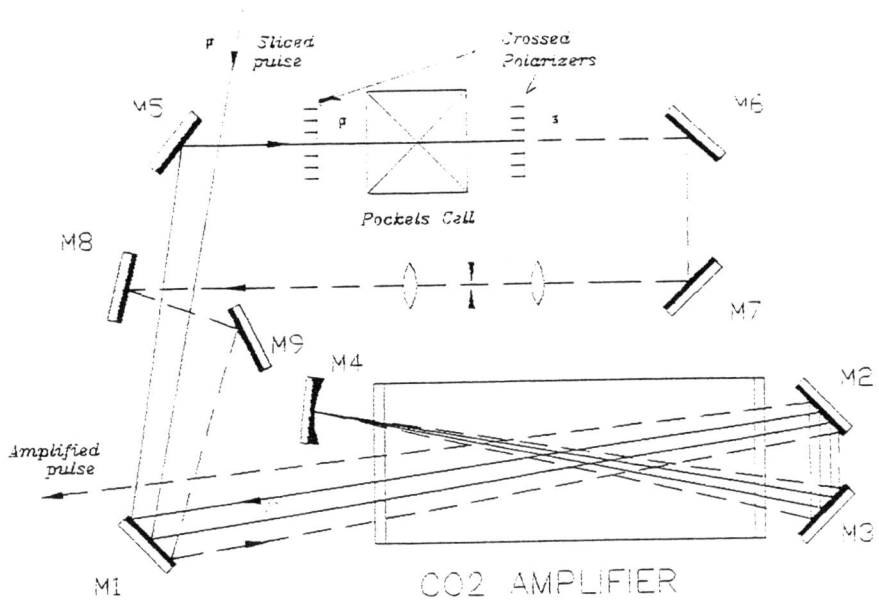

Figure 9: 8-pass optical set up with a polarization rotation.

ACKNOWLEDGMENTS

The author wishes to thank I. Ben-Zvi, A. Fisher, W. Kimura, H. Kirk, N.A. Kurnit, K. Leung, T. Rao and T. Shimada for their helpful discussions and collaboration; J. Fischer and K. Kusche for practical help in modification and maintenance of the CO_2 laser system.

This work was supported by the Department of Energy, Contract No. DE-ACO2-76CH00016 and No. DE-AC06-83ER40128.

PHOTOEMISSION FROM Ag, Cu, AND CsI

T. SRINIVASAN-RAO, J. FISCHER, T. TSANG
BROOKHAVEN NATIONAL LABORATORY, UPTON, N.Y. 11973

ABSTRACT

Photoemission characteristics of three different cathodes, CsI, Ag film and Cu were investigated. CsI, upon irradiation by 213 nm, 10ps laser pulse yields a quantum efficiency of 4% at $0.2\mu J$ input energy. The saturation mechanism observed at higher input energies require further investigation. Ag film, upon irradiation by 630 nm, 300 fs laser emit prompt photoelectrons after absorbing 2 photons. There was no evidence of optical damage of the film up to 10^{11} $w/_{cm}{}^2$. At low intensities, photoemission from Cu is a simple $\nu\text{-}e^-$ interaction, the nonlinearity of the process depending strongly on trace impurities. At higher intensities, there appears to be a change in the emission mechanism.

INTRODUCTION:

In recent years there has been considerable interest [1-3] in using photoemission technique to generate high quality, low emittance and high brightness electron beams for high energy colliders and short wavelength free electron lasers. These applications call for a high quantum efficiency (~1%), rugged, long life-time photocathode material that is capable of operating in 10^{-8} Torr pressure, using a simple, short pulse laser. So far no single photocathode material has met all these requirements simultaneously. Our investigation of CsI was motivated by its capability to meet most of the above requirements. There are also other applications such as the switched power high brightness gun [4] and positron production by $e^-\text{-}\gamma$ and $\gamma\text{-}\gamma$ interaction [5] that require electron bunches of subpicosecond duration. Our investigation of Cu and Ag using femtosecond laser pulses were motivated by these applications. In the subsequent sections the results of experiments with CsI irradiated by 266 nm and 213 nm laser pulse of ~ 10 ps duration and Ag film and Cu irradiated with 630 nm laser pulse of 300 fs are discussed.

MEASUREMENTS ON CsI USING PS LASER PULSES

EXPERIMENTAL ARRANGEMENT:

Preparation of the CsI photocathode was as follows. Bulk CsI was vapor deposited on diamond turned Cu to a thickness of 8000 Å, transported in an evacuated container and the electrodes were assembled together with minimum exposure to air. The anode was a hollow cylinder of 6 mm inner diameter, the opening nearest to the cathode covered with a copper mesh of 70 lines/inch. The electrodes, spaced apart by 1 mm, were then set in the vacuum cell and baked at 100 - 140°C overnight.

The anode can be biased up to a voltage of 10 kV, but normally maintained between 2 to 4 kV to limit the possibility of high voltage breakdown in vacuum.

The laser used to irradiate the CsI was an actively and passively mode-locked Nd-YAG laser operating at 1.06 μm, with ~25 mJ energy in 25 ps duration. Radiation at 0.266 μm and 0.213 μm were obtained using different nonlinear crystals. It was possible to obtain up to 1.5 mJ in 12 ps and 100 μJ in ~10 ps at wavelengths 0.266 μm and 0.213 μm, respectively. The two wavelengths were separated using a fused silica prism and only one of the wavelengths was directed to illuminate the cathode. A small percent of the laser beam was deflected to a calibrated photodiode to monitor the energy of the irradiating beam. The photoemission signal was extracted from the cathode and displayed on the oscilloscope after passing it through a charge sensitive preamplifier and a shaping amplifier when needed. Fig. 1 illustrates the experimental lay-out.

MEASUREMENTS WITH 0.266 μm RADIATION:

The dependence of the charge on the input laser energy was determined by varying the laser energy and observing the corresponding charge emitted and the results are displayed in Fig 2. The slope of this logarithmic plot, an indicator of the number of photons involved in the process, indicates that the photoemission at this wavelength is a single photon process and the quantum efficiency is ~ 2×10^{-6}. Considering the facts that the absorption coefficient of CsI at this wavelength is ~ 1.7 cm^{-1}, [6] the thickness of the CsI in our sample is 8000 Å and the substrate is copper whose quantum efficiency at this wavelength is very close to this measured value, the observed photoelectrons could have originated from the substrate and not from CsI. However, this fact needs to be verified by varying both the thickness and the material of the substrate. The capability of these electrons to tunnel through the CsI layer also needs to be investigated.

MEASUREMENTS WITH 0.213 μm RADIATION:

Fig.2 also illustrates the intensity dependence of the photoelectrons at 0.213 μm radiation. The slope of the plot indicates that photoemission at this wavelength requires two photons. The quantum efficiency at 0.2 μJ incident laser energy is ~4 %. However, as the laser energy approaches ~ 1 μJ, the electron yield appears to saturate and become independent of the incident laser energy. This saturation is unlikely due to space charge effects as evidenced by the voltage dependence data. Fig. 7. Similar effect has been observed on cesium activated GaAs [7] and has been attributed to the formation of plasma and subsequent modification of the work function. The validity of this explanation for CsI is currently under investigation. Measurements at both wavelengths agree with results from other laboratories [8].

MEASUREMENTS WITH SILVER USING FEMTOSECOND LASER PULSES:

The preparation of the silver cathode was similar to that of CsI. 3100 Å thick silver was evaporated on a diamond turned Cu mirror. The cathode was transported in an evacuated cell and the electrodes were assembled and mounted in the vacuum cell. As before, the anode was a hollow cylinder, but with the opening nearest to the cathode covered by a 40 lines/inch nickel mesh. The electrode gap was 1 mm.

The laser used in these experiments was an amplified colliding pulse mode-locked (CPM) laser. The output characteristics of the laser are as follows: wavelength was 630 nm, energy was 1 μJ, pulse duration was 300 fs, focused spot size, FWHM, on the cathode was 100 x 70 μm^2.

Figs. 3 and 4   illustrate the intensity and the voltage dependence of the electron emission respectively. As can be seen from the intensity dependence data, photoelectric emission at this wavelength is a two photon process.   The data taken with the unamplified CPM and the amplified CPM mesh smoothly without any normalization indicating the absence of any other nonlinear mechanism. The maximum current density measured with this cathode was 4000 A/cm^2. The maximum charge extracted was limited by the available laser energy. The measured quantum efficiency was low, however, it has been proven[9] that at grazing incidence, the presence of the electric field of the (p- polarized) laser enhances the quantum efficiency to a level comparable to that of a single photon process, which is ~10^{-5} electrons per incident 2 eV photon for Ag.

The slope of the voltage dependence curve (Fig. 4) is high, implying strong modification of the work function due to Schottky effect. The slopes of these curves for various input laser intensities scale with the input intensity linearly, supporting the evidence for the two photon process.

The pulse duration of the electron bunch was measured using the autocorrelation technique. The laser beam was split into two equal halves, passed through delaying optics, and recombined on the target. One beam was delayed with respect to the other by translating its delaying optics with a fine stepper motor. The electron signal was measured as a function of the delay and displayed in Fig. 5. The autocorrelation trace of the laser pulse is also displayed along side for reference. As can be seen from Fig. 5 the electron pulse duration is comparable to the laser pulse duration. Due to the low signal to noise ratio encountered in these measurements, it is not possible to verify the origin of the slight broadening of the electron pulse length. The silver film was not optically damaged even at intensities of 10^{11} W/cm^2.

MEASUREMENTS WITH COPPER USING FEMTOSECOND LASER PULSES:

The experimental arrangement for these measurements were similar to that for silver. Fig. 6 illustrates the intensity dependence of the photoemitted electrons. Experiments with two different copper samples indicate that nonliniearity of photoemission from copper under visible radiation is significantly affected by trace embedded and surface impurities . The   presence of CuCl could provide a long lived intermediate state [10] and cause a 3 photon process to appear as a 2 photon process and the presence of trace amounts of Si could change the process from two photon to a single photon process. At low laser intensities, the photoemission requires two photons for the sample without trace amounts of Si and one photon for the sample with traces of Si. For a number of Cu samples investigated, as the laser intensity increased beyond 10^{10} W/cm², the number of photoelectrons increased abruptly by thousandfold. The sharp onset of this behavior and the amplitude fluctuation of the laser energy precluded using autocorrelation technique for pulse duration measurements. However, the lower limit on the pulse duration can be estimated. For Ag, the onset of space charge limited photoemission was at 100 fc and the corresponding measured pulse duration was $\simeq$ 250 fs. For copper, even at a charge of 100 pc, the space charge limit is not reached for similar electrode geometry. Hence, the pulse duration of the electrons have to be larger than 250 ps. The time resolution of our measurement system is 1 ns and the oscilloscope traces did not indicate pulses longer than 1 ns. Hence the pulse duration of this bunch must lie between .25 and 1 ns, much longer than the laser pulse duration. Similar behavior from copper with longer pulse duration has been observed previously [11] and attributed to the formation and expansion of plasma from the copper surface. The short pulse behavior is being analyzed at present.

CONCLUSIONS

Photoemission from CsI indicates that quantum efficiencies exceeding 1% could be obtained from rugged materials. The operating wavelength, although high, is not too difficult to achieve. Since CsI is an insulator, the number of available electrons may not be as high as a metal cathode. Hence the capability of CsI to yield short bunch electron charges >> 1 nC has to be investigated in detail. Photoemission measurements on Ag and Cu using femtosecond laser pulses indicate that high current electron pulses in short bursts are achievable with reasonable quantum efficiency. However, further studies, such as effects of electromagnetic fields associated with the laser, non-uniform heating of the electron, plasma production, and emittance characteristics are required before adopting it in other applications.

REFERENCES

1. J. S. Fraser, R. L. Sheffield, E. R. Gray, P. M. Giles, R. W. Springer, and V. A. Loebs, Proceedings of the 1987 Particle Accelerator Conference, Washington, D.C., 1987, p. 1705

2. K. Batchelor, H. Kirk, J. Sheehan, M. Woodle, and K. McDonald, Proceedings of the European Accelerator Conference, Rome, Italy, 1988, p.954

3. J. Madson, CERN, Private Communication

4. R. B. Palmer, Proceedings of the Switched Power Workshop, Shelter Island, U. S. A., 1988, BNL 52211, p. 263

5. R. B. Palmer, Presented at this Workshop

6.   P. Avakian and A. Smakula, Phys.Rev. _120_, 2007, (1960)

7.   Max Zolotorev, SLAC, Private Communication

8.   G. Suburlucq, CERN, Private Communication

9.   J. P. Girardeau-Montaut, C. Girardeau-Montaut, S. D. Moustaizis and C. Fotakis, Submitted for publication in Phys. Rev. Lett.

10. W. F. Krolikowski, Ph. D. Thesis, Stanford University, 1967, p. 231

11. X. J. Wang, T. Tsang, H. Kirk, T. Srinivasan-Rao, J. Fischer, K. Batchelor, P. Russell, and R. C. Fernow, J. Appl. Phys., _72_, 888, (1992)

FIGURE CAPTIONS

1.   Experimental arrangement.

2.   Intensity dependence of photoelectrons from 8000 Å thick CsI on diamond turned Cu. The bias voltage was 3.5 kV across 1 mm gap. o: laser wavelength is 266 nm and spot size is 3 mm diameter. Δ: laser wavelength is 213 nm and the spot size is elliptical with major and minor axes 2.5 and 2 mm respectively.

3.   Intensity dependence of photoelectrons from 3100 Å thick Ag on diamond turned Cu. o: are using unamplified CPM of pulse duration 100 fs Δ: are with amplified CPM of pulse duration 300 fs.

4.   Applied field dependence of photoelectrons from 3100 Å thick Ag on diamond turned Cu with laser energy as a parameter. m is the slope of the line and the ratios listed are m_1/m_2.

5.   Autocorrelation signals of (Δ) electrons and (+) laser. The real width is autocorrelation width/$\sqrt{2}$ for a Gaussian profile.

6.   Intensity dependence of photoelectrons from diamond turned Cu cathode. The bias voltage was 2.5 kV across 2.75 mm gap. The laser pulse duration was 250 fs and the focal spot size was 70 x 100 μm^2 FWHM.

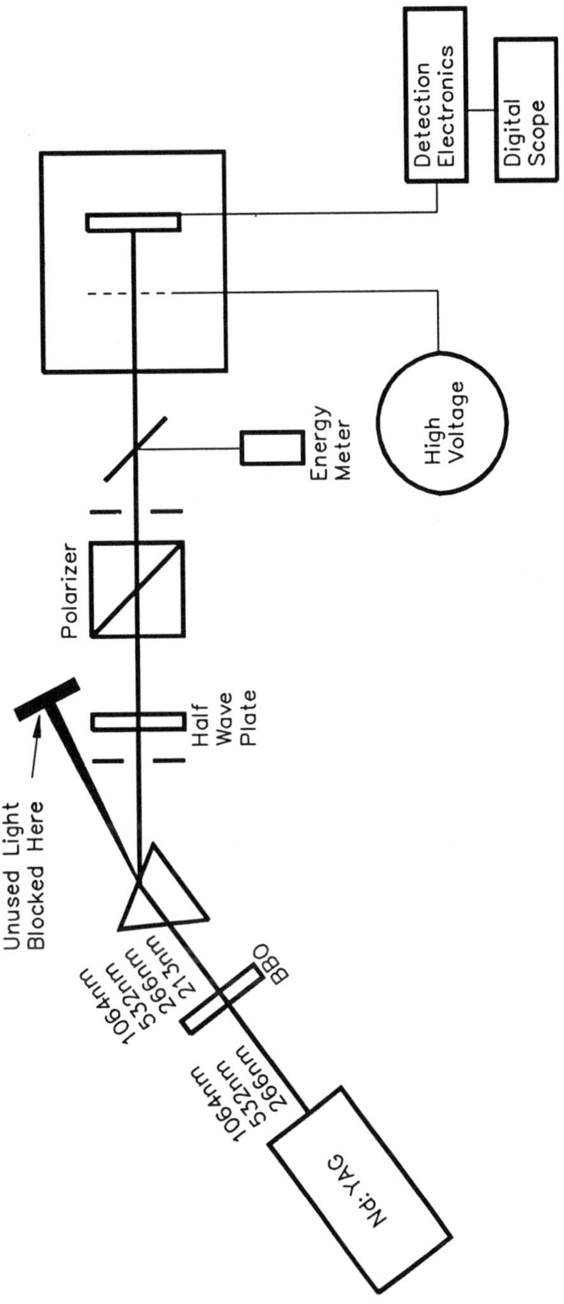

Figure 1. Experimental Arrangement

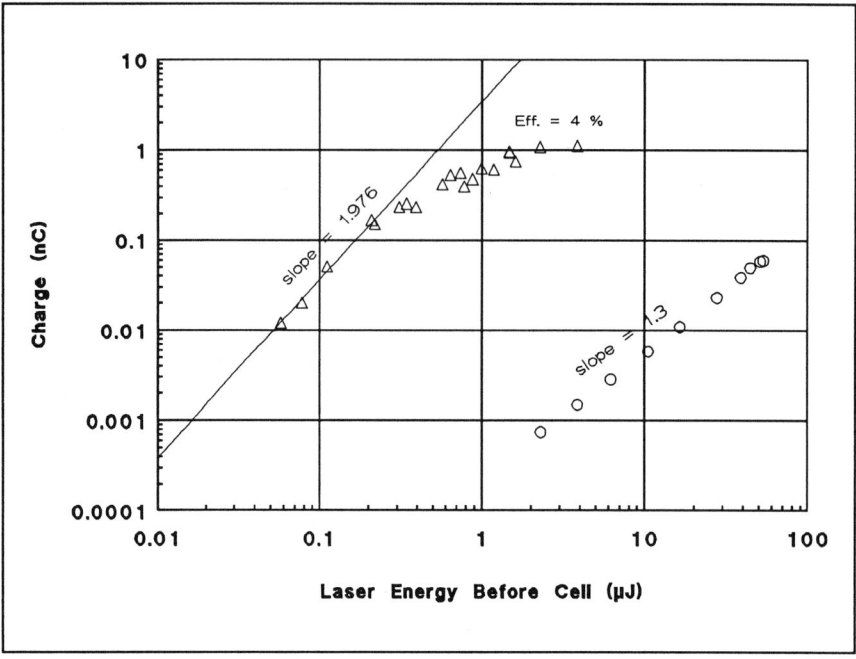

Figure 2.   Intensity dependence of photoelectrons from 8000 Å thick CsI on
diamond turned Cu. The bias voltage was 3.5kV across 1mm gap.
O: laser wavelength is 266nm and spot size is 3mm diameter.
Δ: laser wavelength is 213nm and the spot size is elliptical
with major and minor axes 2.5 and 2 mm respectively.

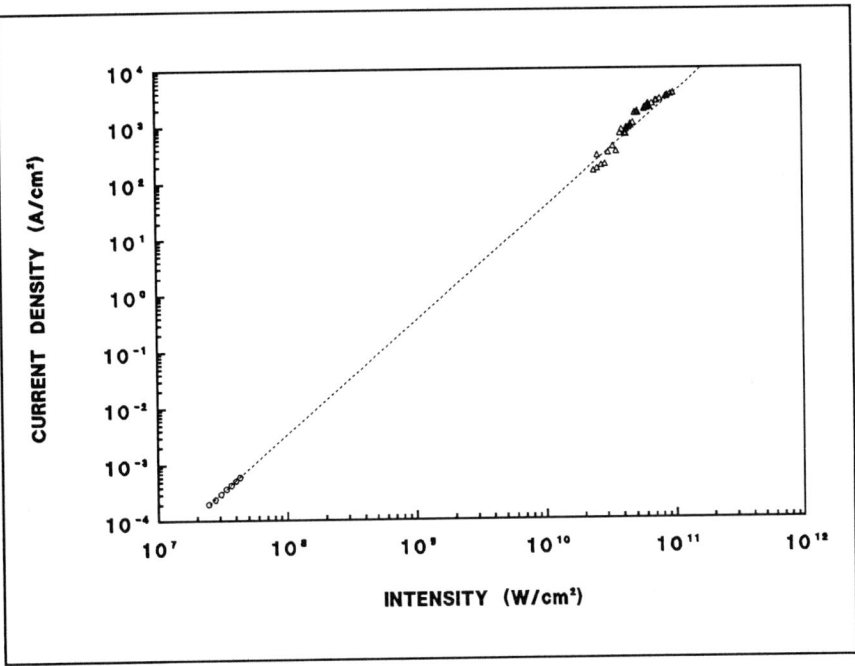

Figure 3. Intensity dependence of photoelectrons from 3100 Å thick Ag on diamond turned Cu.
O:unamplified CPM laser of pulse duration 100 fs
Δ:amplified CPM laser of pulse duration 300 fs

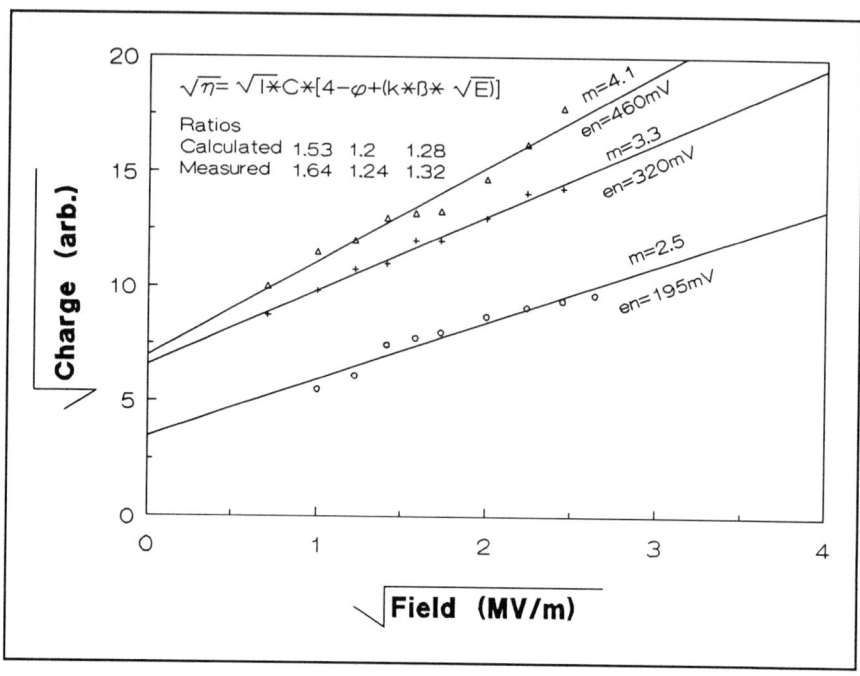

Figure 4.   Applied field dependence of photoelectrons from 3100 Å thick Ag on diamond turned Cu with laser energy as a parameter. m is the slope of the line and the ratios listed are m_1/m_2.

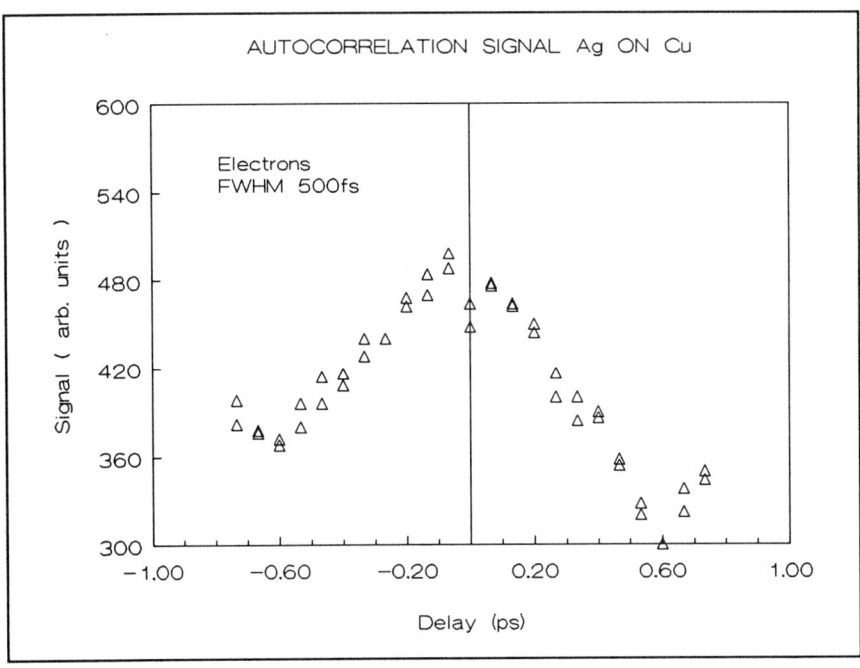

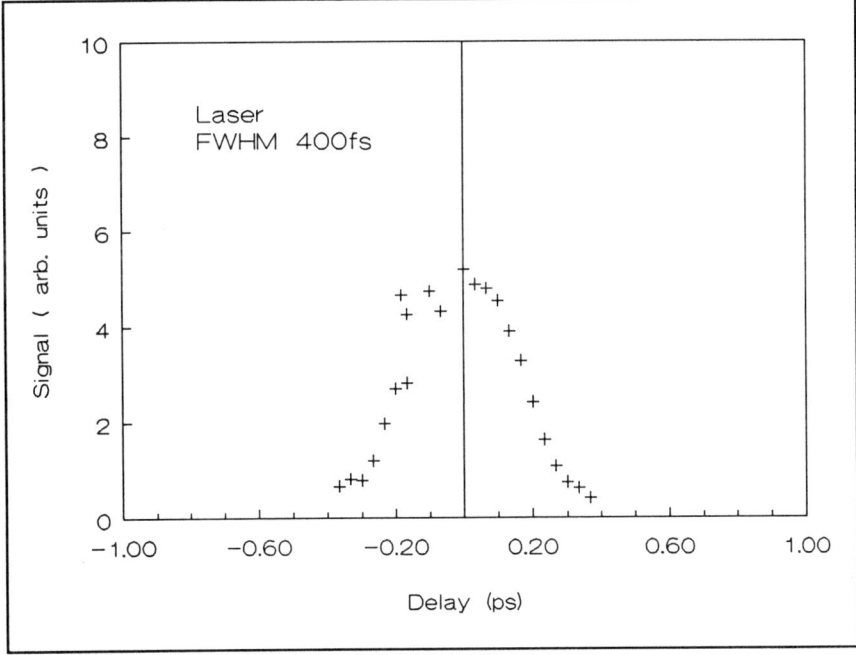

Figure 5.   Autocorrelation signals of (Δ) electrons and (+) laser.

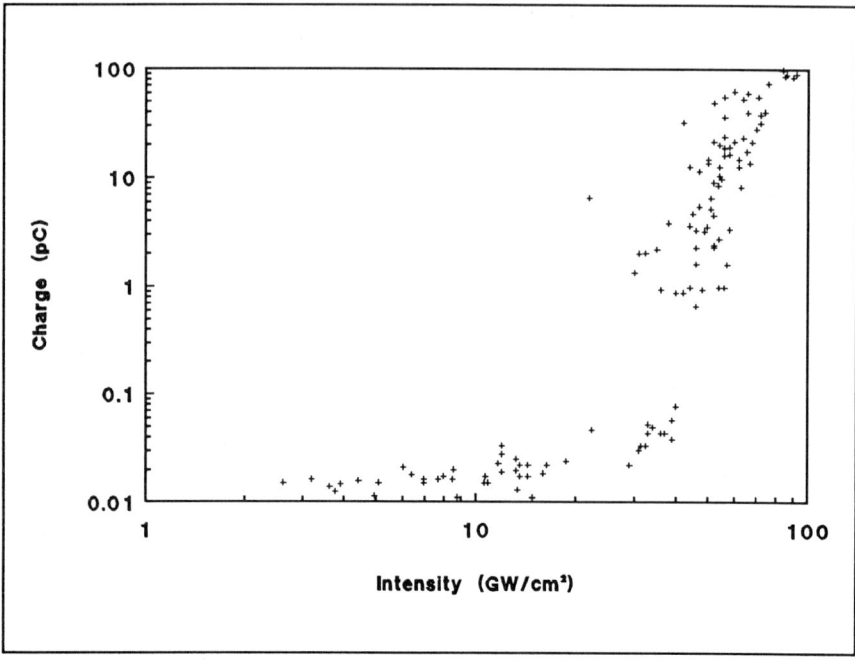

Figure 6.   Intensity dependence of photoelectrons from diamond turned Cu cathode. The bias voltage was 2.5kV across 2.75 mm gap. The laser pulse duration was 250 fs and the focal spot size was 70 x 100 μm^2 FWHM.

PHOTOELECTRON BEAMS FROM THE UCLA RF GUN*

S. Park, N. Barov, S. Hartman, C. Pellegrini, J. Rosenzweig
P. Tran, G. Travish, R. Zhang
Department of Physics, University of California, Los Angeles, CA 90024

P. Davis, G. Hairapetian, C. Joshi, N. Luhmann Jr.
Department of Electrical Engineering
University of California, Los Angeles, CA 90024

ABSTRACT

A high brightness, low emittance photocathode rf gun is starting operation at UCLA as an injector to a 20 MeV linac. This linac will initially be used to drive FELs, plasma wakefield accelerators, and to test plasma lenses. The gun is a $1\frac{1}{2}$ cell π-mode standing wave structure running at 2.856 GHz, and has a copper photocathode. In the initial commissioning of the gun, photoelectron beams of up to 2.5 nC at 4.5 MeV have been produced. We report on the current status of the system, experimental data taken with 50 ps UV laser pulses, and plans for the future.

INTRODUCTION

In this paper we describe the status of the 20 MeV high brightness, compact linear accelerator at UCLA. The ~4 MeV photoinjector gun which will be used to drive a linac system consists of a photocathode rf gun producing a 4 MeV beam, followed by an accelerating structure to raise the energy to about 20 MeV. The beam from the linac can be sent in two beam lines, where experiments utilizing this beam can be performed. For initial measurements, a simplified beam line has been set up (Fig. 1). The experimental program is oriented towards: the development of high brightness beams from photocathodes and rf guns; plasma acceleration and focusing, FELs and other areas[1,2] where small emittance and high brightness are important.

An rf gun with a photocathode is thought to have lower emittance and higher brightness, compared to the more traditional thermionic guns. The lower emittance is due to higher acceleration gradient and optimal phasing of the laser with respect to the rf pulses. The high brightness arises from high charge extraction in a short bunch length, as well as low emittance. Ideally, the spatial and temporal distribution of the photoelectrons at the cathode should be entirely determined by the laser beam profile impinging upon the cathode, so that the electron beam can be completely shaped by means of compression, focusing, and polarization of the laser beam.

In this paper we will describe the characteristics and the present status of the rf gun, the laser illuminating the photocathode, the rf system and controls, the beam line and diagnostics. We will then present some initial measurements

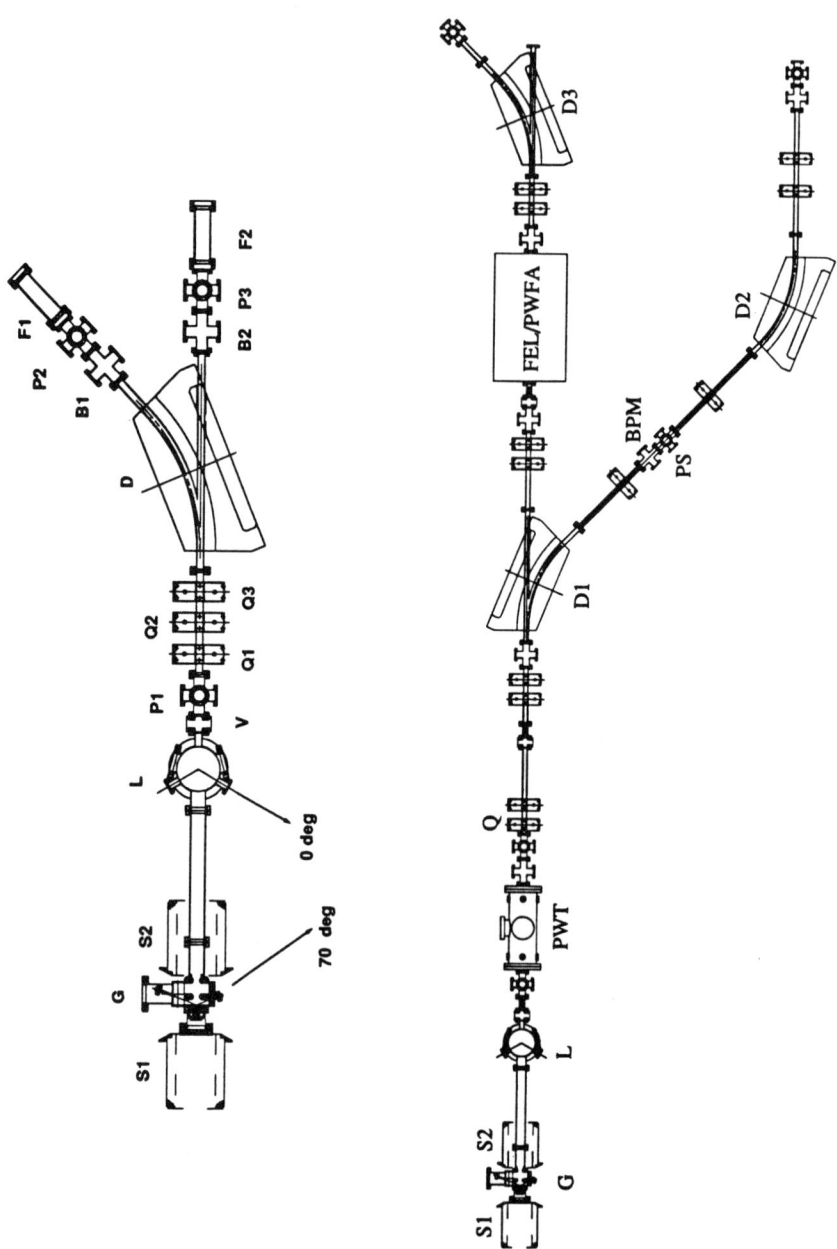

Figure 1: Setup for the quantum efficiency measurements (left) and plan for the future expansion (right). Some components shown here are: gun(G), laser coupling box(L), solenoid(S), dipole(D), quadrupole(Q), BPM(B), phosphor screen(P or PS), and Faraday cup(F). The FEL/PWFA box in the right figure represents the location of the free electron laser or a plasma wakefield accelerator experiment.

on the photoelectron beam produced by the gun when approximately 50 ps long laser pulse irradiated the photocathode.

PHOTOCATHODE RF GUN

The gun design is based on the Brookhaven[3] approach. Optimal coupling of rf power to the cavity, balancing of electric field between the half cell and full cell, and frequency tuning are facilitated by adjustment of cathode position, tuner and rf probe at each cell. The cathode position is controlled through a micrometer and can be finely adjusted. The Q values of the two cells can be determined from the decay of the rf electric field following turning off of the rf. These results show that in the present configuration there is an unbalance between the two cells as seen in Figs. 2(a) and 2(b), which may indicate degradation of the rf coupling to the gun.

The gun can be initially tuned at atmospheric pressure. Once it is installed in the system and pumped down, atmospheric pressure deforms the cavity, resulting in detuning. This is compensated for by passing temperature controlled water through cooling channels of the gun. The rf power reflected from the gun is measured as a function of the gun temperature. With 20 dB attenuation, the input power to the gun is about 100 kW. This level of power is too low for radiation to be of concern, yet sufficiently high for detection through a 50 dB waveguide directional coupler as shown in Fig. 2(c). For minimum reflection, which corresponds to resonance, the temperature is set.

The photocathode is machined from a solid block of OFHC copper to a tolerance of less than 0.001 inch in surface roughness. No active polishing was done. It was baked *in situ* with the temperature no higher than 100°C for about three days. When the cathode is exposed to air for a few hours for beamline modification, conditioning by 5 MW rf and UV laser pulses of up to 200 μJ for a day at 1 Hz seems to restore the surface for photoemission.

LASER AND OTHER OPTICS

The laser system starts with a mode locked Coherent Antares Nd:YAG laser at 1064 nm wavelength. A crystal oscillator at 38.08 MHz drives both the mode locker and gives a seed signal to be multiplied by 75 times to 2.856 GHz for the RF system. The beam at 1064 nm is passed through a 2.2 km optical fiber via self phase modulation for spectral broadening and linearization of the chirp via group velocity dispersion before it enters a Continuum Nd:Glass regenerative amplifier, which is pulsed at 5 Hz by Stanford Research DG-535 pulser, which triggers the RF system simultaneously. The beam from the amplifier can be compressed to less than 4 ps by a grating pair and frequency doubled twice to a wavelength of 266 nm by two KD*P (potassium dihydrogen phosphate) crystals. This UV beam is transported through a maze of radiation shielding and then to the cathode, either via a window at the gun for a 70° angle of incidence or via the laser coupling box for a 2.5° angle of incidence. All the windows for the UV beam passage are made of quartz.

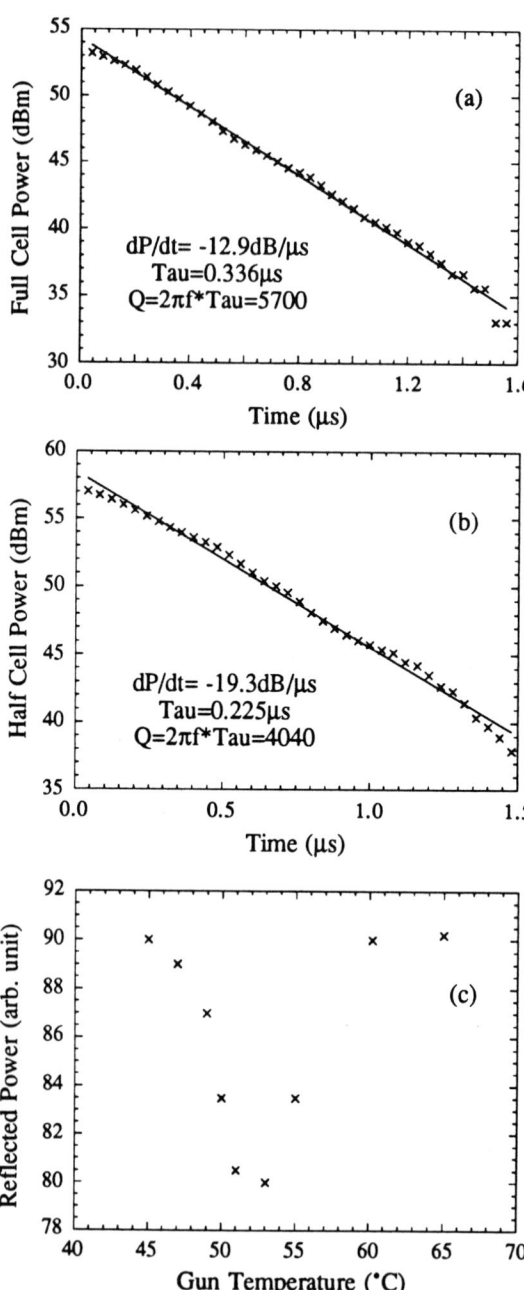

Figure 2: The Q value measurements for full cell (top) and half cell (middle), and the rf power reflected from the gun as a function of gun temperature (bottom).

The beam leaves the laser with a linear polarization. Passing through a half-wave plate, the plane of polarization is rotated by an angle of 2θ if the plate is rotated by θ. Here the quantity of interest is the projection of the wave electric field onto the surface of the cathode. The laser beam spot size on the cathode can be controlled by changing the focal length of the beam. The laser beam has Gaussian distribution in time and space.

While laser timing jitter has been minimized by adding a feedback stabilizer between the laser and mode-locker, the problem of amplitude fluctuation at the final UV output has not been rectified so far. Allowing the amplitude to fluctuate more than 50%, the laser beam energy leaked at a prism is monitored by a photodiode at every shot. This photodiode was calibrated against a pyroelectric detector, since the latter needs too much of beam energy to be a useful online diagnostics. There is a linear correlation between the two. Other diagnostics include autocorrellator, spectrometer, and streak camera.

RF SYSTEM AND CONTROL

The RF system consists of low level RF driver, high power RF, and modulator[4] that drives the klystron. The operating frequency of 2.856 GHz is derived from a crystal oscillator at 38.08 MHz by using a 75 times multiplier. This multiplier is a frequency synthesizer based on a phase-locked loop(PLL). The mode locker provides the rf phase as a reference to the PLL. The PLL output frequency is down converted by a Mod-N counter. The counter output is phase compared with the reference, with the amplified and loop filtered difference driving a voltage controlled oscillator. The stability of this multiplier is as good as the mode locker, and the spectral purity is such that all the other frequencies are at least 70 dB down from the 2.856 GHz center frequency. Concerning the amplitude, the biggest cause of variation is the ambient temperature. If the temperature is stable within 1°C, the output power will remain constant within 0.25%.

A coaxial phase shifter controls phasing of the laser pulse with respect to the rf phase, both of which are monitored on a digital sampling oscilloscope. This phase shifter is followed by a long cable to a pulsed preamplifier and a 1 W cw amplifier, which drives a number of beam position monitors (BPM)[10]. The Pro-Comm preamplifier consists of one 30 dB gain solid state cw amplifier and three stages of 7 dB tuned–cavity triode amplifiers. This amplifier has a rise time of about 1 μs. In order to supply a flat top input signal to the klystron, the amplifier is turned on 2 μs earlier than the modulator trigger, and stays on for about 8 μs. When the drive amplifier is on, with the modulator not yet turned on, the absence of the electron beam in the klystron cavity leads to a gross missmatch of input to the klystron resulting in most of the input power being reflected. Subsequent turn on of the modulator reduces this reflection to less than 10%. Monitoring of the forward and reverse signal between the pulsed amplifier offers an excellent diagnostics of both the low level system and the klystron amplifier. A number of isolators placed at various stages provide protection against excessive reflection of rf power to the previous stage.

About 300 watts of pulsed rf power is fed into a SLAC XK-5 klystron for up to 30 MW of final rf power with 2.5 μs pulse length. This power is divided evenly into two branches by a 3 dB coupler so that half of it drives the photoinjector through a variable attenuator and the other one half drives the PWT Linac[6,7] through an attenuator and a phase shifter. The phase shifter sets the phase relationship between the photoinjector and the PWT linac. Since the acceleration gradient of both the gun and linac is proportional to the square root of the input rf power, stability of klystron output power is compelling in terms of reproducibility for the control of beam energy. With the input rf power being fairly stable in frequency, phase, and amplitude, the dc power from the modulator needs to be controlled in such a manner that the shot to shot variation is minimal.

Timing jitter of the modulator thyratron is less than a few tens of nanoseconds and is not a problem. The pulse forming network (PFN) consists of $N = 20$ stages of $C = 14$ nF capacitors and $L = 0.686$ μH inductors. The characteristic impedance of the PFN is $Z_0 = \sqrt{L/C} = 7.0\Omega$ with a pulse length of $\tau = 2N\sqrt{LC} = 3.92$ μs. Variable inductors are made of 0.375 inch od copper tubing wound to form a multi-turn, 5 inch od helix. Each capacitor with a 50 kV rating has a capacitance variation less than a few percent from specified value. Each tapped inductor is adjusted in such a way that the resonance frequency $\omega = \omega_0/\alpha$, where $\omega_0/2\pi = 1.624$ MHz, $\omega_0 = 1/\sqrt{(0.686\mu H)(14.0 nF)}$, and the actual capacitance $C = (14.0\alpha)$ nF. After this is done for each stage, the PFN is charged up to 5 V and, in place of a thyratron, a switching transistor is triggered at 20 Hz or higher to dump the PFN energy to a 7.00 Ω, 0.25 watt resistor. For a high power test of the modulator, a pair of copper plate electrodes immersed in a tank full of cupric sulfate solution makes a high power resistor. The voltage waveform across that 7 Ω load is monitored while further adjustment of inductance values are made, to produce a flat top pulse at the load. This is to match the impedance of each stage and of overall PFN to that of klystron's pulse transformer primary winding. Since the cathode is pulsed at about 300 kV, a 1% of cathode voltage variation results in 1.5° of rf phase change. For optimal rf waveform out of klystron, a feed–forward system[5] may be introduced at the low level side to minimize ripple in phase and amplitude.

Following the PFN tuning, there is still one more problem to solve: The AC line voltage needs to be regulated. Typically, the three phase 480 V line voltage is lowered by a ganged variac, then stepped up and full wave rectified to produce up to 50 kV of dc, which is the PFN voltage. This line voltage fluctuates by as much as ±5%. The PFN voltage regulation is done as follows. A voltage divider measures the PFN voltage in real time to register 1 volt per 10 kV. A multi-turn potentiometer sets a reference voltage derived from a precision voltage reference. The difference is amplified, dc level shifted, and controls an SCR gate driver by feedback controlling the conduction angle of each of six SCR arrays. In this way, any overshoot or repetition rate dependence of the PFN voltage is avoided. The SCR gate driver is available commercially, and the PFN voltage regulation is better than 0.1% of the set voltage.

Other than voltage control by variacs, there is no active regulation of electric power to the klystron filament. Line voltage fluctuation, as simulated by the change in variac setting, within a few percent has no appreciable effect on the klystron output. The electron emission off the klystron cathode is not temperature limited, but it is space charge limited. Only the cathode voltage is an important parameter. The same is true for the thyratron. The heater and reservoir electric power, and fluctuation thereof, have little effect to the closing characteristics of the device in terms of switching time and impedance at conduction as far as the reservoir voltage is not excessively high to cause self triggering.

Square law crystal detectors are employed for rf power measurements at various stages of interest. Calibrations are made on each detector for output voltage versus rf power. Since detectors, as well as some thermistor probes of powermeters, are not 50Ω devices, higher voltage standing wave ratio (VSWR) can lead to erroneous results, unless enough attenuation is introduced to minimize the VSWR at the rf source. The output voltage V_d of a crystal detector, when terminated by a 50Ω load, and the rf power $P_{\mu w}$ as seen by the detector have been found to have the following relationship. $P_{\mu w}/\text{dBm} = \sum_{i=0}^{n} a_i X^i$, where $X = \log_{10}(V_d/\text{mV})$, and a_i are the coefficient to be determined by a least square fit. Usually, $n = 3$ is sufficient for the power ranging from -10 dBm to 15 dBm.

BEAMLINE AND DIAGNOSTICS

The beamline consists of the rf gun followed by a solenoid focusing magnet and a drift tube, laser coupling box, quadrupole triplet, a dipole bending magnet, and a beam dump, which also serves as a Faraday cup. The solenoid is matched with a bucking coil at the back side of the gun, to provide a field free region on the surface of the cathode. The solenoid can produce up to 3 kG of axial field. There are a number of diagnostics placed along the beamline, which will be described later.

The entire assembly sits on two laser optics tables measuring 3 feet by 5 feet each.Each element has fiducial marks inscribed at the top and/or midplane. A transit pointed along the beam axis detects any misalignment. There are several six–way vacuum crosses to accommodate diagnostics and pumping ports. In line gate valves separate the beamline into sections so that venting during the installation or removal of control and diagnostics is localized.

The entire setup is covered by layers of lead bricks to provide radiation shielding. This shielding is sufficient for beam energies up to 5 MeV. However, for linac operation at energies of $\simeq$20 MeV, neutron generation becomes significant. It is therefore planned to build a concrete structure of three feet thickness and house the beamline inside, with lead bricks remaining for local shielding.

Measurements of rf and laser beam power have been described in the previous sections. At the beamline, short bunches of photoelectron beams out of the rf gun are focused by a solenoid. When they pass through the laser coupling box, the first diagnostic is a Cherenkov radiator mounted on a pneumatic actuator. The radiation is in visible range, and the image is captured by a streak camera to be stored for further processing. This is for bunch length measurements and the projection of the image on the $y = 0$ plane shows the side view of the bunch. Streak camera measurements are currently underway.

Following the radiator is a phosphor screen for beam profile measurements. The phosphor layer is bound to an aluminum base plate by a chemical agent, and a rectangle is scribed on it so that, when viewed from a right angle to the beam axis, it appears to be a square, providing a reference length for beam size determination. The aluminum piece is a cylindrical object machined at 45° to its axis. It is also mounted on a pneumatic actuator. By floating the aluminum, actuator, and flange electrically, this profile monitor also serves as a Faraday cup. A CCD camera is paired with each phosphor screen at the beamline.

As nondestructive diagnostics, beam position monitors(BPMs)[10] are placed adjacent to each phosphor screen. The probe part of the BPM consists of four copper strips running parallel to the beam inside a beam pipe, and the downstream end of each strip is terminated to the ground. These are basically pickups of the electric fields of the electron bunches. A hybrid produces sum and difference from a pair of strips, left/right or top/bottom. Through mixers, these signals are down converted by 2.856 GHz to low frequencies, which are subsequently amplified and integrated. The sum represents the total bunch charge, whereas the difference indicates the transverse position of the bunch with respect to the geometric axis of the BPM. Steering magnets are used to match the beam axis to the BPM and the magnetic center of quadrupole magnets.

Faraday cups are placed at the ends of the beamline to measure the total charge of the beam: One is at the end of straight line, the other at the end of branched line after the dipole. In order to capture all the electrons at relativistic energies, a Faraday cup is made of an aluminum block in the shape of circular cylinder. The cylinder is two inches in height along the beam axis, and one inch in diameter. In its simplest form, a Faraday cup is a capacitor, which is charged up by the beam electrons. The capacitance is increased by a coaxial cable running from the cup to a scope. The 1 MΩ oscilloscope input impedance is a bleeder resistor from the circuit point of view. The charge q collected by the cup is given by $q = CV$, where V is the oscilloscope voltage. If the cable is terminated by a 50 Ω load, the current i is $i = V/50$, which is useful only for a dark current measurement within a single rf pulse.

Another diagnostic device for the measurement of photoelectron charge is an integrating current transformer (ICT). It is a capacitively shorted transformer with a core made of thin ribbons of cobalt/molybdenum amorphous alloy. As the beam passes through the probe, the capacitor is charged up, followed by discharge through the primary winding. The secondary is terminated by a parallel connection of a capacitor and a 50 Ω for readout. The response has time delay and stretching of the pulse length, but the amplitude of the response is proportional to the charge of the bunch, which may have a subpicosecond rise time. This probe has been cross calibrated with the BPM and Faraday cup, resulting in good linear relationship among them. This is a nonperturbing and independent source of data for the measurement of photoelectron beam bunch charge. Figure 3 shows

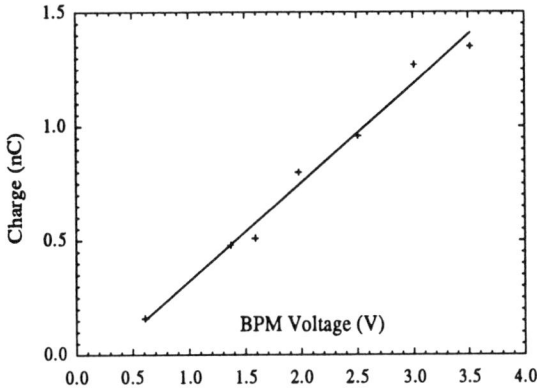

Figure 3: Calibration of the BPM using the ICT as a reference.

the bunch charge as measured by the ICT plotted against the BPM output.

INITIAL ELECTRON BEAM MEASUREMENTS

With the present system configuration[8], we can change the power level in the gun, and the rf phase with respect to the laser pulse, which set the cathode acceleration gradient. Temporal profile of the rf pulse remains the same as long as the PFN tuning remains unchanged. All the other rf parameters such as coupling, frequency, and cavity Q value are not subject to change during the photoelectron beam measurements.

Unlike the rf drive power, the laser beam parameters can be completely changed with relative ease. The angle of incidence, wave polarization, pulse length, and intensity have all been varied individually to study their effect on photoelectron generation. The solenoid focusing was extensively used to maximize the charge collection. In this section we report some of the results obtained at a pulse duration of 50 ps.

Given the large electric field present in the gun, an electron beam (dark current) is produced even when the cathode is not illuminated. Propagating the dark current through the beam line produces an energy spectrum as shown in Fig. 4. Each data point represents one rf pulse. Using a dipole magnet and a Faraday cup as a single channel spectrometer, the distribution would have been vertical somewhere between 2.9 and 3.1 MeV, if there was no shot to shot fluctuation of the rf power. The dark current increases strongly with rf power (Fig. 5). The temporal profile of the dark current depends on the rf power to the gun and so does the duration of it. For the case of 8.3 MW, 24 nC over 2 μs corresponds to 12 mA of average current and the charge per rf period is only 4 pC, whereas the charge of a photoelectron bunch is about 1 nC over a small fraction of an rf period.

Fig. 6 shows two images on phosphor screen captured in a sequence; one with a laser beam and one without. The first one has contribution from the dark

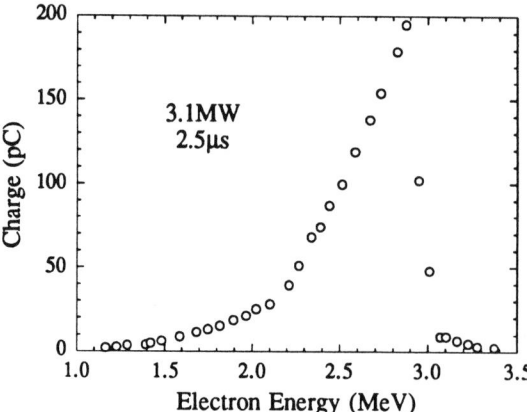

Figure 4: Energy spectrum of dark current.

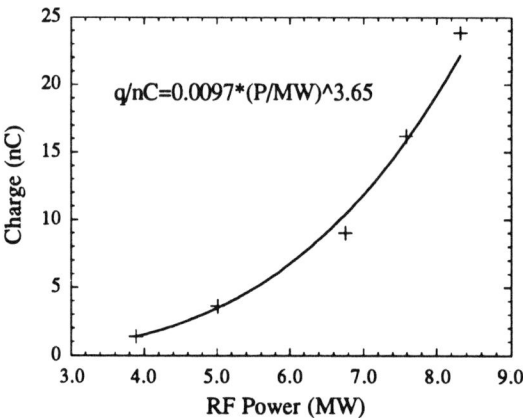

Figure 5: The rf power dependence of dark current.

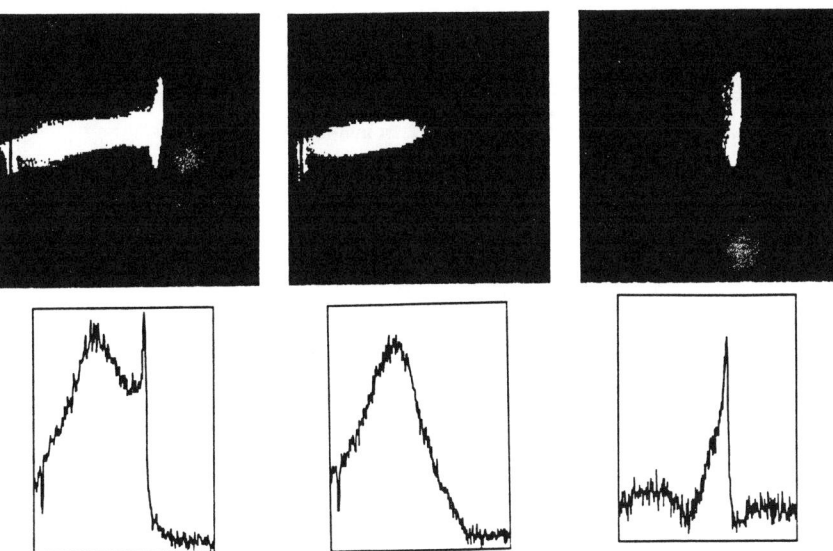

Figure 6: Image(top) and corresponding distribution(bottom) of photoelectron + dark current (left, *a*), dark current only (center, *b*), and photoelectrons (right, $c = a - b$). The horizontal axis represents energy dispersion of the beam.

current and the photocurrent, while the second has only dark current. This one is subtracted from the first to remove the effect of dark current. The horizontal full scale on each frame of Fig. 6 represents about 4% of momentum spread. The vertical axis of the distribution is proportional to the electron density in arbitrary units. A monochromatic beam would have resulted in a delta function distribution. For the frames of image, horizontal is momentum space and vertical is real space. The head and tail of this long pulse beam are subjected to a different rf phase and acceleration. The consequence of this is manifested in high emittance and larger momentum spread of $\Delta p/p \simeq 0.5\%$.

A preliminary emittance measurement with the 50 ps long pulse was done to test the beam transport and diagnostic system. As shown in Fig. 1, the beam out of the gun passes through a quadrupole triplet, a dipole, and then makes a profile on a phosphor screen. The beam transport matrix elements are evaluated by using calibration data of dipole and quadrupole. The beam spot size is measured as a function of current through a quadrupole. A least square fit of this data according to the transport matrix provided a normalized emittance of 12π mm–mrad for this 50 ps, 3.0 MeV photoelectron beam as shown in Fig. 7.

For a given laser beam energy per pulse with the pulse length fixed, one effect of varying the spot size is to change the local power density, and the other is varying the area in which photoelectrons are borne. While electrical conductivity of copper is one of the highest among metals, a characteristic heat conduction time

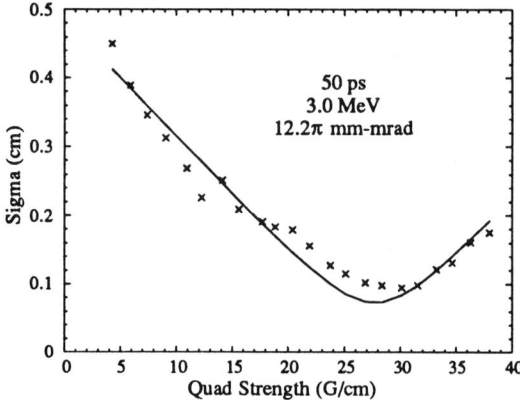

Figure 7: Emittance measurement of 50 ps, 3.0 MeV photoelectron beam.

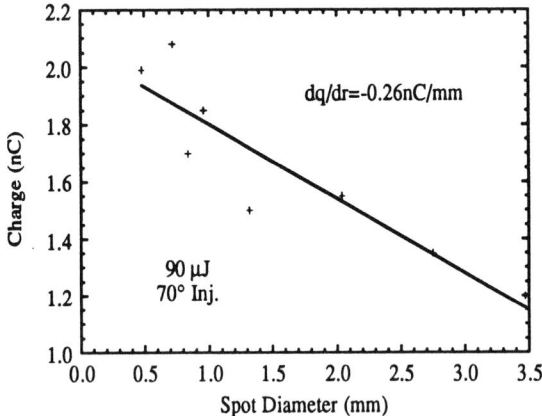

Figure 8: Dependence of photoelectric charge on laser beam spot size on the cathode. Laser energy was kept constant and the angle of incidence was 70°.

is not short enough to conduct away the heat generated on the cathode by laser beam of shorter than 1 ns. Laser beams of pulse lengths much shorter than this time scale develop localized hot spot, which may be of thermionic significance. When the temperature of the lattice exceeds the melting point of copper, it is believed that the laser–rf combination initiates copper plasma that leads to an explosive emission. The probability of this event to occur begins to be appreciable for laser intensity of about 1 GW/cm^2, according to the experiments done at Brookhaven[9].

Our measurements below this intensity limit are shown in Fig. 8. For a 70° angle of incidence, the profile of the laser beam on cathode is elliptical, with its major axis in horizontal direction and the major diameter is $1/\cos 70° = 2.92$ times the minor diameter. When the spot size is small, most of the particles are near the axis so that para–axial beam trajectories are ended at the Faraday cup, leading to higher collection efficiency and thus higher apparent quantum efficiency. As the spot size is made larger, more off–axis particles are lost before they reach the Faraday cup. In this case, a non-zero radial electric field of the rf waves would contribute to this process. Also, small size of the tubing between the window and cathode scrapes off some of the laser energy.

The present status and preliminary experimental results on quantum efficiency can be summarized as follows:

- The rf gun was driven by up to 8.3 MW of rf power. Laser beam pulses of 50 ps, 266 nm with energies up to 100 μJ were used to generate photoelectron beams with up to 2 nC of charge.

- At a given laser energy, more electrons are collected from a smaller spot size. But the efficiency is saturated.

- Preliminary quantum efficiency measurements indicate that the efficiency is that measured by other groups at 0°.

For lower energy spread within a bunch, shorter bunch length is favored. If the pulse length is too short, excessive intensity of the laser may result in explosive emissions and a damaged cathode. The apparent quantum efficiency is based on collected charge. Therefore, the collection efficiency must be understood in terms of beam dynamics, with emphasis on space charge effect and trajectories of off the axis particles. Also, microscopic studies of the cathode and measurements of its thermal response will lead to a better understanding of photoemission characteristics of the cathode.

CONCLUSIONS

We expect to compress the laser pulse to 4 ps in the near future, and perform a complete series of measurements on quantum efficiency at 70° and 2.5°, beam emittance, energy spread and pulse length. Once the gun is completely characterized we will accelerate the beam to 20 MeV and start the experimental program on beam–plasma interaction and FEL.

The linac is expected to be operational in 1993. In the mean time, every effort will be made to achieve the best quality beam in terms of low emittance and high brightness. To that end, theoretical and experimental studies, along with reinforcement of experimental environment, will be continued.

ACKNOWLEDGEMENT

This work has been supported by US Department of Energy under Grant DE-FG03-92ER-40493 and by Office of Naval Research through ONR-SDIO Grant # N00014-90-J-1952.

REFERENCES

[1] J.W. Dodd, *et al.,* Proc. Part. Accel. Conf. (San Francisco, 1991) p. 2751

[2] P. Chen and J. M. Dawson, in *Laser Acceleration of Particles,* C. Joshi and T. Katsouleas, Eds., AIP Conf. Proc. No. 130, p. 201

[3] C. Pellegrini, in *High Gain, High Power Free Electron Laser,* R. Bonifacio, L. De Salvo Souza and C. Pellegrini, Eds., (Elsevier, 1989)
Also, K.T. McDonald, IEEE Trans. Electron Dev. ED-35(1988) 2052.

[4] W. R. Bradford, *et al.,* "Modulators," in *The Stanford Two-Mile Accelerator,* R. B. Neal, Ed., (Benjamin, New York, 1968) pp. 411-462

[5] I. Ben-Zvi, *et al.,* Proc. Part. Accel. Conf. (San Francisco, 1991) p.1323

[6] D. A. Swenson, European Part. Accel. Conf. (Rome, 1988) p. 1418

[7] S. Hartman, *et al,* IEEE Part. Accel. Conf. (San Francisco, 1991) p. 2967

[8] S. Park, *et al.,* Proc. 16th International Linac Conf (Ottawa, Canada, 1992)

[9] X.J. Wang, *et al.,* J.Appl.Phys. **72**, 888 (1992)

[10] J.T. Rogers *et al.,*, BNL Report, unpublished (August, 1991)

IMPROVING THE BEAM QUALITY OF RF GUNS BY CORRECTION OF RF AND SPACE-CHARGE EFFECTS

L. Serafini
INFN and Università di Milano - Via Celoria 16 - 20133 Milano - Italy

ABSTRACT

In this paper we describe two possible strategies to attain ultra-low emittance electron beam generation by laser-driven RF guns. The first one is based on the exploitation of multi-mode resonant cavities to neutralize the emittance degradation induced by RF effects. Accelerating cigar-like (long and thin) electron bunches in multi-mode operated RF guns the space charge induced emittance is strongly decreased at the same time: high charged bunches, as typically requested by future TeV e^-e^+ colliders, can be delivered by the gun at a quite low transverse emittance and good behaviour in the longitudinal phase space, so that they can be magnetically compressed to reach higher peak currents. The second strategy consists in using disk-like electron bunches, produced by very short laser pulses illuminating the photocathode. By means of an analytical study a new regime has been found, where the normalized transverse emittance scales like the inverse of the peak current, provided that the laser pulse intensity distribution is properly shaped in the transverse direction. Preliminary numerical simulations confirm the analytical predictions and show that the minimum emittance achievable is set up, in this new regime, by the wake-field interaction between the bunch and the cathode metallic wall.

INTRODUCTION

The operation of laser-driven RF guns to produce high brightness electron beams has been experimentally demonstrated in the past years at a number of laboratories[1], indicating that such injectors are the most promising sources both for future linac-based Free Electron Lasers in the X-UV domain and electron-positron colliders in the TeV energy range. For both applications intense electron beams are required with very low emittance and low energy spread, in order to meet either the high-gain operation requirements of an FEL or the luminosity specifications of a TeV collider.

The first generation of RF guns being successfully in operation, several techniques and/or strategies have been recently proposed to overcome present limitations on the beam quality (brightness, energy spread, etc), particularly in order to attain the ultra-low emittance domain (less than 1 mm·mrad). Future generation laser-driven injectors shall be designed, in order to meet the extreme requirements coming from future FEL and collider scenarios, according to some of such new techniques, as long as these will be proved to be capable of a significant breakthrough.

This paper is devoted to the presentation of some recently proposed ideas based mainly on a manipulation of the RF field in the gun cavity and/or a proper shaping of the laser pulse illuminating the cathode: other techniques, mainly the emittance correction scheme using a solenoid lens, have been proposed elsewhere[5].

In section 1 we review the main scaling laws for the beam dynamics in a standard RF gun, mainly coming from a previous work by K.J.Kim[2], which have been found in good agreement both with experimental measurements and numerical simulations. Main aim of this section is to give a comprehensive discussion of basic mechanisms limiting the maximum brightness achievable with standard RF guns.

In section 2 we discuss two techniques able to correct the linear correlated emittance produced by RF fields during the acceleration in the gun cavity. The first one is based on the lengthening of the first half-cell in the gun cavity: a better behaviour in the longitudinal phase space is achieved, giving rise to higher compressibility of the beam at the gun exit. The second one consists in the exploitation of an unsymmetrical cavity added downstream the RF gun, able to recover the RF induced emittance down to the minimum value even for injection phases different from the optimum one: the injection phase becomes in this way a free parameter, to be independently optimized according to other requirements on the beam quality.

Section 3 is dedicated to the multi-mode operated RF gun. An analytical study, found in quite good agreement with numerical simulations, shows that the emittance growth induced by RF fields can be cancelled up to fourth-order terms when the RF gun cavity is operated with a harmonic $TM_{012-\pi}$ mode. This idea, recently proposed[11], would allow to accelerate long bunches without any emittance increase due to RF field: this offers the possibility to take advantage from the standard scaling law for the space-charge emittance growth. Uniformly distributed bunches with very low aspect ratio (i.e. radius over length ratio) are known to be characterized by a space-charge emittance growth scaling like the square of the aspect ratio: high charged bunches, as requested by future TeV collider scenarios, can be obtained, in this way at a very low emittance.

A new idea to neutralize the space-charge emittance growth is presented in section 4: a careful inspection of usual scaling laws[2] for the space-charge emittance growth reveals that ultra-short bunches characterized by large aspect ratio are emittance dominated by non linear distortion effects in the transverse phase space. In particular, third order terms in the space charge momentum transfer give rise to a spherical aberration effect which sets up a minimum threshold to the emittance. In order to overcome this limitation, we found that a particular shaping of the laser pulse intensity distribution in the transverse direction is able to neutralize the spherical aberration effect. The scaling law for the emittance in such a new regime is very favourable: it comes out to scale like the inverse of the peak current, when the bunch length is reduced keeping the bunch charge constant. The preliminary results of some simulations are also presented. The scaling law for the space-charge emittance is qualitatively confirmed, but a new relevant effect is observed: the wake-field interaction of the bunch with the metallic wall of the cathode. This effect sets up a new limitation to the minimum emittance achievable with RF guns.

1. BEAM DYNAMICS IN A STANDARD RF GUN

It is well known that the beam brightness achievable by RF guns is limited mainly by two effects:

- the emittance growth due to space charge forces, which can be split into two contributions: the distortion of the transverse phase space distribution caused by non linear transverse components and the phase correlation of the space charge field (i.e. its variation versus the longitudinal position inside the bunch)

- the time (or phase) dependence of RF transverse forces, which produces a transverse momentum at the gun exit strongly correlated to the longitudinal position in the bunch.

The non linear transverse components in the RF field can also give contribution to the emittance blow up; however, a particular shaping of the iris profile[4] allows to minimize such non linear components, making negligible their contribution to the emittance. In the following discussion we will neglect this effect.

The cathode emittance, given by the effective beam temperature at the cathode surface plays also a role: the final emittance at the gun exit is indeed given by the quadratic sum of the total emittance growth (RF plus space-charge) and the cathode emittance. Since the latter scales like the laser spot size times the square root of the photo-electron energy, it can be decreased using laser wavelengths just above the photoemission threshold, in order to cool as much as possible the electrons leaving the cathode surface (however this asks for a step rise quantum efficiency as a function of the wavelength), and/or using a smaller laser spot size.

As shown elsewhere[2], the total rms emittance growth of a gaussian distributed electron bunch, calculated at the gun exit, can be written as:

$$\Delta\varepsilon_{tot}^2 = \left(a_{RF}^2 E_0 \sigma_r^2 \sigma_z^2\right)^2 + \left(a_{sc} \frac{Q\mu_x(A)}{E_0\sigma_z}\right)^2 + \sqrt{2}J(A)a_{RF}^2 a_{sc} Q\sigma_r^2\sigma_z \qquad (1)$$

where E_0 is the peak RF electric field on the cathode surface [MV/m], σ_r and σ_z are the widths of the gaussian bunch current distribution [m], A is the aspect ratio $A \equiv \sigma_r/\sigma_z$, Q the bunch charge [nC], a_{RF} is defined as $a_{RF} \equiv .83k$ ($k=\omega_{RF}/c$ [m^{-1}]) and a_{sc} is given by $a_{sc} \equiv 5.7 \cdot 10^{-6}/\sin\phi_0$ ($\Delta\varepsilon_{tot}$ is in m·rad). The first term on the r.h.s of eq. (1) gives the contribution to the emittance coming from the time dependence of linear RF fields (whose resonant frequency is ω_{RF}), while the second one represents the effect of the space charge forces, where the function $\mu_x(A)$ can be well approximated by $\mu_x(A) = 1/(3A+5)$. The third term gives a contribution coming from the correlation between RF and space charge effects.

The correlation factor J(A) is given by $J(A) = \int_0^\infty (2+x)^{-2}(2+Ax)^{-(3/2)}dx$ .The optimum injection phase ϕ_0, i.e. the phase at which the centre of the laser pulse must strike the cathode surface in order to minimize the RF contribution to the emittance, is specified by the equation

$$\frac{\pi}{2} = \phi_0 + \frac{1}{2\alpha\sin\phi_0} \qquad \left(\alpha \equiv \frac{eE_0}{2kmc^2}\right) \qquad (2)$$

Expression (1) tell us that, once chosen the frequency and field of the RF gun, for a fixed bunch charge some optimum values for σ_r and σ_z can be found which minimizes the emittance growth. An increase of the bunch length causes actually a decrease of the space charge contribution to the emittance but increases at the same time the emittance growth due to RF field. The same holds for the bunch radius σ_r.

Just as an example we show in Fig.1 the behaviour of the quantity $\Delta\varepsilon_{tot}$ corresponding to a bunch charge Q=1 nC and a resonant frequency ν_{RF}=2856 MHz with peak accelerating field at the cathode E_0=100 MV/m .

The equi-level plot for the emittance blow up is drawn in Fig.1 as a function of σ_r and σ_z: the numbers in the plot give the amount of emittance blow up on each line. Two regions of relative minimum are displaied, where $\Delta\varepsilon_{tot}$ ranges below 5 mm·mrad: one for disk-like bunches (large aspect ratio A) and the other one for thin cigar-like bunches (small aspect ratio A). The peak current is of course decreasing as $1/\sigma_z$, hence cigar-like bunches have lower peak currents if the bunch charge is kept constant (as in the present case). As discussed elsewhere[6] - and shown later on in sections 2 and 3 -

the current can be increased via a magnetic compression, applied downstream the RF gun cavity, as long as their longitudinal phase space distributions are not affected by significant non linear distortions.

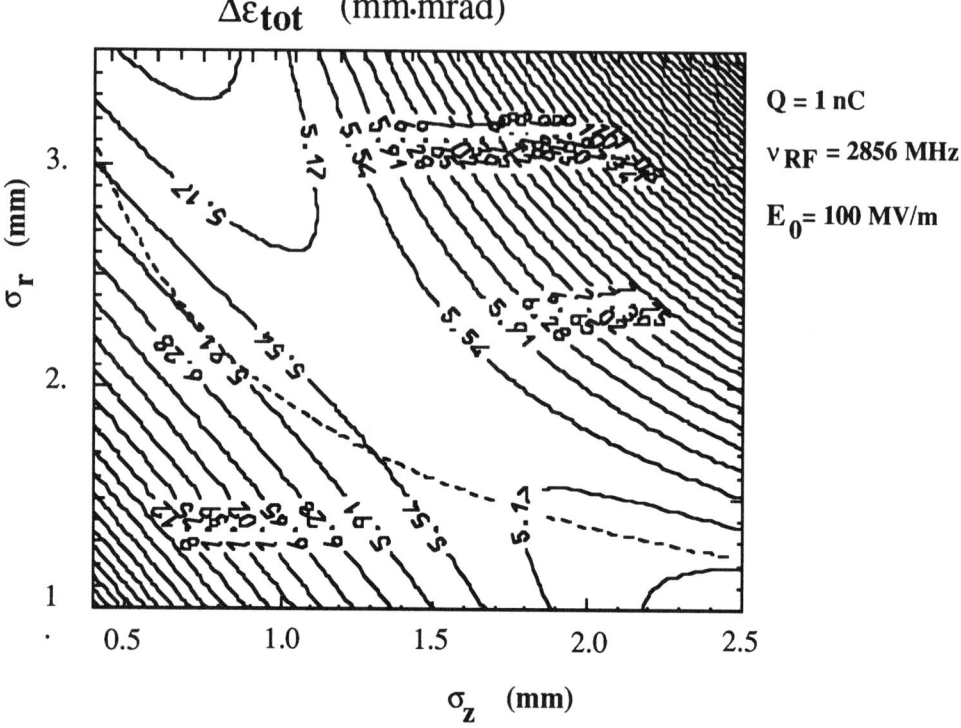

Fig. 1 - Contour plot of $\Delta\varepsilon_{tot}$ (equation 1) as a function of σ_r and σ_z, for fixed Q, ν_{RF} and E_0.

The dashed line splits the plot into two regions: the region below the line is not permitted if the cathode is not able to generate a current density larger than 500 A/cm^2, a typical value for present alkali photo-cathode technology, which can be overcome by metallic cathodes. Cigar-like bunches are in this sense more difficult to generate. As clearly shown in the figure, the upper right corner of the plot is the RF-dominated region for $\Delta\varepsilon_{tot}$ (RF induced blow up is dominant), while the left lower corner is the space-charge dominated region.

We want to stress at this point that in a standard RF gun it is not possible to decrease the beam emittance simply by decreasing the bunch charge density, i.e. using longer and larger bunches in order to keep low the space charge field, since the RF induced emittance blow up sets up and becomes the dominant effect. The possibility to cancel this RF contribution would allow automatically to damp down also the space charge emittance simply by using larger sizes for the bunch (for a given bunch charge).

In particular, it is known that the function $\mu_x(A)$ (equation (1)), governing the space-charge emittance growth, becomes vanishing for a uniform distributed bunch in the limit $A \to 0$, since the space-charge field becomes linear. The use of uniform cigar-like bunches, together with some techniques to damp down the RF emittance growth, will allow, as shown in section 3, to reach ultra-low emittances even with high charged bunches.

It must be noticed that the space-charge term as calculated in ref.2 contains both the contributions due to the correlated emittance (longitudinal variation of the space-charge field inside the bunch) and to the non linear transverse components of the space-charge force.

In section 4 the effect of the two contributions will be separately analyzed, showing that a proper correction of the non linear term will produce a strong damping of the total emittance growth in the domain of ultra-short bunches with very large aspect ratio A.

2. CORRECTION OF THE RF INDUCED EMITTANCE

It is well known that the basic mechanism of the RF induced emittance blow-up consists in the correlation between the exit transverse momentum and the injection phase, as given by the formula (see Appendix A):

$$p_r = \alpha k r \cdot \left(\sin<\phi> + \Delta\phi \cdot \cos<\phi> - \frac{\Delta\phi^2}{2} \sin<\phi> \right) \qquad (3)$$

which gives the well known fan-like shape of the transverse phase space distribution at the gun exit. In this expression r is the radial position (assumed to be constant during the acceleration) of a generic electron of the bunch , whose exit phase ϕ (defined as $\phi \equiv \omega T_f - kL + \phi_0$, L is the gun length, ϕ_0 the injection phase at the cathode, T_f the exit time) is supposed to be slightly distributed around an average exit phase of the bunch $<\phi>$, such that $\phi = <\phi> + \Delta\phi$. The analytical estimation for the exit phase ϕ is reported in Appendix A (it is actually equivalent to the r.h.s. of eq. (2)).

Since the fields (both RF and space charge) are assumed to be axi-symmetrical, the phase space (x,p_x) and (y,p_y) must be identical, hence x and p_x can be substituted in eq. (3) to r and p_r respectively. Taking the standard definition of rms normalized emittance[8] $\varepsilon_x \equiv \sqrt{<x^2><p_x^2> - <xp_x>^2}$, where $< >$ means an average over the phase space distribution, we can calculate the emittance blow up due to the RF field, assuming that the phase distribution is symmetric with respect to $<\phi>$ (i.e. $<\Delta\phi>=<(\Delta\phi)^3>=...=0$) : the result is found to be

$$\varepsilon_x^{RF} = \varepsilon_{min} + \alpha k <x^2> \sqrt{<(\Delta\phi)^2>} \left| \cos<\phi> \right| \qquad (4)$$

where

$$\varepsilon_{min} = \frac{\alpha k <x^2>}{2} \sqrt{<(\Delta\phi)^4> - <(\Delta\phi)^2>^2}$$

In case of a gaussian bunch, recalling that $<x^2>=\sigma_r^2$, $<(\Delta\phi)^4>=3k^4\sigma_z^4$ and $<(\Delta\phi)^2>=k^2\sigma_z^2$, it is easy to verify that ε_{min} gives just the square root of the first term on the r.h.s. of (1).

It is clear from eq. (4) that this contribution has a sharp minimum at $<\phi>=\pi/2$, where the first order term in the (p_r,ϕ) correlation is vanishing: this corresponds to the minimum aperture of the fan in the transverse phase space.

The normalized longitudinal rms emittance ε_z can be calculated by the usual definition $\varepsilon_z \equiv \frac{1}{k}\sqrt{<(\Delta\phi)^2><p_z^2> - <\Delta\phi p_z>^2}$, taking the expression of the exit γ_f as given by (A3), which is equivalent (under the approximation $p_z \approx \gamma_f$) to

$$p_z = 1 + \alpha\cdot[\pi(N+1/2)\sin<\phi> + \cos<\phi>](1-\Delta\phi^2/2) + \tag{5}$$
$$\alpha\cdot[\pi(N+1/2)\cos<\phi> - \sin<\phi>]\Delta\phi$$

The longitudinal emittance ε_z comes out to be, for a uniform distributed bunch of length $\Delta\phi = kL$:

$$\varepsilon_z = \varepsilon_{z\,min} + \frac{\alpha\,\Delta\phi^3}{4\sqrt{2\cdot5!}\,k}\left|\pi(N+1/2)\sin<\phi> + \cos<\phi>\right| \tag{6}$$

where

$$\varepsilon_{z\,min} = \frac{\alpha\,\Delta\phi^4}{2\cdot5!\sqrt{21}\,k}\sqrt{1+\pi^2(N+1/2)^2} \tag{6'}$$

The second order term in $\Delta\phi^2$ of expression (5), representing the non linear correlation in the energy-phase curve, is clearly the main source of longitudinal emittance blow up: it is in fact responsible of the second term in the r.h.s. of (6), which is the dominant contribution. This term vanishes only at an average exit phase defined by

$$\cot<\phi> = -\pi (N+1/2) \tag{7}$$

where the second order term in (5) is vanishing too: the residual minimum emittance ε_{zmin} at such a phase is due to the third order term (in $\Delta\phi^3$) of the exit momentum p_z (not shown in expression (5)). Unfortunately the solution of eq. (7) is close to π, approaching π as the number of cell N becomes larger: it comes out that the average exit phase required to minimize the longitudinal rms emittance blow up is far from the one required ($<\phi> = \pi/2$) to minimize the transverse emittance at the gun exit.

Moreover, it must be recalled that we are neglecting here the uncorrelated longitudinal emittance given by the variation of the energy gain with the particle position off-axis: this is consistent with the choice to use an optimized geometry for the RF gun cavity, able to cancel the non linear transverse components of the RF field. In this case the axial component E_z does not vary versus the radius, as given by expression (A1). As shown later on, for long bunches the uncorrelated longitudinal rms emittance is negligible with respect to the correlated one given by (6).

Since the longitudinal emittance blow up is substantially due to the curvature of the longitudinal phase space, this has a relevant effect on the possibility to increase the peak current of the bunch via a magnetic compression applied downstream the gun exit. This is shown in Fig. 2, where two typical longitudinal phase space distributions, generated by a numerical simulation, are plotted for $<\phi>=90°$ and $<\phi>=110°$, respectively. It is evident that higher injection phases give less curvature in the longitudinal phase space distribution, i.e. a smaller longitudinal emittance.

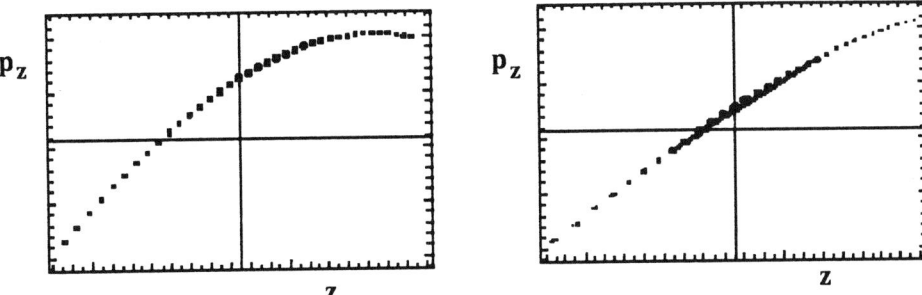

Fig. 2 - Left diagram: longitudinal phase space (z,p_z) distribution, at the gun exit, for the case of minimum transverse emittance (i.e. $<\phi>=90°$) . Right diagram: longitudinal phase space (z,p_z) distribution, at the gun exit, obtained with a higher phase $<\phi>=110°$.

Applying an ideal magnetic compression at the gun exit, defined by the transformation rules $\Delta\phi' = \Delta\phi - \Delta p_z<\Delta\phi p_z>/<\Delta p_z^2>$ and $p_z' = p_z$ ($\Delta p_z \equiv p_z - <p_z>$, $\Delta\phi'$ and p_z' are the phase space variables after the compression), the energy spread $<\Delta p_z^2>$ remains unchanged, while the rms bunch length is reduced down to a minimum value given by $<\Delta\phi'^2> = k^2\varepsilon_z^2/<\Delta p_z^2>$ (the longitudinal phase space is actually uncorrelated after the transformation). We can define therefore a parameter $C \equiv L / L'$ as the ratio between the natural bunch length $L = \Delta\phi/k$ and the minimum length L' (for a uniform distributed bunch $L' = \dfrac{1}{k}\sqrt{12<\Delta\phi'^2>}$). C represents indeed the maximum compressibility of the bunch , i.e. the maximum gain in the peak current. From eq. (5) and (6) we find

$$C = \frac{\sqrt{\dfrac{20}{3}(\pi(N+1/2)\cos<\phi> - \sin<\phi>)^2 + \Delta\phi^2(\pi(N+1/2)\sin<\phi> + \cos<\phi>)^2}}{\Delta\phi\left|\pi(N+1/2)\sin<\phi> + \cos<\phi>\right|} \qquad (8)$$

For a fixed natural bunch length $\Delta\phi$ (C scales like $\Delta\phi^{-1}$ because the longitudinal phase space curvature decreases for smaller $\Delta\phi$) C increases significantly as $<\phi>$ grows above the prescribed $<\phi> = \pi/2$, as evidenced in Fig.3. Hence, looking at eq. (2) it can be deduced that the injection phase ϕ_0 should be pushed towards the peak electric field at 90° RF to get a higher compressibility. That implies a higher transverse emittance at the gun exit, due to the first order term in eq. (4).

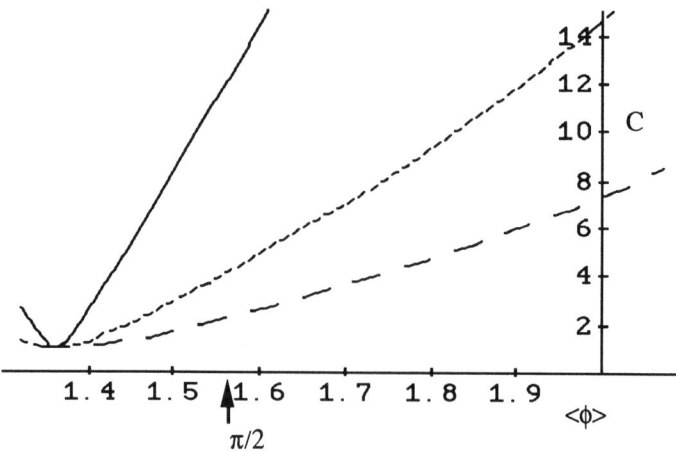

Fig.3 - Compressibility C (i.e. maximum increase of the peak current achievable via a magnetic
compression) as a function of the average exit phase <φ>, for three different bunch lengths
Δφ, namely Δφ=5° RF (solid line), Δφ=15° RF (dotted line) and Δφ=30° RF (dashed line).

To solve this problem two different solutions have been proposed so far: the
lengthening of the first half cell[13] and the correction of the extra-emittance contribution,
due to the first order term, by means of an unsymmetrical cell[9] added downstream the
RF gun cavity.

The lengthening of the first half cell implies an expansion of the RF field in spatial
harmonics: this is reported in Appendix B, where the radial p_r and longitudinal p_z
momentum at the exit of the first cell are calculated.

The divergence at the exit of the first half cell can be found by calculating the ratio
$D = p_r / p_z$ as a function of the parameter δ, giving the relative change of the first half
cell with respect to the standard $\lambda/4$ length:

$$D = \alpha\, k\, r\, \frac{1 + \dfrac{2\delta}{3} - \dfrac{\delta^2}{2}}{1 + \alpha\left(\pi/2 + \dfrac{5\pi\delta}{8} - \pi\delta^2(5/32+\pi^2/48)\right)} \tag{9}$$

A shorter first half cell increases the exit divergence, while the viceversa holds for
longer cells: at $\delta=0.3$ (i.e. a 30% longer cell) the exit divergence is decreased by 6% .

The compressibility can be computed by the same procedure used above for the
ideal $\lambda/4$ cell: since the condition to have a minimum rms transverse emittance is still
$<\phi> = \pi/2$ it is interesting to compute the gain in compressibility at such exit phase, i.e.
the ratio $g \equiv C' / C$, where C' is the compressibility of the longer (or shorter) half-cell
and C is the compressibility given in (8) for the standard half-cell. We found

$$g = \frac{1 + \delta(4/3+\pi^2/8) + \dfrac{5\delta^2\pi^2}{32}}{1 + \dfrac{5\delta}{4} - \delta^2(5/16+\pi^2/24)} \tag{10}$$

The gain g comes out to be less than 1 for $\delta<0$ while grows above 1 for $\delta>0$: that implies a gain in compressibility for longer half-cells. In particular, a half-cell whose length is increased by a factor 1.3 is able to give a compressed current nearly 50 % higher than a standard $\lambda/4$ cell.

It must be stressed, however, that the injection phase is still locked at a fixed value as long as a minimum rms transverse emittance is requested at the gun exit. Moreover, the varied length of the half-cell implies that non linear transverse components will appear in the RF field, as shown in Appendix B: such non linear terms generates an extra emittance growth due to the phase space distortion (at the lowest order, the third order one, which corresponds to a spherical aberration). In case of space-charge dominated emittance, this contribution may be however negligible.

In the following the technique of the unsymmetrical cell, able to recover the extra emittance generated at phases different from the optimum one ($<\phi>=\pi/2$), is described.

Our analysis started from the observation, based on numerical simulations performed for the ARES SC injector design study[6], that an unsymmetrical one-cell cavity is able, under certain conditions, to decrease the rms normalized emittance of a beam produced by a RF gun. The asymmetry here is referred to the fact that the cell is still axi-symmetrical and support TM_{0np} modes, but it is not made up, as usual, by two parts which are symmetric with respect to a central median plane. This cell, as shown later on, is somewhat similar to the first half cell of the RF gun.

The numerical simulations showed that even a standard symmetrical-cell cavity induces some emittance correction when added downstream the RF gun cavity, but with definitely lower efficiency than the unsymmetrical cell, being not able to recover the minimum emittance. Starting from this observation and recalling the results of other studies on higher harmonics[10] we tried to explain the emittance recovery process as due to the higher spatial harmonic content in the RF field of the unsymmetrical cell. This must be compared to the pure first harmonic field typical of a symmetrical cell resonating with the TM_{010} mode, as given in Appendix A. The real field of a single cell $\lambda/2$ cavity is actually a quasi pure first harmonic field: the field in expression (A1) gives actually the field of an infinite array of cells in the $TM_{010-\pi}$ mode. In the same way we will approximate the field of an unsymmetrical single-cell cavity, whose boundary resembles the one of the first half-cell in a RF gun (as shown later on in Fig.5), with the field of infinite array of halve cells resonating in the $TM_{010-\pi}$ mode, whose form factor is given by the function

$$f(x) = \sin(x)\cdot\Sigma_{n=odd}\,[\theta(n\tfrac{\pi}{2}\text{-}x) - \theta((n\text{-}1)\tfrac{\pi}{2}\text{-}x)]$$

where $x=kz$, and $\theta(x)$ is the step-like function as defined in App. A. Fourier analyzing the previous function and keeping terms only up to the 3rd harmonic we found the following expression for the axial field of the unsymmetrical cell

$$E_z^{hc}(z,t) = E_0^{hc}\cdot[\tfrac{1}{2}\sin(kz) + \tfrac{1}{\pi}(\cos(kz)\text{-}\cos(3kz))]\cdot\sin(\omega_{RF}t + \phi_0^{hc}) \qquad (11)$$

which is vanishing at $z=0$ and at $z=\lambda/2$.

As shown elsewhere[9], the momentum change applied by the unsymmetrical cell can be calculated by a two-step approximation, keeping step-like constant the beam divergence through the cell. The momentum change comes out to be dependent on the entrance phase ϕ_0^c (or ϕ_0^{hc}) in the cell. In the following we report the momentum change Δp_r^c produced by a standard symmetrical cell

$$\Delta p_r^c = \alpha kr \left[(0.3 \cos\phi_0^c + 0.2 \sin\phi_0^c) - \frac{1}{\sin\phi} (0.2 \cos^2\phi_0^c + 0.1 \sin\phi_0^c \cos\phi_0^c) \right]$$

(12)

and Δp_r^{hc} by an unsymmetrical cell

$$\Delta p_r^{hc} = \alpha kr \left[(3.6 \cos\phi_0^{hc} + 2.3 \sin\phi_0^{hc}) + \frac{1}{\sin\phi} (0.18 \cos^2\phi_0^{hc} + 0.08 \sin\phi_0^{hc} \cos\phi_0^{hc}) \right]$$

(12')

In the calculation the peak field both for the cell and for the unsymmetrical cell has been taken equal to the gun peak field, i.e. $E_0 = E_0^c = E_0^{hc}$.

Since the emittance increase, for exit phases different from $<\phi>=90°$ RF, is due to the first order term Δp_{1r}^g in the (p_r,ϕ) correlation at the gun exit, as given by eq. (3), we must extract the 1st order terms in $\Delta\phi$ from eq. (12) and (12'), assuming again that the entrance phases ϕ_0^c and ϕ_0^{hc} are slightly distributed around some average phases , $\phi_0^c = <\phi_0^c> + \Delta\phi$ and $\phi_0^{hc} = <\phi_0^{hc}> + \Delta\phi$. The result is found to be:

$$\Delta p_{1r}^c = \alpha kr \left[- 0.3 \sin<\phi_0^c> + 0.2 \cos<\phi_0^c> + \frac{1}{\sin<\phi>} (0.2 \sin(2<\phi_0^c>) - 0.1 \cos(2<\phi_0^c>)) \right]\Delta\phi$$

(13)

and

$$\Delta p_{1r}^{sc} = \alpha kr \left[- 3.6\sin<\phi_0^{hc}> + 2.3\cos<\phi_0^{hc}> - \frac{1}{\sin<\phi>}(0.18\sin(2<\phi_0^{hc}>) - 0.08\cos(2<\phi_0^{hc}>)) \right]\Delta\phi$$

(13')

Once fixed the gun exit phase, the emittance recovery is possible if some values for $<\phi_0^c>$ (or $<\phi_0^{hc}>$) exist such that Δp_{1r}^c (or Δp_{1r}^{hc}) is just equal in amplitude and opposite in sign with respect to the first order term at the gun exit, i.e. $\Delta p_{1r}^g = \alpha k r \cos<\phi>\Delta\phi$. In this way the extra-aperture of the fan in the transverse phase space distribution caused by the first order term can be made vanishing after the acceleration through the cell (or unsymmetrical cell).

For this purpose the two compensation functions $f^c(<\phi_0^c>) = (\Delta p_{1r}^c + \Delta p_{1r}^g)/\Delta p_{1r}^g$ and $f^{hc}(<\phi_0^{hc}>) = (\Delta p_{1r}^{hc} + \Delta p_{1r}^g)/\Delta p_{1r}^g$ are plotted in Fig.4: they give actually the emittance recovery capability of the symmetrical cell and of the unsymmetrical one, respectively. It is clearly shown that only f^{hc} is vanishing for two values of $\langle\phi_0^{hc}\rangle$, while the compensation function f^c for the symmetrical cell never vanishes: this is a qualitative confirmation about the capability of the unsymmetrical cell to recover completely the first order term in the emittance increase, while the symmetrical cell can only partially recover the emittance.

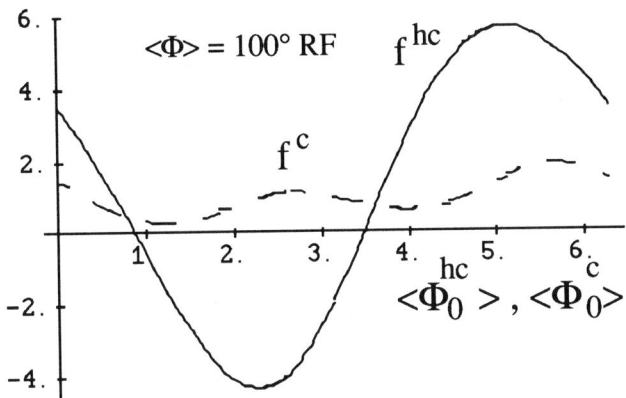

Fig.4 - Compensation functions f^c and f^{hc} versus the injection phase in the symmetrical cell and in the unsymmetrical one, respectively.

The mechanism of emittance correction via the unsymmetrical cell has been proved also by numerical simulations, as shown in Fig.5, where the outline of the

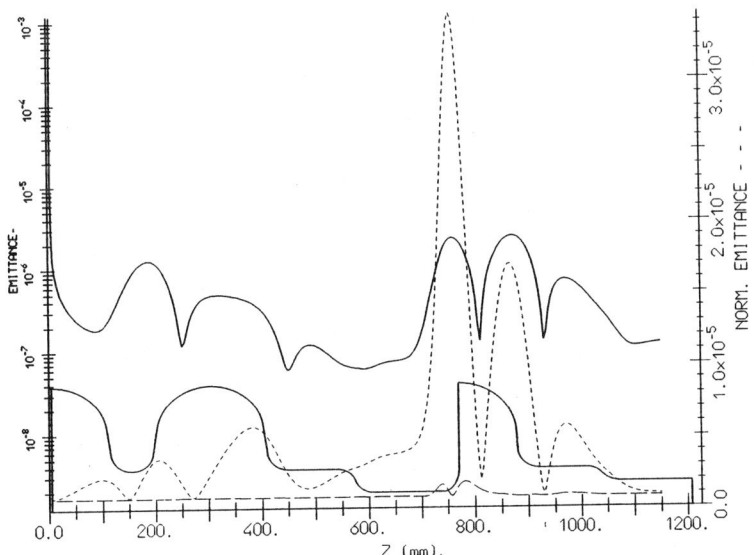

Fig. 5 - Emittance correction by means of an unsymmetrical cell. See text for details

Superconducting RF gun under study for the ARES project is plotted: the gun is formed by one and a half cell together with an uncoupled unsymmetrical cell. The dashed line plotted in the figure shows the behaviour along the acceleration of the rms normalized emittance for a bunch injected 20° RF above the optimum injection phase. It can be seen that the emittance at the gun exit is 5 times higher than the minimum value at 1 mm·mrad, which is fully recovered at the exit of the unsymmetrical cell .

3. NEUTRALIZATION OF THE RF INDUCED EMITTANCE USING MULTI-MODE RF GUNS

In the two previous cases we analyzed the effect of the spatial harmonics onto the transverse and longitudinal emittance: the main result is that the spatial harmonics are capable of correcting a correlated transverse emittance as long as the beam can keep a significant divergence (comparable to the natural divergence in the gun) through the cell. Moreover, if the field amplitude in the cell is comparable to that one in the gun, only the first order emittance term possibly present in the beam can be corrected. It looks that the spatial harmonics play the role just to shift the spatial phase of the first harmonic, but they do not produce any net contribution to the momentum change.

Hence we must search for a real harmonic of the fundamental $TM_{010-\pi}$ mode, i.e. a higher mode of the RF gun cavity whose frequency is an integer multiple of the fundamental one. That is equivalent to the assumption that the RF field can be written as:

$$E_z(z,t) = E_0\cos(kz)\sin(\omega t+\phi_0) + E_n\cos(nkz)\sin(n\omega t+n\phi_0) \tag{14}$$

We repeat the calculation of Appendix A for the longitudinal and transverse momentum, simply summing the separate contribution of the n-th harmonic field. The exit longitudinal momentum comes out to be given by

$$p_z = 1 + \alpha\left[\pi(N+\tfrac{1}{2})\sin\phi + \cos\phi\right] + \alpha_n\left[n\pi(N+\tfrac{1}{2})\sin(n\phi) + \begin{cases} \cos(n\phi) , \text{ n odd} \\ 0 \quad\quad , \text{ n even} \end{cases}\right] \tag{15}$$

and the exit phase ϕ

$$\phi = \frac{1}{2\alpha\sin\phi_0 + 2n\alpha_n\sin(n\phi_0)} + \phi_0 \tag{16}$$

where we have defined the dimensionless parameter α_n for the n-th harmonic in the same way as for the fundamental: $\alpha_n \equiv \dfrac{eE_n}{2nkmc^2}$.

In the following we will study odd harmonics: as shown below, only such harmonics are of interest for the emittance blow up neutralization. Assuming now a symmetric phase distribution over the bunch length, with $\phi = <\phi> + \Delta\phi$, we compute up to 3-rd order the contributions to the longitudinal normalized momentum at the gun exit, for the case $<\phi> = \pi/2$:

$$p_z = \left[1 + \pi(N+1/2)(\alpha + (-1)^{(n-1)/2}n\alpha_n)\right] +$$
$$\left[-\alpha - (-1)^{(n-1)/2}n\alpha_n\right]\Delta\phi +$$
$$\left[(\pi/2)(N+1/2)(-\alpha - (-1)^{(n-1)/2}n^3\alpha_n)\right](\Delta\phi)^2 + \tag{17}$$
$$\left[(1/6)(\alpha + (-1)^{(n-1)/2}n^3\alpha_n)\right](\Delta\phi)^3 + \ldots$$

It is easy to verify that the condition

$$\alpha_n = (-1)^{(n-3)/2}\frac{\alpha}{n^3} \tag{18}$$

which is equivalent to

$$E_n = (-1)^{(n-3)/2}\frac{E_0}{n^2} \tag{18'}$$

gives vanishing third and fourth order terms in the momentum-phase relationship (17).

It is interesting to compute the final momentum for the case of n=3 and n=5, the lowest odd harmonics which can be added to the main accelerating field, given respectively by proper tuned $TM_{012-\pi}$ and $TM_{014-\pi}$ modes of a 1+1/2 cell structure.

3rd harmonic n=3 $E_3 = \dfrac{E_0}{9}$ $p_z = [1+1.5\pi\,\alpha\,(1-\frac{1}{9})] - \alpha\,(1-\frac{1}{9})\,\Delta\phi$

5th harmonic n=5 $E_5 = \dfrac{-E_0}{25}$ $p_z = [1+1.5\pi\,\alpha\,(1+\frac{1}{25})] - \alpha\,(1+\frac{1}{25})\,\Delta\phi$

Therefore, the third harmonic causes a slight (-11%) decrease both in the final average energy (the zero order term in the expression for p_z) and in the energy spread (linear term in $\Delta\phi$) by the same amount, while the fifth harmonic increases both the two terms by a smaller amount (+4%).

The longitudinal rms emittance becomes, for the case of n=3,

$$\varepsilon_z = \frac{\pi\,\alpha\,\Delta\phi^5\,(N+1/2)}{288\sqrt{3}\ k} \tag{19}$$

which gives a very favourable scaling, implying that the longitudinal phase space distribution is free from non linear distortions: the compressibility is in this case much higher than in the standard case (without harmonic). As shown elsewhere[11], this condition is called "straight topping" since the energy-phase relationship exhibits a slanted but straight top, to be compared with the flat-top typical of different type of RF cavity operation with harmonics.

To compute the transverse momentum we simply take into account the presence of the odd n-th harmonic, under the straight-topping condition, $E_n = E_0/n^2$, summing a contribution similar to the one of the fundamental mode:

$$p_r = \alpha kr \cdot \sin\phi + \frac{\alpha kr}{n^2}\sin(n\phi) \tag{20}$$

Again we compute the rms normalized transverse emittance blow up (due to RF field) via the expression

$$\varepsilon_x^{RF} = \alpha k <x^2> \sqrt{ <\left(\sin\phi + \frac{\sin(n\phi)}{n^2}\right)^2> - <\sin\phi + \frac{\sin(n\phi)}{n^2}>^2 } \qquad (21)$$

which gives the following result, taking into account only terms up to $(\Delta\phi)^3$ in the expansion of $\sin\phi$ and $\sin(n\phi)$:

$$\varepsilon_x^2 = [\tfrac{1}{4}(<(\Delta\phi)^4> - <(\Delta\phi)^2>^2)][\sin<\phi> + \sin(n<\phi>)]^2 +$$

$$+ [<(\Delta\phi)^2> - \tfrac{1}{3}<(\Delta\phi)^4>]\cos^2<\phi> +$$

$$+ [\tfrac{1}{n^2}<(\Delta\phi)^2> - \tfrac{1}{3}<(\Delta\phi)^4>]\cos^2(n<\phi>) + \qquad (22)$$

$$+ \tfrac{2}{n}[<(\Delta\phi)^2> - \tfrac{1}{6}<(\Delta\phi)^4> (1+n^2)]\cos<\phi>\cos(n<\phi>)$$

If n=3,7,11,....the previous expression is vanishing at $<\phi>=\pi/2$, meaning that the minimum emittance contains only terms of the order of $<(\Delta\phi)^4>$.
Away from the minimum we have:

$$\varepsilon_x^{RF} = O(<(\Delta\phi)^4>) + \alpha k <x^2> \sqrt{<(\Delta\phi)^2>} \; (\frac{n^2-1}{6}) |<\phi> - \pi/2|^3 \qquad (23)$$

The first term in the r.h.s of (23) gives the minimum value for the emittance at $<\phi>=\pi/2$: this is again the required optimum phase to minimize the emittance blow up, but here the minimum value is at least two order of magnitude lower than the standard value (given by ε_{min} in eq. (4)), since $\Delta\phi$ is usually less than 0.1. Moreover, shifting the average exit phase $<\phi>$ away from $\pi/2$ the emittance blow up is still a factor $(n^2-1)(<\phi>-\pi/2)^2/6$ lower than the corresponding value in a standard RF gun.
    In this respect we can say that the RF induced emittance blow up is neutralized completely, not only at the optimum phase, but even for average exit phases slightly shifted around the optimum one.
    The two relevant effects of the superposition of a third harmonic under the straight topping condition (18) can be summarized as:
-   the minimum of the longitudinal and transverse emittance blow up due to RF effects occur at the same injection phase for both the emittances
-   the transverse emittance blow up is neutralized up to fourth order terms in $\Delta\phi$
-   the longitudinal emittance blow up is neutralized up to fifth order in $\Delta\phi$

A recently proposed geometry for the RF gun cavity, able to support a third harmonic mode $TM_{012-\pi}$ resonating at a frequency triple than the $TM_{010-\pi}$, is shown in Fig.6. The rationale of the design is described elsewhere[11]: here we want to recall that the geometry of the proposed structure was mainly used to verify by means of numerical simulations the validity of the analytical study .

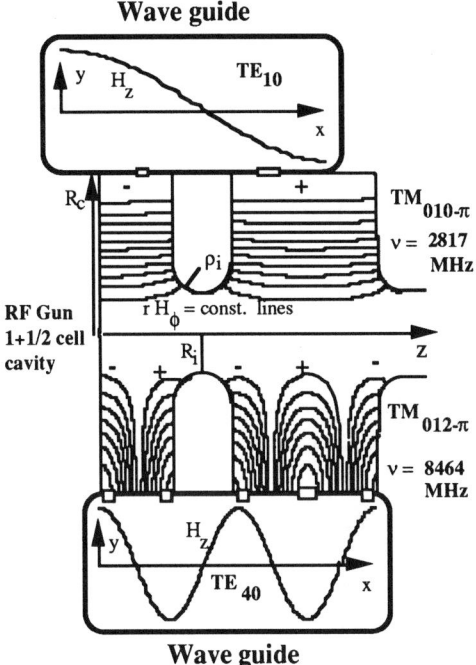

Fig. 6 - Multi-mode RF gun structure lay-out. See text for details.

The neutralization mechanism was indeed proved by simulations, as can be seen in Fig.7. The bunch and field parameter are: E_0=100 MV/m the first harmonic peak field on the cathode, E_3=13 MV/m the third harmonic peak field , σ_ϕ =9° RF , σ_r=2 mm , $<\phi_0>$ = 57° RF the average injection phase, while $<\phi_3>$ = 178 ° RF . The behaviour of the rms normalized emittance ε_x is plotted in the figure (dashed line) versus the average $<z>$ position of the bunch, together with the rms normalized emittance ε_{cs} associated to a central slice of the bunch, centred around $<z>$, carrying 10% of the bunch charge.

The emittance ε_{cs} is clearly much less sensitive to the linear RF field contribution, as can be seen in Fig.7: the strong modulations displaied by the ε_x curve (caused by the alternate opening and closing of the fan-like distribution in the transverse phase space) are in fact much less pronounced in the ε_{cs} curve. The quantity ε_{cs} gives therefore at the exit an estimation of the contribution coming from the non linear terms in the RF field transverse components: in the upper diagram, where no straight-topping has been applied (E_3=0), the difference between ε_x and ε_{cs} is much larger than in the lower one.

The residual rms normalized emittance (2 mm·mrad) is mainly due to the contribution coming from the non linear transverse components in the RF field, as shown by an inspection to the transverse phase space distributions[11].

Under this conditions, the power needed to operate the gun cavity is 5.6 MW on the fundamental mode at 2817 MHz and about 700 kW on the third harmonic mode at 8464 MHz.

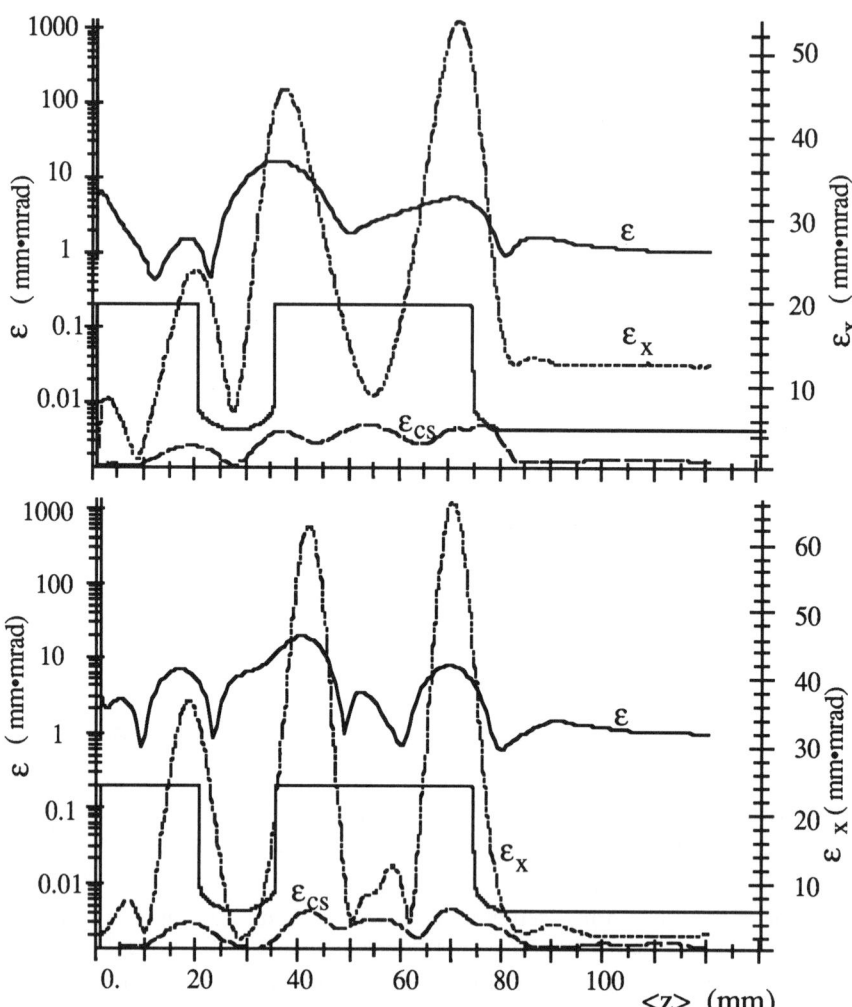

Fig. 7 - Transverse rms emittance as a function of the bunch average position <z> along the acceleration through the multi-mode RF gun: with a straight-topping 3rd harmonic field of amplitude $E_3=14$ MV/m (lower diagram) and without (upper diagram). The dashed line gives the normalized rms transverse emittance ε_x, the solid line gives the rms transverse emittance ε, while the long-dashed line gives the normalized emittance associated to the central bunch slice.

4. NEUTRALIZATION OF THE SPACE-CHARGE INDUCED EMITTANCE USING ULTRA-SHORT LASER PULSES

The multi-mode RF gun structure presented in the previous section can be of great advantage in increasing the quality of long cigar-like bunches that suffer from a relevant RF emittance growth when accelerated throughout a standard rf gun.

Here we analyze the domain of ultra-short bunches, in order to find a technique capable to neutralize the space-charge emittance blow-up.

In Appendix C the transverse component of the electrostatic field of a thin disk-like bunch (characterized by a very large aspect ratio $A \equiv R/L$) is calculated, developing off-axis up to third order the on-axis longitudinal component . Using the same approximation as in ref.2, we compute the total transverse momentum p_r given by the space-charge field during the acceleration through the gun by the formula:

$$p_r = \frac{\pi}{2E_0 \sin\phi_0} E_r^{sc} \tag{24}$$

stating that p_r is proportional to the radial component of the electrostatic field E_r^{sc} produced by the bunch charge at rest in the laboratory frame, divided by the actual RF field at the cathode surface $E_0 \sin\phi_0$ when the bunch is emitted from the cathode.

In our case p_r will be represented by the sum of two contributions, the linear one and the cubic one, i.e $p_r = p_1 \frac{r}{R} + p_3 \frac{r^3}{R^3}$, with p_1 and p_3 functions of the z coordinate inside the bunch.

Since the rms normalized emittance growth due to space-charge forces is defined as $\varepsilon_x^{sc} \equiv \sqrt{<x^2><p_x^2> - <xp_x>^2}$, taking the previous general expression for the space-charge momentum p_r , ε_x^{sc} will be given by $\varepsilon_x^{sc} = \sqrt{\varepsilon_1^2 + \varepsilon_3^2 + \varepsilon_c^2}$, where

$$
\begin{aligned}
\varepsilon_1^2 &\equiv <x^2>2(<p_1^2> - <p_1>^2)\,/\,R^2 \\
\varepsilon_3^2 &\equiv (<x^6><x^2><p_3^2> - <x^4>^2<p_3>^2)\,/\,R^6 \\
\varepsilon_c^2 &\equiv 2<x^2><x^4>(<p_1p_3> - <p_1><p_3>)\,/\,R^4
\end{aligned}
\tag{25}
$$

We split up the emittance into three different terms in order to better enlighten the nature of different effects: indeed, ε_1 represents the linear correlated sp. ch. emittance, being actually given by the z dependence of p_1 (in case p_1 is constant over the bunch length $\varepsilon_1 = 0$). On the other side, ε_3 represents a third order distortion of the transverse phase space distribution and is not vanishing even if p_3 is constant versus z (recalling that for a uniform distribution $<x^2> = R^2/4$, $<x^4> = R^4/8$ and $<x^6> = 5R^6/64$, hence in this case $\varepsilon_3 = p_3R/8$). Finally ε_c represents a correlation term between the two effects.

First we calculate the emittance blow up for a uniform distributed bunch, taking

$$p_1^{uni} = \frac{\pi}{2E_0 \sin\phi_0} E_1^{uni}(z_b)$$

$$p_3^{uni} = \frac{\pi}{2E_0 \sin\phi_0} E_3^{uni}(z_b) \tag{26}$$

with $E_1^{uni}(z_b)$ and $E_3^{uni}(z_b)$ given in Appendix C from (C12) and (C12') respectively. The different emittance terms are given in the following as functions of A, assuming again that $A \gg 1$ (i.e. neglecting terms of the type $O(1/A^3)$):

$$\varepsilon_1^{uni} = \frac{Q}{128\sqrt{5}\varepsilon_0\ E_0\sin\phi_0}\ \frac{1}{A^2R} \tag{27}$$

$$\varepsilon_3^{uni} = \frac{3Q}{1024\varepsilon_0\ E_0\sin\phi_0}\ \frac{1}{R} \tag{27'}$$

$$\varepsilon_x^{sc} = \varepsilon_3^{uni} \tag{27''}$$

It comes out that the total emittance blow up is dominated for ultra-short uniform bunches by the third order term ε_3^{uni}, which represents a spherical aberration[3] in the transverse phase space, and scales unchanged versus A. The emittance is expressed as a function of the bunch charge Q to better display the scaling law for a constant bunch charge: shortening the bunch length by increasing the bunch aspect ratio A gives a bunch current increasing like A, being the current defined as $I \equiv QcA/R$.

A possible cure to the saturation of the emittance blow up, which exhibits a minimum value at large A, is the exploitation of ultra-short bunches with a parabolic distribution in the transverse direction. In Appendix C the optimum distribution that cancel out the third order component at the centre of the bunch is found to be given by $\rho = \rho_0(1-\frac{1}{3}\frac{r^2}{R^2})$. Such a charge density distribution can be achieved using a laser pulse which has a constant intensity profile along its longitudinal direction, and a gaussian clipped profile in the transverse direction. Actually, a typical mode-locked laser used to illuminate the cathode surface has a natural gaussian intensity distribution. Clipping the laser beam at a radius $R = \sqrt{2/3}\ \sigma_r$, where σ_r is the radial width, a radial distribution is obtained having the requested coefficient for the second order term (as r^2) and a really negligible fourth order term (as r^4). The laser clipping may be achieved or by means of a slit (out of the gun) either illuminating a smaller cathode of radius R.

We repeated the emittance calculation using the field behaviour of the optimum parabolic distribution, i.e. taking

$$p_1^{opt} = \frac{\pi}{2E_0\sin\phi_0}\ E_1^{opt}(z_b)$$

$$p_3^{opt} = \frac{\pi}{2E_0\sin\phi_0}\ E_3^{opt}(z_b) \tag{28}$$

with $E_1^{opt}(z_b)$ and $E_3^{opt}(z_b)$ given in Appendix C from (C11) and (C11') respectively. The different emittance terms are in this case

$$\varepsilon_1^{opt} = \frac{7Q}{900\sqrt{5}\varepsilon_0\ E_0\sin\phi_0}\ \frac{1}{AR} \tag{29}$$

$$\varepsilon_3^{opt} = \sqrt{\frac{211}{500}}\ \frac{Q}{256\varepsilon_0\ E_0\sin\phi_0}\ \frac{1}{A^2R} \tag{29'}$$

$$\varepsilon_x^{sc} = \varepsilon_1^{opt} \tag{29''}$$

In this case the scaling law for the emittance blow up becomes really favourable, being dominated by the linear correlated term ε_1^{opt} which scales like A^{-1}. The optimum parabolic distribution cancels out indeed the third order effect as A^{-2}: what is relevant here is the scaling of the emittance like the inverse of the current! Keeping constant the bunch charge and its radius, if one decreases the bunch length (i.e. the time length of the laser pulse) the emittance decreases at the same rate. The behaviour of the two emittances given by (27") for the uniform bunch and (29") for the optimum parabolic one are plotted in Fig.8 as functions of the bunch length L for typical bunch and field parameter values (E_0=100 MV/m , Q=1 nC, R=4 mm, ϕ_0=68 ° RF).

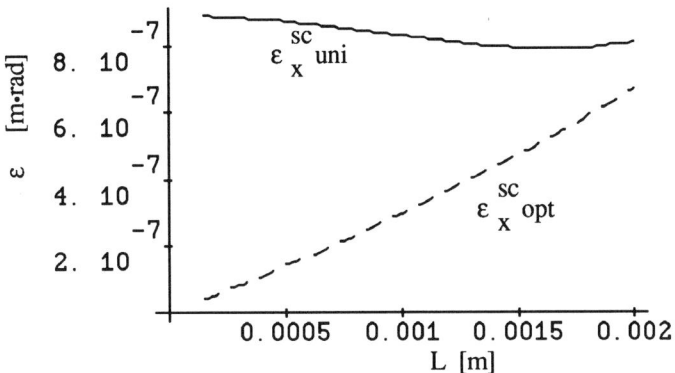

Fig.8 - Space-charge emittance blow up (in m rad) of a 1 nC uniform bunch (ε_x^{sc} uni , solid line) and of an optimum parabolic distributed bunch (ε_x^{sc} opt , dashed line) as functions of the bunch length L.

In order to check the validity of the analytical study prediction, we performed some numerical simulations with the PIC code ITACA[12], using the standard geometry of the BNL RF gun[7] and trying to vary the bunch charge distribution.

We limited the simulations to the region nearby the cathode, reaching just the first half cell exit, mainly for the following reasons:

- the space charge emittance growth takes place just in the first part of the acceleration, when the beam is not yet relativistic: as far as high peak field amplitudes on the cathode are concerned the beam reaches an energy of 1 MeV nearly in the middle of the first half cell

- the simulation of large aspect ratio bunches (short pancakes) asks for mesh steps really very small (typically the mesh step should be less than one tenth of the bunch length) : the capability limits of the code are soon reached, both in terms of core memory and CPU time

- to avoid a mixing between RF and space charge effects (the RF effect are still effective even for ultra-short bunches as far as low emittances are concerned) we concentrated our analysis close to the cathode surface

The peak field at the cathode, the bunch charge and the injection phase were kept fixed throughout all the simulations at the same values as for Fig.8, while the bunch length was taken L=0.6 mm, corresponding to a 2 ps laser pulse. The comparison among different bunch charge distributions was performed selecting four representative distributions:

a) - a radially uniform distribution up to a radius R = 4 mm
b) - a radially gaussian distribution with σ_r = 3.33 mm, clipped at a radius

$$R = 1.2\sigma_r = 4 \text{ mm} \text{ just above the analytical rule } (R = \sqrt{\frac{2}{3}}\sigma_r)$$

c) - a radially gaussian distribution with σ_r = 2.22 mm, clipped at a radius
   R = $1.8\sigma_r$ = 4 mm
d) - a radially gaussian distribution with σ_r = 2.22 mm, not clipped

All distributions are uniform in the longitudinal direction, implying the use of a flat-topped laser pulse. The pseudo-scalar potential r·H_θ is plotted in Fig.9 at a phase ϕ=15° RF (i.e. nearly 15 ps) after the emission from the cathode surface.

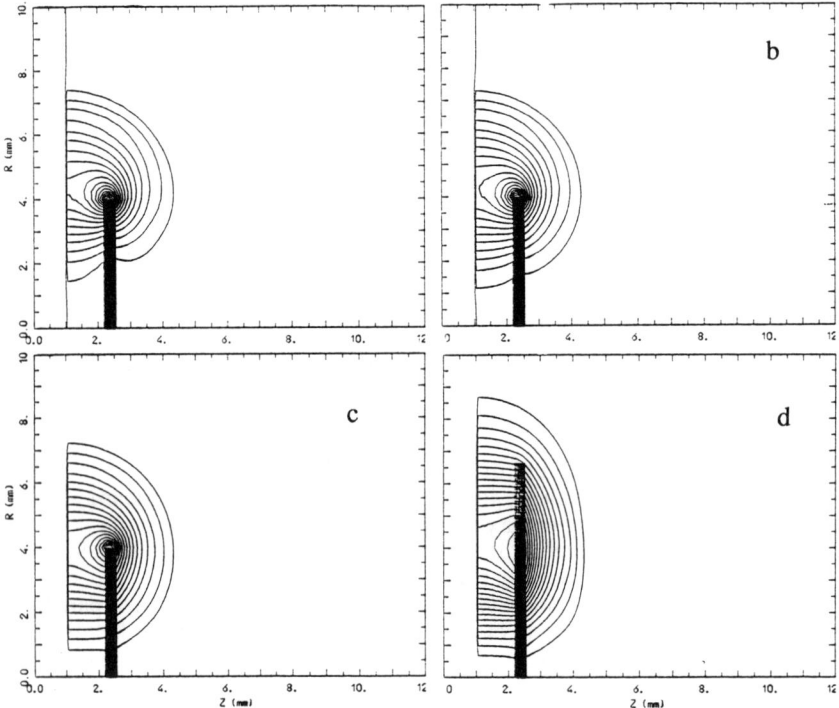

Fig.9 - Electromagnetic field propagation after the photoemission. See text for details.

The electromagnetic perturbation caused by the metallic wall of the cavity is very strong, causing a relevant distortion of the field distribution inside the bunch: that turns out to generate a wake-field like effect by which the bunch core is even focussed (the interaction between the bunch and its image bunch behind the cathode wall is in that regard very similar to the beamstrahlung interaction between an electron and a counter-propagating positron bunch).

The distribution of the radial electric field component as a function of the radius is plotted in Fig. 10 at different longitudinal positions inside the bunch. It can be clearly seen that the uniform distribution has a relevant non linear component with an inner (r<2.5 mm) negative part: the non linear term is progressively corrected as the clip in the gaussian distribution moves outward. The clip at $1.2\sigma_r$ (case b) , corresponding to the analytical prescription, is not sufficient to correct completely the non linearity: the best correction is achieved at $1.8\sigma_r$ clipping, while the natural gaussian distribution gives a relevant non linearity of opposite sign. Unfortunately the metallic wall interaction generates higher order terms which cannot be corrected by this technique and represent a new source of emittance blow up that must be taken into account.

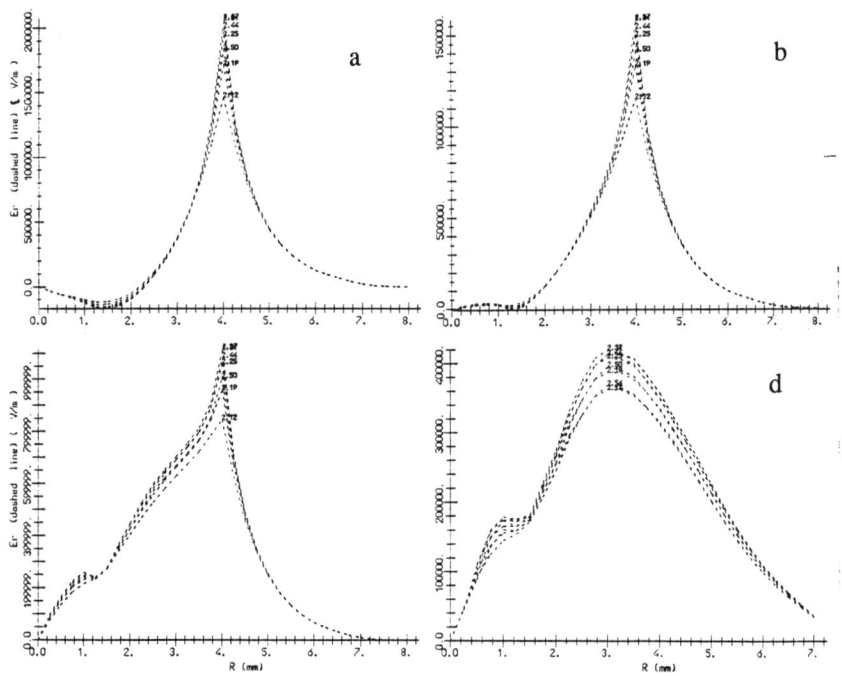

Fig.10 - Radial electric field (inside the bunch) versus r for different charge density distributions.

The final result on the emittance blow up can be understood by looking at the transverse phase space distributions: these are plotted in Fig.11 at a bunch average position z = 30 mm far away from the cathode surface, i.e. when the bunch has just crossed the iris aperture (the first half cell length is 26 mm). The bunch energy is at that time 1.5 MeV. It must be noted that the phase space distributions have been manipulated in order to remove the linear term in the rp_r correlation of the actual distributions: the transformation applied to the phase space is the same as given by a thin lens which makes the beam parallel[3].

The uniform distributed bunch (case a) shows a strong third order term, which is responsible for a 5 mm·mrad emittance growth (corresponding to the rms ellipse area). The natural gaussian bunch (case d) has a little lower emittance, 3 mm·mrad, while the optimum clipped distribution (case c) exhibits the lowest emittance, 1 mm·mrad. The bunch current is for these three cases 500 A. We finally simulated a shorter bunch, 0.3 mm long (1 ps laser pulse), whose phase space is shown in the last diagram (c'). For this case the emittance grows up to 1.2 mm·mrad, but the current is now 1 kA. Since the electrons are emitted cold at the cathode, these emittances must be intended as given only by the space charge and wake field effects, and by the non linear RF emittance growth, which accounts for 0.4 mm·mrad.

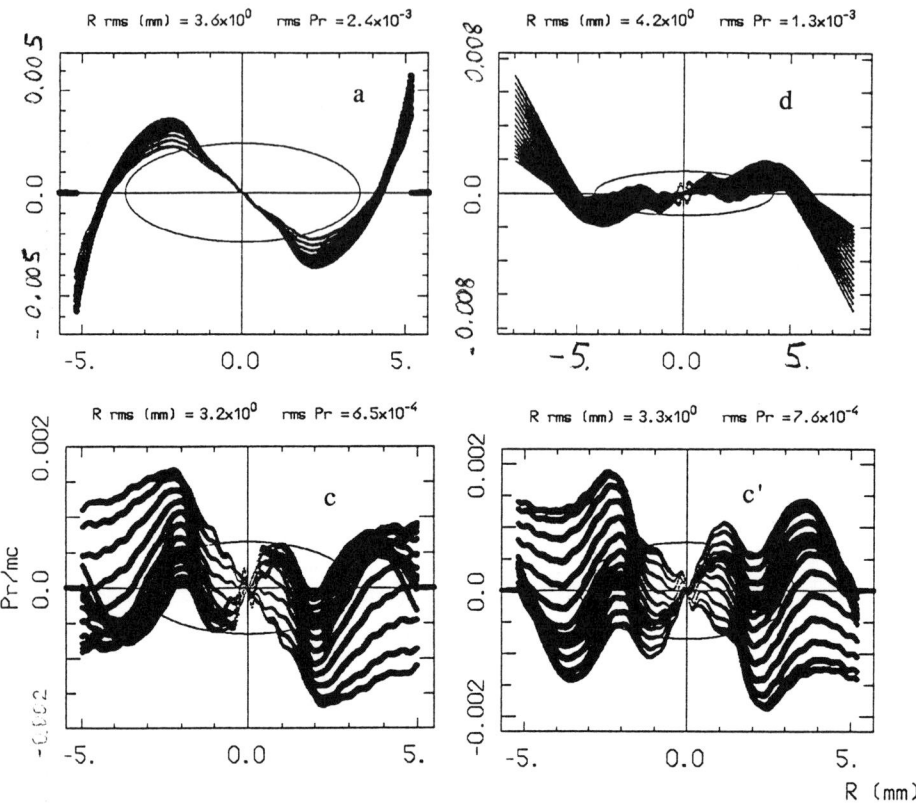

Fig.11 - Transverse phase spaces for three different charge density distributions, carrying a current of 500 A (diagrams a,d and c) and 1 kA (diagram c').

CONCLUSIONS

We have shown, both via an analytical study and by means of some preliminary simulations, that the production of ultra-low emittance beams is feasible by a standard RF gun operated with proper shaped laser pulses. Mode-locked flat-topped 1 ps laser pulses in the 10 µJ energy domain (as typically requested to extract a few nC bunch from a metallic cathode) seem not far beyond the present status of the art. However, further theoretical investigations are needed to better understand the role plaied by wake-field interaction with the cathode wall in deteriorating the beam emittance.

ACKNOWLEDGMENTS

The author is indebted to C.Pagani for valuable suggestions and careful reading of the manuscript. Thanks are due to R.B.Palmer and J.C.Gallardo for enlightening and stimulating discussions on space charge effects in RF guns.

APPENDIX A - ENERGY GAIN AND TRANSVERSE MOMENTUM INSIDE AN IDEAL MULTI-CELL RF GUN CAVITY

We consider an ideal multi-cell structure formed by one $\lambda/4$ half cell plus N $\lambda/2$ cells, with ideal profiling[4] of the iris contour, such that the RF field of the $TM_{010\text{-}\pi}$ mode can be expressed by means of a single first spatial harmonic:

$$E_z = E_0(z) \cos(kz) \sin(\omega t + \phi_0)$$

$$E_r = E_0(z) \frac{kr}{2} \sin(kz) \sin(\omega t + \phi_0) \qquad (A1)$$

$$cB_\theta = E_0(z) \frac{kr}{2} \cos(kz) \cos(\omega t + \phi_0)$$

where $k = \omega/c = 2\pi/\lambda$ and $E_0(z)$ is a step-like function of the type $E_0(z) = \theta((N+1/2)\lambda/2 - z) \, E_0$, with $\theta(x)=1 \; x>0$, $\theta(x)=0 \; x<0$.

A very simple way to compute the exit phase ϕ, final energy γ_f and transverse momentum p_r of an electron accelerated inside the gun is the so-called "impulsive approximation"[11]. It gives same results as in ref.2 as far as the beam at the gun exit can be considered fully relativistic ($\beta \approx 1$), but an easier approach is followed.

Assuming that a strong electric field is applied at the cathode surface, the photo-electrons emerging from the cathode will be quickly accelerated, reaching in a short time (and within a small distance just nearby the cathode surface) a speed quite close to the light speed (i.e. $\beta \approx 1$). One can approximate the RF resonant field, over this time-interval, with a constant electric field of amplitude $E = E_0 \sin\phi_0$.

It is well known that a particle being accelerated from rest by a uniform field (z-oriented) reaches asymptotically a motion described by $z \approx ct - \Delta$, where $\Delta \equiv \dfrac{mc^2}{eE_0\sin\phi_0}$ is the distance travelled by the particle to get a kinetic energy just equal to the rest-mass energy mc^2 (i.e. $\gamma=2$ at $z=\Delta$).

Since in RF guns E_0 is usually very large, Δ turns out to be a small fraction of the RF wavelength λ (with $E_0=100$ MV/m and $\nu_{RF}=2856$ MHz, $\Delta \approx \lambda/18$), hence the non relativistic motion of the electrons is really confined nearby the cathode surface.

Since Δ is just the shift between the electron and the forward travelling wave component of the standing wave RF field, the asymptotic phase shift of the electron with respect to the RF field will be given by $(\phi - \phi_0) = k\Delta$.

Therefore ϕ represents the asymptotic exit phase. Substituting for Δ and recalling the definition of the dimensionless parameter α, $\alpha \equiv eE_0/2m_0c^2k$, one gets the asymptotic exit phase as given by the r.h.s. term of equation (2), i.e.

$$\phi = \frac{1}{2\alpha \sin\phi_0} + \phi_0$$

In order to compute the energy gain we assume now that the electrons are emitted from the cathode surface at $\beta \approx 1$ at a phase ϕ : this is equivalent to looking at the non relativistic part in the electron motion just as a re-definition of the starting conditions.

The total kinetic energy gain ΔT will be given by:

$$\Delta T = eE_0 \int_0^{(N+1/2)\lambda/2} \cos(kz)\sin(\omega t + \phi)dz = \frac{eE_0}{2k} \left[\pi(N + \tfrac{1}{2})\sin\phi + \cos\phi \right]$$

The final value for the relativistic factor γ at the gun exit is therefore:

$$\gamma_f = 1 + \alpha \cdot [\pi(N+1/2)\sin\phi + \cos\phi] \qquad (A2)$$

while during the acceleration holds: $\gamma = 1 + \alpha[(kz)\sin\phi + \sin(kz)\sin(kz+\phi)] \qquad (A3)$

In order to calculate the transverse momentum p_r inside the gun we use the properties of the canonical momentum, defined as usual $\mathbf{P} \equiv \mathbf{p} + e\mathbf{A}$, where $\mathbf{A}$ is the vector potential. A suitable expression for $\mathbf{A}$ is given by:

$$A_z = \frac{E_0(z)}{\omega} \cos(kz)\cos(\omega t + \phi_0)$$

$$A_r = \frac{E_0(z)r}{2c} \sin(kz)\cos(\omega t + \phi_0) \qquad (A4)$$

Assuming that the transverse mechanical momentum p_r^0 of the photoelectrons leaving the cathode is negligible (usually the starting energy is a fraction of eV), from the previous expression it can be deduced that the canonical momentum P_r^0 at the cathode surface is zero at all injection phase ϕ_0 . From the conservation of the canonical momentum we get that P_r must be zero all along the acceleration: the normalized transverse momentum (in unit of m_0c) will be given by $p_r = -eA_r/(m_0c)$.

In particular, recalling that the exit phase ϕ is defined as $\phi \equiv \omega T_f - kL + \phi_0$ and $kL = (N+1/2)\pi$, A_r at the gun exit is $A_r = -\frac{E_0(z)r}{2c}\sin\phi$. Hence the exit transverse momentum is given by:

$$p_r = \alpha \, k \, r \, \sin\phi \qquad (A5)$$

while during the acceleration, according to the impulsive approximation (ϕ_0 must be substituted with ϕ and $\omega t = kz$), p_r reads:

$$p_r = -\alpha \, k \, r \, \sin(kz)\cos(kz+\phi) \qquad (A6)$$

It must be stressed that the radial position r of the electron is indeed varying during the acceleration: in expression (A5) one should substitute the actual exit radial position in place of r. Nevertheless it is usual to approximate the exit radius with the starting position off-axis at the cathode, assuming a constant radius trajectory.

APPENDIX B - RF FIELD EXPANSION IN SPATIAL HARMONICS

We consider here a general form for a standing wave inside a multi-cell cavity resonating on a π mode: the structure is supposed to be axi-symmetrical with a central cell extending between z=-d and z=d, having z=0 as a symmetry (cathode) plane.
We assume also that the half-cell length can be different from $\lambda/4$, setting

$$d = (\lambda/4)(1+\delta) \qquad (B1)$$

with $\lambda = 2\pi/k = 2\pi c/\omega$. A suitable expansion for the field is:

$$E_z(r,z,t) = [a_1 J_0(|k_1|r) \cos(k'z) + a_3 I_0(k_3 r) \cos(3k'r) +$$
$$a_5 I_0(k_5 r) \cos(5k'r) + ...] \sin(\omega t + \phi_0) \qquad (B2)$$

$$E_r(r,z,t) = \frac{k'r}{2} [a_1 \hat{J}_1(|k_1|r) \sin(k'z) + 3a_3 \hat{I}_1(k_3 r) \sin(3k'r) +$$
$$5a_5 \hat{I}_1(k_5 r) \cos(5k'r) + ...] \sin(\omega t + \phi_0) \qquad (B2')$$

$$cB_\theta(r,z,t) = \frac{kr}{2} [a_1 \hat{J}_1(|k_1|r) \cos(k'z) + a_3 \hat{I}_1(k_3 r) \cos(3k'r) +$$
$$a_5 \hat{I}_1(k_5 r) \cos(5k'r) + ...] \cos(\omega t + \phi_0) \qquad (B2'')$$

where $\hat{I}_1(x) \equiv 2\dfrac{I_1(x)}{x}$ and $\hat{J}_1(x) \equiv \hat{I}_1(ix)$, while $k' \equiv \dfrac{k}{(1+\delta)}$ and

$$k_1 = ik\sqrt{1 - \frac{1}{(1+\delta)^2}} \quad ; \quad k_3 = k\sqrt{\frac{9}{(1+\delta)^2} - 1} \quad ; \quad k_5 = k\sqrt{\frac{25}{(1+\delta)^2} - 1}$$

The value of the amplitudes a_1 , a_3 , a_5 ... of each spatial harmonic should be found, in principle, by imposing the boundary conditions: since we are interested in studying a single half-cell, whose length is different from $\lambda/4$, coupled to an ideal multi-cell cavity operated in π mode with $\lambda/2$ cells, we must satisfy some continuity condition at the interface $z_i = (\lambda/4)(1+\delta)$ between the first half cell and the ideal structure ($I_1(x)$ is the first order modified Bessel function).

The first condition is the continuity of the longitudinal derivative of E_z ($\partial E_z/\partial z$) across z_i in order to satisfy the right coupling condition on the π mode: if the ideal multi-cell structure contains a resonant field of amplitude E_0 (as the one given by eq. (A1)) the following condition must hold

$$- a_1 k' + 3a_3 k' - 5a_5 k' + ... = -k E_0 \qquad (B3)$$

Moreover, since the field in the ideal multi-cell structure does not contain any non linear transverse component we must impose that at the interface z_i (just at the iris) all the non linear transverse components of E_r as specified in (B2') must vanish. Recalling that $\hat{I}_1(x) = 1 + x^2/8 + ...$ the condition on the third order components reads

$$- a_1 |k_1|^2 - 3a_3 k_3^2 + 5a_5 k_5^2 + ... = 0 \qquad (B4)$$

One should take into account all the higher order terms, solving finally for a_1 , a_3, a_5 etc. For sake of simplicity we will take into account only the first and third spatial harmonic, hence we solve (B3) and (B4) putting a_5 , a_7 , ... = 0. Substituting the given expression for k_1 and k_3, the solution for a_1 and a_3 comes out to be:

$$a_1 = \frac{E_0(2-\delta)(1+\delta)(4+\delta)}{8} \quad ; \quad a_3 = \frac{-E_0 \delta(1+\delta)(2+\delta)}{24} \qquad (B5)$$

The field E_z at the cathode (z=0) will be given by : $E_c = E_0(1+\delta)(6-2\delta-\delta^2)/6$. At $\delta=0.3$ (i.e. for a cell lengthening of 30%) the field increases by a factor 1.15 .

Substituting a_1 and a_3 given by eq. (B5) into the field expansion (B2), (B2') and (B2'') we get the RF field components in the half-cell: to compute the transverse and longitudinal momentum imparted from the field to an electron leaving the cathode at z=0 and being accelerated through the half-cell we will follow again the same approximation used in Appendix A. The electron is assumed to start from the cathode at v=c with a starting phase ϕ equal to the exit phase, as given by eq. (2), except that we must take into account the phase shift due to the variation in the cell length: here the exit phase will be given by $\phi = \dfrac{1}{2\alpha\sin\phi_0} + \phi_0 + \dfrac{\delta\pi}{2}$. The transverse momentum p_r (in unit of m_0c) at the half-cell exit can be calculated by:

$$p_r = \frac{\alpha k r a_1}{E_0} \int_0^{\pi(1+\delta)/2} \left[\frac{1}{1+\delta}\sin(\frac{x}{1+\delta})\sin(x+\phi) - \cos(\frac{x}{1+\delta})\cos(x+\phi) \right] dx +$$

$$\frac{\alpha k r a_3}{E_0} \int_0^{\pi(1+\delta)/2} \left[\frac{3}{1+\delta}\sin(\frac{3x}{1+\delta})\sin(x+\phi) - \cos(\frac{3x}{1+\delta})\cos(x+\phi) \right] dx$$

The result becomes, up to second order in δ,

$$p_r = \alpha k r \left(1 + \frac{2\delta}{3} - \frac{\delta^2}{2}\right) \sin(\phi) \tag{B6}$$

A similar calculation for the longitudinal momentum,

$$p_z = 1 + \frac{2\alpha a_1}{E_0} \int_0^{\pi(1+\delta)/2} \cos(\frac{x}{1+\delta})\sin(x+\phi)dx +$$

$$\frac{2\alpha a_3}{E_0} \int_0^{\pi(1+\delta)/2} \cos(\frac{3x}{1+\delta})\sin(x+\phi)dx$$

gives

$$p_z = 1 + \alpha [1 + \delta (4/3 + \pi^2/8) + 5\delta^2\pi^2/32] \cos(\phi) + \alpha [\pi/2 + 5\pi\delta/8 + \pi\delta^2(5/32 - \pi^2/48)] \sin(\phi) \tag{B7}$$

It must be noticed that the dimensionless parameter α is now given by $\alpha \equiv eE_c/(2m_0c^2k)$, where E_c is the field at the cathode given above. If the first half-cell is longer than the usual $\lambda/4$, the parameter α is increased, giving a most favourable condition for the damping of the space-charge emittance blow-up.

The normalized transverse rms emittance ε_x due to RF field displays again a minimum at $<\phi> = \pi/2$, where the first order correlation (p_r,ϕ) is vanishing, as can be inferred by (B6). Since the 2nd order term in $\Delta\phi^2$ is changed by a factor$\left(1 + \frac{2\delta}{3} - \frac{\delta^2}{2} \right)$, it follows that the minimum emittance will change by the same amount. Since the space charge emittance scales like E_c^{-1}, the change in the total emittance growth will depend on how much the two contributions (RF and space charge) share the total emittance blow up: for a space-charge dominated emittance a lengthening of the first half cell is, from this point of view, recommended since the growth of the field at the cathode damps more efficiently the space-charge term.

APPENDIX C - SPACE-CHARGE FIELD EXPANSION

We develop here the calculation of the radial component E_r for the electrostatic field of an axi-symmetric charge distribution, looking in particular for the linear and third order terms in the field.

Under the assumption of cylindrical symmetry the electrostatic field components is of the form:

$$E_z = E_0(z) + c_2(z)r^2 + c_4(z)r^4 + \dots \tag{C1}$$

$$E_r = c_1(z)r + c_3(z)r^3 + c_5(z)r^5 + \dots \tag{C1'}$$

where $E_0(z)$ is the field distribution on axis and the coefficients c_i are only function of the z coordinate.

We take a charge density distribution of the form

$$\rho\,(r,z) = \rho_0(z)\,[1 + \rho_2 r^2] \tag{C2}$$

where ρ_0 is dependent only on z and ρ_2 is a constant. Such a distribution represents a variable charge density along the longitudinal position z inside the bunch, with a constant parabolic radial dependence. Later on it is shown that a gaussian distributed bunch, radially clipped in a suitable way, can be represented with a good approximation by such a charge distribution. The uniform distributed bunch is of course obtained putting $\rho_0 = $ const. and $\rho_2 = 0$.

Expressing the Gauss theorem $\nabla \cdot \mathbf{E} = \rho / \varepsilon_0$ in terms of the components given in (C1) and (C1') it is found

$$2c_1 + 4r^2 c_3 + E_0' + c_2' r^2 = \frac{\rho_0(z)\,[1 + \rho_2 r^2]}{\varepsilon_0}$$

where $E_0' \equiv \dfrac{dE_0}{dz}$, $E_0'' \equiv \dfrac{d^2E_0}{dz^2}$, $E_0''' \equiv \dfrac{d^3E_0}{dz^3}$, $\rho_0' \equiv \dfrac{d\rho_0}{dz}$ and $\rho_0'' \equiv \dfrac{d^2\rho_0}{dz^2}$

Equating terms of same order on the l.h.s. and on the r.h.s we get

$$c_1 = -(1/2)E_0' + \frac{\rho_0}{2\varepsilon_0} \tag{C3}$$

$$4c_3 + c_2' = \frac{\rho_0 \rho_2}{\varepsilon_0} \tag{C3'}$$

From $\nabla \times E = 0$ it can be found that $c_1' = c_2$. Taking the derivative with respect to z of this expression and substituting in (C3') we find

$$c_3 = \frac{\rho_0 \rho_2}{4\varepsilon_0} + \frac{E_0'''}{16} - \frac{\rho_0'}{16\varepsilon_0} \qquad (C4)$$

Moreover, from $\nabla \times \nabla \times E = 0$ we get $\nabla^2 E = \nabla \rho / \varepsilon_0$, hence $\nabla^2 E_z = \frac{\rho_0'}{\varepsilon_0}$,

which corresponds to $E_0'' + 4c_2 = \frac{\rho_0'}{\varepsilon_0}$, from which it follows

$$c_2 = \frac{\rho_0'}{4\varepsilon_0} - \frac{E_0''}{4} \qquad (C5)$$

The field components E_z and E_r can be therefore expanded off axis up to third order, as given by

$$E_z (r,z) = E_0(z) + [\frac{\rho_0'}{\varepsilon_0} - E_0''] \frac{r^2}{4}$$

$$(C6)$$

$$E_r (r,z) = [\frac{\rho_0}{\varepsilon_0} - E_0'] \frac{r}{2} + [\frac{4\rho_0 \rho_2}{\varepsilon_0} + E_0''' - \frac{\rho_0''}{\varepsilon_0}] \frac{r^3}{16}$$

In order to minimize the emittance blow up due to third order components in the space-charge field, one would like to make vanishing the coefficient c_3. In the following we study the possibility to achieve such a condition using uniform distributed bunches with either very high aspect ratio (short pancakes) or very low (long cigars). The simplest choice, in order to get $c_3=0$, is indeed to put $\rho_0'' = 0$ and trying to compensate the third derivative of the axis field E_0''' with a proper choice of ρ_2. It will be proved that in the case of high aspect ratio bunches E_0''' becomes nearly a constant (against z), hence it can be compensated exactly. The choice of uniform bunches ($\rho_0 = $ const) must be intended as a seek of simplicity both from the point of view of the calculations and from the point of view of the requirements on the laser pulse characteristics.

In the following we take therefore a uniform-parabolic charge distribution as given by (C2) with $\rho_0 = $ const .

The field on axis produced by such a distribution , extending from z=0 up to z=L and with radius R, is

$$E_0(z) = \frac{\rho_0}{\varepsilon_0} \{ 2z-L + \sqrt{R^2+(z-L)^2} - \sqrt{R^2+z^2} +$$

$$\rho_2 [\frac{2}{3}((L-z)^3-z^3) - R^2\sqrt{R^2+(z-L)^2} + R^2\sqrt{R^2+(z-L)^2} + \qquad (C7)$$

$$\frac{2}{3}(R^2+z^2)^{(3/2)} - \frac{2}{3}(R^2+(z-L)^2)^{(3/2)}] \}$$

Substituting this expression for $E_0(z)$ in the second one of (C6) allows to calculate the two components of E_r, namely the first and the third order one.

Defining $E_r(r,z) = E_1(z)r_b + E_3(z)r_b{}^3$, we get

$$E_1(z_b) = \frac{\rho_0 R}{\varepsilon_0} \left\{ \frac{z_b}{2\sqrt{1+z_b{}^2}} - \frac{z_b-L_b}{2\sqrt{1+(z_b-L_b)^2}} + \right.$$

$$\left. \rho_{2b} \left[z_b{}^2 + (z_b-L_b)^2 - \frac{z_b(2z_b{}^2-1)}{2\sqrt{1+z_b{}^2}} + \frac{(z_b-L_b)(2(z_b-L_b)^2-1)}{2\sqrt{1+(z_b-L_b)^2}} \right] \right\} \qquad (C8)$$

$$E_3(z_b) = \frac{\rho_0 R}{16\varepsilon_0} \left\{ \frac{3z_b}{(1+z_b{}^2)^{(5/2)}} - \frac{3(z_b-L_b)}{(1+(z_b-L_b)^2)^{(5/2)}} + \right.$$

$$\left. \rho_{2b} \left[\frac{9z_b+10z_b{}^3+4z_b{}^5}{(1+z_b{}^2)^{(5/2)}} + \frac{9(z_b-L_b)+10(z_b-L_b)^3+4(z_b-L_b)^5}{(1+(z_b-L_b)^2)^{(5/2)}} \right] \right\} \qquad (C8')$$

after defining the inverse of the aspect ratio $L_b \equiv 1/A$ ($L_b=L/R$), the normalized coordinates $z_b \equiv z/R$, $r_b \equiv r/R$ and the dimensionless parabolic coefficient of the charge distribution $\rho_{2b} \equiv \rho_2 R^2$. Evaluating E_3 at the centre of the bunch ($z_b=L_b/2$) in the limit of large bunch aspect ratio (i.e. for very thin pancake, $L_b\rightarrow 0$), it is found

$$\lim_{L_b->0} \left[E_3\left(\frac{L_b}{2}\right) \right] = \frac{3}{16} (1+3\rho_{2b}) \qquad (C9)$$

Therefore, the optimum value of the parabolic coefficient, able to cancel out the third order component in the transverse field, will be $\rho_{2b} = -1/3$.

Being Q the bunch charge, the charge density coefficient ρ_0 will be given by

$$\rho_0 = \frac{Q}{\pi L R^2 (1+0.5\rho_{2b})} \qquad (C9)$$

Substituting (C9) in (C8) and (C8') and taking for ρ_{2b} the optimum value, we find

$$E_1^{opt}(z_b) = \frac{Q}{5\pi\varepsilon_0 R^2} \left[2z_b - L_b + \frac{z_b\sqrt{1+z_b{}^2}-2z_b{}^2}{L_b} - \frac{(z_b-L_b)\sqrt{1+(z_b-L_b)^2}}{L_b} \right] \qquad (C10)$$

$$E_3^{opt}(z_b) = \frac{Q}{40\pi\varepsilon_0 R^2} \left[\frac{-z_b{}^3(5+2z_b{}^2)}{L_b(1+z_b{}^2)^{(5/2)}} + \frac{(z_b-L_b)^3(5+2(z_b-L_b)^2)}{L_b(1+(z_b-L_b)^2)^{(5/2)}} \right] \qquad (C10')$$

Since for large aspect ratio bunches it holds $L_b \ll 1$ and $z_b \ll 1$, we take the lowest order in z_b and L_b of expressions (C10) and (C10'), finally getting:

$$E_1^{opt}(z_b) = \frac{Q}{10\pi\varepsilon_0 R^2} [2 - 2L_b + L_b^2 + (4 - 3L_b)z_b + (3 - \frac{4}{L_b})z_b^2] \quad (C11)$$

$$E_3^{opt}(z_b) = - \frac{Q}{8\pi\varepsilon_0 R^2} (L_b^2 - 3L_b z_b + 3z_b^2) \quad (C11')$$

By means of a similar evaluation the expression for the first and third order components can be found in case of a non optimized distribution, i.e. for a distribution that is uniform both in the longitudinal and in the transverse direction. Putting $\rho_{2b} = 0$ in (C8) and (C8') and taking again the lowest order in L_b and z_b it is found:

$$E_1^{uni}(z_b) = \frac{Q}{8\pi\varepsilon_0 R^2} (2 - L_b^2 + 3L_b z_b - 3z_b^2) \quad (C12)$$

$$E_3^{uni}(z_b) = - \frac{Q}{8\pi\varepsilon_0 R^2} (\frac{3}{4} - \frac{15}{8}L_b^2 + \frac{45}{8}L_b z_b - \frac{45}{8}z_b^2) \quad (C12')$$

It is interesting to note that the third order component E_3^{uni} of the non optimized uniform distribution has a finite value as L_b tends to zero, while the third order component E_3^{opt} of the optimized parabolic distribution is vanishing as $O(L_b^2)$. That has a relevant effect on the emittance blow up induced by the space-charge field, as discussed in section 4.

REFERENCES

[1] - J.S.Fraser and R.L.Sheffield, IEEE, QE-23, 9, 1987, p.1489.
      C.Travier, NIM A304 (1991), p.285.
[2] - K.J.Kim, NIM A275 (1989), p.201.
[3] - J.C.Gallardo and R.B.Palmer, BNL-43862, CAP 54-ATF-90J.
[4] - K.T.McDonald, "Design of the laser-driven Radio-Frequency electron Gun for the Brookhaven Accelerator Test Facility", Princeton University int. rep., January 17, 1988.
[5] - B.E.Carlsten, NIM A285 (1989), p.313.
[6] - L.Serafini et al., Proc. 2nd EPAC, Nice, June 1990, Ed. Frontieres, p.143.
[7] - K.Batchelor, H.Kirk, J.Sheehan, M.Woodle, K.McDonald, Proc. 1st EPAC, Rome, June 1988, Ed. World Scientific, p.954.
[8] - P.M.Lapostolle, IEEE Trans. NS-18, no.3 (1971), p.1101.
[9] - L.Serafini et al., NIM A318 (1992), p.275.
[10] - C.E.Hess et al., IEEE Trans. NS-32 no.5 (1985), p.2924.
       T.I.Smith, "Intense low emittance linac beams for free electron lasers", Proc. Linac Conf., SLAC PUB-303 (1986).
       T.I.Smith, NIM A250 (1986), p.64.
[11] - L.Serafini et al., INFN rep. INFN/TC-91/07 (1991).
[12] - L.Serafini and C.Pagani, Proc. 1st EPAC, Rome, June 1988, Ed. World Scientific, p.866.
[13] - I.S.Lehrman et al., "Design of a high-brightness, high-duty factor photocathode electron gun", Proc. FEL Int. Conf. 1991, Santa Fe, Aug. 1991.

R.F. PHOTOCATHODE GUNS TRIGGERED BY
SYNCHROTRON LIGHT

G.Dattoli, L.Giannessi

ENEA, Area INN, Dipartimento Sviluppo Tecnologie di Punta,
C.R.E. Frascati, P.O. Box 65, 00044 Frascati (RM), Italy

L.Serafini

INFN, Via Celoria 13, 20133 Milan, Italy

ABSTRACT

The conditions under which synchrotron radiation can be used to trigger a photocathode gun are analysed. We show that such a solution offers in principle the possibility of operating with a pulsed electron beam filling all the buckets of the accelerating radio-frequency field.

INTRODUCTION

In the last few years we have witnessed to a wide diffusion of the radio frequency (RF) photocathode guns technology for those applications requiring a high brightness electron source. An example is given by Free Electron Lasers (FEL) operating at short wavelengths, which demands for high phase space electron density, i.e. low emittances, low energy spread and high current beams. The RF photocathode gun seems to be the best solution for achieving such requirements. However one of the drawbacks of a photocathode gun when compared to a more conventional solution, like the thermoionic injection, is that the latter allows to fill all the buckets of the accelerating field, while in a photocathode driven gun, the maximum repetition rate is limited by the maximum operating frequency of the driving laser. There is however a natural way to overcome the problem. The coherence properties of the laser light are indeed not necessary to trigger the cathode. What is required is enough optical power to extract the necessary charge from the cathode surface. A possible source of radiation might be the synchrotron light emitted by the bunch of electrons itself, while traversing a magnetic structure like a wiggler or an undulator[1]. The advantages offered by this scheme are evident: the light hitting the cathode is shaped by the electron bunch, the limits on the repetition rate due to the driving laser are overcame[*], and the problem of triggering the light with the R.F. is also simplified since the

---

(*)There are obviously other mechanisms limiting the repetition rate operation of an R.F. Gun, like the R.F. power at disposal, or the heat dissipation in the gun.

system is self locked in phase and the time jitter is suppressed. A further advantage consist in the wavelength tunability of synchrotron radiation, that allows to optimize the photons energy to the work function of the cathode material. In principle, new materials with work functions at higher energies might be used.

SYNCHROTRON RADIATION AND
RF PHOTOCATHODE TRIGGERING

In fig. 1 it is shown a schematic layout of the synchrotron radiation triggered photocathode (SRPT). We have essentially a closed loop with the conversion of electrons in photons in a wiggler/undulator and of photons in electrons on the cathode. We denote by $\eta_{e \to p}$ the conversion efficiency of electrons to photons in the undulator, and by $\eta_{p \to e}$ the quantum efficiency of the photocathode. The process of SRPT is self sustained when

$$\eta_{e \to p}\eta_{p \to e} > 1 \qquad (1)$$

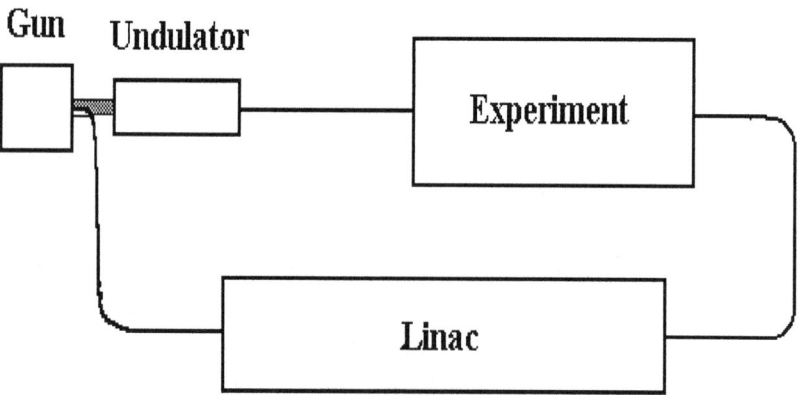

Fig. 1. Layout of the SRPT device.

The number of photons radiated per second by an electron beam with current I and energy $E = \gamma mc^2$, moving in a planar undulator with parameter K, N periods and period length λ_u, is given by

$$N_p = \frac{\alpha N K^2}{2n} \frac{I}{e} F(n,K)$$ (2)

where n is the harmonic index and

$$F(n,K) = 4n^2 \chi \left[J_{\frac{n+1}{2}}(n\chi) - J_{\frac{n-1}{2}}(n\chi) \right]^2, \quad \chi = \frac{1}{4} \frac{K^2}{1+K^2/2}$$ (3)

The photons are emitted in a solid angle

$$\Delta\Omega_i = \pi(K/\gamma)^2$$ (4)

at the wavelengths

$$\lambda_n = \frac{\lambda_u}{2\gamma^2 n}(1+K^2/2) \quad , \quad \omega_n = \frac{2\pi c}{\lambda_n}$$ (5)

in a frequency interval

$$\Delta\omega_i = \frac{\omega_n}{2Nn}$$ (6)

The electron to photons conversion efficiency is then given by[1]

$$\eta_{e\to p} = \frac{\alpha\pi N K^2}{2n} F(n,K)$$ (7)

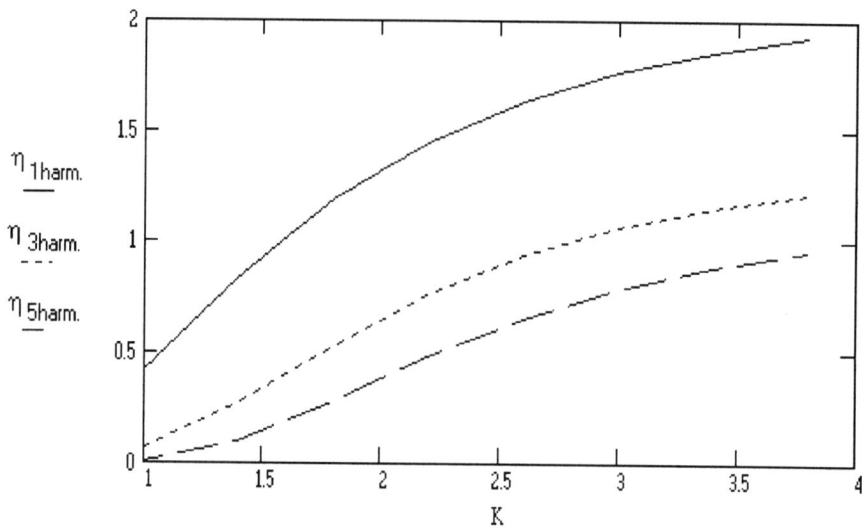

Fig. 2. Efficiency e->p versus K

In fig. 2 we show the plot of the efficiency vs. K for the first three odd harmonics emitted from an undulator of 100 periods. As a practical example let us consider the case of a high efficiency photocathode material like the Cs_3Sb with a work function of approximately 2eV. At 540 nm we can expect a quite large quantum efficiency, of the order of 10%. Considering a wiggler with parameter K=4.5 and assuming that $\lambda = 540$nm is the wavelength of the first resonant harmonic, the condition (1) is a constraint on the number of periods of the wiggler to insure a positive "gain" in the time behaviour of the circulating electrons/photons flux. The threshold is reached for 500 periods. The global length of the magnetic structure depends on the period length, which is a function of the electron energy. At 400 Mev[**], with the specified parameters, the period is 6 cm long and the wiggler length is on the order of 30m. It is clear that SRPT may become a realistic device in a technological context where high remanent magnetic materials, together with high quantum efficiency cathode materials have been developed. On the other hand there are mechanisms enhancing the amount of synchrotron light emitted by electrons, as the free electron laser, or the harmonic generation process due to the FEL action. The distinguishing feature of the emission in these processes consist in the fact that the e-beam is longitudinally

---

[**] The electron energy must be considered as an external parameter and cannot be determined on the basis of the SRPT operation. The value of 400 Mev is only indicative and is given to add physical insight to the example considered.

"bunched", in the sense that the harmonic content of the longitudinal charge or current distribution at the wavelength of resonance, is strongly enhanced by the interaction with an optical electromagnetic wave.

RF PHOTOCATHODE AND HARMONIC GENERATION

As to the harmonic generation process, we can estimate the number of photons emitted by a beam with some harmonic content at the wavelength λ_n necessary for photoemission from the cathode. The harmonic content can be specified through the nth Fourier coefficient of the longitudinal charge distribution. The number of photons emitted from a bunched beam of current I is given by[2]

$$N_b = N_i \frac{I \Delta t}{e} \frac{\Delta \Omega}{\Delta \Omega_i} \frac{\Delta \omega}{\Delta \omega_i} |\rho_n|^2 \tag{8}$$

where Δt is the macropulse duration, $\Delta \omega$ and $\Delta \Omega$ are the frequency and angle intervals where the photons are emitted and $\Delta \omega_i$ and $\Delta \Omega_i$ are their counterpart in the case of incoherent emission. Assuming the periodicity over all the microbunch, we have that $\Delta \omega \Delta t \approx 2\pi$, while the order of magnitude of the solid angle $\Delta \Omega$ can be determined evaluating the correction due to the interference between microbunches as a function of the azimuthal angle. At the wavelength λ_n, the source of radiation can be thought as a series of $L = \Delta t c / \lambda_n$ source points equally spaced of λ_n (see fig. 3).

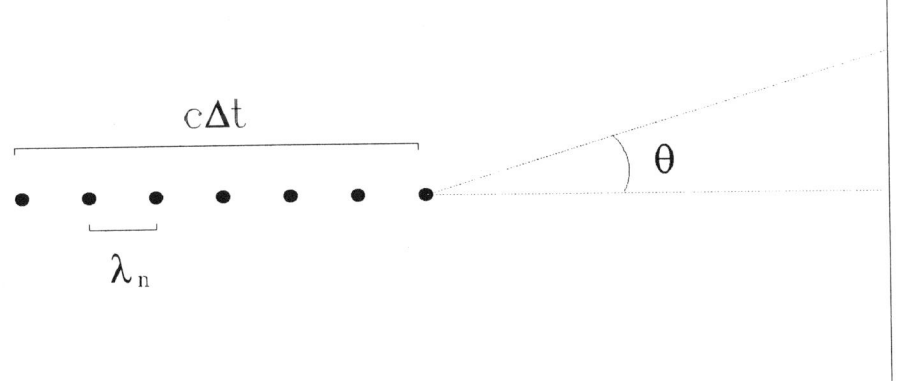

Fig. 3. Geometry of emission.

The interference between source points will introduce the following axially symmetric dependence from the angle θ

$$\frac{\sin^2(\pi(L+1)\cos(\theta))}{\sin^2(\pi\cos(\theta))} \tag{9}$$

the first minimum occurs at

$$\theta_b = \sqrt{2/L} \tag{10}$$

corresponding to a solid angle $\Delta\Omega_b \approx 2\pi/L$.

The cone of emission will be determined by the superposition of the bunched term and the unbunched term (4). A rough estimate of the ratio $\Delta\Omega/\Delta\Omega_i$ is then given by

$$\frac{\Delta\Omega}{\Delta\Omega_i} \approx \frac{\Delta\Omega_b}{\Delta\Omega_i + \Delta\Omega_b} \tag{11}$$

Assuming e.g. $\Delta t \approx 10ps$ and $\lambda_n \approx 540nm$ as typical values we have $\Delta\Omega/\Delta\Omega_i \sim 1$ since the bunched term is larger then the unbunched one. The efficiency for the bunched beam emission can be written then as

$$\eta_{e\rightarrow p,b} = \eta_i |\rho|^2 \frac{I}{ec} nN\lambda_n \tag{11}$$

the threshold condition (1) is now a function of the circulating current and of the bunching coefficient. In figs. 4 and 5 it is shown the plot of the necessary number of periods to have positive gain from the bunched beam SRPT operation respectively on the first and on the third harmonic. We have assumed a threshold current of 10 A and bunching coefficients $\rho \sim 0.01$ for the first and $\rho \sim 0.001$ for the third harmonic.

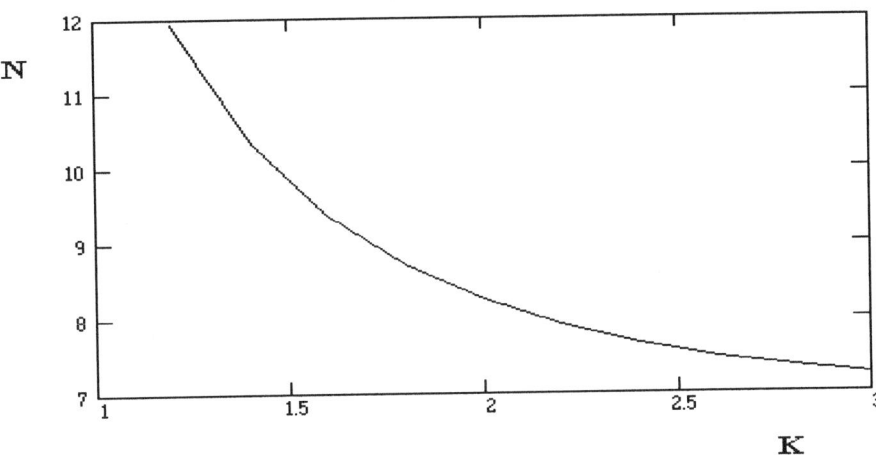

Fig. 4. Threshold periods number vs. K. First harmonic operation.

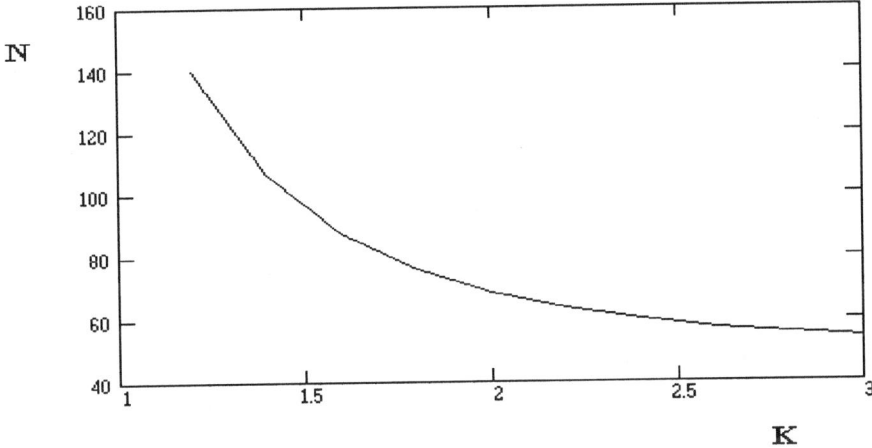

Fig. 5. Threshold periods number vs. K. Third harmonic operation.

The small number of undulator periods required suggests also the possibility of operating a metallic cathode, with a much lower quantum efficiency, but with a longer life time and less contamination problems. In recent tests at Brookhaven[3]it has been measured the quantum efficiency of metallic cathodes for short pulses (10 ps) at a photon energy of 4.66eV. The quantum efficiency measured for Samarium cathodes was 7.2710^{-4}. As an example, we consider a threshold current of 100A, and $\rho \sim 0.01$ for the first harmonic. The plot of the threshold number of periods vs. K is shown in fig 6.

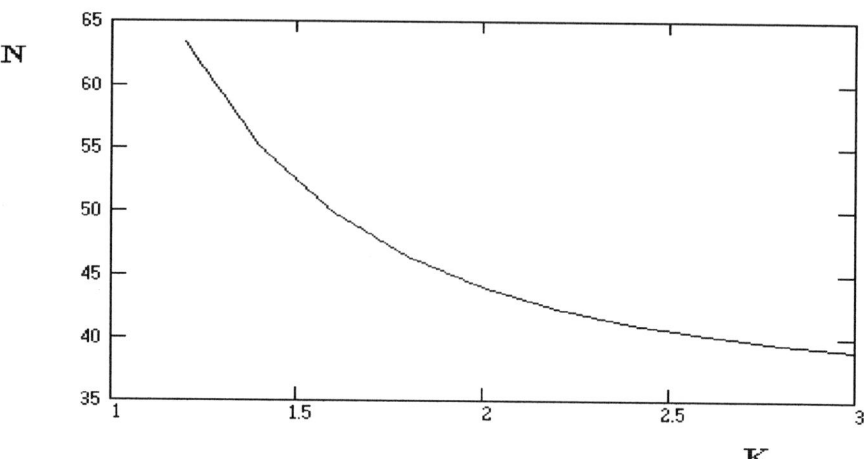

Fig. 6. Threshold periods number vs K, metallic photocathode.

THE STARTUP

A question that naturally arise is, how do we start the SRPT operation ? There are different ways to approach this problem. As we have already mentioned, the SRPT scheme consist essentially in a closed loop, where we switch electrons in photons and vice versa with some "gain" defined as the product of the conversion efficiencies $\eta_{p \to e}$ and $\eta_{e \to p}$. An important parameter is then the time taken by the system to produce a self generated light pulse on the photocathode itself. We will indicate this time interval as the round trip time T_{rt}. The operation at low repetition rate can be primed by an external laser pulse hitting the cathode like in conventional photocathodes. The pulse provided will fill up one of the buckets of the accelerating field, and the system will operate at a frequency $1/T_{rt}$. The resulting frequency is of the order of few MHz, depending on the optical path followed by the electrons/photons bunch along the loop. It is evident that this repetition rate can be increased simply by sending another priming pulse on a different bucket of the accelerating field, and filling of electrons another radio frequency cycle. Another possibility is that of driving the cathode with an external laser with a pulse duration T_{rt}, properly modulated at the r.f. frequency with a Pockels cell. Assuming that inequality (1) is satisfied, then the system should generate a self sustained growth of the circulating current, and modulating longitudinally the train of electrons/photons as a function of the frequency response of the gain function.

References

[1] G.Dattoli, L.Giannessi, A.Renieri, A.Marino, L.Serafini, To be published in Appl. Phys. B

[2] F.De Martini "Coherent Harmonics Generation in Free Electron Lasers" in Laser Handbook 6, p.195, ed. by W.B.Colson, C.Pellegrini and A.Renieri (1990)

[3] T.Srinivasan-Rao, J.Fisher, T.Tsang, BNL Report 45318, October 1990, to be published in J. Applied Physics)

Asymmetric Emittance Rf Photocathode Source for Linear Collider Applications

J.B. Rosenzweig and J. Smolin
UCLA Department of Physics
405 Hilgard Ave.
Los Angeles, CA 90024

November 13, 1992

Abstract

Laser driven rf photocathodes represent a recent advance in high-brightness electron beam sources. We investigate here a variation on these devices, that obtained by using a ribbon laser pulse to illuminate the cathode, yielding a flat beam ($\sigma_x \gg \sigma_y$) which has asymmetric emittances at the cathode proportional to the beam size each transverse dimension. The flat-beam geometry mitigates space charge forces which lead to intensity dependent transverse and longitudinal emittance growth, thus limiting the beam brightness. The fundamental limit on achievable emittance and brightness is set by the transverse momentum distribution and peak current density of the photo-electrons (photon energy and cathode material dependent effects) and appears to allow, taking into account space charge and rf effects, normalized emittances $\epsilon_x < 1 \times 10^{-4}$ m-rad and $\epsilon_y < 10^{-6}$ m-rad, with $Q = 5$ nC and $\sigma_z = 1$ mm. These source emittances are adequate for superconducting linear collider applications, and could preclude the use of a damping ring for the electrons in these schemes.

Introduction

The rapid development of rf photocathodes as high-brightness, low emittance electron sources has been spurred on by their potential for use as injectors for linacs which drive free-electron lasers (FELs) and linear colliders[1]. For linear collider applications it is natural to consider sources which give asymmetric ($\epsilon_x \gg \epsilon_y$) emittances, as this asymmetry is necessary for the final focus, which must produce flat beams ($\sigma_x^* \gg \sigma_y^*$). Collisions of flat beams is an almost universal feature of current linear collider designs, as beam flatness can be used to diminish the effects of the coherent beam-beam focusing, or disruption[2]. The transverse acceleration produced by the beam-beam interaction can cause the colliding beam particles to radiate photons (beamstrahlung), which induces an undesirable spread in collision energies. Detector backgrounds arising from coherent production of e^+e^- pairs by beamstrahlung photons in the intense electromagnetic fields of the oncoming beam is another source of serious beam-beam related problems[2] at the interaction region.

At the lowest energies, as the electron beam leaves the cathode, it is the self-fields of the beam (space-charge fields) which are problematic, as they can give rise to non-uniform defocusing of the beam, which produces transverse emittance growth. The same scheme can be applied for mitigating the strength of the beam's fields here as is applied at the final focus – flattening the beam profile while keeping the cross-sectional area of the beam constant. This not only reduces the effects of space-charge, but also naturally produces an asymmetric emittance.

A preliminary analysis of the prospects for this scheme are presented here. It appears that the demands for the electron emittances of a superconducting linear collider[3], which does not utilize such small beams at collision as a normal-conducting machine, can be met using a flat-beam rf photocathode, thus eliminating the need for an electron damping ring. This device can also be considered for supplying the electron beam emittances demanded by the CLIC design[4], which are not too different than those of a superconducting machine. In addition, a recently proposed far infrared FEL scheme[5], which uses a wave-guide with a few mm height to eliminate slippage between the beam and optical pulse, requires very small beam heights (a few mm) at low energy (< 15 MeV), due to the aperture restriction of the wave-guide. Asymmetric emittances from an rf photocathode source, combined with the natural vertical focusing of a planar wiggler, may allow the small beam height combined with high peak current needed for this application.

Inherent Emittance and Charge Limitations in Photocathodes

The inherent spread in transverse energy of photo-electrons emitted from the cathode is a complicated function of the photon energy, the applied electric fields, and the electronic structure and work function of the cathode material. It is best to take a characteristic energy, typically the difference between the photon energy and the work function, as the transverse temperature of the beam at emission. We take this temperature to be approximately $kT_\perp = 0.5$ eV for the purposes of this paper. The normalized rms horizontal and vertical thermal emittances of the beam at the cathode are simply[6]

$$\epsilon_{n(x,y)}^{th} = \sqrt{\frac{kT_\perp}{mc^2}}\sigma_{x,y} \simeq \sigma_{x,y}(\text{mm}) \times 10^{-6} \text{ m} \cdot \text{rad}.$$

Thus, to take the example of a superconducting linear collider design, to achieve emittances of $\sigma_{y,x} \leq 1,50$ mm-mrad respectively (with a number of particles in a bunch $N_b = 3 \times 10^{10}$), one would illuminate the cathode with a ribbon laser pulse of dimensions less than or equal to $\sigma_{y,x} = 1,50$ mm. Since σ_x is not small compared to typical L or S-band cavity dimensions, it is certain that the beam sizes must be even smaller than dictated by this inherent limit.

The peak current density derived from the cathode is in most cases not limited by physics of the laser-surface interaction, as will be seen below, but by the longitudinal space charge effects near the cathode. Since short pulse beams can usually be considered to be infinitesimally thin at very low energy ($\sigma_z \simeq v_z\sigma_t \ll \sigma_{x,y}$), the maximum decelerating electric force at the trailing end of the beam pulse is approximately, including the contribution from image charges,

$$eE_z \simeq 4\pi e^2 \Sigma_b = \frac{2N_b r_e m_e c^2}{\sigma_x \sigma_y}$$

where N_b is the number of electrons per pulse. If this field is as large as the rf accelerating field $eE_{rf}\sin(\phi_0)$ (ϕ_0 is the initial phase of the rf), then one has a completely space-charge limited flow, analogous to the Child-Langmuir limit for gaps. In practice, the decelerating force must be kept much smaller than the applied accelerating field. We take the practical limit here to be a factor four times smaller than given above, as even at longitudinal space-charge fields much lower than the limit the bunch may lengthen significantly due to differential acceleration of the front and back of the beam, and we are interested in preserving short pulses. Thus the peak surface density allowed is

$$\frac{N_b}{\sigma_x \sigma_y} \leq \frac{eE_{rf}\sin(\phi_0)}{2r_e m_e c^2}$$

and for $eE_{rf} = 80$ MeV/m, $N_b \simeq 2.8 \times 10^{10}\sigma_x\sigma_y(\text{mm}^2)$. For the present example we have $N_b \leq 1.4 \times 10^{12}$, and we can afford to make our initial beam sizes, and thus emittances, smaller by large factors in each plane and still run at the desired N_b; the constraint due to longitudinal space charge can be restated as $\sigma_x \sigma_y \geq 2.7$ mm^2.

There is a final constraint, on the number of liberated electrons per unit area yet to be examined, which is due to cathode surface damage from the laser, as alluded to above. The integrated laser intensity limit has been found experimentally[7] to be about $W = 100\mu J/cm^2$, which for $E_\gamma = 4.66$ eV photons (frequency quadrupled YAG) and a quantum efficiency of $\eta = 1.4 \times 10^{-4}$ (for room temperature copper) gives a limit of $N_b/\sigma_x \sigma_y \leq 2\pi\eta W/E_\gamma = 1.1 \times 10^{11}$ mm^{-2}, which is nearly an order of magnitude larger than due to longitudinal space charge. Thus one is most likely to have limits on the beam intensity due to longitudinal space-charge effects and not peak laser intensity considerations.

Transverse Emittance Growth

The blowup of transverse emittance by phase dependent rf focusing and by transverse space-charge forces has been analyzed by K.J. Kim for round beams[8]. The emittance increase due to phase dependent focusing from the rf in a flat beam, assuming negligible transverse electron motion inside the gun, can be written as a simple extension of Kim's results,

$$\epsilon_{n(x,y)}^{rf} = \frac{eE_{rf}}{\sqrt{2}m_e c^2}(\Delta\phi)^2\sigma_{x,y}^2,$$

where $\Delta\phi = 2\pi\sigma_z/\lambda_{rf}$. It is apparent from this expression that the larger of the two emittances will be affected by this limit first. For our example, with a pulse length of 1 mm, we now take a peak accelerating field of 80 MeV/m in a 1.3 GHz rf gun, and find that the emittance is rf limited when $\sigma_x \geq 12$ mm ($\epsilon_x \geq 12$ mm-mrad), smaller than our maximum beam size due to thermal effects. The rf focusing effects can be ameliorated by running at a lower rf frequency, or by shortening the bunch. In addition, it has been proposed by Serafini et al. to remove the phase space correlations by use of an independent rf cavity after the gun[9]. This scheme may allow total cancellation of the rf derived emittance growth, but is very dependent on the beam particle dynamics inside of the gun. A less sensitive method for cancellation of the phase dependence of the rf focusing using a higher harmonic of the rf ("flat-topping") has also been proposed by Serafini et al.[10] This scheme has an additional advantage, in that a linear energy-phase correlation can be put on the beam in this way, allowing subsequent compression of the pulse length.

Calculation of space-charge driven emittance growth is properly done by by self-consistent simulations, an example of which is shown below. It is instructive, however, to attempt an estimate (good to within a factor of approximately two) of the space-charge contribution to the emittance. We begin by considering the vertical forces of the beam in its rest frame. If the beam is wide in this frame

$(\sigma_x \gg \sigma_y, \sigma_x \gg \gamma\sigma_z)$, then we can exploit Sacherer's result for evolution of the mixed second moment of the vertical distribution function

$$\langle y \frac{dp_y}{dt} \rangle = \frac{Ne^2 \sigma_y}{\sqrt{2\pi}\sigma_x(\sigma_z + \sigma_y)}$$

Simulations indicate that the majority of the emittance growth occurs before the beam particles have a chance to move a significant distance transversely, and thus the time integral of this mixed moment gives a measure of the emittance growth. For an electron distribution subject to a momentum kick of this type, the emittance due to the kick is approximately equal to this moment normalized to $m_e c$

$$\epsilon_{y,n}^{sc} = \frac{1}{m_e c}\sqrt{\langle y^2 \rangle \langle (p_y)^2 \rangle - \langle yp_y \rangle^2} = u(\sigma_y/\sigma_z)\frac{|\langle yp_y \rangle|}{m_e c}.$$

where $u(\sigma_y/\sigma_z)$ is a geometric factor close to to unity for a bi-Gaussian distributions arbitrary aspect ratio.

To convert the electric force in the rest frame to an electromagnetic force in the laboratory frame, we note that the beam length $\sigma_z = \beta c \sigma_t$ in the laboratory frame is expanded by a factor of γ in the beam frame. The vertical electric force obtained in this geometry is converted to the laboratory electromagnetic force by dividing by a factor of γ^2, which is derived from the partial cancellation of the electrostatic repulsion by the magnetic self-focusing of the beam as it picks up velocity. The emittance growth can be most simply written by choosing β as the independent variable to obtain an integral with a nonsingular argument, and integrating from the approximate point, taking into account the effects of the image charges, $z = \sigma_y$ where the space-charge force changes from longitudinal to transverse, to relativistic energies,

$$\epsilon_{y,n}^{sc} \simeq \frac{N_b r_e \sigma_y}{\sqrt{2\pi}\sigma_x W} \int_{\sqrt{2W\sigma_y}}^{1} \frac{d\beta}{(\sigma_y/\sigma_z)\sqrt{1 - \beta^2} + \beta}$$

where $W = eE_{acc}/mc^2 \sin(\phi_0)$. This expression has been evaluated numerically and fit to simple functions, giving

$$\epsilon_{y,n}^{sc} \simeq \frac{2N_b r_e}{7\sigma_x W} \exp\left(-3\sqrt{W\sigma_y}\right)\sqrt{\frac{\sigma_y}{\sigma_z}},$$

We have not attempted as yet to estimate the space-charge induced emittance growth in the horizontal dimension. A rule of thumb in ellipsoidally symmetric charge distributions is that the field at the edge of the beam along any of the symmetry axes is approximately equal to any of the others. Thus we can estimate

$$\epsilon_{x,n}^{sc} \simeq \epsilon_{y,n}^{sc}\frac{\sigma_x}{\sigma_y} \simeq \frac{2N_b r_e}{7\sigma_y W} \exp\left(-3\sqrt{W\sigma_y}\right)\sqrt{\frac{\sigma_y}{\sigma_z}},$$

This implies that raising σ_x to minimize $\epsilon_{x,n}^{sc}$ will cause an inversely proportional rise in $\epsilon_{y,n}^{sc}$, and full constraint of the beam brightness is implied in the preceding equations.

For an example relevant to superconducting linear colliders we now take $Q = 5$ nC the rf frequency to be $f_{rf} = 476$ MHz, the acceleratating field $E_{rf} = 60$ MV/m and the horizontal beam size $\sigma_x = 20$ mm – approximately where the rf contribution to emittance growth asserts itself – and $\sigma_{y,z} = 0.1, 1.25$ mm (note the constraint $\sigma_x \sigma_y \geq N_b/2.8 \times 10^{10}$ mm^2 is satisfied). The total emittances for this example are now, $\epsilon_{nx} = \sqrt{(\epsilon_{nx}^{th})^2 + (\epsilon_{nx}^{rf})^2 + (\epsilon_{nx}^{sc})^2} = 250$ mm-mrad, and $\epsilon_{ny} \leq \sqrt{(\epsilon_{ny}^{th})^2 + (\epsilon_{ny}^{rf})^2 + (\epsilon_{ny}^{sc})^2} = 1.4$ mm-mrad, both reasonably near our design tolerances, especially the vertical emittance, which is critical for achieving high luminosity. The $\leq$ sign is used to indicate that the space-charge and rf derived emittance growth is not independent – correlations should make the total emittance smaller than the sum of squares.

The simulation program PARMELA (the photo-injector version originating from K. McDonald) has been modified to accept an asymmetric initial distribution, and to perform a point-by-point calculation of the space charge forces. The rf forces are calculated using a Fourier decomposition of electromagnetic fields derived from the axisymmetric code SUPERFISH. The results of a simulation using our example input parameters is presented in Fig. 1, which shows the evolution of the emittances in a one-half π-mode cell gun. The emittance growth is explosive, all occuring in the first two centimeters of propagation. The agreement with our analytical estimate of the emittances is quite good for this case, with the horizontalto vertical emittance scaling nearly exactly as estimated. An exploration of the dependence of the emittance on the full parameter space is currently underway, using PARMELA as well as particle-in-cell codes with 2-D Cartesian ("slab") geometry. Figure 2 shows a preliminary result of this study, in which the results of PARMELA for the final y emittance are compared with the theoretical estimare presented here. Again the agreement is quite good in all cases, whihc span a large range of $N_b, W, \sigma_{x,y,z}$. In the future we hope to perform fully 3-D particle-in-cell simulations of the self-consistent electron beam dynamics in these sources.

Some Practical Considerations

The model we have used thus far ignores many dynamical aspects of the problem which can only be calculated accurately by computer simulation. For example, it should be noted that our model for space charge driven emittance growth is based on the assumption that the beam size does not grow significantly after leaving the cathode – this growth will aggravate the space-charge driven emittance blowup. This assumption may not be justified when a thermal beam is quite small, as the beam's 'depth of focus' is approximately proportional to the beam size at the cathode. Thus it may be necessary to provide transverse focusing in such a

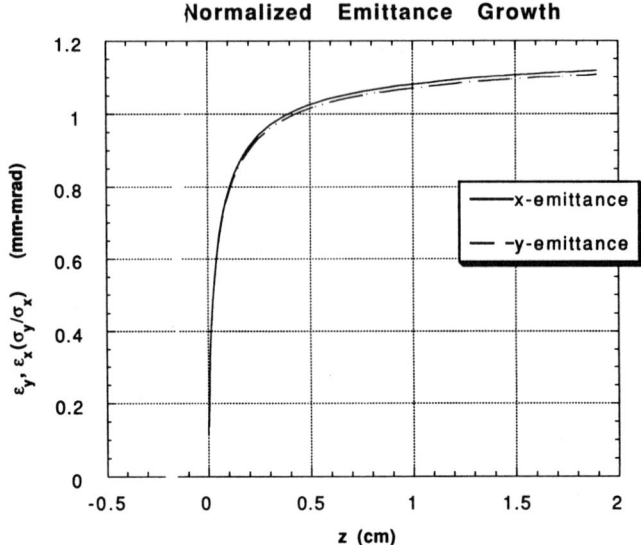

Figure 1: Emittance evolution in 476 MHz rf gun, $Q = 5$ nC, $E_{rf} = 60$ MV/m, $\sigma_{y,x,z} = 0.1, 20, 1.25$ mm, calculated with PARMELA. Note that the ratio of horizontal to vertical emittance is approximately equal to σ_x/σ_y.

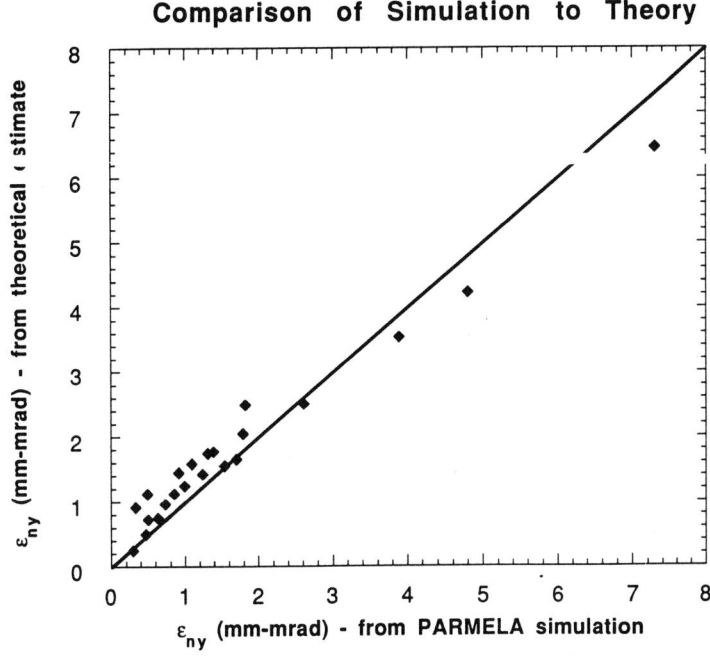

Figure 2: Comparison of the theoretical emittance estimate to the that calculated by PARMELA in 476 MHz rf gun for a range of $N_b = 5 - 10$ nC, $E_{rf} = 20 - 100$ MV/m, and $\sigma_y/\sigma_x = 0.01 - 0.001$, $\sigma_x = 12$ mm, and $\sigma_y = 0.3 - 15$ mm.

device. Also, electrons which are created off-axis also traverse longer path lengths after encountering a transverse focusing element. It is possible to eliminate the pulse lengthening associated with this effect by shaping the laser pulse to liberate the off-axis electrons at slightly earlier times[11]. Since the beam in our scheme is wide in the horizontal dimension, it may be necessary to employ this technique to preserve picosecond pulse lengths.

Thus far, we have only discussed the single bunch beam emittance and peak current requirements for superconducting rf linear colliders. The duty cycle required by such a collider is typically[3] slightly less than a percent. At these high average power levels it may be necessary to use a cryogenic, or superconducting cavity for the rf gun. The superconducting option will almost certainly preclude accelerating fields on the cathode of more than 60 MeV/m. In this case, problems associated with space-charge induced emittance growth will be harder to limit.

In order to exploit the space-charge ameliorating effects of beam flatness to the maximum it may be desirable to operate at lower rf frequencies, to avoid emittance blowup due to rf focusing. Alternatively, one may attempt to control the rf derived emittance growth with "flat-topping" of the rf wave-form, or by using an elliptically shaped cavity (with horizontal dimension larger than vertical dimension).

Conclusions

Further theoretical and computational modeling of rf photocathodes in the flat-beam limit must be done to improve the level of understanding of this scheme. In addition, it is necessary to undertake experimental development of these devices. Tests which are scaled to smaller beam sizes and a shorter rf wavelength can be done at the UCLA compact linac source[12]. Also, it should be noted that the rf parameters given for the superconducting linear collider source are similar to those of the rf photocathode presently under development at for the Argonne Wake-field Accelerator injector[13]. Further tests at that facility would be valuable for judging the suitability of this type of electron source for linear colliders. In addition, further work is necessary for understanding the potential role that rf flat-topping can play in controlling emittance growth in these devices.

References

[1] R. Sheffield, "Progress in Photoinjectors for Linacs", Proceedings of the 1990 Linear Accelerator Conference, 269 (Los Alamos, 1991), and references within.

[2] P. Chen and V. Telnov, *Phys. Rev. Lett.* **63**, 1796 (1989); P. Chen and K. Yokoya, *Phys. Rev. D* **38**, 987 (1988); and "Disruption effects from the interaction of flat e^+e^- beams" submitted to *Phys. Rev. D*.

[3] For an introduction to the literature on superconducting linear colliders, see the Proceedings of the 1st International TESLA Workshop, ed. by H. Padamsee (Cornell, 1990), in particular the articles by U. Amaldi and J. Rosenzweig, and references within.

[4] G. Guignard, "Status of CERN Linear Collider Studies", Proceedings of the 1990 Linear Accelerator Conference, 8 (Los Alamos, 1991).

[5] S.K. Ride, R.H. Pantell and J. Feinstein, "Reducing Slip in a Far-Infrared Free Electron Laser Using a Parallel Plate Waveguide," *Appl. Phys. Lett.* **57**, 1283 (1990).

[6] M. Reiser, "Physics of High Brightness Sources" in *Advanced Accelerator Concepts*, Ed. C. Joshi, AIP Conf. Proc. **193**, 311 (1989).

[7] T. Srinivasan-Rao, J. Fischer, and T. Tsang, "Photoemission Studies on Metals Using Picosecond UV Laser Pulses", *J. Appl. Phys.* **69**, 3291 (1991).

[8] K.J. Kim, "Rf and Space-Charge Effects in Laser-Driven Rf Electron Guns, *Nucl. Instr. Meth. A* **275**, 201 (1989).

[9] L. Serafini, P. Michelato, C. Pagani, A. Peretti, "Design of the SC laser driven injector for ARES", *Proceedings of the 2nd EPAC*, Nice 1990, Editions Frontieres.

[10] L. Serafini, M. Ferrario, C. Pagani, R. Rivolta, "TOPGUN: A new way to increase the beam brightness of rf guns", INFN/TC-91/07, Milan, 1991.

[11] J. Norem and C. Ho, private communication.

[12] S. Hartman, *et al.*, "The UCLA Compact Linac for FEL and Plasma Wakefield Experiments," these proceedings.

[13] J. Simpson, "Wake-field Accelerators", Proceedings of the 1990 Linear Accelerator Conference, 805 (Los Alamos, 1991).

HIGH GRADIENT ACCELERATION IN A 17 GHz PHOTOCATHODE RF GUN

S. C. Chen, B. G. Danly, J. Gonichon, C. L. Lin
R. J. Temkin, S. Trotz, and J. S. Wurtele
Plasma Fusion Center, Massachusetts Institute of Technology

ABSTRACT

A 17 GHz RF acceleration experiment is being constructed at MIT. The goal is to study particle acceleration at high field gradients and to generate high quality electron beams for potential applications in next generation linear colliders and free electron lasers. The RF gun has a $1\frac{1}{2}$ cell cavity with a peak accelerating gradient of about 250 MV/m. The anticipated beam parameters, when operating with a photoemission cathode, are: energy 2 MeV, normalized emittance 0.43π mm-mrad, energy spread 0.18%, bunch charge 0.1 nC, and bunch length 0.39 ps.

The detailed experimental setup, including RF cavity, RF transport and coupling, vacuum system, laser and timing system are described.

1   INTRODUCTION

To meet the stringent requirements set by future applications such as high-energy linear colliders and next generation free electron lasers, efforts have been made recently to create novel electron beam sources. A worldwide review of major RF gun projects can be found in Travier.[1] While the operating frequencies of existing systems range from 500 MHz to 3 GHz, a 17.136 GHz photocathode RF gun has been designed and is under construction at MIT.[2] This RF gun design is basically scaled from the 2.856 GHz RF gun currently under testing in Brookhaven National Laboratory (BNL).[3,4] However, many changes have been made to accomodate the features involved in high frequency operation. Despite technical difficulties and unresolved physics issues associated with high frequencies, the 17.136 GHz operation is very attractive. It

*This research is supported by DOE under Grant DE-FG02-91-ER40648

allows us to achieve a high accelerating gradient, to make the system compact, and to obtain high brightness beams.

In this paper, the experimental design of the 17 GHz photocathode RF gun is presented in detail. A general layout of the experiment is shown in Fig.1. It consists of three parts: (1) the RF gun cavity and the transport line (including the power source and the vacuum system), (2) the laser and timing system, and (3) the beam transport and diagnostic line. Each of these subjects is described successively in sections 2, 3, and 4. Section 5 summarizes the status of the experiment.

2 RF CAVITY AND TRANSPORT LINE

2.1 RF CAVITY AND WAVEGUIDE COUPLING

One of the most interesting phenomena to be studied in this experiment is RF breakdown. Although the breakdown limit at 17 GHz has yet to be studied experimentally, the breakdown threshold should be around 800 MV/m if one extrapolates the data obtained at lower frequencies as measured by Wang.[5] To obtain as high an accelerating gradient as possible without RF breakdown, the cavity geometry must be designed to minimize the ratio of the peak surface field to the field at the cathode. Therefore, our cavity geometry is directly scaled from the BNL design which has already achieved this goal.[3] To keep the cavity surface field strength below the predicted breakdown threshold, the peak accelerating gradient is chosen to be 250 MV/m, corresponding to a peak surface field around 300 MV/m.

The accelerating field profile is also determined by cavity geometry. The beam dynamics and the interplay between time-dependent RF forces, space-charge forces, and nonlinear RF forces have been studied using two computer codes. The first is the particle-in-cell code MAGIC and the second is the particle-pushing code PARMELA. Both codes incorporate the scaled cavity geometry. The laser pulse length, RF phase for photoemission, cathode radius, and current density have been optimized and the results were reported by Lin et. al.[2] The design parameters are summarized in Table 1.

The $1\frac{1}{2}$ cell RF cavity is electroformed from OFHC copper. A second cavity has been constructed by means of a three-piece design instead of electroforming. A side-wall waveguide coupling scheme is used to feed the RF power into the cavity. The experimental arrangement is shown in Fig.2. Both the cavity and waveguide are under vacuum. There are over one thousand holes in the coupler wall to facilitate pumping. An RF monitor and a frequency tuner are installed in each cell of the cavity. The offset in the angle of monitor and tuner is due to space limitations. The tuners are mounted on vacuum motion feedthroughs. The monitors consist of an SMA cable with an RF pick-up loop

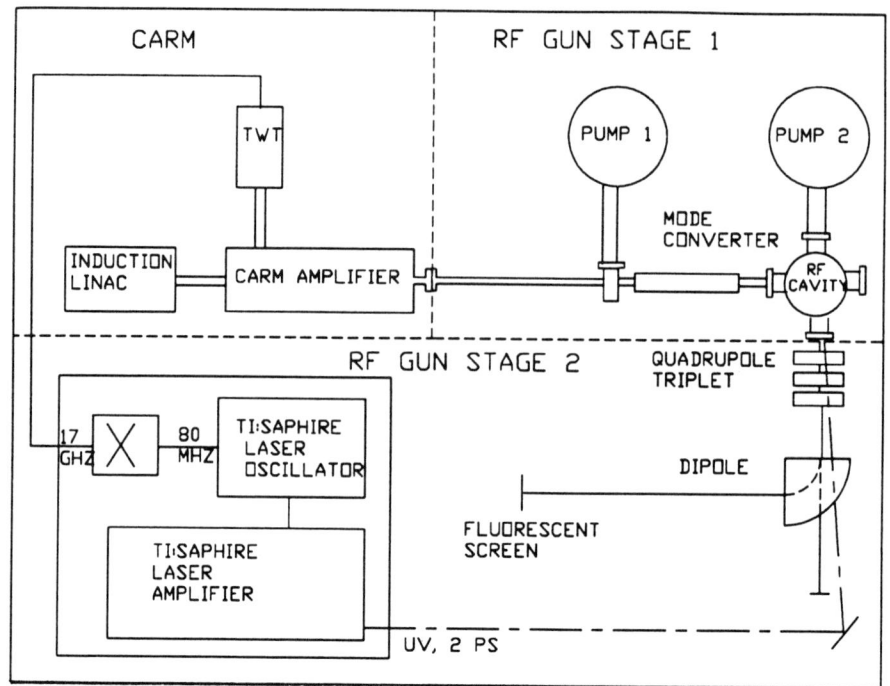

Figure 1: Schematic of the 17 GHz photocathode RF gun experiment.

Table 1: Design Parameters.

Peak accelerating gradient	250MV/m
Laser pulse length	1.4ps
Final bunch length	0.39ps
RF phase for laser pulse	12°
Current density	$6.7kA/cm^2$
Cathode radius	0.525mm
Bunch charge	0.1 nC
Emittance	0.43π mm-mrad
Energy spread	0.18%
Current	258A
Brightness	$1.43 \times 10^{14} \frac{A}{(mrad)^2}$

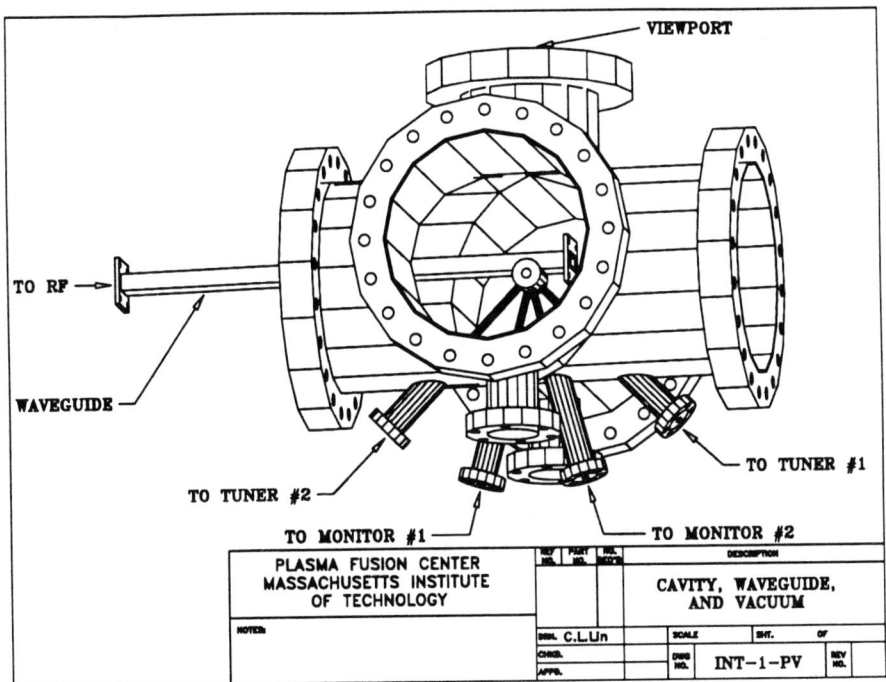

Figure 2: Vacuum chamber.

attached to its end to intersect magnetic flux. All enter the cavity through holes drilled in the cavity wall and can be accessed from outside the vacuum chamber.

The TE_{10} waveguide mode is coupled to the cavity through two rectangular apertures, one on each cell of the cavity, to excite the desired π mode resonance. The waveguide is shorted at a multiple of one-quarter wavelength away from the apertures to maximize the incident field strengths on the apertures. An intensive study of this waveguide side-wall coupling scheme has been conducted both theoretically and experimentally. Theoretically, the cavity's two cells can be modeled as two simple harmonic oscillators driven by two dipole moments that represent the effects of two apertures. The two dipole moments not only excite fields in the two cells, but also excite fields in the waveguide. By imposing proper boundary conditions on the apertures one can solve the two unknown dipole moments and hence the excited fields in two cells and the waveguide in terms of the aperture and cavity geometries.

The theoretical results of the above analysis are in good agreement with experimental results. The ratio of reflected to incident power was measured using a network analyzer. Fig. 3 shows the reflected power as a function of frequency. In this example, the two cavities have not been tuned. There are two clearly visible resonances, each of which corresponds to absorption of most

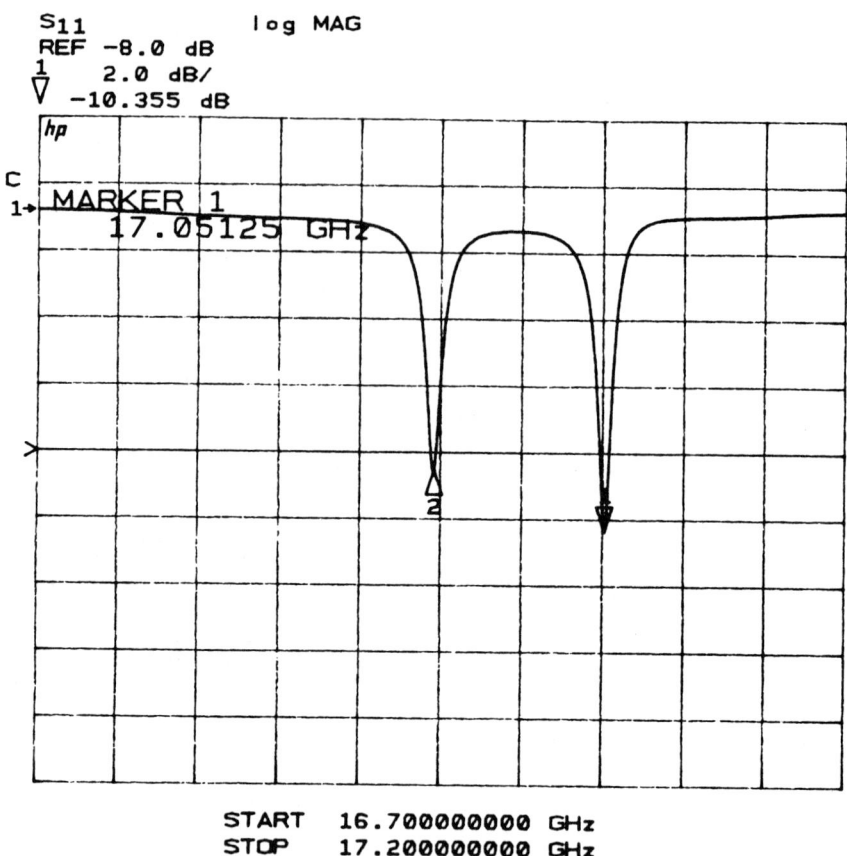

Figure 3: Reflected power as a function of frequency. (Untuned)

of the incident power in one cavity. The width and location of the resonances shows that the two resonances are about 100 MHz apart and that each cavity has a Q value of about 1000. Each cavity absorbs over 80% of the incident power. Fig. 4 shows the reflected power as a function of frequency for the two cavities after tuning. In this case, the two resonances have merged into a single dip corresponding to power absorption of over 90%.

2.2   RF SOURCE AND TRANSPORT

The power source used to feed the RF cavity is a Cyclotron AutoResonance Maser (CARM) under development at MIT.[6] The CARM will deliver 5-10 MW peak in a pulse of 30 ns at a repetition rate of 10 Hz. The RF exits the CARM in a circularly polarized TE_{11} mode and must be converted to the TE_{10} mode

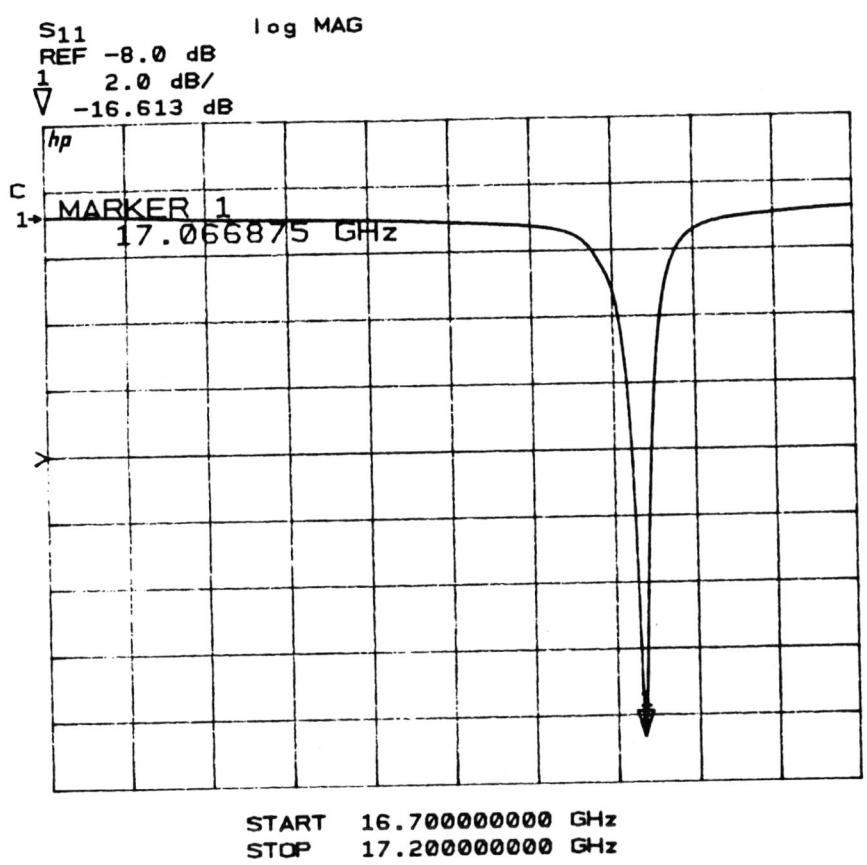

Figure 4: Reflected power as a function of frequency. (Tuned)

RF LINE

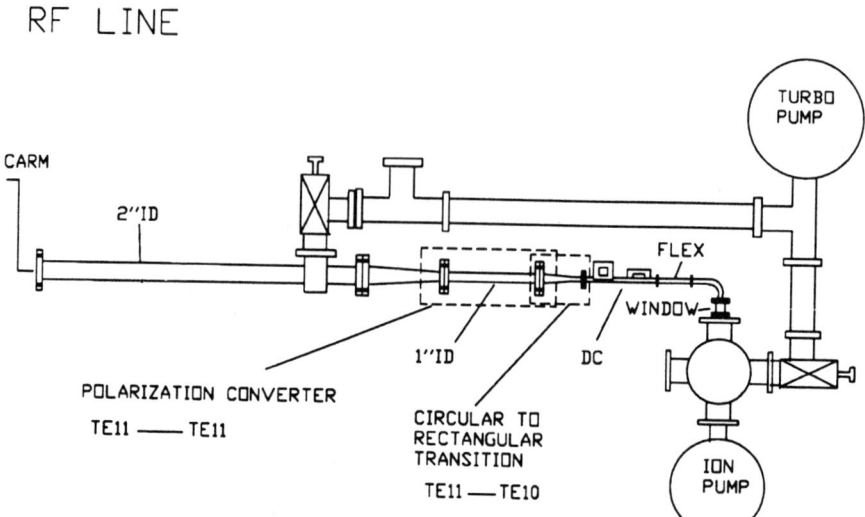

Figure 5: RF transport.

in the rectangular waveguide that couples to the RF cavity. The RF transport line, whose schematic diagram is shown in Fig. 5, has been designed and constructed. A 2″ ID, 6′ long circular waveguide is connected at one end to the output window of the CARM and at the other end to a tapered circular waveguide that reduces the ID from 2″ to 1″. The mode converter follows the taper and is composed of two parts. First, a circular piece of waveguide whose cross section has been slightly deformed to be elliptical in the center converts the rotating polarization of the TE_{11} mode to linear. Second, the TE_{11} linearly polarized mode in the circular waveguide is converted, through a circular to rectangular transition, to TE_{10} in the rectangular waveguide. Between this piece and the waveguide coupler, the construction is modular and involves a dual directional coupler, a high-vacuum RF window, a flexible waveguide, and (optionally) an arc sensor. The total length between the CARM and the cavity is more than 10 feet. This length provides transit time isolation between the RF gun cavity and the CARM amplifier. All components of the RF line have been assembled.

2.3 VACUUM

The vacuum system should be able to provide 10^{-9} Torr inside the RF gun cavity and at least 10^{-5} in the RF transport line to avoid breakdown. The pumping scheme for the cavity and the coupling waveguide has been discussed in Section 2.1. In addition, three hundred holes have been drilled around the long 2" ID circular waveguide to evacuate the RF transport line. The diameter of each hole is small enough (1.6 mm e.g. $\frac{\lambda}{10}$) to be below cutoff. The conductance for each hole is 0.07 l/sec, leading to a total conductance of 21 l/sec, which is higher than the conductance of the 2" ID waveguide (8 l/sec).

3 LASER AND TIMING SYSTEM

3.1 LASER

The design of the laser system is driven by the need to create an electron bunch of the proper size and shape at the optimal time to be accelerated in the RF cavity. The design parameters are summarized in Table 2, and are explained as follows: The wavelength of the light is chosen to maximize the efficiency of the laser and the photoelectric effect for copper. The laser repetition rate matches the duty cycle of the CARM amplifier. The output energy of the laser is calculated to yield an electron bunch of 0.1 nC. This output energy must be stable to limit current fluctuations. The phase jitter of the laser measures the uncertainty in the time between the laser pulse arrival at the cavity relative to the "reference signal" which is derived from the Ti:Sapphire oscillator. The timing jitter is the jitter between the "master trigger signal" and the laser pulse arrival time. These timing figures involve very different timescales. (See section 3.2). The beam should diverge as little as possible in order to provide a small spot size at the photocathode. The beam must accurately hit the cathode which is 1 mm in diameter and is located over eighty feet from the laser and, therefore, should have a low pointing error. Lastly, the mode-lock frequency of the laser oscillator is equal to an integer subharmonic of the 17 GHz source.

An Argon-Ion pumped Ti:Sapphire laser produces a continuous train of microjoule pulses which enter a pulsed Ti:Sapphire laser amplifier. The amplifier is pumped by a separate Nd:YAG laser which provides nearly 1 J/pulse. When triggered, the amplifier optically selects one of the pulses in the CW train and amplifies it to milijoule levels. The amplified IR pulse is frequency tripled into the ultraviolet by an LBO/BBO combination and is directed into the RF cavity. The manner in which the laser system is to be integrated into the experiment is illustrated in Fig. 6.

Table 2: Parameters of the laser system.

Wavelength	220-280 nm
Repetition rate	Single pulse and 0-10 Hz (adjustable)
Final output energy (per pulse)	0-200 μJ (adjustable)
Energy output fluctuation	$\leq \pm 10$ %
Pulse Length	<2 ps
Phase Jitter	<1 ps
Timing Jitter	<3 ns
Polarization	> 99 %
Beam Divergence	0.5 to 1 mrad
Beam Pointing Error	< 10 μrad
Mode-Lock Frequency	82 MHz

3.2 TIMING

With a proper and stable time delay between the firing of the CARM amplifier and the firing of the laser amplifier, the laser pulse should arrive at the cathode surface when the RF gun cavity is filled with the microwaves from the CARM. The 30 ns RF pulse at 17.136 GHz fills up the cavity after about 17 ns. The laser pulse should reach the filled cavity at the optimal microwave phase. As shown in Lin et. al,[2] the electron beam quality is strongly dependent on the RF phase of photoemission. In this experiment, the phase jitter is designed to be less than 1 ps.

In order to satisfy the timing relationships described above, the system will incorporate a highly stable CW Ti:Sapphire laser which serves as the system clock. The repetition rate of the laser is 82 MHz. This frequency is defined by the round-trip time of light traveling between the ends of the laser cavity. The laser is designed to minimize length fluctuations and therefore maintain the 82 MHz value very precisely. However, to account for slight variations in the RF gun cavities' resonance frequency, the mode locker in the laser oscillator is made tunable from 82 to 84.8 MHz. The laser pulse width and power are kept constant through regenerative mode locking, in which the laser cavity resonance (round-trip time) provides the mode locker timing signal. The cavity mirrors are mounted on Invar tubes to minimize length variations.

Since the laser modelocker serves as the system clock, a photodiode is used to convert the optical signal into an electrical signal. A photodiode monitors a sampled portion of the laser output. The resulting signal approximates a series of delta functions at 82 MHz. This signal enters a solid state frequency multiplier which converts it to a 17 GHz sine wave using a x3 and x4 multiplier followed by a comb generator. The multiplier circuit is heavily padded with amplifiers and bandpass filters at all stages. Very little sideband noise is introduced by this process. The resulting 17 GHz signal is used as the seed signal for the CARM amplifier chain. Since any changes in the 82 MHz rate will be multiplied by 204 at the high frequency output of the multiplier, the stability of the Ti:Sapphire laser is essential to ensure that the frequency of microwaves entering the cavity from the CARM matches the resonance of the cavity. The cavity has a Q of about 1000 and should tolerate the measured frequency instability of the laser after multiplication. If the multiplied signal is not stable enough, external modelocking of the Ti:Sapphire laser will be used to provide a master clock for the system.

4   BEAM LINE AND DIAGNOSTICS

The purpose of the beam transport (and diagnostic) line is to bend the paths of electrons exiting the RF cavity away from the laser window, and to measure the electron beam parameters. The quantities to measure are energy, energy spread, transverse emittance, charge, and bunch length.

The bending magnet alters the electron path by an angle of 90 degrees and forms a point-to-point imaging system that can be used for energy spread measurement. Fig. 1 contains a schematic of the beam line including the quadrupole triplet and a bending magnet. A very small beam size is produced before the bending magnet (object plane), with the proper quadrupole setting, and the bending magnet produces a spot on a fluorescent screen in the image plane. The fluorescent spot is observed with a video camera. The position of the spot gives the energy and its horizontal thickness the energy spread.

The quadrupole triplet can also be used to measure the emittance: the bending magnet is switched off and the spot produced by the electrons on another screen positioned along the RF gun axis is observed. The gradient (K) of one of the quadrupoles is varied in order to vary the spot size on the screen. A least-square analysis of the curve σ_{xy} vs. K gives the transverse emittance.

The program TRACE3d, which uses the transport formalism, was used to obtain a preliminary design. Simulations of the same line with the program PARMELA show that the resolution of the spectrometer should be better than 0.1%.

The charge will be measured with a Faraday cup. Several methods for measuring the bunch length are under investigation.

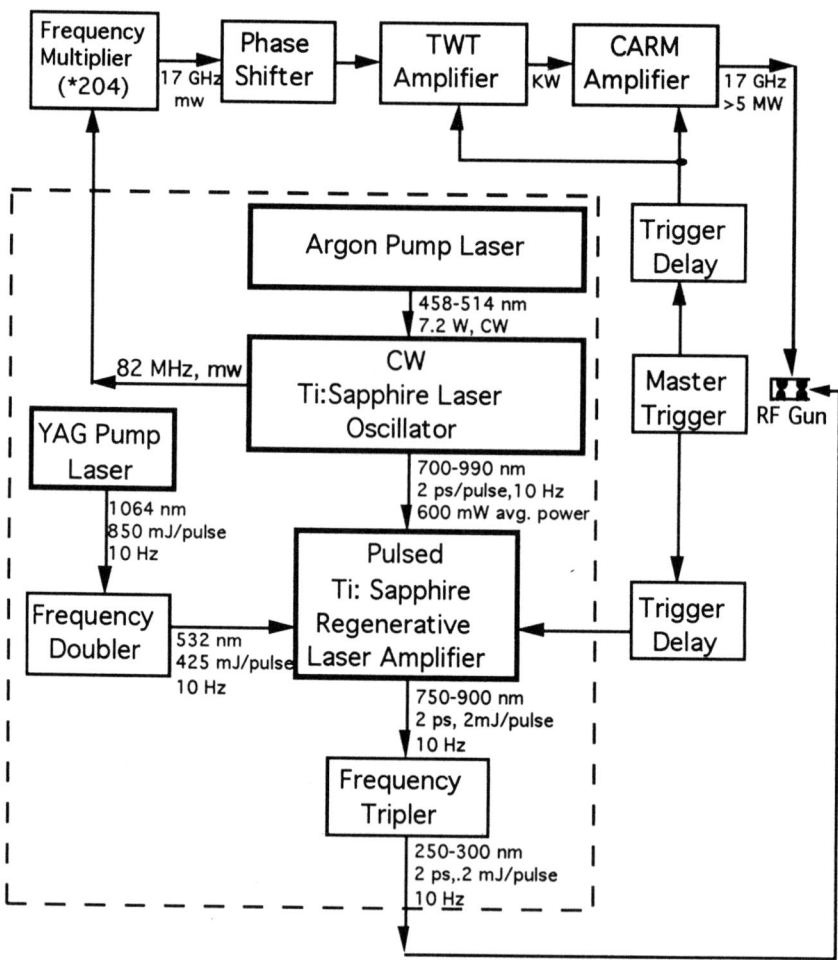

Figure 6: Laser system.

5 SUMMARY

A high gradient RF acceleration experiment is under construction at MIT. The accelerating cavity is a copper RF gun structure with an operating frequency of 17.136 GHz. The designed peak accelerating gradient on axis is 250 MV/m. The accelerating structure and the RF transport line have been constructed and are currently under testing. The first stage experiment involves powering the structure with the high power 17 GHz output from a CARM amplifier. The goal of the initial experiment is to study field emission and RF breakdown, and to condition the cavity for even higher gradient operation.

The following systems will be integrated with the 17 GHz RF gun system in the second stage of the experiment. A UV laser system and the related timing system are being tested to generate picosecond electron bunches through photoemission from the cavity wall. Successful acceleration of these bunches under high field gradient will provide high brightness electron beams[2] suitable for applications in next generation linear colliders and in short wavelength free electron lasers.

REFERENCES

[1] C. Travier, "Review of Microwave Guns," submitted to *Particle Accelerators,* (July, 1991)

[2] C. Lin *et. al., Proc. 1991 IEEE Particle Accelerator Conf.*, p. 2026 (May, 1991)

[3] K. McDonald, "Design of the Laser-Driven RF Electron Gun for BNL Accelerator Test Facility," *IEEE Trans. Electron Devices,* vol. 35, no. 11, p. 2052- 2059 (Nov. 1988)

[4] K. Batchlor *et. al.,* "Design and Modeling of a 5 MeV Radio Frequency Electron Gun," BNL-41766 (1988)

[5] J. W. Wang, "RF Properties of Periodic Accelerating Structures for Linear Colliders," Ph.D. Thesis, Stanford University (1989)

[6] W.L. Menninger *et. al., Proc. 1991 IEEE Particle Accelerator Conf.*, p. 754 (May, 1991)

AN ELECTRON INJECTOR BASED ON A HIGH POWER
X-BAND TWT AMPLIFIER

L. Schächter, J. A. Nation, G. S. Kerslick and J.D. Ivers

**Laboratory of Plasma Studies and School of Electrical Engineering
Cornell University, Ithaca, New York 14853**

ABSTRACT

Theoretical investigation of the interaction of electrons with an electromagnetic wave in a microwave amplifier indicates that nearly 50% of the electrons are in fact accelerated in the amplification process. These fast electrons are phase correlated at the output of an amplifier and they can be further accelerated. For a beam pulse of $100nsec$ and an X-band amplifier, a train of about 1000 bunches can be achieved. Several schemes were considered. Here we present a uniform amplifier, a drift tube (were the slow electrons are dumped) and an accelerator section. With an initial current of $1200A$, and an input power of $20kW$ we calculated electrons with energies of $6MeV$ in buckets of $20°$ corresponding to about 1×10^{10} particles per bunch and an instantaneous current of more than $270A$; the total system length was $1.1m$.

INTRODUCTION

In addition to its role as an RF source in an acceleration system, a TWT can be utilized as the central part of an injection system either of an accelerator or free electron laser. The idea is not just to use the RF output power but also part of the electrons which were bunched in the interaction process. In order to evaluate the potential of such a device let us consider a $100nsec$ electron pulse which is interacting in a traveling wave tube with a $10GHz$ wave. In the interaction process, on average, some of the kinetic energy of the electrons is transmitted to the wave. However, there are a few electrons which in the bunching process are accelerated and their energy is actually significantly larger than the input electron energy; these electrons are basically in anti-phase with the wave. In each period of the wave ($0.1nsec$) the electrons form a bunch of fast electrons on a background of the majority of slow electrons. If these fast electrons are selected then the TWT can generate a train of 1000 such bunches during the beam duration. This is to be compared with tens of pulses with the present laser driven photocathode based injectors. In addition, one should bear in mind that these bunches can be further accelerated by the output RF from the amplifier if both the wave and the bunches are launched into an acceleration section.

TWT BASED INJECTOR

In order to simulate the operation of such an injector without considering reflections from dicontinuities we examine a dielectric loaded waveguide (radius $R = 1.587cm$) since the transition sections can be readily designed, tested and manufactured. The injector consists - see Fig. 1 - of three sections: (i) the first is the amplifier section which is $45cm$ long. This section has a uniform part ($35cm$ long) which is designed such that at $9GHz$ the phase velocity of the TM_{01} mode is $0.92c$ and in the second part ($10cm$) the inner radius of the dielectric ($\epsilon = 2.1$) increases linearly until it reaches the external wall. (ii) The second section is a drift tube which is well above cutoff such that both the RF and the beam can propagate. In this region, the slow electrons are dumped. In the simulation, this was done by assuming 6 filters each one of them dumps the "slow" half of the incident electrons. In the next subsection we shall consider the implementation of this filtering process. (iii) The third section is the acceleration region which is totally $50cm$ long of which $30cm$ is uniform with internal radius, designed such that the phase velocity at $9GHz$ is $1.0c$. On both ends there are tapered sections each of which is $10cm$ long to eliminate reflections of the electromagnetic wave.

It is assumed that $20kW$ of power at $9GHz$ are injected at the input. The beam current is $1200A$ and the electron beam energy is $0.8MeV$. Variations in the dielectric geometry were

706   © 1993 American Institute of Physics

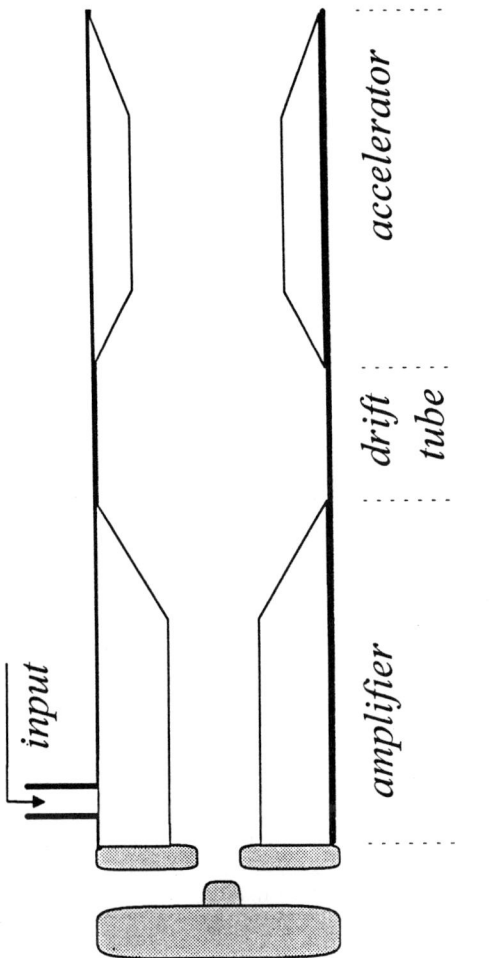

Fig. 1: Schematics of the basic configuration.

fully accounted for in the calculation its result we present now. The equations which describe both the amplifier and the accelerator systems are presented in this volume[1]. Next we present the results of a self-consistent solution of these equations.

Figures 2.a and 2.b illustrate the phase space at various points along the device. We start at $z = 0$ with a uniform distribution in phase-space. At $z = 0.1d$ we observe that the electrons in phase with the wave are decelerated whereas these in anti-phase are accelerated; the process continues along the amplifier as indicated in the next two frames ($z = 0.2d, 0.3d$). As the beam approaches the end of the amplifier section ($z = 0.3d$), we observe that a fraction of the electrons gained a substantial amount of energy ($3 < \gamma_i < 4$). The phase plot indicates that there is a strong correlation between the phase and the energy of these electrons which suggests that they form in real space a well defined bunch. The fifth frame in Fig. 1 shows the phase space of the electrons after they cross the first filter. All the electrons with γ below 3 have been dumped to the wall. Note that in the pipe the phase velocity is larger than c and there is no significant acceleration in this region - see frames $z = 0.4d$ and $z = 0.5d$. The phase space at $z = 0.6d$ shows the distribution in the tapered section at the entrance to the acceleration section. Clearly the maximum energy is more or less like at $z = 0.4d$ only that now after the filtering process a well defined the bunch is formed. The next three frames $z = 0.7, 0.8$ and $0.9d$ shows how the bunch is accelerated in the acceleration section where energies of up to $6MeV$ are achieved. Finally, the phase space after the tapered section is presented and we clearly see the phase slippage due to the velocity mismatch; nevertheless the wave keeps accelerating the electrons even in this section. It is important to note at this point that the main bunch occupies now $20° - 30°$ of the total phase. The average current at the output is $18.75A$ therefore the instantaneous current associated with the bunch is about $270A$ and it contains more than 1×10^{10} electrons.

An alternate way to examine the process in this device is to consider the power carried by all the electrons in a range of γ's which is: $\gamma < \bar{\gamma} < \gamma + 0.2$. Figures 3.a-c illustrate this distribution at exactly the same locations as in the phase space plots above. The first frame shows the process in the amplifier section where the initial power ($\simeq 1GW$), which initially was concentrated in a couple of beamlets corresponding to $\bar{\gamma} = 2.4$ and 2.6, is spread in the bunching process between $\bar{\gamma} = 1.4$ and 3.4. The filtering and the first part of the acceleration process is illustrated in Fig. 3.b. As the slow electrons are dumped to the wall a significant amount of the initial power is lost. In fact only about 10% of the initial power is now carried by the electrons; twice as much power is carried by the RF. The third frame (Fig. 3.c) shows the power distribution associated with the acceleration process in the last 30% of the device. From this picture it is clear that at the output we may expect that the beamlet which consists of electrons in the energy range between $\gamma = 11.8$ and 12.0 will carry a power of about $40MW$.

MAGNETIC FILTER

The presence of the slow electrons in the acceleration section is undesired. Even if the energy separation in the amplifier section is large, problems of noise, debunching and excitation of undesired modes may ultimately diminish the efficiency of the interaction. In addition the system appears to be much less sensitive to current or voltage variation. In the former case it is very important that the slow electrons have a typical velocity of at the most $0.5c$. A 10% variation in the current may result in a substantial number of these electrons having a velocity greater thn $0.5c$ and the beam loading in the accelerator becomes substantial.

The energy selection in the dumping process is very important, therefore we have to consider a resonant process[3] such that we can externally control the electrons to be dumped. For this purpose we shall consider the motion of an electron in a magnetic field which has two components

$$B_x = \delta B cos(kz) \quad , \quad B_z = B_0 \ , \tag{1}$$

where $k = 2\pi/L$ and L is the periodicity of the wiggler; B_0 is assumed to be uniform. We shall ignore here the electromagnetic wave and the waveguide pipe (tacitly assuming that if an

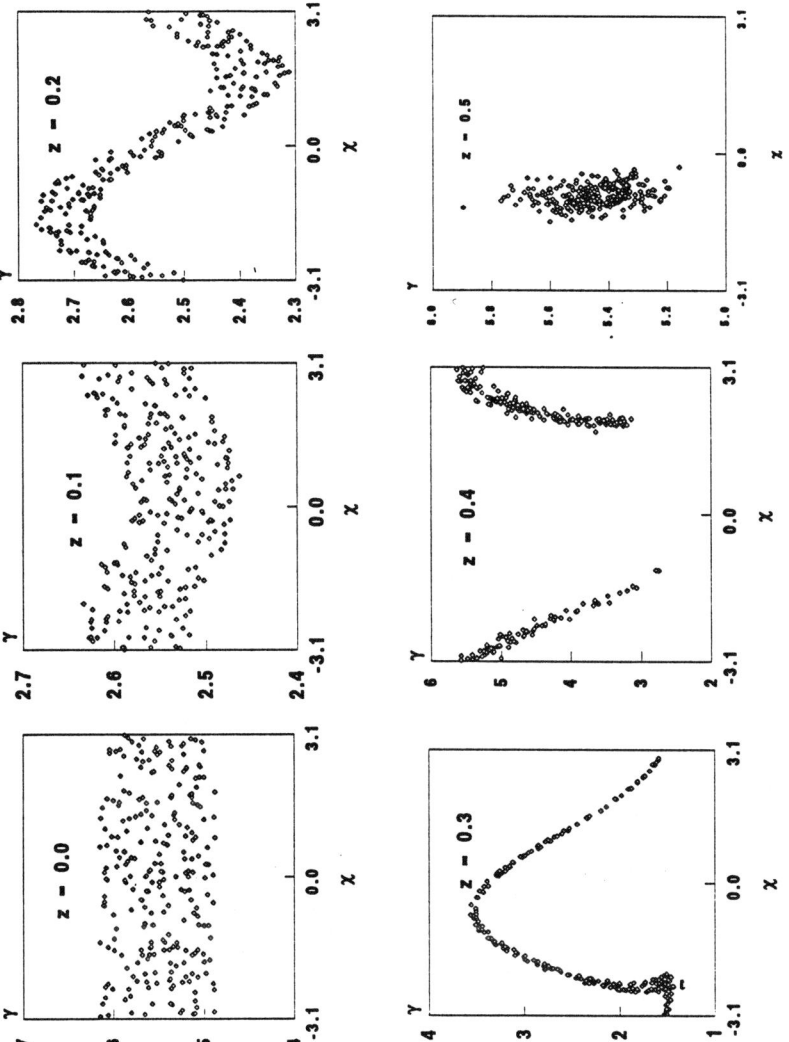

Fig. 2.a: The phase-space plots for the first half of the interaction region.

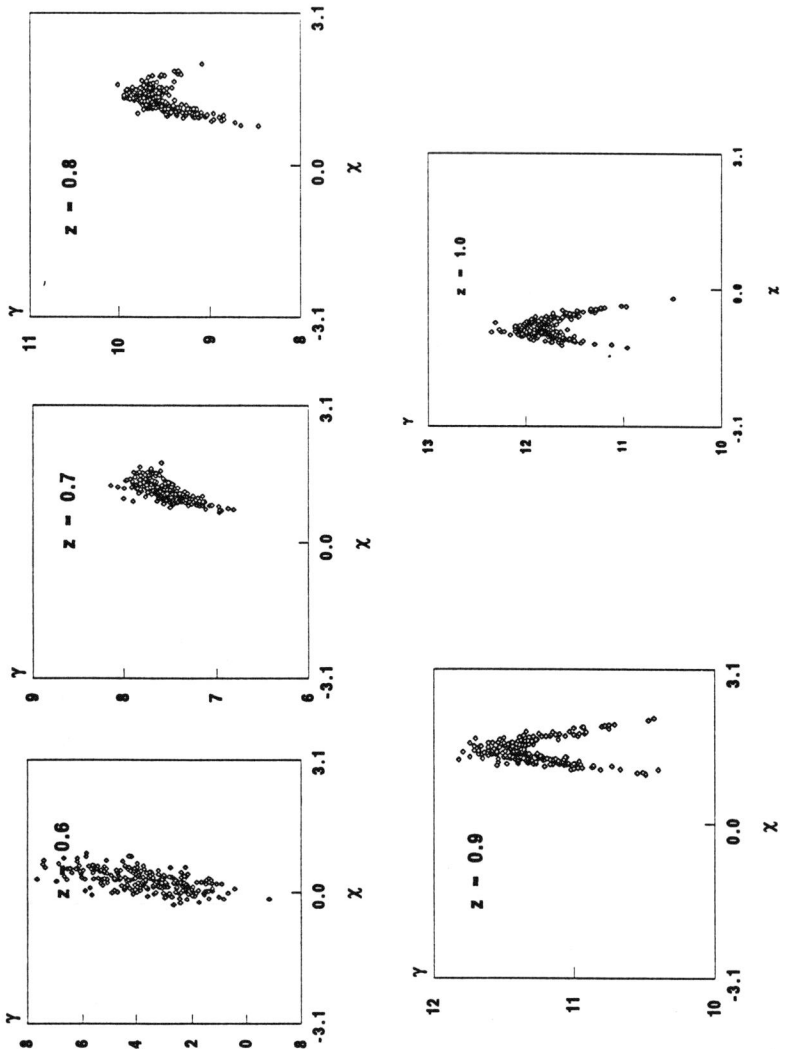

Fig. 2.b: The phase-space plots for the second half of the interaction region.

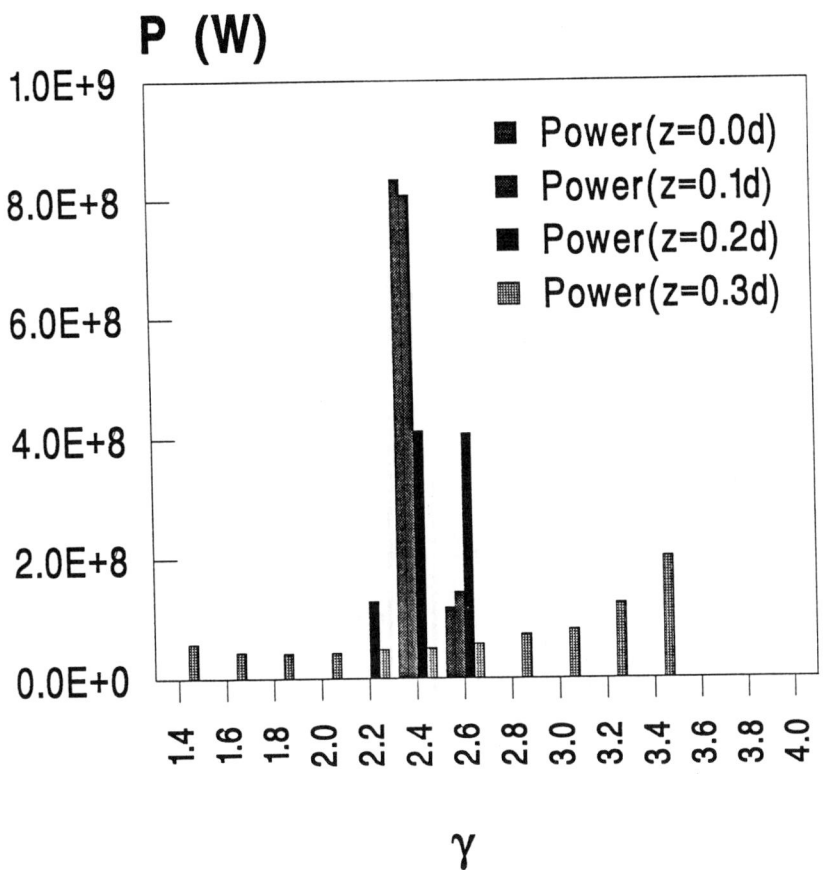

Fig. 3a: Power carried by each beamlet in the amplifier region.

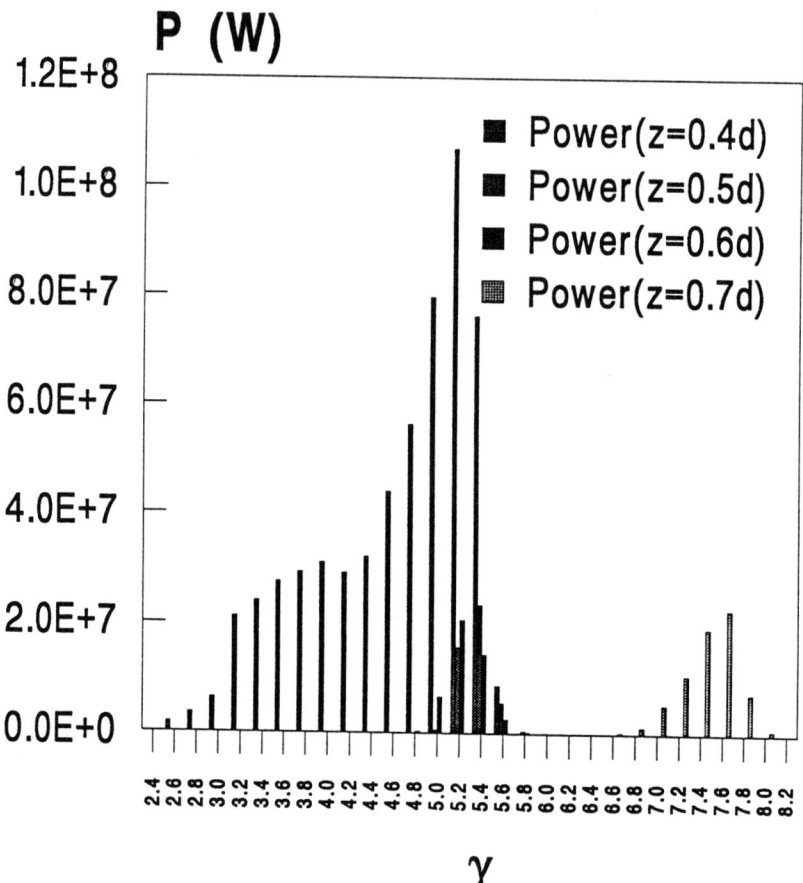

Fig. 3b: Power carried by beamlets in the drift tube region.

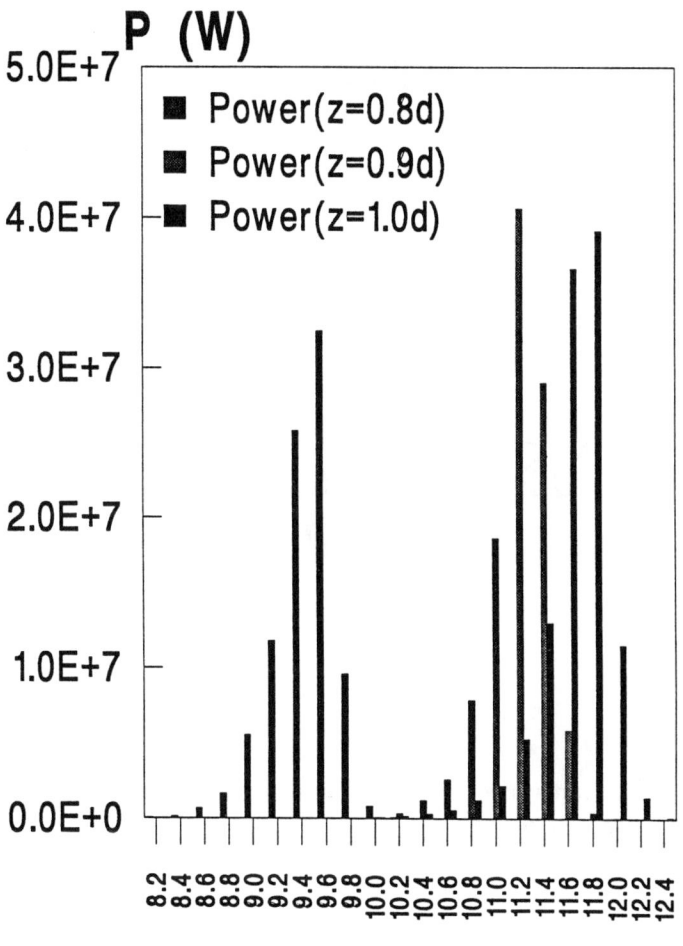

Fig. 3c: Power carried by beamlets in the last 20% of the interaction.

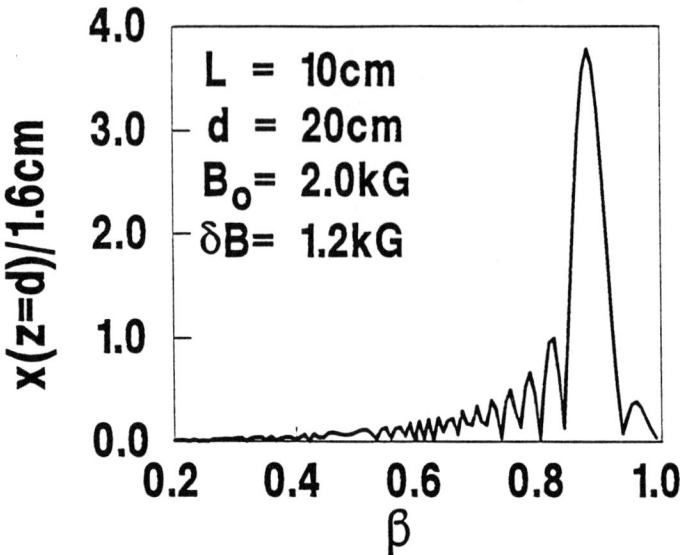

Fig. 4: The transverse normalized position of an electron which at the input has an initial longitudinal velocity βc.

electron hits the pipe it is absorbed there). The problem we wish to consider is the following: if an electron starts at $z = 0$ and on axis, i.e. $r = 0$, with only a longitudinal component of the velocity $V = \beta c$, how should one choose the various parameters such that after a distance $d = nL$ ($n = 1, 2, 3..$) the electrons in certain range of energies will hit the wall.

In the absence of an electric field and bearing in mind the low currents we deal with, the energy of the electrons is expected to remain constant. It is convenient to define the two cyclotron frequencies: $\Omega_0 = eB_0/m\gamma$ and $\omega_0 = e\delta B/m\gamma$ with $\gamma = [1 - \beta^2]^{-1/2}$. The three equations of motion can be readily simplified to two components i.e.

$$\left[\frac{d^2}{dt^2} + \Omega_0^2\right] V_x = \omega_0 \Omega_0 V_z \cos[kz(t)] \quad , \tag{2}$$

$$\frac{d}{dt} V_z = -\frac{\omega_0}{\Omega_0} \cos[kz(t)] \frac{d}{dt} V_x \quad . \tag{3}$$

For a first estimate we shall ignore the <u>average</u> variation of V_z in time therefore $z \approx \beta ct$. Again it is convenient to define $\omega = kc\beta$. We can now integrate the equations of motion and calculate the transverse location of an electron after $\tau = d/c\beta$. The result is

$$x(\tau) = c\beta\tau \frac{\Omega_0\omega_0}{\Omega_0^2 - \omega^2} \left[sinc(\frac{\tau}{2}(\omega + \Omega_0^2/\omega)) - sinc(\omega\tau)\right] \quad . \tag{4}$$

In Fig. 4 we present the normalized transverse location of the electron $X = x(\tau)/1.6cm$. We observe the effect of the resonance at

$$\Omega_0 = \omega \quad ==> \quad \gamma\beta = eB_0L/2\pi mc \quad . \tag{5}$$

All the electrons with β between 0.85 and 0.95 have been dumped to the wall and the fast ones are unaffected. The height of the curve is linearly proportional to the wiggler amplitude.

In this calculation we ignored transverse variation in the magnetic field which will increase the effect because the intensity of the wiggler increases off axis. In addition we have not considered the effect of higher harmonics of the wiggler field. These harmonics will tend to filter the lower energy electrons. For example the second harmonic will filter the electrons with half the momentum filtered by the fundamental harmonic.

SUMMARY
Less than 50% of the electrons interacting in a TWT are accelerated in the amplification process. Their phase is correlated and therefore it is suggested that we take advantage of this fact and construct an injector based on a TWT amplifier. For a beam pulse of $100nsec$ and an X-band amplifier a train of about 1000 bunches can be achieved. We presented a uniform amplifier, drift tube and accelerator section. With an initial current of $1200A$, input power of $20kW$ we calculated a train of electrons with energies of $6MeV$ in buckets of $20°$ each containes about 1×10^{10} electrons and an instantaneous current of more than $270A$; the total length of the system was $1.1m$.

This work was supported by the United States Department of Energy and by AFOSR.

REFERENCES
1. L. Schachter and J.A. Nation; "Narrow Pass-Band High Power TWT Amplifier" in this volume.

2. L. Schachter and J.A. Nation; Proceedings of SPIE: "Intense Microwave and Particle Beams III", Editor Howard E. Brandt, 20-24 January 1992, Los Angeles p.470-480.

3. G. Bekefi *et. al.*, Phys. Fluids B **3**(7), 1755(1991).

THE STUDY OF A DIODE WITH A FERRO-ELECTRIC CATHODE

L. Schächter, J.D. Ivers, J.A. Nation and G.S. Kerslick

Laboratory of Plasma Studies and School of Electrical Engineering
Cornell University, Ithaca NY

ABSTRACT

Recent experiments indicate that current densities of more than $30 A/cm^2$ can be achieved from a cathode made of ferroelectric ceramic, when applying a field across the ceramic of order $0.1 MV/m$. This current exceeds the Child-Langmuir current by two orders of magnitude. The current in the diode varies linearly with the applied voltage, provided that the latter is positive. We show that the ferroelectric material plays a crucial role in the emission process. When a voltage is applied to the ferroelectric, the electric field in the material assisted by the geometry of the gridded cathode cause an enhanced electron extraction. The electrons form in the gap a cloud which can be considered as an extension of the cathode. As a positive voltage is applied to the anode, electrons can be readily transferred through the diode gap. The qualitative and quantitative results of the theory are in good accordance with the experiment.

INTRODUCTION

In the recent years a renewed interest in ferroelectric ceramics for generation of electron beams has been initiated by H. Riege and his collaborators at CERN[1]. High intensity electron beams, $100 A/cm^2$, were produced for a pulse duration of $10 - 100 \mu sec$ containing between $1 nC$ up to several μC of charge. At the Lebedev Institute, A.S. Airapetov et. al.[2] measured current densities of $400 A/cm^2$ in a diode of $0.3 - 1.0 mm$ gap, with an extraction (gap) voltage of $1.6 kV$ and a pulse duration of $160 nsec$. We have examined the operation of a similar device[3][4] but with a diode gap of $2 - 10 mm$ wide, extraction voltage lower than $250V$ $(200 - 600 nsec)$ and a $100 nsec$ long switching voltage. The maximum current density measured was $70 A/cm^2$ which is two orders of magnitude above the Child-Langmuir limit.

The basic mechanism is what we call externally controlled field emission namely, the electrons extraction is due to an electric field which is generated <u>behind</u> the surface of the cathode and, at least in the regime our system was operated, is almost unaffected by the anode voltage. General speaking the device consists of a ferroelectric slab which has a very nonlinear response to an applied voltage and if we were to determine a characteristic dielectric coefficient this would have been larger than 3000. The slab has a uniform electrode on its back side and a gridded and grounded electrode on the front which faces the diode. This is the cathode. The anode consists of a uniform carbon plate. A voltage can be applied to the back electrode of the ferroelectric. If no such voltage is applied, the system behaves like a regular diode and practically for the anode voltages we are interested in, the current is zero. When the ferroelectric is pulsed, a substantial amount of current is measured.

REVIEW OF EXPERIMENTAL RESULTS

The experimental setup is shown schematically in Fig.1. A $1 mm$ thick, $2.5 cm$ diameter ferroelectric disk is coated with a thin uniform silver layer on the back and a gridded silver layer on its front surface; in both cases the thickness is $1 \mu m$. The silver strips are $200 \mu m$ wide and are separated by a similar distance. The effective emission surface is approximately $A \approx 1\ cm^2$. For these experiments we used Lead Zirconate Titanate as the ferroelectric sample. The sample is mounted as a load on a 10Ω transmission line which generates a $150 ns$,$1 - 3 kV$ pulse. A positive pulse is applied to the back electrode of the ferroelectric and the grid is grounded. A planar carbon anode is located at a distance of $g \approx 1 - 10 mm$ from the grid. The anode is

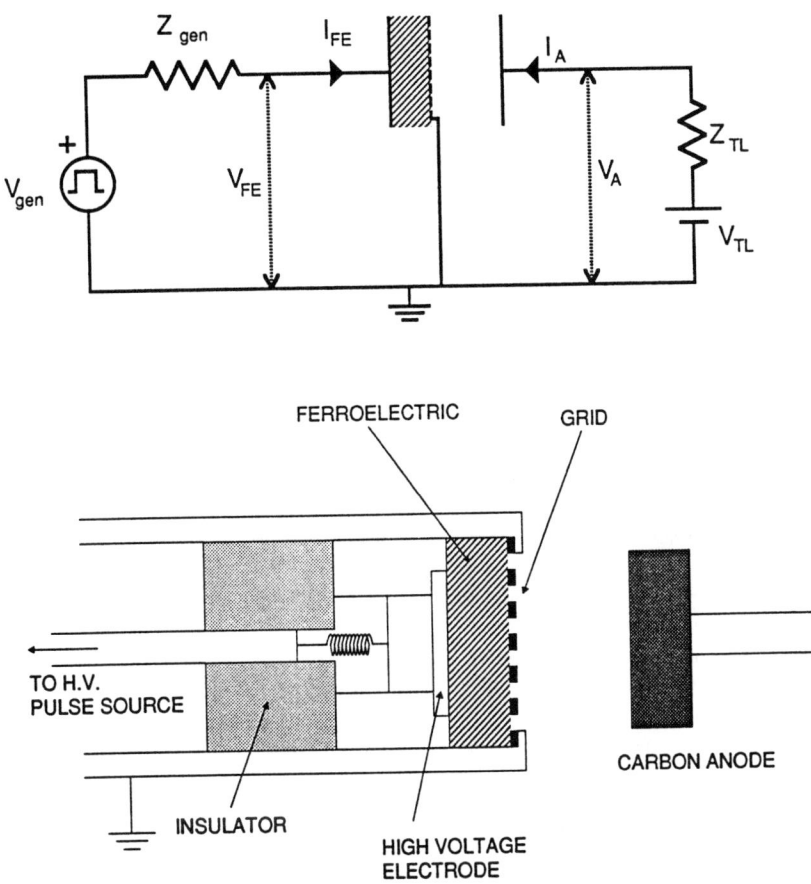

Fig. 1: Schematic experimental set-up.

maintained at a positive potential, V_{TL}, by a charged transmission line whose characteristic impedance can be varied between $R_{TL} = 12.5\Omega$ and 50Ω. The length of the anode voltage pulse is determined by the length of this line, which like the impedance, can be varied.

Three quantities are directly measured: (1) the current charging the ferroelectric capacitor (I_{FE}), (2) the voltage across the ferroelectric (V_{FE}) and (3) the discharging current of the transmission line (the anode current - I_{AN}). The upper frame in Fig.2 illustrates typical anode current results for three shots corresponding to three different voltages ($V_{TL} \approx 100$, 300 and $500V$) applied to a 25Ω cable when the gap was $4mm$ wide. From a large number of such shots we have determined the $I - V$ characteristics of the diode. Fig. 3 illustrates the diode characteristic curve as measured for several cables impedances and lengths. The current into the ferroelectric can be integrated to give the charge and the hysteresis curve describing the ferroelectric capacitor characteristics during the (current) pulse shown in Fig.2, is illustrated in Fig.4 . We have measured the characteristics of the ferroelectric for much slower variation in time ($60Hz$) using the circuit in Fig. 5. Next the experimental data accumulated is summarized.

(a) In our regime of operation, the voltage on the ferroelectric is not affected by the current in the gap - see Fig.2.

(b) The anode current continues even after the signal from the generator pulsing the ferroelectric is effectively off; its length is determined by the length of the cable as could be observed in Fig.2. We shall explain this result in the next section.

(c) The shape of the anode current for low values ($I_A < 15A$) is regular and, as illustrated in Fig.2 (upper frame), it is relatively uniform; at higher levels the shape is somewhat peaked.

(d) At this stage of the experiments it appears that there is no significant difference in the operation of the device when the ferroelectric is pulsed with a positive or negative voltage.

(e) The $Q - V$ curve of the ferroelectric capacitor for slow variations in the voltage and charge reveal a hysteresis characteristic. This curve may be approximately described by:

$$D = \epsilon_0 \left[E + E_0 tanh \left((\epsilon - 1)\frac{E}{E_0} + tanh^{-1}(\frac{P}{\epsilon_0 E_0}) \right) \right] \quad . \tag{1}$$

Based on the hysteresis state of experimental data we determined the parameters of this curve to be: $E_0 \approx 1. \times 10^{10}V/m$, $\epsilon \approx 3000$ and $P_0 \approx 10^{-1}C/m^2$ where $P = \pm P_0$.

(f) Throughout our experiments I_{FE} is larger by at least one order of magnitude ($\simeq 100A$) than the anode current ($I_A \approx 0 - 25A$).

(g) The $I - V$ curve of the gap appears to be linear (see Fig.3) and for a given voltage, the current exceeds by two orders of magnitude the Child-Langmuir limit. If the diode gap is relatively narrow ($< 10mm$), we found that the slope of this curve is strongly influenced by the current flowing into the ferroelectric. In the next section we shall present a model to describe these effects.

(h) For narrow diode gap ($< 2mm$) current was measured at the anode even for zero applied transmission line voltage. This current seems to be strongly dependent on the current charging the ferroelectric capacitor.

DEFINITION OF MODEL

For a complete description of the various processes in a diode with ferro-electric cathode, it is necessary to have an understanding of what happens: (1) in the ferro-electric, (2) at the grid surface and (3) in the gap. For each one of theses regions, a relatively simple model has been constructed. When all are combined together, they reveal a reasonable agreement with the experiment.

The ferroelectric material plays the key role in this device. It is described within the framework of a Weiss-like model for the constitutive relations. Let us denote by Q_p the amount

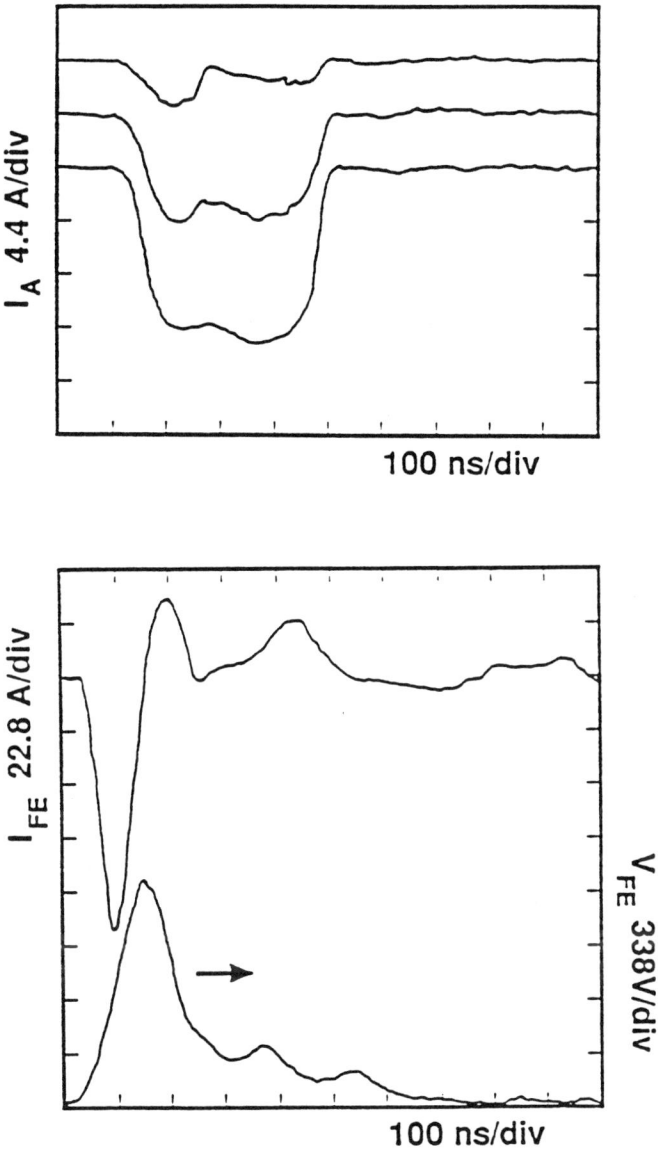

Fig. 2: One set of three shots corresponding to $V_{TL} = 100$, 300 and 500V for a 4mm wide gap; $R_{TL} = 25\Omega$ and the transmission line pulse is 400ns. In the upper frame the anode currents are illustrated. In the lower frame the charging currents (I_{FE}) and the voltage on the ferroelectric (V_{FE}) are presented. Notice that the voltage pulse on the ferroelectric is only about 150ns long.

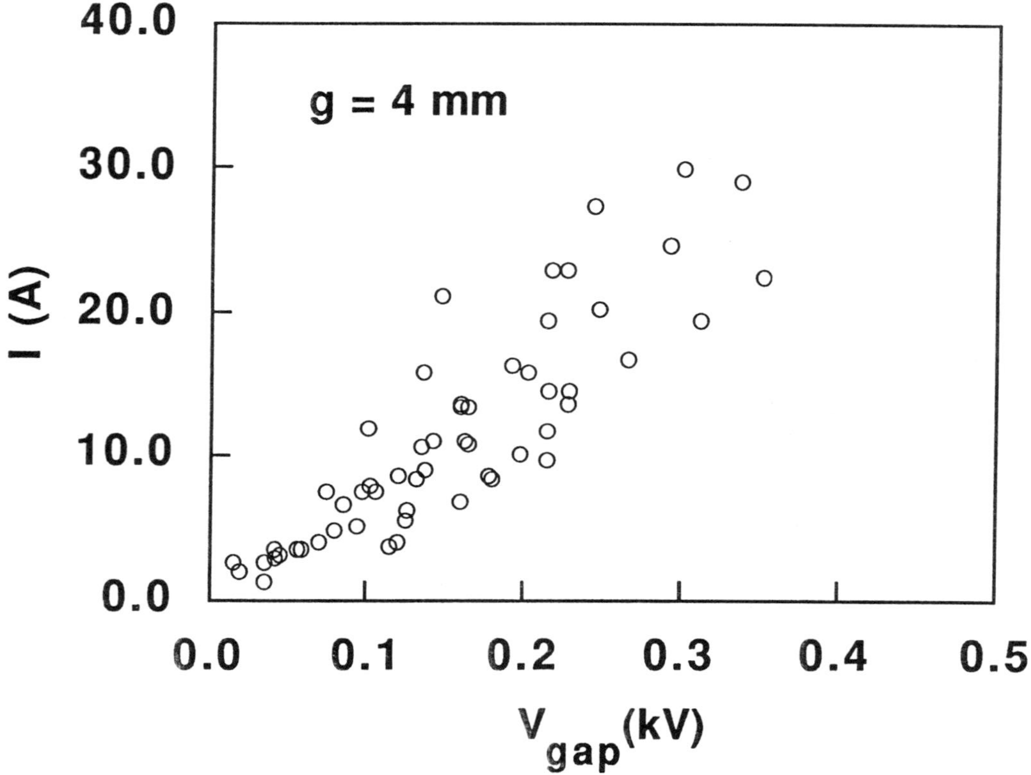

Fig. 3: The $I - V$ characteristic of a diode with a $4mm$ wide gap. Similar results were measured for $2, 6, 8$ and $10mm$ gaps.

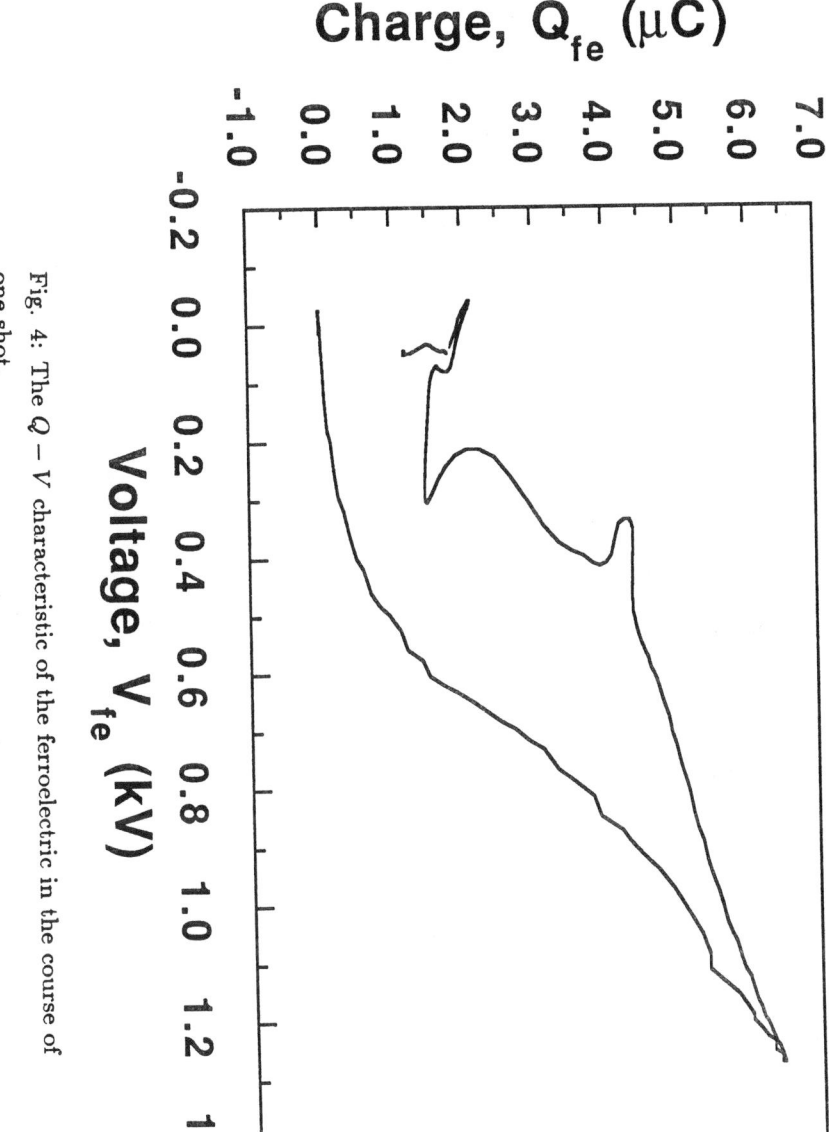

Fig. 4: The $Q - V$ characteristic of the ferroelectric in the course of one shot.

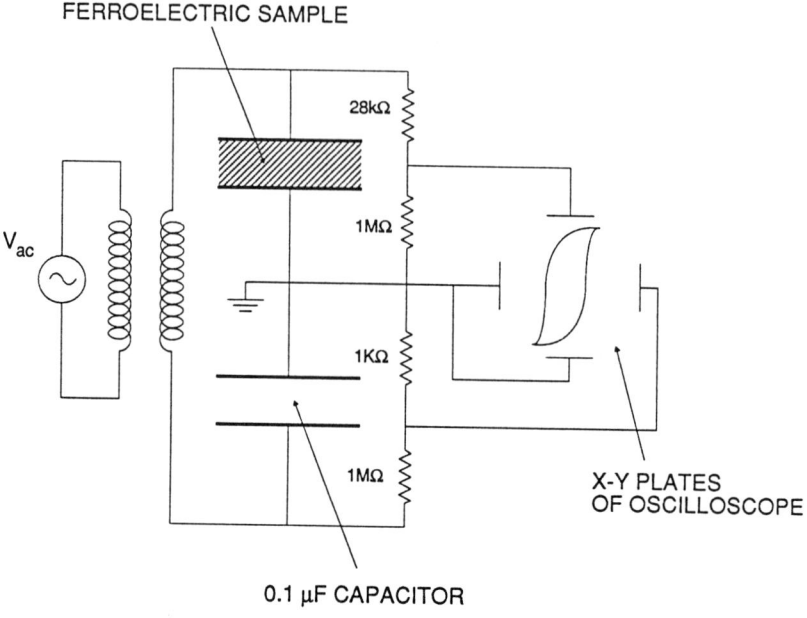

FERROELECTRIC SAMPLE

V_{ac}

28kΩ

1MΩ

1KΩ

1MΩ

0.1 μF CAPACITOR

X-Y PLATES
OF OSCILLOSCOPE

Fig. 5: Schematics of the hysteresis curve at $60Hz$ and the circuit used for this measurement.

of charge which represents the polarization field on the back electrode of the ferroelectric. When an external field is applied, this charge varies as

$$\frac{dQ_p}{dt} + 2\Omega Q_p cosh\left(\frac{V_{FE} + Q_p/C_p}{V_1}\right) = 2\Omega Q_0 sinh\left(\frac{V_{FE} + Q_p/C_p}{V_1}\right) . \tag{2}$$

where V_{FE} is the voltage on the capacitor, $V_1 = k_B T d/p$ is a characteristic voltage which is determined by the material properties; $C_p = \epsilon_1 A/d$, $Q_0 = P_1 A$ are also characteristics of the material and p is the dipole moment of the atom or molecule of interest. The transition rate from one polarization to another is denoted here by Ω.

The ceramic can affect the gap only because the interface electrode is gridded, otherwise no coupling would occur between the two regions. The metallic grid strips are grounded but in between the potential is nonzero. Consequently the average potential on this electrode is nonzero. It can affect significantly the diode gap, when the anode voltage is relatively low, but it does not seem to have a significant influence on the ferroelectric capacitor. In this study, the electrical role of the grid is represented by its own capacitance. In order to determine its capacitance we ignore (for the moment) the nonlinear behavior of the ferroelectric and solve an electrostatic problem using a linear dielectric coefficient, ϵ_{FE} instead. The capacitance was calculated to be

$$C_{grid} = \epsilon_0 A \left[0.5(\epsilon_{FE} + 1)\frac{\pi^3}{1.2L}\right] . \tag{3}$$

The direct result of the voltage applied on the ferroelectric is the change in the polarization field accompanied by electron injection into the gap. The electrons form a cloud which decays exponentially from the cathode. We found the decay parameter to be

$$\alpha \approx \frac{\epsilon_{FE}}{d} , \tag{4}$$

where d is the ferroelectric thickness. It is clear from this representation that if ϵ_{FE} decreases when a voltage is applied on the ferroelectric, the decay parameters decreases and the cloud expands into the gap.

We do not have to know the details of the exact distribution of the electrons in order to estimate their impact on the the potential distribution in the gap. For this purpose we assume that the plane of zero potential is actually on the cathode. From the fact that practically no current was measured for zero anode voltage ($g > 2mm$) we deduce that the electric field near the anode behaves as if no electrons were injected into the gap. This can be understood since the cloud and the grid electrons form all together a "distributed cathode" which neutralizes the positive charge on the back of the ferroelectric. Bearing in mind that the anode potential V_{AN}, is known we consider a solution of the $1D$ Poisson equation which satisfies the boundary conditions mentioned above:

$$\phi(z) = V_{AN}\frac{z}{g} + \Phi\frac{z}{g}\frac{(g-z)^2}{g^2} . \tag{5}$$

The unknown amplitude Φ is determined by substituting Eq.(5) in the Poisson equation and integrating the resulting expression over the entire length of the diode. The source term in the Poisson equation is then proportional to the charge in the gap and therefore so is Φ; explicitly $\Phi = g|Q_{gap}|/3\epsilon_0 A$ where A is the diode surface.

From the potential in Eq.(5) we deduce that the electrons in the cloud are in a dynamic equilibrium in which they oscillate. In this state ($V_{AN} = 0$), the average velocity of the electrons is zero. However, this is because half of the electrons are moving towards the anode whereas the other half move toward the cathode. At any point these two flows are equal. We can estimate the average kinetic energy, $mc^2(\gamma_0 - 1)$, of the electrons by averaging the expression for the energy conservation over the gap spacing. Using Eq.(5) we found that γ_0 reads:

$$\gamma_0 = 1 + \frac{1}{36}\bar{Q} \qquad \bar{Q} \equiv \frac{eQ_{gap}g}{\epsilon_0 Amc^2} \ . \tag{6}$$

As we may have expected, the average kinetic energy of the electrons increases linearly with the total amount of charge in the gap (Q_{gap}). To complete the description of the equilibrium state, we denote the average particle density in the cloud with $\tilde{n}$ and the lowest order estimate of this quantity is just the total number of particles divided by the effective gap volume i.e. $\tilde{n} \approx Q_{gap}/egA$.

When a positive anode voltage, V_{AN}, is applied a net current flows and the ratio of the two determines the gap resistance, R_{gap}:

$$R_{gap} \equiv \frac{V_{AN}}{I_{AN}} = \eta\frac{1}{36}\frac{g^2}{A}\gamma_0^2\sqrt{\frac{\gamma_0+1}{\gamma_0-1}} \ . \tag{7}$$

Where A is the diode surface and $\eta = 377\Omega$. One can easily see that R_{gap} has a minimum as a function of γ_0. This minimum occurs at $\gamma_0 = 1.28$ and for $g = 4mm$ and $R = 5mm$ it corresponds to $Q_{gap}^{opt} = 0.9\mu C$ and $R_{gap}^{min} \approx 10\Omega$. Which agrees well with the experimental data. The linear behavior of the $I - V$ curve in the gap as demonstrated experimentally in Figs.(2-3) is represented theoretically by Eq.(7).

SIMULATION OF THE SYSTEM'S DYNAMICS

Based on the models mentioned above we have constructed an equivalent circuit which is illustrated in Fig.6 There are three features in this system we wish to reemphasize: (1) the non-linear character of the ferroelectric capacitor, (2) the coupling between the two sections of the circuit is through the grid which allows (3) the gap resistance to be determined by the amount of charge in the gap, which in turn is a fraction the charge on the ferroelectric capacitor. The exact value of this fraction is determined by the detailed emission process and it is left as a parameter of the problem i.e. $Q_{gap} = \nu Q_{FE} < Q_{FE}$.

The system of equations which describes the device operation has been solved numerically. The parameters in the following examples are: $\Omega = 11 \times 10^6 rad/sec$, $V_1 = 800V$, $\epsilon_r = 7200$, $d = 1mm$, $g = 4mm$ $A = 0.785cm^2$, $L = 0.4mm$, $R_{TL} = 25\Omega$. The generator pulse is given by $V_{gen}(kV) = 2.0\tau exp(-\tau + 1)$ where $\tau = t/100ns$. In order to understand the details of the functional behavior of the current let us first examine, the hysteresis curve in Fig.(7) which is to be compared with the experimental result in Fig.(4). The large increase in the current see Fig. 8, occurs in the first $50ns$ when the voltage is below $600V$ This is about half a way along the lower hysteresis curve. On the other half of this curve the capacitor is still charging but there are clear indications of saturation and the charging current is dropping until it vanishes as the voltage pulse reaches maximum. From this point on the system follows the upper hysteresis curve. The capacitor discharges at this stage. There is an increasing negative current up to about $270ns$ which corresponds to a voltage of about $800V$ which is somewhat less than half way down the hysteresis curve. In the final stage the discharging current is decreasing and it approaches zero. This latter stage is accompanied, in the experiment, by piezoelectric effects which have not been considered in the present study and except this stage the theory is in very good agreement with the experiment. According to the present model, the gap resistance as formulated in Eq.(7) is illustrated in Fig. 9 and it is relatively constant long after the pulse on the ferroelectric has decayed. This is a direct result of the fact that the ferroelectric follows the upper hysteresis curve in the "recovery process". The duration of the anode current, Fig. 10, is controlled by the voltage applied on the anode.

DISCUSSION AND CONCLUSIONS

In this study we have examined the operation of a diode with a ferroelectric cathode. The voltage applied on the back of the ferroelectric is entirely responsible for the emission process.

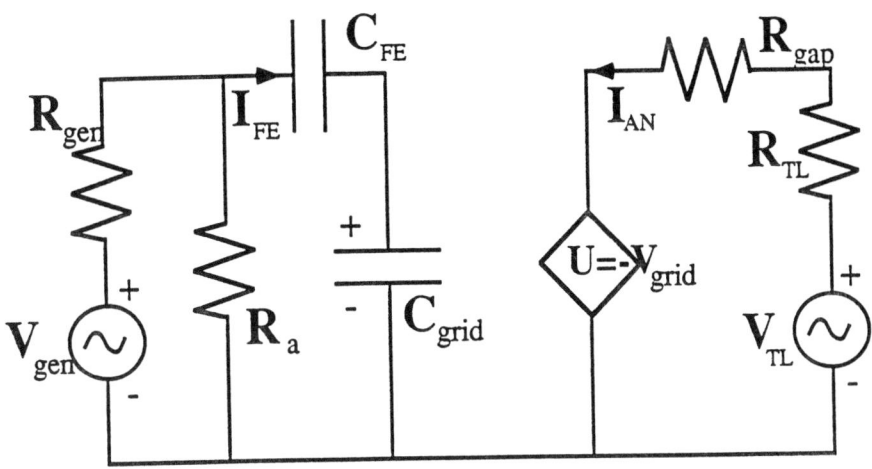

Fig. 6: The equivalent circuit of the system.

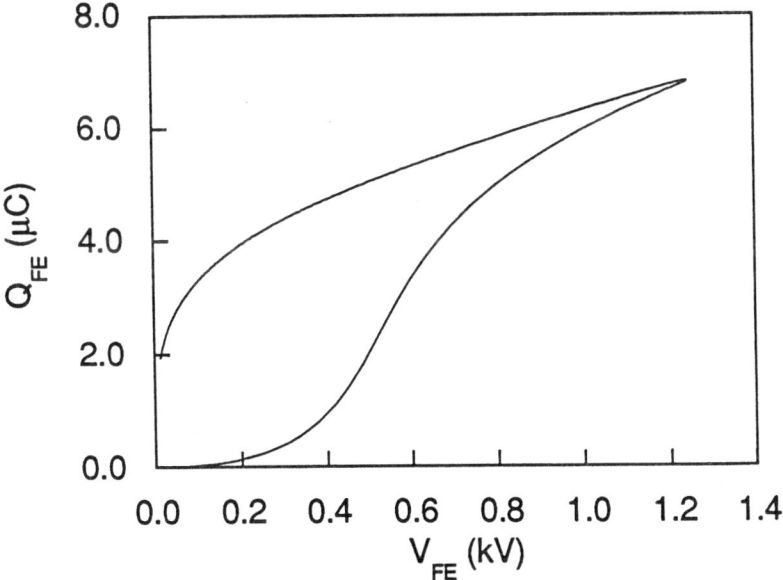

Fig. 7: The hysteresis curve followed by the ceramic during the pulse.

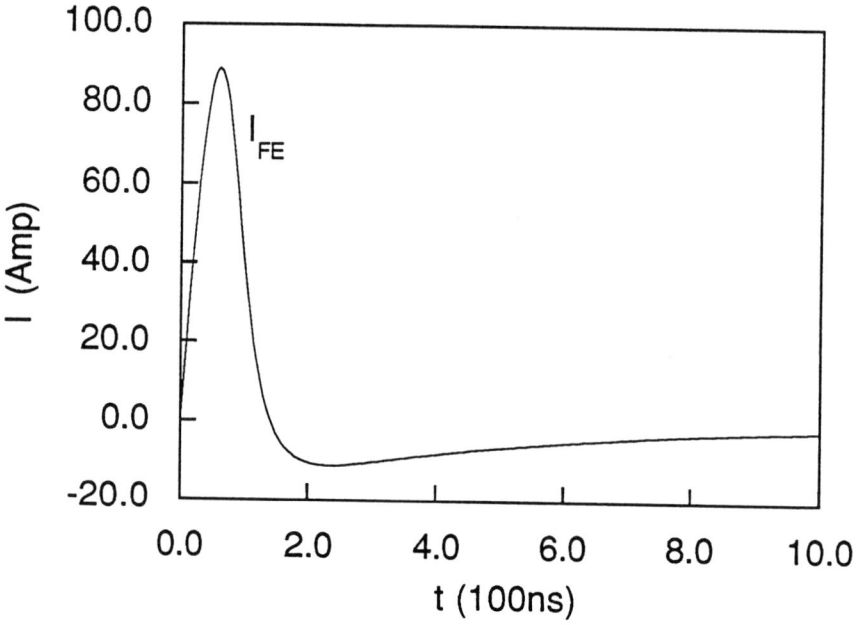

Fig. 8: The current flow into the ferroelectric capacitor.

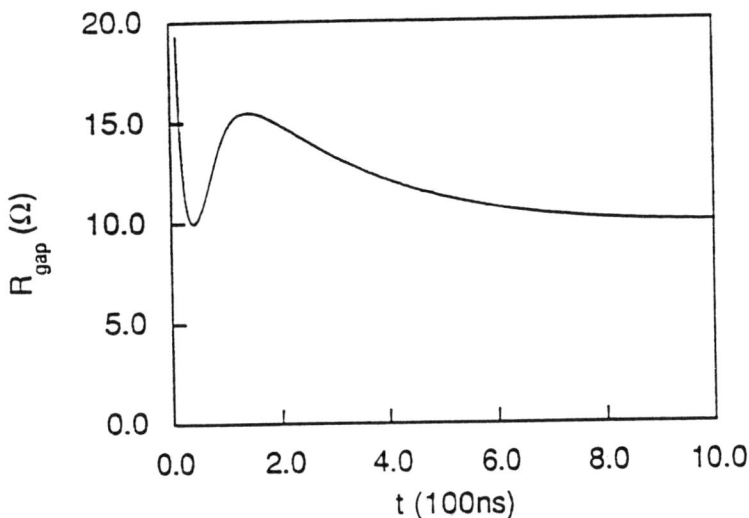

Fig. 9: The resistance induced by the ferroelectric in the gap as a function of time.

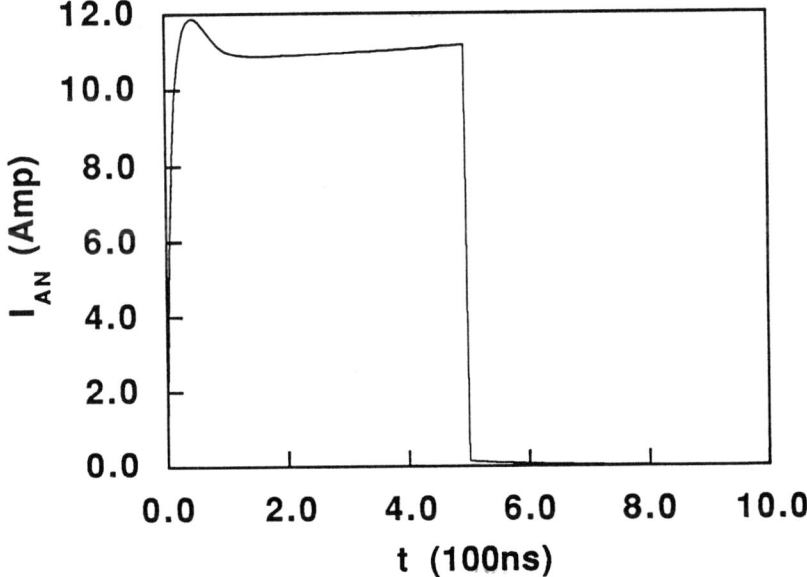

Fig. 10: The anode current is controlled by the transmission line voltage.

From this perspective electron emission from this device is similar to thermionic emission or photo-emission in the sense that it is not the diode voltage which is actually extracting the electrons from the material. However this is a controlled field emission device since the electrons are extracted from the material by an electric field applied to the <u>back</u> of the ferroelectric cathode.

The role of the cathode geometry in this device is far more important than in regular field emission devices. The gridded electrode permits the coupling between the ferroelectric region and the diode gap. Without this coupling the system would operate like a regular field emission diode and for the parameters of interest practically no current would flow. The special geometry of the electrode permits the "penetration" of the potentials space harmonics from the ferroelectric region into the gap. This local field can be very high and it can easily extract electrons from the metallic grid.

The electrons extracted from the metallic grid are not free electrons. Since the system remains neutral in the process, these electrons form a cloud in the gap. As a voltage is being applied on the anode, electrons can readily traverse the gap giving rise to a significant amount of current. This current is linearly dependent on the anode voltage.

We conclude with a quantitative comparison of theory and experiment: (1) Experiment: for $g = 4mm$, $V_{TL} = 300V$, $R_{TL} = 25\Omega$ and $V_{gen} = 1900V$ a current of $8.8A$ was measured. (2) Child-Langmuir current for this gap is of order of $30mA$ which is 200 hundred times smaller. (3) The proposed model ($\nu = 0.5$) predicts a current which varies along the pulse between 8.4 to $9.0A$ in agreement with the experimental data. For another measurement, $V_{TL} = 500V$ (the other parameters remain the same), $14Amps$ of current were measured and our theory predicts a current which varies between 13.5 and $14.7Amps$.

This work was supported by the United States Department of Energy.

REFERENCES

1. H. Riege; CERN-PS 89/42 (AR) 1989.

2. A.S. Airapetov, G.A. Gevorgian, I.I. Ivanchik, A.N. Lebedev, I.V. Levshin, N.A. Tikhomirova and A.L. Feoktistov in Particles Acceleration Meeting San Francisco 1991.

3. J.A. Nation, J.D. Ivers, L. Schachter and G.S. Kerslick in Particles Acceleration Meeting San Francisco 1991.

4. J.D. Ivers, L. Schachter, J.A. Nation, G.S. Kerslick and R. Advani; submitted for publication in J. Appl. Phys. .

HIGH BRIGHTNESS HOLLOW CATHODE ELECTRON BEAM SOURCE

G. Kirkman, N. Reinhardt, B. Jiang
Integrated Applied Physics Inc.
50 Thayer Road
Waltham, MA 02154

M.A. Gundersen, T.Y. Hsu, R.L. Liou
University of Southern California
Los Angeles, CA 90089-0484

R.J. Temkin
Massachusetts Institute of Technology
Cambridge, MA 02139

ABSTRACT

An optically initiated hollow cathode electron source has been demonstrated to produce high brightness electron beams in a low pressure background gas. Three phases of beam production have been identified, first a 3A, 20nsec, 45keV, 10^{10}A/m^{2}rad^2 initial beam, second a 70A, 100nsec,20keV, 10^{11}A/m^{2}rad^2 hollow cathode produced beam and third a 200A, steady state, 500eV superemissive cathode produced beam. A steady state hollow cathode produced beam of ~100mA has also been observed. An experimental characterization of the electron beam source is presented.

INTRODUCTION

Hollow cathode discharges are known to be sources of pulsed[1] and continuous[2] high current electron beams. Under certain conditions the electron beams can be produced with a low transverse energy spread[3] and a high beam brightness can be achieved. The pseudospark[4] is a hollow cathode discharge that produces a high current axial electron beam and has demonstrated high brightness[5] when using a multiple gap structure. In this work a single gap optically initiated pseudospark discharge is demonstrated to produce a high brightness beam similar to the multigap structure. This work has measured a beam brightness >10^{11}A/m^{2}rad^2 at 20keV and 70A peak current. This value of brightness is exceptionally high[6] for this simple source and is applicable to accelerator applications including microwave power generation for next generation colliders[7]. The beams are produced and propagate in a low pressure gas making this source suitable for plasma based microwave sources and plasma accelerators.

The pseudospark is a low pressure gas discharge occurring between flat electrodes each having an axial hole[8]. The discharge can be initiated optically by UV[9] or IR[10] light from a laser or flashlamp[11]. Theoretical modeling of the discharge formation[12,13] and electron beam production[14] has been reported. In the present work electron beam production is studied experimentally showing several phases of beam production including a high brightness initial phase lasting 20-50nsec, a higher current similarly bright hollow cathode phase of 50-100nsec and a lower energy superemissive cathode phase producing a pulselength up to several μsec. The current and duration of each phase are dependent on the external circuit and gas pressure thus operation can be optimized to enhance the desired phase of the beam. The initial phase beam energy is about equal to the applied voltage (up to 45kV in this work), the second hollow cathode beam is produced during the voltage collapse of the discharge and therefore has an energy less than the applied voltage and the final long pulse phase has an energy corresponding to the burning voltage of the superemissive cathode (SEC)[15] discharge which is about 500V. It is also possible to extend the hollow cathode phase of the discharge to produce a steady state beam of ~100mA at 400V for 100msec. The experiments described below have characterized the beam quality of the first two phases and have separately identified each component.

OPTICALLY TRIGGERED PSEUDOSPARK

The optically triggered hollow cathode beam source shown in figure 1, consists of two hollow electrodes with apertures on axis, a window allows light into the hollow cathode space for triggering and the anode is connected to a diagnostic port by a vacuum flange.A single gap structure was fabricated using brazed ceramic-metal techniques. The device is based on a 3 cm diameter electrode, 4 cm diameter ceramic insulator and mounts to vacuum and diagnostic system using standard 2.75 inch high vacuum flanges. The electrodes have molybdenum faces brazed to copper supports making up the hollow electrode structure. The electrodes and flanges are welded to a kovar ring which has been brazed to a metalized surface on the Al_2O_3 ceramic insulator body. An external capacitor or pulse forming network is connected to the device and charged by a dc supply. Light from the UV flashlamp initiates the discharge by providing initial electrons through photoemission[16]. The discharge conducts a current determined by the external circuit, a portion of this current produces the electron beam which propagates past the anode into the diagnostic region. All diagnostic sections are mounted on high vacuum flanges to allow for easy interchange of diagnostic sections and sure high vacuum sealing.The system is pumped by a diffusion pump to a base pressure of 10^{-7}torr and then filled to the desired pressure with argon of 99.999% purity the filling pressure is measured by a capacitance manometer with a 0 to 10torr range.

EXPERIMENT

An experiment was set up with the capability of operating up to 45kV with various external circuits with capacitance ranging from 0.3nF to 100μF. Measurement capabilities include time resolved beam current, current density, gap voltage, beam energy, emittance, brightness, beam profile, and dependence of these values on applied voltage, gas pressure and external circuit.

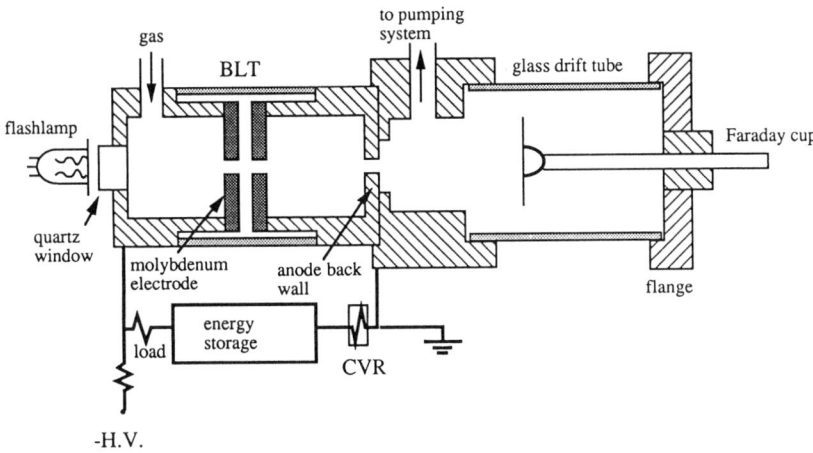

Figure 1. Schematic diagram of hollow cathode beam source experiment.

The beam energy is calculated by measuring its deflection in a magnetic field. The beam passes through a drift tube and is deflected by a transverse magnetic field applied by a solenoid, there is a further drift region after which the beam is observed on a phosphor screen. With no magnetic field the radial current distribution can be observed while with the magnetic field the energy is measured and the energy distribution can be estimated from the deformation of the beam shape. For beam emittance measurements the magnetic field is removed and a "pepper pot" emittance mask is used with the phosphor screen and camera.

EXPERIMENTAL RESULTS

Time resolved faraday cup measurements of the electron beam current indicate three phases of beam production. When using only the self capacitance of the electrode structure charged to 30kV a single electron beam pulse of 20nsec duration and 2A peak current is observed. Adding an external capacitance of 30nF charged to 30kV and limiting the total current to 3.8kA with a 4ohm resistor a second phase of beam current is observed during the formation of the discharge. This beam has a larger current and is strongly dependent on gas pressure and external circuit parameters. Using a PFN to produce a µsec duration pulse the SEC phase of the beam has been observed with a current up to 200A produced with an efficiency of ~5%. When a current limiting resistor of 1000 ohm is used to extend the pulse duration a steady state hollow cathode discharge (HCD) phase of the beam is observed with 500mA current lasting for a time determined by the external circuit.

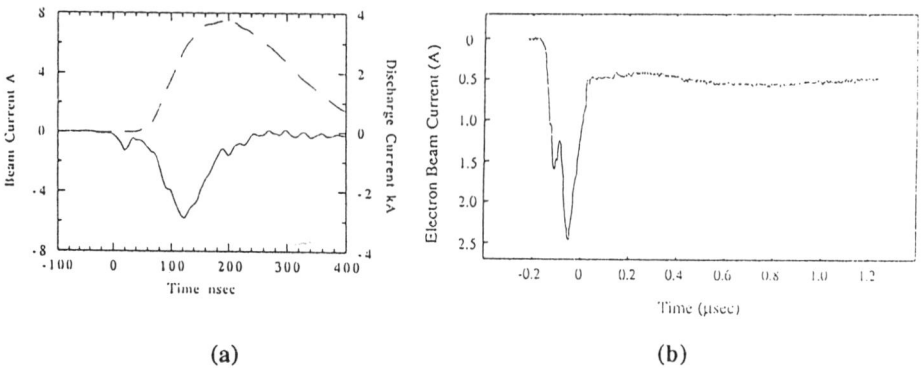

(a)   (b)

Figure 2. Three phases of electron beam production in the hollow cathode source. a) First two phases initial 20nsec is due to self capacitance of the structure and is present even when using no external capacitance. b) Three phases of the beam including the third steady state phase of 0.5A.

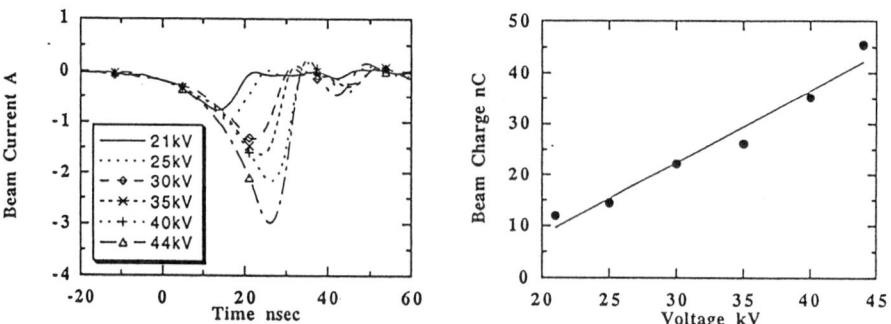

Figure 3. Beam current and total beam charge dependence on voltage during the first phase of beam production.

With no external capacitance the beam current in the first phase has been measured as a function of applied voltage. The beam current pulse shape shows an exponential current rise and a peak value which increases exponentially with applied voltage. The total charge in the beam increases linearly with applied voltage. Beam current and total charge produced at 21 to 44kV is shown in figure 3.

The beam current in the second phase has been measured as a function of gas pressure and applied voltage. The beam current and beam fraction (beam current/discharge current) increases with decreasing gas pressure and increasing applied voltage. A beam current of 70A, 1000A/cm^2 and 2% beam fraction has been obtained at 20kV and 40mtorr argon gas pressure. Dependence on pressure and voltage are shown in figure 4.

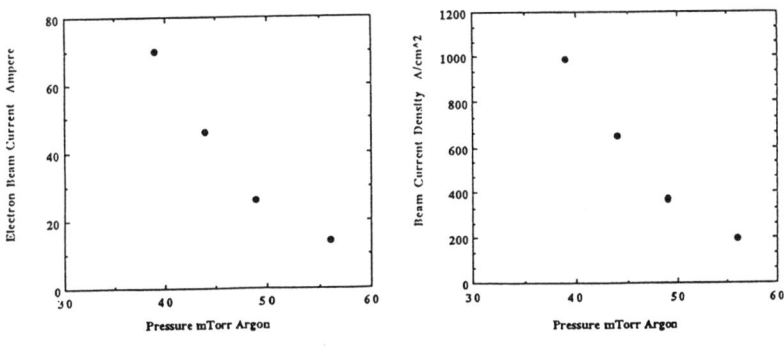

Figure 4. Beam current density and total current dependence on pressure during the second phase of beam production.

The beam energy is measured by deflection in a magnetic field and found to have a component approximately equal to the applied voltage and a component with much lower energy corresponding to the discharge voltage of about 500V. The beam deflection is measured using the phosphor screen. The phosphor has a threshold energy of about 6keV and therefore filters out beam electrons below this voltage. The energy was calculated by approximating the vXB force to be perpendicular to the drift tube which is correct when the beam is not deflected much in the region of interaction with the magnetic field (Larmor radius >> length of interaction region). In this approximation[17] the beam energy is given by $E=0.5\ (e^2/m_e)\ (BLD/x)^2$ where B is the transverse magnetic field, L is the length of the region of interaction with the field, D is the drift distance from the interaction region to the phosphor screen and x is the displacement on the screen, e and m_e are electron charge and mass. At an applied voltage of 20kV the beam displacement is 0.75cm giving an estimated energy of 17keV. The energy has been estimated for several applied voltages and increases linearly with applied voltage.

Separation of the lower energy beam component has been observed by the flourescence produced in the background gas while passing through a known magnetic field. The magnetic field is 80 gauss and the gas pressure is 20mtorr argon. At an applied voltage of 18kV and discharge current of 100A the larmor radius of the low energy beam is 0.9cm yielding an estimated energy of 460eV. The beam energy has been measured at several values of discharge current and increases with increasing current.

The beam density in phase space was measured using the pepperpot emittance diagnostic. In our implementation of this method the beam passes through a pepper pot mask with an array of 300μm diameter holes each separated by 1mm then drifts freely 18.7cm to a phosphor screen where it is photographed. The raw data is used to generate a plot of $x'=v_x/v_z$ against x. The emittance ε_0 is defined as $1/\pi$ times the area of the ellipse in x'-x space that encloses >90% of the beam electrons. The normalized emittance is $\varepsilon_n=\beta\gamma\varepsilon_0$ and the normalized beam brightness is $B_n = 2I/(\pi\varepsilon_n)^2$ The energy threshold of the phosphor screen limits our emittance measurement to the first two components of the electron beam, the first component is measured separately using a very small external capacitance such that the second component is insignificant, the second component can not be produced without the first but during this measurement a large external capacitance is used giving 50A of beam current such that the first component is insignificant.

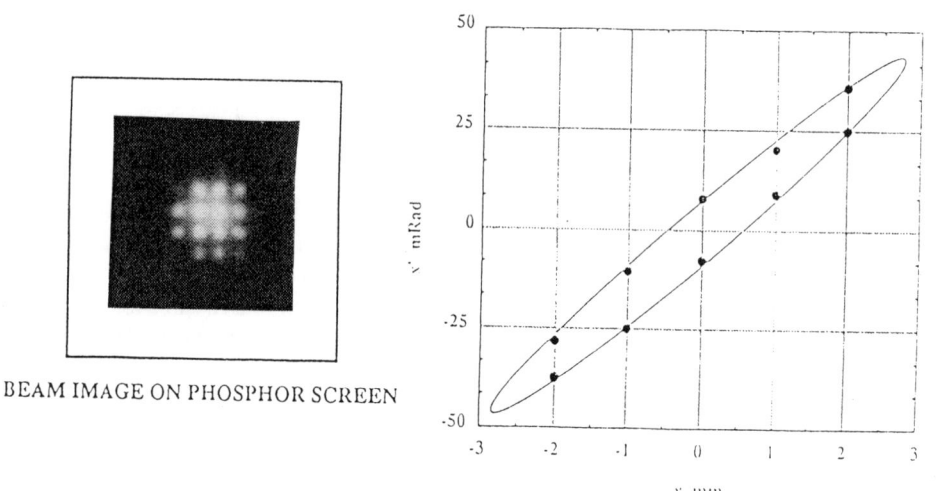

BEAM IMAGE ON PHOSPHOR SCREEN

Figure 5. Beam emittance measurement for the second hollow cathode beam phase.

Using this measurement the emittance of the first component is estimated to be 9π·mm·mrad at 25keV and 1A yielding a normalized emittance and brightness of 2.7π·mm·mrad and 1.4×10^{10}A/m²Rad² respectively. The second component has emittance 25π·mm·mrad at 15keV and 50A, a normalized emittance and brightness of 6π·mm·mrad and 1.6×10^{11} A/m^{2}Rad^2 respectively.

SUPEREMISSIVE CATHODE PRODUCED BEAM

A capacitor discharge and PFN circuit have been used to produce a pulsed electron beam during the superemissive cathode (SEC) phase of the pseudospark/BLT discharge.

With a 40 nF capacitor as the energy storage unit, a 5 ohm load resistor, an electron beam of 140 A peak current, 150 ns full width half maximum (FWHM) was measured 9 cm downstream at 50 mTorr argon gas and 15 kV applied voltage (figure 6b). A charge transfer efficiency, defined as the ratio of total charge of the electron beam to total charge of the discharge current, of 4.7% is obtained. The beam current extended well into the falling portion of the discharge current indicating a SEC beam, as distinct from a HC beam. These electrons are injected from the cathode fall region of the BLT during the high-current conduction phase. The beam current terminates shortly after the maximum of the discharge current, which is a result of fast voltage decay across the BLT during conduction. With a lower load resistor of 2.75 ohm, the maximum discharge current increases to 2.6 kA and the beam maximum increase to 200 A. The charge transfer efficiency also increases to 7%. The magnitude of load resistance is chosen to prevent reversal of discharge current.

Figure 7a shows another BLT-produced electron beam with a 0.2 µF capacitor for the energy transfer. With a 5 ohm load resistor, the discharge current has a much longer falling tail than before but the beam current still terminates 600 ns after the maximum of the discharge pulse. The charge transfer efficiency is 2.4% at applied voltage of 15 kV and argon pressure of 50 mTorr. As before, the beam current and the charge transfer efficiency increase when a lower load resistor of 2.75 ohm is used.

A linear increase of beam current with voltage is observed within our operating voltage range (10 kV to 25 kV). This suggests that the voltage holdoff capability of the BLT, thus the beam current, can be scaled by a multiple-gap structure. Extrapolating the linear relation, it is feasible to generate an electron beam with 1 kA peak current at applied voltage of 100 kV. The beam current also increase with decreasing gas pressure from 60 mTorr to 30 mTorr. Further study of beam current at lower pressure is done by a differential pumping scheme. A simple differential pumping scheme is arranged by positioning the gas pumping outlet at the diagnostic port. The differential pumping effect comes from the center holes of the electrodes and the anode back wall. At drift tube gas pressure of 17 mTorr and applied voltage of 15 kV, an electron beam current of 230 A peak current, 800 ns FWHM was measured as compared with beam current of 180 A peak current, 500ns FWHM at 50 mTorr argon pressure (Figure 7b). The increase of beam current and duration with differential pumping is explained by the reduction of collisions of beam electrons with neutral gas and plasma in the drift region.

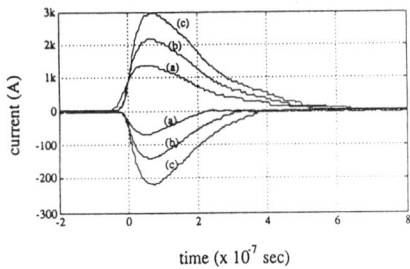

Figure 6. The discharge currents (positive) and the beam currents (negative) at 50 mTorr argon, applied voltages of (a) 10 kV, (b) 15 kV, and (c) 20 kV, using a 40 nF charging capacitor and a 5 ohm load resistor.

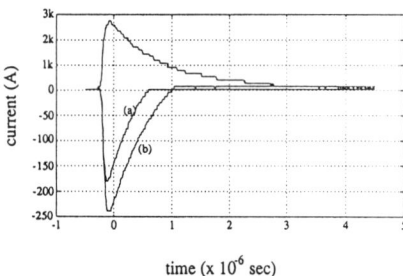

Figure 7. The discharge current (positive) and the beam currents (negative) using a 200 nF charging capacitor and a 5 ohm load resistor at 15 kV applied voltage when the drift tube pressures are (a) 50 and (b) 17 mTorr.

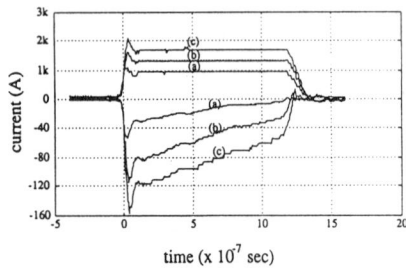

Figure 8. The discharge currents (positive) and the beam currents (negative) at 50 mTorr argon, applied voltages of (a) 10 kV, (b) 15 kV, and (c) 20 kV, using a PFN of characteristic impedance of 5 ohm for transferring energy.

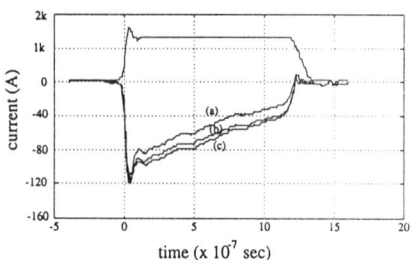

Figure 9. The discharge current (positive) and the beam currents (negative) at 15 kV applied voltage and argon pressures of (a) 50, (b) 40, and (c) 35 mTorr when a PFN of characteristic impedance of 5 ohm is used.

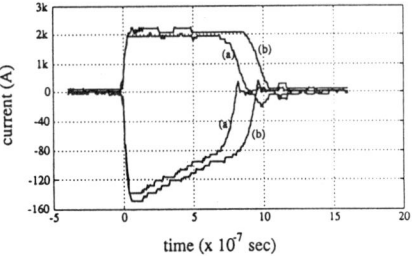

Figure 10. The discharge currents (positive) and the beam currents (negative) at 15 kV applied voltage and 50 mTorr argon when PFN's of characteristic impedances of (a) 4 ohm and (b) 2.75 ohm are used.

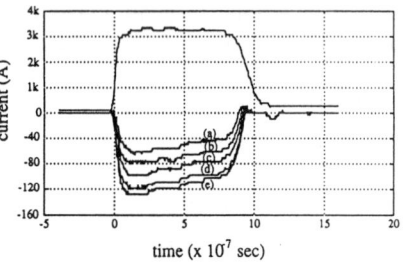

Figure 11. The discharge current (positive) and the beam currents (negative) at 20 kV applied voltage and 6 mTorr drift tube pressure, using a PFN for energy transfer and guiding magnetic fields of (a) 0, (b) 100, (c) 200, (d) 300, and (e) 400Gauss

The beam characteristics have been studied with several different pulse forming network (PFN) circuitries. The first PFN has characteristic impedance of 5 ohm and generates a pulse of 1.3 μs to a matched load. Figure 8 shows the beam pulses and the discharge currents at different applied voltages. The beam has about the same duration as the discharge current but in a decaying way. At 15 kV applied voltage and 50 mTorr argon pressure, the charge transfer efficiency is 3.7% . The magnitude of beam pulse also increases with increasing applied voltage and with decreasing gas pressure, as shown in figure 8 and figure 9. Notice that there is an obvious transition of beam pulse from the HC transient phase beam to the SEC conduction phase beam. As gas pressure decreases, the HC beam does not change too much while the SEC beam increases with a relatively larger proportion. This is consistent with the fact that the lower energy electrons in the SEC beam is affected more by the collisions with bulk plasma and neutral gas.

Two PFN's with characteristic impedances of 4 ohm and 2.75 ohm are used to generate pulses of 750 ns and 1 ms into matched loads respectively. As shown in Figure 10, the SEC beams generated by the BLT with these two PFN's have a relatively lower decaying rate than that with the previous PFN which has a higher characteristic impedance and a lower discharge current. The charge transfer efficiencies also increase to 5.3% and 4.7% respectively. The transition of the beam pulses from the HC component to the SEC component are not so obvious as in the case of PFN with higher impedance because of the increase of the SEC beams. The increase of SEC beam seems to relate to lower load resistance and higher discharge current which might indicate a higher voltage drop in the cathode fall, but further study is needed.

The beam current decreases when the drift tube pressure is < 12 mTorr and is explained by the beam divergence resulting from the space charge effect. To decrease the divergence, a solenoid is used to generate the guiding longitudinal magnetic field. A magnetic field of 200 Gauss is measured at dc current supply of 20 A. Figure 11 shows the increase of beam current with the increase of guiding field at 6 mTorr drift tube pressure and 20 kV applied voltage. With our experiment setup, the guiding field is limited to 400 Gauss and the drift tube pressure can not be lower than 4 mTorr while remaining operating pressure in the BLT hollow cathode.

STEADY STATE HOLLOW CATHODE PRODUCED BEAM

A variable pulse-length electron beam source based on the hollow cathode discharge mode of operation of the back-lighted thyratron has been developed. Long-pulse electron beam generation was achieved by modification of circuit parameters that control the hollow cathode discharge. Two different electric circuit have been adapted in this study; one with an RC discharge (100 μF and 500 Ω), the other a pulse forming network with combination of 100 μF capacitors and 5 H inductors. In the first circuit, an electron beam with a base-to-base duration of ~ 125 msec and peak current of 75 mA has been measured at 5 cm behind anode in a 800 mTorr H_2 discharge. In the pulse forming network circuit, a beam current with FWHM of ~ 200 msec and beam current of 70 mA was measured at 2 cm behind anode in a ~ 1 Torr H_2 discharge. In both circuit, a differential pumping was used to reduce the gas pressure in the drift region. A discussion and preliminary result of the possible DC electron beam operation is included.

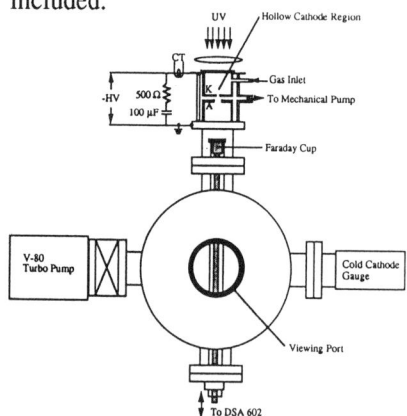

Fig. 12 Experimental setup for long pulse electron beam generation. Electric circuit shown is for capacitive discharge study. Gas will be flowing into the BLT through the cathode side and then being pumped out with both the mechanical pump and turbo pump. The cathode is at negative high potential.

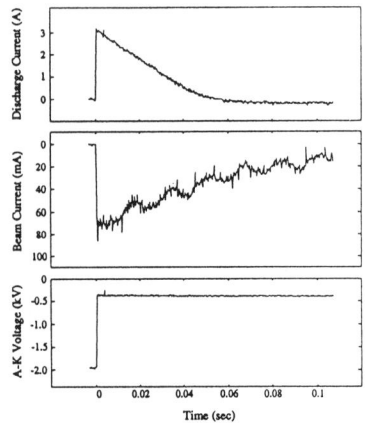

Fig. 13 The electrical data from an RC discharge circuit. The discharge current is measured with a Pearson # 411 current transformer (CT) which has a droop of 0.0009 %/msec. The electron beam current is found to follow the shape of discharge current.

Fig. 12 shows a schematic drawing of the experimental setup. The HCR is a copper cylinder, with a cathode on one end and a quartz window on the other. The HCR is 3 cm in both diameter and length. The cathode central hole size and the thickness of the electrodes are both 3 mm. The anode central hole size and cathode-anode gap are 5 mm. The electron beam emission was initiated with UV light on the back of cathode near the central hole through a quartz window. Cathode current was monitored by Pearson current transformer (CT) with a droop of 0.0009 %/msec. A V-80 turbo pump provides the differential pumping in the drift chamber region. A SD-200 mechanical pump was used to reduce the gas pressure between the anode cathode gap. The purpose is to reduce the possibility of gap closure during the discharge and also reduce the chamber gas pressure. Pressure in the drift chamber was measured with a cold cathode gauge. Cathode voltage was monitored through Tektronix P6015 1000X high voltage probe with a rise time of ~ 5 nsec. The electron beam current is measured with brass Faraday cup positioned along the discharge axis.

Two different discharge circuit were used in this study; one with an RC discharge, the other a pulse forming network with combination of 100 μF capacitors and 5 H inductors. Due to the lack of a DC power supply in the range of several kV and several A output, a capacitor with 100 μF was charged to -2 kV as a main energy source for the initial test. A ballast resistor of 500 W was connected in series with the discharge circuit. The large resistor was chosen to both limit the discharge current into the HCD and increase the time constant. In the former circuit, a current transformer was used to measure the discharge current. In the latter circuit, a current viewing resistor was used.

BEAM GENERATION WITH CAPACITOR DISCHARGE

Our first attempt is to create a long electron beam pulse with duration in the order of 100 msec. For this purpose a simple RC discharge with an 100 μF capacitor and 500 Ω resistor was used. Due to the limitation on the capacitor voltage rating, the circuit was normally operated at ~ -2 kV. Fig. 13 shows the discharge current, electron beam current and the voltage across the anode-cathode gap. The discharge current is overdamped with a peak current of ~ 3 A. The pulse length when measured from base-to-base is ~ 125 msec. The RC time constant of discharge current measured from the data is ~ 30 msec, which is shorter than that calculated from the circuit parameter (i.e. 50 msec). This discrepancy may well be resulting from the droop of current transformer which is 0.0009 %/msec. The data shown therefore is the response of high-pass filter with corner frequency of the order of 1 Hz.

The electron beam is self-extracted with no extraction voltage nor axial magnetic filed. With a pressure of ~ 800 mTorr H_2 in the gas inlet, the pressure in the drift chamber can be pumped down to ~ 8 mTorr. A peak beam current of 75 mA was measured at the peak of discharge current which gives a beam to discharge current ratio of 2.5 %. The Faraday cup is at 5 cm behind anode. The FWHM of the electron beam is ~ 50 msec. The cathode-anode voltage drops from initial applied voltage of 2 kV to ~ 300 V and remains at that value for the rest of the beam current. A transverse magnetic field was used to deflect this electron beam. Faraday cup signal completely disappears under the influence of this test field.

BEAM GENERATION WITH PULSE FORMING NETWORK

The preliminary results of the capacitive discharge show that the electron beam can take the shape of discharge current. Since in most of the applications a constant-amplitude electron beam is desirable. Therefore we constructed a pulse forming network (PFN) with combinations of 100 μF capacitors and 5 H inductors. Fig. 14 shows the electrical circuit of the PFN.

Fig. 15 shows the data with the PFN circuit when Faraday cup was at 7 cm behind anode hole. The gas pressure required for triggering BLT was about 1.2 Torr H_2 in the HCR, which is higher than that of the previous RC discharge circuit. The exact reason for this increase is not clear. This higher chamber pressure (~ 30 mTorr) may be accountable for the decrease in electron beam current. The peak discharge current is ~ 3 A and with a FWHM of ~ 200 msec. The anode cathode voltage still maintains at ~ 400 V through the discharge. The peak beam current is ~ 7 mA.

It is found that the beam current collected by the Faraday cup changes significantly at different positions behind anode while maintaining the same shape. Fig. 16 shows the results with Faraday cup at different locations behind anode hole. A constant ratio was found between any two traces. The results suggests that the electron beam is diverging during the propagation since the number of ionization is small in the distance of several cm. The ratio between the beam current and discharge current is ~ 2.5 % when Faraday cup is 2 cm behind anode.

A test of transverse magnetic field was performed to verify that the current measured with the Faraday cup is from beam electrons not the plasma electrons. A different strength of magnetic field was used. Fig. 17 shows the electron beam current when different transverse magnetic field was applied. The electron beam current decreases gradually with increasing magnetic field which is also an indication of beam divergence. Beam current completely disappears with field ~ 100 Gauss.

It is also found possible to operate BLT in its hollow cathode discharge mode for generation of continuous electron beam with a discharge current up to 100 mA and anode cathode voltage at ~ 500 V. The electron beam is found to increase with decreasing pressure in the HCR in the range of 700 mTorr to 1 Torr. At 700 mTorr in HCR the chamber pressure is ~ 4 mTorr. This DC beam current increases dramatically when the Faraday cup is moved closer to anode. At 1 cm behind anode the beam to discharge current ratio is more than 20 %.

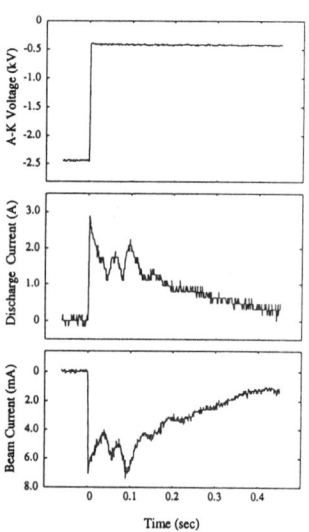

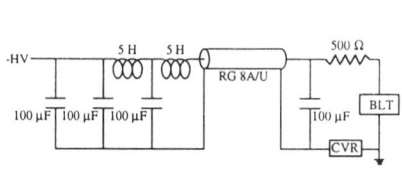

Fig. 14 Pulse Forming Network (PFN) for generating constant electron beam current. The previous capacitive circuit in Fig. 12 was replaced with this PFN. The inductors are copper coil wired on an ion core with an adjustable gap.

Fig. 15 Data obtained with the pulse forming network as shown in Fig. 14. The initial applied voltage is -2.5 kV. The pressure in the HCR is ~ 1.2 Torr H_2. With different pumping, the chamber pressure is ~ 30 mTorr.

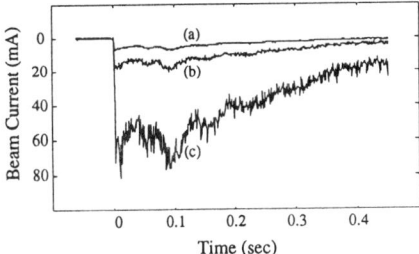

Fig. 16 Electron beam current at different positions behind anode, a) 7 cm, b) 5 cm and c) 2 cm. The ratios between any two curves are found to be constant.

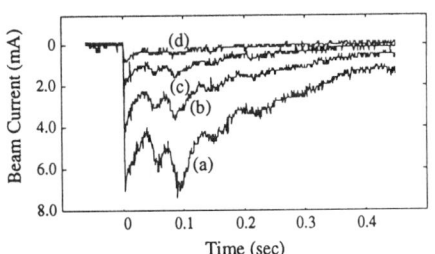

Fig. 17 The electron beam current measured at 7 cm behind anode. Different strength of magnetic were used to deflect the electron beam, a) no magnetic field, b) ~ 35 Gauss, c) ~ 60 Gauss, and d) ~ 80 Gauss. The interaction region is ~ 1.5 cm.

DISCUSSION

Three distinct phases of electron beam production have been observed in the optically triggered hollow cathode discharge. The first phase is produced during the discharge of the self capacitance of the electrode structure. These electrons appear to have energy equal to the applied voltage and a low emittance. The total charge in the beam increases linearly with voltage. The estimated capacitance of the electrode structure is 1.6pF and approximately 40% of the charge applied to the electrode structure is produced as beam current.The second phase is produced during the hollow cathode phase of the discharge and can have an order of magnitude larger current than the first phase. These electrons are produced while the voltage is rapidly falling and therefore have an energy lower than the applied voltage and some spread in energy. The emittance of this beam is larger but the much higher current gives a larger brightness.

The measured brightness of the first two phase of the beam verifies that this optically triggered discharge is a source of high brightness electron beams and agrees with the previous measurement in a multiple gap structure verifying that a single gap structure can produce similar low emittance beams as multiple gap structures when operated at the same voltage.

The third phase of the beam is produced after the hollow cathode discharge when the superemissive discharge has formed drawing high current from the front surface of the cathode which has been heated during the hollow cathode phase. This beam has energy corresponding to the burning voltage of the discharge and a current that is a few percent of the total discharge current. The emittance of this beam could not be measured in our experiments. The pulse length of this beam is determined by the external circuit and appears to be extendable to many msec or more.

The electron beams observed in this device are applicable for accelerator applications. The first phase of the beam could be useful as an injector for a high energy accelerator or plasma based accelerator, the second phase has a current and brightness suitable for microwave generation in plasma filled devices while the third phase if post accelerated could have a variety of applications including electron beam ion traps, electron beam machining and welding.

ACKNOWLEDGEMENT

This work was supported by the U. S. Department of Energy through the SBIR program and Lawerence Livermore Laboratory. The work at U.S.C was supported by the Strategic Defense Initiative and LLNL

REFERENCES

[1] M. Farvre, H. Chuaqui, E. Wyndham and P. Choi, IEEE Trans. Plasma Sci. **PS-20**, 53 (1992).

[2] J. J. Rocca, J. D. Meyer, M.R. Farrell and G.J. Collins, J. Appl. Phys. **56**,790,(1984).

[3] A.I. Herscovitch, V.J. Kovarik and K. Prelec, J. Appl. Phys. **67**,671, (1990).

[4] J. Christiansen and C. Schulheiss, Z. Phys. A, **290**,35 (1979).

[5] K.K. Jain, E Boggasch, M. Reiser and M. J. Rhee, Phys. Fluids **B2**, 2487 (1990).

[6] C. W. Roberson and P. Sprangle, Phys. Fluids, **B1**, 3 (1989).

[7] G. Caryotakis, Proc. 1991 Part. Accel. Conf. IEEE Press, p. 2928 (1991).

[8] J. Christiansen, in Physics and Applications of Pseudosparks, edited by M. Gundersen and G. Schaefer NATO ASI Series B: Vol. 219 Plenum, New York (1990).

[9] G. Kirkman, and M. A, Gundersen, Appl. Phys. Lett. **49**, 494 (1986).

[10] H.H. Chuaqui, M. Farve, E. Wyndham, L. Arroyo, and P. Choi, IEEE Trans. Plasma Sci. **PS-17**, 766 (1989).

[11] G. Kirkman, W. Hartmann and M. Gundersen, Appl. Phys. Lett. **52**, 613 (1988).

[12] J.P. Boeuf and L.C. Pitchford, IEEE Trans,. Plasma Sci. **PS-19**, 286 (1991).

[13] P. Choi ,H. Chuaqui, J. Lunney, R. Reichle, A. J. Davies, and K. Mittag, IEEE Trans,. Plasma Sci. **PS-17**, 770, (1989).

[14] H. Pak and M. Kushner, Appl. Phys. Lett. **57**, 1619, 1990.

[15] W. Hartmann, M. A. Gundersen, Phys. Rev. Lett. **60**, 2371 (1988).

[16] C.G. Braun, W. Hartmann, V. Dominic, G. Kirkman, M. Gundersen and G. McDuff, IEEE Trans. Electron Devices, **ED-35**, 559 (1988).

[17] "An Introduction to the Physics of Intense Charged Particle Beams," R. B. Miller, Plenum Press, New York, 1982.

PERFORMANCE OF THE APEX 40-MEV PHOTOINJECTOR-DRIVEN LINEAR ACCELERATOR

P.G. O'Shea, S.C. Bender, B.E. Carlsten, J.W. Early, D.W. Feldman, R.B. Feldman,
K.F. McKenna, R.L. Martineau, M.J. Schmitt, W.E. Stein, M.D. Wilke, T.J. Zaugg
Mail Stop J579, Los Alamos National Laboratory, Los Alamos, NM 87545 USA

ABSTRACT

Since the mid-1980s, Scientists at Los Alamos National Laboratory have been developing photocathode rf guns for high-brightness electron-beam applications, such as free-electron lasers (FELs). The technology has matured to the point where we now have a routinely operating 40-MeV linac and FEL that uses a photocathode as its electron source. In this paper, we describe the APEX accelerator's performance, with an emphasis on the photocathode's unique features.

INTRODUCTION

Our work on photocathode sources was motivated by the need to develop high-brightness, high-current electron beams for FEL applications.[1] How well an FEL performs at short wavelengths in the emittance dominated regime, depends greatly on the emittance and current; that is, on the brightness of the electron beam delivered to the FEL wiggler. In this paper, we discuss two fundamental issues. First, we discuss how an electron beam is generated and first accelerated in the photoinjector. Second, we discuss the transport of the electron beam through the linac, while minimizing emittance growth. Table 1 lists the nominal design specifications for the APEX linac; Figure 1 shows the layout for both the linac and FEL.

rf Frequency	1.3 GHz
Energy	40 MeV
Micropulse current	300 A
Micropulse length	15 ps
Macropulse current	0.1 A
Macropulse length	100 μs
Macropulse rep. rate	1 Hz
Energy spread	< 0.3%
Emittance	< 13 π mm-mrad

Table 1 APEX Linac Design Specifications

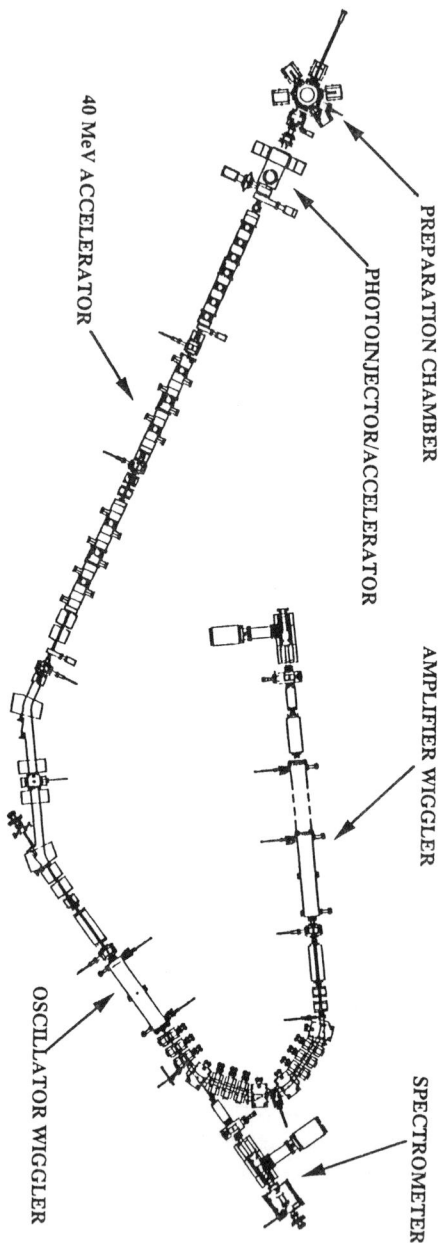

Figure 1 APEX accelerator and FEL layout

PHOTOINJECTOR

The photocathode material used at APEX is CsK_2Sb. This material was chosen because of its high quantum efficiency for frequency-doubled Nd:YLF laser light. While this cathode material makes the drive-laser design for long pulse trains straightforward, it imposes stringent vacuum requirements. For example, the quantum efficiency of CsK_2Sb can easily be degraded by the adsorption of water vapor onto the surface of the cathode. As a result, the vacuum must be better than 1 x 10^{-9} torr[2] Figure 2 shows the effective run time for each of the cathodes used during a recent period. Effective run time is defined as the time at which the electron beam's micropulse current remains above 100 A. The typical starting quantum efficiency for these cathodes is approximately 5% and their active diameter is approximately 12 mm. The actual electron emission diameter is reduced by an aperture in the drive-laser propagation path. Therefore, the emission diameter is governed by the desire to minimize the emittance for a given micropulse charge. In this regard, a balance must be struck between space-charge-induced emittance growth when the diameter is too small and the nonlinear rf-field-induced growth when the cathode diameter is too large.[3] The preferred diameter for a micropulse charge of 1 nC is 6 mm, which rises to 10 mm at 5 nC. The drive-laser spatial mode is gaussian, with truncation at the 36% point, and the emission diameter is defined as the diameter at which the spatial mode is truncated.

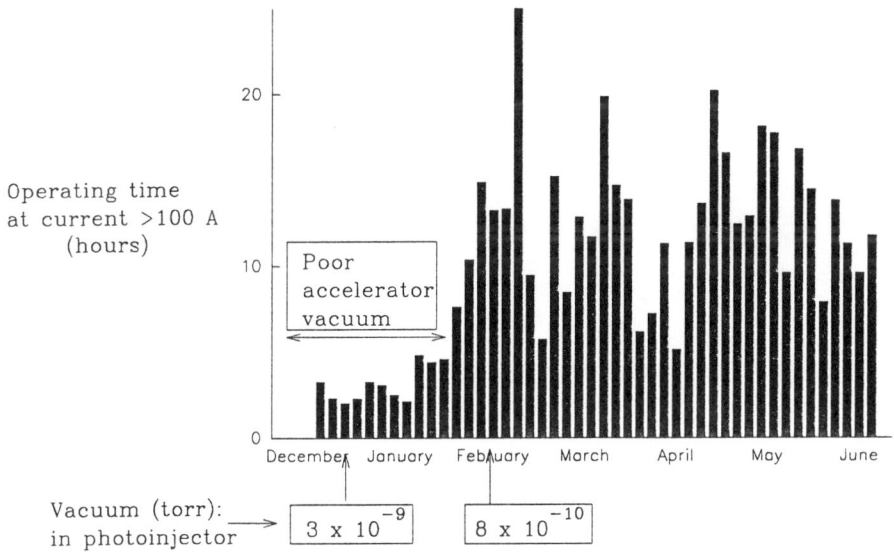

Figure 2 Cathode life-time in the operating accelerator. The times were calculated by extrapolating the measured quantume efficiency decay using an exponential decay model

FIELD EMISSION

The peak electric field at the surface of the cathode is approximately 30 MV/m during normal operation. When the photoinjector was first operated, the background signal from field emission was a significant fraction of the photoelectron current. Because the field emission electrons are emitted over a large phase angle relative to the photoelectrons, their energy spreads and emittance can differ significantly from that of the photoelectrons. At low-duty factor, the field emission electrons create an undesirable background signal on beam diagnostics. At high duty, however, the field emission electrons may radioactivate the accelerator components excessively, and in some extreme cases, physically damage the components. The first APEX cathode had a machined surface whose features were 4 μm high. The radii of curvature at the tips of these features was about 1 μm. Such surface features can result in field enhancement factors of 10 or more over the macroscopic field at the tips of the these features. To solve this problem, we polished the cathode surface to a mirror-finish with a surface feature height of 2 nm rms. Because the curvature radius for the surface features was much greater than the height, we reduced the field enhancement to a very low value. Figure 3 shows the field emission current versus the cathode electric field for both the polished and the unpolished cathode. Polishing the cathode reduced the field emission by a factor of 50. Our analysis of the images produced by the field emission current on our insertable view screens revealed that the residual emission emanated largely from a tiny scratch on the otherwise mirror surface.

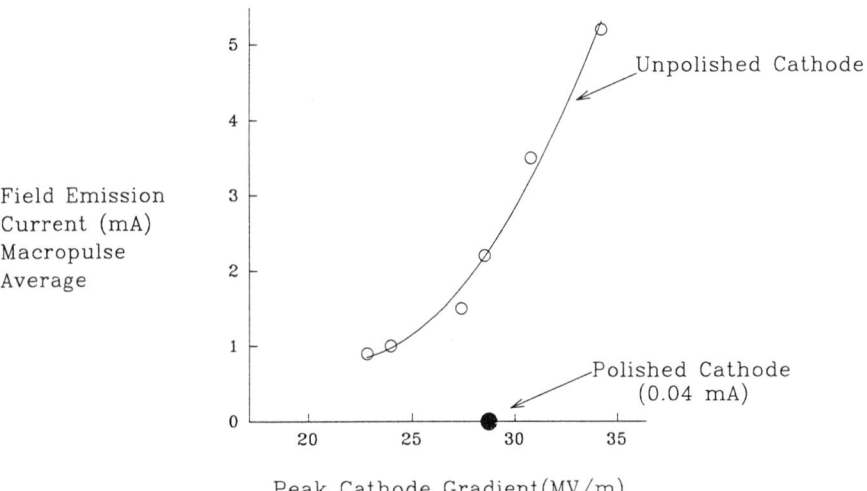

Figure 3 Field emission current versus cathode electric field. the field emission current is averages over the macropulse. Polishing the cathode reduced the emission by approximately a factor of 50.

PHOTOELECTRON EMISSION

Table 2 lists the drive laser's specifications; Figure 4 shows its general layout. Early et. al. provide a detailed description of the laser.[4]

Type	Doubled Nd-YLF
Wavelength	527 nm
Micropulse width	7-15 ps
Micropulse rep. rate	21.7 MHz
Micropulse energy	12 μJ (5-6 μJ at cathode)
Macropulse length	0-200 μs
Phase jitter	< 1 ps
Amplitude jitter	< 1%

Table 2 APEX drive-laser performance

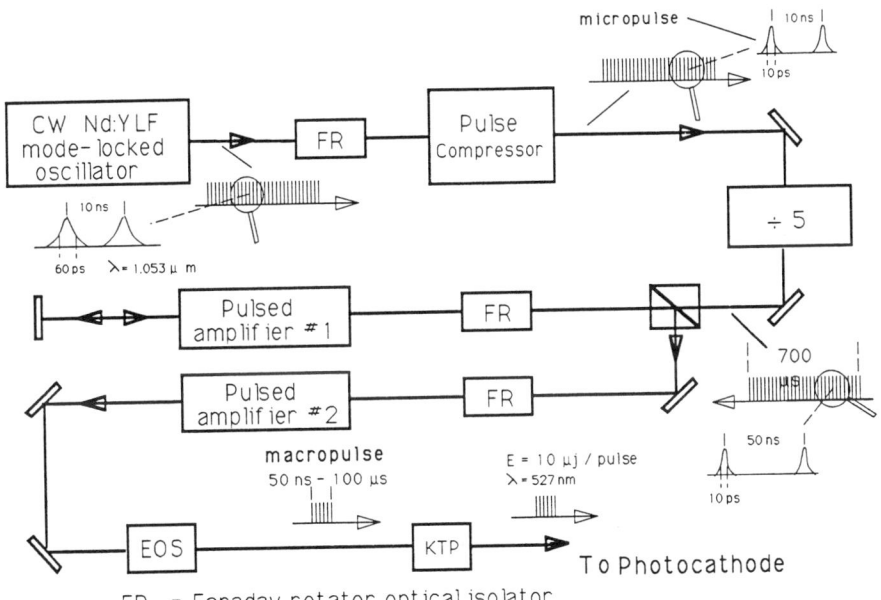

FR   = Faraday rotator optical isolator
EOS = Electro-optic shutter
KTP = Potassium titanyl phosphate frequecy doubling crystal
÷ 5 = Fast pockels cell shutter transmitting every fifth pulse

Figure 4 The APEX photocathode drive-laser

We have studied the photo-emission current as a function of both drive-laser power density on the cathode and injection phase relative to the rf. Figure 4 shows the photo-emission characteristic curves for our cathode. In generating these curves, we held the cathode's diameter at a constant 3 mm and set the drive laser's pulse length at 10 ps. For each drive-laser power, we varied the injection phase and measured the extracted charge using a wall-current monitor located immediately downstream from the photoinjector and approximately 60 cm from the photocathode surface.

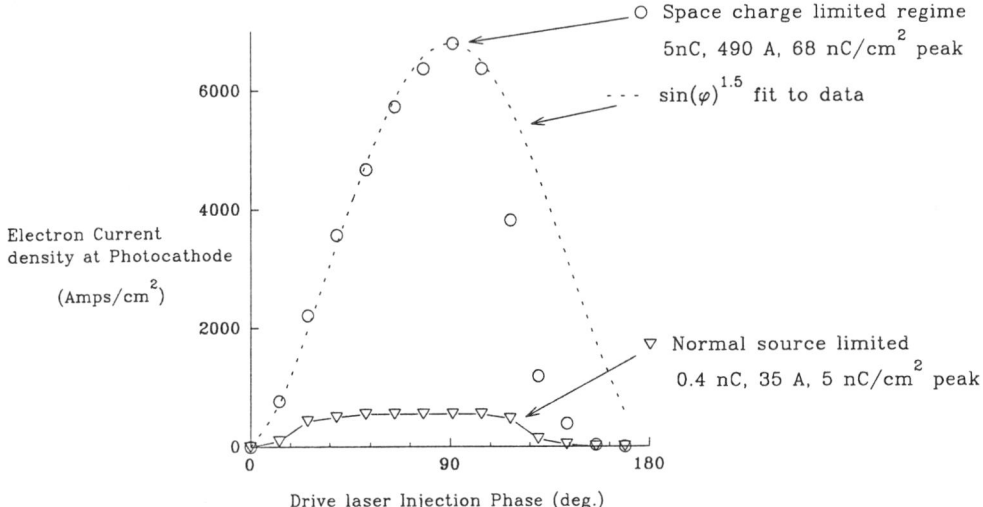

Figure 5 Photoemission current density at the cathode versus drive laser injection phase in the field limitied regime (drive laser power density = 18 MW/cm^2), and source limited regime (drive laser power density = 1.3 MW/cm^2). The cathode radius was 1.5 mm and the drive laser pulse width was 10 ps.

The electron current at the cathode surface is not measured directly. Because the electron emission from the cathode is prompt, the temporal length of the electron pulse near the cathode is less than or equal to the drive-laser pulse length. We measure the charge for each micropulse that reaches the wall-current monitor and estimate the current at the cathode by dividing the total charge per micropulse by the FWHM pulse width of the drive laser. To determine the value of τ, we use the streak camera technique (Hamamatsu C1587, 1-ps resolution).[5]

The variation drive-laser phase and power have their analogs in thermionic cathodes. Phase is analogous to the cathode voltage and power is analogous to cathode temperature. (Note that increasing the drive-laser power does not increase the emission energy of an electron, only the number of electrons emitted.). The electric field, as seen by the photoelectrons at the cathode, will vary as the sine of the injection phase angle.

$$E(\phi) = E_m \sin(\phi) \tag{1}$$

The cases shown in figure 5 illustrate the source-limited regime (low drive-laser power) and the space-charge-limited regime (high drive-laser power). In the case of the low drive-laser power, the curve makes the transition from space-charge-limited (0 to 20°) to the source limited (flat top) and then back to space-charge-limited again. This case corresponds to the normal operating drive-laser power and current density at APEX. PARMELA simulations have shown that the optimum drive-laser phase angle is in between 20° and 30° [3]

In the case of the high drive-laser power, (18 MW/cm^2 or 18 μJ/cm^2 per micropulse) the emission is always in the space-charge-limited regime. We have observed current densities as high as 6800 A/ cm^2. A fit to the data of the form

$$J(\phi) = J_m \sin(\phi)^{1.5} \tag{2}$$

tracked the data well for $\phi < 100°$, and indicated that the electron emission is governed by some form of Child-Langmuir relationship. The reduction of the current below the Child-Langmuir value from $\phi > 100°$ is caused by transit time effects in the six-cell photoinjector. Electrons emitted from the cathode at large phase angles experience decelerating rf fields in the downstream cells and many never reach the wall-current monitor.

BEAM MEASUREMENTS AT HIGH ENERGY

Following the photoinjector, the beam is accelerated to 40 MeV using three side-coupled $\pi/2$-standing wave structures referred to as tanks B, C, D. Each tank has an accelerating cell numbered 15, 13, and 15, respectively, and each is separately powered by a TH2095 klystron. Between each tank, we have insertable view screens from which the electron-beam profile is imaged using optical transition radiation (OTR). We use OTR because its emission is immediate and allows submicropulse resolution when used in conjunction with a fast streak camera. Electron-beam micropulse current is determined from wall-current charge measurements and OTR streak-camera micropulse measurements.

Figure 5 shows the electron beam's micropulse temporal length as a function of micropulse charge. The drive-laser pulse width was held constant at 10 ps. For charges less than 1 nC, the electron-beam pulse width is less than the drive-laser pulse width because of rf bunching in the linac. For higher charges, space-charge debunching dominates to produce longer micropulses.

We measured emittances at two locations immediately before and after the 60° bend, using the quadrupole-scan technique.[6] We measured the electron beam's spot FWHM size as a function of the strength of a preceding quadrupole magnet and

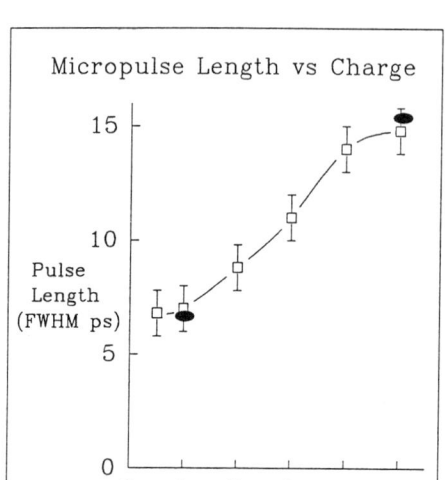
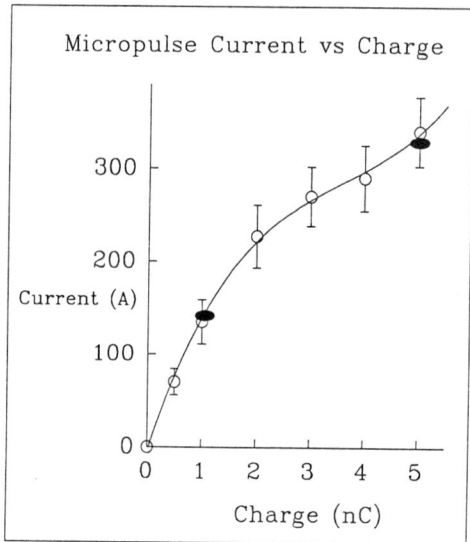

● = PARMELA prediction

Figure 6 Electron beam micropulse temporal length and current versus micropulse charge

determined the emittance by fitting the data to a theoretical curve. The emittance numbers quoted in this paper are normalized root-mean-square values determined by converting FWHM values to rms using a gaussian profile assumption. Figure 7 shows the emittance as a function of micropulse current measured before the 60° bend, which was measured at an electron-beam energy of 36 MeV. The numbers plotted are the geometric mean of two orthogonal measurements averaged over 440 micropulses. The unnormalized emittance at 135 A is 0.046 π mm-mrad. This emittance is sufficient for FEL lasing at wavelengths as short as 160 nm. The measurement of such low emittances at high charge and current from a photoinjector linac verifies the solenoidal-emittance compensation scheme proposed by Carlsten.[3] The PARMELA curve without wakefields corresponds to the emittance when the electron beam is steered to minimize wakefield effects. The PARMELA curve with wakefields corresponds to the electron beam being steered so as to be centered on the screens between each accelerator tank. Because of dipole rf fields in the side-coupled accelerator tanks, the electron beam experiences transverse kicks that induce head-to-tail transverse-wakefield kicks in the micropulse and hence emittance growth. The trajectory required to minimize the wakefields is not intuitively obvious to the accelerator operator. In order to assist in determining the optimum trajectory we have installed our streak camera to look at the electron micropulse temporal and spatial profile on a screen at the end of the linac. By this means we are able to directly observe the effects of the wakefields on the electron beam as shown in figure 8. The operator can easily choose a trajectory that minimizes the wakefields and optimizes the emittance. A detailed knowledge of the optimum trajectory is not required.

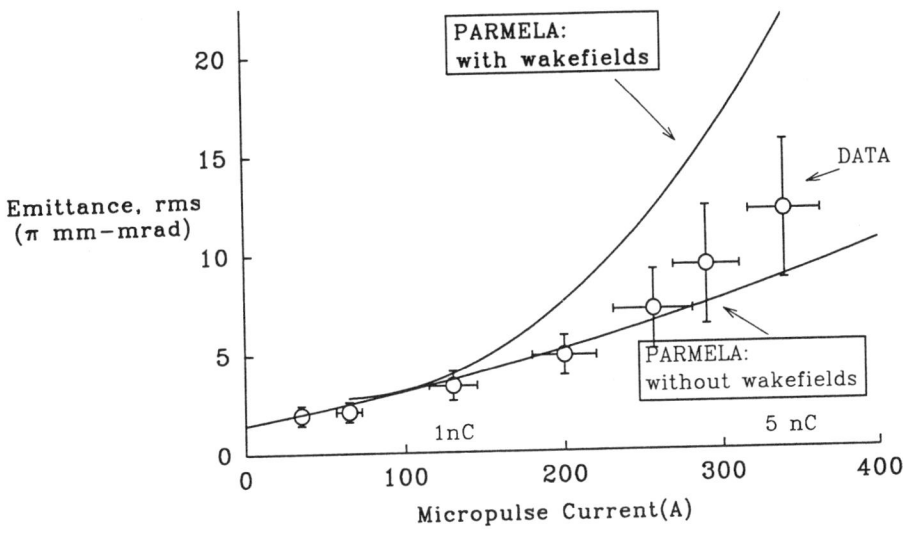

Figure 7 Normalized, rms emittance versus micropulse current.

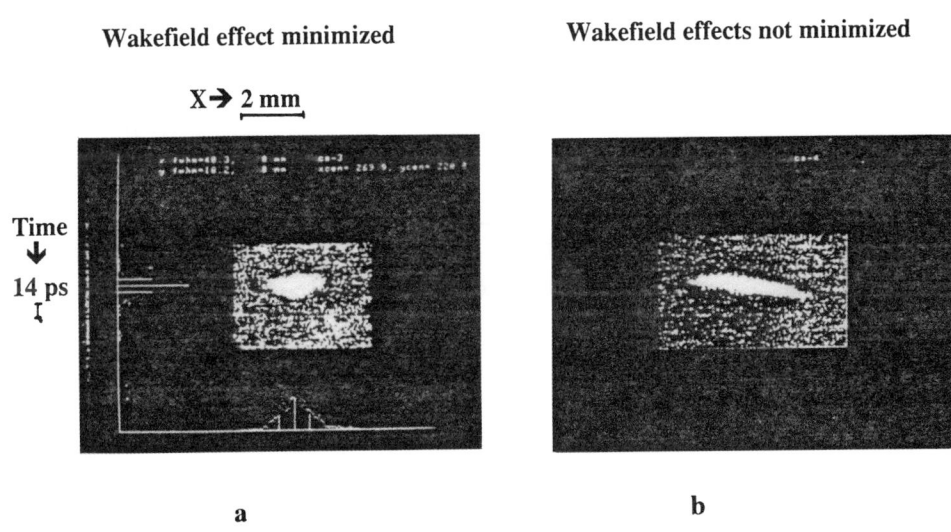

Figure 8 Direct observation of transverse wakefield kicks on electron micropulses. a) head-to-tail kick evident, b) head to tail kick removed by optimum choice of steering.

The normalized brightness of the beam is shown in figure 9 as a function of current. The brightness is defined as

$$B = 2I/\varepsilon^2 \; , \tag{3}$$

where I is the micropulse current and ε is the rms emittance. The factor of π in the definition of emittance is explicitly used to determine the brightness. The maximum brightness observed is to date 3×10^{13} A/(m-rad)^2, at a current of 135 A and an electron-beam energy of 36 MeV. We measured the energy spread at 0.24% FWHM averaged over 220 micropulses.

O'Shea et. al. provide a detailed description of the APEX FEL.[7]

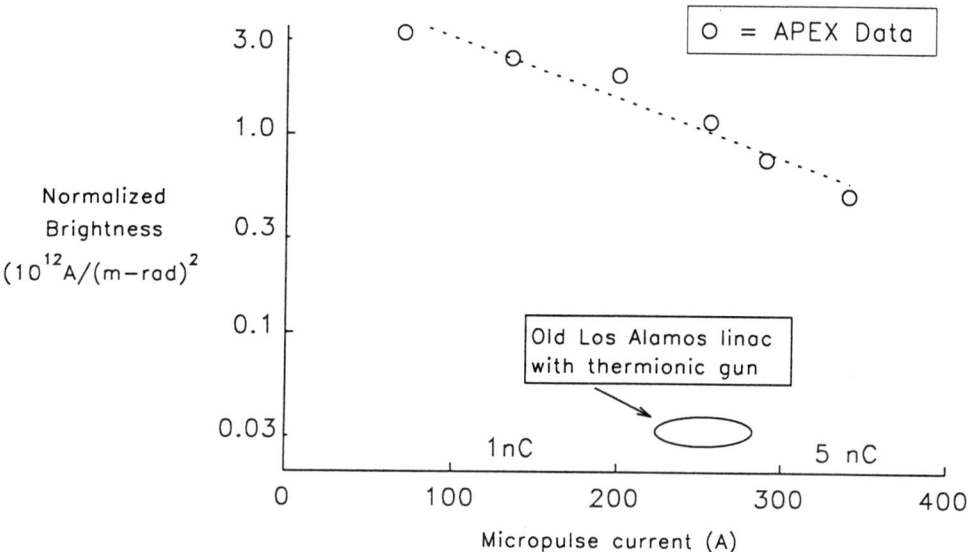

Figure 9 Normalized brightness of the APEX electron beam at 36 MeV versus current. The brightness from our previous accelerator with thermionic injector and 20 MeV energy is shown for comparison.

CONCLUSION

We have shown that photoinjectors can be integrated into linear accelerator systems to produce extremely high-brightness electron beams. Photoinjectors are now a viable and practical alternative to thermionic sources. In the Los Alamos case, replacing the old thermionic gun and subharmonic bunchers with a photoinjector has increased the electron-beam brightness by a factor of 20. We believe that high-quantum-efficiency cathodes are the only practical choice for high-average-current-electron accelerators because the drive-laser power is prohibitively high for low-quantum efficiency sources. The present disadvantage of high-quantum-efficiency sources is the relatively short cathode lifetime. We have made significant progress in understanding the cathode decay processes and in improving the lifetime. We continue to work on more advanced cathode designs that will lead us to both high quantum efficiency and long lifetimes. Based on our experience with the APEX photoinjector, a new 20-MeV photoinjector has been constructed as part of the Advanced FEL (AFEL) project at Los Alamos.[8] The electron beam from the AFEL accelerator should be significantly brighter than that of APEX. In addition, we will use the experience gained at APEX to design and build the high-average power, APLE free-electron laser.

ACKNOWLEDGMENTS

The successful operation of APEX was made possible through the efforts of a large number of dedicated people. The following people deserve special recognition for contributions to the project; N. Okamoto, P. Schafstall, J. Barton, L. Connor, P. Ortega, S. Apgar, C. Webb, J. Lujan, M. Feind and R. Martinez.

REFERENCES

Work supported and funded by the US Department of Defense, Strategic Defense Initiative Organization and Army Strategic Defense Command, under the auspices of the US Department of Energy.

REFERENCES

1. R.L. Sheffield, E.R. Gray, and J.S. Fraser "The Los Alamos Photoinjector Program," *Nucl. Instrum. Methods* **A272**, 222 (1988).

2. P.G. O'Shea, "The Los Alamos High-Brightness Photoinjector," in *High-Brightness Beams for Advanced Accelerator Applications*, W.W. Destler and S.K. Guharay editors, Published by the American Institute of Physics (New York), p. 182 (1992).

3. B.E. Carlsten, L.M. Young, M.E. Jones, L.E. Thode, A.H. Lumpkin, D.W. Feldman, R.B. Feldman, B. Blind, M.J. Browman, and P.G. O'Shea , "Design and Analysis of Experimental Performance of the Los Alamos HIBAF facility Accelerator using the INEX Computer Model," *IEEE J. Quantum Electron . 27*, 2580 (1991).

4.   J.W. Early, J.Barton, R. Wenzel, D. Remelius, and G. Busch, "The Los Alamos FEL Photoinjector Drive Laser," *IEEE J. Quantum Electron.*. 27, 2645 (1991).

5.   A.H. Lumpkin, "The Next Generation of RF FEL Diagnostics," *Nucl. Instr. Meth.* **A304**, 3130 (1991).

6.   M. Ross, N. Phinney, G. Quickfall, H. Shoaee, and J.C. Sheppard, "Automated Emittance measurements in the SLC," *Proc. 1987 IEEE Particle Accelerator Conference*, **CH2387-9**, (1988).

7.   P.G. O'Shea et. al., "Initial results from the Los Alamos Photoinjector-Driven Free-Electron Laser," *Nucl. Instrum. Methods* **A318,** 52 (1992).

8.   K.C.D. Chan, R.H. Kraus, J. Ledford, K. L. Meier, R.E. Meyer, D. Nguyen, R.L. Sheffield, F.L. Sigler, L.M. Young, T.S. Wang, W.L. Wilson and R.L. Wood, "Los Alamos Advanced Free-Electron Laser," *Nucl. Instrum. Methods* **A318**, 148 (1992).

REPORT ON HIGH QUANTUM EFFICIENCY PHOTOCATHODE AT MILANO

P. Michelato

INFN and University of Milano, Lab. LASA, Via F.lli Cervi 201, 20090 Segrate.

ABSTRACT

R&D activity on high quantum efficiency alkali antimonide photocathode is in progress at Milano, in the context of the ARES program. Inside a preliminary preparation chamber, Cs_3Sb layers with quantum efficiency up to 9 % (at $\lambda = 543.5$ nm) and lifetime of some days has been recently produced on copper, stainless steel and niobium, using a reproducible deposition procedure adapted to the material of the different substrata.

A more sophisticated preparation chamber with an improved vacuum system and the possibility to transfer cathode to surface analysis equipments under UHV conditions is close to the final assembling.

INTRODUCTION

Photocathode operation in laser driven RF guns of the superconducting type [1,2] asks for high quantum efficiency (QE) and fast (few picoseconds) photochatodes, able to deliver high current density bunches (hundreds of ampères) [3]. In the context of the ARES program [4], we started at Milano a R&D activity devoted to the production of alkali antimonide large emitting surfaces (Φ=30mm) on different substrata. The extremely high reactivity of alkali metal and alkali antimonide compounds with the UHV system residual gases is the main problem for the life time of such layers [5,6], while the photoemission characteristics are strongly dependent on a proper matching of the preparation parameters (temperature, deposition rate and sequence, ecc) with the substratum composition and surface status.

A simplified prototype of the photocathode preparation system we are designing has been assembled in order to investigate alkali antimonide photocathode preparation processes together with the reaction between the photoemissive surface and the reactive gases: e.g. the activation process and the poisoning of the cathode.

THE PHOTOCATHODE PREPARATION SYSTEM PROTOTYPE

The preparation system prototype is based on a standard UHV 6 ways cross chamber (8" Conflat) with a base pressure in the 10^{-10} mbar range.

On one way of the cross the cathode holder is located together with the integrated heating and temperature monitoring system. The cathode is screwed on the top of a stainless steel tube and a 650 W halogen lamp, located inside the tube, heats the cathode either during the substratum cleaning (600 °C), or during the cathode preparation (120 ÷ 150 °C). Cathode temperature is read by two thermocouple located respectively at the top of the cathode holder and on the cathode itself. Cathode temperature is stable within few degrees. A schematic view of the cathode holder and heating system is shown in fig. 1.

Alkali metal dispensers (produced by SAES GETTERS [7]), instead of ampoules with pure alkali metal, are used in this preparation system for the production of alkali metal vapour. Use of dispensers gives a better control of the alkali metal throughput and reduces the amount of desorbed gases produced by the dispenser heating (smaller size of the dispenser).

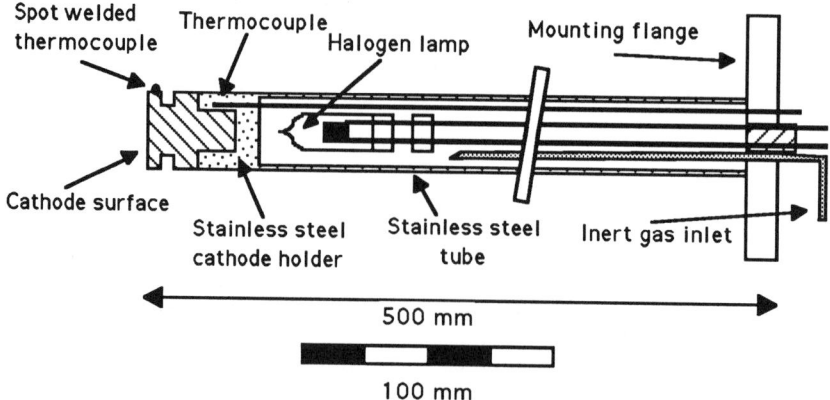

Fig.1 - A schematic view of the cathode holder and heater.

The production of the alkali metal in the dispenser is based on a high temperature (≥ 600 °C) chemical reaction between an alkali metal cromate with a reducing agent. The throughput of alkali metal is controlled by varying the temperature of the dispenser changing the current flowing through. Figure 2 shows the deposition rate vs. current of a 50 mm length cesium dispenser located at a distance of 40 mm from the microbalance sensor.

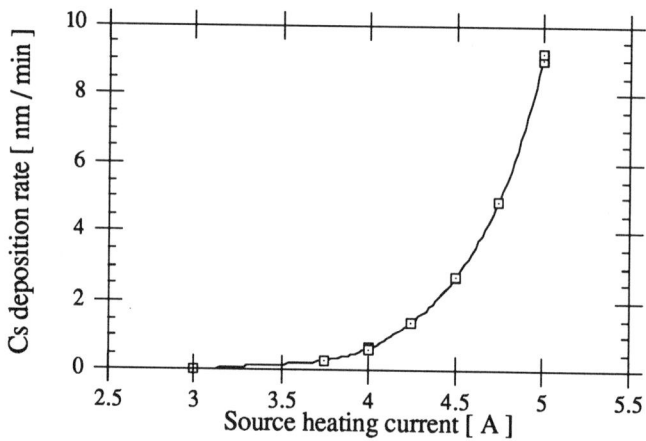

Fig. 2 - Deposition rate vs. heating current of a 50 mm cesium dispenser.

The alkali metal and antimony (especially developed by SAES GETTERS for this purpose) dispensers are assembled on the second way of the double cross in a stair structure able to hold up to 5 different sources. The holding system is moved by a linear transfer mechanism to place the source in front of the cathode or the quartz microbalance, respectively for cathode preparation or source calibration.

On the same port an UHV leak valve is located for the injection of the gas used for the cathode activation (at present O_2).

The other ways of the cross chamber are used for the other ancillary items: bakeable quartz microbalance sensor, anode connection, ionisation gauge, laser viewport and quadrupol mass spectrometer.

The He-Ne green laser (1.5 mW) illuminates the photoemissive surface also during the cathode preparation, while a shutter intercepting the laser beam or a look-in amplifier are used to subtract the eventual dark current for a correct measurement of the photoemitted one.

EXPERIMENTAL RESULTS

Since the first chamber operation, in April 1992, we prepared a number of cathodes on different substrata: stainless steel (SS), copper and niobium.

The best results so far obtained on each substratum are summarized in Table I. All data refer to stable cathodes, with a life time (50 % of the original QE) greater than 2 days. All the three cathodes presented a very uniform QE over the cathode surface (Φ=30mm), as shown in Fig. 3, where the typical data (referred to the cathode prepared on SS) are given. For the same SS cathode the time evolution of the QE is given in Fig.4. In two cases (Nb and Cu) no significant increasing in QE has been observed after the oxidation process (activation).

Tab.I - Cs_3Sb cathode preparation best results.

substratum	QE [%]	activ. effect	stability	uniformity
SS	7.2	Good	Stable	Good
Nb	7.9	Poor	Stable	Good
Cu	9.0	Poor	Stable	Good

In some cases a slowly but significant increase in photocathode QE (more then 50 %) has been observed on high QE non oxidized cathode during the first day of operation. Moreover these emitters are not sensitive to a subsequent activation with oxygen. This effect could be due to a slowly reaction with the residual water vapour that is one of the component of the UHV residual atmosphere. Investigations to correlate water vapour and oxygen partial pressures with the cathode QE enhancement are under way.

To investigate the dependence of the cathode performance on substratum composition and preparation procedure, we assembled at the top of the cathode holder, 3 different metal samples respectively of Cu, Nb and SS (Φ =10mm). Using a compromise between the preparation procedures specific for each substratum (i.e. temperature, Sb thickness, Cs deposition rate, etc), we produced, at the same time, 3 stable cathodes, QE being respectively: 1.6 %, 3.8 % and 0.9 %.

A very preliminary surface analysis on the samples shows that the cathode prepared on copper presents a diffusion of the substratum material into the cesium antimonide layer stronger than that observed in the cases of Nb and SS.

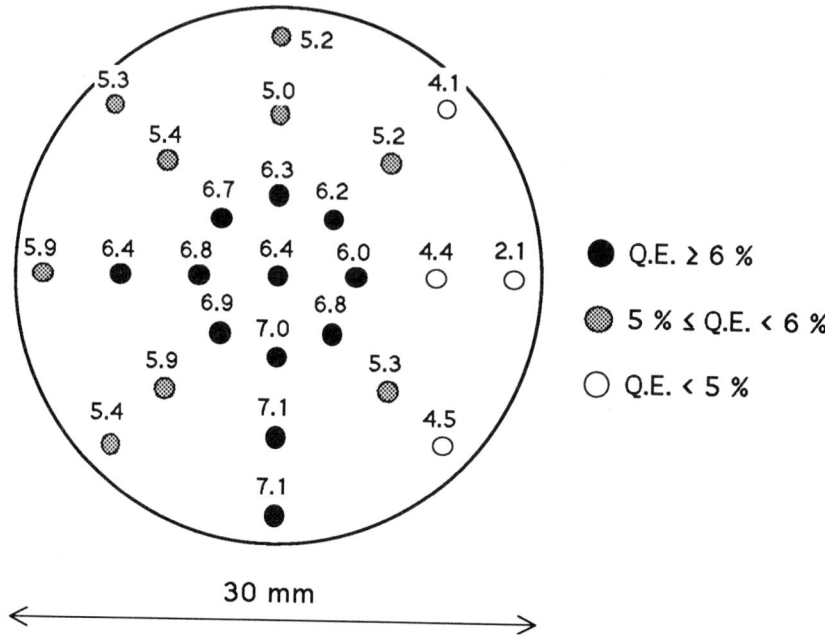

Fig. 3 - Quantum efficiency distribution on the surface of a $\Phi = 30$ mm Cs_3Sb photocathode, prepared on a stainless steel substratum.

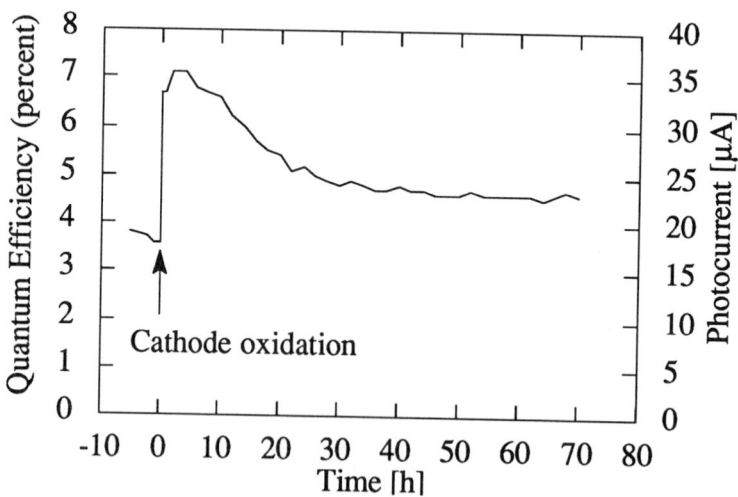

Fig. 4 - Typical time behaviour of a "stable" photocathode

FUTURE DEVELOPMENTS AND CONCLUSIONS

Because we believe that the strategy to widely apply surface analysis techniques to investigate substrata and photocathodes is fundamental to better understand the mechanisms related to cathode preparation, activation and damaging, a new advanced photocathode preparation system has been designed and it will be in operation by October 1992. The advanced preparation chamber is provided with a special device for intervacuum [8] ($< 10^{-9}$ mbar) cathode transfer, so that the cathodes will be analyzed before and during degradation. Moreover the new chamber is pumped by a combination of NEG (not evaporable getter) and sputter ion pumps in order to have a selective pumping effect on different gases.

A schematic drawing of the preparation system is presented in fig. 5.

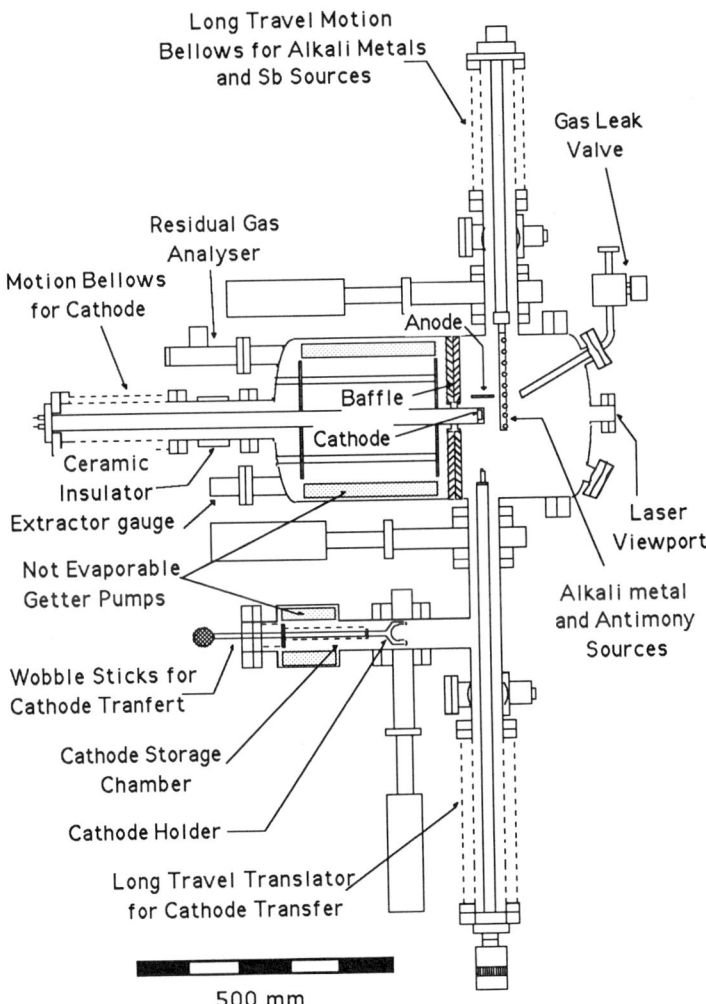

Fig. 5 - Sketch of the advanced photocathode preparation system.

REFERENCES

1. C. Pagani, L. Serafini, S. Tazzari and A. Peretti, The TESLA Injector: a Preliminary Proposal, Proc. of the 1st Int TESLA Workshop, Cornell, July 1990.
2. A. Michalke and H. Piel, C. Sinclair, P. Michelato, Experimental Verification of a Photoemission Source with Superconducting Cavity, Proc. HEACC 92, Hamburg, July 20-24, 1992.
3. P. E. Oettinger, R. E. Shefer, D. L. Birx and M. C. Green, Photoelectron Sources: Selection and Analysis, Nucl. Instr. Methods, A272 (1988) 264.
4. R. Boni et al., Status of the ARES R&D Program, Proc. 1991 IEEE Particle Accelerator Conference, San Francisco, May 6-9, 1991.
5. A. H. Sommer, Photoemissive Materials, Elsevier Sci. Publ., Princeton, 1968.
6. R. W. Decker, Decay of S·20 Photocathode Sensitivity Due to Ambient Gases, Proc. of 4th Symp. on Photo-Electronic Image Devices, London, Sept.16-20, 1968, Adv. in Electr. and Electron Physics, 28 A, Accademic Press, 1969, 357.
7. SAES Getters SpA, Alkali Metal Dispenser, Product Description A830915, Milano, Italy 1983.
8. J. P. Hobson and E. V. Kornelsen, UHV Techique for Intervacuum Sample Transfer, J. Vac. Sci. Technol., Vol 16, No. 2, Mar./Apr 1979.

FEASIBILITY OF A PHOTOEMISSION SOURCE WITH SUPERCONDUCTING RF CAVITY

Achim Michalke

Universität Wuppertal, 5600 Wuppertal 1, Germany

ABSTRACT

Photoemission RF guns have been developed as sources of high brightness electron beams in the recent past. But equipped with normalconducting RF cavities, they are limited to pulsed operation with duty cycles typically below 10^{-3} due to the high RF field required at the photocathode. This limitation could be overcome by using a superconducting acceleration cavity, in principle enabling continuous operation with the same gradient, that means the same beam quality. But the sensitivity of the superconducting cavity imposes severe constraints on the photocathode operation. On the way towards a superconducting prototype gun we have set up experiments to investigate the mutual interferences of these two components and to demonstrate the feasibility of this injector scheme. I'd like to present the results which we have obtained and to discuss their consequences on the design of a prototype source.

INTRODUCTION

In the recent few years, photoemission RF guns have been demonstrated to be reliable sources of electron beams with extreme brightness above 10^{10} A/m²rad² and small energy spread.[1] For this purpose the RF cavities of the source have to be operated at high acceleration gradients between 25 MV/m and 100 MV/m.[2] Normalconducting cavities can withstand such high fields only in pulsed operation with low duty cycles, which keep the average RF power losses in the surface at an acceptable level. This fact is not a limitation as long as the successive accelerator has to be operated with low duty cycle, too, for the same reason. On the other side, nowadays reliable accelerators for bright beams with continuous operation are available,[3] based on superconducting RF cavities. The applications which require bright electron beams (like UV-FELs and linear e^+e^--colliders) would of course profit from a high duty cycle. Thus there is a need for electron sources providing in continuous operation an electron beam with a quality typical for photoemission RF sources.

The most direct approach would be to replace the normalconducting acceleration cavities by superconducting ones. But superconducting cavities impose totally different requirements to their environment than normalconducting ones and also have completely different effects on it. Thus the design of normalconducting photoemission RF guns must not simply be adopted, but the construction has to be reconsidered in detail. The resulting problems can be divided into two categories: One part requires the development of constructive solutions based on knowledge which is available or can be extrapolated from former experiments with superconducting cavities or with photocathodes. The remaining questions concern the interaction of both systems and require experimental investigations because no related information is available.

COMPONENTS AND CAPABILITIES

For normalconducting guns a variety of photocathodes spreading from rugged materials with low quantum efficiency to sensitive materials with high quantum efficiency have been used.[4] In the superconducting high brightness gun the light power absorbed and dissipated in the cathode must be kept as low as possible, because it has to be carried off at cryogenic temperature. Thus the photocathode must have a quantum efficiency as high as possible, especially for high current operation. In addition, its work function should be significantly lower than the work function of the wall material, e.g. 4.5 eV for niobium, to avoid parasitic emission induced by stray light. Therefore a photocathode of alkali antimonide type seems to be the best solution. Their high quantum efficiency up to 10% or more at low work function of 2 eV must be paid by their instability and extreme chemical activity, requiring permanent handling in clean ultrahigh vacuum.

Photocathodes of this type have already been used in normalconducting RF guns [5] where they showed a limited lifetime at high efficiency level. Three effects were known to take part in this decay during operation: Reaction with residual gas components, bombardment with accelerated ions, and thermal desorption of photocathode compounds. The achieved lifetimes of typically a few hours severely limit the performance of the gun. But inside a superconducting cavity, all three processes should be strongly suppressed due to the cryogenic environment and the consequent high vacuum. Therefore we expect a significantly increased lifetime of the cathodes and relaxed service requirements for the source. Whether this effect will really come up and how strong it will be has to be determined in an experiment.

Alkali antimonide photocathodes are produced as thin layers by evaporating the components onto a metallic substrate. Their preparation must be done in a separate preparation chamber as usual. In a superconducting system, the preparation chamber cannot sit just besides the cavity, but must be located outside the cryostat. Therefore the transfer system must be able to carry the photocathode over a longer distance and insert it into the cold cavity. In addition, the stem carrying the cathode must not simply have an electrical contact to the cavity wall: The RF losses in the contact zone would be undefined and much too high for a superconducting system. Thus the cathode stem must be isolated against the adjoining cavity in the high field region, and the contact zone (which is necessary for the DC current) must be shielded against the RF field by means of a choke.

The cavity should be two or three cells following the general shape of super-conducting accelerator cavities with at least the first cell shortened to match the low initial β of the electrons. But compared with normalconducting cells there is much more freedom to adapt the shape, especially in the iris region, to the requirements of beam dynamics without the dominating demand for maximum shunt impedance. The optimum frequency of the superconducting cavity will be around 500 MHz enabling operation at 4.2 K with relaxed requirements on the laser pulse length and synchronisation jitter.

The material of the superconducting cavity should be bulk niobium. Using the standard surface preparation techniques it should be possible to operate the bare cavity reliably at a cathode surface gradient of 25 MV/m – this corresponds to an acceleration gradient of 15 MV/m for standard accelerator cells. Taking into account the lower geometry factor for shortened cells they should achieve a quality factor of about 2×10^9 at 4.2 K. Remark that these values are rather conservative compared with other projects on superconducting accelerators like TESLA. The maximum gradient achieved in RF guns with alkali antimonide cathodes is also about 25 MV/m,[5] thus the superconducting approach does not mean an additional limitation on the gradient.

The above mentioned performances are valid only for bare cavities; they could be severely lowered by an inserted alkali antimonide layer. Various interference effects between cavity and cathode are imaginable: Material could desorb from the photocathode and condense on the cavity surface, causing parasitic field emission (at high electric field areas) or defects and enhanced losses (at high magnetic field areas). Accelerating gradient and quality factor of the cavity could be permanently lowered by the presence or operation of the photocathode. Although this effect should not be significant in the cryogenic environment, experimental investigation of its real size is necessary. Even if the cavity surface is not contaminated, the photocathode itself could cause RF losses or field emission on its surface. First evidence for strong field emission has been obtained at the Los Alamos experiment,[6] but polishing of the substrate surface seems to be an effective cure. Due to its high resistivity, RF losses at the cathode could be significant in spite of its thinness and its high geometry factor. Also here experimental investigation is required; possibly the RF losses will disappear in very thin layers due to the proximity effect.

EXPERIMENTAL SETUP

We have set up an experiment at Wuppertal University to investigate the topics mentioned above [7] which can not be extrapolated from former meas-urements or simulated numerically. This setup consists of a superconducting niobium cavity inside a helium bath cryostat, a photocathode preparation chamber besides the cryostat together with photocathode transfer system, and laser and optical components to illuminate and observe the cathodes. Alkali antimonide photocathodes can be produced in the preparation chamber, transferred to the cavity, and operated there. The arrangement of these components is shown in Figure 1. In this setup we intend to measure in detail the mutual interferences between superconducting cavity and photocathode during operation as well as the behaviour of photocathodes at high fields and low temperatures. Compared to a prototype gun, setup and operation are greatly facilitated by dropping the requirement of generating a definite beam.

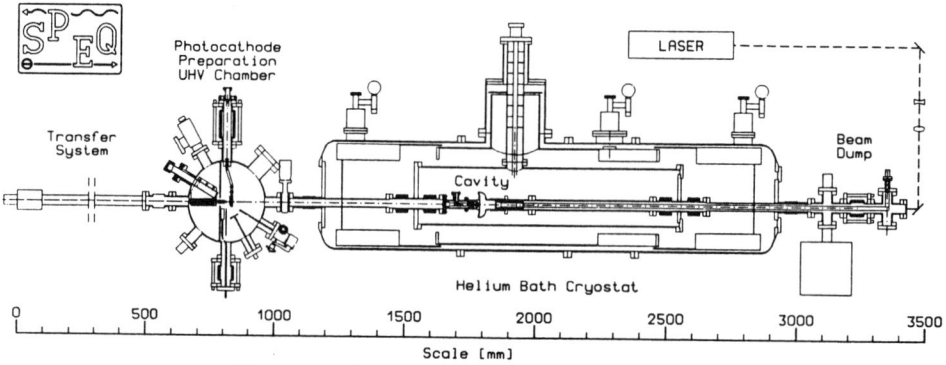

Figure 1: Experimental Setup Overview

The heart of this experiment is the superconducting cavity of bulk niobium, where photocathodes can be inserted and operated. For simplicity of handling and operation, its frequency has been chosen near 3 GHz. Its shape, shown in Figure 2, is roughly like a standard 3 GHz accelerator cell cut half and closed with a plane end plate, preserving the field distribution of the basic mode. Maximum magnetic field is located around the equator, maximum electric field at the iris and in the center of the end plate. Its geometry factor is about half the size of a standard cell; thus a quality factor above 5×10^9 should be possible at temperatures below 2 K. Important parameters of the cavity, measured or calculated by the URMEL computer code, are listed in Table 1. A single-cell shape is chosen instead of a two- or three-cell shape like for a prototype source, because it is not intended to put energy into the electron beam.

The photocathode is located in the center of the end plate, where high electric and low magnetic fields are present. It is evaporated on top of a cylindrical stem beared in a sapphire ring to provide good thermal contact and electrical isolation. Substrate material is niobium as well, to enable high quality measurements with a clean stem and to investigate the occurence of proximity effect with thin photocathode layers. The diameter of the photocathode area is 6 mm, its partial geometry factor about 8 MΩ. This value is large enough not to load the cavity inadequately with a high resistance layer, but small enough to measure the surface resistance of a low resistance layer.

Cathode stem and surrounding vacuum tube form a coaxial line which can extract RF power from the cavity. Therefore it is blocked with a coaxial bandpass filter properly tuned to the main cavity which increases the external

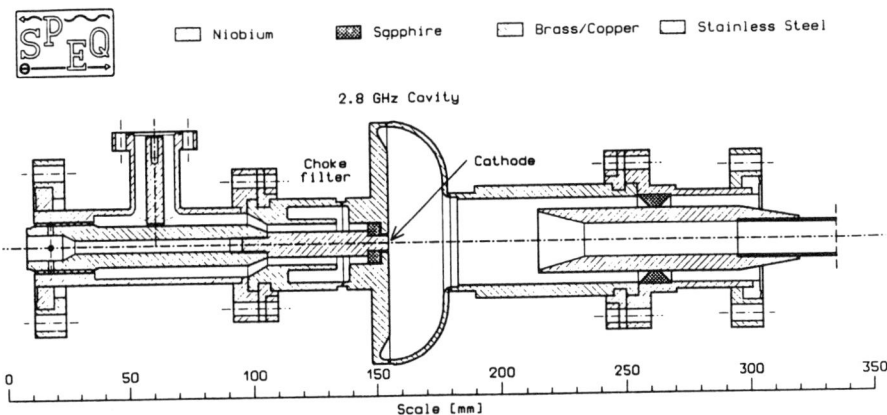

Figure 2: Superconducting 2.83 GHz Cavity

Table 1: Important Geometrical Cavity Parameters

Cavity Parameter		calculated	measured
Basic Mode Frequency f_0	[MHz]	2792 ± 1	2831 ± 1
Total Geometry Factor G_0	[Ω]	150 ± 1	140 ± 10
Beam Tube Cutoff Frequency $f_c^{\text{TM-01}}$	[MHz]	6750	
First HOM Frequency f_{1-1}	[MHz]	4004	
Cathode Geometry Factor G_C	[MΩ]	8.0 ± 0.5	
Cathode Field Gauge Factor E_C^2/W	[$\frac{(\text{MV/m})^2}{\text{W}}$]	5700 ± 200	
$\vec{E}$ Field Ratio Peak/Cathode E_p/E_C		1.1 ± 0.1	(at the Iris)
Peak $\vec{H}$ Field Ratio H_p/E_C	[$\frac{\text{kA/m}}{\text{MV/m}}$]	1.84 ± 0.04	
Filter Area Geometry Factor G_F	[kΩ]	360^*	
Sapphire Stored Energy Portion W_S/W		$2.5 \times 10^{-5*}$	
Peak $\vec{H}$ Field Ratio Filter/Cavity H_F/H_p		0.023^*	

* depending on gap geometry, but internally consistent

quality of this line from 5×10^4 to above 1×10^9. The field strength in the filter strongly depends on the size and exact geometry of the small gap between photocathode and cavity end plate. Assuming a factor 40 lower fields in the filter cavity than in the main cavity, its partial geometry factor is in the order of 360 kΩ. To avoid heat load to the helium bath, the filter is also made from superconducting niobium. Due to its low $\tan\delta$ and the low energy portion stored in it, the sapphire ring should not significantly contribute to the losses.

The photocathodes are prepared in a separate preparation chamber connected to the cryostat. Antimony metal and alkali metals are evaporated on top of the stem using the standard procedures.[8] For alkali metal production dispenser sources are used, which are extremely small and easy to handle. The stem can be heated to provide the adequate temperature for preparation or cleaning. Photocathodes can be tested inside the preparation chamber by means of a DC anode. A basic pressure of 2×10^{-9} mbar has already been achieved in the chamber, consisting of hydrogen with a few percent of methane, carbon monoxide and noble gases. The chamber can be sealed off with a shutter to avoid alkali metal contamination of the cavity area during preparation.

The cathode is illuminated by a laser beam coming through the beam tube from outside the cryostat. This beam can hit the cathode inside the cavity as well as in the preparation chamber passing through the cavity. On the same optical path, supported by a small telescope, the cathode and its surrounding can be visually inspected from outside. Actually, we use a constant wave HeNe-Laser with 0.5 mW power at 543 nm. The continuous laser has the disadvantage that only about a quarter of the illumination time leads to electrons which leave the cavity: Half of the time the electric field at the cathode has wrong polarity, and half of the emitted electrons are accelerated back and strike the cathode with considerable energy. These problems can be avoided by using a synchronized pulse laser with a pulse length not larger than about 60° of the RF phase. The installation of such a laser is scheduled for a later stage of this experiment.

EXPERIMENTAL RESULTS

The experiment described above has been built up recently and was first operated in December 1991. During the recent six months, five tests could be performed and in total six cesium antimonide photocathodes inserted and operated in the cavity.

A quite reliable process to produce Cs_3Sb photocathodes has been established in the preparation chamber. It showed up that preparation on a niobium substrate is extremely difficult if the substrate is not heated to 600°C in advance. We explain this effect by the tight Nb_2O_5 layer covering the niobium which is removed at higher temperatures in UHV. In contrary, on copper photocathodes can be prepared without preceding bakeout. On a copper-plated niobium stem we prepared photocathodes up to 6% quantum efficiency (DC measured in the preparation chamber), while 3–4% are quite reproducible. These layers are not yet activated with oxygen or water, which is known to increase quantum efficiency by a significant amount.

The superconducting cavity has shown a quality factor up to 4×10^8 at 1.9 K. This is far below the expected value, but quite sufficient for most of our purposes. Without cathode stem, a gradient up to $E_C = 21$ MV/m at the cathode location could be achieved, limited by field emission loading. The cathode stem turned out to be badly cooled, getting normalconducting at gradients above 1 MV/m. For this reason in presence of the (clean) stem only gradients up to 7 MV/m could be achieved.

Because we use a continuous laser we get a large fraction of electrons being backaccelerated in the cavity and hitting the cathode. They produce considerable secondary emission which is also extracted. But the dependence of the extracted current from acceleration gradient (Figure 3) is quite conform with the numerical simulation data, which enables us to extrapolate the true photoemitted current as well as the secondary emission coefficient. This true photocurrent does in no case match the expected one quarter of the DC photocurrent measured in the preparation chamber, but was typically quite higher. Probably this effect can be explained at least partially by an increase of quantum efficiency due to the low temperature. Spectral photocurrent measurements could help to decide, because spectral sensitivity is known to get steeper near the threshold at low temperatures.

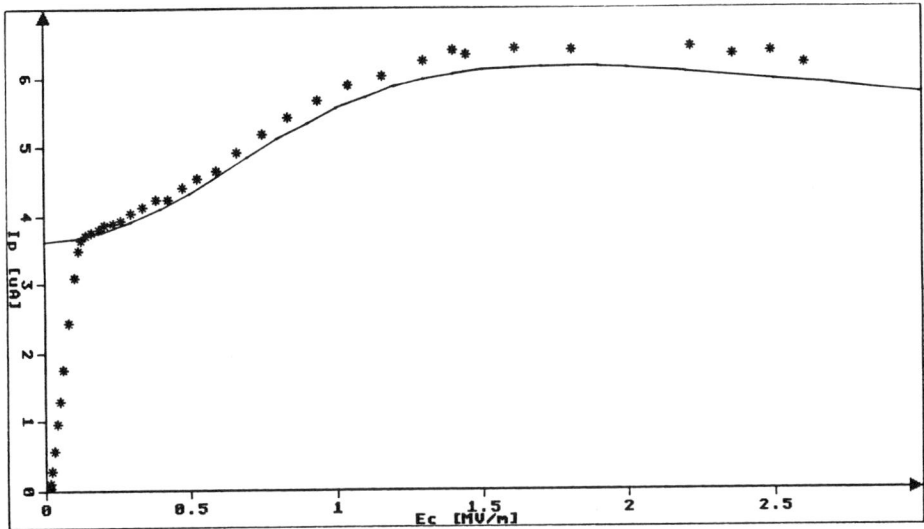

Figure 3: Dependence of RF Extracted Current I_P on Cathode Field E_C: Asterisks are measured data (Cs_3Sb layer 5/2), line is theoretical calculation with secondary emission coefficient $\delta_{max} = 5.8$ at 750 eV

Starting at values between 2 MV/m and 5 MV/m all photocathodes showed strong field emission in the RF field. During increase of the gradient the current is instable, and the emitter sites are processed, but once achieved maximum gradient the dark currents tend to be stable and follow Fowler-Nordheim behaviour (Figure 4). This behaviour suggests tiny emission spots instead of global field emission, and according to the experience obtained at Los Alamos (ref) polishing the stem top should help a lot. During the occurence of field emission currents, we could not observe any glowing spots at the cathode nor any other light emission, except for very high gradients. In contrary to this, in the cavity without cathode stem we could always observe brightly glowing tiny spots near the cathode bore during field emission. Thus, the field emission mechanism seems to be a bit different, probably due to the low work function.

In continuous operation with photoemission (below field emission level) as well as with field emission the cathode layers change their properties significantly on a scale of about 100 s or 1 mC charge. The efficiency can vary in both

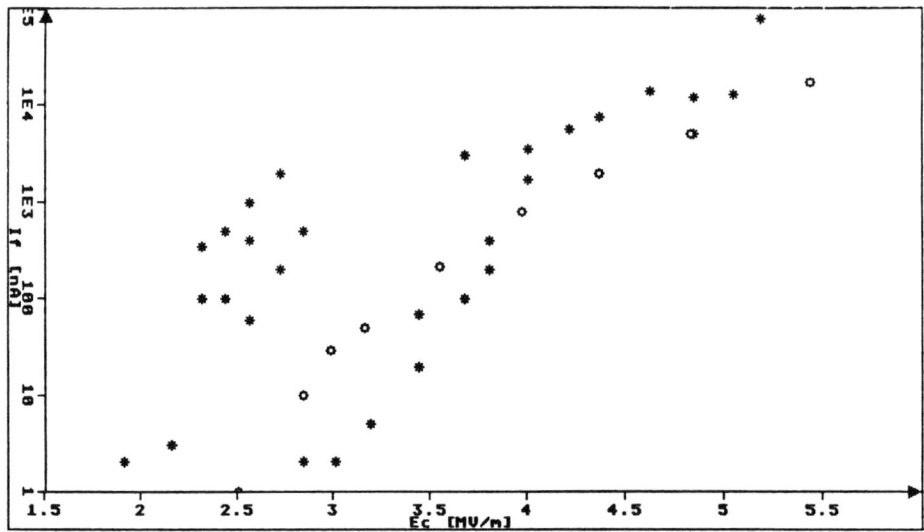

Figure 4: Dependence of RF Field Emission Current I_F of Photocathodes (Cs$_3$Sb layer 4/2) on Cathode Field E_C: Asterisks are measured during field increase, circles during field decrease

directions, even change the direction during operation. We attribute this variation to the bombardment of backaccelerated electrons, which is known to alter the photoemissive properties,[9] rather than to the emission process itself. The illumination with a synchronized pulsed laser could help to decide this problem, avoiding backaccelerated electrons.

The presence of a Cs$_3$Sb photocathode did always significantly load the cavity quality to a value between 5×10^6 and 1×10^7. Taking a partial geometry factor of $G_C = 8$ MΩ for the cathode area, one obtains a surface resistance between 1 and 2 Ω for the Cs$_3$Sb layers. Until now all photocathodes inserted had a thickness of 300 nm or more. The aim is therefore to prepare good layers of thickness below 50 nm on a niobium substrate which should significantly reduce the surface resistance, perhaps even several orders of magnitude by proximity effect.

After removal of the photocathodes, in no case an increase of cavity losses could be detected within our accuracy level. In addition, no decrease of field emission onset has been observed, in contrary, in one case we even observed a significant increase of the onset gradient from 10 MV/m to 17 MV/m. Thus no evidence for a cavity contamination due to presence or operation of the Cs$_3$Sb photocathodes could be found. Nevertheless, experiments have to be repeated with significantly higher photoemission currents and no backaccelerated electrons, requiring a higher-power pulsed laser.

CONCLUSIONS

Several photocathodes from Cs_3Sb have been operated inside a superconducting cavity. The illumination power was several orders of magnitude lower than in a realistic RF gun, and the laser was not pulsed, leading to a considerable amount of backaccelerated electrons. Under these conditions, no permanent degradation of the cavity quality or field emission onset level could be observed. But the photocathodes themselves have high surface resistance and show field emission at relatively low gradients. There is hope to suppress these effects by preparing thinner layers on polished substrates. The photocathode layers also showed significant changes of their performances during operation. This is at least partially attributed to the backaccelerated electrons, and the real meaning of this effect for the operation of the RF gun can only be investigated with a pulsed laser of higher power. From the experience on normalconducting photoemission guns, no severe limitation is expected from this effect.

Conclusively, according to our results a photoemission RF gun with superconducting cavity is feasible using alkali antimonide cathodes. The only observed effect limiting its performance is the strong field emission at the cathode, which has to be suppressed at least up to 20 MV/m.

ACKNOWLEDGEMENT

This project was initiated and guided by Prof. Dr. H. Piel. Most of the experimental work was done by the students P. vom Stein and S. Heinze. Important contributions came from our collaborants C. K. Sinclair at CEBAF and the group of C. Pagani at INFN Milano. I have to thank all of them for their collaboration. We got also support from DESY Hamburg and the companies Siemens Bergisch Gladbach and SAES Milano. Part of this work was funded by the Federal Minister for Research and Technology of Germany.

References

[1] Christian Travier. Review of microwave guns. LAL/SERA/91–286, Laboratoire de l'Accelerateur Lineaire, Orsay.

[2] J.-M. Dolique and M. Coacolo. Relativistic acceleration and retardation effects on photoemission of intense electron short pulses, in RF–FEL photoinjectors. In [10], p. 233.

[3] D. Proch, editor. *Proceedings of the 5th Workshop on RF Superconductivity.* DESY M-92-01.

[4] P. E. Oettinger, R. E. Shefer, D. L. Birx, and M. C. Green. Photoelectron sources: Selection and analysis. *Nucl. Instr. Meth.*, A272:264, 1988.

[5] P. G. O'Shea et al. Performance of the photoinjector accelerator for the Los Alamos free–electron laser. In [10], p. 2754.

[6] Alex H. Lumpkin. Observations on field–emission electrons from the Los Alamos FEL photoinjector. In [10], p. 1967.

[7] A. Michalke, H. Piel, C. K. Sinclair, P. Michelato, C. Pagani, L. Serafini, and M. Peiniger. A proposed superconducting photoemission source of high brightness. In [3], p. 734. DESY M-92-01.

[8] A. H. Sommer. *Photoemissive Materials.* Elsevier, Princeton, 1968.

[9] Kiyoshi Miyake. Effect of electron bombardment on secondary and photoelectron emission of cesium-antimonide. *J. Phys. Soc. Japan*, 10:164, 1955.

[10] Loretta Lizama and Joe Chew, editors. *1991 IEEE Particle Accelerator Conference Record*, San Francisco, May 1991.

POSSIBLE APPLICATION OF TRANSIENT ELECTROMAGNETIC PULSES TO HIGH BRIGHTNESS e-GUNS

N. A. Kurnit, P. K. Benicewicz, and A. J. Taylor
Chemical and Laser Sciences Division
Los Alamos National Laboratory
Los Alamos, NM 87545

ABSTRACT

A number of groups have recently demonstrated the production of freely propagating, focusable pulses of terahertz radiation, consisting of essentially a single subpicosecond cycle of a baseband electromagnetic field. We discuss the possible application of these techniques to the production of strong fields at photocathode surfaces, in a manner analogous to radial-line switched-power concepts. Experimental status in production of these pulses in our laboratory and elsewhere is reviewed, and recent progress in development of short-pulse solid-state lasers useful for this technology is summarized.

INTRODUCTION

In the last several years there has been a growing interest in the generation and application of freely propagating ultrashort-duration electromagnetic pulses. Initially, these were generated at low power by using subpicosecond laser pulses to illuminate small area photoconductors coupled to dipole antennas.[1,2] Focusing of this radiation using optical elements has been demonstrated.[3]

A significant increase in the magnitude of field strengths that could be produced with these techniques was made with the realization that a large-area planar photoconductor could serve as a source of directionally radiated electromagnetic pulses.[4,5] The photocurrent distribution produced by a short-duration optical pulse acts as a phased antenna array that generates transient radiated fields both in the direction of the reflected light and into the photoconductor in a direction given by a generalized Snell's law with respect to the direction of the incident laser pulse (Fig. 1). The field radiated by the accelerating charges produces a pulse with a duration

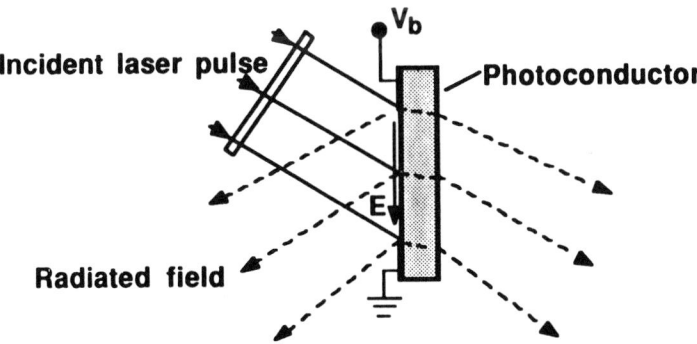

Fig. 1. Illumination of a biased photoconductor by a short laser pulse produces directional electromagnetic pulses radiated in the direction of the reflected light and into the forward direction.

determined by the risetime of the current with a magnitude that can be comparable (but opposite in sign) to the applied bias field. Optical fluences of typically 0.5 mJ/cm^2 are required to switch a large fraction of the bias voltage. The radiated field is confined to a beamwidth given approximately by the diffraction angle corresponding to the median radiated frequency (~terahertz) and the illuminated radiating aperture.

The radiated pulse also contains a lobe of opposite polarity arising from the fall of the current, as well as contributions from reflections at the boundary of the photoconductor. Changes in shape also result from dispersion and from diffraction of different frequency components during propagation into the far field. An example of the time dependence observed in our laboratory with the setup shown in Fig. 2 is displayed in Fig. 3. The detection process uses the charge collected when a fast, radiation damaged silicon-on-sapphire photoconductor is gated on by the simultaneous presence of the transient e-m pulse and a portion of the laser pulse.

APPLICATION TO PARTICLE ACCELERATORS

It is interesting to speculate as to whether such transient pulses may have any applications to particle accelerators. Because of the relatively low efficiency of creating a short laser pulse and then of converting this into a terahertz pulse, it is doubtful that this could be useful as a primary accelerating mechanism. However, there may be some specialized uses, such as proposed for the application of switched-power radial line transformers[6] to e-guns, that could be worth exploring.

We consider first the simple geometry in Fig. 4 in which the forward propagating e-m pulse is pictured focused by a lens to a diffraction-limited spot. (In actuality, a parabolic mirror would probably be used as the focusing element to avoid chromatic aberration caused by the dispersion of different frequency components.) An electron bunch that passes through the focal region during the time that the accelerating field is primarily in accelerating phase receives an impulse $\Delta p = \int eE(r(t),t) \, dt$, where $E(r(t),t)$ is the electric field at the position $r(t)$ of the electron at time t. For slowly moving electrons, the magnetic field of the pulse can curve the electron's trajectory into the direction of propagation of the e-m pulse, where it can continue to gain energy over

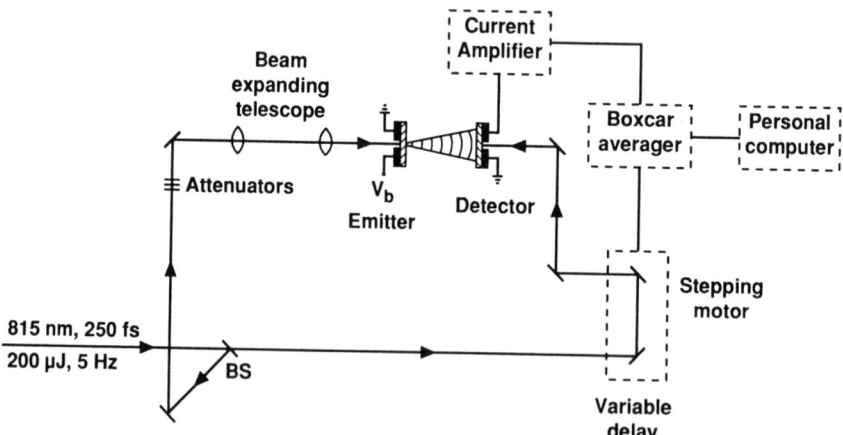

Fig. 2. Schematic of setup for producing and detecting terahertz transient electromagnetic pulses. For the results presented in Fig. 3, a 5-mm-aperture GaAs emitter was illuminated by ~175 μJ from a Ti:sapphire laser and the terahertz signal detected by a radiation-damaged silicon-on-sapphire detector placed 5 cm away.

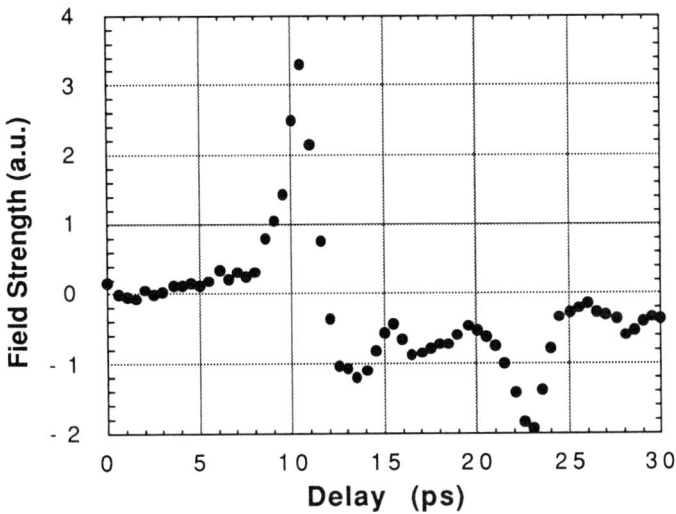

Fig 3. Detected electromagnetic pulse observed by varying the optical delay of the probe beam in Fig. 2. The bias field was 9 kV/cm, and the peak field measured at the detector was estimated to be 300 V/cm.

some finite slippage length, and the analogous situation has been studied as a mechanism for acceleration of electrons in focused laser fields.[7-9] In order to pass through the focal region before they can experience significant reversal of the laser field, the electrons must either gain or initially have near- relativistic velocities. In the latter case, it is expected on the basis of very general superposition principles that the overall interaction leads to a relatively small net acceleration, unless some near-field structures are present to shield part of the interaction.[10] We assume in the following that either a cathode or a near-field aperture are present to allow net acceleration.

The radiation from an aperture of diameter D is expected to be focusable to a diffraction-limited focal spot of diameter $d \sim 2.44 f \lambda_{em} / D$, where λ_{em} is the wavelength corresponding to the central frequency of the radiated pulse. This focal diameter is on the order of 1 mm for a 10-cm radiating aperture and a lens of the same

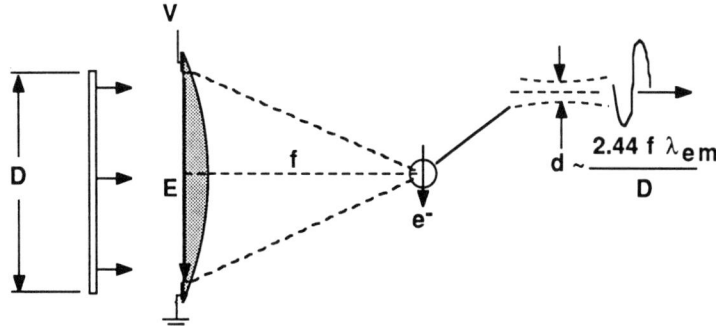

Fig. 4. Focusing of electromagnetic pulse can produce large fraction of bias voltage across small focal volume, possibly allowing acceleration of electron crossing focal region.

focal length, f=10 cm. (In practice, the aperture would be square or rectangular, so that the crystal can be uniformly biased; this does not not qualitatively change the results.) The intensity at the focus is increased by the ratio of $(D/d)^2$, so the field at the focus is larger by D/d, a factor of 100 in the present example. The voltage across the focal region is the same as the voltage at the crystal, $E_{rad}D$. Since GaAs can hold off fields in excess of 100 kV/cm, particularly if pulse-charged, it may be possible to achieve accelerating voltages of 500 kV across the mm focal region in the present example (500 MV/m) if the radiated field can reach half the bias field. Possible limitations on the voltage that can be attained are considered below.

We have neglected the deflecting force of the magnetic field of the pulse in the above discussion, which is certainly not valid.[7-9] However, we now consider a generalization of this geometry in which a symmetrical array of such pulse generators surrounds the focal region (Fig. 5(a)), or, taking this one step further, a biased toroidal semiconductor shaped like a lens on its inner surface is uniformly excited by a converging laser field to produce a converging electromagnetic pulse (Fig 5(b)). This situation is now directly analogous to a switched-power radial line transformer, except that the field is now focused in both the radial and longitudinal direction, giving a higher field increase at the center than can be attained with a single parallel-plate radial line. The magnetic field of this symmetrically converging pulse cancels on-axis, as is evident from the fact that there is no preferred direction for deflection of the electrons. Off-axis electrons experience a time-varying deflection force, which produces a focusing during the convergence of the pulse, and a defocusing once the pulse has passed through the center, provided the electric field does not reverse sign during this time. Whether the acceleration can be useful in light of this, or whether the focusing properties might themselves be useful in other applications, can only be determined by more detailed calculations. It is interesting to note however, that the use of the electric fields from an array of four radial line segments has been proposed as a means to obtain large focusing forces.[11]

The electric field will add coherently over a small cylindrical region at the center of the structure to a value that is $\sim 2\pi$ that of the single f1 lens shown in Fig. 4. The diameter of this high-intensity region, in which the field might reach 3 GV/m under the above assumptions, is $\sim 1/2\pi$ that of the f1 focus, or ~ 160 μm. The longitudinal dimension of the focal region remains unchanged. The axial electrons may experience an accelerating voltage of as much as 3 MV.

The configuration of Fig. 5(b) may be somewhat difficult to construct, and the focusing properties are not optimal because of the frequency dispersion of the semiconductor material. In order to alleviate both of these problems, a nearly equivalent focusing geometry can be achieved by biasing a flat circular disc of semiconductor between an inner and outer radius and focusing the forward-emitted pulse by means of a paraboloidal reflector, as shown in Fig. 6. None of these figures address the problem of providing a low inductance connection for pulse-biasing the structure without intercepting a significant portion of the either the optical or radiated pulse.

POSSIBLE LIMITATIONS

The above estimate of the achievable accelerating field may be optimistic. Limitations on the attainable radiated field strength arise both from the maximum voltage that can be applied without breakdown and from the saturation in electron velocities at high field strength, leading to a decreased effective electron mobility. Some saturation of the radiated field has been observed in InP at bias fields of 5-6 kV/cm at low optical fluence, and perhaps a small amount in GaAs at 4 kV/cm with

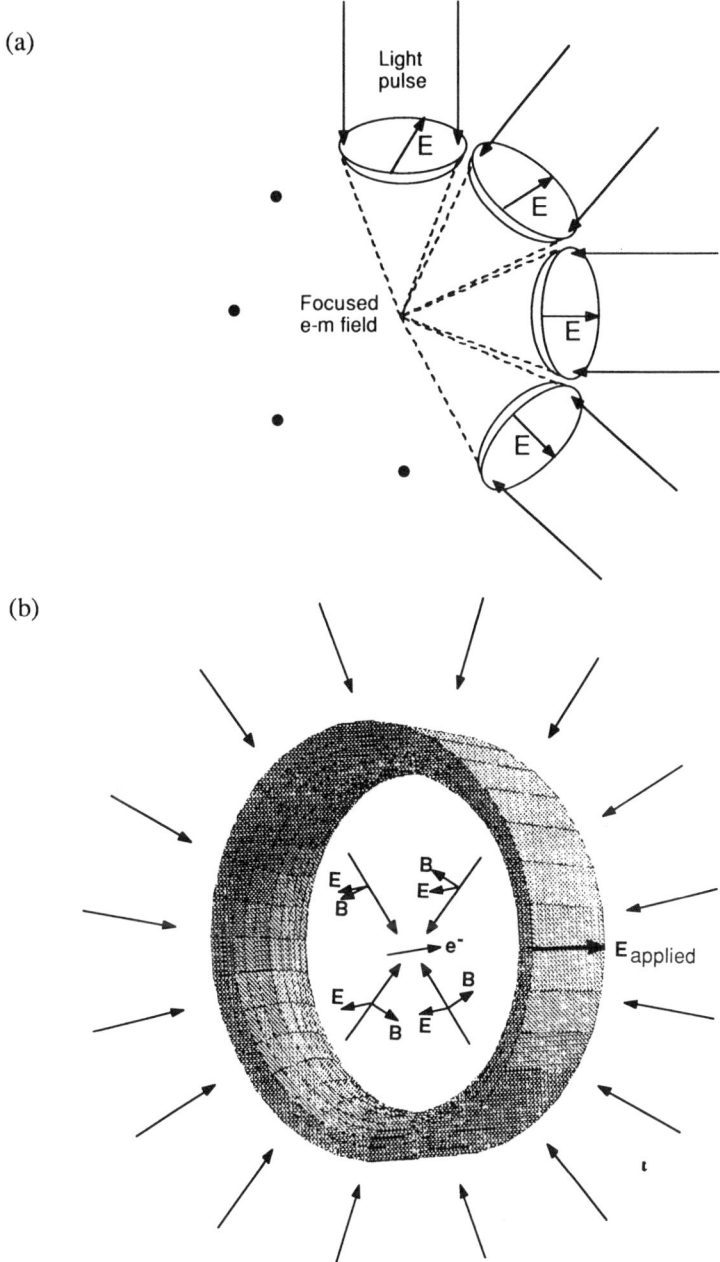

Fig. 5. A number of biased semiconductors can be arrayed around a common focal volume to produce an increased field strength provided they are closely synchronized. These can be individual, as shown in (a) (although they would more likely be square or rectangular than circular as depicted), or they may be fashioned in a continuous toroidal ring, as shown in (b).

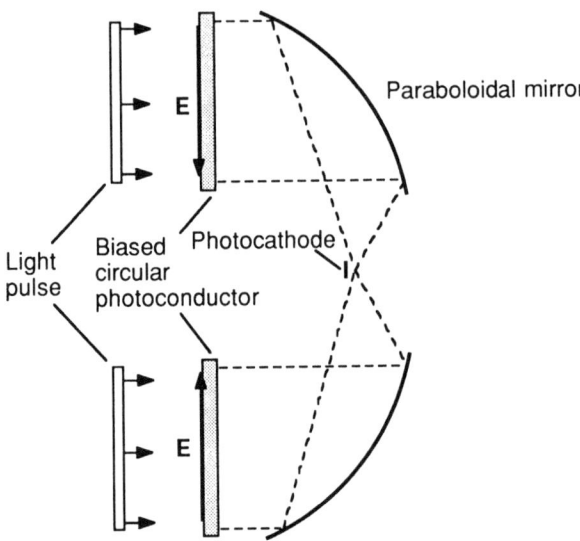

Fig. 6. A flat circular photoconductor may be combined with a paraboloidal reflector to produce a focal geometry similar to Fig. 5.

saturating optical fluences.[5] Earlier work on generation of voltage pulses by switching with 0.53-nm radiation experienced degradation of switching efficiency above 3.2 kV/cm.[12] However, more recent work on switching small structures of GaAs grown by low-temperature molecular-beam epitaxy using 620-nm light has reached 70% switching efficiency at bias fields of 130 kV/cm. At the even longer wavelengths available to us with Ti:sapphire (Ti:Al$_2$O$_3$) lasers, there is expected to be an electron velocity overshoot phenomenon that should help produce high currents on the subpicosecond time scale.[14] A recent paper has in fact reported the generation of greater than 100 nJ of terahertz radiation using 9 mJ of 770-nm light from a Ti:sapphire laser.[15] Bias fields of up to 70 kV/cm were applied to GaAs with no saturation of the emitted radiation observed as a function of bias field, at least up to 50 kV/cm. An alternate technique has also recently been proposed for attaining high field strengths.[16] There is therefore some reason to hope that field strengths of interest can be reached.

AVAILABILITY OF LASER SOURCES

The technique of of chirped pulse amplification[17] has been used recently to produce high-power, ultrashort optical pulses with broadband laser materials such as Ti:sapphire,[18,19] alexandrite,[20] and Cr:LiSrAlF$_6$ (LISAF).[21] These materials lase in the wavelength range aroun 800 nm and therefore yield ideal optical switching pulses for GaAS and InP photoconductors. Pulses from such systems have produced energies approaching 0.5 J in pulsewidths of 100 fs duration with typical repetition rates of 1-10 Hz. Scaling of the output energy to the 10-J or higher level is proceeding; G. Mourou has discussed the prospects for for some of these higher-energy systems at this conference.

Shown in Fig. 7 is the version of a high-brightness solid-state laser system under development in our laboratory. A Kerr-lens mode-locked Ti:sapphire oscillator[22]

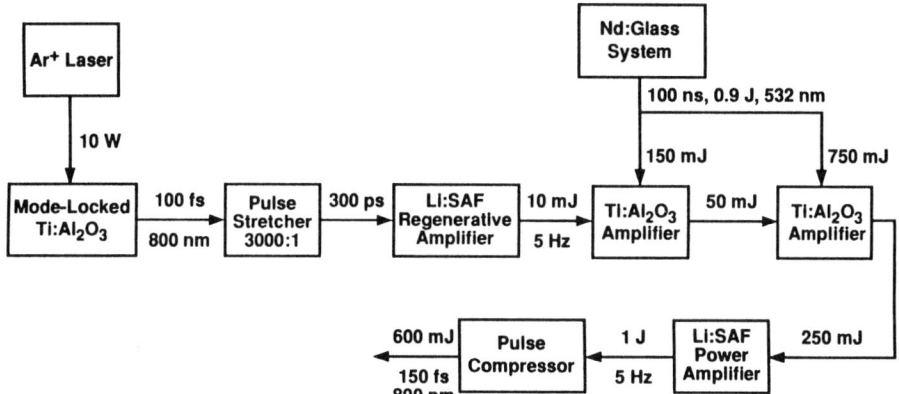

Fig. 7. Solid-state ultrashort-pulse laser system presently under construction at Los Alamos. All parts have been built and tested; system integration is underway.

generates low energy, 100-fs pulses at 810 nm. This has been shown to be a very effective method to generate a stable, ultrashort-pulse mode-locked train, and can be synchronized to an rf source using a mode-locked pump laser.[23] The pulses are stretched to a 300-ps temporal duration by a diffraction-grating stretcher.[24] Amplification to the 10 mJ level occurs in a regenerative amplifier with a flashlamp-pumped 4-mm-diameter LiSAF rod as the gain medium. Subsequent amplification is underway in two multiply-passed Ti:sapphire rods. The pump source for these amplifiers is a frequency-doubled Nd:glass oscillator/Nd:YAG amplifier system that produces 1-J, 100-ns pulses art 532 nm, thus raising the damage threshold above that encountered with conventional 10-ns-duration pump sources. A final, flashlamp-pumped LiSAF amplifier should raise the output energy to the 1-J level, yielding a pulse-compressed 150-fs output of 600 mJ at a 3 Hz repetition rate.

ACKNOWLEDGMENTS

We are grateful to D. H. Auston, X.-C. Zhang, and D. C. MacPherson for help in setting up initial terahertz-pulse experiments at Los Alamos. One of us (NAK) would like to acknowledge helpful discussions with J. R. Fontana, H. G. Kirk, and R. B. Palmer during the course of this workshop.

This work was performed under the auspices of the US Department of Energy.

REFERENCES

1. G. Mourou, C. V. Stancampiano, A. Antonetti, and A. Orzag, Appl. Phys. Lett. 39, 295 (1981).
2. P. R. Smith, D. H. Auston, and M. C. Nuss, IEEE J. Quantum Electron. 24, 255 (1988) and references therein.
3. Ch. Fattinger and D. Grischkowsky, Appl. Phys. Lett. 53, 1480 (1988).
4. B. B. Hu, J. T. Darrow, X.-C. Zhang, D. H. Auston, and P. R. Smith, Appl. Phys. Lett. 56, 886 (1990); J. T. Darrow, B.B. Hu, X.-C. Zhang, and D. H. Auston, Opt. Lett. 15, 323 (1990).
5. J. T. Darrow, X.-C. Zhang, D. H. Auston, and J. D. Morse, IEEE J. Quantum Electron. QE 28, 1607 (1992).

6. W. Willis, *Proc. Workshop On Laser Acceleration of Particles,* Malibu, CA, 1985, C. Joshi and T. Katsouleas, eds., AIP Conf. Proc. <u>130</u>, 421. See also: *Proceedings of the Switched Power Workshop,* Shelter Island, New York, Oct. 16-21, 1988, edited by R. C. Fernow, BNL 52211.

7. M. J. Feldman and R. Y. Chiao, Phys. Rev. A <u>4</u>, 352 (1971).

8. P. K. Kaw and R. M. Karlsrud, The Physics of Fluids <u>16</u>, 321 (1973).

9. J. K. McIver, Jr. and M. J. Lubin, J. Appl. Phys. <u>45</u>, 1682 (1974).

10. R. B. Palmer, Particle Accelerators <u>11</u>, 81 (1980) and references therein.

11. S. H. Aronson and R. C. Fernow, *Proceedings of the Switched Power Workshop,* Shelter Island, New York, Oct. 16-21, 1988, edited by R. C. Fernow, BNL 52211, p.184.

12. C. H. Lee, Appl. Phys. Lett. <u>30</u>, 84 (1977).

13. T. Motet, J. Nees, S. Williamson, and G. Mourou, Appl. Phys. Lett. <u>59</u>, 1455 (1991).

14. G. M. Wysin, D. L. Smith, and A. Redondo, Phys. Rev. B. <u>38</u>, 12514 (1988).

15. D. R. Dykaar, *Proceedings of the Conference on Ultrafast Phenomena VIII,* June 8-12, 1992, paper MC15.

16. L. Xu, X.-C. Zhang and D. H. Auston, submitted to Appl. Phys. Lett.

17. D. Strickland and G. Mourou, Opt. Commun. <u>56</u>, 219 (1985)

18. J. D. Kmetec, J. J. Macklin, and J. F. Young, Opt. Lett. <u>16</u>, 1001 (1991).

19. A. Sullivan, H. Hamster, H. C. Kapteyn, S. Gordon, W. E. White, H. Nathel, R. J. Blair, and R. W. Falcone, Opt. Lett. <u>16</u>, 1406 (1991).

20. M. Pessot, J. Squire, G. Mourou, and D. J. Harter, Opt. Lett. <u>14</u>, 1107 (1989).

21. W. E. White, J. R. Hunter, L. Van Woerkom, T. Ditmire and M. D. Perry, Opt. Lett. <u>17</u>, 1067 (1992).

22. D. Spence, P. N. Kean, and W. Sibbett, Opt. Lett. <u>16</u>, 42 (1991).

23. Ch. Spielmann, F. Krausz, T. Brabec, E. Wintner, and A. J. Schmidt, Opt. Lett. <u>16</u>, 1180 (1991)

24. O. E. Martinez, IEEE J. Quantum Electron. <u>QE-23</u>, 59 (1987).

EXPERIMENTAL INVESTIGATION OF A PSEUDOSPARK DEVICE AS A HIGH-BRIGHTNESS ELECTRON-BEAM SOURCE

B. N. Ding[a] and M. J. Rhee
Laboratory for Plasma Research and Department of Electrical Engineering
University of Maryland, College Park, Maryland 20742

ABSTRACT

A pseudospark as a high-brightness electron-beam source is experimentally investigated. The breakdown voltage of a pseudospark device is measured for a wide range of gas pressure and anode-cathode gap distance. The current and emittance of the electron beam are measured for breakdown voltages up to 40 kV and storage capacitance up to 4.7 nF. The post-acceleration of the beam by an induction linac is studied by constructing the time-resolved energy spectrum of the beam.

INTRODUCTION

Because of its interesting capability of producing a high-quality electron beam, the pseudospark discharge has been a subject of extensive experimental and theoretical investigations.[1-8] The high-brightness electron beam[5-8] produced by the pseudospark has potential applications in such diverse areas as advanced accelerators and free-electron lasers.

In this work, we report experimental studies of a pseudospark device and the electron beam produced by the device. The breakdown voltage characteristic is determined. The scaling studies of the current and the rms emittance of the electron beam are carried out for wide ranges of experimental parameters. The time-resolved energy spectrum of the beam is constructed by analyzing measured currents passed through biased electrode. The post-acceleration of the electron beam is studied by measuring the time-resolved energy spectrum.

EXPERIMENTAL SETUP

In this section, we describe the main experimental setup, which is used for the experimental measurements described in the following sections. As shown in Fig. 1, the discharge chamber consists of a planar cathode with a hollow cavity, a number of sets of intermediate electrodes and insulators (only two sets are shown in Fig. 1), and a planar anode. The hollow cavity is a 2.54-cm diam and 4-cm long cylindrical cavity in which a trigger electrode made of 6.35-mm (1/4 in.) diam semirigid coaxial cable is inserted. All the electrodes are made of brass and the insulators are of plexiglas. A 3.2-mm (1/8 in.) diam center hole is present through the entire electrode system. The storage capacitor used in this experiment is a low-inductance type door knob capacitor. Rogowski coil is built

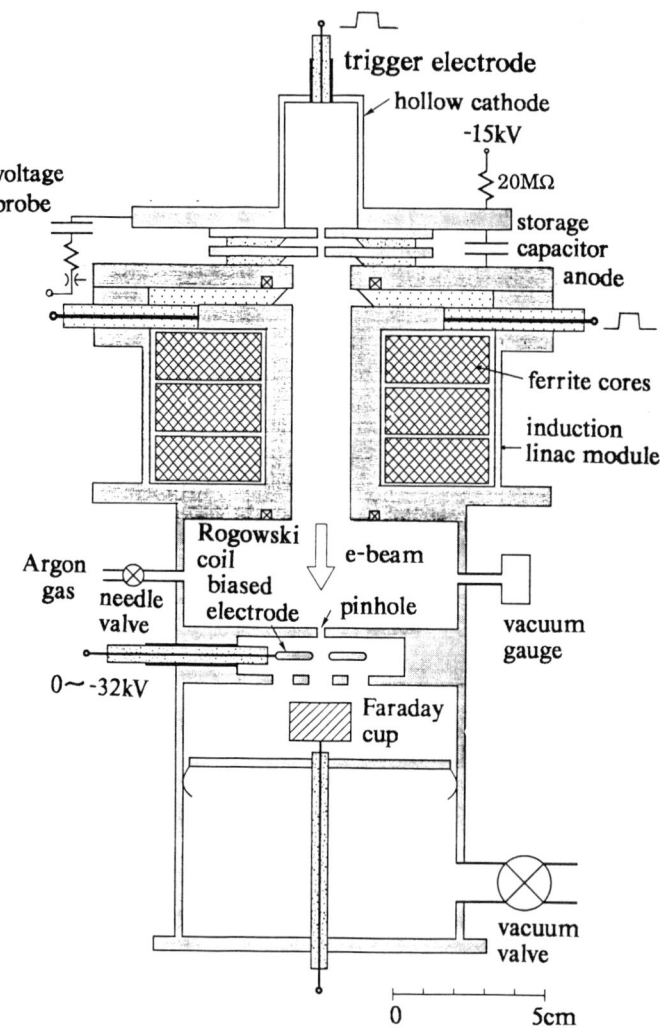

Figure 1: Experimental setup.

into the anode flange to monitor the electron beam current. Placed immediately downstream from the anode is an induction linac module, which is used for the post-acceleration experiment described in a later section. A diagnostic chamber, which is placed below the linac module, can accommodate various diagnostics such as a time-resolved energy spectrometer system (shown in Fig. 1) and an emittance meter (not shown). The vacuum pump system used consists of a mechanical pump and an oil diffusion pump. Argon gas is used throughout the experiments. The pressure is measured by a capacitance-manometer type vacuum gauge.

BREAKDOWN VOLTAGE CHARACTERISTIC

Initially the chamber is evacuated by the oil diffusion pump down to 10^{-5} Torr. The cathode side of the chamber is then charged to a given voltage ranging from zero to -50 kV. The charging voltage is measured by an electrostatic voltmeter. The argon gas pressure in the chamber is then increased at a very slow rate, $dp/dt \approx 1$ mTorr/sec, so that virtually no pressure gradient exists in the chamber. The pressure at which the self-breakdown occurs for the given voltage is determined by reading the last digital display of the vacuum gauge immediately before the breakdown takes place. The same procedure is repeated for various voltages and anode-cathode gap distances. The anode-cathode gap distance is varied by employing a different number of intermediate electrode sets (each set is 6.4 mm thick).

A plot of the measured gas pressure vs. the breakdown voltage and the anode-cathode gap distance is shown in Fig. 2. In order to determine the functional dependence, the data are least-squares fitted numerically[9] to a regression plane given by a simple two variable function,

$$v(p, d) = \alpha/(p^\beta d)^\delta. \tag{1}$$

The coefficients are found to be $\alpha = 0.1865 \pm 0.0019$, $\beta = 1.9952 \pm 0.0064$, and $\delta = 2.226 \pm 0.016$, respectively. The resultant regression plane is depicted by solid lines in Fig. 2. It is interesting to note that the value of β is very close to 2 (well within the error range). This suggests that the breakdown voltage may be described by a function only of the product $p^2 d$. This is a striking contrast to Paschen's law[10] which describes the breakdown voltage only as a function of pd.

ELECTRON BEAM CURRENT

The electron beam current is measured[11] by the Rogowski coil built into the anode flange and a Faraday cup. The Rogowski coil is a 20-turn coil wound around a toroid of 17.5-mm major radius and $1.4 \times 1.4 = 1.96$ mm^2 cross sectional area, and is molded into an axisymmetric groove milled into the anode flange. The rise time of the Rogowski coil may be approximated by $L/Z_0 = 0.18$ ns, where L is the inductance of the coil and Z_0 is the characteristic impedance of the transmission line. The induced voltage signal in the coil is then integrated

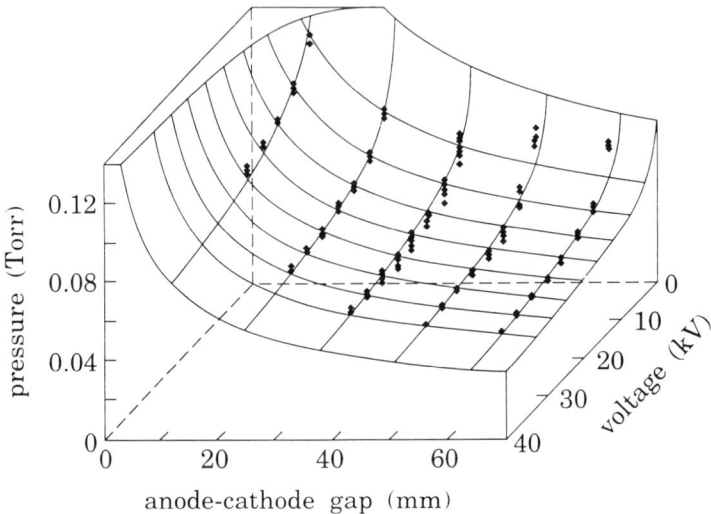

Figure 2: Measured pressures and the least-squares-fit regression plane.

by a passive RC integrating circuit with an RC time constant of 0.75 μs. The Faraday cup consists of a 2-cm diam graphite charge collector and 16 1/8-W carbon resistors (not shown in Fig. 1) which connect radially from the charge collector to the ground. The total resistance measured by a bridge is found to be 192 mΩ. The current waveforms measured for breakdown voltages up to -40 kV and storage capacitances up to 4.7 nF are plotted in Fig. 3. The results are least-squares fitted numerically to a regression plane given by

$$I_p = (\alpha + \beta C)\exp[-(V - V_p)^2/2\sigma^2]\text{ kA}, .\qquad(2)$$

The coefficients are found to be $\alpha = 0.21$ kA, $\beta = 0.84$ kA/nF, $V_0 = 35$ kV, $\sigma = 9.1$ kV, and I_p, C, and V are measured in kA, nF, and kV, respectively. Interesting features of the results are that the current is nearly proportional to the storage capacitance and the current peaks at 35 kV.

RMS EMITTANCE OF ELECTRON BEAM

The root-mean-square (rms) emittance is defined[12] as

$$\epsilon_{rms} = \langle x^2 \rangle \langle x'^2 \rangle - \langle xx' \rangle^{2^{1/2}},\qquad(3)$$

where x' is the gradient of the particle trajectory given by $x' = dx/dz = p_x/p_z$, and the angular brackets denote average values over two-dimensional trace space as $\langle \phi \rangle = I^{-1}\int \phi\rho_2(x, x')dxdx'$, where ρ_2 is the projected density in two-dimensional

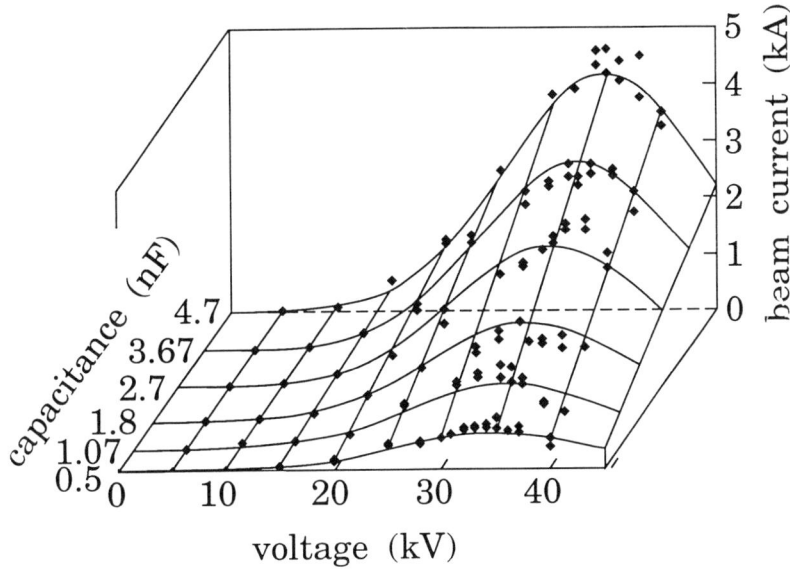

Figure 3: Measured peak current and the least-squares-fit regression plane.

trace space, and I is the total current given by $I = \int \rho_2 dx dx'$. The measurement of the rms emittance rely upon the experimental determination of the two-dimensional trace-space distribution.

It is very reasonable to assume that the beam produced in this experiment is axisymmetric and of Maxwellian transverse velocity distribution, as in Ref. 13. This allows us to use the simple slit-hole type emittance meter, whose results can be easily analyzed to find the rms emittance.[13] The emittance meter consists of a slit plate and a detector film. The slit plate is a series of parallel thin slits of 20 μm width and 1 or 2 mm spacing constructed from 0.4 mm thick stainless steel plate. A 50-μm (2 mil) thick radiachromic film used as a beam detector is placed 10 mm downstream of the slit plane. The emittance of the beam is measured at a fixed position, 5 or 10 cm downstream of the anode. The radiachromic film is exposured to anywhere from 1 to 50 consecutive pulses (depending on beam intensity) so as to produce an appropriate density profile, with the peak value of optical density not exceeding 0.5. This ensures that the measured optical density distribution is linear to the beam intensity.

The density profile is obtained by scanning the exposed film using an optical microdensitometer, and a typical profile is shown in Fig. 4. From the profile we find empirical functions $\alpha(r)$, $\beta(r)$, and $\sigma(r)$ as functions of radial position r, where $\alpha(r)$ is the mean diverging angle, $\beta(r)$ represent the peak values, and $\sigma(r)$ is the rms width of the individual distributions. Numerical integrations are then performed using the empirical functions α, β, and σ to find ρ_2 and

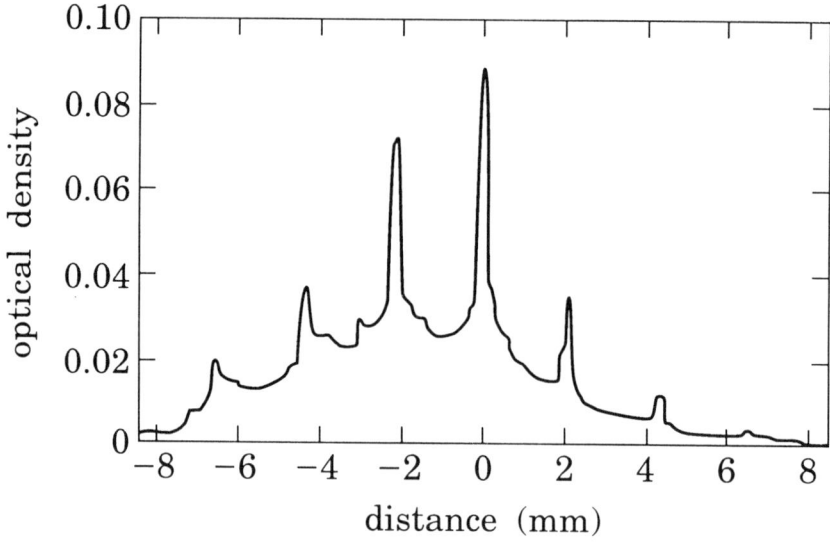

Figure 4: A typical optical density profile of exposed radiachromic film.

the moments $\langle x^2 \rangle, \langle x'^2 \rangle$, and $\langle xx' \rangle$ (See Ref. 13 for details). Several isodensity contours of the resultant $\rho_2(x, x')$ are constructed in x-x' space as shown in Fig. 5 and corresponding value of the effective emittance (four times the rms emittance) is found to be

$$\epsilon_{\text{eff}} = 4\epsilon_{\text{rms}} = 75 \text{ mm mrad.} \qquad (4)$$

Measurements of the emittance are repeated for various storage capacitances up to 4.7 nF and the breakdown voltages up to 40 kV and results are summarized in Fig. 6. It is observed from Fig. 6 that the values of the emittance varies from 60 to 110 mm mrad and no clear dependence of the emittance on either the storage capacitance or the breakdown voltage is found.

TIME-RESOLVED ENERGY SPECTRUM

Experimental setup shown in Fig. 1 is used in this experiment, except that the induction linac is removed. The cathode is charged to -15 kV with respect to the grounded anode through a 20 MΩ charging resistor. Argon gas is then slowly filled at a slow flow rate through a needle valve in the upper chamber while the pinhole allows the gas to leak into the downstream chamber maintaining a constant differential pressure in the upper chamber and vacuum in the downstream chamber. The pressure used in this experiment is 60 mTorr, which gives reproducible beam generation. By applying +20-kV pulse to the trigger electrode, a flashover discharge on the surface of the Teflon insulator at the end of the coaxial

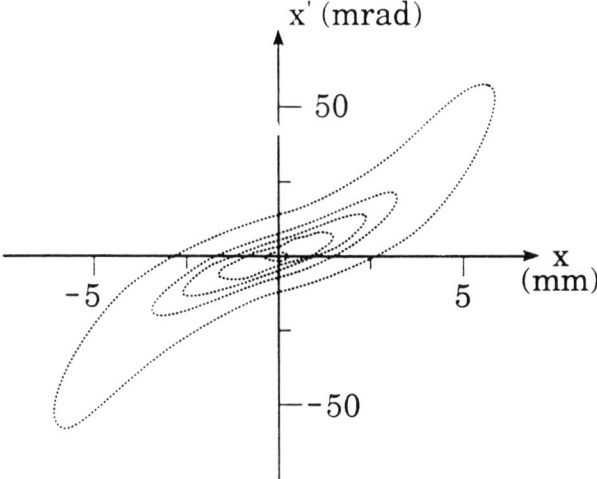

Figure 5: The emittance plot in x-x' trace space. The isodensity contours correspond to 0.1, 20, 40, 60, and 80% of the peak value.

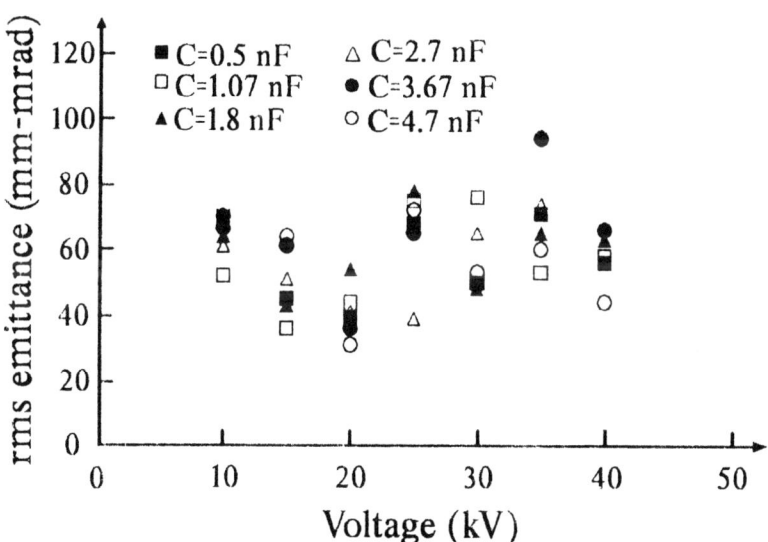

Figure 6: The effective emittance measured for various breakdown voltage and
storage capacitance.

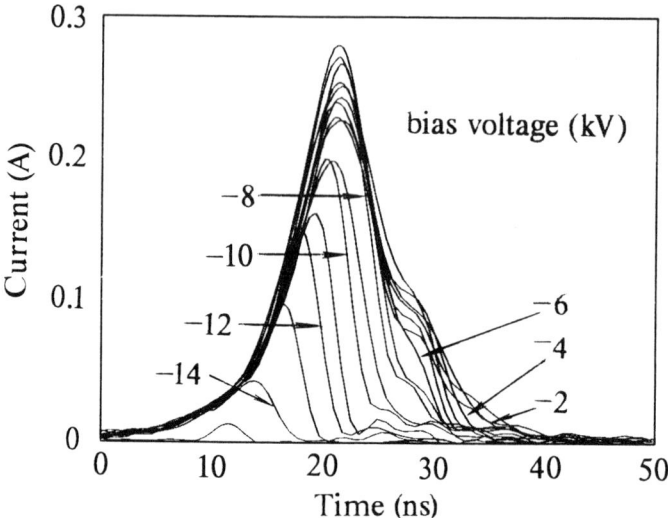

Figure 7: Family of current waveforms measured by a Faraday cup with various bias voltages from 0 to -16 kV with 1 kV step.

trigger electrode triggers the main discharge within a jitter time of <1 ns. The voltage of the cathode during the discharge is measured by a homemade compensated probe and recorded by a digital oscilloscope.[14] The electron beam generated by this discharge is sampled by the 0.5-mm diam pinhole, allowing the sampled beamlet to propagate in the vacuum cavity of the discriminator. Only electrons of energy higher than that corresponding to a given bias voltage can pass through the center hole of the biased electrode and arrive at the Faraday cup. The bias voltages are varied from 0 to -16 kV in 1 kV increments, and the corresponding Faraday cup current signals are recorded by the digital oscilloscope. A typical family of current waveforms is shown in Fig. 7.

The time-resolved energy spectrum is constructed from the family of current waveforms. One may represent the electron beam current passed through the biased electrode, $I(\epsilon_b, t)$, as

$$I(\epsilon_b, t) = \int_{\epsilon_b}^{\infty} i(\epsilon, t) d\epsilon, \tag{5}$$

where $\epsilon_b = eV_b$, V_b is bias voltage, and $i(\epsilon, t)$ is energy resolved current density that is related to the total current by $I(t) = edN/dt = \int_0^{\infty} i(\epsilon, t) d\epsilon$. The difference between two current waveforms, measured at bias voltages V_b and $V_b + \delta V_b$, is the current consisting of electrons with energy between $\epsilon_b = eV_b$ and $\epsilon_b + \delta\epsilon_b = e(V_b + \delta V_b)$. Thus, the time-resolved energy spectrum at the average energy $\epsilon_b + \frac{1}{2}\delta\epsilon_b$ as a function of time is then given by $d^2N/dtd\epsilon = i(\epsilon_b + \frac{1}{2}\delta\epsilon_b, t)/e$, which can be

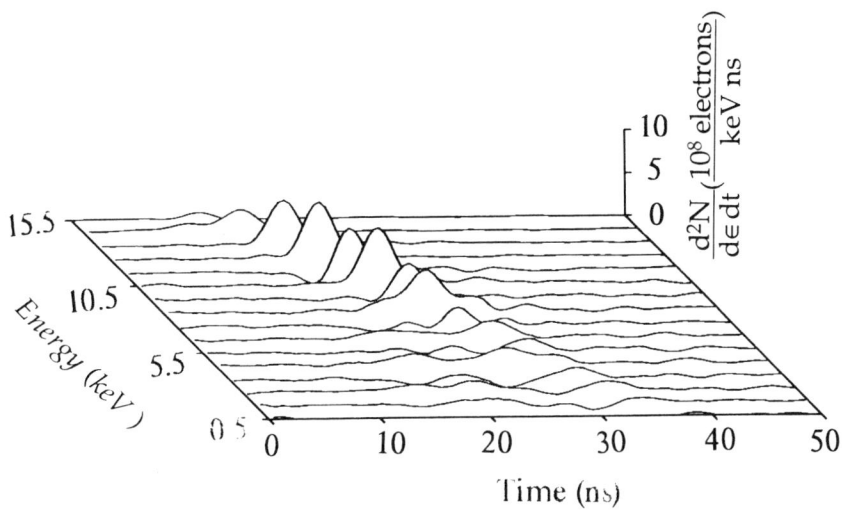

Figure 8: The time-resolved energy spectrum for the pseudospark-produced electron beam.

determined in terms of an experimentally obtained differential current waveform as

$$\frac{d^2N}{dtd\epsilon}\Big|_{\epsilon_b+\frac{1}{2}\delta\epsilon_b} = \frac{I(\epsilon_b,t) - I(\epsilon_b + \delta\epsilon_b,t)}{e\delta\epsilon_b}. \tag{6}$$

Sixteen differential current waveforms are obtained from the family of digitized waveforms shown in Fig. 5. The differences between two adjacent digital waveforms, $I(eV_i,t) - I(eV_{i+1},t)$, where V_i's are the bias voltages ranging from zero to 16 kV in 1 kV increments, are computed conveniently by using a personal computer. The complete time-resolved energy spectrum is then constructed by plotting the resultant differential waveforms as functions of energy as shown in Fig. 8.

POST-ACCELERATION

In this experiment, an induction linac module shown in Fig. 1 that is driven by a 25-Ω Blumlein type modulator, is attached to the downstream side of the anode. The linac module is terminated by a matched load 25 Ω to minimize the beam loading[15] effect. This is done by two 50-Ω cables (not shown in Fig. 1.), one of which is conveniently used for monitoring the voltage across the accelerating gap.

The electron beam generated by this discharge propagates through the induction linac in which the beam is further accelerated. The induction linac is powered by the Blumlein modulator that also triggers the pseudospark. Thus, the beam

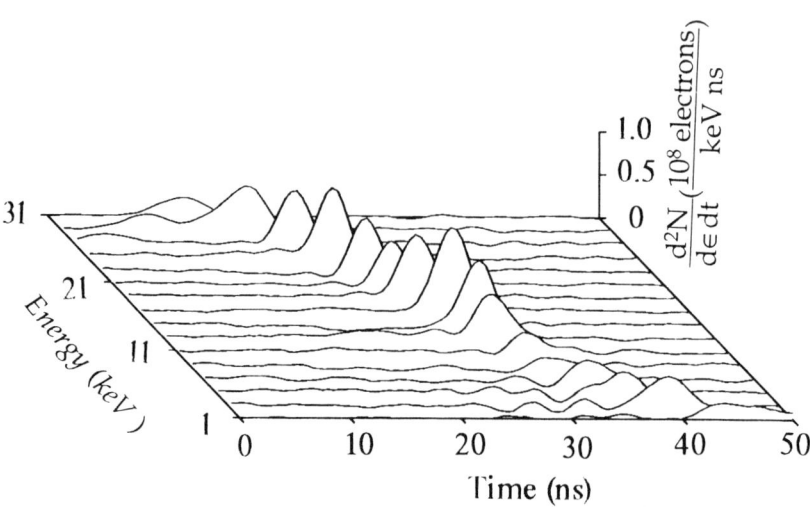

Figure 9: The time-resolved energy spectrum for the post-accelerated pseudospark-produced electron beam.

generation and accelerating voltage are in good synchronization. For this experiment the bias voltages are varied from zero to 32 kV in 2 kV and the time-resolved energy spectrum is constructed by analyzing the Faraday cup current signals as described in the previous section. The resulting spectrum is plotted in Fig. 9.

It is found that both time-resolved energy spectra shown in Figs. 8 and 9 have relatively narrow spreads: instantaneous energy spreads of <1.5 keV and temporal spreads of <2 ns. Energy of these ridge-like distributions of the spectra decrease monotonically from their peak values to zero. The time at which the intensity of each differential waveform attains its peak value for both cases are plotted in the energy-time space as shown in Fig. 10. It is observed that the plots of peak intensity points for the beam before the post-acceleration approximately follow the cathode voltage waveform. This indicates that the source of the electrons is at the cathode or inside the hollow cavity and these electrons are accelerated by the full instantaneous voltage between cathode and anode. The same plots for the post-accelerated beam are favorably compared with a curve that is the sum of the cathode voltage waveform and the accelerating voltage waveform of the induction linac. It is observed, however, that the beam energies are ~10% higher than the sum of the two voltage waveforms. This discrepancy will be further investigated.

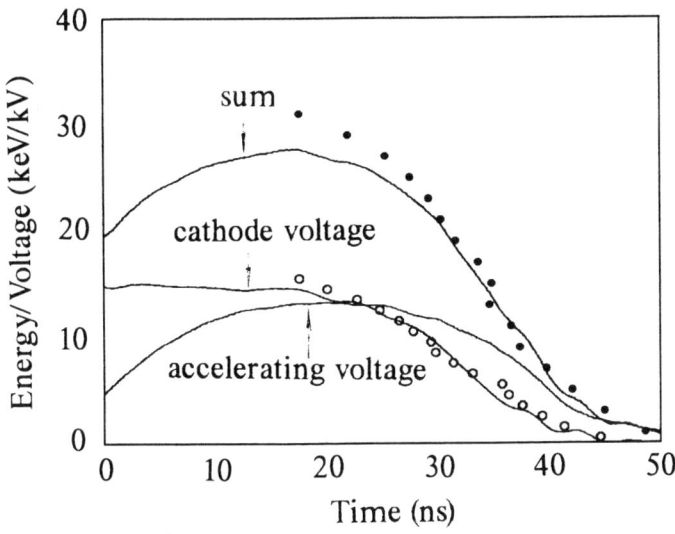

Figure 10: Comparison of spectra with voltage waveforms.

CONCLUSIONS

The breakdown voltages of a pseudospark device, the electron beam currents produced in the device, and rms emittance of the electron beam are measured for a wide range of experimental parameters. The measured data are analyzed by a numerical least- squares-fit method. It is found that the breakdown voltage is a function of the product of gas pressure squared and the anode- cathode gap distance; the electron beam current increases linearly with the capacitance and increases up to 35 kV and then decreases; the range of the effective emittance is found to be from 60 to 110 mm mrad. The time-resolved energy spectrum of the electron beam shows relatively narrow instantaneous energy spreads, and peaks of the distribution projected onto energy-time space approximately follow the voltage waveform of the cathode. The electron beam is further accelerated by an induction linac. The time-resolved energy spectrum of the beam shows that the beam is accelerated by full accelerating voltage of the linac.

ACKNOWLEDGMENT

This work was supported by the U. S. Department of Energy.

REFERENCES

[a]Permanent address: China Academy of Engineering Physics, P.O. Box 523-56, Chengdu, Sichuan, China.

1. J. Christiansen and C. Schultheiss, Z. Physik A <u>290</u>, 35 (1979).

2. D. Bloess, I. Kamber, H. Riege, G. Bittner, V. Bruckner, J. Christiansen, K. Frank, W. Hartmann, N. Lieser, Ch. Schultheiss, R. Seebock, and W. Steudtner, Nucl. Instrum. Methods <u>205</u>, 173 (1983).

3. G. F. Kirkman and M. A. Gundersen, Appl. Phys. Lett. <u>49</u>, 494 (1986).

4. H. Riege and E. Boggasch, IEEE Trans. Plasma Sci. <u>PS-17</u>, 775 (1989).

5. P. Choi, H. H. Chuaqui, M. Favre, and E. S. Wyndham, IEEE Trans. Plasma Sci. <u>PS-15</u>, 428 (1987).

6. M. Hobel, J. Geerk, G. Linker, and C. Schultheiss, Appl. Phys. Lett. <u>56</u>, 973 (1990).

7. E. Boggasch and M. J. Rhee, Appl. Phys. Lett. <u>56</u>, 1746 (1990).

8. K. K. Jain, E. Boggasch, M. Reiser, and M. J. Rhee, Phys. Fluids B <u>2</u>, 2487 (1990).

9. P. R. Bevington, *Data Reduction and Error Analysis for the Physical Sciences*, (McGraw-Hill, New York, 1969), p. 180.

10. F. Paschen, Annalen der Physik und Chemie <u>37</u>, 69 (1889).

11. K. K. Jain, B. N. Ding, and M. J. Rhee, in *Proceedings of the 1991 Particle Accelerator Conference*, edited by L. Lizama and J. Chew (IEEE, New York, 1991) p. 1972.

12. J. D. Lawson, *The Physics of Charged Particle Beams*, 2nd ed. (Clarendon, Oxford, 1988), p. 192.

13. M. J. Rhee and R. F. Schneider, Part. Accel. <u>20</u>, 133 (1986).

14. Tektronix 2440 (500 Msamples/sec).

15. M. J. Rhee and B. N. Ding, in *Proceedings of the 1991 Particle Accelerator Conference*, edited by L. Lizama and J. Chew (IEEE, New York, 1991) p. 2999.

A Novel High Brilliance Electron Source

W. R. Donaldson and A. C. Melissinos

LABORATORY FOR LASER ENERGETICS
University of Rochester
250 East River Road
Rochester, NY 14623-1299

I. INTRODUCTION

One of the most promising applications of laser switched acceleration is in the operation of a very low emittance electron source. In many current applications of high-energy-electron beams the brilliance of the source is a determining factor in the performance of the system. Recent advances in producing short laser pulses have made it advantageous to use a laser to photoemit electrons in ultrashort pulses. Such "rf laser guns" are now becoming widely used but necessitate substantial microwave power in the accelerating cavity. We are interested in demonstrating the acceleration of a laser-produced electron beam in a superconducting cavity. In this arrangement the power needed to establish the same electric field in the cavity is 10^{-4}–10^{-5} of that in a comparable copper cavity. The source could be operated in a continuous mode if the thermal loading due to the optical beam proves to be small. Another advantage is that a low-temperature cathode could possibly provide a beam with smaller transverse emittance than a room temperature source. The emittance of the beam along a particular axis is defined in terms of the corresponding volume in phase space, i.e., $\varepsilon_x = \pi(\Delta x)(\Delta x')$ where $\pm\Delta x$ is the average displacement and $\pm\Delta x'$ the average slope of the beam at any position z along the trajectory. During normal transport of a beam, the emittance is invariant. If the beam is being accelerated, the maximum slope decreases since $\Delta x' = \Delta p_x/p_z$; it is thus advantageous to define an invariant emittance

$$\varepsilon_x(\text{inv}) = \gamma\pi\Delta x\Delta x',$$

where $\gamma = E/mc^2$. The emittance determines the minimum size to which a beam can be focused and is a measure of "beam quality." If the electrons are originally produced over an area of radius r, one finds that even after acceleration the invariant emittance is of order $\varepsilon(\text{inv}) \simeq r$. Thus, it is important to produce the electrons over as small a spot as possible. In a laser-driven photocathode the size of the UV spot determines the area of electron production. A high electric field at the cathode is needed to maintain this small emittance but also to reduce space-charge effects; namely, unless the emitted electrons are accelerated away from the cathode, their negative charge produces a retarding field that limits further emission and increases beam emittance. As a reference point we mention that in our recent experiments using laser-switched pulses to accelerate electrons[1-3] the achieved brilliance was B = 0.6 × 10^{20} e/s–cm² with an invariant emittance of order $\varepsilon(\text{inv})$ ~10^{-4} m·rad. Superconducting cavities have been successfully used for particle acceleration, and correspondingly, laser pulses have been used to produce electron beams. However, the combination of a laser-driven-electron source in a superconducting cavity has not been realized as yet. Similar work is being done by a group in Wuppertal, Germany.[4] The primary difference is that we intend to look at photoemission from the superconducting surface directly.

II. ELECTRON PHOTOEMISSION

Photoemission from metal cathodes by short UV pulses has been a subject of renewed research. For instance, a group at Brookhaven National Laboratory[5] has obtained results indicating that the quantum efficiency for most metals is of order 10^{-4}, a result corroborated by our own work.[1-3]

Efficiencies approaching 0.04 have been reported with Cesiated cathodes that require special treatment in ultrahigh vacuum. This means the optically induced thermal loading will be very low. Since the use of a special cathode in a superconducting cavity may be difficult, we are attempting to produce the photoelectrons directly from the niobium surface. We must however assure that the absorption of the laser power does not drive the cavity normal even if locally the temperature is raised above its critical value, T_c. To limit the optical energy deposited in the niobium, the laser system will initially be operated at low repetition rates of about 10 Hz. An accelerator with such a low repetition rate can be useful as a injector for testbed demonstrations of advanced accelerator concepts. The primary application we are aiming for is electron scattering experiments.

To investigate these effects we have constructed a cryostat suitable for measuring photoemission from a Nb surface at 4°K. The photoelectrons are being collected on an accelerating grid, which has 70% transmission for the laser pulse. We have observed electron production using a 50-μJ UV pulse at $\lambda = 263$ nm, incident on Nb and Au surfaces. These results yielded an efficiency η $= 5 \times 10^{-6}$. The low efficiency is due in part to the space charge limit and in part to the vacuum that was $\simeq 10^{-6}$ Torr. At such pressure, surface contamination is important and is known to degrade the photoemission efficiency.

An interesting alternative to the niobium photocathode might be to use $YBa_2CuO_{7-\delta}$ (YBCO), a high temperature superconductor, as the photocathode. The advantage would be that, if operated at 4K, the YBCO could withstand about 80°K localized heating without going normal. This reduces the to possibility that the stored cavity energy would be dumped into the walls causing a thermal runaway. This would entail drilling a small hole into the niobium so that a thin film of YBCO deposited on a crystalline substrate could be mounted flush with the surface of the niobium. We will test the YBCO off line to see if reasonable amounts of photoemission can be obtained from it.

The laser pulse is derived by quadrupling an infrared pulse, $\lambda = 1.054$ μm, from a Nd:YLF laser system. This system, which was used in our previous studies of the laser-switched linac[1-3] produces short pulses by amplifying a chirped pulse in two stages and then compressing it in time.[6] In this way IR pulses of energy $E \sim 0.1$ J and width $\Delta t \sim 1.4$ ps can be delivered at a repetition rate of ~0.25 Hz. Quadrupling efficiencies are presently at the 1% level but can be improved by proper beam tuning.

III. THE SUPERCONDUCTING ACCELERATING CAVITY

The cavity, which serves also as the cathode, is one-half of an S-band accelerating structure and has been modified to include a "dimple" at the center of its back plane as shown in Fig. 1. The electric field at this perturbation is extremely high (of order 100 MeV/m), and this should enhance photoemission and minimize space charge effects. The dimpled cavity will be used because it is already available from previous experiments done at Cornell. The high electric field at the emitting surface will enhance the photoemission but the perturbation of the cavity modes may degrade the emittance. Eventually a redesigned cavity without the dimple will be used.

The cavity can be operated either in its lowest mode ($TM_{0,1,0}$) at $f = 2.805$ GHz or at the next higher mode ($TM_{0,2,0}$) at $f = 5.780$ GHz. Figure 2(a) shows a plot of the electric field as a function of the axial distance from the endwall for the fundamental mode; Fig. 2(b) indicates the electron energy as a function of the axial distance and reaches 1.2 MeV. In the fundamental mode without perturbations, the defocusing effects of the accelerating field are not severe and we expect to maintain the original emittance. The superconducting cavity has been constructed at the Laboratory for Nuclear Science of Cornell University as part of our collaboration on this work with Professor L. Hand. Our group has had experience with superconducting cavities,[7] and the facilities to carry out this research are available. When high-level power (~100 W) is input in the superconducting cavity in order to reach the field shown in Fig. 2(a), x-ray emission can be substantial and therefore the apparatus has to be properly shielded.

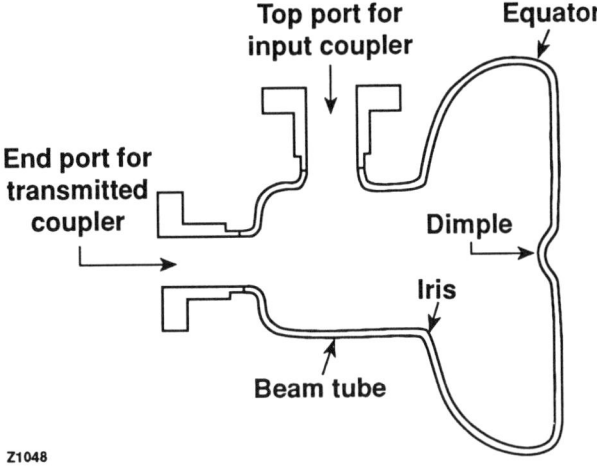

Fig. 1   Superconducting cavity that can serve as an electron source.

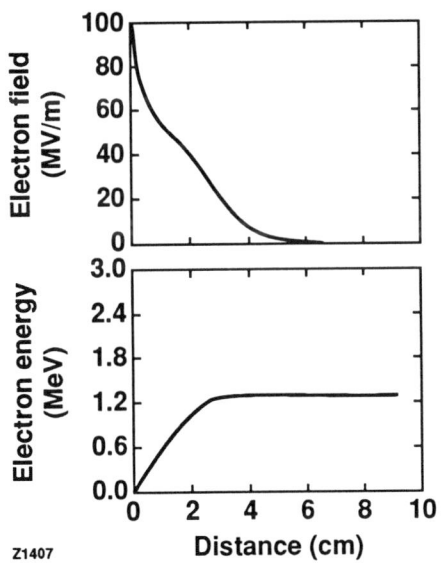

Fig. 2   (a) Accelerating field as a function of axial distance. (b) Electron energy as a function of axial distance for the fundamental mode of the cavity.

IV. DIAGNOSTICS

The cavity shown in Fig. 1 has only two access ports. The vertical port will serve as the microwave input and is equipped with a tunable coupler. The on-axis port will be used to bring in the laser pulse <u>and</u> to extract the electrons. We have had some experience with bringing a laser beam through the cavity aperture and concluded that special instrumentation is required. Since the axial direction must be left clear for the electron beam, the laser pulse will have to be directed to the cavity by reflection at a thin beam splitter such that the 1-MeV electrons can freely traverse through the splitter.

Measuring the total electron current poses no special problem, a normal Faraday cup and integrator being adequate. For low-current intensities, we have in the past successfully used a microchannel-plate (MCP) amplifier. A greater challenge is the measurement of the emittance of the beam: since the cavity is superconducting, we cannot use magnetic focusing in its immediate vicinity, which imposes an increase in the distance between the cavity and the final detector. In this case a horizontal cryostat will be needed.

In spite of the eventual desirability of the horizontal cryostat, we intend to begin this program with a vertical dewar that is presently available. We can measure the beam spot size without focusing and also measure the electron energy by deflection in a weak magnetic field. These preliminary results will be compared to the predicted beam properties, and at that point, further diagnostics will be introduced. Depending on the energy achieved, a Cerenkov radiator can be useful, as well as differential pick-up electrodes to give beam position and current. A problem that must be resolved is the synchronization of the laser pulse with the phase of the microwaves. This can be achieved by dividing down the microwave frequency to the 50 MHz of the laser oscillator mode locking circuit; the two frequencies must be exact multiples of one another and the mode locker tuned accordingly. Our aim is to produce a beam with the following characteristics:

Energy	1.2 MeV
Normalized emittance	10^{-6} m·rad
Energy spread	0.01
Electrons/pulse	10^8
Pulse length	2 ps

V. PRODUCTION OF A TUNABLE SOURCE OF VUV PULSES BY LASER BACKSCATTERING FROM AN ELECTRON BEAM

We assume that we will achieve an electron beam with the characteristic outlined in Sec. IV. If we then scatter a UV pulse from it, the energy of the backscattered photon is given by

$$\omega = \frac{4\gamma^2 \omega_0}{1 + 2\gamma^2(1 - \cos\theta)} \ .$$

For E_e = 1.2 MeV, we have γ = 2.4 and λ_0 = 263 nm. Thus, in the forward direction we can reach wavelengths λ = 10 nm or $\omega \sim$ 100 eV.

To estimate the yield of VUV photons, we take the cross section for Compton scattering at this energy $\sigma_c = 5 \times 10^{-24}$ cm^2; the photon density depends on how tightly the electron beam is focused. Assuming that the laser spot is smaller than the electron beam cross section, the number of VUV photons is given by

$$N_U = \frac{N_e}{A_e} \sigma_c N_L \ ,$$

with

N_e	=	Number of electrons
A_e	=	Area of electron beam
N_L	=	Number of photons in laser pulse.

For a 0.1-J pulse of UV, $N_L = 2 \times 10^{17}$, we take $N_e = 10^8$/pulse and $A_e = \pi\sigma^2$ with $\sigma = 100$ μm to obtain $N_U = 3 \times 10^4$. This is an energy region where it is difficult to obtain well-collimated short probes of radiation such as needed in surface studies.

VI. ACKNOWLEDGMENT

This work was supported in part by the U.S. Department of Energy under contract DE-AC02-76ER13065. This work was also supported by the U.S. Department of Energy Office of Inertial Confinement Fusion under agreement No. DE-FC03-85DP40200 and by the Laser Fusion Feasibility Project at the Laboratory for Laser Energetics, which is sponsored by the New York State Energy Research and Development Authority and the University of Rochester.

VII. REFERENCES

1.   C. Bamber, W. Donaldson, T. Juhasz, L. Kingsley, and A. C. Melissinos, "Radial Compression of Picosecond Electrical Pulses," Part. Accel. **23**, 255 (1988).

2.   Final Technical Report AFOSR-87-0328t1991; see also C. Bamber, W. Donaldson, L. Kingsley, E. Lincke, and A. C. Melissinos, "The Laser Switched Linac Project," University of Rochester Report UR-1207 (ER-13065-657).

3.   C. Bamber, "UA Pulsed Power Electron Accelerator Using Laser Driven Photoconductive Switches," Ph.D. thesis, University of Rochester, 1991 (unpublished).

4.   H. Chaloupka, H. Heinrichs, A. Michalke, H. Piel, C. K. Sinclair, F. Ebeling, T. Weiland, U. Klein, and H. P. Vogel "A Proposed Superconducting Photoemission Source of High Brightness," Nucl. Instrum. & Methods Phys. Res. **A285**, 327 (1989).

5.   J. Fischer and T. Srinivasan-Rao, "UV Photoemission Studies of Metal Photocathodes for Particle Accelerators," Brookhaven National Laboratory Report BNL-42151 (1988).

6.   P. Maine, D. Strickland, P. Bado, M. Pessot, and G. Mourou, "Generation of Ultrahigh Peak Power Pulses by Chirped Pulse Amplification," IEEE J. Quantum Electron. **24**, 398 (1988).

7.   C. E. Reece, P. J. Reiner, and A. C. Melissinos, "Parametric Converters for Detection of Small Harmonic Displacements," Nucl. Instrum. & Methods Phys. Res. **A245**, 299 (1986).

LASER DIAGNOSTICS
for PICOSECOND e-BEAMS

I. Pogorelsky

STI Optronics and Brookhaven National Laboratory

I. Ben-Zvi

Brookhaven National Laboratory

Upton, NY 11973

ABSTRACT

We propose a novel approach to the synchronization and spatial alignment of picosecond e-bunch/laser pulse. The method is based on refraction and reflection of a laser beam on a plasma column created by relativistic electrons trevelling through a gas or solid optical material. The technique may be used in laser accelerators and for general subpicosecond e-beam diagnostics.

INTRODUCTION

A number of laser acceleration proof-of-principle experiments are being initiated or are already under way in several laboratories all over the world. Most of the schemes are based upon tightly focused laser and e-beam interaction on a picosecond time scale. One of the examples is the Inverse Čerenkov Laser Accelerator (IČLA) experiment at the Accelerator Test Facility at Brookhaven National Laboratory where 6-ps e-bunches will be accelerated by >10 MeV along 20 cm by interacting with a 30-ps CO_2 laser pulse[1]. The laser beam will be focused by an axicon mirror onto a 200 μm diameter (FWHM) e-beam in a hydrogen cell.

Interactions on such small time- and space-scale require specific monitoring methods for laser and e-beam alignment and synchronization. A combination of conventional visualizing techniques (phosphor screen, IR TV viewer) allows e-beam/laser spatial alignment to within $\approx$100 μm. But picosecond e-beam/laser synchronization may be a problem which has not been addressed yet.

We discuss here a method of e-beam monitoring using refraction and total reflection of a laser beam off a plasma column created by a relativistic e-beam propagating through an optically transparent medium. In the following paragraphs we consider refractive index radial gradients created in a medium along the e-beam path, oblique incident laser beam reflection on such gradients, and how this effect may be used for picosecond e-beam diagnostics. Our quantitative estimates are based on realistic parameters of the ATF 50 MeV e-beam and 10.6 μm CO_2 laser beam propagating in H_2 or Ge. The results may be easily extended to other cases of e-beam and laser parameters as well as to various other media.

REFRACTIVE INDEX GRADIENTS IN e-BEAMS

An e-beam with a radial density profile N(r,t) propagating in any medium or vacuum will create a negative lens due to a refractive index n drop according to the equation

$$n\left(r,t\right) = n_o \left(1 - N\left(r,t\right)/N_c\right)^{1/2},$$

which follows from the relations

$$n^2 = n_o^2 \left(1 - \frac{\omega_p^2}{\omega^2}\right) \quad and \quad \omega_p^2 = \frac{4\pi N e^2}{m^*},$$

where

   $\omega_p =$ plasma frequency;
   $\omega =$ radiation frequency;
   $N_c =$ critical electron concentration when $\omega_p = \omega$;
   $m^* =$ effective electron mass.

Let us estimate the change in the refractive index

$$\Delta n(r,t) = n_o - n(r,t),$$

within the ATF e-beam for the following parameters: electron energy $E_e = 50$ MeV, electron bunch charge Q = 0.1 nC, pulse width $\tau = 6$ psec (FWHM), radius $r_o = 100\mu$m (FWHM). For the CO_2 laser 10.6 μm radiation, $N_c = 10^{21} cm^{-3}$ for 50 MeV electrons with a relativistic factor $\gamma = 100$. The total number of electrons in the bunch $Q/e = 6 \times 10^8$. The peak density is

$$N(0, t_p) = \frac{Q}{e\tau c\pi r_o^2} = 10^{13} \ cm^{-3}.$$

For the case of $\frac{\omega_p^2}{\omega^2} < 1$, the drop of the refractive index Δn obeys an approximate relation

$$\frac{\Delta n(r,t)}{n_o} \approx \frac{1}{2}\frac{\omega_p(r,t)^2}{\omega^2} = \frac{N(r,t)}{2N_c},$$

and even at the peak relativistic electron concentration, this ratio is very small,

$$\frac{\Delta n(0, t_p)}{n_o} \approx 5 \times 10^{-9}.$$

However, the secondary electron production along the e-beam path will greatly enhance this effect. First of all, in the case of thermal electrons, the critical electron density is 100 times smaller, $N_c = 10^{19} \ cm^{-3}$. The secondary electron density produced at the end of the pulse is

$$N_s(r) = \alpha N(r, t_p)\tau c,$$

where α is a multiplication factor astimated according to the formula

$$\alpha = S\rho/\tilde{E},$$

where

   $S =$ stopping power;
   $\rho =$ medium density;
   $\tilde{E} =$ average excitation energy.

Let us consider two examples characteristic of the conditions for the IČLA and typical IR optical materials:

$\underline{H_2}$: S = 5.7 MeV cm^2/g, $\tilde{E}$ = 19.2 eV,[2] ρ = 8.4 × 10^{-5} g/cm^3. That gives us α = 25 $cm^{-1}atm^{-1}$ and

$$\Delta n = 2.2 \times 10^{-6} \ atm^{-1},$$

along the e-beam axis. Extrapolating this result to other cases of potential interest, we note that Δn increases with the atomic number of the gas (e.g. it will be 2 times higher for air and 3 times higher for Xe), and it is also proportional to the gas pressure P and laser frequency ω_L^2.

For $\underline{Ge, ZnSe}$: S = 5.1 MeV cm^2/g, $\tilde{E}$ = 350 eV,[2] α = 8×10^4 cm^{-1}, then,

$$\frac{\Delta n}{n_o} \approx 10^{-2}.$$

n_o = 4 for Ge and n_o = 2.4 for ZnSe. This result is valid for relatively thin slabs when multiple scattering may be neglected. The limits imposed by multiple scattering are considered below.

The case of oblique incidence of the laser radiation onto a nonuniform secondary electron plasma column is treated below. We will show that such refractive index changes may result in appreciable laser beam distortions.

LASER BEAM REFLECTION INDUCED BY AN e-BEAM

Let us consider wave propagation through an isotropic inhomogeneous medium (particularly a plasma). It is known that in the case of oblique incidence total internal reflection will take place from the region within the plasma layer where n drops to the value

$$n(r) = n_o sin\beta_o,$$

where β_o = angle of incidence.[3]

This effect is responsible for the reflection of radio waves by the ionosphere, as well as of visible laser beams by laser induced plasmas.[4] We should expect the same effect when a laser beam crosses an e-beam at oblique incidence.

Figure 1 illustrates the effect of total internal reflection expected when a laser beam propagates through a radially inhomogeneous plasma created by an e-beam. In the approximation of geometric optics, that is when

$$\frac{\lambda_o \frac{dn}{dr}}{2\pi n^2} << 1,$$

the trajectory of the laser beam is described by the equation[3]

$$z = \int \frac{sin\beta_o dr}{\sqrt{n^2(r) - sin^2\beta_o}}.$$

Strictly speaking, the geometric optics approximation is not applicable in the vicinity of a "reflection point". Instead, wave equations should be used to calculate the reflection coefficient and phase shift for every particular polarization. But that fact does not appreciably change the condition of total internal reflection and its effects on the beam trajectory.

At a small angle between the laser and e-beam $\theta_o = \frac{\pi}{2} - \beta_o$, the condition for total internal reflection will be

$$\frac{\Delta n}{n_o} > \frac{\theta_o^2}{2} \; or \; \theta_o < \sqrt{\frac{N}{N_c}}.$$

In the case of the ATF e-beam, conditions for the total reflection of the CO_2 laser beam will be fulfilled in 2 atm H_2 when $\theta_o < 3 \; mrad \approx 10 \; min$, and in Ge when $\theta_o < 0.15 \; rad \approx 10^o$ (angle measured inside the material).

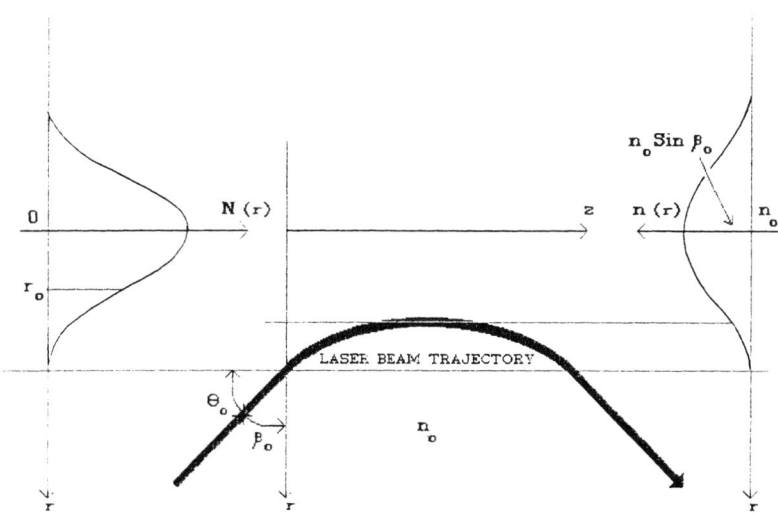

Figure 1: Oblique incidence of electromagnetic wave on an inhomogeneous plasma. $N(r)$ is the e-beam density and $n(r)$ is the refractive index.

LASER AND ELECTRON BEAM ALIGNMENT AND SYNCHRONIZATION

A laser beam reflection, discussed above, may be used for laser and e-beam synchronization within the interaction region. The characteristic time scale for the recombination and ambipolar diffusion of the secondary electron plasma is $\approx 10^{-8}$ sec in solids and $\approx 10^{-6}$ sec in gases. That makes it possible to observe the effect even at relatively long delays of the laser pulse after the passage of the picosecond e-bunch. At that point a precise alignment of the beams may be done. Then laser pulse can be delayed until a reduction and finally disappearance of the effect is observed. This gives the temporal synchronization of the e-beam and laser pulses with the ultimate resolution limited by the width of these pulses.

Methods for implementing this laser beam reflection technique are straightforward and vary according to the optical geometry used to concentrate the laser beam onto the e-beam. For example, a CO_2 laser beam with an initial diameter 1 cm, focused by a f = 1 m lens, will be confined within a 5 mrad angle to the e-beam axis and will be totally reflected by the ATF e-beam in 2 atm of air. Some asymmetry of the initial spatial intensity distribution of the laser beam may be required in order to observe the reflection effect during coaxial illumination. Such an asymmetry may be introduced by a screen blocking a part of the laser beam.

A similar mechanism may be used for alignment and synchronization of an axicon-focused laser beam. In the case of $\theta_o = 20$ mrad (ATF IČLA experiment), 15 atm Xe will be needed to achive the total reflection condition. As mentioned above, we can increase the secondary electron production orders of magnitude by using dense optical materials. That makes it much easier to reach the conditions for the total reflection. When considering solid materials for the e-beam diagnostics, we should take into account multiple scattering of the relativistic electrons causing an expansion of the initially filementary e-beam

$$\Delta r\left(z\right) = \frac{2}{3}\frac{E_S}{E_e\sqrt{L_R}}z^{3/2}.$$

This follows from the equation:[5]

$$\Omega_{1/e} = \frac{E_S}{E_e}\sqrt{\frac{z}{L_R}},$$

where

$\Omega_{1/e}$ = the rms scattering angle;
E_S = multiple scattering constant = 15 MeV;[6]
z = distance inside the scatterer;
L_R = radiation length.

Using the tabulated L_R values[6] for solid materials and the proportionality of L_R to Z^2/A (Z and A are the atomic number and weight of the scatterer),[5] we can predict a plasma column expansion in Ge:

$$\Delta r\left(z\right) \approx 0.2\ z^{3/2},$$

where r and z are measured in cm.

Due to the drop of the electron density from scattering, the condition for total internal reflection will not be fulfilled at a distance L inside the optical slab unless it satisfies the equation

$$\theta_o L^{3/2} < \frac{3}{2}n_o\frac{E_e}{E_S}\sqrt{\frac{\alpha Q L_R}{e\pi N_c}}\qquad (at\ \Delta r >> r_o),$$

where θ_o is measured outside the slab. For the conditions of the ATF IČLA experiment, $L_{Ge} < 2$ mm.

Radiation damage and color center development caused by the e-beam should be considered when choosing an optical material for the e-beam/laser synchronization probe.

REFERENCES

1. W.D. Kimura, I. Pogorelsky, L.C. Steinhauer, S.C. Tidwell, G.H. Kim, and K.P. Kusche, *AIP Proceedings of Advanced Accelerator Concepts Workshop*, Port Jefferson, NY, June 14-20, 1992.

2. M.J. Berger and S.M. Seltzer, *Stopping Powers and Ranges of Electrons and Positrons (2nd Ed.)*, National Bureau of Standards, NBSIR 82-2550-A, 1982.

3. V.L. Ginzburg, *Propagation of Electromagnetic Waves in Plasma*, Gordon and Beach, New York, 1961.

4. H.G. Ahlstrom, *Laser-Plasma Interaction, Proceedings of XXXIV Session of the Summer School*, Les Houches (France), June 30 - July 25, p.1, 1980.

5. V.L. Highland, *Nuclear Instruments and Methods*, v.129, p.497, 1975.

6. Particle Data Group, *Rev. Mod. Phys.*, v.45, p.S35, 1973

ACKNOWLEDGMENTS

Authors wish to thank Wayne Kimura for helpful discussions and enthusiasm to test the proposed method in the course of the upcoming ICLA experiment.

This work was supported by the Department of Energy, Contract No. DE-ACO2-76CH00016 and No. DE-AC06-83ER40128.

A Pulsed-Power Electron Accelerator Using Laser-Driven Photoconductive Switches

C. Bamber, W. R. Donaldson, E. Lincke, and A. C. Melissinos

LABORATORY FOR LASER ENERGETICS
University of Rochester
250 East River Road
Rochester, NY 14623-1299

I. INTRODUCTION

Pulsed-power methods of generating accelerating gradients may prove to be a practical alternative to rf technology. The accelerating gradient is applied to the structure for a short period of time, which may be set by the length of the charge line. A practical device would have a series of accelerating gaps that could be turned on sequentially as the bunch of electrons pass through the structure. For such a device to work effectively, the accelerating gradient should be present only while the bunch is in the gap and should decay to zero as the next gap is turning on. This type of high-power switching must occur in picosecond time scales. A possible candidate is a photoconductive switch that consists of a piece of highly resistive semiconductor in contact with two electrodes.[1] Upon the absorption of photons having energy greater than the band-gap energy, electrons may be promoted out of the valence band into the conduction band, which causes the conductivity of the material to rise. The response times of these switches can be less than a picosecond.[2]

In this work the accelerating structure used is a radial transmission line (RTL) as shown in Fig. 1.

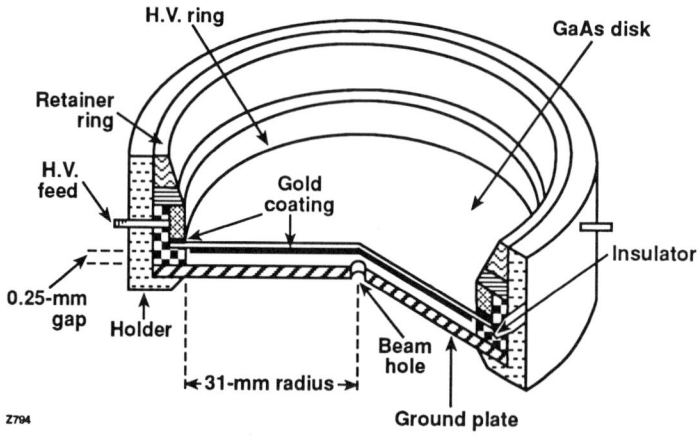

Fig. 1 The RTL boost the accelerating gradient.

This type of geometry was suggested by Willis in 1985.[3] The switch is distributed around the outside edge of the cathode plate with the annular area outside of the switch charged to a negative high voltage. When a short laser pulse illuminates the photoconductive switch, a high voltage pulse is launched toward the center of the RTL. When the pulse arrives at the center of the structure, a bunch of electrons is created photoelectrically and is accelerated toward a hole in the anode where they exit the device. The voltage of the pulse can be much greater when it arrives at the center than it was when applied to the annular charge line. Cassell and Villa[4] showed that the energy gained by a relativistic electron passing through an RTL is given by

$$V \cong 2V_0 \sqrt{\frac{2R}{g + c\tau}} \; ,$$

where V_0 is switched into the edge of the RTL, R is the radius, g is the gap spacing, and τ is the rise time of the pulse.

II. THEORY

A computer simulation of the "radial gain" of an RTL, based upon a numerical integration of the wave equation in cylindrical geometry predicted the electric field enhancements for different RTL geometries and initial rise times. As the pulse propagates in toward the center, the leading edge increases in amplitude. Typical radial gains of factors from 4 to 10 were expected from RTL's within our design range.

The acceleration of electrons with pulsed power is most effective if the length of the accelerating gap is matched to the duration of the electrical pulse. Once the electron beam is relativistic, the gap can be as long as the a light pulse would travel in this interval. In our case, the electron beam starts from rest and so the gap is considerably shorter. In order to predict the energy gained by an electron starting from rest in a pulsed electric field, a numerical simulation of the propagation of electron bunch across the gap was written. It was found that the final electron energy was a function of the starting point of the electron relative to the peak of the electric field (see Fig. 2). In this simulation, a peak field of 10 kV was applied across a gap of 0.5 mm for 30 ps (FWHM). At times much earlier than the peak, the electron energy is insensitive to variations in the starting time, because the electrons are not drawn away from the cathode by the weak fields present. For electrons started within 30 ps of the peak, the final electron energy does show a time delay dependence as they are accelerated to velocities where they can extract larger amounts of energy from the field as it is peaking. If they start even later, the field begins to decrease before they gain significant velocity and so they cannot effectively extract energy from the field. At the smallest gaps possible it is possible to extract the most energy from the voltage pulse. The photoconductive switches used in this experiment have a maximum hold-off voltage in the kilovolt range, and so to get up into the range of a few MV/m gradients the gap should be a millimeter or less. From practical considerations it was found that 0.25 mm was the smallest gap that could be fabricated.

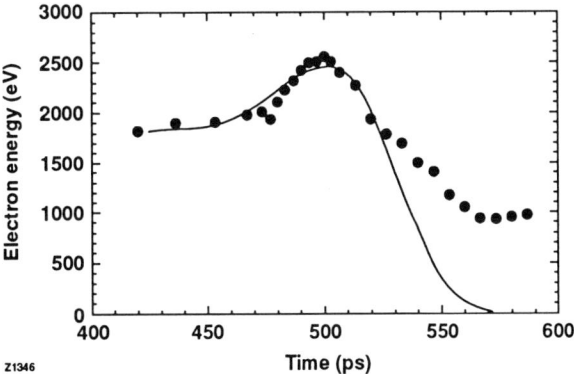

Z1346

Fig. 2 The experimental (points) and theoretical (solid) gain curves match at early time.

III. EXPERIMENTAL

The gap spacings of the RTL ranged from 0.25 mm to 2.0 mm with a radius of 31 mm. The photoconductive switches were fabricated from high resistivity silicon or gallium arsenide. When shielded from light the Si had a resistivity of 7000 Ω cm, and the GaAs had a resistivity of $10^7 \Omega$ cm. The switches were illuminated with light pulses from a Nd:glass laser system operating at a wavelength of 1.053 μm. The laser system consisted of a mode-locked oscillator followed by a regenerative amplifier. Short pulses could be generated by using the chirped pulse amplification and compression technique developed at the University of Rochester's Laboratory for Laser Energetics.[5] Temporally, the pulses had a FWHM of 1.7 ps after compression. The pulses had energies of a few millijoules after extraction from the regenerative amplifier. The pulse passed through a saturable absorber,[6] which attenuated the pedestal and reduced the pulse width to 1.3 ps.

A portion of the light from the laser pulse was split off and passed through 2 BBO doubling crystals to frequency up-convert to the UV (λ = 266 nm). This UV pulse was used to generate an electron bunch photoelectrically. The path length of the UV pulse was adjustable with a translation stage in order to set the time of arrival of the UV pulse to correspond to the time of arrival of the voltage pulse at the center of the RTL. The electrons were emitted from the RTL through the same hole through which the UV entered. The hole had a diameter of 0.5 mm.

In the initial series of experiments the final electron energy was measured, with a calibrated focusing coil, as a function of the time delay between the IR pulse used to fire the photoconductive switch and the UV pulse used to create the bunch of photoelectrons. The results are plotted in Fig. 3. These curves have a series of maxima that are separated in time by 200 ps. This is the round-trip time of the RTL. At the center of an RTL the impedance approaches infinity, so one would expect all of the electrical energy incident to be reflected. At the outside edge of the RTL there is an impedance mismatch at the transition to the switch and charge line. From Fig. 3 we see that the energy of electrons emitted from an RTL with a 0.25-mm gap decreases with each successive reflection much faster than those emitted from an RTL with a 0.5-mm gap. This is consistent with the impedance mismatch at the edge of the 0.25-mm gap being smaller than that of the 0.5-mm gap structure. The impedance of an RTL as a function of radius is given by

$$Z = 377 \Omega \, \frac{g}{2\pi r}.$$

At the outside edge of the RTL with a 0.25-mm gap, the impedance is 0.5 Ω.

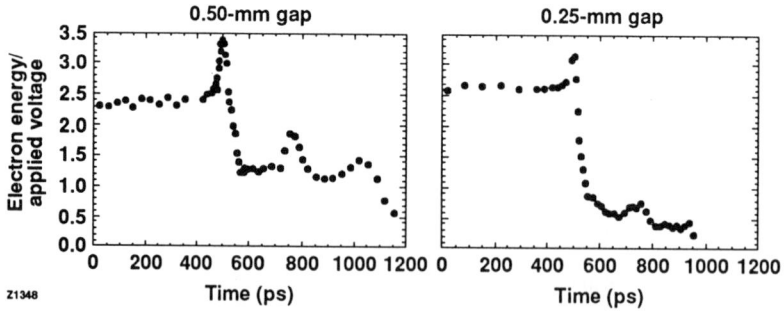

Z1348

Fig. 3 The energy gain versus time delay for two different gaps.

The initial part of the measured curves were fitted by the simulation described earlier. The data is consistent with an electrical pulse width of 36 ps for the RTL with a 0.25-mm gap. The simulation output and data are plotted in Fig. 2. It is possible to estimate the radial gain of the RTL from the best fit parameters in the simulation and an estimate of the impedance mismatch at the switch.

The radial gain is defined as the ratio of the voltage of the pulse at the center of the structure to the voltage of the pulse just after injection at the edge. For the RTL with a 0.25-mm gap and a 31-mm radius, the radial gain is measured to be 5.1±0.4. We expect a gain of 4.7.

The amount of charge per bunch was measured by collecting the charge in the Faraday cup at different optical delay positions. In Fig. 4 this data is displayed for the two geometries described earlier. We find that the maximum electron yields occur when the voltage pulse is present. Apparently the field is enhancing the electron photoelectric production. Yields in the hundreds of fc's are consistent with the relatively low quantum efficiency of the gold cathode. It is interesting to note the behavior of the electron bunch in the time period before the arrival of the first electrical pulse. At this time the agreement between the simulation and the data supports the view that the electron bunch is waiting at the cathode for the pulse to arrive. This simulation does not take into account space charge effects. The amount of charge collected in the Faraday cup decreases as the time between the arrival of the UV pulse and the electrical pulse is increased. A series of measurements were made at various gap spacings in this regime. All of this data shows that the amount of charge collected at the Faraday cup falls off exponentially as the time the electron bunch "waits" as the cathode increases. The characteristic time is of the order of a few hundred picoseconds and is a function of the gap spacing. Loss of charge in this time period can be explained either by reabsorption by the cathode or the anode, or by movement of the electrons into starting positions from which they will not be accelerated into the Faraday cup.

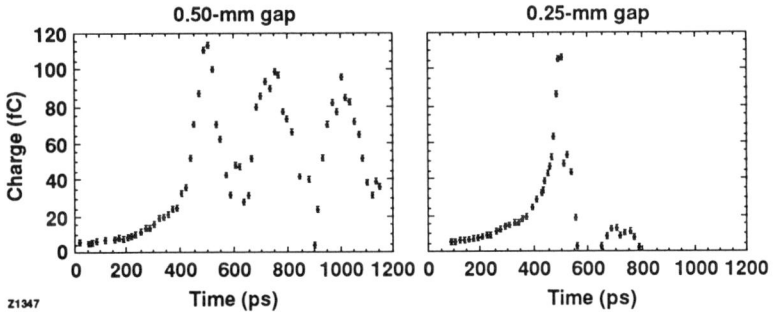

Fig. 4 The charge versus time delay for two different gaps.

IV. CONCLUSIONS

The operation of an RTL-based electron accelerator has been demonstrated. Electrons have been accelerated up to an energy of 11 keV in a gap of 0.25 mm. This represents an average accelerating gradient of 44 MeV/m. Typical electron yields of 100 fc of charge per bunch were generated photoelectrically from the gold cathode surface.

V. ACKNOWLEDGMENT

This work was suppored in part by the United States Air Force Office of Scientific Research under contract AFOSR-87-0328 and the U.S. Department of Energy under contract DE-AC02-76ER13065. This work was also supported by the U.S. Department of Energy Office of Inertial Confinement Fusion under agreement No. DE-FC03-85DP40200 and by the Laser Fusion Feasibility Project at the Laboratory for Laser Energetics, which is sponsored by the New York State Energy Research and Development Authority and the University of Rochester.

VI. REFERENCES

1.   G. Mourou, W. H. Knox, and S. Williamson, "High-Power Switching in Bulk Semiconductors," *Picosecond Optoelectronic Devices: Chapter 7*, edited by C. H. Lee (Academic Press, New York, 1984), pp. 219–248.

2.   J. A. Valdmanis, G. Mourou, and C. W. Gabel, "Subpicosecond Electrical Sampling," IEEE J. Quantum Electron. **QE-19**, 664 (1983).

3.   W. Willis, "Switched Power Linac," in *The Generation of High Fields for Particle Acceleration to Very High Energies* in the *Proceedings of the CAS-ECFA-INFN Workshop*, CERN 85-07 (1985), pp. 166–174.

4.   R. E. Cassell and F. Villa, "Study of a High Gradient Pulsed Linac Structure," Slac-Pub-3804, 1985.

5.   P. Maine, D. Strickland, P. Bado, M. Pessot, and G. Mourou, "Generation of Ultrahigh Peak Power Pulses by Chirped Pulse Amplification," IEEE J. Quantum Electron. **24**, 398 (1988).

6.   Y.-H. Chuang, D. D. Meyerhofer, S. Augst, H. Chen, J. Peatross, and S. Uchida, "Suppression of the Pedestal in a Chirped-Pulse-Amplification Laser," J. Opt. Soc. Am. B **8**, 1226 (1991).

Frontiers of Accelerator Instrumentation

Marc Ross
Stanford Linear Accelerator Center
Stanford, California 94309

I. Abstract

New technology has permitted significant performance improvements of established
instrumentation techniques including beam position and profile monitoring.
Fundamentally new profile monitor strategies are required for the next generation of
accelerators, especially linear colliders (LC). Beams in these machines may be three
orders of magnitude smaller than typical beams in present colliders. In this paper we
review both the present performance levels achieved by conventional systems and present
some new ideas for future colliders.

II. Introduction

The field of accelerator instrumentation is large and growing. Instrumentation has
been the focus of several recent workshops. One reason for this is the broad range of
requirements that new machines place on these systems. This paper addresses mainly
those accelerator instrumentation issues associated with electron-positron linear colliders.

The function of accelerator instrumentation is to provide a measure of subsystem
performance. Through this the accelerator designer may verify component tolerances.
Furthermore, the control system of the accelerator can use information provided by the
diagnostics to perform online corrections through feedback. The next collider generation
will require a substantial effort in the development of the feedback systems.

In parallel with recent improvements in the technology of signal processing and
controls, the next generation of machines will rely heavily on instrumentation and
controls improvements. It is therefore important to identify this aspect of accelerators as
one which can yield significant benefit by relaxing effective tolerances in other
subsystems.[1] Of course, this must be done carefully and may be difficult to prototype and
test accurately.

The challenge of the instrument designer is to develop new technology and also to
improve existing techniques. Through the use of new technology, conventional devices
such as beam-position monitors and video-profile monitors have become more accurate
and free from systematic errors that are harmful to feedback systems. Since a feedback
loop processes and filters information from many sources, it requires an accurate model
of both the accelerator system and its instrumentation.

*Work supported by Department of Energy contract DE–AC03–76SF00515.

The next generation of colliders will require accurate beam-position measuring devices, high-resolution beam-profile monitors, and devices that directly monitor other aspects of the beam such as correlations between energy and longitudinal or transverse position. Present profile monitors project phase space onto one axis, and are therefore not sensitive to correlations. Since smearing effects, such as chromatic filamentation, can irreversibly increase phase-space volume, it is important to accurately monitor the ellipse orientation.[2]

Most work to date has focused on the serious challenge of measuring small-collider, interaction-region spots. However, accurate, durable phase-space monitors will be required throughout the LC in order to maintain the emittance.

III. Beam Position Monitors (BPM)

Present BPM systems can be grouped according to their bandwidth. Colliders, because of their low single-bunch passage rate, probably require high-bandwidth devices. Thermal noise considerations and low bunch spacing presently limit the performance of typical devices to:

$$\frac{V_n}{V_0} = \sqrt{4kTZ_0B}\,\frac{\sqrt{2\pi e}}{\varepsilon a}\,\frac{\sigma^2}{2t_0} \tag{1}$$

Where $\sqrt{4kTZ_0B}$ is the thermal noise, about 25 μV at 25 MHz, a is the signal strength at the electrodes, about 2V ns for 10^{10} particles, ε is the efficiency of the signal collection cables and $2t_0/\sigma^2$ is the fraction of the signal from strip lines of length t_0 in the bandwidth (σ) of the signal processing filter. V_n/V_0 is about 1/5000 for 50 ns interbunch spacing. This is usually adequate since the vacuum chamber can be made to scale with the beam size at least in a coarse manner.

An example of a suitable BPM system is that being built for the Final Focus Test Beam Facility (FFTB)[3] at SLAC. This system has 1 μm resolution near the center of the aperture with intensities of 10^{10} e-, bandwidth of 25 MHz and 16-bit ADC's with noise margin of 3 dB. The absolute accuracy of the BPM is 30 μm. The FFTB BPM's are about 25 mm diameter.

Most BPM systems have significant systematic errors. These can be crudely classified as beam-quality related errors or electrode- and signal-processing errors. Among the former are:

• Beam-shape sensitivity

Beam-shape sensitivity has been studied[4] and can produce an error of about $3/r^2\left(\sigma_x^2 - \sigma_y^2\right)$ for a centered beam, with x and y in the planes formed by the pickup electrodes. Placement of the electrodes at 45° to the beam-coordinate system, or other electrode designs can reduce this error.

• Interference from beam electromagnetic fields

An interesting example of this is the experience[5] in the TRISTAN Main Ring where the wakefields left by the passage of the bunch were sufficient to cause multipactor breakdown near the button electrode.

• Sensitivity to local beam loss

This is certainly the least-understood of the above and will be a significant problem in future high-power machines. The electromagnetic shower generated when a few beam particles strike the vacuum chamber near or just upstream of the monitor can cause a strong secondary-emission signal. This signal, since it is lower-bandwidth than the coupled signal from the beam, can create a large offset error. Guidelines for avoiding this are to 1) recess the strip lines so that they cannot be struck directly by the beam, and 2) provide the appropriate amount of collimator protection upstream.

Other sources of BPM systematic error can be adequately addressed with careful calibration and appropriate data-handling codes. Some of these errors are listed here:

• Electronic offsets and cable imbalance

Non-linear response

In the last ten years, processor speed and digital-signal processing (DSP) techniques have improved enough so that they may now be included in storage-ring feedback loops. In addition to noise-free signal processing, these techniques allow much greater control over the loop-gain and transfer function. Examples of feedback systems using digital signal processing are the tune-control system at LEP, [6] and the proposed longitudinal-feedback system for the SLAC B-factory.[7] An excellent DSP instrument that has been used for development of the SLC damping rings is the Tektronix 3052.[8]

Most LC designs require trains of closely spaced electron and positron bunches. A remaining challenge for LC BPM designs is a system that can, with the accuracy listed above, independently monitor the micro-bunches position. Using the 1.4 ns interbunch spacing of the NLC design, the expected kT noise is six times larger, about 1/1000. Advances in signal processing may help to reduce this. Techniques under development for use with the Fermilab bunched-beam cooling system[9] that use fiber optic signal-delay lines will allow the stripling signal to be repeatedly sampled.[10][11] The use of the fiber optic signal-storage loop effectively lowers the bandwidth considerably.

Narrow-band BPM systems, used in storage rings, can perform much better. Furthermore, frequency-domain techniques allow greater control of electronic offsets. An SSC proposal[12] with 15 KHz bandwidth should have a limiting resolution of 100 nm.

E. Beam-Profile Monitors

The next generation of linear colliders present a serious challenge to the profile-monitor designer. Monitor performance can be characterized by beam-size resolution, (the minimum spot size measurable), and dynamic range, (the accuracy with which the distribution is determined). The first of these has been studied and some proposals for future devices will be outlined here. The second has received less attention but is important in order to gain an understanding of the sources background in high-energy physics-users detectors. Several attempts have been made to find an effective measure of the particle distribution in the beam's tails, especially in e+/e- storage rings.[13]

Most beam-size monitors use scattered radiation produced in the interaction between the beam and a target of some kind. As either the beam size is reduced or the intensity increased, the target may be destroyed in this interaction and must be regenerated before the next measurement. At small beam sizes and high beam current no solid will survive the impact of a single pulse. Gas- or plasma-based ionization monitors are also limited at very high intensities or very small spot sizes, where tunneling ionization, or the ionization of all atoms in the high-field region of the bunch, becomes an important effect. As a result, one must use a target that can be quickly regenerated or another mechanism altogether, such as a laser beam.

A. Review of existing techniques—Wire Scanners and Fluorescent Screens

1. Wires

Present wire scanners are limited to a range of currents and beam sizes because of single-pulse heating due to energy loss in the wire. These limits are roughly given by:

$$\frac{I}{\sigma} \leq 5 \times 10^9 \,/\, 2\mu m \tag{2}$$

for small wires with diameter about the same as the beam size. At SLC, these limits are reached at the IP, and they will be exceeded throughout most of an LC.

The best wire-scanner resolution has been seen with the SLC IP wire scanner. Figure is an SLC interaction-region wire scan with a 4 μm wire.[14]

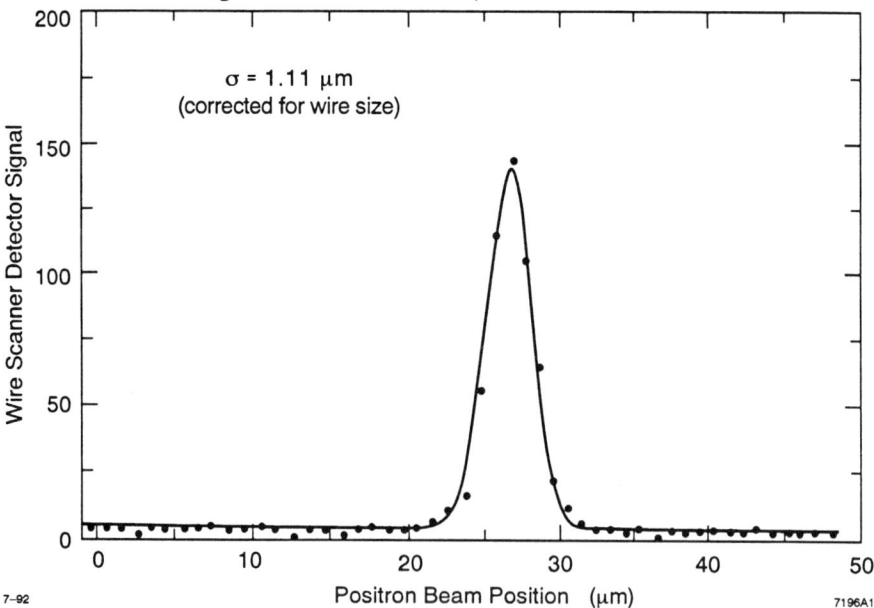

Figure 1. Wire scan results from the SLC interaction-region wire scanner. The reported beam size, after removing the contribution of the finite wire size, is 1.7 μm. This is probably the practical lower limit for this type of device.

Since a scanner samples the beam with a relatively hard process, giving good linearity, systematic errors are generally small. Figure 2 shows a high-resolution wire scan from the SLC. Most high-resolution wire scanner systems use hard, forward, bremstrahlung detectors. These detectors are often very far from the scanner and therefore have limited acceptance and have related systematic errors.

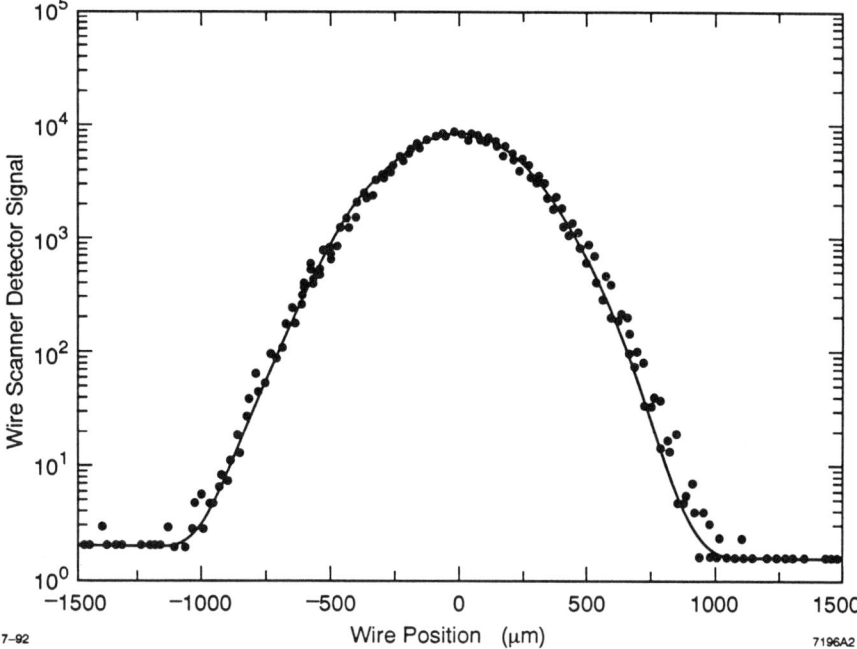

Wire Position   (μm)

Figure 2. High-resolution SLC wire scan using hard-bremstrahlung detector. This device has an 80 dB signal to noise and can effectively measure the beam distribution to ± 5 σ.

2. Fluorescent screens

Fluorescent screens remain a simple, reliable method for measuring beam size. Significant improvements in video technology now allow 16-bit amplitude vertical resolution and spatial resolution beyond other (e.g., optical) practical limits. Accuracy and dynamic range limits are mostly unknown. Excellent radiation hardness has been proven with these devices.

3. Comparison

Wire scanners and video-based profile monitors are in many ways complementary as illustrated in Table 1.

Table 1. Advantage and disadvantage comparison between video based fluorescent screens and wire scanners.

- Disadvantage + Advantage	+/-	Fluorescent Screen		Wire Scanner
Resolution	-	Finer than 20 μm not possible [15]	+	1.5 μm has been achieved and is probably the limit.
Systematic errors	-	Large systematic errors Dominated by optical and phosphor response problems[16]–hard to test	+	Simpler systematic errors Detector non-linearity–relatively easy to test Detector acceptance limits performance of many wire scanners
Dynamic range	-	Camera non-linearity and optical aberrations		
	-	Limited dynamic range without new cooled-CCD technology	+	Wide dynamic range, up to several orders of magnitude
Longevity	-	Phosphor desensitization after prolonged use, motorized systems and frame grabbers can improve	-	Thin wire fragility, wire failure mechanisms not understood
Signal processing	-	Complex frame-grabber data acquisition and background subtraction–quickly improving with new video technology	+	Many possible signal detectors - increases reliability and flexibility
	+	Single-pulse capture at low machine repetition rate	-	Multi-pulse sampling required, difficult to unfold beam jitter
	+	Full two dimensional display	-	Only projections are available. Hard to get detailed information about x-y coupled non-gaussian beams
		Visual presentation, rich intuitive content, real time display		
	-	Image lag and slow scan speed	+	Time resolve closely-spaced bunches (60 ns at SLC)
Radiation resistance	-	Camera and optics radiation sensitivity (<100 KRad without complex optics)[17]	+	Rugged scanner hardware
			-	Scattered particle detector may not be able to operate in high-radiation environments
Operability	-	Invasive–requires the use of kicker magnets for pulse snatching	+	Non-invasive upstream of background suppression collimators. Minimally invasive downstream with pulsed-beam dumpers

4. Wire scanner phase space monitor

An advantage of using narrow forward scattering can be seen when the beam angular divergence or correlation (α) is large compared to $1/\gamma$, the opening angle of the bremsstrahlung. The scan results may be sensitive to the x–x' correlation at the wire, and the steering and position changes at the wire. This may be used to advantage with segmented detectors and allow an estimate of angular divergence. The placement of the device and the optics surrounding it can greatly improve its utility.

5. Wire failure

One significant drawback of a wire scanner is the fragility of the wire. Even well below the threshold for single-pulse breakage listed above, there are other wire-breakage mechanisms. Much work has been done on this problem at the SLC Linac where wire failure has been a problem.

The SLC Linac wire scanners[18] use 40 μm tungsten wire to measure high-intensity (two or three bunches of 3.5 x 10^{10}) 100 μm beams. The single-pulse heating is well below the wire-melting point. The wire is wrapped around cylindrical studs that are set in a ceramic holder. All fractures occur at the point where the wire approaches the stud. Micrographs of the broken wire suggest that high-electric field and a resultant arc to the stud are responsible for these failures. Since the wire and the stud are both in good electrical contact and well-grounded, a strong transient pulse from the beam must cause the arc. About 1 mJ is required. Figure 3 is a micrograph of the broken end of a wire. Similar wire failures have been seen at LEP. In both cases the failure rate has been improved by using only ceramic materials in the support.

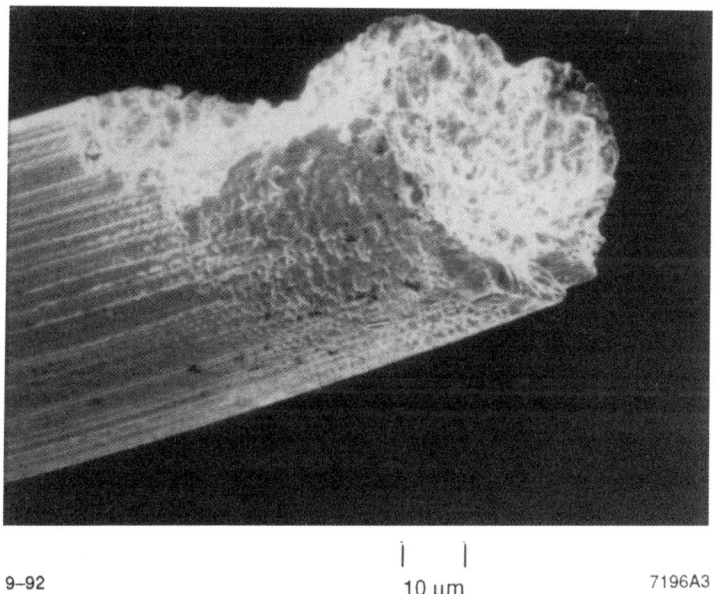

9–92   10 μm   7196A3

Figure 3. Electron microscope picture of a 40 μm tungsten wire break. This wire was installed in an SLC linac wire scanner.

6. New wires for FFTB

Most designs for an LC use large-aspect ratio beams with $\varepsilon_x \sim 100\varepsilon_y$, a new regime for accurate-profile monitors. At several locations in FFTB, during optics tuning, σ_x may be 1000 σ_y. An accurate measurement of σ_y will require novel scanner construction techniques. One proposed technique is to use a radial fan of wires at small angles to one another in order to optimize the angular resolution. Short-stub scanners will probe the beam distribution at several places in x. This data will provide information about the higher-order distortions in the beam matrix.

B. New Techniques

1. Final Focus Test Beam at SLAC

Several novel profile monitors will be tested at SLAC, at the Final Focus Test Beam facility (FFTB). The purpose of the FFTB is to test the optics and tuning tools required for the next generation of linear colliders. This will be done using 10^{10} 50 GeV electrons from the SLC linac. Interaction Point (IP) beam sizes are expected to be $\sigma_x = 0.9$ µm and $\sigma_y = 60$ nm.

2. Ionization Beam Size Monitor

The Orsay/SLAC Ionization Beam-Size Monitor,[19] shown schematically in Figure 4, measures the peak electric field of the beam using the scattered-ion angular distribution and velocity. This device allows the introduction of a controlled amount of gas into a volume surrounding the IP. The periphery of the chamber is equipped with ion detectors, in this case microchannel plates. Low Z He ions will receive a strong impulse from the field of the beam.

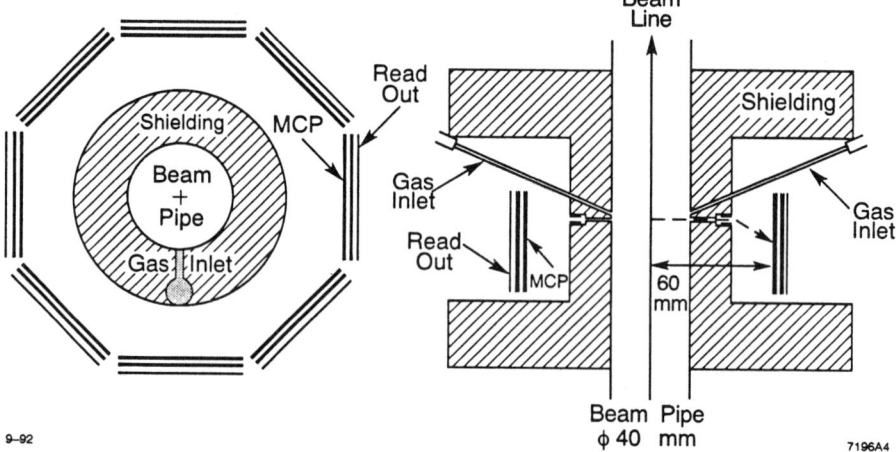

Figure 4. Gas ionization beam size monitor.

At the FFTB the beam will enter an Argon filled volume with a pressure of a few 10–4 mm Hg and ionize through dE/dx. The principle of the device is to measure the maximum velocity of the Argon ions using time of flight. The endpoint of the TOF spectrum is used to derive an estimate of the beam size in the minimum direction (σ_y). To find the beam size in the other plane, He ions are used. Since the He ions are much lighter, they are trapped in the beam, and oscillate with an amplitude that is related to their position with respect to the bunch center at the time they were ionized. The angular distribution of the He ions is related to the aspect ratio of the bunch. This technique does not work very well for positrons, since they are not trapped in the beam. The expected resolution of the device is about 5% for a few (10) pulses at nominal FFTB currents of

1E10. It will function up to a few µm where cross-calibration may be done with respect to the nearby wire scanner.

At an LC, this device will have to use multiply-ionized atoms due to tunneling ionization.

There may be significant background problems with this device since the signal from the ions is small. There are only a few hundred Ar ions detected per beam crossing. With care in beam tuning it can be setup, but under severe conditions it is clear that there may be problems.

3. Laser-Compton scattering profile monitor

The laser-compton profile monitor[20] is another new concept that provides a measure of the beam size using the interaction between the beam and a high-power, standing-wave pattern created by a laser. The beam size is determined by measuring the depth of modulation of the compton-scattered photons as the beam is scanned across the laser-interference fringes. In order to measure a large range of beam sizes (50 nm to a few µm), it is necessary to use several different laser wavelengths. This method has the advantage that the signal is strong, several thousand high-energy photons for typical FFTB parameters, and will therefore be less susceptible to low-level backgrounds. Systematic errors will arise from a poorly-focused laser spot and the resultant large interference zone. The accuracy is expected to be about 20% of the beam size on average. Systematics may also occur when the laser wavelength is changed. The ultimate limits of this device are reached when the beam size becomes comparable to the shortest wavelength laser for which there is enough power and available windows. This is about 130 nm λ and a beam size of about 10 nm, still somewhat larger than the expected σ for some of the proposed LC.

4. Liquid Jet Scanner

The liquid jet scanner [21] will use a small-diameter liquid metal jet delivered from a high pressure nozzle to scan across the beam. A pressure of 95,000 psi is required for a 1 µm-diameter jet of a eutectic alloy about 1 mm long, although the smallest jet seen to date is 4 µm. It may be possible to make jets as small as 0.2 µm. At the smallest orifice size, proper filtering will be difficult.

5. Bremstrahlung edge profile monitor

The principle of operation of this device is a bit more complex.[22] A bremstrahlung beam is produced at the IP using a thin radiator. The x-ray beam is carefully collimated using a nearby collimator and transported through a significant rotation of phase space. The edge is scanned and used to estimate beam size. Diffraction limits the ultimate resolution of this device. Both this device and the Compton-laser scattering scanner rely on long-base line alignment and alignment stability.

C. Synchrotron Light Monitors

Synchrotron light systems have also improved as a result of the development of new technology.[23] An application of synchrotron light at SLC is the use of a fast-gated video camera to monitor the beam profile on successive turns in the damping ring in order to minimize transverse-emittance blowup due to optical mismatch and other errors.[24] Figure 5 shows results from this system.

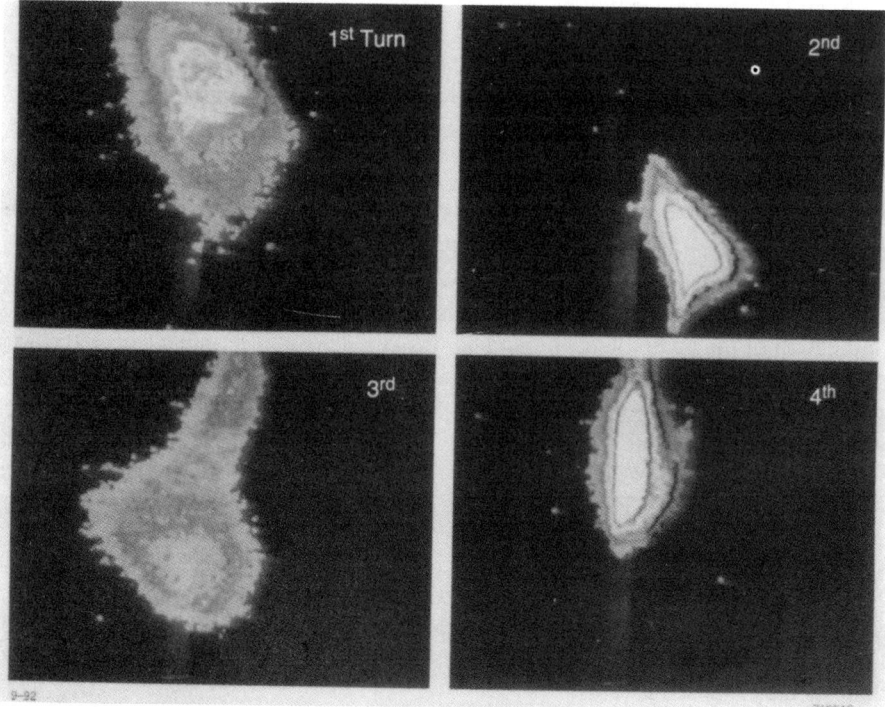

Figure 5. Synchrotron light measurements from the SLC damping ring showing the beam profile on the first four turns.

V Bunch-Length Monitors

1. Energy-Spread Monitors

An example of a non-gaussian beam shape which must be determined in detail is the energy profile of a single collider bunch emerging from the linac. Because of the strong dependence of the energy-spread distribution, especially the extremes of the distribution, on the bunch shape, it is important to measure the shape carefully. Figure 6 shows SLC linac energy-spread wire-scanner data.

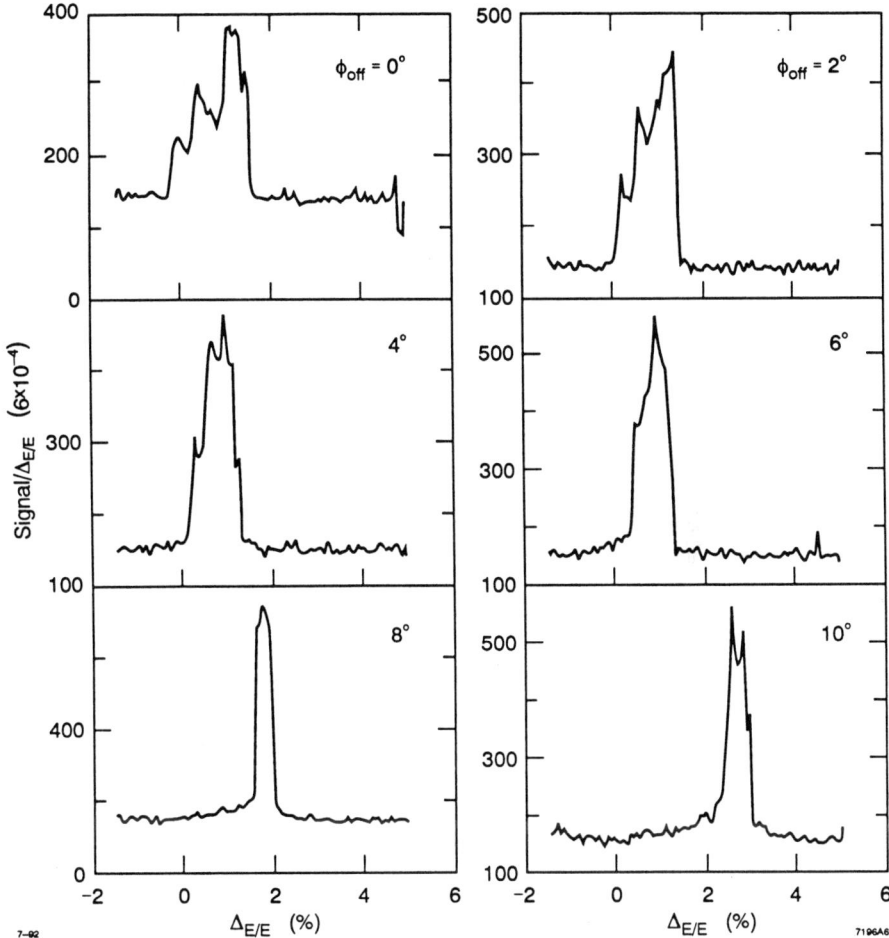

Figure 6. SLC linac energy-spread distributions. These data were produced from wire scans taken as the overall phase of the linac was varied. The data agree with simulations that assume a gaussian, longitudinal bunch distribution, and use a computed higher-order waveguide mode distribution.[25] It is important to measure the energy spread properly in order to predict the performance of downstream optical systems.

2. Coherent Synchrotron Radiation

Coherent synchrotron from a linac beam was first reported in 1989.[26] Coherent synchrotron radiation is the radiation emitted from the bunch as a whole and depends on N^2. In storage rings it is suppressed because the wavelength of the radiation is large compared to the size of the vacuum chamber. The spectrum of coherent radiation can be used as an estimate of the bunch length and for this reason it may be considered a useful beam diagnostic tool. For the nominal operating intensity of $5*10^{10}$, the expected radiated energy is 0.04 µJ or 5 µW at 120 Hz. Experiments done at the Cornell linac[27] with a pneumatic, Golay-cell, thermal-infrared radiation detector.

VI. Conclusion

Table 2 shows a summary of the profile monitor schemes to be tested at the FFTB.

Table 2. Comparison of profile monitor techniques to be used at the FFTB at SLAC.

Device	Gas Ionization	Laser Compton	Fluid jet	Edge scanner	FFTB Wire
	J. Buon	T. Shintake	F. Villa	J. Norem	C. Field
Principle	Field strength	Compton scattering from standing wave	low-melting-point metal	phase space rotation of brems	wire scanner
Min σ at FFTB	60 nm x 1µm	10 nm	jet radius /2 (50 nm)	3 nm	1µm
I limit (at that size)	few 10^{10} (beyond which it operates in a new regime)	none	none	none	5x 10^9
+	not position sensitive; few pulse measure.	large signal	large signal		
-	weak signal	multi-pulse scan difficult alignment	multi-pulse scan	measures integral of beam size, difficult alignment	multi-pulse scan wire failure

[1] F. Bulos, et al., "Beam-Based Alignment and Tuning Procedures for e+ e- Collider Final Focus Systems," Proceedings of the 1991 IEEE Particle Accelerator Conference, p. 3216 (1991).

[2] F. J. Decker et al., "Dispersion and Betatron Matching into the LINAC," Proceedings of the 1991 IEEE Particle Accelerator Conference, p. 905 (1991).

[3] D. L. Burke, "The final focus test beam project," Proceedings of the 1991 IEEE Particle Accelerator Conference, p. 2055 (1991).

[4] R. H. Miller, et al,. "Nonintercepting Emittance Monitor," Proceedings of the 12th International Conference on High-Energy Accelerators, p. 602 (1983).

[5] M. Tejima et al., "Discharge Phenomena in the Button Electrodes of the Beam Position Monitors of the TRISTAN MR," Proceedings of the Second Annual Accelerator Instrumentation Workshop, AIP Conference Proceedings No. 229, p. 287 (1991).

6 H. Schmickler, "'Tune and Chromaticity Measurements in LEP," Proceedings of the Third Annual Accelerator Instrumentation Workshop, AIP Conference Proceedings No. 252, p. 170 (1992).

7 H. Hindi et al., "Down-Sampled Signal Processing for a B Factory Bunch-by-Bunch Feedback," Proceedings of the Third European Particle Accelerator Conference, Berlin (1992).

8 K. J. Cassidy et al., "Development of a Model for Ramping in a Storage Ring," Proceedings of the Third Annual Accelerator Instrumentation Workshop, AIP Conference Proceedings No. 252, p. 144 (1992).

9 G. Jackson et al., "Bunched-Beam Schottky Signal Measurements for the Tevatron Stochastic Cooling System," Proceedings of the Workshop of Advanced Beam Instrumentation, KEK Proceedings 91–92, p. 312 (1991).

10 C. Bovet et al., "Single-Shot Bunch-Length Measurement at LEP by Stochastic Sampling of Synchrotron Light Photons," Proceedings of the Second European Particle Accelerator Conference, p. 762 (1990).

11 G. Jackson, private communication.

12 Don Martin, "Instrumentation Issues at SSC," Proceedings of the Second Annual Accelerator Instrumentation Workshop, AIP Conference Proceedings No. 229, p. 195 (1991).

13 J. T. Seeman, "Beam-Beam Interaction: Luminosity, Tails and Noise," Proceedings of the 12th International Conference on High-Energy Accelerators, p. 212 (1983).

14 R. Fulton et al., "A High-Resolution Wire Scanner for Micron-Size Profile Measurements at the SLC," Nucl. Inst. Meth. A 274, p. 37 (1989).

15 D. P. Russell and K. T. McDonald, "A Beam-Profile Monitor for the BNL Accelerator Test Facility (ATF)," Proceedings of the 1989 IEEE Particle Accelerator Conference, p. 1510 (1989).

16 C. D. Johnson, "Limits to the Resolution of Beam-Size Measurement from Fluorescent Screens due to the Thickness of the Phosphor," SLAC–CN–366 (1988).

17 F. J. Decker, "Beam-Size Measurement at High Radiation Levels," Proceedings of the 1991 IEEE Particle Accelerator Conference, p. 1192 (1991).

18 M. C. Ross et al., "Wire Scanners for Beam-Size and Emittance Measurements at the SLC," Proceedings of the 1991 IEEE Particle Accelerator Conference, p. 1201 (1991).

19 J. Buon et al., "A Beam-Size Monitor for the Final Focus Test Beam," Nucl. Inst. Meth. A 306, p. 93 (1991).

20 T. Shintake, "Proposal of a Nanometer Beam-Size Monitor for e+ e- Linear Colliders," Nucl. Inst. Meth. A 311, p. 453 (1992).

21 F. Villa, "Workshop on Linear Collider Final Focus and Interaction Region," SLAC (1992).

22 J. Norem, "A Beam-Profile Monitor for Small Electron Beams," Rev Sci. Inst. 62 p. 1464 (1991).

23 C. Bovet, "LEP Beam Instrumentation," Proceedings of the Workshop of Advanced Beam Instrumentation, KEK Proceedings 91–92, p. 11 (1991).

[24] F. J. Decker, et al., "Measured Emittance versus Store Time in the SLC Damping Ring," Proceedings of the Third European Particle Accelerator Conference, Berlin (1992).

[25] K. L. Bane et al., "Measurements of Longitudinal Phase Space in the SLC Linac," Proceedings of the Second European Particle Accelerator Conference, p. 1762 (1990).

[26] T. Nakazato, et al., "Observation of Coherent Synchrotron Radiation," Phys. Rev. Lett. **63** p. 1245 (1989).

[27] E. B. Blum et al., "Observation of Coherent Synchrotron Radiation at the Cornell Linac," Nucl. Inst. Meth. A **307,** p. 568 (1991).

Optical Signal Acquisition and Processing in Future Accelerator Diagnostics

Gerald P. Jackson
Fermi National Accelerator Laboratory, MS 341, P.O. Box 500, Batavia IL, 60510

Alizon Elliott
University of Illinois at Chicago, Dept. of Electrical Engineering, Chicago IL, 60607

ABSTRACT

Beam detectors such as striplines and wall current monitors rely on matched electrical networks to transmit and process beam information. Frequency bandwidth, noise immunity, reflections, and signal to noise ratio are considerations that require compromises limiting the quality of the measurement. Recent advances in fiber optics related technologies have made it possible to acquire and process beam signals in the optical domain. This paper describes recent developments in the application of these technologies to accelerator beam diagnostics. The design and construction of an optical notch filter used for a stochastic cooling system is used as an example. Conceptual ideas for future beam detectors are also presented.

BEAM DIAGNOSTIC CONSIDERATIONS

When beam detectors based on coupling to beam generated electric or magnetic fields or vacuum pipe image currents are designed, a number of important decisions must be made. This is especially true with wide frequency bandwidth devices such as beam position monitors[1], Schottky noise detectors[2], and resistive wall monitors[3].

The physical shape of the detector is similar to that of an antenna, detecting some fraction of the frequency spectrum of the beam charge induced fields. The impedance spectrum of the detector determines the signal level and distortion of this signal. Thermal and external electrical noise are considerations which are also included into the design.

Once the signal is acquired, it must be transmitted (usually to a location which is generally inhabitable by humans during beam operations) and processed. Transmission lines suffer from dispersion, attenuation, and multimoding. Dispersion is due to the dielectric material which forms the insulation between the two conductors of the TEM transmission line. The attenuation per unit length of cable is inversely proportional to the surface area of the conductors and increases with frequency, until in the microwave regime one can loose up to 200 dB/km. Multimoding occurs at frequencies at which the distance between the transmission line conductors is approximately a half wavelength. In order to prevent multimoding as higher frequencies are required, physically smaller diameter cables are used, with a correspondingly larger attenuation per unit length.

OPTICAL SIGNAL TRANSMISSION

There are generally two reasons why fiber optic communication of signals is chosen over electrical means. First, there are environments where high voltages and/or large ground current fluctuations make electrical signal transmission hazardous

821

or noisy. Second, because of the very long attenuation length and low dispersion of present glass fibers, long optical transmission and delay lines provide unparalleled performance.

Basic Optical Transmission Schemes

There are presently two basic methods for transmitting signals via optical transmission lines. In the first, the output power of the laser diode is modulated by the signal of interest. Second, an optical modulator after a constant laser source is used to modulate the power. In both cases a photodiode converts power back to electrical current.

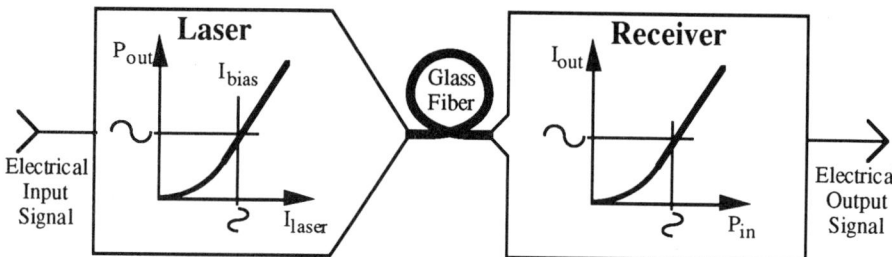

Figure 1: Sketch of an optical signal transmission system in which the laser diode output power is modulated by an input current carrying the signal of interest. Note that the laser chip must provide a bias current to insure linear signal conversion. A photodiode is used to convert the laser power modulation back into an electrical signal.

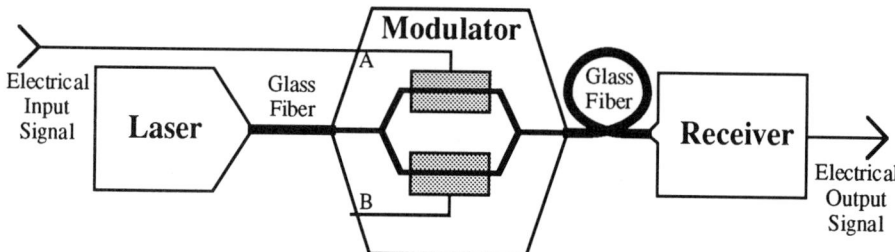

Figure 2: Sketch of an optical signal transmission system in which the constant power from a laser diode is modulated by an external signal. The modulator works by electrically modifying the optical length of a material, such as $LiNbO_3$, through which half of the light traverses, thus inducing partial destructive interference when the light from the two paths is again superimposed.

Figure 1 is a sketch of a standard system in which the laser power is directly modulated[4]. Capable of amplitude modulating the power of the laser in the frequency range of DC to 12 GHz, the vast majority of instrumentation needs are serviced by this type of system. With an attenuation of approximately 0.3 dB/km and negligible dispersion (both due to the doping of the glass and the fact that the carrier frequency of the light is approximately 300 Terahertz), transmissions over very long distances are possible with no signal degradation. The disadvantage of this scheme is the insertion loss of approximately 40 dB (due to the low input impedance of laser diodes and the need to terminate the input signal into 50 Ω), which means high levels of preamplification are needed to avoid signal-to-noise degradations.

In order to attain higher modulation frequencies, it is necessary to use optical modulator technology[5]. In a modulator, the light from an optical fiber is directed into an optical silicon optical channel, where the light is split equally into two separate paths. The two beams are then sent through a material, such as $LiNbO_3$, where the optical length of the path is proportional to the voltage applied across it. The two light signals are then recombined and launched into the output fiber. If the length of both paths are changed together, the output light wave is phase modulated. This is useful in cases where interferometry is being performed in some other part of the optical circuit.

Of more direct interest to instrumentation designers is the mode in which the path length in the two legs are oppositely modified (or in most cases where output phase is not important, just one leg). In this case the power in the output fiber is modulated through partial destructive interference. This technique is actually used in some accelerator RF systems and is called paraphasing[6]. Because of the small size of the modulating material, the capacitance of the electrical junction across it is very small. Therefore, modulation frequencies up to 20 GHz are already available on the market. As apposed to the input to a laser diode, the input impedance of a modulator is very high.

Two applications of optical signal transmission at Fermilab are good examples of the utility of this technology. First, optical signal transmission is used to make impedance measurements of an RF cavity. In the second, fiber optics may be used to transmit Schottky signals across the chord of an accelerator for stochastic cooling.

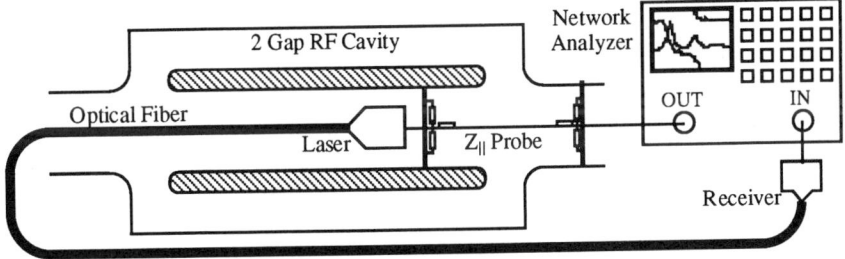

Figure 3: An elegant example of the range of uses for optical fibers in signal transmissions. In order to measure the longitudinal (or even transverse if the wire is moved radially) impedance of a single gap, the signal wire through the other gap would normally invalidate the measurement.

In the first case, because of the fact that most Fermilab RF cavities are of the two gap resonant transmission line variety, where the two gaps are separated by roughly half an RF period, impedance measurements pulling a single wire down the entire length of the cavity can be very confusing[7]. Therefore, it is desirable to measure the impedance of a single gap, and then derive the double gap response independently. The problem is that any S_{21} measurement across an accelerating gap requires a signal cable to cross the other gap, partially shorting it and affecting the measurement. The solution is to transmit the S_{21} response via optical fiber back to the input of the network analyzer[8]. Figure 3 contains a sketch of this solution.

In damper systems, in which measured betatron displacements are corrected by a kicker placed an odd number of 90° downstream, beam position information acquired at the pickup must be transmitted to the kicker in time with the beam. Since the signal in an optical fiber propagates at approximately 2/3 the speed of light, it is not possible to send that signal through the link ahead of a relativistic beam by transmitting along the diameter of a circular accelerator. But, as shown in figure 4, other chords in a wide variety of accelerator shapes are possible.

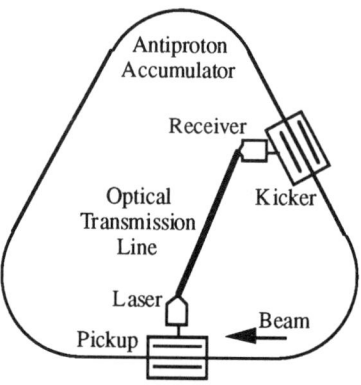

Figure 4: Sketch of a optical transmission system for a stochastic cooling system. In principle, any damper like system where the phase advance between the pickup and the kicker is an odd multiple of 90° could be configured in a similar manner.

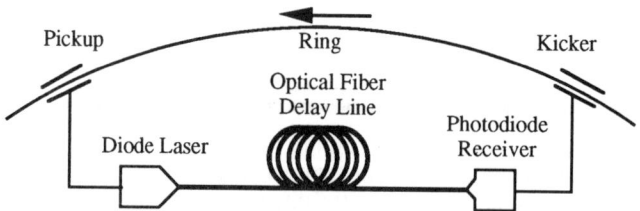

Figure 5: Rough sketch of a damping system where the one turn delay between the pickup and kicker is provided by an optical fiber delay line. In order to suppress beam oscillations or stochastically cool beam, there must be an odd number of quarter wavelengths of phase advance between the pickup and kicker around the ring.

Optical Delay Lines

Designers of RF, stochastic cooling, and feedback systems often need to delay a beam signal for one accelerator revolution in order to correct the beam energy or angle. Stochastic cooling systems often need repetitive notch filters designed to notch or phase shift the beam signal at the revolution harmonics. In both cases, an optical delay line is very useful, especially for accelerators with radii that of the Fermilab Tevatron and beyond.

Figure 5 contains a sketch of a delay line used at Fermilab in the bunched beam stochastic cooling system in the Tevatron[9]. Especially useful in accelerators systems for which cutting a chord across the ring is not a viable option (for instance, at Fermilab the center of the Main Ring and Tevatron is composed of Federally protected wetlands and a very popular prairie grass reclamation project), the pickup and kicker are separated by the machine tune minus enough phase advance to provide an odd number of quarter wavelengths within the feedback loop.

Repetitive Notch Filters

A transverse feedback system whose purpose is to damp injection oscillations, stochastically cool beams, or suppress instabilities often suffers from either power limitations or excessive dynamic range requirements due to power at revolution frequency harmonics caused by the beam passing off-center through the pickup. In order to avoid the first problem, one typically chooses a high power amplifier to drive the kicker that has no DC response. In this way the feedback system will not act as a steering magnet. Though this solution works well in most applications, it can be quite a problem for systems that require bunch-to-bunch isolation. For instance, if a positive kick to a single member of a set of bunches is required, the other bunches receive a negative kick so as to guarantee zero area under the net waveform (as required by a lack of DC response). This residual kick is a source of noise which will eventually cause emittance growth in hadron beams.

Another solution which prevents both the first and second concerns is to suppress the power in the revolution harmonics just after the pickup, before the rest of the system electronics, by means of a repetitive notch filter. Sketched in figure 6, this filter is generated by subtracting from the pickup signal the signal from the previous turn.

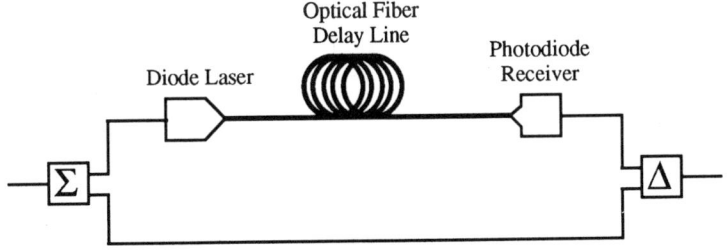

Figure 6: Example of a repetitive notch filter in which an optical fiber is used to generate a very long delay required in large radius accelerators.

The frequency response of such a filter is shown in figures 7 and 8. In the time domain, if both signals are perfectly matched in amplitude and time, the signal net output signal of the filter given a sinusoidal input would be

$$V(t) = V_o e^{i\omega t}\left(1 - e^{-i\omega\tau}\right) \qquad , \qquad (1)$$

where τ is the delay of one leg with respect to the other. In the frequency domain this equation becomes

$$V(\omega) = \left[1 - \cos(\omega\tau)\right] + i\,\sin(\omega\tau) \qquad , \qquad (2)$$

where the real and imaginary parts have been separated. The amplitude and phase of this filter are

$$A(\omega) = \left|\sqrt{2}\,\sin\!\left(\frac{\omega\tau}{2}\right)\right| \qquad , \qquad (3)$$

$$\phi(\omega) = \arctan\!\left[\frac{\sin(\omega\tau)}{1 - \cos(\omega\tau)}\right] \qquad . \qquad (4)$$

As can be seen from the figures, this type of filter has amplitude notches at all frequencies where the delay is exactly one wavelength. In an accelerator, if the delay is exactly one turn every harmonic of the revolution frequency is suppressed. The problem with this type of filter is the slew of the phase response in the passbands. In bunched beam stochastic cooling systems, where the betatron power is occupying half or more of the passband, those particles in the frequency regime beyond 45° are heated instead of cooled.

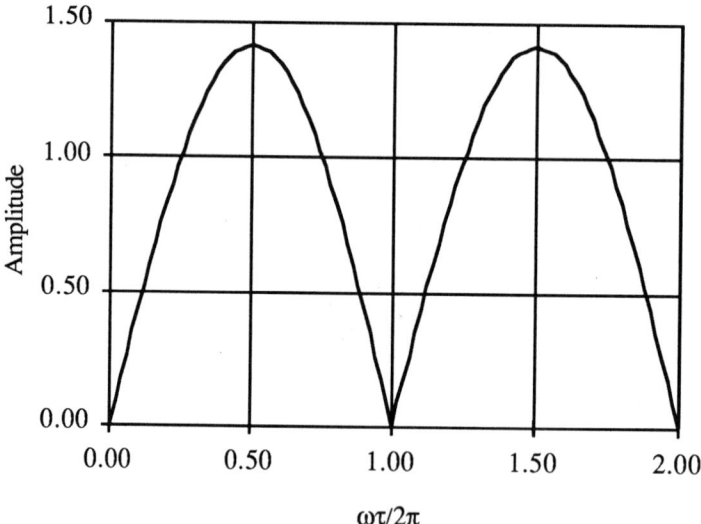

Figure 7:   Amplitude response of a one-turn delay repetitive notch filter.

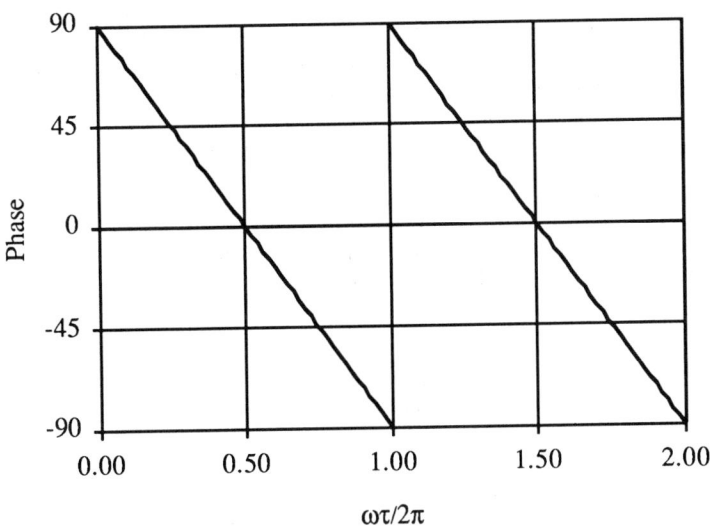

Figure 8: Phase response of a one-turn delay repetitive notch filter.

PHOTON STORAGE RINGS

The technology of optical fibers has quickly moved to provide optical components normally found in the RF and microwave worlds. Devices like directional couplers, attenuators, amplifiers, circulators, and isolators are now all available. With these components it is possible to produce filters and instruments not found in the electrical domain. By far the most powerful fiber geometry found so far for instrumentation and feedback purposes is the fiber optical loop, or photon storage ring.

Loop Notch Filter

In order to reduce the phase slew in the one-turn delay repetitive notch filter, an optical storage ring can be used to subtract the exponential average of all previous turns from the present turn. A sketch of the filter is in figure 9.

The directional couplers transmit some of the light from one path into the other available path in the same direction. Therefore, in the left directional coupler in figure 9, the beam signal launched by the laser is partially inserted into the fiber storage ring in the clockwise direction. The coupling can range anywhere from 3dB to 40 dB, the same range found typically in RF and microwave devices. The difference in the optical domain is that the coupling is exactly uniform over multi-GHz bandwidths, whereas electrical devices have phase and amplitude ripple in their passband which destroy the response of the electrical filters of this type.

Optical amplifiers are the technological breakthrough which really make this filter extraordinary[10]. By directly multiplying photons without the intermediate transformation to electrical current amplification, low nose amplification of the modulated laser power is possible. Gains of up to 40 dB have been attained. There are at present three available technologies: modified silicon laser chips, erbium doped fiber amplifiers, and nonlinear crystal multiplication. The first two are based on the excitation of a metastable state by a pumped source, and generate noise due to the amplification of spontaneous emission of the metastable state before it could be deexcited by stimulated emission. Fortunately, the noise power is spread over Terahertz of bandwidth which is easily removed by an optical bandpass filter. Silicon chip amplifiers are simply diode lasers with the reflecting mirrors at either end of the optical cavity removed. The metastable state is excited by direct electrical stimulation. Erbium doped fibers are pumped with a high power laser of a specific wavelength which excites the required metastable level. At present at Fermilab the silicon amplifier is used, though erbium doped fiber amplifiers are going to be used in the near future.

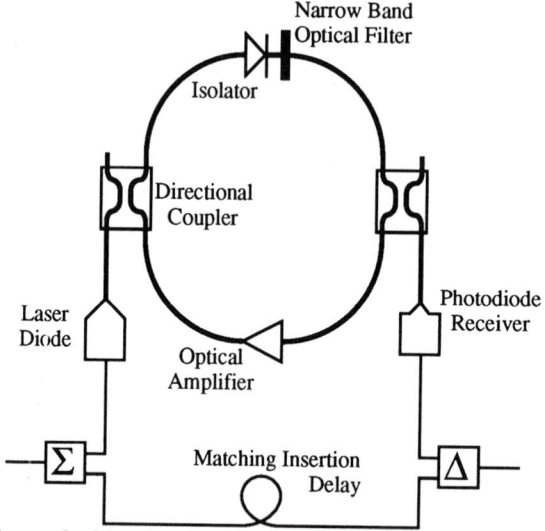

Figure 9: Sketch of a loop notch filter which is being built for the Tevatron bunched beam stochastic cooling system.

The signal in the optical storage ring can be written in the time domain as

$$V(t) = V_0 e^{i\omega t} \sum_{n=1}^{\infty} \alpha^n e^{-in\omega\tau} \quad , \quad (5)$$

were α is the fraction of an inserted signal which survives exactly one turn in the optical storage ring. To write the voltage in the frequency domain, the equality

$$\sum_{n=1}^{\infty} x^n = \frac{x}{1-x} \quad (6)$$

must be employed. Therefore, the voltage in the storage ring in the frequency domain is

$$V(\omega) = V_0 \frac{\alpha e^{-i\omega\tau}}{1 - \alpha e^{-i\omega\tau}} \qquad (7)$$

At the revolution frequency harmonics the exponential factor is unity. Therefore, the gain at these frequencies is $\alpha/(1-\alpha)$. In the full filter, the subtraction needs to be amplitude balanced such that the revolution harmonics are reduced to zero. Therefore, the response of the entire filter is

$$V(\omega) = V_0 \left(1 - \frac{1-\alpha}{\alpha} \frac{\alpha e^{-i\omega\tau}}{1 - \alpha e^{-i\omega\tau}}\right) \qquad (8)$$

By reworking this equation, the compact form

$$V(\omega) = V_0 \frac{1 - e^{-i\omega\tau}}{1 - \alpha e^{-i\omega\tau}} \qquad (9)$$

is attained. Finally, the equation can be written with the real and imaginary parts explicitly separated as

$$V(\omega) = V_0 \frac{\left[(1+\alpha)(1-\cos(\omega\tau))\right] + i\left[(1-\alpha)\sin(\omega\tau)\right]}{1 - \alpha\cos(\omega\tau) + \alpha^2} \qquad (10)$$

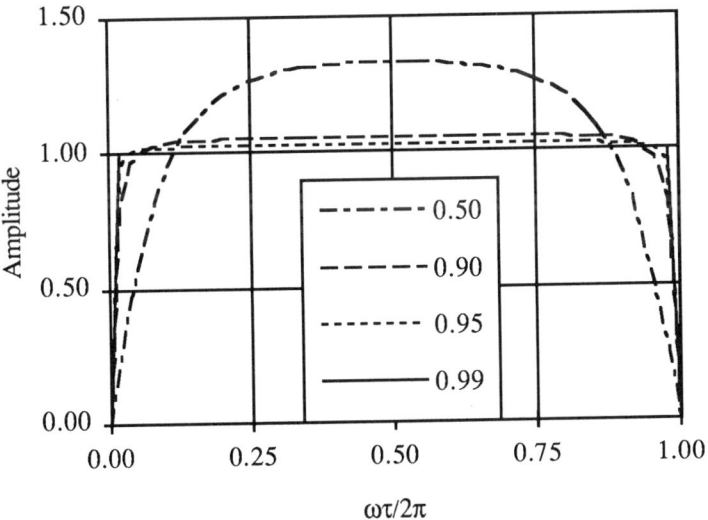

Figure 10: Amplitude response of a loop repetitive notch filter.   The curve for a=0.99 is almost indistinguishable from the value of unity, except for the sharp drops to zero at harmonics of the revolution frequency.

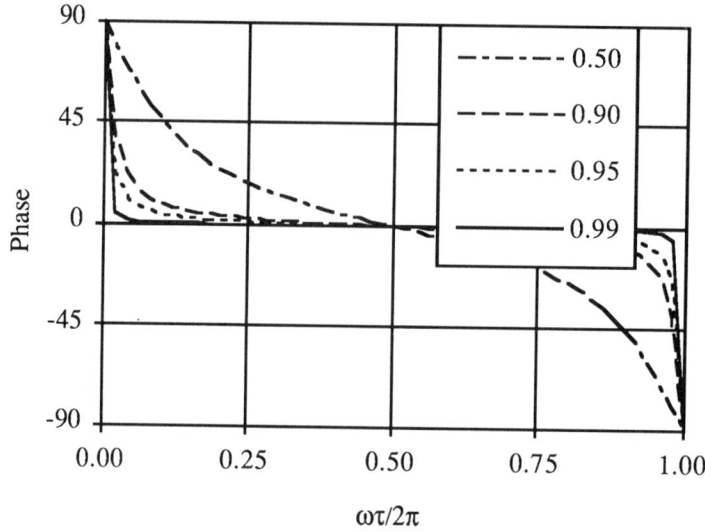

Figure 11: Phase response of a loop repetitive notch filter.

Figures 10 and 11 contain plots of the amplitude and phase of this filter as a function of frequency and the amplitude survival factor α. Note that the response of the filter becomes more like an infinite impulse repetitive notch filter as α goes to unity. The reason for this effect is the fact that the loop filter is acting as an exponential filter with a $1/e$ length of $1/(1-\alpha)$ turns of the loop. Therefore, the signal coupled out of the loop is closer to zero for all non-revolution harmonic frequencies as the averaging gets longer and the filter output is simply the prompt input. In the time domain, when thinking about bunch signals from a beam position detector, the loop filter subtracts the average bunch profile signal (devoid of betatron oscillation induced amplitude modulation) from the prompt signal, therefore subtracting out the closed orbit offset portion of the beam signal without inserting delayed betatron information at the wrong phase (which shows up as the phase slew in the filter response).

Real Time Bunch Length Monitor

An ingenious idea for creating a real time bunch length monitor for an electron beam, created at CERN by Claude Bovet, is also based on a photon storage ring[11]. Shown in figure 12, the basic idea is to generate a histogram with a width proportional to the bunch length by sampling injected synchrotron radiation photons with a fast risetime photodiode sensitive to single photons.

If coupling in and out of the loop is sufficiently small, on average less than one photon per ring revolution is registered at the photodiode. Since the temporal distribution of registered photons comes from the parent distribution of the incident synchrotron radiation, the time spread of photon arrivals is equal to the bunch length. If the circumference of the optical storage ring is much smaller than the spacing of bunches (or the accelerator circumference if bunch gating is employed) real time bunch length monitoring is possible.

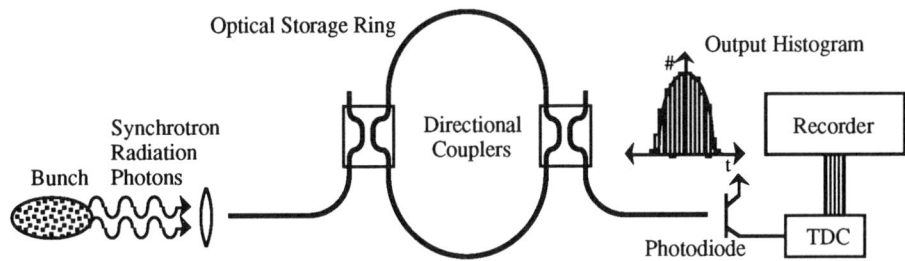

Figure 12: Sketch of a real time bunch length monitor based on a photon storage ring.

Signal Shaping

In many beam signal processing applications the best available technique requires repetitive signals (i.e. are frequency domain based). An excellent example is the AM-PM conversion technique[12] used in many beam position monitoring systems in the world. Unfortunately, sometimes the stimulus for such systems is a single bunch, which acts as an impulse and invalidates the meaning of the processing result.

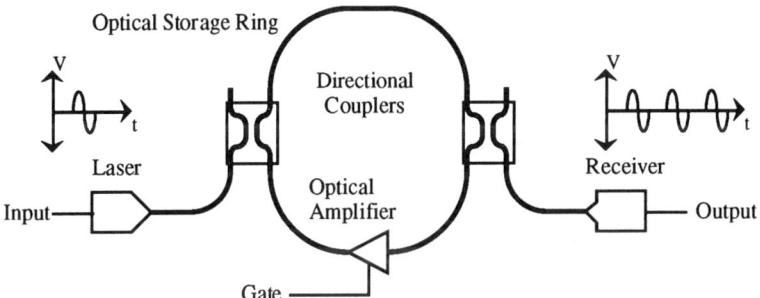

Figure 13: Example of a signal repeater used to make a transient beam signal suitable for processing using an inherently frequency domain method.

There are a number of ways to try to convert an impulse into a signal suitable for frequency domain processing techniques. A very powerful means is to repeat the impulse signal at a regular interval. This can be accomplished by either placing an array of evenly space detectors in the tunnel[13], or repeating the impulse from a single detector by sending the signal into a unity gain loop (see figure 13). Even though in principle electrical signal loops could be used to create repetitive signals, the noise and signal quality features of optical systems makes this a preferred technological platform, especially in large accelerators or those with very low processing frequencies.

OPTICAL DEVICES IN BEAM DETECTORS

Work is just now starting at Fermilab investigating the use of optical components inside electromagnetic beam detectors for direct conversion to optical signal transmission and processing. Especially in the case of wide bandwidth detectors, optical fiber components built into electromagnetic detectors will become a necessity.

Present design concepts fall into two categories. In the first, since the beam acts as a current source and the input impedance of laser diode chips are relatively low, the use of lasers inside low impedance devices such a resistive wall monitors becomes attractive. On the other hand, in pickups which have fixed impedances, local terminations coupled with high input impedance modulators may be the preferred signal encoding method.

In the case of a resistive wall monitor[14] the design goal is to shunt the image currents of the beams through a set of resistors, instead of around the resistors on the inside surface of it's metallic containment vessel. This is accomplished by increasing the inductance of the containment vessel path, raising the impedance to a value much greater than the net resistance across the ceramic beam pipe gap in the frequency range of interest. The present generation of monitors of this type at Fermilab have a useful bandwidth from 3 kHz to 6 GHz. Figure 14 is a sketch of the arrangement of 50Ω resistors crossing over the ceramic gap. The black resistors are replaced by 50Ω cables which are combined to generate a transverse position insensitive response.

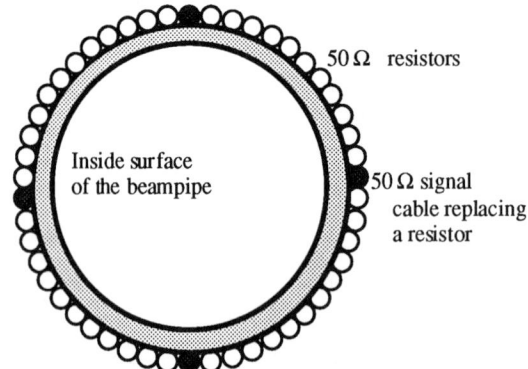

Figure 14: Sketch of an electrically based resistive wall monitor in which 4 50Ω resistors are replaced by 50Ω signal cables which are combined to create a transverse beam position insensitive current monitor.

Clearly, the above detector throws away about 48/52 of the signal, which is dissipated in the non-instrumented resistors. If the resistors were replaced by laser diodes which were all linked together, a 12 GHz bandwidth detector with much less high-μ material needed to generate the required inductance could be produced. In fact, such a device could be built without a ceramic gap, as shown in figure 15, and may be capable of producing better signal-to-noise ratios.

It is also possible to use the high input impedance of optical modulators to encode beam information onto the power from a laser. An excellent example is that of a Schottky detector in which each detector pad is terminated into it's characteristic impedance[15]. Shown in figure 16, beam pickups (directional couplers) of typically

100 Ω transmit signals along 100 Ω transmission lines to properly terminated modulator inputs. Because the propagation velocity in optical fiber is 2/3 the speed of light, the delay between the pickup and the modulator must be increased per stage. The benefit of such an array of pickups is that though the noise power increases linearly with the number of pickups, the signal power goes as the square. Since no impedance matching networks are required to recombine the signal, the deleterious effects stemming from the recombination of slightly different signals (due to misalignment of the array with respect to the beam trajectory) are completely avoided.

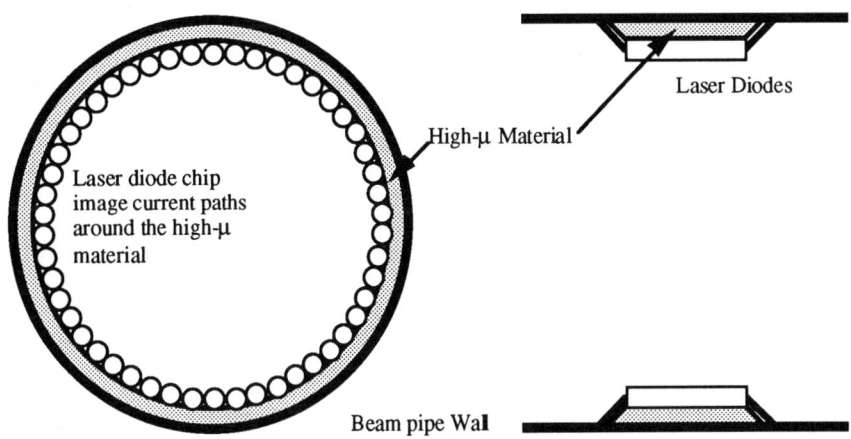

Figure 15: Sketch of another resistive wall monitor in which the laser diodes shunt the image current inside of the high permitivity material through the lasers and modulating the laser power.

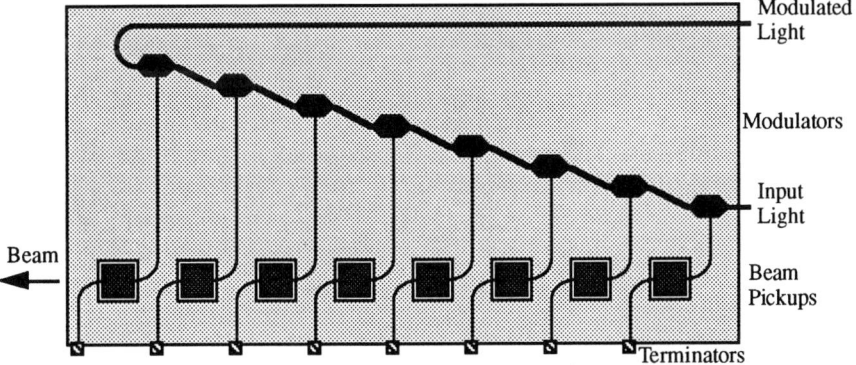

Figure 16: Example of the use of optical modulators as an integral component of an electromagnetic beam detector (in this case, a stochastic cooling pickup) etched on a Teflon-based circuit board. The light shaded surface faces the beam and is a copper ground plane, while the pickup pads, traces, and modulators are on the back surface.

CONCLUSIONS

The present commercial status of optical technology is sufficiently advanced to already provide the means of substantially improving the accuracy, sensitivity, bandwidth, and signal-to-noise of accelerator instrumentation. The rate at which this technology is advancing makes optical fiber transmission and its associated hardware worth monitoring in the future.

ACKNOWLEDGEMENTS

The authors would like to acknowledge the design and fabrication work of R. Pasquinelli and E. Buchanan. Their experience, engineering skills, and careful construction are responsible for the success of the optical fiber based microwave signal transmission and filter projects at Fermilab.

REFERENCES

1. R. Shafer, Acc. Instrum. Workshop, AIP Conf. Proc. 212, 48 (1989).
2. D. Peterson, Acc. Instrum. Workshop, AIP Conf. Proc. 229, 108 (1990).
3. C. Moore, et al., Proc. IEEE Part. Acc. Conf., 1513 (1989).
4. R. Pasquinelli, Acc. Instrum. Workshop, AIP Conf. Proc. 229, 180 (1990).
5. M. Kanda and K. Masterson, Proc. IEEE 80, No. 1, 209 (1992).
6. J.E.. Griffin, et al., IEEE Trans. Nucl. Sci. NS-28, No. 3, 2037 (1981).
7. G. Jackson, Proc. Fermi III Instability Workshop, 245 (1990).
8. P. Colestock, private communication.
9. G. Jackson, et al., Proc. IEEE Part. Acc. Conf., 1758 (1991).
10. C.R. Giles, Jour. Lightwave Tech. 9, No. 2, 147 (1991).
11. C. Bovet, et al., Proc. Europ. Part. Acc. Conf., 762 (1990).
12. R. Shafer, Acc. Instrum. Workshop, AIP Conf. Proc. 212, 48 (1989).
13. D. Briggs, et al., Proc. IEEE Part. Acc. Conf., 1404 (1991).
14. R. Webber, Acc. Instrum. Workshop, AIP Conf. Proc. 212 (1989).
15. D. McGinnis, et. al., Proc. IEEE Part. Acc. Conf., 1389 (1991).

Possible Methods of Measuring the Length of Sub-Picosecond Electron Bunches in the Frequency Domain

Gerald P. Jackson

Fermi National Accelerator Laboratory, MS 341, P.O. Box 500, Batavia IL, 60510

ABSTRACT

The traditional method of measuring extremely short electron beams using streak cameras begins to become problematic and expensive at bunch lengths at a picosecond and below. In this paper a few alternatives, based on the differential measurement of the Fourier spectrum of the longitudinal charge distribution of a bunch, are suggested and evaluated.

FOURIER TRANSFORM OF BUNCH PROFILES

Let us assume that the longitudinal beam current profile is Gaussian in shape and described the equation $(Q=Ne)$

$$I_b(t) = \frac{Q}{\sqrt{2\pi}\sigma_\tau} e^{-\frac{t^2}{2\sigma_\tau^2}} \qquad (1)$$

If Fourier transforms are defined by the equations

$$F(\omega) = \int_{-\infty}^{\infty} f(t)\, e^{i\omega t}\, dt \qquad (2)$$

and

$$f(t) = \frac{1}{2\pi} \int_{-\infty}^{\infty} F(\omega)\, e^{-i\omega t}\, d\omega \qquad , \qquad (3)$$

then the beam current in the frequency domain is described as

$$I_b(\omega) = Q\, e^{-\frac{1}{2}\sigma_\tau^2 \omega^2} \qquad (4)$$

Therefore, in the frequency domain the power spectrum of the bunch shape is also Gaussian, but with an angular frequency width of $\sigma_\omega = 1/\sigma_\tau$. For M bunches each of charge Q_b constrained to occupy buckets of length T_0,

$$I_b(t) = \frac{Q_b}{\sqrt{2\pi}\sigma_\tau} \sum_{m=1}^{M} e^{-\frac{(t-mT_0)^2}{2\sigma_\tau^2}} \qquad (5)$$

The Fourier transform of (5) yields a frequency spectrum of

$$I_b(\omega) = \frac{Q_b}{\sqrt{2\pi}\sigma_\tau} \sum_{m=1}^{M} \int_{(m-\frac{1}{2})T_0}^{(m+\frac{1}{2})T_0} e^{-\frac{(t-mT_0)^2}{2\sigma_\tau^2}+i\omega t}\, dt \quad, \tag{6}$$

which can be reduced to

$$I_b(\omega) = Q_b\, e^{-\frac{1}{2}\sigma_\tau^2\omega^2}\, \mathrm{Re}\left\{\Phi\left(\frac{T_0\sqrt{2}}{4\sigma_\tau}+i\omega\frac{\sigma_\tau}{\sqrt{2}}\right)\right\} \sum_{m=1}^{M} e^{i\omega mT_0} \quad. \tag{7}$$

The summation at the end of equation (7) is a form factor which describes the modulation of the frequency spectrum around harmonics of the bunch passage frequency. Figure 1 contains a plot of this form factor normalized to the number of bunches for M=1, 13, and 84 (cases commonly found at Fermilab).

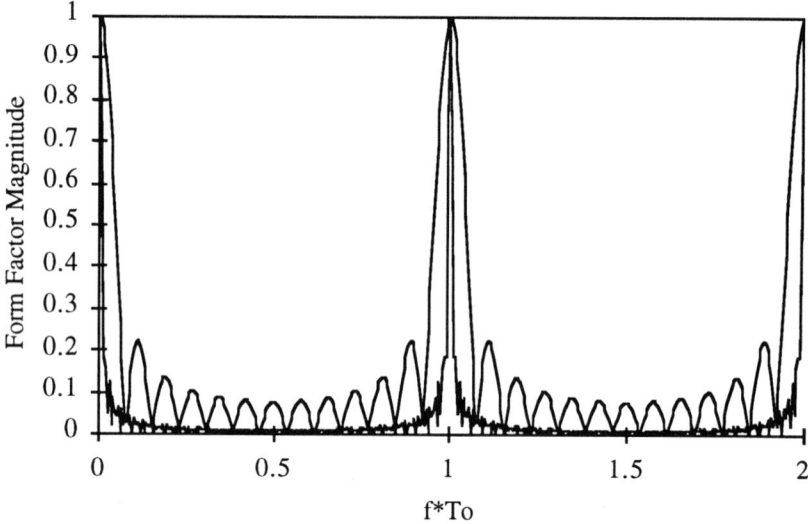

Figure 1: Sum of exp[iωmT$_0$] from m=1 to m=M (the total number of bunches) divided by the number of bunches. The top horizontal line is M=1, the wider modulated curve is M=13, and M=84 is the narrowest curve.

The shape of the curve described by equation (7) for M=1 is almost identical to that of equation (4) in the regime where the rms bunch length σ_τ is less than 1/6th the length of the RF bucket T_0. Figures 2 and 3 compare equations (4) and (7) at the bunch passage frequency and 3 times that frequency, respectively.

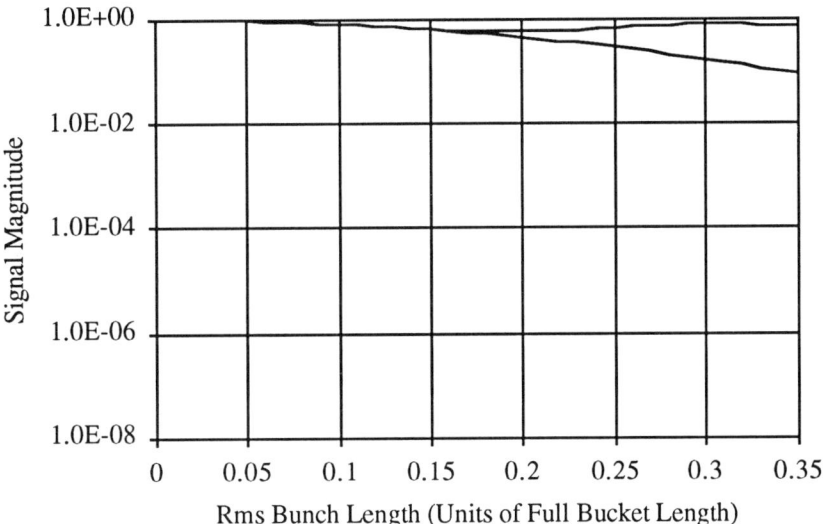

Figure 2: Spectral current density vs σ_τ/T_0 for $\omega=\omega_0$. The lower curve is from equation (4), and upper from equation (7).

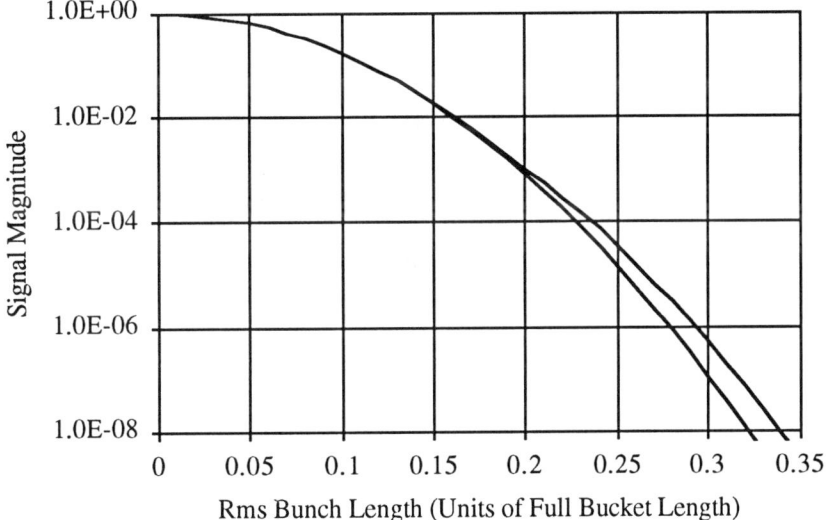

Figure 3: Spectral current density vs σ_τ/T_0 for $\omega=3\omega_0$. The lower curve is from equation (4), and upper from equation (7).

REAL TIME BUNCH LENGTH MONITOR

In order to build a real time bunch length monitor, it is necessary to measure the value of the curve described by equation (4) at two different frequencies. The first frequency is the RF frequency, since the spectral current at this frequency (in units of DC beam current) is relatively insensitive to bunch length (see figure 2) and is independent of the number of bunches in the accelerator. The second frequency is chosen such that $\sigma_\tau \omega \approx 1$, so that the influence of tails is minimized while providing a measurable difference from the current at the lower frequency (see figure 3). Again, the exact frequency of the second line should be a harmonic of the RF frequency to insure that the spectral current (normalized to the DC current in the accelerator) is independent of the number of bunches.

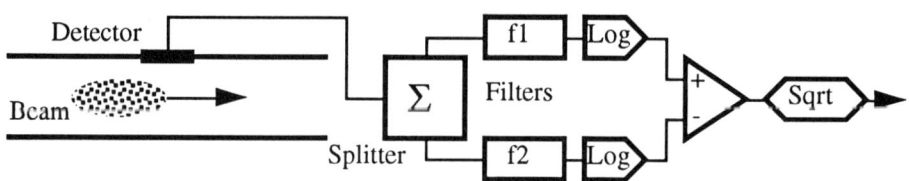

Figure 4: Schematic sketch of a standard real time bunch length monitor based on measuring the ratio of amplitudes at two frequencies of the bunch current spectrum.

As shown in figure 4, one uses a longitudinal, broadband beam current monitor of some type to measure the beam signal. Splitting this signal and filtering each half at the RF fundamental frequency and harmonic N of the RF frequency ω_0, the amplitude difference of two logarithmic detectors[1] yields an intermediate signal

$$S = Ln\left[I_b(\omega_0)\right] - Ln\left[I_b(N\omega_0)\right]$$

or

$$S = \left[-\tfrac{1}{2}\sigma_\tau^2\omega_0^2\right] - \left[-\tfrac{1}{2}N^2\sigma_\tau^2\omega_0^2\right] \tag{8}$$

Solving for σ_τ and performing an analog square root function using an Analog Devices AD532 semiconductor chip, the bunch length is described by the equation

$$\sigma_\tau = \frac{T_0}{2\pi}\sqrt{\frac{2}{(N^2-1)}}\sqrt{S} \tag{9}$$

Therefore, a real time (≈ 10 kHz analog bandwidth using the AD532) bunch length monitor insensitive to beam current and number of bunches as been created. Figure 5 is the result of applying equations (8) and (9) to the data in figures 2 and 3. All that is required is to measure two components of the beam spectral current. Monitors based on this principle are used in most of the Fermilab accelerators[2]. Despite the assumption that the beam profile is Gaussian, remarkably good agreement with the rms size of the actual waveform is generated.

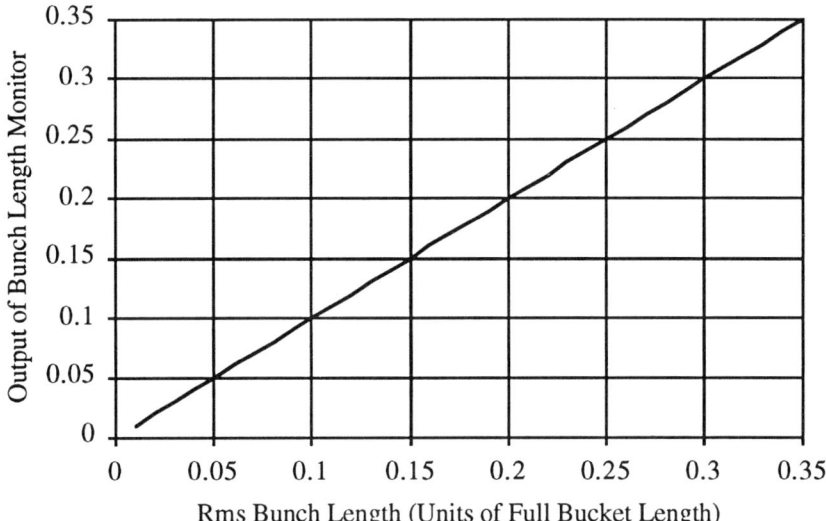

Figure 5: Calculated bunch length vs. input length using equations (8) and (9). Actually, the signals applied to equations (8) and (9) were generated using both equations (4) and (7), and are both plotted (and are completely indistinguishable).

ELECTROMAGNETIC BUNCH LENGTH MONITOR

According to equation (4), to measure the length of a bunch which has a length of one picosecond or less, frequencies on the order of $1/2\pi\sigma_\tau \cong 200$ GHz must be detected. Since structures resonating at such high frequencies are so small, they must usually be attached to the wall of the beam pipe at a port through which electromagnetic energy can be coupled. Figure 6 contains a sketch of a possible geometry for a bunch length monitor.

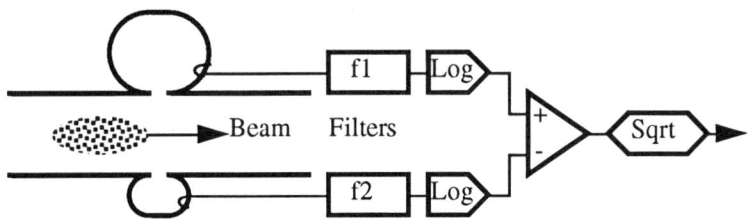

Figure 6: Sketch of a possible bunch length monitor in which two cavities with greatly different resonant frequencies are coupled to the image currents on the inside wall of the vacuum chamber via slots.

The problem with this scheme when employed on bunches from electron guns is the distribution of electromagnetic fields from a non or mildly relativistic charge distribution. The approximate rms width of the longitudinal distribution of image charges on a perfectly conducting wall of radius R from a charge of relativistic energy γ and relativistic velocity β is

$$\sigma_w \cong \frac{R}{\gamma\beta c}$$

(10)

Figure 7 contains a sketch of the electric field lines (and surface charge distribution) generated by a single particle. Because the beam pipe is assumed to be perfectly conducting, the longitudinal component of the electric field lines must be zero at the surface.

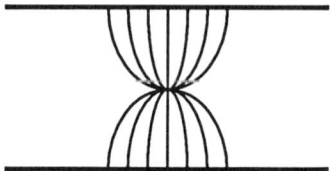

Figure 7: Field and image charge density distribution on the inside surface of an infinite conductivity beam pipe.

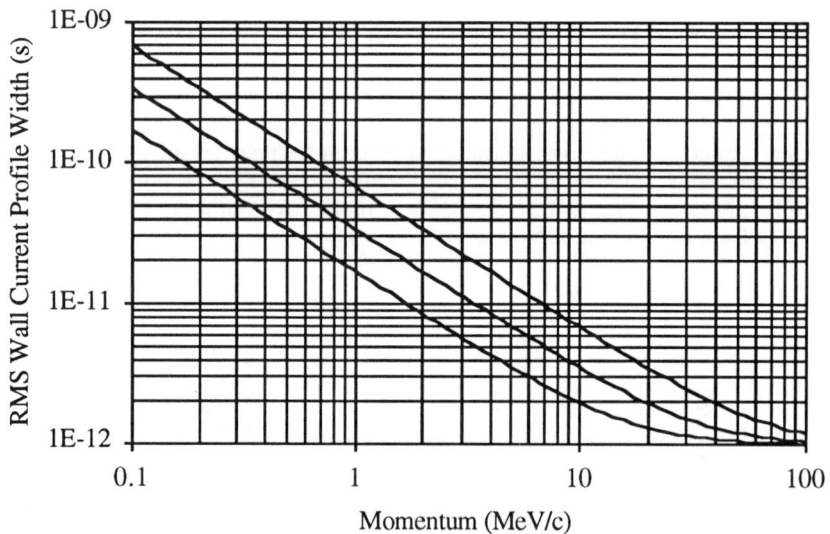

Figure 8: Effective bunch length measured at the beam pipe wall for a 1 psec bunch at different energies. The beam pipe radii were 4 cm (top), 2 cm (middle), and 1 cm (bottom curve).

Using elementary kinematic relationships, the product of energy γ and velocity β can be rewritten as the beam momentum P_0 divided by the mass of the electron m_0. Convoluting over the actual bunch length, the effective bunch length measured by any detectors at the surface of the pipe is

$$\sigma_e = \sqrt{\sigma_\tau^2 + \left(\frac{m_0 R}{c P_0}\right)^2} \quad . \quad (11)$$

Figure 8 contains a plot of this effective signal width as a function of beam momentum assuming a 1 picosecond rms beam width and a typical beam pipe radii of 1, 2 and 4 cm. Below a momentum of 10 MeV/c it becomes very difficult to correct the measurement value to recover the true beam width. Therefore, for RF guns whose output momentum is significantly less than 10 MeV/c, it is necessary to find some other means of measuring a spectral signature of the beam.

TRANSIENT SYNCHROTRON RADIATION SPECTRUM

In the body of a dipole magnet the spectrum of the synchrotron radiation is determined by the time the observer is illuminated by the Lorentz contracted light cone swinging by following the trajectory curvature of the magnet. In the case where the length of the magnet field is short in comparison with this "body" illumination time, frequency spreading of the synchrotron radiation spectrum occurs[3].

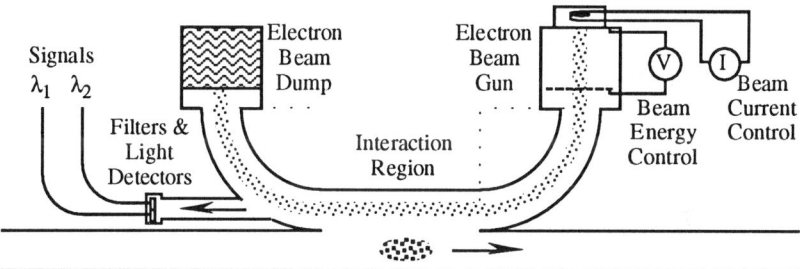

Figure 9: Proposed bunch length monitor using transient synchrotron radiation generated by the sharp beam-beam deflection of the picosecond bunch passing the low energy electron beam.

The transverse beam-beam deflection generated by a picosecond long bunch on a low energy electron beam can be described by an effective dipole field of peak magnitude B_0 and longitudinal position dependence

$$B(z) = B_0 e^{-\frac{z^2}{\sigma_\tau^2}} \quad . \quad (12)$$

The loss of the factor of two in the denominator of the exponential argument comes from the fact that the picosecond long bunch is moving and not a fixed magnet. The spectral density of the light from such a deflection is[4]

$$I(\lambda) = \left(\gamma B_0 c \sigma_\tau\right)^2 \exp\left[-\left(\frac{c\sigma_\tau}{2\gamma^2\lambda}\right)^2\right] \tag{13}$$

The argument of the exponential determines the range of energies required by the electron beam in the monitor to generate light in the wavelength band of available detectors (1-10 µm). Figure 10 is a plot of the energy of the monitor beam required to make the argument of the exponential in equation (13) unity as a function of the length of the picosecond long bunch under test. For sub-picosecond applications a very straightforward gun with an output energy of 5 MeV or less could be used for this monitor. The only outstanding questions are technical: Should one use head-on beam-beam collisions or impact parameters of 2σ of the test bunch? What detector technology is required to measure the fluxes from such a monitor?

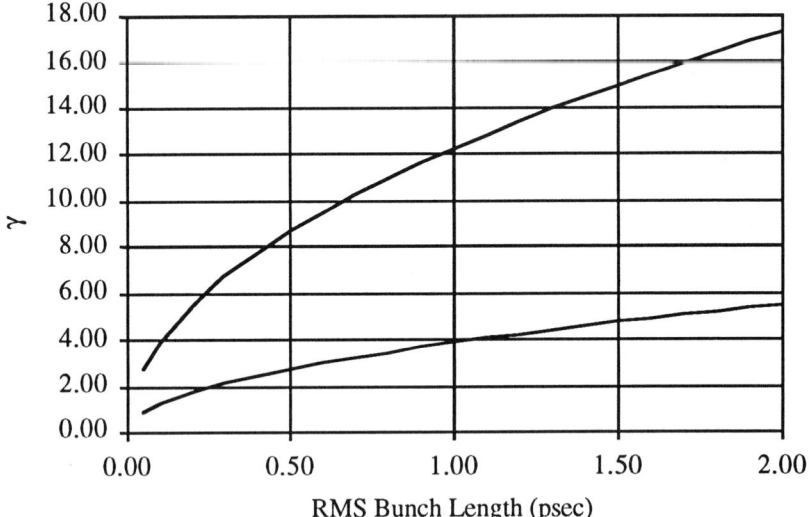

Figure 10: Relativistic energy γ needed to make the argument of the exponential in equation (13) unity for observation wavelengths of 1 µm (top) and 10 µm (bottom curve). The top curve would suggest that a monitor beam of 10 MeV is sufficient to create a useful bunch length detector.

OPTICAL TRANSITION RADIATION

When a relativistic beam traverses a thin foil, optical transition radiation[5] is produced. The frequency spectrum of the forward and backward propagating radiation is identical to that of the beam current[6]. Coupling the electromagnetic radiation from this transition through the metal surface of the foil into two resonant detectors at 200 GHz and 20 GHz (for example), one could construct the same bunch length monitor shown in figure 6 but without the pulse spreading phenomena caused by the low energy nature of some of the beams which need to be measured.

CONCLUSIONS

A variety of techniques have been reviewed for the measurement of picosecond bunch lengths. They have all been based on the highly successful monitors used at Fermilab which provide real time bunch length information. Though they do not provide full pulse shape waveform and the answer can have some other meaning (FWHM, 2 sigma, etc...) for other nonGaussian beam shapes, the availability of a single real time "figure of merit" called the bunch length is a real boon to those tuning and studying accelerator performance.

ACKNOWLEDGMENTS

The author would like to acknowledge the input Alan Hahn, Jamie Rosenzweig, Marc Ross, and Gil Travish. These discussions led to the extrapolation of the Fermilab real time bunch length monitor technology to sub-picosecond bunches.

REFERENCES

1. T. Ieiri, R. Gonzalez, and G. Jackson, Proc. 7th Symposium
   on Acc. Sci. and Tech., Osaka, Japan, 367 (1989).
2. G. Jackson and T. Ieiri, Proc. IEEE Part. Acc. Conf., Chicago, 863 (1989).
3. J.D. Jackson, Classical Electrodynamics (Wiley).
4. R. Coisson, Phys. Rev. A20, 524-28 (1979).
5. V.L. Ginzburg and I.M. Frank, Sov. Phys. JETP 16, 15 (1946).
6. W. Barry, Proc. Workshop Adv. Beam Instr., KEK, 224 (1991).

A High Resolution, Single Bunch, Beam Profile Monitor

J. Norem*

Argonne National Laboratory, Argonne, IL, 60439, USA

INTRODUCTION

Efficient linear colliders require very small beam spots to produce high luminosities with reasonable input power, which limits the number of electrons which can be accelerated to high energies[1]. The small beams, in turn, require high precision and stability in all accelerator components. Producing, monitoring and maintaining beams of the required quality has been, and will continue to be, difficult. A beam monitoring system which could be used to measure beam profile, size and stability at the final focus of a beamline or collider has been developed and is described here.[2] The system uses nonimaging bremsstrahlung optics. The immediate use for this system would be examining the final focus spot at the SLAC/FFTB.[3]

The primary alternatives to this technique are those proposed by P. Chen / J. Buon,[4] which analyses the energy and angular distributions of ion recoils to determine the aspect ratio of the electron bunch, and a method proposed by Shintake,[5] which measures intensity variation of compton backscattered photons as the beam is moved across a pattern of standing waves produced by a laser.

METHOD

The system, Figure 1, consists of a Bremsstrahlung radiator at the focus of the electron beam, a single sided collimator to produce a bremsstrahlung shadow, and a slit and detector system to measure the shape of the shadow edge.[2] The diagnostic slit could be either tilted, as shown, or parallel with the primary collimator. The sharpness of the shadow is inversely proportional to the size of the spot at the bremsstrahlung source. Sweeping magnets and shielding are required to disperse and absorb electron and photon backgrounds. The linear dimensions are not critical. The bremsstrahlung photons would be detected using a Cerenkov counter preceeded by a pair converter.

The optics of the system can be shown by plotting beam phase space at the focus, at the collimator and at the detector downstream, shown in Fig. 2. By moving the detector

*Work supported by the U. S. Department of Energy, Division of High Energy Physics, Contract W-31-109-ENG-38.

slit back and forth it is possible to measure the shadow width, and hence the beam profile. Moving the primary collimator alters the initial x' values, thus the whole phase space density distribution, $\rho(x, x', y, y', t)$, can be measured.

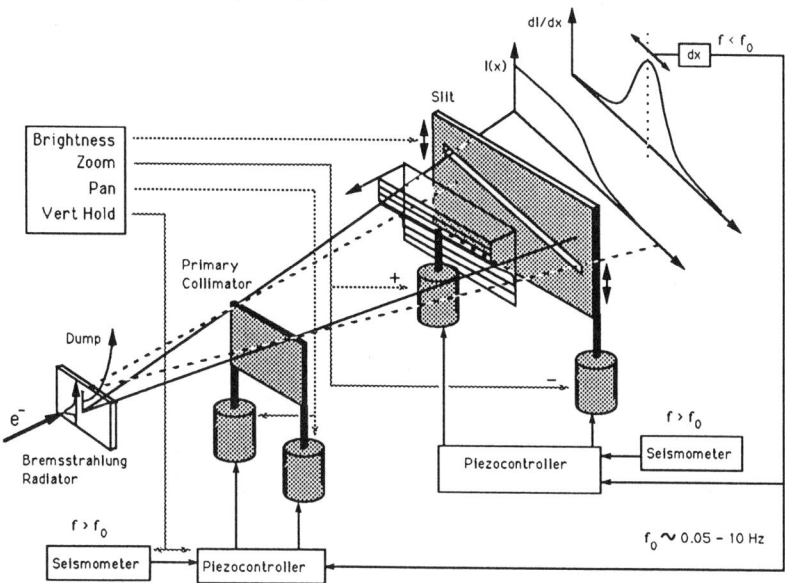

□ Figure 1. The system consists of bremsstrahlung radiator, collimator and slit. Detectors and shielding are not shown.

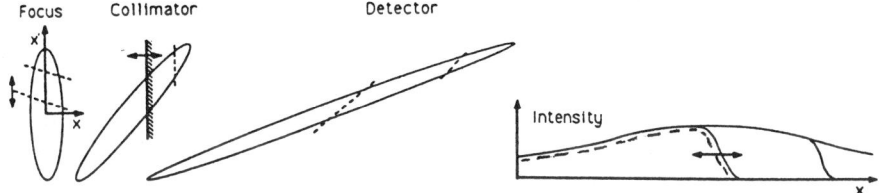

□ Figure 2. Phase space of the system, not to scale. The collimator produces an edge in the phase space distribution which is measured downstream. A projection in x space shows the observed profile.

LIMITATIONS

The ultimate resolution of this system is limited by Fresnel diffraction.[6] This limit can be approximated by considering a virtual slit at the primary collimator location, where the virtual slit width is such that the sagitta is equal to λ, the photon wavelength. If the source to collimator distance is a, the collimator to detector distance is b, and $b >> a$, the expression for the sagitta $\lambda = s^2/2a$ gives the virtual slit width, $s = \sqrt{2a\lambda}$, Fig 3. The angular diffraction width is then $\sim \lambda/s$ and the limiting resolution is

$\sim (\lambda/s)a = \sqrt{\lambda a/2}$, roughly the geometric mean of the beamline dimensions, (1 - 10 m), and the photon wavelength , ($\sim 10^{-16}$ m).

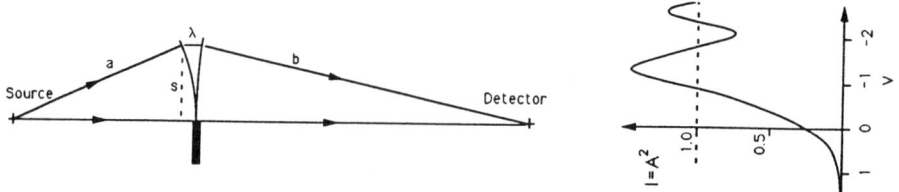

□ Figure 3. Fresnel optics. A point source produces a penumbra of finite width.

The precise calculation the diffraction pattern is done for monochromatic light in most optics books. Integrating the intensity from all paths requires solving the Fresnel integrals, the solution of which generates the Cornu spiral. If the intensity of diffraction pattern on a screen is given by $F(y)$, the width of the pattern is a function of

$$y = v\sqrt{\frac{\lambda a(a + b)}{2b}}, \tag{1}$$

which gives the dimensions of diffraction patterns on a screen in terms of v, the dimensionless variable used to evaluate the Fresnel integrals, and $a = 30$, $b = 30$ m, and λ.[6] The resolution of the system is determined by incoherently adding the diffraction images produced by the bremsstrahlung spectrum as seen by the detector. The acceptance of the detector has been evaluated using EGS4 and and a more specialized monte carlo program which generates a bremsstrahlung spectrum, computes pair production and subsequently evaluates multiple scattering.[7] The detected FFTB spectrum, which depends somewhat on position and angle cuts, is shown in Fig. 4 for minimum detected electron energy of 15 MeV and maximum angle of 2°. This note defines a resolution function as the derivative of this sum of diffraction images and this is shown in Fig 5, for the highest resolution possible with the SLAC line. This curve, which is nongaussian, is the effective shape of a beam at the bremsstrahlung radiator in the limit of a zero width slit at the detector. The width of the resolution function depends on energy, $\delta x, y \sim E_\gamma^{-1/2}$, so high energy photons contribute most to the resolution.

The collimators must have hard edges for optical diffraction, to prevent showers from leaking thru. Since photons are attenuated like e^{-z/L_R}, thick, high Z mirrors (W or Ta), with short radiation length, L_R, should permit very little transmission, $e^{-12} \sim 10^{-5}$. In addition, the body of the primary collimator and slit must be shaded (shielded) to reduce the absorbed heat from showers, and this would significantly reduce the incident and transmitted photon flux.

An option for the bremsstrahlung detector is shown in Fig 6, with pair converter followed by Cherenkov radiator. Sweeping magnets may be required to reduce shower background. The total number of pair produced leptons $n_{e,pair} = n_{e,primary}(l/L_R)\eta(\phi/\sigma_{x',\gamma})$, where n_e is the number of electrons, l/L_R is the thickness of the bremsstrahlung radiator in radiation lengths ,L_R, η is the number of electrons detected for one equivalent full energy photon on the detector, which must be calculated by monte carlo and is roughly 1 - 10 depending on detector, $\phi/\sigma_{x'_\gamma}$ is the acceptance of the detector slit divided by the divergence of the photon beam. In fact, the monte carlos calculate the flux of pair produced leptons of desired angle and energy for a given number of incident beam electrons, integrating over bremsstrahlung angle.

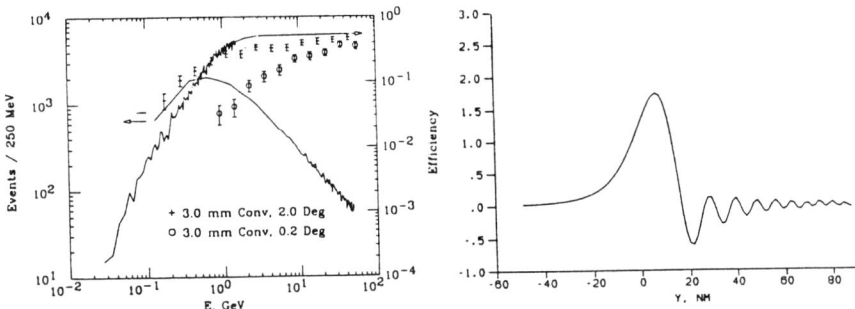

□ Figures 4 and 5. Monte Carlo simulations of the detection efficiency and photon spectrum for different pair acceptance angles, along with the resolution function for ($E_e = 50$ GeV, a=2, b=20m, $\theta_{pair} = 0.2°$).

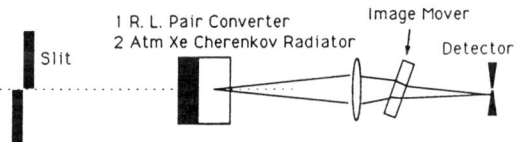

□ Figure 6. One option for a bremsstrahlung detector. Sweeping magnets and distance may be useful in separating the low emittance signal from showers.

The number of detectable photons would then be $n_\gamma \sim (150_{[1/cm]} n_{e,pair} L_{[cm]} \sin^2 \theta_C$, [8] which would yield ~ 2000 photons in one Fresnel half width. It is assumed that Xe gas at 1 atm can be used as the Čerenkov radiator. With a refractive index $n = 1.00071$, the opening angle of radiation is 2.1°, and the minimum detectable electron energy is ~ 12 MeV. This paper assumes that the pair converter is U 3.6 mm thick and the Č radiator is 2 cm thick, and the combined width due to pair production / shower dimensions and Čerenkov optics is $\sim 100\mu$.

It is somewhat difficult to determine the longitudinal position of the waist since this system only sees roughly horizontal slices of the phase space at the focus. The measurement can be made by moving the primary collimator by a large distance $2a\sigma_{y'} \sim 4$ cm, measuring the position of the mean of of the penumbra at two points and extrapolating back to the focus. the error on the focal position is then $\Delta z = \sqrt{2}\delta y/\sigma_{y'} \sim 140\mu \sim \beta_y^*$, where $\delta y = 100$ nm is the measurement error.

Multiple scattering broadens the beam of electrons as they traverse the bremsstrahlung radiator. This effect can be evaluated by Monte Carlo, assuming a scattering angle of $\theta_{ms} = 13.6_{[MeV]}\sqrt{dx/L_R}/p_{e,[MeV]}$, where dx/L_R is a path length element in radiation lengths. In general the multiple scattering correction is small as long as the change in beam size due to multiple scattering, s, is small compared to the unperturbed size.

Self fields will focus the beam and cause it to radiate. At SLAC, magnetic fields $B_\phi \sim \mu_0 cQ/4\sqrt{2}\pi\sigma_x\sigma_z \sim 50T$ will produce deflections, $\theta_{f,max} \sim B_\phi l/B\rho = 0.0002$. The field B_ϕ will cause synchrotron radiation, however the energy seen by the detector can be estimated from $(\theta_{f,max}) \times$ (loss/turn) for electrons in this field, $dE_{[GeV/turn]} = 8.8 \cdot 10^{-5} \times E_{e,[GeV]}^4/\rho_{[m]}$, which is about 0.004 J for $3 \cdot 10^{10}$ electrons. With a critical energy of $E_{c,[GeV]} = 6.6 \cdot 10^{-7} \times E_{e,[GeV]}^2 B_{\phi,[T]} \sim 91$ MeV, most photons are 100 MeV or less. The total energy lost into this radiation will be much smaller than into bremsstrahlung. Total energy for bremsstrahlung is ~ 24 J/pulse, and when the detection efficiency is considered the total energy in synchrotron photons is $\sim 0.1\%$ of the bremsstrahlung energy.

The intense electron beam will ionize the bremsstrahlung radiator and these heavy ions will then be pushed toward the median plane of the electron beam, which will significantly alter the local radiation length. Ionization has been calculated assuming $\sigma_{+,[Mb]} \sim 0.12Z + 0.2$ or Mb for Pb, [4][9] with $\sigma_{++} = \sigma_+/4$, $\sigma_{+++} = \sigma_{++}/4$.[10] Figure 7 shows the result that the Pb is fully ionized, and significant double and triple ionization also occurs. Ion motion has been calculated and is significant. The ions are accelerated to energies of about a keV and then travel toward the median plane of the beam, where they interpenetrate to create what seems to be a very high density for a short time. One effect of this ion motion will be a hole in the radiator, which must be moved for subsequent pulses. For electron pulse trains, a liquid metal jet, of the sort developed by F. Villa [11] could be used as a bremsstrahlung radiator.

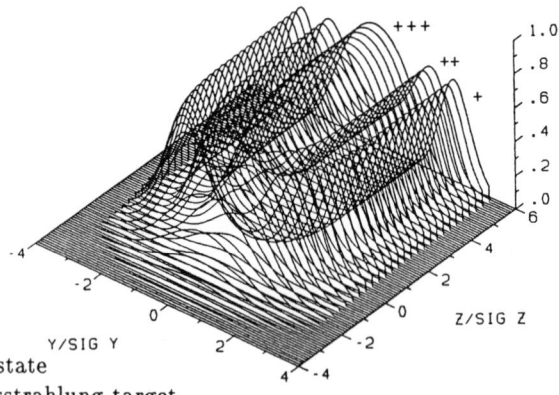

□ Figure 7. The ionization state of atoms in the heavy bremsstrahlung target.

COMPONENTS

It is assumed that high Z materials will be used for the foil, and that these will be locally destroyed by the 1 mJ of ionization energy deposited on every pulse. If a large foil area was inserted and loosely mounted from some point upstream or downstream of the focus, mechanical rastering the foil should be sufficient to provide a new part of the foil for every bunch, while maintaining the position along the beam line. The foil thickness should be less than β^*.

Precision collimators can be made from commercially available metal mirrors, which are described in catalogues with a flatness of $\sigma < \lambda/20 \sim 24$ nm, and surface roughness of better than 1 nm. Synchrotron light sources require microroughness in the range of 0.2 nm and slope errors in the range of 0.1 arc second. Metal mirrors with dimensions of $\sim 5 - 10$ cm, optically polished to a few nm surface roughness, would cost about $1000 - 2000, and take about 10 weeks to fabricate.[12][13]. With glass mirrors, requiring $\sigma < \lambda/200$ adds about 50% to the cost relative to $\lambda/40$.

Since the primary collimators must be thermally and mechanically stable and must be moved with high precision, it is difficult to simultaneously design them to absorb significant beam power. The guard collimators on the other hand should be able to absorb this power, while maintaining alignment only to the level of $\sim 0.1\mu$. Rough alignment can be done with standard techniques using transit and levels. Alignment of the collimator surfaces directly parallel to the beam can be done with optical lenses and prisms, which have angular tolerances of $\pm$ 30 arc seconds, or 0.14 mrad. More precise alignment of the collimators would probably require the beam on target. Slits can be produced using two single collimators, offset to eliminate collisions.

Rough positioning of collimators can be done with a number of commercially available systems, such the the Nanomover sold by Melles Griot, which can set 20 kg loads with ±100 nm resolution over 25 mm.[12] The primary collimator and final slit would have to be more carefully positioned, possibly to tolerances of ±1 nm. It is assumed that the collimators will each be controlled with three actuators, and the precision adjustment of these would be done in real time. Mirrors can be mounted to the structure in a number of ways, a compliant mount, such as pitch, might be desirable.

Since ground vibrations occur at the level of about $\sigma_v \sim 0.035\mu, \sigma_h \sim 0.1\mu$, the beam defining collimators must at least be stabilized against the vertical motion.[15] Although other options are available, The Streckeisen STS-2, while somewhat expensive, is sufficiently sensitive, is linear in amplitude. The velocity and phase response are straightforward for periods from 0.03 to 300 seconds, which spans the low frequency range where ground vibration is largest. An open loop correction system using this system should be able to correct the collimator and/or slit positions to better than 1 nm over the range 0.1 to 10 Hz.

Additional shielding is required for a number of problems: backgrounds in the detectors, heating on the collimators and support frames, activation of the seismometers, which must be installed near the collimators, and minimizing radiation levels outside the shielding. In principle these are problems which can be solved using standard procedures. Considerable spray is produced by the electron beam in the target and this will be a significant source of shower secondaries.

Since the FFTB is being shielded for ~ 2.5 kW of beam power and the bremsstrahlung radiator is anticipated to be 0.1 radiation lengths thick, it is likely that the shielding and cooling of the collimator assemblies must be able to deal with about 250 W of beam power. If low expansion materials are used in structural applications, thermal expansion of components could significantly affect the alignment, critical components must be designed either to absorb power or to avoid the beam spray. It is anticipated that local shielding can protect the majority of the apparatus from the scattered beam however some moveable guard collimators will likely absorb significant heat which will require a total cooling capacity of ~ 0.25 kW.. In a simple test, a piezocrystal was used to measure the vibration excited by the water flow in a 3/8" plastic tube from a 1kW cooling unit. Although large amplitude vibrations were seen from a large number of sources, (movement, pumps etc), the cooling water flowing thru a water fitting caused less than 1 nm of motion.

EXPERIMENTS

A preliminary experiment is being done using the low energy beams of the APS injector at Argonne to help optimize the detector and mechanical systems. This test will look at the beam on the positron production target from downstream of the positron linac. This test should permit optimization of shielding, detectors, mechanical systems and control algorithms.

The first real test should be on the SLAC/FFTB, where beam sizes of 60 nm will be produced. This size is comparable with the spot sizes required by the TESLA design, but somewhat larger than spotsizes required by some collider designs[16] [1]. Other possible applications of the proposed system would be plasma focussing experiments, measurements of beams from nonlinear QED experiments, measurements of drifts and jitter, and multibunch stability.

COLLIDER

If this technique is used for single beams in the final detector of a linear collider it will be necessary to insert a thin radiator of some kind near the center of the large high energy physics detector, Fig 8.[17] Ideally this radiator can be quite small and light, and, since the z position needs to be determined only to some fraction of β^*, the required positioning tolerances are not challenging. While the design of the target holder, collimators and detector would pose significant problems, the most significant interaction with the detector design may be the spray from the bremsstrahlung target.

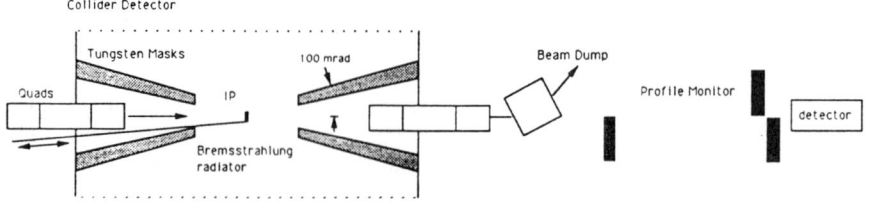

□ Figure 8. Schematic of operation with a collider detector.

While a high Z bremsstrahlung target at the IP would be a copious source of shower secondaries and radiation damage, the magnitude and composition of these showers should be somewhat similar to showers produced by beamstrahlung from e^+/e^- collisions. In normal operation, beamstrahlung will convert a significant fraction, δ, of the total energy of beam bunch into secondaries. Various designs give $\delta = 0.1 - 0.3$ for high frequency, (12GHz), options,[1], and $\delta = 0.01 - 0.1$ for low frequency options, (1.5 GHz), such as TESLA[16]. By comparison a bremsstrahlung target would approximate an energy loss of $\delta < \beta^*/L_R \sim 0.03 - 0.1$. Designs for collider detectors include tungsten masks to prevent low energy beamstrahlung secondaries emitted at angles of less than 100 mrad from entering the detector and these masks should also be useful for bremsstrahlung.

In principle, beamstrahlung from collisions could be used to provide direct images of the colliding bunches using this technique. It may be difficult, however, to get high statistics and high resolution if the collider is run in a mode where the average beamstrahlung photon energy is a small fraction, Υ, of the incident electron energy. Values of Υ vary widely, $\sim 0.01 - 0.6$, from design to design, and should be much lower during tune up, but the method should be useful.

CONCLUSIONS

A single bunch, beam profile monitor using bremsstrahlung should be capable of very high resolution when used with the high energy electron beams associated with linear colliders. The technique uses comparatively simple optics and inexpensive components and seems compatible with operation in linear collider detectors.

REFERENCES

[1] R. B. Palmer, Annu. Rev. Nucl. Part. Sci. **40**, 529, ('90)

[2] J. Norem, Proceedings of the 1988 Linear Accelerator Conf., Newport News, Virginia, (1988), p430

[3] K. Oide, Proc. 1989 Particle Accelerator Conf, Chicago, 1989, p1319

[4] J. Buon, F Couchot, J. Jeanjean, F. Le Diberder, V Lepeltier, H. Nguyen Ngoc, J. Perez-y-Jorba, M. Puzo, and P. Chen, Nucl. Instrum. and Methods., **A306**, (1991), 93

[5] T. Shintake, Nucl. Instrum. and Methods, **A311**, (1992), 453

[6] Jenkins and White, *Fundamentals of Optics*, Fresnel Diffraction, McGraw-Hill (1957)

[7] H. A. Bethe, in Experimental Nuclear Physics Vol 1, E. Segre, ed., John Wiley and Sons Inc. New York, (1953), p166

[8] J. Litt and R. Muenier, Annu. Rev. Nucl. Part. Sci., **23**, 1, ('73)

[9] P. Chen, Contribution to NLT Workshop SLAC, (1988)

[10] B. Rossi, High Energy Particles, Prentice-Hall Inc, Englewood Cliffs N. J., (1952)

[11] F. Villa, (SLAC, private communication, 1991)

[12] Melles Griot, Irvine Calif. 92714

[13] R Lowrey, Rockwell Power Systems, Albuquerque NM (private communication)

[14] J. B. Rosenzweig, P. Schoessow, B. Cole, C. Ho, W. Gai, R. Konecny, S. Mtingwa, J. Norem, M. Rosing, and J. Simpson, Phys. Fluids B **2**, 1376, (1990)

[15] R. E. Ruland, G. E. Fischer, Proceedings of the 2nd International Workshop on Accelerator Alignment, Hamburg, Germany, Sept 10-12, 1990. **and** Stanford Linear Accelerator Laboratory publication SLAC-PUB-5326 (1990)

[16] Proc. of the 1st TESLA Workshop, Cornell U, Ithaca, NY, CLNS 90-1029 (1990), ed H Padamsee.

[17] Schroeder, D. V., Stanford Linear Accelerator Center Report SLAC-371, SLAC, Stanford, CA, (1990)

Summary of the Physics Opportunities Working Group[*]

Pisin Chen

Stanford Linear Accelerator Center
Stanford University, Stanford, California 94309

Kirk T. McDonald

Joseph Henry Laboratories,
Princeton University, Princeton, NJ 08544

ABSTRACT

The Physics Opportunities Working Group was convened with the rather general mandate to explore physics opportunities that may arise as new accelerator technologies and facilities come into play. Five topics were considered during the workshop: QED at critical field strength, novel positron sources, crystal accelerators, suppression of beamstrahlung, and muon colliders. Of particular interest was the sense that a high energy muon collider might be technically feasible and certainly deserves serious study.

INTRODUCTION

A Working Group was convened with the rather general mandate to explore physics opportunities that may arise as new accelerator technologies and facilities come into play. The agenda was set by the interests of the participants,[1] many of whom were inspired to give extemporaneous presentations of ideas they had not expected to discuss, but which ideas had been quietly nurtured awaiting a forum such as this Working Group. The working group considered five topics:

1. QED at critical field strength.

2. Novel positron sources.

3. Crystal accelerators.

4. Suppression of beamstrahlung.

5. Muon colliders.

In the following we attempt to give a flavor of the discussion on each of these topics.

★ Work supported by the Department of Energy, contract DE–AC03–76SF00515 and grant DE–FG02–91ER40671.

QED AT CRITICAL FIELD STRENGTH

K. McDonald reviewed how the combination of low-emittance high-energy electron beams with tabletop teraWatt lasers offers the opportunity to explore QED beyond the critical field strength,

$$E_{\text{crit}} = \frac{m^2 c^3}{e\hbar} \approx 10^{16} \text{ V/cm}, \tag{1}$$

at which the vacuum is unstable against pair creation.[2] A speculative possibility is that a QED phase change occurs that could lead to structure in the $e^+ e^-$ mass spectrum related to the positron peaks reported in heavy-ion collisions at Darmstadt.

A. Varfolomeev recalled that teraWatt lasers could lead to a demonstration of light-by-light scattering using the technique of four-wave mixing. A note discussing this further appears in these proceedings.[3]

A major theme of the Working Group was the production of high-quality, high-energy lepton-lepton collisions in future accelerators. It is anticipated that it will be difficult to maintain a well-defined center-of-mass energy in $e^+ e^-$ collision once the electromagnetic fields of a bunch exceed the QED critical field as observed from the oncoming bunch. The resulting radiation is generally called beamstrahlung.

P. Chen reviewed three aspects of beamstrahlung that will limit the performance of future $e^+ e^-$ colliders. First, the bunches on the average radiate a substantial fraction of their energy once the so-called beamstrahlung parameter

$$\Upsilon = \frac{2\gamma E_{\text{bunch}}}{E_{\text{crit}}} \sim 1. \tag{2}$$

Nevertheless, a good fraction of the bunch particles remain at the initial energy, thanks to the quantum nature of the radiation. For example, with $E_{\text{cm}} = 500$ GeV and $\Upsilon = 0.12$ some 70% of the $e^+ e^-$ collisions still have more than 98% of the nominal center-of-mass energy, although the average energy loss per beam particle is 10%.[4] However, running with Υ much higher than this value would result in a spread of center-of-mass energies rather like that for quark-quark and gluon-gluon scattering at a hadron collider.

Second, once $\Upsilon \gtrsim 1$ there is copious production of $e^+ e^-$ pair by a two-stage process, as discussed earlier.[5] The rates are an extremely rapid function of Υ so the pair creation is negligible for $\Upsilon < 0.2$. Third, the light-by-light scattering processes that lead to $e^+ e^-$ pair creation also lead to gluon-gluon and quark-antiquark pair creation. The most annoying feature of this is radiation of soft-gluon jets ('minijets') which spew low-P_t hadrons into the collision region.[6] While only a few percent of the collisions lead to minijets for $\Upsilon \approx 1$, the number of low-P_t particles in such events is large. Any detector must be effectively blind to these particles, which may compromise the capability for low-P_t physics which has been such a rich source in present $e^+ e^-$ machines.

NOVEL POSITRON SOURCES

The copious production of positrons at critical field strength might be an excellent source of positrons for future linear colliders in a proposal by P. Chen and R. Palmer.[7] In e-laser collisions the QED strong-field processes become prominent once parameter $\Upsilon \gtrsim 1$ where

$$\Upsilon = \frac{E^\star}{E_{\text{crit}}} = \frac{2\gamma E_{\text{laser}}}{E_{\text{crit}}} \tag{3}$$

($E^\star$ is the laser field strength in the electron's rest frame). Pair creation occurs in a two-step process: laser photons collide with high-energy electrons to produce high-energy backscattered photons; then the high-energy photons collide with laser photons to produce e^+e^- pairs.

For $\Upsilon > 1$ the interaction probability approaches unity, and the created positrons and electrons reinteract with the laser to produce an electromagnetic cascade. There are two important aspects in such an approach: First, there exists a threshold at $\Upsilon \sim 1$ in the coherent pair creation process. This helps to accumulate positrons at a low, but finite, energy with a small energy bite. Second, a remarkable fact is that the angular distribution of the positrons closely follows that of the incident electrons, $i.e.$, geometric emittance is preserved. But since the positrons have much lower energy, their invariant emittance is lower than that of the electrons. That is, 'cooling' automatically occurs!

If an optical laser is to be used, the initial electron energy is optimized at 250 GeV. To use 50-GeV electrons, the laser should have $\approx$ 40-nm wavelength. This is strong motivation for a far-UV free-electron-laser program at SLAC.

P. Channell and D. Cline presented two variations on schemes for production of low-energy positrons in p-nucleus interactions. The mechanism is that the proton is absorbed by the nucleus,

$$p + (Z, N) \rightarrow (Z + 1, N) + \gamma, \tag{4}$$

which then decays via positron emission:

$$(Z + 1, N) \rightarrow (Z, N + 1) + e^+. \tag{5}$$

For 5-MeV protons the capture cross sections are of order 300 mb, and the beta-decay times of order 10-100 sec. The positrons emerge with a few-MeV energy.

For high production rates the target must be dense, but for efficient positron extraction the $(Z+1, N)$ nuclei should be in a diffuse medium. Hence the $(Z+1, N)$ nuclei must be separated from the (Z, N) nuclei via chemical or isotope separation. The $(Z + 1, N)$ nuclei could then be stored in a magnetic bottle until they decay.

Small-scale versions of such a scheme are presently implemented as sources for positron emission tomography. Significant R&D is required to achieve production rates of 10^{14}/sec as desired for high-energy physics.

CRYSTAL ACCELERATORS

W. Gabella reviewed suggestions for plasma accelerators in crystals.[8] Here the plasma consists of the conduction electrons of the crystal. In one scheme a acoustic standing wave established by a transducer induces a spatial periodicity to the electron density. Then a pump laser whose wavelength is the same as that of the acoustic wave excites electrons into a plasma oscillation.[9] The crystal must be largely transparent to the laser, so the plasma frequency of the crystal electrons must be less than the laser frequency. A (properly phased) charged-particle beam passing through the crystal can then extract energy from the plasma oscillations.

The use of a crystal is of interest because of channeling, which works best for positively charged particles. In principle very good emittance preservation can be maintained during acceleration.

It appeared that a demonstration of this technique might be possible at the BNL Accelerator Test Facility with 50-MeV electrons and the 10.6-μm CO_2 laser. While negatively charged particles suffer Coulomb collisions after some distance in a crystal, acceleration over a short distance should be feasible. The channeling capture angle was calculated to be 1 mrad at 50 MeV, which is well matched to the emittance of the ATF electron beam. The frequency of the CO_2 laser is $\omega \sim 2 \times 10^{14}$ Hz, so the electron density in the crystal must be $\sim 10^{19}$/cm^3 ($\Rightarrow$ arsenic??). The frequency of the acoustic wave would be about 300 MHz. For a laser intensity near the damage limit, $\sim 10^{13}$ Watts/cm^2 the accelerating gradient would be a few MeV/cm.

Another scheme was presented by S. Bogacz[11] in which a crystal has a periodic strain to form a kind of undulator. Such a crystal could be then used in FEL and inverse FEL configurations, but with characteristic wavelengths much shorter than by other means.

BEAMSTRAHLUNG SUPPRESSION

In this context any additional mechanisms that could suppress the beamstrahlung would be most welcome. A. Sessler reviewed a proposal for beamstrahlung suppression using a plasma at the interaction point.[12] According to Lenz' law we expect charge separation and return currents to be induced in the plasma so as to cancel the **E** and **B** fields of the colliding bunches. For good cancellation the bunch radius must be much larger than the plasma wavelength, which leads to the condition that the plasma charge density must be much larger than that in the e^+e^- bunches. For example, at a Next Linear Collider operating at 500 GeV with bunches of 10^{10} particles each 1 mm long and 100 nm in radius, the electron density is of order 10^{21}/cm^3. Hence plasma density of order 10^{22}/cm^3 are needed. This is a high number, but perhaps achievable. If so, calculation suggests that 90% of the beamstrahlung could be suppressed.

W. Barletta reviewed the possibility of generating such plasma densities by laser excitation.

However, as the energy of the e^+e^- collider rises, the luminosity must increase as s to maintain constant event rates, so the bunch size is likely to shrink. This would require ever higher plasma densities for beamstrahlung suppression, and the scheme becomes difficult for energies of a few TeV or more. In addition, the use of high density plasmas at the interaction point would introduce its own kind of backgrounds such as high energy photons from bremsstrahlung.

J. Rosenzweig reviewed the issue of instabilities in plasma compensation schemes and compared these to alternatives based on 4-beam collisions (two pairs of comoving e^+e^- bunches).[13] The conclusion is that plasma compensation is more stable against dipole (kink) and quadrupole instabilities than 4-beam schemes. The stability of the latter can be improved by tailoring the longitudinal charge-density profile of the bunches.

P. Channell presented a rather speculative "e^+e^- plasma" compensation scheme in which plasma densities exceeding $10^{24}/\text{cm}^3$ might be obtained by beam-strahlung e^+e^- pairs from auxiliary accelerators.

P. Chen reviewed the Landau-Pomeranchuk-Migdal mechanism and other LPM-like effects for suppression of radiation effects in beams. He noted that while the LPM effect is not sufficient in suppressing these raidations, the strong EM field of the opposing beam does help to suppress bremsstrahlung.[14]

It appears that beamstrahlung remains a fundamental limit to performance of e^+e^- colliders in the multi-TeV regime.

MUON COLLIDERS

The limiting effect of beamstrahlung on future colliders makes it timely to reconsider the merits of muon colliders.[15,16] From the definition (2) of the beam-strahlung parameter we see that for muons beams of the same energy and bunch parameters as electrons, Υ will be only 1/205 that for electrons. Without the disruptive effects of beamstrahlung, a muon collider would enjoy all the advantages of a well-defined purely leptonic initial states that has made experimentation at electron colliders crisper than that a hadron colliders.

D. Cline, and also D. Neuffer[16] reviewed the physics potential of muon colliders at the Higgs energy scale, 200 Gev-1 Tev. While production cross sections for quarks, leptons and vector gauge bosons are the same for electron and muon beams, a muon collider has a major advantage over either electrons or hadrons for production of Higgs bosons. Since the latter couple to the mass of the spin-1/2 beam particles, production of Higgs by muons is $(m_\mu/m_e)^2 = 4 \times 10^4$ larger than by electrons. Luminosities of only 10^{30} cm^{-2}sec^{-1} would make a muon collider very competitive for Higgs production.

R. Noble reviewed the prospects for high flux muons sources based on proton accelerators in the 50-200 GeV range.[17] A rapid-cycling proton accelerator, such as that considered for the TRIUMF II upgrade, could provide 10^{15} protons/sec. This would yield some 10^{13} pions/sec into a 1% momentum bite. As the pions decay some 10^{12} muons/sec could be collected into an normalized emittance of $\epsilon_N \approx 10^{-3}$ m-rad. These muons have a total momentum spread of about 40%, so if the muon-momentum spread is limited to 1%, the yield would be 2.5×10^{10} muons/sec, *etc.*

Various participants debated the option of muon production by stopping pions, with the conclusion that further studies are needed. In particular, there are substantial differences between positive- and negative-muon production from stopping pions.

Once a source exists, the muons must be accelerated, cooled, and brought into collision before they decay. Of course, the muons live longer at higher energy according to $\tau = \gamma \times 2.2 \times 10^{-6}$ sec. A useful result is that the lifetime of muons moving in circles under the influence of a fixed magnetic field is independent of the muon energy. In particular, the lifetimes is conveniently expressed in the number of complete revolutions, or 'turns' as

$$\text{muon lifetime in turns} = 300B[\text{Telsa}]. \qquad (6)$$

For example, if 3.3-Tesla magnets are used throughout the accelerator, the muon lifetime is 1000 turns.

D. Neuffer[16] and A. Ruggerio[18] reviewed the luminosity requirements for a muon collider, using 1 TeV $\times$ 1 Tev as an example:

$$\mathcal{L} = \frac{N^+ N^- f}{4\pi \beta^\star \epsilon} \approx 10^{30} \text{ cm}^{-2}\text{s}^{-1}, \qquad (7)$$

where N is the number of muons per bunch, f is the bunch-collision frequency, $\beta^\star_\star$ is the focusing strength, and $\epsilon = \epsilon_N/\gamma$ is the geometric emittance. At 1 TeV the muons have $\gamma = 10^4$, so their lifetime is 20 ms. This suggests the source should cycle at about $\approx$ 50 Hz. The collision region might be either single pass, or multiple pass in a storage ring.

Muon accelerators that include a storage ring to take advantage of the potential 1000 turns of muon lifetime must face a new technical challenge. As the muons decay roughly 1/3 of their energy is dumped into material close to the ring in the form of electromagnetic showers. For example, with 10^{11} 1-TeV muons/sec in the ring, some 10^4 Watts must be dissipated. For a ring of 1000 superconducting magnets, this is 10 Watts per magnet deposited in a localized region of the coil unless precautions are taken.

Possible parameters for a muon accelerator are then:

Single pass: $f = 50$ Hz; $\beta^\star = 5$ μm.

Multiple pass: $f = 50 \times 1000$ turns $= 4 \times 10^4$; $\beta^\star = 5$ mm.

$N^+ = N^- = 10^9$ per cycle.

which leads to a requirement that the geometric emittance be 10^{-10} m-rad, and the normalized emittance be $\epsilon_N = 10^{-6}$ m-rad. This is three orders of magnitude smaller than that expected out of the muon source, so cooling is required, which must be accomplished in approximately 1 msec.

Two possible schemes for muon cooling were presented. D. Neuffer[16] reviewed the method of 'ionization cooling' in which muons pass through an absorber and lose both transverse and longitudinal momentum to ionization, followed by accelerating sections in which the longitudinal momentum only is restored.

A. Ruggerio presented a new scheme based on stochastic cooling, conceived during the Workshop.[18] In this the muon beam is bunched into perhaps 10^9 bunches of 100 particles each. These bunches are transmitted through a series of (≈ 10) arcs separated by accelerating sections. In the arcs stochastic cooling is to be accomplished with very high frequency pickups and kickers. The luminosity is achieved in single pass collisions of each bunch. The invariant emittance of each bunch must be very small ($\lesssim 10^{-16}$ m-rad) in this scheme.

While the ionization-cooling scheme is relatively conservative, the stochastic-cooling scheme provoked lively discussions that continue after the Workshop. There remained very considerable enthusiasm to understand the feasibility of muon accelerators in greater detail, which led to the formation of a new Workshop for that purpose.[19]

HIGHLIGHTS

The major themes developed in each of the five topics explored by the Working Group were:

1. A proposed demonstration of induced light-by-light scattering with a three-beam configuration of a tabletop teraWatt laser.

2. A proposed low-emittance high-yield positron source at SLAC via pair creation by light from a 50-nm free electron laser.

3. A proposed demonstration of a crystal accelerator at the BNL Accelerator Test Facility.

4. Reaffirmation that beamstrahlung is a severe limit to the performance of e^+e^- colliders for $E_{cm} \gtrsim 1$ TeV.

5. A muon collider based on an aggressive cooling scheme that might provide the cleanest high-luminosity source for future high-energy physics.

REFERENCES

1. B. Barletta, A. Bogacz, P. Channell, P. Chen, D. Cline, P. Debenham, B. Gabella, A. Jackson, G. Jackson, K. Kim, N. Kurnit, A. Lebedev, K. McDonald, F. Mills, D. Neuffer, R. Noble, R. Palmer, I. Pogorelsky, J. Rosenzweig, A. Ruggerio, A. Sessler, G. Shuets, J. Simpson, D. Sutter, A. Varfolomeev, X. Wang, J. Wurtele.

2. K.T. McDonald *et al.*, *Study of QED at Critical Field Strength*, in Workshop on beam-Beam and Beam-Radiation Interactions: High Intensity and Nonlinear Effects, ed. by C. Pelligrini, T. Katsouleas and J. Rosenzweig (World Scientific, Singapore, 1992) p. 127.

3. K.T. McDonald, *Induced Scattering of Light by Light*, these Proceedings.

4. P. Chen, *Differential Luminosity under Multiphoton Beamstrahlung*, Phys. Rev. **D46** (1992) 1186.

5. P. Chen and V.L. Telnov, *Coherent Pair Creation in Linear Colliders*, Phys. Rev. Lett. **63** (1989) 1796.

6. M. Drees and R. M. Godbole, *Minijets and Large Hadronic Backgrounds at e^+e^- Supercolliders*, Phys. Rev. Lett. **67** (1991) 1189; J. R. Forshaw and J. K. Storrow, *The Minijet contribution to the Total e^+e^- Cross Section*, Phys. Lett. **B278** (1992) 193; P. Chen, T. Barklow and M.E. Peskin, *Hadronic Production in $\gamma\gamma$ Collisions as a Background for e^+e^- Linear Colliders*, SLAC-PUB/5873 (1992); submitted to *Phys. Rev.* **D**.

7. P. Chen and R. B. Palmer, *Coherent Pair Creation as a Positron Source for Linear Colliders*, these Proceedings.

8. P. Chen and R. J. Noble, *Channeled Particle Acceleration by Plasma Waves in Metals*, NATO ASI Series **B165** (1987) 517.

9. T. Katsouleas *et al.*, *A Side-Injected Plasma Accelerator*, IEEE Trans. Nucl. Sci. **NS-32** (1985) 3554.

10. P. Chen, D. B. Cline, and W. E. Gabella, *Issues Regarding Acceleration in Crystals*, these proceedings.

11. S. A. Bogacz, *Inverse FEL Proton Accelerator via Periodically Modulated Crystal Structure*, these Proceedings.

12. D.H. Whittum *et al.*, *Plasma Suppression of Beamstrahlung*, Part. Accel. **34** (1990) 89.

13. J.B. Rosenzweig, B. Autin and P. Chen, *Instability of Compensated Beam-Beam Collisions*, AIP Conf. Proc. **193** (1989) 324.

14. P. Chen and S. Klein, *The Landau-Pomeranchuk-Migdal Effect and Suppression of Beamstrahlung and Bremsstrahlung in Linear Colliders*, these Proceedings.

15. A. N. Skrinsky, *Accelerator and Instrumentation Prospects of Elementary Particle Physics*, AIP Conf. Proc. **68** (1980) 1056.

16. D. Neuffer, *Principles and Applications of Muon Cooling*, Part. Accel. **14** (1983) 75; *Multi-TeV Muon Colliders*, AIP Conf. Proc. **156** (1987) 201.

17. R.J. Noble, *Particle Production and Survival in Muon Acceleration*, these Proceedings.

18. A.G. Ruggerio, *The Muon Collider*, these Proceedings.

19. Worshop on *Muon Colliders: Particle Physics and Design* (Napa Valley, Dec. 9-11, 1992).

LUMINOSITY ENHANCEMENT BY A SELF-IONIZED PLASMA IN e^+e^- COLLISIONS

P. Chen[1]*, C.-K. Ng[1]* and S. Rajagopalan[2]†

[1]Stanford Linear Accelerator Center, Stanford University, Stanford, CA 94309
[2]Physics Department, University of California, Los Angeles, CA 90024

ABSTRACT

We employ a 2D particle-in-cell code to investigate the focusing of relativistic charged particle beams in plasmas. The intense electric fields generated by electron and positron beams in high energy physics experiments can ionize a gas into a plasma which will then focus the beams. This self-ionization mechanism will enhance the luminosity in e^+e^- collisions by implementing a plasma lens in the interaction region. The dependences of the luminosity enhancement on the plasma lens thickness and plasma density are investigated. The application of plasma lens focusing to SLC is discussed.

1. INTRODUCTION

To achieve high enough luminosities (event rates) for physics studies is one of the important goals for e^+e^- linear colliders. The Stanford Linear Collider (SLC) is the first linear collider for the study of the fundamental electroweak gauge particle, Z, by colliding electron and positron beams at ~ 92 GeV center-of-mass energy. Small spot sizes ($\sim 1\text{-}2\mu$m) can be obtained by the final focusing optics at SLC. Plasma focusing can in principle provide a mechanism to pinch the beams to even smaller size before they collide at the interaction point.

The self-focusing plasma lens has been proposed as a mechanism to increase the luminosity in e^+e^- linear colliders [1]. Conventional quadrupole magnets for final focusing in high energy accelerators have limited focusing strength (a few hundred MG/cm) while plasma lenses are able to produce focusing strength of a few order of magnitudes higher, depending on the plasma density. This self-focusing effect by the plasma on relativistic beams have been verified experimentally at the Argonne National Laboratory[2] and in Japan[3].

Since the proposal of the self-focusing plasma lens, the study of plasma focusing has been distinguished by the overdense regime[4] (i.e. the beam peak density n_b is much smaller than the ambient plasma density n_p) and the underdense regime[5,6] ($n_b > n_p$). In the overdense regime, the beam quantities can be treated as perturbations and hence the plasma dynamics can be well described by linear fluid theory. Although the focusing is strong in this regime, the beam optics are subject to aberrations due to the spatial dependence of the focusing strength. Furthermore, the high plasma density may pose backgrounds to

*Work supported by Department of Energy contract DE-AC03-76SF00515
†Work supported by Department of Energy contract DE-AS03-88ER40381

particle detectors. In the underdense regime, the focusing strength for electron beams is more uniform, and that for positron beams becomes nonlinear.

Recently, it has been proposed that considerable luminosity gain can be achieved using the beams themselves to ionize a gas into a plasma, which subsequently focuses the beams[7]. The major mechanism is the tunneling ionization as a consequence of the distortion of the atomic potential by the strong electric field carried by the intense charged particle beam. This self-ionization mechanism provides an attractive means to implement a plasma lens for final focusing in e^+e^- collisions as it is non-trivial to produce high density plasmas at the center of a complex detector.

Previous studies[1,7] have neglected the effects of ion motion of the plasma. However, because of the strong fields exerted by the beams, the ions are expected to move to neutralize the space charge forces of the beams. On the other hand, the self-ionization process mentioned above is determined spatially by the beam fields, which have to be solved self-consistently when the beams collide. In view of this, simulations are required to give an accurate account of the complicated dynamics of beam-plasma interactions.

The purpose of this work is to simulate beam-plasma interactions by self-ionization mechanism and to investigate the maximal attainable enhancement of luminosity. We will use SLC beams as an example, and our main objective is to determine the plasma density and the plasma lens thickness for maximum luminosity enhancement at SLC. This paper is organized as follows. In the next section, the basic principles of plasma lens focusing are briefly reviewed. In section 3, the self-ionization mechanism by an intense charged particle beam is discussed. In section 4, we employ the 2D particle-in-cell electromagnetic code, CONDOR[8], to simulate beam-plasma interactions and to study the phenomenon of self-focusing. The dependences of the luminosity enhancement on the plasma density and plasma lens thickness at SLC are investigated. Discussions and a summary of our results are given in section 5.

2. PLASMA FOCUSING OF BEAMS

The theory and the literature have thus far[4−6] distinguished between focusing by plasmas of density greater than the beam (overdense lens) and by those of smaller density(underdense lens). This distinction which is made for the purpose of analysis nevertheless may obscure the unifying features of plasma focusing. We will attempt a description of the beam-plasma interaction which describes the focusing aspect of the plasma response. The beams themselves are ultrarelativistic and thus the intra-beam forces are compensated to a factor $1/\gamma^2$, when propagating in vacuum. When the beam impinges a plasma, the initial response of the plasma is to neutralise the electric field of the beam , not necessarily completely. As a consequence the beam particles always experience a net force (intra-beam forces being almost fully balanced) which is a focusing force, due to ions for e^- beams and an excess of electrons for e^+ beams. The magnetic response of the plasma occurs in a much slower fashion and for our purposes is only a correction. The difference between an overdense and an underdense lens for an electron beam is mainly in the optical properties of the lens. In the case of an overdense lens the frequency of oscillation of the plasma electrons is high and the focusing strength follows the beam longitudinal profile with an oscillatory term due to the plasma electron oscillation. For underdense lenses the plasma oscillation has a smaller frequency and in our examples the beam has a length smaller than the plasma wavelength, resulting in the expulsion of plasma electrons from the beam interior and the beam being focused by the plasma ions. In the case of e^+ beams the overdense scenario is similar to the e^- case, but for underdense lenses the plasma electrons oscillate in the field of the beam and provide a focusing force when they spend time in the beam interior.

In general, the β-function of the beam focusing in a plasma is governed by the third order differential equation:

$$\beta''' + 4K\beta' + 2K'\beta = 0 \quad , \tag{1}$$

where $K = K(s)$ is the focusing strength and is in general a function of the longitudinal position of the beam. Let the plasma density be determined by the initial beam density with a ratio η as

$$n_p = \eta n_{b0} \quad . \tag{2}$$

Note that η is also a function of the beam position. Assuming a cylindrically symmetric bi-Gaussian beam-density profile $\rho_b = n_b e^{-r^2/2\sigma_r^2} e^{-z^2/2\sigma_z^2}$, the peak beam density can be derived: $n_b = N/(2\pi)^{3/2}\sigma_r^2\sigma_z$. In terms of the the the initial beam size, $\sigma_{r0}^2 = \beta_0\epsilon_n\gamma^{-1}$, we can write

$$K = \frac{\eta N r_e}{\sqrt{2\pi}\,\beta_0\epsilon_n\sigma_z} \equiv \frac{\eta\zeta}{\beta_0} \quad , \tag{3}$$

where ϵ_n is the normalized emittance. Here, we also introduce the *phase space density* ζ, which measures the beam density in the three-dimensional beam volume of r, r', z.

(a) Focusing by an underdense lens

Assuming that the ions are infinitely heavy, then an underdense plasma reacts to an electron beam by total rarefaction of the plasma electrons inside the beam volume, producing a uniform ion column of charge density en_p. This uniform column produces linear, nearly aberration-free focusing. Simulations have shown that $n_b \sim 2n_p$ is needed to produce linear focusing over most of the bunch[6].

In the underdense plasma regime, the focusing strength K is determined by the density of the plasma, and is essentially constant inside the bulk plasma:

$$K = 2\pi r_e n_p / \gamma \quad . \tag{4}$$

In practice we will choose $\eta \sim 1/2$ in Eq. (3) to ensure the underdense condition. To solve Eq. (1), we first integrate through the δ-function in K' at the start of the lens, and obtain $\Delta\beta'' = -2K\beta_0$. The other two initial conditions are the continuity requirements $\beta' = \beta_0'$ and $\beta = \beta_0$. Also note that $\beta_0'' = 2/\beta_0^*$ just before the lens, where β_0^* is the value at the waist that would be formed in the absence of the lens. The equation of motion is then $\beta'' + 4K\beta = 2/\beta_0^* + 2\zeta$, and we obtain[5]

$$\beta = \frac{\beta_0}{2} + \frac{1}{2K\beta_0^*} + \left(\frac{\beta_0}{2} - \frac{1}{2K\beta_0^*}\right)\cos[\nu(s - s_0)] + \frac{2s_0}{\nu\beta_0^*}\sin[\nu(s - s_0)] \quad , \tag{5}$$

where $\nu = 2\sqrt{K}$. This solution demonstrates an oscillatory behavior without damping effects. We further assume that $\beta_0 = \beta_0^*$ and $s_0 = 0$. To minimize the backgrounds, we look for the next waist at $\sin(\nu s^*) = 0$, then the path length is

$$s^* = \frac{\pi}{\nu} = \frac{\pi}{2\sqrt{K}} \quad . \tag{6}$$

The corresponding β^* is

$$\beta^* = \frac{1}{K\beta_0^*} = \frac{1}{\eta\zeta} \quad . \tag{7}$$

To appreciate the results, we note that beam parameters of $N = 4.0 \times 10^{10}$, $\sigma_z = 0.2$ mm, normalized emittance $\epsilon_n = 4 \times 10^{-3}$cm, and $\beta_0^* = 1$cm give $\zeta = 56$cm^{-1}. If $\eta = 0.5$ is assumed, this will give $\beta_-^* = 0.36$mm, which is a factor ~ 28 reduction in β^*, or about a factor of 5 reduction in beam spot size. The corresponding path length is $s_-^* \simeq 2.10$mm.

For an underdense plasma interacting with a positron beam, the plasma electrons are drawn toward the beam axis by the focusing potential provided by the positron beam. This results in a simultaneous motion of these electrons which is oscillatory and moving with the beam. In each cycle of oscillation the plasma electrons spend a fractional amount of time inside the core of the positron beam, resulting in an effective concentration of negative charges that provides a focusing force. Since the focusing force is nonlinear, the net effect is not simple to describe analytically.

(b) Focusing by an overdense lens

From Ref. 1 the focusing strength of overdense lens (where the plasma response can be treated as a perturbation) at the middle of a Gaussian bunch is

$$K = \frac{r_e N}{(2\pi)^{1/2}\gamma\sigma_r^2\sigma_z}. \tag{8}$$

In a thick lens where the beam size continues to change, then

$$K(s) = \frac{2\pi r_e n_b}{\gamma} = \frac{\zeta}{\beta(s)}.$$
(9)

Inserting Eq. (9) into Eq. (7), we have

$$\beta''' + \frac{2\zeta\beta'}{\beta} = 0.$$
(10)

Integrating twice, the second time after multiplication by β' one finds

$$\frac{\beta'}{2} = \frac{\beta - \beta_0^*}{\beta_0^*} + \zeta\beta ln(\beta_0^*/\beta).$$
(11)

Exact solution to this nonlinear equation is non-trivial. Notice, however, that the very purpose of a plasma lens is to reduce β^*. Thus by definition, $b \equiv \beta^*/\beta_0^* \ll 1$, and the term β/β^* on the right hand side can be neglected in a perturbative approximation. Then the solution is

$$b_0^{b_o} = exp\{-1/\zeta\beta_0^*\},$$
(12)

When restoring the β/β_0^* in Eq. (11) by the value of b_0, we find an improved solution

$$b_1^{b_1} = exp\{-(1 - b_0)/\zeta\beta_0^*\}.$$
(13)

This iterative process can continue to the degree of accuracy that one desires.

3. IONIZATION PROCESSES[7]

There are essentially two ionization mechanisms that can be provided by a high intensity, high energy beam, namely, the collisional ioinzation and the tunneling ionization. In the following, we discuss in details the relative importance of these ionization processes.

(a) Collisional ionization

This is an ionization process in which an individual beam particle ionizes the atom by a virtual photon exchange. The cross section can be estimated via the photo-ionization cross section, using the Weiszacker-Williams spectrum. The ionization cross section in the equivalent photon approximation is given by[9]

$$\sigma_i = \frac{\alpha}{2\pi} \int_{\omega_1}^E [ln(\frac{2E^2}{m\omega}) - 1]\frac{\sigma_\gamma(\omega)}{\omega}d\omega + \frac{2\pi\alpha^2 Z}{m\omega_I},$$
(14)

where

$$\frac{1}{\omega_I} = \frac{1}{Z}\sum_{n=1}^Z \frac{1}{\omega_n}.$$

For hydrogrn atom, $Z = 1$ and $\omega_I = \omega_1 = 13.6\text{eV}$. The spectrum $\sigma_\gamma(\omega)$ can be parametrized from photo-ionization data as

$$\sigma_\gamma(\omega) \approx \left(\frac{27}{\omega[\text{eV}]}\right)^3 \text{Mb} \,. \tag{15}$$

For SLC beams, $E = 45\text{GeV}$ and we find $\sigma_i \sim 0.22\text{Mb}$. The fraction of atoms that can be ionized through this mechanism by an incoming beam with N particles and size σ_r is $R_i = N\sigma_i/4\pi\sigma_r^2$. For the Stanford Linear Collider (SLC) beams, $\sigma_r \sim 1\mu\text{m}$ and $N \sim 10^{10}$, so R_i is only of the order of a few percent, which is far from saturation. One therefore needs to have a gas which is $1/R_i$ times denser to provide the necessary amount of plasma. This is not desirable as the backgrounds become very large. In addition, the nonsaturation of ionization also causes the tail of the beam to encounter a higher concentration of plasma than that seen by the head of the beam. This results in different foci for parts of the beam from different longitudinal positions. Therefore, in this paper, we neglect collisional ionization in our self-ionization process.

(b) Tunneling ionization

There is another ionization mechanism, which relies on the collective field of the beam. When an external electric field is strong enough so that the atomic Coulomb potential is sufficiently distorted, there is a finite probability that the bound state electron can tunnel through the potential barrier and become free. For hydrogen atoms, the ionization probability (per unit time) is given by[10]

$$W = 4\frac{\alpha^5 c}{\lambda_c^2} \frac{mc^2}{eE} \exp\left\{-\frac{2}{3}\frac{\alpha^3}{\lambda_c} \frac{mc^2}{eE}\right\} \,, \tag{16}$$

where E is the external electric field. The coefficient in the exponent is $\frac{2}{3}(\alpha^3/\lambda_c)$ $(mc^2/eE) \simeq 34.1\text{eV}/\text{Å}$. It is interesting to note that the ionization probability is already substantial, long before the exponent reaches a value of the order unity, due to the typical largeness of the non-exponential part. For example, an external field of $3.41\text{eV}/\text{Å}$ would give $W \simeq 1.15 \times 10^{14}\text{sec}^{-1}$. Under this condition, the ionization will be saturated within 10 femto-seconds. In fact, it can be shown for a field strength larger than $eE_{th} = 3.72\text{eV}/\text{Å}$, where the ground state binding energy is above the potential barrier, that, even classically, the electron can escape from the atom. The maximum collective electric field strength in a bi-Gaussian beam can be calculated to be $eE_{max}/mc^2 \simeq r_e N/2\sigma_z\sigma_r$, where r_e is the classical electron radius. A maximum field strength of $3.72\text{eV}/\text{Å}$ corresponds to, for example, a beam of $N = 2.12 \times 10^{10}, \sigma_z = 0.4\ 2\text{m}$, and $\sigma_r = 2.0\mu\text{m}$. This is within the range of the SLC parameters.

In our scheme, both the tunneling ionization and the plasma response rely on the same field strength in the beam, yet the two effects in principle may act against each other. The very nature of the plasma response is to neutralize the space-charge field due to the beam[1]. If the charge neutralization is complete, as in the case of an overdense plasma lens, within a time scale which is comparable to that for the tunneling ionization process, then the beam field would be effectively screened off before the the ionization can be saturated. For an underdense plasma lens the plasma response can never completely neutralize the space-charge, and the problem is less severe. But to ensure that our scheme indeed works let us insist that the time scale involved in the ionization process, τ_i, be much less than the time scale of plasma response to the beam field, τ_p, which in turn should be much less than the beam passage time, τ_b: $\tau_i \ll \tau_p \ll \tau_b$. The last inequality is the known condition for the plasma lens. The first condition, $\tau_i \ll \tau_p$, ensures that the ionization saturates before charge neutralization is effective.

The condition for the saturation of tunneling ionization by an incoming beam is when the integrated ionization probability is close to unity. The function $W(z)$ is an extremely rapid function of z and thus the position z_s where the probability becomes almost unity is not different from the position where the integrated rate is one. Therefore it is reasonable to use the relation

$$1 = \int_{-\infty}^{z_s} W(z)dz/c \quad , \tag{17}$$

where z_s is the position along the beam where the ionization is saturated. Since $W(z)$ is exponentially dependent on $E(z)^{-1}$ of the beam, while $E(z)$ itself follows a Gaussian distribution, we expect that the saturation is predominantly in the small time interval, e.g., $\tau_i \sim W^{-1}(z_s) \sim$ femto-seconds, around the saturation point z_s. On the other hand, as we will see in the following, the matched underdense plasma density to the SLC-like beam density is of the order $n_p \sim 10^{18} \text{cm}^{-3}$, which corresponds to a plasma frequency of $\omega_p \sim 5.6 \times 10^{13} \text{sec}^{-1}$. The typical time scale of plasma perturbation is therefore $\tau_p \sim 2\pi\omega_p^{-1} \sim 100$ femto-second, which is longer than the time scale of tunneling ionization, yet shorter than the beam passage time $\tau_b \sim 1000$ femto-second. Thus for the beam conditions that we anticipate, the three time scales do follow the right ordering.

4. NUMERICAL RESULTS

To give a quantitative account of beam-plasma interactions, we employ the $2\text{-}\frac{1}{2}\text{D}$ particle-in-cell code, CONDOR[8], to simulate the propagation of relativistic electron and positron beams in plasmas. CONDOR is a $2\text{-}\frac{1}{2}\text{D}$, fully electromagnetic particle-in-cell code in which the dynamics of plasma and beam particles are treated self-consistently by the Maxwell equations and the Lorentz force equations. In order to calculate the luminosity enhancement through self-ionization, we have modified CONDOR to include a luminosity calculation and an algorithm for the ionization of gases triggered by the strong electric fields of relativistic charged particle beams.

(a) Beam focusing in plasmas

The geometry of our simulations is in r-z coordinates and hence the beams and the plasma are taken to be cylindrically symmetric. The electron and positron beams are injected from the boundaries in $+\hat{z}$ and $-\hat{z}$ directions respectively. To demonstrate the phenomenon of self-pinching of relativistic beams in plasmas, a pre-formed plasma is assumed in this subsection while the details of self-focusing effects due to tunneling ionization and its consequence on luminosity enhancement will be deferred to a later subsection.

The density profile for a cylindrically symmetric bi-Gaussian beam is

$$\rho_b = n_b e^{-r^2/2\sigma_r^2} e^{-z^2/2\sigma_z^2}, \tag{18}$$

where the peak beam density n_b is related to the number of particles N in the bunch as

$$n_b = N/(2\pi)^{3/2}\sigma_r^2\sigma_z. \tag{19}$$

To simulate the SLC beams, we take $N = 4 \times 10^{10}$, $\sigma_z = 0.2$ mm and $\sigma_r = 2$ μm. The corresponding peak beam density $n_b = 3.17 \times 10^{18}$ cm^{-3}.

We are interested in determining the plasma lens density and thickness for maximum gain of luminosity in e^+e^- collisions, as will be discussed later. Thus, we choose $n_p = 3 \times 10^{18}$ cm^{-3} for the plasma density initially . The simulations run from plasmas which are underdense to overdense plasmas.

The initial conditions for our simulation are as follows. The energies of the electron and positron beams are taken to be 45 GeV which has been achieved at SLC. The emittance of the beams is 4×10^{-10} m-rad and the corresponding $\beta^* = 10$ mm (contrast with 5 mm as the design goal) in the final focusing region before the beams enter the plasma lens. The plasma electrons and ions are assumed to be stationary initially. The plasma lens has a thickness of 4 mm.

In Fig. 1, we show the results from a CONDOR run for the behaviors of the e^+ and e^- beams as they traverse the plasma. The electrons are injected from the left boundary and the positrons from the right. Figs. 1(a)-(d) and Figs. 1(e)-(h) are the snapshots at four different time steps for the distributions of e^- and e^+ beams respectively. From Figs. 1(a)-(b) and (e)-(f), it can be seen that the electrons and positrons are focused gradually by the plasma before they are about to collide in the middle of the lens. The beam fronts are not focused well as the plasma electrons are still responding to the smaller field due to the Gaussian nature of the beams. For a highly relativistic charged particle beam traveling in vacuum, the self-pinching effects are almost balanced by the space charge forces. However, in a plasma, we clearly see that the beams are pinched and focused as a result of the neutralization of the space charge forces of the beams by the plasma electrons or ions. In Figs. 1(c) and (g), the beams collide at smaller beam sizes. This example shows that the outer cores of the beams

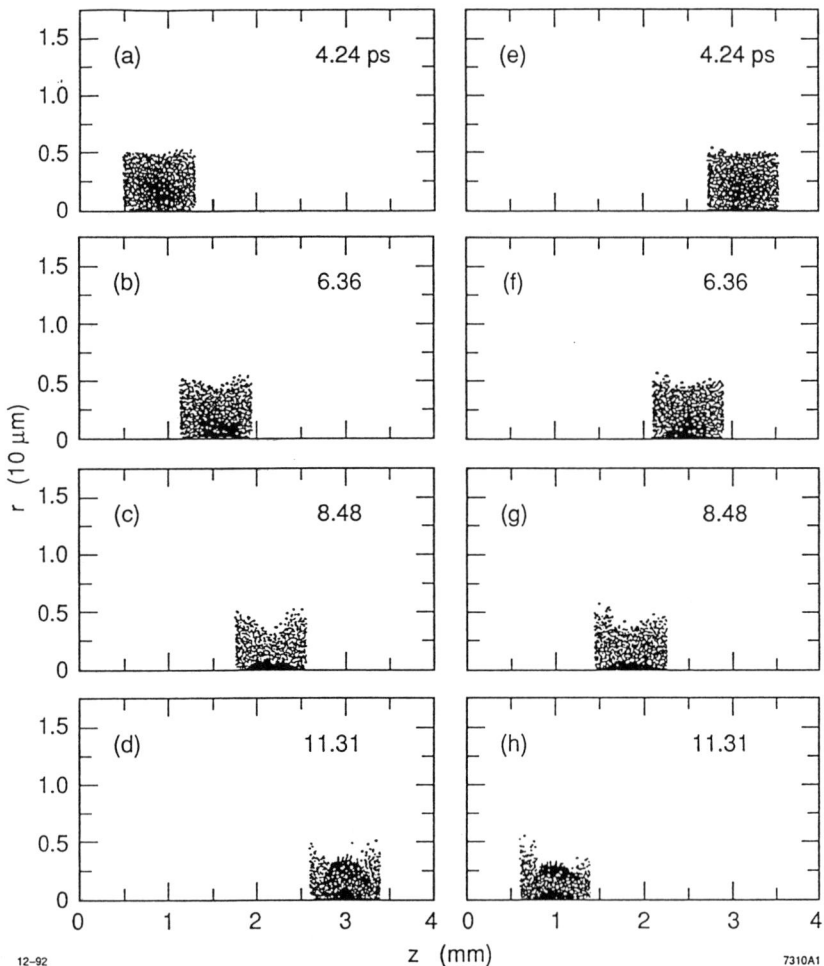

Fig. 1 Collision of e^+ and e^- beams in a pre-formed plasma. (a)-(d) are four
different snapshots for the e^- beam; (e)-(h) are for the corresponding
e^+ beam. The horizontal and vertical axes are in units of mm and 10
μm respectively.

are not focused as well as the central part of the beams. This is due the fact
that the beam density in the outer region is smaller than the plasma density
and is a characteristic of beam focusing in an overdense plasma (note that the
peak beam density is about the same as the plasma density). In Figs. 1(d) and
(h), the two beams start to diverge again after colliding with each other and this
completes the process of e^+e^- collision in the plasma. It is important that the
beams collide at the focal point of the plasma lens for optimal luminosity gain.

(b) Beam focusing by tunneling ionization

As mentioned in the previous section, the major self-ionization process is the tunneling ionization while that due to collisional ionization is at most a few % effect and hence can be neglected. The electric fields of electron and positron beams at SLC are strong enough to ionize a gas into a plasma through the mechanism of tunneling ionization. This mechanism of pinching the beams is attractive since to form a plasma externally in the interaction region is non-trivial. It is important to determine from simulations whether this self-ionization mechanism is able to pinch the beams to achieve enough gain in luminosity for e^+e^- collisions in plasmas lenses. As mentioned before, the maximum of the enhancement factor depends on the focusing strength and the thickness of the plasma lens. By varying the gas density and the lens thickness for fixed beam parameters, it is possible to determine some design parameters for implementing a plasma lens in the final focusing region of SLC.

The criterion for the ionization of a gas element in our system of simulation is given by Eq. (17). For every grid point in our simulation mesh, the ionization rate function W is accumulated at every time step as the beams traverse the gas. The gas element is turned into a plasma element when the integral in Eq. (17) is greater than or equal to 1. The gas will be ionized at a distance behind the beam fronts, which is determined by the beam parameters. In view of this, the beam fronts will never see the plasma and therefore will not be focused at all. Moreover the cores of the beams do not experience substantial focusing because the ionization begins at a radius around 0.2σ and this degrades the final spot size and beam-beam disruption. Therefore, it is expected that the luminosity gain is less than that of colliding e^+e^- beams in a pre-formed plasma.

To study the propagation of electron and positron beams in gases, we choose the gas density and lens thickness to be 6×10^{18} cm^{-3} and 4.0 mm respectively. These parameters of the lens correspond to maximum luminosity enhancement in a gas for the above beam parameters, the result obtained from simulations with varied gas densities and lens thicknesses.

In Fig. 2, we show the simulation results for the behaviors of the e^+ and e^- beams as they traverse a gas of density 6×10^{18} cm^{-3}. Figs. 2(a)-(d) and (e)-(h) are the snapshots at four different time steps for the distributions of e^- and e^+ beams respectively. The description of the focusing of the individual beams and their collision is similar to the case with a pre-formed plasma. The main difference is that a large portion of the front part of the beams is not focused at all during the full passage of the beams through the lens. The electric fields at the front are not strong enough to trigger tunneling ionization of the gas and therefore this part will never be focused. The rear portion of the beams experience the response of the ionised plasma and will be focused(see Figs.2(c)-(d) and (g)-(h)).

In Figs. 3(a) and (b), we show the distributions of the plasma electrons and protons respectively at time step 7.07 psec. This should be compared with the

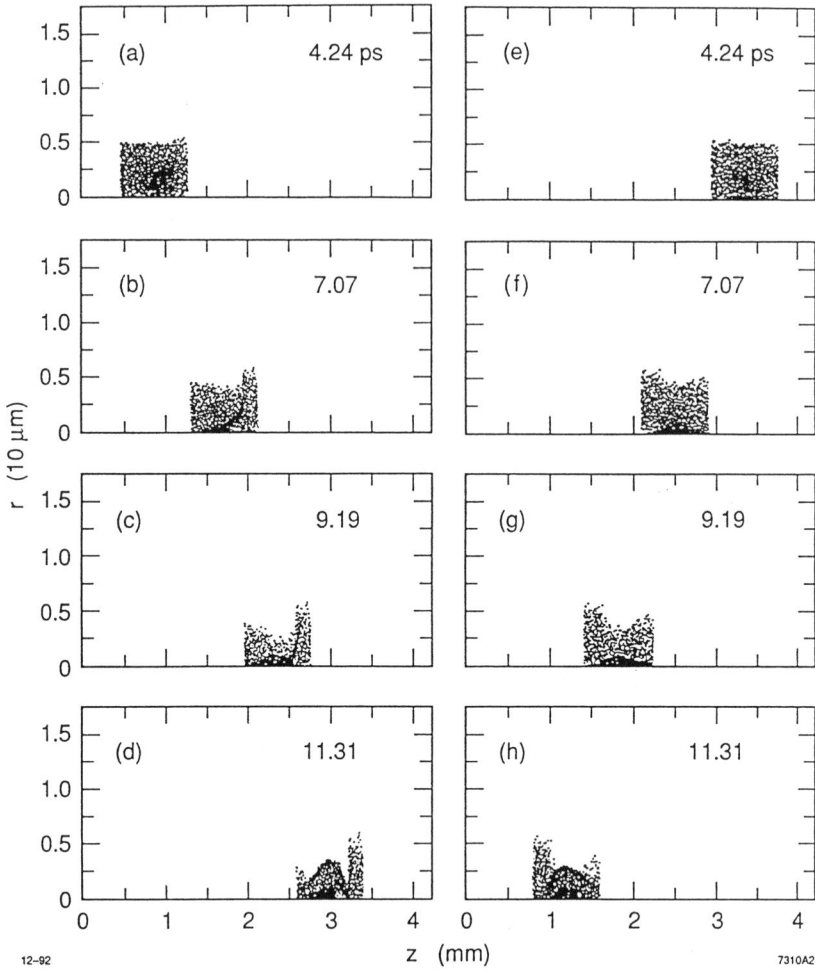

Fig. 2 Collision of e^+ and e^- beams in an ioinzed gas. (a)-(d) are four different snapshots for the e^- beam; (e)-(h) are for the corresponding e^+ beam. The horizontal and vertical axes are in units of mm and 10 μm respectively.

corresponding beam distributions at the same time step in Figs. 2(b) and (f). It can be seen that ionization takes place at a distance of about 0.2 mm behind the beam fronts. The first σ_z of the beams in this example is not focused (remember $\sigma_z = 0.2$ mm). The central core of the gas in the front part following the beams is not ionized because of the nature of the linear rise of the electric fields at small distance from the axis. This will degrade the focusing strength to certain extent for those particles near axis. The motion of the ions is not negligible and results in better focusing at the rear of the beams. This is clearly seen for the electron beam in Fig. 2(b), in which the front part and those particles near the axis are not focused while the rear part is pinched much better (note the banana shape of the particles in the central region). It should also be noted that the positron beam ioinizes the gas farther out from the axis than the electron beam does. This is because the screening of charges is more effective for the electron beam.

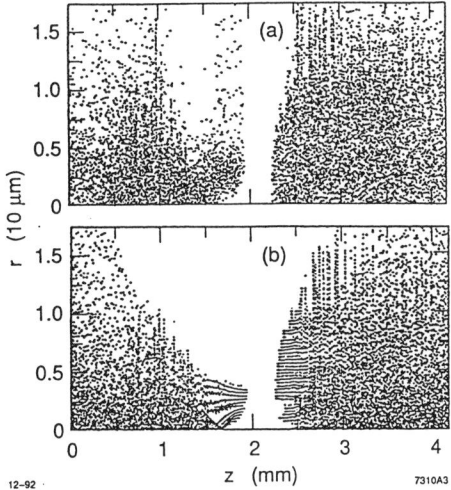

Fig. 3 Distributions of (a) ionized gas electrons; (b) ionized gas protons at snapshot 7.07 psec. The horizontal and vertical axes are in units of mm and 10 μm respectively.

(c) Luminosity enhancement at SLC

Having the optics for the electron and the positron beams described, next we look into the physics of beam-beam interaction inside a plasma. The disruption effect due to the mutual pinching between the colliding e^+e^- beams in vacuum has been studied in details in the past.[11,12] In the situation where the e^+e^- beams collide inside a plasma, the mechanism is modified. When the beams overlap, the total beam current is increased; Therefore we expect an increase of the "return current" induced in the plasma. The return current acts to reduce the magnetic focusing forces and the mutual beam-beam pinching. On the other hand, in the same beam overlapping region the net space charge is reduced. Therefore, we expect a decrease of the space-charge perturbation in the plasma. This helps to reduce the influence of the plasma on the beam-beam disruption.

The overall enhancement on luminosity in our scheme can be estimated as

$$H_D = \frac{H_{D1}H_{D2}}{H_{D0}} \quad , \tag{20}$$

where H_{D1} is the "geometric" enhancement due to the reduction of the beam sizes from the plasma lens, and H_{D0} and H_{D2} are the disruption enhancement due to beam-beam interaction with and without the plasma lens, respectively. Since the plasma-focused e^+e^- beams are different in sizes, the "geometric" enhancement (excluding depth of focus and disruption effects) in luminosity is

$$H_{D1} = \frac{2\sigma_{r0}^{*2}}{\sigma_-^{*2} + \sigma_+^{*2}} \quad . \tag{21}$$

We have shown that the mechanism of self-ionization of a gas can pinch a relativistic e^+ or e^- beam. To ensure that the two beams collide at the optimal position we implemented a luminosity calculation in CONDOR to determine this for the plasma lens. The focal length of the plasma lens is given by

$$s^* = \frac{\pi}{2\sqrt{K}}, \tag{22}$$

where K is the focusing strength of the lens determined by the plasma density n_p and r_e is the classical electron radius. If both e^+ and e^- beams are focused evenly, then the plasma lens thickness should be $l = 2s^*$. With various nonlinear effects, however, the desirable lens thickness cannot be determined analytically. We therefore rely on simulation to determine the optimal lens parameters. In CONDOR simulation, the focal length is determined by optimizing the longitudinal length of the plasma for maximum gain in luminosity.

The luminosity for e^+e^- collisions is defined as the 4-dimensional phase space integral

$$\mathcal{L} = f \int n_1(x,y,z,t) n_2(x,y,z',t) dx\,dy\,dz\,dt, \tag{23}$$

where $z' = z - ct$, n_1 and n_2 are the densities of electrons and positrons as functions of time respectively, and f is the repetition rate of collisions. When e^+ and e^- beams collide in vacuum, luminosity will be enhanced because of disruption, as a consequence of the bending of particle trajectories by the electromagnetic fields of the oncoming beam. To quantify this luminosity enhancement, we define the enhancement factor

$$H_{D0} = \frac{\mathcal{L}}{\mathcal{L}_0}, \tag{24}$$

where $\mathcal{L}_0 = fN^2/4\pi\sigma_r^2$ is the nominal luminosity. When e^+ and e^- beams collide in plasma, the luminosity is further enhanced because of plasma focusing.

In Fig. 4, we show the distributions of the luminosity enhancement factor H_D as a function of the plasma lens thickness L for various gas densities. The number of particles in the beams is 4×10^{10} and 3×10^{10} for the two sets of curves. The gas density varies from 2×10^{18} cm^{-3} to 6×10^{18} cm^{-3}. Let us first consider the case with $N = 4 \times 10^{10}$. We see that for the gas density of 6×10^{18}, we can achieve the maximum gain in luminosity, mainly because of the stronger focusing strength produced by a higher density gas. The optimal lens thickness is found to be about 4.0 mm and this is in good agreement with the theoretical estimate from Eq. (6). It is also seen that the enhancement factor does not change sensitively around this optimal lens thickness. This shows that the enhancement is not very sensitive to the lens thickness or the longitudinal offset of the two beams. For $N = 3 \times 10^{10}$, because the beam fields are weaker, and correspondingly the focusing strength is not as strong as the case with $N = 4 \times 10^{10}$. A longer lens thickness of 5.0 mm is required to achieve the maximum gain in luminosity, and the optimal gain in luminosity is smaller than the case with $N = 4 \times 10^{10}$.

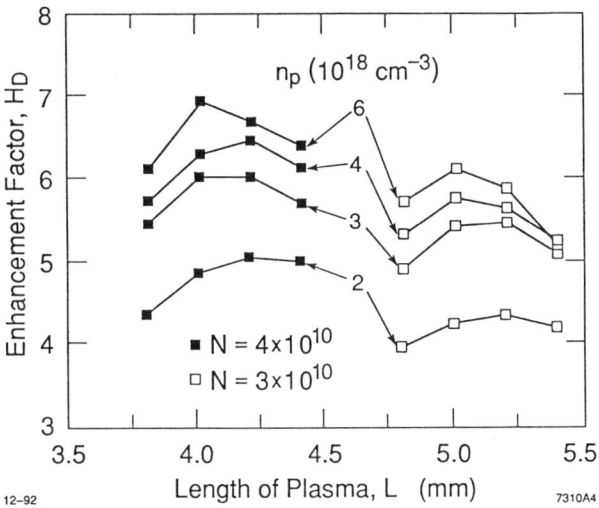

Fig. 4 Luminosity enhancement factor as a function of plasma lens thickness for different plasma density n_p. The beam particle numbers are taken to be 3×10^{10} and 4×10^{10}.

From the previous subsection, we saw that the electrons are focused somewhat better than the positrons. It will be better to have the beams focused with a certain longitudinal offset so that the beams will meet when they have the smallest spot sizes. However, from our simulation, an offest of about 0.67 psec with the positron beam injected later only increases the luminosity gain by a few percent. This is again an illustration of the relatively flat behavior of the distributions near the optimal lens thickness. Furthermore, increasing the gas density beyond 6×10^{18} cm^{-3} does not increase the luminosity gain further. This is expected as in the overdense regime, the focusing will be basically determined by the beam parameters and the aberration effects will increase the spot sizes at the interaction point for higher density gases.

5. CONCLUSIONS

We have used the particle-in-cell code, CONDOR, to study beam-plasma interactions and have demonstrated the focusing of relativistic beams in plasmas. By focusing and colliding e^- and e^+ beams in plasmas, the luminosity in high energy linear collider experiments can be enhanced as a consequence of the self-focusing of the beams.

We have also shown that considerable luminosity enhancement can be obtained by the process of tunneling ionization. A luminosity enhancement factor of 5 to 7 can be achieved by this process for SLC beam parameters of $N = 3 \times 10^{10}$ and $N = 4 \times 10^{10}$, $\sigma_r = 2\mu$m, $\sigma_z = 0.2$mm and an emittance of 4×10^{-10} m-rad. For a repetition rate of 120 sec^{-1}, it will enhance the luminosity to $\sim 3 \times 10^{30}$ cm^{-2}sec^{-1} at SLC. It should be noted that further luminosity enhancement is possible, especially in the core regions, by the process of collisional ionization, which we have neglected in this analysis. The role of impact ioinization complemetary to tunneling ioinzation will play a more important role for the Next Linear Collider (NLC). For example, for NLC beams with $N = 0.65 \times 10^{10}$, $\sigma_x = 300$nm, $\sigma_y = 3$nm, impact ioinzation will be saturated and has to be taken into account for luminosity calculation.

ACKNOWLEDGEMENTS

We are grateful to M. Breidenbach of SLD for earlier discussions which triggered the investigation of beam self-ioinzation. We also thank K. Bane, S. Brandon, K. Eppley and K. Ko for useful discussions.

REFERENCES

1 P. Chen, Particle Accelerator **17**, 121 (1987).

2 J. B. Rosenzweig, et. al., Phys. Fluids **B2**, 1376 (1990).

3 H. Nakanishi et. al., Phys. Rev. Lett. **66**, 1870 (1990).

4 J. B. Rosenzweig and P. Chen, Phys. Rev. **D39**, 2039 (1989).

5 P. Chen, S. Rajagopalan, and J. B. Rosenzweig, Phys. Rev. **D40**, 923 (1989).

6 J. J. Su, T. Katsouleas, J. M. Dawson, and R. Fidele, Phys. Rev. **A41**, 3321 (1990).

7 P. Chen, Phys. Rev. **A39**, 45 (1992).

8 B. Aiminetti, S. Brandon, K. Dyer, J. Moura and D. Nielsen, Jr., *CONDOR User's Guide, Livermore Computing Systems Document*, Lawrence Livermore National Laboratory, Livermore, California, April, 1988.

9 F. LeDiberder, private communication.

10 L. D. Landau and E. M. Lifshitz, *Quantum Mechanics: Non-Relativistic Theory*, 3rd edition, p. 293 (Pergamon Press, 1981).

11 R. Hollebeek, Nucl. Instrum. Methods **184**, 333 (1981).

12 P. Chen and K. Yokoya, Phys. Rev. **D38**, 987 (1988).

Issues Regarding Acceleration in Crystals*

PISIN CHEN

Stanford Linear Accelerator Center
Stanford University, Stanford, California 94309

DAVID B. CLINE AND WILLIAM E. GABELLA

Department of Physics,
University of California, Los Angeles, CA 90024

ABSTRACT

Both self-acceleration and laser-acoustic acceleration in crystals are considered. The conduction electrons in the crystal are treated as a plasma and are the medium through which the acceleration takes place. Self-acceleration is the possible acceleration of part of a bunch due to plasma oscillations driven by the leading part. Laser-acoustic acceleration uses a laser in quasi-resonance with an acoustic wave to pump up the plasma oscillation to accelerate a beam. Self-driven schemes though experimentally simple seem problematic because single bunch densities must be large.

INTRODUCTION

For making dramatically higher gradients in future generations of high energy particle accelerators and for making low energy accelerators more accessible, it would be useful to have a solid state accelerator[1,2] capable of sustaining very high accelerating gradients. The conduction electrons in the solid already make a convenient source of plasma, hence one can invoke all the concepts concerning plasma acceleration and focusing.[3] Also channeling in crystalline solids, the confinement of positively charged particles between planes of atoms (planar channeling), leads to transverse confinement of the beam and preservation of the beam size. These properties encourage the investigation of using crystalline solids as accelerators.

Solid state accelerators were discussed previously by Chen and Noble[1] especially from the point of view of emittance preservation and possible external mechanisms for driving the plasma oscillation. They take as their model that the crystal is just a bag of plasma and it is the plasma interactions, appropriately modified by the crystal structure, that determine focusing and acceleration of a traversing charged particle beam.

---

* Work supported by the Department of Energy, contract DE–AC03–76SF00515.

One reason for the excitement over both plasma and solid state acceleration is that in the cold wave-breaking limit of the plasma-fluid equations, the largest electric fields that can occur are the order of $\sqrt{n_0}$ V/cm when the plasma density n_0 is given in particles per cm^3. So for a metal with conduction electron density around 10^{22}/cm^3, one might expect 100 GeV/cm accelerating gradient. Though practical concerns and instabilities will yield an actual gradient far below this limit.

There are issues that arise about how the crystal structure modifies the physical properties of the free conduction band electrons in the solid when it performs plasma oscillations. One is whether the effective mass or the usual mass of the electron is relevant. For this paper, the plasma oscillations do not carry out such bulk motion that the lattice will modify the behavior of the electron significantly. So, currently it seems appropriate to use the free space rest mass of the electron for the remainder of this paper.

In the following, the self-focusing and self-acceleration of a relativistic beam in a plasma are reviewed, the laser-acoustic accelerator is discussed, and issues concerning radiation in solids are briefly mentioned. Finally, some conclusions as to the best possible approach for testing the ideas of solid-state acceleration are given.

BEAM SELF-ACCELERATION

As a bunched beam enters a plasma, whether the source of that plasma is a solid or not, it very quickly sets up plasma oscillations which in turn can act back on the beam. The principle of self-focusing of a relativistic beam in a plasma is one such consequence.[4] For the discussion presented here, the plasma begins with an unperturbed plasma density of n_0.

In Ref. 4 the focusing fields for a parabolic shaped bunch are calculated. Since this is done in the language of wake fields, it is quite straight forward to use the Panofsky-Wenzel theorem (or Greens theorem) to find the longitudinal fields. The bunch distribution that generates the wake fields is parabolic,

$$\sigma(\mathbf{x}) = \rho_b \left(1 - \frac{r^2}{a^2}\right)\left(1 - \frac{(\zeta + b)^2}{b^2}\right) \quad , \tag{1}$$

with $\rho_b = 3N/(2\pi a^2 b)$ where N is the bunch population. The longitudinal coordinate ζ is the usual co-moving coordinate $z - ct$; the head of the bunch is at $\zeta = 0$ and the tail is at $\zeta = -2b$. Using this distribution and the linearized cold fluid equations for a plasma perturbation, the cylindrically symmetric wake fields are

$$eW_\perp = \frac{8\pi e^2 \rho_b}{k}\left[I_1(kr)K_2(ka) - \frac{r}{ka^2}\right] \times$$

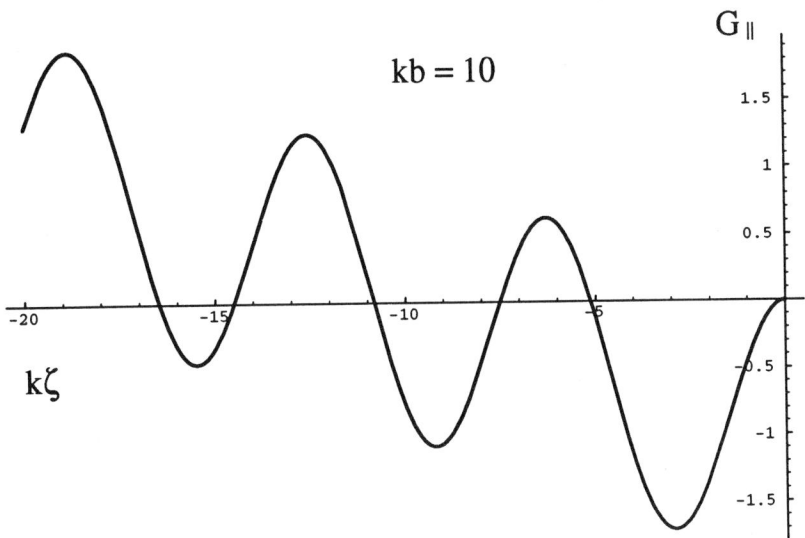

Figure 1. The longitudinal dependence of the longitudinal wake field for $kb = 10$. The middle of the bunch is at $k\zeta = -10$.

$$\left[(1 - \frac{(\zeta + b)^2}{b^2}) + \frac{2}{kb}\sin k\zeta + \frac{2}{k^2 b^2}(1 - \cos k\zeta)\right] \quad , \qquad (2)$$

for the tranverse field on a test particle, and

$$eW_\parallel = \frac{16\pi e^2 \rho_b}{k^2 b}\left[I_0(kr)K_2(ka) + \frac{1}{2}(1 - r^2/a^2) - \frac{2}{k^2 a^2}\right] \times$$
$$\left[\cos k\zeta - (\frac{\zeta + b}{b}) + \frac{1}{kb}\sin k\zeta\right] \quad , \qquad (3)$$

for the longitudinal field on a test particle. The wave number for the plasma oscillation has been used and is $k^2 = \omega^2/c^2 = 4\pi n_0 r_e$, where r_e is the classical electron radius.

For ease of notation and since we are usually considering either the longitudinal or the transverse dependence of the wake fields, the following notation is introduced: $eW_\perp = 16\pi e^2 \rho_b F_\perp(r)G_\perp(\zeta)/k$ and $eW_\parallel = 16\pi e^2 \rho_b F_\parallel(r)G_\parallel(\zeta)/k^2 b$. So, the $F_\perp$ and the $F_\parallel$ are the expressions in the first set of [...] in (2) and (3), and the G's are the terms in the second set of brackets with the left over coefficients.

In Figure 1, $G_\parallel$ is plotted for $kb = 10$ and as a function of the dimensionless parameter $k\zeta$. Notice the rising trend toward the bunch tail ($\zeta = -2b$), this is the accelerating part of the wake field. From the definition of $G_\parallel$, the maximum value it can attain at the tail for $kb > 1$ is about 2.

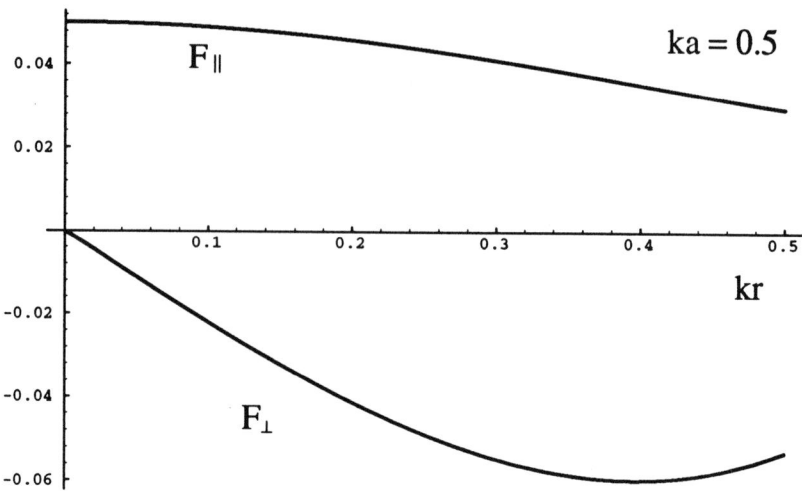

Figure 2. The radial dependence of the longitudinal and the radial wake fields for $ka = 0.5$.

To estimate the maximum gradient in the tail, the r dependence must also be considered. In Figure 2, the $F_\parallel$ and $F_\perp$ are plotted for $ka = 0.5$. Notice that the $F_\parallel$ is relatively flat over the bunch cross-section, at least when $ka < 1$ which seems to be a desirable regime. Note also the linearly rising $F_\perp$ which gives the linear focusing of a plasma lens. To estimate realistic gradients, the $F_\parallel$ is treated as flat and approximately constant, and it is replaced with its value at the origin $F_\parallel(0) \approx K_2(ka) + 1/2 - 2/(ka)^2$.

For the SLAC Final Focus Test Beam[5] (FFTB) the positron bunches are completely determined except for the transverse size. The FFTB has the advantage of a very dense positron beam to excite the plasma oscillations for self-acceleration. This can always be blown up slightly from the values given in Table 1. Replacing ρ_b in (3) and gathering terms, the accelerating gradient can be put in the form

$$eW_\parallel = 24N\frac{r_e}{b^2}mc^2\left[\frac{F_\parallel G_\parallel}{(ka)^2}\right] \quad .\tag{4}$$

The coefficient of the [...] term is about 3.5 MeV/cm and tuning parameters to optimize the [...] term yields a total $eW_\parallel \approx 5.3$ MeV/cm. The maximum occurs around $kb = 6$ and $ka = 1/1000$; the calculated plasma density is $8 \times 10^{14}/\text{cm}^3$, corresponding to Si. The relationships for a cylindrically symmetric parabolic distribution were used to relate RMS values and total lengths: $b = \sqrt{5}\sigma_z$ and

Table 1. SLAC Final Focus Test Beam round beam parameters.

energy	50 GeV
coupled emittance $\gamma\epsilon$	3×10^{-5} m
β_x^*	3 cm
bunch length σ_z	0.5 mm
momentum spread $\Delta p/p$	$\pm 0.3\%$
bunch population	2×10^{10}

$a = \sqrt{6}\sigma_x$ (assumes $\sigma_x = \sigma_y$). Compare the above gradient of 5.3 MeV/cm to the naive estimate $\sqrt{n_0}$ V/cm, or 28 MeV/cm.

At 50 GeV the positrons might traverse 5-10 cm of Si from the point of view of channeling, hence the tail could optimistically gain approximately 50 MeV. Since the bunch spread of ± 150 MeV is not coherent while the energy gain is, there is a possibility of detecting the energy increase. However, the radiative energy loss has so far been ignored.

LASER-ACOUSTIC COUPLING

Besides self-acceleration, it is natural to consider externally driven plasma oscillations as a means for accelerating particles. One such scheme is to couple a side-injected laser to an acoustic mode in the plasma.[6] For the pure plasma case, a standing acoustic wave is set up in the plasma putting a density variation in both the ions and the plasma electrons. A polarized laser is shone on the plasma from the side, with its electric field polarized along the direction of the witness beam and the acoustic wave; see Fig. 3. The electric field of the laser drives a plasma oscillation in the electrons relative to the ions, which are nearly stationary. The time varying charge density gives rise to the electric field along the witness beam.

There are mechanical differences between acoustic waves in solids and gases. In the solid, one naively expects the acoustic wave to be carried predominantly by the ions and couple very little to the conduction electrons. If this is indeed the case, then other techniques for setting up a standing wave in the conduction electron density might have to be investigated, like using a standing wave electric field along the beam direction.

For appropriate parameters, there is an approximate resonance between the acoustic wave and the shaking of the plasma electrons. This leads to growth and then saturation of the amplitude of the plasma oscillations. The dispersion relations and condition for quasi-resonance puts restrictions on the acoustic wave and the laser. That is, the laser is chosen with a frequency just above the plasma frequency, which is the cut-off frequency for propagation of the laser in the plasma.

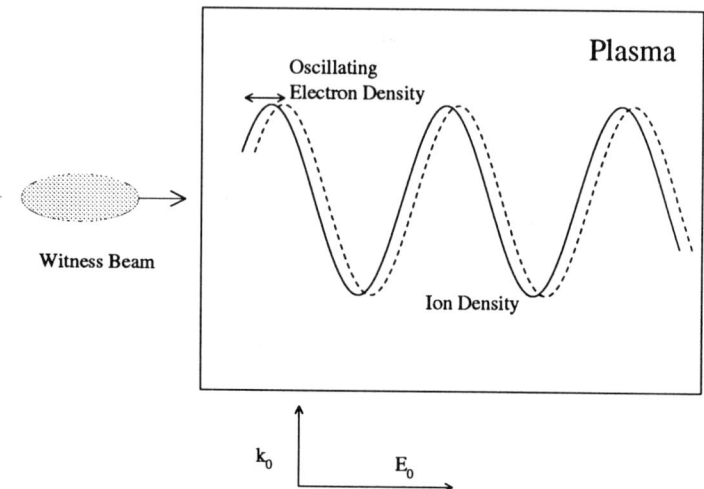

Figure 3. Side-injected laser with electric field polarized in the direction of the witness beam propagation and the acoustic wave. Figure adapted from Ref. 6.

Thus, the wavelength of the laser in the crystal is quite long. To be more definitive, let the frequency of the laser be ω_0, and the wavevector outside the plasma be $\mathbf{k}_0$ and inside be $\mathbf{k}_0'$. The 4-momentum of the acoustic wave is $(\omega_{ac} \approx 0, \mathbf{k}_r)$, and of the plasma wave is $(\omega_p, \mathbf{k}_p)$. The quasi-resonance conditions are

$$\omega_0 \pm \omega_{ac} \approx \omega_p \quad ,$$
$$\mathbf{k}_0' \pm \mathbf{k}_r \approx \mathbf{k}_p \quad . \tag{5}$$

For quasi-resonance then (5) must be satisfied, as well as the dispersion relation for the laser inside and outside the plasma, $ck_0' = \sqrt{\omega_0^2 - \omega_p^2}$, $\omega_0/k_0 = c$, and the dispersion relation for the acoustic wave $\omega_{ac}/k_r = v_s$, where v_s is the speed of sound in the medium. For a gas $v_s \approx 330 \, \text{m/s}$ and for a crystal around 5000 m/s.

For a laser that is just above the cut-off frequency of the plasma, the phase velocity for the plasma oscillation is

$$v_{ph} = \frac{\omega_0 \pm \omega_{ac}}{|k_0 \pm k_r|} \approx \frac{\omega_0}{k_r} \quad . \tag{6}$$

Eqn. (6) can be parameterized as $v_{ph}/c \approx \lambda_r/\lambda_0$, where λ_r is the wavelength of the acoustic wave, and λ_0 is the wavelength of the laser in free space. From this, it is seen that the acoustic wave length must be just less than the free space wavelength of the laser.

Solving the linearized cold fluid equations for the density perturbation to the electrons, Katsouleas *et. al.* in Ref. 6 calculate the accelerating electric field to be $E_z \approx 4\pi e n_1/k_r$, where the electron plasma density is $n = n_0 + \delta n \sin k_r x + n_1(x,t)$, and the velocity is $v_x = v_0 + v_1(x,t)$ where $v_0 = (-eE_0/m\omega_0)\cos\omega_0 = -v_{os}\cos\omega_0 t$ is due to the ponderomotive force of the laser. Using these parameters, the accelerating electric field can be written

$$E_z \approx \varepsilon\sqrt{n_0}\ \frac{V}{cm}\ ,$$
$$\varepsilon = \frac{1}{4}\frac{\delta n}{n_0}\frac{v_{os}}{c} = \frac{1}{4}\frac{\delta n}{n_0}\left(\frac{eE_0}{mc^2}\right)\frac{1}{\omega_0/c}\ . \tag{7}$$

The amplitude $\varepsilon = n_1/n_0$ is found from the solution of the wave equation derived from the cold fluid equations, and δn is the density fluctuation due to the acoustic wave. Ref. 6 gives several theoretical estimates on the saturation of the plasma oscillation growth due to: the wave breaking limit (an effect of the plasma thermal velocity), the detuning from the difference in the laser frequency and the plasma frequency, and the relativistic shift of the plasma frequency. Their simulations show that ε grows linearly with $\omega_p t$ and saturates with a value $\varepsilon \sim 0.10$, in their cases.

Consider for example the Brookhaven Accelerator Test Facility (ATF) RF gun[7] and accompanying CO_2 laser. For this laser operating near the damage limit for material, 10^{13} W/cm^2, and using the laser electric field found from $P_{laser} = E_0^2/(1200\pi)$ for SI units, the $\varepsilon \cong 0.0016$ assuming an acoustic wave with $\delta n/n \sim 0.1$. Since the laser frequency is approximately equal to the plasma frequency, the plasma density is about 10^{19}/cm^3, corresponding to As. This gives an acceleration gradient according to Eqn. (7) of 5.0 MeV/cm. At the upgraded ATF, the electron beam can be either 50 or 10 MeV and would have an energy spread about $\pm 0.4\%$, or ± 0.2 MeV and ± 0.04 MeV respectively. To measure a net energy gain, it must be greater than the energy spread and the energy loss traversing the solid: at 5 MeV/cm this means effective acceleration over more than 0.4 mm or 80 μm, respectively.

RADIATION LOSSES

Relativistic charged particles passing through matter will radiate away some of their energy. For any acceleration scheme to be useful the particle must gain more energy than it radiates. For particles that are not trapped in between crystal planes, this is dominated by bremsstrahlung. For those particles that are confined between crystallographic planes, the radiation is synchrotron-like and is referred to as classical channeling radiation (for ultra-relativistic energies).

BREMSSTRAHLUNG

An estimate of the energy loss of a positron due to bremsstrahlung can be found in the Particle Data Book.[8] For ultrarelativistic positrons the energy loss scale is given by the radiation length,

$$X_0 = \frac{716.4\,A}{Z(Z+1)\ln(287/\sqrt{Z})}\,\frac{\text{gm}}{\text{cm}^2}\ . \tag{8}$$

For Si this is 9.4 cm; a 50 GeV positron loses about 5.3 GeV/cm initially. The solid with the least energy loss is graphite with a radiation length of 18.8 cm giving losses of 2.7 GeV/cm.

For the 50 MeV beam case considered above, it was estimated that the peak gradient from a laser-acoustic accelerator was 5 MeV/cm. For Si, the peak energy loss is about 5.3 MeV/cm, while for graphite it is about 2.7 MeV/cm. This being the case, to overcome the inherent energy spread takes twice the distance than when radiation was ignored, or the 50 MeV beam must be accelerated for a distance of about 0.85 mm not 0.4 mm.

However, at such low energies ionization effects become non-negligible. An estimate of the energy at which radiation loss from ionization and bremsstrahlung are comparable is $E_c = 800/(Z+1.2)$ MeV. For graphite this is about 110 MeV. So, the ionization effects may even dominate the problem of energy loss.

CLASSICAL CHANNELING RADIATION OF POSITRONS

A positron that impinges on a crystallographic direction with a slight enough angle, the critical or channeling angle, will be confined between planes or between strings of atoms. Here we consider the situation when the particle is confined between planes of atoms, so-called planar channeling.[9] The planar channeling angle, or critical angle, below which channeling can occur is given by[10,11]

$$\psi_p = \left[\frac{2\pi Z e^2 n d_p \sqrt{3}\, a_{TF}}{E}\right]^{\frac{1}{2}}\ , \tag{9}$$

where n is the atomic density, d_p is the distance between planes, and a_{TF} is the Thomas-Fermi screening distance. See Table 2 for some estimated channeling angles.

The transverse confining potential for planar channeling defines a quantum system with the number of states[10]

$$n_p^+ = \sqrt{0.3\gamma Z^{2/3} n d_p^3}\ , \tag{10}$$

for a positron of energy $\gamma m c^2$ and a crystal with planar separation d_p and atomic density n. For an electron there are about 1/3 as many states as for a positron.

Table 2. Some properties of crystals.

crystal	$n_0(\text{cm}^{-3})$	$\lambda_p = 1/k$	ψ_p @ 50 GeV	ψ_p @ 50 MeV
Ge	3×10^{13}	0.97 mm		
Si	8×10^{14}	0.19 mm	(110) 39 μrad	0.92 mrad
Bi	3×10^{17}	9.7 μm		
Graphite	3×10^{18}	3.1 μm		
Ni			(110) 47 μrad	1.5 mrad
Cu	8.5×10^{22}	18 nm		
Al	1.7×10^{23}	13 nm		
W			(100) 49 μrad	1.5 mrad

For the Si (110) plane, with $Z = 14$, $n = 0.050$ atoms/Å^3, and $d_p = 1.92\text{Å}$, the number of states is approximately $n_p^+ = 0.33\sqrt{\gamma}$. For a positron with 50 GeV in Si, $n_p^+ = 100$, and for 50 MeV there are about 3 states; for an electron with 50 GeV there are about 33 states, and for 50 MeV there is one. For channeling a positron at high energy, the classical regime is relevant, while for energies around 50 MeV it is clearly a quantum problem.

Kheifets and Knight[12] calculate the classical channeling radiation for electrons and positrons. The radiation intensity averaged over several periods of the orbit is $I = I_0 F_b(x_m)$, where

$$I_0 = \frac{8\pi^2}{3}\gamma^2 r_e^3 mc^2 (Znd_p) \quad . \tag{11}$$

F_b is a universal function of the factor $b = 2\sqrt{3}a_{TF}/d_p$, where a_{TF} is the Thomas-Fermi screening of the atomic potential, and a function of the maximum excursion of the trajectory normalized so that $x_m = 1$ is the edge of the channel, and $x_m = 0$ is the channel center. The factors in (...) in Eqn. (11) are properties of the crystal only. Eqn. (11) is within a factor of 6 or so of what one would calculate assuming the Lenard radiation formula and a parabolic potential, instead of the much better approximation of the channeling potential due to Lindhard[13] that Kheifets and Knight use.

In Table 3, the classical channeling radiation rate is estimated for 4 crystals. Notice the exceptionally high radiation loss rate, at least compared to the above estimate of the acceleration of around a few MeV/cm. However, the scaling with energy is such that at 50 MeV the radiation is negligible—though at this energy it is a quantum problem and is not calculated here though it is presumed small.

Table 3.   Classical channeling radiation rates for a positron with 50 GeV and with a trajectory that goes to 20% of the channel width.

crystal plane	Z	n (atoms/Å^3)	d_p (Å)	a_{TF} (Å)	$I(0.2)/c$ (MeV/cm)
Si (110)	14	0.050	1.92	0.194	210
Al (110)	13	0.060	2.02	0.199	290
Ni (110)	28	0.091	1.76	0.154	1740
W (100)	74	0.055	1.65	0.112	2600

CONCLUSIONS

From the physics mentioned above, there are several ideas, or parameters for an experiment to test acceleration in a crystal that seem possible. Again there is much more work to do on each scenario that one can imagine and on the fundamental physics that might come into play in any given example.

For self-acceleration the beam densities required to create a high accelerating field seem restrictive. Though high energy positrons at the SLAC FFTB would probably channel effectively, their self-acceleration is difficult to observe. Also, for the driven accelerator, the gradients are too small for the energy gain to be measureable compared to the inherent energy spread. However, more work is still needed.

For lower energy beams, 50 MeV, the driven technique seems more realistic. Recall the example of driving the plasma to give a 5 MeV/cm acceleration while the bremsstrahlung losses are about 2.6 MeV/cm. Ionization can be expected to increase and perhaps double this energy loss for a non-channeled particle. Notice also that tuning the plasma does not yield a bigger gradient than 5 MeV/cm because the quasi-resonance conditions and the form for the accelerating gradient taken together do not depend on the plasma parameters. However, this currently seems the best approach for studying acceleration in solids.

REFERENCES

1. P. Chen and R. J. Noble, *Channeled Particle Acceleration by Plasma Waves in Metals*, in *Relativistic Channeling*, edited by Richard A. Carrigan and James A. Ellison, (Plenum Press, New York, 1987).

2. T. Tajima and M. Cavenago, *Crystal X-Ray Accelerator*, Phys. Rev. Lett. **59** (September 1987) 1440.

3. *Proceedings of the 2nd International Workshop on Laser Acceleration of Particles, Malibu, California, 1985*, edited by Chan Joshi and Thomas Katsouleas (AIP, New York, 1985).

4. Pisin Chen, *A Possible Final Focus Mechanism for Linear Colliders*, in Particle Accelerators **20** (1987) 171.

5. *Final Focus Test Beam Project Design Report*, SLAC–Report–376, March 1991.

6. T. Katsouleas *et al.*, *A Side-Injected-Laser Plasma Accelerator*, IEEE Trans. Nucl. Sci., **NS-32**, (October 1985) 3554.

7. K. Batchelor *et al.*, *Performance of the Brookhaven RF Gun*, Nucl. Instrum. Methods **A318** (1992) 372.

8. *Particle Properties Data Booklet*, from Phys. Lett. **B239** (April 1990).

9. *Relativistic Channeling*, edited by Richard A. Carrigan and James A. Ellison, (Plenum Press, New York, 1986).

10. Donald S. Gemmell, *Channeling and Related Effects in the Motion of Charged Particles through Crystals*, Rev. Mod. Phys. **46** (January 1974) 129.

11. M. A. Kumakhov and F. F. Komarov, *Radiation from Charged Particles in Solids*, (AIP, New York, 1989), original in Russian, 1985.

12. S. Kheifets and T. Knight, *Full Average Radiation of Electrons and Positrons Channeled Between the Planes of a Crystal*, J. Appl. Phys. **50** (September 1979) 5937; S. Kheifets and T. Knight, *Radiation Spectra and Angular Distribution of Emitted Quanta for Planar Channelled Particles: Radiation of a Single Particle*, J. Appl. Phys. **51** (July 1980) 3863.

13. J. Lindard, K. Dan. Vidensk. Selsk Mat. Fys. Medd. **34** (1965) 18.

COHERENT PAIR CREATION AS A POSITRON SOURCE FOR LINEAR COLLIDERS*

Pisin Chen and Robert B. Palmer[+]
Stanford Linear Accelerator Center
Stanford University, Stanford, California 94309

ABSTRACT

We propose a positron source for future linear colliders which uses the mechanism of coherent pair creation process from the collision of a high energy electron beam and a monochromatic photon beam. We show that there is a sharp spike in the pair-produced positron energy spectrum at an energy much lower than the primary beam energy. The transverse emittance is "damped", yielding final positrons with lower normalized emittance than the initial electrons. Numerical examples invoking conventional lasers and Free Electron Lasers (FEL) for the photon beams are considered.

INTRODUCTION

A high energy linear collider is a complex system. In the conventional approach, the electron beam, once emitted from the electron gun, is to be pre-accelerated to a certain energy so as to be injected into a damping ring, before eventually be accelerated to the machine energy through the main linac. For the positron beam, there is an additional intermediate step of positron production. This is achieved by bombarding a metalic target by an electron beam at tens of GeV energy (see Fig. 1a), and e^+e^- pairs are produced by the (incoherent) Bethe-Heitler process. Since future linear colliders generally require high beam currents, one potential problem is the melting of the solid target. Furthermore, the damping rings limit the minimum emittance attainable, and are expensive. While there is a good prospect that the electron beam can be produced at very low emittance right from the gun,[1] thus eliminating the need for the electron damping ring, the problem still awaits to be solved for the positron beam.

Recently, it has been suggested[2] that the electron-laser interaction through the single-photon Compton scattering and the subsequent two-photon Breit-Wheeler process can be a potential positron source for linear colliders (see Fig. 1b). An experimental effort is currently underway to test this idea.[3] In this paper we point out that for the purpose of a positron source, it maybe desirable to invoke the multi-photon, or *coherent*, regime of the electrom-photon beam-beam interaction. We show that in this approach the positron yield is high, and the emittance low. When looking at the numerical examples, we find that the laser or FEL technology needed for the photon beam is not too far from reach.

*Work supported by Department of Energy contract DE–AC03–76SF00515.
[+]Also at Brookhaven National Laboratory, Upton, NY.

Figure 1 presents the general form of a feedback controller applied to a dynamic system. This model shows a summing node, from which an error signal is generated, a feedback amplifier with complex gain $A(\omega)$, a second summing node which adds an external driving term $F(\omega)$, and a beam dynamics block with complex transfer function $H(\omega)$. The beam response acts back on the input summing node, closing the feedback loop.

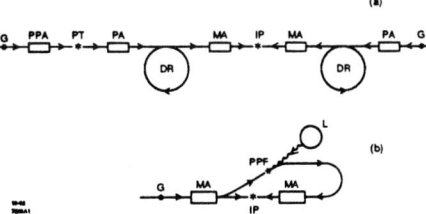

Fig. 1 Conceptual systems of linear colliders; (a) conventional approach, (b) that proposed here. G= gun, PPA= positron production accelereator, PA= pre-accelerator, DR= damping ring, MA= main accelerator, IP= interaction point, L= laser, PPF= Positron production focus.

2. Electron Interaction with Intense EM Waves

For the sake of discussion, we will assume that the monochromatic photon beam is circularly polarized. It is well known that the interaction between a relativistic electron and a monochromatic EM wave can be described by a Lorentz invariant parameter η:[1]

$$\eta \equiv \frac{e\sqrt{-a^2}}{mc} \quad , \tag{2.1}$$

where a is the 4-amplitude of the classical 4-potential of the EM wave. Physically, this parameter measures the ratio of the outcoming angle of the radiated photon, $\theta_c \simeq 1/\gamma$, as a result of the interaction, and the pitch angle, θ_H, of the electron trajectory which is helical in this case:

$$\eta = \frac{\theta_H}{\theta_c} \simeq \gamma \theta_H = \frac{eE}{\omega mc} \quad . \tag{2.2}$$

where ω is the EM wave frequency in the Lab frame. The last expression is identical to eq.(2.1) when a^2 is made explicit. A factor 2 coming from the fact that both E field and B field contribute equally to the bending of the electron trajectory is cancelled by the doubling of the electron oscillation frequency due to the relative electron wave velocities. We see that when $\eta \gg 1$, the radiation cone angle is much smaller than the pitch angle, and the process can be well described by synchrotron radiation; when $\eta \ll 1$, on the other hand, the process is describable by the (single-photon) Compton scattering. In another word, when $\eta \gg 1$, the number of photons absorbed in one physical process becomes large, and therefore the interaction with the photons becomes *coherent*, the physics approaches that of the interaction between the relativistic electron and the calssical EM field in the photon beam. This can be easily appreciated by recalling the well-known Correspondence Principle in quantum mechanics.

Consider a numerical example where the electron beam is at 250 GeV, with normalized emittance $\epsilon_n = 1 \times 10^{-6}$mrad, and the bunch length $\sigma_z \ll 150\mu$m. Consider also a laser at wavelength $\lambda = 350$ nm, energy $J = 15$ Joules, and the pulse length $\sigma_t = 0.5$ psec(150 μm). If this laser is focused with $f/d = 10$, the focused beam will have a radius $r = 2\mu$m, with a depth of focus of 70 μm, and a converging angle $3°$. One can easily varify that maximum EM field in the laser beam is

$$E = \left[\frac{Z_0 J}{\sqrt{2\pi}\sigma_t c \cdot \pi r^2}\right]^{1/2} = 2.43 \times 10^{13} \text{V/m} \quad,$$
$$B = \frac{E}{c} = 0.81 \times 10^9 \text{Gauss} \quad. \tag{2.3}$$

Inserting into eq.(2.2), we find $\eta_{max} = 2.7$. So the interaction is coherent, and is in the multiphoton regime.

In this coherent regime, one may invoke the language which describes the electron interaction with a static EM field. In this case the number of photons emitted by the relativistic electron per unit length of traverse in the transverse static EM field can be described by another Lorentz invariant parameter Υ:[2]

$$\frac{dn_\gamma}{dz} = \frac{5}{2\sqrt{3}}\frac{\alpha \Upsilon}{\lambda_c \gamma}U_0(\Upsilon) \quad, \tag{2.4}$$

where

$$\Upsilon = \gamma\frac{2E}{B_c} \quad; \quad U_0(\Upsilon) \approx \frac{1}{(1 + \Upsilon^{2/3})^{1/2}} \quad. \tag{2.5}$$

Here $B_c = m^2 c^3/e\hbar \approx 4.4 \times 10^{13}$ Gauss is the Schwinger critical field strength. Obviously, Υ does not have any frequency content. We see that Υ and η are related by

$$\Upsilon = \frac{2\gamma\hbar\omega}{mc^2}\eta \quad. \tag{2.6}$$

Thus in the regime of $\eta \gg 1$, a physical process can be either calssical ($\Upsilon \ll 1$) or quantum mechanical ($\Upsilon \gg 1$), depending on whether the photon is much less or more energetic than the electron in the electron rest frame. In the same numerical example discussed above, we find $\Upsilon_{max} = 19$. Thus at the peak of the photon beam field the interaction is deep in the quantum regime.

The photons that are emitted in this process will further interact with the same EM field, and have a finite probability of turning into e^+e^- pairs. This process has been called the *coherent pair creation* in the context of beamstrahlung.[3] In essense, this is the cross channel of the radiation process discussed above.

Let us define an equivalent Lorentz factor $\gamma' \equiv E_\gamma/mc^2$, and introduce an equivalent Lorentz invariant parameter $\Upsilon' = \gamma'2E/B_c$ for the coherent pair creation process: $\gamma \rightarrow e^+e^-$. It can be shown that in the asymptotic limits, the energy spectrum of the pair-produced particle, with fractional energy $x = E_{e+}/E_\gamma$, are given by[3,4]

$$\frac{d^2 n_{e+}}{dx\,dz} = \frac{1}{\sqrt{3}\pi}\frac{\alpha}{\lambda_c \gamma'}\left\{\left[\frac{1-x}{x} + \frac{x}{1-x}\right]K_{2/3}(\xi) + \int_\xi^\infty dz K_{1/3}(z)\right\}$$

$$\approx \frac{\sqrt{3}}{8\pi}\frac{\alpha\Upsilon}{\lambda_c \gamma'}\cdot e^{-\xi}\begin{cases} \sqrt{\pi}(\xi/2)^{1/2}[3 + (1-2x)^2], & \Upsilon' \le 50; \quad (2.7) \\[2ex] 2\Gamma(2/3)(\xi/2)^{1/3}[1 + (1-2x)^2], & \Upsilon' > 50; \end{cases}$$

where

$$\xi \equiv \frac{2}{3\Upsilon'}\frac{1}{x(1-x)}\ .$$

The approximate forms are obtained by taking series and asymptotic expansions of the Bessel functions for $\xi \ll 1$ and $\xi \gg 1$, respectively. Since ξ depends on a combination of Υ' and x, these approximations in principle are not proper in describing different regimes of Υ's. Emperically, however, we find that if we take the upper expression for $Y' \le 50$ and switch to the lower expression for $\Upsilon' > 50$, then the approximation is very good (see Fig. 2). In fact the worst fit occurs only in a very small range around $\Upsilon' \approx 50$ where the error is less then 15%. For all other values of Υ', the error is less than 5%.

In Fig. 2, we see that there exists a *threshold* at $\Upsilon' \sim 1$ in the coherent pair creation process, below which the probability is exponentially suppressed. This fact is one of the major motivations in our proposal for a positron source: In the showering process of successive radiations and pair productions during the electron-laser beam-beam collision, the further branchings essentially ceased when Υ' cascades down to around one.

This helps to accumulate positrons at a certain threshold energy, which in turn helps the yield of positrons within a reasonable energy window (see Fig. 3).

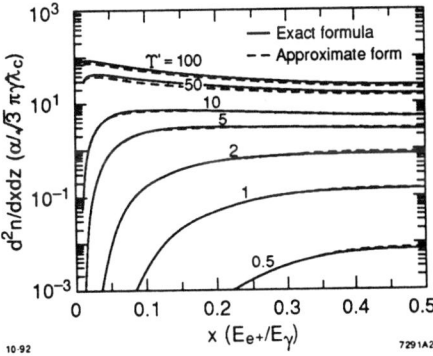

Fig. 2 A comparison of the exact and the approximate coherent pair creation spectrum in Eq.(2.7).

3. Damping of Positron Emittance

Another major advantage of this scheme is the damping of positron emittance. The *rms* angle of the initial electrons is

$$\theta_{in} = \sqrt{\frac{\epsilon_n}{\gamma_{in}\beta^*_{in}}} \quad . \tag{3.1}$$

In the successive radiation-pair creation process, the final state positron outcoming angle is widened.

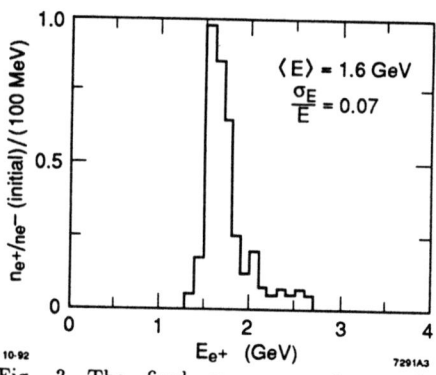

Fig. 3 The final energy spectrum of positrons. The parameters of example A in Table 1 are used in this calculation.

The typical angle of synchrotron radiation is an amount $1/\gamma$ tangential to the instantaneous trajectory of the electron. The pitch angle of the electron, as discussed earlier, is $\theta_H = \eta/\gamma$ at a given γ, where η is independent of the electron energy as long as it is relativistic (or $\gamma \gg 1$). Thus the typical outcoming angle of the photon is about an amount $1/\gamma$ departing from θ_H. For the electron, the difference between the initial state and final state pitch angles due to the energy loss is

$$\Delta\theta_H = \theta_{H1} - \theta_{H2} = \eta\left(\frac{1}{\gamma_2} - \frac{1}{\gamma_2}\right) \quad . \tag{3.2}$$

There is also a contribution to the angular increase from the conservation of momentum:

$$\Delta\theta_\gamma = \frac{1}{\gamma}\frac{E_\gamma}{E_{e_2}} \quad . \tag{3.3}$$

When the photon with energy γ' further turns into a e^+e^- pair, there is an additional transeverse momentum introduced which is of the order $p_\perp \sim mc$. This corresponds to an angle

$$\Delta\theta_{e^+e^-} \simeq 1/\gamma' \tag{3.4}$$

Then the final emittance of the positrons is

$$\epsilon_f = \gamma_{out}\beta^*_{out}\theta^2_{out} \quad , \tag{3.5}$$

where θ_{out} is all the above discussed angular contributions added in quadrature,

$$\theta_{out} = \left[\theta^2_{in} + \Delta\theta^2_H + \Delta\theta^2_\gamma + \Delta\theta^2_{e^+e^-}\right] \quad . \tag{3.6}$$

In the same numerical example that we discussed earlier, we find from a Monte Carlo simulation (see next section) that the *rms* angle increases from

$\theta_{in} = 6.3 \times 10^{-6}$ to $\theta_{out} = 120 \times 10^{-6}$. We see that although the angle is indeed degraded, it nevertheless is rather mild. On the other hand, to match with the input condition, the outcoming β^* remains to be 0.5 mm. The outcoming positron energy, however, is largely reduced due to the cascading process and the threshold effect that we discussed in the previous section. In this case, we find $\langle E \rangle \approx 2$ GeV, or $\langle \gamma \rangle \approx 3400$. This corresponds to an emittance of $\epsilon_{fn} = 3 \times 10^{-8}$ mrad $\ll \epsilon_n = 1 \times 10^{-6}$ mrad! Note that this emittance is comparable to the best one can achieve from the damping ring in one dimension, yet in this case it is in both.

4. Numerical Examples

To demonstrate the possibility of this idea of positron source, we perform Monte Carlo simulations of the electron-photon beam-beam collision process. The cascade of the initial electrons through the successive synchrotron radiation and coherent pair creation processes are tracked, using eq.(2.3) and eq.(2.4) for the radiation, and the approximate formulas in eq.(2.7) for the pair creation spectrum.

The first example studied (A in Table 1) used the parameters defined in earlier sections of this paper. Figure 4 shows the time structure of the parameter η inside the photon beam, which is assumed to be Gaussian, the mean energy of the initial electrons, and the yield of positrons per initial electron, n_{e+}/n_{e-}. We see that before the electrons encounter the maximum field strength in the photon beam, their energies have largely been lost to the synchrotron radiation. On the other hand, the positron yield becomes significant only after η starts to be larger than unity, as we expected. The yield in this example is $n_{e+}/n_{e-} \sim 4.5$, with an rms width equal to 7% of the peak (see Fig. 3). If positrons are accepted in a momentum window of $\Delta p/p = \pm 2.5\%$, then 1.6 positrons are obtained for each initial electron. The final normalized emittance is $\epsilon_{fn} \approx 3 \times 10^{-8}$ mrad.

In Table 1 parameters are given for a number of other examples. In all cases the initial normalized emittance is taken as 1×10^{-6} mrad, the initial and final β^*'s are 0.5 mm, and the apertures used were chosen to set the depth of focus equal to the interaction length. Different initial electron energies, laser pulse lengths and laser frequencies are chosen and in each case a laser power selected to give appoximately 1.5 positrons per initial electron.

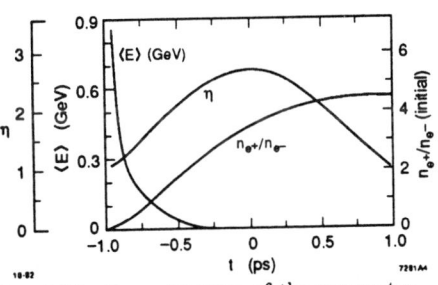

Fig. 4 The time structure of the parameter η inside the photon beam, the mean energy of the initial electrons, and the yield of positrons per initial electron, n_{e+}/n_{e-}.

It is seen that the required number of Joules is roughly inversely proportional to the electron energy, and almost proportional to the EM pulse length. The required EM energies falls even faster than linearly with the wavelength, giving a strong argument in favor of using very short wavelengths. The final emittance is seen to be strongly dependent of η, with values as low as possible favored.

Table 1. Parameters of Examples of Coherent Positron Sources

Electron Beam	A	B	C	D	E	F
E_e[GeV]	250	50	250	50	250	50
ϵ_n[10^{-6}mrad]	1	1	1	1	1	1
β^*[mm]	0.5	0.5	0.5	0.5	0.5	0.5
σ_z max [μm]	150	150	37	37	19	19
Photon Beam						
λ[nm]	350	350	350	350	40	40
J[Joule]	15	75	5	25	0.2	1.5
σ_t[fsec]	500	500	125	125	60	60
f/d	10.4	10.4	5.2	5.2	10.6	10.6
Pair Creation						
η	2.7	6.0	6.2	14	0.9	2.4
Υ	19	9	44	20	54	30
n_{e+}/n_{e-}	4.5	2.5	6.5	3.5	5.6	4.6
$\langle E \rangle$[GeV]	1.6	0.5	0.95	0.28	1.0	0.22
σ_E/E	0.07	0.06	0.11	0.08	0.12	0.08
n_{e+}/n_{e-}[$\Delta p/p = \pm 2.5\%$]	1.6	1.2	1.6	1.3	1.3	1.5
ϵ_{fn}[10^{-6}mrad]	0.03	0.06	0.28	0.69	0.04	0.10

5. Conclusion

The two key ingredients of our positron source are the existence of a thredshold in the coherent pair creation process, and the mild increase of the positron beam divergence at a much lower produced energy. The former helps for the selection of positrons within a narrow energy window, while the latter helps to damp the normalized eimttance of the produced positron beam. These features ensure that the positron beam thus produced has the right qualitative characteristic that meets the demand in future linear colliders.

In the actual Monte Carlo studies that we performed certain approximations have been made. First, at the early stage of the electron- photon beam-beam collision the laser intensity is generally low, and $\eta < 1$. This lies outside the coherent regime of interaction. Yet in the calculations, we apply the coherent formulas to the entire collision. Second, instead of using the general expression for coherent pair creation as in the first line in eq.(2.7), for the ease of computation we employed the approximate forms in eq.(2.7) instead. Furthermore, the transverse variations of the electron and photon beam intensities have been ignored in the calculation. However, since a photon beam would typically follow a Gaussian variation in time, which has a rather rapid rise in intensity, we do not expect the general features of our discussion to have been distorted too much in the calculation. In addition, the approximate coherent pair creation spectrum is actually quite accurate. Nevertheless, a more refined treatment should be persuit in our next effort.

In the numerical examples, it appears that at longer wavelength, e.g., λ = 350 nm, the energy requirements are not too far beyond the reach of convertional lasers. The problem is, such high power lasers tend to have low repetition rates. As a proof-of-princeple experiment, however, this type of lasers can still be very useful. When we turn to much lower wavelengths, it is only natural to consider a Free Electron Laser (FEL). It is believed that such FEL's with $\lambda \sim 40$ nm, energy of a fraction of a Joule, and pulse length around 60 femtosecond are feasible.[5] Certainly, laser and FEL experts will be more competent in assess the technical challenges in meeting the requirements in our scheme.

6. Acknowledgement

We appreciate very helpful discussions with Elan Ben-Zvi on the FEL's.

REFERENCES

1. K. Batchelor et al., *"Performance of the Brookhaven Photocathode RF Gun"*, Nucl. Instr. Meth. **A318**, 372 (1992).

2. J. Spencer, *IEEE Conf. Proc.* No.91CH3038-7 5, 3270 (1991).

3. J. G. Heinrich, C. Lu, K. T. McDonald (Princeton U.), C. Bamber, A. C. Melissinos, D. Meyerhofer, Y. Semertzidis (Rochester U.), P. Chen, J. E. Spencer (SLAC), R. B. Palmer (SLAC & BNL), *Proposal for a Test of QED at Critical Field Strength in Intense Laser High-Energy Electron Collisions at the Stanford Linear Accelerator*, SLAC Proposal E144, Oct. 1991. Additional members of the collaboration at SLAC include D. Burke, T. Barklow, C. Field, J. Frisch, K. Jobe, A. Odian, and D. Walz.

1. V. B. Berestetskii, E. M. Lifshitz and L. P. Pitaevskii, *Quantum Electrodynamics*, Sec. 101, Pergamon Press (1982).

2. K.Yokoya and P. Chen, in *Frontiers of Particle Beams*, Lecture Notes in Physcis **400** (Springer-Verlag, 1992).

3. P. Chen, in *Proc. DPF Summer Study, SNOWMASS '88*, p. 673, World Scientific (1989); P. Chen and V. L. Telnov, *Phys. Rev. Lett.* **63**, 1796 (1989); R. Blankenbecler, S. D. Drell, and N. Kroll, *Phys. Rev.* **D40**, 2462 (1989); M. Jacob and T. T. Wu, *Nucl. Phys.* **B327**, 285 (1989).

4. P. Chen, *Part. Accel.* **30**, 55 (1990).

5. E. Ben-Zvi, private communications, 1992.

An Intense Source of Positrons
Using a Low Energy Proton Beam

Paul J. Channell, AT-7, MS H829
Harry Dreicer, ADRE, MS A104
Los Alamos National Laboratory
Los Alamos, NM 87545

1 Introduction

An intense source of positrons (say $10^{14}e^+$ per second or more) would make it possible to consider linear collider designs for the flavor factories (ϕ and B) now being proposed. Such designs might have the advantage of higher luminosity than circular colliders and would also provide an arena for the further development of linear colliders in preparation for the next generation of TeV linear colliders. An intense source of positrons would also make it possible to be more flexible in the design of the next generation of TeV linear colliders.

The two conventional sources of positrons now used are electron-beam electroproduction of positrons and reactor-produced β^+-unstable isotopes. Electroproduction sources can be fairly efficient (one e^+ per 30-GeV electron used), but suffer from the disadvantages of high capital cost (a multi-GeV electron accelerator) and destruction of the target at higher fluxes. Reactor sources are much more inefficient and have much higher capital costs than electroproduction sources, but might make sense if a reactor already exists at an appropriate site and is principally used for other purposes.

In this paper we will show that it should be possible to produce large fluxes of positrons using intense proton beams (100 mA CW) of modest energy (6 − 10 MeV). The basic idea is to use protons from an accelerator rather than neutrons from a reactor to produce β^+-unstable isotopes by the proton capture reaction. Because the targets of a proton beam can be handled more conveniently and much more quickly than insertions in reactors, we are able to consider a larger range of possible targets with shorter decay times than is reasonable for a reactor. Because an accelerator beam is better defined and more easily controlled than neutrons in a reactor, the efficiency of conversion of beam energy into positrons can be much larger than in a reactor. Because the β^+-unstable isotopes from the target of the proton beam can be quickly transferred into a magnetic trap, after various combinations of chemical and mechanical separation of the β^+-unstable isotopes from the majority target species (in less than a decay time), positron annihilation can be largely avoided and the efficiency of useful positron production can be as high as that of the electroproduction sources.

In the next section, we will discuss the proton beam production of β^+-unstable isotopes and show that one can produce as many as $10^{14}e^+$ per second using 100-mA CW proton beams with energies of $6-10$ MeV. There are a number of possible targets, all with cross sections large enough that approximately one in 10^3 protons will be converted into β^+-unstable isotopes. Of course, once the positrons are produced it is necessary to avoid positron annihilation on background electrons and to capture the positrons in a small enough region of phase space to be useful, i.e., small enough that the positrons can be subsequently accelerated and cooled in a damping ring. Because only a very small fraction ($\approx$ 1 in 10^8) of the target nuclei undergo proton capture, it is essential to separate the β^+-unstable isotopes from the majority target species in order to avoid annihilation. After various combinations of chemical and mechanical separation, the β^+-unstable isotopes can be easily trapped in a magnetic bottle (e.g. a mirror, a cusp, etc.) where they decay into other nuclei and positrons. To produce ultracold positrons one could consider long-time confinement and radiative cooling in a magnetic trap (possibly a mirror or a cusp with electrostatic confinement added), although for large β^+ fluxes the resulting trap sizes are likely to be quite large. In section 3, we will consider a more modest trap, a mirror, which can be analyzed straightforwardly and which is good enough to provide radiative cooling of the multi-MeV positrons by a factor of about 4. The resulting output beam from the mirror could be accelerated to about 5 MeV and then passed through a foil cooling array, with small attendant annihilation, to produce a beam with a normalized emittance of about 2000π mm mrad, i.e., low enough that it can then be accelerated in a reasonable linac to an energy suitable for injection into a damping ring. We will discuss the foil cooling in section 4. In the final section we will summarize our results and briefly discuss variations one might consider.

2 Targets and Beam Requirements

In order to produce β^+-unstable isotopes using proton beams we must have a beam energy larger than the Coulomb barrier but smaller than the threshold energy for neutron production. We have not made an exhaustive study of all possible targets; however, based on the requirements that proton capture yield a β^+-unstable isotope, and that the target isotope be the most abundant form of the element, we have identified five candidate targets; the targets, the proton capture reactions, and the resulting decays are

$$p + {}^{14}N \rightarrow {}^{15}O + \gamma \rightarrow {}^{15}N + e^+,$$

$$p + M^{24}g \rightarrow {}^{25}Al + \gamma \rightarrow {}^{25}Mg + e^+,$$

$$p + {}^{28}Si \rightarrow {}^{29}P + \gamma \rightarrow {}^{29}Si + e^+,$$

$$p + {}^{32}S \rightarrow {}^{33}Cl + \gamma \rightarrow {}^{33}S + e^+,$$

$$p + {}^{40}Ca \rightarrow {}^{41}Sc + \gamma \rightarrow {}^{41}Ca + e^+,$$

where the γ in the intermediate state can stand for one or more prompt or decay gammas. For all of these reactions the (p, n), (p, α), $(p, 2n)$, etc. channels all have Q-values that are negative with magnitudes at least several MeV higher than the Coulomb barrier. The Coulomb barriers (i.e., the minimum proton beam energy required), the decay lifetimes for the second step, and the resulting positron energies for these reactions are shown in Table 1.

Table 1

Target	Coulomb Barrier (MeV)	Decay (sec)	e^+ Energy (MeV)
${}^{14}N$	3.2	124.0	1.74
${}^{24}Mg$	4.6	7.2	3.2
${}^{28}Si$	5.1	4.4	3.95
${}^{32}S$	5.6	2.5	4.5
${}^{40}Ca$	6.5	0.55	5.6

The target thicknesses are limited by the energy loss of the proton beam. For nonrelativistic protons the energy loss formula [1] is approximately

$$\frac{dE}{dz} = \frac{2\pi n Z e^4}{E} (\frac{M}{m}) \ln(\frac{4mE}{\hbar < \omega > M}),$$

where E is the energy of the proton, n is the density of the target, Z is the number of protons in the target nucleus, m is the electron mass, M is the proton mass, and $\hbar < \omega >$ is an average excitation energy of the target atoms. Let us define the loss distance to be

$$L \equiv (\frac{1}{E} \frac{dE}{dz})^{-1}.$$

The computed loss distances for the proton beams (taking the energy to be 3 MeV over the Coulomb barrier) is given in Table 2.

Target	Loss Distance (cm)
N^{14} (100 atm)	0.9
Mg^{24}	0.056
Si^{28}	0.048
S^{32}	0.061
Ca^{40}	0.096

Table 2

The thickness of the target cannot be greater than the loss distances given in this table lest the proton beam energy fall significantly below the Coulomb barrier. To estimate the efficiency of the conversion of protons to positrons, we use a target thickness equal to the proton loss distance and assume the capture cross section is a constant $\sigma = 350$ mb. Of course, the actual cross sections are slightly different from this for each target and vary as the protons slow down. The value we have assumed is typical of the cross section above the Coulomb barrier. The actual cross sections are somewhat higher at the initial beam energy and decrease to slightly less than this value as the beam loses energy. The proton conversion probability then turns out to be about the same in all five cases and is about 8.4×10^{-4}.

Note that the target will be vaporized (if solid) by the beam energy deposition, but this is necessary in any case for subsequent injection into a magnetic trap. One can envision various types of target arrangements, from streams of millimeter-sized pellets fired across the proton beam to flowing jets of gas or prevaporized target material at high pressure.

It is interesting to use the efficiency of conversion to determine the proton current required for the most demanding application, a very high luminosity

B-factory. It would probably be possible to design a linear collider B-factory with a luminosity of $1.3 * 10^{34}$ cm^{-2}sec^{-1} if we had a source of $7 * 10^{14}$ positrons per second [2]. If we assume that all the positrons produced can be trapped, cooled, and extracted, then the above-calculated efficiency can produce $7 * 10^{14}$ positrons per second if the proton beam has a CW current of 130 mA. Using a more realistic overall capture and cooling efficiency of 40%, we see that we need about 325 mA of proton current for the very high luminosity B-factory. Designs based on lower luminosity, of course, require less proton current.

In order to avoid excess annihilation we must separate the β^+-unstable isotopes from the untransmuted isotopes. Because the β^+-unstable isotopes are a different chemical species from the background target material, this should be relatively easy. Consider, for example, the first reaction, transmuting nitrogen to oxygen. It should be possible to devise chemical reactions, possibly followed by mechanical transfer pumping, that will separate these two with high efficiency in the time available (about two minutes). For the other reactions listed the chemical differences are not as great but could be large enough to allow efficient separation techniques.

Of course, other, more exotic, separation techniques can be considered if necessary. For example, because of the relatively large mass differences in the stable and unstable isotopes ($\approx 5\%$), electromagnetic separation of the isotopes should be fast and efficient. If it is necessary to achieve nearly 100% separation, one might consider laser separation by selective ionization of the unstable isotopes; this could be combined with injection into a magnetic trap by ionizing the unstable isotopes as they cross the magnetic bottle and pumping the background un-ionized gas. This technique would, of course, require some laser development, but is probably unnecessary, as ordinary chemical separation should achieve separation efficiencies $greater than 50\%$.

3   Traps

Once the β^+-unstable isotopes have been separated from the background target nuclei, they can be magnetically trapped, say by ionization of gas or vapor in the region of the trap, where they decay and where the positrons are "born" trapped [3]. (Note that having the positrons created in the trap is not a practical alternative for an electroproduction source because of the difficulty of injecting the electron beam into the trap.) It might be possible to make a trap, possibly a spindle cusp with electrostatic fields along the field lines at the field maxima, with very good confinement so that the positrons radiatively cool to very low temperatures. However, such a trap would, in all probability, be quite large and its confinement properties would be untested and thus quite uncertain. In this paper we will consider, instead, a conceptually simpler trap, a magnetic mirror.

Though we can not compute the output distributions from a trap as complex as a cusp, it is straightforward to compute the expected output from a magnetic mirror; as we will see, the performance of a simple mirror is good enough that it should be considered a strong candidate for the positron trap.

3.1   Positron Trapping

In this paper we will use the simplest possible model of a mirror, one in which the particle confinement is determined only by the mirror ratio, i.e., the ratio of maximum magnetic field along a field line to the minimum magnetic field along the line. We will ignore electrostatic fields, though it should be kept in mind that externally imposed electrostatic fields can probably be used to trap positrons for very long times and thus achieve brighter positron beams. We will ignore spatial structure, assuming that the motion along field lines is fast enough that once a particle enters the loss cone it is immediately lost. We have in mind a trap in which the bulk of the plasma is in the low-field region and the end regions occupy only a relatively small volume; the positrons are born from trapped isotopes mostly in the low field region with an energy spectrum given by the Fermi distribution. With these assumptions we can determine whether a particle is trapped or not simply by specifying its position in velocity space.

Our first task is to find the adiabatic invariant for relativistic particles that generalizes the nonrelativistic magnetic moment invariant. This can be computed using the observation that the adiabatic invariant for a periodically oscillating particle is the action integral around the periodic orbit, i.e., the adiabatic invariant, J, is

$$J = \int_0^T \vec{p} \cdot d\vec{q},$$

where T is the period of the oscillation. We let $\vec{p} = \gamma m \vec{v}_\perp$, where $\vec{v}_\perp$ denotes the perpendicular component of velocity. The expression for the period, T, of a particle in a magnetic field is

$$T = \frac{2\pi\gamma mc}{eB},$$

where m is the particle mass, c is the velocity of light, e is the particle charge, and B is the magnitude of the magnetic field. We can now compute J and, multiplying by irrelevant constants to get the right nonrelativistic limit, find that the relativistic magnetic moment is

$$\mu = \frac{\gamma^2 m\vec{v}_\perp^2}{2B}.$$

As the positrons move in the trap, their total energy and their magnetic moment are preserved on time scales short compared with a radiation damping time. Particles with a large enough magnetic moment can not pass over the magnetic field maximum and are trapped. Particles with a small magnetic moment are untrapped and are said to be in the loss cone; they leave the trap almost immediately. In terms of the velocity space of bulk particles, we can compute the location of the loss cone. The Hamiltonian can be expressed as

$$H = \frac{mc^2}{\sqrt{1 - \beta_\parallel^2 - \frac{2\mu B}{\gamma^2 mc^2}}}.$$

where $\vec{\beta} = \vec{v}/c$. In order for a particle in the bulk of the plasma to just reach the magnetic field maximum with no parallel velocity, energy conservation implies

$$\frac{mc^2}{\sqrt{1 - \beta_\parallel^2 - \frac{2\mu B_{\min}}{\gamma^2 mc^2}}} = \frac{mc^2}{\sqrt{1 - \frac{2\mu B_{\max}}{\gamma^2 mc^2}}},$$

where we have used magnetic moment conservation to write the same μ on both sides of this equation. From this equality it is easy to see that a particle will be in the loss cone if

$$\beta_\parallel \geq \sqrt{R-1}\,\beta_\perp,$$

where $R \equiv B_{\max}/B_{\min}$ is the mirror ratio. It is interesting to note that this is the same result one gets for nonrelativistic particles.

A positron "born" in the loss cone will leave the trap almost immediately; a positron not in the loss cone will stay in the trap for some time. We will count all positrons born in the loss cone as lost, because even if they do enter the subsequent accelerating sections it is likely that their longitudinal emittance is too large for them to be useful. Let us estimate the fraction of positrons born in the loss cone. The fraction, F, of velocity space occupied by the loss cone is just the total solid angle of the loss cone divided by 4π. Using the fact that the tangent of the angle of the loss cone is $\beta_\perp/\beta_\| = 1/\sqrt{R-1}$, it is easy to see that this is just

$$F = 1 - \sqrt{1 - \frac{1}{R}},$$

and that this is well approximated by

$$F \simeq \frac{1}{2R}.$$

As an example, for a mirror ratio $R = 4$, 87% of the positrons will be trapped.

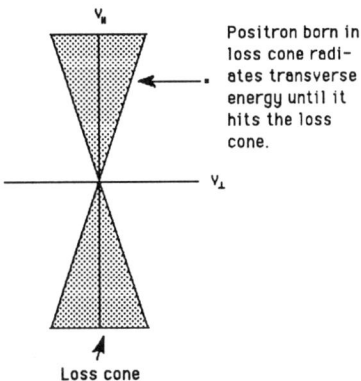

Positron born in loss cone radiates transverse energy until it hits the loss cone.

Loss cone

Figure 1.

Most of the positrons that are trapped will eventually become detrapped either by Coulomb collisions on other positrons and background plasma or by radiating their transverse energy away until they fall into the loss cone. For the relatively energetic positrons in the trap, collisional detrapping is negligible. For example, a 0.5-MeV positron in a trap with a density of 10^{12} cm^{-3} will suffer a collision about every 535 seconds and will undergo about

8 of these collisions before being scattered into the loss cone if the mirror ratio is 4; in other words, collisions are negligible on the isotope decay time scale. Thus we can conclude that most of the trapped positrons will be lost after they radiate away their transverse energy and fall into the loss cone, as depicted in Figure 1.

3.2   Radiative Cooling

To compute the confinement time of the positrons and their output distributions we need to compute the effects of radiation. From reference [1], one can show that the radiated power, P, from a particle moving both parallel and transverse to a magnetic field is given by

$$P = \frac{2\beta_\perp^2 \gamma^2 e^4 B^2}{3m^2 c^3}.$$

Using this and the expression for $\beta_\perp$ in terms of $\beta_\parallel$ and γ, we find that γ satisfies the equation

$$\frac{d\gamma}{dt} = -\frac{\gamma^2}{\tau}(1 - \beta_\parallel^2 - \frac{1}{\gamma^2}),$$

where

$$\tau \equiv \frac{3m^3 c^5}{2e^4 B^2}.$$

If we define a variable

$$\xi \equiv \sqrt{1 - \beta_\parallel^2}\, \gamma,$$

then it is straightforward to see that the solution of the equation is

$$\xi = \frac{\xi_0 + 1 + (\xi_0 - 1)e^{-t/\tau}}{\xi_0 + 1 - (\xi_0 - 1)e^{-t/\tau}},$$

where ξ_0 is the initial value of ξ. This solution shows that ξ decays to a value of 1, which implies that γ approaches $1/\sqrt{1 - \beta_\parallel^2}$, i.e., all the transverse energy radiates away leaving only the longitudinal energy. Of course, before all of the transverse energy is radiated the positron moves into the loss cone and exits the trap.

We can also estimate the confinement time of a particle in the trap. Using the exact solution to the energy equation leads to a complicated solution, so we will approximate the time dependence of ξ by

$$\xi \simeq 1 + (\xi_0 - 1)e^{-t/\tau},$$

in order to get a fairly simple answer. This function approximates the exact solution fairly well for all times. With this time dependence, it is straightforward to find that a particle will stay trapped for a time

$$t_{\text{trapped}} \simeq \frac{\tau}{\sqrt{1 - \beta_\parallel^2}} \left(\ln \left(\sqrt{\frac{1 - \beta_\parallel^2}{1 - \beta_\parallel^2 - \beta_\perp^2}} - 1 \right) \right.$$

$$\left. - \ln \left(\sqrt{\frac{(R - 1)(1 - \beta_\parallel^2)}{1 - \beta_\parallel^2 R}} - 1 \right) \right).$$

Using a mirror ratio of 4 and a maximum positron energy of 1.74 MeV, we have numerically averaged the coefficient of τ over the trapped part of the Fermi distribution and found that the average confinement time is

$$< t_{\text{trapped}} > \simeq 3\tau.$$

Setting this time equal to the isotope decay time allows us to determine the magnetic field. For example, for a 124-sec decay time we find that the magnetic field required is about 3.5 kG. A 1.74-MeV positron has a gyroradius of 2 cm in such a field.

3.3   Output Distribution and Emittance

If we ignore collisions and the particles that are born in the loss cone, the output distribution of the trap can be computed from the initial distribution of positrons and the assumption that the positrons radiate their transverse energy until they hit the loss cone. The output distribution is concentrated along the loss cone in velocity space with a distribution in longitudinal velocity given by the initial longitudinal velocity distribution of trapped positrons.

The full initial distribution of positrons is just the Fermi distribution [4], i.e., the distribution function, $f(\vec{p})$, in momentum is

$$f(\vec{p}) \propto (\chi - \gamma)^2,$$

where χ is the isotope-dependent maximum positron γ and where we have ignored the normalization constant. Converting from momenta, $\vec{p}$, to dimensionless velocities, $\vec{\beta}$, we find that the distribution function in velocity is

$$g(\beta_{\parallel}, \beta_{\perp}) \propto \beta_{\perp} \gamma^5 (\chi - \gamma)^2,$$

where

$$\gamma = \frac{1}{\sqrt{1 - \beta_{\parallel}^2 - \beta_{\perp}^2}}.$$

The distribution function for trapped particles is zero outside the region

$$\frac{|\beta_{\parallel}|}{\sqrt{R-1}} \le \beta_{\perp} \le \sqrt{1 - \frac{1}{\chi^2} - \beta_{\parallel}^2},$$

$$\beta_{\parallel}^2 \le (\frac{R-1}{R})(1 - \frac{1}{\chi^2}).$$

Integrating in $\beta_{\perp}$ over the allowed region for trapped particles we find that the distribution function in $\beta_{\parallel}$, $h(\beta_{\parallel})$, is

$$h(\beta_{\parallel}) \propto \frac{\chi^5}{15} - \frac{2\chi^2}{3D^{3/2}} + \frac{\chi}{D^2} - \frac{2}{5D^{5/2}},$$

where D is defined as

$$D \equiv 1 - \frac{\beta_{\parallel}^2 R}{R-1}.$$

For a particle exiting with a given value of $\beta_{\parallel}$ the transverse velocity is

$$\beta_{\perp} = \frac{|\beta_{\parallel}|}{\sqrt{R-1}},$$

and the total γ is

$$\gamma = (1 - \frac{\beta_{\parallel}^2 R}{R-1})^{-1/2}.$$

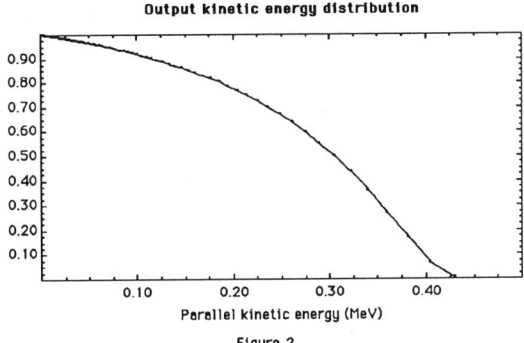

Figure 2.

The distribution function in parallel kinetic energy is shown in Figure 2. For an initial maximum positron energy of 1.74 MeV, the energy of the positron from oxygen decay, the average output energy of the trapped particles is 124 keV. We can estimate the output emittance by noting that all the exiting particles have the same angle, $\theta = \tan^{-1}(1/\sqrt{2(R-1)})$, so that multiplying by the trap radius, r_t, the normalized emittance, ϵ_n, is approximately

$$\epsilon_n \sim \frac{\pi\gamma\beta r_t}{\sqrt{2(R-1)}}.$$

As an example, taking the γ and β appropriate for 124 keV particles and using $R = 4$ and $r_t = 2 - 4$ cm, we find $\epsilon_n \sim 6000 - 12000\pi$ mm mrad, i.e., about a factor of $4 - 8$ less than the emittance produced without a trap.

3.4   Trap Limits

Though the trap parameters seem to be fairly practical, an average field of ~ 3.5 kG, a peak field of ~ 14.0 kG, a radius of a few centimeters, we still have not specified the trap length. Knowing the number of positrons produced in an isotope decay time we know how many positrons have to be contained in the trap. Because we should minimize the trap radius in order to minimize the output emittance, the radius is set by the gyroradius. Thus the trap length is determined when we determine the maximum density that the trap will hold. The maximum trap density allowed is set by two limits, (1) the annihilation rate of the positrons in the trap, and (2) plasma effects.

3.4.1   Annihilation

Because the positron annihilation rate depends on plasma density, the requirement that the probability of annihilation be considerably less than one sets a limit on the plasma density in the trap. The annihilation rate, A, is given in Heitler [5] and, assuming a uniform background of electrons, is given by

$$A = \pi n_e r_0^2 c,$$

where r_0 is the classical radius of the electron, n_e is the electron density, c is the speed of light, and the rate is in $\sec^{-1}$. Assuming a decay time of 124 sec and requiring that the probability of decay be less than one, we find that the electron density must satisfy $n_e < 10^{12}$ cm^{-3}. As an example, let us assume a positron production rate of 10^{14} per second and a trap radius of 4 cm, and measure plasma densities in units of 10^{12} cm^{-3}, i.e., $n_e \equiv \bar{n}_e * 10^{12}$ cm^{-3}; we then find that the plasma length, L, is constrained by

$$L > \frac{2.5}{\bar{n}_e} \text{ meters.}$$

Taking $\bar{n}_e \sim 0.1$ in order to have a 90% probability of survival gives a trap length of ~ 25 meters for this particular isotope in this example. The isotopes with shorter decay times will require, of course, smaller traps.

3.4.2   Plasma Effects

In the trap, it is necessary that the magnetic field pressure be large enough to confine the plasma; this capability is measured by the plasma beta, defined as

$$\beta_p \equiv \frac{8\pi n T}{B^2},$$

where n is the plasma density, T is the plasma temperature, and B is the magnetic field. In principle, there should be a sum over particle species in this definition, but in our case the high temperature of the positrons will cause their pressure to dominate all other species. The condition for equilibrium to be possible is that $\beta_p < 1$. Using the parameters of the example, $B = 3.5$kG, $n_p = 10^{11}$cm^{-3}, and taking the temperature, $T = 0.87$ MeV, i.e., about half the maximum positron energy, we find $\beta_p \approx 0.29$. Thus, equilibrium should be possible, but care must be taken to insure MHD stability for this high a value of beta. To decrease beta, we can either consider designs with greater lengths, or consider other targets with shorter decay times.

There is, of course, the possibility of many instabilities in such a plasma. The most obvious mirror mode, the loss-cone instability [6], is actually helpful in this case because it tends to eject the background ions, along with a roughly equal number of free electrons, and thus reduce the annihilation rate; the low frequency of this mode means that it is unlikely to have much effect on the positrons. There may, of course, be instabilities that affect the positrons directly, but these are likely to be considerably ameliorated by large Larmor radius and wall effects. It thus seems that plasma effects in the trap may be manageable, especially if longer traps are considered, but clearly this must be verified by carrying out experiments.

3.5   Extraction

Controlled extraction of the positrons from the trap is relatively straightforward. By applying longitudinal RF accelerating fields at one end of the trap, we can easily extract the positrons preferentially from that end of the trap in phase with the RF. (To further favor extraction from one end, we could easily weaken the magnetic field on that end.) The depth of the RF bucket required, about 250 keV for the oxygen decay, is not excessive. By continuing the magnetic field in a solenoid configuration one could provide focusing for several cells of RF acceleration until the energy is high enough to allow AG focusing cells to be used.

4   Further Cooling

In the previous sections we have seen that it seems possible to construct a relatively straightforward trap that can put out high currents of positrons with normalized emittances in the range $\epsilon_n \sim 6000 - 12000\pi$ mm mrad. Though these emittances might be small enough to accept into a linac and subsequently a damping ring, it would be easier to design the damping rings if the emittances were lower by another factor of 5 or so. Of course, we can use electrostatic fields to further confine the positrons in the trap and obtain even lower emittances, but it is useful to examine other alternatives as well. In this section we will show that foil cooling should work and that the required foils and their heat loads are reasonable. It is then possible to consider a scheme in which the positron cooling is done in three stages: first, radiative cooling in the magnetic trap decreases the emittances by a factor of 4 or so; then foil cooling in a linac decreases the emittances by another factor of about 5; finally, damping rings reduce the emittances to the required final values.

Foil cooling for applications in accelerators has been proposed and studied before [7], [8], [9], [10]. Briefly, the idea is that a plane foil in the beam

line causes the particles to lose energy parallel to their trajectories; if the
foil is followed immediately by an accelerating gap the particles regain the
lost energy, but only in the direction parallel to the beam axis; thus they
are cooled transversely. If the kinetic energy of the particles is greater than
about three times their rest mass, the particles are cooled longitudinally as
well, though for nuclear and particle- physics applications we do not require
longitudinal cooling. Of course, the positrons are also heated by the scattering because the RMS scattering angle with respect to the direction of
propagation is not zero. The minimum emittance that can be achieved is
thus determined by the balance between these two processes. The performance of the foil cooling system can also be limited by positron annihilation
and by the heat load on the foils. For the examples we consider, we will see
that annihilation and foil heat load are not major concerns.

For the transverse degrees of freedom the equation for either emittance,
ϵ, assuming thin foils is [9]

$$\frac{d\epsilon}{dn} = -\frac{dE}{dz}\frac{\delta}{E}\epsilon + \frac{\beta_f\theta_{rms}^2}{2},$$

where δ is the foil thickness, n is the variable numbering the cooling cell,
β_f is the beta function at the foil location, and θ_{rms} is the RMS scattering
angle in the foil.

For positrons with energies between 5 and 10 MeV the predominant
energy loss mechanism is multiple scattering and the energy loss formula is
approximately [1]

$$\frac{dE}{dz} = \frac{4\pi n Z e^4}{mv^2}\left(\ln((\gamma-1)\sqrt{\frac{\gamma+1}{2}}\frac{mc^2}{\hbar<\omega>}) - \frac{v^2}{c^2}\right),$$

where E is the energy of the proton, n is the density of the target, Z is
the number of protons in the target nucleus, m is the electron mass, M is
the proton mass, and $\hbar<\omega>$ is an average excitation energy of the target
atoms. As an example, if we assume the positrons are at 5 MeV and the
foils are carbon, then the parameters are

$$n \simeq 10^{23} \text{ cm}^{-3},$$

$$Z = 6,$$

$$\hbar<\omega> \simeq 60 \text{ eV},$$

and

$$\gamma \simeq 10.8.$$

We then find that

$$\frac{1}{E}\frac{dE}{dz} \simeq 0.69 \text{ cm}^{-1}.$$

A foil of this type that would induce a 5% energy loss would then be about 0.072 cm thick. If the positron current were 100 μA, the heat deposited would be about 25 W. Cooling such a heat load would be fairly easy. Ignoring for the moment the heating, it would take about 46 foils of this type followed by RF gaps imparting 250 kV each to to the positrons to reduce the emittances by a factor of 10.

The expression for θ_{rms} can also be obtained from reference [1] and is

$$\theta_{\text{rms}} \simeq \frac{16\pi n Z^2 e^4}{\gamma^2 m^2 v^4} \ln(\frac{210}{Z^{1/3}})\delta.$$

Balancing the damping against the heating we see that the minimum emittance that can be obtained is

$$\epsilon_{\text{min}} \simeq \frac{2\beta_f Z(\gamma-1)c^2}{\gamma^2 v^2} \frac{\ln(\frac{210}{Z^{1/3}})}{\left(\ln((\gamma-1)\sqrt{\frac{\gamma+1}{2}}\frac{mc^2}{\hbar<\omega>}) - \frac{v^2}{c^2}\right)}.$$

As an example, if we use the previous parameters, we find

$$\epsilon_{\text{min}} \simeq 0.43\beta.$$

At 5 MeV it should be possible design a transport channel with $\beta_f \simeq 0.5$ cm at the foil positions so that the minimum emittance achievable would be about 2150 π mm mrad, a value small enough to easily accept in a damping ring. Note that the β_f function need not be this small in the initial foils; when the beam is hot the beta function can be larger because heating is less important. The beta function at the foils can be reduced as the beam cools in order to continue the cooling process.

If we double the energy of the positrons to 10 MeV, assuming the same β_f function, then we could obtain emittances about a factor of two lower. However, the heat load on the foils would double, the RF gaps would have to supply twice the acceleration, and it would be harder to achieve this low a β_f function. At even higher energies, bremsstrahlung would begin to compete with scattering as the dominant energy loss mechanism and introduce different scalings. It thus seems that cooling should take place at about 5 to 10 MeV.

To estimate the annihilation probability, P, we use the cross section from reference [5] to find that

$$P \simeq L n Z \pi r_0^2 \frac{1}{\gamma + 1} \left(\frac{\gamma^2 + 4\gamma + 1}{\gamma^2 - 1} \ln(\gamma + \sqrt{\gamma^2 - 1}) - \frac{\gamma + 3}{\sqrt{\gamma^2 - 1}} \right),$$

where L is the total thickness of foils traversed and r_0 is the classical electron radius. Assuming 5-MeV positrons traverse 50 foils each 0.072 cm thick, we find that the total annihilation probability is about 0.14; thus annihilation is not a problem in this example.

5   Summary

We have presented a scheme that would allow the use of low energy proton beams to produce substantial fluxes of useful positrons for nuclear and high-energy physics applications. With a better understanding of the magnetic traps required it might also be possible to produce very high fluxes of cold positrons for solid-state diagnostics applications.

The principal uncertainty in our scheme is estimeated to be the trap physics, which primarily affects the overall trap size. However, even with fairly conservative assumptions on trap performance, it seems that it should be possible to design traps with reasonable overall sizes.

6   Acknowledgments

At various stages of this work we have benefited from conversations with Rich Sheffield, Dave Neuffer, and Fred Wysocki. In particular, Dave Neuffer urged that we consider the foil cooling scheme and Rich Sheffield suggested that we consider RF extraction from the trap.

References

[1] J. D. Jackson, *Classical Electrodynamics, 1st ed.* (John Wiley and Sons, Inc., New York, 1962).

[2] D. B. Cline, Proceedings of the Workshop on Advanced Accelerator Concepts, Madison, 1986, AIP Conference Proceedings #156, 435 (1987).

[3] G. Gibson, W. C. Jordan, E. J. Lauer, Phys. Rev. Lett. 141, 5, Aug. 15, 1960.

[4] J. M. Blatt and V. F. Weisskopf, *Theoretical Nuclear Physics.* (Springer-Verlag, Berlin, 1979).

[5] W. Heitler, *The Quantum Theory of Radiation, 3d ed.* (Oxford University Press, 1954).

[6] N. A. Krall and A. W. Trivelpiece, *Principles of Plasma Physics.* (McGraw-Hill Book Company, New York, 1973).

[7] E. A. Perevedentsev and A. N. Skrinsky, Proc. Second ICFA Workshop on Accelerators and Detectors, 61 (1979).

[8] V. V. Parkhomchuk and A. N. Skrinsky, Proc. 12th Int. Conf. on High-Energy Accelerators, 485 (1983).

[9] D. Neuffer, Particle Accelerators 14, 75 (1983).

[10] D. Neuffer, Proc. 12th Int. Conf. on High-Energy Accelerators, 481 (1983).

INVERSE FEL PROTON ACCELERATOR VIA PERIODICALLY MODULATED CRYSTAL STRUCTURE

S.A. Bogacz

Accelerator Physics Department, Fermi National Accelerator Laboratory[*]

P.O. Box 500, Batavia, IL 60510

ABSTRACT

Presented study explores the idea of using a visible light wave to accelerate relativistic protons via the inverse FEL mechanism. Here, a strain modulated crystal structure - the superlattice, plays the role of a microscopic undulator providing very strong ponderomotive coupling between the beam and the light wave. Purely classical treatment of relativistic protons channeling through a superlattice is performed in a self-consistent fashion involving the Maxwell wave equation for the accelerating electromagnetic field and the relativistic Boltzmann equation for the protons. It yields the accelerating efficiency in terms of the negative gain coefficient for the amplitude of the electromagnetic wave - the rate the energy is extracted from the light by the beam. Presented analytic formalism allows one to find the acceleration rate in a simple closed form, which is further evaluated for a model beam - optical cavity system to verify feasibility of our scheme.

INTRODUCTION

Here we suggest using a solid state superlattice as an undulator in conjunction with an optical pumping cavity to accelerate relativistic protons. Heavier particles (vs electrons or positrons) are more appropriate, because they are not susceptible to emitting photons via spontaneous synchrotron radiation and therefore they are more likely to absorb energy from an electromagnetic wave. The idea of the inverse FEL mechanism employed to accelerate charged particles[1] is not new by itself; what we propose here is to replace a conventional magnetic undulator with a microscopically modulated crystal structure – the superlattice, which assures much stronger ponderomotive coupling to the pumping wave. Furthermore, its microscopic undulator periodicity shifts the wavelength of the pumping optical mode to much shorter waves – it opens a possibility of accelerating the beam with high power laser in the visible region.

Such periodic crystal structures occur naturally in several alloy systems or may be prepared artificially with vapor deposition techniques[2]. We consider a superlattice with an accompanying strain modulation, which is a natural consequence of the two constituents having slightly different lattice spacings (e.g. Si–Ge superlattice). The main idea of using a modulated crystal structure as an undulator is illustrated schematically in Figure 1. A beam of relativistic particles while channeling through the crystal follows a well defined trajectory. Presented treatment will be limited to protons. Figure 1 depicts channeling paths, which for positive particles would lie in low density regions of the crystal. There are two distinct channeling directions: parallel to the superlattice growth direction and also at a 45^0 angle to this direction. The latter are what interest us here because, as one can see from Figure 1, the center of the channeling axis is _modulated by the superlattice periodicity_. This is the essence of the solid state undulator: i.e., the particles are periodically accelerated perpendicular to their flight path as they traverse the channel. The undulator wavelengths typically fall in the range 50–500 Å, far shorter than those of any macroscopic undulator. Furthermore, the electrostatic crystal-fields in-

---

[*]Operated by the Universities Research Association, Inc., under a contract with the U.S. Department of Energy

914   © 1993 American Institute of Physics

volve the line averaged nuclear field and can be two or more orders of magnitude larger than the equivalent fields of macroscopic magnetic undulators (translated into the corresponding electric field in the rest-frame of a relativistic particle). Both of these factors hold the promise of greatly enhanced coupling between the beam and the pumping electromagnetic wave.

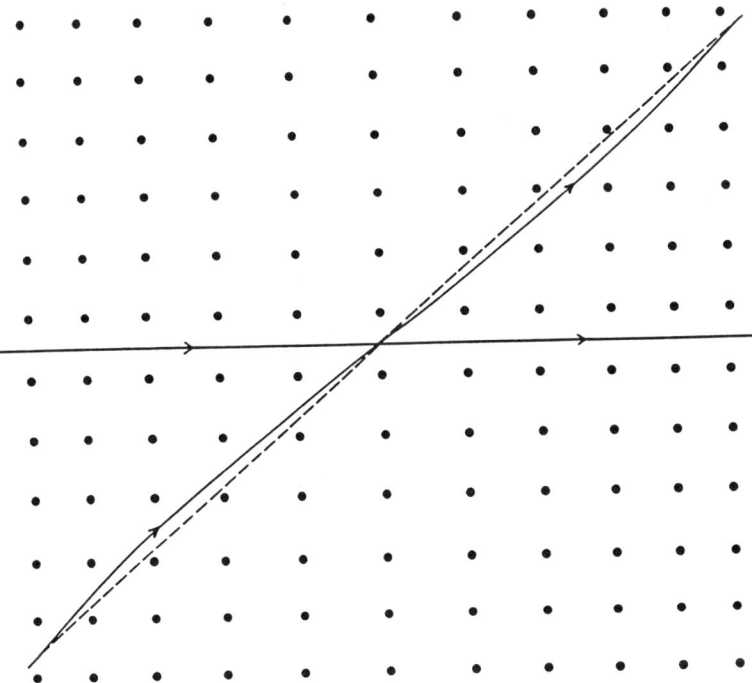

Figure 1   Center of the channeling trajectory for protons in [110] direction in a strain-modulated superlattice. The [100] channeling direction yields no undulator effect.

CHANNELING OF RELATIVISTIC PROTONS THROUGH A STRAIN–MODULATED SUPERLATTICE

Following the spirit of the Vlasov equation[3] we will describe a high intensity proton beam in terms of a classical distribution function, $f(\mathbf{p}, \mathbf{x}, t)$, governed by the relatiyistic Boltzmann equation. The transverse dynamics of relativistic protons propagating in a strain-modulated superlattice is modeled by a harmonic crystal field potential[4] and leads to generation of a transverse current. This couples the Vlasov equation to the Maxwell wave equation. Therefore, presented problem reduces to a self consistent solution of the Vlasov and the wave equations, which will be treated in detail in the next few sections. Finally, a closed analytic expression for the amplitude-gain/loss coefficient will be obtained in the linear approximation for appropriate regime with respect to the cavity length and the momentum spread in the incident beam.

We start with a relativistic Lagrangian describing motion of a proton in an arbitrary electromagnetic field $(\mathbf{A}, \phi)$

$$L = - mc^2 \sqrt{1 - v^2/c^2} + \frac{e}{c} v \cdot A - e\phi . \tag{1}$$

Here A is a vector potential of an electromagnetic field and ϕ is a phenomenological harmonic crystal-field potential, which describes both transverse focusing of the beam and longitudinal modulation of the minimum of the harmonic potential well. More realistic description of the planar channeling is usually given in terms of the Moliere potential[5], which accounts for averaged electric field of ions and electron cloud around each lattice site. For the purpose of our model calculation a harmonic approximation to the focusing potential is quite sufficient and it can be written as follows

$$\phi = \phi_0 + \frac{1}{2} \phi_1 \left(x - x_1 \cos gz \right)^2, \tag{2}$$

where $g = 2\pi/l$ is the strain modulation periodicity and x_1, ϕ_1, ϕ_0 are parameters of the potential. In order to avoid unnecessary complexity we choose to work directly in the laboratory reference frame, and use a space-time description of the problem.

We introduce the canonical momentum of a particle

$$p^\alpha = \frac{\partial L}{\partial \dot{x}^\alpha} = m\gamma v^\alpha + \frac{e}{c} A^\alpha \tag{3}$$

and write the Hamiltonian as

$$H = p^\alpha \dot{x}^\alpha - L = m\gamma c^2 + e\phi , \tag{4}$$

with

$$\gamma(p) = \sqrt{1 + \frac{1}{(mc)^2} \left(p - \frac{e}{c} A \right)^2} . \tag{5}$$

Hamilton's equations of motion are then

$$\dot{x}^\alpha = \frac{\partial H}{\partial p^\alpha} = \frac{1}{m\gamma} \left(p^\alpha - \frac{e}{c} A^\alpha \right) , \tag{6}$$

$$\dot{p}^\alpha = - \frac{\partial H}{\partial x^\alpha} = \frac{e}{mc\gamma} \left(p^\beta - \frac{e}{c} A^\beta \right) \frac{\partial A^\beta}{\partial x^\alpha} - e \frac{\partial \phi}{\partial x^\alpha} . \tag{7}$$

A beam of protons moving along the z axis can be described in terms of a distribution function, $f(p,x,t)$; this distribution function obeys the relativistic Vlasov equation

$$\frac{\partial f}{\partial t} + \frac{1}{m\gamma} \left(p^\alpha - \frac{e}{c} A^\alpha \right) \frac{\partial f}{\partial x^\alpha} \tag{8}$$

$$+ \left[\frac{e}{mc\gamma} \left(p^\beta - \frac{e}{c} A^\beta \right) \frac{\partial A^\beta}{\partial x^\alpha} - e \frac{\partial \phi}{\partial x^\alpha} \right] \frac{\partial f}{\partial p^\alpha} = I(p^\alpha) .$$

Here $I(p^\alpha)$ is a collision integral accounting for various incoherent proton scattering processes. This equation will be treated iteratively and only linear terms in the A-field will be retained. In the 0-th order solution $A = 0$, and the corresponding distribution function $f = f^{(0)}$ is obtained (in the absence of collisions) from the solution of

$$\frac{\partial f^{(0)}}{\partial t} + \frac{p^\alpha}{m\gamma} \frac{\partial f^{(0)}}{\partial x^\alpha} - e \frac{\partial \phi}{\partial x^\alpha} \frac{\partial f^{(0)}}{\partial p^\alpha} = 0 \ . \tag{9}$$

A class of solutions, $f^{(0)}$, describing a beam of protons with a sharply peaked initial momentum distribution, $\Lambda(p_z - p_0)$, $p_x = 0$ can be easily constructed by solving the following set of equations

$$\dot{p}_x = - e \, \phi_1 \left(x - x_1 \cos gz \right), \tag{10}$$

$$\dot{p}_z = - e \, \phi_1 \left(x - x_1 \cos gz \right) x_1 \, g \sin gz \ . \tag{11}$$

Keeping only linear terms in x_1, yields the general solution (including both transient and steady state parts) as follows

$$x = \frac{x_1}{1 - U^2} \cos (gz + \theta) + C \cos k_\beta z \tag{12}$$

and

$$\dot{p}_z = 0 \ ,$$

where $U = g/k_\beta$ and C is an arbitrary constant defining initial amplitude of the transient oscillation. Here, k_β is a focussing strength of the crystal channel given explicitly below

$$k_\beta = \sqrt{\frac{e\phi_1}{p_z v_z}} \ . \tag{13}$$

The remaining phase shift, θ, depends on the relative strength of the two frequencies k_β and g. Its asymptotic behavior can be summarized as follows

$$\theta \to 0 \quad \text{for} \quad g \ll k_\beta \tag{14}$$

$$\theta \to \pi \quad \text{for} \quad k_\beta \ll g \ .$$

It follows from Eqs.(12)–(14) that the initial distribution of canonical momenta in the z direction is conserved during the channeling, while the particles follow the trajectory defined by $x^{(0)}(z) = x$. In principle the trajectory is out of phase with the driving strain modulation (in extreme relativistic regime θ reaches π). This implies that particles which enter the crystal with x coordinates off the channel plane lose energy via scattering mechanisms and asymptotically approach the channel plane[6]. Therefore, the solution, $f^{(0)}$, can be factorized as follows

$$f^{(0)} = n_0 \, \delta(x - x^{(0)}) \, \delta(p_x - p_x^{(0)}) \Lambda(p_z - p_0), \tag{15}$$

where

$$p_x^{(0)} = p_z \frac{\partial x^{(0)}}{\partial z} = - p_z x^{(0)} \frac{gx_1}{1 - U^2} \sin gz \ . \tag{16}$$

Here $\Lambda(p_z - p_0)$ describes an initial momentum distribution and n_0 is a concentration of particles per unit area of the channeling plane. The distribution function, $f^{(1)}$, generated by the A-field obeys the following linearized Vlasov equation

$$\frac{\partial f^{(1)}}{\partial t} + \frac{p_x}{m\gamma} \frac{\partial f^{(1)}}{\partial x} - \frac{1}{m\gamma} \frac{e}{c} A_x \frac{\partial f^{(0)}}{\partial x} + \frac{p_z}{m\gamma} \frac{\partial f^{(1)}}{\partial z} \tag{17}$$

$$+ \frac{ep_x}{mc\gamma} \frac{\partial A_x}{\partial z} \frac{\partial f^{(1)}}{\partial p_x} - e \frac{\partial \phi}{\partial x} \frac{\partial f^{(1)}}{\partial p_x} = \frac{f^{(1)}}{\tau} \; .$$

Here we have used a relaxation time approximation for the collision integral modelling it by the relaxation time, τ. We have also assumed that only the transverse component of the A-field is present and $A_x \equiv A(z, t)$. We seek a solution, $f^{(1)}$, in the following form

$$f^{(1)} = n_0 \, \delta(x - x^{(0)}) \, \delta(p_x - p_x^{(0)}) \, h(z, p_z, t) \; , \tag{18}$$

where h describes bunching of particles due to the presence of the A-field. Substituting Eqs.(15)–(16) and (18) into Eq.(17) leads to the following kinetic equation for h

$$\frac{\partial h}{\partial t} + \frac{p_z}{m\gamma} \frac{\partial h}{\partial z} + \frac{e}{c} \frac{p_z}{m\gamma} \frac{gx_1}{1 - U^2} \frac{\partial A}{\partial z} \frac{\partial \Lambda}{\partial p_z} \sin gz = -\frac{h}{\tau} \; . \tag{19}$$

The inhomogeneous term in the above equation plays the role of a driving force representing acceleration of the particles by the ponderomotive force due to the transverse motion (induced by the crystal field) in the presence of the A-field. The transverse current induced by the fields is given by

$$j_x = e \int_{-\infty}^{\infty} dp_x \int_{-\infty}^{\infty} dp_z \, v_x \, f^{(1)} \; , \tag{20}$$

where

$$v_x = \frac{1}{m\gamma} \left(p_x - \frac{e}{c} A \right) . \tag{21}$$

The current in a single channel can be written to linear accuracy in A as follows:

$$j_x = -n_0 \, e \, \delta(x - x^{(0)}) \int_{-\infty}^{\infty} dp_z \, h(z, p_z, t) \, \frac{p_z}{m\gamma} \frac{gx_1}{1 - U^2} \sin gz. \tag{22}$$

The net transverse current in the crystal is given by a sum of discrete planar currents concentrated at each channeling site of the x direction. According to the Appendix, an infinite array of equally spaced currents is equivalent to a smoothed uniform current distribution given by Eq.(22) but without the factor $\delta(x - x^{(0)})$ and with an associated proton density redefined as $n = n_0/a$, where a is the spacing between adjacent channels. A more general situation where the crystal field potential in each channel is shifted in phase is also discussed in the Appendix. The resulting transverse current couples Eq.(19) to the following wave equation

$$\left(\frac{\partial^2}{\partial z^2} - \frac{1}{c^2}\frac{\partial^2}{\partial t^2}\right)A = \frac{4\pi n e}{c} \int\limits_{-\infty}^{\infty} dp_z \ h \ \frac{p_z}{m\gamma}\frac{g x_1}{1 - U^2} \ \sin\ gz, \tag{23}$$

resulting in a closed system of equations for h and A. Here the A-field can be identified as a sum of the macroscopic driving field and a self consistent electromagnetic field propagating in the crystal structure. We will not attempt to solve this system of partial-differential-integral equations exactly. For the purpose of calculating a linear amplitude–loss/gain coefficient, it is sufficient to confine the solution for the A-field to only the first step of the iteration procedure. The procedure is analogous to the Born approximation in scattering theory[7]. We start with a single plane wave solution of arbitrary ω and k propagating in free space along the z axis in both directions and use it as a 0-th order iteration step

$$A_{\pm}^{(0)} = A_0 \ e^{-i\omega t \pm ikz} \ . \tag{24}$$

Putting $A = A_{\pm}^{(0)}$ in Eq.(19), one can solve it analytically for $h = h_{\pm}^{(1)}$ by constructing a Green's function with the appropriate boundary conditions built in it. We consider a finite crystal undulator extending along the z axis from 0 to L. The bunching function, $h_{+}^{(1)}$, corresponding to the right propagating initial wave, $A_{+}^{(0)}$, vanishes outside the undulator and it is assumed to be continuous at the entry point, z = 0, of the $A_{+}^{(0)}$ wave. Similarly for the left going solution, $A_{-}^{(0)}$, the analogous initial condition is satisfied at its entry point, z = L. The complete set of boundary conditions can be written in the following compact notation

$$h_{\pm}^{(1)}(z_{\pm}) = 0, \tag{25}$$

where

$$z_{+} = 0 \ , \quad z_{-} = L \ . \tag{26}$$

Finally, the solution of Eq.(19) with the above boundary condition (25)–(26) is expressed in the following integral form

$$h_{\pm}^{(1)}(z) = \int\limits_{z_{\pm}}^{z} dz' \ H_{\pm}^{(1)}(z') \ e^{i(z-z')m\gamma\omega/p_z} \ , \tag{27}$$

where

$$H_{\pm}^{(1)}(z) = \pm ikA_{\pm}^{(0)}(z) \sin gz \ Q \frac{\partial \Lambda}{\partial p_z} \tag{28}$$

and

$$Q = \frac{e}{c}\frac{g x_1}{1 - U^2} \ , \tag{29}$$

where in what follows, collisions are modeled by writing $\omega \rightarrow \omega + i/\tau$. Substituting the above solution for $h_{\pm}^{(1)}$ in Eq.(23) reduces it to an inhomogeneous Helmholtz equation for $A_{\pm}^{(0)}$:

$$\left(\frac{\partial^2}{\partial z^2} - \frac{1}{c^2}\frac{\partial^2}{\partial t^2}\right)A_{\pm}^{(1)}(z) = J_{\pm}(z) \ , \tag{30}$$

where the explicit formula for $J_{\pm}$ can be obtained from Eq.(27) and Eqs.(27)–(29). To solve Eq.(30) one may integrate it with the following boundary conditions at the entry points for the right and left propagating solutions

$$A_{\pm}^{(0)}(z_{\pm}) = A_{\pm}^{(1)}(z_{\pm}) \ . \tag{31}$$

and

$$\frac{\partial}{\partial z} A_{\pm}^{(0)}\Big|_{z=z_{\pm}} = \frac{\partial}{\partial z} A_{\pm}^{(1)}\Big|_{z=z_{\pm}} \ . \tag{32}$$

The solution can be written explicitly in terms of the Green's function for the Helmholtz equation

$$A_{\pm}^{(1)}(z) = A_{\pm}^{(0)}(z) + \int_{z_{\pm}}^{z_{\mp}} dz' \ \frac{e^{ik|z-z'|}}{2ki} J_{\pm}(z') \ , \quad k = \omega/c \ , \tag{33}$$

where

$$J_{\pm}(z) = \pm \, ik \int_{-\infty}^{\infty} dp_z \ F(p_z) \sin gz \int_{z_{\pm}}^{z} dz' \sin gz' \ A_{\pm}^{(0)}(z') \ e^{i(z-z')m\gamma\omega/p_z} \ , \tag{34}$$

with the kernel $F(p_z)$ defined as follows

$$F(p_z) = 4\pi n Q^2 \ \frac{p_z}{m\gamma} \ \frac{\partial \Lambda}{\partial p_z} \ . \tag{35}$$

Carrying out the integration in Eq.(34) leads to the required solutions describing the evolution of the right and left propagating waves. To introduce a single pass loss/gain coefficient one should examine the resulting amplitudes after passing through the undulator of length L, i.e., we have to evaluate $A_{\pm}^{(0)}(z=z_{\mp})$ by using Eq.(34). On the other hand, one can model the effect of coupling by adding a small complex part $i\kappa^{\pm} = \alpha^{\pm} + i\beta^{\pm}$ to the k-vector; here α is a gain/loss coefficient and β describes a small shift in the phase velocity of the optical mode. This could be summarized by the following expression

$$A_{\pm}^{(1)}(z) = A_0 \, e^{-i\omega t \pm ikz \pm i\kappa^{\pm}(z-z_{\pm})} \ . \tag{36}$$

Expanding to linear order in $\kappa^{\pm}$ we obtain

$$A_{\pm}^{(1)}(z) = A_{\pm}^{(0)}(z) \left[1 \pm i\kappa^{\pm}(z-z_{\pm}) \right] \ . \tag{37}$$

The above expression evaluated at $z = z_{\pm}$ simplifies to

$$A_{\pm}^{(1)}(z_{\mp}) = A_{\pm}^{(0)}(z_{\mp}) \ (1 \pm i\kappa^{\pm}L \ ) \ . \tag{38}$$

Further comparison of Eqs.(38) and (34) (with $z = z_{\mp}$) allows one to identify two complex loss/gain coefficients $\kappa^{\pm}$, as follows

$$\kappa^+ = \frac{1}{2Li} \int_{-\infty}^{\infty} dp_z\, F(p_z) \int_0^L dz'\, e^{-ikz'} \sin gz' \int_0^{z'} dz''\, \sin gz''\, e^{i(z'-z'')m\gamma\omega/p_z + ikz''} \quad (39)$$

and

$$\kappa^- = \frac{1}{2Li} \int_{-\infty}^{\infty} dp_z\, F(p_z) \int_0^L dz'\, e^{-ikz'} \sin gz' \int_{z'}^L dz''\, \sin gz''\, e^{i(z'-z'')m\gamma\omega/p_z + ikz''} . \quad (40)$$

Apart from a simple integration over z' and z", we have found the complex coefficient, $\kappa^{\pm}$, for waves propagating parallel and antiparallel to the particle beam. The imaginary part of $\kappa^{\pm}$ describes either spontaneous amplification or degradation (depending on its sign) of the optical mode. In the next section, the above integrations will be carried out explicitly and the linear gain/loss coefficients will be calculated in a closed form.

LINEAR LOSS/GAIN COEFFICIENT – ACCELERATION RATE

On carrying out the two spatial integrations in Eqs.(39) and (40), the expressions for the complex gain, $\kappa^{\pm}$, reduce to

$$\kappa^+ = \frac{L}{16} \int_{-\infty}^{\infty} dp_z\, F(p_z) \left\{ \frac{1}{\mu^-} [N(\mu^+) - N(\mu^-) - N(2g) - N(0)] \right. \quad (41)$$

$$+ \frac{1}{\mu^+} [N(\mu^-) - N(\mu^+) + N(0) - N(-2g)]$$

$$- e^{-2ikL} \frac{1}{\mu^-} [N(\nu^-) - N(\nu^+) + N(2k + 2g) - N(-2k)]$$

$$\left. - e^{-2ikL} \frac{1}{\mu^+} [N(\nu^-) - N(\nu^+) + N(2k) - N(2k - 2g)] \right\}$$

and

$$\kappa^- = \frac{L}{16} \int_{-\infty}^{\infty} dp_z\, F(p_z) \left\{ \frac{1}{\nu^-} [N(2g) - N(0) - N(2g - 2k) + N(-2k)] \right. \quad (42)$$

$$- \frac{1}{\nu^+} [N(0) - N(-2g) - N(-2k) + N(-2k - 2g)]$$

$$- e^{-i\nu^- L} \frac{1}{\nu^-} [N(\nu^+) - N(\nu^-) - N(\mu^+) + N(\mu^-)]$$

$$+ e^{-iv^{+}L} \frac{1}{v^{+}} \left[N(v^{+}) - N(v^{-}) - N(\mu^{+}) + N(\mu^{-}) \right] \Bigg\} \ ,$$

where

$$\mu^{\pm} = m\gamma\omega/p_z - k \pm g \quad , \quad v^{\pm} = m\gamma\omega/p_z + k \pm g \tag{43}$$

and

$$N(x) = \frac{e^{ixL/2}}{L/2} \frac{\sin(xL/2)}{xL/2} \ . \tag{44}$$

Using our definition of $\kappa^{\pm}$ the linear gain coefficient is given by $\alpha^{\pm} = - \text{Im } \kappa^{\pm}$ we observe that the function

$$\Gamma(x) \equiv \text{Im} \frac{N(x)}{x} = \left(\frac{\sin(xL/2)}{xL/2} \right)^2 \tag{45}$$

is the characteristic form occurring in diffraction theory, with the principal maximum at $x = 0$. The remaining terms in curly brackets in Eqs.(41) and (42) do not have this feature and their contribution can be neglected relative to the terms containing Q. Therefore, Eqs.(41) and (42) simplify to

$$\alpha^{+} = \frac{L}{16} \int_{-\infty}^{\infty} dp_z \, F(p_z) \left[\Gamma(\mu^{+}) + \Gamma(\mu^{-}) \right] \tag{46}$$

and

$$\alpha^{-} = - \frac{L}{16} \int_{-\infty}^{\infty} dp_z \, F(p_z) \left[\Gamma(v^{+}) + \Gamma(v^{-}) \right] \ . \tag{47}$$

Since the quantities g, k and p_z are positive, for the chosen geometry, only μ^{-} and v^{-} pass through zero. Therefore, the remaining terms, $\Gamma(\mu^{+})$ and $\Gamma(v^{+})$ can be neglected when evaluated far from the maximum compared to the $\Gamma(x) = 1$, $(x = 0)$ term. Using the above argument Eqs.(46) and (47) become

$$\alpha^{+} = \frac{\pi}{4} n \int_{-\infty}^{\infty} dp_z \, Q^2 \frac{p_z}{m\gamma} \frac{\partial \Lambda}{\partial p_z} \, L\Gamma(\mu^{-}) \tag{48}$$

and

$$\alpha^{-} = - \frac{\pi}{4} n \int_{-\infty}^{\infty} dp_z \, Q^2 \frac{p_z}{m\gamma} \frac{\partial \Lambda}{\partial p_z} \, L\Gamma(v^{-}) \ , \tag{49}$$

where the explicit form of the kernel F is used.

Imposing resonant condition, $v^{-} = 0$, in Eq.(49) fixes the wavevector of the optical mode as follows

$$g = m\gamma kc/p_z + k \qquad (50)$$

where

$$p_z = m\gamma\beta c . \qquad (51)$$

Assuming the extreme relativistic limit, $\beta \to 1$, for the proton beam yields the following resonance condition

$$\lambda = 2\ell . \qquad (52)$$

One can summarize the above condition by the following statement: pumping the beam with the electromagnetic wave of *twice the superlattice period* would result in the maximum rate of energy extraction from the wave, given here by α^-, which is the essence of the presented accelerating scheme.

One can notice in passing, that α^+ corresponds to the lasing process, when the forward propagating wave is being amplified by the energy drawn from the beam (the FEL effect). The corresponding resonant condition is given by $\mu^- = 0$, which yields a well known scaling relation between the undulator period and the amplified wavelength

$$\lambda = \frac{\ell}{2\gamma^2}. \qquad (53)$$

From now on, we are interested only in the inverse FEL situation, therefore we will simplify further notation by putting: $\alpha \equiv \alpha^-$ and $\nu \equiv \nu^-$.

One can notice that, apart from a slowly varying function F, the remaining functions occurring in the integrand in Eq.(49), namely, Λ and Γ are sharply peaked functions of momentum characterized by the respective widths:

$$\left(\frac{\Delta p}{p}\right)_\Lambda \text{ and } \left(\frac{\Delta p}{p}\right)_\Gamma .$$

Here the width of the initial momentum distribution, Λ, was introduced in a standard way[8]

$$\left(\frac{\Delta p}{p}\right)_\Lambda = \left[\left(p_z^2 \frac{\partial \Lambda}{\partial p_z}\right)_{max}\right]^{-1/2} . \qquad (54)$$

Similarly the width of Γ is governed by the following simple ratio

$$\left(\frac{\Delta p}{p}\right)_\Gamma = \frac{\ell}{L} . \qquad (55)$$

Now one can compare relative sharpness of both functions; Λ and Γ. Typical value of the relative momentum spread is of the order of 10^{-4}. Assuming superlattice modulation of 500Å and crystal length of 5 cm allows one to evaluate the width of Γ. Both characteristic widths can be summarized as follows

$$\left(\frac{\Delta p}{p}\right)_\Lambda = 10^{-4} , \quad \left(\frac{\Delta p}{p}\right)_\Gamma = 10^{-6} . \qquad (56)$$

The integration in Eq.(49) is carried out assuming that the sharper function, namely Γ, is approximated by the δ-function according to the following asymptotic relationship

$$\lim_{L \gg l} \frac{L}{2\pi} \Gamma(\nu) = \delta(\nu) \ . \tag{57}$$

Applying resonant condition, described by Eqs.(50)–(52), and assuming $\beta \to 1$ reduces the gain/loss coefficient to the following simple expression

$$\alpha = -\frac{\pi}{2} n \frac{Q^2}{m} \gamma l \left(\frac{\Delta p}{p}\right)_\Lambda^{-2} \ . \tag{58}$$

The above final result will serve as a starting point for further feasibility discussion.

THREE WAVE MIXING – PHYSICAL PICTURE

According to the presented model calculation spontaneous bunching of the proton beam channeling through a superlattice and interacting with the electromagnetic wave results in energy flow from the wave to the beam. This particular kind of particle density fluctuation, h, has the form of a propagating plane wave of the same frequency, ω, as the emitted electromagnetic wave. The phase velocity of the moving bunch matches the velocity of protons in the beam. Therefore, the quantity $\gamma m \omega / p_z \equiv k_b$ represents the wavevector of the propagating particle density bunch. Keeping in mind that the periodicity of the undulator represents a static wave with a wavevector g, and that k is the wavevector of the electromagnetic wave, we can analyze our results in the language of three wave mixing.

One can notice, that the resonant denominators appearing in final expressions for the linear gain/loss coefficients, Eq.(43), i.e., $\mu^\pm = m\gamma\omega/p_z - k \pm g$ (forward propagating wave) and $\nu^\pm = m\gamma\omega/p_z + k \pm g$ (backward propagating wave), can be identified with the frequency – wavevector conservation conditions for a three wave mixing process. From this point of view, the FEL or the inverse FEL reduce to an Umklapp process[8] involving: 1) the propagating "bunch" in the proton density, 2) the pumping electromagnetic wave and 3) the static periodic field of the undulator. Matching of frequencies for the two "dynamic modes" assures "energy" conservation. Furthermore "momentum" conservation of all three modes (static and dynamic) yields the $\mu^- = 0$ (FEL), $\nu^- = 0$ (inverse FEL) conditions. The last condition is equivalent to a momentum "recoil", g, between the particle density "bunch" and the electromagnetic wave, i.e., an Umklapp process. One can see immediately that the resonant denominator, ν^-, is responsible for the negative gain (deamplification of the backward propagating wave), where a four momentum (0, g) is transferred from the backward propagating wave to the forward moving proton bunch.

FEASIBILITY ASSESSMENT

In this section we will discuss the feasibility of the proposed scheme by considering (110) planar channeling in a strain modulated Si crystal[9]. We write the undulator period as $l = Nd$, where d = 1.92 Å is the spacing between successive lattice planes and N is the number of such planes. The strain modulation, of course, requires a second component, such as Ge; however, we will use the parameters of Si for convenience.

One can identify the magnetic field of the undulator, which would be equivalent to electric field of the strain modulated crystal lattice (i.e., would result in the same transverse velocity of the channeling particles). The equivalence is given by:

$$B = \phi_1 x_1 \tag{59}$$

According to Ref. 9, the harmonic part of the crystal field potential for (110) planar channeling in Si is $e\phi_1 = 36$ eV $\mathring{A}^{-2}$. Assuming a strain modulation amplitude of $x_1 = 0.1$ $\mathring{A}$ leads to an enormous magnetic field of $B = 1.2 \times 10^6$ Gauss.

Relativistic particles while channeling along the path undergo transverse harmonic oscillations from the crystal field potential, an analog of the betatron oscillations, with the characteristic frequency $\omega_\beta = \sqrt{e\phi_1/m}$. One can see from Eq.(29) that if the angular velocity of a particle traversing the strain modulated path, $\omega = 2\pi v_\parallel/\ell$, approaches $\omega_\beta/\sqrt{\gamma}$ (Doppler shifted betatron frequency), the undulator parameter, Q, has a resonance (U → 1), which would enormously enhance the gain/loss coefficient. However, the excessive growth of the undulator parameter would soon result in a rapid dechanneling of the particles. One can see this easily if Q is rewritten in the following form

$$Q = \frac{e}{c} \frac{v_\perp^{max}}{v_\parallel} \, , \tag{60}$$

where $v_\perp, v_\parallel$ are transverse and longitudinal components of the particle velocity, respectively.

For small values of γ (U ≈ 1), the following simple physical criterion allows one to estimate the maximum value of Q. Dechanneling will occur if the transverse kinetic energy of the particle exceeds the binding energy of the harmonic potential (a particle leaves the channel). If the maximum transverse velocity of a channeling particle is $v_\perp$ and a is the distance between adjacent channels (for (110) channeling in Si a = 5 $\mathring{A}$), the above condition can be written as follows:

$$v_\perp^2 \geq \frac{e\phi_1}{m\gamma} \left(\frac{a}{2}\right)^2. \tag{61}$$

The equality sign in Eq.(61) along with Eq.(60) fix the maximum allowed value of the undulator parameter as

$$Q^{max} = \frac{e}{c} \frac{a}{2} \sqrt{\frac{e\phi_1}{m\gamma c^2}} \frac{\gamma}{\sqrt{\gamma^2 - 1}} \, . \tag{62}$$

The above expression can be evaluated for relativistic protons channeling through our model superlattice as

$$Q^{max} = 7.5 \times \sqrt{\frac{\gamma}{\gamma^2 - 1}} \times 10^{-24} \text{ cm}^{1/2} \text{ g}^{1/2} \, . \tag{63}$$

Now, one can evaluate Eq.(48) assuming only one proton – by assigning n to be an inverse area of the channeling plane per one particle for typical values of the beam concentration, $n = 10^{16}$ cm^{-2}. This way α describes the rate of optical amplitude depletion per one particle – the acceleration rate. Assuming γ of 2, $\ell = 500\mathring{A}$ and

$$\left(\frac{\Delta p}{p}\right)_\Lambda = 10^{-4}$$

yields the following value of the acceleration rate

$$\alpha = 3.53 \times 10^{-4} \text{ cm}^{-1}. \tag{64}$$

The nominal acceleration efficiency in units of eV/cm will, obviously, depend on the energy density of the actual optical cavity, which is left out for further discussion elsewhere.

CONCLUSIONS

The final acceleration rate evaluated numerically in Eq.(64) seems to be quite substantial. Practical feasibility of the presented scheme rests on availability of high density optical mode pumped by a high power laser. Furthermore, there is some radiation damage and thermal heating of the crystal associated with proton channeling. One can apply the beam to a rapidly spinning crystal[10], so that the average power density could be reduced to alleviate this problem.

ACKNOWLEDGEMENT

The author wishes to thank Bob Noble for discussion and his helpful comments on transient vs steady state solution of crystal channeling kinematics.

APPENDIX

We will generalize the treatment given previously, Eqs.(19)–(24), to a situation depicted schematically in Figure 2. Now there is a constant phase shift, ρ, between the crystal field of two neighboring channels. In this case the kinetic equation, Eq.(19), can be modified as follows

$$\left(\frac{\partial}{\partial t} + v_z \frac{\partial}{\partial z}\right) h^j(z, t) = M \frac{\partial A}{\partial z} \sin(gz + j\rho) \ . \tag{65}$$

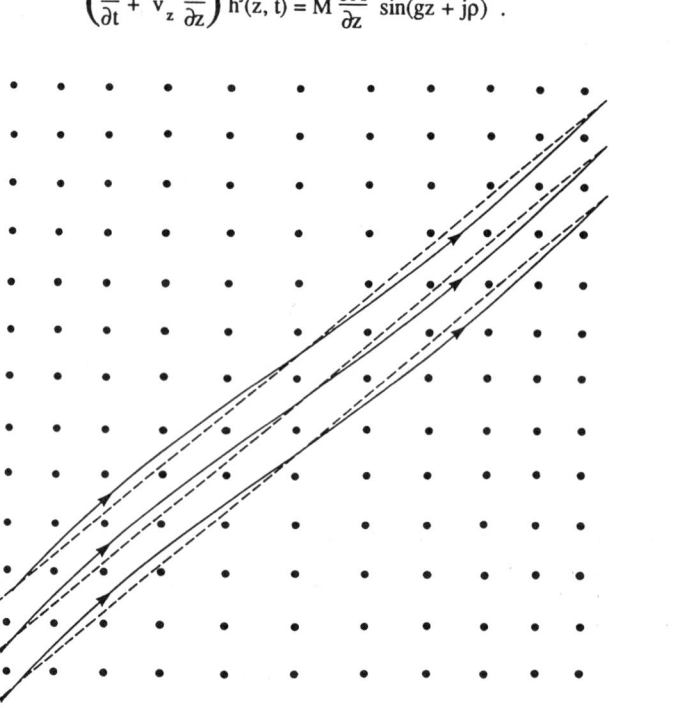

Figure 2   An infinite array of parallel channels in a strain-modulated superlattice. The phase shift in the crystal field potential between two neighboring channels, ρ, is given by $\rho = \frac{a\pi}{\ell}$.

Here j is the index of the channel, which discretizes the variable x as follows: $x = aj$. Furthermore we denote

$$v_z = \frac{p_z}{m\gamma} \qquad (66)$$

and

$$M = v_z Q \frac{\partial \Lambda}{\partial p_z} \,, \qquad (67)$$

Assuming an x-dependent A-field of the form

$$A(x, z, t) = A(z) \, e^{-i\omega t + ik_x x} \,, \qquad (68)$$

we seek a solution for h in the following form

$$h^j(z, t) = h^j(z) \, e^{-i\omega t} \,. \qquad (69)$$

Substituting the above expression in Eq.(65) one can rewrite it as follows

$$\left(v_z \frac{\partial}{\partial z} - i\omega \right) h^j(z) = \frac{M}{2i} \left(e^{igz + i(k_x a + \rho)j} - e^{-igz + i(k_x a - \rho)j} \right) \frac{\partial A}{\partial z} \,. \qquad (70)$$

Therefore the j-dependence of our solution can be written explicitly as

$$h^j(z) = h_+(z) \, e^{i(k_x a + \rho)j} + h_-(z) \, e^{i(k_x a - \rho)j} \,, \qquad (71)$$

where the functions h_+ and h_- are governed by the following pair of equations (equivalent to Eq.(66)

$$\left(v_z \frac{\partial}{\partial z} - i\omega \right) h_+(z) = \frac{M}{2i} e^{igz} \frac{\partial A}{\partial z} \,, \qquad (72)$$

and

$$\left(v_z \frac{\partial}{\partial z} - i\omega \right) h_-(z) = -\frac{M}{2i} e^{-igz} \frac{\partial A}{\partial z} \,. \qquad (73)$$

The net transverse current in the crystal, induced by h_+ and h_-, couples to the A-field through the following wave equation

$$\left(\frac{\partial^2}{\partial x^2} + \frac{\partial^2}{\partial z^2} + k^2 \right) A(x,z) = 4\pi n_0 \int_{-\infty}^{\infty} dp_z \, v_z Q \sum_j \delta(x - ja) \, \sin(gz - j\rho) \, h^j(z) \,, \quad (74)$$

where

$$k^2 = k_x^2 + k_z^2 = (\omega/c)^2 \,.$$

Substituting according to Eq.(67) and representing the δ-function as an integral form,

$$\delta(x) = \int_{-\infty}^{\infty} dk_x \, e^{ik_x x},$$

(74)

one can rewrite Eq.(69) as follows

$$\left(\frac{\partial^2}{\partial z^2} + k_z^2\right) A(z) = 4\pi n_0 \int_{-\infty}^{\infty} dp_z \, v_z \, Q \, \frac{1}{2ia}$$

(75)

$$x \left(h_- \, e^{-igz} - h_+ \, e^{igz} - h_- \, e^{-igz - i(2\rho/a)x} - h_+ \, e^{igz + i(2\rho/a)x} \right).$$

Obviously for a channeling particle x << a and therefore $e^{i(2\rho/a)x} \cong 1$; Eq.(71) then becomes

$$\left(\frac{\partial^2}{\partial z^2} + k_z^2\right) A(z) = 4\pi n_0 \int_{-\infty}^{\infty} dp_z \, v_z \, Q \, \frac{1}{a} \left(h_- + h_+ \right) \sin gz .$$

(76)

Similarly from Eq.(72) one obtains

$$\left(v_z \frac{\partial}{\partial z} - i\omega \right) \left[h_+(z) + h_-(z) \right] = M \frac{\partial A}{\partial z} \sin gz .$$

(77)

One can see immediately that redefining; $h_+ + h_- = h$, in Eqs.(71) and (72) reduces them to the initial pair of equations used previously, namely Eqs.(19) and (22). Therefore, our case (x << a) is equivalent to the one treated before and the same final expression for the linear gain/loss holds even in case of finite phase shift in the crystal potential between neighboring channels.

REFERENCES

1.   R. B. Palmer, Particle Accelerators 11, 81 (1980).

2.   J. A. Ellison, S. T. Picraux, W. R. Allen and W. K. Chu, Phys. Rev. B, 37, 7290 (1988).

3.   S. Krinsky and J. M. Wang, Particle Accelerators, 17, 109 (1985).

4.   B. L. Berman, S. Datz, R. W. Fearick, R. L. Swent, R. H. Pantell, H. Park, J. O. Kephart and R. K. Klein, Nuclear Instruments and Methods in Physics Research B, 2, 90 (1984).

5.   J. A. Ellison, Phys. Rev. B, 18, 5948 (1978).

6.   M. A. Kumakhov, Sov. Phys. JEPT, 4, 781 (1977).

7.   L. D. Landau and E. M. Lifshitz, Electrodynamics of Continuous Media, (Pergamon Press, Oxford, 1960) Ch. XV.

8.   E. M. Lifschitz and L. P. Pitaevskii, Physical Kinetics , (Pergamon Press, Oxford, 1981) Ch. VI.

9.   S. Datz, R. W. Fearick, H. Park, R. H. Pantell, R. L. Swent, J. O. Kephart and B. L. Berman, Nuclear Instruments and Methods in Physics Research B, 2, 74 (1984).

10.   R. H. Pantell and M. J. Alguard, J. Appl. Phys., 50, 798 (1979).

The Landau-Pomeranchuk-Migdal Effect and Suppression of Beamstrahlung and Bremsstrahlung in Linear Colliders

PISIN CHEN

Stanford Linear Accelerator Center
Stanford University, Stanford, California 94309

SPENCER KLEIN

Santa Cruz Institute for Particle Physics
University of California, Santa Cruz, California 95064

ABSTRACT

It is well known that beamstrahlung and bremsstrahlung take place over a finite formation zone distance. If something disturbs the electron during this time, the emission can be suppressed. In this paper, we examine the Landau-Pomeranchuk-Migdal (LPM) effect and other LPM-like effects, such as the longitudinal density (dielectric) suppression and EM field suppression of beamstrahlung and bremsstrahlung in e^+e^- linear colliders. We show that while the LPM effect and the density effect are not sufficient in suppressing these radiations, the strong EM field of the opposing beam does help to suppress bremsstrahlung.

1. Introduction

One of the most important issues in the design of future e^+e^- colliders is the effect of the beam-beam interaction on the physics environment. The single-pass nature of linear colliders necessitates the need for colliding tiny, intense bunches of electrons and positrons in order to achieve the required high luminosity. In this circumstance, these bunches interact strongly with one another, producing large numbers of hard photons, a phenomenon called beamstrahlung,[1] in addition to the conventional bremsstrahlung radiation process. These photons potentially create troublesome backgrounds for experiments on e^+e^- annihilation, and it is highly desirable if they can be suppressed in some way.

⋆ Work supported by the Department of Energy, contract DE–AC03–76SF00515.

929

Bremsstrahlung and beamstrahlung have usually been treated in the context of single particle having an *incoherent* point interaction with another particle or a *coherent* interaction with a collection of target particles, respectively. However, the uncertainty principle puts a lower limit on the size of the interaction; because of the small (especially in the longitudinal direction) momentum transfer between the particles, the interaction must take place in a finite area. For bremsstrahlung in a nuclear medium, this finite size leads to departures from the Bethe-Heitler formula. This paper will consider the effects that this finite size has on beamstrahlung and bremsstrahlung in e^+e^- colliders.

In contrast to bremsstrahlung, beamstrahlung occurs in the situation where the scattering amplitudes between the radiating particle and the target particles within the characteristic length add coherently. Typically for the beam-beam collision in linear colliders there can be over a million target particles involved within the coherence length. The process can therefore be well described in a semi-classical calculation where the target particles are replaced by their collective EM fields.

High energy e^+e^- beams generally follow Gaussian distributions in the three spatial dimensions. In the weak disruption limit, where particle motions are para-axial, it is possible to integrate the radiation process over this volume and derive relation which depend only on averaged, global beam parameters.[2] The overall beamstrahlung intensity is then described by a global *beamstrahlung parameter*,

$$\Upsilon = \gamma \frac{\langle B \rangle}{B_c} = \frac{5}{6} \frac{r_e^2 \gamma N}{\alpha \sigma_z (\sigma_x + \sigma_y)} \quad , \tag{1.1}$$

where $\langle B \rangle$ is the mean electromagnetic field strength of the beam, $B_c = m^2/e \simeq 4.4 \times 10^{13}$ Gauss is the Schwinger critical field, N is the total number of particles in a bunch, $\sigma_x, \sigma_y, \sigma_z$ are the nominal sizes of the Gaussian beam, $\gamma = E_e/m$ is the Lorentz factor of the radiating particle, r_e is the classical electron radius, and α is the fine structure constant. Roughly speaking, for $\Upsilon < 1$, the beamstrahlung spectrum scales as $x^{-2/3}$ for $x \lesssim \Upsilon$, where $x \equiv E_\gamma/E_e$ is the fractional energy of the radiated photon; and decreases exponentially for $x \gtrsim \Upsilon$. When $\Upsilon \gtrsim 1$, the spectrum scales as $x^{-2/3}$ for the entire range of $0 \leq x \leq 1$.

Also relevant to our following discussions is the average particle density of the colliding beams (in the e^+e^- center-of-mass frame). For tri-gaussian distributions, the beam density is

$$n_b = \frac{N}{(2\pi)^{3/2} \sigma_x \sigma_y \sigma_z} \quad . \tag{1.2}$$

To provide a framework for discussion, we will consider here three examples of linear colliders, SLC, a 500 GeV on 500 GeV intermediate collider, and a 5 TeV on 5 TeV super collider. The parameters for the machines are shown in Table 1. Section 2 of this paper will introduce the scale lengths in bremsstrahlung and beamstrahlung. In section 3 we review the physics of the LPM effect and the density suppression in a nuclear medium. Section 4 discusses these two effects in the e^+e^- linear collider environment. We show that both are ineffective in suppressing beamstrahlung and bremsstrahlung. In section 5 we turn to yet another LPM-like mechanism, magnetic suppression. It is shown that magnetic effects can in principle suppress bremsstrahlung effectively. However, they are still not sufficient to suppress beamstrahlung. A conclusion is given in Section 6.

Table 1. Parameters for the 3 linear colliders discussed in the text.

Parameter	SLC	NLC	SuperLC
E_e [GeV]	50	500	5000
N [10^{10}]	4	1.3	0.4
σ_x [nm]	2000	425	26.5
σ_y [nm]	2000	2	0.1
σ_z [μm]	1000	100	15
n_b [cm^{-3}]	6.4×10^{17}	9.7×10^{21}	6.4×10^{24}
Υ	0.001	0.27	70

2. Length Scales in Bremsstrahlung and Beamstrahlung

The classical diagram for bremsstrahlung is presented in Figure 1a; an electron emits a photon, conserving momentum by exchanging a virtual photon with a nearby nucleus. When the electron is of high enough momentum, the longitudinal momentum carried by the virtual photon becomes very small,

$$q_\| = p_e - p'_e - k = \sqrt{E_e^2 - m^2} - \sqrt{E_e'^2 - m^2} - E_\gamma \quad , \qquad (2.1)$$

where p_e, p'_e, E_e, and E'_e are the electron momentum and energy before and after the interaction, respectively, and E_γ is the photon energy. For high energy electrons this simplifies to

$$q_\| \sim \frac{m^2 E_\gamma}{2 E_e(E_e - E_\gamma)} = \frac{m}{2\gamma} \frac{x}{1-x} \equiv \frac{m}{2\gamma} u \quad , \qquad (2.2)$$

where $x = E_\gamma / E_e$ is the fractional energy of the radiated photon. This momentum transfer can be very small. Then, by the uncertainty principle, the virtual photon exchange distance, or the *formation length*, l_f,

$$l_f \sim \frac{\hbar}{q_{\parallel}} = 2\gamma\lambda_c\frac{1}{u} \quad . \tag{2.3}$$

For example, for a 25 GeV electron emitting a 100 MeV photon, $q_{\parallel}$ is only 0.03 eV/c and the formation length l_f is 2μm long. The expression for the formation length is unchanged in the case of e^+e^- scattering.

In the infrared limit, the transverse momentum transfer is essentially absorbed by the final state electron, so the out-going angle of the radiating electron is

$$\theta_{br} \sim \frac{m}{E_e} = \frac{1}{\gamma} \quad . \tag{2.4}$$

This angle is independent of the energy of the radiated photon.

In contrast to the bremsstrahlung process, beamstrahlung occurs due to the bending of the particle trajectory by the external classical EM fields. The overlapping wave-functions between the initial state and the final state contribute maximally to the radiation within a *coherence length*. In terms of Υ, the coherence length is

$$l_c \approx \frac{\gamma\lambda_c}{\Upsilon}\begin{cases} \left(\Upsilon/u\right)^{1/2} \quad , \quad \Upsilon/u \ll 1 \quad ; \\[2ex] \left(\Upsilon/u\right)^{1/3} \quad , \quad \Upsilon/u \gg 1 \quad ; \end{cases} \tag{2.5}$$

Because of the nature of the beamstrahlung spectrum, the part of the spectrum that we are interested will always satisfy the condition $x \ll \Upsilon$. Therefore we will be dealing with the regime where $\Upsilon/u \gg 1$ is always satisfied. So from now on we will simply put

$$l_c \approx \frac{\gamma\lambda_c}{u^{1/3}\Upsilon^{2/3}} \quad . \tag{2.6}$$

The radius of curvature of electron's classical trajectory is $\rho = \gamma^2\lambda_c/\Upsilon$. Thus the corresponding bending angle for the final state electron is

$$\theta_{be} = \frac{l_c}{\rho} = \frac{1}{\gamma}\left(\frac{\Upsilon}{u}\right)^{1/3} \quad . \tag{2.7}$$

We see that the bending angle for radiating a low energy beamstrahlung photon is substantially larger than the typical $\theta_{br} \sim 1/\gamma$ found in bremsstrahlung. The angle is increased due to the large transverse momentum imparted by the electromagnetic field.

3. The LPM Effect and the Density Suppression Effect

In the early 1950's, a group of Russian theorists, led by Landau, Pomeranchuk, Migdal and Feinberg began looking at bremsstrahlung in more detail. They realized that, because of the low longitudinal momentum transfer between the nucleus and the electron, bremsstrahlung is not instantaneous, but occurs over a finite formation zone. During this time, external influences can perturb the electron and suppress the photon emission. When this happens, the traditional Bethe-Heitler formula fails. This can occur in a number of places, most notably in crystals; we shall consider here two examples, due to multiple scattering and due to the dielectric constant of the medium.

Initially, Landau and Pomeranchuk studied suppression by multiple scattering with semiclassical arguments.[6] Later, Migdal presented a fully quantum treatment based on scattering theory[7] Since the LPM theory is unfamiliar to many physicists, we will present here a brief semiclassical derivation, following an article by Feinberg and Pomeranchuk.[8]

The LPM effect appears when one considers that the electron must be undisturbed while it traverses this distance. One factor that can disturb the electron and disrupt the bremsstrahlung is multiple Coulomb scattering. Semiclassically, if the electron multiple scatters by an angle θ_{ns}, greater than the angle made by the bremsstrahlung photon, $\theta_{br} \sim 1/\gamma$, then the bremsstrahlung is suppressed.

The average multiple scattering angle in a nuclear medium is

$$\langle \theta_{ns}^2 \rangle = \left(\frac{E_s}{E_e}\right)^2 \frac{l}{X_0} \quad , \tag{3.1}$$

where $E_s = \sqrt{4\pi/\alpha} \cdot m = 21$ MeV is the characteristic energy, l is the scatterer thickness, and X_0 is the radiation length. The LPM effect becomes important when θ_{ns} is larger than θ_{br}. This occurs for $(E_s/E_e)\sqrt{l/X_0} > m/E_e$. For a fixed electron energy, suppression becomes significant for photon energies below a certain value, given by

$$x < \frac{E_e}{E_{\text{LPM}}} \quad , \tag{3.2}$$

where all of the constants have been lumped into E_{LPM}, given by E_{LPM} [eV] = $m^4 X_0/c\hbar E_s^2 = 7.6 \times 10^{12} X_0$ [cm], about 2.6 TeV in uranium and 4.2 TeV in lead. For example, suppression becomes significant for 250 MeV photons from a 25 GeV electron in uranium. For beamstrahlung, of course, these formulae must be modified.

Finding the magnitude of the suppression is more involved. For low energy photons, the photon spectrum is proportional to $E_\gamma^{-1/2}$, in contrast to the $1/E_\gamma$

Bethe-Heitler spectrum. To go further, Migdal applied scattering theory to the density of wave states to derive detailed formulae. Also of interest is a combined energy-angular distribution; unfortunately the angular aspect of the LPM effect has yet to be worked out.

An analogous effect occurs for pair creation by a high energy photon. As Fig'1. shows, the two processes are closely related. In pair creation, the LPM energy threshold is determined by the lepton with the lower energy. Because of this, the pair creation suppression begins at much higher energies than bremsstrahlung suppression.

Although the LPM effect reduces the divergence of the low energy photon production cross section, it does not eliminate it, since dN/dE_γ still grows as $E_\gamma^{-1/2}$. At low photon energies, another effect removes the divergence. There, the phase shift due to the medium ($\sqrt{\epsilon}k$, where k is the photon wave number) can become significant. In the infrared limit, the contributions to the photon amplitude, $\exp\{i(k \cdot x - \omega t)\}$, from different parts of the electron path through the formation zone can interfere, and photon emission is suppressed.[9,10] This is sometimes known as the longitudinal density effect, and it is related to the dE/dz (transverse) density effect discussed by Fermi. The density effect is significant for photon energies less than $\gamma \omega_p$, where ω_p is the plasma frequency of the medium. For a given material, this occurs at a fixed x, and the suppression factor is[11]

$$F_p = \left(1 + \frac{n r_e \lambda_p^2}{\pi x^2}\right)^{-1} \quad , \tag{3.3}$$

where n is the electron density, and λ_p is the plasma wavelength of the medium. The density effect becomes important for $x = 10^{-4}$ in lead, for example. Below these energies, dN/dE_γ goes as E_γ^2, removing the divergence.

4. LPM and Density Effects in Beam-Beam Interaction

Although the above concepts remain unchanged for e^+e^- linear colliders, most of the details require modification. First, the multiple scattering formulae are modified. Second, the two interacting particles have equal masses, and so divide the momentum transfer equally, halving many of the relevant angles. Third, the relevant formation length changes due to the presence of the electromagnetic field carried by the beam.

There are a number of ways to compute the multiple scattering effect. We will start with the Bhabha cross section. For small angle scattering,

$$\frac{d\sigma}{d\theta} \approx \frac{8\pi r_e^2}{\gamma^2 \theta^3} \quad , \qquad \theta \ll 1 \quad , \tag{4.1}$$

and the average scattering angle is

$$\langle \theta^2 \rangle \approx \frac{\int_0^\infty \theta^2 (d\sigma/d\theta) \cdot d\theta}{\int_0^\infty (d\sigma/d\theta) \cdot d\theta} \quad . \tag{4.2}$$

These integrals are singular; to remove the singularity a minimum momentum cut-off is needed. This is given by the finite size of the beams: $\theta_{min} = \Delta p_{min}/p$. Since $\Delta p_{min} = \hbar/\sigma_x$, $\theta_{min} = \hbar/E_e\sigma_x = \lambda_c/\gamma\sigma_x$. Here, we are assuming that the beams are flat, so that $\sigma_x \gg \sigma_y$. Then, for a single scatter $\langle \theta^2 \rangle = 2\theta^2_{min} \log(1/\theta_{min})$. The total number of scatters is given by $N_s = n_b\sigma_{tot}l$ where n_b is the beam particle density and σ_{ms} is the total integrated cross section, $4\pi(r_e/\gamma\theta_{min})^2$, and l is the applicable length. Then, the total scattering angle is

$$\Theta^2_{ms} = \frac{8\pi r_e^2 n_b l}{\gamma^2} \log \frac{\gamma\sigma_x}{\lambda_c} \quad . \tag{4.3}$$

One way to decide when LPM suppression is important is to calculate a length scale for it. For bremsstrahlung, this length is the distance over which the multiple scattering has accumulated an angle of the order $\theta_{br} = 1/\gamma$. Inserting this condition into Eq.(4.3), we get

$$l_{\mathrm{LPM}}(\mathrm{BR}) = \frac{1}{8\pi r_e^2 n_b} \log^{-1} \frac{\gamma\sigma_x}{\lambda_c} \quad . \tag{4.4}$$

As the beam density rises, this distance gets shorter, indicating that the LPM effect appears at shorter emission length scales. From Table 1 we find that $l_{\mathrm{LPM}}(\mathrm{BR}) \simeq 3 \times 10^4, 1.9$, and 0.003cm, in SLC, NLC, and SuperLC, respectively. E_{LPM} decreases rapidly with energy. Nevertheless, all these lengths are larger than their corresponding bunch lengths, σ_z's, rendering the LPM effect ineffective in linear collider beam-beam interactions.

For beamstrahlung, the distance required to cumulate an angle of θ_{be} through multiple scattering is easily calculated:

$$l_{\mathrm{LPM}}(\mathrm{BE}) = l_{\mathrm{LPM}}(\mathrm{BR}) \cdot \left(\frac{\Upsilon}{u}\right)^{2/3} \quad . \tag{4.5}$$

This length scale is photon-energy dependent. But as we discussed in Sec. 2, the condition $\Upsilon/u > 1$ is generally satisfied in linear colliders. Thus $l_{\mathrm{LPM}}(\mathrm{BE}) > l_{\mathrm{LPM}}(\mathrm{BR}) > \sigma_z$ for all three machines, and we also conclude that the LPM effect cannot suppress beamstrahlung.

The density effect depends on the electron plasma frequency. In colliders, the electrons are relativistic, and hence have increased mass, so, in the center-of-mass frame (also the lab frame), the plasma frequency becomes

$$\omega_p^2 = \frac{4\pi n_b e^2}{\gamma m} = \frac{4\pi c^2 r_e n_b}{\gamma} \quad . \tag{4.6}$$

The appropriate length scale is when $k \cdot x - \omega t$ becomes comparable to 1 for $l = ct$. This occurs when $(1 - \sqrt{\epsilon})\omega l/c = 1$, or giving rise to a density-effect length

$$l_d \simeq \frac{2c\omega}{\omega_p^2} = \frac{1}{2\pi} \frac{\gamma^2 x}{\lambda_c r_e n_b} \quad , \qquad \omega \gg \omega_p \quad . \tag{4.7}$$

When the formation length or the coherence length is longer than this density length, then the density effect comes into play. Since $l_d \sim E_\gamma$, whereas $l_f \sim 1/x$ for small x, the density effect always cuts off the low energy photon spectrum. In the 3 cases that we study, the plasma frequencies are $9.2 \times 10^{-5}, 3.6 \times 10^{-3}$, and 2.9×10^{-2} eV, respectively.

If the bunch is infinitely long, then the density effect would apply for any x which satisfies the condition $l_d \lesssim l_f$, or

$$x \lesssim \frac{\hbar \omega_p}{mc^2} = \left[\frac{4\pi \lambda_c^2 r_e n_b}{\gamma} \right]^{1/2} \quad . \tag{4.8}$$

So in the absence of other suppression effects, the beam density in principle could affect the spectrum up to an energy of $xE_e \lesssim \gamma\omega_p \simeq 9.2$ eV, 3.6 keV, and 0.29 MeV, respectively. However, Eq.(4.7) shows that these energies require bunch lengths of 430, 110, and 130 m! Instead, since the bunch lengths are much shorter than these values, one should equate l_d with σ_z to find the threshold x:

$$x_d \lesssim \frac{2\pi \lambda_c r_e n_b \sigma_z}{\gamma^2} \quad . \tag{4.9}$$

This gives density suppression thresholds of $2.1 \times 10^{-5}, 3.2 \times 10^{-3}$, and 3.2×10^{-2} eV, respectively. All these suppressions are very small considering the very hard spectrums anticipated in linear colliders.

In addition to the changes in dielectric constant due to the real electrons and positrons, there will also be some change in the vacuum polarization due to virtual electron positron pairs in the presence of an external magnetic field.[10] This effect should mainly be important at very high Υ's.

5. Magnetic Suppression

In addition to the sideways kick due to multiple scatterings, which is a incoherent process, there is also the coherent bending of the trajectory. As discussed earlier, this is the same source that give rise to beamstrahlung. The distance associated with the bending of an angle $\sim 1/\gamma$ is

$$l_0(\mathrm{BE}) = \frac{\rho}{\gamma} = \frac{\gamma \lambda_c}{\Upsilon} \quad . \tag{5.1}$$

For SLC, NLC, and SuperLC, $l_0 \simeq 38, 1.4$, and $0.054\ \mu\mathrm{m}$, respectively. These values are about 2 to 3 orders of magnitude smaller than the corresponding bunch lengths.

By comparing l_0 with l_f, we find that bremsstrahlung is suppressed for any

$$x_0(\mathrm{BR}) \lesssim \frac{2\Upsilon}{1 + 2\Upsilon} \quad . \tag{5.2}$$

For $\Upsilon \ll 1$, the spectrum is suppressed roughly up to twice the Υ parameter; whereas for $\Upsilon \gg 1$, essentially the entire spectrum of bremsstrahlung will be suppressed! Thus for bremsstrahlung the dominant source of perturbation comes from the collective EM fields in the beam.

This is in fact not a new effect. Earlier Baier, Katkov, and Strakhovenko[12] studied the suppression of bremsstrahlung in e^+e^- collision due to the presence of a transverse magnetic field. The only difference here is that the coherent bending is due to the collective classical field of the oncoming beam particles, which is locally transverse to the beam propagation. In effect, this puts an upper limit on the electron pathlength that can contribute to bremsstrahlung.[13]

It is also natural to wonder if the angles induced by bremsstrahlung can perturb the beamstrahlung process. It can be shown that for an angular increase of $\theta \sim 1/\gamma$ through bremsstrahlung, it takes about 2 orders of magnitude longer in distance than l_{LPM}. It is evident that the bremsstrahlung cannot suppress beamstrahlung.

6. Conclusions

We have examined the applicability of the LPM effect and density suppression to colliding beam bremsstrahlung and beamstrahlung. We conclude that neither will have measurable effects in SLC and in future colliders. Instead, we show that the collective EM fields in the beam is effective in suppressing bremsstrahlung. In the example of NLC, we find the suppression extends up to $x \sim 0.35$, which is quite significant. As for beamstrahlung, there is no comparable mechanism, as the beam field is exactly the same source that gives rise to the radiation.

We note that a similar effect should in principle also apply to the conversion of photons into e^+e^- pairs, although it would only occur at significantly higher energies.

REFERENCES

1. P. Chen and R. J. Noble, SLAC-PUB-4050 (1986); M. Bell and J. S. Bell, *Part. Accl.* **24**, 1 (1988); R. Blankenbecler and S. D. Drell, *Phys. Rev. Lett.* **61**, 2324 (1988); P. Chen and K. Yokoya, *Phys. Rev. Lett.* **61**, 1101 (1988); M. Jacob and T. T. Wu, *Nucl. Phys.* **B303**, 389 (1988); V. N. Baier, V. M. Katkov, and V. M. Strakhovenko, *Nucl. Phys.* **B328**, 387 (1989).

2. P. Chen, in *Frontiers of Particle Beams*, Lecture Notes in Physics **296**, Springer-Verlag, 1988.

3. K. Yokoya and P. Chen, in *Frontiers of Particle Beams: Intensity Limitations*, Lecture Notes in Physics **400**, Springer-Verlag, 1992.

4. P. Chen and K. Yokoya, *Phys. Rev.* **D38**, 987 (1988).

5. P. Chen, *Phys. Rev.* **D46**, 1186 (1992).

6. L. D. Landau and I.J. Pomeranchuk, *Dokl. Akad. Nauk. SSSR* **92**, 535 (1953); **92** 735 (1953).

7. A.B. Migdal, *Phys. Rev.* **103**, 1811 (1956).

8. E. L. Feinberg and I. Pomeranchuk, *Nuovo Cimento*, Supplement to Vol **3**, 652 (1956).

9. M.L. Ter-Mikaelian, *Dokl. Akad. Nauk. USSR* **94**, 1033 (1954). For a discussion in English, see M.L. Ter-Mikaelian, *High Energy Electromagnetic Processes in Condensed Media*, John Wiley & Sons, 1972.

10. J. Schwinger, W. Y. Tasi, and T. Erber, *Ann. Phys.* **96**, 303 (1976).

11. W. R. Nelson, H. Yirayama and D. W. O. Rogers, SLAC-Report-265, Dec. 1985. Unfortunately, after such a nice discussion they did not include the effect in EGS.

12. V.N. Baier, V.M. Katkov and V.M. Strakhovenko, Institute of Nuclear Physics preprint 87-26 (in Russian); V.N. Baier and V.M Katkov, *Sov. Phys. Doklady* **17**, 1068 (1973); V.M. Katkov and V.M. Strakhovenko, *Sov. J. Nucl. Phys.* **25**, 660 (1977). The latter papers only apply to the case where the transverse momentum is much less than m.

13. P. Chen and S. Klein, work in progress.

SUPPRESSION OF BEAMSTRAHLUNG
BY MEANS OF A PLASMA

Andrew M. Sessler[*]
Lawrence Berkeley Laboratory
University of California
Berkeley, CA 94720
and
David H. Whittum[**]
National Laboratory for High Energy Physics (KEK)
Tsukuba, 1-1 Oho, Ibaraki, 305 Japan

ABSTRACT

It is possible to suppress coherent beam-beam effects at the interaction point of a linear collider by means of charge and current neutralization in a plasma. This concept of "plasma compensation" is reviewed and parameters are noted for two TeV-class linear colliders.

INTRODUCTION

To reach high luminosity in a TeV linear electron-positron collider we must consider beams with spot size of order 0.1 μm or smaller and take into account the coherent processes which occur when these fine, high-current beams collide.[1] One such process is beamstrahlung, the radiation emitted by an electron or positron as it is accelerated in the megagauss field of the two colliding beams.

Beamstrahlung results in an energy loss of 10-30%, as well as an energy spread, in proposed designs. The effect of a large spread in energy, combined with narrowly peaked reaction cross sections, is to reduce the resulting reaction rate, partially negating the effect of increased luminosity. In addition, interactions of beamstrahlung photons with the beams will produce lower-energy background events, including electron-positron pair production.[2] Thus unsuppressed beamstrahlung will be detrimental to the next generation of collider physics experiments.

Equally significant, beamstrahlung has forced designers to consider flat beams 10 nm in width or smaller, and to accept the stringent

[*] Work supported by the Director, Office of Energy Research, Office of High Energy and Nuclear Physics, Division of High Energy Physics, of the U.S. Department of Energy under Contract No. DE-AC03-76SF00098

[**] Supported by the Japan Society for the Promotion of Science, the U.S. National Science Foundation and the National Laboratory for High Energy Physics (KEK).

requirements on emittance and magnet alignment that such beam sizes impose.

It has been proposed to reduce beamstrahlung and other coherent processes by providing a conducting medium—a plasma—at the interaction point, in which return currents will flow and partially cancel the B fields of the high-energy electron beam while totally neutralizing its charge.[3] We find that the key problem with this scheme is the large density required (10^{22}cm^{-3} to 10^{24} cm^{-3}), and attendant difficulties such as background beam-plasma reactions.

The theoretical results of Ref. 3 are presented in the next section. We follow this with application of the concept of plasma compensation to large linear colliders.

THEORY

In Ref. 3 one may find a 1D analytic theory, a 2D MHD theory, and 2D particle simulations. It is shown that all these agree and, therefore, it is adequate to employ the simplest analysis; namely the 1D analytic theory.

The plasma responds instantaneously to the high energy beam. Assuming a preionized channel we have

$$E_z \approx -\frac{1}{c}\frac{\partial A_z}{\partial t} \approx -\frac{L\partial}{c^2 \partial t}\left(I_b + I_p\right), \tag{1}$$

$$\frac{\partial I_p}{\partial t} = \frac{\omega_p^2 a^2}{4} E_z - \nu I_p, \tag{2}$$

where L is a dimensionless inductance on the order of unity which depends on the radial version of the fields. The plasma current is I_p, ν is the collision frequency, a is the beam radius, and ω_p is the plasma frequency.

These equations may be combined to obtain an equation for total current $I_{tot} = I_b + I_p$:

$$\frac{\partial I_{tot}}{\partial t} = \frac{1}{1+\theta}\frac{\partial I_b}{\partial t} - \frac{\nu}{1+\theta}\left(I_{tot} - I_b\right), \tag{3}$$

where

$$\theta = \left(k_p a\right)^2 L / 4 \approx \left(k_p a\right)^2 / 4. \tag{4}$$

This equation determines I_p, the plasma current, given I_b, the beam current. Thus good compensation; that is, I_p very small (compared to I_b) requires θ large. Thus the requirement is simply $k_p a \gg 1$.

The 2D MHD theory can be employed to quantitatively examine plasma compensation. The results, which are displayed in Figs. 1, 2 and 3, are in accord with the 1D results that $k_p a \gg 1$.

APPLICATION

We can readily apply the theory of the previous section to design a plasma-suppressed linear collider. Two examples are given in Table I. It can be seen that the required plasma density is high, just below that characteristic of solids. Creation of such a plasma will be challenging (but shouldn't be impossible). Background events are, of course, a serious issue; it has been (superficially) discussed in Ref. 3.

Note, however, that the parameters of a plasma-suppressed linear collider, even at E_c of $m = 10$ TeV, are greatly relaxed compared to that of ordinary linear colliders. It would appear that the concept merits experimental study and a start in that direction has been made.[4]

REFERENCES

1. Ugo Amaldi, "Introduction to the Next Generation of Linear Colliders," CERN-EP/87-28 (August 1987); R.B. Palmer, "The Interdependence of Parameters for a TeV Linear Collider," SLAC internal report, SLAC-PUB-4295 (April 1987); R.B. Palmer, "Prospects for High Energy e⁺e⁻ Linear Colliders," Annual Reviews of Nuclear and Particle Science 40, 529, Annual Reviews, Inc., Palo Alto (1990).

2. R. Blankenbecler and S.D. Drell, Phys. Rev. Lett. 61, 2324 (14 Nov. 1988); P. Chen, "Disruption, Beamstrahlung and Beamstrahlung Pair Creation," SLAC internal report SLAC-PUB-4822 (Dec. 1988).

3. D.H. Whittum, A.M. Sessler, J.J. Stewart and S.S. Yu, Particle Accelerators 34, 89 (1990).

4. W. Leemans, "Proposed Plasma Lens Experiments Using the ALS Injector," these proceedings.

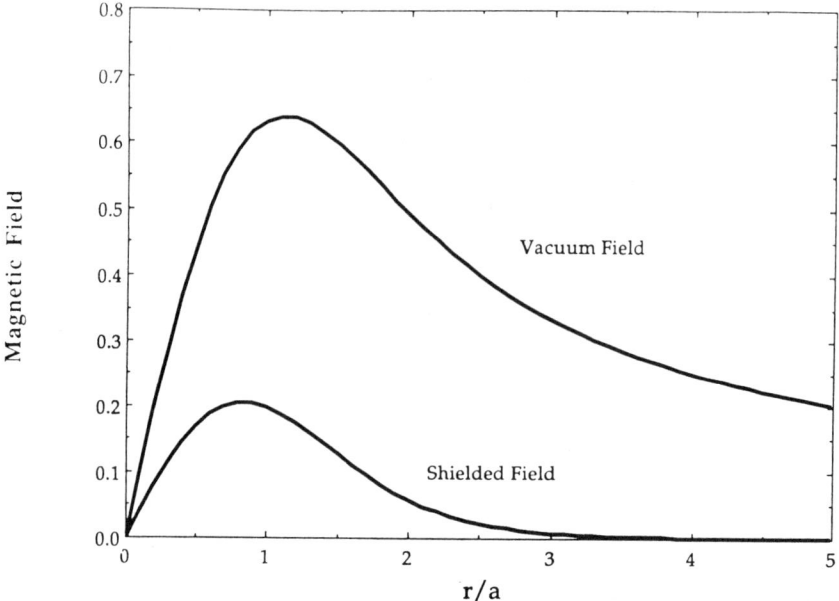

Fig. 1. Example of the radial profile of the shielded magnetic field in the collisionless regime. The radial coordinate is normalized by a, where the radial beam profile is exp $(-r^2/a^2)$ and $k_p a = 2$.

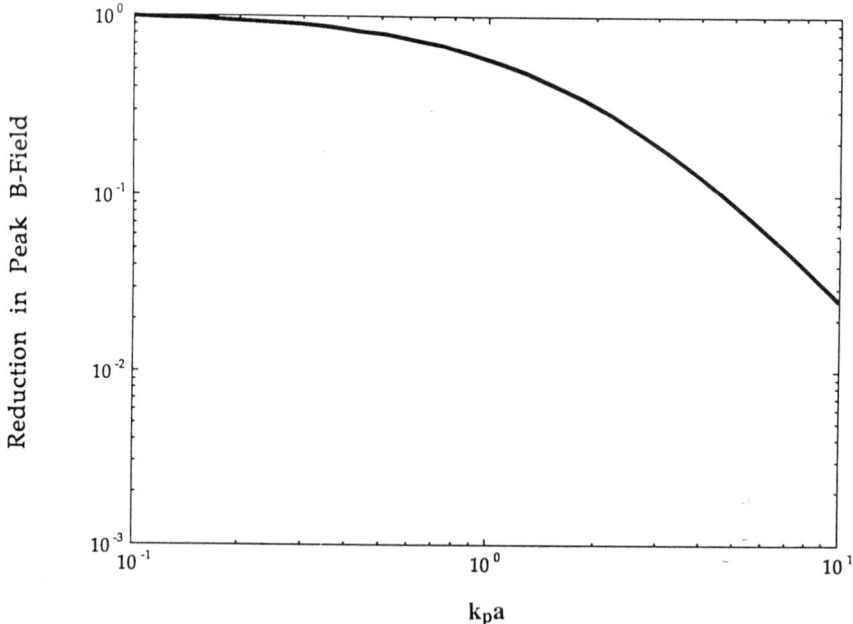

Fig. 2. Reduction in peak B_ϕ field (i.e., maximum as a function of r) as a function of $k_p a$ for a round beam. The peak field value is normalized to the peak field of a gaussian beam in vacuum.

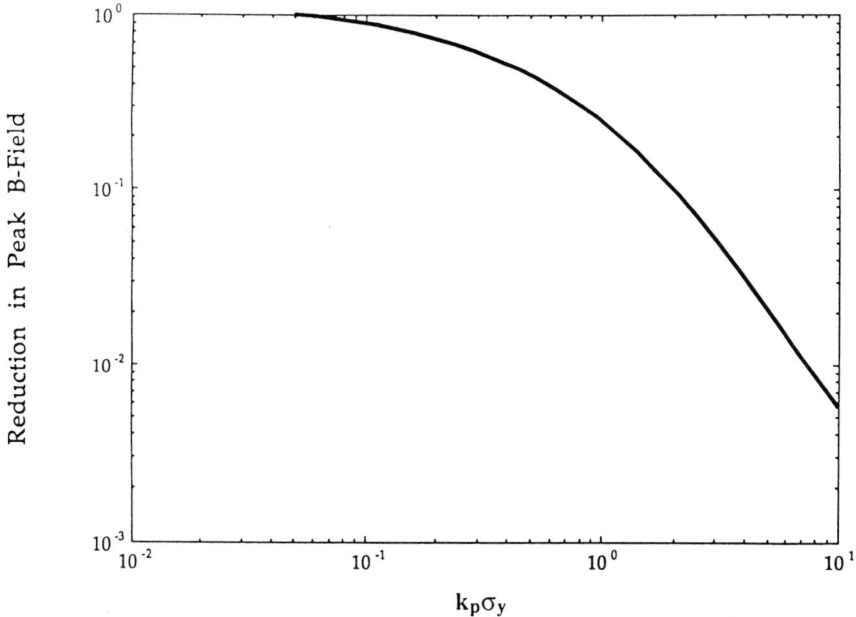

Fig. 3. Reduction in peak B field as a function of $k_p \sigma_y$ for a flat beam.

Table I. Example Collider Parameters

E_c of m (TeV)	1	10
E (TeV) x E (TeV)	1/2 x 1/2	5 x 5
L (cm^{-2}sec^{-1})	1 x 10^{33}	1 x 10^{35}
f (Hz)	500	500
N	5.0 x 10^{10}	5.0 x 10^{10}
n	1	10
R	1.0	1.0
ε_n (m rad)	1 x 10^{-6}	1.0 x 10^{-6}
σ_y (μm)	0.1	0.032
σ_z (mm)	1.0	1.0
I (kA)	1.0	1.0
β^* (cm)	1.0	1.0
n_p (cm^{-2}sec^{-1})	10^{22}	10^{23}

Induced Light-by-Light Scattering Experiment

K.T. McDonald

Joseph Henry Laboratories, Princeton University, Princeton, NJ 08544

Abstract

Past suggestions for a demonstration of light-by-light scattering by a variation of four-wave mixing may be realizable in the near future with tabletop teraWatt lasers.

Introduction

During the Workshop Alexander Varfolomeev pointed out that light-by-light scattering at optical frequencies can be enhanced by use of a third laser beam [1]. Norman Kroll remarked that he had also considered this in the 60's [2]. Here we consider whether a 1-teraWatt laser, such as that built at U. Rochester for SLAC E-144 [3], could be used to perform such an experiment, and conclude that rates are still somewhat low. Perhaps with the recently reported 50-teraWatt lasers [4] the signal can be seen.

The cross section for light-by-light scattering is very small [5]:

$$\sigma = \frac{973}{10125\pi}\alpha^2 r_e^2 \left(\frac{\hbar\omega}{mc^2}\right)^6 \approx 0.03\alpha^2 r_e^2 \left(\frac{\hbar\omega}{mc^2}\right)^6 \approx 7.4 \times 10^{-66} \text{cm}^2 \left(\frac{\hbar\omega}{1 \text{ eV}}\right)^6,$$

where ω is the frequency of the incident photons in the cm frame, and $r_e = e^2/mc^2$ is the classical electron radius.

For example, suppose we collide two laser beams of N photons each at right angles (as would be convenient for the 3-beam experiment discussed below) after focusing them to a spot size of order λ, the laser wavelength. If the laser pulsewidth is τ seconds then the only a fraction $\lambda/c\tau$ of the photons in each beam occupies the interaction volume ($\approx \lambda^3$) at any moment. We may regard the scattering as consisting of $c\tau/\lambda$ successive experiments in which $N\lambda/c\tau$ photons from each beam interact with each other. The total scattering rate would then be

$$\text{Rate} \approx \frac{c\tau}{\lambda}\left(\frac{N\lambda}{c\tau}\right)^2 \frac{\sigma}{\lambda^2} = \frac{N^2\sigma}{\lambda c\tau}.$$

For example, if we have 1 Joule of photons of 1-eV energy ($\lambda = 10^{-4}$ cm) with a pulse length of 1 psec (as for the present Rochester T^3 laser), then the rate is only about 10^{-24} per pulse!

Four-Wave Mixing

The observation of Kroll and Varfolomeev is that when a third laser beam is present and aligned along the direction of a possible final-state photon, the scattering rate is enhanced by the number of photons in the third beam (during each of the subexperiments described above). That is, for N photons in each of the three beams,

$$\text{Rate} \approx \frac{c\tau}{\lambda} \left(\frac{N\lambda}{c\tau} \right)^3 \frac{\sigma}{\lambda^2} = \frac{N^3 \sigma}{(c\tau)^2}.$$

On using the above expression for the cross section, we arrive at the form of Kroll:

$$\text{Rate} = \Gamma \alpha^4 \frac{\lambda_C^5}{\lambda^3 (c\tau)^2} \left(\frac{\mathcal{E}}{mc^2} \right)^3 \approx 10^{-6} \frac{(\mathcal{E}[\text{Joules}])^3}{(\tau[\text{psec}])^2},$$

where $\mathcal{E} = N\hbar\omega$ is the pulse energy, $\lambda_C = \hbar/mc$ is the Compton wavelength of the electron, and the numerical factor Γ is roughly π when the spot size is $\approx \lambda$.

To reach a rate of one scatter per pulse, we would need, for example, 10 Joules in each beam, whose pulselengths have been compressed to 30 fsec.

Configuration of the Laser Beams

It will be highly useful for the fourth photon to have a different frequency from the other three, and to be produced at an angle not along any of the incoming beams. Since $\omega_4 = \omega_1 + \omega_2 - \omega_3$ it is sufficient that $\omega_3 \neq \omega_1$, while we may keep $\omega_1 = \omega_2$ for convenience. In practice, it may be best to choose ω_3 to be the laser frequency, and take $\omega_1 = \omega_2 = 2\omega_3$ by use of a doubling crystal. Then $\omega_4 = 3\omega_3 = 1.5\omega_1$.

To separate beam 4 from the other three, we can make the collision of beams 1 and 2 at some crossing angle less than 180°, and set the pump beam 3 out of the plane of 1 and 2. (Other arrangements are possible as well.) Let θ be the angle between beam 1 and the bisector of the angle between beams 1 and 2. We arrange beam 3 to be in the plane perpendicular to the plane of beams 1 and 2 that contains the bisector. Let θ_3 and θ_4 be the angles between the bisector and beams 3 and 4, respectively. See the Figure.

Under the assumption that $\omega_1 = \omega_2$ energy and momentum conservation then read:

$$2\omega_1 = \omega_3 + \omega_4,$$

$$2\omega_1 \cos\theta = \omega_3 \cos\theta_3 + \omega_4 \cos\theta_4,$$

and

$$\omega_3 \sin\theta_3 = \omega_4 \sin\theta_4.$$

After some algebra we find

$$\cos\theta_3 = \frac{\omega_3 - \omega_1 \sin\theta}{\omega_3 \cos\theta}.$$

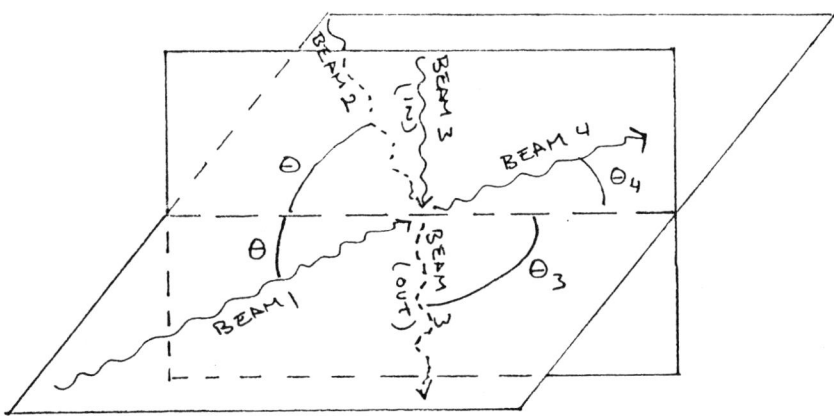

Figure 1: Possible arrangement of the laser beams for the induced light-by-light scattering experiment. Beams 3 and 4 are in the plane perpendicular to the plane of beams 1 and 2 that contains the bisector of the angle 2θ between beams 1 and 2.

A convenient configuration is that $\theta_3 = 90°$, which holds when $\sin\theta = \sqrt{\omega_3/\omega_1}$. In our example where $\omega_3/\omega_1 = 1/2$ we require that $\theta = 45°$, so the angle between beams 1 and 2 is $90°$, as mentioned above. Finally, $\sin\theta_4 = \omega_3/\omega_4 = 1/3$, or $\theta_4 = 19.5°$. The angular separation between photon 4 and the other 3 beams is nearly maximal.

Vacuum Requirements

Background photons might come from residual gas atoms in the scattering chamber vacuum that are ionized by the intense laser beams. Ionization will be probable for any atoms in fields of intensity greater than about 10^{13} Watts/cm^2. The fields at the collision point will be about 10^{21} Watts/cm^2, and the Rayleigh range $\approx \lambda$ for a very strong focus. Then the volume over which ionization is probable is of order $(\sqrt{10^8}\lambda)^3 \approx 1$ cm^3. At 1-atmosphere pressure there are about 3×10^{19} atoms/cm^3, so we would require a vacuum of about 10^{-16} torr. As only about 10^{-10} torr might be achieved in practice, there would be about 10^6 ionized atoms/pulse. There seems a nonzero prospect that several of these atoms would emit radiation at frequency ω_4, for which $\lambda_4 = \lambda_{\text{laser}}/3 = 353$ nm. A narrow interference filter at this wavelength could be used to limit the bandwidth, but if the laser pulse has been compressed to 30 fsec its bandwidth is several percent.

A first step would be to focus the present laser beam in vacuum, and search for ionization photons at 3ω.

References

[1] A.A. Varfolomeev, *Induced Scattering of Light by Light*, Sov. Phys. JETP **23** (1966) 681.

[2] N.M. Kroll, *Parametric Amplification in Spatially Extended Media and Application to the Design of Tuneable Oscillators at Optical Frequencies*, Phys. Rev. **127** (1962) 1207, footnote 9; also, pp. 31-32 of *Quantum Theory of Radiation*, in *Quantum Optics and Electronics* (Les Houches, 1964), ed. by C. DeWitt, A. Blandin, and C. Cohen-Tannoudji (Gordon and Breach, New York).

[3] K.T. McDonald *et al.*, *Study of QED at Critical Field Strength*, in *Workshop on beam-Beam and Beam-Radiation Interactions: High Intensity and Nonlinear Effects*, ed. by c. Pelligrini, T. Katsouleas and J. Rosenzweig (World Scientific, Singapore, 1992) p. 127.

[4] A brief notice on the work of the Limeil group appeared on p. 6 of Lasers & Optronics **11**, No. 11 (Oct. 1992).

[5] See for instance Sec. 127 of V.R. Berestetskii, E.M. Lifshitz and L.P. Pitaevskii, *Quantum Electrodynamics*, 2nd ed., (Pergamon Press, 1982).

PARTICLE PRODUCTION AND SURVIVAL IN MUON ACCELERATION

Robert J. Noble
Fermi National Accelerator Laboratory *
Batavia, Illinois 60510

ABSTRACT

Because of the relative immunity of muons to synchrotron radiation, the idea of using them instead of electrons as probes in high-energy physics experiments has existed for some time, but applications were limited by the short muon lifetime. The production and survival of an adequate supply of low-emittance muons will determine the available luminosity in a high-energy physics collider. In this paper the production of pions by protons, their decay to muons and the survival of muons during acceleration are studied. Based on a combination of the various efficiencies, the number of protons needed at the pion source for every muon required in the final high-energy collider is estimated.

INTRODUCTION

The relative immunity of muons to synchrotron radiation due to their large rest mass ($m_\mu = 105.7\,Mev/c^2$) suggests that they might be used in place of electrons in accelerators for high-energy physics experiments. The idea of using muons as high-energy probes has existed for some time, but applications were limited by the short muon lifetime. Muons have been used in secondary beams as "deep-inelastic" probes of hadron structure. Skrinsky[1] and others have suggested that their role could be extended by accelerating muons after production. Neuffer[2] has recently described the physics of "ionization cooling" of muons for obtaining low-emittance beams needed in high energy colliders.

The short lifetime of muons and the large emittance of initial beams will determine the essential character of any muon acceleration chain: a rapid-cycling facility, the front end of which resembles a "meson factory"[3] in which intense

*Work supported by the U.S. Department of Energy under contract No. DE-AC02-76CHO3000.

proton beams impinging on a stationary target produce copious pion beams that decay into muons for cooling and acceleration. Acceleration to the final collider energy may be in a linear accelerator or rapid-cycling synchrotron, and collisions for physics can be contemplated in either the $\mu^+\mu^-$ or μp channels. If the muons are placed in a storage ring for high-energy experiments, the useful storage time is about one muon lifetime at the collision energy or $2.197 \times 10^{-6}(E_\mu/m_\mu c^2)$ seconds. In any event the production and survival of an adequate supply of low-emittance muons will determine the available luminosity in such machines.

In this paper the production of pions, their decay to muons and the survival of muons during acceleration are studied. The survival of muons during ionization cooling is not discussed since this has been previously considered by Neuffer.[2,4] He showed that the lowest normalized emittance achievable by ionization cooling is approximately $\epsilon_N(rms) = 10^{-2}$ cm rad and is limited by multiple scattering in the absorbers. Cooling of both longitudinal and transverse emittances by a factor of 100 within one muon lifetime at the cooling energy can be achieved with an average energy gradient of about 1 MV/m, so the muon survival ratio during cooling would be $\eta_c \simeq 1/e \simeq 0.37$. This cooling efficiency can be improved by increasing the energy absorption and return per unit length in the cooler which reduces the cooling time.

In the present study a combination of the various efficiencies will suggest that for every muon required in the final high-energy collider, approximately 10^3 protons are needed in the meson factory to produce the initial pions. The basic efficiencies that control the total muon production are the pion production yield $\eta_{p\pi}$, the target efficiency η_t, the pion-muon decay efficiency $\eta_{\pi\mu}$, the cooling survival efficiency η_c and the acceleration survival efficiency η_a. Numerically $\eta_{p\pi}$ and $\eta_{\pi\mu}$ are the smallest coefficients among these and increase almost linearly with increasing momentum spread $\Delta p/p$ of the accepted beams. Significant improvement in muon production could be achieved if initial beams with momentum spreads exceeding $\pm 5\%$ could be captured prior to cooling.

PION PRODUCTION

Muons are not seen in abundance in most high energy processes unless one takes great care to detect them. The exception is charged pion decay $(\pi \rightarrow \mu + \nu_\mu)$ in which muons result from essentially all decays. Pions are produced copiously from proton beams on stationary targets. In this sense muons are tertiary particles from the proton collision.

The calculated particle spectra of Grote et al[5] are useful for estimating pion production on stationary targets for primary proton momenta p_p between 12.5 GeV/c and 800 GeV/c. The spectrum $d^2N_\pi/dp_\pi\,d\Omega$ per interacting proton is relatively insensitive to the target composition. High atomic number targets are preferable however to limit depth of focus problems associated with particle production over long distances. The pion spectrum is very broad in momentum. For forward production ($\theta_\pi = 0°$), the maximum in the π^- spectrum occurs at

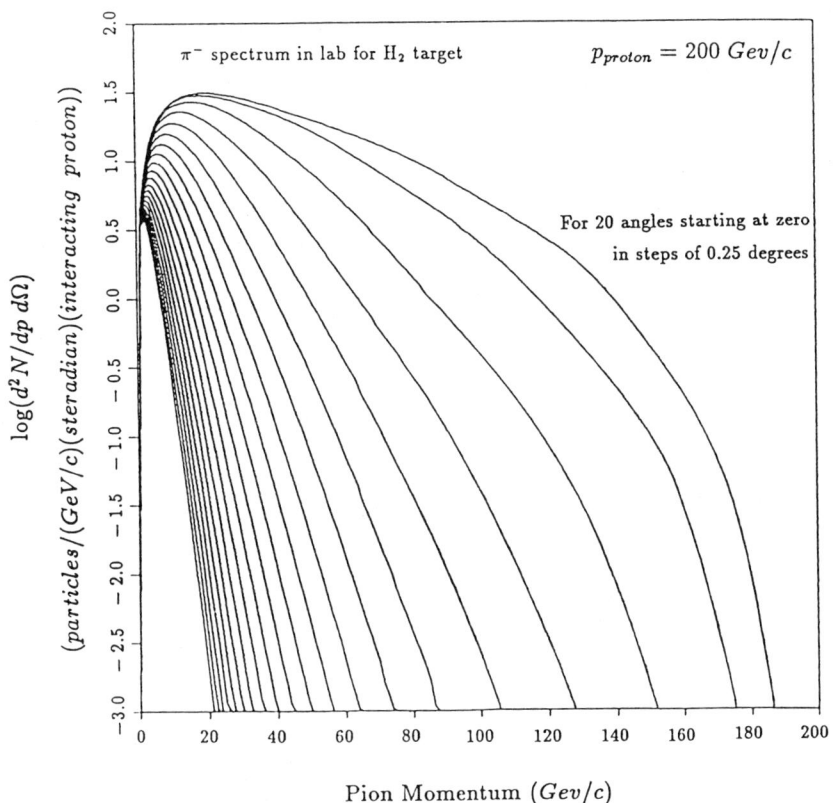

Figure 1: Negative pion spectrum from 200 GeV protons on a hydrogen target (redrawn from Ref. 5).

$p_\pi \simeq p_p/10$ with the position of the maximum decreasing with increasing angle (Figure 1). The π^+ spectrum for $\theta_\pi \simeq 0°$ is relatively flat between $p_p/10$ and $p_p/3$ but is similar to the π^- spectrum at finite angles.

Pion yields are generally increased by accepting larger production angles and momentum spreads. For most pion momenta, the differential production decreases by about an order of magnitude for angles greater than $3.5\, m_\pi c/p_\pi$ relative to forward production. This offsets the advantage of increased solid angle and limits the useful collection angle to about this value. Because of the competing effects of increasing solid angle and decreasing spectral peak at small pion momenta, the maximum pion yield obtained by integrating over angles and a given momentum spread ($\pm\Delta p/p_\pi$) occurs for $p_\pi \simeq p_p/20$. For small momentum spreads, the yield will vary almost linearly with $\Delta p/p_\pi$, but it becomes difficult to transport and capture particle bunches with momentum spreads exceeding several percent.

Tables 1 and 2 show the pion yields $\eta_{p\pi-}$ and $\eta_{p\pi+}$ at different proton momenta calculated from the spectra of Grote et al[5]. These yields are for a maximum accepted production angle of $\theta_{max} = 3.5\, m_\pi c/p_\pi$ and a momentum spread of $\pm 1\%$ around $p_\pi = p_p/20$. The use of an angle proportional to p_π^{-1} in this comparison has the advantage that the yields are quoted at the same normalized transverse emittance for a given beam size at the target. The yields are plotted in Figure 2 and are seen to be nearly constant for proton momenta below 100 GeV/c but then increase steadily. For a momentum spread of $\pm 5\%$, the pion yield $\eta_{p\pi}$ increases from 6% to 10% as the pion momentum increases from 5 GeV/c to 40 GeV/c.

p_p (GeV/c)	p_π (GeV/c)	θ_{max} (mrad)	$dN_{\pi-}/dp_\pi$ ((GeV/c)$^{-1}$)	$\eta_{p\pi-}$ ($\Delta p/p_\pi = \pm 1\%$)
12.5	0.625	784	7.23×10^{-1}	9.04×10^{-3}
19.2	0.960	510	4.63×10^{-1}	8.88×10^{-3}
30.0	1.5	327	2.70×10^{-1}	8.10×10^{-3}
50.0	2.5	196	2.05×10^{-1}	1.02×10^{-2}
70.0	3.5	140	1.48×10^{-1}	1.04×10^{-2}
150	7.5	65.3	9.42×10^{-2}	1.41×10^{-2}
200	10	49.0	7.47×10^{-2}	1.49×10^{-2}
300	15	32.7	5.53×10^{-2}	1.66×10^{-2}
500	25	19.6	3.78×10^{-2}	1.89×10^{-2}
800	40	12.3	2.61×10^{-2}	2.08×10^{-2}

Table 1: Negative pion yields per interacting proton when $\theta_{max} = 3.5\, m_\pi c/p_\pi$.

p_p (GeV/c)	p_π (GeV/c)	θ_{max} (mrad)	$dN_{\pi+}/dp_\pi$ ((GeV/c)$^{-1}$)	$\eta_{p\pi+}$ ($\Delta p/p_\pi = \pm 1\%$)
12.5	0.625	784	6.80×10^{-1}	8.50×10^{-3}
19.2	0.960	510	4.99×10^{-1}	9.58×10^{-3}
30.0	1.5	327	3.35×10^{-1}	1.01×10^{-2}
50.0	2.5	196	2.07×10^{-1}	1.04×10^{-2}
70.0	3.5	140	1.49×10^{-1}	1.05×10^{-2}
150	7.5	65.3	7.89×10^{-2}	1.18×10^{-2}
200	10	49.0	6.31×10^{-2}	1.26×10^{-2}
300	15	32.7	4.49×10^{-2}	1.35×10^{-2}
500	25	19.6	3.04×10^{-2}	1.52×10^{-2}
800	40	12.3	2.11×10^{-2}	1.69×10^{-2}

Table 2: Positive pion yields per interacting proton when $\theta_{max} = 3.5\, m_\pi c/p_\pi$.

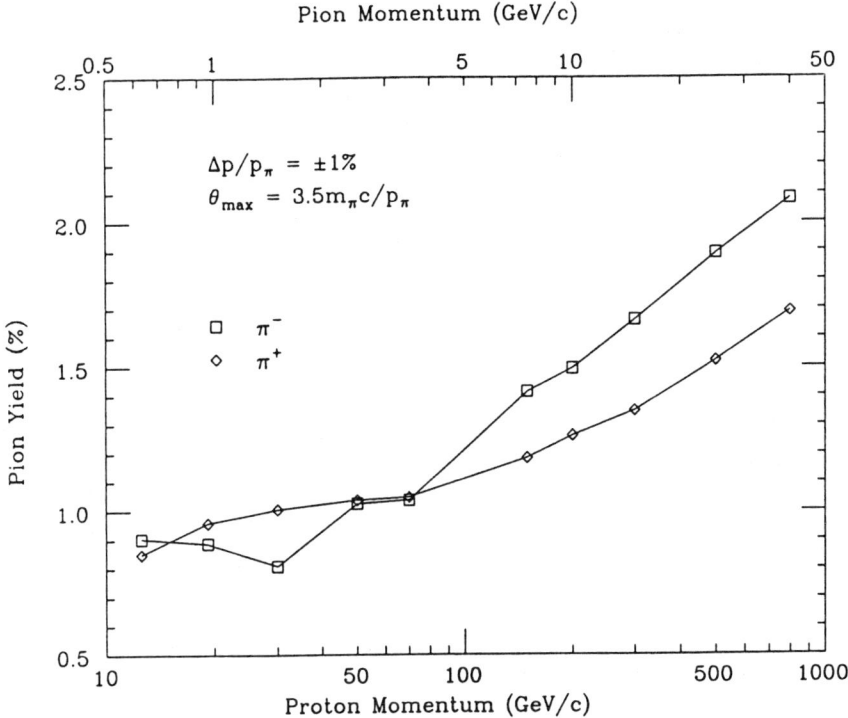

Figure 2: Pion yields per interacting proton.

    The increase in pion yield above 100 GeV/c proton momentum suggests that it is advantageous to collect pions and hence decaying muons at a momentum $p_\pi \simeq p_\mu \geq 5$ GeV/c. Neuffer has shown that ionization cooling of longitudinal emittance (energy spread) is most effective at a muon energy of about 1 Gev with little variation between 0.5 and 5 Gev because of the shape of the high-energy loss curve.[2] Longitudinal emittance cooling can be enhanced by introducing dispersion into the cooler so in fact cooling above 5 GeV is feasible.

    For proton beams with momenta above 100 GeV/c, tungsten targets of length 5 to 10 cm are appropriate for secondary hadronic particle production. These lengths are comparable to the nuclear collision and interaction lengths in tungsten. The target efficiency η_t (the probability that a proton interacts producing a secondary hadronic particle which exits the target) is about 0.4 in such targets. A low emittance pion beam is desirable to reduce both the initial muon emittance and aperture requirements downstream. The proton spot size on the target should be as small as possible without causing target destruction by shock wave depletion.[6] If 95% of the proton and pion beams are within a radius $R(cm)$ at the target, the pion emittance is $\epsilon_N(95\%) \simeq 3.5\ R(cm)$ rad.

The efficient collection of pions emanating from a target at large angles and with a large momentum spread requires metallic lenses similar to those used for antiproton collection.[7] Such lenses are cylindrical conductors carrying a high pulsed current to create an azimuthal magnetic field providing strong linear focussing. Lithium is particularly suitable because it has the least nuclear absorption of any metal. To insure a uniform current distribution and linear focusing field in a metallic lens, the minimum pulse length is chosen to make the skin depth δ comparable to the lens radius, a. Studies of antiproton yields suggest that the lens collection efficiency at the time of maximum field linearity reaches 95% for $\delta/a \simeq 0.4$ and increases slowly for larger ratios. The lens collection efficiency is approximated by unity in the remaining discussion.

PION DECAY TO MUONS

The decay of charged pions to muons involves a two-body final state. The energy spectra of the muon and neutrino are both uniform in the laboratory reference frame with bounds $E_{\pm} = \gamma_{\pi}(E^* \pm \beta_{\pi} cp^*)$, where the decay momentum in the pion rest frame is $p^* = (m_{\pi}^2 - m_{\mu}^2)c/2m_{\pi}$. For pion energies greater than a GeV, β_{π} and β_{μ} are nearly one in the laboratory frame. The muon momentum spectrum is then essentially uniform with upper and lower bounds p_{π} and χp_{π} respectively, where $\chi \equiv (m_{\mu}/m_{\pi})^2 \simeq 0.57$.

For a pion beam with a momentum spread $\pm\epsilon \equiv \pm\Delta p/p_{\pi}$ and $\epsilon \leq (1-\chi)/(1+\chi)$, the muon decay spectrum is illustrated in Figure 3. The muon fractions corresponding to the areas A, B and C in the figure are given by

$$A = \epsilon\chi/(1-\chi) = 1.3\,\epsilon \tag{1}$$

$$B = 1 - \epsilon(1+\chi)/(1-\chi) = 1 - 3.6\,\epsilon \tag{2}$$

$$C = \epsilon/(1-\chi) = 2.3\,\epsilon. \tag{3}$$

The large momentum spread of the resulting muon beam will limit the decay efficiency $\eta_{\pi\mu}$ for obtaining muons from pions.

If the pion decay occurs in a transport line of length approximately equal to the pion decay length ($53.6\,p_{\pi}(GeV/c)$ meters) and momentum acceptance equal to the pion momentum spread, then the decay efficiency $\eta_{\pi\mu} \simeq 0.63\,C = 1.4\,\Delta p/p_{\pi}$. The transport line is assumed to have an aperture adequate to transport all of the resulting muon beam with its increased emittance. The maximum muon angle from pion decay is $\theta_+ \simeq 0.282/\gamma_{\pi}$ radians. If $\bar{\beta}$ is the average betatron focusing function of the line, then the maximum increase in the invariant emittance from the decay of pions to muons is $\Delta\epsilon_N(95\%) \simeq \gamma_{\pi}\bar{\beta}\theta_+^2/2$. For example if $p_{\mu} \simeq p_{\pi} = 10$ GeV/c and $\bar{\beta} = 5$ meters, then $\Delta\epsilon_N(95\%) \simeq 0.28$ cm rad.

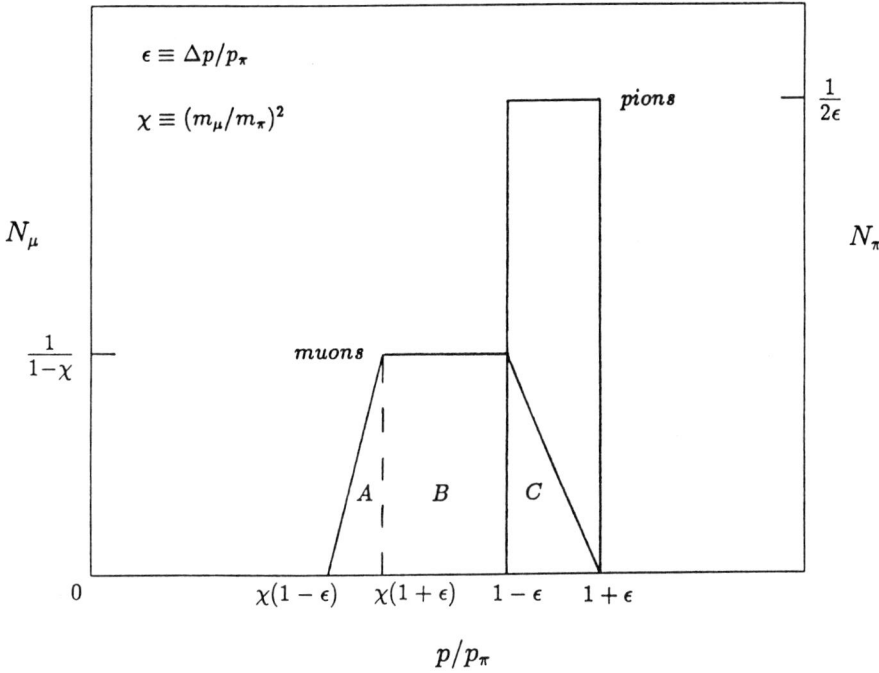

Figure 3: Normalized muon decay spectrum (left) resulting from a uniform pion spectrum (right) when β_π is nearly one. The muon fractions A, B and C are defined in Eqns. (1)-(3).

PARTICLE ACCELERATION AND SURVIVAL

Elementary considerations suggest that the relationship of particle lifetime and acceleration rate is central to the discussion of a rapid-cycling muon facility and is independent of the facility's details. The decay of accelerated particles is described by the equation $dN/N = -dt/\gamma\tau_o = -dz/c\tau_o(\gamma_i + zd\gamma/dz)$ where $\gamma = E/mc^2$, τ_o is the particle lifetime in the rest frame, and $d\gamma/dz$ is the acceleration gradient.

The survival ratio $\eta_a = N_f/N_i$ of unstable particles accelerated from γ_i to γ_f is

$$\eta_a = (\gamma_i/\gamma_f)^{1/(c\tau_o\, d\gamma/dz)}. \qquad (4)$$

The accelerated particle survival in Eqn. (4) as a function of γ_f/γ_i is illustrated in Figure 4 for different normalized acceleration gradients. The normalized acceleration gradient $d\gamma/d(z/c\tau_o)$ is simply the energy gain in units of the rest mass per unit decay length of the particle.

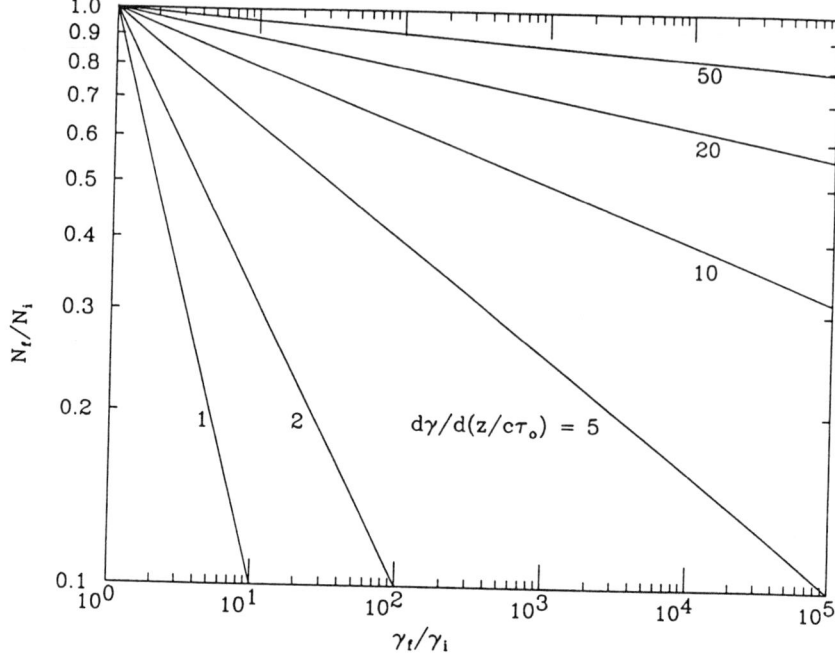

Figure 4: Particle survival as a function of energy ratio for different acceleration gradients.

For muons the range of gradients illustrated in Figure 4 is 0.15 MV/m to 7.5 MV/m. If the initial muon energy is 10 GeV, then acceleration gradients of 1 to 2 MV/m are adequate to insure a particle survival $\eta_a \simeq 0.5$ for final energies between 1 TeV and 10^3 TeV. Muon survival rapidly decreases for lower gradients so this is probably the minimum gradient acceptable for acceleration in a rapid cycling synchrotron complex. Acceleration gradients in a linac are commonly an order of magnitude higher (10 to 20 MV/m), and the muon survival would be greater than 90%.

CONCLUSION

Muon production, emittance cooling and acceleration are inter-related processes which ultimately must be optimized to deliver an adequate supply of low-emittance muons for a high-energy physics experiment. In the present study the approximate survival ratio of high-energy muons to initial protons is the product of the various efficiencies

$$N_\mu/N_p = \eta_{p\pi}\, \eta_t\, \eta_{\pi\mu}\, \eta_c\, \eta_a. \tag{5}$$

The pion production energy will be taken as 10 GeV (200 GeV protons) since anti-proton production near this energy is a well developed technology. The pion momentum spread is assumed to be $\pm\Delta p/p_\pi = \pm 0.05$. From Figure 2 then $\eta_{p\pi} \simeq 0.07$. The target efficiency is $\eta_t = 0.4$. If the pion decay occurs in a transport line of length equal to the pion decay length (536 meters), then $\eta_{\pi\mu} = 0.07$. The ratio of muons surviving emittance cooling will be taken as $\eta_c = 0.5$. If the acceleration to the final energy ($\leq 10^3$ TeV) occurs in a synchrotron complex, then the acceleration survival ratio is assumed to be $\eta_a = 0.5$, and from Eqn. (5) $N_\mu/N_p \simeq 5 \times 10^{-4}$. For acceleration in a linac, η_a is nearly one, and $N_\mu/N_p \simeq 10^{-3}$.

Between 10^3 and 2×10^3 protons are needed in the initial meson factory for every muon required in the high-energy collider assuming the captured pion momentum spread is $\pm 5\%$. Since the ratio N_μ/N_p is approximately proportional to $(\Delta p/p_\pi)^2$ because of $\eta_{p\pi}$ and $\eta_{\pi\mu}$, there can be a significant improvement in muon production if larger momentum spreads can be utilized.

REFERENCES

1. A.N. Skrinsky, *Proc. of the XXth International Conf. on High Energy Physics*, A.I.P. Conf. Proc. **68**, 1056 (1980).

2. D. Neuffer, *Particle Accelerators* **14**, 75 (1983).

3. *Proc. of the Advanced Hadron Facility Accelerator Design Workshop* (Los Alamos, Feb. 22-27, 1988), Los Alamos Report LA-11432-C (Jan. 1989), H.A. Thiessen, editor. See also *The Physics and a Plan for a 45 Gev Facility That Extends the High-Intensity Capability in Nuclear and Particle Physics*, Los Alamos Report LA-10720-MS (May 1986).

4. D. Neuffer, *Advanced Accelerator Concepts*, A.I.P. Conf. Proc. **156**, 201 (1987).

5. H. Grote, R. Hagedorn and J. Ranft, *Particle Spectra* (CERN, Geneva, 1970).

6. C. Hojvat and A. van Ginneken, *Nucl. Inst. Methods* **206**, 67 (1983).

7. G. Dugan et al, *IEEE Trans. Nucl. Sci.* **30**, 3660 (1983). A.J. Lennox, *ibid*, 3663 (1983).

THE MUON COLLIDER*

(Sandro's Snake)

A.G. Ruggiero

Brookhaven National Laboratory
Upton, NY 11973, USA

INTRODUCTION

In the quest for the Higgs bosons, a muon collider may be conceived as the experimental device more affordable and more feasible than electron-positron or very large hadron colliders, like NLC, CLIC, SSC and LHC. Muons have a mass ten times lighter than protons and are therefore easier to be steered on circular trajectories. On the other side their mass is a hundred times heavier than electrons and their motion is considerably less affected by the synchrotron radiation.

Muons are elementary lepton particles, with no internal structure. Like the electrons, they have obvious advantages over the hadron counterpart when they are used as the main projectiles for the production of the Higgs bosons. Moreover, because of their larger mass, they are also better suited than the electrons themselves, due to a considerably larger propagator constant.

Unfortunately, muons do not exist in nature and they have to be produced with the only technique we know these days: impinging an intense beam of protons on a target. This will cause muon production, but with a very large volume of the phase space. Like in the case of the production of antiprotons, in order to make the beam of some use for the subsequent collisions, muons also have to be collected and cooled to a sufficiently high intensity and small beam dimensions, before they can be accelerated and injected in the collider proper.

To make the situation more complicated, there is also the fact that muons are intrinsically unstable particles with a very short lifetime. Accumulation, cooling, acceleration and all other required beam manipulations are then to be executed extremely fast if one requires that a large fraction of the particle beam survives to the collision point.

This paper describes a feasibility study for the design of a muon collider. Recognized the fact that the particle lifetime increases linearly with the energy, we have adopted a scheme where steps of cooling and acceleration are

---

* Work performed under the auspices of the U.S. Department of Energy.
Contribution to the Advanced Accelerator Concepts Workshop, Port Jefferson, NY, 14-20 June 1992.

entwined. We have indeed found convenient to accelerate the beam as fast as possible to increase its chances of survival, and necessary to dilute the action of cooling throughout the entire accelerating process to make it more effective and affordable. All acceleration and cooling steps are executed in a single pass essentially along a curvilinear and open path. We do not believe it is possible to handle the beam otherwise in circular and closed rings, as it has been proposed in the past.[1,2]

The example shown in this paper describes a muon collider at the energy of 250 GeV per beam and a luminosity of $4 \times 10^{28} \text{cm}^{-2}\text{s}^{-1}$. We have adopted an extrapolation of the stochastic cooling method for the reduction of the beam emittance.

PROPOSED SCENARIO

A schematic layout of the muon collider is shown in Figure 1. It is made of three major parts: (1) a high intensity proton source with a target station attached to it for the production of muons; (2) two accelerating sections, one for each beam, with bending dispersed for providing betatron stochastic cooling; and (3) a final collision region which eventually can include a storage and collider ring.

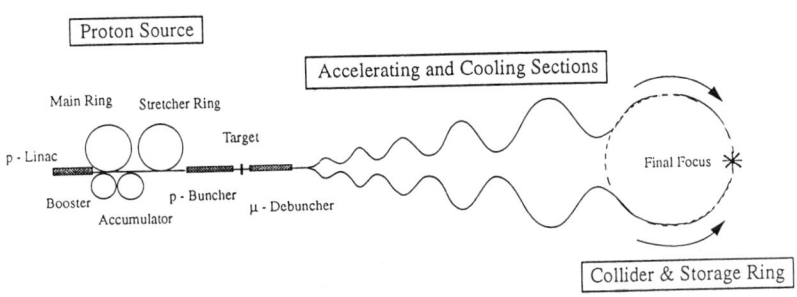

Figure 1: Layout of the Muon Collider (*Sandro's snake*).

THE PROTON SOURCE AND MUON PRODUCTION

Recent studies of hadron facilities (the EHF, for instance[3]) have demonstrated that it is possible to accelerate proton beams to the energy of 30 GeV at high repetition rates for an average output current of $100 \mu A$. This facility can be made of a 1.2 GeV linac, a 9 GeV booster ring with the circumference of 480 m operating at the repetition rate of 50 Hz, an accumulator ring of the same dimension, operating at 25 Hz. If the facility is followed by a Stretcher Ring of the same dimension of the main ring, it is possible to deliver the beam continuously with essentially 100% duty cycle. We shall assume that such proton source is available.

After being slowly extracted from the Stretcher Ring, the proton beam can be rebunched in a sufficiently long traveling wave linac at the frequency $f_b = 3$ GHz (for preparing the time structure of the muon beam to be generated). By entering a second stage of the linac with considerably larger voltage gradient, it is possible to create a mismatch which will make the proton bunches rotate in their own buckets. After a quarter of the oscillation the bunches will present their narrowest length. At the same time the beam is focused to a small spot size at the location of a target for the production of the muon particles. All these processes can be accomplished essentially with no beam losses. The dimensions of the proton bunches can be made small enough to have no consequences on the dimensions of the beam of muons. Each proton bunch impinging the target is made of 2×10^5 particles.

Muons are produced in a cascade as the decay product of π mesons in pairs of μ^+ and μ^-. Large production rates are expected[4]; for instance, the following reference values are customarily taken: a yield of 0.1% by accepting a momentum bite of $\pm 5\%$ and a semi-angular aperture of 50 mrad. Since we require the muon beam coming out of the target to have a reasonably small momentum spread for capture and acceleration, we shall take more conservatively a full momentum bite of only 2%. The production rate increases linearly with the momentum spread and about quadratically with the angular acceptance. Thus, with these adjusted values, we can estimate a pair of production rate of about 2×10^{-4} per proton. There is a continuous streaming of muons of both sign from the target with an average current of approximately 20 nA for each species. At the same time an optimum production energy can be chosen to be about 1 GeV. Since the mass of the muons is about 100 MeV, this corresponds to $\gamma \sim 10$.

The muon beam has the same rf structure f_b of the proton beam, that is 3 GHz, and the same bunch length. There will be about 40 muons of each sign per bunch. The length of the target should match the range for the muon production; we take here for the following estimate a target length of $l = 1$ cm. The resulting muon beam betatron emittance is then $\epsilon = l\theta^2 = 8\pi$mm mrad. The normalized emittance is $\epsilon_{init} = 80\pi$mm mrad.

The two species of muons with opposite electric charge are first separated by a common dipole magnet and then transported by a focusing channel which is to be matched to the beam aspect ratio at the target, that is a value of $\beta_T = 1$ cm. Each of the two beams then undergoes to the same sequence of bending, stochastic cooling and acceleration, until for each of them, the final emittance and energy values are reached.

THE ACCELERATING AND COOLING SECTION

This has the shape of a snake (*Sandro's snake*) with convolutions increasing in size toward the large energy end. We can assume that there M_c of such convolutions, each made of a bending arc followed by a straight accelerating section, as shown in Figure 2. The straight sections are made of traveling-wave rf-cavity structures a'la SLAC for the acceleration of the muon beam; FODO cells for transverse focusing are also provided dispersed. The accelerating rf frequency f_{acc} can also be chosen around 3 GHz, with an effective accelerating gradient W of few tens of MVolt/m. The electric power demand for a continuous mode of operation may be exceedingly too large to be afforded; in this case one might have to resort either to superconducting cavity technology or to the introduction of a duty cycle. As we shall see later, it is indeed possible to re-use the beams over and over in an ultimate large storage ring operating at constant field. During this time the accelerating rf system may be turned off.

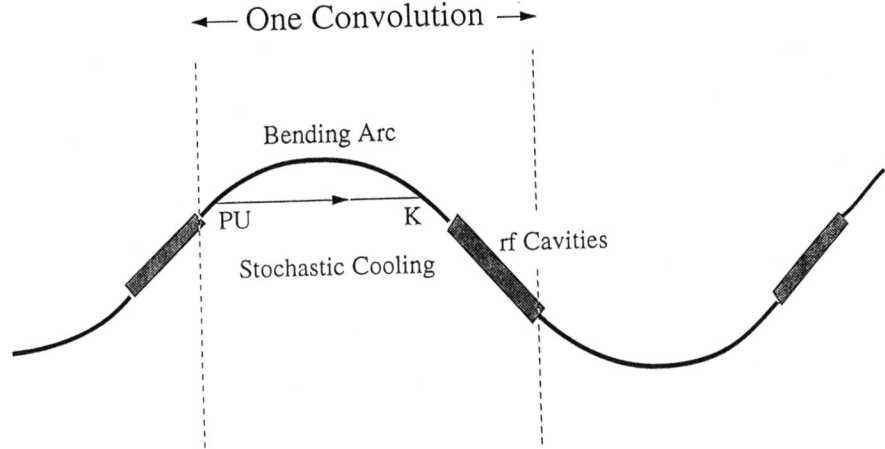

Figure 2: Details of One Convolution.

If the convolutions are labeled in sequential order, $i = 1, 2, \ldots, M_c$, we can then define L_i to be the overall length of the i-th accelerating section; C_i will denote the arc length of the corresponding convolution. We can then easily estimate the overall length of the accelerating and cooling section. As we shall see, a large contribution to the total length of the section is given by the arcs; thus the amount of accelerating gradient is not necessarily an issue. For completeness we shall also denote with E_i the beam kinetic energy and with ϵ_{ni} the normalized betatron emittance at the end of the i-th convolution.

The design of the first straight accelerating section may require special care because the muon beam has still a large momentum spread. Just before entering the first convolution, one will apply bunch rotation at the same accelerating frequency f_{acc} to trade momentum spread with length, as it is done in the antiproton sources of Fermilab and CERN. It is not clear at this moment whether this can be accomplished on a fly, along a straight path with one linac, or whether this will require some sort of circular ring at constant energy (as, for instance, the Debuncher Ring at Fermilab).

The arcs are made of several bending and focusing FODO cells. The bending is provided with dipole magnets operating at a constant field which has the same value B throughout the length of the section. The bending will flip direction from one convolution to the next. The bending angle α_i is only a fraction of π. Other geometries are of course possible, provided they allow convergency of the two beams to the collision point. The convenience of this layout, compared to a complete circular ring, is that one can make use of superconducting magnets without having to cycle them at a too large rate. In the arcs the beam energy is constant, but will vary from arc to arc, and the average bending radius and arc length will also vary accordingly.

Since the beam has essentially the speed of light, the bending is also required to provide enough electric delay for the signal processing of the stochastic cooling. This, as shown in Figures 2, 3 and 4, includes several pickup (PU) stations upstream followed by an equal number of kicker (K) stations downstream. The Schottky signal from the beam travels from the PU's to the K's where it also properly amplified and applied to the beam for the stochastic correction. We assume that a total delay of 10 ns is adequate for signal processing. This value sets a minimum that one can calculate for the arc length and bending. As we shall see later, only a little reduction of betatron emittance is required per convolution. We propose a different method of betatron stochastic cooling, described later, which works effectively at very low beam intensity level and for very short beam bunches.

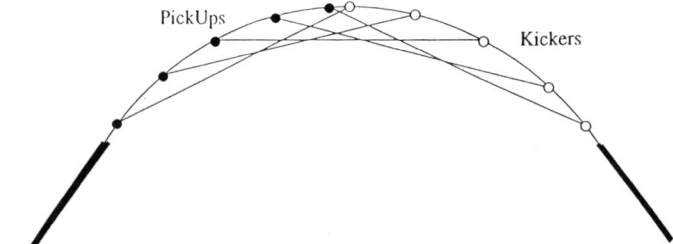

Figure 3: Multiple Single-Pass Stochastic Cooling Steps in the Same Convolution.

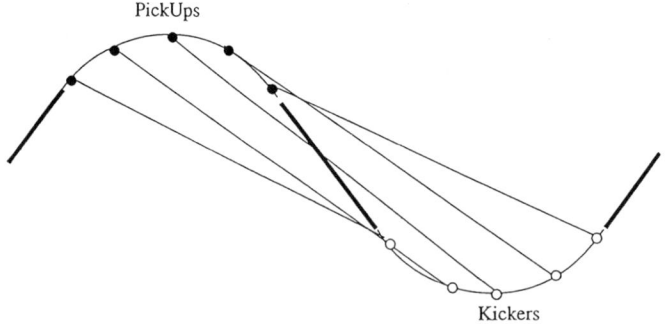

Figure 4: Multiple Single-Pass Stochastic Cooling Steps Shared by Two Consecutive Convolutions.

THE COLLISION REGION

The last convolution of the accelerating and cooling section is of a size large enough to allow the two sides of merge with each other in a straight, head-on collision path as shown in Figure 1. The collision region is the location where the two beams are brought together with a final focus described by β^*. The collision is essentially head-on; if the beam bunches are too close to each other, a small collision angle may be required to avoid beam- beam interaction with subsequent bunches. On the other side the collision angle is to be small enough to avoid any significant reduction of luminosity.

The two beams are *round*, with the same emittance in the two transverse planes; we consider this an advantage since it allows a symmetric focusing arrangement with the same value of β^* in the two planes. Moreover the intensity per bunch, at least in our scenario, is considerably low, so that no serious and disruptive effects are expected from the beam-beam interaction. The large frequency of bunch encounter at the collision point is also an advantage for the experiment setup, which prefers a smoother distribution of the events to be detected in time.

As shown in Figure 1, the last convolution of the two sides can be thought as part of a larger final storage ring. In this case the collision region with the final focus can also be conceived to be an integral, matched part of the storage (and collider) ring. This configuration may be advantageous if the muon beam has acquired enough energy and lifetime to make it survive through several hundreds of revolutions. As the beam circulate and collide until exhaustion in the storage ring, the accelerating section may be temporarily turned off to reduce the electric power demand. This mode of operation is feasible only with a relatively large value of β^*, of few centimeters, if the final focus is to be integral part of the storage ring.

REQUIREMENTS AND GOALS

The major requirement parameters are the final energy E and the luminosity L. For the round beam configuration at collision we have very simply

$$L = \frac{N_+ N_-}{\beta^* \epsilon_n} f_b \gamma \tag{1}$$

where N_+ and N_- are the number of particles per bunch of each species, and ϵ_n is the final normalized betatron emittance, defined as $\epsilon_n = 4\pi\sigma^2/\beta^*$ where σ is the rms beam spot size at the collision point. The requirement on the luminosity versus energy depends eventually on the cross-sections of the μ^+-μ^- collision according to the following scaling

$$L = L_{\text{ref}} \left(\frac{E}{E_{\text{ref}}} \right)^2 \tag{2}$$

where, very likely, if $E_{ref} = 1.25$ TeV then $L_{ref} = 1 \times 10^{30} \text{cm}^{-2} \text{s}^{-1}$. Combining the two equations above yields the following for the normalized emittance

$$\epsilon_n = \frac{N_+ N_- f_b \gamma_{ref}^2}{\beta^* \gamma L_{ref}} \tag{3}$$

It is seen then that the required normalized emittance decreases inversely with the energy at collision. It is obvious that this requirement is considerably smaller, by several orders of magnitude, than the value ϵ_{init} of the beam emittance at the point of production. It is proposed here to recover this difference with stochastic cooling.

Muons are unstable particles and they decay rather fast. At the kinetic energy of 1 GeV the lifetime is of only 21 μs. Due to relativistic effects, the lifetime increases linearly with energy; for instance, it is 21 ms at 1 TeV. It is easy to calculate the survival ratio in every convolution of the collider and the fraction of the beam that survives at the large energy end. It is required that this fraction is large enough, so that eventually the beams can be used again in multiple collision mode in the final storage ring.

In order to make some estimate we shall take here a final energy of 250 GeV (that can be fitted easily in the RHIC tunnel at BNL), that is $\gamma = 2500$. The required luminosity is $4 \times 10^{28} \text{cm}^{-2} \text{s}^{-1}$ and the normalized beam emittance $\epsilon_n = 0.1 \times 10^{-8} \pi$ mm mrad, that is a reduction of eleven orders of magnitude from the value ϵ_{init} at the target production.

STOCHASTIC COOLING

An optimum luminosity configuration, for a given flux of particles entering the collider, which does not impose too stringent requirements on the beam emittance, is a low repetition rate. Unfortunately in this case the number of particles per bunch is too large and it would make impossible the application of a cooling technique like stochastic cooling. This on the contrary, specially in our case, requires a very low number of particles per bunch and thus a considerably higher repetition rate and a final smaller emittance.

Moreover, the bunches are extremely short; a fact which also makes practically impossible the application of stochastic cooling as ordinarily conceived[5] based on the property of *longitudinal mixing*. We shall deviate here from this and consider a *single pass* cooling method which acts on all particles in the same bunch at the same time.

Let us suppose that at the pickup location (PU) in one convolution we measure the center of mass of a bunch particle distribution in either horizontal or vertical plane (denoted by z). The displacement is caused by statistical fluctuations and given as the average over all the particle position, that is

$$\bar{z} = \frac{1}{N} \sum_m z_m \tag{4}$$

where N is the number of particles in the bunch. At the same time, if we denote with β the amplitude lattice function at any desired location, the beam rms emittance ϵ can be defined as

$$\beta\epsilon = \frac{1}{N} \sum_m z_m^2 \tag{5}$$

In average, the relation between these two quantities is the following

$$\bar{z}^2 = \beta\epsilon/N \tag{6}$$

Suppose now that at the Kicker location (K) the position z_m of each particle is corrected by an amount proportional to the average bunch displacement measured at the previous PU location, by a factor g. It is easily seen that

$$(\beta\epsilon)_K = \frac{1}{N} \sum_m (z_m - g\bar{z})^2 \tag{7}$$

$$= (\beta\epsilon)_{PU} - \left(2g - g^2\right)\bar{z}^2$$

So that the emittance reduction during a single step occurring in any one of the convolutions is

$$\frac{\Delta\epsilon}{\epsilon} = -\frac{\left(2g - g^2\right)}{N} \tag{8}$$

The optimum condition, which corresponds to the large emittance reduction, is obtained by setting the gain $g = 1$. In this case the reduction is just inversely proportional to the number N of particles in the bunch. We shall assume that the system parameters are set for this optimum cooling rate. Nevertheless, this mode of operation works only for a *single pass*. To work again in the successive step, one should regenerate the fluctuation signal at the following PU location. In the case of coasting beams, this is usually done with the longitudinal shear of the particle motion due to the difference in speed among particles, and by the fact that the detecting device measures the position of only a longitudinal section of the beam. We propose here to mix the particle relative order by introducing strong octupoles (or other nonlinear devices) to create enough smear in the betatron motion. The octupoles will be placed in the focusing cells FODO cells of both the accelerating and bending sections.

An interesting feature of this method is that the system can be made to work on a narrow bandwidth. Indeed, it is required that all the particles in the same bunch (and bucket) are observed at the same time. Signal overlapping from different bunches is effectively avoided by choosing an electronic bandwidth which matches the bunching frequency f_b, taken in this paper to be 3 GHz.

Denoting with n_s the number of stochastic cooling *single pass* steps, the final beam emittance is given by

$$\epsilon_n = \epsilon_{\text{init}} e^{-(n_s/N)} \tag{9}$$

It is seen that, with $N = 40$ particles per bunch, the final required emittance may be obtained with $n_s \sim 1000$ steps.

A POSSIBLE SOLUTION

There are several ways numbers can be configured together, and the solution we show here is just an example. There may be other more optimal arrangements which need to be found and investigated. In this example we take $M_c = 100$ convolutions.

We begin by taking the same energy gain per convolution (another possibility would be to accelerate faster during the early stages and slower toward the end; one more possibility is just the opposite). This will set the energy gain per convolution to 2.5 GeV which can be achieved over a distance of about 50 meter with an accelerating gradient of 50 MV/m. This alone already require a linear length of 5 kilometers.

We shall take superconducting magnets for the bending arcs. The dipole magnets have a field of 10 Tesla, and we allow for a packing factor of 80%. The arc lengths are adjusted to provide the same total difference of 10 ns between the length of their paths and the length of the associated geometric cords.

We shall assume each arc includes 10 stochastic cooling steps (again, this is just an example of so many possibilities). One possible configuration is sketched in Figure 3 where the steps are entwined with each other in the same arc. Another configuration is shown in Figure 4 where the steps span their function over two consecutive arcs. All these arrangements to work effectively require a considerable amount of transverse mixing from octupole magnets which are placed as often as possible.

The results are shown in Table 1. The total length of the collider is about ten kilometers. At the end, about 98% of the muons have survived. They can then be injected in a storage ring having the dimensions of RHIC, where they can circulate (and collide) for about 400 revolutions, corresponding to their lifetime of 5 ms. Since it takes about 35 μs for the muons to travel the collider, the accelerating rf system can be operated with a duty cycle of less than one percent.

CONCLUSION

We have exposed in this paper the construction in first order approximation of the design of a muon collider. We found this to be a very interesting and appealing project that may be valuable in removing several technical difficulties of an e^+-e^- linear collider and possibly also of the Super Superconducting Collider.

There are still a lot of questions unanswered, and the concepts exposed still need to be carefully evaluated. For instance, there are some questions concerning the muon production: what is the optimum production energy? This may have an impact on the initial beam betatron emittance together to the production angle; what are really the production rates? These questions can be answered with rather simple experiments, for instance, at the BNL facilities.

Table 1 : Collider Parameters versus the Convolution Number

Convolution no.	Kinetic Energy GeV	Arc Length m	Survival Ratio	Total Length km	Norm. Emittance π mm mrad
5	13.5	13.51	0.996	0.284	2.29E+01
10	26.0	20.71	0.994	0.619	6.57E+00
15	38.5	26.81	0.993	0.988	1.88E+00
20	51.0	32.32	0.991	1.386	5.39E-01
25	63.5	37.33	0.990	1.810	1.54E-01
30	76.0	42.02	0.989	2.258	4.42E-02
35	88.5	46.49	0.988	2.730	1.27E-02
40	101.0	50.79	0.988	3.223	3.63E-03
45	113.5	54.90	0.987	3.737	1.04E-03
50	126.0	58.84	0.986	4.271	2.98E-04
55	138.5	62.60	0.986	4.825	8.54E-05
60	151.0	66.23	0.985	5.397	2.45E-05
65	163.5	69.81	0.984	5.987	7.01E-06
70	176.0	73.38	0.984	6.595	2.01E-06
75	188.5	76.71	0.983	7.220	5.76E-07
80	201.0	80.12	0.983	7.863	1.65E-07
85	213.5	83.32	0.982	8.521	4.72E-08
90	226.0	86.50	0.982	9.196	1.35E-08
95	238.5	89.70	0.981	9.886	3.88E-09
100	251.0	92.93	0.981	10.593	1.11E-09

It may be possible to upgrade the scenario to larger energies and to larger luminosities. Indeed the collider could be made longer than described here and one can find an optimum configuration of parameters which makes a more efficient use of the beam intensity. Larger luminosities can be obtained by increasing the muon production rate, for instance, by accepting larger momentum bite. What are the limitations here? Still, increasing the intensity is not enough, as one needs to dilute even more the longitudinal particle distribution to accommodate stochastic cooling.

The idea itself of stochastic cooling in a *single pass* needs more study and careful evaluation of hardware limitations. For instance, what are the effects of thermal noise and Schottky noise on the final beam emittance? The optimum gain regime may be limited by the electronic gain toward the high energy end. We find very intriguing (and challenging) the idea of having to deal, and to measure, an intensity as low as few tens of particles per bunch. Some of the extreme technical conditions of the stochastic cooling performance can be experimentally studied at the Fermilab complex. As the beam dimensions get smaller and smaller we may find more and more difficult to generate particle *mixing* with non linear elements as octupole magnets.

Finally, with the Booster soon completely operational, the AGS complex at BNL will be capable of delivering a proton average intensity of 5 to 10 μA. With the addition of a stretcher ring, a high frequency buncher, a target station and a debuncher for the muons we have then an opportunity to demonstrate experimentally several of the concepts exposed here. We can then later expand from there...

ACKNOWLEDGMENTS

The concepts exposed in this paper stimulated in the Physics Opportunities section of the Workshop on Advanced Accelerator Concepts held in Port Jefferson, New York, on June 11-15, 1992. The author likes to acknowledge and to thank the following people for their comments, critics and very valuable discussions: P. Chen, D. Cline, K. McDonald, D. Neuffer, R. Noble, R. Palmer and A. Sessler.

REFERENCES

1. A.N. Skrinsky, Proc. of the XX International Conf. on High Energy Physics, AIP Conf. Proc. **68**, 1056 (1980).
2. D. Neuffer, Particle Accelerators 14, **75** (1983).
   F . Bradamante, Proposal for a European Hadron Facility, EHF-87-18, 18 May 1987.
3. R.J. Noble, Particle Production and Survival in Muon Accelerator, these proceedings.
4. S. van der Meer, An Introduction to Stochastic Cooling, Physics of Particle Accelerators, AIP Conf. Proc. 153, Volume 2 1628 (1987).

ON SCALING & OPTIMIZATION OF HIGH-INTENSITY, LOW-BEAM-LOSS RF LINACS FOR NEUTRON SOURCE DRIVERS

R. A. Jameson

Los Alamos National Laboratory, Los Alamos, NM 87545

ABSTRACT

Rf linacs providing cw proton beams of 30-250 mA at 800-1600 MeV, and cw deuteron beams of 100-250 mA at 35-40 MeV, are needed as drivers for factory neutron sources applied to radioactive waste transmutation, advanced energy production, materials testing facilities, and spallation neutron sources. The maintenance goals require very low beam loss along the linac. Optimization of such systems is complex; status of beam dynamics aspects presently being investigated is outlined.

INTRODUCTION

The search[1-3] for options to deal with existing radioactive waste and future energy sources with greatly reduced waste streams has resulted in serious consideration of particle accelerator driven transmutation technology, which has the potential to effectively transmute both actinide and fission product waste to shorter-lived, more easily stored and managed forms, and to efficiently produce electricity. The proton linear accelerators needed for the various forms of energy producers and waste burners under consideration are large, high-intensity machines in the 30 – 250 mA cw proton current, 800 – 1600 MeV energy class. The basic layout is shown in Fig. 1. Another application of renewed interest is fusion materials testing using a ~35 MeV deuteron beam on a molten lithium target to produce neutrons. Conceptual designs use modules of up to 250 mA cw deuterons[4].

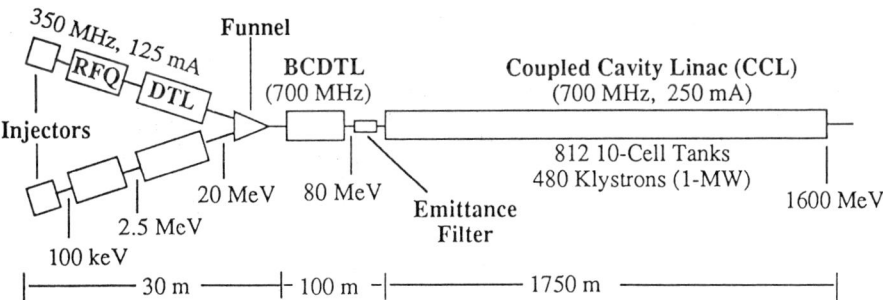

Figure 1. High-intensity proton linear accelerator reference design.

In recent papers[5-8], we have outlined the basic accelerator requirements, presented point designs, discussed the design approach for near-term and longer-range improved options, and the technology issues and technology base. Very briefly summarized, the required room-temperature technology is in hand. An integrated "front-end" funneled system up to at least 40 MeV should be constructed as a testbed for final engineering development and reliability/availability development. Two major reviews, by ERAB[9] and JASON[10], have emphasized these points. We believe that superconducting rf technology may be appropriate in the longer range for energy production systems where efficiency is paramount, and we are initiating studies in this area[6].

A primary design factor for these high-intensity linacs is to insure that beam losses along the linac are kept low enough that maintenance can be done "hands-on", that is, without remote manipulators, over the lifetime of the facility (~40 years). Space-charge in the nonrelativistic ion beam is the main effect that, if not controlled, could lead to stray particles that would cause induced radioactivity along the linac. Typically, fractional losses must be kept below 10^{-5} to 10^{-8}/m, more stringent at higher energy. A basic design strategy is pursued to achieve low beam loss. The radiofrequency-quadrupole (RFQ) is a superb buncher and preaccelerator, well preserving the ion source emittance. The current is kept well below space-charge limits at all times, leading to a funneled system at low energy. Machine transitions (e.g. from RFQ to drift-tube linac (DTL) to coupled-cavity linac (CCL), or where magnet or tank groups change) are minimized in number, made at the lowest energy possible (where beam loss is less important), and always very carefully matched. Attention is then focused on achieving high "rms aperture ratios", the ratio of transverse aperture to rms beam size (at flutter factor maximum) and longitudinal bucket width to rms beam length.

While the resulting designs could proceed with confidence to construction, the scaling and optimization of linacs against the low beam loss requirement is not straightforward or well-codified, even at the rms (large-scale) level. This paper will present some results of ongoing (and therefore incomplete) research toward clarification of the problem at this level, pointing to areas of future work.

Clearly, outlying particle losses at the $10^{-5} - 10^{-8}$ level might have only a small effect on the rms properties of the beam, and thus the total beam size must be constantly kept under observation. There is presently no theory that can predict the total beam size. Indeed, even the complex physics embodied in our modern particle-tracking codes contains approximations limiting their precision. With careful attention[11,12] to numerical aspects, appropriate fitting, and statistical techniques, the numerical behavior of outlying particles observed in full simulations can be predicted better than the code physics deserves. Thus, the total beam size is scrutinized, and, beyond the RFQ, no particle losses are tolerated in an acceptable design when simulated with ≥ ~100,000 particles. An engineering safety factor based on the rms aperture factors is then used, a practice based on experience at LAMPF and many circular accelerators and storage rings. The transverse aperture factor near the end of LAMPF is ~6.3; our current preliminary designs strive for at least 10.

RMS DESCRIPTION OF BEAM/LINAC SYSTEMS

The simplest description of a beam simultaneously matched in the transverse and longitudinal phase-planes is

$$\varepsilon_{tn} = \frac{a^2 \gamma \sigma^t}{n\lambda} \tag{1}$$

$$\varepsilon_{ln} = \frac{(\gamma b)^2 \gamma \sigma^l}{n\lambda} \tag{2}$$

where ε denotes emittance (here total emittance), sub-or-super-t and sub-or-super-l the transverse or longitudinal plane, sub-n the normalization to the canonically preserved emittance during transport or adiabatic acceleration. a and b are the transverse and longitudinal semi-axes, respectively, of an ellipsoidal beam bunch with uniform particle distribution; (the rms radii are $a/Sqrt[5]$ and $b/Sqrt[5]$). n is an integer describing the number of increments of length $\beta\lambda$ in the transverse focusing period, and σ is the phase advance of the particles' oscillating motion over the distance $n\beta\lambda$.

If the potential and kinetic energies between the transverse and longitudinal degrees of freedom are balanced, a third equation, called the equipartitioning equation[13], results:

$$\frac{\varepsilon_{ln}}{\varepsilon_{tn}} = \frac{\sigma^t}{\sigma^l} = \frac{\gamma b}{a} \tag{3}$$

Expanding the phase advance terms:

$$\sigma^{t2} = \sigma_o^{t2} - \frac{l\lambda^3 \hat{k} n^2 (1 - ff)}{a^2 (\gamma b) \gamma^2} \tag{4}$$

$$\sigma^{l2} = \sigma_o^{l2} - \frac{2l\lambda^3 \hat{k} n^2 ff}{a^2 (\gamma b) \gamma^2} \tag{5}$$

where $\hat{k} = (3 Z_o q\, 10^{-6}) / (8\pi\, mc^2)$ for a, b in cm,
and ff = ellipsoid form factor = $\sim a/(3\gamma b)$ for $\sim0.8 < (\gamma b)/a < \sim5$.

$$\sigma_o^{t2} = \frac{1}{8\pi^2}\left[\frac{c\ field\ \beta\ q\ xx\ n^2\lambda^2}{mc^2\ \gamma\ a_o} \right]^2 - \frac{\pi\ q\ e0t\ \sin|\phi_s| n^2\lambda}{mc^2\beta\gamma^3} \tag{6}$$

where $field$ is the quadrupole magnet field, a_o is the bore aperture radius, and xx is a filling factor for the fraction of the period filled with magnets, derived for the smooth approximation. $e0t$ is the accelerating gradient and ϕ_s the synchronous phase angle.

$$\sigma_o^{l2} = -\frac{2\pi\ q\ e0t\ \sin|\phi_s| n^2\lambda}{mc^2\beta\gamma^3} \tag{7}$$

The transverse aperture factors are:

$$tfac_{total} = a_o/(total\ beam\ radius\ at\ flutter\ factor\ maximum) \tag{8}$$
$$= a_o/(a\ Sqrt[psi])$$

$$tfac_{rms} = a_o/(rms\ beam\ radius\ at\ flutter\ factor\ maximum) \tag{9}$$
$$= a_o/(a\ Sqrt[psi/5] = a_o/(rmsr\ Sqrt[psi])$$

The longitudinal aperture factors (justification for this choice is presented later) are:

$$lfac_{total} = |\phi_s|/(total\ beam\ phase\ half\text{-}length) = |\phi_s|/b \tag{10}$$

$$lfac_{rms} = |\phi_s|/(b/Sqrt[5]) = |\phi_s|/rmsl \tag{11}$$

If we were interest in maximum beam brightness or current density only, we can write:

$$\frac{I}{\varepsilon_{in}^{2}}=\frac{\gamma b}{\hat{k}\lambda(1-\mathit{ff})}\left[\frac{\gamma^{2}\sigma_{o}^{\prime2}}{a^{\prime2}n^{2}\lambda^{2}}-\frac{1}{a^{2}}\right] \tag{12}$$

where a' is the divergence; $etn = aa'$, and

$$\frac{I}{a^{2}(\gamma b)}=\frac{\mu_{t}\gamma^{2}\sigma_{o}^{\prime2}}{\lambda^{3}\hat{k}n^{2}(1-\mathit{ff})}=\frac{\mu_{t}\gamma^{2}\sigma_{o}^{\prime2}}{2\lambda^{3}\hat{k}n^{2}\mathit{ff}} \tag{13}$$

Near the space-charge limit, where the beam size stays relatively constant, maximizing these quantities strongly favors higher frequency and strong external focusing. However, these optimizations are in an absolute sense, and not at all the same as the minimization of a and b relative to the machine acceptances, Eqns. (8) – (11), the pertinent aspect of the low beam-loss specification. This seemingly slight shift in emphasis results in the optimum being very difficult to solve for in general.

With only the two matching equations (1) and (2), the two zero-current phase advances, and optionally Eqn. (3), to work with, only 4 – 5 among the variables can be free; the others must be fixed. The coupled equations are nonlinear and represented by polynomials generally of order >8, so it is impossible to find general solutions. The number of variables is high, and many combinations are possible.

In the following, the focus will be on the particular architecture of the practical, baseline linac. A number of special case results are developed as a pathway to gain insight into the general problem and the definition of scaling and optimization criteria.

ARCHITECTURE OF TYPICAL HIGH-INTENSITY LINAC

Typically, the accelerator section of the RFQ, the DTL, and CCLs have been designed using longitudinal focusing from a constant real-estate accelerating gradient and ϕ_S. Sometimes, a ramped $e0t$ or ϕ_S law is used to provide larger acceptance at low energy, and/or to keep the focusing stronger up to higher energy. But for a CCL in particular, the cost of rf power essentially forces constant $e0t$ and ϕ_S for most of the machine. We use this as the baseline choice, and it becomes a very strong architectural feature. Then, in a given linac with frequency and n specified, Eqn. (7) shows that σ_o^l falls as $(\beta\gamma^3)^{1/2}$, and the focusing per unit length, $\sigma_o^l/n\beta\lambda$, is independent of n.

The transverse focusing law in the three accelerator sections has also usually been chosen in a similar way, by using a focusing law based on a slope and intercept from the stability chart without directly accounting for the beam. A constant σ_o^l has been used for the present Los Alamos reference designs. This is appealing because it generally results in a decreasing rms beam size at higher energy, and also because the quad settings do not depend on the beam current. Per unit length, the focusing provided by this choice is $\sigma_o^l/n\beta\lambda = k_2/n\beta\lambda$, where k_2 is a constant < 90° to avoid the envelope resonance. We also used a constant $n = 10$, which minimizes matching requirements between sections.

Typical simulation results for such a linac are shown in Figs. 2–5, for a 700 MHz, 20-1600 MeV, 10-cell per tank CCL reference design with constant $\sigma_o^l = 80°$ over a 13-$\beta\lambda$ period, constant real-estate accelerating gradient = 1 MV/m, nearly constant $\phi_S = -30°$ (ramped from -41.4° – -30° over 20–40 MeV), and constant 2.5 cm bore radius. The input $\varepsilon_{in\ rms} = 0.02$ cm·mrad, and $\varepsilon_{ln\ rms} = 0.04$ cm·mrad. In order to probe deeper, we also compute the phase advances and the energy balance factor. A summary of various quantities for these and other sets of conditions is given in Tables 1 and 2. In most cases, the variation is smooth and monotonic, with most of the

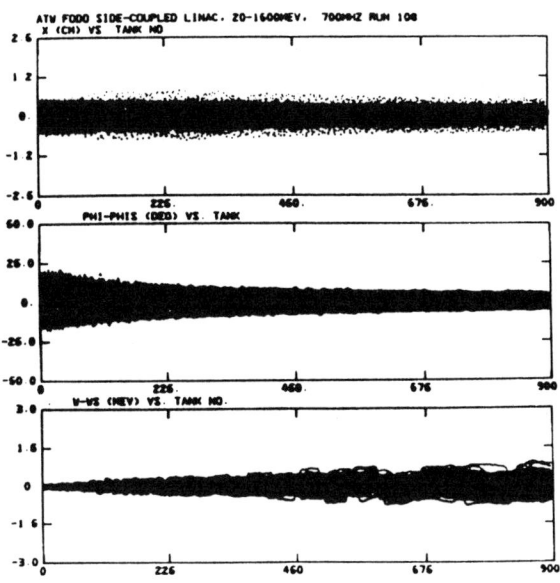

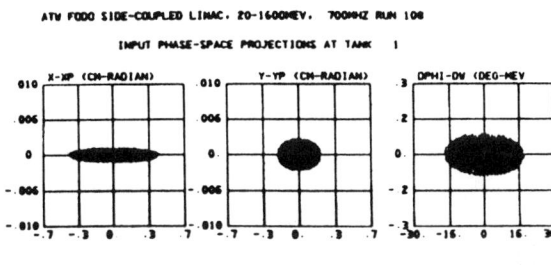

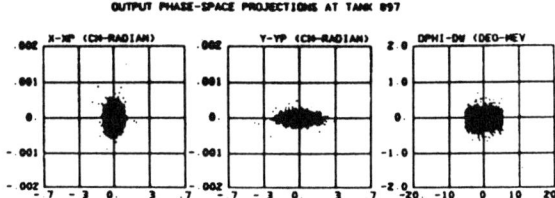

Fig. 2. 700 MHz CCL, current = 140 mA, $\varepsilon_{ln\ rms}$ input = 0.02 cm·mrad. 20 MeV input and 1600 MeV output phase-space projections of x, phi-phis, and w - ws along the linac. These are typical of all cases.

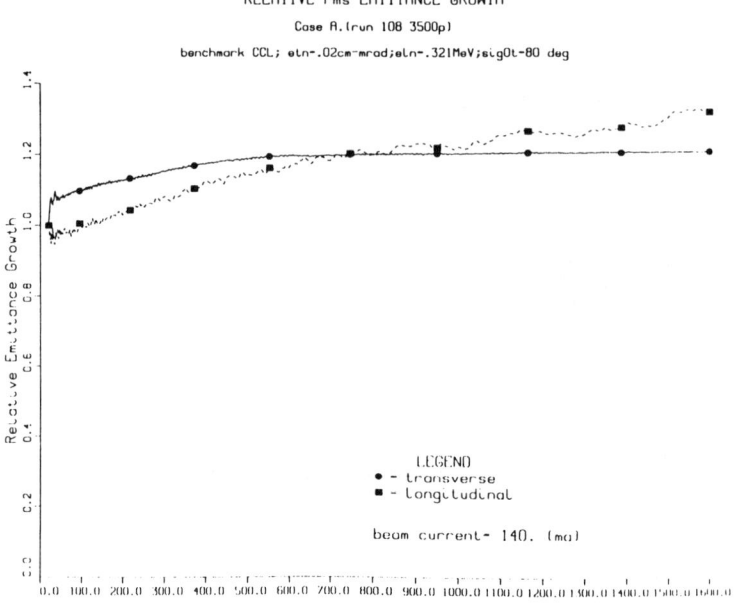

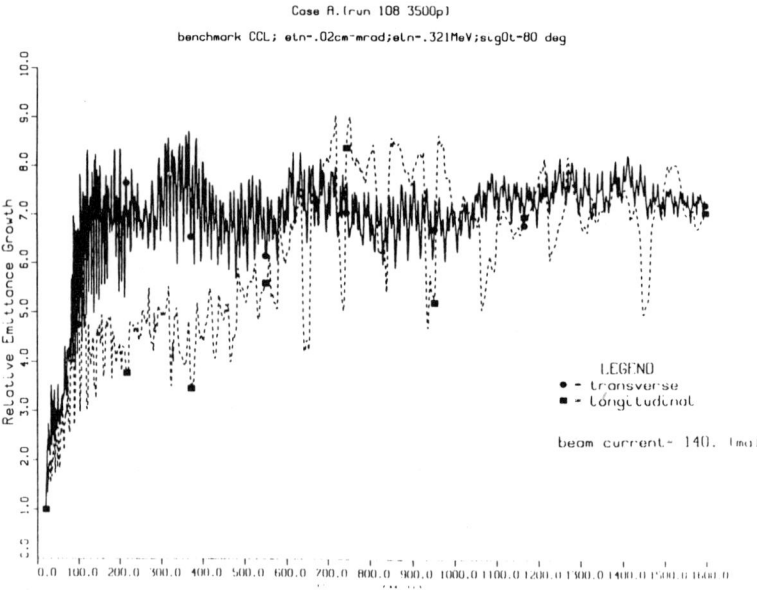

Fig. 3. 700 MHz CCL, current = 140 mA, $\varepsilon_{tn\ rms}$ input = 0.02 cm·mrad. Rms (top) and total (bottom) emittance growth along the linac.

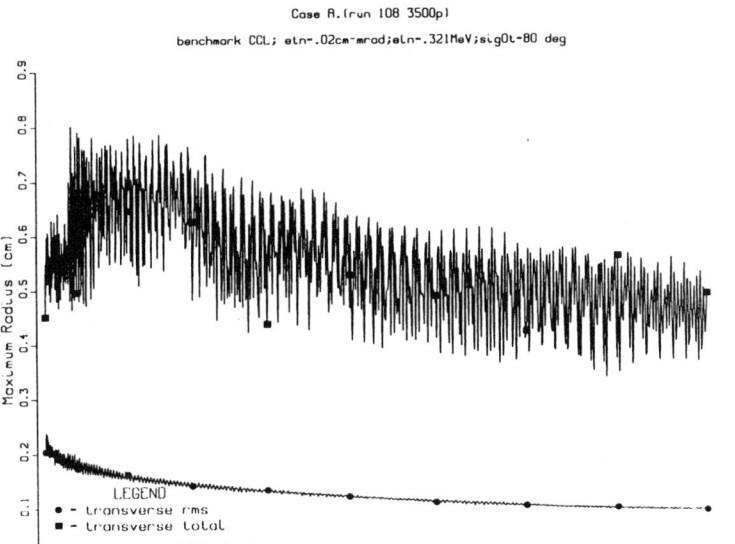

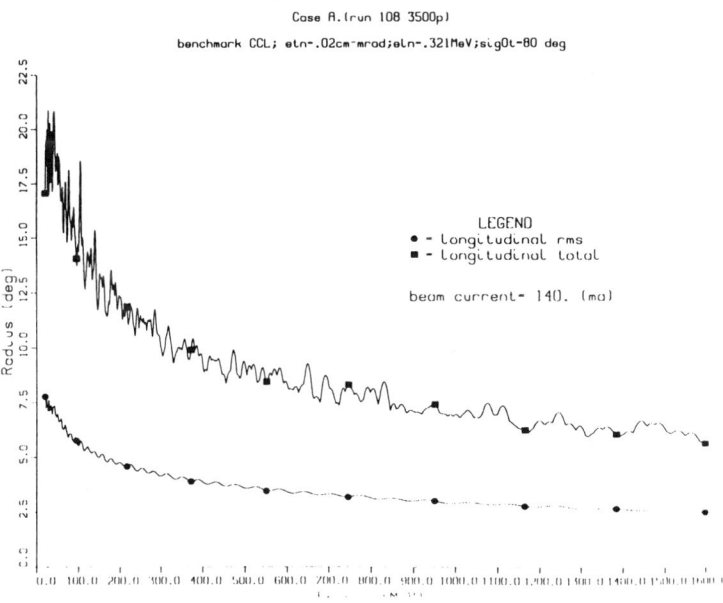

Fig. 4.   700 MHz CCL, current = 140 mA, $\varepsilon_{ln\ rms}$ input = 0.02 cm·mrad.   Top -
maximum transverse total beam radius along the linac, and maximum rms beam radius.
Bottom - longitudinal total and rms beam radius, degrees.

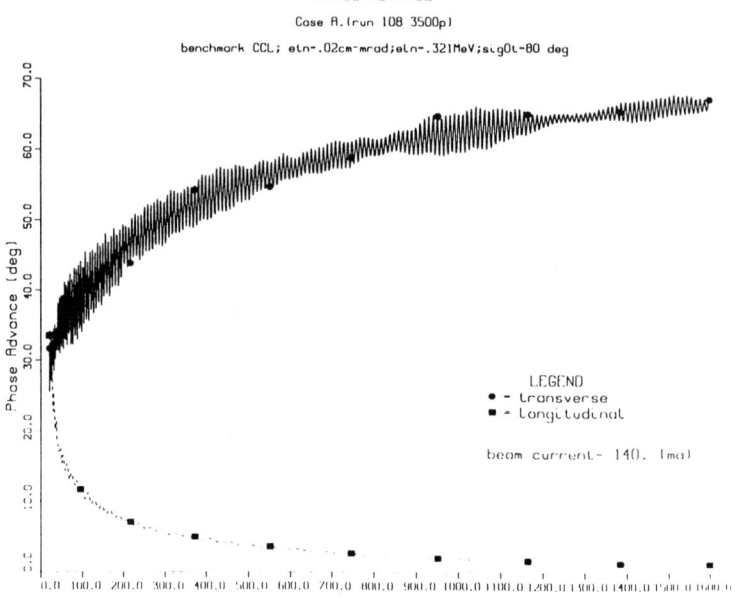

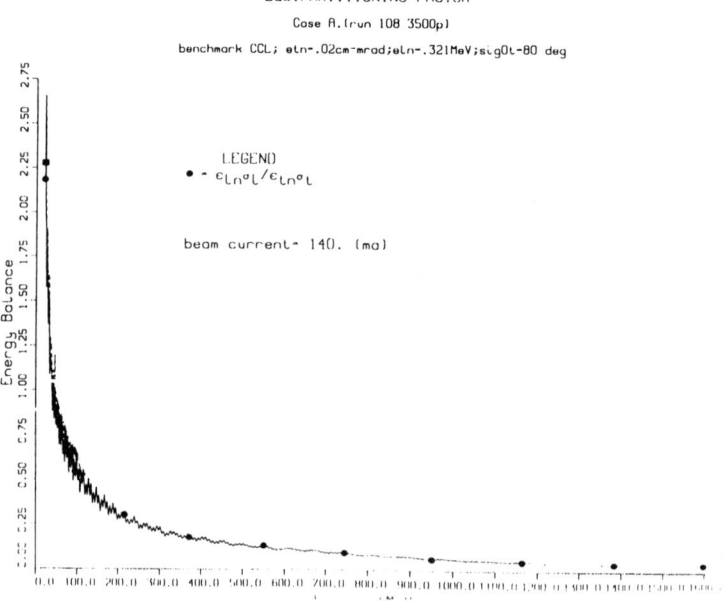

Fig. 5. 700 MHz CCL, current = 140 mA, $\varepsilon_{tn\ rms}$ input = 0.02 cm·mrad. Top - transverse and longitudinal phase advance along the linac. Bottom - energy balance ratios.

Table 1. 700 MHz CCL reference design at 140 mA.

Quantities listed from 20 -> 1600 MeV.

ε_{tn} rms. input	.02			.04			.08		
σ^t	30	->	67	44	->	69	58	->	70
$\sigma^t/\sigma_0{}^t$	.38	->	.84	.56	->	.86	.83	->	1.0
$\sigma_0{}^l$	69	->	6.5	69	->	6.5	69	->	6.5
σ^l	33	->	2	39	->	2	45	->	2
$\sigma^l/\sigma_0{}^l$	.48	->	.3	.56	->	.3	.65	->	.3
σ^l/σ^t	1.1	->	.03	.87	->	.03			
σ^t/σ^l							1.3	->	35
σ_p	86	->	62	75	->	57	58	->	54
ε_{tn} rms	.02	->	.024	.04	->	.041	.08	->	.08
ε_{tn} rms growth		1.2	(1.13)		1.04	(1.05)		1.0	(1.0)
ε_{ln} rms	.04	->	.052	.04	->	.052	.04	->	.052
ε_{ln} rms growth		1.3	(1.34)		1.3	(1.22)		1.3	(1.15)
ε_{ln} rms/ε_{tn} rms	2	->	2.2	1	->	1.3			
ε_{tn} rms/ε_{ln} rms							2	->	1.5
ε_{tn} total	.10	->	.85	.20	->	.70	.40	->	1.2
ε_{tn} total growth		8.5	(3.4)		4	(2)		3	(2)
ε_{ln} total	.20	->	1.6	.20	->	2.8	.20	->	1.2
ε_{ln} total growth		7-8	(4)		14	(7)		6	(3)
ε_{tn} rms·σ^t	.6	->	1.6	1.8	->	3.1	4.6	->	5.6
ε_{ln} rms·σ^l	1.3	->	.34	1.6	->	.10	1.8	->	.10
ε_{ln} rms·$\sigma^l/\varepsilon_{tn}$ rms·σ^t	2.2	->	.10	.9	->	.03	.4	->	.02
max transv rad, rms	.2	->	.1	.24	->	.11	.3	->	.16
max transv rad, total	.45 -> .8 -> .53			.52 -> .9 -> .55			.65 -> 1.1 -> .65		
tot/rms transv rad	2.24	->	5.74	2.24	->	5	2.24	->	4.1
tfac rms	12.5	->	25	10.4	->	23	8.3	->	16
tfac total	5.5 -> 3.1 -> 4.7			4.8 -> 2.8 -> 4.5			3.8 -> 2.3 -> 3.8		
bm long radius, rms, °.	7.8	->	2.5	7.5	->	2.5	7	->	2
bm long radius, total, °.	17 -> 21 -> 6			16.5 -> 22 -> 6			15.7 -> 20 -> 5.5		
tot/rms long rad	2.24	->	2.1	2.24	->	2.4	2.24	->	2.75
lfac rms	3.9	->	12	4	->	12	4.3	->	15
lfac total	1.8 -> 1.4 -> 5			1.8 -> 1.4 -> 5			1.9 -> 1.5 -> 5.5		

Table 2. 700 MHz CCL reference design at 250 mA.

Quantities listed from 20 -> 1600 MeV.

ε_{tn} rms, input	.02	.04	.08
σ^t	22 -> 64	35 -> 66	50 -> 70
$\sigma^t/\sigma_0{}^t$	.28 -> .8	.44 -> .82	.7 -> 1.0
$\sigma_0{}^l$	69 -> 6.5	69 -> 6.5	69 -> 6.5
σ^l	26 -> 2	28 -> 2	35 -> 2
$\sigma^l/\sigma_0{}^l$	.38 -> .31	.4 -> .3	.5 -> .3
σ^l/σ^t	1.2 -> .03	.8 -> .03	
σ^t/σ^l			1.4 -> 35
σ_p	88 -> 66	80 -> 64	65 -> 54
ε_{tn} rms	.02 -> .026	.04 -> .044	.08 -> .08
ε_{tn} rms growth	1.3 (1.13)	1.06 (1.09)	1.0 (1.02)
ε_{ln} rms	.04 -> .08	.04 -> .07	.04 -> .07
ε_{ln} rms growth	2 (2)	1.8 (1.5)	1.8 (1.3)
ε_{ln} rms/ε_{tn} rms	2 -> 3	1 -> 1.6	
ε_{tn} rms/ε_{ln} rms			2 -> .9
ε_{tn} total	.1->1.0->.7	.2 -> 1.2	.4 -> 1.2
ε_{tn} total growth	7 - 10 (3.3)	6 (2)	3 (1.5)
ε_{ln} total	.2 -> 4.2	.2 -> 4.6 -> 3.6	.2 -> 4
ε_{ln} total growth	21 (7)	->23->18 (7)	20 (6)
ε_{tn} rms·σ^t	.44 -> 1.7	1.4 -> 2.9	4 -> 5.6
ε_{ln} rms·σ^l	1.0 -> .16	1.1 -> .14	1.4 -> .14
ε_{ln} rms·σ^l/ε_{tn} rms.σt	2.7 -> .1	.8 -> .04	.35 -> .02
max transv rad, rms	.24 -> .1	.27 -> .13	.31 -> .17
max trans rad, total	.54 -> .83 -> .6	.6 -> 1.2 -> .6	.7 -> 1.1 -> .62
tot/rms transv rad	2.24 -> 6	2.24 -> 4.6	2.24 -> 3.65
tfac rms	10.4 -> 25	9.3 -> 19	8.1 -> 15
tfac total	4.7 -> 3 -> 4.2	4.2 -> 2.1 -> 4.2	3.6 -> 2.3 -> 4
bm long radius, rms, °.	9 -> 3	9 -> 3	8 -> 2.5
bm long radius, total, °.	19.5 -> 27 -> 7.5	18.5 -> 27.5 -> 7.5	18 -> 29 -> 7
tot/rms long rad	2.24 -> 2.5	2.24 -> 2.5	2.24 -> 3
lfac rms	3.3 -> 10	3.3 -> 10	3.8 -> 12
lfac total	4.7 -> 3 -> 4.2	4.2 -> 2.1 -> 4.2	3.6 -> 2.3 -> 4

variation before ~500 MeV, as would be expected because the space charge effects decrease with energy. In some cases, a significantly different value is found at an intermediate energy; this is signified by the notation # -> # -> #.

There are a number of questions to be explored in the following sections:
- What is the cause and importance of the observed increase in beam size/emittance?
- What is the proper bucket width to be used in assessing the longitudinal aperture factor?
- What considerations and results are involved in various scalings (particularly frequency) and optimization of this design?

DISCUSSION OF BEAM SIZE AND EMITTANCE

Some general features are observed:
- The rms beam profiles are smooth, with only small betatron or synchrotron oscillations, indicating that the input beam is rms-matched, and that the beam stays rms-matched along the linac.
- There is an immediate equilibration process occurring in the first few tanks that results in an increase of both emittance and beam size.
- There is little further transverse rms emittance growth, with less at larger transverse emittances. After the initial jump, $\varepsilon_{tn\ rms}$ grows ~ 13, 5, and 0% (at 140 mA) for the $\varepsilon_{tn\ rms}$ = 0.02, 0.04, and 0.08 cases respectively. These values are shown in parentheses in the tables.

The transverse total emittance growth is also less at larger input emittances. After the initial jump, further growth of 3.4, 2, and 2 times is seen for the 140 mA cases, reaching a peak and then leveling off at a level near that of the peak. The total emittance is the behavior of the single outermost particle in the phase plane for that simulation run, through which an ellipse of the rms shape and orientation has been fitted. Thus the total emittance value depends on the statistics of the initial particle selection, and final estimates require sophisticated techniques[11,12]. The convergence criterion on the number of particles (3500) used in the simulation is that the rms emittance has stopped changing significantly. This actually highlights the concern that total emittance and beam size also have to be carefully observed.
- After the initial jump, the longitudinal rms emittance grows steadily along the linac in all cases, with more growth at higher current and lower transverse input emittance. The longitudinal total emittance grows correspondingly, and also tends to level off. The energy where the longitudinal total emittance levels off does not correspond particularly to the energy where the transverse total emittance levels off.
- The phase advances change rapidly during the first few hundred MeV, particularly the longitudinal phase advance, as expected from the decrease of σ_o^l as $\beta\gamma^3$. The combination of rapidly increasing transverse phase advance and rapidly decreasing longitudinal phase advance is a strong feature of this architecture that will be discussed further below. By ~50 MeV, the transverse and longitudinal tunes already differ by a factor of two or more; most of the emittance and size growth occur after this point.
- The energy balance falls rapidly at first, and then more slowly, following the ratio of the tunes.

The main aspect of concern is the behavior of beam size with respect to the transverse aperture and longitudinal acceptance.

In the longitudinal plane, both the rms and total phase width are relatively smooth, and damp as the energy increases. In the following section, we discuss the behavior of the longitudinal acceptance as the beam is accelerated. Taking the acceptance left-edge phase width as a constant 30°, the longitudinal aperture factors, both rms and total, are reduced sharply during a period of initial growth, and then grow as phase damping occurs, to final values that are nearly the same over this range of parameters. There is a predictable increase in beam length with beam current and smaller transverse emittance.

The transverse aperture is 2.5 cm through the entire linac. The maximum rms beam radius gets smaller with energy, as expected from the constant σ_o^t focusing law, and the rms aperture factor exceeds those at LAMPF ($\sim$4 - 6.3 from 100 - 800 MeV).

At low energy, the space-charge effects are larger and additional current reduces the aperture factors more. At 140 mA, a smaller transverse input emittance gives a better aperture factor. If the current were zero, the factor of 4 range in the injected emittance would correspond to a factor of 2 in injected beam size, so it is seen that the space-charge is exerting an influence on the matched beam size. At 250 mA, the space-charge effects are much more dominant over the emittance.

There is a strong early growth in the total beam radius, in the region where the energy balance is changing quickly. The minimum transverse total aperture factors are similar ($\sim$2-3) for all six cases, indicating that space-charge is the dominant effect and is strong even at 140 mA. Overall, however, there seems to be some advantage to smaller input transverse emittance. After the initial growth, the maximum total transverse radius remains approximately constant until the transverse emittance equilibrates, and then the total radius decreases because of the strong $\sigma_o^t = 80^o$ focusing law.

At 1600 MeV, space-charge effects are small, Over this parameter range, there is an advantage in using a smaller injected transverse emittance.

The development of the particle distribution into a core plus halo is very interesting. Note the ratios of total-to-rms transverse beam radius. They start at Sqrt[5] = 2.24, the relationship for the mathematical "Type 8" input distribution, and develop to factors of 5 to 6 for the smaller input emittance cases. As a simple estimate, these factors can be thought of as applying to a Gaussian distribution (although it is known that single or combined Gaussian distributions poorly fit linac beams[12]); for now, it is enough to note that the ratios are large and to compare them to the longitudinal. The total-to-rms longitudinal beam radius ratio stays near that of the uniform initial distribution. We observe that the longitudinal tune depression is relatively low through the entire linac. On Fig. 2, we note that the dphi-dw phase-plane distribution at 1600 MeV has squared up and has quite sharp phase boundaries, and that the dw trajectory of the outermost particle has a square-wave characteristic. This is expected for a strongly depressed tune where space-charge dominates. The focusing forces and space-charge forces essentially cancel inside the bunch and particles drift, reflecting back into the bunch when the nonlinearities at the bunch edge are encountered. Thus we see that strongly space-charge dominated beams can be inherently more free from halos. This may be an important aspect in the consideration of equipartitioned or other alternative tuning architectures discusssed below.

A beam performing in a focusing/accelerating system is characterized by the kinetic and internal space charge field energy of the beam and potential energy from the external fields. The definition of a matched beam is that the beam and its surroundings are in perfect equilibrium, including any nonlinear or time-varying effects. Our design philosophy is to try to achieve matching; our optimization philosophy should start with the matched condition, but is subject to practical constraints.

An imbalance (mismatch) anywhere constitutes a source of free energy[13-15]. There are many possible sources – independent or dependent on beam intensity, linear or nonlinear, static or time-varying. Via interaction with a nonlinearity, a mechanism, or dynamical path, exists for the free energy to convert to (coarse-grained) emittance growth, leading toward a new equilibrium (in the absence of further driving terms). In the case at hand, beam size growth would directly reduce the aperture factors. The evolution can occur through single-particle or collective motion, and be a stable or unstable process. The rate of evolution will vary accordingly.

For example, emittance growth will occur if the beam is rms-mismatched (wrong phase-space ellipse orientation , off-axis, or off-synchronous)[15,16]. If there is a strong enough anisotropy of divergence between the degrees of freedom, an equipartitioning energy transfer between planes may occur through coherent instabilities[14,17-19]. Inter-plane effects will result from mismatch even without instability, through the couplings between planes.

In the CCL simulation, care has been taken to provide matching to the extent that we do understand how to provide rms matching of the coupled input beam envelopes in the transverse and longitudinal planes, and by injecting the beam on-axis and with the centroid at the correct synchronous energy and phase. That this has been accomplished is seen by the smooth profiles of the beam along the machine, without significant betatron or synchrotron rms envelope oscillations.

We know that if the beam density distribution is not matched to the channel, an equilibration will occur in about 1/4 of a plasma period[14,19].

From the work of I. Hofmann[14,18,20], we also have information about certain coherent instabilities that can lead to energy redistribution between planes via emittance growth. This occurs if there is a sufficient initial energy unbalance among the degrees of freedom such that the instability starting thresholds are exceeded; in the present case, two degrees of freedom, the transverse and longitudinal planes, will be discussed. Hofmann found that the thresholds are functions of several ratios, and has prepared mode threshold charts such as shown in Fig. 6. Although the charts are derived for K-V beam distributions, which are unphysical and give rise to these instabilities from the too-abrupt edge, it has been found that they are also quite accurate for physically realized distributions[17-20]. The charts are not easy to use, so the reader must be patient.

The local operating point along the linac is to be plotted as a trajectory from 20 – 1600 MeV on this chart. The required phase advances and emittances are taken from the simulation runs. The first step is to set the ratio $\varepsilon_x/\varepsilon_y$ >1. In the 700 MHz CCL, for the transverse input emittances of 0.02 and 0.04 cm·mrad, the longitudinal emittance is always larger than the transverse, so $\varepsilon_x/\varepsilon_y = \varepsilon_{ln}/\varepsilon_{tn}$. Other simulation runs for increased $\varepsilon_{tn\ rms}$ are also shown. For the transverse input emittance cases of 0.08 cm·mrad, the opposite is true, so $\varepsilon_x/\varepsilon_y = \varepsilon_{tn}/\varepsilon_{ln}$. Then the operating point is plotted according to the x- or y-plane tune depression on the upper or lower chart respectively.

To interpret whether or not an instability will be excited, the emittance ratio at the operating point in question has to be used. Because the emittance may be changing along the trajectory, this is tedious by hand (later it should be computerized). If the

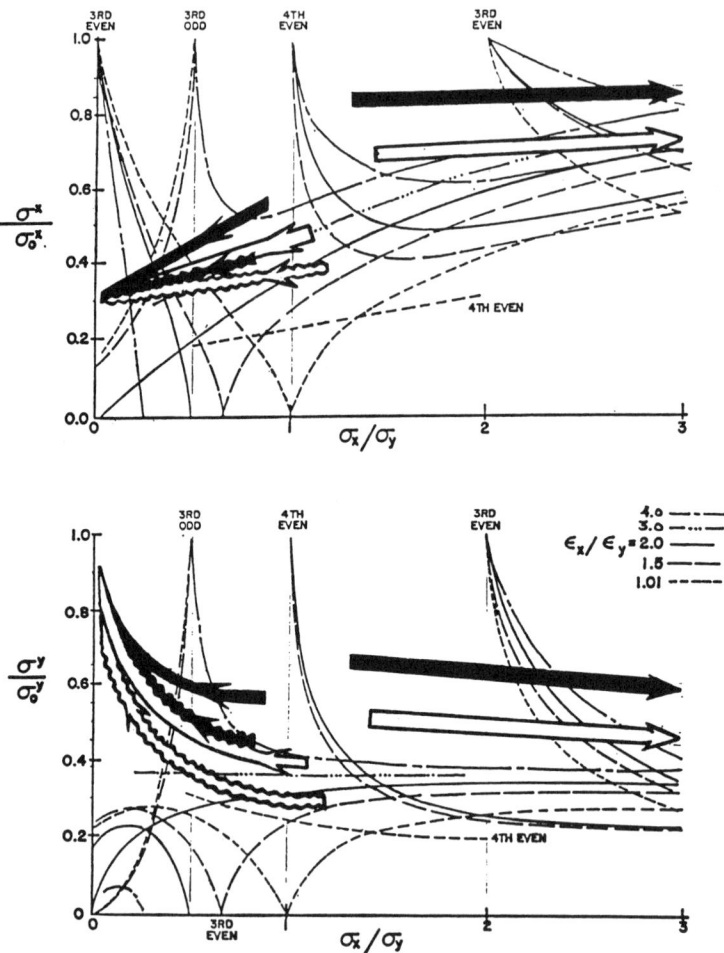

Fig. 6. Operating point trajectories of the 700 MHz CCL on the Hofmann coherent instability mode charts. On the left side of chart, where $\sigma_x/\sigma_y < 1$, the arrows with straight border are for current = 140 mA; arrows with wavy border are for current = 250 mA; the white arrows are for input $\varepsilon_{tn\ rms} = 0.02$ cm·mrad; and the black arrows are for input $\varepsilon_{tn\ rms} = 0.04$ cm·mrad. On the right side of the chart, where $\sigma_x/\sigma_y > 1$, the trajectories are for input $\varepsilon_{tn\ rms} = 0.08$ cm·mrad; the black arrow is for current = 140 mA and the white arrow is for current = 250 mA.

operating point is below a mode threshold corresponding to the emittance ratio on both charts, the instability will be excited.

First, consider the quick rms emittance changes in the first few tanks of the linac. This stems from two different effects. In the input $\varepsilon_{tn\ rms} = 0.04$ cases, the energy balance is approximately equipartitioned at injection. There appears to be an immediate redistribution of the particle density distribution to a distribution more naturally matched to the channel. The plasma phase advance per transverse focusing period (2 tanks), $(\sigma_p{}^2 = 2(\sigma_o{}^t{}^2 - \sigma_t{}^2) - (\sigma_o{}^l{}^2 - \sigma_l{}^2))$, is about 86° during the initial growth. This type of redistribution takes only ~1/4 of a plasma period, so it would be finished in only a few tanks, in agreement with the simulation. (There are 100 tanks between 20 and 94 MeV.)

In the input $\varepsilon_{tn\ rms} = 0.02$ case, Fig. 3 shows that $\varepsilon_{tn\ rms}$ grows abruptly, with a slight decrease in $\varepsilon_{ln\ rms}$. Here, two effects are probably competing; one is the density redistribution that would cause growth in the same direction as above. However, the energy balance, Fig. 5, shows the input beam is also not equipartitioned and has $\sigma^l/\sigma^t > 1$ until the end of the sharp initial growth, when σ^l/σ^t reaches ≤ 1. Thus an equipartitioning transfer from longitudinal to transverse would be predicted, and this effect appears to be stronger than the density redistribution effect.

Looking at the equipartitioning contribution more closely on Fig. 6, we see that for the 0.02 input emittance cases, the initial $\varepsilon_{ln\ rms}/\varepsilon_{tn\ rms} \sim 2$; the initial part of the trajectory is below the 4th-even mode thresholds, and the 250 mA case is also under the 3rd-odd mode threshold. Growth can be seen in $\varepsilon_{tn\ rms}$; growth in $\varepsilon_{ln\ rms}$ is delayed compared to the cases with larger input transverse emittance. The trajectories quickly move out to $\sigma^l/\sigma^t < 1$ and the growth stops. For the 0.04 and 0.08 input emittance cases, the starting points are above all low-order mode thresholds.

Subsequently, in the $\sigma^l/\sigma^t < 1$ region, the 0.02 and 0.04 cases move very quickly (by Tank No. 20 - 30) through the 3rd-odd mode (resonance) region, with no observable effect on the rms emittance, and then the transverse tune depression gets even less and the transverse trajectory moves into the upper left corner as σ^l/σ^t becomes $\ll 1$, and lies above any of the mode thresholds shown. Similarly, the 0.08 input emittance cases move very quickly through the 3rd-even, and then the longitudinal trajectory stays above the thresholds shown (although it may move back under the 3rd-odd again as $\sigma^l/\sigma^t \gg 1$ and $\sigma^l/\sigma_o{}^l \to 0.3$). Most of the emittance growth occurs in this final region where the tunes differ by a factor of two or more. Note that σ_p is reasonably large over the linac, so the growth is occurring over many plasma periods.

The energy balance also indicates that there is no long-term movement toward equipartitioning. In some of the cases, the energy balance is briefly at one, but does not remain there because the tune prescription does not insist on equipartitioning.

Thus, only the small, abrupt initial increase is explained by available methods. The coherent instabilities, as least to the order shown, do not seem to explain the sustained longitudinal emittance growth or the initial swelling in total beam size.

Three possible explanations for the emittance growth are hypothesized.

One is that higher order modes, especially the 5th, may be involved. The tunes are inevitably below the thresholds of many higher-order modes, and remain so for the entire transit of this (relatively long) linac. The original development by Hofmann found that equipartitioning would occur, for strong enough tune depressions, when $\varepsilon_x/\varepsilon_y \gg 1$ and $\sigma_x/\sigma_y > \sim 1.5$. This situation is not present in these cases. Whether transfer can occur when $\varepsilon_x/\varepsilon_y > 1$ but $\sigma_x/\sigma_y \ll 1$ needs further exploration.

The rf nonlinearity across the bunch length may be contributing[21]. However, the rms phase radius is only about $7.5 - 9$ degrees initially, and damps smoothly. This will be tested by running the simulation with linear longitudinal focusing.

The third hypothesis is that a systematic error is occurring in the linear part of the matching (the most powerful source of free energy), because the linac is not a truly periodic (steady-state) system, as assumed in the rms matching procedures. A global look at an RFQ, DTL, or this CCL underscores the point — there is acceleration, and also the tune prescriptions (here constant $\sigma_o{}^t = 80^\circ$, constant real-estate accelerating gradient and ϕ_S) result in the rapidly changing tunes along the linac. A steady-state might never be reached in a particular linac, or a steady-state might be reached for one phenomenon and not another. Acceleration at a rate faster than adiabatic (necessary in a practical linac) means that matching in the steady-state sense is not possible. Paraphrasing Hofmann[4]: "An rms matched beam is [still] (intrinsically) mismatched if the nonlinear field energy term [or a source of free energy] changes rapidly within a coherent oscillation period,..." Investigation into the amount of free energy that may be available from the aperiodic nature of the real linac, and the mechanisms that may allow this free energy to generate emittance growth, must now be explored.

As a first estimate of the acceleration effect with beam current, Eqns. (1) and (2) were expanded to first order in γ and $\sigma_o{}^t$, and solved for the correction necessary in $\sigma_o{}^t$ to cancel the effect of an error in γ:

$$\Delta\sigma^t_{ocorr} = -\frac{\Delta\beta}{\beta}\sigma^t_o\left[3-\left(\frac{\sigma^t}{\sigma^t_o}\right)^2\right]\bigg/\left[6\left(\frac{\sigma^t}{\sigma^t_o}\right)^2\right] \tag{14}$$

which can be large at strong tune depressions; e.g., with $\sigma_o{}^t = 80^\circ$ and tune depressions of 0.4 in each plane, $\Delta\sigma_o{}^t{}_{corr}$ $,^\circ = \sim 237\ (\Delta\beta/\beta)$. On the other hand, the first term of Eqn.(6) for computing $\sigma_o{}^t$ is $\sim\beta/\gamma$ and the error in computing $\sigma_o{}^t$ is only $\sim\beta(\Delta\beta/\beta)$ for small β. Further theoretical and simulation work is in progress.

ACCEPTANCE WIDTH UNDER ACCELERATION CONDITIONS

In order to use the concept of longitudinal aperture factors as scaling and optimization criteria for low beam loss, the behavior of the accelerator longitudinal acceptance, or "bucket", must be known. Once again, at present, we are in the situation of having no accurate and convenient theoretical description, because of the complex role of space-charge for high-intensity beams, and because acceleration causes the system to depart greatly from the strict periodicity upon which present theory is based, The accelerator must be viewed as a transient system, not steady-state.

In this section, simulation methods are introduced that allow the acceptance behavior of a linac to be studied. The behavior of the 700 MHz CCL is investigated, because it represents a practical design and we need to get acquainted with a real case. Later, systematic studies of accelerating channels set up under more controlled conditions need to be done, to determine characteristics at different acceleration rates as a function of current, and so on.

Theoretical Framework

The available theoretical framework was explored by Nishikawa[22] and Gluckstern[23] many years ago. Their development is couched in terms of the smooth approximation envelope equations used in this paper. Their point-of-view was different from what ours must be. They were concerned with the "current limit" of the accelerating channel, i.e., how much total current would be transmitted if the input were saturated, or flooded, with beam, without regard for the quality of the output beam, or concern that a large fraction of the injected beam would obviously be lost under such circumstances. Our objective is thus considerably different – we want to accelerate and transmit a high-intensity beam with negligible beam loss, hence requiring a large aperture factor.

Our requirement has two major meanings: one, that particles are not actually lost from the bucket. Lost particles are not accelerated further, and can become over-focused by the quadrupoles and eventually strike the transverse aperture.

The second concern is that the longitudinal phase-space is highly nonlinear except very near the synchronous phase and energy. Particles in outlying regions experience very different phase advances (there is a wide "tune spread"); thus over time (distance), the phase-space becomes filled with a diffuse halo through nonlinear filamentation processes, even without space-charge. Through the rf gap and other nonlinearities, the longitudinal motion is coupled to the transverse phase-plane, and halos can form that could eventually lead to transverse beam loss.

Gluckstern lists several other factors that make simple analysis difficult, including the introduction of acceleration, asymmetries in the longitudinal focusing force and transverse beam cross-section in an alternating-gradient focusing system, non-uniform charge distribution, rapidly changing image forces, and coupling between planes.

The approach leading to Eqn. (2) indicates that without acceleration, the phase-stable region would shrink under the influence of space sharge, from $-|\phi_s| \leq \phi \leq 2|\phi_s|$ to $-|\phi_s|/(\sigma^l/\sigma_o^l)^2 \leq \phi \leq 2|\phi_s|/(\sigma^l/\sigma_o^l)^2$, with the assumption that the beam fills this entire phase width. This would be a very large shrinkage, e.g. to 16% if the longitudinal tune depression were 0.4, as it is in the CCLs under study. If the beam is allowed to fill the whole $3|\phi_s|/(\sigma^l/\sigma_o^l)^2$ width, a current limit can be calculated.

Even without acceleration, the above limit can be exceeded by several times. Gluckstern and Nishikawa develop some aspects that are not applicable to our case, such as the offset of the bunch center from the synchronous angle if the bunch were to fill the whole bucket, or particles traveling "outside" the ellipsoid that see weakened repulsive forces and could consititue additional transmitted current. In this latter case, we would have to carefully explore the halo aspects of such "outside" particles. Gluckstern discusses the difficulty of finding self-consistent models that can be treated analytically, and indicates the utility of local use of the envelope equations with various apriori prescriptions, as we have been doing. He points out that numerical computations show that acceleration results in an "effective injection energy (when some sort of dynamic stability is established) that tends to extremely large values".

Nishikawa considerably extends the analysis, treating first the zero-current case and assuming a particular variation of the synchronous $\beta_s\gamma_s$ along the linac. He shows the well-known result that the amplitude of the particle motion damps as $(\beta_s\gamma_s)^{(-3/4)}$ along the linac (for constant accelerating gradient). Adding space-charge, if the tune depression is constant, the oscillation period is increased by $1/(\sigma^l/\sigma_o^l)$, and the damping is still as $(\beta_s\gamma_s)^{(-3/4)}$. Next, he lets the tune depression vary in a prescribed way, in the form $(1 - (\sigma^l/\sigma_o^l)^2) = \mu^l = \mu^{l\prime}(\beta_s\gamma_s)^n$. Deriving a formula to estimate n, he demonstrates reasonable agreement with an example simulation in which μ^l starts at 0.54 and rises to ~0.9. The simulation approach apparently used the envelope equations above, which cannot account for emittance growth.

Again discussing the case where the beam completely fills the phase acceptance, the (pessimistic) estimate of the bucket shrinkage given above indicates that the maximum space-charge limited current would occur at $\mu^l = 1/3$. By more carefully considering the fact that the center of such a beam would no longer be at ϕ_s, the current limit is shown to occur at $\mu^l = 0.4$, implying less shrinkage of the bucket and a higher current limit.

Including the exterior potential of the bunch, he finds a self-consistent limiting maximum current without acceleration at $\mu^l = 1$, corresponding to a stationary distribution, again for the case where the beam fills the entire phase width of the bucket.

He also develops the effect of a uniformly charged cylindrical bunch, considered more realistic than an ellipsoidal bunch under strong space-charge conditions. We can prove this now with our modern simulations; in Fig. 2, note that the 1600 MeV dphi-dw phase plane projection has squared up and has very sharp phase boundaries; the cylindrical bunch is a feature of our case as well. Although we endeavor to keep the bunch size considerably smaller than the bucket, the bunch becomes cylindrical.

Considering acceleration, again his emphasis is on limiting current and particles that may start as unstable but later are captured as the bucket energy height enlarges with acceleration. A conjecture is made that unstable particles may be lost during the first half-period of the phase oscillation of stable particles. During this half-period, the particle velocity would typically about double, and this velocity would be used in the computation of an approximate current limit.

These models for a saturated system are valuable as general guidance, but are not directly useful to the low-beam-loss case. To get some initial insight, we turn to numerical simulation.

Simulation study of longitudinal acceptance

A general method for studying acceptance has been developed. It is applied here to the specific 700 MHz CCL design, which has generic features (constant accelerating gradient, constant σ_o^t and n) similar to the accelerating section of conventional RFQs and DTLs. Later, systematic studies of acceptance as a function of accelerating rate and under other conditions will be made.

The simulations begin by transporting the design "core" beam through the linac as usual, computing and storing the space-charge fields at each cell. Then a grid of particles uniformly covering an area of ($\Delta\phi$, Δw) phase-space is injected at a given cell, e.g. the beginning of the linac. These particles are "test particles"; they are assigned no current and do not contribute to the space-charge fields previously computed. They are only acted upon by these space-charge forces and the external forces as they traverse the linac. The grid particles outside the acceptance will be lost, either by falling behind in energy by more than a prescribed amount, or by striking the transverse aperture. By referring the particles that have survived (a distance beyond which the picture stops changing) back to the injection grid, the acceptance from that point is determined.

Fig. 7 summarizes the outer contours of the acceptances from 0 – 250 mA from the 20, 24, 50, 104, 404, and 783 MeV starting points. Beyond a few hundred MeV, there is no difference in the acceptance from 0 – 250 mA, even though the longitudinal tune depression is strong throughout the entire CCL. These plots are in terms of $\Delta w_s/w_s$; this ratio falls as energy increases, but the absolute energy height of the bucket increases and thus particles that become trapped tend to stay trapped.

Fig. 8 summarizes the 0 mA and 140 mA contours at the several injection points along the CCL. The introduction of beam current is seen to produce "intrusions" into the zero-current bucket contour. Later, we plan to study the onset and development of these contours in detail.

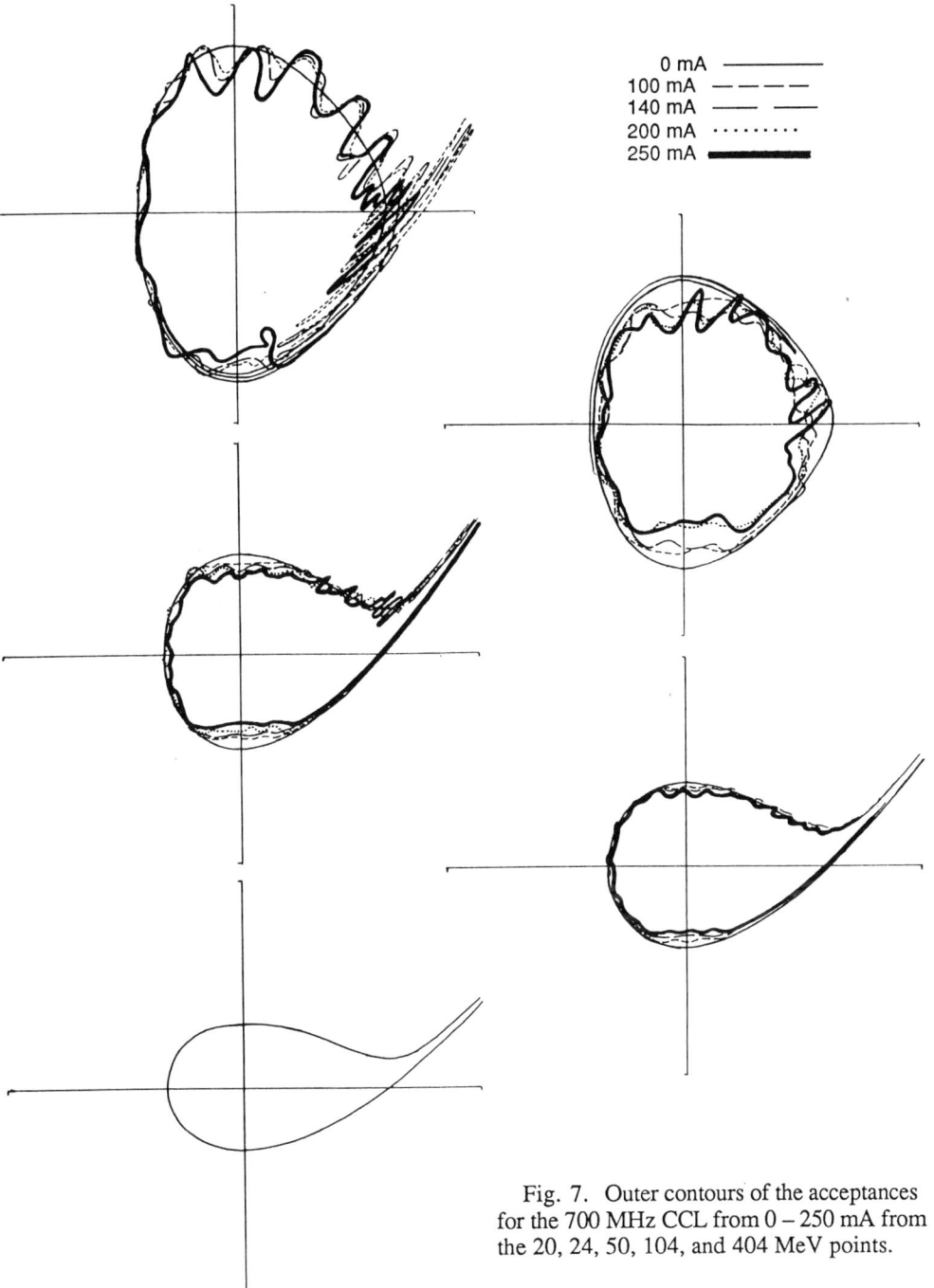

Fig. 7. Outer contours of the acceptances for the 700 MHz CCL from 0 – 250 mA from the 20, 24, 50, 104, and 404 MeV points.

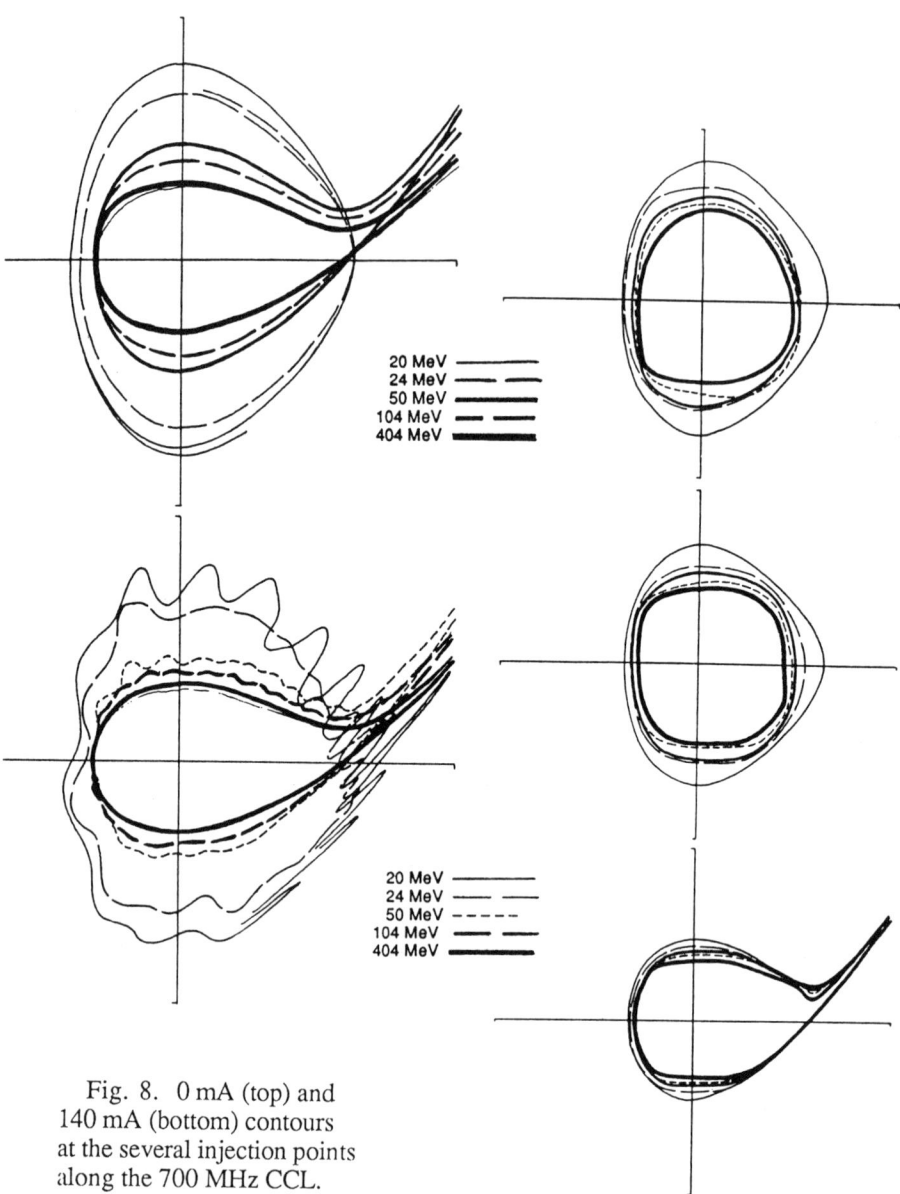

Fig. 8.   0 mA (top) and 140 mA (bottom) contours at the several injection points along the 700 MHz CCL.

Fig. 9.   Acceptance contours drawn by eye through points of deepest intrusion. Top, acceptance from 20 MeV, middle – from 24 MeV, bottom – from 50 MeV. Beam currents, from outer contour to inner, are 0, 100, 140, 200, and 250 mA.

In this CCL, the synchronous phase is ramped from -41.4° to -31° over 20 – 40 MeV, and then slowly changes to -30° by around 450 MeV, remaining at ~-30° to 1600 MeV. Figs. 7 and 8 indicate that the effective phase width remains fairly closely at $-/\phi_s/$ $\leq \phi \leq 2/\phi_s/$ over the full current range. The bucket phase width follows the ϕ_s ramp in, and remains close to $3/\phi_s/$, especially above 50 MeV for all currents. The left edge is close to $-/\phi_s/$ in all cases.

For the present, we will use $-/\phi_s/ \leq \phi \leq 2/\phi_s/$ as the effective bucket width, and compute the longitudinal aperture factor on the $-/\phi_s/$ side.

A few additional observations on the simulation results are now outlined:

• Most of the "action" in this CCL occurs before 100 MeV. Injecting at 20 MeV, the acceleration effect is such that there is no "fish" characteristic to the acceptance. From 50 MeV, the acceptance has the usual fish shape.

• Little shift in the bunch centroid from ϕ_s is observed.

• The damping as $(\beta_s \gamma_s)^n$ of the top and bottom edges was checked for the 50 MeV (after ϕ_s has stopped changing) – 1205 MeV acceptances as a function of beam current. The damping remains nearly $n = 3/2$ as expected, as indicated in the {current, n} pairs: {{0,1.5},{100,1.53},{140,1.58},{200,1.61},{250,1.62}}.

• Although the phase width of the acceptance is not much affected, the energy height at the low energy end is steadily reduced as current is increased, and the total acceptance area shrinks. Fig. 9 summarizes contours drawn by eye through all the points of deepest intrusion into the acceptance area. The result from 20 MeV is not easy to interpret, while from 50 MeV, the left edge is seen to follow the ϕ_s ramp, and there is a clear shrinkage and flattening of the top and bottom edges. The contours from 24 MeV are most interesting; here a progression toward a square shape is quite apparent as beam current increases. This is consistent with the increasing cancellation of the focusing force by the space-charge force at higher μ^l (σ^l/σ_o^l is about 0.45 and 0.35 ($\mu^l \sim 0.8$ and 0.88) for 140 mA and 250 mA respectively). While there is a good probability that particles outside these contours could be transmitted, a more systematic investigation of this effect may help establish an approximate theory including acceleration in the future.

SCALING AND OPTIMIZATION

Optimization and scaling of Eqns. (8-11) is not possible in general because of the high polynomial order (>8), nonlinearily and coupling of Eqns. (1), (2), and optionally (3). Unfortunately, the desired operating regime is not at either extreme of zero beam current or at the space-charge limit, but where the emittance and space-charge contributions to the beam envelope are roughly equal and the coupling and nonlinearity are at their worst. Some of the issues are addressed in the following preliminary study of choices between linacs of different frequency, or within one linac as a function of energy.

As mentioned earlier, only a few variables are free; the others must be fixed. As in any scaling problem, the choice of what is to be fixed (either as constant, or with prescribed variation) during the scaling is crucial to the result. Presently there are no established criteria for scaling a high-intensity linac intended to have low beam loss, so it is necessary to develop one or more that will be "fair" and practical across the scaling range. A large variety of options are available; only an outline of the approach and conclusions from more detailed discussion[24-26] can be given here.

Table 3. Zero-current variable dependence as frequency( )

	5, E.P. general	5, E.P.	5.	4.b, E.P. general	4.b, E.P.	4.a, E.P. general	4.a, E.P.	4.a.	3, E.P.	2.b, E.P.	2.a, E.P.	1, E.P. general	1.
tfac0	$m/2 - 1$	$-1/2$	$m/2 - 1$	$m-1$	0	$-1 + (m+1/2)/2$	$-1/4$	0	$-3/4$	$-3/2$	$1/2$	$m-1/2$	$-1/2$
lfac0	$(m+1)/2 - 1$	0	$(m+1/2)/2 - 1$	$m-1$	0	$-1 + (m+1/2)/2$	$-1/4$	$-1/4$	$-3/4$	0	0	-1	-1
a	$-m/2$	$-1/2$	$-m/2$	$-m$	-1	$-(m+1/2)/2$	$-3/4$	-1	$-1/4$	$1/2$	$-3/2$	$-(m+1/2)$	$-1/2$
b	$-(m+1)/2$	-1	$-(m+1/2)/2$	$-m$	-1	$-(m+1/2)/2$	$-3/4$	$-3/4$	$-1/4$	-1	-1	0	0
ε_{tn}	$-m$	-1	$-m$	$1/2 - 2m$	$-3/2$	$-m$	k/f	k/f	0	0	-2	$-m$	0
ε_{ln}	$-(m+1/2)$	$-3/2$	$-m$	$1/2 - 2m$	$-3/2$	$-m$	$2k/f$	$2k/f$	0	$-3/2$	$-3/2$	$1/2$	$1/2$
σ_o^t	0	0	0	$1/2$	$-1/2$	$-1/2$	$-1/2$	0	$-1/2$	-2	0	0	0
c	1	1	1	0	0	0	0	0	0	0	0	0	0
$\sigma_o^l/n\beta\lambda$	0	0	0	$1/2$	$1/2$	$1/2$	$1/2$	1	$1/2$	-1	1	$m+1$	1

Frequency Scaling

It is useful to establish the boundary condition behavior at zero current. If the input emittances are described by functions that do not depend on beam current[*] (e.g., constant), then when $I \neq 0$, a and b will increase, because $\sigma^t \leq \sigma_o^t$ and $\sigma^l \leq \sigma_o^l$. The zero-current equations are thus helpful as a guide to maximum aperture factors.

When $I \neq 0$, the equations are strongly coupled. Many feasible cases would not be energy balanced, and emittance growth would probably occur. If we want to invoke equipartitioning to help avoid emittance growth, we want the solutions to converge smoothly at $I = 0$, so the zero-current form of Eqn. (3) is useful when used simultaneously with the zero-current form of Eqns. (1) and (2).

Table 3 summarizes a number of examples

---

[*]   In future work, it may be of interest to explore relationships in which the emittances are also prescribed or free functions of current. It may be important to the optimization that in strongly space-charge dominated beams, emittance redistribution occurs mainly in the divergence, rather than in the physical size.

of how the zero-current aperture factors and other quantities vary with frequency, for various fixed conditions on the other variables (shown as shaded boxes in each column). The general cases with e_{tn} & e_{ln} (4.a.), or a & b (4.b.), having equal functional variations with frequency are equivalent in the resulting frequency variation of the zero-current aperture factors; e.g., when $tfac0$ and $lfac0$ vary as f^0, a & b vary as $f^{(-1)}$ and ε_{tn} & ε_{ln} vary as $f^{(-3/2)}$.

As tentative criteria for frequency comparisons, assume that both the transverse aperture and the longitudinal acceptance vary as $f^{(-1)}$, that the input ε_{tn} and/or ε_{ln} can be varied between f^0 and $f^{(-1)}$ that the smallest frequency variation of zero-current $tfac0$ and $lfac0$ is desired, and that $tfac0$ and $lfac0$ should have as close to the same variation with frequency as possible. From Table 3:

1.) Case 4.a. with equipartitioning seems the best, with emittances varying as $f^{(-1)}$, resulting in equal $tfac0$ and $lfac0$ variations as $f^{(-1/4)}$. $\sigma_o{}^t$ varies as $f^{(-1/2)}$, and the transverse focusing per unit length as $f^{(-3/2)}$.

2.) Case 4.a. without equipartitioning has constant $tfac0$, and $tfac0/lfac0$ at $f^{(1/4)}$. There could be more emittance growth than with equipartitioning.

3.) Case 5, without equipartitioning, was an apparently good choice for the point-design frequency comparison. Constant transverse focusing per unit length results in $tfac0/lfac0 \sim f^{(-1/4)}$ when the emittances vary as $f^{(-1)}$; however, $tfac0 \sim f^{(-1/2)}$. The performance of this case was studied above; emittance/beam size growth was seen.

4.) Case 5 with equipartitioning can also give $tfac0 \sim f^{(-1/2)}$, with $lfac0$ constant.

In future work, items 1, 2 and 4 should be studied in detailed simulations.

At the space charge limit, σ^t and σ^l equal zero and the external focusing is exactly cancelled by the space-charge forces. Eqns. (4) and (5) show, for a spherical beam:

$$a^3 = \frac{2\hat{Ik}\lambda(n\beta\lambda)^2}{(\beta\gamma)^2\left(2\sigma_o^{t2} + \sigma_o^{l2}\right)} = \frac{2\hat{Ik}n^2\lambda^3}{\gamma^2\left(2\sigma_o^{t2} + \hat{k}_3n\lambda\right)} \tag{15}$$

For either constant $\sigma_o{}^t$ or constant $\sigma_o{}^l$ and transverse focusing per unit length, the aperture factor dependences on frequency vary between $f^{(-1/3)}$ to f^0 as frequency increases, so direct comparison seems reasonable.

Now add the requirement that the beam also be equipartitioned, so that Eqns. $(1-3)$ have be to solved simultaneously. For frequency comparison, assume a fixed current and energy. Set $(\sigma_o{}^l)^2 = k_3\lambda$. $\sigma_o{}^t$ can be prescribed, or can be a variable to satisfy conditions on the emittance, beam size, or phase advance quantities in Eqns. (1)-(3).

This general problem, with its many variables, nonlinearity, coupling, and characteristic of involving very high-order polynomial expressions, is being studied in an effort to find out more about it and perhaps to find solutions that would be helpful. A general solution for a from the simultaneous solutions of Eqns.(1)-(3) is:

$$a = 3e_{tn}{}^2\left[\left(\frac{\varepsilon_{ln}}{\varepsilon_{tn}}\right)^2 - \left(\frac{\sigma_o^t}{\sigma_o^l}\right)^2\right] \bigg/ \left[I\lambda\hat{k}\left[2\left(\frac{\sigma_o^t}{\sigma_o^l}\right)^2 + 1 - 3\left(\frac{\varepsilon_{ln}}{\varepsilon_{tn}}\right)\right]\right] \tag{16}$$

which shows that restrictions apply to the $\varepsilon_{ln}/\varepsilon_{tn}$ and $\sigma_o{}^t/\sigma_o{}^l$ ratios[24] to keep a positive.

If the equipartitioning ratios of Eqn. (3) are set to a constant, k_1, and if n, ε_{tn}, $\sigma_o{}^t$, and λ are also fixed (leaving $\sigma_o{}^l$ and ε_{ln} free), information can be obtained about the beam size as a function of k_1. It can be shown that the beam radii have no maximum or minimum where the derivative with respect to k_1 equals zero. The study of a longitudinal aperture factor in which the bucket phase width was assumed to vary as $3/\phi_s/(1 - \mu^l)$ yielded information that is important to further work. It was found that this aperture factor has a maximum within the allowed range of k_l. From numerical simulations, it was observed that this maximum appeared to occur at the same place in the range of k_l, for different choices of the other variables. A change of variable in the equations to a range variable proved that the peak occurs very close to 3/4 of the range, but not exactly except at one limit. However, the solution kernel was found, and the form of departure from 3/4 was made clear. The longitudinal tune depression at the $lfac$ maximum is also equal to 3/4 with the same departures.

By assuming that the peak occurs exactly at 3/4, an analytic expression for this type of longitudinal aperture factor can be written:

$$lfac_{max} = 2\sqrt{5}\beta|\phi_s| \left[\frac{3 \, \gamma^5 \sigma_o'^2}{I \, n^2 \hat{k} \, k_1 \left(6k_1^2 + 3k_1 - 1 \right)} \right]^{1/3} \tag{17}$$

Above, we found that the phase width does not shrink in this fashion, although some effective shrinkage to be sure of avoiding "intrusion" regions was noticed. A more refined model for bucket width may result in a form similar to that used in the derivation of Eqn. (17), in which case it would become directly useful.

Summarizing, Eqn. (17) was obtained under equipartitioned conditions. The wavelength was fixed (it can be used as a parameter). The form of the solution kernel may be of value in the search for other kernels.

The equipartitioned case with $(\sigma_o^l)^2 = k_3 \lambda$ and constant I, $\beta\gamma$, σ_o', n and ε_{tn} has also been studied extensively. Defining a transverse aperture factor as proportional to λ divided by the maximum rms beam size, numerical simulations typically show a variation with respect to frequency, with beam current as a parameter, as shown in Fig. 10. For non-zero current, there will be an optimum frequency where $tfac$ is maximized. The order of the polynomials describing the curves is very high, and from one side of the peaks to the other, the significant coefficients slide from the low-order to the high-order terms. No accurate approximation for the direct curves has yet been found.

It has been found numerically that for a wide range of input conditions, the $tfac$ peak occurs when the transverse tune depression, σ^t/σ_o^t, is near $1/\text{Sqrt}[5]$. Similarly, a longitudinal aperture factor defined to include the $(1 - \mu^l)$ shrinkage peaks as a function of frequency near $\sigma^l/\sigma_o^l = 0.75$, as above. It is clear that the peaks occur at a point where the beam emittance and space-charge have an equal influence in some sense, and the appropriate normalization and solution kernel are now being sought.

We observe from Fig. 10:

• At zero current, low frequency gives a higher $tfac$ for a given ε_{tn}.

• In this problem, σ^t/σ_o^t, σ^l/σ_o^l and $\varepsilon_{ln}/\varepsilon_{tn}$ increase with frequency, i.e. a given ε_{tn} means that if more current is added, the frequency where $tfac$ peaks, at $\sigma^t/\sigma_o^t \sim 1/\text{Sqrt}[5]$, will be higher.

• Within a given frequency range, depending on the current, $tfac$ can vary monotonically downward with frequency, have a peak, or vary monotonically upward. The $tfac$ peak for a given current may not lie in the frequency range of interest. The change in $tfac$ over the frequency range of interest may not be very great.

Assuming that $\sigma^t/\sigma_o^t = 1/\text{Sqrt}[5]$, Eqns (1) and (4) can be used to solve for the wavelength where $tfac$ is maximum:

$$\lambda_{\text{tfacmax}} = \varepsilon_{tn} \left(\frac{\gamma \, \sigma_o^t}{n} \right)^{1/3} \left(\left[12 \left(\frac{\varepsilon_{ln}}{\varepsilon_{tn}} \right)^2 \right] \middle/ \left[5^{1/4} I \, \hat{k} \left(3 \frac{\varepsilon_{ln}}{\varepsilon_{tn}} - 1 \right) \right] \right)^{2/3} \tag{18}$$

and $tfac = \beta\lambda/k_5 a$, where k_5 is a constant:

$$tfac_{\text{max}} = \frac{\beta}{k_5} \left(\frac{\gamma \, \sigma_o^t}{n} \right)^{2/3} \left(\left[12 \left(\frac{\varepsilon_{ln}}{\varepsilon_{tn}} \right)^2 \right] \middle/ \left[5^{1/4} I \, \hat{k} \left(3 \frac{\varepsilon_{ln}}{\varepsilon_{tn}} - 1 \right) \right] \right)^{1/3} \tag{19}$$

This gives information at a specified ε_{ln}; however, because ε_{ln} is a function of the input conditions, it is not a general solution.

It appears that the rationale followed in this section does provide some basis for choosing a meaningful condition for comparing linacs at different frequencies; more work is needed.

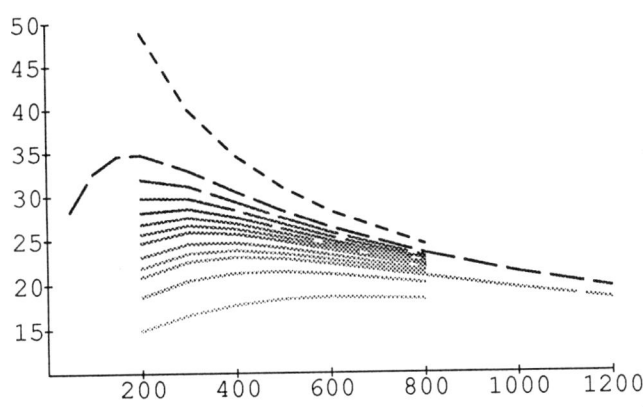

Fig. 10. Transverse aperture factor vs. frequency (MHz), at 20 MeV in a DTL with $\sigma_o^t = 70^\circ$, $e0t = 2$ MV/m, $\phi_s = -30^\circ$, aperture $= \beta\lambda/9$, and $\varepsilon_{tn} = 0.1$ cm·mrad (ε_{tn} rms $= 0.02$ cm·mrad), with equipartitioning. Top line is at 0 mA; aperture factor decreases for increasing currents 100, 150, 200, 250, 300, 350, 400, 500, 600, 700, 1000, 2000 mA.

Choice of Tune During Acceleration

Again checking the behavior at zero-current, for ε_{tn}, σ_o^t, n and λ constant, $a \sim \gamma^{(-1/2)}$ and $b \sim \beta^{(1/4)}\gamma^{(-3/4)}$. When $(\sigma_o^t/n\beta\lambda)$, σ_o^t, ε_{tn}, ε_{ln}, and λ are constant, $a \sim (\beta\gamma)^{(-1/2)}$, and $b \sim (\beta/\gamma)^{(3/4)}$. So the transverse beam radius decreases a little more with constant transverse focusing per unit length during acceleration.

Checks of the smooth approximation envelope equations against the full simulation results show good agreement when the emittance growth in the full simulation is used as local input to the envelope equations. Thus, the envelope equations were used to explore different tune strategies along the CCL. The constant $\sigma_o^t = 80^\circ$ transverse tune is a function of the external forces only, and does not involve the beam. Other tunes to be explored involve the beam itself in the tunes, so the external forces have to be calculated self-consistently. This is easy to do locally with the envelope equations, but emittance growth cannot be accounted for. Therefore, both transverse and longitudinal emittances were assumed to remain constant, and we use the trajectory of a tune recipe on the Hofmann Chart to make conjectures about what might happen to the emittance. Later, the more promising strategies must be studied in full simulations. In an earlier code, now in disrepair, the beam was repeatedly passed through each period of the linac until the desired (matched) tune was achieved, using an iterative, nonlinear optimization procedure – future work will rebuild this code.

Five 140 mA cases were run for the 700 MHz CCL and the operating point trajectories plotted on the Hofmann Chart, Fig. 11. In all cases, the longitudinal conditions are determined by the 1 MV/m accelerating gradient and a constant ϕ_s = -30°, and input $\varepsilon_{in\ rms}$ = 0.0407 cm·mrad. $\varepsilon_{in\ rms}$ = 0.02 cm·mrad except as noted. Our argument in this particular study is that the longitudinal conditions are largely fixed by the economics of rf power costs, and therefore, the operating trajectory can only be changed by adjusting the quadrupole focusing. The values of certain variables at 20 and 1600 MeV are listed in Table 4.

Table 4. Five CCL tune strategies; values at 20 and 1600 MeV

	Energy	σ^t/σ^t	σ^l/σ_o^l	σ^t/σ_o^t	σ_o^t	rms radius	rms tfac
Case 1 –	20	1.1	.53	.41	80	.18	14
Constant $\sigma_o^t = 80°$	1600	.1	.08	.91	80	.09	28
Case 2 –	20	.51	.42	.5	109	.16	16
Constant $\sigma^t/\sigma_o^t = 0.5$	1600	.18	.22	.5	15	.24	10.4
Case 3 – Equipartitioned	20	.49	.43	.52	117	.14	18
	1600	.49	.30	.37	11	.34	7.4
Case 4 – Equipartitioned,	20	.98	.59	.59	71	.24	10.4
$\varepsilon_{in\ rms} = 0.04$ cm·mrad	1600	.98	.44	.44	6.7	.57	4.4
Case 5 – Const. $\sigma^t = 40°$	20	.87	.5	.44	90.3	.18	13.6
	1600	.02	.1	.84	48	.11	23

Case 1 – Constant $\sigma_o^t = 80°$
These are the same conditions as above for the full simulation shown in Fig. 2-5, but with the emittances held constant. Without the emittance growth, the longitudinal trajectory drives further down toward the origin, and the transverse tune depression even further toward 1.

Case 2 – Constant $\sigma^t/\sigma_o^t = 0.5$
The beam participates in setting the tune. This could be considered a problem if the beam current were to be varied often during startup or operation. However, with electromagnetic quadrupoles and a computer control system, and assuming we will finally be successful in predicting the quad settings desired as a function of current, this should not be a problem. For this tune, the quads are weakened along the machine to

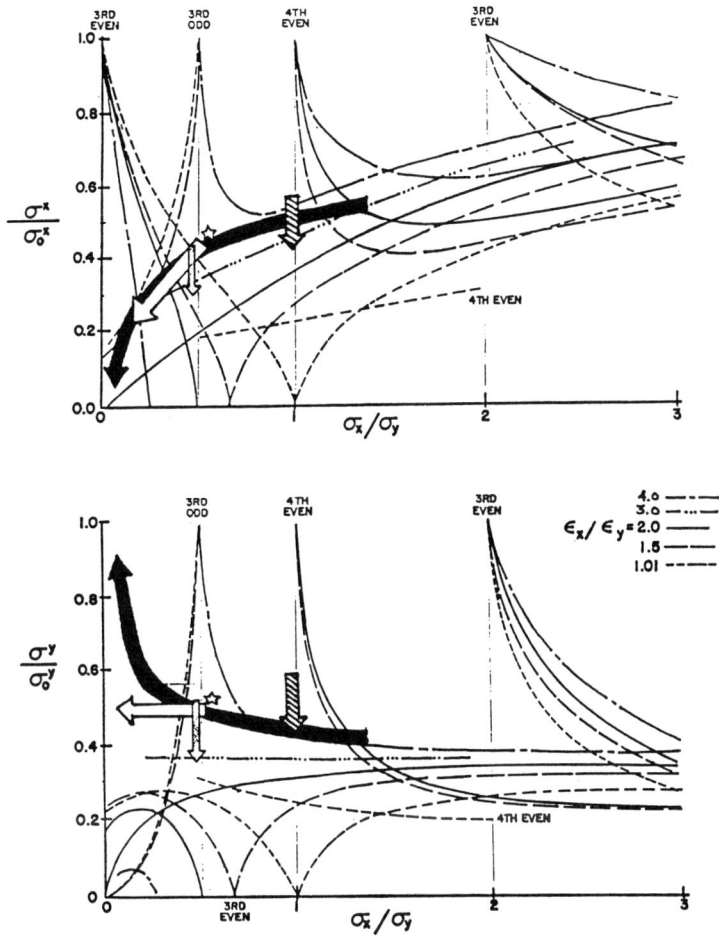

Fig. 11   Operating point trajectories on the Hofmann coherent instability mode chart for the 700 MHz CCL with four transverse tune strategies. The arrows run from 20 -> 1600 MeV. Case 1, black arrows;   Case 2, white arrows;   Case 3, narrow, double-hatched vertical arrows;   Case 4, single-hatched, broad vertical arrows.

keep the transverse tune depression constant. The operating trajectory on the Hofmann Chart is not so different from Case 1, so similar emittance performance might be expected.   There is an additional problem, in that the parameters require $\sigma_o{}^t$ to be above 90° initially, because in that region the tune of some particles will resonate with the alternating gradient period of the quadrupoles. If there were no emittance growth, the design would look reasonable in terms of the transverse rms aperture factor, which ranges from 16 - 10.5.

Case 3 – Equipartitioned, $\varepsilon_{tn\ rms} = 0.02$ cm·mrad

Assuring energy balance within the beam should remove the possibility for exciting the coherent instabilities identified in Fig. 11. For this case, $\varepsilon_{ln\ rms}/\varepsilon_{tn\ rms} = 0.0407/0.02 = 2.035$, so $\sigma^l/\sigma^t = 0.49$. The trajectory moves vertically downward.

It can be conjectured that the strongly depressed transverse tune in both this case and Case 4 may cause less halo for either or both of two reasons: one, that the beam is kept equipartitioned; and two, that the Debye sheath restricts the extent of the halo, as apparently happens in the longitudinal plane simulations in Fig. 2.

However, there could be several problems. The entire trajectory lies under the 3rd-odd mode boundaries; while Hofmann's work to date indicates that operation in this region should not result in strong excitation, there could be significant excitation over a system this long. The points $\{\sigma^l/\sigma^t, \sigma^l/\sigma_o^l\} = \{0.5, 1.0\}$ and $\{\sigma^l/\sigma^t, \sigma^t/\sigma_o^t\} = \{0.5, 1.0\}$ represent a resonance in single particle motion; the resonance is smeared out by the coherent action of space-charge. Thus, the 3rd-odd mode could act like a sustained resonant effect for this tune. Also, σ_o^t is above 90° for some distance. Finally, there remain the unknown effects discussed earlier, that may derive from the aperiodic nature of the linac, manifested through higher-order mode interactions or some other mechanism.

Case 4 – Equipartitioned, $\varepsilon_{tn\ rms} = 0.04$ cm·mrad

For an emittance ratio near one, the Hofmann Chart shows a large area free of modes, and indicates that there should be no excitation of the lower order modes shown if the tune depressions are kept above ~0.3. The initial σ_o^t is below 90°. There is, however, a price to be paid in the transverse rms aperture factor, which falls from about 10.4 to 4.4. If there were no growth of total emittance, the total aperture factor would be ~4.4Sqrt[5] = 2.8. The equipartitioning requirement with the given weak longitudinal focusing results in a strong reduction in quad strengths to keep the transverse tune near that of the longitudinal, resulting in a larger matched beam, even without any emittance growth. Recall that in the $\varepsilon_{tn\ rms} = 0.04$ case of Table 1, $tfac_{rms}$ increased from 10.4 to 23. The $tfac_{total}$ varied from 4.8 -> 2.8 -> 4.5, and there would certainly be particles even farther out. Thus, the two cases might be somewhat equivalent in $tfac_{total}$, the most important aspect for low-beam-loss, if in fact there were no growth in the equipartitioned case.

Case 5 – Constant $\sigma^t = 40°$

This case is quite similar to Case 1. It may merit further exploration when the full simulation tools are ready, because the author has noted, in past studies, important differences in the amount of halo formation between Cases 1 and 5 under certain input conditions.

Design Target ?

We note in comparing Cases 3 and 4 that a smaller transverse emittance results in larger aperture factor, assuming no emittance growth. So, a good design target may be an intermediate equipartitioned case with $\varepsilon_{ln\ rms}/\varepsilon_{tn\ rms}$ somewhat less than two, in order to have σ^l/σ^t a little larger than 0.5, to the right of the 3rd-odd mode threshold. The sensitivity of the magnet settings and of halo formation to tune depression would be explored. The initial point on the trajectory of this design target, to be investigated in future studies, is shown as a star on Fig. 11.

Earlier Work on Tune Strategy

Ref. [27] explored constant $\sigma_o{}^t$ vs. constant-length – constant-strength (cl-cs) quadrupole tuning in terms of minimizing the maximum beam size in the initial part of the FMIT DTL. The issue of insuring insensitivity to mismatch errors that could be encountered in operation was introduced — another complication. Some of the questions raised by this work should be revisited using our slowly growing insight into emittance growth, equipartitioning and mismatch.

ACKNOWLEDGEMENT

The able assistance of G.P. Boicourt in making the simulation studies is gratefully acknowledged and appreciated.

REFERENCES

[1] "The 2nd International Symposium on Advanced Nuclear Energy Research - Evolution by Accelerators, January 24-26, 1990, Mito, Japan, Proceedings by Japan Atomic Energy Research Institute, Tokai, Ibaraki, Japan.

[2] C.D. Bowman, et. al., "Nuclear Energy Generation and Waste Transmutation Using an Accelerator-Driven Intense Thermal Neutron Source", LA-UR-91-2601, Los Alamos National Laboratory.

[3] R.A. Jameson, Compiler, "Specialist Meeting on Accelerator-Driven Transmutation Technology for Radwaste and Other Applications", 24-28 June 1991, Saltsjöbaden, Stockholm, Sweden, spons. by Los Alamos National Laboratory and Swedish National Board for Spent Nuclear Fuel (SKN), LA-12005-C, SKN Rpt. No. 54

[4] G.P. Lawrence, et al., "High-Performance Deuterium-Lithium Neutron Source for Fusion Materials and Technology Testing", 1989 Particle Accelerator Conference, Chicago, Illinois, March 20-23, 1989.

[5] G.P. Lawrence, High-Power Proton Linac for Transmuting the Long-Lived Fission Products in Nuclear Waste", LA-UR-91-1335, IEEE Particle Accelerator Conference, May 1991, San Francisco, CA.

[6] G.P. Lawrence, R.A. Jameson, S.O. Schriber, "Accelerator Technology for Los Alamos Nuclear-Waste-Transmutation and Energy Production Concepts", LA-UR-91-2797, Proc. ICNES '91, to be published in Fusion Technology.

[7] R.A. Jameson, G.P. Lawrence & C.D. Bowman, "Accelerator-Driven Transmutation Technology for Incinerating Radwaste and for Advanced Application to Power Production", 2nd european conference on accelerators in applied research and technology (ecaart), 3-7 September 1991, Frankfurt-am-Main, Germany, LA-UR-91-2687.

[8] R.A. Jameson, G.P. Lawrence, S.O. Schriber, "Accelerator-Driven Transmutation Technology for Energy Production and Nuclear Waste Treatment", Third European Particle Accelerator Conference, Berlin Technical University, 24-28 March 1992, LA-UR-92-865, Los Alamos National Laboratory.

[9] "Accelerator Production of Tritium (APT), February 1990, A Report of the Energy Research Advisory Board to the USDOE, DOE/S-0074.

[10] S. Drell, Chrmn, "Accelerator Production of Tritium (APT)", JASON, The MITRE Corp., JSR-92-310, January 1992.

[11] G.P. Boicourt and R. A. Jameson, "A Statistical Approach to the Estimation of Beam Spill", Proc. 10th Linear Accelerator Conf., Montauk, New York September 10-14, 1979, Brookhaven National Laboratory report BNL 51134, p. 238, September 1980.

[12] G.P. Boicourt, "A Probability Function To Fit Radial Distributions in PARMILA Simulation Beams", 1983 Particle Accelerator Conf., IEEE Trans. Nucl. Sci, Vol. NS-30, No. 4, August 1983, p. 2534.

[13] P. Channell, "Mismatch Emittance Growth", Lectures, Notes compiled by W. Lysenko, Los Alamos National Laboratory, AT-Division, unpublished, 1991.

[14] I. Hofmann, "Generalized Equations for Emittance and Field Energy of High-Current Beams in Periodic Focusing", LA-11132-MS, Los Alamos National Laboratory, December 1987.

[15] M. Reiser, "Emittance Growth in Mismatched Particle Beams", 1991 Particle Accelerator Conf, IEEE Conf Record 91CH3038-7, p. 2497.

[16] R. A. Jameson, R. S. Mills, O. R. Sander, "Report on Foreign Travel - Switzerland", LASL Office Memo AT-DO-351(U)MP-9, Dec. 28, 1978; and R. A. Jameson, ""Emittance Growth in the New CERN Linac - Transverse Plane Comparison between Experimental Results and Computer Simulation", LASL Office Memo AT-DO-377(U), Jan. 15, 1979; and R. A. Jameson, "CERN Linac Tests, LASL Office Memo MP-9/AT-DO-(U), Mar. 1, 1979; and R. A. Jameson, "CERN Linac Tests" LASL Office Memo AT-DO-514(U), Apr. 26, 1979.

[17] R. A. Jameson, "Beam-Intensity Limitations in Linear Accelerators," (Invited), Proc. 1981 Particle Accelerator Conf., Washington, DC, March 11-13, 1981, IEEE Trans. Nucl. Sci. 28, p. 2408, June 1981, and, R.A. Jameson, "Equipartitioning in Linear Accelerators", 1981 Linear Accelerator Conf., LA-9234-C, Los Alamos National Laboratory, p. 125 .

[18] I. Hofmann,, "Emittance Growth of Beams Close to the Space Charge Limit", 1981 PAC, IEEE Trans. Nucl. Sci., Vol. NS-28, No. 3, June 1981, p 2399, and, I. Hofmann, I. Bozsik,"Computer Simulation of Longitudinal-Transverse Space Charge Effects in Bunched Beams", 1981 Linac Conf., October 1981, LA-9234-C, Los Alamos National Laboratory, Feb. 1982, p. 116.

[19] T.P. Wangler, F.W. Guy, "The Influence of Equipartitioning on the Emittance of Intense Charged-Particle Beams", 1986 Linear Accelerator Conference, Stanford Linear Accelerator Center, SLAC-303 CONF-860629, pp. 336-345.

[20] I. Hofmann, "Transport and Focusing of High-Intensity Unneutralized Beams",Advances in Electronics and Electron Physics, Supplement 13C, Applied Charged PArticle Optics, Part C: Very-High-Density Beams, Edited by A. Septier, Academic Press, 1983.

[21] K. Mittag, H.G. Hereward, "On Space Charge Instabilities and Emittance Growth in RF Linacs", Proc. 1981 Linear Accelerator Conf., LA-9234-C, Los Alamos National Laboratory, p. 120.

[22] T. Nishikawa, "Analytic Approach to the Space Charge Effect on the Longitudinal Phase Motions in Linear Accelerators", SJC-A-67-1, Working Group for Construction of Proton Synchrotron, Institute for Nuclear Study, University of Tokyo, Tokyo, Japan

[23] R.L. Gluckstern, "Space Charge Effects", in "Linear Accelerators", Ed. Lapostolle & Septier, North-Holland Pub., 1970, pp. 828-830.

[24] R.A. Jameson, Principal Investigator, "Scaling and Optimization in High-Intensity Linear Accelerators", LA-CP-91-272, Los Alamos National Laboratory, July 1991.

[25] R.A. Jameson, Principal Investigator, "Progress Toward Scaling and Optimization Criteria for High-Intensity, Low-Beam-Loss RF Linacs", LA-CP-92-221, Los Alamos National Laboratory, July 1992.

[26] R.A. Jameson, Principal Investigator, "Conceptual Design Aspects of a Deuteron Linac for Materials Testing Neutron Source", LA-CP-92-249, Los Alamos National Laboratory, July 1992.

[27] G.P. Boicourt, R.A. Jameson, "A Study of the Variation of Maximum Beam Size With Quadrupole Gradient in the FMIT Drift Tube Linac", 1981 PAC, IEEE Trans. Nucl. Sci., Vol. NS-28, No. 3, June 1981.

PROPOSAL FOR THE EXPERIMENTAL DEMONSTRATION OF THE COHERENT RADIATION*

A.G. Ruggiero

Brookhaven National Laboratory
Upton, NY 11973, USA

INTRODUCTION

It is of great importance to provide an experimental demonstration of the *coherence* of the radiation of electromagnetic waves from a short bunch of electrons performing oscillations in a direction transverse to the main direction of motion.

It is known that electrons circulating in a storage ring lose energy to *synchrotron radiation*.[1] Another method to stimulate radiation is to let an electron bunch travel through a *wiggler* or an *ondulator* device.[2] In either case the spectrum usually peaks in correspondence of wavelengths considerably smaller the length of the electron bunch; in this situation the power radiated is then linearly proportional to the number N of electrons in the bunch, as it is customarily observed.[3] Nevertheless, it may be possible conceiving a situation where the bunch length is considerably smaller or at least comparable to the wavelength of the peak of the radiation spectrum. If this is the case, it is then speculated[4] that the power radiated is proportional to the square N^2 of the number of particles in the bunch. This effect, which we can call a *coherent effect*, is of course very important since it would help to enhance the amount of the power radiated with a lower electron intensity.[5]

Unfortunately the *coherent effect* is difficult to observe since it is not easy to create beam and trajectory parameters which yield a bunch of length smaller than the radiation wavelength.[6,7] We thus propose here an experiment which has the two goals: (i) to generate an experimental situation where electrons in a bunch of length ℓ radiate electromagnetic power at a variable wavelength λ, and (ii) to observe with measurements the total amount of radiation versus λ and the beam intensity N. This experiment can be executed at the Accelerator Test Facility either at Brookhaven[8] or at Argonne National Laboratory[9].

---

* Work performed under the auspices of the U.S. Department of Energy.
Contribution to the Advanced Accelerator Concepts Workshop, Port Jefferson, NY, 14-20 June 1992.

EXPERIMENTAL APPARATUS

The experimental apparatus is simple and shown schematically in Fig. 1. It takes at the start a short bunch of electrons at the relatively low energy of 4 MeV. It is important that the electron bunch is as short as possible. Beam intensity is not very crucial; actually a lower intensity is preferable to control easily both the bunch length and the transverse emittance. The experiment needs to make only relative measurements of the electromagnetic power being radiated and the beam intensity is to be large enough to allow easy detection of the radiation. We assume that the bunch length is 6 psec and the normalized transverse emittance is $4 \times 10^{-6} \pi \cdot$mm$\cdot$mrad. A beam intensity with a peak value of 0.1 n Coulomb may be required.

At the exit of the electron gun there is a conventional wiggler magnet with a field variable up to 1 kG and a period of 40 cm. The wiggler is 4 m long and thus is made of ten periods. The devices can be made of conventional magnets with quadrupoles included for focusing. A magnet gap of few cm is adequate for letting the beam through. Previous experiments[6,7] employed conventional bending magnets for the production of synchrotron radiation. The use of a wiggler, as proposed here, is more advantageous since it provides a better defined and controlled radiation wavelength.

Leaving the wiggler, the electron bunch enters a bending magnet which deflects the beam away toward a collector for energy recovery, or toward a dump underground. At the same time the radiation traveling in a straight line will leave the bending magnet and enter a radiation detector. The bending magnet has a field of 1.0 kG and a length of 0.2 m for a bending angle of 20°. It may be useful that a data gathering is attached to the radiation detector for the collection of the experimental results and subsequent data analysis. At the same time it will be important to gather information on the bunch length, transverse dimensions and intensity. The electron beam energy and the wiggler field are also to be recorded.

THE ELECTROMAGNETIC RADIATION

The electrons going through the wiggler magnet perform transverse oscillations with a period given by the period of the wiggler $2\ell_w$ and an amplitude which depends on the wiggler field B_w and the magnetic rigidity $(B\rho)$ of the particles according to the following formula

$$a = \frac{(B\rho)}{B_w} \left[1 - \sqrt{1 - (\ell_w B_w / 2B\rho)^2} \right]. \qquad (1)$$

With a field of 250 Gauss one derives an amplitude of oscillations of 8.4 mm.

In the limit each electron is performing an infinitely long sequence of oscillations, the spectrum of the radiation is monochromatic with a frequency corresponding to the wavelength

$$\lambda_r = \ell_w / 2\gamma^2 \qquad (2)$$

which, with the values of the parameters taken above, gives 2.6 mm. This is about the assumed bunch length $\ell_e = 2$ mm. The only way to increase the radiation wavelength is by increasing also the wiggler period or lowering the beam energy. But it is clear that there are conflicting requirements and that it is not indeed easy to fulfill the *coherence regime* obtained with $\ell_e \gg \lambda_r$.

Since the electrons are performing in the experimental set up only ten oscillations, the radiation will have a wider spectrum with a bandwidth around the critical frequency of about 10%. Moreover the radiation has spatial distribution with an angular aperture of about $1/\gamma \sim 100$ mrad.

The total energy radiated for the case the bunch is made of a single electron is given by

$$W_0 = (56.3 \text{ keV}) \frac{E^4}{(B\rho)} n B_w \arcsin \left[\frac{\ell_w B_w}{2(B\rho)} \right] \qquad (3)$$

where $(B\rho)$ is the particle rigidity in Gauss·meter, E the kinetic energy in GeV, n the number of periods in the wiggler, $2\ell_w$ the period length in meter and B_w the wiggler field in Gauss.

In the approximation of a long bunch with N electrons the total energy radiated by the bunch is

$$W_L = N W_0 \qquad (4)$$

whereas in the limit of a very short bunch, because of the *coherence effect* the total power radiated is

$$W_s = N^2 W_0 \qquad (5)$$

As an example, let us take $B_w = 250$ Gauss and $N = 1 \times 10^9$ electrons, then

$$W_0 = 40 \ \mu eV$$

$$W_L = 40 \text{ keV}$$

$$W_s = 40 \text{ TeV}$$

At the repetition rate of 40 MHz the cw power for the short bunch case is 260 W, easily detectable. In the case of long bunches the power level is nine orders of magnitude lower and thus more difficult to measure.

MEASUREMENTS

It is thus conceived that our proposal in a "yes" or "no" experiment. If any radiation will be observed in significant amount it will be caused by *coherence effects*; if no radiation is measured the same effects can be questioned.

To perform a complete set of measurements to demonstrate further the enhancement due to *coherence effects* the following parameters will be varied:

 a) The wiggler field B_w. This will vary the amplitude of the oscillation and the total power being radiated. According to Eq. (1) the amplitude is linearly proportional to B_w, and according to Eq. (3) the power radiated increases quadratically with B_w. Too large values of the field are not useful since the amplitude of the oscillation gets too large and it will make the source spot size of the radiation also too large to be detected.

b) The number N of electrons. Depending on whether the *coherence effects* are present, the power radiated will have either one of the two dependencies shown by Eqs. (4 and 5). Moreover, in the case the bunch length is just about equal to the radiation wavelength, one can expect a mixture of N and N^2 dependencies. A plot of measurements of power versus N is then very useful.

c) The beam energy. For a fixed length of the wiggler period, the only way to vary the wavelength of the radiation is to change the beam energy according to Eq. (2). By increasing the beam energy, the power radiated of course increases, but the wavelength will correspondingly decrease and the dependence with the number N of particles should become more linear.

In summary, it is important to have the possibility of varying and measuring the wiggler field B_w, the beam intensity N and the beam energy γ. It is assumed that the control and reading of the last two parameters is provided by the Accelerator Test Facility itself; the control and measurements of the wiggler field will be included in the design of the wiggler itself.

TRANSVERSE FOCUSING

An important requirement for this experiment is to keep the transverse beam bunch size to a dimension smaller or comparable to the radiation wavelength. At the energy of 4 MeV the betatron emittance is 0.4 πmm·mrad, and one will need a strong focusing system to keep the bunch size to around a millimeter. For this purpose, quadrupoles with alternating gradients (QF and QD) in a FODO cell arrangement are interplaced with the wiggler magnets. The quadrupoles have a 5 cm length and a maximum gradient of 100 G/cm and are separated by 0.4 m, the wiggler period. This focusing system provides a $\beta_{max} \sim 1.3$ m at the center of each quadrupole.

Before entering the wiggler, the electron bunch will be sufficiently focused to match the transport in the wiggler itself. Care is taken to avoid introducing dispersion and other chromatic effects that may lead to a bunch length increase. Leaving the wiggler, the electron bunch is deflected by the dipole magnet which is then followed by a suitable focusing elements for the beam disposal.

THE RADIATION DETECTOR

The most important part of the experiment is the measurement of the radiation lost by the electrons. The radiation has a wavelength in the few millimeters range and it can first be transported by a waveguide attached to the end of the wiggler on one side of the magnet which deflects the electrons away. In order for the radiation to propagate freely through the waveguide this can be made of rectangular cross-section, 4 cm wide and 1 cm high, which will serve also as a filter to lower frequency noise radiation. The waveguide is 50 cm long and terminated at the receiving end with an adapter; if the adapter includes a resistive load, the square of the current flowing through the load is a direct measurement of the power being radiated. Other methods of measurement include directional couplers attached to the waveguide and followed by a frequency analyzer.

COMPONENTS AND COST

Table I gives a list of the components required for the execution of the experiment and the corresponding cost estimate. The total technical cost is 135 K$ to which the labor of the equivalent two scientific man/year is to be added at 65 K$ each; finally a 35 K$ is added for travel, computing and other administrative tasks. The total cost of the experiment is thus 300 K$. To this one should add overhead, contingency and escalation. In order to procure the several parts, one might to have wait two years for the completion of the experiment in which case the spending profile can be split in 200 K$ for the first year and 100 K$ for the second. Once in place, the experiment itself should take only about 10 weeks run of the Accelerator Test Facility in order to gather all the necessary data.

REFERENCES

1. J. Schwinger, Phys. Rev., 75, No. 12 (1949), p. 1912.
2. J.M.J. Madey, J. Appl. Phys. 42, 1906 (1971).
3. J.D. Jackson, Classical Electrodynamics, Chapter 14, Second Edition, John Wiley & Sons.
4. F.C. Michel, Phy. Rev. Letters, 48, No. 9, (1982), p. 580.
5. A.G. Ruggiero, AD/AP-35 and 41. BNL Internal Reports (1992). See also contribution on this subject in these conference proceedings.
6. T. Nakazato, et al., Phy. Rev. Letters, 63, No. 12 (1989), p. 1245.
7. E.B. Blum, et al., NIM A307, p. 568 (1991).
8. I. Ben-Zvi, 1991 Particle Accelerator Converence, IEEE 91CH3038-7, page 550.
9. P. Schoessow, et al., Proceedings of EPAC 90, Nice, France, June 1990, Vol. 1, p. 606.

Table I Components and Cost

Wiggler	20	Dipoles		
		length	20 cm	
		B_{max}	1 kG	
		Aperture	8×2 cm^2 ($H \times V$)	50 K$
Focusing	11	Quadrupoles	(QF or QD)	
		length	5 cm	
		G_{max}	100 G/cm	
		Bore Radius	4 cm	
	3	quadrupole triplet for matching		
		between RF gun and wiggler		20
	3	quads triplet after deflecting dipole		
Deflecting Dipole	1	length	0.2 m	2.5
		field	1 kG (similar to wiggler)	
Vacuum Chamber		stainless steel or aluminum		
		2 mm thick 8×2 cm^2 ($H \times V$)		
		length	6 m straight	2
Waveguide		0.5 m long copper 4×2 cm^2		0.5
		attached with flanges to v.c.		
Radiation Detectors				10
Vacuum Valves	2			
Pumps	2	(10^{-7} torr)	}	10
Power Supply for Dipoles				10
Quadrupoles				10
Data Gathering/Computer				<u>20</u>
				135 K$
2-One man year (scientific)				130
Travel, computing, etc.				<u>35</u>
				300 K$

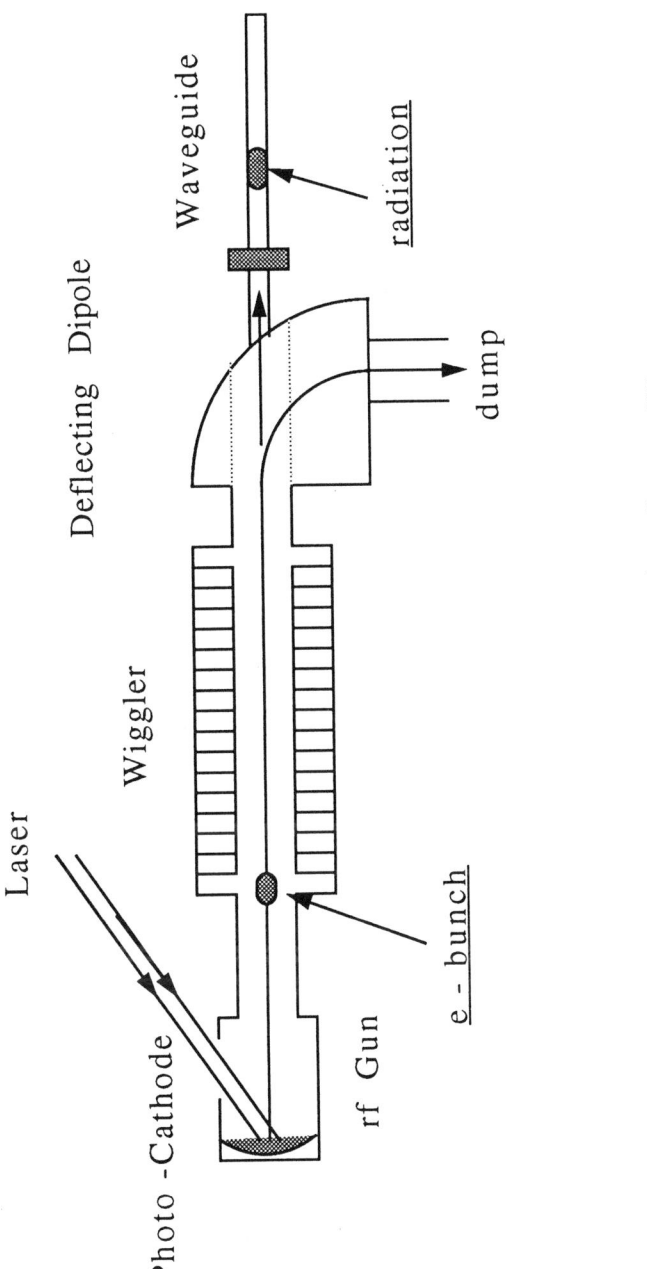

Fig. 1   Experimental   Set - Up